Pronunciations of Selected Microorganisms

Absidia (ab-sid′ē-ă)

Acanthamoeba (ă-kan-thă-mē′bă)

Acetobacter (a-sē′tō-bak-ter)

Acidanus (as-i-dā′nus)

Acinetobacter (as-i-nē′tō-bak′ter)

Acremonium (ak′rĕ-mō′nē-ŭm)

Actinomyces israelii (ak′ti-nō-mī′sēz is-rā′el-ē-ē)

Actinomycetes (ak′ti-nō-mī-sē′tēz)

Agrobacterium tumefaciens (ag′rō-bak-ti′rē-um tū′me-fāsh-enz)

Alternaria (al-ter-nā′rē-ă)

Amanita muscaria (am-ă-nī′tă mus-ka′rē-ă)

Amanita phalloides (am-ă-nī′tă fal-ōy′dēz)

Amoeba (am-ē′bă)

Aquaspirillum magnetotacticum (ă-kwă-spī′ril-ŭm mag-ne-tō-tak′ti-kŭm)

Aquifex (ăk′wē-feks)

Aspergillus oryzae (as-per-jil′ŭs o′ri-zē)

Azomonas (ā-zō-mō′nas)

Azospirillum (ā-zō-spī′ril-ŭm)

Azotobacter (ā-zō-tō-bak′ter)

Babesia (bă-bē′zē-ă)

Bacillus anthracis (ba-sil′ŭs an-thrā′sis)

Bacillus cereus (ba-sil′ŭs se′rē-ŭs)

Bacillus licheniformis (ba-sil′ŭs lī-ken-i-for′mis)

Bacillus polymyxa (ba-sil′ŭs po-lē-miks′ă)

Bacillus popilliae (ba-sil′ŭs pop-pil′ē-ē)

Bacillus sphaericus (ba-sil′ŭs sfe′ri-kŭs)

Bacillus stearothermophilus (ba-sil′ŭs ste-rō-ther-ma′fil-ŭs)

Bacillus subtilis (ba-sil′ŭs sŭt′i-lis)

Bacillus thuringiensis (ba-sil′ŭs thur-in-jē-en′sis)

Bacteroides fragilis (bak-tur-oy′dēz fra′ji-lis)

Balantidium coli (bal-an-tid′ē-ŭm kō′lī)

Bartonella bacilliformis (bar-tō-nel′ă ba-sil′li-for′mis)

Bartonella henselae (bar-tō-nel′ă hen′sel-ī)

Bartonella quintana (bar-tō-nel′ă kwin′ta-nă)

Bdellovibrio (del-lō-vib′rē-ō)

Beggiatoa (bej′jē-a-tō′ă)

Blastomyces dermatitidis (blas-tō-mī′sēz der-mă-tit′i-dis)

Bordetella pertussis (bōr-dĕ-tel′ă per-tus′is)

Borrelia burgdorferi (bō-rēl′ē-ă burg-dōr′fer-ē)

Borrelia recurrentis (bō-rēl′ē-ă re-kur-ren′tis)

Botryococcus braunii (bot′rē-ō-kok′ŭs brow′nē-ē)

Brucella abortus (broo-sel′lă a-bort′us)

Brucella canis (broo-sel′lă kā′nis)

Brucella melitensis (broo-sel′lă me-li-ten′sis)

Brucella suis (broo-sel′lă soo′is)

Burkholderia (burk-hol-der′ē-ă)

Campylobacter jejuni (kam′pi-lō-bak′ter jē-jū′nē)

Candida albicans (kan′did-ă al′bi-kanz)

Caulobacter (kaw′lō-bak-ter)

Cephalosporium (sef′ă-lō-spor′ē-ŭm)

Chlamydia pneumoniae (kla-mid′ē-ă noo-mō′nē-ī)

Chlamydia psittaci (kla-mid′ē-ă sit′ă-sē)

Chlamydia trachomatis (kla-mid′ē-ă tra-kō′ma-tis)

Chondrus crispus (kon′drŭs krisp′ŭs)

Chromatium buderi (krō-ma′tē-ŭm bū′de-rī)

Citrobacter (sit′rō-bak-ter)

Cladophialophora carrionii (klă-dŏf′ē-ă-lof′ŏ-rā kar-rē-on′ē-ī)

Claviceps purpurea (klav′i-seps per′pŭ-rē′ă)

Clostridium botulinum (klos-trid′ē-ŭm bo-tū-li′nŭm)

Clostridium difficile (klos-trid′ē-ŭm di-fi′sēl)

Clostridium perfringens (klos-trid′ē-ŭm per-frin′jens)

Clostridium tetani (klos-trid′ē-ŭm te′tan-ē)

Coccidioides immitis (kok-sid-ē-oy′dēz im′mi-tis)

Codium (kō′dē-ŭm)

Corynebacterium diphtheriae (kŏ-rī′nē-bak-tēr′ē-ŭm dif-thi′rē-ē)

Coxiella burnetii (kok-sē-el′ă ber-ne′tē-ē)

Cryptococcus neoformans (krip-tō-kok′ŭs nē-ō-for′manz)

Cryptosporidium parvum (krip-tō-spō-rid′ē-ŭm par′vŭm)

Cyclospora cayetanensis (sī-klō-spōr′ă kī-ē-tan-en′sis)

Cytomegalovirus (sī-tō-meg′ă-lō-vī′rŭs)

Cytophaga (sī-tof′ă-gă)

Deinococcus radiodurans (dī-nō-kok′ŭs rā-dē-ō-dur′anz)

Desulfovibrio (dē′sul-fō-vib-rē-ō)

Dictyostelium (dik-tē-os-tē′lē-ŭm)

Didinium (dī-di′nē-ŭm)

Diplococcus pneumoniae (dip′lō-kok′ŭs nū-mō′nē-ī)

Echinococcus granulosus (ĕ-kī′nō-kok′ŭs gra-nū-lō′sŭs)

Edwardsiella (ed′ward-sē-el′ă)

Ehrlichia chaffeensis (er-lik′ē-ă chaf-ē-en′sis)

Ehrlichia equi (er-lik′ē-ă ē′kwi)

Entamoeba histolytica (ent-ă-mē′bă his-tō-li′ti-kă)

Enterobacter (en′ter-ō-bak′ter)

Enterobius vermicularis (en-ter-ō′bī-ŭs ver-mi-kū-lar′is)

Enterococcus faecalis (en′ter-ō-kok′ŭs fē-kă′lis)

Enterococcus faecium (en′ter-ō-kok′ŭs fē-sē′ŭm)

Epidermophyton floccosum (ep′i-der-mof′i-ton flŏk′ŏ-sŭm)

Epulopiscium fishelsoni (ep′yoo-lō-pis′sē-ŭm fish-el-sō′nī)

Escherichia coli (esh-ĕ-rik′ē-ă kō′lī)

Euglena granulata (yū-glān′ă gran-yū-lă′tă)

Eupenicillium (yū-pen-i-sil′ē-ŭm)

Exophiala (ek-sō-fī′ă-lă)

Fasciola gigantica (fa-sē′ō-lă ji-gan′ti-kă)

Fasciola hepatica (fa-sē′ō-lă he-pa′ti-kă)

Ferroplasma acidarmarnus (fe′rō-plaz-ma a-sid′ar-mar-nŭs)

Fonsecaea compacta (fon-sē-sē′ă kom-pak′ta)

Pronunciations of Selected Microorganisms continues on back endsheets

The Microbiology Place

Log on.

Explore.

Succeed.

To help you succeed in microbiology, your professor has arranged for you to enjoy access to great media resources, on an interactive CD-ROM and on The Microbiology Place website. You'll find that these resources that accompany your textbook will enhance your course materials.

home   search   tech support   feedback   aw.com   credits

the microbiology place

1: History and Scope of Microbiology

Microbiology by Robert Bauman

Home > 1: History and Scope of Microbiology > Welcome

Welcome
Objectives
Pre-Tests
Practice Tests
Review Question Answers
Web Links
Microbe Review
Flashcards
Videos
Careers
Research News
Research Navigator

Chapter 1: History and Scope of Microbiology
Welcome

Welcome to The Microbiology Place web site for *Microbiology*, by Robert Bauman. From this site, you may sample a rich selection of practice quizzes, flashcards, videos, web links, and other study and instructional materials supporting the Bauman textbook. To begin, select a chapter from the pull-down menu above, and then click on any feature to the left.

Among the resources you will find:

- *Objectives* help you focus on the most important concepts and topics to master in each chapter.
- *Pre-Tests* are multiple-choice questions intended to assess your prior knowledge and/or pre-conceptions of topics relevant to the chapter you are about to begin studying.
- *Practice Tests* are intended to help you assess your mastery of chapter topics. There are 30-40 practice test questions per chapter, consisting of true/false, multiple-choice, and fill-in-the-blank questions.

What your system needs to use these media resources:

WINDOWS™
266 MHz Intel Pentium processor or greater
Windows 98 or higher
64 MB Minimum RAM
800 X 600 screen resolution
Browser: Internet Explorer 5.0 or higher; Netscape 4.7, 6.2, or higher
Plug-Ins: Shockwave Player 8, Flash Player 5, QuickTime 4
Internet Connection: 56k modem minimum for website
NOTE: Use of Netscape 6.0 is not recommended due to a known compatibility issue between Netscape 6.0 and the Flash and Shockwave plug-ins.

MACINTOSH™
266 MHz Minimum CPU
OS 8.6 or higher
64 MB Minimum RAM
800 x 600 screen resolution, thousands of colors
Browser: OS 8-9: Netscape 4.7, 6.2, or higher; (Web only: Internet Explorer 5 or higher); OS X: Internet Explorer 5.2 or higher
Plug-Ins: Shockwave Player 8, Flash Player 5, QuickTime 4
Internet Connection: 56k modem minimum for website
NOTE: Use of Netscape 6.0 is not recommended due to a known compatibility issue between Netscape 6.0 and the Flash and Shockwave plug-ins. Also, Mac OS X users should run this CD-ROM on Internet Explorer only due to known compatibility issues between OS X and Netscape.

Got technical questions?

For technical support, please visit www.aw.com/techsupport and complete the appropriate online form. Technical support is available Monday-Friday, 9 a.m. to 6 p.m. Eastern Time (US and Canada).

Here's your personal ticket to success:

How to log on to www.microbiologyplace.com:

1. Go to www.microbiologyplace.com.
2. Click *Microbiology* by Robert W. Bauman.
3. Click "Register."
4. Scratch off the silver foil coating below to reveal your pre-assigned access code.
5. Enter your pre-assigned access code exactly as it appears below.
6. Complete the online registration form to create your own personal user Login Name and Password.
7. Once your personal Login Name and Password are confirmed by email, go back to www.microbiologyplace.com, type in your new Login Name and Password, and click "Log In."

Your Access Code is:

If there is no silver foil covering the access code above, the code may no longer be valid. In that case, you need to either:
- Purchase a new student access kit at your campus bookstore.
- Purchase access online using a major credit card. Go to www.microbiologyplace.com, click *Microbiology* and click "Buy Now."

Microbiology

ROBERT W. BAUMAN, Ph.D.

Amarillo College

Contributions by:

Elizabeth Machunis-Masuoka, Ph.D.

University of Virginia

Ian Tizard, Ph.D.

Texas A&M University

PEARSON

Benjamin
Cummings

San Francisco Boston New York
Cape Town Hong Kong London Madrid Mexico City
Montreal Munich Paris Singapore Sydney Tokyo Toronto

Publisher *Daryl Fox*
Executive Editor *Serina Beauparlant*
Project Editor *Barbara Yien*
Managing Editor *Wendy Earl*
Production Editor *Sharon Montooth*
Text Development Editor *Laura Bonazzoli*
Art Development Editor *Laura Southworth*
Principal Artist *Kenneth X. Probst*
Artists *Darwen Hennings, Cassio Lynm, Precision Graphics, J.B. Woolsey & Associates*
Copyeditor *Alan Titche*
Art and Photo Coordinator *Kelly Murphy*
Photo Researchers *Diane Austin, Kathleen Olson*
Text Designer *tani hasegawa*
Cover Designer *Lillian Carr*
Contributing Editors *Claire Brassert, Kay Ueno, Amy Teeling*
Market Development Manager *Cheryl Cechvala*
Executive Marketing Manager *Lauren Harp*
Publishing Assistants *Ryan Shaw, Ziki Dekel*
Manufacturing Buyer *Stacey Weinberger*
Compositor *GTS Graphics*

Cover Image: T4 bacteriophages bursting from an *E. coli* bacterium. Kenneth X. Probst
Photograph by Lee D. Simon, Photo Researchers, Inc.

Text, photography, and illustration credits appear following the Appendices.

Library of Congress Cataloging-in-Publication Data

Bauman, Robert W.
    Microbiology/Robert W. Bauman ; contributions by Elizabeth Machunis-Masuoka, Ian
    Tizard.
        p. cm.
    Includes index.
    ISBN 0-8053-7590-2
    1. Microbiology. I. Machunis-Masuoka, Elizabeth. II. Tizard, Ian R. III. Title.

QR41.2.B38 2003
579—dc21   2003051619

ISBN 0-8053-7590-2

2 3 4 5 6 7 8 9 10—DOW—07 06 05 04

www.aw.com/bc

I dedicate this labor of love to my students,
who urged me to write this book, and to my family,
who stuck with me to the end.

Robert W. Bauman

ABOUT THE AUTHOR

Robert W. Bauman is a professor of biology and chairman of the Department of Biological Sciences at Amarillo College in Amarillo, Texas. He teaches microbiology and human anatomy and physiology. He received an M.A. degree in botany from the University of Texas at Austin, and a Ph.D. in biology from Stanford University. His research interests have included the morphology and ecology of freshwater algae, the cell biology of marine algae (particularly the deposition of cell walls and intercellular communication), and environmentally triggered chromogenesis in butterflies. He is a member of the American Society of Microbiology (ASM), Texas Community College Teacher's Association (TCCTA), American Association for the Advancement of Science (AAAS), and The Lepidopterist's Society. When he is not writing books, he enjoys spending time with his family: gardening, hiking, camping, playing games (with lots of laughter), and reading classics out loud by a crackling fire.

ABOUT THE ILLUSTRATION TEAM

Kenneth Xavier Probst, the principal artist, received his M.A. in Medical and Biological Illustration from the Johns Hopkins University School of Medicine and has been a staff scientific illustrator for Benjamin Cummings since 2000. In addition to his work as an educator, fine artist, graphic and industrial designer, Kenneth has received recognition for his work as a medical illustrator, including Certificates of Merit from the Association of Medical Illustrators and a Vesalius Trust Research Grant.

Laura Southworth is a freelance art development editor and scientific illustrator in Forestville, California. She began her career while obtaining a B.S. degree in biology from California Polytechnic State University, San Luis Obispo. For over twenty years she has been designing and creating scientific art for publications in many disciplines, including biology, human anatomy and physiology, entomology, microbiology, and molecular biology.

Richard A. Robison photographed many of the unique micrographs found in this text. He is an Associate Professor of microbiology at Brigham Young University in Provo, Utah. He earned a Ph.D. in microbiology from BYU in 1988, with graduate minors in chemistry and biochemistry. His current research interests center around frank bacterial pathogens of humans and animals. Dr. Robison is director of a biosafety level-3 facility and collaborates with scientists at several national and private laboratories. Photographing the microscopic world has been a favorite hobby of his for over 20 years.

PREFACE

I have taught microbiology to undergraduates for over 15 years, and have witnessed firsthand how students struggle with the same topics and concepts year after year. Increasingly, I also find that students come into class with various levels of preparation—while some have strong science backgrounds, others lack a foundation in chemistry and/or biology—making it a challenge to decide how to gear the course. In setting out to create this textbook, my goal was to write an introductory text that helped address these concerns and that was also engaging and enjoyable to read, weaving in the many exciting and interesting microbiology-related topics that appear daily in the news and in research journals. I wished to highlight the positive effects of microorganisms in our lives, along with the medically important microorganisms that cause disease. Lastly, I aimed to write a text that explains complex topics (especially metabolism, genetics, and immunology) in a way beginning students can understand, while still presenting a thorough, comprehensive, and accurate overview of microbiology.

To achieve these goals, a great deal of thought and research, and the feedback of literally hundreds of students and instructors (listed at the end of this preface), went into the development of this textbook. From the beginning I was determined *not* to write a book that would be just like every other microbiology textbook, but instead one that improved upon conventional explanations and illustrations in substantive and effective ways. To this end, I have collaborated with educators, students, editors, and top scientific illustrators in the field to develop this text. The following sections provide an outline of the book's main themes and features.

A Groundbreaking Art Program That Engages as it Teaches

This book's stunning art program features a number of innovations and pedagogical tools designed to help students understand the material. **Half-micrograph, half-illustration figures (Figures 1 and 2)** depict microbial structures in a three-dimensional, visually striking way that draws students' interest while also teaching them what these structures really look like. **Orientation figures (Figure 3)** remind students of "the big picture". **Step-by-step diagrams (Figure 3)** break complex processes into smaller, more manageable pieces; the step numbers are also color-coded to correspond to step numbers in the text. **Figure legend questions (Figure 4)** engage students by asking them to apply what they've learned in the text to answer questions about topics featured in the art. **Consistent use of icons and colors (Figure 4)** from figure to figure makes it easier for students to recognize structures and processes throughout the book. **Ruler bars** and TEM / SEM / LM designations provide readers added context with which to study the art.

Half-micrograph, half-illustration figures help students visualize microbial structures as they really are.

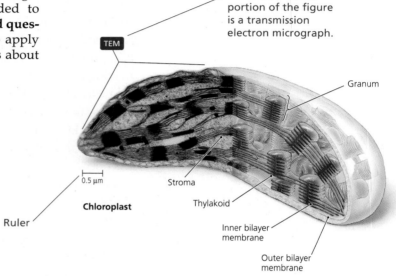

Indicates that this portion of the figure is a transmission electron micrograph.

TEM

Granum

0.5 µm

Stroma

Thylakoid

Inner bilayer membrane

Outer bilayer membrane

Chloroplast

Ruler

▲ *Figure 1*

0.5 µm

Ribosome

Cytoplasm

Nucleoid

Glycocalyx

Cell wall

Cell membrane

Inclusions

Flagellum

TEM

◀ *Figure 2*

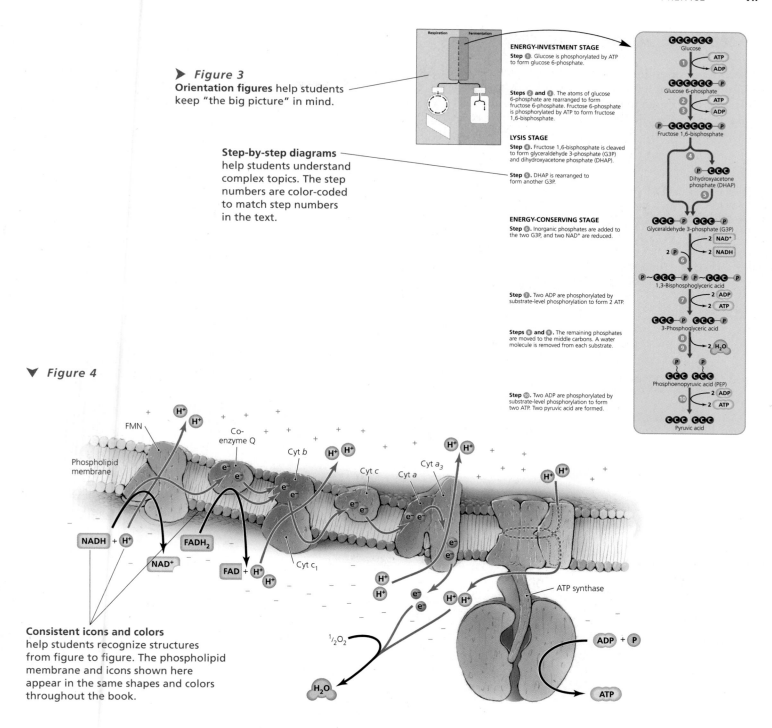

▶ Figure 3
Orientation figures help students keep "the big picture" in mind.

Step-by-step diagrams help students understand complex topics. The step numbers are color-coded to match step numbers in the text.

ENERGY-INVESTMENT STAGE
Step ❶. Glucose is phosphorylated by ATP to form glucose 6-phosphate.

Steps ❷ and ❸. The atoms of glucose 6-phosphate are rearranged to form fructose 6-phosphate. Fructose 6-phosphate is phosphorylated by ATP to form fructose 1,6-bisphosphate.

LYSIS STAGE
Step ❹. Fructose 1,6-bisphosphate is cleaved to form glyceraldehyde 3-phosphate (G3P) and dihydroxyacetone phosphate (DHAP).

Step ❺. DHAP is rearranged to form another G3P.

ENERGY-CONSERVING STAGE
Step ❻. Inorganic phosphates are added to the two G3P, and two NAD^+ are reduced.

Step ❼. Two ADP are phosphorylated by substrate-level phosphorylation to form 2 ATP.

Steps ❽ and ❾. The remaining phosphates are moved to the middle carbons. A water molecule is removed from each substrate.

Step ❿. Two ADP are phosphorylated by substrate-level phosphorylation to form two ATP. Two pyruvic acid are formed.

▼ Figure 4

FMN
Co-enzyme Q
Cyt b
Cyt c
Cyt a
Cyt a_3
H^+
Phospholipid membrane
e^-
NADH + H^+
NAD$^+$
FADH$_2$
FAD + H^+
H^+
Cyt c_1
ATP synthase
$\frac{1}{2}O_2$
H_2O
ADP + P
ATP

Consistent icons and colors help students recognize structures from figure to figure. The phospholipid membrane and icons shown here appear in the same shapes and colors throughout the book.

▲ Figure 5.21
One possible arrangement of an electron transport chain. Electron transport chains are located in the cytoplasmic membranes of prokaryotes, and in the inner membranes of mitochondria in eukaryotes. The exact types and sequences of carrier molecules in electron transport chains vary among organisms. As electrons move down the chain (red arrows), their energy is used to pump protons (H^+) across the membrane. The protons then flow through ATP synthase (ATPase), which synthesizes ATP. Approximately one molecule of ATP is generated for every two protons that cross the membrane. *Why is it essential that all the electron carriers of an electron transport chain be membrane bound?*

Figure legend questions ask students to apply concepts learned in the text to answer questions about the art. Answers to the questions appear upside-down.

Figure 5.21 The carrier molecules of electron transport chains are membrane bound for at least two reasons: (1) Carrier molecules need to be in fairly close proximity so they can pass electrons between each other. (2) The goal of the chain is the pumping of H^+ across a membrane to create a proton gradient; without a membrane, there could be no gradient.

Exceptional Scientific Accuracy

Exceptional attention has been paid to **scientific accuracy (Figure 5)** throughout the art program. Many proteins, for instance, are drawn to look like their actual shapes based on data from the Protein Data Bank from the Research Collaboratory for Structural Bioinformatics. Microbial structures are drawn from actual micrographs. And, where appropriate, photos are used in place of (or in addition to) illustrations in order to more accurately depict a topic or process **(Figure 6).**

❶ Slide is flooded with crystal violet for 1 min, then rinsed with water.

Result: All cells are stained purple.

LM   5 μm

❷ Slide is flooded with iodine for 1 min, then rinsed with water.

Result: Iodine acts as a mordant; all cells remain purple.

LM   5 μm

❸ Slide is flooded with solution of alcohol and acetone for 10–30 sec, then rinsed with water.

Result: Smear is decolorized; Gram-positive cells remain purple, but Gram-negative cells are now colorless.

LM   5 μm

❹ Slide is flooded with safranin for 1 min, then rinsed with water and blotted dry.

Result: Gram-positive cells remain purple, Gram-negative cells are pink.

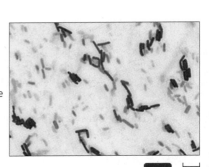

LM   5 μm

▼ *Figure 5*
Scientific accuracy
The reverse transcriptase depicted in this illustration of HIV was based on data from the Protein Data Bank.

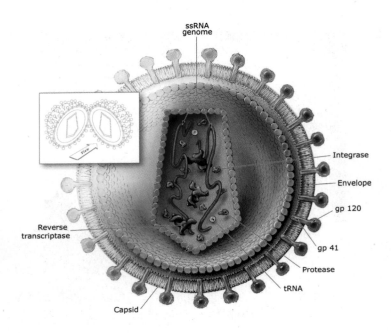

ssRNA genome

Integrase

Envelope

gp 120

gp 41

Protease

tRNA

Reverse transcriptase

Capsid

◄ *Figure 6*
High-quality micrographs
are found throughout the text, and are also available to instructors in electronic format. Here, four same-field micrographs depict the Gram staining procedure.

Coverage of Timely and High-Interest Topics from Both Research and Everyday Life

The text includes several features designed to draw students' interest and bring the wonders of microbiology to life. **Chapter-opening vignettes (Figure 7)** pique the reader's curiosity about the chapter at hand, covering such topics as the use of microbes in genetically modified foods (Ch. 8) and the problem of so-called "superbugs" that are resistant to our most powerful antibiotics (Ch. 10). **Highlight boxes (Figure 8)** focus on fascinating topics in microbiology, such as microbial control in the preparation of sushi (Ch. 9), the medicinal and cosmetic uses of Botox (Ch. 11), and oil-eating bacteria (Ch. 26). These boxes demonstrate the relevance of applied microbiology to everyday life, encouraging interest in the field. Meanwhile, **New Frontiers boxes (Figure 9)** introduce students to cutting-edge research and future challenges related to microbiology. Topics include bioterrorism (Chs. 1 and 26), biofilms (Chs. 3 and 6), genetic modification of mosquitoes to curb the spread of malaria (Ch. 8), and the use of *E. coli* in making blue denim (Ch. 26). A complete list of box topics is included on page xxvi.

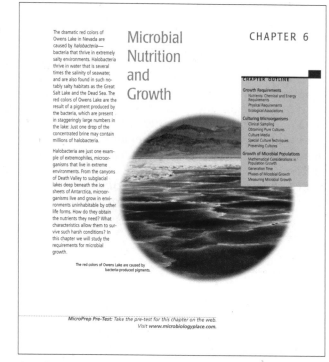

▲ *Figure 7*
Chapter openers include striking photos and interesting vignettes to draw readers in.

Highlight 9.2   Sushi: A Case Study in Microbial Control

Sushi—it either makes you squirm or salivate! Although technically the term refers to rice, it's generally understood to refer to bite-sized slices of raw fish served with rice. Long a staple of Japanese cuisine, sushi has become popular throughout the United States. But isn't eating raw fish dangerous? What methods of microbial control are applied in the preparation of sushi?

It's true that raw fish can contain harmful microorganisms. Such parasites as the kind of roundworms called anasakids are commonly found in fish and can cause gastrointestinal symptoms. Accordingly, before fish can be served raw, the Food and Drug Administration requires that it be frozen at −20°C (−4°F) for 7 days, or at −37°C (−35°F) for 15 hours. Unfortunately, freezing does *not* kill all bacterial or viral pathogens. Consumers should be aware that they always assume some risk of food poisoning caused by such bacteria as *S. aureus*, *E. coli,* and species of *Salmonella* and *Vibrio* whenever they eat any kind of raw seafood, including raw oysters, seviche, or carpaccio.

Even though diners tend to fixate on the safety of raw fish, cooked rice left sitting at room temperature is also vulnerable to the growth of pathogens. To counter this potential problem, sushi rice is prepared with vinegar, which acidifies the rice. At pH values below 4.6, rice becomes too acidic to support the growth of pathogens. Sushi bars can also reduce the risk posed by pathogens by keeping restaurant temperatures cool. Additionally, it is believed that wasabi, the fiery horseradish-like green paste commonly eaten with sushi, contains antimicrobial properties, although its mechanism of action is not well understood.

With these antimicrobial precautions in place, the vast majority of diners consume sushi safely meal after meal (although pregnant women and people with compromised immune systems should avoid all raw seafood). When properly prepared, sushi is beautiful, low in calories, and a source of heart-healthy omega 3-fatty acids—and very delicious.

◀ *Figure 8*
Highlight boxes focus on fascinating topics in microbiology.

New Frontiers 8.1   Building a Better Mosquito

What if scientists could genetically engineer mosquitoes to be incapable of transmitting deadly diseases, such as malaria? Researchers are attempting to do just that. One possible strategy involves using transposons to carry genes into mosquitoes that would boost their production of defensive proteins that destroy pathogens. Mosquitoes naturally produce such proteins, but not in high enough doses to kill off all pathogens. Scientists theorize that if they can genetically engineer and breed mosquitoes that destroy pathogens, these mosquitoes could then be released into the environment to breed with wild mosquitoes, creating subsequent populations that can no longer spread disease.

To date, researchers have successfully used transposons as vectors to insert defense-boosting genes into *Aedes* mosquitoes, which transmit yellow fever. While this is a promising achievement, *Aedes* is only one of over 100 disease-transmitting mosquito species. Recombinant technologies that work on *Aedes* may not necessarily work on, for instance, *Anopheles gambia* mosquitoes, which are the primary carriers of the most deadly form of malaria. In addition, some worry that unleashing transgenic mosquitoes into the wild may have adverse consequences. Given that malaria alone kills over 2 million people each year, however, the quest for a safer mosquito is one that scientists are compelled to continue.

Aedes mosquito

◀ *Figure 9*
New Frontiers boxes highlight cutting-edge research as well as future challenges in microbiology.

Student-Focused Pedagogy

Every page of this textbook was written with students' specific needs in mind. To help students be better prepared in class, **Online "MicroPrep" Pre-tests** are available for each chapter at www.microbiologyplace.com. The pre-tests are cited at the beginning of each chapter, and are designed to be taken before a given chapter is read in order to (1) review relevant topics introduced in previous chapters (or in fundamental chemistry/biology courses), and (2) test student's preconceptions and/or knowledge concerning the chapter's topics. **Learning Objectives (Figure 10)** are integrated throughout each chapter to give students a preview of topics covered. **Integrated pronunciation and etymology guides (Figure 10)** help students pronounce and remember vocabulary. **Critical Thinking questions (Figure 10),** which are both integrated throughout the text and listed at the end of each chapter, encourage students to apply what they have learned to a novel situation or to a case study. **End-of-chapter summaries** and **review questions (Figure 11)** help students review the material.

Given that so many students enrolled in introductory microbiology courses are preparing for careers in the health profession, **Case Studies (Figure 12)** are included in the chapters that discuss pathogens (Chs. 19–25). These boxes ask students to apply material they have learned in the text to clinical scenarios, often featuring a micrograph or clinical photo that the student is asked to interpret.

CHAPTER 5 Microbial Metabolism   **151**

Subsequent reactions further catabolize the glycerol and fatty acid molecules. Glycerol is converted to DHAP, which as one of the substrates of glycolysis is oxidized to pyruvic acid **(see Figure 5.14)**. The fatty acids are degraded in a catabolic process known as **beta-oxidation (Figure 5.24b).**[9] In this process, enzymes repeatedly split off pairs of the hydrogenated carbon atoms that make up a fatty acid and join each pair to coenzyme A to form acetyl-CoA, until the entire fatty acid has been converted to molecules of acetyl-CoA. NADH and FADH₂ are generated during beta-oxidation, and more of these molecules are generated when the acetyl-CoA is utilized in the Krebs cycle to generate ATP **(see Figure 5.19)**. As is the case for the Krebs cycle, the enzymes involved in beta-oxidation are located in the cytoplasm of prokaryotes, and in the mitochondria of eukaryotes.

CRITICAL THINKING

How and where do cells use the molecules of NADH and FADH₂ produced during beta-oxidation?

Protein Catabolism

Learning Objective

✓ Explain how proteins are catabolized for energy and metabolite production.

Some microorganisms, particularly food-spoilage bacteria, pathogenic bacteria, and fungi, normally catabolize proteins as an important source of energy and metabolites. Most other cells catabolize proteins and their constituent amino acids only when carbon sources such as glucose and fat are not available.

Generally, proteins are too large to cross cell membranes, so microorganisms conduct the first step in the process of protein catabolism outside the cell by secreting **proteases** (pro´tē-ās-ez)—enzymes that split proteins into their constituent amino acids (Figure 5.25 on page 152). Once released by the action of proteases, amino acids are transported into the cell, where special enzymes split off amino groups in a reaction called **deamination**. The resulting altered molecules enter the Krebs cycle, and the amino groups are either recycled to synthesize other amino acids or excreted as nitrogenous wastes such as ammonia (NH₃), ammonium ion (NH₄⁺), or trimethylamine oxide (which has the chemical formula (CH₃)₃NO and is abbreviated TMAO). As **Highlight 5.2** on page 152 describes, TMAO plays an interesting role in the production of the odor we describe as "fishy."

Thus far, we have examined the catabolism of carbohydrates, lipids, and proteins. Now we turn our attention to the *synthesis* of these molecules, beginning with the anabolic reactions of photosynthesis.

[9] "Beta" is part of the name of this process because enzymes break the bond at the second carbon atom from the end of a fatty acid, and beta is the second letter in the Greek alphabet.

◀ *Figure 10*
Critical thinking questions are integrated throughout the text, as well as at the end of each chapter.

Learning Objectives throughout each chapter keep the student focused.

Integrated pronunciation guides help students pronounce key terms.

Etymology guides help students retain word definitions and enhance comprehension.

CHAPTER SUMMARY

The Early Years of Microbiology (pp. 2–7)

1. Leeuwenhoek's observations of **microbes** introduced **microorganisms** to the world and earned him the title "Father of Bacteriology and Protozoology." His discoveries were named and classified by Linnaeus in his **taxonomic** system.

2. Relatively large microscopic eukaryotic **fungi** include **molds** and **yeasts**.

3. Animal-like **protozoa** are single-celled eukaryotes. Some cause disease.

4. Plant-like eukaryotic **algae** are important providers of oxygen, serve as food for many marine animals, and make chemicals used in microbiological growth media.

5. Small **prokaryotic bacteria** and **archaea** live in a variety of communities and in most habitats. Even though some cause

4. Koch initiated careful microbiological laboratory techniques in his search for disease agents. **Koch's postulates**, the logical steps he followed to prove the cause of an infectious disease, remain an important part of microbiology today.

5. The procedure for the **Gram stain** was developed in the 1870s and is still used to differentiate bacteria into two categories: Gram-positive and Gram-negative.

6. The investigations of Semmelweis, Lister, Nightingale, and Snow are the foundations upon which **infection control** and **epidemiology** are built.

7. Jenner's use of a cowpox-based vaccine for preventing smallpox began the field of **immunology**. Pasteur significantly advanced the field.

8. Ehrlich's search for "magic bullets"—chemicals that differentially kill microorganisms—laid the foundations for the field

QUESTIONS FOR REVIEW   *(Answers to multiple choice, fill in the blanks, and matching questions are on the web, along with additional review questions. Visit microbiologyplace.com.)*

Multiple Choice

1. Which of the following microorganisms are not eukaryotic?
   a. bacteria
   b. yeasts
   c. molds
   d. protozoa

2. Which microorganisms are used to make microbiological growth media?
   a. bacteria
   b. fungi
   c. algae
   d. protozoa

3. In which habitat would you most likely find archaea?
   a. acidic hot springs
   b. swamp mud
   c. Great Salt Lake
   d. all of the above

4. Of the following scientists, who defended the theory of abiogenesis?
   a. Aristotle

Matching Questions

Match each of the following descriptions with the person it best describes. An answer may be used more than once.

1. ___ Developed smallpox immunization   A. Alexander Fleming
2. ___ First photomicrograph of bacteria   B. Paul Ehrlich
3. ___ Germ theory of disease   C. Louis Pasteur
4. ___ Germs cause disease   D. Antoni van Leeuwenhoek
5. ___ Sought a "magic bullet" to destroy pathogens   E. Carolus Linnaeus
6. ___ Discovered penicillin   F. Turberville Needham
7. ___ Father of Microbiology   G. Eduard Buchner
8. ___ Father of Microbiological Laboratory Techniques   H. Robert Koch
9. ___ Father of Bacteriology   I. Joseph Lister
10. ___ Father of Protozoology   J. Edward Jenner
11. ___ Founder of antiseptic   K. Girolamo Fracastoro

▲ *Figure 11*
End-of-chapter summaries and **review questions** help students master the material.

Case Study 22.2   Is it Athlete's Foot?

A 30-year-old man comes into a community clinic complaining of persistent redness, itching, and peeling skin on the soles of his feet and between his toes. He states that he works out daily and plays on several sports teams. He concluded he has athlete's foot, but every over-the-counter antifungal agent he has tried has failed to cure his condition. He has finally come in because his toenails have begun to thicken, yellow, and detach from the nail bed. His condition is pictured.

1. Given the information here, can athlete's foot be diagnosed definitively? If not, what laboratory tests should be performed to identify the infecting organism?

2. What treatment is likely? Would the recommended treatment be different if two fungal species, and not just one fungal species, were present?

◀ *Figure 12*
Case Studies ask students to apply their knowledge to clinical scenarios.

A Powerful Media and Supplements Package

This text is accompanied by **The Microbiology Place CD-ROM/Website (Figure 13).** The Microbiology Place (www.microbiologyplace.com) hosts the MicroPrep Pre-tests, along with over 40 practice test questions for each chapter of the text, review questions that utilize micrographs and other art from the text, weblinks, electronic flashcards, answers to questions in the text, information about careers in microbiology, and more. It also contains links to **Research Navigator,** a powerful online research tool with access to three exclusive databases of credible and reliable source material, including the *Journal of Applied Microbiology,* the *Annual Review of Microbiology, Science,* and the *New York Times* science pages. Available in web and/or CD format.

Other supplements available with this text include the following:

For the Instructor

Instructor's Resource CD-ROM 0-8053-7657-7
This cross-platform CD-ROM provides PowerPoint® lecture outlines, digital videos of microorganisms, and electronic files of the line art, photographs, and micrographs from the book (approximately 800 images, **Figure 14**). All images are identified by figure number and are available in both PowerPoint® and jpeg formats, enabling instructors to quickly and easily prepare powerful, visually engaging classroom presentations. In addition, several key illustrations are provided in a customizable, step-by-step format—allowing instructors maximum flexibility in deciding how to present complex topics within PowerPoint®.

Instructor's Manual and Test Bank 0-8053-7658-5
The Instructor's Manual contains expanded chapter summaries, a visual guide to the art, tips for teaching the course, a guide to The Microbiology Place Website, and more. The Test Bank contains approximately 1800 multiple-choice, matching, true/false, fill-in-the-blank, and essay questions, with their answers.

Transparency Acetates 0-8053-7661-5
Features approximately 400 key illustrations from the text.

Computerized Test Bank CD-ROM 0-8053-7659-3
This easy-to-use, cross-platform testing program enables instructors to view and edit electronic questions from the Test Bank, create multiple tests, and print them in a variety of formats.

Biological Agents of Disease/HIV Slide Set for Microbiology 0-8053-7671-2
An additional 100 clinical photos of microbes and pathologies, all from sources other than the text.

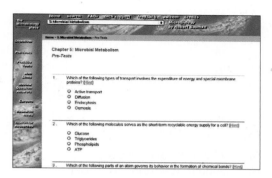

▲ *Figure 13*
The Microbiology Place website/CD-ROM includes MicroPrep Pre-Tests to aid student preparation, along with practice quizzes, web links, flashcards, and more.

▲ *Figure 14*
The Instructor's Resource CD-ROM includes lecture outlines, digital videos of microorganisms, and approximately 800 illustrations and photos from the Bauman text.

CourseCompass™

CourseCompass™ is a nationally hosted online course management system offering preloaded book-specific content, including testing and assessment, weblinks, illustrations, and photos. To view a demonstration, visit www.coursecompass.com.

For the Student

Study Guide 0-8053-7662-3
by Elizabeth Machunis-Masuoka
Students can master key concepts and earn a better grade with the help of the valuable tools found in this study guide. The study guide includes expanded chapter summaries, a guide to The Microbiology Place website, critical thinking questions, and a variety of self-test questions with answers.

Microbiology: A Photographic Atlas for the Laboratory
0-8053-2732-0
by Steve K. Alexander and Dennis Strete
A full-color atlas of approximately 400 high-quality color photographs that demonstrate the results of laboratory procedures and the morphology of important microorganisms.

Laboratory Experiments in Microbiology, Seventh Edition
0-8053-7673-9
by Ted R. Johnson and Christine L. Case
This lab manual includes 57 experiments that demonstrate basic microbiological techniques and applications.

Microbiology Coloring Book 0-06-041925-3
by I. Edward Alcamo and Lawrence M. Elson
This microbiology coloring atlas asks the reader to color a series of figures that convey microbiological principles and processes. An efficient review of all areas pertinent to a microbiology course, it simplifies the learning process and provides visually appealing figures that can be used for future study. The act of coloring helps students remember the structure and processes of microbiology, and how they relate to one another.

Microbiology: A Laboratory Manual, Sixth Edition
0-8053-7648-8
by James Cappuccino and Natalie Sherman
This versatile laboratory manual can be used with any undergraduate microbiology text and course. Known for its brief laboratory activities, minimal equipment requirements, and competitive price, the manual includes a variety of experiments selected to assist in the teaching of basic principles and techniques. Each of the 77 experiments includes an overview, a purpose, an in-depth discussion of the principle involved, easy-to-follow procedures, and lab reports with review and critical thinking questions. Comprehensive introductory material and laboratory safety instructions are provided.

The Microbe Files: Cases in Microbiology for the Undergraduate
by Marjorie K. Cowan
The first of its kind, *The Microbe Files* provides allied health and non-major microbiology students a fascinating series of short cases that help them apply what they have learned in the course. It's available in two versions:
(with answers) 0-8053-4927-8
(without answers) 0-8053-4928-6

Reviewers, Class Testers, and Focus Group Participants

I wish to thank the hundreds of instructors and students who reviewed and/or class-tested early drafts of this textbook. I also wish to thank the faculty and students who participated in focus groups during this book's development. Your comments have informed this book from beginning to end, and I am deeply grateful.

Reviewers and/or Class Testers

Ghava Ahmed
Seton Hall University

Craig Almeida
Stonehill College

Robin Almeida
University of North Carolina—Greensboro

John Amaral
University of San Francisco

Dawn Anderson
Berea College

Gail Baker
LaGuardia Community College

Susan Baxley
Troy State University—Montgomery

Keith Belcher
Austin Peay State University

Clinton Benjamin
Lower Columbia College

Margaret Beucher
Brigham Young University

Benjie Blair
Jacksonville State University

Pat Blaney
Brevard Community College—Cocoa

Susan Bornstein-Forst
Marian College

Kathryn Brooks
Michigan State University

Linda Bruslind
Oregon State University

Dave Bryan
Cincinnati State Technical Community College

Gregory W. Buck
Texas A&M University—Corpus Christi

George Bullerjahn
Bowling Green State University

Jackie Butler
Grayson County Community College

Christie Canady
Middle Georgia College

Jeff Cavanaugh
Troy State University—Montgomery

Elaine Charters
Xavier University

Kathy Clark
College of Southern Idaho

L. Clendaniel
Delaware Technical College—Georgetown Campus

Candice Coffin
Briar Cliff University

Don Collier
Calhoun Community College

Janet Cooper
Rockhurst University

Rhena Cooper
North Idaho College

Elaine Cox
Bossier Parish Community College

Lorraine Cramer
University of North Carolina—Chapel Hill

Victoria Crnekovic
Parkland College

Tucker Crum
Wellesley College

Don Dailey
Austin Peay State University

Roxanne Davenport
Tulsa Community College

Kelley Davis
University of Missouri, Columbia

Louis Debetaz
Angelina College

Paul DeLange
Kettering College of the Medical Arts

Paul Demchick
Barton College

Charles Denny
University of South Carolina—Sumter

Don Deters
Bowling Green State University

Beverly Dixon
California State University—Hayward

Diane Dixon
Southeast Oklahoma State University

Sandra Dubowsky
Allen County Community College

Khrys Duddleston
University of Alaska—Anchorage

Frances Duncan
Pensacola Jr. College

David Elmendorf
University of Central Oklahoma

Debby Filler
Anoka Ramsey Community College

John Friede
Villanova University

Denise Friedman
Hudson Valley Community College

David Fulford
Edinboro University of Pennsylvania

Karen Fulford
Morris Brown College

Fred Funk
Northern Arizona University

Rhonda Gamble
Mineral Area College

Beth Gaydos
San Jose City College

Aiah Gbakima
Morgan State University

David Gilmore
Arkansas State University

Michael Griffin
Angelo State University

Dale Hall
Southwestern Community College

Hershell Hanks
Collin County Community College

Ernie Hannig
University of Texas—Dallas

Ann Hanson
University of Maine

Elaine Hardwick
Minnesota State University—Mankato

Julie Harlass
Montgomery College

Donna Harman
Lubbock Christian University

Randall Harris
William Carey College

Sharon Harris
Columbia Basin Community College

Donna Hazelwood
Dakota State University

Alice Helm
University of Illinois—Urbana-Champaign

Daniel Herman
Grand Valley State University

Elizabeth Hoffman
Ashland Community College

Carolyn Holcroft
Foothill College

James Holda
University of Akron

Georgia Ineichen
Hinds Community College

Janice Ito
Leeward Community College

Bob Iwan
Anoka Ramsey Community College

Ruper Iyer
Wharton County Jr. College

Edythe Jones
Claflin University

Hinrich Kaiser
LaSierra University

Kevin Kaiser
Cape Fear Community College

Phyllis Katz
Endicott College

Karen Keller
Allegany College of Maryland

Karen Kendall-Fite
Columbia State Community College

Karen Kesterson
University of Nevada—Las Vegas
Harry Kestler
Lorain County Community College
Peggy Knittel
Eastern Wyoming College
Dubear Kroening
University of Wisconsin—Fox Valley
Karl Kvistberg
Nicolet College
Carol Lauzon
California State University—Hayward
Michael Lee
Joliet Jr. College
Jettie Lights
Columbus State University
Gayle LoPiccolo
Montgomery College
Candi Lostroh
Simmons College
Bonnie Lustigman
Montclair State University
Dawn Madl
Northeast Wisconsin Technical College
Reza Marvdashti
San Jacinto College, North
Luis Materon
University of Texas—Pan American
Kim Maznicki
Seminole Community College
Donald McGarey
Kennesaw State University
Elizabeth McPherson
University of Tennessee—Knoxville
Robert McReynolds
San Jacinto College, Central
Gina Marie Morris
Frank Phillips College
Quian Moss
Des Moines Community College
Rita Moyes
Texas A&M University
John Myers
Owens Community College
Linda Nazitski
Tunxis Community College
Mary Jane Niles
University of San Francisco
Catherine O'Brien
San Jacinto College, South
Karen Palin
Bates College
Robin Patterson
Butler Community College
Bobbie Pettriess
Wichita State University
Marcia Pierce
Eastern Kentucky University
Laraine Powers
East Tennessee State University
Davis Pritchett
University of Louisiana—Monroe
John Pritchett
Florence Darlington Technical College
Charlie Purvis
Jefferson Community College, Southwest

Kirsten Raines
San Jacinto College, South
Kimberly Raun
Wharton County Jr. College
Sabina Rech
San Jose State University
Thomas Reilly
Ball State University
Clifford Renk
Florida Gulf Coast University
Jean Revie
South Mountain Community College
Fayette Reynolds
Berkshire Community College
Tim Rhoads
South Georgia College
Kathleen Richardson
Portland Community College
Shelly Robertson
University of Wyoming
Jackie Rosby
Milwaukee Area Technical College
Ed Rowland
Ohio University
Michael Ruhl
Vermont Regional Jr. College
Barbara Rundell
College of DuPage
Jerry Sanders
University of Michigan—Flint
Pramila Sen
Houston Community College, Central Campus
Victoria Sharpe
Blinn College—Bryan Campus
Patty Shields
University of Maryland
Brian R. Shmaefsky
Kingwood College
Linda Simpson
University of North Carolina—Charlotte
Luci Nell Simpson
Weatherford College
Cheryl Smith
Pennsylvania College of Technology
Sally Smith
Okefenokee Technical College
Kris Snow
Fox Valley Technical College
Walter Spangenberg
Spokane Falls Community College
Robert Speed
Wallace Community College
Kathy Steinart
Bellevue Community College
Kathleen Steinert
Bellevue Community College
Gail Stewart
Camden County College
John Story
Northwest Arkansas Community College
Julie Sutherland
College of DuPage
Jeffrey Taylor
North Central State College
Monica Tischler
Benedictine University

Stephen Torosian
University of New Hampshire
James Urban
Kansas State University
Margaret Waldman
Allegany College of Maryland
Burton Webb
Indiana University School of Medicine
Larry Weiskirch
Onondaga Community College
S. Williams
Delaware Technical College—Georgetown Campus
Christina Wilson
Southern Illinois University, Edwardsville
T. Yee Wong
University of Memphis
Patricia Yokell
New Hampshire Technical Institute
Kathy Zarilla
Durham Technical Community College

Faculty Focus Group Participants

Bill Boyko
Sinclair Community College
Elizabeth Hoffman
Ashland Community College
Richard Karp
University of Cincinnati
Marcia Pierce
Eastern Kentucky University
Charlie Purvis
Jefferson Community College, Southwest
Galen Renwick
Indiana University, Southeast
Bill Staddon
Eastern Kentucky University

Student Focus Group Participants

Instructor: Pat Johnson
Palm Beach Community College
Sharon Cordero
Palm Beach Community College
Samantha Dill
Palm Beach Community College
Melanie Radd
Palm Beach Community College
Joshua Suttle
Palm Beach Community College
Instructor: Kathryn Brooks
Michigan State University
Traci Klomparens
Michigan State University
Katy Loyd
Michigan State University
Nicole Westover
Michigan State University
Instructor: Joy Voorhees
Capital University
Christina Tompkins
Capital University

Acknowledgements

Writing a textbook was a larger undertaking than I imagined when I began. Though my name is on the cover, this book has truly been a team effort. I am deeply grateful to Daryl Fox, Serina Beauparlant, and Barbara Yien of Benjamin Cummings, and to the team they gathered to make this book happen. Daryl's support for this book never wavered, and his willingness to take risks in support of innovation is much appreciated. Serina and Barbara helped develop the "vision" for this book, constantly coming up with ideas for making it more effective and compelling. As project editor, Barbara also had the Herculean task of coordinating everything and keeping us all on track. Thank you, Barbara—"Mack" and I will miss the daily phone calls. Laura Bonazzoli and Alan Titche edited the manuscript thoroughly and meticulously—credit for my successes is in large part due to them, while any errors or problems are mine alone. Laura Southworth developed many of this book's complex "process" figures and ensured that all the art in the text was styled consistently. The incomparable Ken Probst is responsible for creating this book's amazingly beautiful biological illustrations, including the cover image. Kay Ueno and Amy Teeling provided editorial guidance in the early stages of this book. Elizabeth Machunis-Masuoka, Randall Harris, Temma Al-Mukhtar, Michelle Bauman, Laura Bonazzoli, and Ian Tizard all made valuable writing contributions to the text and its supplements. Wendy Earl, Sharon Montooth, and Kelly Murphy pulled out all the stops to meet the book's aggressive production schedule. Diane Austin and Kathleen Olson did a superb job researching the photos for this text and Rich Robison supplied many of the text's wonderful and unique micrographs. tani hasegawa created the book's elegant design. Thanks also to Tracey Hines, Nan Kemp, Ziki Dekel and Ryan Shaw for their administrative and research assistance. Finally, Cheryl Cechvala, Lauren Harp, and the Benjamin Cummings book reps have done a terrific job of keeping in touch with the professors and students who provided so many wonderful suggestions for this textbook. I am especially grateful to Larry Weiskirch of Onondaga Community College and Mary Jane Niles of the University of San Francisco for their expertise and advice on the immunology chapters.

On the home front, I am grateful for Dr. Nichol Dolby and Art Schneider at Amarillo College; Chris Gilbert, former student, current racquetball partner, and dear friend; Tracey Hays, business consultant and partner in what really matters; Dr. Vance Esler, Larry Latham, Mike Isley, and Dr. Dustin Anderson, lunch buddies—all of them were always supportive and helpful. My "secretarial staff," Michelle, Jennie, Elizabeth, and Jeremy Bauman, were always there to type, file, surf the web, run to the FedEx box, or rub my feet and neck. Many men may be blessed, but I am blessed above all; you, my family, are the best. Finally, thanks to JC, who got me through it all.

Robert W. Bauman
Microbiology
Benjamin Cummings
1301 Sansome Street
San Francisco, CA 94111

BRIEF CONTENTS

1   A Brief History of Microbiology   1

2   The Chemistry of Microbiology   26

3   Cell Structure and Function   56

4   Microscopy, Staining, and Classification   92

5   Microbial Metabolism   125

6   Microbial Nutrition and Growth   167

7   Microbial Genetics   199

8   Recombinant DNA Technology   241

9   Controlling Microbial Growth in the Environment   262

10   Controlling Microbial Growth in the Body: Antimicrobial Drugs   287

11   Characterizing and Classifying Prokaryotes   313

12   Characterizing and Classifying Eukaryotes   342

13   Characterizing and Classifying Viruses, Viroids, and Prions   377

14   Infection, Infectious Diseases, and Epidemiology   404

15   Nonspecific Lines of Defense   437

16   Specific Defense: The Immune Response   462

17   Immunization and Immune Testing   487

18   Hypersensitivities, Autoimmune Diseases, and Immune Deficiencies   509

19   Pathogenic Gram-Positive Cocci and Bacilli   532

20   Pathogenic Gram-Negative Cocci and Bacilli   566

21   Mycoplasmas, Rickettsias, Chlamydias, Spirochetes, and Vibrios   595

22   Pathogenic Fungi   623

23   Parasitic Protozoa, Helminths, and Arthropod Vectors   649

24   Pathogenic DNA Viruses   679

25   Pathogenic RNA Viruses   703

26   Applied and Environmental Microbiology   743

Appendix A   Metabolic Pathways   A-1

Appendix B   Some Mathematical Considerations in Microbiology   B-1

Appendix C   Classification of Bacteria According to *Bergey's Manual*   C-1

Appendix D   Major Microbial Agents of Disease by Body System Affected   D-1

Glossary   G-1

Index   I-1

CONTENTS

CHAPTER 1
A Brief History of Microbiology   1

The Early Years of Microbiology   2
What Does Life Really Look Like?   2
How Can Microbes Be Classified?   3

The Golden Age of Microbiology   7
Is Spontaneous Generation of Microbial Life
    Possible?   7
What Causes Fermentation?   10
What Causes Disease?   13
How Can We Prevent Infection and Disease?   16

The Modern Age of Microbiology   19
What Are the Basic Chemical Reactions of Life?   19
How Do Genes Work?   20
What Roles Do Microorganisms Play in the
    Environment?   21
How Do We Defend Against Disease?   21
What Will the Future Hold?   23

Chapter Summary   23

Questions for Review   24

CHAPTER 2
The Chemistry of Microbiology   26

Atoms   27
Atomic Structure   27
Isotopes   27
Electron Configurations   28

Chemical Bonds   29
Nonpolar Covalent Bonds   30
Polar Covalent Bonds   32
Ionic Bonds   33
Hydrogen Bonds   34

Chemical Reactions   35
Synthesis Reactions   35

Decomposition Reactions   36
Exchange Reactions   36

Water, Acids, Bases, and Salts   37
Water   37
Acids and Bases   37
Salts   39

Organic Macromolecules   39
Functional Groups and Monomers   39
Lipids   40
Carbohydrates   42
Proteins   45
Nucleic Acids   50

Chapter Summary   53

Questions for Review   54

CHAPTER 3
Cell Structure and Function   56

Processes of Life   57

Eukaryotic and Prokaryotic Cells: An Overview   58

External Structures of Prokaryotic Cells   61
Glycocalyces   62
Flagella   62
Fimbriae and Pili   64

Prokaryotic Cell Walls   66
Bacterial Cell Walls   66
Archaeal Cell Walls   68

Prokaryotic Cytoplasmic Membranes   70
Structure   70
Function   71

Cytoplasm of Prokaryotes   74
Cytosol   75
Inclusions   75
Nonmembranous Organelles   75

External Structures of Eukaryotic Cells **76**

   Glycocalyces 77

   Flagella 77

   Cilia 77

Eukaryotic Cell Walls and Cytoplasmic Membranes **77**

Cytoplasm of Eukaryotes **80**

   Nonmembranous Organelles 80

   Membranous Organelles 81

   Endosymbiotic Theory 85

Chapter Summary **88**

Questions for Review **90**

CHAPTER 4
Microscopy, Staining, and Classification 92

Units of Measurement **93**

Microscopy **93**

   General Principles of Microscopy 94

   Light Microscopy 95

   Electron Microscopy 101

   Probe Microscopy 104

Staining **106**

   Preparing Specimens for Staining 106

   Principles of Staining 108

   Simple Stains 108

   Differential Stains 108

   Special Stains 112

   Staining for Electron Microscopy 112

Classification and Identification of Microorganisms **112**

Linnaeus, Whitaker, and Taxonomic Categories **113**

   Domains 117

   Taxonomic and Identifying Characteristics 118

   Taxonomic Keys 121

Chapter Summary **122**

Questions for Review **123**

CHAPTER 5
Microbial Metabolism 125

Basic Chemical Reactions Underlying Metabolism **126**

   Catabolism and Anabolism 126

   Oxidation and Reduction Reactions 127

   ATP Production and Energy Storage 127

   The Roles of Enzymes in Metabolism 128

Carbohydrate Catabolism **136**

   Glycolysis 139

   Alternatives to Glycolysis 139

   Cellular Respiration 141

   Fermentation 148

Other Catabolic Pathways **149**

   Lipid Catabolism 149

   Protein Catabolism 151

Photosynthesis **152**

   Chemicals and Structures 153

   Light-Dependent Reactions 154

   Light-Independent Reactions 156

Other Anabolic Pathways **156**

   Carbohydrate Biosynthesis 157

   Lipid Biosynthesis 158

   Amino Acid Biosynthesis 159

   Nucleotide Biosynthesis 160

Integration and Regulation of Metabolic Functions **160**

Chapter Summary **163**

Questions for Review **164**

CHAPTER 6
Microbial Nutrition and Growth 167

Growth Requirements **168**

   Nutrients: Chemical and Energy Requirements 168

   Physical Requirements 172

   Ecological Associations 177

Culturing Microorganisms **177**

   Clinical Sampling 178

   Obtaining Pure Cultures 178

   Culture Media 181

   Special Culture Techniques 185

   Preserving Cultures 186

Growth of Microbial Populations **186**

   Mathematical Considerations in Population Growth 187

   Generation Time 188

   Phases of Microbial Growth 188

   Measuring Microbial Growth 189

Chapter Summary **195**

Questions for Review **197**

CHAPTER 7
Microbial Genetics   199

The Structure and Replication of Genomes   200
The Structure of Nucleic Acids   200
The Structure of Prokaryotic Genomes   200
The Structure of Eukaryotic Genomes   202
DNA Replication   204

Gene Function   208
The Relationship Between Genotype and Phenotype   208
The Transfer of Genetic Information   208
The Events in Transcription   208
Translation   212
Control of Transcription   218

Mutations of Genes   221
Types of Mutations   221
Effects of Mutations   222
Mutagens   223
Frequency of Mutation   224
DNA Repair   224
Identifying Mutants, Mutagens, and Carcinogens   226

Genetic Recombination and Transfer   229
Horizontal Gene Transfer Among Prokaryotes   229
Transposons and Transposition   232

Chapter Summary   236

Questions for Review   237

CHAPTER 8
Recombinant DNA Technology   241

The Role of Recombinant DNA Technology in Biotechnology   242

The Tools of Recombinant DNA Technology   242
Mutagens   242
The Use of Reverse Transcriptase to Synthesize cDNA   243
Synthetic Nucleic Acids   243
Restriction Enzymes   244
Vectors   245
Gene libraries   246

Techniques of Recombinant DNA Technology   247
Multiplying DNA *in vitro:* The Polymerase Chain Reaction (PCR)   247
Selecting a Clone of Recombinant Cells   248

Separating DNA Molecules: Gel Electrophoresis and the Southern Blot   248
Inserting DNA into Cells   251

Applications of Recombinant DNA Technology   252
Genome Mapping   252
Pharmaceutical and Therapeutic Applications   254
Agricultural Applications   257

The Ethics and Safety of Recombinant DNA Technology   259

Chapter Summary   259

Questions for Review   260

CHAPTER 9
Controlling Microbial Growth in the Environment   262

Basic Principles of Microbial Control   263
Terminology of Microbial Control   263
Microbial Death Rates   264
Action of Antimicrobial Agents   264

The Selection of Microbial Control Methods   265
Factors Affecting the Efficacy of Antimicrobial Methods   265
Methods for Evaluating Disinfectants and Antiseptics   267

Physical Methods of Microbial Control   268
Heat-Related Methods   268
Refrigeration and Freezing   272
Desiccation and Lyophilization   272
Filtration   274
Osmotic Pressure   274
Radiation   275

Chemical Methods of Microbial Control   277
Phenol and Phenolics   278
Alcohols   278
Halogens   279
Oxidizing Agents   279
Surfactants   280
Heavy Metals   280
Aldehydes   281
Gaseous Agents   281
Antimicrobics   282
Development of Resistant Microbes   283

Chapter Summary   283

Questions for Review   285

CHAPTER 10
Controlling Microbial Growth in the Body: Antimicrobial Drugs   287

The History of Antimicrobial Agents   288

Mechanisms of Antimicrobial Action   289
Inhibition of Cell Wall Synthesis   290
Inhibition of Protein Synthesis   292
Disruption of Cytoplasmic Membranes   292
Inhibition of Metabolic Pathways   294
Inhibition of Nucleic Acid Synthesis   295
Prevention of Attachment   295

Clinical Considerations in Prescribing Antimicrobial Drugs   296
Spectrum of Action   297
Efficacy   297
Routes of Administration   300
Safety and Side Effects   300

Resistance to Antimicrobial Drugs   301
The Development of Resistance in Populations   301
Mechanisms of Resistance   302
Multiple Resistance and Cross Resistance   303
Retarding Resistance   304

Chapter Summary   310

Questions for Review   311

CHAPTER 11
Characterizing and Classifying Prokaryotes   313

General Characteristics of Prokaryotic Organisms   314
Morphology of Prokaryotic Cells   314
Reproduction of Prokaryotic Cells   314
Arrangements of Prokaryotic Cells   315
Endospores   316

Modern Prokaryotic Classification   318

Survey of Archaea   319
Extremophiles   320
Methanogens   322

Survey of Bacteria   323
Deeply Branching and Phototrophic Bacteria   323
Low G + C Gram-Positive Bacteria   325
High G + C Gram-Positive Bacteria   328
Gram-Negative Proteobacteria   330
Other Gram-Negative Bacteria   337

Chapter Summary   339

Questions for Review   340

CHAPTER 12
Charaterizing and Classifying Eukaryotes   342

General Characteristics of Eukaryotic Organisms   343
Reproduction in Eukaryotes   343
The Classification of Eukaryotic Organisms   347

Protozoa   348
Distribution of Protozoa   348
Morphology of Protozoa   348
Nutrition of Protozoa   349
Reproduction in Protozoa   349
Classification of Protozoa   350

Fungi   355
The Significance of Fungi   355
Morphology of Fungi   355
Nutrition of Fungi   356
Reproduction in Fungi   357
Classification of Fungi   359
Lichens   363

Algae   366
Distribution of Algae   366
Morphology of Algae   366
Reproduction in Algae   367
Classification of Algae   367

Water Molds and Slime Molds   369
Water Molds   370
Slime Molds   370

Other Eukaryotes of Microbiological Interest: Parasitic Helminths and Vectors   371

Chapter Summary   373

Questions for Review   374

CHAPTER 13
Characterizing and Classifying Viruses, Viroids, and Prions   377

Characteristics of Viruses   378
Genetic Material of Viruses   379
Hosts of Viruses   379
Sizes of Viruses   380

Capsid Morphology   380

Viral Shapes   382

The Viral Envelope   382

Classification of Viruses   384

Viral Replication   384

Lytic Replication of Bacteriophages   385

Lysogeny   388

Replication of Animal Viruses   390

The Role of Viruses in Cancer   394

Culturing Viruses in the Laboratory   395

Culturing Viruses in Whole Organisms   396

Culturing Viruses in Embryonated Chicken Eggs   396

Culturing Viruses in Cell (Tissue) Culture   396

Other Parasitic Particles: Viroids and Prions   397

Characteristics of Viroids   397

Characteristics of Prions   398

Are Viruses Alive?   400

Chapter Summary   400

Questions for Review   402

CHAPTER 14

Infection, Infectious Diseases, and
Epidemiology   404

**Symbiotic Relationships Between Microbes and Their
Hosts   405**

Types of Symbiosis   405

Normal Microbiota in Hosts   406

How Normal Microbiota Become Opportunistic
Pathogens   407

The Movement of Microbes into Hosts: Infection   409

Exposure to Microbes: Contamination and
Infection   409

Portals of Entry   409

The Role of Adhesion in Infection   410

The Nature of Infectious Disease   411

Manifestations of Disease: Symptoms, Signs, and
Syndromes   412

Causation of Disease: Etiology   412

Virulence Factors of Infectious Agents   415

The Stages of Infectious Diseases   418

**The Movement of Pathogens Out of Hosts: Portals
of Exit   419**

Sources of Infectious Diseases in Humans   419

Animal Reservoirs   420

Human Carriers   420

Nonliving Reservoirs   420

Modes of Infectious Disease Transmission   421

Contact Transmission   421

Vehicle Transmission   422

Vector Transmission   423

Classification of Infectious Diseases   423

Epidemiology of Infectious Diseases   424

Frequency of Disease   424

Epidemiological Studies   426

Hospital Epidemiology: Nosocomial Infections   430

Epidemiology and Public Health   432

Chapter Summary   433

Questions for Review   434

CHAPTER 15

Nonspecific Lines of Defense   437

An Overview of the Body's Defenses   438

The Body's First Line of Defense   438

The Role of Skin in Nonspecific Defense   438

The Role of Mucous Membranes in Nonspecific
Defense   439

The Role of Normal Microbiota in Nonspecific
Defense   440

Other First-Line Defenses   441

An Overview of the Body's Second Line of Defense   442

Defense Components of Blood   443

The Body's Second Line of Defense   448

Phagocytosis   448

Extracellular Killing by Leukocytes   450

Nonspecific Chemical Defenses Against Pathogens   450

Inflammation   455

Fever   457

Chapter Summary   459

Questions for Review   460

CHAPTER 16

Specific Defense: The Immune Response   462

Elements of Specific Immunity   463

Antigens   463

The Cells, Tissues, and Organs of the Lymphatic System   465

B Lymphocytes (B Cells) and Antibodies   468

T Lymphocytes (T Cells)   471

Immune System Cytokines   473

The Body's Preparation for a Specific Immune Response   473

Lymphocyte Editing by Clonal Deletion   473

The Roles of the Major Histocompatibility Complex   474

Antigen Processing   475

The Humoral Immune Response   477

B Cell Activation and Clonal Selection   477

Memory B Cells and the Establishment of Immunological Memory   477

The Cell-Mediated Immune Response   480

The Perforin-Granzyme Cytotoxic Pathway   480

The CD95 Cytotoxic Pathway   481

Memory T Cells   481

T Cell Regulation   481

Types of Acquired Immunity   482

Naturally Acquired Active Immunity   482

Naturally Acquired Passive Immunity   483

Artificially Acquired Active Immunity   483

Artificially Acquired Passive Immunity   483

Chapter Summary   483

Questions for Review   485

CHAPTER 17

Immunization and Immune Testing   487

Immunization   488

Brief History of Immunization   488

Active Immunization   489

Passive Immunization   494

Immune Testing   496

Precipitation Tests   496

Agglutination Tests   499

Neutralization Tests   500

The Complement Fixation Test   501

Labeled Antibody Tests   502

Recent Developments in Immune Testing   505

Chapter Summary   506

Questions for Review   507

CHAPTER 18

Hypersensitivities, Autoimmune Diseases, and Immune Deficiencies   509

Hypersensitivities   510

Type I (Immediate) Hypersensitivity   510

Type II (Cytotoxic) Hypersensitivity   515

Type III (Immune-Complex Mediated) Hypersensitivity   518

Type IV (Delayed or Cell-Mediated) Hypersensitivity   520

Autoimmune Diseases   523

Causes of Autoimmune Diseases   523

Single Tissue Autoimmune Diseases   524

Systemic Autoimmune Diseases   525

Immunodeficiency Diseases   526

Primary Immunodeficiency Diseases   527

Acquired Immunodeficiency Diseases   528

Chapter Summary   528

Questions for Review   530

CHAPTER 19

Pathogenic Gram-Positive Cocci and Bacilli   532

Staphylococcus   533

Structure and Physiology   533

Pathogenicity   534

Epidemiology   534

Staphylococcal Diseases   535

Diagnosis, Treatment, and Prevention   537

Streptococcus   537

Group A Streptococcus: *Streptococcus pyogenes*   538

Group B Streptococcus: *Streptococcus agalactiae*   541

Other Beta-Hemolytic Streptococci   542

Alpha-Hemolytic Streptococci: The Viridans Group   542

Streptococcus pneumoniae   542

Enterococcus **544**
  Structure and Physiology 544
  Pathogenesis, Epidemiology, and Diseases 544
  Diagnosis, Treatment, and Prevention 544

Bacillus **545**
  Structure, Physiology, and Pathogenicity 545
  Epidemiology 545
  Disease 545
  Diagnosis, Treatment, and Prevention 546

Clostridium **547**
  Clostridium perfringens 547
  Clostridium difficile 548
  Clostridium botulinum 548
  Clostridium tetani 550

Listeria **551**
  Pathogenesis, Epidemiology, and Disease 551
  Diagnosis, Treatment, and Prevention 553

Corynebacterium **554**
  Pathogenesis, Epidemiology, and Disease 554
  Diagnosis, Treatment, and Prevention 554

Mycobacterium **555**
  Tuberculosis 555
  Leprosy 559
  Other Mycobacterial Infections in AIDS Patients 560

Propionibacterium **560**

Nocardia and Actinomyces **562**
  Nocardia asteroides 562
  Actinomyces 562

Chapter Summary 563

Questions for Review 564

CHAPTER 20
Pathogenic Gram-Negative Cocci and Bacilli 566

Pathogenic Gram-Negative Cocci: *Neisseria* 567
  Structure and Physiology of *Neisseria* 567
  The Gonococcus: *Neisseria gonorrhoeae* 568
  The Meningococcus: *Neisseria meningitidis* 569

Pathogenic, Gram-Negative, Facultatively Anaerobic Bacilli 570
  The Enterobacteriaceae: An Overview 571
  Coliform Opportunistic Enterobacteriaceae 574

  Noncoliform Opportunistic Enterobacteriaceae 576
  Truly Pathogenic Enterobacteriaceae 577
  The Pasteurellaceae 582

Pathogenic, Gram-Negative, Aerobic Bacilli 583
  Bartonella 583
  Brucella 584
  Bordetella 585
  Pseudomonads 587
  Francisella 588
  Legionella 589
  Coxiella 590

Pathogenic, Gram-Negative, Anaerobic Bacilli 591
  Bacteroides 591
  Prevotella 591

Chapter Summary 592

Questions for Review 593

CHAPTER 21
Mycoplasmas, Rickettsias, Chlamydias, Spirochetes, and Vibrios 595

Mycoplasmas 596
  Mycoplasma pneumoniae 597
  Other Mycoplasmas 597

Rickettsias 598
  Rickettsia 598
  Orienta 600
  Ehrlichia 600

Chlamydias 602
  Chlamydia trachomatis 602
  Chlamydia pneumoniae 605
  Chlamydia psittaci 605

Spirochetes 607
  Treponema 607
  Borrelia 610
  Leptospira 613

Pathogenic Gram-Negative Curved Bacilli 614
  Vibrio 614
  Campylobacter jejuni 616
  Helicobacter pylori 617

Chapter Summary 618

Questions for Review 620

CHAPTER 22
Pathogenic Fungi   623

An Overview of Medical Mycology   624
The Epidemiology of Mycoses   624
Categories of Fungal Agents: True Fungal Pathogens and Opportunistic Fungi   624
Clinical Manifestations of Fungal Diseases   626
The Diagnosis of Fungal Infections   627
Antifungal Therapies   627

Systemic Mycoses Caused by Pathogenic Fungi   628
Blastomycosis   628
Coccidioidomycosis   629
Histoplasmosis   630
Paracoccidioidomycosis   631

Systemic Mycoses Caused by Opportunistic Fungi   631
Aspergillosis   632
Candidiasis   633
Cryptococcosis   635
Pneumocystis Pneumonia   636
Zygomycoses   636
The Emergence of Fungal Opportunists in Immunosuppressed Individuals   637

Superficial, Cutaneous, and Subcutaneous Mycoses   637
Superficial Mycoses   638
Cutaneous and Subcutaneous Mycoses   640

Fungal Intoxications and Allergies   643
Mycotoxicoses   643
Mycetismus: Mushroom Poisoning   643
Allergies to Fungi   644

Chapter Summary   645
Questions for Review   646

CHAPTER 23
Parasitic Protozoa, Helminths, and Arthropod Vectors   649

Protozoan Parasites of Humans   650
Ciliates   650
Amoebae   650
Flagellates   653
Apicomplexans   658

Helminthic Parasites of Humans   663
Cestodes   664

Trematodes   666
Nematodes   669

Arthropod Vectors   672
Arachnids   673
Insects   674

Chapter Summary   675
Questions for Review   677

CHAPTER 24
Pathogenic DNA Viruses   679

Poxviridae   680
Smallpox   680
Molluscum Contagiosum   682
Other Poxvirus Infections   683

***Herpesviridae*   683**
Herpes Simplex Infections   684
Varicella-Zoster Virus Infections   687
Epstein-Barr Virus Infections   689
Cytomegalovirus Infections   691
Other Herpesvirus Infections   692

***Papillomaviridae* and *Polyomaviridae*   693**
Papillomavirus Infections   693
Polyomavirus Infections   695

***Adenoviridae*   696**

***Hepadnaviridae*   696**
Hepatitis B Infections   697
The Role of Hepatitis B Virus in Hepatic Cancer   699

***Parvoviridae*   699**

Chapter Summary   700
Questions for Review   701

CHAPTER 25
Pathogenic RNA Viruses   703

Naked, Positive ssRNA Viruses: *Picornaviridae, Caliciviridae, Astroviridae,* and *Hepeviridae*   704
Common Colds Caused by Rhinoviruses   704
Diseases of Enteroviruses   705
Acute Gastroenteritis and Hepatitis E   709

Enveloped, Positive ssRNA Viruses: *Togaviridae, Flaviviridae,* and *Coronaviridae*   709
Diseases of +RNA Arboviruses   710

Rubella   713

Hepatitis C   714

Diseases of Coronaviruses   715

Enveloped, Positive ssRNA Viruses with Reverse Transcriptase: *Retroviridae*   **716**

Oncogenic Retroviruses   717

Acquired Immunodeficiency Syndrome   718

Enveloped, Unsegmented, Negative ssRNA Viruses: *Paramyxoviridae, Rhabdoviridae,* **and** *Filoviridae*   **724**

Measles   725

Diseases of Parainfluenza Virus   727

Mumps   727

Disease of Respiratory Syncytial Virus   728

Rabies   728

Hemorrhagic Fevers   730

Enveloped, Segmented, Negative ssRNA Viruses: *Orthomyxoviridae, Bunyaviridae,* **and** *Arenaviridae*   **731**

Influenza   731

Diseases of Bunyaviruses   735

Diseases of Arenaviruses   736

Naked, Segmented dsRNA Viruses: *Reoviridae*   **737**

Rotaviruses   737

Coltiviruses   737

Chapter Summary   739

Questions for Review   740

CHAPTER 26
Applied and Environmental Microbiology   743

Food Microbiology   744

The Roles of Microorganisms in Food Production   744

The Causes and Prevention of Food Spoilage   748

Food-Borne Illnesses   751

Industrial Microbiology   752

The Roles of Microbes in Industrial Fermentations   752

Industrial Products of Microorganisms   753

Water and Wastewater Treatment   756

Bioremediation   761

The Problem of Acid Mine Drainage   762

Environmental Microbiology   763

Microbial Ecology   763

The Roles of Microorganisms in Biogeochemical Cycles   765

Soil Microbiology   768

Aquatic Microbiology   770

Biological Warfare and Bioterrorism   772

Assessing Microorganisms as Potential Agents of Warfare or Terror   772

Known Microbial Threats   774

Defense Against Bioterrorism   775

The Roles of Recombinant Genetic Technology in Bioterrorism   775

Chapter Summary   776

Questions for Review   778

Appendix A   Metabolic Pathways   A-1

Appendix B   Some Mathematical Considerations in Microbiology   B-1

Appendix C   Classification of Bacteria According to *Bergey's Manual*   C-1

Appendix D   Major Microbial Agents of Disease by Body System Affected   D-1

Glossary   G-1

Index   I-1

BOXES

New Frontiers

1.1 Bioterrorism 22
3.1 Understanding Biofilms 67
4.1 "Smart" Bandages of the Future 110
5.1 Gold-Mining Microbes 128
6.1 Is There Life on Other Planets? 169
6.2 Biofilms, Quorum Sensing, and Disease Prevention 177
7.1 Controlling Disease by Controlling Gene Expression 221
8.1 Building a Better Mosquito 247
8.2 Using the Human Genome to Detect Disease-Causing Microbes 254
8.3 Vaccines on the Menu 255
10.1 The AIDS "Cocktail" and the Problem of Drug Resistance 303
11.1 *Lactobacillus:* A Probiotic to Watch 328
12.1 Amoeba Midwives 353

13.1 High-Pressure Viruses 387
16.1 Attacking Cancer with Lab-Grown T Cells 482
17.1 Listening for Pathogens 506
18.1 Can Pets Help Decrease Children's Allergy Risks? 513
18.2 When Kissing Triggers Allergic Reactions 515
21.1 The "Germ Theory" of Heart Disease 605
23.1 In Search of a Malaria Vaccine 661
24.1 The Importance of Adenoviruses in Research 698
25.1 West Nile Virus and Severe Acute Respiratory Syndrome (SARS) 715
25.2 Nipah Virus: From Pigs to Humans 726
26.1 Making Blue Jeans "Green" 754
26.2 Bacterial Batteries 755
26.3 Could Bioterrorists Manufacture Viruses from Scratch? 776

Highlights

2.1 Raw Oysters and Antacids: a Deadly Mix? 39
2.2 Why Can't We Digest Wood? 47
3.1 Are Viruses Alive? 58
3.2 Unusual Giants 59
4.1 "One of the Most Important Inventions of the Century" 102
4.2 Woese's Taxonomic Revolution 117
5.1 Glowing Bacteria 146
5.2 What's that Fishy Smell? 152
6.1 Some Like it Hot: Extreme Thermophiles 175
7.1 Understanding the Central Dogma: It's (No Longer) Greek to Me 209
8.1 Does Genetically Modified Corn Harm Monarch Butterflies? 258
9.1 Odwalla's Road to Pasteurization 272

9.2 Sushi: A Case Study in Microbial Control 273
9.3 Should You Use Antimicrobial Soaps? 283
10.1 Why Do Microbes Make Antibiotics? 289
10.2 Ciprofloxacin and Anthrax 297
11.1 Botulinum Toxin: A Killer and a Healer 326
12.1 Truffles: The Gourmet Fungi 363
13.1 Prescription Bacteriophages? 381
13.2 Kuru and Cannibalism in the Fore Tribe of Papua New Guinea 400
14.1 A Tale of Two Organisms 406
15.1 What Happens to All That Skin? 440
15.2 The Discovery of Phagocytes 449
15.3 The Fight for Iron 450
16.1 The Loss of Helper T Cells in AIDS Patients 472

17.1 Why Isn't There a Cold Vaccine?   490

17.2 Smallpox: To Vaccinate or Not to Vaccinate?   494

18.1 SCID: "Bubble Boy" Disease   527

19.1 An Outbreak of Group A Streptococci in the Marine Corps   539

20.1 The Tale of "Typhoid Mary"   579

22.1 Does "Killer Mold" Exist?   645

24.1 The "Gift" of Smallpox: Bioterrorism During the French and Indian War   683

24.2 Reye's Syndrome: Can Aspirin Harm Children with Viral Diseases?   691

25.1 Rabies: A Threat from the Wild   730

26.1 Oil-Eating Bacteria to the Rescue   763

Case Studies

14.1 A Deadly Carrier   420

14.2 Unusual Transmission of West Nile Virus   423

14.3 *Legionella* in the Produce Aisle   431

15.1 Evaluating a Complete Blood and Differential Leukocyte Count (CBC) Profile   447

19.1 A Fatal Case of Methicillin-Resistant *S. aureus*   538

19.2 Anthrax in the Mail: The Case of Tom Brokaw's Assistant   547

19.3 The Man with No Feelings   560

20.1 *Neisseria* Nightmare   573

20.2 When "Health Food" Isn't Healthy   584

21.1 Blame It on the Chiggers?   601

21.2 Campylobacter and the Lactovegetarians   617

22.1 What's Ailing the Bird Enthusiast?   631

22.2 Is it Athlete's Foot?   641

23.1 A Protozoan Mystery   660

23.2 A Fluke Disease?   669

24.1 Grandfather's Shingles   689

24.2 A Child with Warts   695

25.1 A Case of AIDS   725

CHAPTER 1

A Brief History of Microbiology

Life as we know it would not exist without microorganisms. Plants depend on microorganisms to help them obtain the nitrogen they need for survival. Animals such as cows and sheep need microorganisms in order to digest the cellulose in their plant-based diets. Our ecosystems rely on microorganisms to enrich soil, degrade wastes, and support life. We use microorganisms to make wine and cheese and to develop vaccines and antibiotics. The human body is home to billions of microorganisms, many of which help keep us healthy. Microorganisms are not only an essential part of our lives, they are quite literally a part of us.

Of course, some microorganisms do cause disease, from the common cold to AIDS. The threats of bioterrorism and new or re-emerging infectious diseases are real. This textbook explores all the roles—both harmful and beneficial—that microorganisms play in our lives, as well as their sophisticated structures and processes. We begin with a look at the history of microbiology, starting with the invention of crude microscopes that revealed, for the first time, the existence of this miraculous, miniature world.

Pond water microorganisms.

CHAPTER OUTLINE

The Early Years of Microbiology
What Does Life Really Look Like?
How Can Microbes Be Classified?

The Golden Age of Microbiology
Is Spontaneous Generation Possible?
What Causes Fermentation?
What Causes Disease?
How Can We Prevent Infection and Disease?

The Modern Age of Microbiology
What Are the Basic Chemical Reactions of Life?
How Do Genes Work?
What Roles Do Microorganisms Play in the Environment?
How Do We Defend Against Disease?
What Will the Future Hold?

MicroPrep Pre-Test: Take the pre-test for this chapter on the web. Visit www.microbiologyplace.com.

Science is the study of nature that proceeds by posing questions about observations. Why are there seasons? What is the function of the nodules at the base of this plant? Why does this bread taste sour? What does plaque from between teeth look like when magnified? Why are so many crows dying this winter? What causes disease?

Many early written records show that people have always asked questions like these. For example, the Greek physician Hippocrates (ca. 460–ca. 377 B.C.) wondered whether there is a link between environment and disease, and the Greek historian Thucydides (ca. 460–ca. 404 B.C.) questioned why he and other survivors of the plague could have intimate contact with victims and not fall ill again. For many centuries the answers to these and other fundamental questions about the nature of life remained largely unanswered. But then, about 350 years ago, the invention of the microscope began to provide some clues.

In this chapter we'll see how one man's determination to answer a fundamental question about the nature of life— What does life really look like?—led to the birth of a new science called *microbiology*. We'll then see how the search for answers to other questions, such as those concerning spontaneous generation, why fermentation occurs, and the cause of disease, prompted advances in this new science. Finally, we'll look briefly at some of the key questions microbiologists are asking today.

The Early Years of Microbiology

The early years of microbiology brought the first observations of microbial life and the initial efforts to organize them into logical classifications.

What Does Life Really Look Like?

Learning Objectives

✓ Describe the world-changing scientific contributions of Leeuwenhoek.

✓ Define microbes in the words of Leeuwenhoek and as we know them today.

Only a few people have changed the world of science forever. We've all heard of Galileo, Newton, and Einstein, but the list also includes Antoni van Leeuwenhoek (lā'věn-huk; 1632–1723), a Dutch tailor, merchant, and lens grinder, and the man who first discovered the microbial world **(Figure 1.1)**.

Leeuwenhoek was born in Delft, the Netherlands, and lived most of his 90 years in the city of his birth. What set Leeuwenhoek apart from other men of his generation was an insatiable curiosity coupled with an almost stubborn desire to do everything for himself. His journey to fame began simply enough, when as a tailor he needed to examine the quality of cloth. Rather than merely buying one of the magnifying lenses already available, he learned to grind glass and made his own **(Figure 1.2)**. Soon he began asking the

▲ *Figure 1.1*

Antoni van Leeuwenhoek, the Father of Protozoology and Bacteriology, who reported the existence of protozoa in 1674 and of bacteria in 1676. *Why did Leeuwenhoek discover protozoa before bacteria?*

Figure 1.1 Protozoa are generally larger than bacteria.

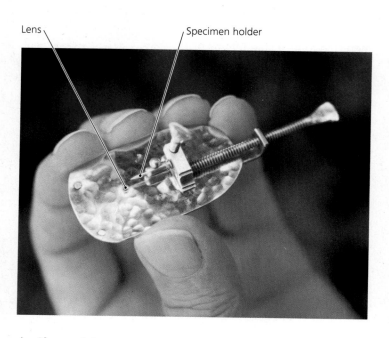

▲ *Figure 1.2*

Leeuwenhoek's microscope. This simple device is little more than a magnifying glass with screws for manipulating the specimen; yet with it, Leeuwenhoek changed the way we see our world. The lens, which is convex on both sides, is about the size of a pinhead. The object to be viewed was mounted either directly on the specimen holder or inside a small glass tube, which was then mounted on the specimen holder.

question "What does it really look like?" of everything in his world: the stinger of a bee, the brain of a fly, the leg of a louse, a drop of blood, flakes of his own skin. To find answers, he spent hours examining, reexamining, and recording every detail of each object he observed.

Making and looking through his simple microscopes, most really no more than magnifying glasses, became the overwhelming passion of his life. His enthusiasm and dedication are evident from the fact that he sometimes personally extracted the metal for his microscope from ore. Further, he often made a new microscope for each specimen, which remained mounted so that he could view it again and again. Then one day, he turned a lens onto a drop of water. We don't know what he expected to see, but certainly he saw more than he had anticipated. As he reported to the Royal Society of London[1] in 1674, he was surprised and delighted by

> *some green streaks, spirally wound serpent-wise, and orderly arranged. . . . Among these there were, besides, very many little animalcules, some were round, while others a bit bigger consisted of an oval. On these last, I saw two little legs near the head, and two little fins at the hind most end of the body. . . . And the motion of most of these animalcules in the water was so swift, and so various, upwards, downwards, and round about, that 'twas wonderful to see.*

Leeuwenhoek had discovered a previously unknown microbial world, which today we know to be populated with tiny animals, fungi, algae, and single-celled protozoa **(Figure 1.3)**. In a later report to the Royal Society, he noted that

> *the number of these animals in the scurf [plaque] of a man's teeth, are so many that I believe they exceed the number of men in a kingdom. . . . I found too many living animals therein, that I guess there might have been in a quantity of matter no bigger than the 1/100 part of a [grain of] sand.*

From the figure accompanying this report and the precise description of the size of these organisms from between his teeth, we know that Leeuwenhoek was reporting the existence of bacteria. By the end of the 19th century, Leeuwenhoek's "beasties," as he sometimes dubbed them, were called **microbes,** and today we also know them as **microorganisms.** Both terms include all organisms that are too small to be seen without a microscope.

Because of the quality of his microscopes, his profound observational skills, his detailed reports over a 50-year period, and his report of the discovery of many types of microorganisms, Antoni van Leeuwenhoek was elected to the Royal Society in 1680. He and Isaac Newton were probably the most famous scientists of their time. Because he changed forever the way we view our world, Leeuwenhoek is known today as the Father of Protozoology and Bacteriology.

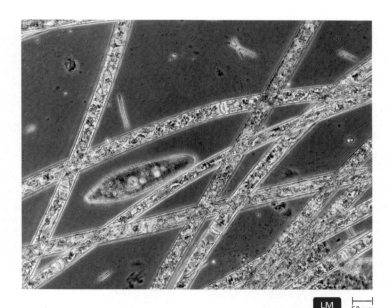

LM   50 μm

▲ *Figure 1.3*
The microbial world. Leeuwenhoek reported seeing a scene very much like this, one full of numerous, fantastic, cavorting creatures.

How Can Microbes Be Classified?

Learning Objectives

✓ List five groups of microorganisms.
✓ Explain why protozoa, algae, and nonmicrobial parasitic worms are studied in microbiology.
✓ Differentiate between prokaryotic and eukaryotic organisms.

Shortly after Leeuwenhoek made his discoveries, the Swedish botanist Carolus Linnaeus (1707–1778) developed a **taxonomic system**—that is, a system for naming plants and animals and grouping similar organisms together. For instance, Linnaeus and other scientists of the period grouped all organisms into either the animal kingdom or the plant kingdom. Today, biologists still use this basic system, but they have modified Linnaeus's scheme by adding categories that more realistically reflect the relationships among organisms. For example, scientists no longer classify yeasts, molds, and mushrooms as plants, but instead as fungi. We examine taxonomic schemes in more detail in Chapter 4.

The microorganisms that Leeuwenhoek described can be grouped into five basic categories: fungi, protozoa, algae, prokaryotes, and small animals. The only microbes not described by Leeuwenhoek are *viruses,*[2] which are too small to

---

[1]The Royal Society of London for the Promotion of Natural Knowledge, granted a royal charter in 1662, is one of the older and more prestigious scientific groups in Europe.
[2]Technically, viruses are not "organisms" because they neither replicate themselves nor carry on the chemical reactions of living things. See Chapter 3 for a fuller discussion of this issue.

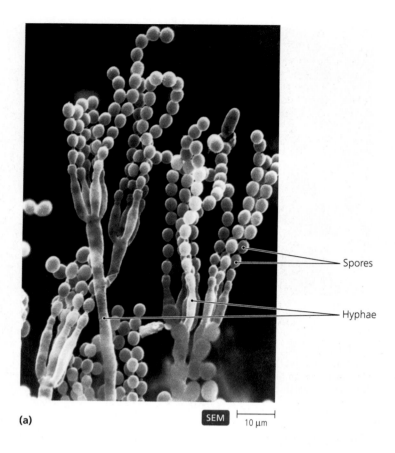

(a)

SEM ⊢———⊣ 10 μm

—— Spores

—— Hyphae

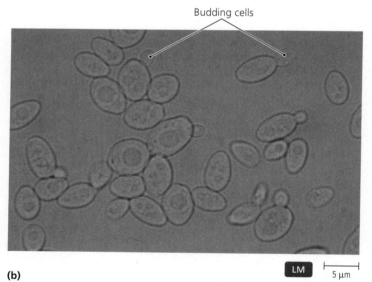

Budding cells

(b)

LM ⊢———⊣ 5 μm

▲ *Figure 1.4*

Fungi. **(a)** The mold *Penicillium chrysogenum*, which produces penicillin, has long filamentous hyphae that intertwine to form its body. It reproduces by spores. **(b)** The yeast *Saccharomyces cerevisiae*. Yeasts are round to oval and typically reproduce by budding.

be seen without an electron microscope. We briefly consider organisms in the first four categories in the following sections.

Fungi

Fungi (fŭn′jī)[3] are organisms whose cells are **eukaryotic;**[4] that is, each of their cells contains a nucleus composed of genetic material surrounded by a distinct membrane. Fungi are different from plants because they obtain their food from other organisms (rather than making it by photosynthesis). They differ from animals by having cell walls.

Microscopic fungi include some molds and yeasts. **Molds** are typically multicellular organisms that grow as long filaments, called *hyphae,* that intertwine to make up the body of the mold. Molds reproduce by sexual and asexual spores, which are cells that produce a new individual without fusing with another cell **(Figure 1.4a).** The cottony growths on cheese, bread, and jams are examples of molds. *Penicillium chrysogenum* (pen-i-sil′ē-ŭm krī-so′jĕn-ŭm) is a mold that produces penicillin.

Yeasts are unicellular and typically oval to round. They reproduce asexually by *budding,* a process in which a daughter cell grows off the mother cell. Some yeasts also produce sexual spores. An example of a useful yeast is *Saccharomyces cerevisiae* (sak-ă-rō-mī′sēz se-ri-vis′ē-ē; **Figure 1.4b**) which causes bread to rise and produces alcohol from sugar. *Candida albicans* (kan′did-ă al′bi-kanz) is a yeast that causes most cases of yeast infections in women.

Fungi and their significance in the environment, in food production, and as agents of human disease are discussed in Chapters 12 and 22.

Protozoa

Protozoa are single-celled eukaryotes that are similar to animals in their nutritional needs and cellular structure. In fact, *protozoa* is Greek for "first animals." Most protozoa are capable of locomotion, and one way scientists categorize protozoa is according to their locomotive structures: *pseudopodia,*[5] *cilia,*[6] or *flagella.*[7] Pseudopodia are extensions of a cell that flow in the direction of travel **(Figure 1.5a).** Cilia are numerous, short, hairlike protrusions of a cell that beat rhythmically to propel the protozoan through its environment **(Figure 1.5b).** Flagella are also extensions of a cell, but are fewer, longer, and more whiplike than cilia **(Figure 1.5c).** Some protozoa, such as the malaria-causing *Plasmodium* (plaz-mō′dē-ŭm), are nonmotile in their mature forms.

Protozoa typically live freely in water, but some live inside animal hosts, where they can cause disease. Most protozoa reproduce asexually, though some are sexual as well. The protozoa are discussed further in Chapters 12 and 23.

---

[3]Plural of the Latin *fungus,* meaning mushroom.
[4]From Greek *eu,* meaning true, and *karyon,* meaning kernel (which in this case refers to the nucleus of a cell).
[5]From Greek *pseudes,* meaning false, and *podos,* meaning foot.
[6]Plural of the Latin *cilium,* meaning eyelid.
[7]Plural of the Latin *flagellum,* meaning whip.

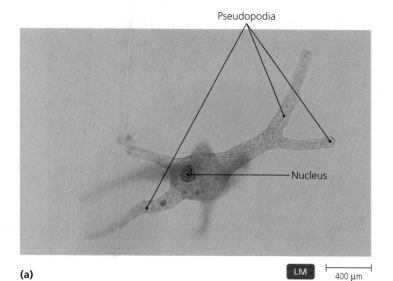

Pseudopodia

Nucleus

(a)

LM | 400 μm

Cilia

(b)

LM | 10 μm

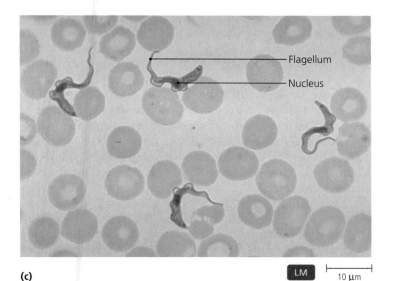

Flagellum

Nucleus

(c)

LM | 10 μm

◀ *Figure 1.5*

Locomotive structures of protozoa. **(a)** Pseudopodia, which are cellular extensions used for locomotion and feeding, as seen in *Amoeba proteus*. **(b)** Cilia, which are short, motile, hairlike extrusions over the entire surface of the cell, as seen in *Paramecium*. **(c)** Flagella, which are whiplike extensions that are less numerous and longer than cilia, as seen in *Trypanosoma brucei*, the protozoan that causes African sleeping sickness. *How do cilia and flagella differ?*

Figure 1.5 Cilia are short, numerous, and cover the cell, whereas flagella are long and relatively few in number.

Algae

Algae[8] are unicellular or multicellular *photosynthetic* organisms (**Figure 1.6** on page 6); that is, like plants they make their own food from carbon dioxide and water using energy from sunlight. They differ from plants in the relative simplicity of their reproductive structures. Algae are categorized on the basis of their pigmentation, their storage products, and the composition of their cell walls.

Large algae, commonly called seaweeds and kelps, are common in the world's oceans. Chemicals from their gelatinous cell walls are used as thickeners and emulsifiers in many food and cosmetic products, as well as in microbiological laboratory media.

Unicellular algae are common in freshwater ponds, streams, and lakes, and in the oceans as well. They are the major food of small aquatic and marine animals and provide most of the world's oxygen as a by-product of photosynthesis. The glasslike cell walls of diatoms provide grit for many polishing compounds. Other aspects of the biology of algae are discussed in Chapter 12.

Prokaryotes

Prokaryotes[9] are unicellular microbes that lack nuclei. There are two kinds of prokaryotes—**bacteria** and **archaea.** Bacterial cell walls are composed of a polysaccharide called *peptidoglycan,* though some bacteria lack cell walls. The cell walls of archaea lack peptidoglycan and instead are composed of other polymers. Members of both groups reproduce asexually. Chapters 4 and 11 examine other differences between bacteria and archaea and Chapters 19–21 discuss pathogenic (disease-causing) bacteria.

Most bacteria and archaea are much smaller than eukaryotic cells (**Figure 1.7** on page 6.) They live singly or in pairs, chains, or clusters in almost every habitat containing sufficient moisture. Archaea are found in extreme environments, such as the highly saline Mono Lake in California, acidic hot springs in Yellowstone National Park, and oxygen-depleted mud at the bottom of swamps.

[8]Plural of the Latin *alga,* meaning seaweed.
[9]From Greek *pro,* meaning before, and *karyon,* meaning kernel.

(a)   LM   10 µm   (b)   LM   10 µm

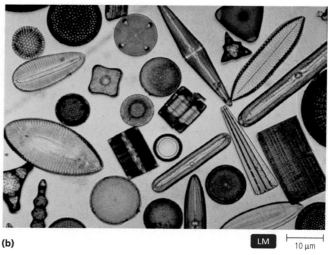

▲ *Figure 1.6*

Algae. **(a)** *Spirogyra* sp. These microscopic algae grow as chains of cells containing helical photosynthetic organs. **(b)** Diatoms. These beautiful algae have glasslike cell walls.

Prokaryotic bacterial cells   Eukaryotic cheek cell

LM   20 µm

▲ *Figure 1.7*

Cells of the bacterium *Streptococcus,* which Gram stain purple, surrounding a (eukaryotic) human cheek cell, which Gram stains pink. Notice the size difference.

Though bacteria often have a poor reputation in our world, the great majority of bacteria do not cause disease in animals, humans, or crops. Indeed, bacteria are beneficial to us in many ways. For example, dead plants and animals are degraded by bacteria (and fungi) to release phosphorus, sulfur, nitrogen, and carbon back into the air, soil, and water to be used by new generations of organisms. Without microbial recyclers, the world would be buried under the petrified corpses of uncountable dead organisms.

CRITICAL THINKING

A few bacteria produce disease because they derive nutrition from human cells and produce toxic wastes. Algae do not cause disease. Why not?

Other Organisms of Importance to Microbiologists

Microbiologists also study parasitic worms, which range in size from microscopic forms **(Figure 1.8)** to adult tapeworms over 7 meters (approximately 23 feet) in length. Even though most of these worms are not microscopic as adults, many of them cause diseases that were studied by early microbiologists. Further, laboratory technicians diagnose infections of parasitic worms by finding microscopic eggs and immature stages in blood, fecal, urine, and lymph specimens. Chapter 23 discusses parasitic worms.

The only type of microbes that remained hidden from Leeuwenhoek and other early microbiologists are viruses, which are much smaller than the smallest prokaryote and are not visible by light microscopy **(Figure 1.9)**. Viruses could not be seen until the electron microscope was invented in 1932. All viruses are acellular (not composed of cells) obligatory parasites composed of small amounts of genetic material (either DNA or RNA) surrounded by a protein coat. The general characteristics of viruses are discussed in Chapter 13, and specific viral pathogens are the subject of Chapters 24 and 25.

Leeuwenhoek first reported the existence of microorganisms in 1674, but microbiology did not develop significantly as a field of study for almost two centuries. There were a number of reasons for this delay. First, Leeuwenhoek was a suspicious and secretive man. Though he built over 400 microscopes, he never trained an apprentice, and he never sold or gave away a microscope. In fact, he never let *anyone*—not his family or such distinguished visitors as the Czar of Russia and the Queen of England—so much as peek through his very best instruments. When Leeuwenhoek died, the secret of creating superior microscopes was lost. It took almost 100 years for scientists to make microscopes of equivalent quality.

Another reason that microbiology was slow to develop as a science is that scientists in the 1700s considered microbes to be curiosities of nature and insignificant to human affairs. But in the late 1800s, scientists began to adopt a new philosophy, one that demanded experimental proof rather than mere acceptance of traditional knowledge. This fresh philosophical foundation, accompanied by improved microscopes, new laboratory techniques, and a drive to answer a series of pivotal questions, propelled microbiology to the forefront as a scientific discipline.

The Golden Age of Microbiology

Learning Objective

✓ List and answer four questions that propelled research in what is called the "Golden Age of Microbiology."

For about 50 years, during what is now called "The Golden Age of Microbiology," scientists and the blossoming field of microbiology were driven by the search for answers to the following four questions:

1. Is spontaneous generation of microbial life possible?
2. What causes fermentation?
3. What causes disease?
4. How can we prevent infection and disease?

Competition among scientists, who were striving to be the first to answer these questions, drove exploration and discovery in microbiology during the late 1800s and early 1900s. These scientists' discoveries and the fields of study they initiated continue to shape the course of microbiological research today.

In the next sections we consider these questions, and how the great scientists accumulated the experimental evidence that answered them.

Is Spontaneous Generation of Microbial Life Possible?

Learning Objectives

✓ Identify the scientists who argued in favor of spontaneous generation.
✓ Compare and contrast the investigations of Redi, Needham, Spallanzani, and Pasteur concerning spontaneous generation.
✓ List four steps in the scientific method of investigation.

A dry lakebed has lain under the relentless North African desert sun for eight long months. The cracks in the baked, parched mud are wider than a man's hand. There is no sign of life anywhere in the scorched terrain. With the

Red blood cell

LM   30 µm

▲ *Figure 1.8*
An immature stage of a parasitic worm in blood.

Bacterium   Virus

TEM   200 nm

▲ *Figure 1.9*
Viruses, which are acellular obligatory parasites, are too small to be seen with a light microscope. This photo is a colorized electron microscope image of viruses that infect bacteria. Notice how small the viruses are compared to the bacterium.

abruptness characteristic of desert storms, rain falls in a torrent, and a raging flood of roiling water and mud crashes down the dry streambed and fills the lake. Within hours, what had been a lifeless, dry mudflat becomes a pool of water teeming with billions of shrimp; by the next day it is home to hundreds of toads. Where did these animals come from?

Flask unsealed

Flask sealed

Flask covered
with gauze

▲ *Figure 1.10*

Redi's experiments. When the flask remained unsealed, maggots covered the meat within a few days. When the flask was sealed, flies were kept away and no maggots appeared on the meat. When the flask opening was covered with gauze, flies were kept away and no maggots appeared on the meat, although a few maggots appeared on top of the gauze.

Aristotle (384–322 B.C.), the most famous Greek scientist, wrote that living things can arise via three processes: through asexual reproduction, through sexual reproduction, or from nonliving matter. The appearance of shrimp and toads in the mud of what so recently was a dry lakebed was seen as an example of the third process, which came to be known as *abiogenesis*[10] or **spontaneous generation.** The theory of spontaneous generation as promulgated by Aristotle was widely accepted for over 1900 years because it seemed to explain a variety of commonly observed phenomena, such as the appearance of maggots on spoiling meat. However, the validity of the theory came under challenge in the 17th century.

Redi's Experiments

In the late 1600s, the Italian physician Francesco Redi (1626–1697) demonstrated by a series of experiments that when decaying meat was kept isolated from flies, maggots never developed, while meat exposed to flies was soon infested **(Figure 1.10)**. As a result of experiments such as these, scientists began to doubt Aristotle's theory and adopt the view that animals only come from other animals.

Needham's Experiments

The debate over spontaneous generation was rekindled when Leeuwenhoek discovered microbes and showed that they appeared after a few days in freshly collected rainwater. Though scientists agreed that larger animals could not arise spontaneously, they disagreed about Leeuwenhoek's "wee animalcules"; surely they did not have parents, did they? They must arise spontaneously.

The proponents of spontaneous generation pointed to the careful demonstrations of British investigator Turberville Needham (1713–1781). He boiled beef gravy and infusions[11] of plant material in vials, which he then tightly sealed with corks. Some days later, Needham observed that the vials were cloudy, and examination revealed an abundance of "microscopical [*sic*] animals of most dimensions." As he explained it, there must be a "life force" that causes inanimate matter to spontaneously come to life, since he had heated the vials sufficiently to kill everything. Needham's experiments so impressed the Royal Society that they elected him a member.

Spallanzani's Experiments

Then, in 1799, the Italian scientist Lazzaro Spallanzani (1729–1799) reported results that contradicted Needham's findings. Spallanzani boiled some infusions for almost an hour and sealed the vials by melting their slender necks closed. His infusions remained clear, unless he broke the seal and exposed the infusion to air, after which they became cloudy with microorganisms. He concluded three things:

1. Needham had either failed to heat his vials sufficiently to kill all microbes, or he had not sealed them tightly enough.

2. Microorganisms exist in the air and can contaminate experiments.

3. Spontaneous generation of microorganisms does not occur; all living things arise from other living things.

---

[10]From Greek *a*, meaning not, *bios*, meaning life, and *genein*, meaning to produce.
[11]Infusions are broths made by steeping plant or animal material in water.

Although Spallanzani's experiments would appear to have settled the controversy once and for all, it proved difficult to dethrone a theory that had held sway for 2000 years, especially when so notable a scientist as Aristotle had propounded it. One of the criticisms of Spallanzani's work was that his sealed vials did not allow enough air for organisms to thrive; another objection was that his prolonged heating destroyed the "life force." The debate continued until the French chemist Louis Pasteur **(Figure 1.11)** conducted experiments that finally laid the theory of spontaneous generation to rest.

Pasteur's Experiments

Louis Pasteur (1822–1895) was an indefatigable worker who pushed himself as hard as he pushed others. As he wrote his sisters, "To *will* is a great thing dear sisters, for Action and Work usually follow Will, and almost always Work is accompanied by Success. These three things, Work, Will, Success, fill human existence. Will opens the door to success both brilliant and happy; Work passes these doors, and at the end of the journey Success comes to crown one's efforts." When his wife complained about his long hours in the laboratory, he replied, "I will lead you to fame."

Pasteur's determination and hard work are apparent in his investigations of spontaneous generation. Like Spallanzani, he boiled infusions long enough to kill everything. But instead of sealing the flasks, he bent their necks into an S-shape, which allowed air to enter while preventing the introduction of dust and microbes into the broth **(Figure 1.12a** on page 10).

Crowded for space and lacking funds, he improvised an incubator in the space under a staircase. Day after day he crawled on hands and knees into this incommodious space and examined his flasks for the cloudiness that would indicate the presence of living organisms. In 1861, he reported that his "swan-necked flasks" remained free of microbes even 18 months later. Since they contained all the nutrients (including air) known to be required by living things, he concluded, "Never will spontaneous generation recover from the mortal blow of this simple experiment."

Pasteur followed this experiment with demonstrations that microbes in the air were the "parents" of Needham's microorganisms. He broke the necks off some flasks, exposing the liquid in them directly to the air, and he carefully tilted others so that the liquid touched the dust that had accumulated in their necks **(Figure 1.12b)**. The next day, all of these flasks were cloudy with microbes. He concluded that the microbes in the liquid were the progeny of microbes that had been on the dust particles.

The Scientific Method

The debate over spontaneous generation led to the development of a generalized **scientific method** by which questions are answered through observations of the outcomes of carefully controlled experiments, instead of by conjecture or

▲ *Figure 1.11*

Louis Pasteur, who disproved spontaneous generation and came to be known as the Father of Microbiology.

according to the opinions of any authority figure. The scientific method, which provides a *framework* for conducting an investigation rather than a rigid set of specific "rules," consists of four basic steps **(Figure 1.13** on page 11):

1 A group of observations leads a scientist to ask a question about some phenomenon.

2 The scientist generates a hypothesis—that is, a potential answer to the question.

3 The scientist designs and conducts an experiment to test the hypothesis.

4 Based on the observed results of the experiment, the scientist either accepts, rejects, or modifies the hypothesis.

As shown in Figure 1.13, the scientist then returns to earlier steps in the method, either modifying hypotheses and then testing them, or repeatedly testing accepted hypotheses, until the evidence for a hypothesis is convincing. Accepted hypotheses that explain many observations and are repeatedly verified by numerous scientists over many years may become *theories* or *laws*.

Note that in order for experiments (and their results) to be accepted as valid by the scientific community, they must include appropriate *control groups*—groups that are treated exactly the same as the other groups in the experiment, except for the one variable that the experiment is designed to test. In Pasteur's experiments on spontaneous generation, for example, his "control flasks" contained a sterile infusion composed of all the nutrients living things need, as well as air made available through the flasks' "swan necks." His "experimental flasks" for testing his hypothesis—that microbes would reach (and subsequently grow in) the infusion through contact with dust particles—were exposed to exactly the same conditions, *plus* contact with the dust in the

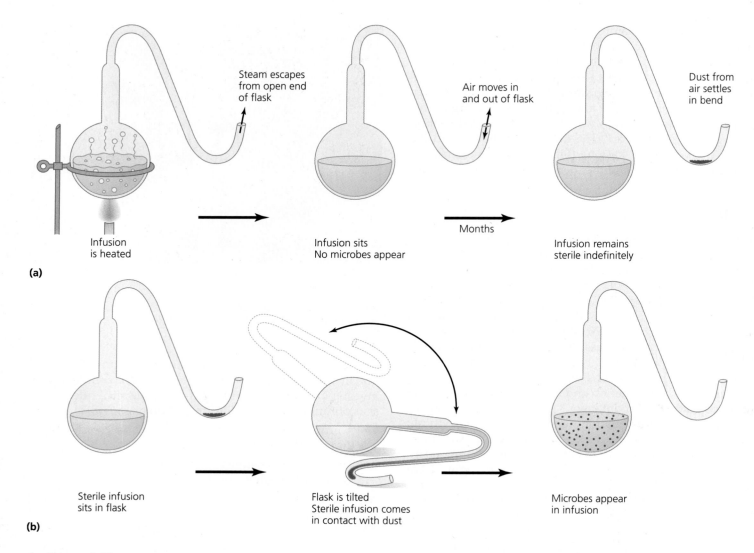

(a)

Steam escapes from open end of flask

Infusion is heated

Air moves in and out of flask

Infusion sits No microbes appear

Months

Dust from air settles in bend

Infusion remains sterile indefinitely

(b)

Sterile infusion sits in flask

Flask is tilted Sterile infusion comes in contact with dust

Microbes appear in infusion

▲ *Figure 1.12*

Pasteur's experiments with "swan-necked flasks." **(a)** As long as the flask remained upright, no microbial growth appeared in the infusion. **(b)** When the flask was tilted so that some dust from the bend in the neck was carried back into the flask, the infusion became cloudy with microbes within a day.

bend in the neck. Because exposure to the dust was the *only* difference between the control and experimental groups, Pasteur was able to conclude that the microbes growing in the infusion arrived on the dust particles.

What Causes Fermentation?

Learning Objectives

✓ Discuss the significance of Pasteur's fermentation experiments to our world today.

✓ Explain why Pasteur is known as the Father of Microbiology.

✓ Identify the scientist whose experiments led to the field of biochemistry and the study of metabolism.

The controversy over spontaneous generation was largely a philosophical exercise among men who conducted research to gain basic scientific knowledge. They had no practical goals in mind—except, perhaps, personal aggrandizement in the form of financial support, honor, and prestige. However, the second question that moved microbial studies forward in the 1800s had tremendous practical applications.

Our story resumes in 19th-century France where spoiled, acidic wine was threatening the livelihood of many grape growers. The initial question was, "Why is the wine spoiled?" but this led to a more fundamental question, "What causes the fermentation of grape juice into wine?" These questions were so important to vintners that they funded research concerning fermentation, hoping scientists could develop methods to promote the production of alcohol and prevent spoilage by acid during fermentation.

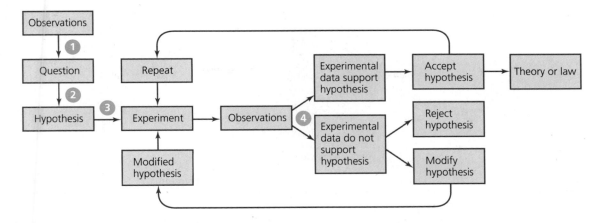

▲ *Figure 1.13*

The scientific method, which forms the framework for scientific research. ① Observations of natural phenomena lead a scientist to ask a question. ② The scientist generates a hypothesis to explain the observations. ③ The scientist designs and conducts an experiment to test the hypothesis. ④ Observations of the experiment's results either support the hypothesis, which is then provisionally accepted, or fail to support the hypothesis. Unsupported hypotheses may be either rejected or modified and retested. A hypothesis that continues to be accepted after repeated testing may be called a theory or a law.

Pasteur's Experiments

Scientists of the 1800s used the word *fermentation* to mean not only the formation of alcohol from sugar, but also other chemical reactions such as the formation of lactic acid, the putrefaction of meat, and the decomposition of waste. Many scientists asserted that air caused fermentation reactions, others insisted that living organisms were responsible.

The debate over the cause of fermentation reactions was linked to the debate over spontaneous generation. Some scientists proposed that the yeasts observed amidst alcohol fermentation products were nonliving globules of chemicals and gases. Others thought that yeasts were alive and were spontaneously generated during fermentation. Still others asserted that yeasts not only were living organisms, but also caused fermentation.

Pasteur conducted a series of careful observations and experiments that answered the question "What causes fermentation?" First, he observed yeast cells growing and budding in grape juice and conducted experiments showing that they arise only from other yeast cells. Then, by sealing some sterile flasks containing grape juice and yeast, and by leaving others open to the air, he demonstrated that yeast could grow with or without oxygen; that is, he discovered that yeasts are *facultative anaerobes*[12]—organisms that can live with or without oxygen. Finally, by introducing bacteria and yeast cells into different flasks of sterile grape juice, he proved that bacteria ferment grape juice to produce acids, and that yeast cells ferment grape juice to produce alcohol (**Figure 1.14** on page 12).

Pasteur's discovery that *anaerobic* bacteria fermented grape juice into acids suggested a method for preventing the spoilage of wine. His name became a household word when

he developed *pasteurization*, a process of heating the grape juice just enough to kill most contaminating bacteria without changing the juice's basic qualities, so that it could then be inoculated with yeast to ensure that alcohol fermentation occurred. Pasteur thus began the field of **industrial microbiology** (or **biotechnology**) in which microbes are intentionally used to manufacture products (**Table 1.1** on page 13; see also Chapter 26). Today pasteurization is used routinely on milk to eliminate pathogens that cause diseases such as bovine tuberculosis and brucellosis; it is also used to eliminate pathogens in juices and other beverages.

These are just a few of the many experiments Pasteur conducted with microbes. While a few of Pasteur's successes can be attributed to the superior microscopes available in the late 1800s, his genius is clearly evident in his carefully designed and straightforward experiments. Because of his many, varied, and significant accomplishments in working with microbes, Pasteur is considered the Father of Microbiology.

Buchner's Experiments

Studies on fermentation began with the idea that fermentation reactions were strictly chemical and did not involve living organisms. This idea was supplanted by Pasteur's work showing that fermentation proceeded only when living cells were present, and that different types of microorganisms growing under varied conditions produced different end products.

In 1897, the German scientist Eduard Buchner (1860–1917) resurrected the chemical explanation by showing

---

[12]From Greek *an*, meaning not, *aer*, meaning air (i.e., oxygen), and *bios*, meaning life.

Observation:

Fermenting
grape juice

Microscopic analysis
shows juice contains
yeasts and bacteria

Hypothesis	Experiment		Observation	Conclusion
	Day 1: Flasks of grape juice are heated sufficiently to kill all microbes	Day 2		
1. Spontaneous fermentation occurs	Flask is sealed		No fermentation; juice remains free of microbes	Reject hypothesis 1
2. Air ferments grape juice	Flask remains open to air via curved neck		No fermentation; juice remains free of microbes	Reject hypothesis 2
3. Bacteria ferment grape juice into alcohol	Juice in flask is inoculated with bacteria and sealed		Bacteria reproduce; acids are produced	Modify hypothesis 3; bacteria ferment grape juice into acids
4. Yeasts ferment grape juice into alcohol	Juice in flask is inoculated with yeast and sealed		Yeasts reproduce; alcohol is produced	Accept hypothesis 4; yeasts ferment grape juice into alcohol

▲ *Figure 1.14*

How Pasteur applied the scientific method in investigating the nature of fermentation.
After observing that fermenting grape juice contained both yeasts and bacteria, Pasteur
hypothesized that these organisms cause fermentation. Upon eliminating the possibility
that fermentation could occur spontaneously or be caused by air (hypotheses 1 and 2),
he concluded that fermentation requires the presence of living cells. The results of
additional experiments (those testing hypotheses 3 and 4) indicated that bacteria
ferment grape juice to produce acids, and that yeasts ferment grape juice to produce
alcohol. *Which of Pasteur's flasks was the control?*

Figure 1.14   The sealed flask that remained free of microorganisms served as the control.

that fermentation does not require living cells **(Figure 1.15).**
Buchner's experiments demonstrated the presence of *enzymes,*
which are cell-produced proteins that promote chemical
reactions. Buchner's work began the field of **biochemistry**
and the study of **metabolism,** a term that refers to the sum of
all chemical reactions within an organism.

CRITICAL THINKING

How might the debate over spontaneous generation have been
different if Buchner had conducted his experiments in 1857 instead of
1897?

Table 1.1   **Some Industrial Uses of Microbes**

Product or Process	Contribution of Microorganism
Foods and Beverages	
Cheese	Flavoring and ripening produced by bacteria and fungi; flavors dependent on the source of milk and the type of microorganism
Alcoholic beverages	Alcohol produced by bacteria or yeast by fermentation of sugars in fruit juice or grain
Soy sauce	Produced by fungal fermentation of soybeans
Vinegar	Produced by bacterial fermentation of sugar
Yogurt	Produced by bacteria growing in skim milk
Sour cream	Produced by bacteria growing in cream
Artificial sweetener	Amino acids synthesized by bacteria from sugar
Bread	Rising of dough produced by action of yeast; sourdough results from bacterial-produced acids
Other Products	
Antibiotics	Produced by bacteria and fungi
Human growth hormone, human insulin	Produced by genetically engineered bacteria
Laundry enzymes	Isolated from bacteria
Vitamins	Isolated from bacteria
Diatomaceous earth (used in polishes and buffing compounds)	Composed of cell walls of microscopic algae
Pest control chemicals	Insect pests killed or inhibited by bacterial pathogens
Drain opener	Protein-digesting and fat-digesting enzymes produced by bacteria

▲ *Figure 1.15*

Buchner's demonstration of acellular fermentation. After he ground yeast cells in water with quartz sand, the filtered liquid was capable of fermenting sterilized sugar cane juice into alcohol. This experiment started the field of biochemistry. *What substances from the crushed yeast cells carried out fermentation?*

What Causes Disease?

Learning Objectives

✓ List at least seven contributions made by Koch to the field of microbiology.

✓ List the four steps that must be taken to prove the cause of an infectious disease.

✓ Describe the contribution of Gram to the field of microbiology.

You are a physician in London, and it is August, 1854. It is past midnight, and you have been visiting patients since before dawn. As you enter the room of your next patient you observe with frustration and despair that this case is like hundreds of others you and your colleagues have attended in the neighborhood over the past month.

A five-year-old boy with a vacant stare lies in bed listlessly. As you watch, he is suddenly gripped by severe abdominal cramps, and his gastrointestinal tract empties in an explosion of watery diarrhea. The voided fluid is clear, colorless, odorless, and streaked with thin flecks of white mucus, reminiscent of water poured off a pot of cooking rice. His anxious mother changes his bedclothes as his father gives him a sip of water, but it is of little use. With a heavy heart you confirm the parents' fear—their child has cholera, and there is nothing you can do. He will likely die before morning. As you despondently turn to go, the question that has haunted you for two months is foremost in your mind: What causes such a disease?

The third question that propelled the advance of microbiology concerned disease, defined generally as any abnormal condition in the body. Prior to the 1800s, disease was attributed to various factors, including evil spirits, sin, imbalances in body fluids, and foul vapors. Although the Italian philosopher Girolamo Fracastoro (1478–1553) conjectured as early as 1546 that "germs[13] of contagion" cause disease, the idea that germs might be invisible living organisms awaited the invention of Leeuwenhoek's microscope 230 years later.

Pasteur's discovery that bacteria are responsible for spoiling wine led naturally to his hypothesis in 1857 that microorganisms are also responsible for diseases. This idea came to be known as the **germ theory of disease.** Since a particular disease is typically accompanied by the same symptoms in all affected individuals, early investigators suspected that diseases such as cholera, tuberculosis, and anthrax are each caused by a specific germ, called a **pathogen.**[14] Today we know that some diseases are genetic and that allergic reactions and environmental toxins cause others, so the germ theory applies only to *infectious*[15] *diseases.*

Just as Pasteur was the chief investigator in disproving spontaneous generation and determining the cause of fermentation, so investigations in **etiology**[16] (the study of causation of disease) were dominated by Robert Koch (1843–1910) **(Figure 1.16).**

Koch's Experiments

Koch was a country doctor in Germany when he began a race with Pasteur to discover the cause of anthrax, which is a potentially fatal disease, primarily of animals, in which toxins produce ulceration of the skin. Anthrax caused untold financial losses to farmers and ranchers in the 1800s, and the disease can be spread to humans. (see **New Frontiers 1.1** on page 22.)

Koch carefully examined the blood of infected animals, and in every case he identified a rod-shaped bacterium[17]

▲ *Figure 1.16*

Robert Koch, who was instrumental in modifying the scientific method to prove that a given pathogen caused a specific disease.

that formed chains. He observed the formation of resting stages (endospores) within the bacterial cells and showed that the spores always produced anthrax when they were injected into mice. This was the first time that a bacterium was proven to cause a disease. As a result of his successful work on anthrax, Koch was able to move to Berlin and was given facilities and funding to continue his research.

Heartened by his success, Koch turned his attention to other diseases. He had been fortunate when he chose anthrax for his initial investigations, because anthrax bacteria are quite large and easily identified with the microscopes of that time. However, most bacteria are very small, and different types exhibit few or no visible differences. Koch puzzled how to distinguish among these bacteria.

He solved the problem by taking specimens (for instance, blood, pus, or sputum) from disease victims and then smearing the specimens onto a solid surface such as a slice of potato or a gelatin medium. He then waited for bacteria and fungi present in the specimen to multiply and form distinct colonies **(Figure 1.17).** Koch hypothesized that each colony consisted of the progeny of a single cell. He then inoculated samples from each colony into laboratory animals to see which caused disease. Koch's method of isolation is a standard technique in microbiological and medical labs to this day, though a gel called *agar,* derived from red seaweed, is used instead of gelatin or potato.

---

[13]From Latin *germen,* meaning sprout.
[14]From Greek *pathos,* meaning disease, and *genein,* meaning to produce.
[15]From Latin *inficere,* meaning to taint (*i.e.,* with a pathogen).
[16]From Greek *aitia,* meaning cause, and *logos,* meaning word or study.
[17]Now known as *Bacillus anthracis*—Latin for "the rod of anthrax."

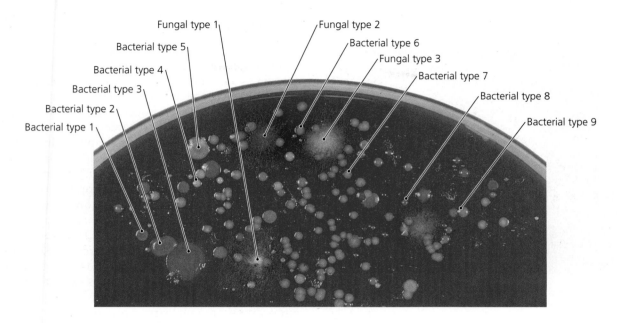

Fungal type 1
Fungal type 2
Bacterial type 5
Bacterial type 6
Bacterial type 4
Fungal type 3
Bacterial type 3
Bacterial type 7
Bacterial type 2
Bacterial type 8
Bacterial type 1
Bacterial type 9

▲ *Figure 1.17*

Bacterial and fungal colonies on a solid surface (agar). Differences in colony size, shape, and color indicate the presence of different species of bacteria and fungi. Such differences allowed Koch to isolate specific types of microbes that could be tested for their ability to cause disease.

Koch and his colleagues are also responsible for many other advances in laboratory microbiology, including the following:

- Simple staining techniques for bacterial cells and flagella
- The first photomicrograph of bacteria
- The first photograph of bacteria in diseased tissue
- Techniques for estimating the number of bacteria in a solution based on the number of colonies that form after inoculation onto a solid surface
- The use of steam to sterilize growth media
- The use of Petri[18] dishes to hold solid growth media
- Aseptic laboratory techniques such as transferring bacteria between media using a platinum wire that had been heat-sterilized in a flame
- Elucidation of bacteria as distinct species

For these achievements, Koch is considered the Father of the Microbiological Laboratory.

CRITICAL THINKING

French microbiologists, led by Pasteur, tried to isolate a single bacterium by diluting liquid media until only a single type of bacterium could be microscopically observed in a sample of the diluted medium. What advantages does Koch's method have over the French method?

Koch's Postulates

After discovering the anthrax bacterium, Koch continued to search for disease agents. In two pivotal scientific publications in 1882 and 1884, he announced that the cause of tuberculosis was a rod-shaped bacterium, *Mycobacterium tuberculosis* (mī-kō-bak-tēr′ē-ŭm tū-ber-kyū-lō′sis). In 1905 he received the Nobel Prize in Physiology or Medicine for this work.

In his publications on tuberculosis, Koch elucidated a series of steps that must be taken to prove the cause of any infectious disease. These steps, now known as **Koch's postulates,** are one of his more important contributions to microbiology. The postulates, which we discuss in more detail in Chapter 14, are the following:

1. The suspected causative agent must be found in every case of the disease and be absent from healthy hosts.
2. The agent must be isolated and grown outside the host.
3. When the agent is introduced to a healthy, susceptible host, the host must get the disease.
4. The same agent must be reisolated from the diseased experimental host.

We use the term *suspected causative agent* because "agent" can refer to a fungus, protozoan, bacterium, virus, or other pathogen. It is "suspected" until the postulates have

---

[18]Named for Richard Petri, Koch's assistant, who invented them in 1887.

Gram-negative   Gram-positive

LM |——| 10 µm

▲ *Figure 1.18*

Results of Gram staining. Gram-positive cells (in this case *Staphylococcus aureus*) are purple; Gram-negative cells (in this case *Escherichia coli*) are pink.

been fulfilled. There are practical and ethical limits in the application of Koch's postulates, but in almost every case they must be satisfied before the cause of an infectious disease is proven.

CRITICAL THINKING

Why aren't Koch's postulates always useful in proving the cause of a given disease? Consider a variety of diseases, such as cholera, pneumonia, Alzheimer's, AIDS, Down's syndrome, and lung cancer.

During microbiology's "golden years," other scientists used Koch's postulates, as well as laboratory techniques introduced by Koch and Pasteur, to discover the causes of most protozoan and bacterial diseases, as well as some viral diseases. For example, Charles Laveran (1845–1922) showed that a protozoan is the cause of malaria, and Edwin Klebs (1834–1913) described the bacterium that causes diphtheria. Dmitri Ivanowski (1864–1920) and Martinus Beijerinck (1851–1931) discovered that a certain disease in tobacco plants is caused by a pathogen that passes through filters with such extremely small pores that bacteria cannot pass through. Beijerinck, recognizing that the pathogen was not bacterial, called it a *filterable virus.* Now such pathogens are simply called *viruses.* As previously noted, viruses couldn't be seen until electron microscopes were invented in 1932. The American physician Walter Reed (1851–1902) proved in 1900 that viruses can cause such diseases as yellow fever in humans. Chapters 13, 24, and 25 deal with the science of *virology.*

A partial list of scientists and the pathogens they discovered is provided in **Table 1.2.**

Gram's Stain

The first of Koch's postulates demands that the suspected agent be found in every case of a given disease, which presupposes that minute microbes can be seen and identified. However, because most microbes are colorless and difficult to see, scientists began to use dyes to stain them and make them more visible under the microscope.

Though Koch reported a simple staining technique in 1877, the Danish scientist Christian Gram (1853–1938) developed a more important staining technique in 1884. His procedure, which involves the application of a series of dyes, leaves some microbes purple and others pink. We now label the first group of cells as *Gram-positive* and the second as *Gram-negative,* and we use the Gram procedure to separate bacteria into these two large groups **(Figure 1.18).**

The **Gram stain** is still the most widely used staining technique. It is one of the first steps carried out in any laboratory where bacteria are being identified, and is one of the procedures you will learn in micro lab. The full procedure is discussed in Chapter 4 (page 109).

How Can We Prevent Infection and Disease?

Learning Objectives

✓ Identify four health care practitioners who did pioneering research in the areas of public health microbiology and epidemiology.

✓ Name two scientists whose work with vaccines began the field of immunology.

✓ Describe the quest for a "magic bullet."

The last great question that drove microbiological research during the "Golden Age" was how to prevent infectious diseases. Though some methods of preventing or limiting disease were discovered even before it was understood that microorganisms caused contagious diseases, great advances occurred only after Pasteur and Koch showed that life comes from life and that microorganisms can cause diseases.

In the mid-1800s, modern principles of hygiene, such as those involving sewage and water treatment, personal cleanliness, and pest control, were not widely practiced. Typically, medical personnel and health care facilities lacked adequate cleanliness. *Nosocomial*[19] *infections*—infections acquired in a health care setting—were rampant. For example, surgical patients frequently succumbed to gangrene acquired while under their doctor's care, and many women who gave birth in hospitals died from puerperal[20] fever. Four health care practitioners who were especially instrumental in changing the way health care is delivered were Semmelweis, Lister, Nightingale, and Snow.

---

[19]From Greek *nosos,* meaning disease, and *komein,* meaning to care for (relating to a hospital).
[20]From Latin *puerperus,* meaning childbirth.

Table 1.2 **Some Notable Scientists of the "Golden Age of Microbiology" and the Agents of Human Disease They Discovered**

Scientist	Year	Disease	Agent
Robert Koch	1876	Anthrax	*Bacillus anthracis* (bacterium)
Albert Neisser	1879	Gonorrhea	*Neisseria gonorrheae* (bacterium)
Charles Laveran	1880	Malaria	*Plasmodium* species (protozoa)
Carl Eberth	1880	Typhoid fever	*Salmonella typhi* (bacterium)
Robert Koch	1882	Tuberculosis	*Mycobacterium tuberculosis* (bacterium)
Edwin Klebs	1883	Diphtheria	*Corynebacterium diphtheriae* (bacterium)
Theodore Escherich	1884	Traveler's diarrhea Bladder infection	*Escherichia coli* (bacterium)
Albert Fraenkel	1884	Pneumonia	*Streptococcus pneumoniae* (bacterium)
Robert Koch	1884	Cholera	*Vibrio cholerae* (bacterium)
David Bruce	1887	Undulant fever (brucellosis)	*Brucella melitensis* (bacterium)
Anton Weichselbaum	1887	Meningococcal meningitis	*Neisseria meningitidis* (bacterium)
A. A. Gartner	1888	Salmonellosis (form of food poisoning)	*Salmonella* species (bacterium)
Shibasaburo Kitasato	1889	Tetanus	*Clostridium tetani* (bacterium)
Dmitri Iwanowski and Martinus Beijerinck	1892 1898	Tobacco mosaic disease	*Tobamovirus* tobacco mosaic virus
William Welsh and George Nuttall	1892	Gas gangrene	*Clostridium perfringens* (bacterium)
Alexandre Yersin and Shibasaburo Kitasato	1894	Bubonic plague	*Yersinia pestis* (bacterium)
Kiyoshi Shiga	1898	Shigellosis (a type of severe diarrhea)	*Shigella dysenteriae* (bacterium)
Walter Reed	1900	Yellow fever	*Flavivirus* yellow fever virus
Robert Forde and Joseph Dutton	1902	African sleeping sickness	*Trypanosoma brucei gambiense* (protozoan)

Semmelweis and Handwashing

Ignaz Semmelweis (1818–1865) was a physician on the obstetric ward of a teaching hospital in Vienna. In about 1848 he observed that women giving birth in the wing where medical students were trained died from puerperal fever at a rate 20 times higher than the mortality rates of either women attended by midwives in an adjoining wing or women who gave birth at home.

Though Pasteur had not yet elaborated his germ theory of disease, Semmelweis hypothesized that medical students carried "cadaver particles" from their autopsy studies into the delivery rooms, and that these "particles" resulted in puerperal fever. Semmelweis gained support for his hypothesis when a doctor who sliced his finger during an autopsy died after showing symptoms similar to those of puerperal fever. Today we know that the primary cause of puerperal fever are bacteria in the genus *Streptococcus* (strep-tō-kok'ŭs; see Figure 1.7), which are usually harmless on the skin or in the mouth but cause severe complications when they enter the blood.

Semmelweis began requiring medical students to wash their hands with chlorinated lime water, a substance long used to eliminate the smell of cadavers. Mortality in the subsequent year dropped from 18.3% to 1.3%. Despite his success, Semmelweis was ridiculed by the director of the hospital and eventually forced to leave. He returned to his native Hungary, where his insistence on handwashing met with general approval when it continued to produce higher patient survival rates.

Though his impressive record made it easier for later doctors to institute changes, Semmelweis was unsuccessful in gaining support for his method from most European doctors. He became severely depressed and was committed to a mental hospital, where he died from an infection of *Streptococcus*, the very organism he had fought for so long.

Lister's Antiseptic Technique

Shortly after Semmelweis was rejected in Vienna, the English physician Joseph Lister (1827–1912) modified and advanced the idea of *antisepsis*[21] in health care settings. As a surgeon, Lister was aware of the dreadful consequences that

---

[21]From Greek *anti,* meaning against, and *sepein,* meaning putrefaction.

resulted from the infection of *wounds. Therefore,* he began *spraying wounds,* surgical incisions, and dressings with carbolic acid (phenol), a chemical that had previously proven effective in reducing odor and decay in sewage. Like Semmelweis, he initially met with some resistance, but when he showed that it reduced deaths among his patients by two-thirds, his method was accepted into common practice. In this manner, Lister vindicated Semmelweis, became the founder of antiseptic surgery, and opened new fields of research into antisepsis and disinfection.

Nightingale and Nursing

Florence Nightingale (1820–1910) **(Figure 1.19)** was a dedicated English nurse who succeeded in introducing cleanliness and other antiseptic techniques into nursing practice. She was instrumental in setting standards of hygiene that saved innumerable lives during the Crimean War of 1854–56. One of her first requisitions in the military hospital was for 200 scrubbing brushes, which she and her assistants used diligently in the squalid wards. She next arranged for each patient's filthy clothes and dressings to be replaced or cleaned at a different location, thus removing many sources of infection. She thoroughly documented statistical comparisons to show that poor food and unsanitary conditions in the hospitals were responsible for the deaths of many soldiers.

After the war, Nightingale returned to England, where she actively exerted political pressure to reform hospitals and implement public health policies. Perhaps her greatest achievements were in nursing education. For example, she founded the Nightingale School for Nurses—the first of its kind in the world.

Snow and Epidemiology

Another English physician, John Snow (1813–1858), also played a key role in setting standards for good public hygiene to prevent the spread of infectious diseases. Snow had been studying the propagation of cholera and suspected that the disease was spread by a contaminating agent in water. In 1854, he mapped the occurrence of cholera cases during an epidemic in London and showed that they centered around a public water supply on Broad Street. When, after his recommendation, the water pump was dismantled, the cholera epidemic abated.

Though Snow did not know the cause of cholera, his careful documentation of the epidemic highlighted the critical need for adequate sewage treatment and a pure water supply. His study was the foundation for two branches of microbiology—**infection control** and **epidemiology,**[22] which is the study of the occurrence, distribution, and spread of disease in humans.

Jenner's Vaccine

In about 1789, the English physician Edward Jenner (1749–1823) tested the hypothesis that a mild disease called

▲ *Figure 1.19*

Florence Nightingale, the founder of modern nursing, was influential in introducing antiseptic technique into nursing practice.

cowpox provided protection against potentially fatal smallpox. After he intentionally inoculated his son with pus collected from a milkmaid's cowpox lesion, the boy developed cowpox, which, of course, he survived. When Jenner then attempted to infect his son with smallpox, he found that the boy had become immune[23] to it. (Note that experiments that intentionally expose human subjects to deadly pathogens are considered unethical today.) In 1798 Jenner reported similar results from additional experiments, demonstrating the validity of the procedure he named *vaccination* after *Vaccinia,*[24] the virus that causes cowpox. Jenner invented vaccination (the term *immunization* is often used synonymously today), established a safe treatment for preventing smallpox, and began the field of **immunology**—the study of the body's specific defenses against pathogens. Chapters 16–18 discuss immunology.

Pasteur later capitalized on Jenner's work by producing weakened strains of various pathogens for use in preventing the serious diseases they cause. In honor of Jenner's work with cowpox, Pasteur used the term *vaccine* to refer to all weakened, protective strains of pathogens. He subsequently developed successful vaccines against fowl cholera, anthrax, and rabies.

Ehrlich's "Magic Bullets"

Gram's discovery that bacteria could be differentiated into two types by staining suggested to the German microbiologist

---

[22]From Greek *epi,* meaning upon, *demos* meaning people, and *logos* meaning word or study.
[23]From Latin *immunis,* meaning free.
[24]From Latin *vacca,* meaning cow.

Paul Ehrlich (1854–1915) that chemicals could be used to kill microorganisms differentially. To investigate this idea, Ehrlich undertook an exhaustive survey of chemicals to find a "magic bullet" that would destroy pathogens while remaining nontoxic to humans. By 1908, he had discovered chemicals active against trypanosomes (the protozoan parasites that cause sleeping sicknesses) and against *Treponema pallidum* (trep-ō-nē′mă pal′li-dŭm), the causative agent of syphilis. His discoveries began the branch of medical microbiology known as **chemotherapy.**

In summary, the Golden Age of Microbiology was a time when researchers proved that living things come from other living things, that microorganisms can cause fermentation and disease, and that certain procedures and chemicals can limit, prevent, and cure infectious diseases. These discoveries were made by scientists who applied the scientific method to biological investigation, and they led to an explosion of knowledge in a number of scientific disciplines **(Figure 1.20).**

The Modern Age of Microbiology

Learning Objective

✓ List four major questions that drive microbiological investigations today.

The vast increase in the number of microbiological investigations and in scientific knowledge during the 1800s opened new fields of science, including disciplines called environmental science, immunology, epidemiology, chemotherapy, and genetic engineering **(Table 1.3,** page 20). Microorganisms played a significant role in the development of these disciplines because microorganisms are relatively easy to grow, take up little space, and are available by the trillions. Much of what has been learned about microbes also applies to other organisms, including humans. In the rest of this text we examine advances made in these branches of microbiology, though it would require thousands of books this size to deal with all that is known.

Once the developing science of microbiology had successfully answered questions about spontaneous generation, fermentation, and disease, additional questions arose in each branch of the new science. Since the early 20th century, microbiologists have worked to answer these new questions. In this section we briefly consider some of the 20th century's overarching questions in both basic and applied research. The chapter concludes with a look at some of the questions that might propel microbiological research for the next 50 years.

What Are the Basic Chemical Reactions of Life?

Biochemistry is the study of metabolism—that is, the chemical reactions that occur in living organisms. Biochemistry

▲ *Figure 1.20*

Some of the many scientific disciplines and applications that arose from the pioneering work of scientists just before and during the "Golden Age of Microbiology."

began with Pasteur's work on fermentation by yeast and bacteria, and with Buchner's discovery of enzymes in yeast extract, but by the early 1900s many scientists thought that the metabolic reactions of microbes had little to do with the metabolism of plants and animals.

In contrast, microbiologists Albert Kluyver (1888–1956) and his student C. B. van Niel (1897–1985) proposed that basic biochemical reactions are shared by all living things, that these reactions are relatively few in number, and that their primary feature is the transfer of hydrogen. In adopting this

Table 1.3   Fields of Microbiology

Disciplines	Subject(s) of Study
Basic Research	
Microbe-Centered	
Bacteriology	Bacteria and archaea
Phycology	Algae
Mycology	Fungi
Protozoology	Protozoa
Parasitology	Parasitic protozoa and parasitic animals
Virology	Viruses
Process-Centered	
Microbial metabolism	Biochemistry: chemical reactions within cells
Microbial genetics	Functions of DNA and RNA
Environmental microbiology	Relationships between microbes, and among microbes, other organisms, and their environment
Applied Microbiology	
Medical Microbiology	
Serology	Antibodies in blood serum, particularly as an indicator of infection
Immunology	Body's defenses against specific diseases
Epidemiology	Frequency, distribution, and spread of disease
Etiology	Causes of disease
Infection control	Hospital hygiene and control of nosocomial infections
Chemotherapy	Development and use of drugs to treat infectious diseases
Applied Environmental Microbiology	
Bioremediation	Use of microbes to remove pollutants
Public health microbiology	Sewage treatment, water purification, and control of insects that spread disease
Agricultural microbiology	Use of microbes to control insect pests
Industrial Microbiology (Biotechnology)	
Food and beverage technology	Reduction or elimination of harmful microbes in food and drink
Pharmaceutical microbiology	Manufacture of vaccines and antibiotics
Recombinant DNA technology	Alteration of genes in microbes to synthesize useful products

view, scientists could use microbes as model systems to answer questions about metabolism in all organisms. Research during the 20th century validated this approach to understanding basic metabolic processes, but scientists have also documented an amazing metabolic diversity. Chapter 5 discusses basic metabolic processes, and Chapter 6 considers metabolic diversity.

Basic biochemical research has many practical applications, including:

- The design of herbicides and pesticides that are specific in their action and have no long-term adverse effects on the environment.

- The diagnosis of illnesses and the monitoring of a patient's responses to treatment. For example, physicians routinely monitor liver disease by measuring blood levels of certain enzymes and products of liver metabolism.

- The treatment of metabolic diseases. One example is treating phenylketonuria, a disease resulting from the inability to properly metabolize the amino acid phenlyalanine, by eliminating foods containing phenylalanine from the diet.

- The design of drugs to treat leukemia, gout, bacterial infections, malaria, herpes, AIDS, asthma, and heart attacks.

CRITICAL THINKING

Albert Kluyver said, "From elephant to . . . bacterium—it is all the same!" What did he mean?

How Do Genes Work?

Genetics, the scientific study of inheritance, started in the mid-1800s as an offshoot of botany, but scientists studying microbes made most of the great advances in this discipline.

Microbial Genetics

While working with the bacterium *Streptococcus pneumoniae* (strep-tō-kok'ŭs nū-mō'nē-ī), Oswald Avery (1877–1955), Colin MacLeod (1909–1972), and Maclyn McCarty (1911–) determined that genes are contained in molecules of DNA. In 1958, George Beadle (1903–1989) and Edward Tatum (1909–1975), working with the bread mold *Neurospora crassa* (nū-ros'pōr-ă kras'ă), established that a gene's activity is related to the function of the specific protein coded by that gene. Other researchers, also working with microbes, determined the exact way in which genetic information is translated into a protein, the rates and mechanisms of genetic mutation, and the methods by which cells control genetic expression. We examine all of these aspects of microbial genetics in Chapter 7.

Over the past 40 years, advances in microbial genetics developed into several new disciplines that are among the faster growing areas of scientific research today, including *molecular biology, recombinant DNA technology,* and *gene therapy.*

Molecular Biology

Molecular biology combines aspects of biochemistry, cell biology, and genetics to explain cell function at the molecular level. Molecular biologists are particularly concerned with genome[25] sequencing. Using techniques perfected on microorganisms, molecular biologists have sequenced the genomes of many organisms, including humans and some of their pathogens. It is hoped that a fuller understanding of the genomes of organisms will result in practical ways to limit disease, repair genetic defects, and enhance agricultural yield.

The American Nobel laureate Linus Pauling (1901–1994) proposed in 1965 that gene sequences could provide a means of understanding possible evolutionary relationships and processes, establishing taxonomic categories that more closely reflect these relationships, and identifying the existence of microbes that have never been cultured in a laboratory. Two examples illustrate such uses of gene sequencing data:

- In the 1970s, Carl Woese (1928–) discovered that significant differences in nucleic acid sequences among organisms clearly reveal that cells belong to one of *three* major groups—bacteria, archaea, or eukaryotes—and not merely two groups (prokaryotes and eukaryotes) as previously thought.

- Scientists showed in 1990 that cat-scratch fever is caused by a bacterium that cannot be cultured. The bacterium was discovered by recognition of the sequence of a portion of its ribonucleic acid that differs from all other known ribonucleic acid sequences.

Recombinant DNA Technology

Molecular biology is applied in **recombinant DNA technology,**[26] commonly called *genetic engineering,* which was first developed using microbial models. Geneticists manipulate genes in microbes, plants, and animals for practical applications. For instance, once scientists have inserted the gene for human blood-clotting factor into *Escherichia coli* (esh-ĕ-rik′ē-ă kō′lī), the bacterium produces the factor in a pure form. This technology is a boon to hemophiliacs, who previously depended on clotting factor isolated from donated blood which was possibly contaminated by life-threatening viral pathogens.

Gene Therapy

An exciting new area of study is the use of recombinant DNA technology for **gene therapy,** a process that involves inserting a missing gene or repairing a defective one in human cells. This procedure uses harmless viruses to insert a desired gene into host cells, where it is incorporated into a chromosome and begins to function normally. Chapter 8 examines recombinant DNA technology and gene therapy in more detail.

What Roles Do Microorganisms Play in the Environment?

Learning Objectives

✓ Identify the field of microbiology that studies the role of microorganisms in the environment.

✓ Name the fastest growing scientific disciplines in microbiology today.

Ever since Koch and Pasteur, most research in microbiology has focused on pure cultures of individual species; however, microorganisms are not alone in the "real world." Instead, they live in natural microbial communities in the soil, water, the human body, and other habitats, and these communities play critical roles in such processes as the production of vitamins and *bioremediation*—the use of living bacteria, fungi, and algae to detoxify polluted environments.

Microbial communities also play an essential role in the decay of dead organisms and the recycling of chemicals such as carbon, nitrogen, and sulfur. Martinus Beijerinck discovered bacteria capable of converting nitrogen gas (N_2) from the air into nitrate (NO_3), the form of nitrogen used by plants, and the Russian microbiologist Sergei Winogradsky (1856–1953) elucidated the role of microorganisms in the recycling of sulfur. Together these two microbiologists developed laboratory techniques for isolating and growing environmentally important microbes. Their discoveries and other important aspects of **environmental microbiology** are discussed in Chapter 26.

Another role of microbes in the environment is the causation of disease. Although most microorganisms are not pathogenic, in this book (particularly in Chapters 19–25), we focus on pathogenic microbes because of the threat they pose to human health. We examine their characteristics and the diseases they cause, as well as the steps we can take to limit their abundance and control their spread in the environment, such as sewage treatment, water purification, disinfection, pasteurization, and sterilization.

CRITICAL THINKING

The ability of farmers around the world to produce crops such as corn, wheat, and rice is often limited by the lack of nitrogen-based fertilizer. How might scientists use Beijerinck's discovery to increase world supplies of grain?

How Do We Defend Against Disease?

Why do some people get sick during the flu season while their close friends and family remain well? The germ theory of disease showed not only that microorganisms can cause diseases, but also that the body can defend itself—otherwise, everyone would be sick most of the time.

---

[25]A genome is the total genetic information of an organism.
[26]Recombinant DNA is DNA composed of genes from more than one organism.

▲ *Figure 1.21*

The bacteriocidal effects of penicillin on a bacterial "lawn" in a Petri dish. The clear ring surrounding the fungus colony, which is producing the antibiotic, is where the penicillin killed the bacteria.

The work of Jenner and Pasteur on vaccines showed that the body can protect itself from repeated diseases by the same organism. The German bacteriologist Emil von Behring (1854–1917) and the Japanese microbiologist Shibasaburo Kitasato (1856–1931), working in Koch's laboratory, reported the existence in the blood of chemicals and cells that fight infection. Their studies developed into the fields of *serology*, the study of blood serum[27]—specifically, the chemicals in the liquid portion of blood that fight disease—and *immunology*, the study of the body's defense against specific pathogens. Chapters 16–18 cover these aspects of microbiology, which are of utmost importance to physicians, nurses, and other health care practitioners.

Ehrlich introduced the idea of a "magic bullet" that would kill pathogens, but it wasn't until Alexander Fleming (1881–1955) discovered penicillin **(Figure 1.21)** in 1929 and Gerhard Domagk (1895–1964) discovered sulfa drugs in 1935 that medical personnel finally had drugs effective against a wide range of bacteria. We study chemotherapy, and some physical and chemical agents used to control microorganisms in the environment, in Chapters 9 and 10.

[27]Latin, meaning whey. Serum is the liquid that remains after blood coagulates.

New Frontiers 1.1   Bioterrorism

At 2 A.M. on October 2, 2001, a 63-year-old photo editor named Bob Stevens was admitted to JFK Medical Center in Florida, exhibiting fever, confusion, and low blood pressure. Within a few hours, he had lost consciousness. The diagnosis, confirmed on October 4, was both startling and unexpected: inhalational anthrax, the first documented case on U.S. soil since 1976. On October 5, Stevens died. At the time, no one yet knew that Stevens had contracted anthrax after handling a piece of mail containing anthrax spores—and that his death was the result of a bioterrorist attack that would ultimately claim four more lives, infect 13 other people, and sound alarms throughout a nation still reeling from the events of September 11.

Bioterrorism—the use of viruses, bacteria, fungi, or their toxins for threatening or harmful purposes—is not new. British soldiers sent blankets containing smallpox to enemy tribes during the French and Indian War, and the use of biological agents in both world wars (such as the dropping of plague-infected fleas in cities to spread disease) has been documented. However, bioterrorism is arguably a greater threat today than ever before. The technology to cultivate biological agents, and to effectively disseminate them for purposes of mass destruction, is more advanced than ever. And the roster of potential biological agents now includes newly discovered viruses such as Ebola and bac-

Skin lesion caused by *Bacillus anthracis*

teria that have been genetically modified to resist antibiotics (including drug-resistant strains of *Bacillus anthracis,* the bacterium that causes anthrax.)

Combating bioterrorism requires a joint effort by microbiologists, epidemiologists, immunologists, molecular biologists, and health care workers in governmental agencies and the private sector. The challenges include (first and foremost) preventing such an attack; detecting that an outbreak has occurred; distinguishing between a natural outbreak and a bioterrorist act; and identifying the agent(s) responsible. Plans must then be made for containing the agent's spread, as well as for treating victims expediently and effectively.

Microbiology plays an important role in addressing all these challenges. For example, scientists are in the process of developing *biosensors,* which are handheld devices powered by genetically altered immune cells from mice and jellyfish that can quickly detect harmful microbial agents in the air. Health care workers must apply their knowledge of microbiology to diagnosing and treating patients. And of course, the development of new vaccines, antibiotics, and germicides to defend against biological attacks will necessarily be informed by microbiological research.

Fortunately, transforming microorganisms into effective weapons for mass destruction remains difficult. Pathogens and biotoxins often lose their virulence or toxicity during storage. Even after deployment, various technical and environmental factors—such as dilution in air and water or deactivation by sunlight—further curb their effectiveness. Nonetheless, the threat of bioterrorism remains a real one. Mounting a defense is critical and has the added benefit of increasing our ability to respond to what is likely an even greater threat: the emergence of new infectious diseases, or the reemergence of naturally occurring ones.

What Will the Future Hold?

The science of microbiology is built on asking and answering questions. What began with the curiosity of a dedicated lens grinder in the Netherlands has come far in the past 350 years and has expanded into disciplines as diverse as immunology, recombinant DNA technology, and bioremediation. However, the adage remains true: *The more questions we answer, the more questions we have.*

What will microbiologists discover next? Among the questions for the next 50 years are the following:

- What is it about the physiology of life forms known only by their nucleic acid sequences that prevents those life forms from being grown in the laboratory?

- Can bacteria and archaea be used in ultraminiature technologies such as living computer circuit boards?

- How can an understanding of microbial communities help us understand communities of larger organisms, including humans?

- What genetic sequences make some microbes pathogenic, and what can we do at a genetic level to defend against these pathogens?

- Does life exist beyond planet Earth, and if so, what are its features?

- How can we reduce the threat of infectious diseases, especially those that can be used by bioterrorists? (See New Frontiers 1.1, page 22.)

CHAPTER SUMMARY

The Early Years of Microbiology (pp. 2–7)

1. Leeuwenhoek's observations of **microbes** introduced **microorganisms** to the world and earned him the title "Father of Bacteriology and Protozoology." His discoveries were named and classified by Linnaeus in his **taxonomic system.**

2. Relatively large microscopic eukaryotic **fungi** include **molds** and **yeasts.**

3. Animal-like **protozoa** are single-celled eukaryotes. Some cause disease.

4. Plant-like eukaryotic **algae** are important providers of oxygen, serve as food for many marine animals, and make chemicals used in microbiological growth media.

5. Small **prokaryotic bacteria** and **archaea** live in a variety of communities and in most habitats. Even though some cause disease, most are beneficial.

6. Viruses, the smallest microbes, are so small they can be seen only by using an electron microscope.

7. Parasitic worms, the largest organisms studied by microbiologists, are often visible without a microscope, although their immature stages are microscopic.

The Golden Age of Microbiology (pp. 7–19)

1. The study of the Golden Age of Microbiology includes a look at the men who proposed or refuted the theory of **spontaneous generation**: Aristotle, Redi, Needham, Spallanzani, and Pasteur (the Father of Microbiology). The **scientific method** that emerged then remains the accepted sequence of study today.

2. The study of fermentation by Pasteur and Buchner led to the fields of **industrial microbiology (biotechnology)** and **biochemistry,** and to the study of **metabolism.**

3. Koch, Pasteur, and others proved that **pathogens** cause infectious diseases, an idea that is known as the **germ theory of disease.** **Etiology** is the study of the causation of diseases.

4. Koch initiated careful microbiological laboratory techniques in his search for disease agents. **Koch's postulates,** the logical steps he followed to prove the cause of an infectious disease, remain an important part of microbiology today.

5. The procedure for the **Gram stain** was developed in the 1870s and is still used to differentiate bacteria into two categories: Gram-positive and Gram-negative.

6. The investigations of Semmelweis, Lister, Nightingale, and Snow are the foundations upon which **infection control** and **epidemiology** are built.

7. Jenner's use of a cowpox-based vaccine for preventing smallpox began the field of **immunology.** Pasteur significantly advanced the field.

8. Ehrlich's search for "magic bullets"—chemicals that differentially kill microorganisms—laid the foundations for the field of **chemotherapy.**

The Modern Age of Microbiology (pp. 19–23)

1. Microbiology in the modern age has focused on answering questions regarding **biochemistry,** which is the study of metabolism; microbial genetics, which is the study of inheritance in microorganisms; and **molecular biology,** which involves investigations of cell function at the molecular level.

2. Scientists have applied knowledge from basic research to answer questions in **recombinant DNA technology** and **gene therapy.**

3. The study of microorganisms in their natural environment is **environmental microbiology.**

4. The discovery of chemicals in the blood that are active against specific pathogens advanced immunology and began the field of serology.

5. Advancements in chemotherapy were made in the 1900s with the discovery of numerous substances, such as penicillin and sulfa drugs, that inhibit pathogens.

QUESTIONS FOR REVIEW

(Answers to multiple choice, fill in the blanks, and matching questions are on the web, along with additional review questions. Visit www.microbiologyplace.com.)

Multiple Choice

1. Which of the following microorganisms are not eukaryotic?
   a. bacteria
   b. yeasts
   c. molds
   d. protozoa

2. Which microorganisms are used to make microbiological growth media?
   a. bacteria
   b. fungi
   c. algae
   d. protozoa

3. In which habitat would you most likely find archaea?
   a. acidic hot springs
   b. swamp mud
   c. Great Salt Lake
   d. all of the above

4. Of the following scientists, who defended the theory of abiogenesis?
   a. Aristotle
   b. Pasteur
   c. Needham
   d. Spallanzani

5. Which of the following scientists hypothesized that a bacterial colony arises from a single bacterial cell?
   a. Antoni van Leeuwenhoek
   b. Louis Pasteur
   c. Robert Koch
   d. Richard Petri

6. Which scientist first hypothesized that medical personnel can infect patients with pathogens?
   a. Edward Jenner
   b. Joseph Lister
   c. John Snow
   d. Ignaz Semmelweis

Fill in the Blanks

Fill in the blanks with the name(s) of the scientist(s) whose investigations led to the following fields of study in microbiology.

1. Environmental microbiology _____ and _____
2. Biochemistry _____ and _____
3. Chemotherapy _____
4. Immunology _____
5. Public health microbiology _____
6. Etiology _____
7. Epidemiology _____
8. Biotechnology _____
9. Food microbiology _____

Matching Questions

Match each of the following descriptions with the person it best describes. An answer may be used more than once.

1. ___ Developed smallpox immunization   A. Alexander Fleming

2. ___ First photomicrograph of bacteria   B. Paul Ehrlich

3. ___ Germ theory of disease   C. Louis Pasteur

4. ___ Germs cause disease   D. Antoni van Leeuwenhoek

5. ___ Sought a "magic bullet" to destroy pathogens   E. Carolus Linnaeus

6. ___ Discovered penicillin   F. Turberville Needham

7. ___ Father of Microbiology   G. Eduard Buchner

8. ___ Father of Microbiological Laboratory Techniques   H. Robert Koch

9. ___ Father of Bacteriology   I. Joseph Lister

10. ___ Father of Protozoology   J. Edward Jenner

11. ___ Founder of antiseptic surgery   K. Girolamo Fracastoro

Short Answer

1. Why was the theory of spontaneous generation a hindrance to the development of the field of microbiology?

2. Discuss the significant difference between the flasks used by Pasteur and Spallanzani. How did Pasteur's investigation settle the dispute about spontaneous generation?

3. List four types of microorganisms from smallest to largest.

4. Defend this statement: "The investigations of Antoni van Leeuwenhoek changed the world forever."

5. Why would a *macro*scopic tapeworm be studied in *micro*biology?

6. Describe what has been called "The Golden Age of Microbiology" with reference to four major questions that propelled scientists during that period.

7. List four major questions that drive microbiological investigations today.

8. Refer to the four steps in the scientific method in describing Pasteur's fermentation experiments.

9. List Koch's postulates, and explain why they are significant.

CRITICAL THINKING

1. If Robert Koch had become interested in a viral disease such as influenza instead of anthrax (caused by a bacterium), how might his list of lifetime accomplishments been different? Why?

2. In 1911, the Polish scientist Kasimir Funk proposed that a limited diet of polished white rice (rice without the husks) caused beriberi, a disease of the central nervous system. Even though history has proven him correct—beriberi is caused by a thiamine deficiency, which in his day resulted from unsophisticated milling techniques that removed the thiamine-rich husks—Funk was criticized by his contemporaries, who told him to find the microbe that caused beriberi. Explain how the prevailing scientific philosophy of the day shaped Funk's detractors' point of view.

3. Haemophilus influenzae does not cause flu, but it received its name because it was once thought to be the cause. Explain how a proper application of Koch's postulates would have prevented this error in nomenclature.

4. Just before winter break in early December, your roommate stocks the refrigerator with a gallon of milk, but both of you leave before opening it. When you return in January, the milk has soured. Your roommate is annoyed because the milk was pasteurized and thus should not have spoiled. Explain why your roommate's position is unreasonable.

5. Design an experiment to prove that microbes don't spontaneously generate in milk.

6. The British General Board of Health concluded in 1855 that the Broad Street cholera epidemic (see page 18) resulted from fermentation of "nocturnal clouds of vapor" from the polluted Thames River. How could an epidemiologist prove or disprove this claim?

CHAPTER 2

The Chemistry of Microbiology

CHAPTER OUTLINE

Atoms
Atomic Structure
Isotopes
Electron Configurations

Chemical Bonds
Nonpolar Covalent Bonds
Polar Covalent Bonds
Ionic Bonds
Hydrogen Bonds

Chemical Reactions
Synthesis Reactions
Decomposition Reactions
Exchange Reactions

Water, Acids, Bases, and Salts
Water
Acids and Bases
Salts

Organic Macromolecules
Functional Groups and Monomers
Lipids
Carbohydrates
Proteins
Nucleic Acids

Is there microbial life on other planets? From telescopic observations and space probes we have identified the chemicals found on our nearest planetary neighbors. Venus, for example, has a thick atmosphere of CO_2, sulfur, and sulfuric acid (H_2SO_4), and its soil contains a significant amount of iron. Microorganisms could potentially live on Venus, since carbon (essential for all life forms) is available in Venus's atmosphere, and we know some microbes use iron and sulfur for energy. To date, probes have found no evidence of life on Venus or any other planet, but the tantalizing possibility remains.

What environmental conditions are needed to support microbial life on Earth? How can we develop treatments against harmful bacteria and viruses? How do we differentiate one microorganism from another? The answers to these questions, and to many other important questions in microbiology, have their basis in chemistry.

Does microbial life exist on Venus?

MicroPrep Pre-Test: *Take the pre-test for this chapter on the web.*
Visit ***www.microbiologyplace.com***

Learning some basic concepts of chemistry will enable you to understand more fully the variety of interactions between microorganisms and their environment—which includes you. If you plan a career in health care, you will find microbial chemistry involved in the diagnosis of disease, the response of the immune system, the growth and identification of pathogenic microorganisms in the laboratory, and the function and selection of antibiotics. Even preserving your own health is aided by an understanding of the fundamentals of chemistry.

In this chapter we study atoms, which are the basic units of chemistry, and we consider how atoms react with one another to form chemical bonds and molecules. Then we examine the three major categories of chemical reactions. The chapter concludes with a look at the molecules of greatest importance to life: water, acids, bases, lipids, carbohydrates, proteins, nucleic acids, and ATP.

Atoms

Learning Objective

✓ Define *matter, atom,* and *element,* and explain how these terms relate to one another.

Matter is defined as anything that takes up space and has mass.[1] The smallest chemical units of matter are **atoms.** Atoms are extremely small, and only the very largest of them can be seen using the most powerful microscopes. Therefore, scientists have developed various models to conceptualize and illustrate the structure of atoms.

Atomic Structure

Learning Objective

✓ Draw and label an atom, showing the parts of the nucleus and orbiting electrons.

In 1913, the Danish physicist Niels Bohr proposed a simple model in which negatively charged subatomic particles called **electrons** orbit a centrally located nucleus like planets in a miniature solar system **(Figure 2.1).** A nucleus is composed of uncharged **neutrons** and positively charged **protons.** (The only exception to this description is the nucleus of a normal hydrogen atom, which is composed of only a single proton and no neutrons.) The number of electrons in an atom typically equals the number of protons, so overall atoms are electrically neutral.

An **element** is matter that is composed of a single type of atom. For example, gold is an element because it consists of only gold atoms. In contrast, the ink in your pen is not an element because it is composed of many different kinds of atoms.

Elements differ from one another in their **atomic number,** which is the number of protons in their nuclei. For example, the atomic numbers of hydrogen, carbon, and oxygen are 1, 6, and 8, respectively, because all hydrogen nu-

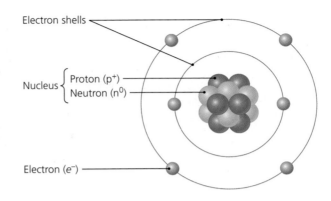

▲ *Figure 2.1*

An example of a Bohr model of atomic structure. This drawing is not to scale; for the electrons to be shown in scale with the greatly magnified nucleus, they would have to occupy orbits located many miles from the nucleus. Put another way, the volume of an entire atom is 100 trillion times the volume of its nucleus.

clei contain a single proton, all carbon nuclei have six protons, and all oxygen nuclei have eight protons.

The **atomic mass** of an atom (sometimes called its *atomic weight*) is the sum of the masses of its protons, neutrons, and electrons. Protons and neutrons each have a mass of approximately 1 *atomic mass unit,*[2] which is also called a *dalton.*[3] An electron is much less massive, with a mass of about 0.00054 dalton. Electrons are often ignored in discussions of atomic mass because their contribution to the overall mass is negligible. Therefore, the sum of the number of protons and neutrons approximates the atomic mass of an atom.

There are 93 naturally occurring elements known;[4] however, organisms typically utilize only about 20 elements, each of which has its own symbol that is derived from its English or Latin name **(Table 2.1** on page 28).

Isotopes

Learning Objective

✓ List at least four ways that radioactive isotopes are useful.

Every atom of an element has the same number of protons, but atoms of a given element can differ in the number of neutrons in their nuclei. Atoms that differ in this way are called **isotopes.** For example, there are three naturally

---

[1]*Mass* and *weight* are sometimes confused. Mass is the quantity of material in something, whereas weight is the effect of gravity on mass. Even though an astronaut is weightless in space, his mass is the same in space as on Earth.
[2]An atomic mass unit (dalton) is 1/597,728,630,000,000,000,000,000, or 1.673×10^{-24} grams.
[3]Named for John Dalton, the British chemist who helped develop atomic theory around 1800.
[4]For many years, scientists thought that there were only 92 naturally occurring elements, but natural plutonium was discovered in Africa in 1997.

Table 2.1 **Common Elements of Life**

Element	Symbol	Atomic Number	Atomic Mass*	Biological Significance
Hydrogen	H	1	1	Component of organic molecules and water; H$^+$ released by acids
Boron	B	5	11	Essential for plant growth
Carbon	C	6	12	Backbone of organic molecules
Nitrogen	N	7	14	Component of amino acids, proteins, and nucleic acids
Oxygen	O	8	16	Component of many organic molecules and water; OH$^-$ released by bases; necessary for aerobic metabolism
Sodium (Natrium)	Na	11	23	Principal cation outside cells
Magnesium	Mg	12	24	Component of many energy-transferring enzymes
Silicon	Si	14	28	Component of cell wall of diatoms
Phosphorus	P	15	31	Component of nucleic acids and ATP
Sulfur	S	16	32	Component of proteins
Chlorine	Cl	17	35	Principal anion outside cells
Potassium (Kalium)	K	19	39	Principal cation inside cells; essential for nerve impulses
Calcium	Ca	20	40	Utilized in many intercellular signaling processes; essential for muscular contraction
Manganese	Mn	25	54	Component of some enzymes
Iron (Ferrum)	Fe	26	56	Component of energy-transferring proteins; transports oxygen in the blood of many animals
Cobalt	Co	27	59	Component of vitamin B$_{12}$
Copper (Cuprum)	Cu	29	64	Component of some enzymes
Zinc	Zn	30	65	Component of some enzymes
Molybdenum	Mo	42	96	Component of some enzymes
Iodine	I	53	127	Component of thyroid hormones

*Rounded to nearest whole number.

occurring isotopes of carbon, each having six protons and six electrons **(Figure 2.2).** Over 95% of carbon atoms also have six neutrons. Because these atoms have six protons and six neutrons, the atomic mass of this isotope is about 12 daltons, and it is known as carbon-12, symbolized as ^{12}C. Atoms of carbon-13 (^{13}C) have seven neutrons per nucleus, and ^{14}C atoms each have eight neutrons.

Unlike the first two isotopes, the nucleus of ^{14}C is unstable due to the ratio of its protons and neutrons. Unstable atomic nuclei release energy and subatomic particles such as neutrons, protons, and electrons in a process called *radioactive decay*. Atoms that undergo radioactive decay are *radioactive isotopes*. Radioactive decay and radioactive isotopes play important roles in microbiological research, medical diagnosis, the treatment of disease, and the complete destruction of contaminating microbes (sterilization) of medical equipment and chemicals.

(a) Carbon –12
6 Protons
6 Neutrons

(b) Carbon –13
6 Protons
7 Neutrons

(c) Carbon –14
6 Protons
8 Neutrons

▲ *Figure 2.2*

Nuclei of the three naturally occurring isotopes of carbon. Each isotope also has six electrons, which are not shown. *What are the atomic number and atomic mass of each of these isotopes?*

Electron Configurations

While the nuclei of atoms determine their identities, nuclei of different atoms almost never come close enough together

Figure 2.2 The atomic number of all three is 6; their atomic masses are 12, 13, and 14, respectively.

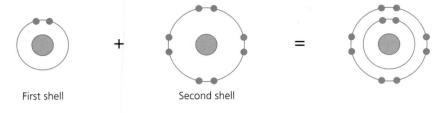

First shell   Second shell

(a) Electron shells of neon: two-dimensional view

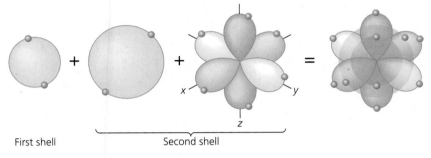

First shell   Second shell

(b) Electron shells of neon: three-dimensional view

▲ *Figure 2.3*

Electron configurations. **(a)** Two-dimensional model (Bohr diagram) of the electron shells of neon. **(b)** Three-dimensional model of the electron shells of neon. In this model, the first shell is a small sphere, whereas the second shell consists of a larger sphere plus three pairs of ellipses that extend from the nucleus at right angles. Larger shells (not shown) are even more complex.

to interact.[5] Typically, only the electrons of atoms interact, which is why it is electrons that determine an atom's *chemical behavior.* Thus, because all of the isotopes of carbon (for example) have the same number of electrons, all these isotopes behave the same way in chemical reactions, even though their nuclei are different.

Scientists know that electrons do not really orbit the nucleus in a two-dimensional circle, as indicated by a Bohr diagram **(Figure 2.3a),** but instead in three-dimensional *electron shells* or *clouds* that assume unique shapes dependent on the energy of the electrons **(Figure 2.3b).** More accurately put, an electron shell depicts the *probable* locations of electrons at a given time; nevertheless, it is simpler and more convenient to draw electron shells as circles.

Each electron shell can hold only a certain, maximum number of electrons. For example, the first shell (the one nearest the nucleus) can accommodate a maximum of two electrons, and the second shell can hold no more than eight electrons. Atoms of hydrogen and helium have one and two electrons, respectively; thus, these two elements have only a single electron shell. A lithium atom, which has three electrons, has two shells.

Atoms with more than 10 electrons require more shells. The third shell holds up to eight electrons when it is the outermost shell, though its capacity increases to 18 when the fourth shell contains two electrons. Atoms of heavier atoms have even more shells, but these atoms do not play significant roles in the processes of life.

It is the electrons in the outermost shell of atoms, called *valence electrons,* that actually interact with other atoms. **Figure 2.4** on page 30 depicts the electron configurations of atoms of some elements important to microbial life. Notice that except for helium, atoms of all elements in a given column have the same number of valence electrons. Helium is placed in the far right-hand column with the other inert gases, because its outer shell is full, though it has two rather than eight valence electrons.

Chemical Bonds

Learning Objectives

✓ Describe the configuration of electrons in a stable atom.
✓ Contrast molecules and compounds.

Outer electron shells are stable when they contain eight electrons (except for the first electron shell, which is stable with only two electrons, since that is its maximum number). When atoms' outer shells are not filled with eight electrons, they either have room for more electrons or have "extra" electrons, depending on whether it is easier for them to gain electrons or lose electrons. For example, an oxygen atom, with six electrons in its outer shell, has two "unfilled spaces" (see Figure 2.4), because it requires less energy for the oxygen atom to gain two electrons than to lose six electrons. A calcium atom, by contrast, has two "extra" electrons in its outer (fourth) shell, because it requires less energy to lose these two electrons than to gain six new ones. When a calcium atom loses two electrons, its third shell, which is then its outer shell, is full and stable with eight electrons.

As previously noted, an atom's outermost electrons are called valence electrons, and thus the outermost shell of an atom is the *valence shell.* An atom's **valence,**[6] defined as its combining capacity, is considered to be positive if its valence shell has extra electrons to give up, and to be negative if its valence shell has spaces to fill. Thus a calcium atom, with two electrons in its valence shell, has a valence of $+2$, whereas an oxygen atom, with two spaces to fill in its valence shell, has a valence of -2.

Atoms combine with one another by either sharing or transferring valence electrons in such a way as to fill their valence shells. Such interactions between atoms are called **chemical bonds.** Two or more atoms held together by chemical bonds form a **molecule.** If a molecule contains atoms of more than one element, it is a **compound.** Two hydrogen atoms bonded together form a hydrogen molecule, which is not a compound because only one element is involved.

---

[5]Except during nuclear reactions, such as occur in nuclear power plants.
[6]From Latin *valentia,* meaning strength.

▲ *Figure 2.4*

Bohr diagrams of the first 20 elements, and their places within the chart known as the periodic table of the elements. Note that the number of valence electrons increases from left to right in each row, and that every element in a column has the same number of valence electrons (with the exception of helium). Heavier atoms have been omitted because most heavy elements are less important to living organisms.

However, two hydrogen atoms bonded to an oxygen atom form a molecule of water (H_2O), which is a compound.

In this section, we discuss the three principal types of chemical bonds: *nonpolar covalent bonds, polar covalent bonds,* and *ionic bonds.* We also consider *hydrogen bonds,* which are weak forces that act with polar covalent bonds to give certain large chemicals their characteristic three-dimensional shapes.

CRITICAL THINKING

Neon (atomic mass 10) and argon (atomic mass 18) are *inert* elements, which means that they very rarely form chemical bonds. Give the electron configuration of their atoms and explain why these elements are inert.

Nonpolar Covalent Bonds

Learning Objective

✓ Contrast nonpolar covalent, polar covalent, and ionic bonds.

A **covalent**[7] **bond** is the sharing of a pair of electrons by two atoms. Consider, for example, what happens when two hydrogen atoms approach one another. Each hydrogen atom consists of a single proton orbited by a single electron. Since

the valence shell of each atom requires two electrons to be filled, each atom shares its single electron with the other, forming a hydrogen molecule in which both atoms have full shells **(Figure 2.5a).** Similarly, two oxygen atoms can share electrons, but they must share *two* pairs of electrons for their valence shells to be full **(Figure 2.5b).** Since two pairs of electrons are involved, oxygen atoms form two covalent bonds, or a *double covalent bond,* with one another.

The attraction of an atom for electrons is called its **electronegativity.** The more electronegative an atom, the greater the pull its nucleus exerts on electrons. Note in **Figure 2.6** on page 32, which displays the electronegativities of atoms of several elements, that electronegativities tend to increase from left to right in the chart. The reason is that elements toward the right of the chart have more protons and thus exert a greater pull on electrons. Electronegativities of elements decrease from top to bottom in the chart because of the increasing distance between the nucleus and the valence shell as elements get larger.

Atoms with equal or nearly equal electronegativities, such as two hydrogen atoms or a hydrogen and a carbon, share electrons equally or nearly equally. In chemistry and

[7]From Latin *co,* meaning with or together, and *valentia,* meaning strength.

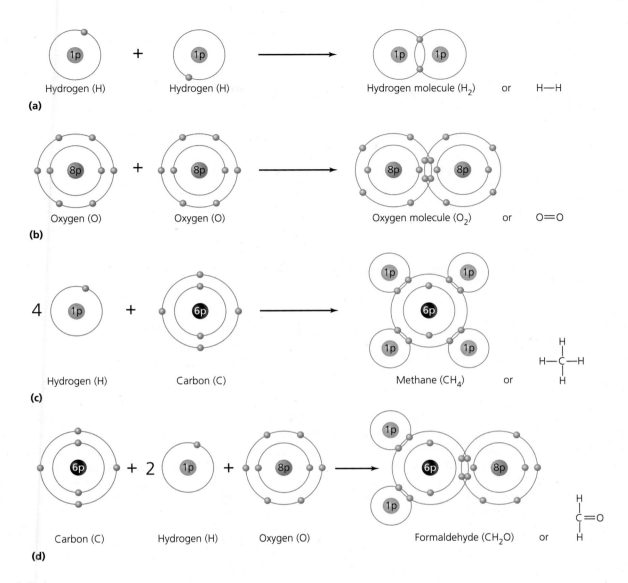

▲ *Figure 2.5*

Four molecules formed by covalent bonds. **(a)** Hydrogen. Each hydrogen atom needs another electron to have a full valence shell. The two atoms share their electrons, forming a covalent bond. **(b)** Oxygen. Oxygen atoms have six electrons in their valence shells; thus, they need two electrons each. When they share with each other, two covalent bonds are formed. Note that the valence electrons of oxygen atoms are in the second shell. **(c)** A methane molecule, which has four single covalent bonds. **(d)** Formaldehyde. The carbon atom forms a double bond with the oxygen atom and single bonds with two hydrogen atoms. *Which of these molecules are also compounds? Why?*

Figure 2.5 Methane and formaldehyde molecules are also compounds because they are composed of more than one element.

physics, "poles" are opposed forces, such as north and south magnetic poles or positive and negative terminals of a battery. In the case of atoms with similar electronegativities, the shared electrons tend to spend an equal amount of time around each nucleus of the pair, and no poles exist; therefore, the bond between them is a **nonpolar covalent bond.** All the covalent bonds illustrated in Figure 2.5 are nonpolar.

A hydrogen molecule can be symbolized a number of ways:

$$H—H \qquad H:H \qquad H_2$$

In the first symbol, the dash represents the chemical bond between the atoms. In the second symbol, the dots represent the electron pair of the covalent bond. These two symbols

	I	II											III	IV	V	VI	VII	Inert gases
	H 2.1																	He 0.0
	Li 1.0	Be 1.5											B 2.0	C 2.5	N 3.0	O 3.5	F 4.0	Ne 0.0
	Na 0.9	Mg 1.2											Al 1.5	Si 1.8	P 2.1	S 2.5	Cl 3.0	Ar 0.0
	K 0.8	Ca 1.0	Sc 1.3	Ti 1.5	V 1.6	Cr 1.6	Mn 1.5	Fe 1.8	Co 1.8	Ni 1.8	Cu 1.9	Zn 1.6	Ga 1.6	Ge 1.8	As 2.0	Se 2.4	Br 2.8	Kr 0.0

▲ *Figure 2.6*

Electronegativity values of selected elements. The values are expressed according to the Pauling scale, named for the Nobel Prize-winning chemist Linus Pauling, who based the scale on bond energies. Pauling chose to compare the electronegativity of each element to that of fluorine, to which he assigned a value of 4.0.

are known as *structural formulas.* In the third symbol, known as a *molecular formula,* the subscript "2" indicates the number of hydrogen atoms that are bonded, not the number of shared electrons. Each of these symbols indicates the same thing—two hydrogen atoms are sharing a pair of electrons.

Many atoms need more than one electron to fill their valence shell. For instance, a carbon atom has four valence electrons and needs to gain four more if it is to have eight in its valence shell **(see Figure 2.5c).** As before, a line in the structural formula represents a covalent bond formed from the sharing of two electrons. Note that two covalent bonds are formed between an oxygen atom and a carbon atom in formaldehyde **(see Figure 2.5d).** This fact is represented by a double line, which indicates that the carbon atom shares four electrons with the oxygen atom.

Carbon atoms are critical to life. Since a carbon atom has four electrons in its valence shell, it has equal tendency to either lose four electrons or gain four electrons. Either event produces a full outer shell. The result is that carbon atoms tend to share electrons and form four covalent bonds with one another, and with many other types of atoms. Each carbon atom in effect acts as a four-way intersection where different components of a molecule can attach. One result of this feature is that carbon atoms can form very large chains that constitute the "backbone" of many biologically important molecules. Carbon chains can be branched or unbranched, and some even close back on themselves to form rings. Compounds that contain carbon and hydrogen atoms are called **organic compounds.** Among the many biologically important organic compounds are proteins and carbohydrates, which are discussed later in the chapter.

CRITICAL THINKING

An article in the local newspaper about gangrene states that the tissue-destroying toxin, lecithinase, is "an organic compound." But many people consider "organic" chemicals to mean something is good. Explain the apparent contradiction.

Polar Covalent Bonds

Learning Objective

✓ Explain the relationship between electronegativity and the polarity of a covalent bond.

If two covalently bound atoms have significantly different electronegativities, their electrons will not be shared equally. Instead, the electron pair will spend more time orbiting the nucleus of the atom with greater electronegativity. This type of bond, in which there is unequal sharing of electrons, is a **polar covalent bond.** An example of a molecule with polar covalent bonds is water **(Figure 2.7a).**

Because oxygen is more electronegative than hydrogen, the electrons spend more time near the oxygen nucleus than near the hydrogen nuclei, and thus the oxygen atom acquires a partial negative charge (symbolized as δ^-). The hydrogen nuclei each have a corresponding partial positive charge (δ^+). The covalent bonds between the oxygen atom and two hydrogen atoms are called polar because they have opposite electrical charges.

Polar covalent bonds can form between many different elements, but the most important polar covalent bonds for life are those that involve hydrogen because they allow hydrogen bonding, which we discuss shortly.

Both nonpolar and polar covalent bonds form angles between atoms such that the distances between electron orbits are maximized. The bond angle for water is shown in **Figure 2.7b.** However, it is more convenient to simply draw molecules as if all the atoms were in one plane, as in **Figure 2.7c.**

CRITICAL THINKING

The deadly poison hydrogen cyanide has the chemical formula H—C≡N. Describe the bonds between carbon and hydrogen, and between carbon and nitrogen, in terms of the number of electrons involved.

Triple covalent bonds are stronger and more difficult to break than single covalent bonds. Explain the reason why by referring to the stability of a valence shell that contains eight electrons.

(a)

(b)

H—O—H

(c)

▲ *Figure 2.7*

Polar covalent bonding in a water molecule. **(a)** A Bohr model of a water molecule, which has two polar covalent bonds. When the electronegativities of two atoms are significantly different, the shared electrons of covalent bonds spend more time around the more electronegative atom, giving it a partial negative charge (δ^-). **(b)** The bond angle in a water molecule. Atoms maximize the distances between electron orbitals in polar and nonpolar covalent bonds. **(c)** A common, convenient way of representing water as a polar molecule without showing bond angles.

Sodium chloride (NaCl)

(a)

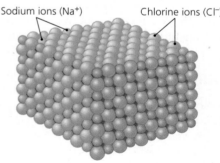

Sodium chloride crystal

(b)

▲ *Figure 2.8*

The interaction of sodium and chlorine to form an ionic bond. **(a)** Because the chlorine atom has a much higher electronegativity than the sodium atom, it "strips" the valence electron from sodium ①, forming a positively charged sodium ion and a negatively charged chlorine ion ②. The ion's opposite charges attract one another, forming an ionic bond ③. **(b)** The arrangement of sodium and chlorine ions in a sodium chloride (NaCl) crystal.

Ionic Bonds

Learning Objective

✓ Define ionization using the terms *cation* and *anion*.

Consider what happens when two atoms with vastly different electronegativities—for example, sodium, with one electron in its valence shell and an electronegativity of 0.9, and chlorine, with seven electrons in its valence shell and an electronegativity of 3.0—come together. Chlorine has such a higher electronegativity that it very strongly attracts sodium's valence electron, and the result is that the sodium loses that electron to chlorine (step ① in **Figure 2.8a**).

Now that the chlorine atom has one more electron than it has protons, it has a full negative charge, and the sodium atom, which has lost an electron, now has a full positive charge (step ❷). An atom or group of atoms that has either a full negative charge or a full positive charge is called an *ion.* Positively charged ions are called **cations,** whereas negatively charged ions are called **anions.**

Because of their opposite charges, cations and anions attract each other and form what is termed an **ionic bond** (step ❸). They form crystalline ionic compounds known as **salts,** such as sodium chloride (NaCl), also known as table salt **(Figure 2.8b),** and potassium chloride (KCl, sodium-free table salt). Ionic bonds differ from covalent bonds in that ions do not share electrons. Instead, the bond is formed from the attraction of opposite electrical charges.

The polar bonds of water molecules interfere with the ionic bonds of salts, causing *dissociation* (also called *ionization*) **(Figure 2.9).** This occurs as the partial negative charge on the oxygen atom of water attracts cations, and the partial positive charge on hydrogen atoms attracts anions. Obviously, the presence of polar bonds interferes with the attraction between the cation and anion.

When cations and anions disassociate from one another and become surrounded by water molecules (are hydrated), they are called **electrolytes** because they can conduct electricity through the solution. Electrolytes are critical for life because they stabilize a variety of compounds, act as electron carriers, and allow electrical gradients to exist within cells. We examine these functions of electrolytes in later chapters.

In nature, chemical bonds range from nonpolar bonds to polar bonds to ionic bonds. The important thing to remember is that electrons are shared between atoms in covalent bonds, and transferred from one atom to another in ionic bonds.

CRITICAL THINKING

According to the chart in Figure 2.6, what type of bond (nonpolar covalent, polar covalent, or ionic) would you expect between chlorine and potassium? Between carbon and nitrogen? Between phosphorus and oxygen? Explain your reasoning in each case.

Hydrogen Bonds

Learning Objective

✓ Describe hydrogen bonds, and discuss their importance in living organisms.

As we have seen, hydrogen atoms bind to oxygen atoms by means of polar covalent bonds, which results in partial positive charges on the hydrogen atoms. Hydrogen atoms form polar covalent bonds with atoms of other elements as well.

The electrical attraction between a partially charged hydrogen atom and a full or partial negative charge on either a different region of the same molecule or another molecule is called a **hydrogen bond (Figure 2.10).** Hydrogen bonds can be likened to weak ionic bonds in that they arise from the

Water surrounds ions in salt crystal

Individual Na⁺ and Cl⁻ ions break away

Hydrated sodium ion   Hydrated chlorine ion

▲ *Figure 2.9*

Dissociation of NaCl in water. When water surrounds the ions in a NaCl crystal, the partial charges on water molecules are attracted to charged ions, and the water molecules hydrate the ions by surrounding them. The partial negative charges on oxygen atoms are attracted to cations (in this case, the sodium ions), and the partial positive charges on hydrogen atoms are attracted to anions (the chlorine ions). Because the ions no longer attract one another, the salt crystal dissolves. Hydrated ions are called electrolytes.

Hydrogen bond

Cytosine   Guanine

▲ *Figure 2.10*

Hydrogen bonds can hold together portions of the same molecule, or hold two different molecules together. In this case, three hydrogen bonds are holding molecules of cytosine and guanine together.

Table 2.2   Characteristics of Chemical Bonds

Type of Bond	Description	Relative Strength
Nonpolar covalent bond	Pair of electrons is nearly equally shared between two atoms	Strong
Polar covalent bond	Electrons spend more time around the more electronegative of two atoms	Strong
Ionic bond	Electrons are stripped from a cation by an anion	Weaker than covalent in aqueous environments
Hydrogen bond	Partial positive charges on hydrogen atoms are attracted to full and partial negative charges on other molecules or other regions of the same molecule	Weaker than ionic

attraction of positive and negative charges. Notice also that although they are a consequence of polar covalent bonds between hydrogen atoms and other, more electronegative atoms, hydrogen bonds themselves are not covalent bonds—they do not involve the sharing of electrons.

As we have seen, covalent bonds are essential for life because they strongly link atoms together to form molecules. Hydrogen bonds, though weaker than covalent bonds, are also essential. The cumulative effect of numerous hydrogen bonds is to stabilize the three-dimensional shapes of large molecules. For example, the familiar double-helix shape of DNA is due to the stabilizing effects of thousands of hydrogen bonds holding the molecule together. Exact shape is critical for the functioning of enzymes, antibodies, intercellular chemical messengers, and the recognition of target cells by pathogens. Further, because hydrogen bonds are weak, they can be overcome when necessary. For example, the two complementary halves of a DNA molecule are held together primarily by hydrogen bonds, and can be separated for DNA replication and other processes (see Figure 7.5 on page 206).

Table 2.2 summarizes the characteristics of chemical bonds.

Chemical Reactions

Learning Objective

✓ Describe three general types of chemical reactions found in living things.

You are already familiar with many consequences of chemical reactions: You add yeast to bread dough and it rises; enzymes in your laundry detergent remove grass stains; and gasoline burned in your car releases energy to speed you on your way. What exactly is happening in these reactions? What is the precise definition of a chemical reaction?

We have discussed how bonds are formed via the sharing of electrons or the attraction of positive and negative charges. Scientists define **chemical reactions** as the making or breaking of such chemical bonds. All chemical reactions begin with **reactants**—the atoms, ions, or molecules that exist at the beginning of a reaction. Similarly, all chemical reactions result in **products**—the atoms, ions, or molecules left after the reaction is complete.

Reactants and products may have very different physical and chemical characteristics. For example, hydrogen and oxygen are gases, and both have very different properties from water. However, the numbers and types of atoms never change in a chemical reaction; atoms are neither destroyed nor created, only rearranged.

Now let's turn our attention to three general categories of chemical reactions that occur in microorganisms: *synthesis, decomposition,* and *exchange reactions.*

Synthesis Reactions

Learning Objectives

✓ Give an example of a synthesis reaction that involves the formation of a water molecule.
✓ Contrast endothermic and exothermic chemical reactions.

Synthesis reactions involve the formation of larger, more complex molecules. Synthesis reactions can be expressed symbolically as:

$$\text{Reactant} + \text{Reactant} \rightarrow \text{Product(s)}$$

The arrow indicates the direction of the reaction and the formation of new chemical bonds. For example, algae make their own glucose (sugar) using the following reaction:

$$6\,H_2O + 6\,CO_2 \rightarrow C_6H_{12}O_6 + 6\,O_2$$

The reaction is read, "Six molecules of water plus six molecules of carbon dioxide yield one molecule of glucose and six molecules of oxygen." Notice that the total number and kind of atoms are the same on both sides of the reaction.

An important type of synthesis reaction is **dehydration synthesis,** in which two smaller molecules are joined together by a covalent bond, and a water molecule is also formed (**Figure 2.11a** on page 36). The word *dehydration* in the name of this type of reaction refers to the fact that one of the products is a water molecule formed when a hydrogen ion (H^+) from one reactant combines with a hydroxyl ion (OH^-) from another reactant.

Synthesis reactions require energy to break bonds in the reactants and to form new bonds to make products. Reactions

(a) Dehydration synthesis

(b) Hydrolysis

▲ *Figure 2.11*

Two types of chemical reactions in living things. **(a)** Dehydration synthesis. In this energy-requiring reaction, a hydroxyl ion (OH⁻) removed from one reactant and a hydrogen ion (H⁺) removed from another reactant combine to form hydrogen hydroxide (HOH), which is water. **(b)** Hydrolysis, an energy-yielding reaction that is the reverse of a dehydration synthesis reaction. *What are the scientific words meaning "energy-requiring" and "energy-yielding"?*

Figure 2.11 Endothermic means "energy requiring," and exothermic means "energy releasing."

that require energy are said to be **endothermic**[8] **reactions** because they trap energy within new molecular bonds. As we will see in Chapter 6, an energy supply for fueling synthesis reactions is one common requirement of all living things.

Taken together, all of the synthesis reactions in an organism are called **anabolism.**

Decomposition Reactions

Learning Objective

✓ Give an example of a decomposition reaction that involves breaking the bonds of a water molecule.

Decomposition reactions are the reverse of synthesis reactions in that they break bonds within larger molecules to form smaller atoms, ions, and molecules. These reactions release energy and are therefore **exothermic.**[9] In general, decomposition reactions can be represented by the following formula:

$$\text{Reactant} \rightarrow \text{Product} + \text{Product}$$

An example of a biologically important decomposition reaction is the aerobic decomposition of glucose to form carbon dioxide and water:

$$C_6H_{12}O_6 + 6\,O_2 \rightarrow 6\,H_2O + 6\,CO_2$$

Note that this reaction is exactly the reverse of the synthesis reaction in algae that we examined previously. Synthesis and decomposition reactions are often reversible in living things.

A common type of decomposition reaction is **hydrolysis,**[10] the reverse of dehydration synthesis **(Figure 2.11b).** In hydrolytic reactions, a covalent bond in a large molecule is broken, and the ionic components of water (H⁺ and OH⁻) are added to the products.

Collectively, all of the decomposition reactions in an organism are called **catabolism.**

Exchange Reactions

Learning Objective

✓ Compare exchange reactions to synthesis and decomposition reactions.

Exchange reactions (also called *transfer reactions*) have features similar to both synthesis and decomposition reactions. For instance, they involve breaking and forming covalent bonds, and they involve both endothermic and exothermic steps. As the name suggests, atoms are moved from one molecule to another. In general, these reactions can be represented as either

$$A + BC \rightarrow AB + C$$

or

$$AB + CD \rightarrow AD + BC$$

An important exchange reaction within organisms is the phosphorylation of glucose:

$$C_6H_{12}O_6 + A\text{-}P\text{-}P\text{-}P \rightarrow C_6H_{12}O_6\text{-}P + A\text{-}P\text{-}P$$

| Glucose | Adenosine triphosphate | Glucose phosphate | Adenosine diphosphate |

The sum of all of the chemical reactions in an organism, both catabolic and anabolic, is called **metabolism.** We examine metabolism in more detail in Chapter 5.

---

[8]From Greek *endon,* meaning within, and *thermos,* meaning heat (energy).
[9]From Greek *exo,* meaning outside, and *thermos,* meaning heat (energy).
[10]From Greek *hydor,* meaning water, and *lysis,* meaning loosing.

Figure 2.12 ▶

The cohesiveness of liquid water. **(a)** Water molecules are cohesive because hydrogen bonds cause them to stick to one another. **(b)** One result of cohesiveness in water is surface tension, which can be strong enough to support the weight of insects known as water striders.

Hydrogen bonds

(a)

(b)

Water, Acids, Bases, and Salts

As previously noted, living things depend on organic compounds, those that contain carbon and hydrogen atoms. Living things also require a variety of **inorganic chemicals,** which typically lack carbon. Such inorganic substances include water, oxygen molecules, metal ions, and many acids, bases, and salts. In this section we examine the characteristics of some of these inorganic substances.

Water

Learning Objective

✓ Describe five qualities of water that make it vital to life.

Water is the most abundant substance in organisms, constituting 50–99% of their mass. Most of the special characteristics that make water vital result from the fact that a water molecule has two polar covalent bonds, which allows hydrogen bonding between water molecules and their neighbors. Among the special properties of water are the following:

1. Water molecules are cohesive; that is, they tend to stick to one another through hydrogen bonding **(Figure 2.12).** This property generates many special characteristics of water, including *surface tension,* which allows water to form a thin layer on the surface of cells. This aqueous layer is necessary for the transport of dissolved materials into and out of a cell.

2. Water is an excellent *solvent;* that is, it dissolves salts and other electrically charged molecules because it is attracted to both positive and negative charges (see Figure 2.9).

3. Water remains a liquid across a wider range of temperatures than other molecules of its size. This is critical because living things require water in liquid form.

4. Water can absorb significant amounts of heat energy without itself changing temperature. Further, when

heated water molecules eventually evaporate, they take much of this absorbed energy with them. These properties moderate temperature fluctuations that would otherwise damage organisms.

5. Water molecules participate in many chemical reactions within cells, both as reactants in hydrolysis and as products of dehydration synthesis.

CRITICAL THINKING

How can hydrogen bonding between water molecules help explain water's ability to absorb large amounts of energy before evaporating?

Acids and Bases

Learning Objective

✓ Contrast acids, bases, and salts, and explain the role of buffers.

As we have seen, the polar bonds of water molecules dissociate salts into their component cations and anions. A similar process occurs with substances known as acids and bases.

An **acid** is a substance that dissociates into one or more hydrogen ions (H^+) and one or more anions (**Figure 2.13a** on page 38). Acids can be inorganic molecules such as hydrochloric acid (HCl) and sulfuric acid (H_2SO_4), or organic molecules such as amino acids and nucleic acids. Familiar organic acids are found in lemon juice, black coffee, and tea. Of course, the anions of organic acids contain carbon, while those of inorganic acids do not.

A **base** is a molecule that binds with H^+ when dissolved in water. Some bases dissociate into cations and *hydroxyl ions* (OH^-) **(Figure 2.13b),** which then combine with hydrogen ions to form water molecules:

$$H^+ + OH^- \rightarrow H_2O$$

Other bases, such as household ammonia (NH_3), directly accept hydrogen ions and become compound ions such as NH_4^+ (ammonium). Another common household base is baking soda (sodium bicarbonate, $NaHCO_3$).

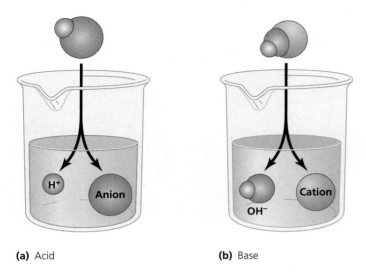

(a) Acid   **(b)** Base

▲ *Figure 2.13*

Acids and bases. **(a)** Acids dissociate in water into hydrogen ions and anions. **(b)** Many bases dissociate into hydroxyl ions and cations.

▲ *Figure 2.14*

The pH scale. Values below 7 are acidic; values above 7 are basic.

Metabolism requires a relatively constant balance of acids and bases because hydrogen ions and hydroxyl ions are involved in many chemical reactions. Further, many complex molecules such as proteins lose their functional shapes when acidity changes. If the concentration of either hydrogen ions or hydroxyl ions deviates too far from normal, metabolism ceases.

The concentration of hydrogen ions in a solution is expressed using a logarithmic **pH scale (Figure 2.14).** The term *pH* comes from *potential hydrogen,* which is the negative of the logarithm of the concentration of hydrogen ions. In this logarithmic scale, it is important to notice that acidity increases as pH values decrease, and that each decrease by a whole number in pH indicates a 10-fold increase in acidity (hydrogen ion concentration). For example, a glass of grapefruit juice, which has a pH of 3.0, contains 10 times as many hydrogen ions as the same volume of tomato juice, which has a pH of 4.0. Similarly, tomato juice is 10,000 times more acidic than pure water, which has a pH of 7.0 (neutral). Water is neutral because it dissociates into one hydrogen cation and one hydroxyl anion:

$$H_2O \rightarrow H^+ + OH^-$$

Alkaline (basic) substances have pH values greater than 7.0. They reduce the number of free hydrogen ions by combining with them. For bases that produce hydroxyl ions, the concentration of hydroxyl ions is inversely related to the concentration of hydrogen ions.

Organisms can tolerate only a certain, relatively narrow pH range. Fluctuations outside an organism's preferred range inhibit its metabolism and may even be fatal. Most organisms contain natural **buffers**—substances, such as proteins, that prevent drastic changes in internal pH. In a laboratory culture, the metabolic activity of microorganisms

can change the pH of microbial growth solutions as nutrients are taken up and wastes are released; therefore, pH buffers are often added to them (see Table 6.6 on page 195). One common buffer used in microbiological media is KH_2PO_4, which exists as either a weak acid or a weak base, depending on the pH of its environment. Under acidic conditions, KH_2PO_4 is a base that combines with H^+, neutralizing the acidic environment; in alkaline conditions, however, KH_2PO_4 acts as an acid, releasing hydrogen ions.

Microorganisms differ in their ability to tolerate various ranges of pH. Most grow best when the pH is between 6.5 and 8.5. Photosynthetic bacteria known as *cyanobacteria* grow well in more basic solutions. Fungi generally tolerate acidic environments better than most prokaryotes, though acid-loving prokaryotes, called *acidophiles,* require acidic conditions. Two pathogenic bacteria that tolerate acidic conditions in the human body, allowing them to grow where other bacteria cannot, are *Propionibacterium acnes* (prō-pē-on-i-bak-tēr′ē-ŭm ak′nēz), which contributes to acne in the skin, which normally has a pH of about 4.0; and *Helicobacter pylori* (hel'ĭ-kō-bak'ter pī'lō-rē),[11] a curved bacterium that has been shown to cause ulcers in the stomach, which can have a pH as low as 1.5 when acid is being actively secreted. **Highlight 2.1** focuses on how the use of antacids may increase the survival rates of certain disease-causing bacteria in the stomach.

Microorganisms can change the pH of their environment by utilizing acids and bases and by producing acidic or basic wastes. For example, fermentative microorganisms form organic acids from the decomposition of sugar, and the bacterium *Thiobacillus* (thī-ō-bă-sil'ŭs) can reduce the pH of its environment to 0.0. Acid produced by this bacterium in mine water dissolves enough uranium and copper from low-grade ore to make some mines profitable.

Scientists measure pH with a pH meter or with test papers impregnated with chemicals (such as litmus or phenol red) that change color in response to pH. In a microbiological laboratory, changes in color of such pH indicators incorporated into microbial growth media are commonly used to distinguish among bacterial genera.

Salts

As we have seen, a salt is a compound that dissociates in water into cations and anions other than H^+ and OH^-. Acids and hydroxyl-yielding bases neutralize each other during exchange reactions that produce water and salt. For instance, milk of magnesia (magnesium hydroxide), is an antacid used to neutralize excess stomach acid. The chemical reaction is

$$Mg(OH)_2 \quad + \quad 2\,HCl \quad \rightarrow \quad MgCl_2 \quad + 2\,H_2O$$

Magnesium hydroxide   hydrochloric acid   magnesium chloride   water

Cations and anions of salts are electrolytes. A cell uses electrolytes to create electrical differences between its inside and outside, to transfer electrons from one location to another, and as important components of many enzymes. Certain organisms also use salts such as calcium carbonate ($CaCO_3$) to provide structure and support for their cells.

CRITICAL THINKING

Why are two molecules of hydrochloric acid neutralized by only one molecule of magnesium hydroxide?

Organic Macromolecules

Inorganic molecules play important roles in an organism's metabolism; however, water excluded, they compose only about 1.5% of its mass. Inorganic molecules are typically too small and too simple to constitute an organism's basic structures, or to perform the complicated chemical reactions required of life. These functions are fulfilled by organic molecules, which are larger and much more complex.

Functional Groups and Monomers

Learning Objective

✓ Define a "functional group" as it relates to organic chemistry.

As we have seen, organic molecules contain carbon and hydrogen atoms, and each carbon atom can form four covalent bonds with other atoms (see Figure 2.5c and d). Carbon atoms that are linked together in branched chains, unbranched chains, and rings provide the basic frameworks of organic molecules.

Atoms of other elements are bound to these carbon frameworks to form an unlimited number of compounds. Besides carbon and hydrogen, the most common elements in organic compounds are oxygen, nitrogen, phosphorus, and sulfur. Other elements, such as iron, copper, molybdenum, manganese, zinc, and iodine, are important in some proteins.

Atoms often appear in certain common arrangements called **functional groups.** For example, $—NH_2$, the amino functional group, is found in all amino acids, and $—OH$, the hydroxyl functional group,[12] is common to all alcohols. When a class of organic molecules is discussed, the letter **R** (for *residue*) designates atoms in the compound that vary from one molecule to another. The symbol R—OH, therefore,

---

[11]The name *pylori* refers to the pylorus, a region of the stomach.
[12]Note that the hydroxyl functional group is not the same thing as the hydroxyl *ion* because the former is covalently bonded to a carbon atom.

Table 2.3 Functional Groups of Organic Molecules, and Some Classes of Compounds in Which They Are Found

Structure	Name	Class of Compounds
—OH	Hydroxyl	Alcohol Monosaccharide Amino acid
R—CH₂—O—CH₂—R'	Ether	Disaccharide Polysaccharide
R—C(=O)—R'	Internal carbonyl (a carbon atom on each side)	Ketone Carbohydrate
R—C(=O)—H	Terminal carbonyl (a carbon atom on only one side)	Aldehyde
R—C(=O)—O—H	Carboxyl	Amino acid Nucleic acid
R—C(H)—NH₂	Amino	Amino acid Protein
R—C(=O)—O—R'	Ester	Fat Wax
R—CH₂—SH	Sulfhydryl	Amino acid Protein
R—CH₂—O—P(OH)(OH)=O	Organic phosphate	Phospholipid Nucleotide ATP

represents the general formula for an alcohol. **Table 2.3** describes some common functional groups of organic molecules.

There is a great variety of organic compounds, but certain basic types are used by all organisms. These molecules, known as *macromolecules* because they are very large, are lipids, carbohydrates, proteins, and nucleic acids. Proteins, carbohydrates, and nucleic acids are composed of simpler subunits known as **monomers,**[13] which are basic building blocks. The monomers of these macromolecules are joined together to form chains of monomers called **polymers.**[14] Some macromolecular polymers are composed of hundreds of thousands of monomers.

Lipids

Learning Objectives

✓ Describe the structure of a triglyceride molecule, and compare it to that of a phospholipid.

✓ Distinguish among saturated, unsaturated, and polyunsaturated fatty acids.

Lipids are a diverse group of organic macromolecules not composed of monomers. They have one common trait—they are **hydrophobic;**[15] that is, they are insoluble in water. Lipids have little or no affinity for water because they are composed almost entirely of carbon and hydrogen atoms linked by nonpolar covalent bonds. Because these bonds are nonpolar, they have no attraction to the polar bonds of water molecules. To look at it another way, the polar water molecules are attracted to each other and exclude the nonpolar lipid molecules. There are four major groups of lipids in cells: fats, phospholipids, waxes, and steroids.

Fats

Organisms make **fats** in dehydration synthesis reactions that form *esters* between three chainlike fatty acids and an alcohol named *glycerol* **(Figure 2.15a).** Fats are also called *triglycerides* because they contain three fatty acid molecules linked to a molecule of glycerol.

The three fatty acids in a fat molecule may be identical or different from one another, but each usually has 12 to 20 carbon atoms. An important difference among fatty acids is the presence and location of double bonds between the carbon atoms **(Figure 2.15b).** When the carbon atoms are linked solely by single bonds, every carbon atom, with the exception of the terminal ones, is covalently linked to two hydrogen atoms. Such a fatty acid is **saturated** with hydrogen. In contrast, **unsaturated fatty acids** contain at least one double bond between adjacent carbon atoms, and therefore contain

[13]From Greek *mono*, meaning one, and *meris*, meaning part.
[14]From Greek *poly*, meaning many, and *meris*, meaning part.
[15]From Greek *hydor*, meaning water, and *phobos*, meaning fear.

Glycerol + 3 fatty acids

(a)

Saturated fatty acid

Unsaturated fatty acid

(b)

▲ *Figure 2.15*

Fats (triglycerides). **(a)** Fats are made in dehydration synthesis reactions that form ester bonds between a glycerol molecule and three fatty acids. **(b)** Saturated fatty acids have only single bonds between their carbon atoms, whereas unsaturated fatty acids have double bonds between carbon atoms. Scientists often use abbreviated diagrams of fatty acids in which each angle represents a carbon atom, and hydrogen atoms are omitted, as seen in part (a) of this figure. Compare these abbreviated diagrams to the "full" diagrams shown in Table 2.4. *According to Table 2.4 on page 42, which fatty acids are shown in Figure 2.15a?*

Figure 2.15 Stearic acid, palmitic acid, and oleic acid.

at least one carbon atom bound to only a single hydrogen atom. If several double bonds exist in even one fatty acid of a molecule of fat, then it is a **polyunsaturated** fat.

Saturated fats (composed of saturated fatty acids), like those found in animals, are usually solid at room temperature because their fatty acids can pack close together. Unsaturated fatty acids, by contrast, are bent at every double bond, and so cannot pack tightly; they remain liquid at room temperature. Most fats in plants are unsaturated or polyunsaturated. **Table 2.4** on page 42 compares the structures and melting points of four common fatty acids.

Fats contain an abundance of energy stored in their carbon–carbon covalent bonds. Indeed, the primary role of fats in organisms is to store energy. As we see in Chapter 5, fats can be catabolized to provide energy for movement, synthesis, and transport.

Phospholipids

Phospholipids are similar to fats, but they contain two fatty acid chains instead of three. In phospholipids, the third carbon atom of glycerol is linked to a phosphate (PO_4) functional group instead of to a fatty acid (**Figure 2.16a** on page 43). Like fats, different phospholipids contain different fatty acids. Small organic groups linked to the phosphate group provide additional variety.

Part of a phospholipid molecule is attracted to water; that is, whereas the fatty acid "tail" portion of the molecule is nonpolar and thus hydrophobic, the phospholipid "head" is polar and is thus **hydrophilic**[16] **(Figure 2.16b)**. As a result, phospholipids placed in a watery environment will always self-assemble into forms that keep the fatty acid tails away from water. Two such forms are possible **(Figure 2.16c)**: micelles, which have a single layer of phospholipid and resemble balls, and phospholipid bilayers, which resemble a two-ply bag.

The fatty acids tails, which are hydrophobic, congregate in the water-free interior of micelles and bilayers. The polar phosphate heads orient toward the water because they are hydrophilic. Phospholipid bilayers make up the outer membranes of all cells, as well as the internal membranes of plant, fungal, and animal cells.

Waxes

Waxes contain one long-chain fatty acid linked covalently to a long-chain alcohol by an ester bond. Waxes do not have a hydrophilic head; thus, they are completely water insoluble. Certain microorganisms, such as *Mycobacterium tuberculosis* (mī′kō-bak-tēr′ē-ŭm tū-ber-kyū-lō′sis), are surrounded by a waxy wall, making them resistant to drying. Some marine

---

[16]From Greek *philos,* meaning love.

Table 2.4 Common Fatty Acids in Fats and Cell Membranes

Number of Carbon Atoms: Double Bonds	Type of Fatty Acid	Structure and Formula	Common Name	Melting Point (°C)
16 : 0	Saturated	$CH_3(CH_2)_{14}COOH$	Palmitic acid	63
18 : 0	Saturated	$CH_3(CH_2)_{16}COOH$	Stearic acid	70
18 : 1	Monounsaturated	$CH_3(CH_2)_7CH=CH(CH_2)_7COOH$	Oleic acid	16
18 : 2	Polyunsaturated	$CH_3(CH_2)_4(CH=CHCH_2)_2(CH_2)_6COOH$	Linoleic acid	−5

microbes use waxes instead of fats as energy storage molecules.

Steroids

A final group of lipids are **steroids.** Steroids consist of four rings (containing five or six carbon atoms) that are fused to one another and attached to various side chains and functional groups (**Figure 2.17a** on page 44). Steroids play many roles in human metabolism. Some act as hormones; another steroid, *cholesterol,* is familiar to you as an undesirable component of food. However, cholesterol is also an essential part of the phospholipid bilayer membrane surrounding all animal cells. Cells of fungi, plants, and one group of bacteria (mycoplasmas) have similar sterol molecules in their membranes. Sterols, which are steroids with an —OH functional group, interfere with the tight packing of the fatty acid chains of phospholipids (**Figure 2.17b**). This keeps the membranes fluid and flexible at low temperatures. Without

steroids such as cholesterol, the membranes of cells would become stiff and inflexible in the cold.

CRITICAL THINKING

We have seen that it is important that biological membranes remain flexible. Most bacteria lack sterols in their membranes and instead incorporate unsaturated phospholipids in the membranes to resist tight packing. Reexamine Table 2.4. Which fatty acid might best protect the membranes of an ice-dwelling bacterium?

Carbohydrates

Learning Objective

✓ Discuss the roles of carbohydrates in living systems.

Carbohydrates are organic molecules composed solely of atoms of carbon, hydrogen, and oxygen. Most carbohydrate

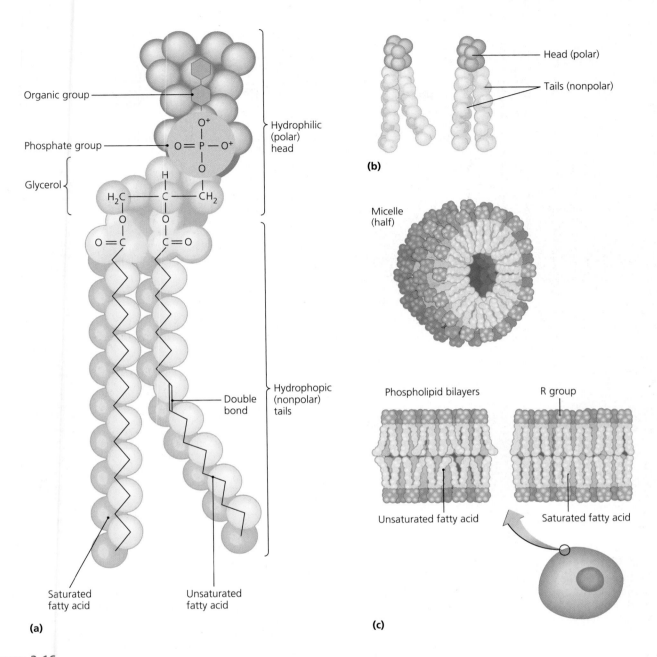

Organic group

Phosphate group

Glycerol

Hydrophilic (polar) head

Saturated fatty acid

Unsaturated fatty acid

Double bond

Hydrophopic (nonpolar) tails

Saturated fatty acid

Unsaturated fatty acid

(a)

Head (polar)

Tails (nonpolar)

(b)

Micelle (half)

Phospholipid bilayers

R group

Unsaturated fatty acid

Saturated fatty acid

(c)

▲ *Figure 2.16*

Phospholipids. **(a)** A phospholipid is composed of a hydrophilic (polar) "head," which is composed of glycerol and a phosphate group, and two hydrophobic (nonpolar) fatty acid "tails." **(b)** The symbols commonly used to represent phospholipids. **(c)** In water, phospholipids self-assemble into spherical micelles or bilayers. Note that phospholipids containing unsaturated fatty acids do not pack together as tightly as those containing saturated fatty acids.

compounds contain an equal number of oxygen and carbon atoms, and twice as many hydrogen atoms as carbon atoms, so the general formula for a carbohydrate is $(CH_2O)_n$, where n indicates the number of CH_2O units.

Carbohydrates play many important roles in organisms. Large carbohydrates such as starch and glycogen are used for the long-term storage of chemical energy, while a smaller carbohydrate molecule—glucose—serves as a ready energy source in most cells. Carbohydrates also form part of the backbones of DNA and RNA, and others are converted routinely into amino acids. Additionally, polymers of carbohydrate form the cell walls of fungi, plants, algae, and prokaryotes, and are involved in intercellular interactions between animal cells. For example, specific carbohydrates found on the surfaces of white blood cells determine which cells interact in immune responses against pathogens.

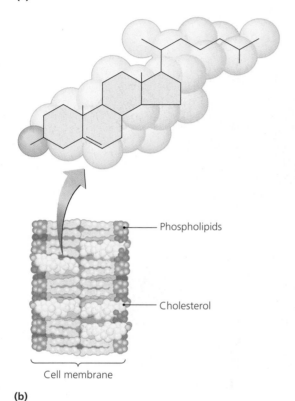

(a)

Phospholipids

Cholesterol

Cell membrane

(b)

▲ *Figure 2.17*

Steroids. **(a)** Steroids are lipids characterized by four "fused" carbon rings. **(b)** The steroid cholesterol functions in animal and protozoan cell membranes to prevent packing of phospholipids, which keeps the membranes fluid at low temperatures.

Monosaccharides

The simplest carbohydrates are **monosaccharides**[17] or simple sugars **(Figure 2.18).** The general names for the classes of monosaccharides are formed from a prefix indicating the number of carbon atoms, and the suffix *-ose.* For example, *pentoses* are sugars with five carbon atoms, and *hexoses* are sugars with six carbon atoms. Pentoses and hexoses are particularly important in the metabolism of cells. For example, deoxyribose, which is the sugar component of DNA, is a pentose. Glucose is a hexose and the primary energy molecule of cells, and fructose is a hexose found in fruit.

---

[17]From Greek *sakcharon,* meaning sugar.

(a) Glucose

α configuration   β configuration

(b) Fructose

acetyl group

(c) N-acetylglucosamine

▲ *Figure 2.18*

Monosaccharides (simple sugars). Although simple sugars may exist as either linear molecules (at left) or rings (at right), energy dynamics generally favor ring forms. **(a)** Glucose, a hexose, is the primary energy source for cellular metabolism and an important monomer in many larger carbohydrates. Alpha and beta ring configurations exist. **(b)** Fructose, or fruit sugar, a hexose. **(c)** N-acetylglucosamine (NAG), a monomer in many bacterial cell walls.

Monosaccharides may exist as linear molecules, but due to energy dynamics they usually take cyclic (ring) forms. In some cases, more than one cyclic structure may exist. For example, glucose can assume an alpha (α) configuration or a beta (β) configuration (see Figure 2.18a). As we will see, these configurations play important roles in the formation of different polymers.

Disaccharides

When two monosaccharide molecules are linked together via dehydration synthesis, the result is a **disaccharide.** For

(a) Dehydration synthesis of sucrose

(b) Hydrolysis of sucrose

▲ *Figure 2.19*

Disaccharides. **(a)** Formation of the disaccharide sucrose via dehydration synthesis. **(b)** Breakdown of sucrose via hydrolysis.

example, the linkage of two hexoses, glucose and fructose, forms sucrose (table sugar) and a molecule of water **(Figure 2.19a).** Other disaccharides include maltose (malt sugar) and lactose (milk sugar). Disaccharides can be broken down via hydrolysis into their constituent monosaccharides **(Figure 2.19b).**

Polysaccharides

Polysaccharides are polymers composed of tens or thousands of monosaccharides that have been covalently linked in dehydration synthesis reactions. Even polysaccharides that contain only glucose monomers can be quite diverse because they can differ according to their monosaccharide monomer configurations (either alpha or beta) and their shapes (either branched or unbranched). Cellulose, the main constituent of the cell walls of plants and green algae, is a long unbranched molecule that contains only β-monomers of glucose linked between carbons 1 and 4 of alternating monomers; such bonds are termed β-1,4 bonds **(Figure 2.20a on page 46).** Amylose, a starch (a storage compound in plants), has only α-1,4 bonds and is unbranched **(Figure 2.20b);** amylopectin, another plant starch, contains mostly α-1,4 monomers, but also has a few α-1,6 bonds and is branched **(Figure 2.20c).** Glycogen, a storage molecule formed in the liver and muscle cells of animals, is a highly branched molecule with both α-1,4 and α-1,6 bonds **(Figure 2.20d). Highlight 2.2** on page 47 explores the roles of these bond types, and other factors as well, in the digestibility of cellulose, the most abundant polysaccharide on Earth.

The cell walls of bacteria are composed of *peptidoglycan,* which is made of polysaccharides and amino acids (see Figure 3.14 on page 70). Polysaccharides may also be linked to lipids to form lipopolysaccharides, which can form cell markers such as those involved in the ABO blood typing system in humans.

Proteins

Learning Objectives

✓ Describe five general functions of proteins in organisms.
✓ Sketch and label four levels of protein structure.

The most complex organic compounds are **proteins,** which are composed mostly of carbon, hydrogen, oxygen, nitrogen, and sulfur. Proteins perform many functions in cells, including:

- Structure. Proteins are structural components in cell walls, in membranes, and within cells themselves. Proteins are also the primary structural material of hair, nails, the outer cells of skin, muscle, and flagella and cilia (which act to move microorganisms through their environment).

- Enzymatic catalysis. *Catalysts* are chemicals that enhance the speed or likelihood of a chemical reaction. Catalysts in cells, which are typically proteins, are called *enzymes.*

- Regulation. Some proteins regulate cell function by stimulating or hindering either the action of other proteins or the expression of genes. Hormones are an example of a regulatory protein.

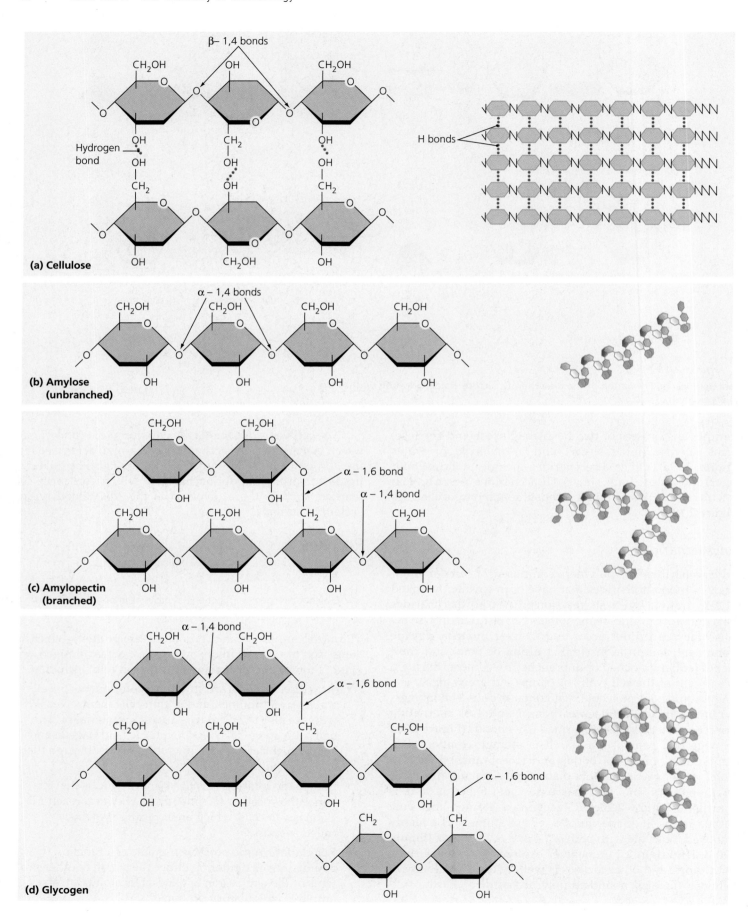

(a) Cellulose

(b) Amylose (unbranched)

(c) Amylopectin (branched)

(d) Glycogen

◀ *Figure 2.20*

Polysaccharides. All four polysaccharides shown here are composed solely of glucose but differ in the configuration of the glucose monomers and the amount of branching. **(a)** Cellulose, the major structural material in plants, is unbranched and contains only β-1,4 bonds. **(b)** Amylose is an unbranched plant starch with only α-1,4 bonds. **(c)** Amylopectin is a branched plant starch with mostly α-1,4 bonds but also a few α-1,6 branches. **(d)** Glycogen, a highly branched storage molecule in animals, is composed of glucose monomers linked by α-1,4 or α-1,6 bonds.

- Transportation. As we will see in Chapter 3, certain proteins act as channels and "pumps" that move substances into or out of cells.
- Defense and offense. *Antibodies* and *complement* are examples of proteins that defend your body against microorganisms, and some bacteria produce proteins called *bacteriocins* that kill other bacteria.

A protein's function is dependent on its shape, which is determined by the molecular structures of its constituents parts.

Amino Acids

Proteins are polymers composed of monomers called **amino acids.** Amino acids contain a basic amino group (—NH₂), an acidic carboxyl group (—COOH), and a hydrogen atom, all attached to the same carbon atom, which is known as the *α-carbon* **(Figure 2.21)**. A fourth bond attaches the α-carbon to a side group (—R) that varies among different amino acids. The side group may be a single hydrogen atom, various chains, or various complex ring structures. There are hundreds of amino acids, but most organisms use only 21 amino acids to build proteins. The different side groups affect the

| Highlight 2.2 | **Why Can't We Digest Wood?** |

Why is it that termites can digest wood, but humans cannot? The answer obviously involves biochemistry, but it involves bacteria as well.

Wood consists mainly of cellulose, a constituent of plant cell walls. A comparison of the structures of cellulose, plant starches, and animal glycogen in Figure 2.20 reveals that the glucose monomers of cellulose are joined with covalent bonds called β-1,4 bonds, but the monomers in starches and glycogen are linked with α-1,4 bonds and α-1,6 bonds. Animals possess enzymes that recognize and hydrolyze alpha bonds, but they lack *cellulase*, the enzyme that recognizes the shape of β-1,4 bonds. Therefore, animals (including humans) cannot digest cellulose.

How, then, do termites survive on wood, and herbivores such as horses and cattle thrive on grass, when their diets are composed primarily of cellulose? Their guts provide an ideal environment for billions of cellulose-digesting bacteria, which secrete cellulase. This enzyme hydrolyzes cellulose into glucose, which the animals can then absorb.

way amino acids interact with one another within a given protein, as well as how the protein interacts with other molecules. A change in an amino acid's side group may seriously interfere with a protein's normal function.

Because amino acids contain both an acidic carboxyl group and a basic amino group, they have both positive and negative charges and are easily soluble in water. The French physicist Jean Baptiste Biot (1774–1862) discovered that organic molecules such as amino acids and simple sugars, when dissolved in water, bend light rays as they pass through the solution. Some molecules, known as

▲ *Figure 2.21*

Amino acids. **(a)** The basic structure of amino acids. The central carbon (called the α-carbon) is attached to an amino group, a hydrogen atom, a carboxyl group, and a side group (—R group) that varies among amino acids. **(b)** Some selected amino acids, with their side groups highlighted. Note that each amino acid has a distinctive three-letter abbreviation.

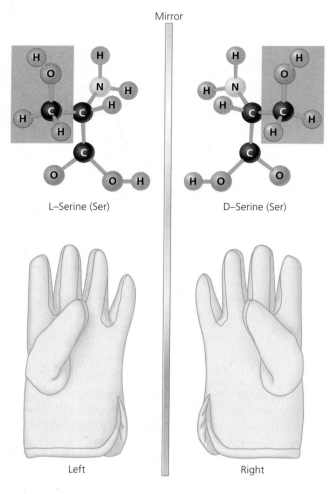

Mirror

L–Serine (Ser)   D–Serine (Ser)

Left   Right

▲ *Figure 2.22*

Stereoisomers, molecules that are mirror images of one another. When dissolved in water, D isomers bend light clockwise, and L forms bend light counterclockwise. Just as a right-handed glove does not fit a left hand, so a D stereoisomer cannot be substituted for an L stereoisomer in metabolic reactions.

D *forms,*[18] bend light rays clockwise; other molecules bend light rays counterclockwise and are known as L *forms.*[19]

Many organic molecules exist as both D and L forms that are *stereoisomers* of one another; that is; they have the same atoms and functional groups but are mirror images of each other **(Figure 2.22).** Amino acids in proteins are almost always L forms—except for glycine, which does not have a stereoisomer. Interestingly, organisms almost always use D sugars in metabolism and polysaccharides. The rare stereoisomers—D amino acids and L sugars—do exist in some bacterial cell walls and in some antibiotics.

CRITICAL THINKING

Why isn't there a stereoisomer of glycine?

Peptide Bonds

Cells link amino acids together in chains that somewhat resemble beads on a necklace. By a dehydration synthesis reaction, a covalent bond is formed between the carbon of the carboxyl group of one amino acid, and the nitrogen of the amino group of the next amino acid in the chain **(Figure 2.23).** As we study in detail in Chapter 7, cells follow the organism's genetic instructions to link amino acids together in precise sequences.

Scientists refer to covalent bonds between amino acids by the special name: **peptide**[20] **bonds.** A molecule composed of two amino acids linked together by a single peptide bond is called a dipeptide (see Figure 2.23); longer chains of amino acids are called *polypeptides.*

Protein Structure

Proteins are unbranched polypeptides composed of hundreds to thousands of amino acids linked together in specific patterns as determined by genes. The structure of a protein molecule is directly related to its function; therefore, an understanding of protein structure is critical to understanding certain specific chemical reactions (see Chapter 5), the action of antibiotics (see Chapter 10), and specific defense against pathogens (see Chapter 16). Every protein has at least three levels of structure, and some proteins have four levels.

1. Primary structure. The primary structure of a protein is its sequence of amino acids **(Figure 2.24a).** Cells use many different types of amino acids in proteins, though not every protein contains all types. The primary structures of proteins vary widely in length and amino acid sequence.

   A change in a single amino acid can drastically affect a protein's overall structure and function, though this is not always the case. For instance, the replacement of valine by alanine in position 136 of the primary structure of a particular sheep brain protein, called cellular prion (prē′on) protein, may result in a disease called *scrapie.* The altered protein is believed to have spread into cows, causing *mad cow disease,* and from cows into humans, causing *new variant Creutzfeldt-Jakob* (kroytsfelt-yah-kŭp) *disease.*[21] However, numerous substitutions can be made in other, noncritical regions of cellular prion protein, with no ill effects.

2. Secondary structure. Ionic and hydrogen bonds, and hydrophobic and hydrophilic characteristics, cause many polypeptide chains to fold into either coils, which are called α-*helices,* or accordionlike structures called β-*pleated sheets* **(Figure 2.24b).** Proteins are typically

---

[18]From Latin *dexter,* meaning on the right.
[19]From Latin *laevus,* meaning on the left.
[20]From *peptone,* the name given to short chains of amino acids resulting from the partial digestion of protein.
[21]Named for the two German neurobiologists who first described the disease.

The linkage of amino acids by peptide bonds via a dehydration reaction. In this reaction, removal of a hydroxyl group from amino acid 1 and a hydrogen atom from amino acid 2 results in the production of a dipeptide—which is two amino acids linked by a single peptide bond—and a molecule of water.

(a) Primary structure

(b) Secondary structure

▲ *Figure 2.24*

Levels of protein structure. **(a)** A protein's primary structure is the sequence of amino acids in a polypeptide. **(b)** Secondary structure arises as a result of interactions, such as hydrogen bonding, between regions of the polypeptide. Secondary structure takes two basic shapes: α-helices and β-pleated sheets. **(c)** The more complex tertiary structure is a three-dimensional shape defined by further hydrogen bonding as well as covalent bonding between sulfur atoms of adjoining cysteine. **(d)** Those proteins that are composed of more than one polypeptide chain have a quaternary structure.

(c) Tertiary structure

(d) Quaternary structure

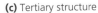

composed of both α-helices and β-pleated sheets linked by short sequences of amino acids that do not show such secondary structure. Because of its primary structure, the protein that causes new variant Creutzfeldt-Jakob disease has β-pleated sheets in locations where the normal protein has α-helices.

3. Tertiary structure. Polypeptides further fold into complex three-dimensional shapes that are not repetitive like α-helices and β-pleated sheets **(Figure 2.24c)**, but are uniquely designed to accomplish the function of the protein. Scientists are only beginning to understand the interactions that determine tertiary structure, but it is clear that covalent bonds between —R groups of amino acids, hydrogen bonds, ionic bonds, and other molecular interactions are important. For instance, nonpolar (hydrophobic) side chains fold into the interior of molecules, away from the presence of water.

   Some proteins form strong covalent bonds between sulfur atoms of cysteine molecules that are brought into proximity by the folding of the polypeptide. These *disulfide bridges* are critical in maintaining tertiary structure of many proteins.

4. Quaternary structure. Some proteins are composed of two or more polypeptide chains linked together by disulfide bridges or other bonds. The overall shape of such a protein may be globular **(Figure 2.24d)** or fibrous (threadlike).

Organisms may further modify proteins by combining them with other organic or inorganic molecules. For instance, *glycoproteins* are proteins covalently bound with carbohydrates, *lipoproteins* are proteins bonded with lipids, *metalloproteins* contain metallic ions, and *nucleoproteins* are proteins bonded with nucleic acids.

Since protein function is determined by protein shape, anything that severely interrupts shape also disrupts function. As we have seen, shape and function can be altered by amino acid substitution. Additionally, physical and chemical factors such as heat, changes in pH, and salt concentration can interfere with hydrogen and ionic bonding, which in turn disrupts the three-dimensional structure of proteins. This process is called **denaturation.** Denaturation can be perma-

nent or temporary (if the denatured protein is able to return to its original shape again).

Nucleic Acids

Learning Objectives

✓ Describe the basic structure of a nucleotide.
✓ Compare and contrast DNA and RNA.
✓ Contrast the structures of ATP, ADP, and AMP.

The nucleic acids **deoxyribonucleic acid (DNA)** and **ribonucleic acid (RNA)** are vital as the genetic material of cells and viruses. Moreover, RNA, acting as an enzyme, binds amino acids together to form polypeptides. DNA and RNA are both unbranched macromolecular polymers that differ primarily in the structures of their monomers, which we discuss next.

Nucleotides

Each monomer of nucleic acids, a **nucleotide,** consists of three parts **(Figure 2.25a):** (1) phosphoric acid, which at the pH of cells is ionized and called *phosphate* (PO_4^{2-}); (2) a pentose sugar, either deoxyribose or ribose **(Figure 2.25b);** and (3) one of five cyclic (ring-shaped) nitrogenous bases: **adenine (A), guanine (G), cytosine (C), thymine (T),** or **uracil (U) (Figure 2.25c).** Adenine and guanine are double-ringed molecules of a class called *purines,* whereas cytosine, thymine, and uracil have single rings and are *pyrimidines.* DNA contains A, G, C, and T bases, whereas RNA contains A, G, C, and U bases. As their names suggest, DNA nu-

Figure 2.25 ▸

Nucleotides. **(a)** The basic structure of nucleotides, each of which is composed of a phosphate, a pentose sugar, and a nitrogenous base. **(b)** The pentose sugars deoxyribose, which is found in deoxyribonucleic acid (DNA), and ribose, which is found in ribonucleic acid (RNA). **(c)** The nitrogenous bases, which are either the double-ringed purines adenosine or guanine, or the single-ringed pyrimidines cytosine, thymine, or uracil.

(a) Phosphate group / Pentose sugar / Purine or pyrimidine / Nitrogenous base

(b) Deoxyribose / Ribose

(c) Purines / Pyrimidines
Adenine (A) (DNA and RNA)
Cytosine (C) (DNA and RNA)
Guanine (G) (DNA and RNA)
Thymine (T) (DNA)
Uracil (U) (RNA)

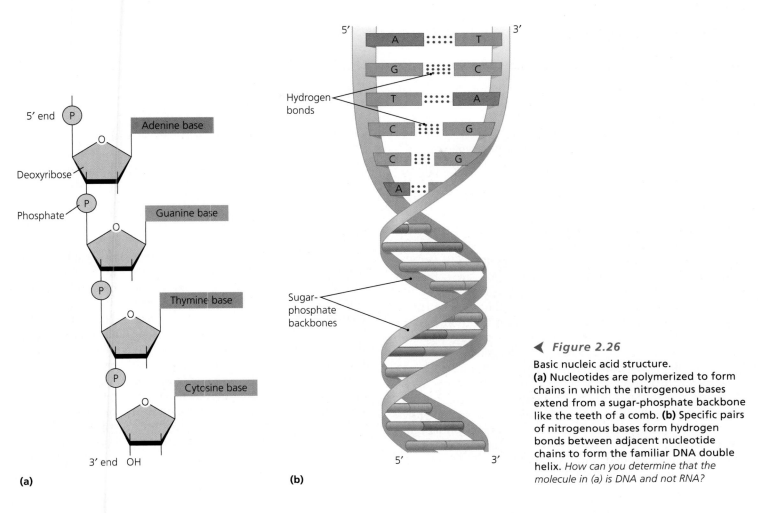

(a)

5' end — P

Adenine base

Deoxyribose

Phosphate — P

Guanine base

P

Thymine base

P

Cytosine base

3' end   OH

(b)

Hydrogen bonds

Sugar-phosphate backbones

◄ *Figure 2.26*

**Basic nucleic acid structure.
(a)** Nucleotides are polymerized to form chains in which the nitrogenous bases extend from a sugar-phosphate backbone like the teeth of a comb. **(b)** Specific pairs of nitrogenous bases form hydrogen bonds between adjacent nucleotide chains to form the familiar DNA double helix. *How can you determine that the molecule in (a) is DNA and not RNA?*

Figure 2.26 It is DNA because its nucleotides have deoxyribose sugar and because some of them have thymine bases (not uracil, as in RNA).

cleotides contain deoxyribose, and RNA nucleotides contain ribose.

Each nucleotide is also named for the base it contains. Thus a nucleotide made with uracil is a uracil RNA nucleotide (or uracil *ribonucleotide*), and a nucleotide made with adenine and ribose is an adenine ribonucleotide. Likewise, a nucleotide composed of adenine and deoxyribose is an adenine DNA nucleotide (or adenine *deoxyribonucleotide*).

CRITICAL THINKING

A textbook states that only five nucleotide bases are found in cells, but a laboratory worker correctly reports that she has isolated eight different nucleotides. Explain why both are correct.

Nucleic Acid Structure

Nucleic acids, like polysaccharides and proteins, are polymers. They are composed of nucleotides linked by covalent bonds between the phosphate of one nucleotide and the sugar of the next. Polymerization results in a linear spine composed of alternating sugars and phosphates, with bases extending from it rather like the teeth of a comb **(Figure**

2.26a). The ends of a chain of nucleotides are different. At one end, called the 5' end[22] (five prime end), carbon 5' of the sugar is attached to a phosphate group. At the other end (3' end), carbon 3' of the sugar is not attached to a phosphate group.

The atoms of the bases in nucleotides are arranged in such a manner that hydrogen bonds readily form between specific bases of two adjacent nucleic acid chains. Three hydrogen bonds form between an adjacent pair composed of cytosine (C) and guanine (G), whereas two hydrogen bonds form between an adjacent pair composed of adenine (A) and thymine (T) in DNA **(Figure 2.26b),** or between an adjacent pair composed of adenine (A) and uracil (U) in RNA. Hydrogen bonds do not readily form between other combinations of nucleotide bases; for example, adenine does not readily pair with cytosine, guanine, or another adenine nucleotide.

In cells and most viruses that use DNA as a genome, DNA molecules are double stranded. The two strands of

---

[22]Carbon atoms in organic molecules are commonly identified by numbers. In a nucleotide, carbon atoms 1, 2, 3, etc. belong to the base, and carbon atoms 1', 2', 3', etc. belong to the sugar.

Table 2.5 Comparison of Nucleic Acids

Characteristic	DNA	RNA
Sugar	Deoxyribose	Ribose
Purine nucleotides	A and G	A and G
Pyrimidine nucleotides	T and C	U and C
Number of strands	Double stranded in cells and in most DNA viruses; single stranded in parvoviruses	Single stranded in cells and in most RNA viruses; double stranded in reoviruses
Function	Genetic material of all cells and DNA viruses	Protein synthesis in all cells; genetic material of RNA viruses

DNA are complementary to one another; that is, the specificity of nucleotide base pairing ensures that opposite strands are composed of complementary nucleotides. For instance, if one strand has the sequence AATGCT, then its complement has TTACGA.

The two strands are also *antiparallel*; that is, they run in opposite directions. One strand runs from the 3' end to the 5' end, while its complement runs in the opposite direction from its 5' end to its 3' end. Though hydrogen bonds are relatively weak bonds, thousands of them exist at normal temperatures, forming a stable, double-stranded DNA molecule, which looks much like a ladder: The two deoxyribose-phosphate chains are the side rails, and base pairs form the rungs. Hydrogen bonding also twists the phosphate-deoxyribose backbones into a helix (see Figure 2.26b). Thus, typical DNA is a double helix. As you will see in Chapter 24, parvoviruses use single-stranded DNA, which is an exception to this rule.

Nucleic Acid Function

DNA is the genetic material of all organisms and of many viruses; it carries instructions for the synthesis of RNA molecules and proteins. By controlling the synthesis of enzymes and regulatory proteins, DNA controls the synthesis of all other molecules in an organism. Genetic instructions are carried in the sequence of nucleotides that make up the nucleic acid. Even though only four kinds of bases are found in DNA (A, T, G, and C), they can be sequenced in distinctive patterns that create genetic diversity and code for an infinite number of proteins, just as an alphabet of only four letters could spell a very large number of words. Cells replicate their DNA molecules and pass copies to their descendents, ensuring that each has the instructions necessary for life.

Ribonucleic acids play several roles in the synthesis of proteins, including catalyzing the synthesis of proteins. RNA molecules also function as structural components of ribosomes, and in place of DNA as the genome of RNA viruses.

Table 2.5 compares and contrasts RNA and DNA. We examine the synthesis and function of DNA and RNA in detail in Chapter 7.

ATP

Phosphate in nucleotides and other molecules is a highly reactive functional group and can form covalent bonds with other phosphate groups to make diphosphate and triphosphate molecules. Such molecules made from ribose nucleotides are important in many metabolic reactions. The names of these molecules indicate the nucleotide base and the number of phosphate groups they contain. Thus, cells make adenosine monophosphate (AMP) from the nitrogenous base adenine, ribose sugar, and one phosphate group; adenosine diphosphate (ADP), which has two phosphate groups; and **adenosine triphosphate** (ă-den′ō-sēn trī-fos′ fāt) or **ATP,** which has three phosphate groups **(Figure 2.27).**

ATP is the principal, short-term, recyclable energy supply for cells. When the phosphate bonds of ATP are broken,

▲ *Figure 2.27*

Adenosine triphosphate (ATP), the primary short-term, recyclable energy supply for cells. Energy is stored in "high-energy" bonds between the phosphate groups. *What is the relationship between AMP and adenine ribonucleotide?*

a significant amount of energy is released; in fact, more energy is released from phosphate bonds than is released from most other covalent bonds. For this reason, the phosphate–phosphate bonds of ATP are known as *high-energy bonds*, and to show these specialized bonds, ATP is often symbolized as

$$A-\textcircled{P}\sim\textcircled{P}\sim\textcircled{P}$$

Energy is released when ATP is converted to ADP, and when phosphate is removed from ADP to form AMP, though the latter reaction is not as common in cells. Energy released from the phosphate bonds of ATP is used for important life-sustaining activities, such as synthesis reactions, locomotion, and transportation of substances into and out of cells.

Cells also use ATP as a structural molecule in the formation of *coenzymes.* As we will see in Chapter 5, coenzymes such as *flavin adenine dinucleotide, nicotinamide nucleotide,* and *coenzyme* A function in many metabolic reactions.

A cell's supply of ATP is limited; therefore, an important part of cellular metabolism is replenishing ATP stores. We discuss the important ATP-generating reactions in Chapter 5.

CHAPTER SUMMARY

Atoms (pp. 27–29)

1. **Atoms** contain negatively charged **electrons** spinning around a nucleus composed of uncharged **neutrons** and positively charged **protons.**

2. An **element** is matter composed of a single type of atom.

3. The number of protons in the nucleus of an atom is its **atomic number.** The sum of the masses of its protons, neutrons, and electrons is an atom's **atomic mass,** which is estimated by adding the number of neutrons and protons (because electrons have little mass).

4. **Isotopes** are atoms of an element that differ only in the numbers of neutrons they contain.

Chemical Bonds (pp. 29–35)

1. The region of space occupied by electrons is an electron shell. The number of electrons in the outermost or **valence shell** of an atom determines the atom's reactivity. Most valence shells hold a maximum of eight electrons. Sharing or transferring valence electrons to fill a valence shell results in **chemical bonds.**

2. A chemical bond results when two atoms share a pair of electrons. The **electronegativities** of each of the atoms, which is the strength of their attraction for electrons, determines whether the bond between them will be a **nonpolar covalent bond** (equal sharing of electrons), a **polar covalent bond** (unequal sharing of electrons), or an **ionic bond** (giving up of electrons from one atom to another).

3. If a **molecule** contains atoms of more than one element, it is a **compound. Organic compounds** are those that contain carbon and hydrogen atoms.

4. An **anion** is an atom with an extra electron and thus a negative charge. A **cation** has lost an electron and thus has a positive charge. Ionic bonding between the two types of ions makes **salt.** When salts dissolve in water, their ions are called **electrolytes.**

5. **Hydrogen bonds** are relatively weak but important chemical bonds. They hold molecules in specific shapes and confer unique properties to water molecules.

Chemical Reactions (pp. 35–36)

1. **Chemical reactions** result from the making or breaking of chemical bonds in a process in which **reactants** are changed into **products.**

2. **Synthesis reactions** form larger, more complex molecules. In **dehydration synthesis,** a molecule of water is removed from the reactants as the larger molecule is formed. **Endothermic reactions** require energy. **Anabolism** is the sum of all synthesis reactions in an organism.

3. **Decomposition reactions** break larger molecules into smaller molecules and are **exothermic** because they release energy. **Hydrolysis** is a decomposition reaction that uses water as one of the reactants. The sum of all decomposition reactions in an organism is called **catabolism.**

4. **Exchange reactions** involve exchanging atoms between reactants.

5. **Metabolism** is the sum of all anabolic and catabolic chemical reactions in an organism.

Water, Acids, Bases, and Salts (pp. 37–39)

1. **Inorganic** molecules typically lack carbon.

2. Water is a vital inorganic compound because of its properties as a solvent, its liquidity, its great capacity to absorb heat, and its participation in chemical reactions.

3. **Acids** release hydrogen ions. **Bases** release hydroxyl anions. The relative strength of each is assessed on a logarithmic **pH scale,** which measures the hydrogen ion concentration in a substance.

4. **Buffers** are substances that prevent drastic changes in pH.

Organic Macromolecules (pp. 39–53)

1. Certain groups of atoms in common arrangements, called **functional groups,** are found in organic macromolecules. **Monomers** are simple subunits that can be covalently linked to form chainlike **polymers.**

2. **Lipids,** which include fats, phospholipids, waxes, and steroids, are **hydrophobic** (insoluble in water) macromolecules.

3. **Fat** molecules are formed from a glycerol and three chainlike fatty acids. **Saturated fatty acids** contain more hydrogen in their structural formulas than **unsaturated fatty acids,** which contain double bonds between some carbon atoms. If several double bonds exist in the fatty acids of a molecule of fat, it is a **polyunsaturated** fat.

4. **Phospholipids** contain two fatty acid chains and a phosphate functional group. The phospholipid head is **hydrophilic,** whereas the fatty acid portion of the molecule is hydrophobic.

5. Waxes contain a long-chain fatty acid covalently linked to a long-chain alcohol. Waxes, which are water insoluble, are components of cell walls and are sometimes used as energy storage molecules.

6. **Steroid** lipids such as cholesterol help maintain the structural integrity of membranes as temperature fluctuates.

7. **Carbohydrates** such as **monosaccharides, disaccharides,** and **polysaccharides** serve as energy sources, structural molecules, and recognition sites during intercellular interactions.

8. **Proteins** are structural components of cells, enzymatic catalysts, regulators of various activities, molecules involved in the transportation of substances, and molecules involved in an organism's defense. They are composed of **amino acids** linked by **peptide bonds,** and they possess primary, secondary, tertiary, and (sometimes) quaternary structures that affect their function. **Denaturation** of a protein disrupts its structure and subsequently its function.

9. **Deoxyribonucleic acid (DNA)** and **ribonucleic acid (RNA)** are unbranched macromolecular polymers of nucleotides, each composed either of deoxyribose or ribose sugar, phosphoric acid, and a nitrogenous base. Five different bases exist: **adenine, guanine, cytosine, thymine,** and **uracil.** DNA contains A, G, C, and T nucleotides. RNA uses U nucleotides instead of T nucleotides.

10. The structure of nucleic acids allows for genetic diversity, correct copying of genes for their passage on to the next generation, and the accurate synthesis of proteins.

11. **Adenosine triphosphate (ATP),** which is related to adenine nucleotide, is the most important short-term energy storage molecule in cells. It is also incorporated into the structure of many coenzymes.

QUESTIONS FOR REVIEW
(Answers to multiple choice and fill in the blanks are on the web, along with additional review questions. Visit www.microbiologyplace.com.)

Multiple Choice

1. Which of the following structures have no electrical charge?
   a. protons
   b. neutrons
   c. electrons
   d. ions

2. The atomic mass of an atom most closely approximates the sum of the masses of all its
   a. protons.
   b. electrons.
   c. isotopes.
   d. protons and neutrons.

3. One isotope of iodine differs from another in
   a. the number of protons.
   b. the number of electrons.
   c. the number of neutrons.
   d. atomic number.

4. Which of the following is *not* an organic compound?
   a. monosaccharide
   b. protein
   c. water
   d. steroid

5. Which of the following terms most correctly describes the bonds in a molecule of water?
   a. nonpolar covalent bond
   b. polar covalent bond
   c. ionic bond
   d. hydrogen bond

6. In water, cations and anions of salts disassociate from one another and become surrounded by water molecules. In this state, the ions are also called
   a. electrically negative.
   b. ionically bonded.
   c. electrolytes.
   d. hydrogen bonds.

7. Which of the following can be most accurately described as a decomposition reaction?
   a. $C_6H_{12}O_6 + 6 O_2 \rightarrow 6 H_2O + 6 CO_2$
   b. glucose + ATP $\rightarrow$ glucose phosphate + ADP
   c. $6 H_2O + 6 CO_2 \rightarrow C_6H_{12}O_6 + 6 O_2$
   d. $A + BC \rightarrow AB + C$

8. Which of the following statements about a carbonated cola beverage with a pH of 2.9 is true?
   a. It has a relatively high concentration of hydrogen ions.
   b. It has a relatively low concentration of hydrogen ions.
   c. It has equal amounts of hydroxyl and hydrogen ions.
   d. Cola is a buffered solution.

9. Proteins are polymers of
   a. amino acids.
   b. fatty acids.
   c. nucleic acids.
   d. monosaccharides.

10. Which of the following are hydrophobic organic molecules?
    a. proteins
    b. carbohydrates
    c. lipids
    d. nucleic acids

Fill in the Blanks

1. The outermost electron shell of an atom is known as the _____ shell.
2. The type of chemical bond between atoms with nearly equal electronegativities is called a _____ _____ bond.
3. The principal short-term energy storage molecule in cells is _____.
4. The most common long-term energy storage molecule is _____.
5. Groups of atoms such as NH_2 or OH that appear in certain common arrangements are called _____.
6. The reverse of dehydration synthesis is _____.
7. Reactions that release energy are called _____ reactions.
8. All chemical reactions begin with reactants and result in new molecules called _____.

Short Answer

1. List three main types of chemical bonds, and give an example of each.
2. Name five properties of water that are vital to life.
3. Describe the difference(s) among saturated fatty acids, unsaturated fatty acids, and polyunsaturated fatty acids.
4. What is the difference between atomic oxygen and molecular oxygen?
5. Explain how the polarity of water molecules makes water an excellent solvent.

CRITICAL THINKING

1. Anthrax is caused by a bacterium, *Bacillus anthracis*, a pathogen that avoids defenses against disease by synthesizing an outer glycoprotein covering made from D-glutamic acid. This covering is not digestible by white blood cells that normally engulf bacteria. Why is the covering indigestible?

2. Dehydrogenation is a chemical reaction in which a saturated fat is converted to an unsaturated fat. Explain why the name for this reaction is an appropriate one.

3. Two freshmen disagree about an aspect of chemistry. The nursing major insists that H^+ is the symbol for a hydrogen ion. The physics major insists that H^+ is the symbol for a proton. How can you help them resolve their disagreement?

4. When an egg white is heated, it changes from liquid to solid. When gelatin is cooled, it changes from liquid to solid. Both gelatin and egg white are proteins. From what you have learned about proteins, why can the gelatin be changed back to liquid but the cooked egg cannot?

5. When amino acids are synthesized in a test tube, D and L forms occur in equal amounts. However, cells use only L forms in their proteins. Occasionally, meteorites are found to contain amino acids. Based on these facts, how could NASA scientists determine whether the amino acids recovered from space are evidence of Earth-like extraterrestrial life or of non-metabolic processes?

CHAPTER 3

Cell Structure and Function

CHAPTER OUTLINE

Processes of Life

Eukaryotic and Prokaryotic Cells:
An Overview

External Structures of Prokaryotic
Cells
    Glycocalyces
    Flagella
    Fimbriae and Pili

Prokaryotic Cell Walls
    Bacterial Cell Walls
    Archaeal Cell Walls

Prokaryotic Cytoplasmic
Membranes
    Structure
    Function

Cytoplasm of Prokaryotes
    Cytosol
    Inclusions
    Nonmembranous Organelles

External Structures of Eukaryotic
Cells
    Glycocalyces
    Flagella
    Cilia

Eukaryotic Cell Walls and
Cytoplasmic Membranes

Cytoplasm of Eukaryotes
    Nonmembranous Organelles
    Membranous Organelles
    Endosymbiotic Theory

Can a microbe be a magnet? The answer is yes, if it is a magneto-bacterium.

Magnetobacteria are microorganisms with an unusual feature: cellular structures called *magnetosomes.* Magnetosomes are stored deposits (also called inclusions) of the mineral magnetite. These deposits allow magnetobacteria to respond to the lines of the Earth's magnetic field, much like a compass. In the Southern Hemisphere, magnetobacteria exist as south-seeking varieties; in the Northern Hemisphere, they exist as north-seeking varieties.

How do these bacteria benefit from magnetosomes? Magneto-bacteria prefer environments with little or no oxygen, such as those that exist far below the surfaces of land and sea. The magnetosomes are drawn toward the underground magnetic poles, helping magnetobacteria move toward the oxygen-scarce regions that are their favored homes.

Aquaspirillum magnetotacticum, a magnetobacterium. The dark circles make up the magnetosome.

MicroPrep Pre-Test: *Take the pre-test for this chapter on the web. Visit* **www.microbiologyplace.com.**

All living things—including our bodies and the bacterial, protozoan, and fungal pathogens that attack us—are composed of living cells. If we want to understand disease and its treatment, therefore, we must first understand the life of cells. How pathogens attack our cells, how our bodies defend themselves, how current medical treatments assist our bodies in recovering—all of these activities have their basis in the biology of our, and our pathogens', cells.

In this chapter, we will examine cells and the structures within cells. We will discuss similarities and differences between the two major kinds of cells, eukaryotes and prokaryotes. The differences are particularly important because they allow researchers to develop treatments that inhibit or kill pathogens without adversely affecting a patient's own cells. We will also learn about cellular structures that allow pathogens to evade the body's defenses and cause disease.

Processes of Life

Learning Objective

✓ Describe the four major processes of living cells.

As we discussed in Chapter 1, microbiology is the study of particularly small living things. What we have not yet discussed is the question of how we define life. Scientists once thought that living things were composed of special organic chemicals, such as glucose and amino acids, that carried a "life force" found only in living organisms. These organic chemicals were thought to be formed only by living things and to be very different from the inorganic chemicals of nonliving things.

The idea that organic chemicals could come only from living organisms had to be abandoned in 1828, when Friedrich Wöhler (1800–1882) synthesized urea, an organic molecule, using only inorganic substrates in his laboratory. Today we know that all living things contain both organic and inorganic chemicals, and that many organic chemicals can be made from inorganic chemicals by laboratory processes. If organic chemicals can be made even in the absence of life,

what is the difference between a living thing and a nonliving thing? What is life?

At first this may seem a simple question. After all, you can usually tell when something is alive. However, defining "life" itself is difficult, so biologists generally avoid setting a definition, preferring instead to describe characteristics common to all living things. Biologists can agree that all living things share at least four processes of life: growth, reproduction, responsiveness, and metabolism.

- **Growth.** Living things can grow; that is, they can increase in size.

- **Reproduction.** Organisms normally have the ability to reproduce themselves. Reproduction means that they increase in number, producing more organisms like themselves. Reproduction may be accomplished asexually (alone) or sexually with a mate. Note that reproduction is an increase in number, whereas growth is an increase in size. Growth and reproduction often occur simultaneously. We consider several methods of reproduction when we examine microorganisms in detail in Chapters 11–13.

- **Responsiveness.** All living things respond to their environment. They have the ability to change internal and/or external properties in reaction to changing conditions around or within them. Many organisms also have the ability to move toward or away from environmental stimuli—a response called *taxis*.

- **Metabolism.** Metabolism can be defined as the ability of organisms to take in nutrients from outside themselves and use the nutrients in a series of controlled chemical reactions to provide the energy and structures needed to grow, reproduce, and be responsive. Metabolism is a unique process of living things; nonliving things cannot metabolize. Cells store metabolic energy in the chemical bonds of *adenosine triphosphate* (ă-den'ō-sēn trī-fos'fāt) or *ATP*. The major processes of microbial metabolism, including the generation of ATP, are discussed in Chapters 5–7.

Table 3.1 shows how these characteristics, along with cell structure, relate to various kinds of microbes.

Table 3.1	Characteristics of Life and Their Distribution in Microbes		
Characteristic	Prokaryotes	Eukaryotes	Viruses
Growth: increase in size	Occurs in all	Occurs in all	Does not occur
Reproduction: increase in number	Occurs in all	Occurs in all	Occurs only inside a host cell
Responsiveness: ability to react to environmental stimuli	Occurs in all	Occurs in all	Occurs in some viruses as a reaction to host cells
Metabolism: controlled chemical reactions of organisms	Occurs in all	Occurs in all	Uses host cell's metabolism
Cellular structure: membrane-bound structure capable of all of the above functions	Present in all	Present in all	Lacks cytoplasmic membrane or cellular structure

Highlight 3.1   Are Viruses Alive?

An ongoing discussion in microbiology concerns the nature of viruses. Are they alive?

Viruses are noncellular; that is, they lack cell membranes, cell walls, and most other cellular components. However, they do contain genetic material in the form of either DNA or RNA. (No virus has both DNA and RNA.) While many viruses use DNA molecules for their genes, others such as HIV and poliovirus use RNA for genes instead of DNA. No cells use RNA molecules for their genes.

Influenzaviruses   **TEM**   100 nm

Viruses have some other characteristics of living cells. For instance, some demonstrate responsiveness to their environment when they inject their genetic material into susceptible host cells. However, viruses lack most characteristics of life: They are unable to grow, reproduce, or metabolize outside of a host cell, although once they enter a cell they take control of the cell's metabolism and cause it to make more viruses. This takeover typically leads to the death of the cell and results in disease in an organism.

CRITICAL THINKING

In your opinion, are viruses alive? What factors led you to your conclusion?

Organisms may not exhibit these processes at all times. For instance, in some organisms, reproduction may be postponed or curtailed by age or disease, or in humans at least, by choice. Likewise, the rate of metabolism may be reduced, as occurs in a seed, a hibernating animal, or a bacterial endospore,[1] and growth often stops when an animal reaches a certain size. However, microorganisms typically grow, reproduce, respond, and metabolize as long as conditions are suitable. The proper conditions for the metabolism and growth of various types of microorganisms are discussed in Chapter 6.

A group of pathogens essential to the debate regarding "What constitutes life?" are the viruses. **Highlight 3.1** discusses the question of whether or not viruses should be considered living things.

Eukaryotic and Prokaryotic Cells: An Overview

Learning Objective

✓ Compare and contrast prokaryotic and eukaryotic cells.

In the 1800s, two German biologists, Theodor Schwann (1804–1881) and Matthias Schleiden (1810–1882) developed the theory that all living things are composed of cells. *Cells* are living entities, surrounded by a membrane, that are capable of growing, reproducing, responding, and metabolizing. The smallest living things are single celled microorganisms.

There are many different kinds of cells **(Figure 3.1).** Some cells are free-living, independent organisms; others live together in colonies or form the bodies of multicellular organisms. Cells also exist in various sizes, from the smallest

bacteria to bird eggs, which are the largest of cells. All cells are described as either *prokaryotes* (prō-kar'ē-ōts) or *eukaryotes* (yū-kar'ē-ōts).

The distinctive structural feature of **prokaryotes** is not what they have, but what they lack. The light microscope reveals that prokaryotes do not have a membrane surrounding their genetic material (DNA); that is, they do not have a nucleus **(Figure 3.2a** on page 60). The word *prokaryote* comes from Greek words meaning "before nucleus." Moreover, electron microscopy has revealed that prokaryotes also lack various types of internal membrane-bound structures present in eukaryotic cells. Not only are prokaryotes simpler than eukaryotes; they are also typically smaller—approximately 1.0 μm in diameter, as compared to 10–100 μm for eukaryotic cells **(Figure 3.3** on page 61). (See **Highlight 3.2** on page 59 for some interesting exceptions concerning the size of prokaryotes.)

There are two types of prokaryotic cells: *bacteria* and *archaea*. These cell types fundamentally differ in the nucleotide base sequences of their ribosomal RNA, in the type of lipids in their cytoplasmic membranes, and in the chemistry of their cell walls. Chapter 11 discusses these differences in more detail.

Eukaryotes have a membrane surrounding their DNA, forming a nucleus (see Figure 3.2b). Indeed, the term *eukaryote* comes from Greek words meaning "true nucleus." Besides the nuclear membrane, eukaryotes have numerous other internal membranes that compartmentalize cellular functions. These membrane-bound compartments are **organelles,** specialized structures that act like tiny organs to carry on the various functions of the cell. Organelles and their functions are discussed later in this chapter. The cells of algae, protozoa, fungi, animals, and plants are eukaryotic. Eukaryotes are usually larger and more complex than prokaryotes.

---

[1]Endospores are resting stages, produced by some bacteria, that are tolerant of environmental extremes.

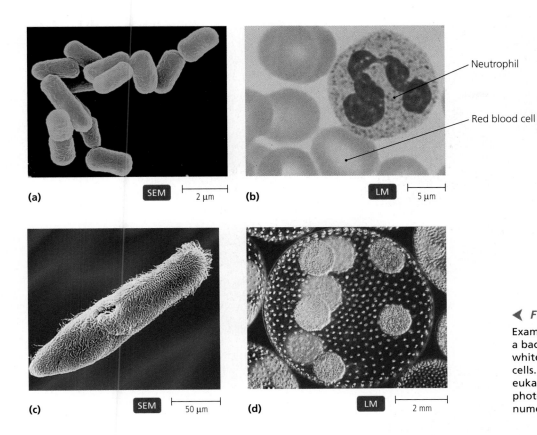

(a) SEM   2 μm

(b) LM   5 μm

— Neutrophil

— Red blood cell

(c) SEM   50 μm

(d) LM   2 mm

◀ *Figure 3.1*

Examples of types of cells. **(a)** *Escherichia coli*, a bacterial cell. **(b)** A neutrophil, a defensive white blood cell, amidst human red blood cells. **(c)** *Paramecium*, a single-celled eukaryote. **(d)** *Volvox*, a colonial, photosynthetic eukaryote composed of numerous small cells and daughter colonies.

Highlight 3.2   **Unusual Giants**

Most prokaryotes are small compared to eukaryotes. The dimensions of a typical bacterial cell—for example, a cell of *Escherichia coli*—are $1.0 \ \mu m \times 2.0 \ \mu m$. Typical eukaryotic cells are often thousands of micrometers in diameter.

For many years biologists thought that the small size of prokaryotes was a necessary result of the nature of their cells. Without internal compartments, any prokaryotes that were as large as eukaryotes could not isolate and control the metabolic reactions of life; nor could they efficiently distribute nutrients and eliminate wastes.

Paramecium
Epulopiscium fishelsoni
LM   250 μm

In 1985 scientists located a new unicellular organism from the intestines of a surgeonfish. Its cells were 0.6 mm (600 μm) long, a size visible to the unaided eye. Microscopic examination revealed the presence of a fine covering of what appeared to be cilia, which are a feature of eukaryotic cells only. The new organism was named *Epulopiscium fishelsoni*,[a] and it was assumed that such a large ciliated cell was eukaryotic. A surprise came when these cells were examined with an electron microscope. They contained no nuclei or other eukaryotic organelles, and the "cilia" more closely resembled short bacterial flagella. *Epulopiscium*, as it turned out, is indeed a giant prokaryote. (Compare the size of *Epulopiscium* with that of *Paramecium*, a eukaryote, in the box photo.)

It is not, however, the largest prokaryote. In 1997, Heide Shultz, a graduate student of marine microbiology in Bremen, Germany, discovered an even larger bacterium, named *Thiomargarita namibiensis*,[b] which lives in chains of 2–50 cells in ocean sediments off the coast of Namibia, Africa. The spherical cells of *Thiomargarita* can grow to 750 μm in diameter, or about the size of the period at the end of this sentence. Such large cells have volumes over 66 times those of the largest *Epulopiscium*. If a single cell of *Thiomargarita* were the size of an African elephant, then *Epulopiscium* would be the size of a cow, and *Escherichia* would be the size of an ant.

[a]*Epulopiscium* in Latin means "guest at a banquet of fish."

[b]This name means "sulfur pearl of Namibia," because granules of sulfur stored in the cytoplasm cause the cells to glisten white, like a string of pearls.

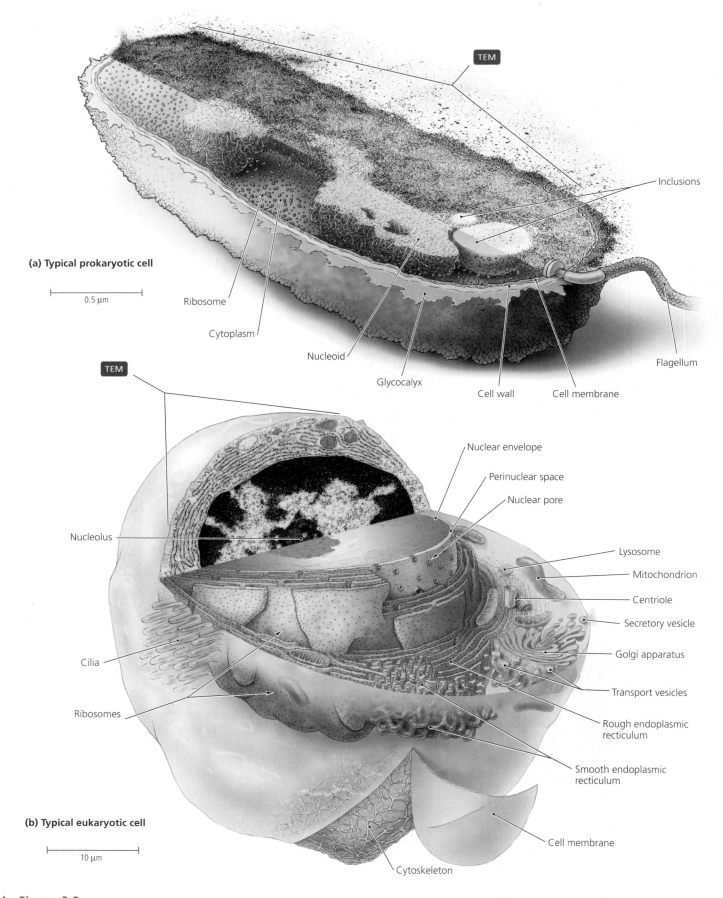

(a) Typical prokaryotic cell

0.5 μm

Ribosome

Cytoplasm

Nucleoid

Glycocalyx

Cell wall

Cell membrane

Inclusions

Flagellum

TEM

TEM

Nucleolus

Cilia

Ribosomes

(b) Typical eukaryotic cell

10 μm

Cytoskeleton

Nuclear envelope

Perinuclear space

Nuclear pore

Lysosome

Mitochondrion

Centriole

Secretory vesicle

Golgi apparatus

Transport vesicles

Rough endoplasmic recticulum

Smooth endoplasmic recticulum

Cell membrane

▲ *Figure 3.2*

Comparison of cells. **(a)** A prokaryote. **(b)** A eukaryote. Note differences in magnification.
What major difference between prokaryotes and eukaryotes was visible to early microscopists?

Figure 3.2 Eukaryotes contain nuclei, which are visible with light microscopy, whereas prokaryotes lack nuclei.

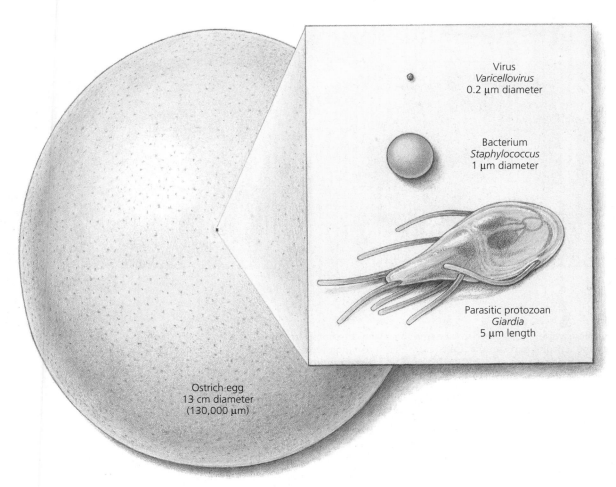

Virus
Varicellovirus
0.2 µm diameter

Bacterium
Staphylococcus
1 µm diameter

Parasitic protozoan
Giardia
5 µm length

Ostrich egg
13 cm diameter
(130,000 µm)

*At this scale, the inset box on the egg would actually be too small to be visible. (Width of box would be about 0.006 mm.)

▲ *Figure 3.3*

Approximate size of various types of cells. Note that *Staphylococcus*, a prokaryote, is smaller than *Giardia*, a eukaryote. A chicken pox virus is shown only for comparison; viruses are not cellular.

Although there are many kinds of cells, they all share the characteristic processes of life as previously described, as well as certain physical features. In this chapter, we will distinguish between prokaryotic and eukaryotic "versions" of physical features common to cells, including (1) external structures, (2) the cell wall, (3) the cytoplasmic membrane, and (4) the cytoplasm. We will also discuss features unique to each type. Further details of prokaryotes and eukaryotes, their classification, and their ability to cause disease are discussed in Chapters 11, 12, and 19–23.

External Structures of Prokaryotic Cells

Many cells have special external features that enable them to respond to other cells and their environment. In prokaryotes, these features include glycocalyces, flagella, fimbriae, and pili.

Glycocalyx
(Capsule)

Glycocalyx
(Slime layer)

(a)   TEM   200 nm

(b)   TEM   200 nm

▲ *Figure 3.4*

Glycocalyces. **(a)** Micrograph of *Streptococcus pneumoniae*, the common cause of pneumonia, showing a prominent capsule. **(b)** *Bacteroides*, a common fecal bacterium, has a slime layer surrounding the cell. *What advantage does a glycocalyx provide a cell?*

Figure 3.4 A glycocalyx provides protection from drying and from being devoured; it may also help attach cells to one another and to surfaces in the environment.

Glycocalyces

Learning Objectives

✓ Describe the composition, function, and relevance to human health of glycocalyces.
✓ Distinguish between capsules and slime layers.

Some cells have a gelatinous, sticky substance that surrounds the outside of the cell. This substance is known as a **glycocalyx** (plural, *glycocalyces*), which literally means "sugar cup." The glycocalyx may be composed of polysaccharides, polypeptides, or both. These chemicals are produced inside the cell and extruded onto the cell's surface.

When the glycocalyx of a prokaryote is composed of organized repeating units of organic chemicals firmly attached to the cell surface, the glycocalyx is called a **capsule (Figure 3.4a).** A loose, water-soluble glycocalyx is called a **slime layer (Figure 3.4b).** Capsules and slime layers protect cells from desiccation (drying).

The presence of a glycocalyx is a feature of numerous pathogenic prokaryotes. Their glycocalyces play an important role in the ability of these cells both to survive and to cause disease. Slime layers are often *viscous* (sticky), providing one means by which prokaryotes attach to surfaces. For example, they enable oral bacteria to colonize the teeth, where they produce acid and cause decay. Since the chemicals in many capsules are similar to those normally found in the body, they may prevent bacteria from being recognized and devoured by defensive cells of the host. For example,

the capsules of *Streptococcus pneumoniae* (strep-tō-kok-ŭs nū-mō′nē-ē) and *Klebsiella pneumoniae* (kleb-sē-el′ă nū-mō′nē-ē) enable these prokaryotes to avoid destruction by defensive cells in the respiratory tract and cause pneumonia. Unencapsulated strains of these bacteria do not cause disease because the body's defensive cells destroy them.

Flagella

Learning Objectives

✓ Discuss the structure and function of prokaryotic flagella.
✓ List and describe four prokaryotic flagellar arrangements.

A cell's motility may enable it to flee from a harmful environment or move toward a favorable environment such as one where food or light is available. The structures responsible for this microbial movement are flagella. **Flagella** (singular: *flagellum*) are long, whiplike structures that extend beyond the surface of the cell and glycocalyx and propel the cell through its environment.

Structure

Prokaryotic flagella are composed of three parts: a long, thin *filament,* a *hook,* and a *basal body* **(Figure 3.5).** The filament is the whiplike shaft, about 20 nm in diameter, that extends out into the cell's environment. It is composed of many identical globular molecules of a protein called *flagellin*. Molecules of flagellin are arranged in chains and form a helix around a

(a)

◀ *Figure 3.5*

**Proximal structure of prokaryotic flagella.
(a)** Detail of flagellar structure of a Gram-positive cell. **(b)** Detail of the flagellum of a Gram-negative prokaryote. *How do flagella of Gram-positive prokaryotes differ from those of Gram-negative prokaryotes?*

Figure 3.5 Flagella of Gram-positive cells have a single pair of rings in the basal body, which function to attach the flagellum to the cytoplasmic membrane. The flagella of Gram-negative cells have two pairs of rings; one pair anchors the flagellum to the cytoplasmic membrane, the other pair to the cell wall.

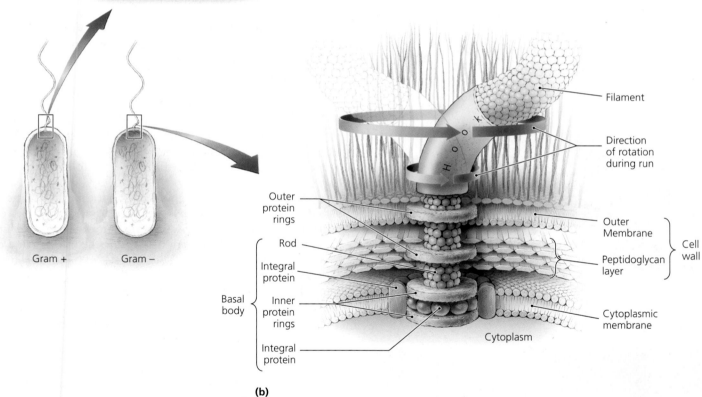

(b)

Arrangement

Prokaryotes may have one of several flagellar arrangements (**Figure 3.6** on page 64). Cells with a single flagellum are **monotrichous.**[2] **Lophotrichous**[3] cells have a tuft of flagella at only one end of the cell, whereas **amphitrichous**[4] cells have flagella at both ends. Flagella that cover the surface of

hollow core. No membrane covers the filament of prokaryotic flagella. At its base the filament inserts into a curved structure, the hook, which is composed of a different protein. The basal body, which is composed of still different proteins, anchors the filament and hook to the cell wall by means of a rod and a series of either two or four rings. Together the hook, rod, and rings allow the filament to rotate 360°. Differences in the proteins associated with prokaryotic flagella vary enough to allow classification of prokaryotic species into groups (strains) called *serovars.*

[2]From Greek *monos,* meaning one, and *trichos,* meaning a hair.
[3]From Greek *lophos,* meaning tuft.
[4]From Greek *amphi,* meaning both.

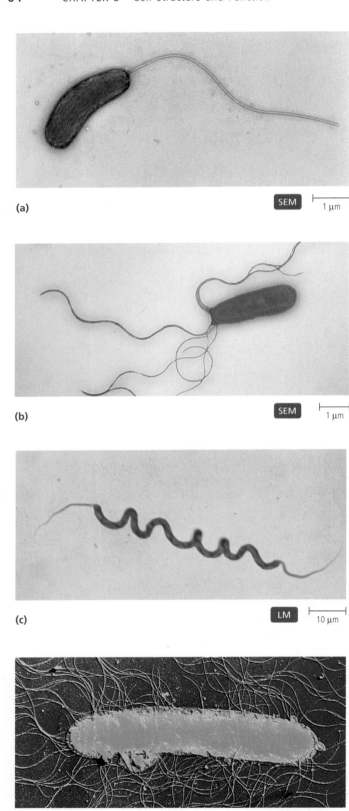

(a) SEM 1 μm

(b) SEM 1 μm

(c) LM 10 μm

(d) SEM 1 μm

▲ *Figure 3.6*

Micrographs of basic arrangements of bacterial flagella.
(a) Monotrichous. **(b)** Lophotrichous. **(c)** Amphitrichous.
(d) Peritrichous.

the cell are termed **peritrichous,**[5] in contrast to **polar** flagella, which are only at the ends.

Some spiral-shaped bacteria, called *spirochetes* (spī′rō-kēts),[6] have amphitrichous flagella that spiral tightly around the cell instead of protruding into the surrounding medium. These flagella, called **endoflagella,** form an **axial filament** that wraps around the cell between its cell membrane and an outer membrane found in spirochetes. **(Figure 3.7).** Rotation of endoflagella evidently causes the axial filament to rotate around the cell, causing the spirochete to "corkscrew" through its medium. *Treponema pallidum* (trep-ō-nē′mă pal′li-dŭm), the agent of syphilis, and *Borrelia burgdorferi* (bō-rē′lē-a bŭrg-dor′fer-ē), the cause of Lyme disease, are notable spirochetes. Some scientists think the corkscrew motility of these pathogens allows them to invade viscous human tissues.

Function

Although the precise mechanism by which prokaryotic flagella move is not completely understood, we do know that they rotate 360° like the shaft of an electric motor rather than whipping from side to side. This rotation propels the bacterium through the environment. Rotation of a flagellum may be either clockwise or counterclockwise and can be reversed by the cell. If more than one flagellum is present, the flagella align and rotate together as a unit.

Prokaryotes move with a series of "runs" punctuated by "tumbles" **(Figure 3.8).** Runs are movements of a cell in a single direction for some time; tumbles are abrupt, random, changes in direction. Both runs and tumbles occur in response to stimuli.

Receptors for light or chemicals on the surface of the cell send signals to the flagella, which then adjust their speed and direction of rotation. A prokaryote can position itself in a more favorable environment by varying the number and duration of runs and tumbles. The presence of favorable stimuli increases the number of runs and decreases the number of tumbles; as a result, the cell tends to move toward an attractant. Unfavorable stimuli increase the number of tumbles, which increases the likelihood that it will move in another direction, away from a repellant.

Movement in response to a stimulus is termed **taxis.** The stimulus may be either light **(phototaxis)** or a chemical **(chemotaxis).** Movement toward a favorable stimulus is *positive taxis*, whereas movement away from an unfavorable stimulus is *negative taxis*. For example, movement toward a nutrient would be *positive chemotaxis*.

Fimbriae and Pili

Learning Objective

✓ Compare and contrast the structures and functions of fimbriae, pili, and flagella.

---

[5]From Greek *peri,* meaning around.
[6]From Greek *speira,* meaning coil, and *chaeta,* meaning hair

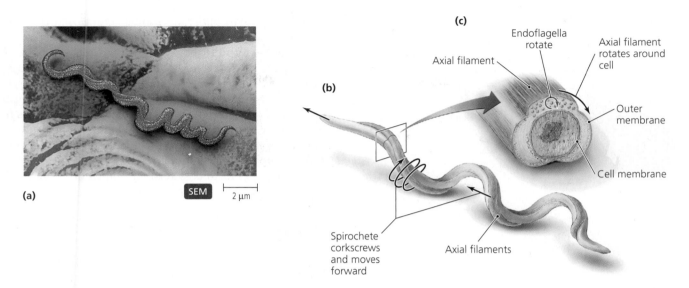

▲ *Figure 3.7*

Axial filament. **(a)** Scanning electron micrograph of *Treponema pallidum*, a spirochete.
(b) Diagram of axial filament wrapped around a spirochete. **(c)** Cross-section of the
spirochete, which reveals that the axial filament is composed of endoflagella.

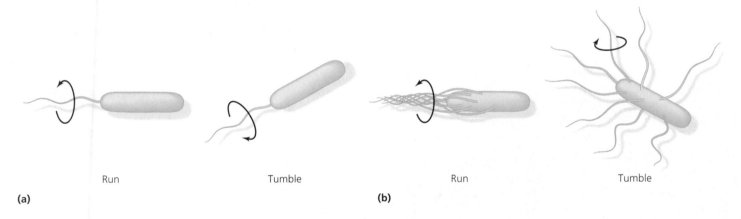

▲ *Figure 3.8*

Patterns of bacterial motility. **(a)** Motion of a monotrichous bacterium. **(b)** Motion of a
peritrichous bacterium. In peritrichous bacteria, runs occur when all of the flagella
rotate counterclockwise and become bundled. Tumbles occur when the flagella rotate
clockwise, become unbundled, and the cell spins randomly. *What triggers a bacterial
flagellum to rotate counterclockwise, producing a run?*

Figure 3.8 Favorable environmental conditions induce runs.

Some prokaryotes have nonmotile extensions called fimbriae
and pili. Some Gram-negative bacteria (defined shortly) use
fimbriae—sticky, proteinaceous, bristlelike projections—to
adhere to one another and to substances in the environment.
There may be hundreds of fimbriae per cell. They are
usually shorter than flagella (**Figure 3.9** on page 66). An example of a prokaryote with fimbriae is *Neisseria gonorrhoeae*
(nī-sē'rē-ă go-nōr-rē'ē), the bacterium that causes gonorrhea.
Pathogens must be able to adhere to their hosts if they are to
survive and cause disease. This bacterium is able to colonize
the mucous membrane of the reproductive tract by attaching with fimbriae. *Neisseria* cells that lack fimbriae are nonpathogenic.

Fimbriae also serve an important function in **biofilms,**
slimy masses of bacteria adhering to a substrate by means of
fimbriae and glycocalyces. It has been estimated that about
99% of bacteria in nature exist in biofilms. Biofilms are a
topic of intense interest to researchers because of the roles
they play in human diseases and in industry. See the **New
Frontiers** box on page 67.

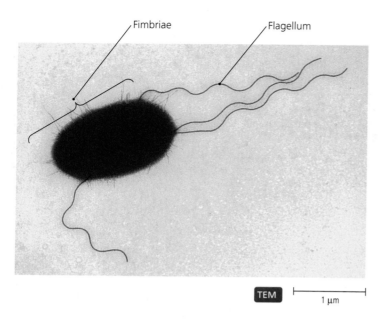

Fimbriae   Flagellum

TEM   1 µm

▲ **Figure 3.9**

Fimbriae. *Escherichia coli* have fimbriae and flagella.

Fimbriae   Conjugation pilus

TEM   1 µm

▲ **Figure 3.10**

Pilus. Two *Escherichia coli* cells are connected by a conjugation (sex) pilus. *How are pili different from prokaryotic flagella?*

Figure 3.10 Prokaryotic flagella are flexible structures that rotate to propel the cell; pili are hollow, tubes that mediate transfer of DNA from one cell to another.

Pili (singular: *pilus*) are tubules composed of a protein called *pilin*. Pili are longer than fimbriae, but usually shorter than flagella. Typically only one to ten pili are present per cell in bacteria that have them. Bacteria use pili to move across a substrate or towards another bacterium via a process that appears to be similar to the use of a grappling hook. The bacterium extrudes a pilus, which attaches to the substrate or to another bacterium, then the bacterium pulls itself along the pilus toward the attachment point. So-called **conjugation pili** mediate the transfer of DNA from one cell to the other via a process termed *conjugation* **(Figure 3.10)**, which is discussed in detail in Chapter 7.

Prokaryotic Cell Walls

The cells of most prokaryotes are surrounded by a **cell wall** that provides structure and shape to the cell and protects it from osmotic forces. In addition, a cell wall assists some cells in attaching to other cells or in eluding antimicrobial drugs. Note that animal cells do not have walls, a difference that plays a key role in treatment of many bacterial diseases with certain types of antibiotics. For example, penicillin attacks the cell wall of bacteria but is harmless to human cells, which lack walls.

Recall that there are two basic types of prokaryotes: bacteria and archaea. One difference between these prokaryotes involves the chemistry of their cell walls. Scientists have studied bacterial walls more extensively than those of archaea, so we will concentrate our discussion on bacterial cell walls.

Bacterial Cell Walls

Learning Objectives

✓ Compare and contrast the cell walls of Gram-positive and Gram-negative prokaryotes in terms of structure and Gram staining.

✓ Describe the clinical implications of the structure of the Gram-negative cell wall.

✓ Compare and contrast the cell walls of acid-fast bacteria with typical Gram-positive cell walls.

Most bacteria have a cell wall, composed of **peptidoglycan,** a complex polysaccharide. Peptidoglycan is composed of two regularly alternating sugars, called *N-acetylglucosamine (NAG)* and *N-acetylmuramic acid (NAM),* which are structurally similar to glucose **(Figure 3.11)**. NAG and NAM are covalently linked in long chains in which NAG alternates with NAM **(Figure 3.12** on page 68). These chains are the "glycan" portions of peptidoglycan.

Chains of NAG and NAM are attached to other chains by crossbridges of four amino acids (tetrapeptides). These crossbridges are the "peptido" portion of peptidoglycan. Depending on the bacterium, these tetrapeptide bridges are either covalently bonded to one another or are held together by *short connecting* chains of amino acids as shown in Figure 3.12. Peptidoglycan covers the entire surface of the cell.

New Frontiers 3.1   Understanding Biofilms

They form plaque on teeth; they are the slime on rocks in rivers and streams; and they can cause disease, clog drains, or aid in hazardous-waste cleanup. They are **biofilms**—organized, layered systems of bacteria attached to a surface. Understanding biofilms holds the key to many important clinical and industrial applications.

Recent research has demonstrated that bacteria in biofilms behave in significantly different ways from individual, free-floating bac-teria. For example, as a free-floating cell, the soil bacterium *Pseudomonas putida* propels itself through water with its flagella; however, once it becomes part of a biofilm, it turns off the genes for flagellar proteins and starts synthesizing pili instead. In addition, the genes for antibiotic resistance in *P. putida* are more active within cells in a biofilm than within a free-floating cell.

Biofilm on a tooth surface   SEM |—————| 5 µm

Bacteria in a biofilm also communicate via chemical messages that help them organize and form three-dimensional structures. The architecture of a biofilm provides protection that free-floating bacteria lack. For example, the lower concentrations of oxygen found in the interior of biofilms thwart the effectiveness of some antibiotics. Moreover, the presence of multiple bacterial species in most biofilms increases the likelihood that some bacteria within the biofilm community will remain resistant to a given antibiotic.

The more we understand biofilms, the more readily we can reduce their harmful effects or put them to good use. Biofilms account for two-thirds of bacterial infections in humans, including gum disease and the serious lung infections suffered by cystic fibrosis patients. Biofilms are also the culprits in many industrial problems, including corroded pipes and clogged water filters, which cause millions of dollars of damage each year. Fortunately, not all biofilms are detrimental; some have shown potential as aids in preventing and controlling certain kinds of pollution.

1. "Slimebuster," *Nature*, 408: 284–286.

2. The Center for Biofilm Engineering at Montana State University, Bozeman: www.erc.montana.edu

The two basic types of bacterial cell walls are **Gram-positive** cell walls and **Gram-negative** cell walls, and bacterial cells are described as being either "Gram-positive" or "Gram-negative" depending on the kind of wall they have. Gram-positive and Gram-negative cells are distinguished by the use of the Gram staining procedure, which is described in Chapter 4. The Gram staining procedure was invented long before the structure and chemical nature of prokaryotic cell walls were known.

Gram-Positive Cell Walls

Gram-positive cell walls have a relatively thick layer of peptidoglycan that also contains unique polysaccharides called *teichoic* (tī-kō′ĭk)[7] *acids*. Some teichoic acids are covalently linked to lipids, forming *lipoteichoic acids* that anchor the peptidoglycan to the cell membrane (**Figure 3.13a** on page 69). The thick cell wall of a Gram-positive organism retains the crystal violet dye used in the Gram staining procedure, so the stained cells appear purple under magnification.

Some additional chemicals are associated with the walls of some Gram-positive cells. For example, species of *Mycobacterium* (mī′kō-bak-tēr′ē-ŭm), which include the causative agents of tuberculosis and leprosy, have walls with up to 60% mycolic acid, a waxy lipid. Mycolic acid helps these cells survive desiccation and makes them

▲ *Figure 3.11*

Comparison of the structures of glucose, NAG, and NAM. **(a)** Glucose. **(b)** N-acetylglucosamine (NAG) and N-acetylmuramic acid (NAM) molecules linked as in peptidoglycan. Blue shading indicates the differences between glucose and the other two sugars. Orange boxes highlight the difference between NAG and NAM.

difficult to stain with regular water-based dyes. Special staining procedures such as the acid-fast stain have been developed for these Gram-positive cells that contain large amounts of waxy lipids so they are called *acid-fast bacteria* (see Chapter 4).

[7]From Greek *teichos*, meaning wall.

Figure 3.12 ▶

Structure of peptidoglycan.
Peptidoglycan is composed of chains
of NAG and NAM linked by
tetrapeptide crossbridges and
connecting chains of amino acids to
form a relatively rigid structure. The
amino acids of the crossbridges differ
among bacterial species.

Sugar
backbone

Tetrapeptide
(amino acid)
crossbridge

Connecting chains
of amino acids

Gram-Negative Cell Walls

Gram-negative cell walls have only a thin layer of peptidoglycan **(Figure 3.13b),** but outside this layer is a bilayer membrane composed of phospholipids, channel proteins (called *porins*), and **lipopolysaccharide (LPS).** LPS is a union of lipid with sugar. The lipid portion of LPS is known as **lipid A.** The erroneous idea that lipid A is *inside* Gram-negative cells led to the use of the term *endotoxin*[8] for this chemical. Lipid A is released from dead cells when the cell wall disintegrates, and it may trigger fever, vasodilation, inflammation, shock, and blood clotting in humans. Because killing large numbers of Gram-negative cells with antimicrobial drugs releases large amounts of lipid A, which might threaten the patient more than the live bacteria, any internal infection by Gram-negative bacteria is cause for concern.

The Gram-negative cell wall can also be an impediment to the treatment of disease. The outer membrane may prevent the movement of penicillin to the underlying peptidoglycan, thus rendering the drug ineffectual against many Gram-negative pathogens.

Between the cell membrane and the outer membrane of Gram-negative organisms is a **periplasmic space** (see Figure 3.13b). The periplasmic space contains the peptidoglycan and *periplasm*, the name given to the gel between the membranes of Gram-negative cells. Periplasm contains water,

[8]From Greek *endo*, meaning inside, and *toxikon*, meaning poison.

nutrients, and substances secreted by the cell, such as digestive enzymes and proteins involved in specific transport. The enzymes function to catabolize large nutrient molecules into smaller molecules that can be absorbed or transported into the cell.

Because the cell walls of Gram-positive and Gram-negative organisms differ, the Gram stain is an important diagnostic tool. After the Gram staining procedure Gram-negative cells appear pink, and Gram-positive cells appear purple.

Bacteria without Cell Walls

A few bacteria, such as *Mycoplasma pneumoniae* (mī′kō-plaz-mă nū-mō′nē-ē), lack cell walls entirely. In the past, these bacteria were often mistaken for viruses because of their small size and lack of walls. However, they do have other features of prokaryotic cells, including prokaryotic ribosomes (discussed later in the chapter).

Archaeal Cell Walls

Learning Objective

✓ Contrast types of archaeal cell walls with one another and with bacterial cell walls.

The archaea have walls containing a variety of specialized polysaccharides and proteins, but no peptidoglycan. The absence of peptidoglycan is one reason scientists classify archaea in a separate group from bacteria.

(a) Gram-positive bacteria

Teichoic acid

Peptidoglycan layer
(cell wall)

Cell membrane

Lipoteichoic acid

Integral
protein

Porin

Polysaccharide ⎤
⎥ LPS layer
Lipid A ⎦

Porin
(sectioned)

Outer
membrane
of cell wall

Periplasmic space

Peptidoglycan
layer of cell wall

Cell membrane

Phospholipid layers

(b) Gram-negative bacteria

Integral
proteins

Nevertheless, archaeal cells share some characteristics with bacteria. Like Gram-positive bacteria, Gram-positive archaea have a thick wall and Gram stain purple. Gram-negative archaeal cells have a layer of protein covering the wall, rather than a lipid membrane as seen in Gram-negative bacteria, but they do stain pink.

CRITICAL THINKING

After a man infected with the bacterium *Escherichia coli* was treated with the correct antibiotic for this pathogen, the bacterium was no longer found in the man's blood, but his symptoms of fever and inflammation worsened. What caused the man's response to the treatment? Why was his condition worsened by the treatment?

▲ *Figure 3.13*

Comparison of cell walls of Gram-positive and Gram-negative bacteria. **(a)** The Gram-positive cell wall has a thick layer of peptidoglycan and teichoic acids that anchor the wall to the cell membrane. **(b)** The Gram-negative cell wall has a thin layer of peptidoglycan and an outer membrane composed of lipopolysaccharide (LPS), phospholipids, and proteins. *What effects can lipid A have on human physiology?*

Figure 3.13 Lipid A can cause shock, blood clotting, and fever in humans.

Prokaryotic Cytoplasmic Membranes

Beneath the glycocalyx and the cell wall is the **cytoplasmic membrane.** The cytoplasmic membrane may also be referred to as the *cell membrane* or the *plasma membrane.*

Structure

Learning Objectives

✓ Diagram a phospholipid bilayer, and explain its significance in reference to a cytoplasmic membrane.

✓ Explain the fluid mosaic model of membrane structure.

Cytoplasmic membranes are composed of lipids and associated proteins. Bacterial membranes contain phospholipids **(Figure 3.14);** by contrast, archaeal membranes are composed of lipids that lack phosphate groups and have branched hydrocarbons linked to glycerol by ether linkages instead of ester linkages (see Table 2.4). Some archaea—particularly those that thrive in very hot water—have a single layer of lipid composed of two glycerol groups linked with branched hydrocarbon chains. Some bacterial membranes also contain

sterol-like molecules, called *hopanoids,* that help stabilize the membrane. Because scientists are most familiar with bacterial cytoplasmic membranes, the remainder of our discussion will concentrate on these structures.

The structure of a bacterial cytoplasmic membrane is referred to as a **phospholipid bilayer** (see Figure 3.14). As discussed in Chapter 2, a phospholipid molecule is bipolar; that is, the two ends of the molecule are different. The phosphate heads of each phospholipid molecule are *hydrophilic*[9] that is, they are attracted to water at the two surfaces of the membrane. The hydrocarbon tails of each phospholipid molecule are *hydrophobic*[10] and huddle together with other tails in the interior of the membrane, away from water. Phospholipids placed in a watery environment naturally form a bilayer because of their bipolar nature.

About half of a bacterial cytoplasmic membrane is composed of proteins, many of which are *integral proteins* inserted between the phospholipids. Some integral proteins

---

[9]From Greek *hydro,* meaning water, and *philos,* meaning love.
[10]From Greek *hydro,* meaning water, and *phobos,* meaning fear.

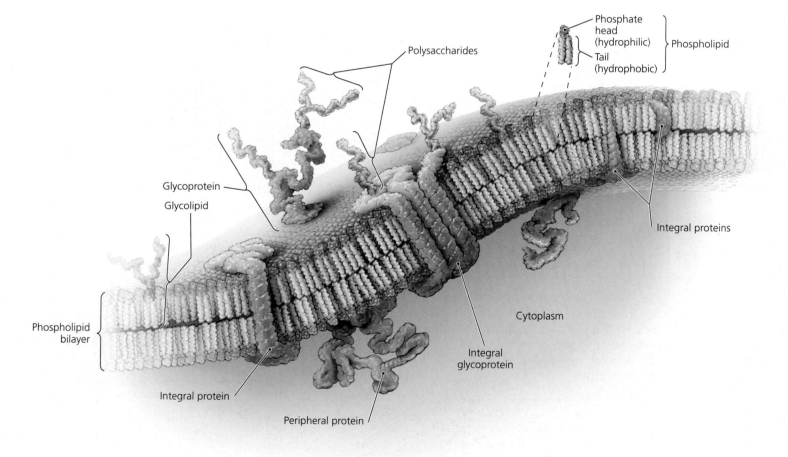

▲ *Figure 3.14*

The structure of a prokaryotic cytoplasmic membrane: a phospholipid bilayer.

penetrate the entire bilayer; others are found in only half the bilayer. *Peripheral proteins* are loosely attached to the membrane on one side or the other. Some of the membrane proteins have polysaccharide groups attached to them. Proteins chemically bound to polysaccharides are called *glycoproteins*. Proteins of cell membranes may act as recognition proteins, enzymes, receptors, carriers, or channels.

The **fluid mosaic model** describes our current understanding of membrane structure. The term *mosaic* indicates that the membrane proteins are arranged in a way that resembles the tiles in a mosaic, and *fluid* indicates that the proteins and lipids are free to flow laterally within the membrane.

Function

Learning Objectives

✓ Describe the functions of the cytoplasmic membrane as they relate to permeability.

✓ Compare and contrast the passive and active processes by which materials cross the cytoplasmic membrane.

✓ Define osmosis, and distinguish among isotonic, hypertonic, and hypotonic solutions.

The cytoplasmic membrane does more than separate the contents of the cell from the outside environment. In all prokaryotic cells, the cytoplasmic membrane controls the passage of substances into and out of the cell. Nutrients are brought into the cell, and wastes are removed. The membrane also functions in energy production, and to harvest light energy in photosynthetic prokaryotes. Energy production and photosynthesis are discussed in Chapter 5.

In its function of controlling the contents of the cell, the cytoplasmic membrane is **selectively permeable;** that is, it allows some substances to cross it while preventing the crossing of others. How does the membrane exert control over the contents of the cell, and the substances that move across it?

A phospholipid bilayer is naturally impermeable to most substances. Large molecules cannot cross through it; ions and molecules with an electrical charge are repelled by it; and hydrophilic substances cannot easily cross its hydrophobic interior. However, cytoplasmic membranes, unlike plain phospholipid bilayers in a scientist's test tube, contain proteins, and these proteins allow substances to cross the membrane by functioning as pores, channels, or carriers.

Movement across the cytoplasmic membrane occurs either by passive or active processes. Passive processes do not require the expenditure of a cell's ATP energy store, whereas active processes require the expenditure of ATP, either directly or indirectly. Active and passive processes will be discussed shortly, but first you must understand another feature of selectively permeable cytoplasmic membranes: their ability to maintain a *concentration gradient*.

Membranes enable a cell to concentrate chemicals on one side of the membrane or the other. The difference in concentration of a chemical on the two sides of a membrane is its **concentration gradient** (also known as a *chemical gradient*).

▲ *Figure 3.15*

Electrical potential of a cell membrane. The electrical potential exists across a membrane because there are more negative charges inside the cell than outside it.

Because many of the substances that have concentration gradients across cell membranes are electrically charged chemicals, a corresponding **electrical gradient,** or voltage, exists across the membrane **(Figure 3.15).** For example, a greater concentration of negatively charged proteins exists inside the membrane, while positively charged sodium ions are more concentrated outside the membrane. One result of the segregation of electrical charges by a membrane is that the interior of a cell is usually electrically negative compared to the exterior. This tends to repel negatively charged chemicals and attract positively charged substances into the cell.

Together the chemical and electrical gradients are called the **electrochemical gradient.** A cytoplasmic membrane can use the energy inherent in its electrochemical gradient to transport substances into or out of a cell.

Passive Processes

In passive processes, the electrochemical gradient provides a source of energy; the cell does not expend its ATP energy reserve. Passive processes include diffusion, facilitated diffusion, and osmosis.

Diffusion   **Diffusion** is the net movement of a chemical down its concentration gradient—that is, from an area of higher concentration to an area of lower concentration. It requires no energy output by the cell, a common feature of all passive processes. In fact, diffusion occurs even in the absence of cells or their membranes. In the case of diffusion into or out of cells, only chemicals that are small or lipid soluble can diffuse through the lipid portion of the membrane **(Figure 3.16a)**. For example, oxygen, carbon dioxide, alcohol, and fatty acids can freely diffuse through the cytoplasmic membrane, but molecules such as glucose and proteins cannot.

Facilitated Diffusion   The phospholipid bilayer blocks the movement of large or electrically charged molecules, so they cannot cross the membrane unless there is a pathway for diffusion. As we have seen, cell membranes contain integral proteins. Some of these proteins act as channels or carriers to allow certain molecules to diffuse into or out of the cell. This process is called **facilitated diffusion** because the proteins facilitate the process by providing a pathway for diffusion. The cell expends no energy in facilitated diffusion; the electrochemical gradient provides all of the energy necessary.

Channel proteins, also known as *permeases*, can be either nonspecific or specific. Nonspecific channels, which are common in prokaryotes, allow the passage of a wide range of chemicals that have the right size or electrical charge **(Figure 3.16b)**. Specific channel proteins, which are more common among eukaryotic cells, carry only specific substrates **(Figure 3.16c)**. A specific carrier has a binding site that is selective for one substance.

CRITICAL THINKING

A scientist, who is studying passive movement of chemicals across the cell membrane of *Salmonella typhi* (sal′mō-nel′ă tī ′fē), measures the rate at which two chemicals diffuse into a cell as a function of external concentration. The results are shown in the figure. Chemical A diffuses into the cell more rapidly than does B at lower external concentrations, but the rate levels off as the external concentration increases. The rate of diffusion of chemical B continues to increase as the external concentration increases.

1. How can you explain the differences in the diffusion rates of chemicals A and B?

2. Why does the diffusion rate of chemical A taper off?

3. How could the cell increase the diffusion rate of chemical A?

4. How could the cell increase the diffusion rate of chemical B?

Osmosis   When discussing simple and facilitated diffusion, we considered a solution in terms of the solutes (dissolved materials) it contains, because it is those solutes that move in and out of the cell. In contrast, with osmosis it is useful to consider the concentration of the solvent, which in organisms is always water. Osmosis is the special name given to the diffusion of water across a selectively permeable membrane—that is, across a membrane that is permeable to water molecules, but not to all solutes that are present, such as proteins, amino acids, salts, or glucose **(Figure 3.16d)**. Because they cannot freely penetrate the membrane, such solutes cannot diffuse no matter how unequal their concentrations on either side of the membrane may be. Instead, what diffuses is the water, which crosses from the side of the membrane that contains a higher concentration of water (lower concentration of solute) to the side that contains a lower concentration of water

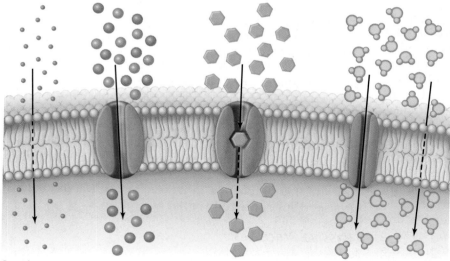

Extracellular fluid

Cytoplasm

(a) Diffusion through the phospholipid bilayer

(b) Facilitated diffusion through a nonspecific channel protein

(c) Facilitated diffusion through a channel protein specific for one chemical; binding of substrate causes shape change in channel protein

(d) Osmosis, the diffusion of water through a specific channel protein or through a phospholipid bilayer

◀ *Figure 3.16*

Passive processes of movement across a cell membrane. Passive processes always involve movement down an electrochemical gradient. **(a)** Diffusion. **(b)** Facilitated diffusion through a nonspecific channel protein. **(c)** Facilitated diffusion through a specific channel protein. **(d)** Osmosis through a nonspecific channel protein or through a phospholipid bilayer.

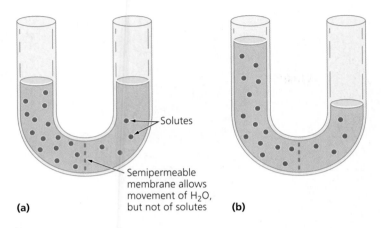

Solutes

Semipermeable membrane allows movement of H$_2$O, but not of solutes

(a)　　　　　　**(b)**

▲ *Figure 3.17*

Osmosis, the diffusion of water across a selectively permeable membrane. **(a)** A membrane separates two solutions of different concentrations in a U-shaped tube. The membrane is permeable to water, but not to the solute. **(b)** After time has passed, water has moved down its concentration gradient until water pressure prevented the osmosis of any additional water. *Which side of the tube more closely represents a living cell?*

Figure 3.17 The left-hand side represents the cell, because cells are typically hypertonic to their environment.

(higher concentration of solute). Osmosis continues until equilibrium is reached, or until the pressure of water is equal to the force of osmosis **(Figure 3.17).**

We commonly classify solutions according to their concentrations of solutes and water. When solutions on either side of a selectively permeable membrane have the same concentration of solutes and water, the two solutions are said to be **isotonic.**[11] In an isotonic situation, neither side of a selectively permeable membrane will experience a net loss or gain of water **(Figure 3.18a on page 74).**

When the concentrations of solutions are unequal, they are named according to their relative concentration of *solutes.* The solution with the higher concentration of solutes is said to be **hypertonic**[12] to the other. The solution with a lower concentration of solutes is **hypotonic**[13] in comparison. Note that the terms *hypertonic* and *hypotonic* refer to the concentration of solute, even though osmosis refers to the movement of the *solvent,* which, in cells, is water. The terms *isotonic, hypertonic,* and *hypotonic* are relative. For example, a glass of tap water is isotonic to another glass of the same water, but it is hypertonic compared to distilled water, and hypotonic when compared to seawater. In biology, the three terms are traditionally used relative to the interior of cells.

Obviously, a higher concentration of solutes necessarily means a lower concentration of water; that is, a hypertonic solution has a lower concentration of water than does a hypotonic solution. Like other chemicals, water moves down its concentration gradient from a hypotonic solution into a

hypertonic solution. A cell placed in a hypertonic solution will therefore lose water and shrivel **(Figure 3.18b).**

On the other hand, water will diffuse into a cell placed in a hypotonic solution because the cell has a higher solutes-to-water concentration. As water moves into the cell, water pressure against its cytoplasmic membrane increases, and the cell expands **(Figure 3.18c).** One function of a cell wall is to resist further osmosis and prevent cells from bursting.

It is useful to compare solutions to the concentration of solutes in a patient's blood cells. Isotonic saline solutions administered to a patient have the same percent dissolved solute (in this case, salt) as do the patient's blood cells. Thus, the patient's intracellular and extracellular environments remain in equilibrium when an isotonic saline solution is administered. However, if the patient is infused with a hypertonic solution, water will move out of the patient's cells, and the cells will shrivel, a condition called *plasmolysis.* Conversely, if a patient is infused with a hypotonic solution, water will move into the patient's cells, and they will swell and possibly burst.

CRITICAL THINKING

Solutions hypertonic to bacteria and fungi are used for food preservation. For instance, jams and jellies are hypertonic with sugar, and pickles are hypertonic with salt. How do hypertonic solutions kill bacteria and fungi that would otherwise spoil these foods?

Active Processes

As stated previously, active processes require the cell to expend energy to move materials across the cytoplasmic membrane against their electrochemical gradient. This is analogous to moving water uphill. Energy stored in molecules of ATP is used for active transport. As we will see, ATP may be utilized directly during transport, or indirectly at some other site and at some other time. Active processes in prokaryotes include *active transport* by means of carrier proteins and a special process termed *group translocation.*

Active Transport　Like facilitated diffusion, **active transport** utilizes transmembrane carrier proteins; however, the functioning of active transport carrier proteins require the cell to expend ATP to transport molecules across the membrane. Some carrier proteins are referred to as *gated channels* or *ports* because they are controlled. When the cell is in need of a substance, the carrier protein becomes functional (the gate "opens"). At other times the gate is "closed."

If only one substance is transported at a time, the carrier protein is called a *uniport* **(Figure 3.19a).** In contrast, *antiports* simultaneously transport two chemicals, but in opposite directions; that is, one substance is transported into the cell at the same time that a second substance is transferred out of

---

[11]From Greek *isos,* meaning equal, and *tonos,* meaning tone.
[12]From Greek *hyper,* meaning more or over.
[13]From Greek *hypo,* meaning less or under.

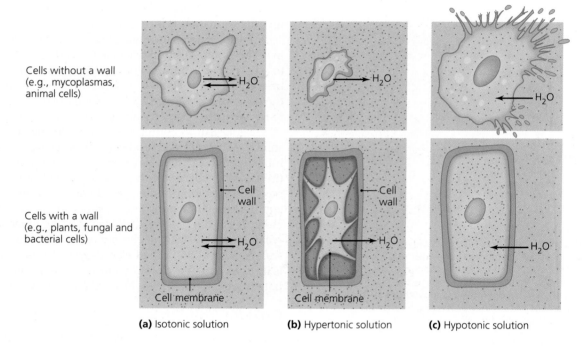

Cells without a wall
(e.g., mycoplasmas,
animal cells)

Cells with a wall
(e.g., plants, fungal and
bacterial cells)

Cell wall

Cell wall

Cell membrane

Cell membrane

(a) Isotonic solution   **(b)** Hypertonic solution   **(c)** Hypotonic solution

▲ *Figure 3.18*

Effects of isotonic, hypertonic, and hypotonic solutions on cells. **(a)** Cells in isotonic
solutions experience no net movement of water. **(b)** Cells in hypertonic solutions shrink
due to the net movement of water out of the cell. **(c)** Cells in hypotonic solutions
undergo a net gain of water. Animal cells burst because they lack a cell wall; in cells
with a cell wall, the pressure of water pushing against the interior of the wall
eventually stops the movement of water into the cell.

the cell **(Figure 3.19b)**. In other types of active transport, two substances move together in the same direction across the membrane by means of a single carrier protein. Such carrier proteins are known as *symports* **(Figure 3.19c)**.

In all cases, active transport moves substances against their electrochemical gradient. Typically, the carrier protein breaks down ATP into ADP and inorganic phosphate during transport. ATP releases energy that is used to move the chemical against its electrochemical gradient across the membrane.

With symports and antiports, one chemical's electrochemical gradient may provide the energy needed to transport the second chemical. For example, H^+ moving into a cell down its electrochemical gradient by facilitated diffusion provides energy to carry glucose into the cell, against the glucose gradient. The two processes are linked by a symport. However, ATP is still utilized for transport because the H^+ gradient was previously established by the active pumping of H^+ to the outside of the cell by an ATP-dependent H^+ uniport. The use of ATP is thus separated in time and space from the active transport of glucose, but ATP was still expended.

Group Translocation   **Group translocation** is an active process that occurs only in some prokaryotes. In group translocation, the substance being actively transported across the membrane is chemically changed during transport **(Figure 3.20)**. The membrane is impermeable to the altered substance, trapping it inside the cell. Group translocation is very efficient at bringing substances into a cell. It can operate efficiently even if the external concentration of the chemical being transported is as low as 1 part per million (ppm).

One well-studied example of group translocation is the accumulation of glucose inside a bacterial cell. As glucose is transported across the bacterial cell membrane, it is phosphorylated; that is, a phosphate group is added to the glucose. The glucose is changed into glucose-6-phosphate, a sugar that cannot cross back out, but can be utilized in the ATP-producing metabolism of the cell. Other carbohydrates, fatty acids, purines, and pyrimidines are also brought into bacterial cells by group translocation. A summary of prokaryotic transport processes is shown in **Table 3.2** on page 76.

Cytoplasm of Prokaryotes

Cytoplasm is the general term used to describe the semiliquid, gelatinous material inside a cell. Cytoplasm is semitransparent, fluid, elastic, and aqueous. It is composed of cytosol, inclusions, ribosomes, and a cytoskeleton.

Figure 3.19 ➤
Mechanisms of active transport.
(a) Via a uniport. **(b)** Via an antiport. **(c)** Via a uniport coupled with a symport. *What is the usual source of energy for active transport?*

Figure 3.19 ATP is the usual source of energy for active transport processes.

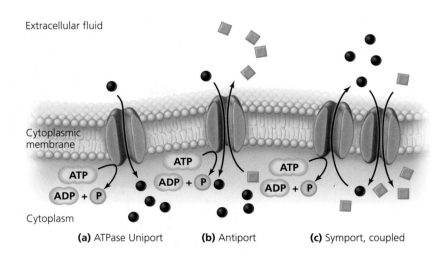

(a) ATPase Uniport   **(b)** Antiport   **(c)** Symport, coupled

▼ **Figure 3.20**

Group translocation. This process involves a chemical change in a substance as it is being transported. This figure depicts glucose being transported into a bacterial cell via group translocation.

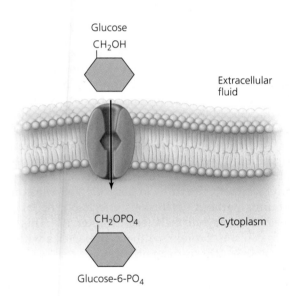

Cytosol

Learning Objective

✓ Describe cytoplasm and its basic contents.

The liquid portion of the cytoplasm is called **cytosol.** It is mostly water, but it also contains dissolved and suspended substances, including ions, carbohydrates, proteins (mostly enzymes), lipids, and wastes. The cytosol of prokaryotes also contains the cell's DNA in a region called the **nucleoid.** Most prokaryotes have a single, circular chromosome that is not surrounded by a nuclear membrane. *Vibrio cholerae* (vib′rē-ō kol′er-ē), the bacterium that causes cholera, is an unusual prokaryote in that it has two chromosomes.

The cytosol is the site of some chemical reactions. For example, enzymes within the cytosol function to produce amino acids.

Inclusions

Deposits, called **inclusions,** are often found within the cytosol of prokaryotes. Inclusions may include reserve deposits of lipids, starch, or compounds containing nitrogen, phosphate, or sulfur (**Figure 3.21** on page 76). Such chemicals may be taken in and stored in the cytosol when nutrients are in abundance and then utilized when nutrients are scarce. The presence of inclusions is diagnostic for several pathogenic bacteria.

Many aquatic cyanobacteria (blue-green photosynthetic prokaryotes) contain inclusions called *gas vesicles* that store gases in protein sacs. The gases buoy the cells to the surface and into the light needed for photosynthesis. Other interesting inclusions are the small crystals of magnetite stored by *magnetobacteria,* featured at the beginning of this chapter.

Nonmembranous Organelles

Learning Objective

✓ Describe the structure and function of ribosomes and the cytoskeleton.

As previously noted, prokaryotes lack the membrane-bound organelles found in eukaryotic cells. However, two types of *nonmembranous organelles* are found in direct contact with the cytosol in prokaryotic cytoplasm. Some scientists do not consider them to be true organelles because they lack a membrane, but in this text they are considered a type of organelle. Nonmembranous organelles in prokaryotes include ribosomes and the cytoskeleton.

Ribosomes

Ribosomes are the sites of protein synthesis in cells. Prokaryotic cells have thousands of ribosomes in their cytoplasm, which gives cytoplasm a grainy appearance. The approximate size of ribosomes—and indeed other cellular

Table 3.2   Transport Processes Across Prokaryotic Cytoplasmic Membranes

	Description	Substances Transported
Passive Transport Processes	Processes require no use of energy by the cell; the electrochemical gradient provides energy.	
Diffusion	Molecules move down their electrochemical gradient through the phospholipid bilayer of the membrane.	Oxygen, carbon dioxide, lipid-soluble chemicals
Facilitated diffusion	Molecules move down their electrochemical gradient through channels or carrier proteins.	Glucose, fructose, urea, some vitamins
Osmosis	Water molecules move down their concentration gradient across a selectively permeable membrane.	Water
Active Transport Processes	Cell expends energy in the form of ATP to move a substance against its electrochemical gradient. Some active processes are linked to a passive process and use the energy of the passive process to carry a substance across the membrane.	
Active transport	ATP-dependent carrier proteins bring substances into cell.	Na^+, K^+, Ca^{2+}, H^+, Cl^-
Group translocation	The substance is chemically altered during transport; found only in some prokaryotes.	Glucose, mannose, fructose

Inclusion

LM   0.5 μm

▲ *Figure 3.21*

Cellular inclusions. Shown here are sulfur globules in the purple sulfur bacterium *Chromatium buderi*.

structures—is expressed in Svedbergs (S),[14] and is determined by their sedimentation rate—the rate at which they move to the bottom of a test tube during centrifugation. As you might expect, large, compact, heavy particles sediment faster than small, loosely packed, or light ones, and are assigned a higher number. Prokaryotic ribosomes are 70S: in contrast, the larger ribosomes of eukaryotes are 80S.

All ribosomes are composed of two subunits, each of which is composed of protein and a type of RNA called **ribosomal RNA (rRNA).** The subunits of prokaryotic 70S ribosomes are a smaller 30S subunit and a larger 50S subunit; the 30S subunit contains protein and a single rRNA molecule, whereas the 50S subunit has two rRNA molecules. Because sedimentation rates are dependent not only on mass and size but also on shape, the sedimentation rates of subunits do not add up to the sedimentation rate of a whole ribosome.

Many antibiotics act on prokaryotic 70S ribosomes or their subunits without deleterious effects on the larger 80S ribosomes of eukaryotic cells (see Chapter 10). This is why antibiotics can stop protein synthesis in bacteria without affecting protein synthesis in a patient.

Cytoskeleton

Most cells contain an internal network of fibers called a **cytoskeleton,** which appears to play a role in forming a cell's basic shape. Prokaryotes were long thought to lack cytoskeletons, but recent research has revealed that rod-shaped prokaryotes have a simple one **(Figure 3.22).** Spherical prokaryotes appear to lack cytoskeletons.

To this point, we have discussed basic features of bacterial and archaeal prokaryotic cells. Chapter 11 discusses the classification of prokaryotic organisms in more detail. Next we turn our attention to eukaryotic cells.

External Structures of Eukaryotic Cells

Eukaryotic cells have many external structural similarities to prokaryotic cells, and some uniquely eukaryotic features as well. In this section we discuss the structure and function of eukaryotic glycocalyces, flagella, and cilia.

---

[14]Svedberg units are named for Nobel Prize winner and inventor of the ultracentrifuge, Theodor Svedberg.

LM   ⊢——⊣ 3 μm

▲ *Figure 3.22*

The simple helical cytoskeleton of the rod-shaped prokaryote *Bacillus subtilis*. This cytoskeleton, which is composed of only a single protein, has been stained with a fluorescent dye.

Glycocalyces

Learning Objective

✓ Describe the composition, function, and importance of eukaryotic glycocalyces.

Animal and protozoan cells lack cell walls, but they have sticky carbohydrate glycocalyces that are anchored to their cytoplasmic membranes by covalently bonding to membrane proteins and lipids. The functions of eukaryotic glycocalyces, which are never as structurally organized as prokaryotic capsules, include helping to anchor animal cells to each other, strengthening the cell surface, providing some protection against dehydration, and functioning in cell-to-cell recognition and communication. Glycocalyces are absent in eukaryotic cells that have cell walls.

Flagella

Learning Objective

✓ Compare and contrast the structure and function of prokaryotic and eukaryotic flagella.

Structure and Arrangement

Flagella of eukaryotes (**Figure 3.23a** on page 78) differ structurally from flagella of prokaryotes. The shaft of a eukaryotic flagellum is composed of molecules of a globular protein called *tubulin* arranged in chains to form hollow *microtubules* (**Figure 3.23c**). Nine pairs of microtubules surround two microtubules in the center. This "9 + 2" arrangement of microtubules is common to all flagellated eukaryotic cells, whether they are found in protozoa, algae, animals, or plants. The filaments of eukaryotic flagella are anchored to the cell by a basal body, but no hook connects the two parts, as in prokaryotes. The basal body has *triplets* of microtubules instead of pairs, and there are no microtubules in the center, so it has a "9 + 0" arrangement of microtubules.

Another difference between the flagella of eukaryotes and prokaryotes is that eukaryotic flagella are surrounded by an extension of the cell membrane and are filled with cytosol. Thus, the flagella of eukaryotes are inside the cell, not extensions outside the cell. Eukaryotic flagella may be single or multiple and are generally found at one pole of the cell.

Function

The flagella of eukaryotes also move differently from those of prokaryotes. Rather than rotating like prokaryotic flagella, those of eukaryotes undulate rhythmically (**Figure 3.24a** on page 79). Some eukaryotic flagella push the cell through the medium (as occurs in animal sperm), whereas others pull the cell through the medium (as occurs in many protozoa). Positive and negative phototaxis and chemotaxis are seen in eukaryotic cells, but such cells do not move in runs and tumbles.

Cilia

Learning Objectives

✓ Describe the structure and function of cilia.
✓ Compare and contrast eukaryotic cilia and flagella.

Some eukaryotic cells move by means of hairlike structures called **cilia,** which cover the surface of the cell and are shorter and more numerous than flagella (see **Figure 3.23b**). No prokaryotic cells have cilia. Like flagella, cilia are composed primarily of tubulin microtubules, which are arranged in a "9 + 2" arrangement of pairs in their shafts and a "9 + 0" arrangement of triplets in their basal bodies (see **Figure 3.23c**).

A single cell may have hundreds or even thousands of cilia. Cilia beat rhythmically, much like a swimmer doing a butterfly stroke (**Figure 3.24b).** Coordinated beating of cilia propels single-celled eukaryotes through their environment. Cilia are also used within multicellular eukaryotes to move substances in the local environment past the surface of the cell. For example, such movement of cilia helps cleanse the human respiratory tract of dust and microorganisms.

Eukaryotic Cell Walls and Cytoplasmic Membranes

Learning Objectives

✓ Compare and contrast prokaryotic and eukaryotic cell walls and cytoplasmic membranes.
✓ Contrast exocytosis and endocytosis.
✓ Describe the role of pseudopodia in eukaryotic cells.

The eukaryotic cells of fungi, algae, and plants have cell walls. Recall that glycocalyces are absent from eukaryotes with cell walls; instead, the cell wall takes on one of the functions of a glycocalyx by providing protection from the environment. The wall also provides shape and support against osmotic pressure. Eukaryotic cell walls are composed of various polysaccharides, but not the peptidoglycan seen in the walls of most bacteria.

(a)

Nucleus   Flagellum

(b)

Cilia   Macronucleus

LM   150 µm

TEM   200 nm

(c)

Cell membrane

Cytosol

Central pair microtubules

Microtubules (doublet)

"9+2" arrangement

Cell membrane

Portion cut away to show transition area from doublets to triplets and the end of central microtubules.

Basal body

Microtubules (triplet)

"9+0" arrangement

Figure 3.23 ➤

Eukaryotic flagella and cilia. **(a)** Light micrograph of the pathogen *Leishmania donovani;* each cell possesses a single flagellum. **(b)** Light micrograph of the human pathogen *Balantidium coli,* which is covered with cilia. **(c)** Details of the arrangement of microtubules of eukaryotic flagella and cilia. Both flagella and cilia have the same internal structure. *How do eukaryotic flagella differ from cilia?*

Figure 3.23 Flagella are longer and less numerous than cilia.

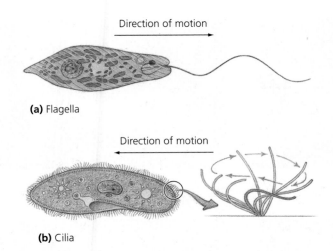

(a) Flagella

Direction of motion

(b) Cilia

Direction of motion

▲ *Figure 3.24*

Movement of eukaryotic flagella and cilia. **(a)** Eukaryotic flagella undulate in waves that begin at one end and traverse the length of the flagellum. **(b)** Cilia move with a power stroke followed by a return stroke. In the power stroke a cilium is stiff; it relaxes during the return stroke. *How is the movement of eukaryotic flagella different from that of prokaryotic flagella?*

Figure 3.24 Eukaryotic flagella undulate in a wave that moves down the flagellum; the flagella of prokaryotes rotate about the basal body.

Cell wall

Cytoplasmic membrane

TEM   1 μm

▲ *Figure 3.25*

A eukaryotic cell wall, as seen in the red alga *Gelidium*. This cell wall is composed of layers of the polysaccharide called agar. *What is the function of a cell wall?*

Figure 3.25 The cell wall provides support, protection, and resistance to osmotic forces.

The walls of plant cells are composed of *cellulose*, a polysaccharide that is familiar to you as either paper or dietary fiber. Fungi also have walls of polysaccharides, including cellulose, *chitin*, and/or *glucomannan*. The walls of algae **(Figure 3.25)** are composed of a variety of polysaccharides and other chemicals, depending on the type of alga. These chemicals include cellulose, *agar*, *carrageenan*, *silicates*, *algin*, calcium carbonate, or a combination of these substances. Fungi and algae are discussed in more detail in Chapter 12.

All eukaryotic cells have cytoplasmic membranes **(Figure 3.26).** (Botantists call the cytoplasmic membrane of an algal or plant cell a *plasmalemma*.) A eukaryotic cytoplasmic membrane, like those of bacteria, is a fluid mosaic of phospholipids and proteins, which act as recognition molecules, enzymes, receptors, carriers, or channels. Additionally, within multicellular animals some proteins serve to anchor cells to each other.

Unlike most bacterial membranes, eukaryotic membranes contain steroid lipids (*sterols*), such as cholesterol in animal cells, that help maintain membrane fluidity. Paradoxically, at high temperatures sterols stabilize a phospholipid bilayer by making it less fluid, but at low temperatures sterols have the opposite effect—they prevent phospholipid packing, making the membrane more fluid.

Like its prokaryotic counterpart, a eukaryotic cytoplasmic membrane controls the movement of materials into and out of the cell. Eukaryotic cytoplasmic membranes use both

Cytoplasmic membrane

Intercellular matrix

Cytoplasmic membrane

TEM   35 μm

▲ *Figure 3.26*

Eukaryotic cytoplasmic membrane. Note that this micrograph depicts the cytoplasmic membranes of two adjoining cells.

passive processes (diffusion, facilitated diffusion, and osmosis; see Figure 3.16) and active transport (see Figure 3.19). Eukaryotic membranes do not perform group translocation, which occurs only in some prokaryotes, but many perform another type of active transport—**endocytosis (Figure 3.27** on page 80), which involves physical manipulation of the cytoplasmic membrane around the cytoskeleton. Endocytosis occurs when the membrane distends to form **pseudopodia** (false feet) that surround a substance, bringing it into the cell. Endocytosis is termed **phagocytosis** if a solid is brought into the cell, and **pinocytosis** if only liquid is brought into the cell. Nutrients brought into a cell by endocytosis are then enclosed in a *food vesicle.* Vesicles and digestion of the nutrients they contain are discussed in more detail shortly. The process of phagocytosis is more fully discussed in Chapter 15 as it relates to the defense of the body against disease.

Exocytosis, another eukaryotic process, is the reverse of endocytosis in that it enables substances to be exported from the cell. Not all eukaryotic cells can perform endocytosis or exocytosis. Some eukaryotes also use pseudopodia as a means of locomotion. The cell extends a pseudopod and then the cytoplasm streams into it, a process called *amoeboid action* **(Figure 3.28).**

Table 3.3 lists some of the features of endocytosis and exocytosis.

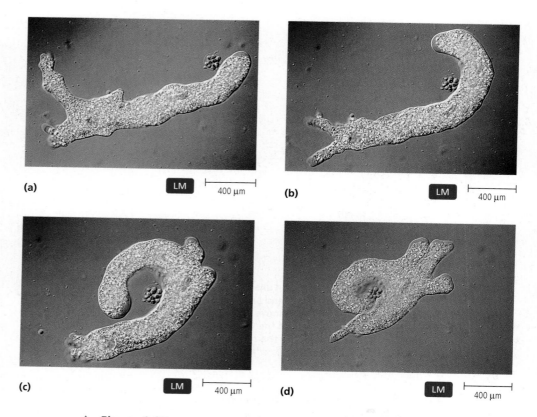

(a) LM 400 μm
(b) LM 400 μm
(c) LM 400 μm
(d) LM 400 μm

▲ **Figure 3.27**

Endocytosis. Pseudopodia extend to surround solid and/or liquid nutrients, which become incorporated into a food vesicle inside the cytoplasm. *What is the difference between phagocytosis and pinocytosis?*

Figure 3.27 Phagocytosis is endocytosis of a solid; pinocytosis is endocytosis of a liquid.

Cytoplasm of Eukaryotes

Learning Objectives

✓ Compare and contrast the cytoplasm of prokaryotes and eukaryotes.
✓ Identify nonmembranous and membranous organelles.

The cytoplasm of eukaryotic cells is more complex than that of either bacteria or archaea. The most distinctive difference is the presence of numerous membranous organelles in eukaryotes. However, before we discuss these membranous organelles, we will briefly consider the nonmembranous organelles in eukaryotes.

Nonmembranous Organelles

Learning Objectives

✓ Describe the structure and function of ribosomes, cytoskeletons, and centrioles.
✓ Compare and contrast the ribosomes of prokaryotes and eukaryotes.
✓ List and describe the three filaments of a eukaryotic cytoskeleton.

Here we discuss three nonmembranous organelles found in eukaryotes: ribosomes and cytoskeleton (both of which are also present in prokaryotes), and centrioles (which are present only in certain kinds of eukaryotic cells).

Ribosomes

The cytosol of eukaryotes, like that of prokaryotes, is a semitransparent fluid composed primarily of water containing dissolved and suspended proteins, ions, carbohydrates, lipids, and wastes. Within the cytosol of eukaryotic cells are ribosomes that are larger than prokaryotic ribosomes; instead of 70S ribosomes, eukaryotic ribosomes are 80S and are composed of 60S and 40S subunits. (Chapter 10 discusses many antibiotics that inhibit protein synthesis by 70S prokaryotic ribosomes without adversely affecting protein synthesis by the 80S ribosomes in eukaryotic cells.) In addition to the ribosomes found within the cytosol, many eukaryotic ribosomes are attached to the membranes of the *endoplasmic reticulum* (discussed shortly).

Table 3.3   Active Transport Processes Found Only in Eukaryotes: Endocytosis and Exocytosis

	Description	Substances Transported
Endocytosis: phagocytosis and pinocytosis	Substances are surrounded by pseudopodia and brought into the cell. Phagocytosis involves solid substances; pinocytosis involves liquids.	Bacteria, viruses, aged and dead cells; liquid nutrients in extracellular solutions
Exocytosis	Vesicles containing substances are fused with cytoplasmic membrane, dumping their contents to the outside.	Wastes, secretions

Pseudopodia

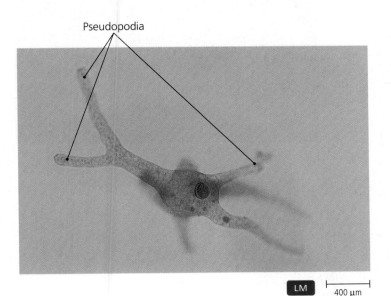

LM |———| 400 μm

▲ *Figure 3.28*

Amoeboid action. Pseudopodia move a cell to a new location. This micrograph is of a cell caught in motion. Pseudopodia protrude from the surface, cytoplasm flows into them, and the cell progresses across the surface.

Cytoskeleton

Eukaryotic cells contain an extensive cytoskeleton composed of an internal network of fibers and tubules. The eukaryotic cytoskeleton acts to anchor organelles, functions in cytoplasmic streaming and in movement of organelles within the cytosol, enables contraction of the cell, moves the cell membrane during endocytosis and amoeboid action, and provides the basic shape of many cells.

The eukaryotic cytoskeleton is made up of *tubulin microtubules* (also found in flagella, cilia, and centrioles), thinner *microfilaments* composed of *actin*, and *intermediate filaments* composed of various proteins (**Figure 3.29** on page 82).

Centrioles and Centrosome

Animal cells and some fungal cells contain two **centrioles,** which lie at right angles to each other near the nucleus, in a region of the cytoplasm called the **centrosome (Figure 3.30** on page 82). Plants, algae, fungi, and prokaryotes lack both centrioles and centrosomes. Centrioles are composed of nine

triplets of microtubules arranged in a way that resembles the "9 + 0" arrangement seen at the base of eukaryotic flagella and cilia.

Centrioles appear to play a role in *mitosis* (nuclear division), *cytokinesis* (cell division), and in the formation of flagella and cilia. However, because many eukaryotic cells that lack centrioles, such as brown algal sperm and numerous one-celled algae, are still able to form flagella and undergo mitosis and cytokinesis, the function of centrioles is the subject of ongoing research.

Membranous Organelles

Learning Objectives

✓ Discuss the function of each of the following membranous organelles: *nucleus, ER, Golgi body, lysosome, peroxisome, vesicle, vacuole, mitochondrion,* and *chloroplast.*

✓ Label the structures associated with each of the membranous organelles.

Eukaryotic cells contain a variety of organelles that are surrounded by phospholipid bilayer membranes similar to the cytoplasmic membrane. These membranous organelles include the nucleus, endoplasmic reticulum, Golgi body, lysosomes, peroxisomes, vacuoles, vesicles, mitochondria, and chloroplasts. Prokaryotic cells lack these structures.

Nucleus

The **nucleus** is usually spherical to ovoid and is often the largest organelle in a cell[15] (**Figure 3.31** on page 83). Some cells have a single nucleus; others are multinucleate, while still others lose their nuclei. The nucleus is often referred to as "the control center of the cell" because it contains most of the cell's genetic instructions in the form of DNA. Cells that lose their nuclei, such as mammalian red blood cells, can survive for only a few months.

Just as the semiliquid portion of the cell is called cytoplasm, the semiliquid matrix of the nucleus is called **nucleoplasm.** Within the nucleoplasm may be one or more **nucleoli** (singular: *nucleolus*), which are specialized

---

[15]Historically, the nucleus was not considered an organelle because it is large and not considered part of the cytoplasm.

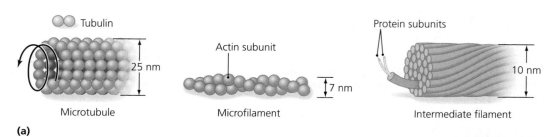

(a)

▲ *Figure 3.29*

Eukaryotic cytoskeleton. The cytoskeleton of eukaryotic cells serves to anchor organelles, provides a "track" for the movement of organelles throughout the cell, and provides shape to animal cells. Their cytoskeletons are composed of microtubules, microfilaments, and intermediate filaments. **(a)** Artist's rendition of cytoskeleton filaments. **(b)** Various elements of the cytoskeleton shown here have been stained with different fluorescent dyes.

(b)

(a)

(b)

▲ *Figure 3.30*

Centrosome. A centrosome is composed of two centrioles at right angles to one another; each centriole has nine triplets of microtubules. **(a)** Transmission electron micrographs of centrosomes and centrioles. **(b)** Artist's rendition of a centrosome. *How do centrioles compare with the basal body and shafts of eukaryotic flagella and cilia (see Figure 3.23)?*

Figure 3.30 Centrioles have the same "9 + 0" arrangement of microtubules that is found in the basal bodies of eukaryotic cilia and flagella.

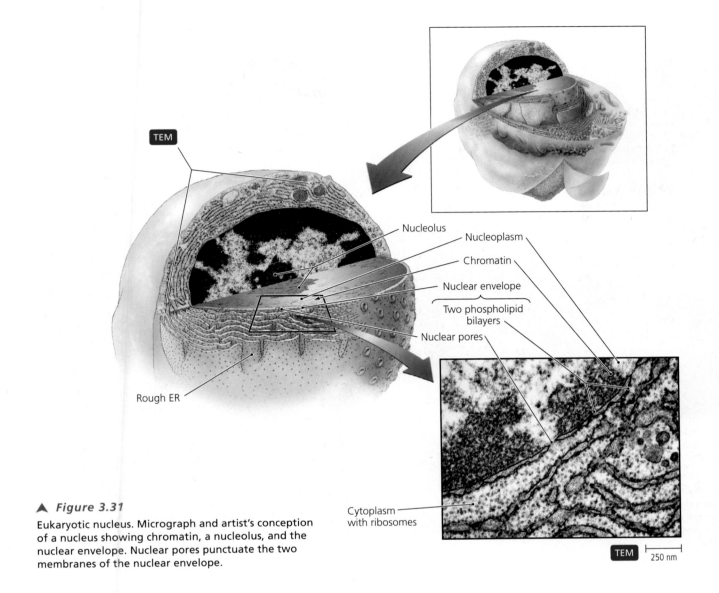

▲ *Figure 3.31*
Eukaryotic nucleus. Micrograph and artist's conception of a nucleus showing chromatin, a nucleolus, and the nuclear envelope. Nuclear pores punctuate the two membranes of the nuclear envelope.

regions where RNA is synthesized. The nucleoplasm also contains **chromatin,** which is a threadlike mass of DNA associated with special proteins called *histones* that play a role in packaging nuclear DNA. During mitosis, the process of eukaryotic nuclear division, chromatin becomes visible as *chromosomes*. Mitosis is discussed in more detail in Chapter 12.

Surrounding the nucleus is a double membrane called the **nuclear envelope,** which is composed of two phospholipid bilayers, for a total of four phospholipid layers (see **Figure 3.31**). The nuclear envelope contains **nuclear pores** that function to control the import and export of substances through the envelope.

Endoplasmic Reticulum

Continuous with the outer membrane of the nuclear envelope is a netlike arrangement of hollow tubules called **endoplasmic reticulum (ER)** (**Figure 3.32** on page 84). The ER traverses the cytoplasm of eukaryotic cells. Endoplasmic

reticulum functions as a transport system and is found in two forms: **smooth endoplasmic reticulum (SER)** and **rough endoplasmic reticulum (RER).** SER plays a role in lipid synthesis as well as transport. Rough endoplasmic reticulum is rough because ribosomes adhere to its outer surface. Proteins produced by ribosomes on the RER are inserted into the lumen (central canal) of the RER and transported throughout the cell.

Golgi Body

A **Golgi**[16] **body** is like the shipping department of a cell: it receives, processes, and packages large molecules for export from the cell (**Figure 3.33** on page 84). The Golgi body packages secretions in sacs called **secretory vesicles,** which then fuse with the cytoplasmic membrane before dumping their

---

[16]Golgi was an Italian histologist who first described the organelle in 1898. This organelle is also known as a Golgi complex or Golgi apparatus, and in plants and algae as a dictyosome.

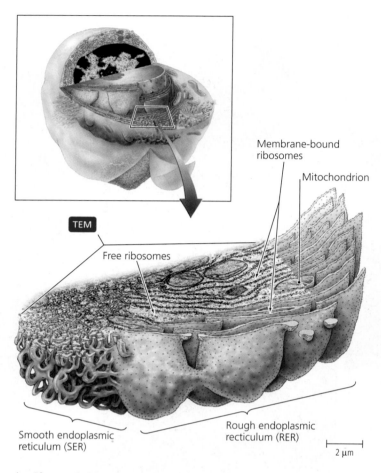

Membrane-bound ribosomes

Mitochondrion

TEM

Free ribosomes

Smooth endoplasmic reticulum (SER)

Rough endoplasmic recticulum (RER)

2 μm

▲ *Figure 3.32*

Endoplasmic reticulum. ER functions in transport throughout the cell. Ribosomes are on the surface of rough ER; smooth ER lacks ribosomes.

Secretory vesicles

TEM

Vesicles arriving from ER

0.5 μm

▲ *Figure 3.33*

Golgi body. A Golgi body is composed of flattened sacs. Proteins synthesized by ribosomes on RER are transported via vesicles to the Golgi. The Golgi then modifies the proteins and sends them via secretory vesicles to the cytoplasmic membrane, where they can be secreted from the cell by exocytosis.

contents outside the cell via exocytosis. Golgi bodies are composed of a series of flattened hollow sacs that are circumscribed by a phospholipid bilayer.

Lysosomes, Peroxisomes, Vacuoles, and Vesicles

Lysosomes, peroxisomes, vacuoles, and vesicles are membranous sacs that function to store and transfer chemicals within eukaryotic cells. Both **vesicle** and **vacuole** are general terms for such sacs. Large vacuoles are found in plant and algal cells that store starch, lipids, and other substances in the center of the cell. Often a central vacuole is so large that the rest of the cytoplasm is pressed against the cell wall in a thin layer **(Figure 3.34).**

Lysosomes, which are found in animal cells, contain catabolic enzymes that damage the cell if they are released from their packaging into the cytosol. The enzymes are used during the self-destruction of old, damaged, and diseased cells, and to digest nutrients that have been phagocytized. For example, the digestive enzymes in lysosomes are

utilized by white blood cells to destroy phagocytized pathogens **(Figure 3.35).**

Peroxisomes are vesicles that contain *oxidase* and *catalase*, which are enzymes that degrade poisonous metabolic wastes (such as free radicals and hydrogen peroxide) resulting from some oxygen-dependent reactions. Peroxisomes are found in all eukaryotic cells but are especially prominent in the kidney and liver cells of mammals.

Mitochondria

Mitochondria are spherical to elongated structures found in most eukaryotic cells **(Figure 3.36** on page 86). Like nuclei, they have two membranes, each composed of a phos-

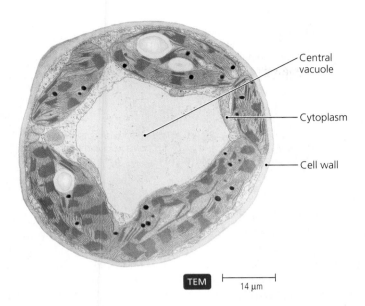

▲ Figure 3.34

Vacuole. The large central vacuole of a plant cell, which constitutes a storehouse for the cell, presses the cytoplasm against the cell wall.

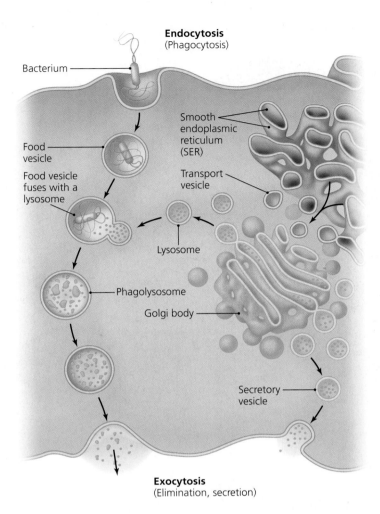

▲ Figure 3.35

The roles of vesicles in the destruction of a phagocytized pathogen within a white blood cell. Before endocytosis, vesicles from the SER deliver digestive enzymes to the Golgi body, which then packages them into lysosomes. During endocytosis, phagocytized particles (in this case, a bacterium) are enclosed within a food vesicle, which then fuses with a lysosome to form a phagolysosome. Once digestion within the phagolysosome is complete, the resulting wastes can be expelled from the cell via exocytosis.

pholipid bilayer. The inner membrane is folded into numerous **cristae** that increase the inner membrane's surface area. Mitochondria are often called the "powerhouses of the cell" because their cristae produce most of the ATP in the cell. The chemical reactions that produce ATP are discussed in Chapter 5.

The interior matrix of a mitochondrion contains small ("prokaryotic") 70S ribosomes and a circular molecule of DNA. This DNA contains genes for some RNA molecules and for a few mitochondrial polypeptides that are manufactured by mitochondrial ribosomes; however, most mitochondrial proteins are coded by nuclear DNA and synthesized by cytoplasmic ribosomes.

Chloroplasts

Chloroplasts are light-harvesting structures found in photosynthetic eukaryotes (**Figure 3.37** on page 86). Like mitochondria and the nucleus, chloroplasts have two phospholipid bilayer membranes and DNA. Further, like mitochondria, chloroplasts can synthesize a few polypeptides with their own 70S ribosomes. The pigments of chloroplasts gather light energy to produce ATP and form sugar from carbon dioxide. Numerous membranous sacs called *thylakoids* form an extensive surface area for the biochemical and photochemical reactions of chloroplasts. The fluid between the thylakoids and the inner membrane is called the *stroma*.

Photosynthetic prokaryotes lack chloroplasts and instead have infoldings of their cytoplasmic membranes called *photosynthetic lamellae*. The details of photosynthesis are discussed in Chapter 5.

The functions of the nonmembranous and membranous organelles, and their distribution among prokaryotic and eukaryotic cells, are summarized in **Table 3.4** on page 87.

Endosymbiotic Theory

Learning Objectives

✓ Describe the endosymbiotic theory of the origin of mitochondria, chloroplasts, and eukaryotic cells.
✓ List evidence for and against the endosymbiotic theory.

Mitochondria and chloroplasts are semiautonomous; that is, they divide independently of the cell but remain dependent

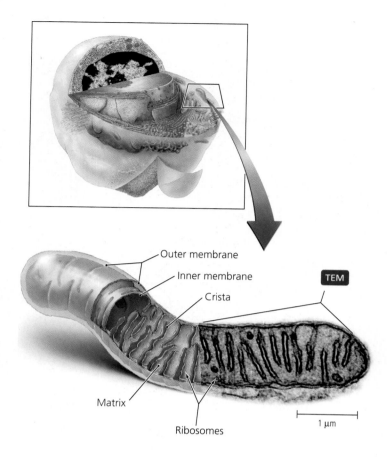

◄ **Figure 3.36**

Mitochondria. Note the double membrane. The inner membrane is folded into cristae that increase its surface area. *What is the importance of the increased surface area of the inner membrane that results from having cristae?*

Figure 3.36 The chemicals involved in aerobic ATP production are located on the inner membranes of mitochondria. Increased surface area provides more space for more chemicals.

Outer membrane

Inner membrane

TEM

Crista

Matrix

Ribosomes

1 μm

▼ **Figure 3.37**

Chloroplasts. Chloroplasts have an ornate internal structure designed to harvest light energy for photosynthesis.

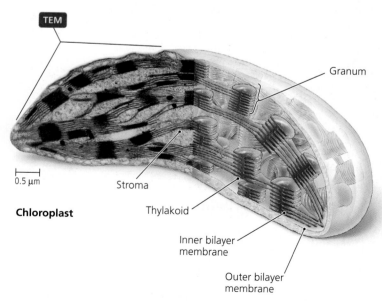

TEM

Granum

Chloroplast

Stroma

Thylakoid

Inner bilayer membrane

Outer bilayer membrane

0.5 μm

on the cell for most of their proteins. As we have seen, both mitochondria and chloroplasts contain a small amount of DNA and 70S ribosomes, and each can produce a few polypeptides with its own ribosomes. The presence of circular DNA, 70S ribosomes, and two bilipid membranes in these semiautonomous organelles led scientist Lynn Margulis (1938–) to propose the **endosymbiotic**[17] **theory** for the formation of eukaryotic cells, which suggests that eukaryotes formed from the phagocytosis of small aerobic[18] prokaryotes by larger anaerobic prokaryotes. The smaller prokaryotes were not destroyed by the larger cells, but instead became internal parasites that remained surrounded by a vesicular membrane of the host.

According to the theory, the parasites eventually lost the ability to exist independently, but they retained a portion of their DNA, some ribosomes, and their cytoplasmic membranes. During the same time, the larger cell became dependent on the parasites for aerobic ATP production. According to the theory, the aerobic prokaryotes eventually evolved into mitochondria. Margulis described a similar scenario for the origin of chloroplasts from phagocytized photosynthetic prokaryotes. The theory provides an explanation for the presence of 70S ribosomes and circular DNA within mitochondria and chloroplasts, and it accounts for the presence of their two membranes.

The endosymbiotic theory is not universally accepted, however, partly because it does not explain all of the facts. For example, the theory provides no explanation for the two membranes of the nuclear envelope; nor does it explain why only a few polypeptides of mitochondria and chloroplasts are made in the organelles while the bulk of their proteins come from nuclear DNA and cytoplasmic ribosomes.

Table 3.5 summarizes features of prokaryotic and eukaryotic cells.

CRITICAL THINKING

Eukaryotic cells are almost always larger than prokaryotic cells. What structures might allow for their larger size?

[17]From Greek *endo*, meaning inside, and *symbiosis*, meaning to live with.
[18]Aerobic means requiring oxygen; anaerobic is the opposite.

Table 3.4   Nonmembranous and Membranous Organelles of Cells

	General Function	Prokaryotes	Eukaryotes
Nonmembranous Organelles			
Ribosomes	Protein synthesis	Present in all	Present in all
Cytoskeleton	Shape in prokaryotes; support, cytoplasmic streaming, and endocytosis in eukaryotes	Present in some	Present in all
Centrioles	Appear to play a role in mitosis, cytokinesis, and flagella and cilia formation in animal cells	Absent in all	Present in animals
Membranous Organelles	Sequester chemical reactions within the cell		
Nucleus	"Control center" of the cell	Absent in all	Present in all
Endoplasmic reticulum	Transport within the cell, lipid synthesis	Absent in all	Present in all
Golgi bodies	Exocytosis, secretion	Absent in all	Present in some
Lysosomes	Breakdown of nutrients, self-destruction of damaged or aged cells	Absent in all	Present in some
Peroxisomes	Neutralization of toxins	Absent in all	Present in some
Vacuoles	Storage	Absent in all	Present in some
Vesicles	Storage, digestion, transport	Absent in all	Present in all
Mitochondria	Aerobic ATP production	Absent in all	Present in most
Chloroplasts	Photosynthesis	Absent in all	Present in plants and algae

Table 3.5   Comparison of Prokaryotic and Eukaryotic Cells

Characteristic	Prokaryotes	Eukaryotes
Size	Commonly 0.2–2.0 μm in diameter	Commonly 10–100 μm in diameter
Nucleus	Absent	Present
Membrane-bound organelles	Absent	Present; include ER, Golgi, lysosomes, mitochondria, and chloroplasts
Glycocalyx	Present as organized capsule or as unorganized slime layer	Present surrounding some animal cells
Motility	Some have simple flagella composed of a filament, hook, and basal body that rotates	Some have complex flagella or cilia composed of a "9 + 2" arrangement of microtubules that undulate; others move by means of amoeboid action
Flagella	Present	Present
Cilia	Absent	Present
Fimbriae and pili	Present	Absent
Cell wall	Present in most; in bacteria, typically composed of peptidoglycan	Present in most; composed of cellulose, algin, agar, carrageenan, silicate, glucomannan, or chitin
Cytoplasmic membrane	Lacks carbohydrates; usually lacks sterols	Contains glycolipids, glycoproteins, and sterols
Cytosol	Present	Present
Inclusions	Present	Present
Ribosomes	Smaller (70S)	Larger (80S) in cytoplasm, but 70S in mitochondria and chloroplasts
Chromosomes	Commonly single, circular, and lacking histones	Typically more than one, linear, and containing histones

CHAPTER SUMMARY

Processes of Life (pp. 57–58)

1. All living things have some common features, including **growth,** an increase in size; **reproduction,** an increase in number; **responsiveness,** reactions to environmental stimuli; **metabolism,** controlled chemical reactions in an organism; and cellular structure.

2. Viruses do not grow, self reproduce, or metabolize.

Eukaryotic and Prokaryotic Cells: An Overview (pp. 58–61)

1. All cells can be classified as either **prokaryotic** or **eukaryotic.** Prokaryotic cells lack a nucleus and membrane-bound organelles. Bacteria and archaea are prokaryotic. Eukaryotes have internal, membrane-bound **organelles,** including nuclei. Animals, plants, algae, fungi, and protozoa are eukaryotic.

External Structures of Prokaryotic Cells (pp. 61–66)

1. Cells share common structural features. These include external structures, cell walls, cytoplasmic membranes, and cytoplasm. The external structures of prokaryotic cells include glycocalyces, flagella, fimbriae, and pili.

2. **Glycocalyces** are sticky external sheaths of cells. They may be loosely attached **slime layers** or firmly attached **capsules.** Glycocalyces prevent cells from drying out. Capsules protect cells from phagocytosis by other cells, and slime layers enable cells to stick to each other and to surfaces in their environment.

3. A **flagellum** is a long, whiplike protrusion of some cells composed of a basal body, hook, and filament. Flagella allow cells to move toward favorable conditions such as nutrients or light, or move away from unfavorable stimuli such as poisons.

4. The flagella of prokaryotes may be single (**monotrichous**), grouped at one end of the cell (**lophotrichous**), be at both ends of the cell (**amphitrichous**), or cover the cell (**peritrichous**). **Endoflagella,** which are special flagella of spirochetes, form **axial filaments,** located in the periplasmic space.

5. **Taxis** is movement that may be either a positive response or a negative response to light (**phototaxis**) or chemicals (**chemotaxis**).

6. **Fimbriae** are nonmotile extensions of some bacterial cells that function along with glycocalyces to adhere cells to one another and to environmental surfaces. A mass of such bacteria on a surface is termed a **biofilm.**

7. **Pili,** which are also called **conjugation pili,** are hollow, nonmotile tubes of protein that connect some prokaryotic cells. They mediate the movement of DNA from one cell to another. Not all bacteria have fimbriae or pili.

Prokaryotic Cell Walls (pp. 66–69)

1. Most prokaryotic cells have **cell walls** that provide shape and support against osmotic pressure. Cell walls are composed of polysaccharide chains.

2. The cell walls of bacteria are composed of a large interconnected molecule of **peptidoglycan.** Peptidoglycan is composed of alternating sugar molecules called N-acetylglucosamine (NAG) and N-acetylmuramic acid (NAM). **Gram-positive** cells have thick layers of peptidoglycan. **Gram-negative** cells have thin layers of peptidoglycan and an external wall membrane with a **periplasmic space** between. This wall membrane contains **lipopolysaccharide (LPS),** which contains **lipid A.** During an infection with Gram-negative bacteria, lipid A can accumulate in the blood, causing shock, fever, and blood clotting. Acid-fast bacteria have waxy lipids in their cell walls.

3. Cell walls of archaea lack peptidoglycan.

Prokaryotic Cytoplasmic Membranes (pp. 70–74)

1. A **cytoplasmic membrane** is typically composed of lipid molecules arranged in a double layer configuration called a **phospholipid bilayer.** Proteins associated with the membrane vary in location and function and are able to flow laterally within the membrane. The **fluid mosaic model** is descriptive of the current understanding of the membrane.

2. Archaea do not have phospholipid membranes, and some have a single layer of lipid instead of a bilayer.

3. The **selectively permeable** cytoplasmic membrane prevents the passage of some substances while allowing other substances to pass through protein pores or channels, sometimes requiring carrier molecules. The relative concentrations inside and outside the cell of chemicals (**concentration gradients**) and of electrical charges (**electrical gradients**) create an **electrochemical gradient** across the membrane. The electrochemical gradient has a predictable effect on the passage of substances through the membrane.

4. Passive processes that move chemicals across the cytoplasmic membrane require no energy expenditure by the cell. Molecular size and concentration gradients determine the rate of simple **diffusion. Facilitated diffusion** depends on the electrochemical gradient and carriers within the membrane that allow certain substances to pass through the membrane. **Osmosis** specifically refers to the diffusion of water molecules across a selectively permeable membrane.

5. The concentrations of solutions can be compared. **Hypertonic** solutions have a higher concentration of solutes than **hypotonic** solutions, which have a lower concentration of solutes. Two **isotonic solutions** have the same concentrations of solutes.

6. Active transport processes require cell energy from ATP. **Active transport** moves a substance against its electrochemical gradient via carrier proteins. These carriers may move two substances in the same direction at once (*symports*) or move substances in opposite directions (*antiports*). **Group translocation** occurs in prokaryotes; during it the substance being transported is chemically altered in transit.

Cytoplasm of Prokaryotes (pp. 74–76)

1. **Cytoplasm** is composed of the liquid **cytosol** inside a cell plus nonmembranous organelles and inclusions. **Inclusions** in the cytosol are deposits of various substances.

2. Both prokaryotic and eukaryotic cells contain nonmembranous organelles.

3. The **nucleoid** is the nuclear region in prokaryotic cytosol. It has no nuclear envelope and usually contains a single circular molecule of DNA.

4. Inclusions include reserve deposits of lipids, starch, or compounds containing nitrogen, phosphate, or sulfur. Inclusions called gas vesicles store gases.

5. **Ribosomes,** composed of protein and **ribosomal RNA (rRNA),** are nonmembranous organelles, found in both prokaryotes and eukaryotes, that function to make proteins. The 70S ribosomes of prokaryotes are smaller than the 80S ribosomes of eukaryotes.

6. The **cytoskeleton** is a network of fibrils that appears to provide the basic shape of rod-shaped prokaryotes.

External Structures of Eukaryotic Cells (pp. 76–77)

1. Eukaryotic animal and some protozoan cells lack cell walls but have glycocalyces that prevent desiccation, provide support, and enable cells to stick together.

2. Some eukaryotic cells have long, whiplike flagella that differ from the flagella of prokaryotes. They have no hook and the basal bodies and shafts are composed arrangements of microtubules. Further, eukaryotic flagella are surrounded by the cytoplasmic membrane.

3. Some eukaryotic cells are covered with **cilia,** which have the same structure as eukaryotic flagella but are much shorter and more numerous.

Eukaryotic Cell Walls and Cytoplasmic Membranes (pp. 77–80)

1. Fungal, plant, algal, and some protozoan cells have cell walls composed of polysaccharides. Cell walls provide support, shape, and protection from osmotic forces.

2. Fungal cell walls are composed of chitin or other polysaccharides. Plant cell walls are composed of cellulose. Algal cell walls contain agar, carrageenan, algin, cellulose, or other chemicals.

3. Eukaryotic cytoplasmic membranes contain steroids such as cholesterol, which act to strengthen and solidify the membranes when temperatures rise and provide fluidity when temperatures fall.

4. Some eukaryotic cells transport substances into the cytoplasm via **endocytosis,** which is an active process requiring the expenditure of energy by the cell. In endocytosis, **pseudopodia**—moveable extensions of the cytoplasm and membrane of the cell—surround a substance and move it into the cell. When solids are brought into the cell, endocytosis is called **phagocytosis;** the incorporation of liquids by endocytosis is called **pinocytosis.**

5. **Exocytosis** is the active export of substances out of a cell.

Cytoplasm of Eukaryotes (pp. 80–87)

1. Eukaryotic cytoplasm is characterized by membranous organelles, particularly a nucleus. It also contains nonmembranous organelles and cytosol.

2. The 80S ribosomes of eukaryotic cells are composed of 60S and 40S subunits. They are found free in the cytosol and attached to endoplasmic reticulum. The ribosomes within mitochondria and chloroplasts are 70S.

3. The eukaryotic cytoskeleton is composed of microtubules, intermediate filaments, and microfilaments. It provides an infrastructure and aids in movement of cytoplasm and organelles.

4. **Centrioles,** which are nonmembranous organelles in animal cells only, are found in a region of the cytoplasm called the **centrosome** and are composed of triplets of microtubules in a "9 + 0" arrangement. Centrioles appear to function in the formation of flagella and cilia and in cell division.

5. The **nucleus,** a membranous structure in eukaryotic cells, contains **nucleoplasm** in which are found one or more **nucleoli** and **chromatin.** Chromatin consists of the multiple strands of DNA and associated histone proteins that become obvious as chromosomes during mitosis. **Nuclear pores** penetrate the four phospholipid layers of the **nuclear envelope** (membrane).

6. The **endoplasmic reticulum** functions as a transport system. It can be **rough ER (RER),** which has ribosomes on its surface, or **smooth ER (SER),** which lacks ribosomes.

7. The **Golgi body** is a series of flattened hollow sacs surrounded by phospholipid bilayers. It packages large molecules destined for export from the cell in **secretory vesicles,** which release these molecules from the cell via exocytosis.

8. **Vesicles** and **vacuoles** are general terms for membranous sacs that store or carry substances. More specifically, **lysosomes** of animal cells contain digestive enzymes, and **peroxisomes** contain enzymes that neutralize poisonous free radicals and hydrogen peroxide.

9. Four phospholipid layers surround **mitochondria,** site of production of most ATP in a eukaryotic cell. The inner bilayer is folded into **cristae,** which greatly increase the surface area available for chemicals that generate ATP.

10. Photosynthetic eukaryotes possess **chloroplasts,** which are organelles containing membranous thylakoids that provide increased surface area for photosynthetic reactions.

11. The **endosymbiotic theory** has been suggested to explain why mitochondria and chloroplasts have 70S ribosomes, circular DNA, and two membranes. The theory states that the ancestors of these organelles were prokaryotic cells that were internalized by other prokaryotes and then lost the ability to exist outside their host—thus forming early eukaryotes.

QUESTIONS FOR REVIEW

(Answers to multiple choice and matching questions are on the web, along with additional review questions. Visit www.microbiologyplace.com.)

Multiple Choice Questions

1. A cell may allow a large or charged chemical to move across the cytoplasmic membrane, down the chemical's electrochemical gradient, in a process called
   a. active transport.
   b. facilitated diffusion.
   c. endocytosis.
   d. pinocytosis.

2. Which of the following statements concerning growth and reproduction is *not* true?
   a. Growth and reproduction may occur simultaneously in living organisms.
   b. A living organism must reproduce to be considered alive.
   c. Living things may stop growing and reproducing, yet still be alive.
   d. Normally, living organisms have the ability to grow and reproduce themselves.

3. A "9 + 2" arrangement of microtubules is seen in
   a. prokaryotic flagella.
   b. nucleoids.
   c. eukaryotic flagella.
   d. Golgi bodies.

4. Which of the following is most associated with diffusion?
   a. symports
   b. antiports
   c. carrier proteins
   d. endocytosis

5. Which of the following is *not* associated with prokaryotic organisms?
   a. nucleoid
   b. glycocalyx
   c. cilia
   d. circular DNA

6. Which of the following is true of Svedbergs?
   a. They are not exact but are useful for comparisons.
   b. They are abbreviated "sv."
   c. They are prokaryotic in nature but exhibit some eukaryotic characteristics.
   d. They are an expression of sedimentation rate during high-speed centrifugation.

7. Which of the following statements is true?
   a. The cell walls of bacteria are composed of peptidoglycan.
   b. Peptidoglycan is a fatty acid.
   c. Gram-positive cell walls have a relatively thin layer of peptidoglycan anchored to the cytoplasmic membrane by teichoic acids.
   d. Peptidoglycan is found mainly in the cell walls of fungi, algae, and plants.

8. Which of the following is *not* a function of a glycocalyx?
   a. It forms pseudopodia for faster mobility of an organism.
   b. It can protect a bacterial cell from drying out.
   c. It hides the bacterial cell from other cells.
   d. It allows a bacterium to stick to a host.

9. Bacterial flagella
   a. are anchored to the cell by a basal body.
   b. are composed of many identical globular proteins arranged around a central core.
   c. are surrounded by an extension of the cytoplasmic membrane.
   d. are composed of tubulin in hollow microtubules in a "9 + 2" arrangement.

10. Which cellular structure is important in classifying a bacterial species as Gram-positive or Gram-negative?
    a. flagella
    b. cell wall
    c. cilia
    d. glycocalyx

11. A cell is moving uric acid across the cell membrane against its electrochemical gradient. Which of the following statements is true?
    a. The exterior of the cell is probably electrically negative compared to the interior of the cell.
    b. The acid probably moves by a passive means such as facilitated diffusion.
    c. The acid moves by an active process such as active transport.
    d. The movement of the acid requires phagocytosis.

12. Gram-positive cells
    a. have a thick cell wall, which retains crystal violet dye.
    b. contain teichoic acids in their cell walls.
    c. appear purple under magnification.
    d. all of the above

Matching Questions

1. Match the structures on the left with the descriptions on the right. A letter may be used more than once or not at all, and more than one letter may be correct for each blank.

   ___ Glycocalyx   A. Bristlelike projections found in quantities of 100 or more

   ___ Flagella   B. Long whip

   ___ Axial filaments   C. Responsible for conjugation

   ___ Cilia   D. "Sugar cup" composed of polysaccharides and/or polypeptides

   ___ Fimbriae   E. Short, numerous, nonmotile projections

   ___ Pili   F. Responsible for motility of spirochetes

      G. Not used for motility

      H. Made of tubulin in eukaryotes

      I. Made of flagellin in prokaryotes

2. Match the term on the left with its description on the right. Only one description is intended for each term.

   ___ Ribosome   A. Site of protein synthesis

   ___ Cytoskeleton   B. Contains enzymes to neutralize hydrogen peroxide

   ___ Centriole   C. Functions as the transport system within a eukaryotic cell

___ Nucleus

___ Mitochondrion

___ Chloroplast

___ ER

___ Golgi body

___ Peroxisome

D. Allows contraction of the cell

E. Site of most DNA in eukaryotes

F. Contains microtubules in "9 + 0" arrangement

G. Light-harvesting organelle

H. Packages large molecules for export from the cell

I. Its internal membranes are sites for ATP production

Essay Questions

1. A local newspaper writer has contacted you, an educated microbiology student from a respected college. He wants to obtain scientific information for an article he is writing about "life" and poses the following query: "What is the difference between a living thing and a nonliving thing?" Knowing that he will edit your material to fit the article, give an intelligent, scientific response.

2. Define cytosol.

3. The term *fluid mosaic* has been used in describing the cytoplasmic membrane. How does each word of that phrase accurately describe our current understanding of a cell membrane?

4. Describe (or draw) an example of diffusion down a concentration gradient.

5. Sketch, name, and describe four flagellar arrangements in prokaryotes.

6. What is the difference between growth and reproduction?

7. Compare bacterial cells and algal cells, giving at least four similarities and four differences.

8. Contrast a cell of *Streptococcus pyogenes* (a bacterium) with the unicellular protozoan *Entamoeba histolytica*, listing at least eight differences.

9. Differentiate among pili, fimbriae, prokaryotic flagella, eukaryotic flagella, and cilia using sketches and descriptive labels.

10. Can nonliving things metabolize? Explain your answer.

11. Contrast prokaryotic and eukaryotic cells by filling in the following table.

Characteristic	Prokaryotes	Eukaryotes
Size		
Presence of nucleus		
Presence of membrane-bound organelles		
Structure of flagella		
Chemicals in cell walls		
Type of ribosomes		
Structure of chromosomes		

12. What is the function of glycocalyces and fimbriae in forming a biofilm?

13. What factors may prevent a molecule from moving across a cell membrane?

14. Compare and contrast three types of passive transport across a cell membrane.

15. Contrast the following active processes for transporting materials into or out of a cell: active transport, group translocation, endocytosis, exocytosis.

16. Contrast symports and antiports.

17. Describe the endosymbiotic theory. What evidence supports the theory? Which features of eukaryotic cells are not explained by the theory?

CRITICAL THINKING

1. A scientist develops a chemical that prevents Golgi bodies from functioning. Contrast the specific effects the chemical would have on human cells versus bacterial cells.

2. Methylene blue binds to DNA. What structures in a yeast cell would be stained by this dye?

3. A new chemotherapeutic drug kills bacteria, but not humans. Discuss the possible ways the drug may act selectively on bacterial cells.

4. Some bacterial toxins cause cells lining the digestive tract to secrete ions, making the contents of the tract hypertonic. What effect does this have on a patient's water balance?

5. A researcher carefully inserts an electrode into a plant cell. He determines that the electrical charge across the cytoplasmic membrane is −70 millivolts. Then he slips the electrode deeper into the cell across another membrane and measures an electrical charge of −90 millivolts compared to the outside. What relatively large organelle is surrounded by the second membrane? Explain your answer.

6. Does Figure 3.2a on page 5 best represent a Gram-positive or a Gram-negative cell? Defend your answer.

7. An electron micrograph of a newly discovered cell shows long projections with a basal body in the cell wall. What kind of projections are these? How might this cell behave in its environment because of the presence of this structure?

8. An entry in a recent scientific journal reports positive phototaxis in a newly described species. What condition could you create in the lab to encourage the growth of these organisms?

9. A medical microbiological lab report indicates that sample X contained a biofilm, and that one species in the biofilm was identified as *Neisseria gonorrhoeae*. Is this strain of *Neisseria* likely to be pathogenic? Why or why not?

CHAPTER 4

Microscopy, Staining, and Classification

CHAPTER OUTLINE

Units of Measurement

Microscopy
General Principles of Microscopy
Light Microscopy
Electron Microscopy
Probe Microscopy

Staining
Preparing Specimens for Staining
Principles of Staining
Simple Stains
Differential Stains
Special Stains
Staining for Electron Microscopy

Classification and Identification of Microorganisms
Linnaeus, Whitaker, and Taxonomic Categories
Domains
Taxonomic and Identifying Characteristics
Taxonomic Keys

When we want to weigh ourselves, we just step on a bathroom scale. But what can we use to measure the mass of something as delicate as the cell wall of a microorganism? The answer is an interference microscope, which uses a split beam of light to form vertical light and dark bands across a specimen (see the photo). Where light rays have traveled through a specimen and slowed, the pattern of light and dark bands is shifted, and the amount of shift is directly proportional to the change in speed. That change in speed can then be correlated with the specimen's density; the denser the specimen, the slower light travels through it. In this way, we can "weigh" a cell wall—or a nucleus or a chloroplast, or even a single chromosome.

This chapter explores the illuminating world of microscopes, and the amazing lessons about microbial life that we continue to learn from them.

A piece of cell wall from the marine red alga *Griffithsia pacifica*, as seen through an interference microscope. Its mass is 0.65 picogram per square micrometer. (A picogram equals 10^{-12} gram.)

MicroPrep Pre-Test: *Take the pre-test for this chapter on the web.*
Visit ***www.microbiologyplace.com***

Either the well was very deep, or she fell very slowly, for she had plenty of time as she went down to look about her, and to wonder what was going to happen next. . . ."Curiouser and curiouser!" cried Alice. . . .
—*Alice's Adventures in Wonderland* by Lewis Carroll

Like Alice falling into Wonderland or traveling through a looking glass, scientists have entered the marvelous microbial world through advances in microscopy. With the invention of new laboratory techniques and the construction of new instruments, biologists are still discovering "curiouser and curiouser" wonders about the microbial world. In this chapter we will discuss some of the techniques microbiologists use to enter that world. We begin with a discussion of metric units as they relate to measuring the size of microbes. We then examine the instruments and staining techniques used in microbiology. Finally, we consider the classification schemes used to categorize the inhabitants of the microbial wonderland.

Units of Measurement

Learning Objectives

✓ Identify the two primary metric units used to measure the diameters of microbes.

✓ List the metric units of length in order, from meter to nanometer.

Microorganisms are small. This may seem an obvious statement, but it is one that can be taken for granted. Exactly how small are they? How can we measure the width and length of microbes?

Typically, a unit of measurement is smaller than the object being measured. For example, we measure a person's height in feet or inches, not in miles. Likewise, the diameter of a dime is measured in fractions of an inch, not in feet. So, measuring the size of a microbe requires units that are smaller than even the smallest interval on a ruler marked with English units (typically 1/16 inch). Even smaller units, such as 1/64 inch or 1/128 inch, become quite cumbersome

and very difficult to use when we are dealing with microorganisms.

So that they can work with units that are simpler and in standard use the world over, scientists use metric units of measurement. Unlike the English system, the metric system is a decimal system, so each unit is one tenth the size of the next largest unit. Using the metric system, even extremely small units are much easier to use than the fractions involved in the English system.

The unit of length in the metric system is the *meter (m)*, which is slightly longer than a yard. One tenth of a meter is a *decimeter (dm)*, and one hundredth of a meter is a *centimeter (cm)*, which is equivalent to about a third of an inch. One tenth of a centimeter is a *millimeter (mm)*, which is the thickness of a dime. A millimeter is still too large to measure the size of most microorganisms, but in the metric system we continue to divide by multiples of 10 until we have a unit appropriate for use. Thus, one thousandth of a millimeter is a *micrometer (µm)*, which is small enough to be useful in measuring the size of cells. One thousandth of a micrometer is a *nanometer (nm)*, a unit used to measure the smallest cellular organelles and viruses. A nanometer is one billionth of a meter.

Table 4.1 presents these metric units and some English equivalents. Refer to Figure 3.3 on page 61 for a visual size comparison of a typical eukaryotic cell, prokaryotic cell, and virus particle.

Microscopy

Learning Objective

✓ Define microscopy.

Microscopy[1] refers to the use of light or electrons to magnify objects. The science of microbiology began when Leeuwenhoek used primitive microscopes to observe and report the

---

[1]From Greek *micro,* meaning small, and *skopein,* meaning to view.

Table 4.1 Metric Units of Length

Metric Unit (abbreviation)	Meaning of Prefix	Metric Equivalent	U.S. Equivalent	Representative Microbiological Application of the Unit
meter (m)	—[a]	1 m	39.37 in (about a yard)	Length of pork tapeworm, *Taenia solium* (e.g., 1.8 m–8.0 m)
decimeter (dm)	1/10	0.1 m = 10^{-1} m	3.94 in	—[b]
centimeter (cm)	1/100	0.01 m = 10^{-2} m	0.39 in; 1 in = 2.54 cm	Diameter of a mushroom cap (e.g., 12 cm)
millimeter (mm)	1/1000	0.001 m = 10^{-3} m	—	Diameter of a bacterial colony (e.g., 2.3 mm); length of a tick (e.g., 5.7 mm)
micrometer (µm)	1/1,000,000	0.000001 m = 10^{-6} m	—	Diameter of white blood cells (e.g., 5 µm–25 µm)
nanometer (nm)	1/1,000,000,000	0.000000001 m = 10^{-9} m	—	Diameter of a poliovirus (e.g., 25 nm)

[a]The meter is the standard metric unit of length.
[b]Decimeters are rarely used; centimeters are much more commonly used units.

▲ *Figure 4.1*

The electromagnetic spectrum. Visible light comprises a narrow band of wavelengths of radiation. Visible and ultraviolet light are used in microscopy.

existence of microorganisms. Since that time, scientists and engineers have developed a variety of light and electron microscopes.

General Principles of Microscopy

Learning Objectives

✓ Explain the relevance of electromagnetic radiation to microscopy.
✓ Define *empty magnification.*
✓ List and explain two factors that determine resolving power.
✓ Discuss the relationship between contrast and staining in microscopy.

General principles involved in both light and electron microscopy include the wavelength of radiation, the magnification of an image, the resolving power of the instrument, and contrast in the specimen.

Wavelength of Radiation

Visible light is one part of a spectrum of electromagnetic radiation that includes X-rays, microwaves, and radio waves **(Figure 4.1).** Note that beams of radiation may be referred to as either rays or waves. These various forms of radiation differ in **wavelength**—the distance between two corresponding parts of a wave. The human eye discriminates among different wavelengths of visible light and sends patterns of nerve impulses to the brain, which interprets the impulses as different colors. For example, we see wavelengths of 400 nm as violet and wavelengths of 650 nm as red. White light, composed of many colors (wavelengths), has an average wavelength of 550 nm.

In Chapter 2 we learned that electrons are negatively charged particles that orbit the nuclei of atoms. Besides being particulate, moving electrons act as waves with wavelengths dependent upon the voltage of an electron beam. For example, the wavelength of electrons at 10,000 volts (V) is 0.01 nm, whereas the wavelength of electrons at 1,000,000 V is 0.001 nm. As we will see, using radiation of smaller wavelengths results in enhanced microscopy.

Magnification

Magnification is the apparent increase in size of an object. It is indicated by a number and "×", which is read "times." For example, the *Paramecium* in Figure 4.15c is magnified 2000× or 2000 times. Magnification results when a beam of radiation *refracts* (bends) as it passes through a lens. Curved glass lenses refract light, and magnetic fields (magnetic lenses) refract electron beams. Let's consider the magnifying power of a glass lens that is convex on both sides.

A lens refracts light because it is *optically dense* compared to the surrounding medium (such as air); that is, light travels more slowly through the lens than through air. This can be compared to a car moving from a paved road onto a dirt road at an angle. As the right front tire leaves the pavement, it has less traction and slows down. Since the other wheels continue at their original speed, the car will veer toward the dirt, and the line of travel will bend to the right. Likewise, the leading edge of a light beam slows as it enters glass, and the beam bends **(Figure 4.2a).** Light also bends as it leaves the glass and reenters the air.

Because of its curvature, a lens refracts light rays that pass through its periphery more than light rays that pass

(a)

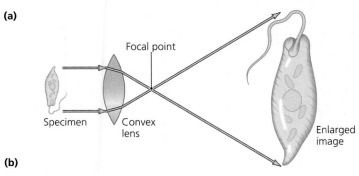

Focal point

Specimen
Convex
lens

Enlarged
image

(b)

▲ *Figure 4.2*

Light refraction and image magnification by a convex glass lens. **(a)** Light passing through a lens refracts (bends) because light rays slow down as they enter the glass, and light at the leading edge of a beam that strikes the glass at an angle slows first. **(b)** A convex lens focuses light on a focal point. The image is enlarged as light rays pass the focal point and spread apart.

through its center. The lens focuses light rays on a *focal point.* Importantly, for the purpose of microscopy, light rays spread apart as they travel past the focal point and produce an enlarged image **(Figure 4.2b).** The degree to which the image is enlarged depends on the thickness of the lens, its curvature, and the speed of light through its substance.

Microscopists could combine lenses to obtain an image magnified millions of times, but the image would be faint and blurry. Such magnification is said to be *empty magnification.* The properties that determine the clarity of an image, which in turn determines the useful magnification of a microscope, are *resolution* and *contrast.*

Resolution

Resolution, which is also called *resolving power,* is the ability to distinguish between objects that are close together. An optometrist's eye chart is a test of resolution at a distance of 20 feet (6.1 m). Leeuwenhoek's microscopes had a resolving power of about 1 μm; that is, he could distinguish between objects if they were more than about 1 μm apart, and objects closer together than 1 μm appeared as a single object. The closer together two objects are still distinguishable as separate objects, the better the resolution. Modern microscopes have five-fold better resolution than Leeuwenhoek's; they can distinguish between objects as close together as 0.2 μm. **Figure 4.3** on page 96 illustrates the size of various objects

that can be resolved by the unaided human eye and by various types of microscopes.

Why do modern microscopes have better resolution than Leeuwenhoek's microscopes? A principle of microscopy is that resolution distance is dependent on (1) the wavelength of the light or electron beam and (2) the **numerical aperture** of the lens, which refers to the ability of a lens to gather light.

Resolution distance is calculated using the following formula:

$$\text{resolution distance} = \frac{0.61 \times \text{wavelength}}{\text{numerical aperture}}$$

The resolution of today's microscopes is greater than that of Leeuwenhoek's microscopes because modern microscopes use shorter wavelength radiation, such as blue light or electron beams, and because they have lenses with larger numerical apertures.

Contrast

Contrast refers to differences in intensity between two objects, or between an object and its background. Contrast is important in determining resolution. For example, although you can easily distinguish between two golf balls lying side by side on a putting green 15 m away, at that distance it is much more difficult to distinguish between them if they are lying on a white towel.

Most microorganisms are colorless and have very little contrast whether one uses light or electrons. One way to increase the contrast between microorganisms and their background is to stain them. Stains and staining techniques are covered later in the chapter. As we will see, the use of light that is in *phase*—that is, in which all of the waves' crests and troughs are aligned—can also enhance contrast.

Light Microscopy

Learning Objectives

✓ Contrast simple and compound microscopes.
✓ Compare and contrast bright-field microscopy, dark-field microscopy, and phase microscopy.
✓ Compare and contrast fluorescent and confocal microscopes.

Several classes of microscopes use various types of light to examine microscopic specimens. The most common microscopes are *bright-field microscopes,* in which the background (or *field*) is illuminated. In *dark-field microscopes,* the specimen is made to appear light against a dark background. *Phase microscopes* use the alignment or misalignment of light waves to achieve the desired contrast between a living specimen and its background. *Fluorescent microscopes* use invisible ultraviolet light to cause specimens to radiate visible light, a phenomenon called *fluorescence.* Microscopes that use lasers to illuminate fluorescent chemicals in a thin plane of a specimen are called *confocal microscopes.* Next we examine each of these kinds of light microscope in turn.

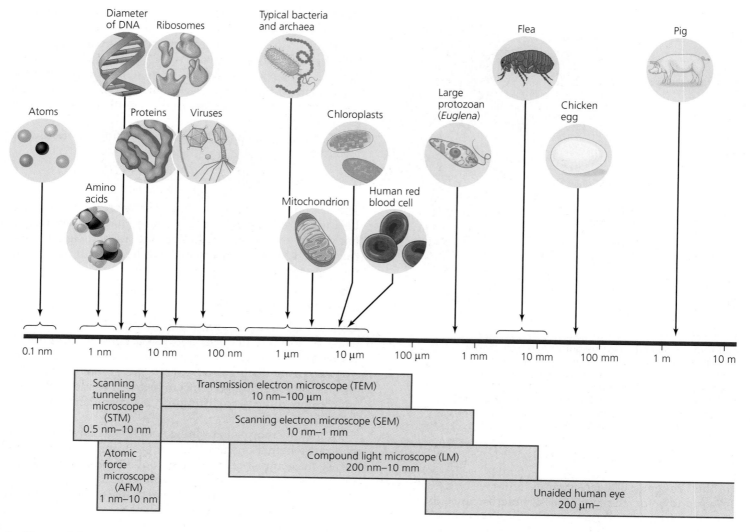

▲ *Figure 4.3*

The limits of resolution (and some representative objects within those ranges) of the human eye and of various types of microscopes.

Bright-Field Microscopes

There are two basic types of bright-field microscopes: *simple microscopes* and *compound microscopes*.

Simple Microscopes   As we learned in Chapter 1, Leeuwenhoek first reported his observations of microorganisms using a simple microscope in 1673. A **simple microscope,** which contains a single magnifying lens, is more similar to a magnifying glass than to a modern microscope (see Figure 1.2). Though Leeuwenhoek did not invent the microscope, he was the finest lens grinder of his day and produced microscopes of exceptional quality. They were capable of approximately 300× magnification and achieved excellent clarity, far surpassing other microscopes of his time.

Compound Microscopes   Simple microscopes have been replaced in modern laboratories by compound microscopes. A **compound microscope** uses a series of lenses for magnification **(Figure 4.4a).** Many scientists, including Galileo Galilei, made compound microscopes as early as 1590, but it was not until about 1830 that scientists developed compound microscopes that exceeded the clarity and magnification of Leeuwenhoek's simple microscope.

In a basic compound microscope, magnification is achieved as light rays pass through a specimen and into an **objective lens,** which is the lens immediately above the object being magnified **(Figure 4.4b).** An objective lens is really a series of lenses that not only create a magnified image, but are also engineered to reduce aberrations in the shape and color of the image. Most light microscopes used in biology have three or four objective lenses mounted on a **revolving**

Ocular lens
Remagnifies the image formed by
the objective lens

Body
Transmits the image from the
objective lens to the ocular lens
using prisms

Arm

Objective lenses
Primary lenses that
magnify the specimen

Stage
Holds the microscope
slide in position

Condenser
Focuses light
through specimen

Diaphragm
Controls the amount of
light entering the condenser

Illuminator
Light source

Coarse focusing knob
Move the stage up and
down to focus the image

Fine focusing knob

Base

Line of vision

Ocular lens

Path of light

Prism

Body

Objective
lenses

Specimen

Condenser
lenses

Illuminator

(a)   (b)

▲ *Figure 4.4*

A bright field, compound light microscope. **(a)** The parts of a compound microscope,
which uses a series of lenses to produce an image up to 2000× magnification.
(b) The path of light in a compound microscope; light travels from bottom to top.
Why can't light microscopes produce clear images that are magnified 10,000×?

*Figure 4.4 While it is possible for a light microscope to produce an image that is magnified
10,000×, magnification above 2000× is empty magnification because pairs of objects in the
specimen are too close together to resolve with even the shortest-wavelength (blue) light.*

nosepiece. The objective lenses on a typical microscope are:
scanning objective lens (4×), *low power objective lens* (10×), *high
power lens* or *high dry objective lens* (40×), and *oil immersion
objective lens* (100×).

Not only does an oil immersion lens increase magnification, it also increases resolution. As we have seen, light refracts as it travels from air into glass and also from glass into
air; therefore, some of the light passing out of a glass slide is
bent so much that it bypasses the lens (**Figure 4.5a** on page
98). Placing *immersion oil* between the slide and an *oil immersion objective lens* enables the lens to capture this light because light travels through immersion oil at the same speed
as through glass. Since light is traveling at a uniform speed
through the slide, the immersion oil, and the glass lens, it
does not refract (**Figure 4.5b**). Immersion oil increases the
numerical aperture, which increases resolution, because
more light rays are gathered into the lens to produce the image. Obviously, the space between the slide and the lens can

be filled with oil only if the distance between the lens and
the specimen, called the *working distance,* is small.

An objective lens bends the light rays, which then pass
up through the hollow body of the microscope and through
one or two **ocular lenses,** which are the lenses closest to the
eyes. Microscopes with a single ocular lens are *monocular,*
and those with two are *binocular.* Ocular lenses magnify
the image created by the objective lens, typically another
10 times (10×).

The **total magnification** of a compound microscope is
determined by multiplying the magnification of the objective lens by the magnification of the ocular lens. Thus, total
magnification using a 10× ocular lens and a 10× low power
objective lens is 100×. Using the same ocular and a 100× oil
immersion objective produces 1000× magnification. Some
light microscopes, using higher magnification oil immersion
objective lenses and ocular lenses, can achieve 2000× magnification, but this is the limit of useful magnification for

(a) Without immersion oil

(b) With immersion oil

▲ *Figure 4.5*

The effect of immersion oil on resolution. **(a)** Without immersion oil, light is refracted as it moves from the cover glass into the air. Part of the scattered light misses the objective lens. **(b)** With immersion oil. Because light travels through the oil at the same speed as it does through glass, no light is refracted as it leaves the specimen, and more light enters the lens, which increases resolution.

light microscopes because their resolution is restricted by the wavelength of visible light.

Modern compound microscopes also have a **condenser lens** (or lenses), which directs light through the specimen, as well as one or more mirrors or prisms that deflect the path of the light rays from an objective lens to the ocular lens (see Figure 4.4b). Some microscopes have mirrors or prisms that direct light to a camera through a special tube. Photographs of a microscopic image are called **micrographs.**

Dark-Field Microscopes

Pale objects are best observed with **dark-field microscopes.** These microscopes utilize a *dark-field stop* in the condenser that prevents light from directly entering the objective lens **(Figure 4.7).** Instead, light rays are reflected inside the condenser so that they pass into the slide at such an oblique angle that they miss the objective lens. Only those light rays that are scattered by the specimen enter the objective lens and are seen, so the specimen appears light against a dark background (see Figure 4.6b). This increases contrast and enables observation of more details than are visible in bright-field microscopy. Dark-field microscopes are especially useful for examining small or colorless cells.

Phase Microscopes

Scientists use **phase microscopes** to examine living microorganisms or specimens that would be damaged or altered by attaching them to slides or staining them. Basically, phase microscopes treat one set of light rays differently from another set of light rays.

Light rays are said to be *in phase* when their crests and troughs are aligned, and *out of phase* when their crests and troughs are not aligned **(Figure 4.8a** on page 100). Light rays that are in phase reinforce one another, producing a brighter image, and light rays that are out of phase interfere with one another and produce a darker image. Light rays passing through a specimen naturally slow down and are shifted about 1/4 wavelength out of phase. A special filter called a *phase plate,* which is mounted in the phase objective lens, retards these rays another 1/4 wavelength, so that they are 1/2 wavelength out of phase with their neighbors. When the phase microscope lens brings the two sets of rays together, troughs of one wave interfere with the crests of the other—because they are out of phase **(Figure 4.8b)**—and contrast is created. There are two types of phase microscopes: phase-contrast and differential interference contrast microscopes.

Phase-Contrast Microscopes The simplest phase microscopes, **phase-contrast microscopes,** produce sharply defined images in which fine structures can be seen in living cells (see Figure 4.6c). These microscopes are particularly useful for observing cilia and flagella.

Differential Interference Contrast Microscopes Differential **interference contrast microscopes** (also called Nomarski[2] microscopes) create phase interference patterns. They also use prisms that split light beams into their component wavelengths (colors). This significantly increases

---

[2]After the French physicist Georges Nomarski, who invented the differential interference contrast microscope.

(a)   LM   20 μm

(b)   LM   20 μm

Nucleus

(c)   LM   20 μm

Bacteria

(d)   LM   20 μm

◀ *Figure 4.6*

Four kinds of micrographs, in this case of bacteria and a human cheek cell. **(a)** Bright-field microscopy reveals some internal structures. **(b)** Dark-field microscopy increases contrast between some internal structures and between the edges of the cell and the surrounding medium. **(c)** Phase-contrast microscopy provides greater resolution of internal structures. **(d)** Differential interference contrast (Nomarski) microscopy produces a three-dimensional effect.

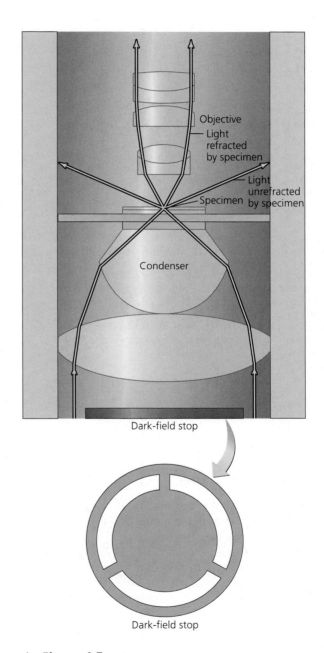

Objective

Light refracted by specimen

Light unrefracted by specimen

Specimen

Condenser

Dark-field stop

Dark-field stop

▲ *Figure 4.7*

The light path in a dark-field microscope. A dark-field stop prevents light from entering the specimen directly; only light rays that are scattered by the specimen reach the objective lens and can be seen. The resulting high-contrast image—a brightly lit specimen against a dark background—enhances resolution.

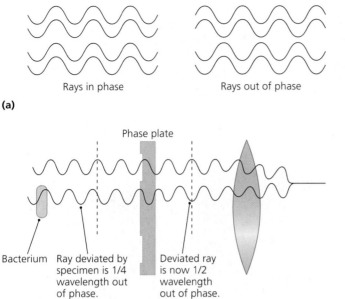

(a)

Rays in phase   Rays out of phase

Phase plate

Bacterium | Ray deviated by specimen is 1/4 wavelength out of phase. | Deviated ray is now 1/2 wavelength out of phase.

(b)

▲ *Figure 4.8*

Principles of phase microscopy. **(a)** Light rays that are in phase are aligned; their crests and troughs reinforce one another to produce a brighter image. Rays that are out of phase interfere with one another and produce a darker image. **(b)** Together, the regions of a specimen and a phase plate built into a phase objective lens slow some of the light rays such that they are 1/2 wavelength out of phase. These rays interfere with light rays that bypass the specimen, which produces contrast between the field and various regions of the specimen (see Figure 4.6c and d).

contrast and gives the image a dramatic three-dimensional or shadowed appearance, almost as if light were striking the specimen from one side (see Figure 4.6d). This technique also produces unnatural colors, which enhance contrast.

Fluorescent Microscopes

Molecules that absorb energy from invisible radiation (such as ultraviolet light) and then radiate the energy back as a longer, visible wavelength are said to be *fluorescent*. **Fluorescent microscopes** use an ultraviolet (UV) light source to fluoresce objects. UV light increases resolution because it has a shorter wavelength than visible light, and contrast is improved because fluorescing structures are visible against a black background.

Some cells—for example, the pathogen *Pseudomonas aeruginosa* (soo-dō-mō′nas ā-rū-ji-nō′să)—and some cellular molecules (such as chlorophyll in photosynthetic organisms) are naturally fluorescent. Other cells and cellular structures can be stained with fluorescent dyes. When these dyes are bombarded with ultraviolet light, they emit visible light and show up as bright orange, green, yellow, or other colors against a black background (see Figure 3.30 on page 82).

(a)   LM   10 μm

(b)   LM   10 μm

▲ *Figure 4.9*

Fluorescent microscopy, in which fluorescent chemicals absorb invisible short-wavelength radiation and emit visible (longer wavelength) radiation. **(a)** When viewed under normal illumination, *Mycobacterium tuberculosis* cells stained with the fluorescent dye auramine O are invisible amid the mucus and debris in a sputum smear. **(b)** When the same smear is viewed under UV light, the bacteria fluoresce and are clearly visible.

Some fluorescent dyes are specific for certain cells. For example, the dye fluorescein isothiocyanate attaches to cells of *Bacillus anthracis* (ba-sil′ŭs an-thrā′sis), the causative agent of anthrax, and appears apple green when viewed in a fluorescent microscope. Another dye, auramine O, which fluoresces yellow, stains *Mycobacterium tuberculosis* (mī′kō-bak-tēr′ē-ŭm tū-ber-kyū-lō′sis; **Figure 4.9**).

Fluorescent microscopy is also used in a process called *immunofluorescence*. First, fluorescent dyes are covalently linked to Y-shaped immune system proteins called *antibodies* **(Figure 4.10a).** Given the opportunity, these dye-tagged antibodies will bind specifically to complementary-shaped *antigens,* which are molecules (or portions of molecules) that are present, for example, on the surface of microbial cells. When viewed under UV light, a microbial specimen that has bound dye-tagged antibodies becomes visible **(Figure 4.10b).** In addition to identifying pathogens, including those that cause syphilis, rabies, and Lyme disease, scientists use immunofluorescence to locate and make visible a variety of proteins of interest.

Confocal Microscopes

Even though **confocal**[3] **microscopes** also use fluorescent dyes or fluorescent antibodies, these microscopes use ultraviolet lasers to illuminate the fluorescent chemicals in only a single plane that is no thicker than 1.0 μm; the rest of the specimen remains dark and out of focus. Visible light emitted by the dyes passes through a pinhole aperture, which helps eliminate blurring that can occur with other types of microscopes and increases resolution by up to 40%. Each image from a confocal microscope is thus an "optical slice" through the specimen, as if it had been thinly cut. Once individual images are digitized, a computer is used to construct a three-dimensional representation, which can be rotated and viewed from any direction (**Figure 4.11** on page 102). Confocal microscopes have been particularly useful for examining the relationships among various organisms within complex microbial communities called *biofilms.* Regular light microscopy cannot produce clear images within a living biofilm, and removing surface layers from a biofilm would change the dynamics of the community.

Electron Microscopy

Learning Objective

✓ Contrast transmission electron microscopes with scanning electron microscopes in terms of how they work, the images they produce, and the advantages of each.

Even with the most expensive phase microscope using the best oil immersion lens with the highest numerical aperture, resolution is still limited by the wavelength of visible light. Because the shortest visible radiation (violet) has a wavelength of about 400 nm, structures closer together than about 200 nm cannot be distinguished using a light microscope. By contrast, electrons traveling as waves have wavelengths between 0.01 nm and 0.001 nm, which is one ten-thousandth to one hundred-thousandth the wavelength of visible light. The resolving power of electron microscopes is therefore much greater than that of light microscopes, and with greater resolving power comes the possibility of

[3]From *coinciding focal* points of (laser) light.

(a)

(b) LM 10 μm

▲ *Figure 4.10*

Immunofluorescence. **(a)** After a fluorescent dye is covalently linked to an antibody, the dye-antibody combination binds to the antibody's target, making the target visible under fluorescent microscopy. **(b)** Immunofluorescent staining of *Treponema pallidum,* the causative agent of syphilis. The bacteria are brightly-colored against a dark background.

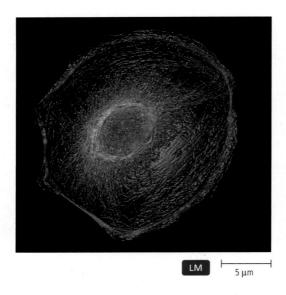

LM ⊢——————⊣ 5 µm

▲ *Figure 4.11*

Confocal microscopy, in which a three-dimensional computer-generated image (here, of a mouse cell) is created from many individual "optical slices."

greater magnification. **Highlight 4.1** describes the invention of the electron microscope in the 1930s.

Generally, electron microscopes magnify objects $10,000\times$ to $100,000\times$, though millions of times magnification with good resolution is possible. Electron microscopes provide detailed views of the smallest bacteria, viruses, internal cellular structures, and even molecules and large atoms. Cellular structures that can be seen only by using electron microscopy are referred to as a cell's *ultrastructure*. Ultrastructural details cannot be made visible by light microscopy because they are too small to be resolved.

There are two general types of electron microscopes: *transmission electron microscopes* and *scanning electron microscopes*.

Transmission Electron Microscopes

A **transmission electron microscope (TEM)** generates a beam of electrons that ultimately produces an image on a fluorescent screen **(Figure 4.12a).** The path of electrons is similar to the path of light in a light microscope. From their source, the electrons pass through the specimen, through magnetic fields (instead of glass lenses) that manipulate and focus the beam, and then onto a fluorescent screen that absorbs electrons, thereby changing some of their energy into visible light **(Figure 4.12b).** Dense areas of the specimen block electrons from striking the screen, resulting in a dark area on the screen. In regions where the specimen is less dense, the screen fluoresces more brightly. As with light microscopy, contrast and resolution can be enhanced through the use of electron-dense stains, which are discussed later. The brightness of each region of the screen corresponds to the number of electrons striking it. Therefore, the image on the screen is composed of light and dark areas much like a photographic negative.

The screen can be folded out of the way to enable the electrons to strike a photographic film, located in the base of the microscope. Prints made from the film are called *transmission electron micrographs* or *TEM images*. Such images can be colorized to emphasize certain features **(Figure 4.13).**

Matter, including air, absorbs electrons, so the column of a transmission electron microscope must be a vacuum, and the specimen must be very thin. Before thicker specimens such as whole cells can be examined, they must be dehydrated, embedded in plastic, and cut to a thickness of about 100 nm with a diamond or glass knife mounted in a slicing machine called an *ultramicrotome*. Such thin sections are placed on a small copper grid and inserted into the microscope through an air lock.

Because the vacuum and slicing of the specimens are required, electron microscopes cannot be used to study living organisms. Unfortunately dehydration and sectioning can introduce shrinkage, distortion, and other *artifacts*, which

---

Highlight 4.1   "One of the Most Important Inventions of the Century"

By the beginning of the 20th century, microscopists had reached an impasse. They could not increase the useful magnification of microscopes because resolution is dependent on the wavelength of light. In 1931, Ernst Ruska, an engineering student at the Technical University in Berlin, posited that electrons, which had been shown to sometimes act like waves with very short wavelengths, might be focused on a specimen, providing the means for greater resolution and greater useful magnification. But one major obstacle to this idea remained: the lack of a lens that could focus electron waves. So Ruska set out to produce one.

In 1931, Ruska built the first magnetic lens that could focus a beam of electrons, much as a glass lens focuses light waves. In 1933, using a series of magnetic lenses, he created the first electron microscope. This prototype was capable of magnification of only a few

hundred times, and it would be another decade before a working electron microscope would become a reality.

Still Ruska's surpassing idea—the principle that electrons could be used for microscopy—opened the door for modern microbiologists to study the ultrastructure of cells. Calling the electron microscope "one of the most important inventions of the century," the Nobel Society awarded Ernst Ruska the Nobel Prize for physics more than 50 years later, in 1986.

CRITICAL THINKING

Electron microscopy reveals the ultrastructure of a cell. What cellular structures are considered ultrastructural?

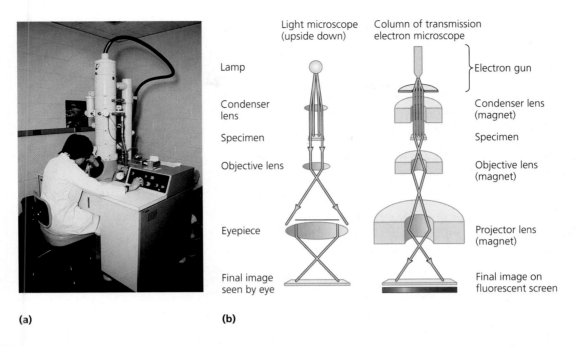

Light microscope (upside down)

Column of transmission electron microscope

Lamp

Condenser lens

Specimen

Objective lens

Eyepiece

Final image seen by eye

Electron gun

Condenser lens (magnet)

Specimen

Objective lens (magnet)

Projector lens (magnet)

Final image on fluorescent screen

(a) **(b)**

▲ *Figure 4.12*

A transmission electron microscope (TEM). **(a)** TEMs are much larger than light microscopes. **(b)** The path of electrons through a TEM (at right), as compared to the path of light through a light microscope (at left, drawn upside down to facilitate the comparison). *Why must air be evacuated from the column of an electron microscope?*

Figure 4.12 Air would absorb electrons, so there would be no radiation to produce an image.

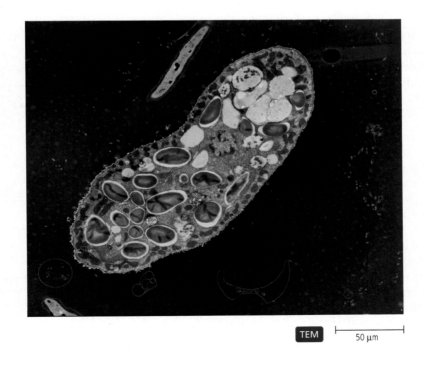

TEM ⊢——⊣ 50 µm

▲ *Figure 4.13*

Transmission electron image. A transmission electron micrograph reveals much internal detail not visible by light microscopy.

are structures that appear in a TEM image but are not present in natural specimens.

Scanning Electron Microscopes

A **scanning electron microscope (SEM)** also uses magnetic fields within a vacuum tube to manipulate a beam of electrons, called primary electrons **(Figure 4.14a)**. However, rather than passing electrons through a specimen, the SEM rapidly focuses them back and forth across the specimen's surface, which has previously been coated with a metal such as platinum or gold **(Figure 4.14b)**. The primary electrons knock electrons off the surface of the specimen, and these scattered secondary electrons pass through a detector and a photomultiplier, producing an amplified signal that is displayed on a monitor. Typically, scanning microscopes are used to magnify up to 10,000× with a resolution of about 20 nm.

One advantage of scanning microscopy over transmission microscopy is that whole specimens can be observed, because sectioning is not required. Scanning electron micrographs can be beautifully realistic and three-dimensional **(Figure 4.15)**. Two disadvantages of a scanning electron microscope are that it magnifies only the external surface of a specimen, and, like TEM, it requires a vacuum and thus can examine only dead organisms.

Probe Microscopy

Learning Objective

✓ Describe two variations of probe microscopes.

A relatively recent advance in microscopy utilizes miniscule, pointed, electronic probes to magnify more than 100,000,000 times. There are two variations of probe microscopes: *scanning tunneling microscopes* and *atomic force microscopes*.

Scanning Tunneling Microscopes

A **scanning tunneling microscope** passes a tungsten probe, sharpened to end in a single atom, back and forth across and slightly above the surface of a specimen. Rather than scattering a beam of electrons into a detector, as in scanning electron microscopy, a scanning tunneling microscope measures the flow of electrons to and from the probe and the specimen's surface. The amount of electron flow, called a *tunneling current,* is directly proportional to the distance from the probe to the specimen's surface. A scanning tunneling microscope can measure distances as small as 0.01 nm and reveal details on the surface of a specimen at the atomic level **(Figure 4.16a** on page 106). A requirement for scanning tunneling microscopy is that the specimen be electrically conductive.

(a)

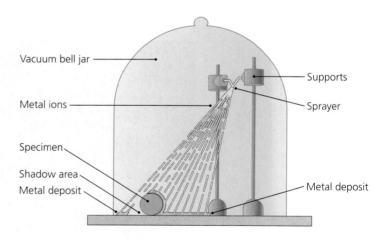

(b)

▲ *Figure 4.14*

Scanning electron microscope (SEM). **(a)** The SEM uses magnetic lenses to focus a beam of primary electrons, which are scanned across the metal-coated surface of a specimen. Secondary electrons, knocked off the surface of the specimen by the primary electrons, are collected by a detector, and their signal is amplified and displayed on a monitor. **(b)** The coating of a specimen with a thin layer of metal ions in preparation for its examination in the SEM.

(a) Diatom
SEM   10 μm

(b) *Aspergillus*
SEM   50 μm

(c) *Paramecium*
SEM   25 μm

(d) *Streptococcus*
SEM   1 μm

▲ *Figure 4.15*
SEM images.

Atomic Force Microscopes

An **atomic force microscope** also uses a pointed probe, but it traverses the tip of the probe lightly on the surface of the specimen, rather than at a distance. This might be likened to the way a person reads Braille. Deflection of a laser beam aimed at the probe's tip measures vertical movements, which when translated by a computer reveals the atomic topography.

Unlike tunneling microscopes, atomic force microscopes can magnify specimens that do not conduct elec-

trons. They can also magnify living specimens, because neither an electron beam nor a vacuum is required (see Figure 4.16b). These microscopes have been used to magnify the surfaces of bacteria, viruses, proteins, and amino acids. Recent studies using atomic force microscopes have provided insights into the molecular folding of lipopolysaccharides on the surfaces of Gram-negative pathogenic bacteria.

Table 4.2 on page 107 summarizes the features of the various types of microscopes.

▶ *Figure 4.16*

Probe microscopy. **(a)** Scanning tunneling microscopes reveal surface detail; in this case, three turns of a DNA double helix. **(b)** Plasmid DNA being digested by an enzyme, viewed through atomic force microscopy.

(a) STM ⊢ 1 nm   **(b)** AFM ⊢ 15 nm

Spread culture in thin film over slide

Air dry

Pass slide through flame to fix it

(a)

(b)

▲ *Figure 4.17*

Preparing a specimen for staining. **(a)** Microorganisms are spread in liquid across the surface of a slide using a circular motion. After drying in the air, the smear is passed through the flame of a Bunsen burner to fix the cells to the glass. Alternatively, chemical fixation can be used. **(b)** A properly prepared smear lacks clumps and is thinner toward the edges than in the middle. *Why must a smear be fixed to the slide?*

Staining

Earlier we discussed the difficulty of resolving two distant white golf balls viewed against a white background. If the balls were painted black, they could be distinguished more readily from the background, and from one another. This illustrates why staining increases contrast and resolution.

Most microorganisms are colorless and difficult to view with bright-field microscopes. Microscopists use stains to make microorganisms and their parts more visible because stains increase contrast between structures, and between a specimen and its background. Electron microscopy also requires the treatment of specimens with stains or coatings to enhance contrast.

In this section we examine how scientists prepare specimens for staining and how stains work, and we consider five kinds of stains used for light microscopy. We conclude with a look at staining for electron microscopy.

Preparing Specimens for Staining

Learning Objective

✓ Explain the purposes of a smear, heat fixation, and chemical fixation in the preparation of a specimen for microscopic viewing.

Many investigations of microorganisms, especially those seeking to identify pathogens, begin with light microscopic observation of stained specimens. **Staining** simply means coloring specimens with stains, which are also called dyes.

Before microbiologists stain microorganisms, they must place them on and then firmly attach them to a microscope slide **(Figure 4.17)**. Typically, this involves making a *smear* and *fixing* it to the slide. If the organisms are growing in a liquid, a small drop is spread across the surface of the slide. If the organisms are growing on a solid surface, such as an agar plate, then they are mixed into a small drop of water on

Table 4.2   Comparison of Types of Microscopes

Type of Microscope	Typical Image	Description of Image	Special Features	Typical Uses
Light Microscopes		Useful magnification 1× to 2000×; resolution to 200 nm	Use visible light; shorter, blue wavelengths provide better resolution	
Bright-Field		Colored or clear specimen against bright background	Simple to use; relatively inexpensive; stained specimens often required	Observation of killed stained specimens and naturally colored live ones; also used to count microorganisms
Dark-Field		Bright specimen against dark background	Use a special filter in the condenser that prevents light from directly passing through a specimen; only light scattered by the specimen is visible	Observation of living, colorless, unstained organisms
Phase-Contrast		Specimen has light and dark areas	Use a special condenser that splits a polarized light beam into two beams, one of which passes through the specimen, and one of which bypasses the specimen; the beams are then rejoined before entering the oculars; contrast in the image results from the interactions of the two beams	Observation of internal structures of living microbes
Differential Interference Contrast (Nomarski)		Image appears three-dimensional	Use two separate beams instead of a split beam; false color and a three-dimensional effect result from interactions of light beams and lenses; no staining required	Observation of internal structures of living microbes
Fluorescent		Brightly colored fluorescent structures against dark background	An ultraviolet light source causes fluorescent natural chemicals or dyes to emit visible light	Localization of specific chemicals or structures; used as an accurate and quick diagnostic tool for detection of pathogens
Confocal		Single plane of structures or cells that have been specifically stained with fluorescent dyes	Use a laser to fluoresce only one plane of the specimen at a time	Detailed observation of structures of cells within communities
Electron Microscopes		Typical magnification 1000× to 100,000×; resolution to 0.001 nm	Use electrons traveling as waves with short wavelengths; require specimens to be in a vacuum, so cannot be used to examine living microbes	
Transmission		Monotone, two-dimensional, highly magnified images; may be color-enhanced	Produce two-dimensional image of ultrastructure of cells	Observation of internal ultrastructural detail of cells and observation of viruses and small bacteria
Scanning		Monotone, three-dimensional, surface images; may be color-enhanced	Produce three-dimensional view of the surface of microbes and cellular structures	Observation of the surface details of structures
Probe Microscopes		Magnification greater than 100,000,000× with resolving power greater than that of electron microscopes	Use microscopic probes that move over the surface of a specimen	
Scanning Tunneling		Individual molecules and atoms visible	Measures the flow of electrical current between the tip of a probe and the specimen to produce an image of the surface at atomic level	Observation of the surface of objects; provide extremely fine detail, high magnification, and great resolution
Atomic Force		Individual molecules and atoms visible	Measure the deflection of a laser beam aimed at the tip of a probe that travels across the surface of the specimen	Observation of living specimens at the molecular and atomic levels

▲ *Figure 4.18*

Simple stains, which are simply and quickly performed, increase contrast, and allow determination of size, shape, and arrangement of cells. **(a)** Unstained *Staphylococcus aureus.* **(b)** *Staphylococcus aureus* stained with methyl blue. **(c)** *Escherichia coli* stained with crystal violet. Note that all cells, no matter their type, stain the same color with a simple stain because only one dye is used.

the slide. Either way, the thin film of organisms on the slide is called a **smear.**

The smear is air dried and then attached or fixed to the surface of the slide using either heat or chemicals. Robert Koch developed **heat fixation** more than a hundred years ago. In heat fixation, the slide is gently heated by passing the slide, smear up, through the flame of a Bunsen burner. Alternately, **chemical fixation** involves applying methyl alcohol (methanol) or formalin (a solution of formaldehyde in water) to the smear for one minute. Desiccation (drying) and fixation kill the microorganisms, attach them firmly to the slide, and generally preserve their shape and size. It is important to smear and fix specimens properly so that they are not lost during staining.

Specimens prepared for electron microscopy are also dried, because water vapor from a wet specimen would stop the electron beam. As we have seen, transmission electron microscopy requires that the desiccated sample also be sliced very thin, generally before staining.

Principles of Staining

Learning Objective

✓ Describe the uses of acidic and basic dyes, mentioning ionic bonding and pH.

Dyes used as microbiological stains for light microscopy are usually salts. As we saw in Chapter 2, a salt is composed of a positively charged *cation* and a negatively charged *anion.* At least one of the two ions in the molecular makeup of dyes is colored; this colored portion of a dye is known as the *chromophore.* Chromophores typically bind to chemicals via covalent, ionic, or hydrogen bonds. For example, methylene blue chloride is composed of a cationic chromophore, methylene blue, and a chloride anion. Since methylene blue is positively charged, it ionically bonds to negatively charged molecules in cells, including DNA and many proteins. In

contrast, anionic dyes, for example, eosin, bind to positively charged molecules such as some amino acids.

Anionic chromophores are also called **acidic dyes** because they stain alkaline structures and work best in acidic (low pH) environments. Positively charged, cationic chromophores are called **basic dyes** because they combine with and stain acidic structures; further, they work best under basic (higher pH) conditions. In microbiology, basic dyes are used more commonly than acidic dyes because most cells are negatively charged. Acidic dyes are used in negative staining, which is discussed shortly.

Some stains do not form bonds with cellular chemicals, but rather function because of their solubility characteristics. For example, Sudan Black selectively stains membranes because it is lipid soluble and accumulates in phospholipid bilayers.

Simple Stains

Learning Objective

✓ Describe the simple, Gram, acid-fast, and endospore staining procedures.

Simple stains are composed of a single basic dye such as crystal violet, safranin, or methylene blue. They are "simple" because they involve no more than soaking the smear in the dye for 30–60 seconds and then rinsing off the slide with water. (A properly fixed specimen will remain attached to the slide despite this treatment.) After carefully blotting the slide dry, the microbiologist observes the smear under the microscope. Simple stains are used to determine size, shape, and arrangement of cells **(Figure 4.18).**

Differential Stains

Most stains used in microbiology are **differential stains,** which use more than one dye so that different cells, chemicals,

Figure 4.19 ▶

The Gram staining procedure. A specimen is smeared and fixed to a slide. The classical procedure consists of four steps. Gram-positive cells (in this case, *Bacillus cereus*) remain purple throughout the procedure; Gram-negative cells (here, *Escherichia coli*) end up pink.

or structures can be distinguished when microscopically examined. Common differential stains are the *Gram stain*, the *acid-fast stain*, and the *endospore stain*.

Gram Stain

In 1884, the Danish scientist Hans Christian Gram developed the most frequently used differential stain, which now bears his name. The **Gram stain** differentiates between two large groups of microorganisms: purple-staining Gram-positive cells, and pink-staining Gram-negative cells. Recall from Chapter 3 that these cells differ significantly in the chemical and physical structures of their cell walls (see Figure 3.13 on page 69). Typically, a Gram stain is the first step a medical laboratory technologist performs to identify certain kinds of pathogens.

Let's examine the Gram staining procedure as it was originally developed, and as it is typically performed today, over a hundred years later. For the purposes of our discussion, we'll assume that a smear has been made on a slide and heat fixed, and that the smear contains both Gram-positive and Gram-negative colorless bacteria. The classical Gram staining procedure has the following four steps **(Figure 4.19):**

1 Flood the smear with the basic dye crystal violet for 1 min, and then rinse with water. Crystal violet, which is called the **primary stain,** colors all cells.

2 Flood the smear with an iodine solution for 1 min, and then rinse with water. Iodine is a **mordant,** a substance that binds to a dye and makes it less soluble. After this step, all cells remain purple.

3 Rinse the smear with a solution of ethanol and acetone for 10–30 sec, and then rinse with water. This solution, which acts as a **decolorizing agent,** breaks down the thin cell wall of Gram-negative cells, allowing the stain and mordant to be washed away; these cells are now colorless. Gram-positive cells, with their thicker cell walls, remain purple.

4 Flood the smear with safranin for 1 min, and then rinse with water. This red **counterstain** provides a contrasting color to the primary stain. Although all types of cells may absorb safranin, the resulting pink color is masked by the darker purple dye already in Gram-positive cells. After this step, Gram-negative cells now appear pink, whereas Gram-positive cells remain purple.

After the final step, the slide is blotted dry in preparation for microscopy.

1 Slide is flooded with crystal violet for 1 min, then rinsed with water.

Result: All cells are stained purple.

LM   5 μm

2 Slide is flooded with iodine for 1 min, then rinsed with water.

Result: Iodine acts as a mordant; all cells remain purple.

LM   5 μm

3 Slide is flooded with solution of alcohol and acetone for 10–30 sec, then rinsed with water.

Result: Smear is decolorized; Gram-positive cells remain purple, but Gram-negative cells are now colorless.

LM   5 μm

4 Slide is flooded with safranin for 1 min, then rinsed with water and blotted dry.

Result: Gram-positive cells remain purple, Gram-negative cells are pink.

LM   5 μm

(a) 24-hour culture   `LM` ⊢———⊣ 10 µm

(b) 72-hour culture   `LM` ⊢———⊣ 10 µm

▲ *Figure 4.20*

Effects of time on Gram staining of *Staphylococcus aureus,* a Gram-positive organism.
(a) Cells grown for 24 hours stain uniformly purple. **(b)** In this 72-hour culture, older,
presumably less healthy cells are bleached by the decolorizer, lose their purple color,
and stain pink. This creates a false Gram-negative reaction in many cells.

The Gram procedure works best with young cells. Older
Gram-positive cells bleach more easily than younger cells
and can therefore stain pink, which makes them appear to
be Gram-negative cells **(Figure 4.20)**. Therefore, smears for
Gram-staining should come from freshly grown bacteria.

Microscopists have developed minor variations on
Gram's original procedure. For example, 95% ethanol may
be used to decolorize instead of Gram's ethanol-acetone
mixture. In a three-step variation, safranin dissolved in
ethanol simultaneously decolorizes and counterstains.

New Frontiers 4.1 focuses on a miniature sensor that is
being developed to differentiate between Gram-positive
and Gram-negative bacteria.

Acid-Fast Stain

The **acid-fast stain** is another important differential stain because it stains cells of the genera *Mycobacterium* and *Nocardia*
(nō-kar′dē-ă), which cause many human diseases, including
tuberculosis, leprosy, and other lung and skin infections.
Cells of these bacteria have large amounts of waxy lipid in

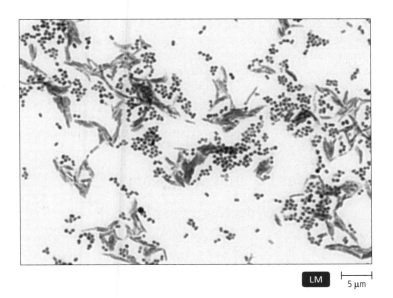

LM   5 µm

▲ *Figure 4.21*

Ziehl-Neelsen acid-fast stain. Acid-fast cells such as these rod-shaped *Mycobacterium bovis* cells stain pink or red. Non-acid-fast cells—in this case, *Streptococcus pyogenes*—stain blue. *Why isn't the Gram stain utilized to stain* Mycobacterium?

Figure 4.21 Cell walls of Mycobacterium are composed of waxy materials that repel the water-based dyes of the Gram stain.

their cell walls, so they do not readily stain with simple water-based dyes such as methylene blue and crystal violet, or with the Gram stain.

Modern microbiological laboratories commonly use a variation of the acid-fast stain developed by Franz Ziehl (1857–1926) and Friedrich Neelsen (1854–1894) in 1883. Their procedure is as follows:

1. Cover the smear with a small piece of tissue paper to retain the dye during the procedure.

2. Flood the slide with the red primary stain, carbolfuchsin, for several minutes while warming it over steaming water. In this procedure, heat is used to drive the stain through the waxy wall and into the cell, where it remains trapped.

3. Remove the tissue paper, cool the slide, and then decolorize the smear by rinsing it with a solution of hydrochloric acid (pH < 1.0). The bleaching action of hydrochloric acid removes color from both non-acid-fast cells and the background. Acid-fast cells retain their red color because the acid cannot penetrate the waxy wall. The name of the procedure is derived from this step; that is, the cells are colorfast in acid.

4. Counterstain with methylene blue, which stains only bleached, non-acid-fast cells.

The Ziehl-Neelsen acid-fast staining procedure results in pink acid-fast cells, which can be differentiated from blue

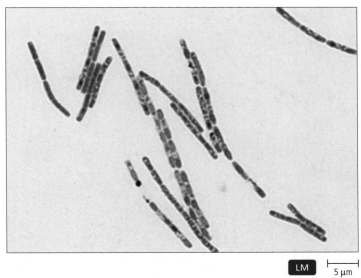

LM   5 µm

▲ *Figure 4.22*

Schaeffer-Fulton endospore stain of *Bacillus anthracis*. The nearly impermeable spore wall retains the green dye during decolorization. Vegetative cells, which lack spores, pick up the counterstain and appear red. *Why don't the spores stain red as well?*

Figure 4.22 Heat from steam is used to drive the green primary stain into the endospores. Counterstaining is performed at room temperature, and the thick, impermeable walls of the endospores resist the counterstain.

non-acid-fast cells, including human cells and tissue (**Figure 4.21**). The presence of *acid-fast rods (AFRs)* in sputum is indicative of mycobacterial infection.

CRITICAL THINKING

In what ways are the Gram stain and the acid-fast staining procedures similar? In the acid-fast procedure, what takes the place of Gram's iodine mordant?

Endospore Stain

Some bacteria—notably those of the genera *Bacillus* and *Clostridium* (klos-trid'ē-ŭm), which contain species that cause such diseases as anthrax, gangrene, and tetanus—produce **endospores.** These dormant, highly resistant cells form inside the cytoplasm of the bacteria and can survive environmental extremes such as desiccation, heat, and harmful chemicals. Endospores cannot be stained by normal staining procedures because their walls are practically impermeable to all chemicals. The **Schaeffer-Fulton endospore stain** uses heat to drive the primary stain, *malachite green*, into the endospore. After cooling, the slide is decolorized with water and counterstained with safranin. This staining procedure results in green-stained endospores and red-colored vegetative cells (**Figure 4.22**).

Background   Capsules   Rods

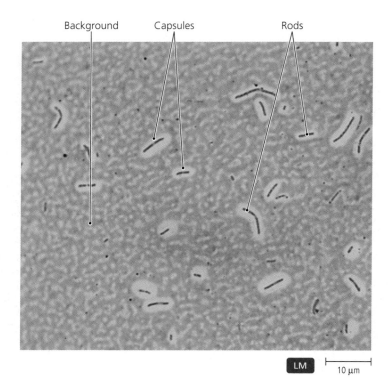

LM   10 µm

▲ *Figure 4.23*

Negative (capsule) stain of *Klebsiella pneumoniae*.
Notice that the acidic dye stains the background and
does not penetrate the capsule. The cells have been
counterstained with a different (basic) dye.

Special Stains

Special stains are simple stains designed to reveal special
microbial structures. There are three types of special stains:
negative stains, flagellar stains, and *fluorescent stains* (which we
already discussed in the section on fluorescent microscopy).

Negative (Capsule) Stain

Most dyes used to stain bacterial cells, such as crystal violet,
methylene blue, malachite green, and safranin, are basic
dyes. These dyes stain cells by attaching to negatively
charged molecules within them.

Acidic dyes, by contrast, are repulsed by the negative
charges on the surface of cells and therefore do not stain
them. Such stains are called **negative stains** because they
stain the background and leave cells colorless. Eosin and ni-
grosin are examples of acidic dyes used for negative staining. A counterstain may be added to color the cells.

Negative stains are used primarily to reveal the presence of negatively charged bacterial capsules. Therefore,
they are also called **capsule stains.** Encapsulated cells appear to have a halo surrounding them **(Figure 4.23).**

Flagellar Stain

Bacterial flagella are extremely thin and thus normally invisible with light microscopy, but their presence, number,

and arrangement are important in the identification of some
species, including some pathogens. Flagellar stains, such as
pararosanaline and carbolfuchsin, and mordants, such as
tannic acid and potassium alum, are applied in a series of
steps. These molecules bind to the flagella, increase their diameter, and change their color, all of which increases contrast and makes them visible **(Figure 4.24).**

Stains used for light microscopy are summarized in
Table 4.3 on page 114.

Staining for Electron Microscopy

Learning Objective

✓ Explain how stains used for electron microscopy differ from those
   used for light microscopy.

Laboratory technicians increase contrast and resolution for
transmission electron microscopy by using stains, just as
they do for light microscopy. However, stains used for transmission electron microscopy are not colored dyes, but instead chemicals containing atoms of heavy metals such as
lead, osmium, tungsten, and uranium, which absorb electrons. Electron-dense stains may bind to molecules within
specimens, or they may stain the background. The latter
type of negative staining is used to provide contrast for extremely small specimens such as viruses and molecules.

Stains for electron microscopy can be general, in that
they stain most objects to some degree, or they may be
highly specific. For example, osmium tetraoxide (OsO_4) has
an affinity for lipids and is thus used to enhance the contrast
of membranes. Electron-dense stains can also be linked to
antibodies to provide an even greater degree of staining
specificity because antibodies bind only to their specific target molecules.

Classification and Identification of Microorganisms

Learning Objective

✓ Discuss the purposes of classification and identification of
   organisms.

Biologists classify organisms for several reasons: to bring a
sense of order and organization to the variety and diversity
of living things, to enhance communication, to make predictions about the structure and function of similar organisms,
and to uncover and understand potential evolutionary con-
nections. They sort organisms on the basis of mutual similarities into nonoverlapping groups called **taxa.**[4] As we
discussed in Chapter 1, **taxonomy**[5] is the science of classifying and naming organisms. Taxonomy consists of *classification,* which is assigning organisms to taxa based upon

---

[4]From Greek *taxis,* meaning order.
[5]From *taxis* and Greek *nomos,* meaning rule.

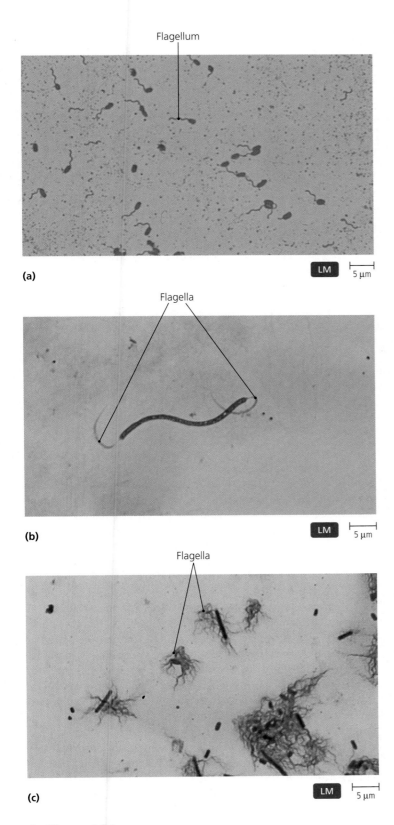

Flagellum

(a)   LM   5 μm

Flagella

(b)   LM   5 μm

Flagella

(c)   LM   5 μm

▲ *Figure 4.24*

Flagellar stain. Various bacteria have different numbers and arrangements of flagella, features that are important in identifying some species. **(a)** *Pseudomonas aeruginosa* has a single flagellum. **(b)** *Spirullum volutans* has flagella at each end. **(c)** *Proteus vulgaris* is covered with many flagella.

similarities; *nomenclature,* which is concerned with the rules of naming organisms; and *identification,* which is the practical science of determining that an isolated individual or population belongs to a particular taxon. In this book we concentrate on classification and identification.

Since all members of any given taxon share certain common features, taxonomy enables scientists to both organize large amounts of information about organisms and make predictions based on knowledge of similar organisms. For example, if one member of a taxon is important in recycling nitrogen in the environment, it is likely that others in the group will play a similar ecological role. Likewise, a clinician might suggest a treatment against one pathogen based on what has been effective against another pathogen in the same taxon.

Identification of organisms is an essential part of taxonomy because it enables scientists to communicate effectively and be confident that they are discussing the same organism. Further, identification is often essential for treatment of groups of diseases such as meningitis and pneumonia, which can be caused by pathogens as different as fungi, bacteria, and viruses.

In this section we examine the historical basis of taxonomy, consider modern advances in this field, and briefly consider various taxonomic methods. This chapter also presents a general overview of the taxonomy of prokaryotes (which are considered in greater detail in Chapter 11); of animals, protozoa, fungi, and algae (Chapter 12); and of viruses, viroids, and prions (Chapter 13).

Linnaeus, Whitaker and Taxonomic Categories

Learning Objectives

✓ Discuss the difficulties in defining species of microorganisms.
✓ List the hierarchy of taxa from general to specific.
✓ Define binomial nomenclature.
✓ Describe a few modifications of the Linnaean system of taxonomy.

Our current system of taxonomy began in 1758 with the publication of *Systema Naturae* by the Swedish botanist Carolus Linnaeus. Until his time, the names of organisms were often strings of descriptive terms that varied from country to country and from one scientist to another. Linnaeus provided a system that standardized the naming and classification of organisms based on characteristics they have in common. He grouped similar organisms that can successfully interbreed into categories called **species.**

The definition of species as "a group of organisms that interbreed to produce viable offspring" works relatively well for more complex, sexually reproducing organisms, but it is not satisfactory for asexual organisms, including most microorganisms. As a result, for asexual organisms many scientists define a species as a collection of *strains*—populations of cells that arose from a single cell—that share many stable properties and differ from other strains. Not surprisingly, this definition sometimes results in disagreements and inconsistencies in the classification of microbial life.

Table 4.3   Stains Used for Light Microscopy

Type of Stain	Examples	Results	Typical Images	Representative Uses
Simple Stain (uses a single dye)	Crystal violet Methylene blue	Uniform purple stain Uniform blue stain		Reveals size, morphology, and arrangement of cells
Differential Stains (use two or more dyes to differentiate between cells or structures)	Gram stain	Gram-positive cells are purple; gram-negative cells are pink		Differentiates between Gram-positive and Gram-negative bacteria, which is typically the first step in their identification
	Ziehl-Neelsen acid-fast stain	Pink to red acid-fast cells and blue non-acid-fast cells		Distinguishes the genera *Mycobacterium* and *Nocardia* from other bacteria
	Schaeffer-Fulton endospore stain	Green endospores and pink to red vegetative cells		Highlights the presence of endospores produced by species in the genera *Bacillus* and *Clostridium*
Special Stains	Negative stain for capsules	Background is dark, cells unstained or stained with simple stain		Reveals bacterial capsules
	Flagellar stain	Bacterial flagella become visible		Allows determination of number and location of bacterial flagella

In Linnaeus's system, which forms the basis of modern taxonomy, similar species are grouped into **genera,**[6] and similar genera into still larger taxonomic categories. That is, genera sharing common features are grouped together to form **families;** similar families are grouped into **orders;** orders are grouped into **classes;** classes into **phyla,**[7] and phyla into **kingdoms (Figure 4.25).**

All these categories, including species and genera, are taxa, which are hierarchical; that is, each successive taxon has a broader description than the preceding one, and each taxon includes all the taxa beneath it. The rules of nomenclature require that all taxa have Latin or Latinized names, in part because the language of science during Linnaeus's time was Latin, and in part because using Latin ensures that no country or ethnic group has priority in the language of taxonomy. The name *Chondrus crispus* (kon'drŭs krisp'ŭs) describes the exact same algal species all over the world, despite the fact that in England its common name is Irish moss, in Ireland it is carragheen, in North America it is curly moss, and it isn't really a moss at all!

When new microscopic, genetic, or biochemical techniques identify new or more detailed characteristics of organisms, a taxon may be split into two or more taxa. Alternatively, several taxa may be lumped together into a single taxon. For example, the genus "*Diplococcus*" has been united (synonymized) with the genus *Streptococcus* (strep-tō-kok'ŭs), and the name *Diplococcus pneumoniae* has been changed to reflect this synonymy—to *Streptococcus pneumoniae.*

Linnaeus assigned each species a descriptive name consisting of its genus name and a **specific epithet.** The genus name is always a noun, and it is written first and capitalized. The specific epithet always contains only lowercase letters and is usually an adjective. Both names, together called a *binomial,* are either printed in italics or underlined. Because the Linnaean system assigns two names to every organism, it is said to use **binomial**[8] **nomenclature.**

Consider the following examples of binomials. *Enterococcus faecalis* (en'ter-ō-kok'ŭs fē-kǎ'lis)[9] is a fecal bacterium. Whereas *Enterococcus faecium* (fē-sē'ŭm) is in the same genus, it is a different species because of certain differing characteristics, so it is given a different specific epithet. Humans are classified as *Homo sapiens* (hō'mō sā'pē-enz); the genus name means "man" and the specific epithet means "wise." The genus *Homo* contains no other living species, but scientists have assigned some fossil remains to *Homo neanderthalensis* (nē-an'der-thol-en'sis).

Note that even though binomials are often descriptive of an organism, sometimes they can be misleading. For instance, *Haemophilus influenzae* (hē-mof'i-lŭs in-flū-en'zē) does not cause influenza. In some cases, binomials honor

---

[6]Plural of Latin *genus,* meaning race or birth.
[7]Plural of *phylum,* from Greek *phyllon,* meaning tribe. The term *phylum* is used for animals and bacteria; the corresponding taxon in mycology and botany is called a *division.*
[8]From Greek *bi,* meaning two, and *nomos,* meaning rule.
[9]From Greek *enteron,* meaning intestine, and *kokkos,* meaning berry.

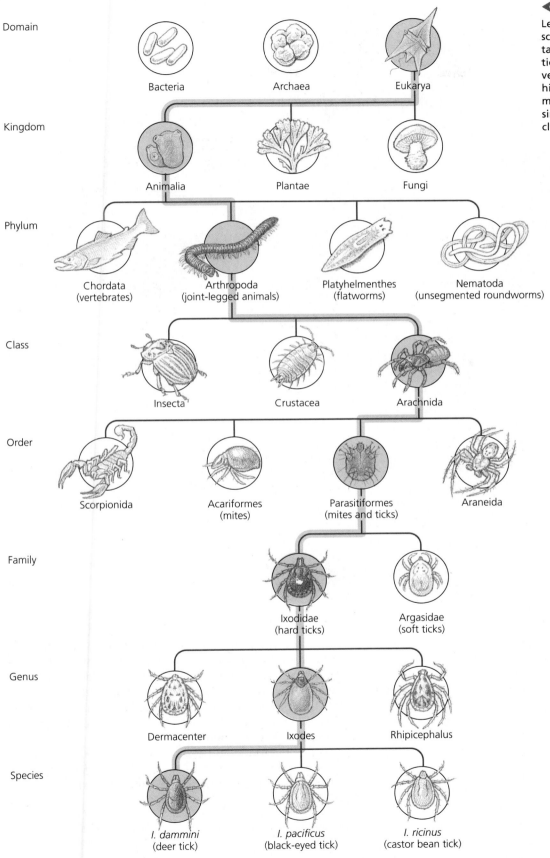

Levels in a Linnaean taxonomic scheme. Illustrated here are the taxonomic categories for the deer tick, *Ixodes dammini,* the primary vector of Lyme disease. Notice that higher taxonomic categories are more inclusive than lower ones. To simplify the diagram, only selected classification possibilities are shown.

Domain

Bacteria   Archaea   Eukarya

Kingdom

Animalia   Plantae   Fungi

Phylum

Chordata
(vertebrates)   Arthropoda
(joint-legged animals)   Platyhelmenthes
(flatworms)   Nematoda
(unsegmented roundworms)

Class

Insecta   Crustacea   Arachnida

Order

Scorpionida   Acariformes
(mites)   Parasitiformes
(mites and ticks)   Araneida

Family

Ixodidae
(hard ticks)   Argasidae
(soft ticks)

Genus

Dermacenter   Ixodes   Rhipicephalus

Species

I. dammini
(deer tick)   *I. pacificus*
(black-eyed tick)   *I. ricinus*
(castor bean tick)

Figure 4.26 ➤
Whittaker's five-kingdom taxonomic scheme.

people. Examples include *Pasteurella haemolytica* (pas-ter-el′ ă hē-mō-lit′i-kă), a bacterium named after the microbiologist Louis Pasteur; *Escherichia coli* (esh-ĕ-rik′ē-ă kō-lī), a bacterium named after the physician Theodor Escherich; and *Isabella abottae* (iz-a-bel′ah ab′ot-tē), a marine alga named after the phycologist and taxonomist Isabella Abbott.

CRITICAL THINKING

Examine the binomials of the species discussed in this section. What do the genus names and specific epithets indicate about the organisms?

Most scientists still use the Linnaean system today, though significant modifications have been adopted. For example, scientists sometimes use additional categories, such as tribes, sections, subfamilies, and subspecies.[10] Further,

Linnaeus divided all organisms into only two kingdoms (Plantae and Animalia), and he did not know of the existence of viruses and smaller pathogens. As scientists learned more about organisms, they adopted taxonomic schemes to reflect these advances in knowledge. For example, in 1959, Robert Whittaker (1920–1980) proposed a widely accepted taxonomic approach based on five kingdoms: Animalia, Plantae, Fungi, Protista,[11] and Prokaryotae (also called Monera) **(Figure 4.26).** Though widely accepted, this scheme groups together in a single kingdom (Protista) such obviously

---

[10]Subspecies, which are also called *varieties* or *strains,* differ only slightly from each other. Not all taxonomists agree on how much difference between two microbes constitutes a strain versus a species.
[11]The Kingdom Protista consists of all algae, slime molds, and single-celled animal-like organisms.

Woese's Taxonomic Revolution

Carl Woese labored for years to understand the taxonomic relationships among prokaryotic cells. Morphology (shape) and biochemical tests did not provide enough information to classify prokaryotes fully, so for over a decade Woese painstakingly sequenced the nucleotides of prokaryotic ribosomal RNA (rRNA) in an effort to unravel the relationships among these organisms.

Then, in 1976, he sequenced rRNA from an odd group of prokaryotes that produce methane gas as a metabolic waste. Woese was surprised when their rRNA did not contain nucleotide sequences characteristic of bacteria. At first he thought he had made a mistake in handling the samples, but repeated testing showed that methanogens, as they are called, were not like other prokaryotic or eukaryotic organisms. They were something new to science, a third branch of life. Woese had discovered the Archaea, ushering microbiologists into a new and curious wonderland of microbes.

Ribosomal nucleotide sequences suggest that there may be at least 50 kingdoms of Bacteria and three kingdoms of Archaea. Further, scientists examining substances such as human saliva, water,

soil, and rock regularly discover novel nucleotide sequences. When these sequences are compared to known sequences stored in a computer database, they cannot be associated with any previously identified organism. This suggests that many curious new forms of microbial life have never been grown in a laboratory and still await discovery.

CRITICAL THINKING

In 1999, microbiologists at Stanford University announced the discovery of 31 new species of bacteria that thrive between the teeth and gums of humans. The bacteria could be not be grown in the researcher's laboratories, nor were any of them ever observed via any kind of microscopy.

If they couldn't culture them or see them, how could the researchers know they had discovered 31 new species? If they couldn't examine the cells for the presence of a nucleus, how did they determine that the organisms were prokaryotes and not eukaryotes?

---

disparate organisms as massive brown seaweeds (kelps) and unicellular protozoa. Further, because it does not address the taxonomy of viruses nor all the differences between microorganisms, scientists have proposed other taxonomic schemes that have from four to more than 50 kingdoms.

Sometimes students are upset that taxonomists do not agree about all the taxonomic categories or the species they contain, but it must be realized that classification of organisms reflects the state of our current knowledge and theories. Taxonomists change their schemes to accommodate new information, and not every expert agrees with every proposed modification.

Another significant development in taxonomy is a shift in its basic goal—from Linnaeus's goal of classifying and naming organisms as a means of cataloging them, to the more modern goal of understanding the relationships among groups of organisms. Linnaeus based his taxnomic scheme primarily on organisms' structural similarities, and whereas such terms as *genus* and *family* may suggest the existence of some common lineage, he and his contemporaries thought of species as divinely created and generally immutable entities. However, when Charles Darwin (1809–1882) propounded his theory of the evolution of species by natural selection (a century after Linnaeus published his pivotal work on the taxonomy of plants), taxonomists came to consider that common ancestry largely explains the similarities among organisms in the various taxa. Today, most taxonomists agree that a major goal of modern taxonomy is to reflect a *phylogenetic*[12] *hierarchy;* that is, that the ways in which organisms are grouped reflect their derivation from common ancestors. The spatial arrangement of kingdoms

seen in Figure 4.26 (and the placement of smaller taxa within them, as well) reflects this phylogenetic approach to taxonomy.

Taxonomists' efforts to classify organisms according to their ancestry has resulted in reduced emphasis on comparisons of physical and chemical traits, and in greater emphasis on comparisons of their genetic material. Such work has led to a proposal to add a new, most inclusive taxon: the *domain*.

Domains

Learning Objective

✓ List and describe the three domains proposed by Carl Woese.

Spurred primarily by the work of Carl Woese (1928–) and his collaborators, taxonomists are now comparing the nucleotide sequences of the smaller ribosomal RNA (rRNA) subunits of both prokaryotes and eukaryotes. Because these rRNA molecules are present in all cells and are crucial to protein synthesis, changes in their nucleotide sequences are presumably very rare. By sequencing the nucleotides in rRNA, Woese and his coworkers discovered three basic type of ribosomes **(Highlight 4.2)**, which led them to propose a new classification scheme in which a new taxon, called a **domain,** contains (or in some schemes replaces) the Linnaean taxon of kingdom. The three domains identified by Woese— **Eukarya, Bacteria,** and **Archaea**—are based on three basic types of cells as determined by ribosomal nucleotide sequences.

Domain Eukarya includes all eukaryotic cells, all of which contain eukaryotic rRNA sequences. The domains Bacteria and Archaea include all prokaryotic cells. They contain bacterial and archaeal rRNA sequences, respectively,

---

[12]From Greek *phyllon,* meaning tribe, and Latin *genus,* meaning birth (*i.e.,* origin of a group).

which differ significantly from one another and from those in eukaryotic cells. In addition to differences in rRNA sequences, cells of the three domains differ in many other characteristics, including the lipids in their cell membranes, transfer RNA (tRNA) molecules, and sensitivity to antibiotics. The taxonomy of organisms within the three domains is discussed in Chapters 11 and 12, which cover prokaryotes and eukaryotes, respectively.

Ribosomal nucleotide sequences have also given microbiologists a new way to define prokaryotic species. Some scientists propose that prokaryotes whose rRNA sequence differs from that of other prokaryotes by more than 3% be classified as a distinct species. While this definition has the advantage of being precise, not all taxonomists agree with it.

Taxonomic and Identifying Characteristics

Learning Objective

✓ Describe five procedures used by taxonomists to identify and classify microorganisms.

The criteria and laboratory techniques used for classifying and identifying microorganisms are quite numerous and include macroscopic and microscopic examination of physical characteristics, differential staining characteristics, growth characteristics, microorganisms' interactions with antibodies, microorganisms' susceptibilities to viruses, nucleic acid analysis, biochemical tests, and organisms' environmental requirements, including the temperature and pH ranges of their various types of habitats. Clearly, then, microbial taxonomy is too broad a subject to cover in one chapter, and thus the details of the criteria for the classification of major groups are provided in subsequent chapters.

It is important to note that even though scientists may use a given technique to either classify or identify microorganisms, the criteria used for identifying a particular organism are not always the same as those that were used to classify it. For example, even though laboratory technologists distinguish the genus *Escherichia* from other bacterial genera by its inability to utilize citric acid (citrate) as a sole carbon source, this identifying characteristic was not vital in the classification of *Escherichia*.

Bergey's Manual of Determinative Bacteriology, first published in 1923 and now in its 9th edition (1994), contains information used for the identification of prokaryotes. *Bergey's Manual of Systematic Bacteriology* (2nd edition, 2000) is a similar reference work that is used for classification. Its scheme is based on ribosomal RNA sequences, which presumably reflect phylogenetic (evolutionary) relationships. Each of these volumes is known as "Bergey's Manual."

Linnaeus did not know of the existence of viruses and thus did not include them in his original taxonomic hierarchy; nor are viruses assigned to any of the five kingdoms or three domains because viruses are acellular and generally lack rRNA. Virologists do classify viruses into families and genera, but higher taxa are poorly defined for viruses. Viral taxonomy is discussed further in Chapter 13.

With this background, let's turn now to some brief discussions of five types of information microbiologists commonly use to distinguish among microorganisms: physical characteristics, biochemical tests, serological tests, phage typing, and analysis of nucleic acids.

Physical Characteristics

Many physical characteristics are used to identify microorganisms. Scientists can usually identify protozoa, fungi, algae, and parasitic worms based solely upon their *morphology* (shape). As we see in Chapter 6, lab technologists can also use the physical appearance of a bacterial colony[13] to help identify microorganisms. As we have discussed, stains are used to view the size and shape of individual bacterial cells and to show the presence or absence of identifying features such as endospores and flagella.

Linnaeus categorized prokaryotic cells into two genera based upon two prevalent shapes. He classified spherical prokaryotes in the genus "*Coccus*,"[14] and he placed rod-shaped cells in the genus *Bacillus*.[15] However, subsequent studies have revealed that there are vast differences between many of the thousands of spherical and rod-shaped prokaryotes, and thus physical characteristics alone are not sufficient for the classification of prokaryotes. Instead, taxonomists rely primarily on genetic differences as revealed by metabolic dissimilarities, and more and more frequently on rRNA sequences.

CRITICAL THINKING

Why is the genus name "*Coccus*" placed within quotation marks, but the genus name *Bacillus* is not?

Biochemical Tests

Microbiologists distinguish many prokaryotes that are similar in microscopic appearances and staining characteristics on the basis of differences in their ability to utilize or produce certain chemicals. Biochemical tests include procedures that determine an organism's ability to ferment various carbohydrates; utilize various substrates such as specific amino acids, starch, citrate, and gelatin; or produce waste products such as hydrogen sulfide (H_2S) gas **(Figure 4.27)**. Differences in fatty acid composition of bacteria are also used to distinguish between bacteria. Obviously, biochemical tests can be used to identify only those microbes that can be grown under laboratory conditions.

Laboratory technicians utilize biochemical tests to identify pathogens, allowing physicians to prescribe appropriate treatments. Many tests require that the microorganisms be *cultured* (grown) for 12 to 24 hours, though this time can be greatly reduced by the use of rapid identification tools. Such

---

[13]A group of bacteria that has arisen from a single cell grown on a solid laboratory medium.
[14]From Greek *kokkos*, meaning berry. This genus name has been supplanted by many genera, including *Staphylococcus*, *Micrococcus*, and *Streptococcus*.
[15]From Latin *bacillum*, meaning small rod.

Gas bubble   Inverted tubes
to trap gas

Acid with gas   Inert   Uninoculated
control

(a)

Hydrogen
sulfide produced

No hydrogen
sulfide

(b)

▲ *Figure 4.27*
Two biochemical tests for identifying bacteria. **(a)** A
carbohydrate utilization test. At left is a tube in which
the bacteria have metabolized a particular carbohydrate
to produce acid (which changes the color of a pH
indicator, phenol red, to red) and gas, as indicated by
the gas bubble. (Other species of bacteria produce acid
but no gas.) At center is a tube inoculated with bacteria
that are "inert" with respect to this test (they do not
metabolize the carbohydrate). At right is a control tube
into which no bacteria were inoculated. **(b)** A hydrogen
sulfide (H_2S) test. Bacteria that produce H_2S are
identified by the black precipitate formed by the
reaction of the H_2S with iron present in the medium.

(a) MicroScan instrument

Wells

(b) MicroScan® panel

▲ *Figure 4.28*
One tool for the rapid identification of bacteria, the
automated MicroScan® system. **(a)** The MicroScan®
instrument. **(b)** A MicroScan® panel, a plate containing
numerous wells, each the site of a particular
biochemical test. The instrument ascertains the identity
of the organism by reading the pattern of colors in the
wells after the biochemical tests have been performed.

tools exist for many groups of medically important
pathogens, such as Gram-negative bacteria in the family En-
terobacteraceae, Gram-positive bacteria, yeasts, and filamentous fungi. Automated systems for identifying
pathogens, such as the one shown in **Figure 4.28a,** read the
results of a whole battery of biochemical tests performed in
a plastic plate containing numerous small wells **(Figure
4.28b).** A color change in a well indicates the presence of a
particular metabolic reaction, and the machine reads the

Negative result   Positive result

(a)

Negative result   Positive result

(b)

▲ *Figure 4.29*

An agglutination test, one type of serological test. **(a)** In a positive agglutination test, visible clumps are formed by the binding of antibodies to their target antigens present on cells. **(b)** The processes involved in agglutination tests. In a negative result, antibody binding cannot occur because its specific target is not present; in a positive result, specific binding does occur. Note that agglutination occurs because each antibody molecule can bind to two antigen molecules.

pattern of colors in the plate to ascertain the identity of the pathogen.

Serological Tests

In the narrowest sense, *serology* is the study of serum, the liquid portion of blood after the clotting factors have been removed, and an important site of antibodies. In its most practical application, serology is the study of antigen-antibody reactions in laboratory settings. As we have seen, antibodies are immune system proteins that bind very specifically to target antigens. Antibodies and antigens are discussed in detail in Chapter 16; in this section we briefly consider the use of serological testing to identify microorganisms.

Many microorganisms are *antigenic*; that is, within a host organism they trigger an immune response that results in the production of antibodies. Suppose, for example, that a scientist injects a sample of *Borrelia burgdorferi* (bō-rē′lē-ă bŭrg-dōr′fer-ē), the bacterium that causes Lyme disease, into a rabbit. The bacterium has many surface proteins and carbohydrates that are antigenic because they are foreign to the rabbit. The rabbit responds to these foreign antigens by producing antibodies against them. These antibodies can be isolated from the rabbit's serum and concentrated into a solution known as an **antiserum.** Antisera bind to the antigens that triggered their production.

In a procedure called an **agglutination test,** antiserum is mixed with a sample that potentially contains its target cells. If the antigenic cells are present, antibodies in the antiserum will clump (*agglutinate*) the antigen **(Figure 4.29).** Other antigens, and therefore other organisms, remain unaffected because antibodies are highly specific for their targets.

Antisera can be used to distinguish among species, and even among strains of the same species. For example, a particularly pathogenic strain of *Escherichia coli* was classified and is identified by the presence of both antigen number 157 on its cell wall (designated O157) and antigen number 7 on

its flagella (designated H7). This pathogen, known as *E. coli* O157:H7, has caused several deaths in the United States over the past few years. Other serological tests, such as *enzyme-linked immunosorbent assay (ELISA)* and *Western blotting,* are discussed in Chapter 17.

Phage Typing

Bacteriophages (or simply **phages**) are viruses that infect and usually destroy bacterial cells. Just as antibodies are specific for their target antigens, phages are specific for the hosts they can infect. **Phage typing,** like serological testing, works because of such specificity. One bacterial strain may be susceptible to a particular phage while a related strain is not.

In phage typing, a technician spreads a solution containing the bacterium to be identified across a solid surface of growth medium and then adds small drops of solutions containing different types of bacteriophage. Wherever a specific phage is able to infect and kill bacteria, the resulting lack of bacterial growth produces within the bacterial lawn a clear area called a **plaque (Figure 4.30).** A microbiologist can identify an unknown bacterium by comparing the phages that form plaques with known phage-bacteria interactions.

Analysis of Nucleic Acids

As we have discussed, the sequence of nucleotides in nucleic acid molecules provides a powerful tool for the classification and identification of microbes. In many cases, nucleic acid analysis has confirmed classical taxonomic hierarchies. In other cases, as in Woese's discovery of domains, curious new organisms and relationships that were not obvious from classical methodologies have come to light. Techniques of nucleotide sequencing and comparison, such as *DNA base composition, polymerase chain reaction (PCR),* and *nucleic acid hybridization,* are best understood after we have discussed microbial genetics; therefore, these techniques are discussed in Chapter 8.

▲ *Figure 4.30*

Phage typing. Drops containing type A bacteriophages were added to this plate after its entire surface was inoculated with an unknown strain of *Salmonella.* After 12 hours of bacterial growth, clear zones, called plaques, developed where the phages killed bacteria. Given the great specificity of phage A for infecting and killing its host, the strain of bacterium can be identified as *Salmonella typhi* type A.

▼ *Figure 4.31*

Use of a dichotomous taxonomic key. The example presented here involves identification of the genera of potentially pathogenic, intestinal bacteria. **(a)** A sample key. To use it, choose the one statement in a pair that applies to the organism to be identified, and then either refer to another key (as indicated by words) or go to the appropriate place within this key (as indicated by a number). **(b)** A flowchart that shows the various paths that might be followed in using the key presented in part a. Highlighted is the path taken when the bacterium in question is *Escherichia.*

1a. Gram-positive cells.................................. Gram-positive bacteria
1b. Gram-negative cells................................ 2

2a. Rod-shaped cells..................................... 3
2b. Non-rod-shaped cells............................. Cocci and pleomorphic bacteria

3a. Can tolerate oxygen................................ 4
3b. Cannot tolerate oxygen........................... Obligate anaerobes

4a. Ferments lactose.................................... 5
4b. Cannot ferment lactose......................... Non-lactose fermenters

5a. Can use citric acid as a sole carbon source... 6
5b. Cannot use citric acid alone................... 8

6a. Produces hydrogen sulfide gas............... *Salmonella*
6b. Does not produce hydrogen sulfide gas.. 7

7a. Produces acetoin.................................... *Enterobacter*
7b. Does not produce acetoin....................... *Citrobacter*

8a. Produces gas from glucose..................... *Escherichia*
8b. Does not produce gas from glucose....... *Shigella*

(a)

Determination of the proportion of a cell's DNA that is guanine and cytosine, a quantity referred to as the cell's *GC ratio,* has also become a part of prokaryotic taxonomy. Scientists express the ratio as follows:

$$\frac{G + C}{A + T + G + C} \times 100\%$$

GC ratios vary from 20% to 80% among prokaryotes, and often (but not always) organisms that share other characteristics have similar GC ratios. Organisms that were once thought to be closely related but have widely different GC ratios are invariably not as closely related as thought.

Taxonomic Keys

As we have seen, taxonomists, medical clinicians, and researchers can use a wide variety of information, including morphology, chemical characteristics, and results from biochemical, serological, and phage typing tests, in their efforts to identify microorganisms, including pathogens. But how can all these characteristics and results be organized so that they can be used efficiently to identify an unknown organism? All this information is often arranged in **dichotomous keys,** which contain a series of paired statements worded so that only one of two "either/or" choices applies to any particular organism **(Figure 4.31)**. Based on which of the two

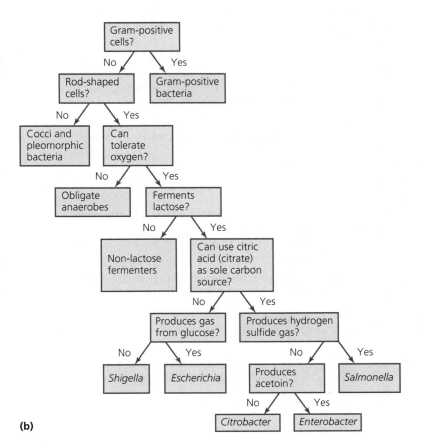

(b)

statements applies, the key either directs the user to another pair of statements, or it provides the name of the organism in question. Note that more than one key can be created to enable the identification of a given set of organisms, but all such keys involve mutually exclusive, "either/or" choices that send the user along a path that leads to the identity of the unknown organism.

CHAPTER SUMMARY

Units of Measurement (p. 93)

1. The metric system is a decimal system in which each unit is one tenth the size of the next largest unit.

2. The basic unit of length in the metric system is the **meter.**

Microscopy (pp. 93–105)

1. **Microscopy** refers to the passage of light or electrons of various **wavelengths** through lenses to **magnify** objects and provide **resolution** and contrast so that those objects can be viewed and studied.

2. **Immersion oil** is used to fill the space between the specimen and a lens to reduce light refraction and thus increase the **numerical aperture** and resolution.

3. Staining techniques and polarized light may be used to enhance **contrast** between the object and its background.

4. **Simple microscopes** contain a single magnifying lens, whereas **compound microscopes** use a series of lenses for magnification.

5. The lens closest to the object being magnified is the **objective lens,** several of which are mounted on a **revolving nosepiece.** The lenses closest to the eyes are **ocular lenses. Condenser lenses** lie beneath the stage and direct light though the slide.

6. The magnifications of the objective lens and the ocular lens are multiplied together to give **total magnification.**

7. A photograph of a microscopic image is a **micrograph.**

8. **Dark-field microscopes** provide a dark background for small or colorless specimens.

9. **Phase microscopes,** such as **phase-contrast** and **differential interference contrast microscopes,** cause light rays that pass through a specimen to be out of phase with light rays that pass through the field, producing contrast.

10. **Fluorescent microscopes** use ultraviolet light and fluorescent dyes to fluoresce specimens and enhance contrast.

11. A **confocal microscope** uses fluorescent dyes in conjunction with computers to provide three-dimensional images of a specimen.

12. A **transmission electron microscope (TEM)** provides an image produced by the transmission of electrons through a thinly sliced, dehydrated specimen.

13. A **scanning electron microscope (SEM)** provides a three-dimensional image by scattering electrons from the surface of a specimen.

14. Minuscule electronic probes are used in **scanning tunneling microscopes** and in **atomic force microscopes** to reveal details at the atomic level.

Staining (pp. 106–112)

1. **Staining** organisms with dyes for light microscopy involves making a **smear** or thin film of the specimens on a slide and then either passing the slide through a flame **(heat fixation)** or applying a chemical **(chemical fixation)** to attach the specimens to the slide. **Acidic dyes** or **basic dyes** are used to stain different portions of an organism to aid viewing and identification.

2. **Simple stains** involve the simple process of soaking the smear with one dye and then rinsing with water. **Differential stains** such as the **Gram stain, acid-fast stain,** and **endospore stain** use more than one dye to differentiate different cells, chemicals, or structures.

3. The Gram stain procedure includes use of a **primary stain,** a **mordant,** a **decolorizing agent,** and a **counterstain** that results in either purple (Gram-positive) or pink (Gram-negative) organisms, depending on the chemical structures of their cell walls.

4. The acid-fast stain is used to differentiate cells with waxy cell walls. **Endospores** are stained by the **Schaeffer-Fulton endospore stain** procedure.

5. Stains that stain the background and leave the cells colorless are called **negative** (or **capsule**) **stains.**

Classification and Identification of Microorganisms (pp. 112–122)

1. **Taxa** are nonoverlapping groups of organisms that are studied and named in **taxonomy.** Carolus Linnaeus invented a system of taxonomy, grouping similar interbreeding organisms into **species,** species into **genera,** genera into **families,** families into **orders,** orders into **classes,** classes into **phyla,** and phyla into **kingdoms.**

2. Linnaeus gave each species a descriptive name consisting of a genus name and **specific epithet.** This practice of naming organisms with two names is called **binomial nomenclature.**

3. Carl Woese proposed the existence of three taxonomic **domains** based on three cell types revealed by rRNA sequencing: **Eukarya, Bacteria,** and **Archaea.**

4. Taxonomists rely primarily on genetic differences revealed by metabolic dissimilarities to classify organisms. Species or strains within species may be distinguished by using **antisera, agglutination tests,** nucleic acid analysis, or **phage typing** with **bacteriophages,** in which unknown bacteria are identi-fied by observing **plaques** (regions of a bacterial lawn where the phage has killed bacterial cells).

5. Microbiologists use **dichotomous keys,** which involve step-wise choices between paired characteristics, to assist them in identifying microbes.

QUESTIONS FOR REVIEW

(Answers to multiple choice and fill in the blanks are on the web, along with additional review questions. Visit www.microbiologyplace.com.)

Multiple Choice

1. Which of the following is smallest?
   a. decimeter
   b. millimeter
   c. nanometer
   d. micrometer

2. A nanometer is _____ than a micrometer.
   a. 10 times larger
   b. 10 times smaller
   c. 1000 times larger
   d. 1000 times smaller

3. Resolution is best described as
   a. the ability to view something that is small.
   b. the ability to magnify a specimen.
   c. the ability to distinguish between two adjacent objects.
   d. the difference between two waves of electromagnetic radiation.

4. Curved glass lenses _____ light.
   a. refract
   b. bend
   c. magnify
   d. both a and b

5. Which of the following factors is important in making an image appear larger?
   a. the thickness of the lens
   b. the curvature of the lens
   c. the speed of the light passing through the lens
   d. all of the above

6. Which of the following is different between light microscopy and transmission electron microscopy?
   a. magnification
   b. resolution
   c. wavelengths
   d. all of the above

7. Which of the following types of microscopes produces a three-dimensional image with a shadowed appearance?
   a. simple microscope
   b. Nomarski microscope
   c. fluorescent microscope
   d. transmission electron microscope

8. Which of the following microscopes combines the greatest magnification with the best resolution?
   a. confocal microscope
   b. probe microscope
   c. TEM
   d. SEM

9. Negative stains such as eosin are also called
   a. capsule stains.
   b. endospore stains.
   c. simple stains.
   d. acid-fast stains.

10. In the binomial system of nomenclature, which term is always written in lowercase letters?
    a. kingdom
    b. domain
    c. genus
    d. specific epithet

Fill in the Blanks

1. If an objective magnifies 40×, and each binocular lens magnifies 15×, the total magnification of the object being viewed is _____.

2. The type of fixation developed by Koch for bacteria is _____.

3. Immersion oil _____ (increases/decreases) the numerical aperture, which _____ (increases/decreases) resolution because _____ (more/fewer) light rays are involved.

4. _____ refers to differences in intensity between two objects.

5. Cationic chromophores such as methylene blue ionically bond to _____ (positively/negatively) charged chemicals such as DNA and proteins.

Short Answer

1. Explain how the principle "electrons travel as waves" applies to microscopy.

2. Critique the following definition of magnification given by a student on a microbiology test: "Magnification makes things bigger."

3. Why can electron microscopes magnify only dead organisms?

4. Put the following substances in the order they are used in a Gram stain: counterstain, decolorizing agent, mordant, primary stain.

5. Why is Latin used in taxonomic nomenclature?

6. Give three characteristics of a "specific epithet."

7. How does the study of the nucleotide sequences of ribosomal RNA fit into a discussion of taxonomy?

8. An atomic force microscope can magnify a living cell, whereas electron microscopes and scanning tunneling microscopes cannot. What requirement of electron and scanning tunneling microscopes precludes the imaging of living specimens?

CRITICAL THINKING

1. Miki came home from microbiology lab with very green fingers and a bad grade. When asked about this, she replied that she was doing a Gram stain but it never worked the way the book said it should. Boone overheard the conversation and said that she must have used the wrong chemicals. What dye was she probably using, and what structure does that chemical normally stain?

2. Why is the definition of species as "successfully interbreeding organisms" not satisfactory for most microorganisms?

3. With the exception of the discovery of new organisms, is it logical to assume that taxonomy as we currently know it will stay the same? Why or why not?

4. A novice microbiology student incorrectly explains that immersion oil increases the magnification of his microscope. What is the function of immersion oil?

The next time you bite into a delicious piece of chocolate, consider this: Microbial metabolism played a key role in how that chocolate came to taste so good.

Chocolate comes from cacao seeds, found inside the pods of *Theobroma cacao* trees. After the pods are split open, the seeds and the surrounding pulp are scooped out and placed in heaps on top of plantain or banana leaves. The heaps are then covered and left to ferment for 2–7 days. Fermentation occurs as microorganisms—including yeast and several kinds of bacteria—grow on the fleshy, sugary pulp. Bathed in fermenting pulp, the cacao seeds (which start off tasting bitter) begin to develop the flavors and colors that we associate with chocolate. After fermentation, the seeds are dried, roasted, and then processed further by chocolate manufacturers before becoming one of our favorite desserts.

In this chapter we will learn about many metabolic processes of microorganisms, including fermentation.

Microbial Metabolism

CHAPTER 5

CHAPTER OUTLINE

Basic Chemical Reactions Underlying Metabolism
Catabolism and Anabolism
Oxidation and Reduction Reactions
ATP Production and Energy Storage
The Roles of Enzymes in Metabolism

Carbohydrate Catabolism
Glycolysis
Alternatives to Glycolysis
Cellular Respiration
Fermentation

Other Catabolic Pathways
Lipid Catabolism
Protein Catabolism

Photosynthesis
Chemicals and Structures
Light-Dependent Reactions
Light-Independent Reactions

Other Anabolic Pathways
Carbohydrate Biosynthesis
Lipid Biosynthesis
Amino Acid Biosynthesis
Nucleotide Biosynthesis

Integration and Regulation of Metabolic Functions

A worker prepares cacao seeds for fermentation.

MicroPrep Pre-Test: Take the pre-test for this chapter on the web.
Visit: ***www.microbiologyplace.com***.

How do pathogens acquire energy and nutrients

at the expense of a patient's health? How does grape juice turn into wine, and how does yeast cause bread to rise? How do disinfectants, antiseptics, and antimicrobial drugs work? When laboratory personnel perform biochemical tests to identify unknown microorganisms and help diagnose disease, what exactly are they doing?

The answers to all of these questions require an understanding of microbial **metabolism,**[1] the collection of controlled biochemical reactions that takes place within the cells of an organism. While it is true that metabolism in its entirety is complex, consisting of thousands of chemical reactions and control mechanisms, the reactions are nevertheless elegantly logical and can be understood in a simplified form. In this chapter we will concern ourselves only with the central metabolic pathways and energy metabolism.

Your study of metabolism will be manageable if you keep in mind that the ultimate function of metabolism is to reproduce the organism, and that metabolic processes are guided by the following eight elementary statements:

1. Every cell acquires *nutrients,* which are the chemicals necessary for metabolism.

2. Metabolism requires energy from light or from the *catabolism* (kă-tab′ō-lizm), or breakdown, of acquired nutrients.

3. Energy is stored in the chemical bonds of *adenosine triphosphate (ATP).*

4. Using *enzymes,* cells catabolize nutrient molecules to form elementary building blocks called *precursor metabolites.*

5. Using these precursor metabolites, energy from ATP, and other enzymes, cells construct larger building blocks in *anabolic* (an-ă-bol′ik), or biosynthetic, reactions.

6. Cells use enzymes and additional energy from ATP to link building blocks together to form macromolecules in *polymerization* reactions.

7. Cells grow by assembling macromolecules into cellular structures such as ribosomes, membranes, and cell walls.

8. Cells typically reproduce once they have doubled in size.

We will discuss each aspect of metabolism in the chapters that most directly apply. For instance, we discussed the first step of metabolism—the active and passive transport of nutrients into cells—in Chapter 3. In this chapter we will examine the importance of enzymes in catabolic and anabolic reactions, study the three ways that ATP molecules are synthesized, and show that catabolic and anabolic reactions are linked. We will also examine the catabolism of nutrient molecules; the anabolic reactions involved in the synthesis of carbohydrates, lipids, amino acids, and nucleotides; and a

[1]From Greek *metabole,* meaning change.

▲ *Figure 5.1*
Metabolism is composed of catabolic and anabolic reactions. Some energy released in catabolism is stored in ATP molecules, but most is lost as heat. Anabolic reactions require energy, typically provided by ATP. There is some heat loss in anabolism as well. The products of catabolism provide many of the building blocks for anabolic reactions; other anabolic building blocks must be acquired directly as nutrients.

few ways that cells control their metabolic activities. Genetic control of metabolism and the polymerization of DNA, RNA, and proteins are discussed in Chapter 7, and the specifics of cell division are covered in Chapters 11 and 12.

Basic Chemical Reactions Underlying Metabolism

In the following sections we will examine the basic concepts of catabolism, anabolism, and a special class of reactions called *oxidation-reduction reactions.* The latter involve the transfer of electrons between molecules. Then we will turn our attention briefly to the synthesis of ATP and energy storage before we discuss the organic catalysts called *enzymes,* which make metabolism possible.

Catabolism and Anabolism

Learning Objective

✓ Distinguish among metabolism, anabolism, and catabolism.

Metabolism, which is all of the chemical reactions in an organism, can be divided into two major classes of reactions: **catabolism** and **anabolism (Figure 5.1).** A series of reactions is called a *pathway.* Cells have *catabolic pathways,* which break larger molecules into smaller products, and *anabolic pathways,* which synthesize large molecules from the smaller products of catabolism. Even though catabolic and anabolic pathways are intimately linked in cells, it is often useful to study the two types of pathways as if they were separate.

When catabolic pathways break down large molecules, they release energy; that is, catabolic pathways are *exergonic* (ek-ser-gon'ik). Cells store some of this released energy in the bonds of ATP, though much of it is lost as heat. Another result of the breakdown of large molecules by catabolic pathways is the production of numerous smaller molecules, some of which are **precursor metabolites** of anabolism. Some organisms, such as *Escherichia coli* (esh-ĕ-rik'ē-ă kō'lī), can synthesize everything in their cells from these precursor metabolites; other organisms must acquire some anabolic building blocks as nutrients. Note that catabolic *pathways,* but not necessarily *individual* catabolic *reactions,* produce ATP and metabolites; a given catabolic pathway may produce ATP, or metabolites, or both. An example of a catabolic pathway is the breakdown of lipids into glycerol and fatty acids.

Anabolic pathways are functionally the opposite of catabolic pathways in that they synthesize macromolecules and cellular structures. Because building anything requires energy, anabolic pathways are *endergonic* (en-der-gon'ik); that is, they require more energy than they release. The energy required for anabolic pathways usually comes from ATP molecules produced during catabolism. An example of an anabolic pathway is the synthesis of lipids for cell membranes from glycerol and fatty acids.

To summarize, then, a cell's metabolism involves both catabolic pathways that break down macromolecules to supply molecular building blocks and energy in the form of ATP, and anabolic pathways that use the building blocks and ATP to synthesize macromolecules needed for growth and reproduction.

Oxidation and Reduction Reactions

Learning Objective

✓ Contrast reduction and oxidation reactions.

Many metabolic reactions involve the transfer of electrons from a molecule that donates an electron (called an *electron donor*) to a molecule that accepts an electron (called an *electron acceptor*). Such electron transfers are called **oxidation-reduction reactions** or **redox reactions.** The reactions in which electrons are accepted are *reduction reactions,* whereas the reactions in which electrons are donated are *oxidation reactions* **(Figure 5.2).** Electron acceptors are said to be *reduced* because their gain in electrons reduces their overall electrical charge (that is, they are more negatively charged). Molecules that donate electrons are said to be *oxidized* because frequently their electrons are donated to oxygen atoms.

Reduction and oxidation reactions are always coupled (as represented in Figure 5.2) because every electron that is gained by one molecule must be donated by some other molecule. A molecule may be reduced by gaining either a simple electron or an electron that is part of a hydrogen atom—which, as we saw in Chapter 2, is composed of one

▲ **Figure 5.2**

Oxidation-reduction or redox reactions. When electrons are transferred from donor molecules to acceptor molecules, donors become oxidized and acceptors become reduced. *Why are acceptor molecules said to be reduced when they are gaining electrons?*

Figure 5.2 "Reduction" refers to the overall electrical charge on a molecule. Because electrons have a negative charge, the gain of an electron reduces the molecule's overall charge.

proton and one electron. (**New Frontiers 5.1** on page 128 describes an interesting example of how some prokaryotes are able to reduce gold dissolved in solution.) In contrast, a molecule may be oxidized in one of three ways: by losing a simple electron, by losing a hydrogen atom, or by gaining an oxygen atom. Because biological oxidations often involve the loss of hydrogen atoms, such reactions are also called *dehydrogenation* (dē-hī'drō-jen-ā'shŭn) *reactions.*

Electrons rarely exist freely in cytoplasm; instead, they orbit atomic nuclei. Therefore, cells use electron carrier molecules to carry electrons (often in hydrogen atoms) from one location in a cell to another. Three important electron carrier molecules, which are derived from vitamins, are **nicotinamide adenine dinucleotide (NAD$^+$), nicotinamide adenine dinucleotide phosphate (NADP$^+$),** and **flavine adenine dinucleotide (FAD).** Cells use each of these molecules in specific metabolic pathways to carry pairs of electrons. One of the electrons carried by either NAD$^+$ or NADP$^+$ is part of a hydrogen atom, forming NADH and NADPH. FAD carries two electrons as hydrogen atoms (FADH$_2$). Many metabolic pathways, including those that synthesize ATP, require such electron carrier molecules.

ATP Production and Energy Storage

Learning Objective

✓ Compare and contrast the three types of ATP phosphorylation.

Nutrients contain energy, but that energy is spread throughout their chemical bonds and generally is not concentrated enough for use in anabolic reactions. During catabolism, organisms release energy from nutrients that can then be concentrated and stored in high-energy phosphate bonds of molecules such as ATP. This happens by a general process called *phosphorylation* (fos' fōr-i-lā'shŭn), in which inorganic phosphate (PO$_4{}^{2-}$) is added to a substrate. For example,

New Frontiers 5.1   Gold-Mining Microbes

Gold, as found in nature, exists in two forms: gold-ore deposits, usually found near the Earth's crust, and gold dissolved in solution, as found in thermal springs and in seawater. Dissolved gold (which is gold in its oxidized form) is largely useless to humans; it cannot be converted easily or inexpensively into the valuable objects that we produce from solid gold (which is gold in its reduced form). Even though gold in either form is toxic when ingested by most living things, scientists have recently discovered that certain bacteria and archaea can metabolize dissolved gold. When placed in a solution containing gold, these microorganisms reduce the dissolved gold and shed flecks of solid gold as metabolic waste.

Solid gold is gold in its reduced form.

Scientists have long known that some prokaryotes can transfer electrons from an electron donor (commonly hydrogen) to metals such as iron and uranium, in the process reducing these metals. This knowledge led a team of researchers from the Uni-

versity of Massachusetts, Amherst, headed by microbiologist Derek Lovley, to hypothesize that such microorganisms might also be able to transfer electrons to gold in its dissolved form, thereby reducing it and precipitating solid gold. Focusing on microorganisms known for their ability to reduce iron, Lovley's team found that although not all iron-reducing microbes could also reduce gold, some could, including *Pyrobaculum islandicum* and *Pyrococcus furiosus* (archaea), and *Thermotoga maritima* and *Shewanella algae* (bacteria). Lovley's research thus suggests that microorganisms may play a role in the formation of some gold-ore deposits.

Entrepreneurial minds may wonder if this research also has practical—that is, potentially profitable—applications. While it is true that a great deal of dissolved gold is found in thermal springs and oceans, the gold is very dilute—only minute amounts are present in very large volumes of water. Moreover, were someone to perfect a way of using microorganisms to convert dissolved gold to great quantities of solid gold, they would be wise to keep it to themselves: So much solid gold could become available that its market value would plunge dramatically.

Reference: Kashefi, K., J. M. Tor, K. P. Nevin, and D. R. Lovley. 2001. Reductive Precipitation of Gold by Assimilatory Fe (III)-Reducing Bacteria and Archaea. *Applied and Environmental Microbiology.* 67(7):3275–9.

cells phosphorylate adenosine diphosphate (ADP), which has two phosphate groups, to form adenosine triphosphate ATP, which has three phosphate groups **(Figure 5.3).**

As we will examine in the following sections, cells phosphorylate ADP to form ATP in three specific ways:

- *Substrate-level phosphorylation* (see page 139), which involves the transfer of phosphate to ADP from another phosphorylated organic compound
- *Oxidative phosphorylation* (see page 147), in which energy from redox reactions of *respiration* (described shortly) is used to attach inorganic phosphate to ADP
- *Photophosphorylation* (see page 154), in which light energy is used to phosphorylate ADP with inorganic phosphate

We will investigate each of these in more detail as we proceed through the chapter.

In summary, after ADP is phosphorylated to produce ATP, anabolic pathways use some energy of ATP by breaking a phosphate bond (which re-forms ADP). Thus the cyclical interconversion of ADP and ATP functions somewhat like rechargeable batteries: ATP molecules store energy from light (in photosynthetic organisms) and from catabolic reactions and then release stored energy to drive cellular processes (including anabolic reactions, active transport, and movement). Reformed ADP molecules can be "recharged" to ATP again and again (Figure 5.3).

The Roles of Enzymes in Metabolism

Learning Objectives

✓ Draw a table listing the six basic types of enzymes, their activities, and an example of each.

✓ Describe the components of a holoenzyme, and contrast protein and RNA enzymes.

✓ Define *activation energy, enzyme, apoenzyme, cofactor, coenzyme, active site,* and *substrate,* and describe their roles in enzyme activity.

✓ Describe how temperature, pH, substrate concentration, and competitive and noncompetitive inhibition affect enzyme activity.

As we saw in Chapter 2, reactions occur when chemical bonds are broken or formed between atoms. In catabolic reactions, a bond must be destabilized before it will break, whereas in anabolic reactions reactants must collide with sufficient energy before bonds will form between them. In anabolism, increasing either the concentrations of reactants or ambient temperatures will increase the number of collisions and produce more chemical reactions; however, in living organisms, neither reactant concentration nor temperature is usually high enough to ensure that bonds will form. Therefore, the chemical reactions of life depend upon *catalysts,* which are chemicals that increase the likelihood of a reaction but are not permanently changed in the process. Organic catalysts are known as **enzymes.**

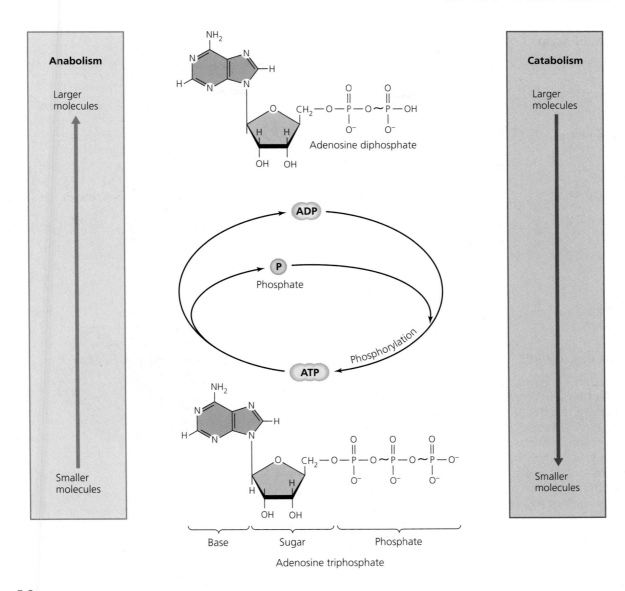

▲ *Figure 5.3*

Phosphorylation of ADP to form ATP. Cells add phosphate via a high energy bond to ADP, making ATP. During anabolism, ATP gives up its energy and phosphate to become ADP, which is then recycled to ATP.

Naming and Classifying Enzymes

The names of enzymes usually end with the suffix "-ase," and the name of each enzyme often incorporates the name of that enzyme's **substrate,** which is the molecule the enzyme acts upon. Based on their mode of action, enzymes can be grouped into six basic categories:

1. *Hydrolases* catabolize molecules by adding water in a decomposition process known as *hydrolysis.* Hydrolases are used primarily in the depolymerization of macromolecules.

2. *Isomerases*[2] rearrange the atoms within a molecule but do not add or remove anything (so they are neither catabolic nor anabolic).

3. *Ligases* or *polymerases* join two molecules together (and are thus anabolic). They often use energy supplied by ATP.

4. *Lyases* split large molecules (and are thus catabolic) without using water in the process.

5. *Oxidoreductases* remove electrons from (oxidize) or add electrons to (reduce) various substrates. They are used in both catabolic and anabolic pathways.

6. *Transferases* transfer functional groups, such as an amino group (NH_2), a phosphate group, or a two-carbon (acetyl) group, between molecules. Transferases can be anabolic.

---

[2]An isomer is a compound with the same molecular formula as another molecule, but with a different arrangement of atoms.

Table 5.1   **Enzyme Classification Based on Reaction Types**

Class	Type of Reaction Catalyzed	Example
Hydrolase	Hydrolysis (catabolic)	Lipase—breaks down lipid molecules
Isomerase	Rearrangement of atoms within a molecule (neither catabolic nor anabolic)	Phosphoglucoisomerase—converts glucose 6-phosphate into fructose 6-phosphate during glycolysis
Ligase or polymerase	Joining two or more chemicals together (anabolic)	Acetyl-CoA synthetase—combines acetate and coenzyme A to form acetyl-CoA for the Krebs cycle
Lyase	Splitting a chemical into smaller parts without using water (catabolic)	Fructose 1,6-bisphosphate aldolase—splits fructose 1,6-bisphosphate into G3P and DHAP during glycolysis
Oxidoreductase	Transfer of electrons or hydrogen atoms from one molecule to another	Lactic acid dehydrogenase—oxidizes lactic acid to form pyruvic acid during fermentation
Transferase	Moving a functional group from one molecule to another (may be anabolic)	Hexokinase—transfers phosphate from ATP to glucose in the first step of glycolysis

Table 5.1 summarizes these types of enzymes and gives examples of each.

The Makeup of Enzymes

Many enzymes are composed entirely of protein, but others are composed of protein portions, called **apoenzymes** (ap'ō-en-zīms), that are inactive if they are not bound to one or more nonprotein substances called cofactors. **Cofactors** are either inorganic ions (such as iron, magnesium, zinc, or copper ions) or certain organic molecules called coenzymes. All **coenzymes** (ko'-en'zīms) are either vitamins (or contain vitamins), which are organic molecules that are required for metabolism but cannot be synthesized by certain organisms (especially mammals). Some apoenzymes bind with inorganic cofactors, some bind with coenzymes, and some bind with both. The binding of an apoenzyme and its cofactor(s) forms an active enzyme, called a **holoenzyme** (hol-ō-en'zīm; **Figure 5.4**).

Table 5.2 lists several examples of inorganic cofactors and organic cofactors (coenzymes). Note that three important coenzymes are the electron carriers NAD^+, $NADP^+$, and FAD, which, as we have seen, carry electrons in hydrogen atoms from place to place within cells. We will examine more closely the roles of these coenzymes in the generation of ATP later in the chapter.

Not all enzymes are proteinaceous; some are RNA molecules called **ribozymes.** In eukaryotes, ribozymes process other RNA molecules by removing sections of RNA and splicing the remaining pieces together. Recently, researchers have discovered that the functional core of a ribosome is a ribozyme; therefore, given that ribosomes make all proteins, ribosomal enzymes make protein enzymes.

Enzyme Activity

Within cells, enzymes catalyze reactions by lowering the **activation energy,** which is the amount of energy needed to trigger a chemical reaction **(Figure 5.5).** Whereas heat can provide energy to trigger reactions, the temperatures

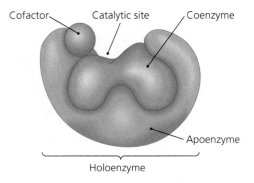

▲ *Figure 5.4*

Makeup of a protein enzyme. The combination of a proteinaceous apoenzyme with one or more cofactors forms a holoenzyme, which is the active form of the enzyme. The label "cofactor" represents either an inorganic ion or a coenzyme, which is an organic cofactor derived from a vitamin. The apoenzyme is inactive unless it is bound to its cofactors. *Name four metal ions that can act as cofactors.*

Figure 5.4 Iron, magnesium, zinc and copper ions can act as cofactors.

needed to reach activation energy for most metabolic reactions are often too high to allow cells to survive, so enzymes are needed if metabolism is to occur. This is true regardless of whether the enzyme is a protein or RNA, or whether the chemical reaction is anabolic or catabolic.

The activity of enzymes depends on the closeness of fit between the functional sites of an enzyme and its substrate. The shape of an enzyme's functional site, called its **active site,** is complementary to the shape of the substrate. Generally, the shapes and locations of only a few amino acids or nucleotides determines the shape of an enzyme's active site. A change in a single component—for instance, through mutation—can render an enzyme less effective or even completely nonfunctional.

Table 5.2   Representative Cofactors of Enzymes

Cofactors	Example of Use in Enzymatic Activity	Substance Transferred in Enzymatic Activity	Vitamin Source (of Coenzyme)
Inorganic (metal ion)			
Magnesium (Mg^{2+})	Forms bond with ADP during phosphorylation	Phosphate	None
Organic (coenzymes)			
Nicotinamide adenine dinucleotide (NAD^+)	Carrier of reducing power	Two electrons and a hydrogen ion	Niacin
Nicotinamide adenine dinucleotide phosphate ($NADP^+$)	Carrier of reducing power	Two electrons and a hydrogen ion	Niacin
Flavine adenine dinucleotide (FAD)	Carrier of reducing power	Two hydrogen atoms	Riboflavin
Tetrahydrofolate	Used in synthesis of nucleotides and some amino acids	One-carbon molecule	Folic acid
Coenzyme A	Formation of acetyl-CoA in Krebs cycle and beta-oxidation	Two-carbon molecule	Pantothenic acid
Pyridoxal phosphate	Transaminations in the synthesis of amino acids	Amine group	Pyridoxine
Thiamine pyrophosphate	Decarboxylation of pyruvic acid	Aldehyde group (CHO)	Thiamine

▲ *Figure 5.5*

The effect of enzymes on chemical reactions. Enzymes catalyze reactions by lowering the activation energy— that is, the energy needed to trigger the reaction.

This *enzyme-substrate specificity*, which is critical to enzyme activity, has been likened to the fit between a lock and key. This analogy is not completely apt because enzymes change shape slightly when they bind to their substrate, almost as if a lock could grasp its key once it has been inserted. This latter description of enzyme-substrate specificity is called the *induced fit model* (**Figure 5.6** on page 132).

In some cases, several different enzymes possess active sites that are complementary to various portions of a single substrate molecule. For example, an important precursor metabolite called phosphoenolpyruvic acid (PEP) is the substrate for at least five enzymes; depending on the enzyme involved, various products are produced from PEP. For instance, in a catabolic pathway PEP is converted to pyruvic acid, whereas in an anabolic pathway PEP is converted to the amino acid phenylalanine.

Although the exact ways that enzymes lower activation energy are not known, it appears that several mechanisms are involved. Some enzymes appear to bring reactants into sufficiently close proximity to enable a bond to form, whereas other enzymes change the shape of a reactant, inducing a bond to be broken. In any case, enzymes increase the likelihood that bonds will form or break.

The activity of enzymes is believed to follow the process illustrated in **Figure 5.7** on page 133, which depicts the catabolic lysis of a molecule called fructose 1,6-bisphosphate:

1 An enzyme associates with a specific substrate molecule having a shape that is complementary to that enzyme's active site.

2 The enzyme and its substrate bind to form a temporary intermediate compound called an *enzyme-substrate complex*. The binding of the substrate induces the enzyme to fit the shape of the substrate even more closely.

3 Bonds within the substrate are broken, forming two

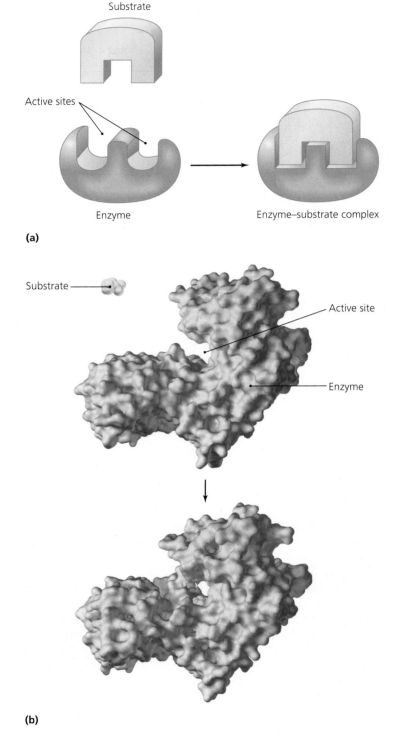

(a)

(b)

▲ *Figure 5.6*

(a) The induced-fit model of enzyme-substrate interaction. The enzyme's active site is generally complementary to the shape of its substrate, but a perfect fit between them does not occur until the substrate and enzyme bind to form an enzyme-substrate complex. **(b)** Space-filling models of an enzyme binding to a substrate.

(and in some other reactions, more than two) products. (Note that in an anabolic reaction, instead of the breakage of a bond, two reactants are linked together to form a single product.)

④ The enzyme disassociates from the newly formed molecules, which diffuse away from the site of the reaction, and the enzyme resumes its original configuration and is ready to associate with another substrate molecule.

Many factors influence the rate of enzymatic reactions, including temperature, pH, enzyme and substrate concentrations, and the presence of inhibitors.

Temperature   As mentioned, higher temperatures tend to increase the rate of most chemical reactions because molecules are moving faster and collide more frequently, which encourages bonds to form or break. However, this is not entirely true of enzymatic reactions, because the active sites of enzymes change shape as temperature changes. If the temperature rises too high or falls too low, an enzyme is often no longer able to achieve a fit with its substrate.

Each enzyme has an optimal temperature for its activity (**Figure 5.8a** on page 134). The optimum temperature for the enzymes in the human body is about 37°C, which is normal body temperature. Part of the reason certain pathogens can cause disease in humans is that the optimal temperature for the enzymes in those microorganisms is also 37°C. The enzymes of some other microorganisms, however, function best at much higher temperatures; this is the case for *hyperthermophiles*, organisms that grow best at temperatures above 80°C.

If temperature rises beyond a certain critical point, the noncovalent bonds within an enzyme (such as the hydrogen bonds between amino acids) will break, and the enzyme will **denature** (**Figure 5.9** on page 134). Denatured enzymes lose their specific three-dimensional structure, so they are no longer functional. Denaturation is said to be *permanent* when an enzyme cannot regain its original three-dimensional structure once conditions return to normal, much like the irreversible solidification of the protein albumin when egg whites are cooked and then cooled. In other cases denaturation is *reversible*—the denatured enzyme's noncovalent bonds re-form upon the return of normal conditions.

CRITICAL THINKING

Explain why thermophiles do not cause disease in humans.

pH   Extremes of pH also denature enzymes when ions released from acids and bases interfere with hydrogen bonding and distort and disrupt an enzyme's secondary and tertiary structures. Therefore, each enzyme has an optimal pH (see Figure 5.8b).

Changing the pH provides a way to control the growth of unwanted microorganisms by denaturing their proteins. For example, vinegar (acetic acid, pH 3.0) acts as a preserva-

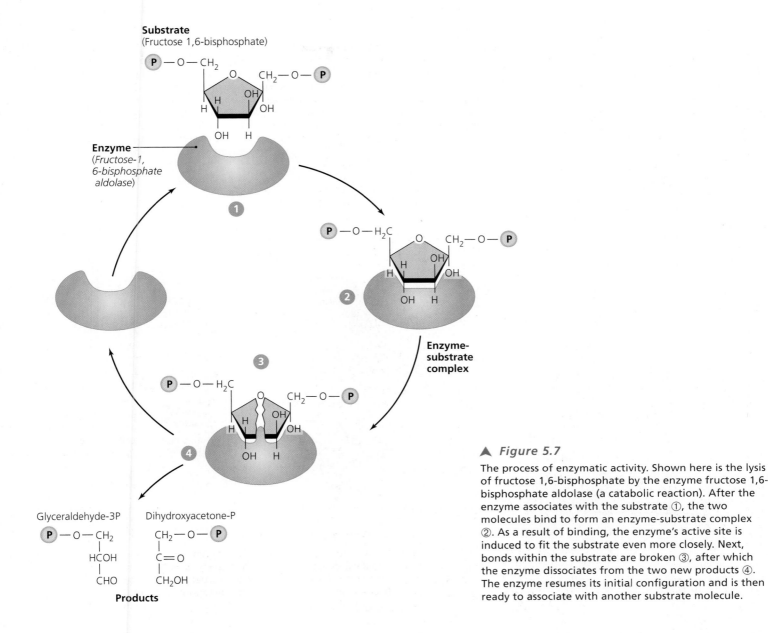

Substrate
(Fructose 1,6-bisphosphate)

Enzyme
(*Fructose-1,*
6-bisphosphate
aldolase)

Enzyme-substrate complex

Glyceraldehyde-3P

Dihydroxyacetone-P

Products

▲ *Figure 5.7*

The process of enzymatic activity. Shown here is the lysis of fructose 1,6-bisphosphate by the enzyme fructose 1,6-bisphosphate aldolase (a catabolic reaction). After the enzyme associates with the substrate ①, the two molecules bind to form an enzyme-substrate complex ②. As a result of binding, the enzyme's active site is induced to fit the substrate even more closely. Next, bonds within the substrate are broken ③, after which the enzyme dissociates from the two new products ④. The enzyme resumes its initial configuration and is then ready to associate with another substrate molecule.

tive in dill pickles, and ammonia (pH 11.5) can be used as a disinfectant.

CRITICAL THINKING

In addition to extremes in temperature and pH, other chemical and physical agents, including ionizing radiation, alcohol, enzymes, and heavy-metal ions, denature proteins. For example, the first antimicrobial drug, salvarsan, contained the heavy metal arsenic and was used to inhibit the enzymes of the bacterium *Treponema pallidum,* the causative agent of syphilis.

Given that both human and bacterial enzymes are denatured by heavy metals, how was salvarsan used to treat syphilis without poisoning the patient? Why is syphilis no longer treated with arsenic-containing compounds?

Enzyme and Substrate Concentration   Another factor that determines the rate of enzymatic activity within cells is the concentration of substrate present (see Figure 5.8c). As substrate concentration increases, enzymatic activity increases as more and more enzyme active sites bind more and more substrate molecules. Eventually, when all enzyme active sites have bound substrate, the enzymes have reached their saturation point, and the addition of more substrate will not increase the rate of enzymatic activity.

Obviously, the rate of enzymatic activity is also affected by the concentration of enzyme within cells. In fact, one way that organisms regulate their metabolism is by controlling the quantity and timing of enzyme synthesis. In other words, many enzymes are produced in the amounts and at the times they are needed to maintain metabolic activity. Chapter 7 discusses the role of genetic mechanisms in the regulation of enzyme synthesis. Additionally, eukaryotic

(a) Temperature

(b) pH

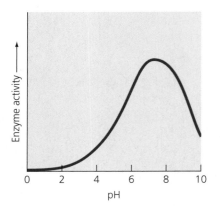

(c) Substrate concentration

◀ *Figure 5.8*

The effects of temperature, pH, and substrate concentration on enzyme activity.
(a) Rising temperature enhances enzymatic activity to a point, but above some optimal temperature the enzyme denatures and loses function. **(b)** Enzymes typically have some optimal pH, at which point enzymatic activity reaches a maximum. **(c)** At lower substrate concentrations, enzyme activity increases as the substrate concentration increases and as more and more active sites are utilized. At the substrate concentration at which all active sites are utilized, termed the saturation point, enzymatic activity reaches a maximum, and any additional increase in substrate concentration has no effect on enzyme activity. *What is the optimal pH of the enzyme shown in part (b)?*

Figure 5.8 The enzyme's optimal pH is approximately 7.

Functional protein   Denatured protein

▲ *Figure 5.9*

Denaturation of protein enzymes. Breakage of noncovalent bonds (such as hydrogen bonds) causes the protein to lose its secondary and tertiary structure and become denatured; as a result, the enzyme is no longer functional.

cells control some enzymatic activities by compartmentalizing enzymes inside membranes so that certain metabolic reactions proceed physically separated from the rest of the cell. For example, white blood cells catabolize phagocytized pathogens using enzymes packaged within lysosomes.

Inhibitors   Enzymatic activity can be influenced by a variety of inhibitory substances that block an enzyme's active site. Enzymatic inhibitors, which may be either competitive or noncompetitive, do not denature enzymes.

Competitive inhibitors are shaped such that they fit into an enzyme's active site and thus prevent the normal

substrate from binding **(Figure 5.10a).** However, such inhibitors do not undergo a chemical reaction to form products. Competitive inhibitors can bind permanently or reversibly to an active site. Permanent binding results in permanent loss of enzymatic activity; reversible competition can be overcome by an increase in the concentration of substrate molecules, which increases the likelihood that active sites will be filled with substrate instead of inhibitor **(Figure 5.10b).**

A good example of competitive inhibition is the action of sulfanilamide (found in sulfa drugs), which has a shape similar to that of para-aminobenzoic acid (PABA).

Sulfanilamide has great affinity for the active site of an enzyme required in the conversion of PABA

Sulfanilamide   PABA

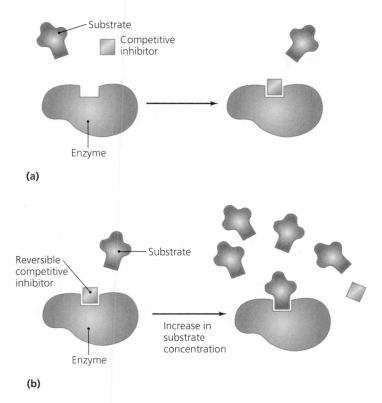

(a)

(b)

▲ *Figure 5.10*

Competitive inhibition of enzyme activity. **(a)** Inhibitory molecules, which are similar in shape to substrate molecules, compete for and block active sites.
(b) Reversible inhibition can be overcome by an increase in substrate concentration.

into the B vitamin folic acid, which is essential for DNA synthesis. Once sulfanilamide is bound to the enzyme, it stays bound. As a result, it prevents synthesis of folic acid. Sulfanilamide effectively inhibits bacteria that make folic acid from PABA without harming people because humans lack the necessary enzymes; they must acquire folic acid in their diets.

Noncompetitive inhibitors do not bind to the active site but instead prevent enzymatic activity by binding to an *allosteric* (al-ō-stār'ik) *site* located elsewhere on the enzyme. Binding at an allosteric site alters the shape of the active site so that substrate cannot be bound. Allosteric control of enzyme activity can take two forms: inhibitory and excitatory. *Allosteric (noncompetitive) inhibition* halts enzymatic activity in the manner just described **(Figure 5.11a)**. In *excitatory allosteric control,* the binding of certain activator molecules (such as a heavy-metal ion cofactor) to an allosteric site causes a change in shape of the active site, which activates an otherwise inactive enzyme **(Figure 5.11b)**. Some enzymes have several allosteric sites, both inhibitory and excitatory, which allows their function to be closely regulated.

Cells often control the action of enzymes through **feedback inhibition** (also called **negative feedback** or *end-product inhibition*). Allosteric feedback inhibition functions in

much the way a thermostat controls a heater. As the room gets warmer, a sensor inside the thermostat changes shape and sends an electrical signal that turns off the flame or electrical coil in the heater. Similarly, in metabolic feedback inhibition, the end-product of a series of reactions is an allosteric inhibitor of an enzyme in an earlier part of the pathway (**Figure 5.12a** on page 136). Because the product of each reaction in the pathway is the substrate for the next reaction, inhibition of the first enzyme in the series inhibits the entire pathway, thereby saving the cell energy. For example, in *Escherichia coli*, the presence of the amino acid isoleucine allosterically inhibits the first enzyme in the metabolic pathway that produces isoleucine. In this manner, the bacterium prevents the accumulation of isoleucine (and intermediate products) when the amino acid is available from the environment. When environmental isoleucine is depleted, the first metabolic enzyme is no longer inhibited, and isoleucine production resumes.

Feedback inhibition can occur in even more complex ways. For instance, even though the first step in the synthesis of the amino acids tyrosine, phenylalanine, and tryptophan is the same—the linkage of phosphoenol pyruvic acid (PEP) and erythrose 4-phosphate to form 3-deoxy-arabino-heptulosonic acid 7-phosphate (DAHAP)—three different

(a) Allosteric inhibition

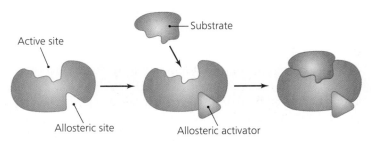

(b) Allosteric activation

▲ *Figure 5.11*

Allosteric control of enzyme activity. **(a)** Allosteric (noncompetitive) inhibition results from a change in the shape of the active site when an inhibitor binds to an allosteric site. **(b)** Allosteric activation results when the binding of an activator molecule to an allosteric site causes a change in the active site that makes it capable of binding substrate.

(a)

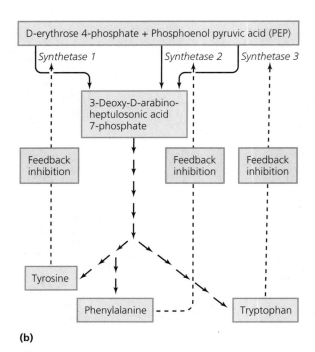

(b)

◀ *Figure 5.12*

Feedback inhibition. **(a)** The end-product of a metabolic pathway allosterically inhibits the initial step, shutting down the pathway. **(b)** In this example, each of the three end-products (tyrosine, phenylalanine, and tryptophan) inhibits a different synthetase enzyme. An excess of all three end-products is needed to completely hinder the synthesis of 3-deoxy-arabino-heptulosonic acid 7-phosphate.

synthetase enzymes are involved **(Figure 5.12b).** In other words, enzymatic reactions convert DAHAP into the three amino acids via three different pathways. As shown, each of the end-product amino acids inhibits only one of the three synthetase enzymes. Therefore, an excess of all three amino acids is needed to completely hinder synthesis of DAHAP and thereby stop the production of these amino acids.

To this point we have viewed the concept of metabolism as a collection of chemical reactions (pathways) that can be categorized as either catabolic (breaking down) or anabolic (building up). Because enzymes are required to lower the activation energy of these reactions, we examined these catalysts in some detail.

Energy is also critical to metabolism, so we examined redox reactions as a means of transferring energy within cells. We saw, for example, that redox reactions and carrier molecules are used to transfer energy from catabolic pathways to ATP, a molecule that stores energy in cells.

We will now consider how cells acquire and utilize metabolites, which are used to synthesize the macromolecules necessary for growth and, eventually, reproduction—the ultimate goal of metabolism. We will also consider in more detail the phosphorylation of ADP to make ATP.

Carbohydrate Catabolism

Learning Objective

✓ In general terms, describe the three stages of aerobic glucose catabolism (glycolysis, the Krebs cycle, and the electron transport chain), including their substrates, products, and net energy production.

Organisms oxidize carbohydrates as their primary energy source for anabolic reactions. They use glucose most commonly, though other sugars, amino acids, and fats are also utilized, often by first converting them into glucose. Glucose is catabolized via one of two processes: either via *cellular*

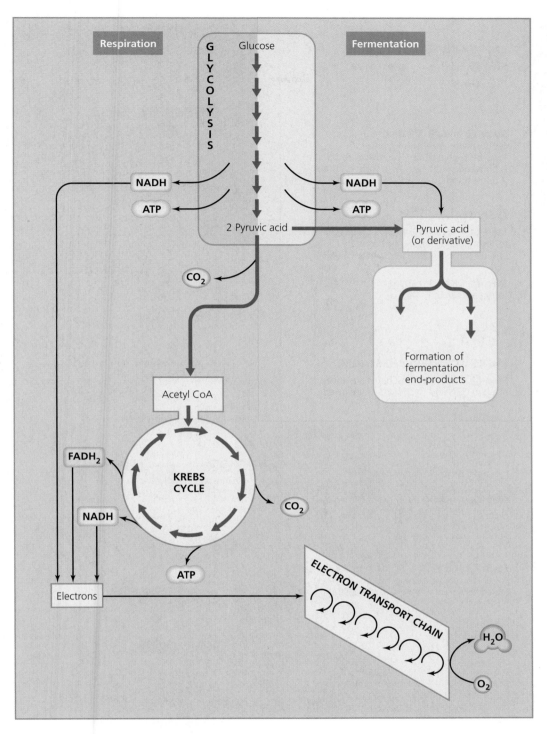

Summary of glucose catabolism. Glucose catabolism begins with glycolysis, which forms pyruvic acid and two molecules of both ATP and NADH. Two pathways branch from pyruvic acid: respiration and fermentation. In aerobic respiration, the Krebs cycle and the electron transport chain completely oxidize pyruvic acid to CO_2 and H_2O, in the process synthesizing many molecules of ATP. Fermentation results in the incomplete oxidation of pyruvic acid to form organic fermentation products.

respiration[3]—a process that involves the complete breakdown of glucose to carbon dioxide and water—or via *fermentation*, which results in organic waste products.

As shown in **Figure 5.13,** both cellular respiration and fermentation begin with *glycolysis* (glī-kol'i-sis), a process

that catabolizes a single molecule of glucose to two molecules of pyruvic acid (also called pyruvate) and results in a small amount of ATP production. Respiration then continues via the *Krebs cycle* and the *electron transport chain,* which results in a significant amount of ATP production. Fermentation involves the conversion of pyruvic acid into other organic compounds. Because it lacks the Krebs cycle and electron transport chain, fermentation results in the production of much less ATP than does respiration.

---

[3]Cellular respiration is often referred to simply as *respiration,* which should not be confused with breathing, also called respiration.

ENERGY-INVESTMENT STAGE

Step ①. Glucose is phosphorylated by ATP to form glucose 6-phosphate.

Steps ② and ③. The atoms of glucose 6-phosphate are rearranged to form fructose 6-phosphate. Fructose 6-phosphate is phosphorylated by ATP to form fructose 1,6-bisphosphate.

LYSIS STAGE

Step ④. Fructose 1,6-bisphosphate is cleaved to form glyceraldehyde 3-phosphate (G3P) and dihydroxyacetone phosphate (DHAP).

Step ⑤. DHAP is rearranged to form another G3P.

ENERGY-CONSERVING STAGE

Step ⑥. Inorganic phosphates are added to the two G3P, and two NAD^+ are reduced.

Step ⑦. Two ADP are phosphorylated by substrate-level phosphorylation to form 2 ATP.

Steps ⑧ and ⑨. The remaining phosphates are moved to the middle carbons. A water molecule is removed from each substrate.

Step ⑩. Two ADP are phosphorylated by substrate-level phosphorylation to form two ATP. Two pyruvic acid are formed.

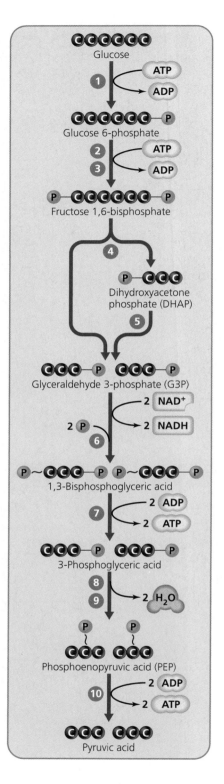

▲ *Figure 5.14*

Glycolysis (also known as the Embden-Meyerhof pathway), in which glucose is cleaved and ultimately transformed into two molecules of pyruvic acid. Four ATPs are formed and two ATPs are used, so a net gain of two ATPs results. Two molecules of NAD^+ are reduced to NADH.

The following is a simplified discussion of glucose catabolism. Details of the substrates and enzymes involved are provided in the Appendix on page A-1. To help understand the basic reactions in each of the pathways of glucose catabolism, pay special attention to three things: the number of carbon atoms in each of the intermediate products, the relative numbers of ATP molecules produced in each pathway, and the changes in the coenzymes NAD^+ and FAD as they are reduced and then oxidized back to their original forms.

Glycolysis

Glycolysis,[4] also called the *Embden-Meyerhof pathway* after the scientists who discovered it, is the first step in the catabolism of glucose via both respiration and fermentation. Glycolysis occurs in most cells. In general, as its name implies, glycolysis involves the splitting of a six-carbon glucose molecule into two three-carbon sugar molecules. When these three-carbon molecules are oxidized to pyruvic acid, some of the energy released is stored in molecules of ATP.

Glycolysis, which occurs in the cytoplasm, can be divided into three stages involving a total of 10 steps **(Figure 5.14),** each of which is catalyzed by its own enzyme. The three stages of glycolysis are:

1. Energy-investment stage (steps **1**–**3**). As with money, one must invest before a profit can be made. In this case, the energy in two molecules of ATP is invested to phosphorylate a six-carbon glucose molecule and rearrange its atoms to form fructose 1,6-bisphosphate.

2. Lysis stage (steps **4** and **5**). Fructose 1,6-bisphosphate is cleaved into glyceraldehyde 3-phosphate (G3P)[5] and dihydroxyacetone phosphate (DHAP). Each of these compounds contains three carbon atoms and is freely convertible into the other.

3. Energy-conserving stage (steps **6**–**10**). G3P is oxidized to pyruvic acid, yielding two ATP molecules. DHAP is converted to G3P and also oxidized to pyruvic acid, yielding another two ATP molecules, for a total of four ATP molecules.

Our study of glycolysis provides our first opportunity to study *substrate-level phosphorylation* (see steps **1**, **7**, and **10** in Figure 5.14). Let's examine this important process more closely by considering the 10th and final step of glycolysis.

Each of the two phosphoenolpyruvic acid (PEP) molecules produced in step **9** of glycolysis is a three-carbon compound containing a high-energy phosphate bond. In the presence of a specific holoenzyme (which requires a Mg^{2+} cofactor), the high-energy phosphate in PEP (one substrate) is transferred to an ADP molecule (a second substrate) to form ATP **(Figure 5.15);** the direct transfer of the phosphate between the two substrates is the reason the process is called **substrate-level phosphorylation.** A variety of substrate-

▲ *Figure 5.15*

Substrate-level phosphorylation, in which high-energy phosphate bonds are transferred from one substrate to another. *What role does Mg^{2+} play in this reaction?*

Figure 5.15 Mg^{2+} is a cofactor of the enzyme.

level phosphorylations occur in metabolism. As you might expect, each type has its own enzyme that recognizes both its substrate molecule and ADP.

In glycolysis, two ATP molecules are invested by substrate-level phosphorylation to prime glucose for lysis, and four molecules of ATP are produced, also by substrate-level phosphorylation. Therefore, a net gain of two ATP molecules occurs for each molecule of glucose that is oxidized to pyruvic acid. Glycolysis also yields two molecules of NADH.

Alternatives to Glycolysis

Learning Objective

✓ Compare the pentose phosphate pathway and the Entner-Doudoroff pathway with glycolysis in terms of energy production and products.

The initial part of the catabolism of glucose can also proceed via two alternate pathways: the pentose phosphate pathway and the Entner-Doudoroff pathway. Though they yield fewer molecules of ATP than glycolysis, these alternate pathways reduce coenzymes and yield different substrate metabolites that are needed in anabolic pathways. Next we briefly examine each of these alternate pathways.

Pentose Phosphate Pathway

The **pentose phosphate pathway** is named for the phosphorylated pentose (five-carbon) sugars—ribulose, xylulose, and ribose—that are formed from glucose 6-phosphate by enzymes in the pathway **(Figure 5.16** on page 140). The pentose phosphate pathway is primarily used for the production of precursor metabolites used in anabolic reactions, including the synthesis of nucleotides for nucleic acids, of certain amino acids, and of glucose by *photosynthesis*

---

[4]From Greek *glykys,* meaning sweet, and *lysein,* meaning to loosen.
[5]G3P is also known as phosphoglyceraldehyde or PGAL.

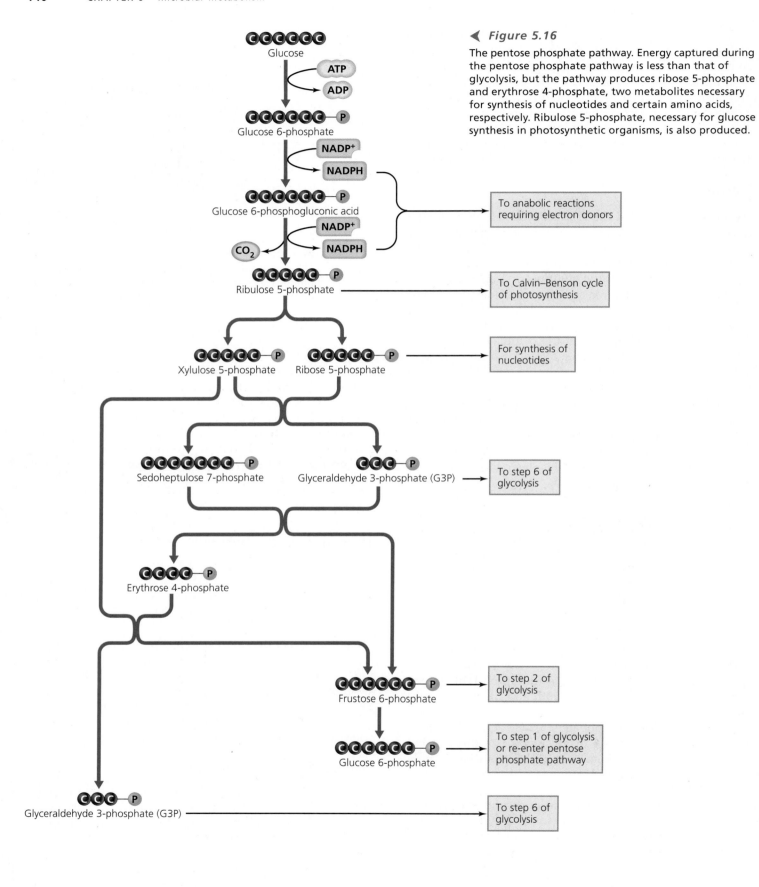

◄ *Figure 5.16*

The pentose phosphate pathway. Energy captured during the pentose phosphate pathway is less than that of glycolysis, but the pathway produces ribose 5-phosphate and erythrose 4-phosphate, two metabolites necessary for synthesis of nucleotides and certain amino acids, respectively. Ribulose 5-phosphate, necessary for glucose synthesis in photosynthetic organisms, is also produced.

(described in a later section). The pathway also reduces two molecules of $NADP^+$ to NADPH and nets a single molecule of ATP from each molecule of glucose. NADPH is a necessary coenzyme for anabolic enzymes that synthesize DNA nucleotides, steroids, and fatty acids.

CRITICAL THINKING

Examine the biosynthetic pathway for the production of the amino acids tryptophan, tyrosine, and phenylalanine in Figure 5.12b. Where do the initial reactants (PEP and erythrose 4-phosphate) originate?

Entner-Doudoroff Pathway

Most bacteria use glycolysis and the pentose phosphate pathway, but a few substitute the **Entner-Doudoroff pathway (Figure 5.17)** for glycolysis. This pathway, named for its discoverers, is a series of reactions that catabolize glucose to pyruvic acid using different enzymes from those used in either glycolysis or the pentose phosphate pathway.

Among organisms, only a very few bacteria use the Entner-Doudoroff pathway. These include the Gram-negative *Pseudomonas aeruginosa* (soo-dō-mō′nas ā-rŭ-ji-nō′să), and the Gram-positive bacterium *Enterococcus faecalis* (en-te-rō-kok′kus fē-kā′lis). Like the pentose phosphate pathway, the Entner-Doudoroff pathway nets only a single molecule of ATP for each molecule of glucose, but it does yield precursor metabolites and NADPH. The latter is unavailable from glycolysis.

CRITICAL THINKING

Even though *Pseudomonas aeruginosa* and *Enterococcus faecalis* usually grow harmlessly in the body, they can cause disease. Because these bacteria use the Entner-Doudoroff pathway instead of glycolysis to catabolize glucose, investigators sometimes use clinical tests that provide evidence of the Entner-Doudoroff pathway to identify the presence of these potential pathogens.

Suppose you were able to identify the presence of any specific organic compound. Name a substrate molecule you would find in *Pseudomonas* and *Enterococcus* cells, but not in human cells.

Cellular Respiration

Learning Objectives

✓ Discuss the roles of acetyl-CoA, the Krebs cycle, and electron transport in carbohydrate catabolism.

✓ Contrast electron transport in aerobic and anaerobic respiration.

✓ Identify four classes of carriers in electron transport chains.

✓ Describe the role of chemiosmosis in oxidative phosphorylation of ATP.

After glucose has been oxidized via glycolysis or one of the alternate pathways, a cell uses the resultant pyruvic acid molecules to complete either cellular respiration or fermentation (which we will discuss in a later section). Our topic here—**cellular respiration**—is a metabolic process that involves the complete oxidation of substrate molecules and

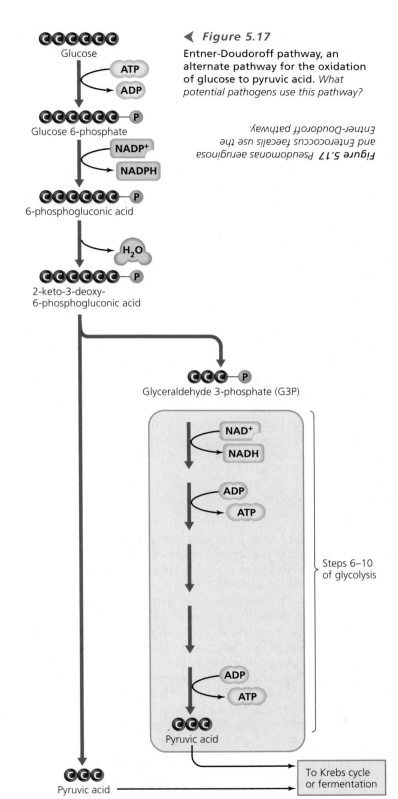

◀ *Figure 5.17*

Entner-Doudoroff pathway, an alternate pathway for the oxidation of glucose to pyruvic acid. *What potential pathogens use this pathway?*

Figure 5.17 Pseudomonas aeruginosa and Enterococcus faecalis use the Entner-Doudoroff pathway.

Glucose

ATP
ADP

Glucose 6-phosphate

$NADP^+$
NADPH

6-phosphogluconic acid

H_2O

2-keto-3-deoxy-6-phosphogluconic acid

Glyceraldehyde 3-phosphate (G3P)

NAD^+
NADH

ADP
ATP

ADP
ATP

Pyruvic acid

Steps 6–10 of glycolysis

Pyruvic acid

To Krebs cycle or fermentation

Figure 5.18 ❯

Formation of acetyl-CoA. The responsible enzyme acts in a stepwise manner to ① remove CO_2 from pyruvic acid, ② attach the remaining two-carbon acetate to coenzyme A, and ③ reduce a molecule of NAD^+ to NADH.

then production of ATP by a series of redox reactions. The three stages of cellular respiration are: (1) synthesis of acetyl-CoA, (2) the Krebs cycle, and (3) a final series of redox reactions, which collectively constitute an electron transport chain.

Synthesis of Acetyl-CoA

Before pyruvic acid (generated by glycolysis, the pentose phosphate pathway, and the Entner-Doudoroff pathway) can enter the Krebs cycle for respiration, it must first be converted to *acetyl-coenzyme A* or **acetyl-CoA** (as'e-til kō-ā'; see Figure 5.13). Enzymes remove one carbon from pyruvic acid as CO_2 and join the remaining two-carbon acetate to *coenzyme-A* with a high-energy bond **(Figure 5.18)**. The removal of CO_2, called *decarboxylation*, requires a coenzyme derived from the vitamin thiamine. One molecule of NADH is also produced during this reaction.

    Recall that two molecules of pyruvic acid were derived from each molecule of glucose. Therefore, at this stage, two molecules of acetyl-CoA, two molecules of CO_2, and two molecules of NADH are produced.

The Krebs Cycle

At this point in the catabolism of a molecule of glucose, a great amount of energy remains in the bonds of acetyl-CoA. The **Krebs cycle**[6] is a series of eight enzymatically catalyzed reactions that transfer much of this stored energy to the coenzymes NAD^+ and FAD. The two carbons in acetate are oxidized, and the coenzymes are reduced. The Krebs cycle, which occurs in the cytoplasm in prokaryotes and in the matrix of mitochondria in eukaryotes, is diagrammed in **Figure 5.19** and presented in more detail in the Appendix on page A-6. It is also known as the *tricarboxylic acid (TCA) cycles,* because many of its compounds have three carboxyl (—COOH) groups, and as the *citric acid cycle,* for the first compound formed in the cycle.

    There are five types of reactions in the Krebs cycle:

1. Anabolism of citric acid (step ❶)
2. Isomerization reactions (steps ❷, ❼, and ❽)
3. Redox reactions (steps ❸, ❹, ❻, and ❽)
4. Decarboxylations (steps ❸ and ❹)
5. Substrate-level phosphorylation (step ❺)

---

[6]Named for biochemist Sir Hans Krebs (1900–1981), who elucidated its reactions in the 1940s.

    In the first step of the Krebs cycle, the splitting of the high-energy bond between acetate and coenzyme A releases enough energy to enable the binding of the freed two-carbon acetate to a four-carbon compound called oxaloacetic acid, forming the six-carbon compound citric acid.

    As you study **Figure 5.19,** notice that after isomerization (step ❷), the decarboxylations of the Krebs cycle release two molecules of CO_2 for each acetyl-CoA that enters (steps ❸ and ❹). Thus, for every two carbon atoms that enter the cycle, two are lost to the environment. At this juncture in the respiration of a molecule of glucose, six carbon atoms have been lost to the environment: two as CO_2 molecules produced in decarboxylation of two molecules of pyruvic acid to form two acetyl-CoA molecules, and four in CO_2 molecules produced in decarboxylations in the *two* turns through the Krebs cycle. (One molecule of acetyl-CoA enters the cycle at a time).

    A small amount of ATP is also produced in the Krebs cycle. For every two molecules of acetyl-CoA that pass through the Krebs cycle, two molecules of ATP are generated by substrate-level phosphorylation (step ❺). A molecule of guanosine triphosphate (GTP), which is similar to ATP, serves as an intermediary in this process.

    Redox reactions reduce FAD to $FADH_2$ (step ❻) and NAD^+ to NADH (steps ❸, ❹, and ❽), so that for every two molecules of acetyl-CoA that move through the cycle, six molecules of NADH and two of $FADH_2$ are formed. In the Krebs cycle, little energy is captured directly in high-energy phosphate bonds, but much energy is transferred via electrons to NADH and $FADH_2$. These coenzymes are the most important molecules of respiration because they carry a large amount energy that is subsequently used to phosphorylate ADP to ATP.

CRITICAL THINKING

We have examined the total ATP, NADH, and $FADH_2$ production in the Krebs cycle for each molecule of glucose coming through Embden-Meyerhof glycolysis. How many of each of these molecules would be produced if the Entner-Doudoroff pathway were used instead of glycolysis?

Electron Transport

Some scientists estimate that each day an average human synthesizes his or her own weight in ATP molecules and uses them for metabolism, responsiveness, growth, and cell reproduction. ATP turnover in prokaryotes is relatively as

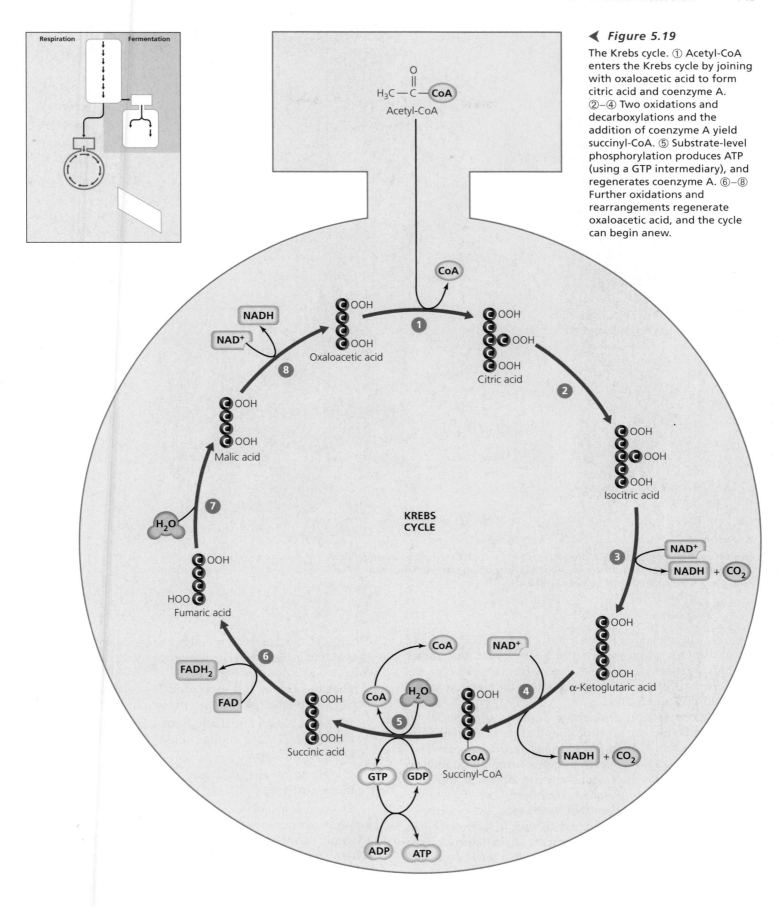

◀ *Figure 5.19*

The Krebs cycle. ① Acetyl-CoA enters the Krebs cycle by joining with oxaloacetic acid to form citric acid and coenzyme A. ②–④ Two oxidations and decarboxylations and the addition of coenzyme A yield succinyl-CoA. ⑤ Substrate-level phosphorylation produces ATP (using a GTP intermediary), and regenerates coenzyme A. ⑥–⑧ Further oxidations and rearrangements regenerate oxaloacetic acid, and the cycle can begin anew.

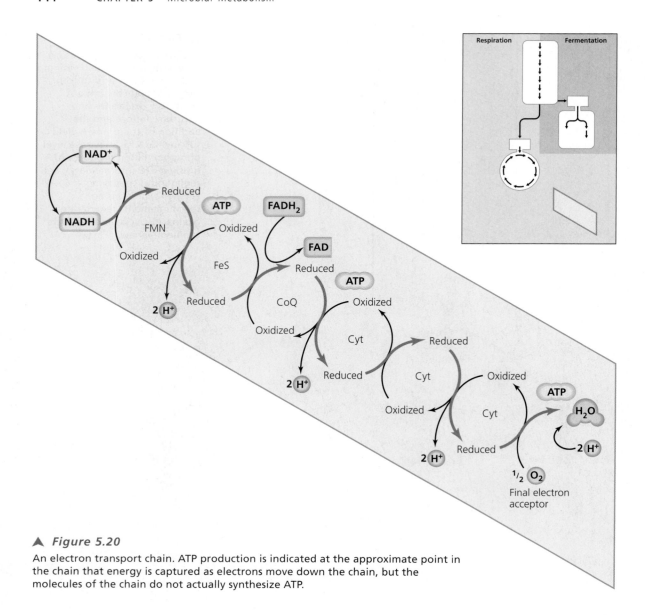

▲ *Figure 5.20*

An electron transport chain. ATP production is indicated at the approximate point in the chain that energy is captured as electrons move down the chain, but the molecules of the chain do not actually synthesize ATP.

copious. The most significant production of ATP does not occur through glycolysis or the Krebs cycle, but rather through the stepwise release of energy from a series of redox reactions between molecules known as an **electron transport chain (Figure 5.20).**

An electron transport chain consists of a series of membrane-bound carrier molecules that pass electrons from one to another and ultimately to a *final electron acceptor.* Typically, as we have seen, electrons come from the catabolism of an organic molecule such as glucose; however, microorganisms called *chemolithotrophs* (kem'ō-lith'-ō-trōfs) acquire electrons from inorganic sources such as H_2, NO_2^-, or Fe^{2+}. (Chemolithotrophs are discussed further in Chapter 6.) In any case, electrons are passed down the chain like buckets in a fire brigade to the final acceptor. As with a bucket brigade, the final step of electron transport is irreversible. Energy from the electrons is used to actively transport (pump)

protons (H^+) across the membrane, establishing a *proton gradient* that generates ATP via a process called *chemiosmosis,* which we will discuss shortly.

To avoid getting lost in the details of electron transport, keep the following critical concepts in mind:

- Electrons pass sequentially from one membrane-bound carrier molecule to another, and eventually to a final acceptor molecule.

- The electrons' energy is used to pump protons across the membrane.

Electron transport chains are located in the inner mitochondrial membranes (cristae) of eukaryotes and in the cytoplasmic membrane of prokaryotes **(Figure 5.21).** Though NADH and FADH$_2$ donate electrons as hydrogen atoms (electrons and protons), many carrier molecules pass only the electrons down the chain. There are four categories of carrier molecules in electron transport chains.

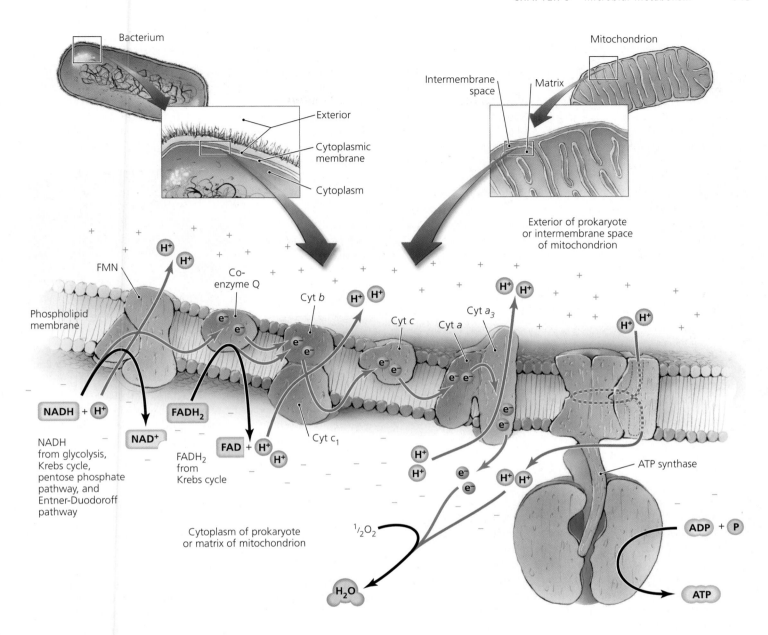

▲ Figure 5.21

One possible arrangement of an electron transport chain. Electron transport chains are located in the cytoplasmic membranes of prokaryotes, and in the inner membranes of mitochondria in eukaryotes. The exact types and sequences of carrier molecules in electron transport chains vary among organisms. As electrons move down the chain (red arrows), their energy is used to pump protons (H^+) across the membrane. The protons then flow through ATP synthase (ATPase), which synthesizes ATP. Approximately one molecule of ATP is generated for every two protons that cross the membrane. *Why is it essential that all the electron carriers of an electron transport chain be membrane bound?*

Figure 5.21   The carrier molecules of electron transport chains are membrane bound for at least two reasons: (1) Carrier molecules need to be in fairly close proximity so they can pass electrons between each other. (2) The goal of the chain is the pumping of H^+ across a membrane to create a proton gradient; without a membrane, there could be no gradient.

Highlight 5.1   Glowing Bacteria

Bacteria in the genus *Photobacterium* possess an interesting electron transport chain that generates light instead of ATP. These organisms can switch the flow of electrons from a standard electron transport chain (composed of cytochromes, proton pumps, and O_2 as the final electron acceptor) to an alternate chain. Whereas the standard chain establishes a proton gradient that is used to synthesize ATP, the alternate chain uncouples electron transport and ATP; instead of transferring electrons to proton pumps, the alternate chain shunts the electrons to the coenzyme flavin mononucleotide (FMN). Then, in the presence of an enzyme called luciferase and a long-chain hydrocarbon, the alternate chain emits light as it transfers electrons to O_2 (see the box diagram). The exact mechanism of bioluminescence is not known, but both FMN and the hydrocarbon are oxidized as oxygen is reduced.

Interestingly, free-living *Photobacterium* are usually not bioluminescent. It is primarily when these bacteria colonize the tissues of marine animals such as squid and fish that they use their light-generating pathway. The animals gain from the association because the light produced serves as an attractant for mates and a warning against predators. One species, the "flashlight fish" (*Photoblepharon palpebratus*), has a special organ near its mouth that is specially adapted for the growth of luminescent bacteria. Enough light is generated from millions of bacteria that the fish can navigate over coral reefs at night and attract prey to their light. The light organ even has a membrane that descends like an eyelid to control the amount of light emitted.

Photoblepharon palpebratus

It is not clear what the bacteria gain from this association. Presumably, the protection and nutrients the bacteria gain from the fish make up for the enormous metabolic cost the bacteria incur in the form of lost ATP synthesis.

They are:

1. *Flavoproteins* are integral membrane proteins, many of which contain flavin, a coenzyme derived from riboflavin (vitamin B_2). One form of flavin is flavin mononucleotide (FMN), which is the initial carrier molecule of electron transport chains of mitochondria. The familiar FAD is a coenzyme for other flavoproteins. Like all carrier molecules in the electron transport chain, flavoproteins alternate between the reduced and oxidized states.

2. *Ubiquinones* (yū-bik′wi-nōns) are lipid-soluble, nonprotein carriers that are so named because they are ubiquitous in all cells. Ubiquinones are derived from vitamin K. In mitochondria, the ubiquinone is called *coenzyme Q*.

3. *Metal-containing proteins* are a mixed group of integral proteins with a wide-ranging number of iron, sulfur, and copper atoms that can alternate between the reduced and oxidized states. Iron-sulfur proteins occur in various places in electron transport chains of various organisms. Copper proteins are found only in electron transport chains involved in photosynthesis (discussed shortly).

4. *Cytochromes* (sī′tō-krōms) are integral proteins associated with *heme,* which is the same iron-containing, nonprotein, pigmented molecule found in the hemoglobin of blood. Iron can alternate between a reduced (Fe^{2+}) state and an oxidized (Fe^{3+}) state. Cytochromes are identified by letters and numbers based on the order in which they were identified, so their sequence in electron transport chains does not always seem logical.

The carrier molecules in electron transport chains are diverse—bacteria typically have different carrier molecules arranged in different sequences than do archaea or the mitochondria of eukaryotes. Some prokaryotes, including *E. coli,* can even vary their carrier molecules under different environmental conditions. Even among bacteria, the makeup of carrier molecules can be variable. For example, the pathogens *Neisseria* (nī-se′rē-ă) and *Pseudomonas* contain two cytochromes, *a* and a_3—together called *cytochrome oxidase*—that oxidize cytochrome *c;* such bacteria are said to be *oxidase positive.* In contrast, other bacterial pathogens, such as *Escherichia, Salmonella,* and *Proteus* (prō′tē-ŭs), lack cytochrome oxidase and are thus considered to be *oxidase negative.*

Electrons carried by NADH enter the transport chain at a flavoprotein, and those carried by $FADH_2$ are introduced via a ubiquinone. This explains why more molecules of ATP are generated from NADH than from $FADH_2$. Researchers do not agree on which carrier molecules are the actual proton pumps, nor on the number of protons that are pumped. Figure 5.21 shows one possibility.

In some organisms, the final electron acceptors are oxygen atoms, which, with the addition of hydrogen ions, generate H_2O; these organisms conduct **aerobic[7] respiration** and are called *aerobes*. Other organisms, called *anaerobes*,[8] use other inorganic molecules (or rarely an organic molecule) instead of oxygen as the final electron acceptor and perform **anaerobic respiration.** The anaerobic bacterium *Desulfovibrio* (dē′sul-fō-vib-rē-ō), for example, reduces sulfate (SO_4^{2-}) to hydrogen sulfide gas (H_2S), whereas other anaerobes in the genera *Bacillus* (ba-sil′ŭs) and *Pseudomonas* utilize nitrate (NO_3^-) to produce nitrite ions (NO_2^-), nitrous oxide (N_2O), or nitrogen gas (N_2). Some prokaryotes—particularly archaea called methanogens—reduce carbonate (CO_3^{2-}) to methane gas (CH_4). **Highlight 5.1** describes an unusual bacterial electron transport system that produces light instead of ATP.

Laboratory technologists test for products of anaerobic respiration, such as nitrite, to aid in identification of some species of bacteria. As discussed more fully in Chapter 26, anaerobic respiration is also critical for the recycling of nitrogen and sulfur in nature.

CRITICAL THINKING

Cyanide is a potent poison because it irreversibly blocks cytochrome a_3. What effect would its action have on the rest of the electron transport chain? What would be the redox state (reduced or oxidized) of coenzyme Q in the presence of cyanide?

In summary, glycolysis, the pentose phosphate pathway, the Entner-Doudoroff pathway, and the Krebs cycle strip electrons, which carry energy, from glucose molecules and transfer them to molecules of NADH and $FADH_2$. In turn, NADH and $FADH_2$ pass the electrons to an electron transport chain. As the electrons move down the electron transport chain, proton pumps use the electrons' energy to actively transport protons (hydrogen ions) across the membrane.

Recall, however, that the significance of electron transport is not merely that it pumps protons across a membrane, but that it ultimately results in the synthesis of ATP. We turn now to the process by which cells synthesize ATP using the concentration gradient produced by the large concentration of protons on one side of a membrane.

Chemiosmosis

Chemiosmosis is a general term for the use of ion gradients to generate ATP; that is, ATP is synthesized utilizing energy released by the flow of ions down their electrochemical gradient across a membrane. The term should not be confused with osmosis of water. To understand chemiosmosis, we need to review several concepts from Chapter 3 concerning diffusion and phospholipid membranes.

Recall that chemicals diffuse from areas of high concentration to areas of low concentration, and toward an electrical charge opposite their own. We call the composite of differences in concentration and charge an *electrochemical gradient.* Chemicals diffuse down their electrochemical gradients. Recall as well that membranes of cells and organelles are impermeable to most chemicals unless a specific protein channel allows their passage across the membrane. A membrane maintains an electrochemical gradient by keeping one or more chemicals in a higher concentration on one side. The blockage of diffusion creates potential energy, like water behind a dam.

Chemiosmosis uses the potential energy of an electrochemical gradient to phosphorylate ADP into ATP. Even though chemiosmosis is a general principle with relevance to both *oxidative phosphorylation* and *photophosphorylation,* here we consider it as it relates to oxidative phosphorylation.

As we have seen, cells use the energy released in the redox reactions of electron transport chains to actively transport protons (H^+) across a membrane. Theoretically, an electron transport chain pumps three pairs of protons for each pair of electrons contributed by NADH, and pumps two pairs of protons for each electron pair delivered by $FADH_2$. This difference results from the fact that $FADH_2$ delivers electrons farther down the chain than does NADH; therefore, energy carried by $FADH_2$ is used to transport one-third fewer protons (see Figure 5.20). Because lipid bilayers are impermeable to protons, the transport of protons to one side of the membrane creates an electrochemical gradient known as a **proton gradient,** which has potential energy known as a *proton motive force.*

Hydrogen ions, propelled by the proton motive force, flow down their electrochemical gradient through protein channels, called **ATP synthases (ATPases),** that phosphorylate molecules of ADP to ATP (see Figure 5.21). Such phosphorylation is called **oxidative phosphorylation** because the proton gradient is created by the oxidation of components of an electron transport chain.

In the past, scientists attempted to calculate the exact number of ATP molecules synthesized per pair of electrons that travel down an electron transport chain. However, it is now apparent that phosphorylation and oxidation are not directly coupled. In other words, chemiosmosis does not require exact constant relationships among the number of molecules of NADH and $FADH_2$ reduced, the number of electrons that move down an electron transport chain, and the number of molecules of ATP that are synthesized. Additionally, cells use proton gradients for other cellular processes, including active transport and bacterial flagellar motion, so not every transported electron results in ATP production.

Nevertheless, about 34 molecules of ADP per molecule of glucose are oxidatively phosphorylated to ATP via chemiosmosis: three from each of the 10 molecules of NADH generated from glycolysis, the synthesis of acetyl-CoA, and the Krebs cycle, and two from each of the two

---

[7]From Greek *aer,* meaning air (that is, oxygen), and *bios,* meaning life.
[8]The Greek prefix *an* means not.

molecules of $FADH_2$ generated in the Krebs cycle. Given that glycolysis produces a net two molecules of ATP by substrate-level phosphorylation, and that the Krebs cycle produces two more, the complete aerobic oxidation of one molecule of glucose by a prokaryote can theoretically yield a net total of 38 molecules of ATP **(Table 5.3)**. The theoretical net maximum for eukaryotic cells is generally given as 36 molecules of ATP because the energy from two ATP molecules is required to transport NADH generated in glycolysis into the mitochondria.

CRITICAL THINKING

Suppose you could insert a tiny pH probe into the space between mitochondrial membranes. Would the pH be above or below 7.0? Why?

Fermentation

Learning Objectives

✓ Describe fermentation, and contrast it with respiration.

✓ Identify three useful end-products of fermentation, and explain how fermentation reactions are used in the identification of bacteria.

✓ Discuss the use of biochemical tests for metabolic enzymes and products in the identification of bacteria.

Sometimes cells cannot completely oxidize glucose by cellular respiration. For instance, they may lack sufficient final electron acceptors, as is the case, for example, for an aerobic bacterium in the anaerobic environment of the colon. Electrons cannot flow down an electron transport chain unless oxidized carrier molecules are available to receive them. Our bucket brigade analogy can help clarify this point.

Suppose the last person in the brigade did not throw the water, but instead held onto two full buckets. What would happen? The entire brigade would soon consist of firefighters holding full buckets of water. The analogous situation occurs in an electron transport chain: All the carrier molecules are forced to remain in their reduced states when there is not a final electron acceptor. Without the movement of electrons down the chain, protons cannot be transported, the proton motive force is lost, and oxidative phosphorylation of ADP to ATP ceases. Without sufficient ATP, a cell is unable to anabolize, grow, or divide.

ATP could be synthesized in glycolysis and the Krebs cycle by substrate-level phosphorylation. After all, together these pathways produce four molecules of ATP per molecule of glucose. However, careful consideration reveals that glycolysis and the Krebs cycle require a continual supply of oxidized NAD^+ molecules (see Figures 5.14 and 5.19). In respiration, electron transport produces the required NAD^+, but without a final electron acceptor, this source of NAD^+ ceases to be available. A cell in such a predicament must use an alternate source of NAD^+ provided by alternative metabolic pathways, called *fermentation pathways*.

▲ *Figure 5.22*

Fermentation. In the simplest fermentation reaction, NADH reduces pyruvic acid (from glycolysis) to form lactic acid. Another simple fermentation pathway involves a decarboxylation reaction and reduction to form ethanol.

In the vernacular, fermentation refers to the production of alcohol from sugar, but in microbiology, fermentation has an expanded meaning. **Fermentation** is the partial oxidation of sugar (or other metabolites) to release energy using an organic molecule as an electron acceptor rather than an electron transport chain. In other words, fermentation pathways are metabolic reactions that oxidize NADH to NAD^+ while reducing organic molecules, which are the final electron acceptors. In contrast, as previously noted, aerobic respiration reduces oxygen, and anaerobic respiration reduces other inorganic chemicals such as sulfate and nitrate or (rarely) an organic molecule. **Figure 5.22** illustrates two common fermentation pathways that reduce pyruvic acid to lactic acid and ethanol, oxidizing NADH in the process.

Table 5.3	Summary of Prokaryotic Aerobic Respiration of One Molecule of Glucose				
Pathway	ATP Produced	ATP Used	NADH Produced	FADH$_2$ Produced	Comments
Glycolysis	4	2	2	0	Two molecules of water produced
Synthesis of acetyl-CoA and Krebs cycle	2	0	8	2	Two molecules of water used, six molecules of CO$_2$ produced
Electron transport chain	34	0	0	0	Ten molecules of NADH, two molecules of FADH$_2$, and six molecules of O$_2$ used, six molecules of water produced
Total	40	2			
Net total	**38**				

The essential function of fermentation is the regeneration of NAD$^+$ for glycolysis, so that ADP molecules can be phosphorylated to ATP. Even though fermentation pathways are not as energetically efficient as respiration because much of the potential energy stored in glucose remains in the bonds of fermentation products, the major benefit of fermentation is that it allows ATP production to continue in the absence of cellular respiration. Microorganisms with fermentation pathways can colonize anaerobic environments, which are unavailable to strictly aerobic organisms. **Table 5.4** on page 150 compares fermentation to aerobic and anaerobic respiration with respect to four crucial aspects of these processes.

Microorganisms produce a variety of fermentation products depending on the enzymes and substrates available to each. Though fermentation products are wastes to the cells that make them, many are useful to humans, including ethanol (drinking alcohol), acetic acid (vinegar), and lactic acid (used in the production of cheese, sauerkraut, and pickles) (**Figure 5.23** on page 150).

Other fermentation products are harmful to human health and industry. For example, fermentation products of the bacterium *Clostridium perfringens* (klos-tri'dē-um per-frin'jens) are involved in the necrosis (death) of muscle tissue associated with gangrene. Also, as you may recall from Chapter 1, Pasteur discovered that bacterial contaminants in grape juice fermented the sugar into unwanted products such as acetic acid and lactic acid, which spoiled the wine.

Laboratory personnel routinely use the detection of fermentation products in the identification of microbes. For example, *Proteus* ferments glucose but not lactose, whereas *Escherichia* and *Enterobacter* ferment both. Further, glucose fermentation by *Escherichia* produces mixed acids (acetic, lactic, succinic, and formic), while *Enterobacter* produces 2,3-butanediol. Common fermentation tests contain a carbohydrate and a pH indicator, which is a molecule that changes color as the pH changes. An organism that utilizes the carbohydrate causes a change in pH, causing the pH indicator to change color.

In this section on carbohydrate catabolism, we have spent some time examining glycolysis, alternatives to glycolysis, the Krebs cycle, and electron transport because these pathways are central to metabolism. They generate all of the precursor metabolites and most of the ATP needed for anabolism. We have seen that some ATP is generated in respiration by substrate-level phosphorylation (in both glycolysis and the Krebs cycle), and that most ATP is generated by oxidative phosphorylation via chemiosmosis utilizing the reducing power of NADH and FADH$_2$. We also saw that some microorganisms use fermentation to provide an alternate source of NAD$^+$ when conditions are not suitable for respiration.

Thus far we have concentrated on the catabolism of glucose as a representative carbohydrate, but microorganisms can also use other molecules as energy sources. In the next section we will examine catabolic pathways that utilize lipids and proteins.

Other Catabolic Pathways

Lipid and protein molecules contain abundant energy in their chemical bonds and can also be converted into precursor metabolites. These molecules are first catabolized to produce their constituent monomers, which serve as substrates in glycolysis and the Krebs cycle.

Lipid Catabolism

Learning Objective

✓ Explain how lipids are catabolized for energy and metabolite production.

The most common lipids involved in ATP and metabolite production are fats, which, as we saw in Chapter 2, consist of glycerol and fatty acids. In the first step of fat catabolism, enzymes called *lipases* hydrolyze the bonds attaching the glycerol to the fatty acid chains (**Figure 5.24a** on page 151).

Table 5.4 Comparison of Aerobic Respiration, Anaerobic Respiration, and Fermentation

	Aerobic Respiration	Anaerobic Respiration	Fermentation
Oxygen required	Yes	No	No
Type of phosphorylation	Substrate-level and oxidative	Substrate-level and oxidative	Substrate-level
Final electron (hydrogen) acceptor	Oxygen	NO_3^-, SO_4^{2-}, or CO_3^{2-}	Organic molecules
Potential molecules of ATP produced	36–38	2–36	2

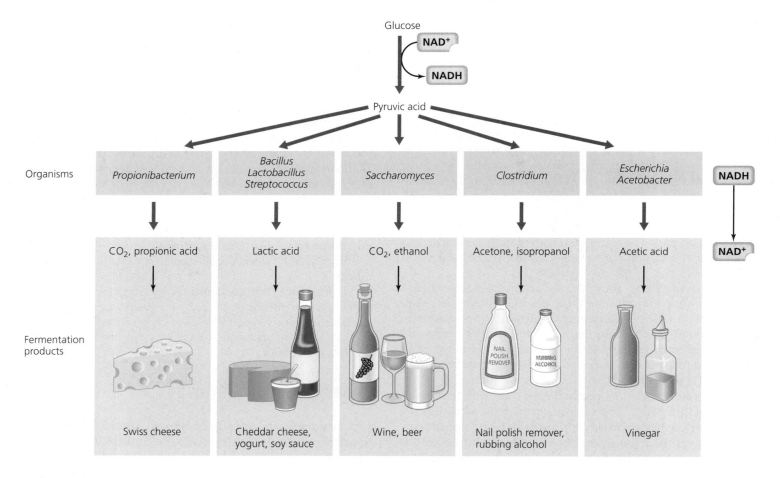

▲ *Figure 5.23*

Representative fermentation products and the organisms that produce them. All of the organisms are bacteria except *Saccharomyces,* which is a yeast. *Why does Swiss cheese have holes, but cheddar does not?*

Figure 5.23 Swiss cheese is a product of Propionibacterium, which ferments pyruvic acid to produce CO₂; bubbles of this gas cause the holes in the cheese. Cheddar is a product of Lactobacillus, which does not produce gas.

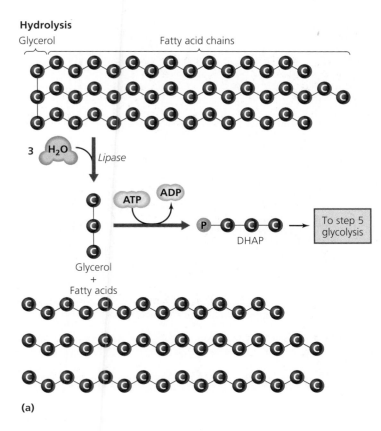

Hydrolysis

Glycerol

Fatty acid chains

To step 5 glycolysis

DHAP

Glycerol
+
Fatty acids

(a)

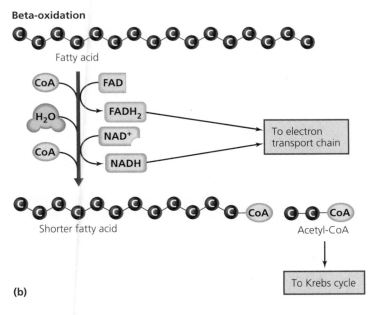

Beta-oxidation

Fatty acid

To electron transport chain

Shorter fatty acid

Acetyl-CoA

To Krebs cycle

(b)

▲ *Figure 5.24*

Catabolism of a fat molecule. **(a)** Lipase breaks fats into glycerol and three fatty acids by hydrolysis. Glycerol is converted to DHAP, which can be catabolized via glycolysis and the Krebs cycle. **(b)** Fatty acids are catabolized via beta-oxidation reactions that produce molecules of acetyl-CoA and reduced coenzymes (NADH and FADH₂).

Subsequent reactions further catabolize the glycerol and fatty acid molecules. Glycerol is converted to DHAP, which as one of the substrates of glycolysis is oxidized to pyruvic acid **(see Figure 5.14).** The fatty acids are degraded in a catabolic process known as **beta-oxidation (Figure 5.24b).**[9] In this process, enzymes repeatedly split off pairs of the hydrogenated carbon atoms that make up a fatty acid and join each pair to coenzyme A to form acetyl-CoA, until the entire fatty acid has been converted to molecules of acetyl-CoA. NADH and FADH₂ are generated during beta-oxidation, and more of these molecules are generated when the acetyl-CoA is utilized in the Krebs cycle to generate ATP **(see Figure 5.19).** As is the case for the Krebs cycle, the enzymes involved in beta-oxidation are located in the cytoplasm of prokaryotes, and in the mitochondria of eukaryotes.

CRITICAL THINKING

How and where do cells use the molecules of NADH and FADH₂ produced during beta-oxidation?

Protein Catabolism

Learning Objective

✓ Explain how proteins are catabolized for energy and metabolite production.

Some microorganisms, particularly food-spoilage bacteria, pathogenic bacteria, and fungi, normally catabolize proteins as an important source of energy and metabolites. Most other cells catabolize proteins and their constituent amino acids only when carbon sources such as glucose and fat are not available.

Generally, proteins are too large to cross cell membranes, so microorganisms conduct the first step in the process of protein catabolism outside the cell by secreting **proteases** (prō′ tē-ās-ez)—enzymes that split proteins into their constituent amino acids (**Figure 5.25** on page 152). Once released by the action of proteases, amino acids are transported into the cell, where special enzymes split off amino groups in a reaction called **deamination.** The resulting altered molecules enter the Krebs cycle, and the amino groups are either recycled to synthesize other amino acids or excreted as nitrogenous wastes such as ammonia (NH_3), ammonium ion (NH_4^+), or trimethylamine oxide (which has the chemical formula (CH_3)$_3$NO and is abbreviated TMAO). As **Highlight 5.2** on page 152 describes, TMAO plays an interesting role in the production of the odor we describe as "fishy."

Thus far, we have examined the catabolism of carbohydrates, lipids, and proteins. Now we turn our attention to the *synthesis* of these molecules, beginning with the anabolic reactions of photosynthesis.

---

[9]"Beta" is part of the name of this process because enzymes break the bond at the second carbon atom from the end of a fatty acid, and beta is the second letter in the Greek alphabet.

Figure 5.25 ➤

Protein catabolism in microbes. Hydrolysis of proteins by secreted proteases releases amino acids, which are deaminated after uptake to produce molecules used as substrates in the Krebs cycle. R indicates the side group, which varies among amino acids.

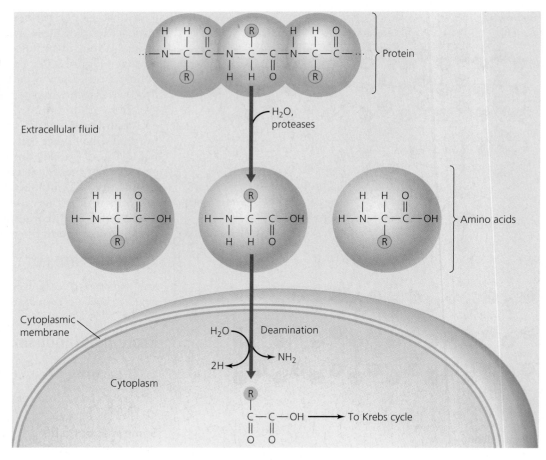

Highlight 5.2   What's That Fishy Smell?

Ever wonder why fish at the supermarket sometimes have that "fishy" smell? The reason relates to bacterial metabolism.

Trimethylamine oxide (TMAO) is a nitrogenous waste product of fish metabolism used to dispose of excess nitrogen (as amine groups) produced in the catabolism of amino acids. TMAO is an odorless substance that does not affect the smell, appearance, or taste of fresh fish. However, some bacteria use TMAO as a final electron acceptor in anaerobic respiration, reducing TMAO to trimethylamine (TMA), a compound with a very definite—and, for most people, unpleasant—"fishy" odor. A human nose is able to detect even a few molecules of TMA. Because bacterial degradation of fish and the production of TMA begin as soon as a fish is dead, a fish that smells fresh probably is fresh, or was frozen while still fresh.

Photosynthesis

Learning Objective

✓ Define photosynthesis.

Many organisms use only organic molecules as a source of energy and metabolites, but where do they acquire organic molecules? Ultimately every food chain begins with anabolic pathways in organisms that synthesize their own organic molecules from inorganic carbon dioxide. Most of these organisms capture light energy from the sun and use it to drive the synthesis of carbohydrates from CO_2 and H_2O by a process called **photosynthesis.** Cyanobacteria, purple sulfur bacteria, green sulfur bacteria, green nonsulfur bacteria, purple nonsulfur bacteria, a few protozoa, algae, and green plants are photosynthetic.

CRITICAL THINKING

Photosynthetic organisms are rarely pathogenic. Why?

▼ *Figure 5.26*

Photosynthetic structures in a prokaryote. **(a)** Chlorophyll molecules are grouped together in thylakoid membranes to form photosystems. Chlorophyll contains a light-absorbing active site that is connected to a long hydrocarbon tail. The chlorophyll shown here is bacteriochlorophyll *a*; other chlorophyll types differ in the side chains protruding from the central ring. **(b)** Prokaryotic thylakoids are infoldings of the cytoplasmic membrane.

Active site

Mg^{2+}

Tail (carbon chain)

Chlorophyll active sites (tails not shown)

Chlorophyll

Thylakoid membrane

Photosystem embedded in membrane (sectioned)

(a)

(b)   TEM   0.2 μm

Chemicals and Structures

Learning Objective

✓ Compare and contrast the basic chemicals and structures involved in photosynthesis in prokaryotes and eukaryotes.

Photosynthetic organisms capture light energy with pigment molecules, the most important of which are **chlorophylls.** Chlorophyll molecules are composed of a hydrocarbon tail attached to a light-absorbing *active site* centered around a magnesium ion (Mg^{2+}) **(Figure 5.26a).** Active sites are structurally similar to the cytochrome molecules found in electron transport chains, except chlorophylls use Mg^{2+} rather then Fe^{2+}. Chlorophylls, typically designated with letters—for example, chlorophyll *a*, chlorophyll *b*, and bacteriochlorophyll *a*—vary slightly in the lengths and structures of their hydrocarbon tails and in the atoms that extend from their active sites. Green plants, algae, photosynthetic protozoa, and cyanobacteria principally use chlorophyll *a*, whereas green and purple bacteria use bacteriochlorophylls.

The slight structural differences among chlorophylls cause them to absorb light of different wavelengths. For example, chlorophyll *a* from algae best absorbs light with wavelengths of about 425 nm and 660 nm (violet and red), whereas bacteriochlorophyll *a* from purple bacteria best absorbs light with wavelengths of about 350 nm and 880 nm (ultraviolet and infrared). Because they best use light with differing wavelengths, algae and purple bacteria successfully occupy different ecological niches.

Cells arrange numerous molecules of chlorophyll and other pigments within a protein matrix to form light-harvesting matrices called **photosystems** that are embedded in cellular membranes called **thylakoids.** Thylakoids of photosynthetic prokaryotes are invaginations of the cytoplasmic membranes **(Figure 5.26b).** The thylakoids of eukaryotes appear to be formed from infoldings of the inner membranes of chloroplasts, though thylakoid membranes and the inner membranes are not connected in mature chloroplasts (see Figure 3.37). Thylakoids of chloroplasts are arranged in stacks called *grana.* An outer chloroplast membrane surrounds the grana, and the space between the outer membrane and the thylakoid membrane is known as the *stroma.*

There are two types of photosystems, named photosystem I (PS I) and photosystem II (PS II), in the order of their discovery. Photosystems absorb light energy and use redox reactions to store the energy in molecules of ATP and NADPH. Because they depend upon light energy, these reactions of photosynthesis are classified as **light-dependent reactions.** Photosynthesis also involves **light-independent reactions** that actually synthesize glucose from carbon dioxide and water. Historically, these reactions were called *light* and *dark reactions,* but this older terminology implies that light-independent reactions occur only in the dark, which is not the case. We first consider the light-dependent reactions of the two photosystems.

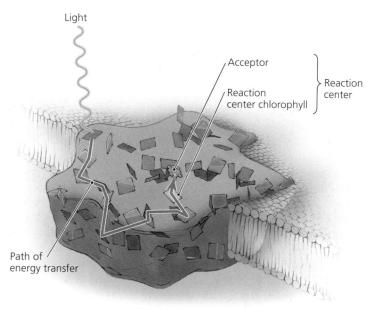

Light

Acceptor

Reaction center chlorophyll

Reaction center

Path of energy transfer

Photosystem I

▲ *Figure 5.27*

Reaction center of a photosystem. Light energy absorbed by pigments anywhere in the photosystem is transferred to the reaction center chloroplast, where it is used to excite electrons for delivery to the reaction center's electron acceptor.

Light-Dependent Reactions

Learning Objectives

✓ Describe the components and function of the two photosystems, PS I and PS II.

✓ Contrast cyclic and noncyclic photophosphorylation.

The pigments of photosystem I absorb light energy and transfer it to a neighboring molecule within the photosystem until the energy eventually arrives at a special chlorophyll molecule called the **reaction center chlorophyll** (Figure 5.27). Light energy from hundreds of such transfers excites electrons in the reaction center chlorophyll, which passes its excited electrons to the reaction center's electron acceptor, which is the initial carrier of an electron transport chain. As electrons move down the chain, their energy is used to pump protons across the membrane, creating a proton motive force. In prokaryotes, protons are pumped out of the cell; in eukaryotes they are pumped from the stroma into the interior of the thylakoids.

The proton motive force is used in chemiosmosis, in which, you should recall, protons flow down their electrochemical gradient through ATPases, which generate ATP. In photosynthesis, this process is called *photophosphorylation,* which can be either cyclic or noncyclic.

Figure 5.28 ▶

The light-dependent reactions of photosynthesis: cyclic and noncyclic photophosphorylation. The red arrows indicate the flow of electrons excited by light energy. **(a)** In cyclic photophosphorylation, electrons excited by light striking photosystem I travel from its reaction center, down an electron transport chain, and then return to the reaction center. The proton gradient produced drives the phosphorylation of ADP by ATP synthase (chemiosmosis). **(b)** In noncyclic photophosphorylation, light striking photosystem II excites electrons that are passed down an electron transport chain to photosystem I, simultaneously establishing a proton gradient that is used to phosphorylate ADP to ATP. Light energy collected by photosystem I further excites the electrons, which are used to reduce $NADP^+$ to NADPH via an electron transport chain. In oxygenic organisms, new electrons are provided by the lysis of H_2O (bottom left).

Cyclic Photophosphorylation

Electrons moving from one carrier molecule to another in a thylakoid must eventually pass to a final electron acceptor. In **cyclic photophosphorylation,** which occurs in all photosynthetic organisms, the final electron acceptor is the original reaction center chlorophyll that donated the electrons **(Figure 5.28a).** In other words, when light energy excites electrons in PS I, they pass down an electron transport chain and return to PS I. The energy from the electrons is used to establish a proton gradient that drives the phosphorylation of ADP to ATP by chemiosmosis.

Noncyclic Photophosphorylation

Some photosynthetic bacteria and all plants and algae also utilize **noncyclic photophosphorylation.** (Exceptions are green and purple sulfur bacteria.) Noncyclic photophosphorylation, which requires both PS I and PS II, not only generates molecules of ATP, but also reduces molecules of coenzyme $NADP^+$ to NADPH **(Figure 5.28b).**

When light energy excites electrons of PS II, they are passed to PS I through an electron transport chain. (Note that photosystem II occurs first in the pathway, and photosystem I second, a result of the fact that the photosystems were named for the order in which they were discovered, not the order in which they operate.) PS I further energizes the electrons with additional light energy and transfers them through an electron transport chain to $NADP^+$, which is thereby reduced to NADPH. Hydrogen ions in NADPH come from the stroma or cytosol. NADPH subsequently participates in the synthesis of glucose in the light-independent reactions, which we will examine in the next section.

In noncyclic photophosphorylation, a cell must constantly replenish electrons to the reaction center of photosystem II. *Oxygenic* (oxygen-producing) organisms, such as algae, green plants, and cyanobacteria, derive electrons from the dissociation of H_2O (see Figure 5.28b). In these organisms, two molecules of water give up their electrons, producing molecular oxygen (O_2) as a waste product during

(a) Cyclic photophosphorylation

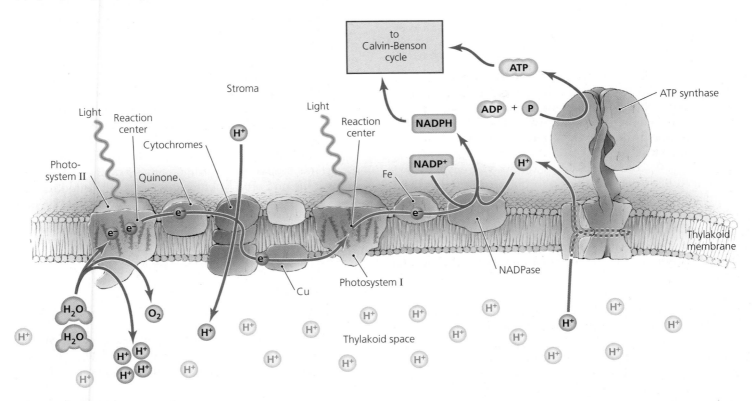

(b) Noncyclic photophosphorylation

photosynthesis. *Anoxygenic* photosynthetic bacteria get electrons from inorganic compounds such as H_2S, which results in a nonoxygen waste such as sulfur.

Table 5.5 on page 156 compares photophosphorylation to substrate-level and oxidative phosphorylation.

CRITICAL THINKING

We have seen that of the two ways that ATP is generated via chemiosmosis—photophosphorylation and oxidative phosphorylation—the former can be cyclical, but the latter is never cyclical. Why can't oxidative phosphorylation be cyclical; that is, why aren't electrons passed back to the molecules that donated them?

Table 5.5	A Comparison of the Three Types of Phosphorylation			
	Source of Phosphate	Source of Energy	Location in Eukaryotic Cell	Location in Prokaryotic Cell
Substrate-level phosphorylation	Organic molecule	High-energy bond between donor and phosphate	Cytosol and mitochondrial matrix	Cytosol
Oxidative phosphorylation	Inorganic phosphate (PO_4^{2-})	Proton motive force	Inner membrane of mitochondrion	Cell membrane
Photophosphorylation	Inorganic phosphate (PO_4^{2-})	Proton motive force	Thylakoid of chloroplast	Thylakoid of cell membrane

To this point, we have examined the use of photosynthetic pigments and thylakoid structure to harvest light energy to produce both ATP and reducing power in the form of NADPH. Next we examine the light-independent reactions of photosynthesis.

Light-Independent Reactions

Learning Objectives

✓ Contrast the light-dependent and light-independent reactions of photosynthesis.

✓ Describe the reactants and products of the Calvin-Benson cycle.

Light-independent reactions of photosynthesis do not require light directly; instead, they use ATP and NADPH generated by the light-dependent reactions. The key reaction of the light-independent pathway of photosynthesis is **carbon fixation** by the **Calvin-Benson cycle,**[10] which involves the attachment of molecules of CO_2 to molecules of a five-carbon organic compound called ribulose 1,5-bisphosphate (RuBP). RuBP is derived from phosphorylation of a precursor metabolite produced by the pentose phosphate pathway (see Figure 5.16).

The Calvin-Benson cycle is reminiscent of the Krebs cycle in that the substrates of the cycle are regenerated. It is helpful to notice the number of carbon atoms during each part of the three steps of the Calvin-Benson cycle **(Figure 5.29):**

1 Fixation of CO_2. Enzymes attach three molecules of carbon dioxide (3 carbon atoms) to three molecules of RuBP (15 carbon atoms), which are split to form six molecules of 3-phosphoglyceric acid (18 carbon atoms).

2 Reduction. NADPH reduces the six molecules of 3-phosphoglyceric acid to form six molecules of glyceraldehyde 3-phosphate (G3P) (18 C). These reactions require six molecules each of ATP and NADPH generated by the light-dependent reactions.

3 Regeneration of RuBP. The cell regenerates three molecules of RuBP (15 C) from five molecules of G3P (15 C); it also uses the remaining molecule of glyceraldehyde 3-

phosphate to synthesize glucose by reversing the reactions of glycolysis.

In summary, ATP and NADPH from the light-dependent reactions drive the synthesis of glucose from CO_2 in the light-independent reactions of the Calvin-Benson cycle. For every three molecules of CO_2 that enter the Calvin-Benson cycle, a molecule of glyceraldehyde 3-phosphate (G3P) leaves. Glycolysis is subsequently reversed to anabolically synthesize glucose 6-phosphate from two molecules of G3P. A more detailed account of the Calvin-Benson cycle is given in the Appendix on page A-7.

The processes of oxygenic photosynthesis and aerobic respiration complement one another to complete both a carbon cycle and an oxygen cycle. During the synthesis of glucose in oxygenic photosynthesis, water and carbon dioxide are used, and oxygen is released as a waste product; in aerobic respiration, oxygen serves as the final electron acceptor in the oxidation of glucose to carbon dioxide and water.

In this section we examined an essential anabolic pathway for life on Earth—photosynthesis, which produces glucose. Next we will consider anabolic pathways involved in the synthesis of other organic molecules.

Other Anabolic Pathways

Learning Objective

✓ Define amphibolic reaction.

Anabolic reactions are synthesis reactions. As such, they require energy and a source of metabolites. Energy for anabolism is provided by ATP generated in the catabolic reactions of aerobic respiration, anaerobic respiration, and fermentation, and by the initial redox reactions of photosynthesis. Glycolysis, the Krebs cycle, and the pentose phosphate pathway provide 12 basic precursor metabolites from which all macromolecules and cellular structures can be made. Some microorganisms, such as *E. coli*, can synthesize all 12 precursors (**Table 5.6** on page 160); other organisms, such as humans, must acquire some precursors in their diets.

Many anabolic pathways are the reversal of the catabolic pathways we have discussed; therefore, much of the

[10]Named for the men who elucidated its pathways.

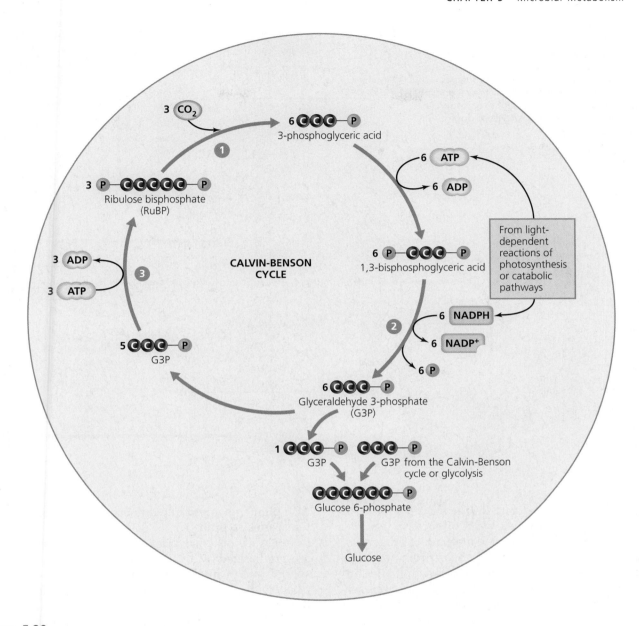

▲ Figure 5.29

Simplified diagram of the Calvin-Benson cycle ①. Three molecules of RuBP combine with three molecules of CO₂ ②. The resulting molecules are reduced to form six molecules of G3P ③. Five molecules of G3P are converted to three molecules of RuBP, which completes the cycle. Two turns of the cycle also yield two molecules of G3P, which are polymerized to synthesize glucose 6-phosphate.

material that follows has *in a sense* been discussed in the sections on catabolism. Reactions that can proceed in either direction—toward catabolism or toward anabolism—are said to be **amphibolic.** The following sections discuss the synthesis of carbohydrates, lipids, amino acids, and nucleotides. The polymerizations of amino acids into proteins, and of nucleotides into RNA and DNA, are closely linked to genetics, so they are covered in Chapter 7.

Carbohydrate Biosynthesis

Learning Objective

✓ Describe the biosynthesis of carbohydrates.

As we have seen, anabolism begins in photosynthetic organisms with carbon fixation by the enzymes of the Calvin-Benson cycle to form molecules of G3P. Enzymes use G3P as the starting point for synthesizing sugars, complex

Figure 5.30 ➤

The role of gluconeogenesis in the biosynthesis of complex carbohydrates. Complex carbohydrates are synthesized from simple sugar molecules such as glucose, glucose-6 phosphate, and fructose-6 phosphate. Starch and cellulose are found in algae; glycogen is found in animals and protozoa; and peptidoglycan is found in bacteria.

polysaccharides such as starch, cellulose for cell walls in algae, and peptidoglycan for cell walls of bacteria. Animals and protozoa synthesize the storage molecule glycogen.

Some cells are able to synthesize sugars from noncarbohydrate precursors such as amino acids, glycerol, and fatty acids by pathways collectively called *gluconeogenesis*[11] (glū'kō-nē-ō-jen'ě-sis; **Figure 5.30**). Most of the reactions of gluconeogenesis are amphibolic, using enzymes of glycolysis in reverse, but four of the reactions require unique enzymes. Gluconeogenesis is highly endergonic and can proceed only if there is an adequate supply of energy.

Lipid Biosynthesis

Learning Objective

✓ Describe the biosynthesis of lipids.

As we saw in Chapter 2, lipids are a diverse group of organic molecules that function as energy-storage compounds and as components of membranes. *Carotenoids*, which are red-

dish pigments found in many bacterial and plant photosystems, are also lipids.

Because of their variety, it is not surprising that lipids are synthesized by a variety of routes. For example, fats are synthesized in anabolic reactions that are the reverse of their catabolism—they polymerize glycerol and three fatty acids **(Figure 5.31)**. Glycerol is derived from G3P generated by the Calvin-Benson cycle and glycolysis; the fatty acids are produced by the linkage of two-carbon acetyl-CoA molecules to one another by a sequence of endergonic reactions that effectively reverse the catabolic reactions of beta-oxidation. Other lipids, such as steroids, are synthesized in complex pathways involving polymerizations and isomerizations of sugar and amino acid metabolites.

Mycobacterium tuberculosis (mī'kō-bak-tēr'ē-ŭm tū-ber-kyū-lō-sis), the pathogen that causes tuberculosis, makes copious amounts of a waxy lipid called *mycolic acid*, which is incorporated into its cell wall. The arduous, energy-intensive process of making long lipid chains explains why this organism grows slowly, and why tuberculosis requires a long course of treatment with antimicrobial drugs.

[11]From Greek *glykys*, meaning sweet, *neo*, meaning new, and *genesis*, meaning generate.

Figure 5.31 ➤

Biosynthesis of fat, a lipid. A fat molecule is synthesized from glycerol and three molecules of fatty acid, the precursors of which are produced in glycolysis.

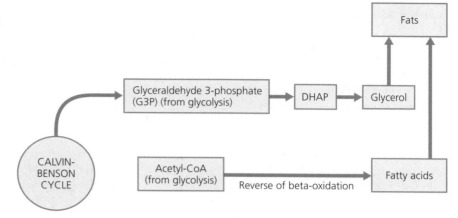

Figure 5.32

The synthesis of amino acids via amination and transamination. **(a)** In amination, an amine group from ammonia is added to a precursor metabolite. In this example, oxaloacetic acid is converted into the amino acid aspartic acid. **(b)** In transamination, the amine group is derived from an existing amino acid. In the example shown here, the amino acid glutamic acid donates its amine group to oxaloacetic acid, which is converted into aspartic acid. Note that transamination is a reversible reaction.

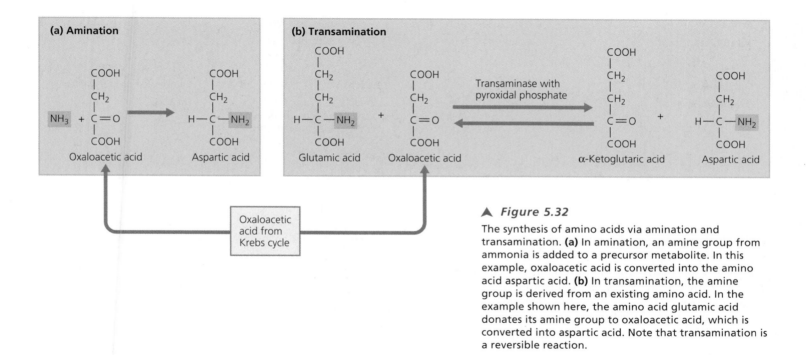

Amino Acid Biosynthesis

Learning Objective

✓ Describe the biosynthesis of amino acids.

Amino acids are synthesized from precursor metabolites derived from glycolysis, the Krebs cycle, and the pentose phosphate pathway, and from other amino acids. Some organisms (such as *E. coli,* and most plants and algae) synthesize all their amino acids from precursor metabolites. Other organisms, including humans, must acquire so-called *essential amino acids* in their diets because they cannot synthesize them. One extreme example is *Lactobacillus* (lak-tō-

bǎ-sil'ŭs), a bacterium that ferments milk and produces some cheeses. This microorganism cannot synthesize any of its amino acids; it acquires all of them by catabolizing proteins in its environment.

Precursor metabolites are converted to amino acids by the addition of an amine group. This process is called **amination** when the amine group comes from ammonia (NH_3); an example is the formation of aspartic acid from NH_3 and the Krebs cycle intermediate oxaloacetic acid **(Figure 5.32a).** Amination reactions are the reverse of the catabolic deamination reactions we discussed previously. More commonly, however, a cell moves the amine group from one amino acid and adds it to a metabolite, producing a different amino

Table 5.6 The Twelve Precursor Metabolites

	Pathway that Generates the Metabolite	Examples of Macromolecule Synthesized from Metabolite[a]	Examples of Functional Use
Glucose 6-phosphate	Glycolysis	Lipopolysaccharide	Outer membrane of cell wall
Fructose 6-phosphate	Glycolysis	Peptidoglycan	Cell wall
Glyceraldehyde-3-phosphate (G3P)	Glycolysis	Glycerol portion of lipids	Fats—energy storage
Phosphoglyceric acid	Glycolysis	Amino acids: cysteine, glycine, and serine	Enzymes
Phosphoenolpyruvic acid (PEP)	Glycolysis	Amino acids: phenylalanine, tryptophan, and tyrosine	Enzymes
Pyruvic acid	Glycolysis	Amino acids: alanine, leucine, and valine	Enzymes
Ribose 5-phosphate	Pentose phosphate pathway	DNA, RNA, amino acid and histidine	Genome, enzymes
Erythrose 4-phosphate	Pentose phosphate pathway	Amino acids: phenylalanine, tryptophan, and tyrosine	Enzymes
Acetyl-CoA	Krebs cycle	Fatty acid portion of lipids	Cell membrane
α-Ketogluteric acid	Krebs cycle	Amino acids: arginine, glutamic acid, glutamine, and proline	Enzymes
Succinyl-CoA	Krebs cycle	Heme	Cytochrome electron carrier
Oxaloacetate	Krebs cycle	Amino acids: aspartic acid, asparagine, isoleucine, lysine, methionine, and threonine	Enzymes

[a]Examples given apply to the bacterium *E. coli.*

acid. This process is called **transamination** because the amine group is transferred from one amino acid to another **(Figure 5.32b)**. All transamination enzymes use a coenzyme, *pyridoxal phosphate,* which is derived from vitamin B_6.

Ribozymes of ribosomes polymerize amino acids into proteins. This energy-demanding process is examined in Chapter 7 because it is intimately linked with genetics.

CRITICAL THINKING

What class of enzyme is involved in amination reactions? What class of enzyme catalyzes transaminations?

Nucleotide Biosynthesis

Learning Objective

✓ Describe the biosynthesis of nucleotides.

Recall from Chapter 2 that the building blocks of nucleic acids are nucleotides, each of which consists of a five-carbon sugar, a phosphate group, and a purine or pyrimidine base (see Figure 2.24a). Nucleotides are produced from precursor metabolites of glycolysis and the Krebs cycle **(Figure 5.33):**

- The five-carbon sugars—ribose in RNA and deoxyribose in DNA—are derived from ribose-5 phosphate from the pentose phosphate pathway.
- The phosphate group is derived ultimately from ATP.
- Purines and pyrimidines are synthesized in a series of ATP-requiring reactions from the amino acids glutamine and aspartic acid derived from Krebs cycle intermediates, ribose-5 phosphate and *folic acid* (which is a vitamin for humans but can be synthesized by many bacteria and protozoa).

The anabolic reactions by which polymerases polymerize nucleotides to form DNA and RNA are discussed in Chapter 7.

Integration and Regulation of Metabolic Functions

Learning Objectives

✓ Describe interrelationships between catabolism and anabolism in terms of ATP and substrates.
✓ Discuss regulation of metabolic activity.

As we have seen, catabolic and anabolic reactions interact with one another in several ways. First, ATP molecules

Figure 5.33 ➤
The biosynthesis of nucleotides.

produced by catabolism are used to drive anabolic reactions. Second, precursor metabolites produced by catabolic pathways are used as substrates for anabolic reactions. Additionally, most metabolic pathways are amphibolic; they function as part of either catabolism or anabolism as needed.

Cells regulate metabolism in a variety of ways to maximize efficiency in growth and reproductive rate. Among the mechanisms involved are the following:

- Cells synthesize or degrade channel and transport proteins to increase or decrease the concentration of chemicals in the cytosol and organelles.

- Cells often synthesize the enzymes needed to catabolize a particular substrate only when that substrate is available. For instance, the enzymes of beta-oxidation are not produced when there are no fatty acids to catabolize.

- If two energy sources are available, cells catabolize the more energy efficient of the two. For example, a bacterium growing in the presence of both glucose and lactose will produce enzymes only for the transport and catabolism of glucose. Once the supply of glucose is depleted, lactose-utilizing proteins are produced.

- Cells synthesize the metabolites they need, but they typically cease synthesis if a metabolite is available as a nutrient. For instance, bacteria grown with an excess of aspartic acid will cease the amination of oxaloacetic

acid (see Figure 5.32).

- Eukaryotic cells keep metabolic processes from interfering with each other by isolating particular enzymes within membrane-bounded organelles. For example, proteases sequestered within lysosomes digest phagocytized proteins without destroying vital proteins in the cytosol.

- Cells use inhibitory and excitatory allosteric sites on enzymes to control the activity of enzymes (see Figure 5.11).

- Feedback inhibition slows or stops anabolic pathways when the product is in abundance (see Figure 5.12).

- Cells regulate catabolic and anabolic pathways that use the same substrate molecules by requiring different coenzymes for each. For instance, NADH is used almost exclusively with catabolic enzymes, whereas NADPH is typically used for anabolism.

Note that these regulatory mechanisms are generally of two types: *control of gene expression,* in which cells control the amount and timing of protein (enzyme) production, and *control of metabolic expression,* in which cells control the activity of proteins (enzymes) once they have been produced.

Numerous interrelationships among the metabolic pathways discussed in this chapter are diagrammed schematically in **Figure 5.34** on page 162.

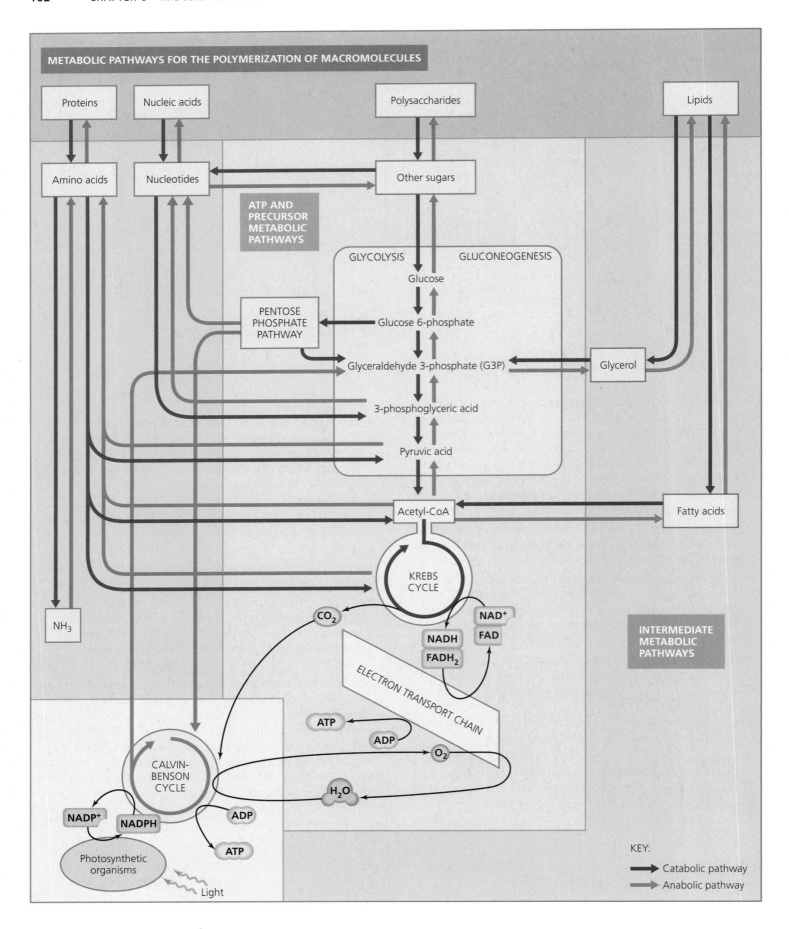

◀ *Figure 5.34*

Integration of cellular metabolism. Cells possess three major categories of metabolic pathways: pathways for the polymerization of macromolecules (proteins, nucleic acids, polysaccharides, and lipids), intermediate pathways, and ATP and precursor pathways (glycolysis, Krebs cycle, the pentose phosphate pathway, and the Entner-Doudoroff pathway [not shown]). Cells of photosynthetic organisms also have the Calvin-Benson cycle

CHAPTER SUMMARY

Basic Chemical Reactions Underlying Metabolism (pp. 126–136)

1. **Metabolism** is the sum of complex biochemical reactions within the cells of an organism, including **catabolism,** which breaks down molecules and releases energy, and **anabolism,** which synthesizes molecules and uses energy.

2. **Precursor metabolites,** often produced in catabolic reactions, are used to synthesize all other organic compounds.

3. **Reduction** reactions are those in which the electrical charge on a chemical is made more negative by the addition of electrons. The molecule that donates an electron is **oxidized.** If the electron is part of a hydrogen atom, an oxidation reaction is also called dehydrogenation. Oxidation and reduction reactions always occur in pairs called **oxidation-reduction (redox) reactions**.

4. Three important electron carrier molecules are **nicotinamide adenine dinucleotide (NAD^+), nicotinamide adenine dinucleotide phosphate ($NADP^+$),** and **flavine adenine dinucleotide (FAD).**

5. Phosphorylation is the addition of phosphate to a molecule. Three types of phosphorylation form ATP: **Substrate-level phosphorylation** involves the transfer of phosphate to ADP from another phosphorylated organic compound. In **oxidative phosphorylation,** energy from redox reactions of respiration is used to attach inorganic phosphate (PO_4^{2-}) to ADP. **Photophosphorylation** is the phosphorylation of ADP with inorganic phosphate using energy from light.

6. Catalysts increase the rates of chemical reactions and are not permanently changed in the process. **Enzymes,** which are organic catalysts, are often named for their **substrates,** the molecules upon which they act. Enzymes can be classified according to their mode of action into hydrolases, isomerases, ligases (polymerases), lyases, oxidoreductases, or transferases.

7. **Apoenzymes** are the protein portions of protein enzymes that may require one or more **cofactors** such as inorganic ions or organic cofactors called **coenzymes.** The combination of both apoenzyme and its cofactors is a **holoenzyme.** RNA molecules functioning as enzymes are called **ribozymes.**

8. **Activation energy** is the amount of energy required to initiate a chemical reaction.

9. Substrates fit into the specifically shaped **active sites** of the enzymes that catalyze their reactions.

10. Enzymes may be **denatured** by physical and chemical factors such as heat and pH. Denaturation may be reversible or permanent.

11. Enzyme activity proceeds at a rate proportional to the concentration of substrate molecules until all the active sites are filled to saturation.

12. **Competitive inhibitors** block active sites and thereby block enzyme activity. In **allosteric inhibition,** noncompetitive inhibitors attach to an allosteric site on an enzyme, altering the active site so that it is no longer functional.

13. **Feedback inhibition (negative feedback)** occurs when the final product of a series of reactions is an allosteric inhibitor of some previous step in the series. Thus, accumulation of the end-product "feeds back" into the series a signal that stops the process.

Carbohydrate Catabolism (pp. 136–149)

1. **Glycolysis** involves the splitting of a glucose molecule in a three-stage, 10-step process that ultimately results in two molecules of pyruvic acid and a net gain of two ATP and two NADH molecules.

2. The **pentose phosphate** and **Entner-Doudoroff pathways** are alternative means for the catabolism of glucose that yield fewer ATP molecules than does glycolysis. However, they produce precursor metabolites not produced in glycolysis.

3. **Cellular respiration** is a metabolic process that involves the complete oxidation of substrate molecules and the production of ATP following a series of redox reactions.

4. Two carbons from pyruvic acid join coenzyme A to form acetyl-coenzyme A (**acetyl-CoA**), which then enters the **Krebs cycle,** a series of eight enzymatic steps that transfer energy from acetyl-CoA to coenzymes NAD^+ and FAD.

5. An **electron transport chain** is a series of redox reactions that pass electrons from one membrane-bound carrier to another, and then to a final electron acceptor. The energy from these electrons is used to pump protons across the membrane.

6. The four classes of carrier molecules in electron transport systems are flavoproteins, ubiquinones, metal-containing proteins, and cytochromes.

7. Aerobes use oxygen atoms as final electron acceptors of their electron transport chains in a process known as **aerobic respiration,** whereas anaerobes use other inorganic molecules (such as NO_3^-, SO_4^{2-}, and CO_3^{2-}, or rarely an organic molecule) as the final electron acceptor in **anaerobic respiration.**

8. In **chemiosmosis,** ions flow down their electrochemical gradient across a membrane through **ATP synthase (ATPase)** to synthesize ATP.

9. A **proton gradient** is an electrochemical gradient of hydrogen ions across a membrane. It has potential energy known as a proton motive force.

10. Oxidative phosphorylation and photophosphorylation use chemiosmosis.

11. **Fermentation** is the partial oxidation of sugar to release energy using an organic molecule rather than an electron transport chain as the final electron acceptor. End-products of fermentation, which are often useful to humans and aid in identification of microbes in the laboratory, include acids, alcohols, and gases.

Other Catabolic Pathways (pp. 149–152)

1. Lipids and proteins can be catabolized into smaller molecules, which can be used as substrates for glycolysis and the Krebs cycle.

2. **Beta-oxidation** is a catabolic process in which enzymes split pairs of hydrogenated carbon atoms from a fatty acid and join them to coenzyme A to form acetyl-CoA.

3. **Proteases** secreted by microorganisms digest proteins outside the microbe's cell wall. The resulting amino acids are moved into the cell and used in anabolism, or **deaminated** and catabolized for energy.

Photosynthesis (pp. 152–156)

1. **Photosynthesis** is a process in which light energy is captured by pigment molecules called **chlorophylls** (bacteriochlorophylls in some bacteria) and transferred to ATP and metabolites. **Photosystems** are networks of light-absorbing chlorophyll molecules and other pigments held within a protein matrix on membranes called **thylakoids.**

2. The redox reactions of photosynthesis are classified as **light-dependent reactions** and **light-independent reactions.**

3. A **reaction center** chlorophyll is a special chlorophyll molecule in a photosystem in which electrons are excited by light energy

and passed to an acceptor molecule of an electron transport chain.

4. In **cyclic photophosphorylation,** the electrons return to the original reaction center after passing down the electron transport chain.

5. In **noncyclic photophosphorylation,** photosystem II works with photosystem I, and the electrons are used to reduce $NADP^+$ to NADPH. In oxygenic photosynthesis, cyanobacteria, algae, and green plants replenish electrons to the reaction center by dissociation of H_2O molecules, resulting in the release of O_2 molecules. Anoxygenic bacteria derive electrons from inorganic compounds such as H_2S, producing waste such as sulfur.

6. In the light-independent pathway of photosynthesis, **carbon fixation** occurs in the **Calvin-Benson cycle,** in which CO_2 is reduced to produce glucose.

Other Anabolic Pathways (pp. 156–160)

1. **Amphibolic reactions** are metabolic reactions that are reversible.

2. Some cells are able to synthesize glucose from amino acids, glycerol, and fatty acids via a process called gluconeogenesis.

3. **Amination** reactions involve adding an amine group from ammonia to a metabolite to make an amino acid. **Transamination** occurs when an amine group is transferred from one amino acid to another.

4. Nucleotides are synthesized from precursor metabolites produced by glycolysis, the Krebs cycle, and the pentose phosphate pathway.

Integration and Regulation of Metabolic Functions (pp. 160–163)

1. Cells regulate metabolism by control of gene expression or metabolic expression. They control the latter in a variety of ways, including synthesizing or degrading channel proteins and enzymes, sequestering reactions in membrane-bounded organelles (seen only in eukaryotes), and feedback inhibition.

QUESTIONS FOR REVIEW

(Answers to multiple choice, matching questions, and fill in the blanks are on the web, along with additional review questions. Visit www.microbiologyplace.com.)

Multiple Choice

For each of the phrases in Questions 1–7, indicate the type of metabolism referred to, using the following choices:
  a. anabolism only
  b. both anabolism and catabolism (amphibolic)
  c. catabolism only

1. Breaks a large molecule into smaller ones

2. Includes dehydration synthesis reactions

3. Is exergonic

4. Is endergonic

5. Involves the production of cell membrane constituents

6. Includes hydrolytic reactions

7. Includes metabolism

8. Redox reactions
  a. transfer energy.
  b. transfer electrons.
  c. involve oxidation and reduction.
  d. are involved in all of the above.

9. A reduced molecule
  a. has gained electrons.
  b. has become more positive in charge.

c. has lost electrons.

d. is an electron donor.

10. Activation energy
    a. is the amount of energy required during an activity such as flagellar motion.
    b. requires the addition of nutrients in the presence of water.
    c. is usually lowered by the action of organic catalysts.
    d. results from the movement of molecules.

11. Coenzymes
    a. are types of apoenzymes.
    b. are proteins.
    c. are inorganic cofactors.
    d. are organic cofactors.

12. Which of the following statements best describes ribozymes?
    a. Ribozymes are proteins that aid in the production of ribosomes.
    b. Ribozymes are nucleic acids that produce ribose sugars.
    c. Ribozymes store enzymes in ribosomes.
    d. Ribozymes process RNA molecules in eukaryotes.

13. Which of the following does *not* affect the function of enzymes?
    a. ubiquinone
    b. substrate concentration
    c. temperature
    d. competitive inhibitors

14. Most oxidation reactions in bacteria involve the
    a. removal of hydrogen ions and electrons.
    b. removal of oxygen.
    c. addition of hydrogen ions and electrons.
    d. addition of hydrogen ions.

15. Under ideal conditions, the fermentation of one glucose molecule by a bacterium allows a net gain of how many ATP molecules?
    a. 2
    b. 4
    c. 38
    d. 0

16. Under ideal conditions, the complete aerobic oxidation of one molecule of glucose by a bacterium allows a net gain of how many ATP molecules?
    a. 2
    b. 4
    c. 38
    d. 0

17. Which of the following statements is *not* true of the Entner-Doudoroff pathway?
    a. It is a series of reactions that synthesizes glucose.
    b. Its products are sometimes used to determine the presence of *Pseudomonas*.
    c. It is a pathway of chemical reactions that catabolizes glucose.
    d. It is an alternative pathway to glycolysis.

18. Reactions involved in the light-independent reactions of photosynthesis constitute the
    a. Krebs cycle.
    b. Entner-Doudoroff pathway.
    c. Calvin-Benson cycle.
    d. pentose phosphate pathway.

Matching Questions

19. ___ Occurs when energy from a compound containing phosphate reacts with ADP to form ATP

20. ___ Involves formation of ATP via reduction of coenzymes in the electron transport chain

21. ___ Begins with glycolysis

22. ___ Occurs when all active sites on substrate molecules are filled

a. Saturation

b. Oxidative phosphorylation

c. Substrate-level phosphorylation

d. Photophosphorylation

e. Carbohydrate catabolism

Fill in the Blanks

23. The final electron acceptor in cyclic photophosphorylation is _____.

24. Two ATP molecules are used to initiate glycolysis. Enzymes generate molecules of ATP for each molecule of glucose that undergoes glycolysis. Thus a net gain of _____ molecules of ATP is produced in glycolysis.

25. The initial catabolism of glucose occurs by glycolysis and/or the _____ and _____ pathways.

26. _____ is a cyclic series of eight reactions involved in the catabolism of acetyl-CoA that yields eight molecules of NADH and two molecules of $FADH_2$.

27. The final electron *acceptor* in aerobic respiration is _____.

28. Three common electron acceptors in anaerobic respiration are _____, _____, and _____.

29. Chemolithotrophs *acquire* electrons from (organic/inorganic) _____ compounds.

30. Complete the following chart:

Category of Enzymes	Description
	Catabolyzes substrate by adding water
Isomerase	
Ligase/polymerase	
	Moves functional groups such as an acetyl group
	Adds or removes electrons
Lyase	

Short Answer

1. Why are enzymes necessary for anabolic reactions to occur in living organisms?

2. How do organisms control the rate of metabolic activities in their cells?

3. How does a noncompetitive inhibitor at an allosteric site affect a whole pathway of enzymatic reactions?

4. Explain the mechanism of negative feedback with respect to enzyme action.

5. Facultative anaerobes can live under either aerobic or anaerobic conditions. What metabolic pathways allow these organisms to continue to harvest energy from sugar molecules in the absence of oxygen?

6. How does oxidation of a molecule occur without oxygen?

7. List at least four groups of microorganisms that are photosynthetic.

8. Why do we breathe oxygen and give off carbon dioxide?

9. Why do cyanobacteria and algae take in carbon dioxide and give off oxygen?

10. What happens to the carbon atoms in sugar catabolized by *E. coli*?

11. How do yeast cells make alcohol and cause bread to rise?

12. Where specifically does the most significant production of ATP occur in prokaryotic and eukaryotic cells?

13. Why are vitamins essential metabolic factors for microbial metabolism?

14. A laboratory scientist notices that a certain bacterium does not utilize lactose when glucose is available in its environment. Describe a cellular regulatory mechanism that would explain this observation.

CRITICAL THINKING

1. Why does an organism that uses glycolysis and the Krebs cycle also need the pentose phosphate pathway?

2. Describe how bacterial fermentation causes milk to sour.

3. *Giardia lamblia* and *Entamoeba histolytica* are protozoa that live in the colons of mammals and can cause life-threatening diarrhea. Interestingly, these microbes lack mitochondria. What kind of pathway must they have for carbohydrate catabolism?

4. Two cultures of a facultative anaerobe are grown in the same type of medium, but one is exposed to air and the other is maintained under anaerobic conditions. Which of the two cultures will contain more cells at the end of a week? Why?

5. What is the maximum number of molecules of ATP that can be generated by a bacterium after the complete aerobic oxidation of a fat molecule containing three 12-carbon chains? (Assume that all the available energy released during catabolism goes to ATP production.)

6. In terms of its effects on metabolism, why is a fever over 40°C often life threatening?

7. Using the categories defined in the text on page 129, classify each of the enzymes of glycolysis (see pages A-2 and A-3), the Krebs cycle (see page A-6), and transamination (see Figure 5.32 on page 159).

8. How are photophosphorylation and oxidative phosphorylation similar? How are they different?

9. Members of the pathogenic bacterial genus *Haemophilus* require NAD^+ and heme from their environment. For what purpose does *Haemophilus* use these growth factors?

10. Compare and contrast aerobic respiration, anaerobic respiration, and fermentation.

11. Scientists estimate that up to one-third of Earth's biomass is composed of methanogenic prokaryotes in ocean sediments (*Science* 295:2067–2070). Describe the metabolism of these organisms.

12. A young student was troubled by the idea that a bacterium is able to control its diverse and complex metabolic activities, even though it lacks a brain. How would you explain their metabolic control?

13. If a bacterium uses beta-oxidation to catabolize a molecule of the fatty acid arachidic acid, which contains 20 carbon atoms, how many acetyl-CoA molecules will be generated?

CHAPTER 6

Microbial Nutrition and Growth

The dramatic red colors of Owens Lake in Nevada are caused by *halobacteria*—bacteria that thrive in extremely salty environments. Halobacteria thrive in water that is several times the salinity of seawater, and are also found in such notably salty habitats as the Great Salt Lake and the Dead Sea. The red colors of Owens Lake are the result of a pigment produced by the bacteria, which are present in staggeringly large numbers in the lake: Just one drop of the concentrated brine may contain millions of halobacteria.

Halobacteria are just one example of extremophiles, microorganisms that live in extreme environments. From the canyons of Death Valley to subglacial lakes deep beneath the ice sheets of Antarctica, microorganisms live and grow in environments uninhabitable by other life forms. How do they obtain the nutrients they need? What characteristics allow them to survive such harsh conditions? In this chapter we will study the requirements for microbial growth.

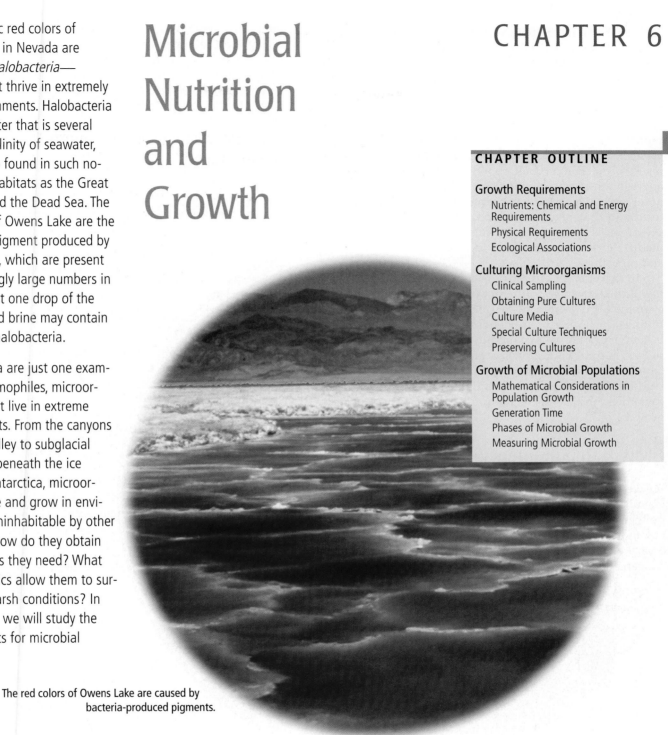

The red colors of Owens Lake are caused by bacteria-produced pigments.

CHAPTER OUTLINE

Growth Requirements
Nutrients: Chemical and Energy Requirements
Physical Requirements
Ecological Associations

Culturing Microorganisms
Clinical Sampling
Obtaining Pure Cultures
Culture Media
Special Culture Techniques
Preserving Cultures

Growth of Microbial Populations
Mathematical Considerations in Population Growth
Generation Time
Phases of Microbial Growth
Measuring Microbial Growth

MicroPrep Pre-Test: Take the pre-test for this chapter on the web.
Visit **www.microbiologyplace.com.**

We saw in Chapter 5 that metabolism—the collection of controlled chemical reactions within cells—is a major characteristic of all living things. Recall as well that the ultimate outcome of metabolic activity is reproduction, an increase in the number of individual cells or organisms. When speaking of the reproductive activities of microbes in general, and of bacteria in particular, microbiologists typically use the term *growth,* referring to an increase in a *population* of microbes rather than to an increase in size. The result of such microbial growth is a discrete **colony,** an aggregation of cells arising from a single parent cell. Put another way, the *reproduction* of individual microorganisms results in the *growth* of a colony. Further, the common expression "The microorganisms *grow* in salt-containing media" is widely understood to mean that they metabolize and reproduce.

In this chapter we consider the characteristics of microbial growth from two different but related perspectives: We examine the requirements of microbes living in their natural settings, including their chemical, physical, and energy requirements, and we explore how microbiologists try to duplicate natural conditions to grow microorganisms in the laboratory so that they can be transported, identified, and studied. We conclude by examining bacterial population dynamics and some techniques for measuring bacterial growth.

Growth Requirements

As we discussed in Chapter 5, organisms use a variety of chemicals—called **nutrients**—for their energy needs and to build organic molecules and cellular structures. The most common of these nutrients are compounds containing necessary elements such as carbon, oxygen, nitrogen, and hydrogen (Table 6.1 on page 170). Like all organisms, microbes obtain nutrients from a variety of sources in their environment, and they must bring nutrients into their cells by the passive and active transport processes we discussed in Chapter 3. When they acquire their nutrients by living in or on another organism, they may cause disease as they interfere with their hosts' metabolism and nutrition.

Nutrients: Chemical and Energy Requirements

Learning Objectives

✓ Describe the roles of carbon, hydrogen, oxygen, nitrogen, trace elements, and vitamins in microbial growth and reproduction.

✓ Compare the four basic categories of organisms based on their carbon and energy sources.

✓ Distinguish among anaerobes, aerobes, aerotolerant anaerobes, facultative anaerobes, microaerophiles, and capnophiles.

✓ Explain how oxygen can be fatal to organisms by discussing singlet oxygen, superoxide radical, peroxide anion, and hydroxyl radical, and describe how organisms protect themselves from toxic forms of oxygen.

We begin our examination of microbial growth requirements by considering three things all cells need for metabolism: a carbon source, a source of energy, and a source of electrons or hydrogen atoms.

Sources of Carbon, Energy, and Electrons

Organisms can be categorized into two broad groups based on their source of carbon. Organisms that utilize an inorganic carbon source (that is, carbon dioxide) as their sole source of carbon are called *autotrophs* (aw'tō-trōfs), so-named because they "feed themselves."[1] More precisely, autotrophs make organic compounds from CO_2 and thus need

_____

[1]From Greek *auto,* meaning self, and *trophe,* meaning nutrition.

	Energy Source	
	Light *(photo-)*	**Chemical compounds *(chemo-)***
Carbon dioxide *(auto-)*	***Photoautotrophs*** • Plants, algae, and cyanobacteria use H_2O to reduce CO_2, producing O_2 as a byproduct • Photosynthetic green sulfur and purple sulfur bacteria do not use H_2O nor produce O_2	***Chemoautotrophs*** • Hydrogen, sulfur, and nitrifying bacteria
Organic compounds *(hetero-)*	***Photoheterotrophs*** • Green nonsulfur and purple nonsulfur bacteria	***Chemoheterotrophs*** • Aerobic respiration: most animals, fungi, and protozoa, and many bacteria • Anaerobic respiration: some animals, protozoa, and bacteria • Fermentation: some bacteria and yeasts

(Left axis label: Carbon Source)

◀ *Figure 6.1*

The four basic groups of organisms based on their carbon and energy sources. Additionally, every organism utilizes electrons from either organic molecules (organotrophs) or inorganic molecules (lithotrophs) in redox reactions.

New Frontiers 6.1   Is There Life on Other Planets?

We are accustomed to thinking of a biological world based upon photosynthesis—photosynthetic organisms provide a suitable carbon source for animals, fungi, protozoa, and most bacteria, which are then food for larger organisms, including humans. However, based on evidence from samples within the earth's crust, many living things are lithotrophic chemoautotrophic prokaryotes that acquire all of their nutritional and energy needs from inorganic chemicals such as hydrogen, sulfur, and ferrous iron (Fe^{2+}). Such microbes also fill the role of primary producers and support entire communities of animals on the bottoms of the oceans, where no light penetrates. Some scientists believe that organisms of this type were the first living things on Earth, and that they are the only things that could live on other planets in the solar system.

Additionally, scientists have recently discovered novel microorganisms living in hydrothermal waters far beneath the mountain ranges of Idaho. These chemoautolithotrophic microorganisms, 90% of which are archaea, live in an environment completely devoid of sunlight. Instead of obtaining energy from light or organic matter, they use hydrogen as an energy source. Because the habitat of these microorganisms is very similar to what we know of the environment on Mars, some scientists speculate that microbial life may be possible on the Red Planet.

Mars

---

not feed on organic compounds from other organisms to acquire carbon. In contrast, organisms called *heterotrophs*[2] (het'er-ō-trōfs) catabolize reduced organic molecules (such as proteins, carbohydrates, amino acids, and fatty acids) they acquire from other organisms.

Organisms can also be categorized according to whether they use chemicals or light as a source of energy for such cellular processes as anabolism, intracellular transport, and motility. Organisms that acquire energy from redox reactions involving inorganic and organic chemicals are called *chemotrophs* (kēm'ō-trōfs). (Recall from Chapter 5 that these reactions are either aerobic respiration, anaerobic respiration, or fermentation, depending on the final electron acceptor). Organisms that use light as their energy source are called *phototrophs*[3] (fō'tō-trōfs).

Based on their carbon and energy sources, then, most organisms can be categorized into one of four basic groups: **photoautotrophs, chemoautotrophs, photoheterotrophs,** and **chemoheterotrophs (Figure 6.1).** Plants, some protozoa, and algae are photoautotrophs and animals, fungi, and other protozoa are chemoheterotrophs. Prokaryotes are most diverse, with members in all four groups.

Additionally, the cells of all organisms require electrons or hydrogen atoms. Hydrogen is the most common chemical element in cells, and it is so common in organic molecules and water that it is never a *limiting nutrient*; that is, metabolism is never interrupted by a lack of hydrogen. Hydrogen is essential for hydrogen bonding and in electron transfer. Organisms that acquire electrons (typically as part of hydrogen atoms) from the same organic molecules that provide them carbon and energy are called **organotrophs** (ōr'găn-ō-trōfs); alternatively, organisms that acquire electrons or hydrogen atoms

from inorganic sources (such as H_2, NO^{2-}, H_2S, and Fe^{2+}) are called **lithotrophs**[4] (lith'ō-trōfs). **New Frontiers 6.1** focuses on lithotrophic chemoautotrophs (also called chemolithoautotrophs), which are prokaryotic organisms that acquire carbon, energy, and electrons from inorganic chemicals.

CRITICAL THINKING

The filamentous bacterium *Beggiatoa* gets its carbon from carbon dioxide and electrons and energy from hydrogen sulfide. What is its nutritional classification?

Not all organisms are easy to classify. For instance, *Euglena granulata* typically uses light energy and gets its carbon from carbon dioxide. However, when it is cultured on a suitable medium in the dark, this microbe utilizes energy and carbon solely from organic compounds. What is the nutritional classification of *Euglena*?

Oxygen Requirements

As we learned in Chapter 5, oxygen is essential for **obligate aerobes** because it serves as the final electron acceptor of electron transport chains, which produce most of the ATP in these organisms. By contrast, oxygen is a deadly poison for **obligate anaerobes.** How can oxygen be a fatal toxin for one group of organisms and yet be essential for another?

The key to understanding this apparent incongruity is understanding that neither gaseous molecular oxygen (O_2) of the atmosphere nor the oxygen that is covalently bound in compounds such as carbohydrates and water is poisonous.

---

[2]From Greek *hetero*, meaning other, and *trophe*, meaning nutrition.
[3]From Greek *photos*, meaning light, and *trophe*, meaning nutrition.
[4]From Greek *lithos*, meaning rock, and *trophe*, meaning nutrition.

Table 6.1   Elements Required by Microorganisms

Element	Proportion of Prokaryotic Cells' Dry Weight	FUNCTION
Carbon	50%	Basic structural component of organic compounds
Oxygen	20%	Component of many organic and inorganic compounds; O_2 is the final electron acceptor in aerobic respiration
Nitrogen	14%	Constituent of amino acids and nucleotides
Hydrogen	8%	Constituent of organic compounds; electrons of hydrogen atoms are used in redox reactions
Phosphorus	3%	As phosphate (PO_4^{2-}), constituent of nucleic acids, phospholipids, some coenzymes, and ATP
Sodium	1%	Cation used to maintain electrical balance of cells
Sulfur	1%	Constituent of the amino acids cysteine and methionine; disulfide bonds are important for tertiary structure of proteins
Potassium	1%	Inorganic cation involved in maintaining electrical balance inside cells; cofactor for some enzymes, including some involved in protein synthesis
Chlorine	0.5%	Principal inorganic anion involved in maintaining electrical balance of cells
Magnesium	0.5%	Component of chlorophyll; cofactor for some enzymes
Calcium	0.5%	Responsible in part for the resistance of bacterial endospores; cofactor for some enzymes; constituent of some algal cell walls and protozoan shells
Iron	0.25%	Constituent of cytochromes in electron transport chains; cofactor for some enzymes
Trace elements (cobalt, copper, manganese, molybdenum, nickel, selenium, silicon, tungsten, vanadium, and zinc)	0.25%	Cofactors of some enzymes and constituents of algal cell walls

Rather, the forms of oxygen that are toxic are those that are highly reactive. They are toxic for the same reason that oxygen is the final electron acceptor for aerobes: They are excellent oxidizing agents, so they steal electrons from other compounds, which in turn steal electrons from still other compounds. The resulting chain of vigorous oxidations causes irreparable damage to cells by oxidizing important compounds, including proteins and lipids.

There are four toxic forms of oxygen:

1. **Singlet oxygen ($^1O_2$)** Singlet oxygen is molecular oxygen with electrons that have been boosted to a higher energy state, typically during aerobic metabolism. Singlet oxygen is a very reactive oxidizing agent. Phagocytic cells, such as certain human white blood cells, use it to oxidize pathogens. Because singlet oxygen is also photochemically produced by the reaction of oxygen and light, phototrophic microorganisms often contain pigments called **carotenoids** (ka-rot'e-noyds) that prevent toxicity by removing the excess energy of singlet oxygen.

2. **Superoxide radical (O_2^-)** A few superoxide radicals form during the incomplete reduction of O_2 during electron transport in aerobes and during metabolism by anaerobes in the presence of oxygen. Superoxide radicals are so reactive and toxic that aerobic organisms

must produce enzymes called *superoxide dismutases* (dis′myu-tās-es) to detoxify them. These enzymes, which have active sites that contain metal ions—Zn^{2+}, Mn^{2+}, Fe^{2+}, Ni^{2+}, or Cu^{2+}, depending on the organism—combine two superoxide radicals and two protons to form hydrogen peroxide (H_2O_2) and molecular oxygen (O_2):

$$2O_2^- + 2H^+ \rightarrow H_2O_2 + O_2$$

The reason that anaerobes are susceptible to oxygen is that they lack superoxide dismutase; they die as a result of the oxidizing reactions of superoxide radicals formed in the presence of oxygen.

3. **Peroxide anion (O_2^{2-})** Hydrogen peroxide formed during reactions catalyzed by superoxide dismutase (and during other metabolic reactions) contains peroxide anion, another highly reactive oxidant. It is peroxide anion that makes hydrogen peroxide an effective antimicrobial agent. Aerobes contain either catalase or peroxidase, enzymes that detoxify peroxide anion.

   Catalase converts hydrogen peroxide to water and molecular oxygen:

$$2\,H_2O_2 \rightarrow 2\,H_2O + O_2$$

A simple test for catalase involves adding a sample from a bacterial colony to a drop of hydrogen peroxide. The production of bubbles of oxygen indicates the presence of catalase **(Figure 6.2).**

Peroxidase breaks down hydrogen peroxide without forming oxygen, using a reducing agent such as the coenzyme NADH:

$$H_2O_2 + NADH + H^+ \rightarrow 2\,H_2O + NAD^+$$

Obligate anaerobes either lack both catalase and peroxidase or have only a small amount of them, so they are susceptible to the toxic action of hydrogen peroxide.

4. **Hydroxyl radical (OH·)** Hydroxyl radicals result from ionizing radiation and from the incomplete reduction of hydrogen peroxide:

$$H_2O_2 + e^- + H^+ \rightarrow H_2O + OH·$$

Hydroxyl radicals are the most reactive of the four toxic forms of oxygen, but because hydrogen peroxide does not accumulate in aerobic cells (due to the action of catalase and peroxidase), the threat of hydroxyl radical is virtually eliminated in aerobic cells.

Besides the enzymes superoxide dismutase, catalase, and peroxidase, aerobes use other antioxidants, such as vitamins C and E, to protect themselves against toxic oxygen products. These antioxidants provide electrons that reduce toxic forms of oxygen.

Not all organisms are either strict **aerobes** or **anaerobes;** many organisms can live in various oxygen concentrations between these two extremes. For example, some aerobic organisms can maintain life via fermentation or anaerobic respiration, though their metabolic efficiency is often reduced in the absence of oxygen. Such organisms are called

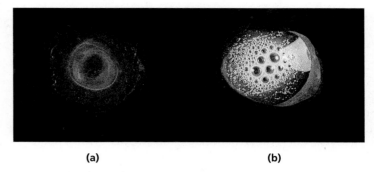

(a)   (b)

▲ *Figure 6.2*

Catalase test. The enzyme catalase converts hydrogen peroxide into water and oxygen, the latter of which can be seen as visible bubbles. *Enterococcus faecalis* **(a)** is catalase negative, whereas *Staphylococcus epidermidis* **(b)** is catalase positive.

facultative anaerobes. *Escherichia coli* (esh-ĕ-rik′ē-ă kō′lī) is an example of a facultatively anaerobic bacterium.

Aerotolerant anaerobes do not use aerobic metabolism, but they tolerate oxygen by having some of the enzymes that detoxify oxygen's poisonous forms. The lactobacilli that transform cucumbers into pickles, and milk into cheese, are aerotolerant. These organisms can be grown without the special conditions required by strict anaerobes, which will be discussed shortly.

Microaerophiles, such as the ulcer-causing pathogen *Helicobacter pylori*[5] (hel′ĭ-kō-bak′ter pī′lo-rē), are aerobes that require oxygen levels of 2% to 10%. This concentration of oxygen is found in the stomach. Microaerophiles are damaged by the 21% concentration of oxygen in the atmosphere, presumably because they have limited ability to detoxify hydrogen peroxide and superoxide radicals.

Microbial groups contain members with each of the five types of oxygen requirement. Algae, most fungi and protozoa, and many prokaryotes are obligate aerobes. A few yeasts and numerous prokaryotes are facultative anaerobes. Many prokaryotes and a few protozoa are aerotolerant, microaerophilic, or obligate anaerobes. The oxygen requirement of an organism can be identified by growing it in a medium that contains an oxygen gradient from top to bottom **(Figure 6.3).**

Nitrogen Requirements

Another essential element is nitrogen, which is contained in many organic compounds, including the amine group of amino acids and as part of nucleotide bases. Nitrogen makes up about 14% of the dry weight of microbial cells.

Nitrogen is often a growth-limiting nutrient for many organisms; that is, their anabolism ceases because they do not have sufficient nitrogen to build proteins and nucleotides.

---

[5]From semi-helical shape of the cell, and *pylorus*, which is the distal portion of the stomach.

Oxygen
concentration
High

Loose-fitting cap

(a) Obligate aerobes   **(b)** Obligate anaerobes   **(c)** Facultative anaerobes   **(d)** Aerotolerant anaerobes

Low

▲ *Figure 6.3*

Using a liquid growth medium to identify the oxygen requirements of organisms. The surface of thioglycollate medium, which is exposed to atmospheric oxygen, is aerobic. Oxygen concentration decreases with depth; the bottom of the tube is anaerobic. **(a)** Obligate aerobes cannot survive below the depth to which oxygen penetrates the medium. **(b)** Obligate anaerobes cannot tolerate any oxygen. **(c)** Facultative anaerobes can grow with or without oxygen, but their ability to use aerobic respiration pathways enhances their growth near the surface. **(d)** Aerotolerant aerobes can grow equally well with or without oxygen; their growth is relatively evenly distributed throughout the medium. *Where in such a test tube would the growth zone be for a microaerophilic aerobe?*

Figure 6.3 Microaerophiles would be found slightly below the surface, but neither directly at the surface nor in the depths of the tube.

Organisms acquire nitrogen from organic and inorganic nutrients. For example, most photosynthetic organisms can reduce nitrate (NO_3^-) to ammonium (NH_4^+), which can then be used for biosynthesis. In addition, all cells recycle nitrogen from their amino acids and nucleotides.

Though nitrogen constitutes about 79% of the atmosphere, relatively few organisms can utilize nitrogen gas. A few bacteria, notably many cyanobacteria and *Rhizobium* (rī-zō'bē-ŭm), reduce nitrogen gas (N_2) to ammonia (NH_3) via a process called **nitrogen fixation.** Nitrogen fixation is essential for life on Earth because nitrogen-fixers provide nitrogen in a useable form to other organisms. In Chapter 11 we discuss nitrogen-fixing prokaryotes, as well as nitrifying prokaryotes—those that oxidize nitrogenous compounds to acquire electrons for electron transport.

Other Chemical Requirements

Together, carbon, hydrogen, oxygen, and nitrogen make up more than 95% of the dry weight of cells; phosphorus, sulfur, calcium, manganese, magnesium, copper, iron, and a few other elements constitute the rest. Phosphorus is a component of phospholipid membranes, DNA, RNA, ATP, and some proteins. Sulfur is a component of sulfur-containing amino acids, which bind to one another via disulfide bonds

that are critical to the tertiary structure of proteins, and in vitamins such as thiamine (B_1) and biotin.

Other elements are called **trace elements** because they are required in very small ("trace") amounts (see Table 6.1 on page 170). For example, a few atoms of selenium dissolved out of the walls of glass test tubes provide the total requirement for the growth of green algae in a laboratory. Other trace elements are usually found in sufficient quantities dissolved in water. For this reason, tap water can sometimes be used instead of distilled or deionized water to grow microorganisms in the laboratory.

Some microorganisms—for example, algae and photosynthetic bacteria—are lithotrophic photoautotrophs; that is, they can synthesize all of their metabolic and structural needs from inorganic nutrients. These organisms have every enzyme and cofactor they need to produce all their cellular components. Most organisms, however, require small amounts of certain organic chemicals that they cannot synthesize, in addition to those that provide carbon and energy. These organic chemicals are called **growth factors** **(Table 6.2).** For example, vitamins are growth factors for some microorganisms. Recall that vitamins constitute all or part of many coenzymes and cannot be synthesized by some organisms. (Note, however, that vitamins are not growth factors for microorganisms that can manufacture them, such as *E. coli.*) Growth factors for various microbes include certain amino acids, purines, pyrimidines, cholesterol, NADH, and heme.

CRITICAL THINKING

Given that *Haemophilus ducreyi* is a chemoheterotrophic pathogen that requires heme as a growth factor, deduce how this bacterium phosphorylates most of its ADP to form ATP. Defend your answer.

Physical Requirements

Learning Objective

✓ Explain how extremes of temperature, pH, and osmotic and hydrostatic pressure limit microbial growth.

In addition to chemical nutrients, organisms have physical requirements for growth, including specific conditions of temperature, pH, osmolarity, and pressure.

Temperature

Temperature plays an important role in microbial life through its effects on the three-dimensional configurations of biological molecules. Recall that to function properly, proteins require a specific three-dimensional shape that is determined in part by temperature-sensitive hydrogen bonds, which are more likely to form at lower temperatures, and more likely to break at higher temperatures. When the hydrogen bonds break, the proteins denature and lose function. Additionally, lipids, such as those that are components of the membranes of cells and organelles, are also temperature

Table 6.2 Some Growth Factors of Microorganisms and Their Functions

Growth Factor	Function
Amino acids	Components of proteins
Cholesterol	Used by mycoplasmas (bacteria) for cell membranes
Heme	Functional portion of cytochromes in electron transport system
NADH	Electron carrier
Niacin (nicotinic acid)	Precursor of NAD^+ and $NADP^+$
Pantothenic acid	Component of coenzyme A
Para-aminobenzoic acid	Precursor of folic acid, which is involved in metabolism of one-carbon compounds and nucleic acid synthesis
Purines, pyrimidines	Components of nucleic acids
Pyroxidine (vitamin B_6)	Utilized in transamination syntheses of amino acids
Riboflavin (vitamin B_2)	Precursor of FAD
Thiamine (vitamin B_1)	Utilized in some decarboxylation reactions

(a)

5°C 25°C 35°C

(b)

◀ *Figure 6.4*

The effects of temperature on microbial growth. (a) Minimum, optimum, and maximum temperatures, as determined from a graph showing growth rate plotted against temperature. **(b)** Growth of *Staphylococcus aureus* on blood agar after 24 hours of incubation at three different temperatures. *If microorganisms can survive at temperatures lower than their minimum growth temperature, then why is it so-named?*

Figure 6.4 The minimum growth temperature is defined as the lowest temperature that supports metabolism. Many organisms can survive at low temperatures without actively metabolizing.

sensitive. If the temperature is too low, membranes become rigid and fragile; if the temperature is too high, the lipids become too fluid, and the membrane cannot contain the cell or organelle.

Thus, because temperature plays an important role in the three-dimensional structure of many types of biological molecules, different temperatures have different effects on the survival and growth of microbes **(Figure 6.4).** The lowest temperature at which an organism is able to conduct metabolism is called the *minimum growth temperature.* Note, however, that many microbes, particularly bacteria, survive temperatures far below this temperature, despite the fact that cell membranes are less fluid and transport processes are too slow to support metabolic activity. The highest temperature at which an organism continues to metabolize is called the *maximum growth temperature;* when the temperature exceeds this value, the organism's proteins are permanently

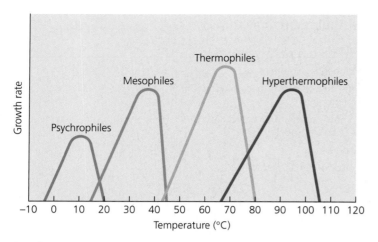

▲ *Figure 6.5*

Four categories of microbes based on temperature ranges for growth. *Categorize the bacterium* Vibrio marinus, *which has an optimum growth temperature near 10°C.*

Figure 6.5 Vibrio marinus is a psychrophile.

denatured, and it dies. The temperature at which an organism's metabolic activities produce the highest growth rate is the **optimum growth temperature.** Each organism thus survives over a *temperature range,* within which its growth and metabolism are supported.

CRITICAL THINKING

Examine the graph in Figure 6.4. Note that the growth rate increases slowly until the optimum is reached, and then it declines steeply over higher temperatures. In other words, organisms tolerate a wider range of temperatures below their optimal temperature than they do above the optimum. Explain this observation.

Based on their preferred temperature ranges—the temperatures within which their metabolic activity and growth are best supported—microbes can be categorized into four overlapping groups **(Figure 6.5). Psychrophiles**[6] (sī'krō-fīls) grow best at temperatures below about 15°C and can even continue to grow at temperatures below 0°C. They die at temperatures much above 20°C. In nature, psychrophilic algae, fungi, and bacteria live in snowfields, ice, and cold water **(Figure 6.6).** They do not cause disease in humans because they cannot survive at body temperature; some do cause food spoilage in refrigerators. Psychrophiles present unique challenges to laboratory investigations, because they must be kept at cold temperatures. For example, microscope stages must be refrigerated, and the air temperatures needed to maintain living psychrophiles are uncomfortably cold for lab personnel.

Mesophiles[7] (mez'ō-fīls) are organisms that grow best in temperatures ranging from 20°C to about 40°C (see Figure 6.5), though they can survive at higher and lower

temperatures. Because normal body temperature is approximately 37°C, human pathogens are mesophiles. *Thermoduric*[8] *organisms* are mesophiles that can survive brief periods at higher temperatures. Inadequate heating during pasteurization and canning can result in food spoilage by thermoduric mesophiles.

Thermophiles[9] (ther'mō-fīls) grow at temperatures above 45°C in habitats such as compost piles and hot springs. Some members of the Archaea, called **hyperthermophiles,** grow in water above 80°C; others live at temperatures above 100°C[10]. The current record-holder is *Pyrodictium,* a bacterium of submarine hot springs, which grows naturally in water at 113°C and can survive up to an hour in an autoclave at 121°C. Thermophiles and hyperthermophiles do not cause disease because they "freeze" at body temperature. **Highlight 6.1** discusses the biology of these interesting organisms.

CRITICAL THINKING

Over 100 years ago, doctors infected syphilis victims with malaria parasites to induce a high fever. Surprisingly, such treatment often cured the syphilis infection. Explain how this could occur.

pH

Organisms are sensitive to changes in acidity because hydrogen ions and hydroxyl ions interfere with hydrogen bonding within the molecules of proteins and nucleic acids; as a result, organisms have ranges of acidity that they prefer and can tolerate. Recall from Chapter 2 that pH is a measure of the concentration of hydrogen ions in a solution; that is, it is a measure of the acidity or alkalinity of a substance. A pH below 7.0 is acidic; the lower the pH value, the more acidic a substance is. Alkaline (or basic) pH values are higher than 7.0.

Most bacteria and protozoa, including most pathogens, grow best in a narrow range around a neutral pH—that is, between pH 6.5 and pH 7.5, which is also the pH range of most tissues and organs in the human body; such microbes are thus called **neutrophiles** (nū'trō-fīls). By contrast, other bacteria and many fungi are **acidophiles** (ă-sid'ō-phīls), organisms that grow best in acidic habitats. One example of acidophilic microbes are the chemoautotrophic bacteria that live in mines and in water that runs off from mine tailings (waste rock), habitats that have pHs as low as 0.0. These bacteria oxidize sulfur to sulfuric acid, further lowering the pH of their environment. Whereas *obligate acidophiles* require an acidic environment and die if the pH approaches 7.0, *acid-tolerant microbes* merely survive in acid without preferring it.

---

[6]From Greek *psuchros,* meaning cold, and *philos,* meaning love.
[7]From Greek *mesos,* meaning middle, and *philos,* meaning love.
[8]From Greek *therme,* meaning hot. The organisms are so-named because of their ability to endure or tolerate heat.
[9]From Greek *therme,* meaning hot, and *philos,* meaning love.
[10]Water can remain a liquid above 100°C if it has a high salt content or is under pressure, such as occurs in geysers or deep ocean troughs.

(a)   (b)

LM   100 μm

▲ *Figure 6.6*

An example of a psychrophile. **(a)** The alga *Chlamydomonas nivalis* colors this summertime snowbank in the Sierra Nevada mountains pink. **(b)** Microscopic view of the red-pigmented spores of *C. nivalis.*

Highlight 6.1   Some Like it Hot: Extreme Thermophiles

Extreme thermophiles, also called hyperthermophiles, live in extremely high-temperature environments, even those in excess of 100°C. Their heat-stable lipids, nucleic acids, and proteins allow them to survive—indeed thrive—at temperatures that would destroy the organic molecules of mesophiles. Hyperthermophilic organisms present unique challenges to our understanding of organic chemistry: Why don't their lipid membranes melt? How do their DNA molecules remain intact? What prevents their proteins from denaturing at such high temperatures?

The cells and chemicals of hyperthermophiles have special features that enable them to remain intact at high temperatures:

- Their cytoplasmic membranes do not contain fatty acids, which would melt at such high environmental temperatures. Instead, the membranes are monolayers (instead of bilayers) composed of hydrocarbon chains containing 40 carbon atoms bonded to glycerol phosphate. This accounts, at least in part, for their stability at high temperatures.

- The nucleic acids of extreme thermophiles appear to be stabilized by the presence of enzymes that fold DNA into unique heat-stable supercoils, by high concentrations of potassium ion, and by heat-stable proteins that bind to and stabilize DNA.

- The enzymes of thermophiles are also heat stable; they contain more hydrophobic amino acids than their counterparts in mesophiles, and form additional bonds between neighboring amino acids.

As we will see in Chapters 8 and 26, enzymes from hyperthermophiles play important roles in genetic engineering and industry. For example, DNA polymerases isolated from *Thermus aquaticus*, a thermophile found in hot springs at Yellowstone National Park, and from *Thermococcus litoralis*, a deep ocean bacterium, are used in the *polymerase chain reaction (PCR)* technique, a method of creating unlimited amounts of identical DNA strands. Additionally, enzymes from thermophiles and extreme thermophiles catalyze reactions in industrial processes that optimally operate between 50°C and 100°C.

Thermophilic bacteria, which can be distinguished by their orange-colored carotenoids, surround the Grand Prismatic Spring in Yellowstone National Park.

Many organisms produce acidic waste products that accumulate in their environment until eventually they inhibit further growth. For example, many cheeses are acidic due to lactic acid produced by fermenting bacteria and fungi. The low pH of these cheeses then acts as a preservative by preventing any further microbial growth. Other acidic foods, such as sauerkraut and dill pickles, are also kept from spoiling because most organisms cannot tolerate their low pH.

The normal acidity of certain regions of the body inhibits microbial growth and retards many kinds of infection. At one site, the vaginas of adult women, acidity results from the fermentation of carbohydrates by normal resident bacteria. If the growth of these normal residents is disrupted—for instance, by antibiotic therapy—the resulting higher pH may allow yeasts to grow and cause a yeast infection. Another site, the stomach, is inhospitable to most microbes because of the normal production of stomach acid. However, the acid-tolerant bacterium *Helicobacter pylori* neutralizes stomach acid by secreting bicarbonate and urease, an enzyme that converts urea to ammonia, which is alkaline. The growth of *Helicobacter* in certain portions of the digestive tract is the cause of most ulcers.

Alkaline conditions also inhibit the growth of most microbes, but **alkalinophiles** live in alkaline soils and water up to pH 11.5. For example, *Vibrio cholerae* (vib′rē-ō kol′er-ī), the causative agent of cholera, grows best outside of the body in water at pH 9.0.

Physical Effects of Water

Microorganisms require water; they must be in a moist environment if they are to be metabolically active. Water is not only needed to dissolve enzymes and nutrients but is also an important reactant in many metabolic reactions. Even though most cells die in the absence of water, some microorganisms—for example, the bacterium *Mycobacterium tuberculosis* (mī′kō-bak-tēr′ē-ŭm tū-ber-kyū-lō′sis)—have cell walls that retain water, allowing them to survive for months under dry conditions. Additionally, the spores and cysts of some other single-celled microbes can cease most metabolic activity for years; these cells are in essence in a state of suspended animation because they neither grow nor reproduce in their dry condition.

We now consider the physical effects of water on microbes by examining two topics: osmotic pressure and hydrostatic pressure.

Osmotic Pressure As we saw in Chapter 3, *osmosis* is the diffusion of water across a semipermeable membrane and is driven by unequal solute concentrations on the two sides of such a membrane. The *osmotic pressure* of a solution is the pressure exerted on a semipermeable membrane by a solution containing solutes (dissolved material) that cannot freely cross the membrane. Osmotic pressure is related to

the concentration of dissolved molecules and ions in a solution. Solutions with greater concentrations of such solutes are *hypertonic* relative to those with a lower solute concentration, which are *hypotonic*.

Osmotic pressure can have dire effects on cells. For example, a cell placed in fresh water (a hypotonic solution relative to the cell's cytoplasm) gains water from its environment and swells to the limit of its cell wall. Cells that lack a cell wall—animal cells and some bacterial, fungal, and protozoan cells—will swell until they burst in hypotonic solutions. By contrast, a cell placed in seawater, which is a solution containing about 3.5% solutes and thus hypertonic to most cells, loses water into the surrounding saltwater. Such a cell can die from **plasmolysis** (plaz-mol′i-sis), or shriveling of its cytoplasm. Osmotic pressure accounts for the preserving action of salt in jerky and salted fish, and of sugar in jellies, preserves, and honey. In those foods, the salt and sugar are solutes that draw water out of any microbial cells that are present, preventing the growth and reproduction of the microbes.

Osmotic pressure restricts organisms to certain environments. Some microbes, called **obligate halophiles,**[11] are adapted to growth under high osmotic pressure such as exists in the Great Salt Lake and smaller salt ponds. They may grow in up to 30% salt and will burst if placed in freshwater. Other microbes are *facultative halophiles;* that is, though they do not require high salt concentrations, they can tolerate them. One potential bacterial pathogen, *Staphylococcus aureus* (staf′i-lō-kok′ŭs o′rē-ŭs), can tolerate up to 20% salt, which allows it to colonize the surface of the skin—an environment that is too salty for most microbes. *S. aureus* causes a number of different skin and mucous membrane diseases ranging from pimples, styes, and boils to life-threatening scalded skin and toxic shock syndromes. These diseases are covered more fully in Chapter 19.

Hydrostatic Pressure Water exerts pressure in proportion to its depth. For every additional 10 m of depth, water pressure increases 1 atmosphere (atm). Therefore, the pressure at 100 m below the surface is 10 atm—ten times greater than at the surface. Obviously, the pressure in deep ocean basins and trenches, which are thousands of meters below the surface, is tremendous. Organisms that live under such extreme pressure are called **barophiles**[12] (bar′ō-fīls). Their membranes and enzymes do not merely tolerate pressure, but are dependent on pressure to maintain their three-dimensional, functional shapes. Thus barophiles brought to the surface quickly die because their proteins denature. Obviously, barophiles cannot cause diseases in humans, plants, or animals that do not live at great depths.

---

[11]From Greek *halos,* meaning salt, and *philos,* meaning love.
[12]From Greek *baros,* meaning weight, and *philos,* meaning love.

Frontiers 6.2   **Biofilms, Quorum Sensing, and Disease Prevention**

⌐t many microorganisms become more harmful when they are
◻iofilm, scientists are seeking ways to prevent biofilms from
the first place. Researchers have learned that drugs that block
⌐rs, thereby disrupting communication among bacterial cells,
will inhibit biofilm formation and thus prevent disease. Such receptor-
blocking drugs have successfully blocked biofilm formation and prevented disease in mice and are being considered for use in humans.

Another possible approach to preventing biofilm formation involves artificially amplifying quorum sensing while bacteria are still
relatively few in number. Some pathogens hide within capsules and
in blood clots, and only after they have multiplied significantly and
formed a biofilm do they emerge and become "visible" to the
immune system. With amplified quorum sensing, bacteria might
produce certain proteins earlier in an infection cycle than normal,
thus "revealing" themselves sooner to the immune system, which
may then eliminate them before they can form biofilms and cause
disease.

Ecological Associations

Learning Objective

✓ Describe how quorum sensing can lead to formation of a biofilm.

Organisms don't live in a purely chemical/physical (abiotic)
environment—they live in association with other individuals of their own and different species. The relationships
between organisms can be viewed as falling along a
continuum stretching from causing harm to providing
benefits.

Relationships in which one organism harms or even kills
another organism are considered *antagonistic relationships.* As
discussed in Chapter 1, viruses are especially clear examples
of antagonistic microbes; they require a cell in which to replicate themselves and almost always kill their cellular hosts.
Beneficial relationships take at least two forms: synergistic relationships and symbiotic relationships. In *synergistic relationships,* the individual members of an association cooperate
such that each receives benefits that exceed those that would
result if each lived by itself, even though each member could
live separately. In *symbiotic relationships,* organisms live in
such close nutritional or physical contact that they become
interdependent, such that the members rarely (if ever) live
outside the relationship. Symbiotic relationships are discussed in greater detail in Chapter 14, particularly as they
relate to the production of disease.

Biofilms are examples of complex relationships among
numerous individuals, which are often different species,
that attach as a group to surfaces and display metabolic
and structural traits different from those expressed by any
of the microorganisms alone. Some scientists estimate that
65% of bacterial diseases are caused by biofilms.

Biofilms often form as a result of a process called **quorum
sensing,** in which bacteria respond to the density of nearby
bacteria. The bacteria secrete molecules into their environment that act as signals; the cells also possess receptors for
these signal molecules. When bacterial densities are low, the
concentration of signaling molecules remains low, and few
receptors bind signal molecules. However, when the density
of bacteria increases, the concentration of signal molecules
also increases, such that more and more receptors become
occupied. Once the binding of signal molecules exceeds a
certain threshold amount, the expresion of previously
suppressed genes is triggered, and the result is that the
bacteria have new characteristics, such as the ability to
form biofilms.

Dental plaque on teeth is an example of a biofilm. The
ability of some strains of *Salmonella enterica* (sal'mŏ-nel'ă enter'i-kă), *Staphylococcus aureus,* and *Pseudomonas aeruginosa*
(soo-dō-mō'nas ā-rū-ji-nō'să) to cause disease in humans is
also linked to their ability to form biofilms. **New Frontiers 6.2**
describes research into preventing disease by manipulating
the formation of biofilms.

Culturing Microorganisms

As we saw in Chapter 1, the second of Koch's postulates for
demonstrating that a certain agent causes a specific disease
requires that microorganisms be isolated and cultivated.
Medical laboratory personnel must also grow pathogens as
a step in the diagnosis of many diseases. To cultivate or *culture* microorganisms, a sample called an **inoculum** (plural:
inocula) is introduced into a collection of nutrients called a
medium. Microorganisms that grow from an inoculum are
also called a *culture;* thus, the word **culture** can refer to the
act of cultivating microorganisms or to the microorganisms
that are cultivated.

Cultures can be grown in liquid media called **broths,** or
on the surface of solid media. Cultures that are visible on
the surface of solid media are called colonies. Bacterial and
fungal colonies often have distinctive characteristics—
including color, size, shape, elevation, texture, and appearance of the colony's margin (edge)—that taken together aid
in identifying the microbial species that formed the colony
(Figure 6.7).

Microbiologists obtain inocula from a variety of sources.
Environmental specimens are taken from such sources as ponds,
streams, soil, and air; environmental sampling is discussed in

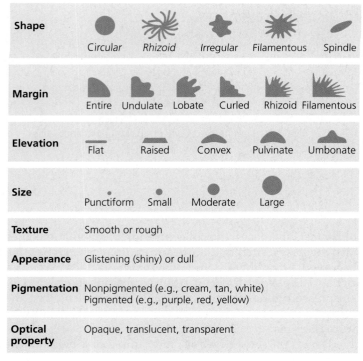

Shape	Circular	Rhizoid	Irregular	Filamentous	Spindle
Margin	Entire	Undulate	Lobate	Curled	Rhizoid	Filamentous
Elevation	Flat	Raised	Convex	Pulvinate	Umbonate
Size	Punctiform	Small	Moderate	Large
Texture	Smooth or rough				
Appearance	Glistening (shiny) or dull				
Pigmentation	Nonpigmented (e.g., cream, tan, white) Pigmented (e.g., purple, red, yellow)				
Optical property	Opaque, translucent, transparent				

(a)

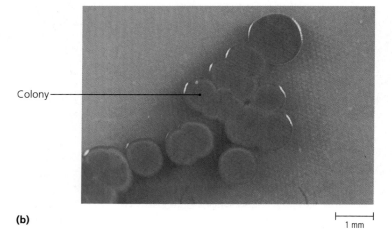

Colony

1 mm

(b)

▲ *Figure 6.7*

Characteristics of bacterial colonies. **(a)** Shape, margin, elevation (side view), size, texture, appearance, pigmentation (color), and optical properties are described by a variety of terms. **(b)** *Serratia marcescens* growing on an agar surface. These colonies are circular, entire, convex, small, smooth, shiny, red, and opaque.

some detail in Chapter 26. *Clinical specimens* are taken from patients and handled in ways that facilitate the examination of or testing for the presence of microorganisms. Finally, another source of inocula is the cultures originally grown from environmental or clinical specimens and maintained

in storage in the laboratory. Next we briefly examine clinical sampling.

Clinical Sampling

Learning Objective

✓ Describe methods for collecting clinical specimens from the skin and from the respiratory, reproductive, and urinary tracts.

Diagnosis and treatment of disease often depend on the isolation and correct identification of pathogens. Physicians and other health care professionals must properly obtain samples from their patients, and must then transport them quickly and correctly to a microbiology laboratory for culture and identification. They must take care to prevent the contamination of samples with microorganisms from the environment or other regions of the patient's body, and they must prevent infecting themselves with pathogens while sampling. In this regard, the Centers for Disease Control and Prevention (CDC) has established a set of guidelines, called *universal precautions*, to protect health care professionals from contamination by pathogens.

In clinical microbiology, a **clinical specimen** is a sample of human material, such as feces, saliva, cerebrospinal fluid, or blood, that is examined or tested for the presence of microorganisms. As summarized in **Table 6.3,** health care professionals collect clinical specimens using a variety of techniques and equipment. Specimens must be properly labeled and promptly transported to a microbiological laboratory to both avoid death of the pathogens and minimize the growth of normal organisms. Clinical specimens are often transported in special *transport media* that are chemically formulated to maintain the relative abundance of different microbial species or to maintain an anaerobic environment.

Obtaining Pure Cultures

Learning Objective

✓ Describe the two most common methods by which microorganisms can be isolated for culture.

Clinical specimens are collected in order to identify a suspected pathogen, but they also contain *normal microbiota,* which are microorganisms associated with a certain area of the body (see Table 14.2) without causing diseases. As a result, the suspected pathogen in a specimen must be isolated from the normal microbiota in culture. Scientists use several techniques to isolate organisms in **pure cultures,** that is, cultures composed of cells arising from a single progenitor. The word *axenic*[13] (ā-zen′ik) is also used to refer to a pure culture. The progenitor from which a particular pure culture is derived may be either a single cell or a group of related cells; therefore, the progenitor is termed a **colony-forming unit (CFU).**

---

[13]From Greek *a,* meaning no, and *xenos,* meaning stranger.

Table 6.3   **Clinical Specimens and the Methods Used to Collect Them**

Type or Location of Specimen	Collection Method
Skin, accessible mucous membrane (including eye, outer ear, nose, throat, vagina, cervix, urethra) or open wounds	Sterile swab brushed across the surface; care should be taken not to contact neighboring tissues
Blood	Needle aspiration from vein, anticoagulants are included in the specimen transfer tube
Cerebrospinal fluid	Needle aspiration from subarachnoid space of spinal column
Stomach	Intubation, which involves inserting a tube into the stomach, often via a nostril
Urine	In aseptic collection, a catheter is inserted into the bladder through the urethra; in the "clean catch" method, initial urination washes the urethra, and the specimen is midstream urine
Lungs	Collection of sputum either dislodged by coughing or acquired via a catheter
Diseased tissue	Surgical removal (biopsy)

Figure 6.8 ❯

The streak plate method of isolation. **(a)** An inoculum is spread across the surface of an agar plate in a sequential pattern of streaks (as indicated by the numbers and arrows). The loop is sterilized between streaks. In streaks 2, 3, and 4, bacteria are picked up from the previous streak, diluting the number of cells each time. **(b)** A streak plate showing colonies of *Neisseria gonorrhoea.*

(a)

(b)

Of course, in all microbiological procedures care must be taken to reduce the chance of contamination as occurs when instruments or air currents carry foreign microbes into culture vessels. All media, vessels, and instruments must be **sterile**—that is, free of any microbial contaminants. Sterilization and *aseptic techniques,* which are designed to limit contamination, are discussed in Chapter 9. Now we examine two common isolation techniques: streak plates and pour plates.

Streak Plates

The most commonly used isolation technique in microbiological laboratories is the **streak plate** method. In this technique, a sterile inoculating loop (or sometimes a needle) is used to spread an inoculum across the surface of a solid medium in *Petri dishes,* which are clear, flat culture dishes with loose-fitting lids. The loop is used to lightly streak a set pattern that gradually dilutes the sample to a point that CFUs are isolated from one another **(Figure 6.8a).** After an appropriate period of time called **incubation,** colonies develop from each isolate **(Figure 6.8b).** The various types of organisms present are distinguished from one another by differences in colonial characteristics (see Figure 6.9b). Samples from each variety can then be inoculated in new media to establish axenic cultures.

CRITICAL THINKING

Using the terms in Figure 6.7a, describe the shape, margin, pigmentation, and optical properties of bacterial colonies 3 and 5 seen in Figure 6.9b.

Pour Plates

In the **pour plate** technique, CFUs are separated from one another using a series of dilutions. There are various ways to perform pour plate isolations. In one method, an initial 1-milliliter sample is mixed into 9.0 ml of medium in a test tube. After mixing, a new sample from this medium is then used to inoculate a second tube of liquid medium. The process is repeated to establish a series of dilutions **(Figure 6.9a)**. Samples from the more diluted media are mixed in Petri dishes with sterile, warm medium containing *agar*—a gelling agent derived from the cell walls of red algae. After the agar cools and solidifies, and the Petri plates are incubated, isolated colonies—colonies that are separate and distinct from all others—form in the dishes from CFUs that have been separated via the dilution series **(Figure 6.9b)**. One difference between this method and the streak plate technique is that colonies form both at and below the surface of the medium. As before, pure cultures can be established from distinct colonies.

Isolation techniques work well only if a relatively large number of CFUs of the organism of interest are present in the initial sample, and if the medium used supports the growth of that microbe. As discussed later, special media and enriching techniques can be used to increase the likelihood of success.

(a)

◀ *Figure 6.9*

The pour plate method of isolation. **(a)** After an initial sample is diluted through a series of transfers, the final dilutions are mixed with warm agar in Petri plates. Individual CFUs form colonies in and on the agar. **(b)** A portion of a plate showing the results of isolation.

(b)

Other Isolation Techniques

Streak plates and pour plates are used primarily to establish pure cultures of bacteria, but they can also be used for some fungi, particularly yeasts. Protozoa and motile unicellular algae are not usually cultured on solid media because they do not remain in one location to form colonies. Instead, they are isolated through a series of dilutions, but remain in broth culture media. In cases of fairly large microorganisms such as the protists *Euglena* and *Amoeba*, hollow tubes with small diameters called *micropipettes* can be used to pick up a single cell, which is then used to establish a culture.

Culture Media

Learning Objective

✓ Describe six types of general culture media available for bacterial culture.

Culturing microorganisms can be an exacting science. Although some microbes, such as *E. coli*, are not particular about their nutritional needs and can be grown in a variety of media, others, such as *Neisseria gonorrheae* (nī-sē′rē-ă go-nor-re′-ē′) and *Haemophilus influenzae* (hē-mof′i-lŭs in-flū-en′zē), require specific nutrients, including specific growth factors. The majority of prokaryotes have never been successfully grown in any culture medium, in part because scientists have concentrated their efforts on culturing pathogens and commercially important species. However, some pathogens, such as the syphilis bacterium *Treponema pallidum* (trep-ō-nē′mă pal′li-dŭm), have never been cultured in any laboratory medium, despite over a century of effort.

A variety of media are available for microbiological cultures, and more are developed each year to support the needs of food, water, industrial, and clinical microbiologists. Most media are available from commercial sources and come in powdered forms that require only the addition of water to make broths. A common medium, for example, is *nutrient broth,* which contains powdered beef extract and peptones (short chains of amino acids produced by enzymatic digestion of protein) dissolved in water. For some purposes broths are adequate, but if solid media are needed, dissolving about 1.5% agar into warm broth, pouring the liquid mixture into an appropriate vessel, and allowing it to cool provides a solid surface to support colonial growth. Media made solid by the addition of agar to a broth have the word *agar* in their names; thus *nutrient agar* is nutrient broth to which 1.5% agar has been added.

Agar, a complex polysaccharide, is a useful compound in microbiology for several reasons:

- Most microbes cannot digest agar; therefore, agar media remain solid even when bacteria and fungi are growing on them.

- Powdered agar dissolves in water at 100°C, a temperature at which most nutrients remain undamaged.

- Agar solidifies at temperatures below 40°C, so temperature-sensitive, sterile nutrients such as vitamins and

◀ *Figure 6.10*
Slant tube containing solid media (in this case, acetamide agar).

Slant

Butt

blood can be added without detriment, and liquid agar can be poured over most bacterial cells without harming them. The latter technique plays a role in the pour plate isolation technique.

- Solid agar does not melt below 100°C; thus, it can be used to culture many thermophiles.

As we have seen, still-warm liquid agar media can be poured into Petri dishes, which once the agar solidifies are then called **Petri plates.** When warm agar media are poured into test tubes that are then placed at an angle and left to cool until the agar solidifies, the result is **slant tubes,** or **slants (Figure 6.10).** The slanted surface provides a larger surface area for aerobic microbial growth while the butt of the tube remains almost anaerobic. If the tubes are kept vertical until the agar solidifies, they are called *deeps.*

Next we examine six types of general culture media: defined media, complex media, selective media, differential media, anaerobic media, and transport media. It is important to note that these types of media are not mutually exclusive categories; that is, in some cases a given medium can be classified into more than one category.

Defined Media

If microorganisms are to grow and multiply in culture, the medium must provide essential nutrients (including an appropriate energy source for chemotrophs), water, an appropriate oxygen level, and the required physical conditions (such as the correct pH and suitable osmotic pressure and temperature). A **defined medium** (also called a **synthetic medium**) is one in which the exact chemical composition is known (**Table 6.4** on page 182). Relatively simple defined media containing inorganic salts and a source of CO_2 (such as sodium bicarbonate) are available for chemolithotrophs and phototrophs, particularly cyanobacteria and algae. Chemoheterotrophs also require organic molecules such as glucose, amino acids, and vitamins, which supply carbon and energy or are vital growth factors.

Organisms that require a relatively large number of growth factors are termed *fastidious.* Such organisms may be used as living assays for the presence of growth factors. For example, a scientist needing to know if a sample contains a certain vitamin could inoculate the sample with a fastidious

Table 6.4	Ingredients of a Representative Defined (Synthetic) Medium for Culturing *E. coli*
Glucose	1.0 g
Na_2HPO_4	16.4 g
KH_2PO_4	1.5 g
$(NH_4)_2PO_4$	2.0 g
$MgSO_4 \cdot 7H_2O$	0.2 g
$CaCl_2$	0.01 g
$FeSO_4 \cdot 7H_2O$	0.005 g
Distilled or deionized water	Enough to bring volume to 1 L

organism that requires the vitamin. If the microbe grows, the vitamin is present in the sample. The amount of growth provides an estimate of the amount of vitamin present, scant growth indicating a small amount.

Complex Media

For most clinical cultures, defined media are unnecessarily troublesome to prepare. Most chemoheterotrophs, including pathogens, are routinely grown on **complex media** that contain nutrients released by the partial digestion of yeast, beef, soy, or proteins such as casein from milk. The exact chemical composition of a complex medium is unknown because partial digestion releases many different chemicals in a variety of concentrations.

Complex media have advantages over defined media. Because a complex medium contains a variety of nutrients, including growth factors, it can support a wider variety of different microorganisms. Complex media are also used to culture organisms whose exact nutritional needs are unknown. Nutrient broth, Trypticase™ soy agar, and MacConkey agar are some common complex media. Blood is often added to complex media to provide additional growth factors, such as NADH and heme. Such a fortified medium is said to be *enriched* and can support the growth of many fastidious microorganisms.

Selective Media

Selective media typically contain substances that either favor the growth of particular microorganisms or inhibit the growth of unwanted ones. Eosin, methylene blue, and crystal violet dyes as well as bile salts are included in media to inhibit the growth of Gram-positive bacteria without affecting Gram-negatives. A high concentration of NaCl (table salt) in a medium selects for halophiles and for salt-tolerant bacteria such as the pathogen *S. aureus*. Sabouraud dextrose agar has a slightly low pH, which by inhibiting the growth of bacteria is selective for fungi **(Figure 6.11)**.

A medium can also become a selective medium when a single crucial nutrient is left out of it. For example, leaving glucose out of Trypticase™ soy agar makes the resulting medium selective for organisms that can meet all their carbon requirements by catabolizing amino acids.

Fungal colonies   Bacterial colonies

▲ *Figure 6.11*

An example of the use of a selective medium. After inoculation with a diluted soil sample, the presence of a slightly acidic pH in Sabouraud dextrose agar (left) makes it selective for fungi by inhibiting the growth of bacteria. At right for comparison is a nutrient agar plate inoculated with an identical sample.

CRITICAL THINKING

Examine the ingredients of MacConkey agar as listed in Table 6.5 on page 184. Does this medium select for Gram-positive or Gram-negative bacteria? Explain your reasoning.

The sole carbon source in citrate medium is citric acid (citrate). Why might a laboratory microbiologist use this medium?

Differential Media

Differential media are formulated such that either the presence of visible changes in the medium or differences in the appearances of colonies helps microbiologists differentiate among different kinds of bacteria growing on the medium. Such media take advantage of the fact that different bacteria utilize the ingredients of any given medium in different ways. One example of the use of a differential medium involves the differences in organisms' utilization of the red blood cells in blood agar **(Figure 6.12)**. *Streptococcus pneumoniae* (strep-tō-kok′ŭs nu-mō′nē-ī) partially

▲ *Figure 6.12*

The use of blood agar as a differential medium. *Streptococcus pyogenes* (left) completely uses red blood cells, producing a clear zone termed beta-hemolysis. *Streptococcus pneumoniae* (middle) partially uses red blood cells, producing a discoloration termed alpha-hemolysis. *Enterococcus faecalis* (right) does not use red blood cells; the lack of any change in the medium around colonies is termed gamma-hemolysis even though no red cells are hemolyzed.

Escherichia coli   *Alcaligenes faecalis*   Uninoculated control

Carbohydrate utilization tube

▲ *Figure 6.13*

The use of carbohydrate utilization tubes as differential media. Each tube contains a single kind of simple carbohydrate (a sugar) as a carbon source, and the dye phenol red as a pH indicator. *Escherichia coli* fermented the sugar, and the acid produced lowered the pH enough to cause the phenol red to turn yellow. This bacterium also produced gas, which is visible as a small bubble in the Durham tube. *Alcaligenes faecalis* did not ferment this carbohydrate; because no acid was produced, the color of the medium did not change.

digests (lyses) red blood cells, producing around its colonies a greenish-brown discoloration denoted *alpha-hemolysis.* By contrast, *Streptococcus pyogenes* (pī-oj′en-ēz) completely digests red blood cells, producing around its colonies clear zones termed *beta-hemolysis. Enterococcus faecalis* (en′ter-ō-kok′ŭs fē-kǎ′lis) does not digest red blood cells, so the agar appears unchanged, a reaction called *gamma-hemolysis* even though no lysis occurs. In some differential media, such as carbohydrate utilization broth tubes, a pH-sensitive dye changes color when bacteria metabolizing sugars produce acid waste products **(Figure 6.13).** Some common differential complex media are described in **Table 6.5** on page 184.

Many media are both selective and differential; that is, they enhance the growth of certain species that can then be distinguished from other species by variations in their effect on the medium or by the color of colonies they produce. For example, bile salts and crystal violet in MacConkey agar both inhibit the growth of Gram-positive bacteria and differentiate between lactose-fermenting and non-lactose-fermenting Gram-negative bacteria **(Figure 6.14).**

CRITICAL THINKING

Using as many of the following terms that apply—selective, differential, broth, solid, defined, complex—categorize each of the media listed in Tables 6.4 and 6.5.

Anaerobic Media

Obligate anaerobes require special culture conditions in that their cells must be protected from free oxygen. Anaerobes can be introduced with a straight inoculating wire into the anoxic (oxygen-free) depths of solid media to form a *stab culture,* but special media called **reducing media** provide better anaerobic culturing conditions. These media contain compounds, such as sodium thioglycollate, that chemically combine with free oxygen and remove it from the medium. Heat is used to drive the absorbed oxygen from the thioglycollate immediately before such a medium is inoculated.

The use of Petri plates presents special problems for the culture of anaerobes because each dish has a loose-fitting lid that allows the entry of air. For the culture of anaerobes, inoculated Petri plates are placed in sealable containers containing reducing chemicals **(Figure 6.15a).** Of course, the air-tight lids of anaerobic culture vessels must be sealed so that oxygen cannot enter. Only anaerobes that can tolerate exposure to oxygen can be cultured by this method because inoculation and transfer occur outside of the anaerobic environment. Laboratories that routinely study strict anaerobes have larger anaerobic glove boxes, which are transparent

Table 6.5   Representative Differential Complex Media

Medium and Ingredients		Use	Interpretation of Results
MacConkey Medium		For the culture and differentiation of enteric bacteria based on their ability to ferment lactose	Lactose fermenters produce red to pink colonies; nonfermenters form colorless or transparent colonies
Peptone	20.0 g		
Agar	12.0 g		
Lactose	10.0 g		
Bile salts	5.0 g		
NaCl	5.0 g		
Neutral red	0.075 g		
Crystal violet	0.001 g		
Water to bring volume to 1 L			
Eosin Methylene Blue (EMB) Agar		For the isolation, culture, and differentiation of Gram-negative enteric bacteria	Lactose-fermenting bacteria appear with a green metallic sheen; nonfermenters appear as colorless or light purple colonies
Agar	13.5 g		
Pancreatic digest of casein	10.0 g		
Lactose	5.0 g		
Sucrose	5.0 g		
K_2HPO_4	2.0 g		
Eosin Y	0.4 g		
Methylene blue	0.065 g		
Water to bring volume to 1 L			
Triple Sugar Iron Agar		For the differentiation of enteric Gram-negative bacteria based on their fermentation of glucose, sucrose, and lactose and the production of H_2S gas	Red slant/red butt: no fermentation; yellow slant/red butt: glucose fermentation only; yellow slant/yellow butt: glucose and lactose and/or sucrose fermentation; if the butt turns black, H_2S is produced
Peptone	20.0 g		
Agar	12.0 g		
Lactose	10.0 g		
Sucrose	10.0 g		
NaCl	5.0 g		
Beef extract	3.0 g		
Yeast extract	3.0 g		
Glucose	1.0 g		
Ferric citrate	0.3 g		
$Na_2S_2O_3$	0.3 g		
Phenol red	0.025 g		
Water to bring volume to 1 L			
Blood Agar		For culture of fastidious microorganisms and differentiation of hemolytic microorganisms	Partial digestion of blood: alpha-hemolysis; complete digestion of blood: beta-hemolysis; no digestion of blood: gamma-hemolysis
Agar	15.0 g		
Pancreatic digest of casein	15.0 g		
Papaic digest of soybean meal	5.0 g		
NaCl	5.0 g		
Sterile blood	50.0 ml		
Water to bring volume to 950.0 ml (Blood is added to medium after autoclaving and cooling.)			

Figure 6.14 ➤

Use of MacConkey agar as a selective and differential medium. **(a)** Whereas both the Gram-positive *Staphylococcus aureus* (*Sa*) and the Gram-negative *Escherichia coli* (*Ec*) grow on nutrient agar, MacConkey agar **(b)** selects for Gram-negative bacteria and inhibits Gram-positive bacteria. **(c)** MacConkey agar also differentiates between Gram-negative bacteria on their ability to ferment lactose. The colonies of the lactose-fermenting *E. coli* are easily distinguished from those of the non-lactose-fermenting *Shigella flexneri* (*Sf*).

(a)   (b)   (c)

Clamp

Air-tight lid

Chamber

$2H_2 + O_2 \rightarrow 2H_2O$

Palladium pellets to catalyze reaction removing O_2

H_2 CO_2 O_2

Envelope containing chemicals to release CO_2 and H_2

Methylene blue (anaerobic indicator)

Petri plates

(a)

(b)

▲ *Figure 6.15*

Anaerobic culture equipment. **(a)** The Gas-Pak anaerobic system utilizes chemicals to create an anaerobic environment inside a sealable, airtight jar. Methylene blue, which turns colorless in the absence of oxygen, indicates when the environment within the jar is anaerobic. **(b)** An anaerobic glove box, which provides an environment for the transfer and growth of strict anaerobes.

airtight chambers with special airtight rubber gloves, chemicals that remove oxygen, and air locks **(Figure 6.15b).** These chambers allow scientists to manipulate equipment and anaerobic cultures in an oxygen-free environment.

Transport Media

Hospital personnel use special **transport media** to carry clinical specimens of feces, urine, saliva, sputum, blood, and other bodily fluids in such a way as to ensure that people are not infected and that the specimens are not contaminated. Speed in transporting clinical specimens to the laboratory is extremely important because pathogens often do not survive outside the body as long as normal microbiota. Stool and other specimens are transported in buffered media designed to maintain the ratios among different microorganisms. Anaerobic clinical specimens can be transported for less than an hour in syringes from which the needles have been removed, but longer transport times require that the specimens be injected into anaerobic transport media.

Special Culture Techniques

Learning Objective

✓ Discuss the use of special culture methods including animal and cell culture, low-oxygen culture, and enrichment culture.

Not all organisms can be grown under the culture conditions we have discussed. Scientists have developed other techniques to culture many of these organisms.

Animal and Cell Culture

Microbiologists have developed animal and cell culture techniques for growing microbes for which artificial media are inadequate. The causative agents of leprosy and syphilis, for example, must be grown in animals because all attempts to grow them using standard culture techniques have been unsuccessful. *Mycobacterium leprae* (mī'kō-bak-tēr'ē-ŭm lep'-rī) is cultured in armadillos, whose internal conditions (including a relatively low body temperature) provide the conditions this microbe prefers. Rabbits meet the culture needs for *Treponema pallidum,* the bacterium that causes syphilis. Because viruses and small bacteria called rickettsias and chlamydias are obligate intracellular parasites—that is, they grow and reproduce only within living cells—bird eggs and cultures of living cells are used to culture these organisms.

Low-Oxygen Culture

As we have discussed, many types of organisms prefer oxygen conditions that are intermediate between strictly aerobic and anaerobic environments. *Carbon dioxide incubators,* machines that electronically monitor and control CO_2 levels,

Figure 6.16 ▶

A candle jar. The burning candle consumes oxygen and releases carbon dioxide until it extinguishes itself; the resulting environment is suitable for growing microaerophiles and capnophiles.

provide atmospheres that mimic the environments of the intestinal tract, the respiratory tract, and other body tissues and thus are useful for culturing these kinds of organisms. Smaller and much less expensive alternatives to CO_2 incubators are *candle jars* (**Figure 6.16**). In these simple but effective devices, culture plates are sealed in a jar along with a lit candle; the flame consumes much of the O_2, replacing it with CO_2. The candle eventually extinguishes itself, creating an environment that is ideal for aerotolerant anaerobes, microaerophiles, and **capnophiles** which are organisms such as *Neisseria gonorrhoeae* that grow best with a relatively high concentration of carbon dioxide in addition to low oxygen levels. Remaining oxygen in the jar prevents the growth of strict anaerobes. The use of packets of chemicals that remove oxygen from the jar has largely replaced candles in modern microbiology labs.

Enrichment Culture

Bacteria that are present in small numbers may be overlooked on a streak plate or overwhelmed by faster-growing, more abundant strains. This is especially true of organisms in soil and fecal samples that contain a wide variety of microbial species. To isolate potentially important microbes that might otherwise be overlooked, microbiologists enhance the growth of less abundant organisms by a variety of techniques.

In the late 1800s, the Dutch microbiologist Martinus Beijerinck (1851–1931) introduced the most common of these methods, called simply **enrichment culture.** Enrichment cultures use a selective medium and are designed to increase very small numbers of a chosen microbe to observable levels. For example, suppose a microbiologist specializing in environmental clean up wanted to isolate an organism capable of digesting crude oil to have on hand should it be required to clean an oil-soaked beach. Even though a sample of beach sand might contain a few such organisms, it would also likely contain many millions of

unwanted common bacteria. To isolate oil-utilizing microbes, the scientist would inoculate a sample of sand into a tube of selective medium containing oil as the sole carbon source, and then incubate it. Then a small amount of the culture would be transferred into a new tube of the same medium, to be incubated again. After a series of such enrichment transfers, any remaining bacteria will be oil-utilizing organisms. Different species could be isolated by either streak-plate or pour-plate methods.

Cold enrichment is another technique used to enrich a culture with cold-tolerant species, such as *Vibrio cholerae,* the bacteria that cause cholera. Stool specimens or water samples suspected of containing the bacterium are incubated in a refrigerator instead of at 37°C. Cold enrichment works because *Vibrio* cells are much less sensitive to cold than are more common fecal bacteria such as *E. coli;* therefore, *Vibrio* continues to grow in the cold while the other species are inhibited. The result of cold enrichment is a culture with a greater percentage of *Vibrio* cells than the original sample; the *Vibrio* cells can then be isolated by other methods.

CRITICAL THINKING

Beijerinck used the concept of enrichment culture to isolate aerobic and anaerobic nitrogen-fixing bacteria, sulfate-reducing bacteria, and sulfur-oxidizing bacteria. What kind of selective media could he have used for isolating each of these four classes of microbes?

Preserving Cultures

Learning Objective

✓ Contrast refrigeration, deep freezing, and lyophilization as methods for preserving cultures of microbes.

Because cells metabolize, storing living cells necessitates slowing their metabolism to prevent the excessive accumulation of waste products and the exhaustion of all nutrients in a medium. **Refrigeration** is often the best technique for storing bacterial cultures for short periods of time.

Deep-freezing and lyophilization are used for long-term storage of bacterial cultures. **Deep-freezing** involves freezing the cells at temperatures from −50°C to −95°C. Deep frozen cultures can be restored years later by thawing them and placing a sample in an appropriate medium.

Lyophilization (lī-of′i-li-zā′shŭn) (freeze drying) involves removing water from a frozen culture using an intense vacuum. Under these conditions, ice directly becomes a gas (sublimates) and is removed from cells without permanently damaging cellular structures and chemicals. Lyophilized cultures can last for decades and are revived by adding lyophilized cells to liquid culture media.

Growth of Microbial Populations

Most unicellular microorganisms reproduce by *binary fission,* a process in which a cell grows to twice its normal size and

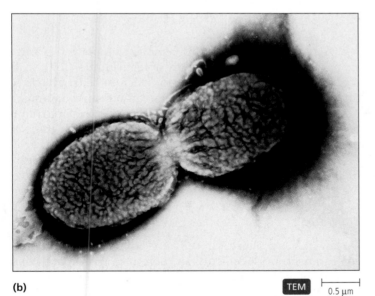

(a)

(b)   TEM   0.5 μm

▲ *Figure 6.17*

Binary fission. **(a)** The events in binary fission. ① The cell replicates its chromosome; the duplicated DNA molecules are attached to the cytoplasmic membrane. ② The cell elongates, and growth between attachment sites pushes the chromosomes apart. ③ The cell forms a new cell wall (a septum) across the midline. ④ The septum is completed; daughter cells may remain attached (as shown here) or may completely separate. ⑤ The process repeats. All the cells may divide in parallel planes and remain attached to form a chain, or they may divide in different planes to form a cluster (as shown here). **(b)** Transmission electron micrograph of a bacterial cell undergoing binary fission.

divides in half to produce two equally sized daughter cells. Binary fission generally involves four steps as illustrated in **Figure 6.17** for a prokaryotic cell.

① The cell replicates its chromosome. The duplicated chromosomes are attached to the cytoplasmic membrane. (In eukaryotic cells chromosomes are attached to microtubules.)

② The cell elongates and growth between attachment sites pushes the chromosomes apart. (Eukaryotic cells segregate their chromosomes by *mitosis,* a process described in Chapter 12.)

③ The cell forms a new cell membrane and wall (septum) across the midline.

④ When the septum is completed, the daughter cells may remain attached as shown in the figure, or they may completely separate. When the cells remain attached, further binary fissions in parallel planes produces a chain. When further divisions are in different planes, the cells become a cluster (as shown in the figure).

⑤ The process repeats.

Other reproductive strategies of prokaryotes and eukaryotes are discussed in Chapters 11 and 12. Here we consider only the growth of populations by binary fission, using bacterial cultures as examples. We begin with a brief discussion of the mathematics of population growth.

Mathematical Considerations in Population Growth

Learning Objective

✓ Describe logarithmic growth.

With binary fission, any given cell divides to form two cells; then each of these new cells divides in two, to make four, and then four become eight, and so on. This type of growth, called **logarithmic growth** or **exponential growth,** produces very different results from simple addition, or *arithmetic growth*. We can compare these two types of growth by considering what would happen over time to two identical hypothetical populations, as shown in **Figure 6.18.** In this case we assume that a population of hypothetical species A increases by adding one new cell every 20 minutes, whereas the cells in hypothetical species B divide by binary fission every 20 minutes. After 20 minutes, each population, which started with but a single cell, would have two cells; after 40 minutes, species A would have three cells, while species B would have four cells. At this point there is little difference in the growth of the two populations, but after 2 hours, the arithmetically growing species A would have only seven cells, whereas the logarithmically growing species B would have increased to 64 cells. Clearly, logarithmic growth can increase a population's size dramatically—after only 7 hours, species B will have reproduced over 2 million cells!

The number of cells arising from a single cell reproducing by binary fission is calculated as 2^n, where n is the number of generations, in other words, multiply two times itself n number of times. To calculate the total number of cells in a population, we multiply the original number of cells by $2n$. If, for example, species B had begun with three cells instead of one, then after 2 hours it would have 192 cells ($3 \times 2^6 = 3 \times 64 = 192$).

A visible culture of bacteria may consist of trillions of cells, so microbiologists use scientific notation to deal with the huge numbers involved. One advantage of scientific notation is that large numbers are expressed as powers of 10, making them easier to read and write. For example, consider our culture of species B. After 10 hours (30 generations) it would have 1,073,741,824 (2^{30}) cells. This large number can be rounded off and expressed more succinctly in scientific notation as 1.07×10^9. After 30 more generations, scientific notation would be the only practical way to express the huge number of cells in the culture, which would be 1.15×10^{18} (2^{60}). Such a number, if written out (the digits 115 followed by 16 zeroes), would be impractically large. Scientific notation is discussed in more detail in Appendix B.

Generation Time

Learning Objective

✓ Explain what is meant by the generation time of bacteria.

The time required for a bacterial cell to grow and divide is its **generation time.** Viewed another way, generation time is also the time required for a population of cells to double in number. Generation times vary among populations and are dependent on chemical and physical conditions. Under optimal conditions, some bacteria (such as *E. coli* and *S. aureus*) have a generation time of 20 minutes or less. For this reason, food contaminated with only a few of these organisms can cause food poisoning if not properly refrigerated and cooked. Most bacteria have a generation time of 1–3 hours, though some slow-growing species such as *Mycobacterium leprae* require more than 10 days before they double. The math required to calculate generation time for a population is presented in Appendix B.

Phases of Microbial Growth

Learning Objectives

✓ Draw and label a bacterial growth curve.
✓ Describe what occurs at each phase of a population's growth.

A graph that plots the numbers of organisms in a growing population over time is known as a **growth curve.** When drawn using an arithmetic scale on the y-axis, a plot of exponential growth presents two problems **(Figure 6.19a):** It is difficult or impossible to distinguish numbers in early generations from the baseline, and as the population grows it becomes impossible to accommodate the graph on a single page.

(a) Species A   **(b)** Species B

▲ *Figure 6.18*

A comparison of arithmetic and logarithmic growth. Given two initial populations consisting of a single cell each, after 2 hours the arithmetically growing species A will have seven cells, whereas the logarithmically growing species B will have 64 cells.

The solution to these problems is to replace the arithmetic scale on the y-axis with a logarithmic (log) scale **(Figure 6.19b).** Such a log scale, in which each division is 10 times larger than the preceding one, can accommodate small numbers at the lower end of the graph, and very large numbers at the upper end. This kind of graph is *semilogarithmic,* because only one axis uses a log scale.

When bacteria are inoculated into a liquid medium, there are four distinct phases to a population's growth curve: the lag, log, stationary, and death phases **(Figure 6.20).**

Lag Phase

During the **lag phase** the cells are adjusting to their new environment; most cells do not reproduce immediately, but instead actively synthesize enzymes to utilize novel nutrients in the medium. For example, bacteria inoculated from a medium containing glucose as a carbon source into a medium containing lactose must synthesize two types of proteins: membrane proteins to transport lactose into the cell, and the enzyme lactase to catabolize the lactose. The lag phase can last less than an hour or for days depending on the species and the chemical and physical conditions of the medium.

Log Phase

Eventually, the bacteria synthesize the necessary chemicals for conducting metabolism in their new environment, and

(a)

(b)

▲ *Figure 6.19*

Two growth curves of logarithmic growth. The generation time for this *E. coli* population is 20 minutes. **(a)** An arithmetic graph. Using an arithmetic scale for the *y*-axis makes it difficult to ascertain actual numbers of cells near the beginning, and impossible to plot points after only a short time. **(b)** A semilogarithmic graph. Using a logarithmic scale for the *y*-axis solves both of these problems. Note that a plot of logarithmic population growth using a logarithmic scale produces a straight line.

▲ *Figure 6.20*

A typical microbial growth curve, including the four phases of population growth. *Why do cells trail behind their optimum reproductive potential during the lag phase?*

Figure 6.20 During lag phase, cells are synthesizing the metabolic machinery and chemicals required for optimal reproduction.

they then enter a phase of rapid chromosome replication, growth, and reproduction. This is the **log phase,** so-named because the population increases logarithmically, and the reproductive rate reaches a constant as DNA and protein syntheses are maximized. Because the metabolic rate of individual cells is at a maximum during log phase, this phase is sometimes preferred for industrial and laboratory purposes. Populations in log phase are more susceptible to antimicrobial drugs that interfere with metabolism, such as erythromycin, and to drugs that interfere with the formation of cell structures, such as the inhibition of cell wall synthesis by penicillin. Further, populations in log phase are preferred for Gram staining because most cells' walls are intact—an important characteristic for correct staining.

Stationary Phase

If bacterial growth continued at the exponential rate of the log phase, bacteria would soon overwhelm the earth. This does not occur because as nutrients are depleted and wastes accumulate, the rate of reproduction decreases. Eventually, the number of dying cells equals the number of cells being produced, and the size of the population becomes stationary—hence the name **stationary phase.** During this phase the metabolic rate of surviving cells declines.

The onset of the stationary phase can be postponed indefinitely (and thus exponential growth can be maintained indefinitely) by a special apparatus called a *chemostat*, which continually removes wastes (along with old medium and some cells) and adds fresh medium. Chemostats are used in industrial fermentation processes.

Death Phase

If nutrients are not added and wastes are not removed, a population reaches a point at which cells die at a faster rate than they are produced. Such a culture has entered the **death phase** (or *decline phase*). Bear in mind that during the death phase, some cells remain alive and continue metabolizing and reproducing, but the number of dying cells exceeds the number of new cells produced, so that eventually the population decreases to a fraction of its previous abundance. In some cases, all the cells die, while in others a few survivors may remain indefinitely. The latter case is especially true for cultures of bacteria that can develop resting structures called *endospores* (see Chapter 11).

Measuring Microbial Growth

Learning Objective

✓ Contrast direct and indirect methods of measuring bacterial growth.

We have discussed the concepts of population growth and have seen that large numbers result from logarithmic growth, but we have not discussed practical methods of determining the size of a microbial population. Because of each cell's small size and incredible rate of reproduction, it is not possible to actually count every one in a population. (For one thing, they grow so rapidly that their number changes during the count.) Therefore, laboratory personnel must estimate the number of

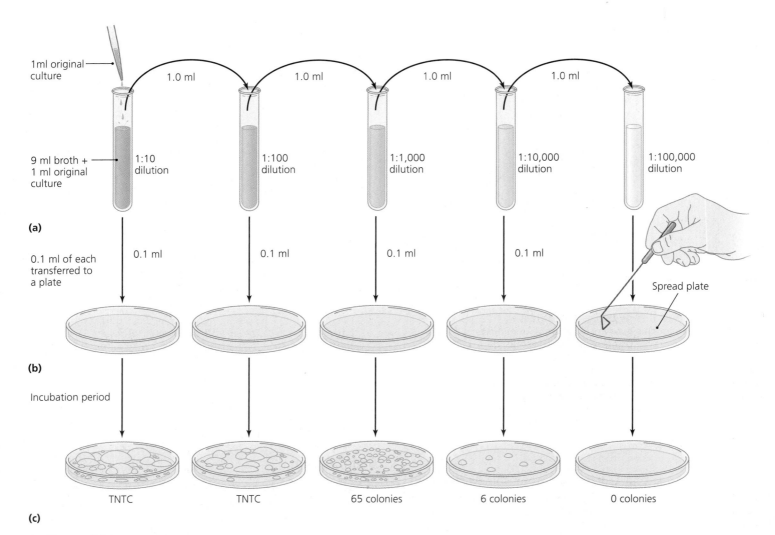

1ml original culture

1.0 ml   1.0 ml   1.0 ml   1.0 ml

9 ml broth + 1 ml original culture

1:10 dilution   1:100 dilution   1:1,000 dilution   1:10,000 dilution   1:100,000 dilution

(a)

0.1 ml of each transferred to a plate

0.1 ml   0.1 ml   0.1 ml   0.1 ml

Spread plate

(b)

Incubation period

TNTC   TNTC   65 colonies   6 colonies   0 colonies

(c)

▲ *Figure 6.21*

A viable plate count for estimating microbial population size. **(a)** Serial dilutions. A series of 10-fold dilutions is made. **(b)** Plating. A 0.1-ml sample from each dilution is poured onto a plate and spread with a sterile rod. Alternatively, 0.1 ml of each dilution can be mixed with melted agar medium and poured into plates. **(c)** Counting. Plates are examined after incubation. Some plates may contain so much growth that colonies are too numerous to count (TNTC). The number of colonies is multiplied by the reciprocal of the dilution to estimate the concentration of bacteria in original culture—in this case, 65 colonies × 10,000 = 650,000 bacteria/ml.

cells in a population by counting the number in a small, representative sample and then multiplying to estimate the number in the whole specimen. For example, if there are 25 cells in a microliter (μl) sample of urine, then there are approximately 25 million cells in a liter of urine.

Estimating the number of microorganisms in a sample is useful for determining such things as the severity of urinary tract infections, the effectiveness of pasteurization and other methods of food preservation, the degree of fecal contamination of water supplies, and the effectiveness of particular disinfectants and antibiotics.

Microbiologists use either direct or indirect methods to estimate the number of cells. We begin with direct methods of measuring bacterial growth.

Direct Methods

Among the many direct techniques are viable plate counts, membrane filtration, microscopic counts, the use of electronic counters, and the most probable number method.

Viable Plate Counts   What if the number of cells in even a very small sample is still too great to count? If, for example, a l-milliliter sample of milk containing 20,000 bacterial cells were plated on a Petri plate, there would be too many colonies to count. In such cases, microbiologists make a series of dilutions and count the number of colonies resulting on a spread or pour plate from each dilution. Scientists count the colonies on plates with 30–300 colonies and multiply the number by the reciprocal of the dilution to estimate the number of bacteria per ml of the original culture. This method is called a **viable plate count (Figure 6.21).**

(a)

Sample to be filtered

Membrane filter retains cells

To vacuum

Membrane transferred to culture medium

Incubation

Colonies

(b)

LM 2.5 μm

(c)

◀ *Figure 6.22*

The use of membrane filtration to estimate microbial population size. **(a)** After all the bacteria in a given volume of sample are trapped on a membrane filter, the filter is transferred onto an appropriate medium and incubated. The microbial population is estimated by multiplying the number of colonies counted by the volume of sample filtered. **(b)** Bacteria trapped on the surface of a membrane filter. **(c)** Colonies growing on a solid medium after being transferred from the membrane filter. Scientists use a super-imposed grid to help them count the colonies. *If the colonies in (c) resulted from filtering 2.5 liters of stream water, what is the minimum number of bacteria per liter in the stream?*

$$\textit{Figure 6.22} \quad \frac{25 \; colonies}{2.5 \; L} = 10 \; colonies/L$$

When a plate has fewer than 30 colonies, it is not used to estimate the number of bacteria in the original sample because the chance of underestimating the population increases when the number of colonies is small. Recall that the number of colonies on a plate indicates the number of *colony-forming units* that were inoculated onto the plate. This number differs from the actual number of cells when the colony forming units are composed of more than one cell. In such cases, a viable plate count underestimates the number of cells present in the sample.

The accuracy of a viable plate count is also dependent on the homogeneity of the dilutions, the ability of the bacteria to grow on the medium used, the number of cell deaths, and the growth phase of the sample population. Thoroughly mixing each dilution, inoculating multiple plates per dilution, and using log-phase cultures minimize errors.

Membrane Filtration   Viable plate counts allow scientists to estimate the number of microorganisms when the population is very large, but if the population density is very small—as is the case, for example, for fecal bacteria in a stream or lake—microbes are more accurately counted by **membrane filtration (Figure 6.22).** In this method, a large sample (perhaps as large as several liters) is poured (or drawn under a vacuum) through a membrane filter with pores small enough to trap the cells. The membrane is then transferred onto a solid medium, and the colonies present after incubation are counted. In this case, the number of colonies is equal to the number of CFUs in the original large sample.

CRITICAL THINKING

Viable plate counts are used to estimate population size when the density of microorganisms is high, whereas membrane filtration is used when the density is low. Why is a viable plate count appropriate when the density is high, but not when the density is low?

(a)

(b)

(c)

▲ *Figure 6.23*

The use of a cell counter for estimating microbial numbers.
(a) The counter is a glass slide with an etched grid that is
exactly 0.02 mm lower than the bottom of the cover slip. A
bacterial suspension placed next to the cover slip through a
pipette moves under the cover slip and over the grid by
capillary action. **(b)** View of a 1 mm² portion of the grid
through the microscope. Each square millimeter of the grid
has 25 large squares, each of which is divided into 16 small
squares. **(c)** Enlarged view of one large square containing
15 cells. The number of bacteria in several large squares is
counted and averaged. The calculations involved in
estimating the number of bacteria per milliliter (cm³) of
suspension are described in the text.

Microscopic Counts   Microbiologists can also count microorganisms directly through a microscope rather than inoculating them onto the surface of a solid medium. In this method, particularly suitable for stained prokaryotes and relatively large eukaryotes, a sample is placed on a *cell counter* (also called a *Petroff-Hauser counting chamber*), which is a glass slide composed of an etched grid positioned beneath a glass cover slip **(Figure 6.23)**. Because the cover slip is 0.02 mm above the grid, the volume of bacterial suspension over a 1 mm² portion of the grid is 1 mm × 1 mm × 0.02 mm = 0.02 mm³. Each 1 mm² grid contains 25 large squares; a microbiologist counts the number of bacteria in several of the large squares and then calculates the mean number of bacteria per square. The number of bacteria per milliliter (cm³) can be calculated as follows:

$$\text{mean no. of bacteria per square} \times 25 \text{ squares}$$
$$= \text{no. of bacteria per } 0.02 \text{ mm}^3$$

$$\text{no. of bacteria per } 0.02 \text{ mm}^3 \times 50 = \text{no. of bacteria per mm}^3$$

$$\text{no. of bacteria per mm}^3 \times 1000 = \text{no. of bacteria per cm}^3 \text{ (ml)}$$

This means that one only needs to multiply the mean number of bacteria per square by 1,250,000 (25 × 50 × 1000) to calculate the number of bacteria per ml of bacterial suspension.

Direct microscopic counts are advantageous when there are more than 10,000,000 cells per ml or when a speedy estimate of population size is required. However, direct counts can be problematic because it is often difficult to differentiate between living and dead cells, and difficult to count motile microorganisms.

Electronic Counters   A *Coulter counter* is a device that directly counts cells as they interrupt an electrical current flowing across a narrow tube held in front of an electronic detector. This device is useful for counting the larger cells of yeast, algae, and protozoa; it is less useful for bacterial counts because of debris in the media and the presence of filaments and clumps of cells.

Flow cytometry is a variation of counting with a Coulter counter. A cytometer uses a light-sensitive detector to record changes in light transmission through the tube as cells pass. Scientists use this technique to distinguish among cells that have been differentially stained with fluorescent dyes or tagged with fluorescent antibodies. They can count bacteria in a solution and even count host cells that contain fluorescently stained intracellular parasites.

Most Probable Number   The **most probable number (MPN) method** is a statistical estimating technique based on the fact that the more bacteria in a sample, the more dilutions are required to reduce their number to zero.

Let's consider an example of the use of the MPN method to estimate the number of fecal bacteria contaminating a

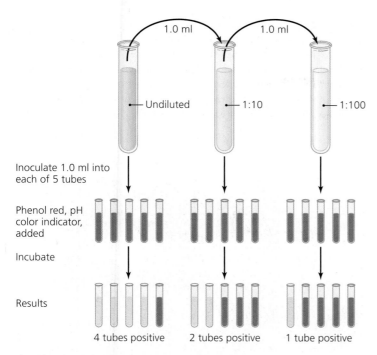

1.0 ml   1.0 ml

Undiluted   1:10   1:100

Inoculate 1.0 ml into each of 5 tubes

Phenol red, pH color indicator, added

Incubate

Results

4 tubes positive   2 tubes positive   1 tube positive

▲ *Figure 6.24*

The most probable number (MPN) method for estimating microbial numbers. Typically, sets of five test tubes are used for each of three dilutions. After incubation, the number of tubes showing growth in each set is used to enter an MPN table (see Table 6.6), which provides an estimate of the number of cells per 100 ml of liquid. *If the results were 5, 3, 1, what would be the most probable number of microorganisms in the original broth?*

Figure 6.24 The MPN is 110 ml.

stream. A researcher inoculates a set of test tubes of a broth medium with a sample of stream water. Even though the more tubes that are used, the more accurate is the MPN method, accuracy must be balanced against the time and cost involved in inoculating and incubating numerous tubes. Typically, a set of five tubes is inoculated.

The researcher also inoculates a set of five tubes with a 1:10 dilution, and another set of five tubes with a 1:100 dilution of stream water. Thus, there are 15 test tubes—the first set of five tubes inoculated with undiluted sample, the second set with a 1:10 dilution, and the third set with a 1:100 dilution **(Figure 6.24)**.

After incubation for 48 hours, the researcher counts the number of test tubes in each set that show growth as determined by some method, such as cloudiness of the broth, gas production, or pH changes. This generates three numbers, in this case—4, 2, 1—that are compared to the numbers in an MPN table (**Table 6.6** on page 195). How statisticians develop MPN tables is beyond the scope of our discussion, but they are constructed in such a way that they accurately estimate

the number of cells in a culture 95% of the time. Thus, when growth occurs in four of the undiluted broth tubes, two of the 1:10 tubes, and only one of the 1:100 tubes (4,2,1), the MPN table estimates that there were 26 bacteria/100 ml of stream water.

The most probable number method is useful for counting microorganisms that will not grow on solid media, when bacterial counts are required routinely, and when samples of wastewater, drinking water, and food samples contain too few organisms to use a viable plate count. The MPN method is also used to count algal cells because algae seldom form distinct colonies on solid media.

Indirect Methods

It is not always necessary to count microorganisms to estimate population size or density. Industrial and research microbiologists use indirect methods that measure such variables as metabolic activity, dry weight, and turbidity instead of counting microorganisms, colonies, or MPN tubes.

Metabolic Activity   Under standard temperature conditions, the rate at which a population of cells utilizes nutrients and produces wastes is dependent on their number. Once they establish the metabolic rate of a microorganism, scientists can indirectly estimate the number of cells in a culture by measuring changes in such things as nutrient utilization, waste production, or pH.

Dry Weight   The abundance of some microorganisms, particularly filamentous microorganisms, is difficult to measure by direct methods. Instead, these organisms are filtered from their culture medium, dried, and weighed. The *dry weight method* is suitable for broth cultures, but growth cannot be followed over time because the organisms are killed during the process.

Turbidity   As bacteria reproduce in a broth culture, the broth often becomes *turbid* (cloudy) **(Figure 6.25a)**. Generally, the greater the bacterial population, the more turbid a broth will be. An indirect method for estimating the growth of a microbial population involves measuring changes in turbidity using a device called a *spectrophotometer* **(Figure 6.25b)**.

A spectrophotometer measures the amount of light transmitted through a culture under standardized conditions **(Figure 6.25c)**. The greater the concentration of bacteria within a broth, the more light will be absorbed and scattered, and the less light will pass through and strike a light-sensitive detector. Generally, transmission is inversely proportional to the population size; that is, the larger the population grows, the less light will reach the detector.

Scales on the gauge of a spectrophotometer report *percentage of transmission* and *absorbance*. These are two ways of looking at the same things, for example, 25% transmission is the same thing as 75% absorbance. Direct counts must be calibrated with transmission and absorbance readings to provide estimates of population size. Once these values are determined, spectrophotometry provides estimates of population size more quickly than any direct method.

The benefits of measuring turbidity to estimate population growth include ease of use and speed. However, the technique is useful only if the concentration of cells exceeds 1 million per milliliter; densities below this value generally do not produce turbidity. Further, the technique is accurate only if the cells are suspended uniformly in the medium. If they form either a *pellicle* (a film of cells at the surface) or a *sediment* (an accumulation of cells at the bottom), their number will be underestimated. Further, spectrophotometry does not distinguish between living and dead cells.

(a)

(b)

(c)

Figure 6.25 ▶

Turbidity and the use of spectrophotometry in indirectly measuring population size. **(a)** Turbidity (right), an increased optical density or cloudiness of a solution. **(b)** A spectrophotometer. **(c)** The principle of spectrophotometry. After passing a light beam through an uninoculated sample of the culture medium, the scale is set at 100% transmission. In an inoculated sample, the microbial cells absorb and scatter light, reducing the amount reaching the detector. The percentage of light transmitted is inversely proportional to population density.

Table 6.6   Most Probable Number Table

Number Out of 5 Tubes Giving Positive Results in Three Dilutions			Most Probable Number of Bacteria per 100ml	Number Out of 5 Tubes Giving Positive Results in Three Dilutions			Most Probable Number of Bacteria per 100ml
0	0	0	1.8	4	3	0	27
0	0	1	2	4	3	1	33
0	1	0	2	4	4	0	34
0	2	0	4	5	0	0	23
1	0	0	2	5	0	1	30
1	0	1	4	5	0	2	40
1	1	0	4	5	1	0	30
1	1	1	6	5	1	1	50
1	2	0	6	5	1	2	60
2	0	0	4	5	2	0	50
2	0	1	7	5	2	1	70
2	1	0	7	5	2	2	90
2	1	1	9	5	3	0	80
2	2	0	9	5	3	1	110
2	3	0	12	5	3	2	140
3	0	0	8	5	3	3	170
3	0	1	11	5	4	0	130
3	1	0	11	5	4	1	170
3	1	1	14	5	4	2	220
3	2	0	14	5	4	3	280
3	2	1	17	5	4	4	350
4	0	0	13	5	5	0	240
4	0	1	17	5	5	1	300
4	1	0	17	5	5	2	500
4	1	1	21	5	5	3	900
4	1	2	26	5	5	4	1,600
4	2	0	22	5	5	5	$\geq 1,600$
4	2	1	26				

CHAPTER SUMMARY

Growth Requirements (pp. 168–177)

1. A **colony,** which is a visible population of microorganisms living in one place, grows in size as the number of cells increases.

2. Chemical **nutrients** such as carbon, hydrogen, oxygen, and nitrogen are required for the growth of microbial populations.

3. **Photoautotrophs** use carbon dioxide as a carbon source and light energy from the environment to make their own food; **chemoautotrophs** use carbon dioxide as a carbon source but catabolize organic molecules for energy. **Photoheterotrophs** are photosynthetic organisms that acquire energy from light and acquire nutrients via catabolism of organic compounds; **chemoheterotrophs** use organic compounds for both energy and carbon. **Organotrophs** acquire electrons from organic sources, whereas **lithotrophs** acquire electrons from inorganic sources.

4. **Obligate aerobes** require oxygen as the final electron acceptor of the electron transport chain, whereas **obligate anaerobes** cannot tolerate oxygen and use an electron acceptor other than oxygen.

5. The four toxic forms of oxygen are **singlet oxygen ($^1O_2$),** which is neutralized by pigments called **carotenoids; superoxide radicals (O_2^-),** which are detoxified by superoxide dismutase; **peroxide anion (O_2^{2-}),** which is detoxified by catalase or peroxidase; and **hydroxyl radicals (OH·),** the most reactive of the toxic forms of oxygen.

6. Microbes are described in terms of their oxygen requirements and limitations as strict **aerobes,** which require oxygen; strict **anaerobes,** which cannot tolerate oxygen; as **facultative anaerobes,** which can live with or without oxygen; as **aerotolerant anaerobes,** which prefer anaerobic conditions, but can tolerate exposure to low levels of oxygen; or as **microaerophiles,** which require low levels of oxygen. A **capnophile** grows best with high CO_2 levels in addition to low oxygen levels.

7. Nitrogen, acquired from organic or inorganic sources, is an essential element for microorganisms. Some bacteria can reduce nitrogen gas into a more usable form via a process called **nitrogen fixation.**

8. In addition to the main elements found in microbes, very small amounts of **trace elements** are required. Vitamins are among the **growth factors,** which are organic chemicals required in small amounts for metabolism.

9. Though microbes survive within the limits imposed by a minimum growth temperature and a maximum growth temperature, an organism's metabolic activities produce the highest growth rate at the **optimum growth temperature.**

10. Microbes are described in terms of their temperature requirements as (from coldest to warmest) **psychrophiles, mesophiles, thermophiles,** or **hyperthermophiles.**

11. **Neutrophiles** grow best at neutral pH, **acidophiles** grow best in acidic surroundings, and **alkalinophiles** live in alkaline habitats.

12. Osmotic pressure can cause cells to die from either swelling and bursting or shriveling **(plasmolysis).** The cell walls of some microorganisms protect them from osmotic shock. **Obligate halophiles** require high osmotic pressure, whereas facultative halophiles do not require but can tolerate such conditions.

13. **Barophiles,** organisms that normally live under the extreme hydrostatic pressure at great depth below the surface of a body of water, often cannot live at the pressure found at the surface.

14. **Quorum sensing** is the process by which bacteria respond to changes in microbial density by utilizing signal and receptor molecules. **Biofilms,** which are communities of cells attached to surfaces, are one result of quorum sensing.

Culturing Microorganisms (pp. 177–186)

1. Microbiologists culture microorganisms by transferring an **inoculum** from a clinical or environmental **specimen** into a **medium** such as **broth** or solid media. The microorganisms grow into a **culture.** On solid surfaces, cultures are seen as colonies.

2. **Pure cultures** (axenic cultures) contain cells of only one species and are derived from a **colony-forming unit (CFU)** composed of a single cell or group of related cells. To obtain pure cultures, **sterile** equipment and use of aseptic techniques are critical.

3. The **streak plate** method allows CFUs to be isolated, incubated, and observed. The **pour plate** technique isolates CFUs via a series of dilutions.

4. Petri dishes that are filled with solid media are called **Petri plates. Slant tubes (slants)** are test tubes containing agar media that solidified while the tube was resting at an angle.

5. A **defined medium** (also known as **synthetic medium**) provides exact known amounts of nutrients for the growth of a particular microbe. **Complex media** contain a variety of growth factors. **Selective media** either inhibit the growth of unwanted microorganisms or favor the growth of particular microbes. Microbiologists use **differential media** to distinguish among groups of bacteria. **Reducing media** provide conditions conducive to culturing anaerobes. **Transport media** are designed to move specimens safely from one location to another while maintaining the relative abundance of organisms and preventing contamination of the specimen or environment.

6. Special culture techniques include the use of animal and cell cultures, low-oxygen cultures, **enrichment cultures,** and **cold enrichment cultures.**

7. Cultures can be preserved in the short term by **refrigeration,** and in the long term by **deep-freezing** and **lyophilization.**

Growth of Microbial Populations (pp. 186–195)

1. Bacteria grow by **logarithmic** or **exponential growth.**

2. A population of bacteria doubles during its **generation time**—the time also required for a bacterial cell to grow and divide.

3. A graph that plots the number of bacteria growing in a population over time is called a **growth curve.** When bacteria are grown in a broth and the growth curve is plotted on a semi-logarithmic scale, the population's growth curve has four phases. In the **lag phase,** the organisms are adjusting to their environment. In the **log phase,** the population is most actively growing. In the **stationary phase,** new organisms are being produced at the same rate at which they are dying. In the **death phase,** the organisms are dying more quickly than they can be replaced by new organisms.

4. Direct methods for estimating population size include **viable plate counts, membrane filtration,** microscopic counts, electronic counters, flow cytometry, and the **most probable number (MPN) method.**

5. Indirect methods include measurements of metabolic activity, dry weight, and turbidity.

QUESTIONS FOR REVIEW

(Answers to Multiple Choice and Fill in the Blanks are on the Web, along with additional review questions. Visit www.microbiologyplace.com.)

Multiple Choice

1. Which of the following would be found growing in a Petri plate?
   a. an anaerobe
   b. a colony on an agar surface
   c. viruses on an agar surface
   d. barophiles

2. Which of the following terms best describes an organism that cannot exist in the presence of oxygen?
   a. obligate aerobe
   b. facultative aerobe
   c. obligate anaerobe
   d. facultative anaerobe

3. Superoxide dismutase
   a. causes hydrogen peroxide to become toxic.
   b. detoxifies superoxide radicals.
   c. neutralizes singlet oxygen.
   d. is missing in aerobes.

4. The most reactive of the four toxic forms of oxygen is
   a. the hydroxyl radical.
   b. the peroxide anion.
   c. the superoxide radical.
   d. singlet oxygen.

5. Microaerophiles that grow best with a high concentration of carbon dioxide in addition to a low level of oxygen are called
   a. aerotolerant.
   b. capnophiles.
   c. facultative anerobes.
   d. fastidious.

6. Which of the following is *not* a growth factor for various microbes?
   a. cholesterol
   b. water
   c. vitamins
   d. heme

7. Organisms that may be found thriving in icy waters are described as
   a. barophiles.
   b. thermophiles.
   c. mesophiles.
   d. psychrophiles.

8. Barophiles
   a. cannot cause diseases in humans.
   b. live at normal barometric pressure.
   c. die if put under high pressure.
   d. thrive in warm air.

9. This statement, "In the laboratory, a sterile inoculating loop is moved across the agar surface in a culture dish, thinning a sample and isolating individuals," describes which of the following:
   a. broth culture
   b. pour plate
   c. streak plate
   d. dilution plate

10. In a defined medium,
    a. the exact chemical composition of the medium is known.
    b. agar is available for microbial nutrition.
    c. blood may be included.
    d. organic chemicals are excluded.

11. Which of the following is most useful in representing population growth on a graph?
    a. logarithmic reproduction of the growth curve
    b. a semilogarithmic graph using a log scale on the *x*-axis
    c. an arithmetic graph of the lag phase followed by a logarithmic section for the log, stationary, and death phases
    d. none of the above would best represent a population growth curve

12. Which of the following methods is best for counting fecal bacteria from a stream to determine the safety of the water for drinking?
    a. dry weight
    b. turbidity
    c. viable plate counts
    d. membrane filtration

13. A Coulter counter is
    a. a statistical estimation using five dilution tubes and a table of numbers to estimate the number of bacteria per milliliter.
    b. an indirect method of counting microorganisms.
    c. a device that directly counts microbes as they pass through a tube in front of an electronic detector.
    d. a device that directly counts microbes that are differentially stained with fluorescent dyes.

14. Lyophilization is best described as
    a. freeze-drying.
    b. deep-freezing.
    c. refrigeration.
    d. pickling.

15. Quorum sensing is
    a. the ability to respond to changes in population density.
    b. a characteristic of most bacteria.
    c. dependent on direct contact among cells.
    d. associated with colonies on an agar plate.

Fill in the Blanks

1. All cells require a source of _____, _____, and _____,

2. A toxic form of oxygen, _____ oxygen, is molecular oxygen with electrons that have been boosted to a higher energy state.

3. All cells recycle the essential element _____ from amino acids and nucleotides.

4. _____ are small organic molecules that are required in minute amounts for metabolism.

5. The lowest temperature at which a microbe continues to metabolize is called its _____.

6. Cells that shrink in hypertonic solutions such as saltwater are responding to _____ pressure.

7. Obligate _____ may exist in salt ponds because of their ability to withstand high osmotic pressure.

Short Answer

1. High temperature affects the shape of particular molecules. How does this affect the life of a microbe?

2. Support or refute the following statement: Microbes cannot tolerate the low pH of the human stomach.

3. Explain quorum sensing, and describe how it is related to biofilm formation.

4. Why must media, vessels, and instruments be sterilized before they are used for microbiological procedures?

5. Why is agar used in microbiology?

6. What is the difference between complex media and defined media?

7. Draw and label the four distinct phases of a bacterial growth curve. Describe what is happening within the culture as it passes through the phases.

8. If there are 47 cells in a μl of sewage, how many cells are there in a liter?

9. List three indirect methods of counting microbes.

10. List five direct methods of counting microbes.

11. Explain the differences among photoautotrophs, chemoautotrophs, photoheterotrophs, chemoheterotrophs, organotrophs, and lithotrophs.

CRITICAL THINKING

1. A microbiologist describes an organism as a chemoheterotrophic, aerotolerant, mesophilic, facultatively halophilic bacillus. Describe the organism's metabolic and structural features in plain English.

2. Pasteurization is a technique that uses temperatures of about 72°C to neutralize potential pathogens in foods. What effect does this temperature have on the enzymes and cellular metabolism of pathogens? Why does the heat of pasteurization kill some microorganisms yet fail to affect thermophiles?

3. If two cultures of a facultative anaerobe were grown under identical conditions except that one was exposed to oxygen and the other was completely deprived of oxygen, what differences would you expect to see between the dry weights of the cultures? Why?

CHAPTER 7

Microbial Genetics

Do genes hold the secrets to life? If we can identify all the genes in an organism, and determine the functions of each, can we explain all of biological behavior? What genes do organisms have in common, and what genes make them unique? What genes cause certain microorganisms to be harmful or even deadly, and how can we develop drugs or techniques to target a pathogen's genes?

These are the kinds of questions that drive the dynamic world of genetic research. We now have complete genome maps, or genetic blueprints, of hundreds of viruses, bacteria, and other organisms. We even have a working draft of the human genome, which our knowledge of microbial genetics is helping us to analyze. For example, by studying the genes of the *Escherichia coli* bacterium and then identifying which genes we share, we can determine the roles these same genes play in humans. Genetically speaking, you may have much more in common with microorganisms than you think!

CHAPTER OUTLINE

The Structure and Replication of Genomes
  The Structure of Nucleic Acids
  The Structure of Prokaryotic Genomes
  The Structure of Eukaryotic Genomes
  DNA Replication

Gene Function
  The Relationship Between Genotype and Phenotype
  The Transfer of Genetic Information
  The Events in Transcription
  Translation
  Control of Transcription

Mutations of Genes
  Types of Mutations
  Effects of Mutations
  Mutagens
  Frequency of Mutation
  DNA Repair
  Identifying Mutants, Mutagens, and Carcinogens

Genetic Recombination and Transfer
  Horizontal Gene Transfer Among Prokaryotes
  Transposons and Transposition

Ribosomes on two strands of messenger RNA (mRNA). mRNA carry genetic information from chromosomes to ribosomes, enabling protein synthesis. The synthesis of polypeptides is visible in the strand to the right.

MicroPrep Pre-Test: Take the pre-test for this chapter on the web. Visit www.microbiologyplace.com.

Genetics is the study of inheritance and inheritable traits as expressed in an organism's genetic material. Geneticists study many aspects of inheritance, including the physical structure and function of genetic material, mutations, and the transfer of genetic material among organisms. In this chapter, we will examine these topics as they apply to microorganisms, the study of which have formed much of the basis of our understanding of human, animal, and plant genetics.

The Structure and Replication of Genomes

Learning Objective

✓ Compare and contrast the genomes of prokaryotes and eukaryotes.

The **genome** (je'nōm) of a cell or virus is its entire genetic complement, including both its **genes**—specific sequences of nucleotides that code for polypeptides or RNA molecules—and nucleotide sequences that connect genes to one another. The genomes of cells and DNA viruses are composed solely of molecules of deoxyribonucleic acid (DNA), whereas RNA viruses use ribonucleic acid instead. We will examine the genomes of viruses in more detail in Chapter 13. The remainder of this chapter focuses on *bacterial* genomes—their structure, replication, function, mutation, and repair, and how they compare and contrast with eukaryotic genomes and with the genomes of archaea. We begin by examining the structure of nucleic acids.

The Structure of Nucleic Acids

Learning Objective

✓ Describe the structure of DNA, and discuss how it facilitates the ability of DNA to act as genetic material.

As we studied in Chapter 2, nucleic acids are polymers of nucleotides, each of which contains a pentose sugar (deoxyribose in DNA, ribose in RNA), a phosphate, and one of five nitrogenous bases (guanine, cytosine, adenine, thymine, or uracil). These bases hydrogen bond in specific ways called **base pairs (bp):** In DNA, the complementary bases guanine and cytosine bond to one another with three hydrogen bonds **(Figure 7.1a),** and the complementary bases adenine and thymine bond to one another with two hydrogen bonds **(Figure 7.1b).** In RNA, uracil (not thymine) bonds with adenine **(Figure 7.1c).**

Deoxyribonucleotides are linked through their sugars and phosphates to form the two backbones of a helical, double-stranded DNA (dsDNA) molecule **(Figure 7.1d).** The carbon atoms of deoxyribose are numbered 1′ (pronounced "one prime") through 5′. One end of a DNA strand is called the 5′ end because it terminates in a phosphate group attached to a 5′ carbon; the opposite (3′) end terminates with a hydroxyl group bound to a 3′ carbon of deoxyribose. The two strands are oriented in opposite directions to each other; one strand runs in a 3′ to 5′ direction, while the other runs in a 5′ to 3′ direction. Scientists say the two strands are *antiparallel.* The base pairs extend into the middle of the molecule in a way reminiscent of the steps of a spiral staircase.

The lengths of DNA molecules are not usually given in metric units; instead the length of a DNA molecule is expressed in base pairs. For example, the genome of *Mycoplasma genitalium* (mī'kō-plaz-mă jen-ē-tal'ē-ŭm) is 580,070 bp long, making it the smallest known cellular genome.

The structure of DNA helps explain its ability to act as genetic material. First, the linear sequence of nucleotides carries the instructions for the synthesis of polypeptides and RNA molecules—in much the way a sequence of letters carries information used to form words and sentences. Secondly, the complementary structure of the two strands allows a cell to make exact copies to pass to its progeny. We will examine the genetic code and DNA replication shortly.

CRITICAL THINKING

The chromosome of *Mycobacterium tuberculosis* is 4,411,529 bp long. A scientist who isolates and counts the number of nucleotides in its DNA molecule discovers that there are 2,893,963 molecules of guanine. How many molecules of the other three nucleotides are in the original DNA?

The amount of DNA in a genome can be extraordinary, as some examples will illustrate. The bacterium *Escherichia coli* (esh-ĕ-rik'ē-ă kō'lī) is approximately 2 μm long and 1 μm in diameter, but its genome consists primarily of a 4.6 × 10^6 bp DNA molecule that is about 1600 μm long—800 times longer than the cell. The human genome has about 3 billion base pairs in 46 nuclear DNA molecules plus a mitochondrial DNA molecule, and it would be 1.6 meters (1,600,000 μm) long if all 47 molecules were laid end to end. Most of the human genome is packed into a nucleus that is typically only 5 μm in diameter. This is like packing 10 miles of thread into a golf ball! To understand how cells package such prodigious amounts of DNA into such small spaces, we must first understand that prokaryotes and eukaryotes package DNA in different ways. We begin by examining the structure of prokaryotic genomes.

The Structure of Prokaryotic Genomes

The DNA in prokaryotic genomes is found in two structures: chromosomes and plasmids.

Prokaryotic Chromosomes

Prokaryotic cells package the main portion of their DNA, along with associated molecules of protein and RNA, as one or two **chromosomes.**[1] Scientists have studied the genes of

---

[1]From Greek *chroma*, meaning color (because they typically stain darkly), and *soma*, meaning body.

(a) **G–C base pair (DNA and RNA)**

(b) **A–T base pair (DNA)**

(c) **A–U base pair (RNA)**

(d)

▲ *Figure 7.1*

The structure of nucleic acids, which are polymers of nucleotides consisting of a pentose sugar, a phosphate, and a nitrogenous base. **(a)** Base pairing between the complementary bases guanine (G) and cytosine (C) formed by three hydrogen bonds, found in both DNA and RNA. **(b)** Base pairing between the complementary bases adenine (A) and thymine (T) formed by two hydrogen bonds, found in DNA only. **(c)** Base pairing between adenine and uracil (U), found in RNA only. Notice the structural similarities between thymine and uracil. **(d)** Double-stranded DNA, which consists of antiparallel strands of nucleotides held to one another by the hydrogen bonding between complementary bases. *What structures do DNA nucleotides and RNA nucleotides have in common?*

Figure 7.1 Both DNA and RNA nucleotides are composed of a pentose sugar, a phosphate, and a nitrogenous base.

bacteria more than those of archaea, so the remainder of our discussion deals with bacterial genomes.

A typical bacterial chromosome (**Figure 7.2a** on page 202) consists of a circular molecule of DNA localized in a region of the cell called the **nucleoid.** No membrane surrounds the nucleoid, though the chromosome is packed in such a way that a distinct boundary is visible between the nucleoid and the rest of the cytosol. Chromosomal DNA is folded into loops that are 50,000–100,000 bp long **(Figure 7.2b)** held in place by molecules of protein and RNA. The entire chromosome is further folded and coiled like a skein of yarn into a compact mass.

For many years scientists thought that all prokaryotes had only a single circular chromosome, but we now know that some bacterial species contain two chromosomes, and that at least one member of such a pair may be linear. *Agrobacterium tumefaciens* (ag'rō-bak-ti'rē-um tū' me-fāsh-enz), a

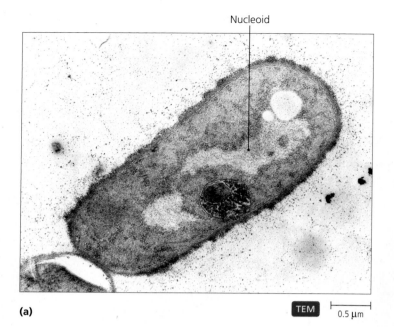

Nucleoid

(a) TEM 0.5 μm

Plasmids Bacterium Chromosome

(b) SEM 0.5 μm

▲ *Figure 7.2*

Bacterial genome. **(a)** Bacterial chromosomes are packaged in a region of the cytosol called the nucleoid, which is not surrounded by a membrane. **(b)** The packing of a circular bacterial chromosome into loops, as seen after the cell was gently broken open to release the chromosome. Extrachromosomal DNA in the form of plasmids is also visible.

bacterium used to transfer genes into plants, is an example of a prokaryote with two chromosomes, one circular and one linear. Another example of a prokaryote with two chro-

mosomes is *Vibrio cholerae* (vib'rē-ō kol'er-ă), the bacterium that causes cholera.

Plasmids

Learning Objective

✓ Describe the structure and function of plasmids.

In addition to chromosomes, many bacterial cells contain one or more **plasmids,** which are small, circular molecules of DNA that replicate independently of the chromosome. Plasmids are circular and usually 1% to 5% the size of a bacterial chromosome (see Figure 7.2b), ranging in size from a few thousand bp to a few million bp. Each plasmid carries information required for its own replication, and often for one or more cellular traits. Typically, genes carried on plasmids are not essential for normal metabolism, for growth, or for cellular reproduction, but *plasmid* genes can confer advantages to the cells that carry them.

Researchers have identified many types of plasmids (sometimes also called *factors*), including the following:

- *Fertility (F) factors* carry instructions for *conjugation*, a process involved in transferring genes from one bacterial cell to another. We will consider conjugation in more detail near the end of this chapter.

- *Resistance (R) factors* carry genes for resistance to one or more antimicrobial drugs, heavy metals, or toxins. By processes we will discuss shortly, certain cells can transfer resistance factors to other cells, which then acquire resistance to the same antimicrobial chemicals. One example of the effects of an R factor involves gonorrhea, which is no longer treated with penicillin because cells of *Neisseria gonorrhoeae* (nī-sē'rē-ă gon-ō-rē'ē) acquired an R factor for resistance to penicillin from strains of *Streptococcus.*

- *Bacteriocin factors* carry genes for proteinaceous toxins called *bacteriocins*, which kill bacterial cells of the same or similar species that lack the factor. In this way a bacterium containing the plasmid can kill its competitors.

- *Virulence plasmids* carry instructions for structures, enzymes, or toxins that enable a bacterium to become pathogenic. For example, *E. coli*, a normal resident of the human gastrointestinal tract, causes diarrhea only when it carries plasmids that code for certain toxins.

- So-called *cryptic plasmids* have been observed in photographs, but scientists have as yet been unable to identify their functions.

Now that we have examined the structure of prokaryotic genomes, we turn to the structure of eukaryotic genomes.

The Structure of Eukaryotic Genomes

Eukaryotic genomes consist of both nuclear and extranuclear DNA.

(a) Nucleosomes

(b) Chromatin fiber

(c) Euchromatin and heterochromatin

(d) Highly condensed, duplicated chromosome of dividing nucleus

▲ *Figure 7.3*

Eukaryotic nuclear chromosomal packaging. **(a)** Histones stabilize and package DNA to form nucleosomes connected by linker DNA. **(b)** Nucleosomes clump to form chromatin fibers. **(c)** Chromatin fibers fold and are organized into active euchromatin and inactive heterochromatin. **(d)** During nuclear division (mitosis), duplicated chromatin fully condenses into a mitotic chromosome that is visible by light microscopy. If the nucleosomes were actually the size shown in **(a)**, the chromosome in **(d)** would be 20 m (about 65 feet) long.

Nuclear Chromosomes

Learning Objective

✓ Compare and contrast prokaryotic and eukaryotic chromosomes.

Typically, eukaryotic cells have more than one nuclear chromosome in their genomes, though one species of Australian ant has a single chromosome per nucleus, and some eukaryotic cells such as mammalian red blood cells lose their chromosomes as they mature. Eukaryotic chromosomes differ from their typical prokaryotic counterparts in that they are linear (rather than circular) and are sequestered within a nu-

cleus. As we saw in Chapter 3, the nucleus is an organelle surrounded by two membranes, which together are called the *nuclear envelope.* Given that a typical eukaryotic cell must package substantially more DNA than its prokaryotic counterpart, it is not surprising that nuclear chromosomes are more elaborate than those of prokaryotes.

The structures involved in the packaging of eukaryotic chromosomes are depicted in **Figure 7.3.** Eukaryotic chromosomes are composed of DNA and globular proteins called **histones.** DNA, which has an overall negative electrical charge, wraps around the positively charged histones to

Table 7.1   Characteristics of Microbial Genomes

	Bacteria	Archaea	Eukarya
Number of chromosomes	One or rarely two	One	With one exception, two or more
Plasmids present?	In some cells; frequently more than one per cell	In some cells	In some fungi and protozoa
Type of nucleic acid	Circular or linear dsDNA	Circular dsDNA	Linear dsDNA in nucleus; circular dsDNA in mitochondria, chloroplasts, and plasmids
Location of DNA	In nucleoid and in plasmids in cytosol	In nucleoid and in plasmids in cytosol	In nuclei, and in mitochondria, chloroplasts, and plasmids in cytosol
Histones present?	No, though chromosome is associated with a small amount of nonhistone protein	Yes	Yes

form 10-nm-diameter beads called **nucleosomes (Figure 7.3a).** Nucleosomes clump with other proteins to form **chromatin fibers** that are about 30 nm in diameter **(Figure 7.3b).** Except during *mitosis* (nuclear division), chromatin fibers are dispersed throughout the nucleus and are too thin to be resolved without the extremely high magnification of electron microscopes. In regions of the chromosome where genes are active, the chromatin fibers are loosely packed to form *euchromatin* (yū-krō′mă-tin); inactive DNA is more tightly packed and is called *heterochromatin* (het′er-ō-krō′mă-tin; **Figure 7.3c).**

Prior to *mitosis* (nuclear division), a cell replicates its chromosomes and then condenses them into pairs of chromosomes visible by light microscopy **(Figure 7.3d).** (One molecule of each pair is destined for each daughter nucleus. Chapter 12 discusses mitosis in more detail.) The net result is that each DNA molecule has been packaged as a mitotic chromosome that is 50,000× shorter than its extended length.

Extranuclear DNA of Eukaryotes

Not all of the DNA of a eukaryotic genome is contained in its nuclear chromosomes; most eukaryotic cells also have mitochondria, and plant, algal, and some protozoan cells have chloroplasts that contain their own DNA. The DNA of mitochondria and chloroplasts is circular and resembles the circular chromosomes of prokaryotes. Genes located on these "prokaryotic" chromosomes code for about 5% of the RNA and polypeptides required for the organelle's replication and function; nuclear DNA codes for the remaining 95% of polypeptides and RNA molecules required for mitochondrial and chloroplast function. Recall from Chapter 2 that some proteins have a quarternary structure formed from the association of individual polypeptides. Interestingly, polypeptides coded by mitochondrial or chloroplast chromosomes do not alone constitute functional proteins. Rather, they become functional only when associated with polypeptides coded by nuclear chromosomes.

In addition to the extranuclear DNA in their mitochondria, some fungi and protozoa carry plasmids. For instance, most strains of the yeast *Saccharomyces cerevisiae* (sak-ah-rō-mī′sēz se-ri-vis′ē-ē) contain about 70 copies of a plasmid known as a *2-μm circle.* Each 2-μm circle is about 6300 bp long and has four protein-encoding genes that are involved solely in replicating the plasmid and confer no other traits to the cell.

In summary, the genome of a prokaryotic cell consists of both chromosomal DNA, which is usually in a single circular chromosome, and all extrachromosomal DNA in the form of plasmids that is present. By contrast, a eukaryotic genome consists of nuclear chromosomal DNA in one or more linear chromosomes, plus all the extranuclear DNA in mitochondria, chloroplasts, and any plasmids that are present. The genomes of prokaryotes and eukaryotes are compared and contrasted in **Table 7.1.**

CRITICAL THINKING

In Chapter 3 we learned that the endosymbiotic theory proposes that mitochondria and chloroplasts evolved from prokaryotes living within other prokaryotes. What aspects of the eukaryotic genome support this theory? What aspects do not support the theory?

DNA Replication

Learning Objectives

✓ Describe the replication of DNA as a semiconservative process.
✓ Compare and contrast the synthesis of leading and lagging strands in DNA replication

DNA replication, an anabolic polymerization process, allows a cell to pass copies of its genome to its descendants. As discussed in Chapter 2, all polymerization processes require monomers (building blocks) and energy. *Triphosphate deoxyribonucleotides*—nucleotides with three phosphate groups linked together by two high-energy bonds—serve both functions in DNA replication. In other words, the building blocks of DNA carry within themselves the energy required for DNA synthesis **(Figure 7.4).** The structure of guanosine triphosphate nucleotide (dGTP), shown in Figure 7.4a, differs from that of cytidine triphosphate (dCTP), thymidine triphosphate (dTTP), and adenosine triphosphate (dATP)

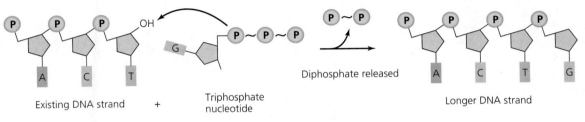

Guanosine triphosphate deoxyribonucleotide (dGTP)

Guanine nucleotide (dGMP)

Guanine base

Deoxyribose

(a)

(b)

Existing DNA strand + Triphosphate nucleotide

Diphosphate released

Longer DNA strand

◄ *Figure 7.4*

The dual role of triphosphate deoxyribonucleotides as building blocks and energy sources in DNA synthesis. **(a)** Guanosine triphosphate deoxyribonucleotide (dGTP), like all the triphosphate monomers of DNA, is a nucleotide to which two additional phosphate groups are attached. **(b)** The energy required for DNA polymerization (the addition of nucleotide building blocks to a DNA strand) is carried in each triphosphate nucleotide in the high-energy bonds between phosphate groups. *What is the difference between dGTP and guanosine triphosphate ribonucleotide (rGTP)?*

Figure 7.4 This molecule (dGTP) contains deoxyribose; rGTP contains ribose.

only in the kind of base present; dATP has a structure similar to that of the energy-storage molecule ATP, except that ATP is an RNA nucleotide (see Figure 2.26).

The key to DNA replication is the complementary structure of the two strands: Adenine and guanine in one strand bond with thymine and cytosine, respectively, in the other. DNA replication is a simple concept—a cell separates the two original strands and uses each as a template for the synthesis of a new complementary strand. Biologists say that DNA replication is *semiconservative* because each daughter DNA molecule is composed of one original strand and one new strand. The following sections examine DNA replication in more detail.

Initial Processes in DNA Replication

DNA replication (**Figure 7.5** on page 206) begins at a specific sequence of nucleotides called an *origin* (not shown). First, a cell removes chromosomal proteins, exposing the DNA helix. Next, enzymes called *DNA helicases* locally "unzip" the DNA molecule by breaking the hydrogen bonds between complementary nucleotide bases, which exposes the nucleotides in a *replication fork* (Figure 7.5a). Other protein molecules stabilize the single strands so that they do not rejoin while replication proceeds.

After the helicases untwist and separate the strands, a molecule of an enzyme called *DNA polymerase* (po-lim′er-ās) binds to each strand. Scientists have identified five kinds of prokaryotic DNA polymerase, and six kinds from eukary-

otes. These eleven enzymes vary in their specific functions, but all of them share one important feature—they synthesize DNA by adding nucleotides only to a hydroxyl group at the 3′ end of a nucleic acid. Because of the latter feature, DNA polymerases replicate DNA in only one direction—5′ to 3′—like a jeweler stringing pearls to make a necklace, adding them one at a time, always moving from one end of the string to the other.

Because the two template strands are antiparallel, cells synthesize new strands in two different ways. One strand, called the *leading strand,* is synthesized continuously as a single long chain of nucleotides. The other strand, called the *lagging strand,* is synthesized in short segments that are later joined. We will consider synthesis of the leading strand before examining replication of the lagging strand, even though the two processes occur simultaneously.

Synthesis of the Leading Strand

The cell synthesizes the leading strand toward the replication fork in the following series of five steps, the first three of which are shown in Figure 7.5b:

1. An enzyme called *primase* synthesizes a short RNA molecule that is complementary to the template DNA strand. This *RNA primer* provides a 3′ hydroxyl group required by DNA polymerase.

2. Triphosphate deoxyribonucleotides form hydrogen bonds with their complements in the parental strand.

(a) Initial processes

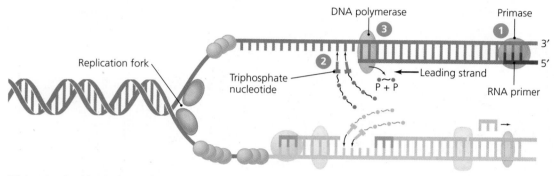

(b) Synthesis of leading strand

(c) Synthesis of lagging strand

▲ *Figure 7.5*

DNA replication. **(a)** Initial processes. The cell removes histones and other proteins from the DNA molecule. Helicases unzip the double helix—breaking hydrogen bonds between complementary base pairs—to form a replication fork. **(b)** Continuous synthesis of the leading strand. DNA synthesis always moves in the 5′ to 3′ direction, so the leading strand is synthesized toward the replication fork. The numbers refer to the steps in the process, which are described in the text. (Steps 4 and 5, the proofreading function of DNA polymerase and the replacement of the RNA primer with DNA, are not shown.) **(c)** Discontinuous synthesis of the lagging strand, which proceeds moving away from the replication fork. Actual Okazaki fragments are about 1000 nucleotides long. *Why is DNA replication termed "semiconservative"?*

❸ Using the energy in the high-energy bonds of the triphosphate deoxyribonucleotides, DNA polymerase covalently joins them by dehydration synthesis to form the leading strand. A typical DNA polymerase can add about 500–1000 nucleotides per second to a new strand.

❹ DNA polymerase also performs a proofreading function (not shown). About one out of every 100,000 nucleotides is mismatched with its template; for instance, a guanine might become incorrectly paired with a thymine. DNA polymerase recognizes most such errors and removes the incorrect nucleotides before proceeding with synthesis. Because of this proofreading function, only about one error remains for every billion (10^9) base pairs replicated.

Figure 7.5 "Semiconservative" refers to the fact that each of the daughter molecules retains one parental strand and has one new strand; in other words, each is half new and half old.

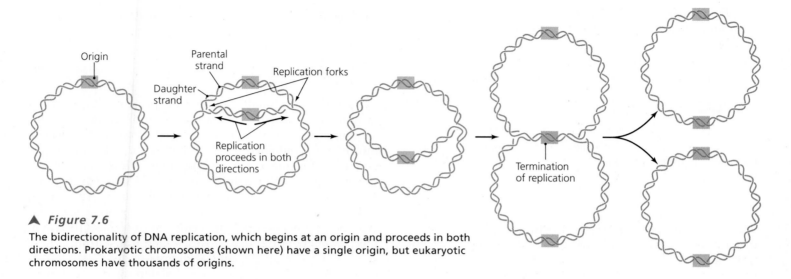

▲ *Figure 7.6*

The bidirectionality of DNA replication, which begins at an origin and proceeds in both directions. Prokaryotic chromosomes (shown here) have a single origin, but eukaryotic chromosomes have thousands of origins.

5 Another enzyme replaces the RNA primer with DNA (not shown).

Synthesis of the Lagging Strand

Because DNA polymerase adds nucleotides only in a 5' to 3' direction, it moves away from the replication fork as it synthesizes the lagging strand. As a result, the lagging strand is synthesized discontinuously and always lags behind the process occurring in the leading strand. The steps in the synthesis of the lagging strand are as follows (Figure 7.5c):

1 As with the leading strand, primase synthesizes RNA primers.

2 Nucleotides pair up with their complements in the template—adenine with thymine, and cytosine with guanine.

3 DNA polymerase joins neighboring nucleotides and proofreads. In contrast to synthesis of the leading strand, however, the lagging strand is synthesized in discontinuous segments called *Okazaki fragments,* named for the Japanese scientist Reiji Okazaki (1930–1975), who first identified them. Each Okazaki fragment requires a new RNA primer and consists of about 1000 nucleotides.

4 Another type of DNA polymerase replaces the RNA primers of Okazaki fragments with DNA and further proofreads the daughter strand.

5 *DNA ligase* seals the gaps between adjacent Okazaki fragments to form a continuous DNA strand.

In summary, synthesis of the leading strand proceeds continuously toward the replication fork from a single RNA primer at the origin, following the helicases and replication fork down the DNA. The lagging strand is synthesized away from the replication fork, discontinuously as a series of Okazaki fragments, each of which begins with its own RNA primer. All the primers are eventually replaced with DNA nucleotides, and ligase joins the Okazaki fragments.

DNA replication is *semiconservative;* each daughter molecule is composed of one parental strand and one daughter strand. This replication process, which produces double-stranded daughter molecules with a nucleotide sequence identical to that in the original double helix, ensures that the integrity of an organism's genome is maintained each time it is copied.

Other Characteristics of DNA Replication

DNA replication is usually *bidirectional;* that is, DNA synthesis proceeds in both directions from the origin. In prokaryotes, the process of replication typically proceeds from a single origin, so it involves two sets of enzymes, two replication forks, two leading strands, and two lagging strands **(Figure 7.6).** By contrast, the large size of eukaryotic chromosomes necessitates thousands of origins per molecule, each generating two replication forks; otherwise, the replication of eukaryotic chromosomes would take days instead of hours.

DNA replication is further complicated by **methylation** of the daughter strands, in which a cell adds a methyl group ($-CH_3$) to one or two bases that are part of specific nucleotide sequences. Plant and animal cells methylate cytosine bases exclusively, whereas prokaryotes typically methylate adenine bases and only rarely a cytosine base.

Methylation can play a role in a variety of cellular processes, including the following:

• *Control of genetic expression.* In some cases, genes that are methylated are "turned off" and are not transcribed, whereas in other cases methylated genes are "turned on" and are transcribed.

• *Initiation of DNA replication.* In many prokaryotes, methylated nucleotide sequences play a role in initiating DNA replication.

• *Protection against viral infection.* Methylation at specific sites in the nucleotide sequence enables cells to distinguish their DNA from viral DNA, which lacks methylation. The cells can then selectively degrade the viral DNA.

- *Repair of DNA.* The role of methylation in some DNA repair mechanisms is discussed on page 225.

CRITICAL THINKING

We have seen that the hydrogen bonds between complementary nucleotides are crucial to the structure of dsDNA because they hold the two strands together. Why couldn't the two strands be effectively linked by covalent bonds?

We have examined the physical structure of cellular genes—the specific sequences of DNA nucleotides—and how cells replicate their genes. Now we will consider how genes function, and how cells control genetic expression.

Gene Function

The first topic we must consider if we are to understand gene function is the relationship between an organism's genotype and its phenotype.

The Relationship Between Genotype and Phenotype

Learning Objective

✓ Explain how the genotype of an organism determines its phenotype.

The **genotype**[2] (jen′ō-tīp) of an organism is the actual set of genes in its genome. A genotype differs from a genome in that a genome also includes nucleotides that are not part of genes, such as the nucleotide sequences that link genes together. At the molecular level, the genotype consists of all the series of DNA nucleotides that carry instructions for an organism's life. **Phenotype**[3] (fē′nō-tīp) refers to the physical features and functional traits of an organism, including characteristics such as structures, morphology, and metabolism. For example, the shape of a cell, the presence and location of flagella, the enzymes and cytochromes of electron transport chains, and membrane receptors that trigger chemotaxis are all phenotypic traits.

Genotype determines phenotype by specifying what kinds of RNA and which structural, enzymatic, and regulatory protein molecules are produced. Though genes do not code *directly* for such molecules as phospholipids or for behaviors such as chemotaxis, ultimately phenotypic traits result from the actions of RNA and protein molecules that are themselves coded by DNA.

Not all genes are active at all times; that is, the information of a genotype is not always expressed as a phenotype. For example, *E. coli* activates genes for lactose catabolism only when it detects lactose in its environment.

The Transfer of Genetic Information

Learning Objective

✓ State the central dogma of genetics, and explain the roles of DNA and RNA in polypeptide synthesis.

Cells must continually synthesize proteins required for growth, reproduction, metabolism, and regulation. This synthesis requires that they accurately transfer the genetic information contained in DNA nucleotide sequences to the amino acid sequences of polypeptides. However, cells do not transfer the information coded in DNA directly, but first make an RNA copy of the gene. In this copying process, called **transcription**,[4] the information is copied as RNA nucleotide sequences; RNA molecules in ribosomes then synthesize polypeptides in a process called **translation**.[5] These processes make up the **central dogma** of genetics: DNA is transcribed to RNA, which is translated to form polypeptides. **Highlight 7.1** presents an analogy of the central dogma. There are a few exceptions to the central dogma. For example, some RNA viruses transcribe DNA from an RNA template—a process that is the reverse of cellular transcription.

In the following sections, we will examine the processes of transcription and translation.

The Events in Transcription

Learning Objective

✓ Describe three steps in RNA transcription, mentioning the following: DNA, RNA polymerase, promoter, 5′ to 3′ direction, and terminator.

Cells transcribe three types of RNA from DNA:

- **messenger RNA (mRNA)** molecules, which carry genetic information from chromosomes to ribosomes;

- **ribosomal RNA (rRNA)** molecules, which combine with ribosomal polypeptides to form ribosomes—the organelles that synthesize polypeptides; and

- **transfer RNA (tRNA)** molecules, which deliver amino acids to the ribosomes.

We will more closely examine the functions of each type of RNA shortly.

Whereas transcription occurs in the nucleoid region of the cytoplasm in prokaryotes, in eukaryotes it occurs in the nucleus, mitochondria, and chloroplasts. In both prokaryotes and eukaryotes, the steps of RNA transcription are similar: *initiation of transcription, elongation of the RNA transcript,* and *termination of transcription.* The events of transcription in bacteria are depicted in **Figure 7.7** on page 210.

---

[2]From Greek *genos,* meaning race, and *typos,* meaning type.
[3]From Greek *phainein,* meaning to show.
[4]From Latin *trans,* meaning across, and *scribere,* meaning to write—that is, to transfer in writing.
[5]From Latin *translatus,* meaning transferred.

Highlight 7.1   Understanding the Central Dogma: It's (No Longer) Greek to Me

Suppose you were trying to understand the following message (which is a portion of the oath of Hippocrates, written in the Greek alphabet):

ΔιαιτημασιτεχρησομαιεπωΦελεινκαμνοντωνκατα δυναμινκαικρισινεμηνεπιδηληειδεκαιαδικιηειρζειν.

If the Greek alphabet is foreign to you, you might have the Greek characters *transcribed* into the familiar English alphabet as a first step in understanding the message:

Diaiteimasi te chreisomai ep ophelein kamnonton kata dunamin kai krisin emein epi deileisei de kai adikiei eirzein

Then you could begin the process of having the Greek words, now expressed in English letters, *translated* into English words:

I will prescribe treatment to the best of my ability and judgment to help the sick and never for a harmful or illicit purpose

To a ribosome in a cell, DNA would be like a foreign language written in a foreign alphabet. Thus, a cell must use processes analogous to those just described: It must first *transcribe* the "foreign alphabet" of DNA nucleotides (genes) into the more "familiar alphabet" of RNA nucleotides; then it must *translate* the message formed by these "letters" into the "words" (amino acids) that make up the "message" (a polypeptide). In this way a genotype can be expressed as a phenotype.

Initiation of Transcription

RNA polymerase—the enzyme that synthesizes RNA—moves by an unknown mechanism along a DNA helix until it reaches a specific nucleotide sequence called a **promoter,** which is located near the beginning of a gene and initiates transcription (Figure 7.7a). In bacteria, a portion of RNA polymerase called the *sigma factor* is necessary for recognition of the promoter. In contrast, separate protein *transcription factors* assist in binding eukaryotic RNA polymerase to promoters. Once it adheres to a promoter, RNA polymerase unwinds and unzips the DNA in the promoter region; and then continues moving along the DNA, unzipping the double helix as it moves.

RNA polymerases do not adhere equally strongly to all promoter sequences. The greater the attraction between RNA polymerase and a promoter, the more likely that transcription will proceed. One way that cells control the relative amounts of transcribed RNA is by using different promoter sequences for different genes. Ultimately, variation in promoters affects the amounts and kinds of polypeptides produced.

Elongation of the RNA Transcript

RNA transcription does not actually begin in the promoter region, but at a spot 10 nucleotides away. There, triphosphate ribonucleotides (rATP, rUTP, rGTP, and rCTP) align opposite their complements in the open DNA. RNA polymerase links together two adjacent ribonucleotide molecules using energy from the phosphate bonds of the first ribonucleotide. The enzyme then moves down the DNA strand, elongating RNA by repeating the process (Figure 7.7b). During transcription of the first 10 ribonucleotides, RNA polymerase releases its sigma factor, which causes RNA polymerase to adhere tightly to the DNA. Once this occurs, transcription proceeds to completion. The released sigma factor joins with another molecule of RNA polymerase to locate another promoter and initiate another transcription.

Many molecules of RNA polymerase may concurrently transcribe the same gene (**Figure 7.8** on page 211). In this way, a cell simultaneously produces numerous identical copies of RNA from a single gene—much in the way many identical prints can be made from a single photographic negative.

Like DNA polymerase, RNA polymerase links nucleotides in the 5' to 3' direction only; however, RNA polymerase differs from DNA polymerase in the following ways:

- RNA polymerase unwinds and opens DNA by itself; helicase is not required.
- RNA polymerase is slower than DNA polymerase, proceeding at a rate of about 50 nucleotides per second.
- RNA polymerase incorporates ribonucleotides instead of deoxyribonucleotides.
- Uracil nucleotides are incorporated instead of thymine nucleotides.
- RNA polymerase lacks a proofreading function, leaving a base pair error about every 10,000 nucleotides.

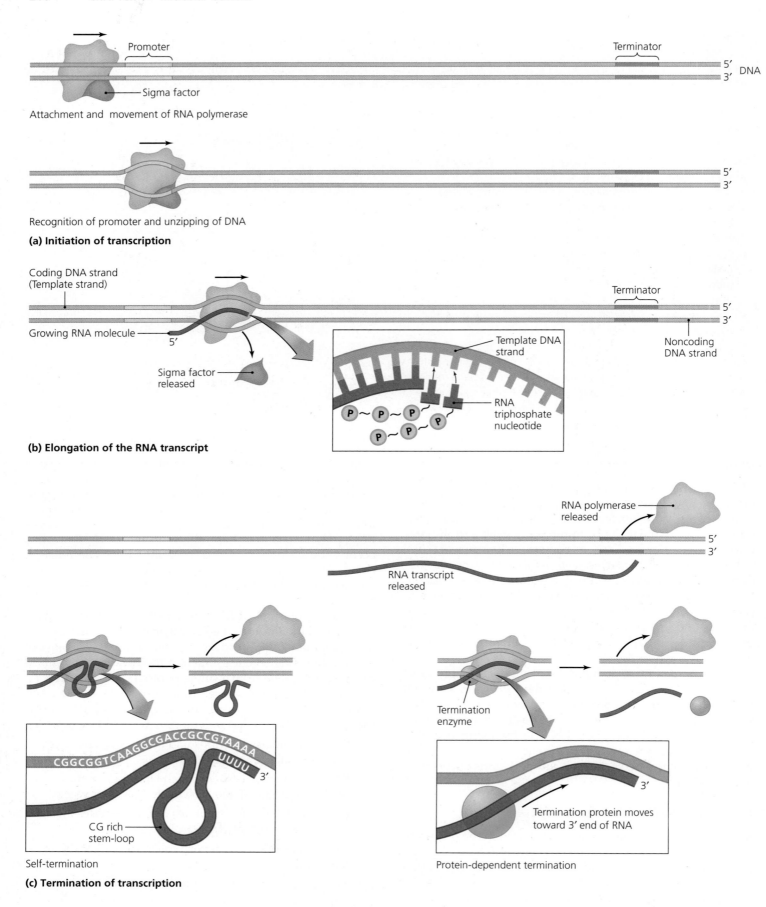

(a) Initiation of transcription

Promoter

Sigma factor

Attachment and movement of RNA polymerase

Recognition of promoter and unzipping of DNA

Terminator

5′
3′ DNA

Coding DNA strand
(Template strand)

Growing RNA molecule

5′

Sigma factor released

Template DNA strand

RNA triphosphate nucleotide

Terminator

5′
3′

Noncoding DNA strand

(b) Elongation of the RNA transcript

RNA polymerase released

RNA transcript released

5′
3′

Termination enzyme

CGGCGGTCAAGGCGACCGCCGTAAAA
UUUU
3′

CG rich stem-loop

Self-termination

Termination protein moves toward 3′ end of RNA

3′

Protein-dependent termination

(c) Termination of transcription

◀ *Figure 7.7*

The events in the transcription of RNA in bacteria. The helical dsDNA molecule is depicted as straight, parallel strands for clarity. **(a)** Initiation of transcription. RNA polymerase attaches nonspecifically to DNA and travels down its length until it recognizes a promoter sequence. Sigma factor enhances promoter recognition in bacteria. Upon recognition of the promoter, RNA polymerase unzips the DNA molecule beginning at the promoter. **(b)** Elongation of the RNA transcript. After triphosphate ribonucleotides align with their DNA complements, RNA polymerase links them together, synthesizing RNA. The triphosphate ribonucleotides also provide the energy required for RNA synthesis. **(c)** Termination of transcription, which is effected by the release of RNA polymerase. In self-termination, the transcription of DNA terminator sequences cause the RNA to fold, loosening the grip of RNA polymerase on the DNA; in enzyme-dependent termination, a termination enzyme pushes between RNA polymerase and the DNA, releasing the polymerase. *What is the difference between a promoter sequence and an origin?*

Figure 7.7 A promoter is a DNA sequence that initiates transcription; an origin is a point where DNA replication begins.

▲ *Figure 7.8*

Concurrent RNA transcription. Once an RNA polymerase molecule has cleared the promoter, another molecule can recognize the promoter and initiate transcription. In this manner, multiple copies of RNA are transcribed simultaneously.

CRITICAL THINKING

On average, RNA polymerase makes one error for every 10,000 nucleotides it incorporates in RNA. By contrast, only one base pair error remains for every billion base pairs during DNA replication. Explain why the accuracy of RNA transcription is not as critical as the accuracy of DNA replication.

Termination of Transcription

Transcription terminates when RNA polymerase and the transcribed RNA are released from DNA (see Figure 7.7c). As stated previously, once RNA polymerase loses its sigma factor, it becomes tightly associated with the DNA molecule

and cannot be removed easily; therefore, the termination of transcription is complicated. Scientists have elucidated two types of termination processes in bacteria: those that are self-terminating and those that depend on the action of an additional termination enzyme.

Self-Termination   Self-termination occurs when RNA polymerase transcribes a **terminator** sequence of DNA composed of two symmetrical series: one that is very rich in guanine and cytosine bases, followed by a region rich in adenine bases (see Figure 7.7c, lower left). RNA polymerase slows down during transcription of the GC rich portion of the terminator because the three hydrogen bonds between each guanine and cytosine base pair makes unwinding the DNA helix more difficult. This pause in transcription provides enough time for the RNA molecule to form hydrogen bonds between its own symmetrical sequences, forming a stem and loop structure that puts tension on the union of RNA polymerase and the DNA. When RNA polymerase transcribes the adenine rich portion of the terminator, the relatively few hydrogen bonds between the adenine bases of DNA and the uracil bases of RNA cannot withstand the tension, and the RNA transcript breaks away from the DNA, releasing RNA polymerase.

Enzyme-Dependent Termination   The second type of termination depends on a termination protein that binds to a specific RNA sequence near the end of an RNA transcript. The protein moves toward the 3′ end, pushing between RNA polymerase and the DNA strand and forcing them apart; this releases RNA polymerase and the RNA transcript (see Figure 7.7c, lower right).

Now that we have discussed how cells use DNA as the genetic material, maintain the integrity of their genomes through semiconservative replication, and transcribe RNA from DNA genes, we turn to the process of translation and the role of each type of RNA.

Translation

Learning Objectives

✓ Describe the genetic code in general, and identify the relationship between codons and amino acids.

✓ Describe the translation of polypeptides, identifying the roles of the three types of RNA.

Translation is the process whereby ribosomes use the genetic information of nucleotide sequences to synthesize polypeptides composed of specific amino acid sequences. As we studied in Chapter 2, some proteins are simple polypeptides, whereas other proteins are composed of several polypeptides bound together in a quaternary structure.

Ribosomes can be thought of as "polypeptide factories," so consider the following analogy between translation and a hypothetical automobile factory. Boxcars deliver preformed parts to the factory at the correct times and in the correct order to manufacture one of a large variety of automobile models, depending on instructions from corporate headquarters delivered by special courier. Similarly, molecules of tRNA (the boxcars) deliver preformed amino acids (the parts) to a ribosome (the factory), which can manufacture an infinite variety of polypeptides (the car models) by assembling amino acids (the parts) in the correct order according to the instructions from DNA (corporate headquarters) delivered via mRNA (the special courier).

How do ribosomes interpret the nucleotide sequence of mRNA to determine the correct order in which to assemble amino acids? To answer this question, we will consider the genetic code, examine in more detail the RNA molecules that participate in translation, and describe the specific steps of translation.

The Genetic Code

When geneticists in the early 20th century began to consider that DNA might be the genetic molecule, they were confronted with a problem: How can four kinds of DNA nucleotides (adenine, thymine, guanine, and cytosine) specify the 21 different amino acids commonly found in proteins? If each nucleotide coded for one amino acid, only four amino acid could be specified. Even if pairs of nucleotides served as the code—for instance, if AA, AT, and TA each specified a different amino acid—only 16 amino acids (that is, 4^2) could be accommodated. Eventually scientists showed that genes are composed of sequences of three nucleotides that specify amino acids. For example, the DNA nucleotide sequence TTT specifies the amino acid lysine, and TTA codes for asparagine. There are 64 possible arrangements of the four nucleotides in triplets (4^3)—more than enough to specify 21 amino acids.

These examples are DNA triplets, but ribosomes do not directly access genetic information on a DNA molecule. Instead, molecules of mRNA carry the code to the ribosomes; therefore, scientists define the genetic code **(Figure 7.9)** as triplets of mRNA nucleotides called **codons** (kō′ donz) that code for specific amino acids. AAA is a codon for lysine, and AAU is a codon for asparagine.

In most cases, 61 codons specify amino acids and three codons—UAA, UAG, and UGA—instruct ribosomes to stop translating; though, under some conditions, UGA codes for the 21st amino acid, selenocysteine. Codon AUG also has a dual function, acting as both a start signal and coding for an amino acid. For eukaryotic and archaeal proteins, AUG codes for methionine, while in prokaryotes, mitochondria, and chloroplasts it codes for N-formylmethionine (fMet), which is similar in structure to methionine:

Methionine   N-formylmethionine

As you examine the genetic code, notice that it is redundant; that is, more than one codon is associated with all the amino acids except methionine and tryptophan. With most redundant codons, the first two nucleotides determine the amino acid, and the third nucleotide is inconsequential. For example, the codons GUU, GUC, GUA, and GUG all specify the amino acid valine.

Interestingly, the genetic code is nearly universal; that is, with few exceptions, ribosomes in archaeal, bacterial, plant, fungal, protozoan, and animal cells use the same genetic code. Some exceptions are listed in **Table 7.2.**

Figure 7.9

The genetic code, the complete set of mRNA codons and the amino acids for which they code. AUG is both the start codon and specifies methionine (Met) in eukaryotes and N-formylmethionine (fMet) in prokaryotes, mitochondria, and chloroplasts. Three codons (UAA, UAG, and UGA) are stop codons that do not typically specify amino acids.

Table 7.2   Some Exceptions to the Genetic Code

Codon	Usual Use	Alternate Use
AUA	Codes for isoleucine	Codes for methionine in mitochondria
UAG	STOP	Codes for glutamine in some protozoa and algae and for pyrrolysine (22nd amino acid) in some prokaryotes
CGG	Codes for arginine	Codes for tryptophan in plant mitochondria
UGA	STOP, selenocysteine	Codes for tryptophan in mitochondria and mycoplasmas (type of bacteria)

CRITICAL THINKING

A scientist isolates a molecule of mRNA with the following base sequence: AUGUACGACAUAUGCAUA. What is the sequence of amino acids in the polypeptide synthesized by a prokaryotic ribosome from this message? What would be different if the message were translated in a mitochondrion instead?

Participants in Translation

As we have discussed, transcription produces messenger RNA, transfer RNA, and ribosomal RNA—each of which is involved in translation. We now discuss each kind of RNA in turn.

Messenger RNA   Messenger RNA carries genetic information (in the form of RNA nucleotide sequences) from a chromosome to ribosomes. In prokaryotes a basic mRNA molecule contains sequences of nucleotides that are recognized by ribosomes: an AUG start codon, sequential codons for other amino acids in the polypeptide, and at least one of the three stop codons. A single molecule of prokaryotic mRNA often contains start codons and instructions for more than one polypeptide arranged in series (**Figure 7.10** on page 214). Because both transcription and the subsequent events of translation occur in the cytosol of prokaryotes, prokaryotic ribosomes can begin translation before transcription is finished.

Eukaryotic mRNA differs from prokaryotic mRNA in several ways:

- Newly synthesized eukaryotic mRNA is called *pre-messenger RNA* because it contains noncoding

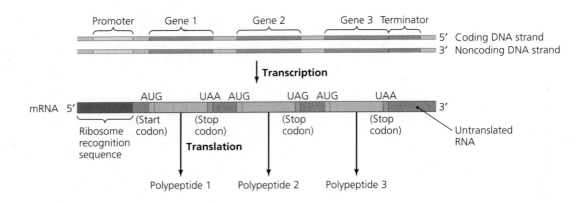

Figure 7.10 ▶

Prokaryotic mRNA. Prokaryotes typically code for several related polypeptides via a single mRNA molecule. The mRNA molecule shown here has transcripts of three genes coding three polypeptides. The transcript of each gene begins with a start codon and ends with a stop codon.

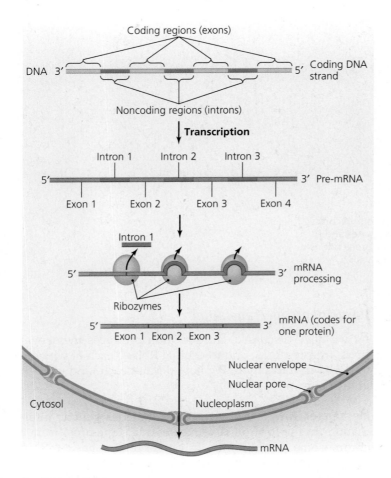

▲ **Figure 7.11**

Eukaryotic mRNA. Within the nucleus, transcription produces pre-mRNA, which contains coding exons and noncoding introns. Ribozymes process pre-mRNA by removing introns and splicing together exons to form a molecule that codes for a single polypeptide. Eukaryotic mRNA then moves from the nucleus to the cytoplasm.

sequences called **introns,** which are removed to make functional mRNA containing only coding regions called **exons (Figure 7.11).** The "in" in *intron* refers to *intervening* sequences (that is, they lie between coding regions), whereas the "ex" in *exon* refers to the fact that

these coding regions are *expressed.* Small RNA molecules in the nucleus act as ribozymes (ribosomal enzymes) to process pre-mRNA into mRNA; that is, they remove the introns and connect the exons to produce a functional mRNA molecule.

- A molecule of eukaryotic mRNA contains instructions for only one polypeptide.
- Eukaryotic mRNA is not translated until it is fully transcribed and processed and has left the nucleus because eukaryotic ribosomes are located in the cytoplasm. In other words, transcription and translation of a molecule of eukaryotic mRNA do not occur simultaneously.

Transfer RNA A transfer RNA (tRNA) molecule is a sequence of about 75 ribonucleotides that curves back on itself to form three main hairpin loops held in place by hydrogen bonding between complementary nucleotides. Although transfer RNA molecules can be modeled simplistically by a cloverleaf structure **(Figure 7.12a),** their three-dimensional shape is more complex. For simplicity, tRNA will be represented in subsequent figures by an icon shaped like the illustration in **Figure 7.12b.**

A molecule of tRNA transfers the correct amino acid to a ribosome during polypeptide synthesis. To this end, a tRNA has an *acceptor stem* for a specific amino acid at its 3' end, and an **anticodon** (an-tē-kō′ don) triplet in its bottom loop (see Figure 7.12b). Specific enzymes in the cytoplasm *charge* each tRNA molecule; that is, they attach the appropriate amino acid to the acceptor stem. The existence of only one specific enzyme for each amino acid ensures that every tRNA molecule carries only one specific amino acid.

Anticodons are complementary to mRNA codons, and each acceptor stem is designed to carry one particular amino acid, which varies with the tRNA. In other words, each transfer RNA carries a specific amino acid and recognizes mRNA codons only for that amino acid. A tRNA molecule is designated by a superscript abbreviation of its amino acid. For example, tRNA[Phe] carries phenylalanine, and tRNA[Ser] transfers serine.

The fact that 62 codons specify the 21 amino acids used by cells does not mean that there must be 62 anticodons on

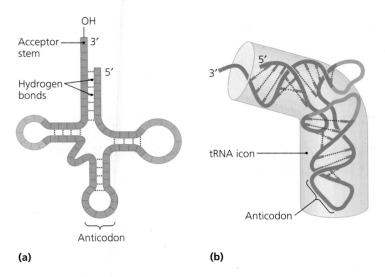

(a)

(b)

▲ *Figure 7.12*

Transfer RNA. (a) A two-dimensional "cloverleaf" representation of tRNA showing three hairpin loops held in place by intramolecular hydrogen bonding. **(b)** A three-dimensional drawing of the same tRNA. A specific amino acid attaches to the 3' acceptor stem; the anticodon is a nucleotide triplet that is complementary to the mRNA codon for that amino acid.

62 different types of tRNA, because many tRNA molecules recognize more than one codon. *E. coli*, for example, has only about 40 different tRNAs. The variability in codon recognition by tRNA is due to "wobble" of the anticodon's third nucleotide. Wobble, which is a change of angle from the normal axis of the molecule, allows the third nucleotide to hydrogen bond to a nucleotide other than its usual complement. For example, a guanine nucleotide in the third position normally bonds to cytosine, but it can wobble and also pair with uracil; whether a codon has cytosine or uracil in the third position makes no difference, because the same tRNA recognizes either nucleotide in the third position. For example, the codons UUC and UUU both specify the amino acid phenylalanine because the anticodon AAG recognizes both of them. Similarly, UCC and UCU code for serine, and UAC and UAU code for tyrosine. This redundancy in the genetic code helps protect cells against the effects of errors in replication and transcription.

CRITICAL THINKING

We have seen that wobble makes the genetic code redundant in the third position for C and U. After reexamining the genetic code in Figure 7.9, state what other nucleotides in the second or third position appear to accommodate anticodon wobbling.

Ribosomes and Ribosomal RNA   Prokaryotic ribosomes, which are also called 70S ribosomes based on their sedimentation rate in an ultracentrifuge, are extremely complex associations of ribosomal RNAs and polypeptides. Each ribosome is composed of two subunits: 50S and 30S **(Figure**

(a)

(b)

(c)

▲ *Figure 7.13*

Ribosomal structure. (a) The 70S prokaryotic ribosome, which is composed of polypeptides and three rRNA molecules arranged in 50S and 30S subunits. **(b)** The 80S eukaryotic ribosome, which is composed of molecules of rRNA and polypeptides, arranged in 60S and 40S subunits. **(c)** Transfer RNA binding sites in a ribosome, which is presented schematically for simplicity. The P site holds the transfer RNA attached to the growing polypeptide, whereas the A site accepts the tRNA carrying the next amino acid in the polypeptide. Transfer RNAs exit from the E site.

7.13a). The 50S subunit is in turn composed of two rRNA molecules (23S and 5S) and 33 different polypeptides, whereas the 30S subunit consists of one molecule of 16S rRNA and 21 ribosomal polypeptides. The ribosomes of mitochondria and chloroplasts are also 70S ribosomes composed of the similar subunits and polypeptides.

In contrast, both the cytosol and the rough endoplasmic reticulum (RER) of eukaryotic cells have 80S ribosomes composed of 60S and 40S subunits **(Figure 7.13b).** These subunits contain larger molecules of rRNA and more polypeptides than the corresponding prokaryotic subunits, though researchers do not agree on their exact number. The term *eukaryotic ribosome* is understood to mean only the 80S

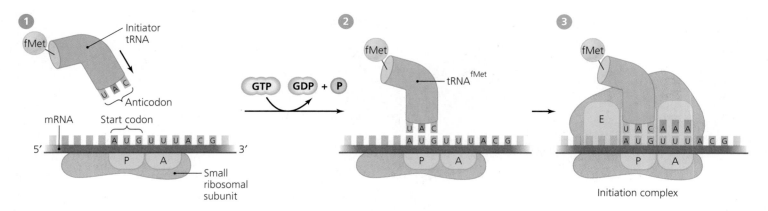

▲ *Figure 7.14*

The initiation of translation in prokaryotes. ① The smaller ribosomal subunit attaches to mRNA at a ribosomal recognition sequence containing a start codon (AUG). ② The anticodon of tRNA^{fMet} aligns with the start codon on the mRNA; energy from GTP (not shown) is used to bind the tRNA in place. ③ The larger ribosomal subunit attaches to form an initiation complex—a complete ribosome attached to mRNA.

ribosomes of the cytosol and RER. Since the ribosomes of mitochondria and chloroplasts are 70S, they are called *prokaryotic ribosomes* even though they are in eukaryotic cells.

The structural differences between prokaryotic and eukaryotic ribosomes play a crucial role in the efficacy and safety of antimicrobial drugs. Because erythromycin, for example, binds only to the 23S rRNA found exclusively in prokaryotic ribosomes, it has no effect on eukaryotic ribosomes, and thus no deleterious effect on the patient. Chapter 10 discusses antimicrobial drugs in more detail.

The smaller subunit of a ribosome is shaped to accommodate three codons at one time—that is, nine nucleotide bases of a molecule of mRNA. Each ribosome also has three tRNA binding sites that are named for their function (**Figure 7.13c**):

- The **A site** accommodates a tRNA delivering an *amino acid.*
- The **P site** holds a tRNA and the growing *polypeptide.*
- Discharged tRNAs *exit* from the **E site.**

We will now examine translation, the process whereby ribosomes actually synthesize polypeptides using amino acids delivered by tRNAs and following the genetic instructions of mRNA.

Stages of Translation

Researchers divide translation into three stages: *initiation, elongation,* and *termination.* All three stages require additional protein factors that assist the ribosomes. Initiation and elongation also require energy provided by molecules of the ribonucleotide GTP, which are free in the cytosol (that is, they are not part of an RNA molecule).

Initiation   During initiation, mRNA, the two ribosomal subunits, several protein factors, and tRNA^{fMet} (in prokary-

otes) or tRNA^{Met} (in eukaryotes) form an *initiation complex.* The events of initiation in a bacterium are as follows (**Figure 7.14**):

① The smaller ribosomal subunit attaches to mRNA at a ribosomal recognition sequence, with the start codon at its P site.

② tRNA^{fMet} (whose anticodon is complementary to the start codon) attaches to the ribosome's P site; GTP supplies the energy required for binding.

③ The larger ribosomal subunit attaches to form a complete initiation complex.

Elongation   Synthesis of a polypeptide is a cyclical process that involves the sequential addition of amino acids to a polypeptide chain growing at the P site. **Figure 7.15** illustrates several cycles of the process. The steps of each cycle occur as follows:

① The transfer RNA whose anticodon matches the next codon—in this case, phenylalanine (Phe)—delivers its amino acid to the A site. Another protein called *elongation factor* (not shown) escorts the tRNA along with a molecule of GTP. Energy from GTP (not shown) is used to stabilize each tRNA as it is added to the A site.

② A ribozyme in the larger ribosomal subunit forms a peptide bond by dehydration synthesis between the terminal amino acid of the polypeptide chain (in this case, N-formylmethionine) and the newly introduced amino acid. The polypeptide is now attached to the tRNA occupying the A site.

③ Using energy supplied by more GTP, the ribosome moves one codon down the mRNA. This transfers each tRNA to the adjacent binding site; that is, the first tRNA moves from the P site to the E site, and the second tRNA

▲ *Figure 7.15*

The elongation stage of translation. Transfer RNAs sequentially deliver amino acids as directed by the codons of the mRNA. Ribosomal RNA in the large ribosomal subunit catalyzes a peptide bond between the amino acid at the A site and the growing polypeptide at the P site. The steps in the process are described in the text. *Which component macromolecule of a ribosome—protein or rRNA—is the enzyme that actually links amino acids to form a polypeptide?*

Figure 7.15 A ribozyme (RNA) is the active enzyme in protein synthesis.

(with the attached polypeptide) moves to the vacated P site.

④ The ribosome releases the "empty" tRNA from the E site. In the cytosol, the appropriate enzyme recharges it with another molecule of its specific amino acid.

⑤ The cycle repeats, each time adding another amino acid (in this case, threonine, then alanine, and then glutamine).

As elongation proceeds, ribosomal movement exposes the start codon, allowing another ribosome to attach behind the first one. In this way, one ribosome after another attaches at the start codon and begins to translate identical polypeptide molecules from the same message. Such a group of ribosomes, called a *polyribosome*, resembles beads on a string (**Figure 7.16** on page 218).

Termination   Termination does not involve tRNA; instead, proteins called *release factors* halt elongation. Though the exact mechanism is unknown, it appears that release factors somehow recognize stop codons and modify the larger ribosomal subunit in such a way as to activate another of its ribozymes, which severs the polypeptide from the final tRNA (resident at the P site). The ribosome then dissociates into its subunits.

CRITICAL THINKING

If a scientist synthesizes a DNA molecule with the nucleotide base sequence TACGGGGGGAGGGGGAGGGGGA and then uses it for transcription and translation, what would be the amino acid sequence of the product? (Refer to Figure 7.9 for the genetic code.)

(a) Direction of transcription

(b) TEM | 50 nm

▲ *Figure 7.16*

A polyribosome in a prokaryotic cell. As a ribosome moves down the mRNA, the start codon (AUG) becomes available to another ribosome. In this manner, numerous identical polypeptides are translated simultaneously from a single mRNA molecule.

Table 7.3   Comparison of Genetic Processes

Process	Purpose	Beginning Point	Ending Point
Replication	To duplicate the cell's genome	Origin	Origin or the end of the DNA molecule
Transcription	To synthesize RNA	Promoter	Terminator
Translation	To synthesize polypeptides	AUG start codon	UAA, UAG, or UGA stop codons

The processes we have examined thus far—how a cell replicates DNA, transcribes RNA, and translates RNA into polypeptides—are summarized in **Table 7.3.** Next we examine the way cells control the process of transcription.

Control of Transcription

Learning Objectives

✓ Explain the operon model of transcriptional control in prokaryotes.

✓ Contrast the regulation of an inducible operon with that of a repressible operon, and give an example of each.

About 75% of genes are expressed at all times; that is, they are constantly transcribed and translated and play a persistent role in the phenotype. These genes code for RNAs and polypeptides that are needed in large amounts by the cell—for example, integral proteins of the cytoplasmic membrane, structural proteins of ribosomes, and enzymes of glycolysis.

Other genes are regulated so that the polypeptides they encode are synthesized only when a cell has need of them. Protein synthesis requires a large amount of energy, which can be conserved if a cell forgoes production of unneeded polypeptides. For example, when lactose is missing in its environment, a bacterial cell saves energy by not synthesizing proteins whose only functions are to import or catabolize lactose. Prokaryotic and eukaryotic cells regulate protein

synthesis in many ways. Cells most typically stop synthesis by stopping transcription. In a few cases, cells stop translation directly.

Proteins coded by regulated genes include a few structural proteins and many enzymes. An example of the former is found in *Trypanosoma brucei* (tri-pan'ō-sō' mă brus'ē) (the protozoan that causes African sleeping sickness), which has many genes for surface glycoproteins. Each protozoan activates only one of the genes at a time so that a population of cells contains members with a wide variety of glycoproteins. This is advantageous to *Trypanosoma* because when its host's immune system recognizes and attacks cells with one kind of glycoprotein, cells with other types may remain unrecognized and continue to reproduce.

Since much of our knowledge of regulatory mechanisms has come from the study of microorganisms such as *E. coli,* we will examine two types of enzyme regulation in this bacterium: *induction* and *repression.* But before we examine induction and repression, we must first consider a special arrangement of prokaryotic genes that plays a role in gene expression.

The Nature of Prokaryotic Operons

In 1961, Francois Jacob (1920–) and Jacques Monod (1910–1976) proposed the *operon model* to explain gene

▲ Figure 7.17

An operon, which consists of genes, their promoter, and a contiguous operator. The genes code for enzymes and structures such as channel and carrier proteins. A nearby regulatory gene codes for a protein that controls the operon.

regulation in bacteria. An **operon** consists of a promoter, an adjacent regulatory element called an **operator** where a *repressor protein* binds to stop transcription, and a series of genes, which code for enzymes and structures such as channel proteins **(Figure 7.17)**. Operons are either repressed (turned off) or induced (turned on) by proteins coded by a regulatory gene (located elsewhere). **Inducible operons** are not usually transcribed and must be activated by *inducers*. **Repressible operons** operate in reverse fashion—they are transcribed continually until deactivated by *repressors*. To

clarify these concepts, let's examine some inducible and repressible operons found in *E. coli*.

The Lactose Operon, an Inducible Operon

The *lactose (lac) operon* of *E. coli* is an inducible operon and the first operon whose structure and action were elucidated. It includes a promoter, an operator, and three genes that encode for proteins involved in the catabolism of lactose **(Figure 7.18a)**. The operon is controlled by a regulatory gene that is constantly transcribed and translated to produce a

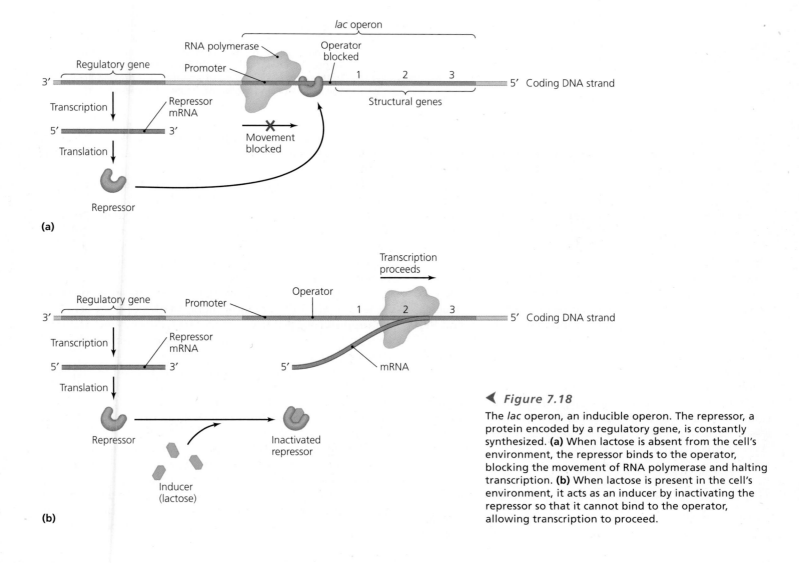

◀ Figure 7.18

The *lac* operon, an inducible operon. The repressor, a protein encoded by a regulatory gene, is constantly synthesized. **(a)** When lactose is absent from the cell's environment, the repressor binds to the operator, blocking the movement of RNA polymerase and halting transcription. **(b)** When lactose is present in the cell's environment, it acts as an inducer by inactivating the repressor so that it cannot bind to the operator, allowing transcription to proceed.

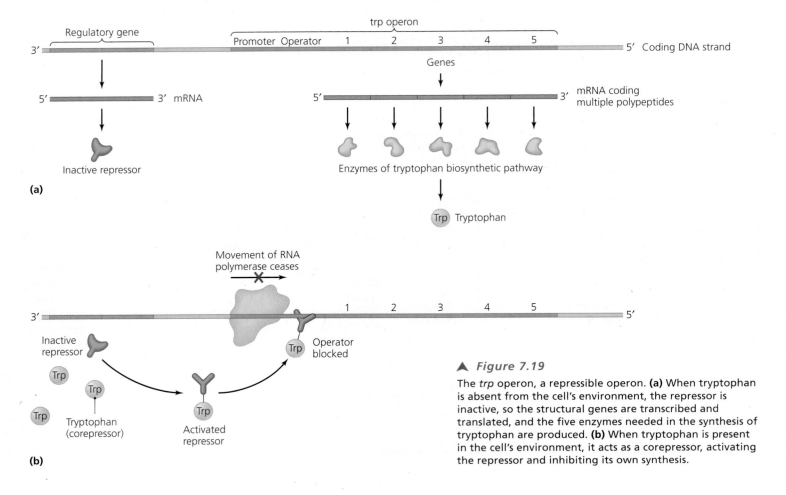

▲ Figure 7.19

The *trp* operon, a repressible operon. **(a)** When tryptophan is absent from the cell's environment, the repressor is inactive, so the structural genes are transcribed and translated, and the five enzymes needed in the synthesis of tryptophan are produced. **(b)** When tryptophan is present in the cell's environment, it acts as a corepressor, activating the repressor and inhibiting its own synthesis.

repressor protein that attaches to DNA at the *lac* operator. This repressor prevents RNA polymerase from moving beyond the promoter, stopping synthesis of mRNA. Thus, the *lac* operon is usually inactive.

Whenever lactose becomes available in an *E. coli* cell's environment, it acts as an inducer and changes the quaternary structure of the repressor so that it is inactivated and can no longer attach to DNA **(Figure 7.18b)**. This absence of binding allows transcription of the three structural genes to proceed—the operon has been induced and has become active. Ribosomes translate the newly synthesized mRNA to produce enzymes that catabolize lactose. Once the lactose supply has been depleted, there is no more inducer, and the repressor once again becomes active, suppressing transcription and translation of the *lac* operon. In this manner, *E. coli* cells conserve energy by synthesizing enzymes for the catabolism of lactose only when lactose is available to them.

Such inducible operons are often involved in controlling catabolic pathways whose polypeptides are not needed unless a particular nutrient is available. A different situation occurs with anabolic pathways such as those that synthesize amino acids such as tryptophan.

The Tryptophan Operon, a Repressible Operon

E. coli normally synthesizes the amino acids it needs for polypeptide synthesis; however, it can save energy by using amino acids available in its environment. In such cases, *E. coli* represses the genes for a given amino acid's synthetic pathway.

The *tryptophan (trp) operon*, which consists of a promoter, an operator, and five genes that code for the enzymes involved in the synthesis of tryptophan, is an example of such a repressible operon. Just as with the *lac* operon, a regulatory gene codes for a repressor molecule that is constantly synthesized. In contrast to inducible operons, however, the repressor of repressible operons is normally *inactive*. Thus, in the case of the repressible *trp* operon, whenever tryptophan is not present in the environment, the *trp* operon is active: The appropriate mRNA is transcribed, the enzymes for tryptophan synthesis are translated, and tryptophan is produced **(Figure 7.19a)**.

When tryptophan is available, it activates the repressor by binding to it. The activated repressor then binds to the operator, halting the movement of RNA polymerase and halting transcription **(Figure 7.19b)**. In other words, tryptophan acts as a *corepressor* of its own synthesis.

Biochemical analyses have revealed numerous variations of repressor-operator regulatory control. For example, scientists have discovered operons that use dual regulatory proteins, multiple operons controlled by a single repressor, and operons with multiple operators. Furthermore, some operons are not merely on-off systems, but can be fine-tuned

Controlling Disease by Controlling Gene Expression

Recent advances in our understanding of transcriptional control may provide new techniques for treating and even preventing diseases caused by numerous bacteria.

For example, many strains of the bacterium *Staphylococcus* that cause food poisoning and a variety of diseases (including impetigo, ear infections, and toxic shock syndrome) are resistant to penicillin. One protein that confers such resistance is an enzyme called beta-lactamase, which deactivates molecules of penicillin by cleaving them. Normally, the beta-lactamase gene is repressed, but in the presence of penicillin the gene is induced and beta-lactamase is synthesized.

New research has provided scientists a greater understanding of how penicillin induces the beta-lactamase gene in these bacteria. Penicillin binds to a receptor protein on the bacterial cell membrane, which causes the receptor to become enzymatic. The newly formed enzyme breaks down the repressor of the beta-lactamase gene. In the absence of repressor, the gene is transcribed, beta-lactamase is produced, and penicillin is deactivated. Now that this metabolic path-

way for penicillin resistance is understood, it may be possible to design drugs that block the receptor's action and leave *Staphylococcus* susceptible to penicillin.

Staphylococcus aureus.   SEM   1 μm

Table 7.4	The Roles of Operons in the Regulation of Transcription	
Type of Regulation	Type of Metabolic Pathway Regulated	Regulating Condition
Inducible operons	Catabolic pathways	Presence of substrate of pathway
Repressible operons	Anabolic pathways	Presence of product of pathway

so that transcription rates vary with the concentration of corepressors.

Table 7.4 summarizes how the basic characteristics of inducible and repressible operons relate to the regulation of transcription. **New Frontiers 7.1** examines potential practical aspects of transcriptional control in health care.

Mutations of Genes

Learning Objective

✓ Define mutation.

The phenotype of a cell is dependent upon both the integrity and accurate control of its genes; however, the nucleotide sequences of genes are not always accurately maintained. A **mutation** is a change in the nucleotide base sequence of a genome, particularly its genes. Mutations of genes are almost always deleterious, though a few make no difference to the organism. Very rarely a mutation could lead to a protein having a novel property that improves the ability of an

organism and its descendants to survive and reproduce. Mutations in asexual organisms are generally passed on to the organism's progeny, but mutations in multicellular sexual organisms are passed to offspring only if the mutation occurs in gametes or gamete-producing cells.

Types of Mutations

Learning Objective

✓ Define and describe three types of point mutations.

Mutations range from large changes in an organism's genome, such as the loss or gain of an entire chromosome, to the most common type of mutation—**point mutations**—in which just one or a few nucleotide base pairs are affected. Point mutations include base pair **substitutions, insertions,** and **deletions.** These types of mutations can be demonstrated by the following analogy. Suppose that the DNA code was represented by the letters THECATATEELK. Grouping the letters into triplets (like codons) yields THE CAT ATE ELK. The substitution of a single letter could either change the meaning of the sentence, as in THE RAT ATE ELK, or result in a meaningless phrase, such as THE CAT RTE ELK. Insertion or deletion of a letter produces more serious changes, such as TRH ECA TAT EEL K or TEC ATA TEE LK. Insertions and deletions are also called **frameshift mutations** because nucleotide triplets subsequent to the mutation are displaced, creating new sequences of codons that result in vastly altered polypeptide sequences. Frameshift mutations affect proteins much more seriously than mere substitutions because a frame shift affects all codons subsequent to the mutation.

Mutations can also involve inversion (THE ACT ATE KLE), duplication (THE CAT CAT ATE ELK ELK), or

Figure 7.20 ▶
The effects of the various types of point mutations. Base-pair substitutions can result in silent mutations **(a)**, missense mutations **(b)**, or nonsense mutations **(c)**. Frameshift insertions **(d)** and frameshift deletions **(e)** usually result in severe missense or nonsense mutations because all codons downstream from the mutation are altered.

Normal

transposition (THE ELK ATE CAT). Such mutations and even larger deletions and insertions are *gross mutations*.

Effects of Mutations

Learning Objective

✓ List three effects of mutations.

Some base-pair substitutions produce **silent mutations** because the substitution does not change the amino acid sequence due to the redundancy of the genetic code **(Figure 7.20a).** For example, when the DNA triplet AAA is changed to AAG, the mRNA codon will be changed from UUU to UUC; however, because both codons specify the amino acid phenylalanine, there is no change in the phenotype—the mutation is silent because it affects the genotype only.

Of greater concern are substitutions that change a codon for one amino acid into a codon for a different amino acid. A change in a nucleotide sequence resulting in a codon that specifies a different amino acid is called a **missense mutation (Figure 7.20b);** what gets transcribed and translated makes sense, but not the right sense. The effect of missense mutations depends on where in the protein the different amino acid occurs. When the different amino is in a critical region of a protein, the protein becomes nonfunctional; however, when the different amino acid is in a less important region, the mutation has no adverse effect.

A third type of mutation occurs when a base-pair substitution changes an amino acid codon into a stop codon. This is called a **nonsense mutation (Figure 7.20c).** Nearly all nonsense mutations result in nonfunctional proteins.

Frameshift mutations (that is, insertions or deletions) typically result in drastic missense and nonsense mutations **(Figure 7.20d, e),** except when the insertion or deletion is very close to the end of a gene.

Table 7.5 summarizes the types and effects of point mutations.

CRITICAL THINKING

Find the codon UGU in Figure 7.9. What DNA nucleotide triplet codes for this codon? Identify a base-pair substitution that would produce a silent mutation at this codon. Identify a base-pair substitution that would result in a missense mutation at this codon. Identify a base-pair substitution that would produce a nonsense mutation at this codon.

(a) Silent mutation

(b) Missense mutation

(c) Nonsense mutation

(d) Frameshift insertion

(e) Frameshift deletion

Table 7.5   The Types of Point Mutations and Their Effects

Type of Point Mutation	Description	Effects
Substitution	Mismatching of nucleotides or replacement of one base pair by another	*Silent mutation* if change results in redundant codon, as amino acid sequence in polypeptide is not changed. *Missense mutation* if change results in codon for a different amino acid; effect depends on location of different amino acid in polypeptide. *Nonsense mutation* if codon for an amino acid is changed to a stop codon.
Frameshift (insertion)	Addition of one or a few nucleotide pairs creates new sequence of codons	Missense and nonsense mutations
Frameshift (deletion)	Removal of one or a few nucleotide pairs creates new sequence of codons	Missense and nonsense mutations

Mutagens

Learning Objectives

✓ Discuss how different types of radiation cause mutations in a genome.

✓ Describe three kinds of chemical mutagens and their effects.

Mutations occur naturally during the life of an organism. These *spontaneous mutations* result from errors in replication and repair as well as from *recombination* in which relatively long stretches of DNA move among chromosomes, plasmids, and viruses, introducing frameshift mutations. For example, we have seen that mismatched base pairing during DNA replication results in one error in every billion (10^9) base pairs. Since an average gene has 10^3 base pairs, about one of every 10^6 (million) genes contains an error. Further, though cells have repair mechanisms to reduce the effect of mutations, the repair process itself can introduce additional errors. Mutations can also be induced by many physical or chemical agents called **mutagens** (myū′tă-jenz), which include radiation and several types of DNA-altering chemicals.

Radiation

In the 1920s, Hermann Muller (1890–1967) discovered that *X-rays* increased phenotypic variability in fruit flies by causing mutations. *Gamma rays* also damage DNA. As discussed in Chapter 9, X-rays and gamma rays are *ionizing radiation;* that is, they energize electrons in atoms, causing some of the electrons to escape from their atoms. These free electrons strike other atoms, producing ions that can react with the structure of DNA, producing mutations. More seriously, electrons and ions can break the covalent bonds between the sugars and phosphates of a DNA backbone, causing physical breaks in chromosomes and complete loss of cellular control.

Nonionizing radiation in the form of *ultraviolet (UV) light* is also mutagenic because it causes adjacent thymine bases to covalently bond to one another, forming **thymine dimers (Figure 7.21).** The presence of dimers prevents hydrogen

▲ *Figure 7.21*

A thymine dimer. Ultraviolet light causes adjacent thymine bases to covalently bond to each other, which prevents hydrogen bonding with adenine bases in the complementary strand. The resulting distortion of the sugar-phosphate backbone prevents proper replication and transcription.

bonding with the adenine nucleotides in the complementary strand, distorts the sugar-phosphate backbone, and prevents proper replication and transcription. Cells have several methods of repairing thymine dimers (discussed shortly).

Chemical Mutagens

Here we consider three of the many basic types of mutagenic chemicals.

Nucleotide Analogs   **Nucleotide analogs** are compounds that are structurally similar to normal nucleotides (**Figure 7.22a** on page 224). When nucleotide analogs are available to replicating cells, they may be incorporated into DNA in place of normal nucleotides, where their structural differences either inhibit nucleic acid polymerases or result in mismatched base pairing. When, for example, thymine is replaced by 5′-bromouracil, it pairs with the wrong complement—it pairs with guanine rather than adenine, resulting in a point mutation (**Figure 7.22b**).

Nucleotide analogs make potent antiviral and anticancer drugs because viruses and cancer cells typically

Normal nucleotide

Thymine

Analog

5'-bromouracil

(a)

Normal DNA

Incorporation of 5'-bromouracil (bU) in place of thymine

First replication cycle: bU pairs with guanine

Second replication cycle: bU pairs with guanine

Base-pair substitution in daughter DNA molecules

(b)

▲ *Figure 7.22*

The structure and effects of nucleotide analogs. **(a)** The structures of thymine and its nucleotide analog, 5'-bromouracil (bU). **(b)** When 5'-bromouracil is incorporated into DNA, the result is a point mutation. A single replication cycle results in one normal DNA molecule and one mutated molecule in which bU pairs with guanine. After a second replication cycle, a complete base-pair substitution has occurred in one of the four DNA molecules: CG has been substituted for TA.

replicate faster than normal cells. The use of nucleotide analogs as antimicrobial agents is discussed in Chapter 10.

Nucleotide-Altering Chemicals   Some chemical mutagens alter the structure of nucleotides. For example, *nitrous acid* (H—O—N=O) removes the amine group of adenine, converting it into a guanine analog. When a cell replicates DNA containing this analog, an AT base pair is changed to a GC base pair in one daughter molecule—a base-pair-substitution mutation.

Another group of nucleotide-altering chemicals are *aflatoxins,* which are produced by *Aspergillus* (as-per-jil'ŭs) molds growing on grains and nuts. Aflatoxins convert guanine nucleotides into thymine nucleotides, so that a GC base pair is converted to a TA base pair, resulting in missense mutations and possibly cancer.

Frameshift Mutagens   Still other mutagenic chemical agents insert or delete nucleotide base pairs, resulting in frameshift mutations. Examples of frameshift mutagens are *benzopyrene,* which is found in smoke; *ethidium bromide,* which is used to stain DNA; and *acridine,* one of a class of dyes commonly used as mutagens in genetic research. These chemicals are exactly the right size to slip between adjoining nucleotides in DNA, producing a bulge in the molecule **(Figure 7.23).** When DNA polymerase copies the misshapen strands, one or more base pairs may be inserted or deleted in the daughter strand.

Frequency of Mutation

Learning Objective

✓ Discuss the relative frequency of deleterious and useful mutations.

Mutations are rare events. If they were not, organisms could not live or effectively reproduce themselves. As we have seen, about one of every million (10^6) genes contains an error. Mutagens typically increase the mutation rate by a factor of 10–1000 times; that is, mutagens induce an error in one of every 10^3 to 10^5 genes.

Most mutations are deleterious because they code for nonfunctional proteins or stop transcription entirely. Cells without functional proteins cannot metabolize, therefore many deleterious mutations are removed from the population when the cells die. Rarely, however, a cell acquires a beneficial mutation that allows it to survive, reproduce, and pass the mutation to its descendents. This change in gene frequency is the basis of evolution. For example, a bacterium might randomly acquire a mutation that confers resistance to an antibiotic. In an environment containing the antibiotic, cells without resistance die, but the mutated cell survives and reproduces. As long as the antibiotic is present, cells with such a mutation have an advantage over cells without the mutation.

DNA Repair

Learning Objective

✓ Describe light and dark repair of thymine dimers, base excision repair, mismatch repair, and the SOS response.

We have seen that although a mutation might rarely convey an advantage, most mutations are deleterious. To respond to the dangers mutations pose, cells have numerous methods for repairing damaged DNA, including light and dark repair of thymine dimers, base excision repair, mismatch repair, and an SOS response.

◀ *Figure 7.23*

The action of a frameshift mutagen. When DNA polymerase passes the bulge caused by the insertion of acridine between the two DNA strands, it incorrectly synthesizes a daughter strand, which contains either a deletion or an insertion mutation.

Repair of Thymine Dimers

Thymine dimers resulting from exposure to UV light constitute the most common type of mutation. Many prokaryotes contain *DNA photolyase,* an enzyme that is activated by visible light to break the bonds between adjoining thymine nucleotides, reversing the mutation and restoring the original DNA sequence (**Figure 7.24a** on page 226). This so-called **light repair** mechanism is advantageous for the prokaryote, but it presents a difficulty to scientists studying UV-induced mutations—they must keep such bacterial strains in the dark, or the mutants will revert to their previous form.

So-called **dark repair** involves a different repair enzyme—one that doesn't require light. Dark repair enzymes cut the damaged section of DNA from the molecule, creating a gap that is repaired by DNA polymerase and DNA ligase **(Figure 7.24b).** Though called dark repair, this mechanism operates either in light or in the dark.

Base-Excision Repair

Sometimes DNA polymerase incorporates an incorrect nucleotide during DNA replication. If the proofreading function of DNA polymerase does not repair the error, cells may use another enzyme system in a process called **base-excision repair.** This enzyme system excises the erroneous base, and then DNA polymerase fills in the gap **(Figure 7.24c).** A typical excision repair is the replacement of uracil, which is not supposed to be in DNA, with thymine.

Mismatch Repair

A similar repair mechanism is called **mismatch repair.** Mismatch repair enzymes scan newly synthesized DNA looking for mismatched bases, which they remove and replace **(Figure 7.24d).** How does the mismatch repair system determine which strand to repair? If it chose randomly, 50% of the time it would choose the wrong strand and introduce mutations.

Mismatch repair enzymes can distinguish between a new DNA strand and an old strand because old strands are methylated. Recognition of an error as far as 1000 base pairs away from an unmethylated portion of DNA triggers the mismatch repair enzymes. Once a new DNA strand is methylated, mismatch repair enzymes cannot correct any errors that remain.

Figure 7.24 ▶

DNA repair mechanisms. **(a)** Light repair of thymine dimers. A light-activated enzyme breaks thymine-to-thymine bonds. **(b)** Dark repair of thymine dimers. After the repair enzyme removes the entire damaged section from one strand of DNA, DNA polymerase and DNA ligase repair the breach. This mechanism is called dark repair because it does not require light; in fact, it operates in either light or darkness. **(c)** Base-excision repair, in which enzymes remove a segment with an incorrect base and DNA polymerase fills the gap. **(d)** Mismatch repair, which involves total excision of an incorrect nucleotide.

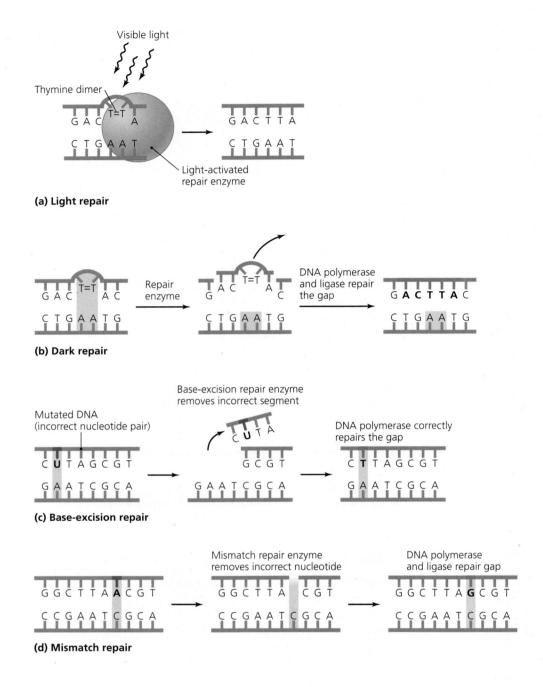

(a) Light repair

(b) Dark repair

(c) Base-excision repair

(d) Mismatch repair

SOS Response

Sometimes damage to DNA is so extreme that regular repair mechanisms cannot cope with the damage. In such cases, bacteria resort to what geneticists call an **SOS response** involving a variety of processes, such as the production of a novel DNA polymerase capable of copying less-than-perfect DNA. This polymerase replicates DNA with little regard to the base sequence of the template strand. Of course, this introduces many new and potentially fatal mutations, but presumably SOS repair allows a few offspring of these bacteria to survive.

Identifying Mutants, Mutagens, and Carcinogens

Learning Objectives

✓ Contrast the positive and negative selection techniques for isolating mutants.

✓ Describe the Ames test, and discuss its use in discovering carcinogens.

If a cell does not successfully repair a mutation, it and its descendents are called **mutants.** In contrast, cells normally found in nature (in the wild) are called **wild type** cells.

Scientists distinguish mutants from wild type cells by observing or testing for altered phenotypes. Since mutations are rare, and nonfatal mutations are even rarer, mutants can easily be "lost in the crowd." Therefore, researchers have developed methods to recognize mutants amidst their wild type neighbors.

Positive Selection

Positive selection involves selecting a mutant by eliminating wild type phenotypes **(Figure 7.25)**. Assume, for example, that researchers want to isolate bacterial mutants that are resistant to penicillin from a liquid culture. To do so they spread the liquid medium, which contains mostly penicillin-sensitive cells, but also the few penicillin-resistant mutants, onto a plate containing medium that includes penicillin; only the penicillin-resistant mutants multiply on this medium and produce visible colonies **(Figure 7.25a)**.

When a mutagenic agent is added to the liquid culture, it increases the number of mutants **(Figure 7.25b)**. While the researchers are isolating mutants, they can also determine the rate of mutation by comparing the number of mutant colonies formed after use of the mutagen with the number formed before treatment. The rate of mutation can be calculated as follows.

$$\frac{\begin{matrix}\text{number of colonies} \\ \text{seen with use of mutagen}\end{matrix} - \begin{matrix}\text{number of colonies} \\ \text{seen without use of mutagen}\end{matrix}}{\text{number of colonies seen without the use of mutagen}} \times 100\%$$

Negative (Indirect) Selection

An organism with nutritional requirements that differ from those of its wild-type phenotype is known as an *auxotroph* (awk'sō-trōf).[6] For example, a mutant bacterium that has lost the ability to synthesize tryptophan is auxotrophic for this amino acid—it must acquire tryptophan from the environment. Obviously, if a researcher attempts to grow tryptophan auxotrophs on media lacking tryptophan, the bacteria will be unable to synthesize all its proteins and will die. Therefore, to isolate such auxotrophs we must use a technique called **negative (indirect) selection**. The process by which a researcher uses negative selection to culture a tryptophan auxotroph is as follows **(Figure 7.26** on page 228):

1 The researcher inoculates a sample of a bacterial suspension containing potential mutants onto a plate containing complete media (including tryptophan). The sample is diluted such that the plate receives only about 100 cells.

2 Both auxotrophs and wild-type cells reproduce and form colonies on the plate, but the colonies are indistinguishable.

3 The researcher transfers cells from all the colonies on the plate to a sterile velvet pad by pressing the pad onto the plate.

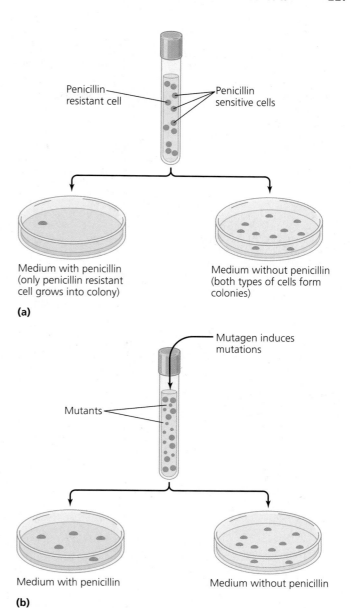

Medium with penicillin (only penicillin resistant cell grows into colony)

Medium without penicillin (both types of cells form colonies)

(a)

Medium with penicillin

Medium without penicillin

(b)

▲ *Figure 7.25*

Positive selection of mutants. Only mutants that are resistant to penicillin can survive on the plate containing the antibiotic. **(a)** Normally a population includes very few mutants. **(b)** Introduction of a mutagen increases the number of mutants and thus the number of colonies that grow in the presence of penicillin. *What is the rate of mutation induced by the mutagen in (b)?*

Figure 7.25 $\frac{5-1}{1} \times 100\% = 400\%$

4 The researcher inoculates two new plates—one containing tryptophan, the other lacking tryptophan—by pressing the pad onto each of them. This technique is called *replica plating*.

5 After the plates have incubated for several hours, the researcher compares the two replica plates. Tryptophan

---

[6]From Greek *auxein*, meaning to increase, and *trophe*, meaning nutrition.

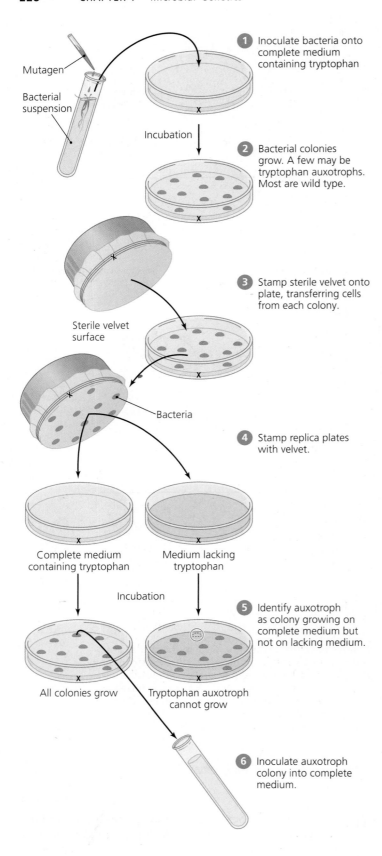

1. Inoculate bacteria onto complete medium containing tryptophan

Mutagen

Bacterial suspension

Incubation

2. Bacterial colonies grow. A few may be tryptophan auxotrophs. Most are wild type.

Sterile velvet surface

3. Stamp sterile velvet onto plate, transferring cells from each colony.

Bacteria

4. Stamp replica plates with velvet.

Complete medium containing tryptophan

Medium lacking tryptophan

Incubation

5. Identify auxotroph as colony growing on complete medium but not on lacking medium.

All colonies grow

Tryptophan auxotroph cannot grow

6. Inoculate auxotroph colony into complete medium.

◄ **Figure 7.26**

The use of negative (indirect) selection to isolate a tryptophan auxotroph. The plates are marked (in this case, with an X) so that their orientation can be maintained throughout the procedure. Researchers may have to inoculate hundreds of such plates to identify a single mutant.

auxotrophs growing on medium containing tryptophan are revealed by the absence of a corresponding colony on the plate lacking tryptophan.

6. The researcher takes cells of the auxotroph colony from the replica plate and inoculates them into a complete medium. The auxotroph is now isolated.

The Ames Test for Identifying Carcinogens

Numerous chemicals in food, the workplace, and the environment in general have been suspected of being **carcinogenic** (kar′si-nō-jen′ik) mutagens; that is, of causing mutations that result in cancer. Because animal tests to prove that they are indeed carcinogenic are expensive and time-consuming, researchers use a fast and inexpensive method for screening mutagens called an **Ames test,** which is named for its inventor, Bruce Ames (1928–). Interpretation of an Ames test includes the assumption that any given mutagen is a potential carcinogen.

An Ames test uses mutant *Salmonella* bacteria possessing a point mutation that prevents the synthesis of the amino acid histidine; in other words, they are histidine auxotrophs, indicated by the abbreviation his^-. To perform the test, an investigator mixes his^- mutants with liver extract and the substance suspected to be a mutagen **(Figure 7.27).** The presence of liver extract simulates the conditions in the body under which liver enzymes can turn innocuous chemicals into mutagens. The researcher then spreads the treated bacteria on a solid medium lacking histidine. If the suspected substance does in fact cause mutations, some of the mutations will reverse the effect of the original mutation, producing revertant cells (designated his^+) that have regained the ability to synthesize histidine and thus can survive on a medium lacking histidine. Thus, the presence of colonies during an Ames test reveals that the suspected substance is mutagenic in *Salmonella*.

Given that DNA in all cells is very similar, the ability of a chemical to cause mutations in *Salmonella* indicates that it is likely to cause mutations in humans as well. Note, however, that a positive Ames test does not prove that the chemical is a carcinogen; it only shows that the substance or a derivative of it is mutagenic in *Salmonella*. To prove that the substance can in fact cause cancer, scientists must test it in laboratory animals, a process that usually is warranted only if a chemical is mutagenic in *Salmonella*. Ames testing reduces the cost and time that would be required to assay every chemical for carcinogenicity in animals.

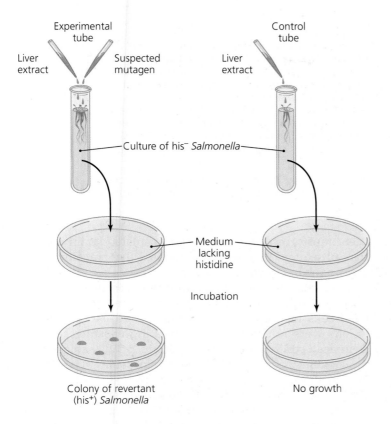

▲ Figure 7.27

The Ames test. A mixture containing his⁻ *Salmonella* mutants, rat liver extract, and the suspected mutagen is inoculated onto a plate lacking histidine. Colonies will form only if a mutagen reverses the his⁻ mutation, producing revertant his⁺ organisms with the ability to synthesize histidine. A control tube that lacks the suspected mutagen demonstrates that reversion did not occur in the absence of the mutagen. *What is the purpose of liver extract in an Ames test?*

Figure 7.27 Liver extract simulates conditions in the body by providing enzymes that may degrade harmless substances into mutagens.

Genetic Recombination and Transfer

Learning Objective

✓ Define genetic recombination.

Genetic recombination refers to the exchange between two DNA molecules of segments that are composed of identical or nearly identical nucleotide sequences called *homologous sequences*. Scientists have discovered a number of molecular mechanisms for genetic recombination, one of which is illustrated simply in **Figure 7.28a** on page 230. In this type of recombination, enzymes nick one strand of DNA at the homologous sequence, and another enzyme inserts the nicked strand into the second DNA molecule. Ligase then reconnects the strands in new combinations, and the molecules resolve themselves into novel molecules. Such DNA molecules that contain new arrangements of nucleotide sequences (and the cells that contain them) are called **recombinants.** Scientists first observed recombinants in eukaryotes during *crossing over*—a process in which portions of homologous chromosomes are recombined during the formation of gametes **(Figure 7.28b)**. Chapter 12 discusses crossing over and the formation of gametes in more detail.

Horizontal Gene Transfer Among Prokaryotes

Learning Objectives

✓ Contrast vertical gene transfer with horizontal gene transfer.
✓ Explain the roles of an F factor, F⁺ cells, and Hfr cells in bacterial conjugation.
✓ Describe the structures and actions of simple and complex transposons.
✓ Compare and contrast crossing over, transformation, transduction, and conjugation.

As we have discussed, both prokaryotes and eukaryotes replicate their genomes and supply copies to their descendants. This is known as *vertical gene transfer*—the passing of genes to the next generation. In addition, many prokaryotes can acquire genes from other microbes of the same generation—a process termed **horizontal gene transfer.** In horizontal gene transfer, a **donor cell** contributes part of its genome to a **recipient cell,** which may be a different species or even different genus from the donor. Typically, the recipient cell inserts part of the donor's DNA into its own chromosome, becoming a recombinant cell. Cellular enzymes then usually degrade the remaining unincorporated DNA. Horizontal gene transfer is a rare event, typically occurring in less than 1% of a population.

Here we consider the three types of horizontal gene transfer: *transformation, transduction,* and *bacterial conjugation.*

Transformation

In **transformation,** a recipient cell takes up DNA from the environment, such DNA as might be released by dead organisms. Frederick Griffith (1879–1941) discovered this process in 1928 while attempting to develop a vaccine for pneumonia caused by *Streptococcus pneumoniae* (strep-tō-kok'ŭs nū-mō'nē-ē), though the organism was called *Diplococcus pneumoniae* (dip'lō-kok'ŭs nū-mō'nē-ē) at that time. Griffith worked with two strains of *Streptococcus.* Cells of the first strain have a protective capsule that enables them to escape a body's defensive white blood cells; thus, these encapsulated cells cause deadly pneumonia when injected into mice **(Figure 7.29a** on page 231). (They are called *strain S* because they form *smooth* colonies on an agar surface.) The application of heat kills these encapsulated cells and renders them harmless when injected into mice **(Figure 7.29b).** In contrast, cells of the second strain (called *strain R* because they form *rough* colonies) are mutants that cannot make a capsule. The unencapsulated cells of strain R do not cause

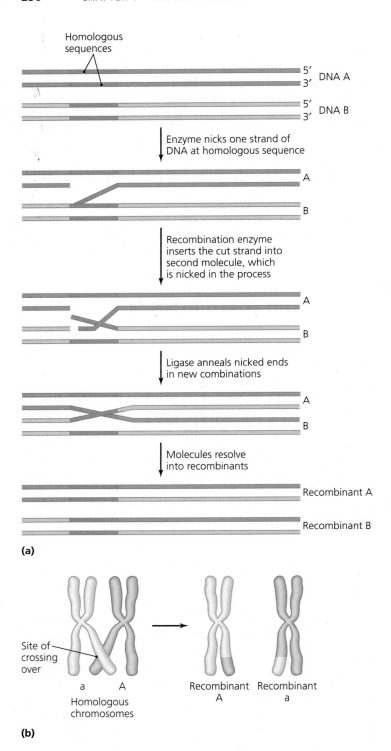

Homologous sequences

5′ } DNA A
3′

5′ } DNA B
3′

↓ Enzyme nicks one strand of DNA at homologous sequence

A

B

↓ Recombination enzyme inserts the cut strand into second molecule, which is nicked in the process

A

B

↓ Ligase anneals nicked ends in new combinations

A

B

↓ Molecules resolve into recombinants

Recombinant A

Recombinant B

(a)

Site of crossing over

a   A
Homologous chromosomes

Recombinant   Recombinant
A   a

(b)

◄ **Figure 7.28**

Genetic recombination. **(a)** Simplified depiction of one type of recombination between two DNA molecules. After an enzyme nicks one strand (here, strand A), a recombination enzyme rearranges the strands, and ligase seals the gaps to form recombinant molecules. **(b)** Crossing over, a process by which recombination routinely occurs between homologous chromosomes in eukaryotes. *What is the function of ligase during DNA replication?*

Figure 7.28 Ligase functions to anneal Okazaki fragments during replication of the lagging strand.

been transformed into deadly, encapsulated strain S bacteria. Subsequent investigations showed that transformation also occurs *in vitro* **(Figure 7.29e)**.

The fact that the living encapsulated cells retrieved at the end of this experiment outnumbered the dead encapsulated cells injected at the beginning indicated that strain R cells were not merely appropriating capsules released from dead strain S cells. Instead, strain R cells had acquired the capability of producing their own capsules by assimilating capsule-coding genes released from strain S cells. In 1944, Oswald Avery, Colin MacLeod, and Maclyn McCarty extracted various chemicals from S cells and determined that the transforming agent was DNA. This discovery was one of the conclusive pieces of proof that DNA is the genetic material of cells.

Cells that have the ability to take up DNA from their environment are said to be **competent.** Competence results from alterations in the cell wall and cytoplasmic membrane that allow DNA to enter the cell. Natural competency occurs in only a few types of bacteria, including pathogens in the genera *Streptococcus, Haemophilus* (hē-mof′i-lŭs), *Neisseria, Bacillus* (ba-sil′ŭs), *Staphylococcus* (staf′i-lō-kok′ŭs), and *Pseudomonas* (soo-dō-mō′ nas). Scientists can also generate competency artificially in *Escherichia* and other bacteria by manipulating the temperature and salt content of the medium. Because competent cells take up DNA from any donor genome, competency and transformation are important tools in genetic engineering.

Transduction

A second method of horizontal gene transfer, called **transduction,** involves the transfer of DNA from one cell to another via a replicating virus. Transduction can occur either between prokaryotic cells or between eukaryotic cells; it is limited only by the availability of a virus capable of infecting both donor and recipient cells. Here we will consider transduction in bacteria.

A virus that infects bacteria is called a **bacteriophage** or **phage** (fāj).[7] The process by which a phage participates in transduction is depicted in **Figure 7.30** on page 232. To repli-

disease **(Figure 7.29c)** because a mouse's defensive white blood cells quickly devour them.

Griffith discovered that when he injected both heat-killed strain S and living strain R into a mouse, the mouse died from pneumonia, even though neither of the injected strains was harmful when administered alone **(Figure 7.29d).** Further, and most significantly, Griffith isolated numerous living, *encapsulated* cells from the dead mouse. He realized that harmless, unencapsulated strain R bacteria had

---

[7]From Greek *phagein*, meaning to eat.

▲ *Figure 7.29*

Transformation in *Streptococcus pneumoniae*. Griffith's observations revealed that **(a)** encapsulated strain S killed mice, **(b)** heating renders strain S harmless to mice, and **(c)** unencapsulated strain R did not harm mice. In Griffith's experiment **(d)**, a mouse injected concurrently with killed strain S and live strain R (each harmless) died and was found to contain numerous living, encapsulated bacteria. **(e)** A demonstration that transformation of R cells to S cells also occurs *in vitro*.

cate, a bacteriophage attaches to a bacterial host cell and injects its genome into the cell ❶. Phage enzymes, translated by cellular ribosomes, degrade the cell's DNA ❷. The phage genome now controls the cell's functions and directs it to synthesize new phage DNA and phage proteins. Normally, phage proteins assemble around phage DNA to form new phage particles, but some phages mistakenly incorporate remaining fragments of bacterial DNA to form **transducing phages** ❸. Eventually the host cell lyses, releasing daughter and transducing phages. Transduction occurs when a transducing phage injects donor DNA into a new host cell (the recipient) ❹. The recipient host cell incorporates the donated DNA into its chromosome by recombination ❺.

In *generalized transduction*, the transducing phage carries a random DNA segment from a donor host cell's chromosome or plasmids to a recipient host cell. The transduction is not limited to a particular DNA sequence. In *specialized transduction*, only certain host sequences are transferred (along with phage DNA). In nature, specialized transduction is important in transferring genes encoding for certain bacterial toxins—including those responsible for diphtheria, scarlet fever, and the bloody, life-threatening diarrhea

caused by *E. coli* O157:H7—into cells that would otherwise be harmless. Chapter 8 discusses the use of specialized transduction to intentionally insert genes into cells.

Bacterial Conjugation

A third method of genetic transfer in bacteria is **conjugation**.[8] Unlike the donor cells in transformation and transduction, a donor cell in conjugation remains alive. Further, conjugation requires physical contact between donor and recipient cells. Scientists discovered conjugation between cells of *E. coli*, and it is best understood in this species. Thus, the remainder of our discussion will focus on conjugation in this bacterium.

Conjugation is mediated by **conjugation pili** (pīlī, singular: *pilus*), also called *sex pili*,[9] which are proteinaceous, rodlike structures extending from the surface of a cell. The gene coding for conjugation pili is located on a plasmid

---

[8]From Latin *conjugatus*, meaning yoked together.
[9]Bacterial conjugation may resemble intercourse, but because no exchange of gametes occurs, it is not true sex.

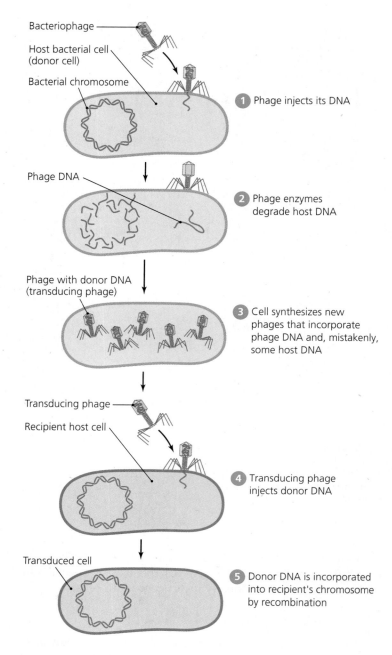

Bacteriophage

Host bacterial cell (donor cell)

Bacterial chromosome

1 Phage injects its DNA

Phage DNA

2 Phage enzymes degrade host DNA

Phage with donor DNA (transducing phage)

3 Cell synthesizes new phages that incorporate phage DNA and, mistakenly, some host DNA

Transducing phage

Recipient host cell

4 Transducing phage injects donor DNA

Transduced cell

5 Donor DNA is incorporated into recipient's chromosome by recombination

▲ *Figure 7.30*

Transduction. After a virus called a bacteriophage (phage) attaches to a host bacterial cell, it injects its genome into the cell and directs the cell to synthesize new phages. During assembly of new phages, some host DNA may be incorporated, forming transducing phages, which subsequently carry donor DNA to a recipient host cell.

called an **F (fertility) plasmid** or **F factor.** (Recall that a plasmid is a small, circular, extrachromosomal, molecule of DNA; see Figure 7.2b.) Cells that contain an F plasmid are called F^+ cells, and they serve as donors during conjugation. Recipient cells are F^-; that is, they lack an F plasmid and therefore have no conjugation pili.

The process of bacterial conjugation is illustrated in **Figure 7.31.** First, a sex pilus connects a donor cell (F^+) to a recipient cell (F^-) cell **1**, and the pilus then draws the cells together. After the cells touch, they stabilize their contact, probably via the fusion of cell membranes **2**. The donor then replicates one strand of its F plasmid and transfers it to the recipient **3**. The F^- recipient then synthesizes a complementary strand of F plasmid DNA and thus becomes an F^+ cell **4**.

In some bacterial cells, an F plasmid does not remain independent in the cytosol but instead integrates into the cellular chromosome at a specific DNA sequence. Such cells, which are called **Hfr (high frequency of recombination) cells,** can also conjugate with an F^- cell (**Figure 7.32** on page 234). After the F plasmid has integrated **1** and the Hfr and F^- cells join via a sex pilus **2**, DNA replication begins in the middle of the F plasmid, and the newly synthesized strand crosses to the recipient cell, carrying with it a copy of the donor's chromosome **3**. In most cases, movement of the cells breaks the intercellular connection before an entire donor chromosome is transferred **4**. Because the recipient receives only a portion of the F plasmid, it remains an F^- cell; however, it also acquires some chromosomal genes from the donor. Recombination can integrate the donor DNA into the recipient's chromosome **5**. The recipient is now a recombinant cell that contains its own genes as well as some donor genes.

Because the F plasmid integrates into chromosomes at only a few locations, the order in which genes are transferred is always the same. Scientists have produced maps of the genes on bacterial chromosomes by noting the time, in minutes, required for a particular gene to transfer from donor cells to recipient cells (**Figure 7.33** on page 235).

Conjugation occurs in several species of bacteria, which can be quite promiscuous; that is, conjugation can occur among bacteria of widely varying kinds and even between a bacterium and a yeast cell or between a bacterium and a plant cell. For example, the crown gall bacterium, *Agrobacterium,* transfers some of its genes into the chromosome of its host plant.

The natural transfer of genes by conjugation among diverse organisms heightens some scientists' concerns about the spread of R plasmids among pathogens. These scientists note that antibiotic resistance developed by one pathogen can spread to other pathogens.

Table 7.6 on page 235 summarizes the mechanisms of horizontal gene transfer in bacteria.

Transposons and Transposition

Transposons[10] are segments of DNA, 700–40,000 bp in length, that transpose (move) themselves from one location in a DNA molecule to another location in the same or different molecule. The result of the action of a transposon is

---

[10]From *transposable* elements.

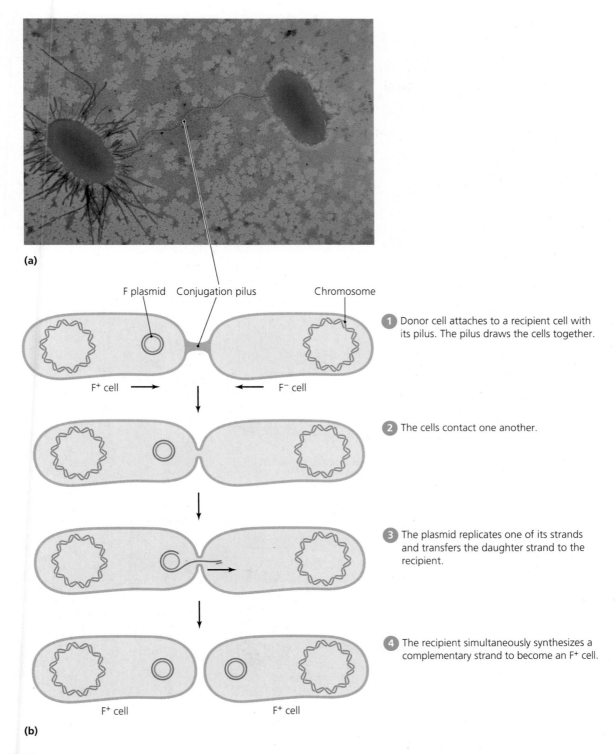

(a)

F plasmid   Conjugation pilus   Chromosome

F⁺ cell →   ← F⁻ cell

1 Donor cell attaches to a recipient cell with its pilus. The pilus draws the cells together.

2 The cells contact one another.

3 The plasmid replicates one of its strands and transfers the daughter strand to the recipient.

4 The recipient simultaneously synthesizes a complementary strand to become an F⁺ cell.

F⁺ cell   F⁺ cell

(b)

▲ *Figure 7.31*
Bacterial conjugation, a process in which a conjugation pilus connecting two cells mediates the transfer of DNA between the cells. **(a)** Two *E. coli* cells are connected by a conjugation pilus. **(b)** Artist's rendition of the process.

termed *transposition;* in effect it is a kind of frameshift insertion (see Figure 7.20d). This "illegitimate" recombination does not need a region of homology, unlike other recombination events.

The American geneticist Barbara McClintock discovered these so-called "jumping genes" through a painstaking analysis of the colors of the kernels of corn (**Figure 7.34** on page 235). She discovered that the genes for kernel color were turned on and off by the insertion of transposons. Subsequent research has shown that transposons are found in many, if not all,

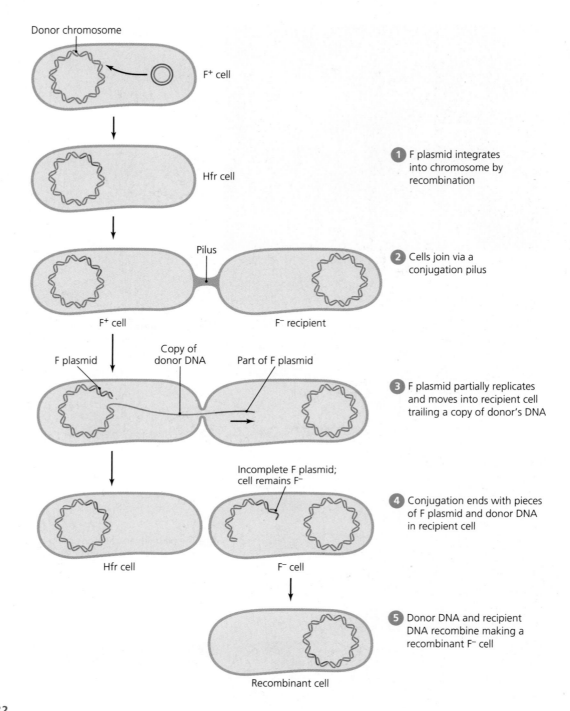

Donor chromosome

F⁺ cell

Hfr cell

1 F plasmid integrates into chromosome by recombination

Pilus

2 Cells join via a conjugation pilus

F⁺ cell

F⁻ recipient

F plasmid

Copy of donor DNA

Part of F plasmid

3 F plasmid partially replicates and moves into recipient cell trailing a copy of donor's DNA

Incomplete F plasmid; cell remains F⁻

4 Conjugation ends with pieces of F plasmid and donor DNA in recipient cell

Hfr cell

F⁻ cell

5 Donor DNA and recipient DNA recombine making a recombinant F⁻ cell

Recombinant cell

▲ *Figure 7.32*

Conjugation involving an Hfr cell, which is formed when an F⁺ cell integrates its F plasmid into its chromosome. Hfr cells donate a partial copy of their DNA and a portion of the F plasmid to a recipient, which is rendered a recombinant cell but remains F⁻.

prokaryotes, eukaryotes, and viruses. Fortunately, while transposons are common, transposition and the frameshift mutations it causes are relatively rare occurrences.

Transposons vary in their nucleotide sequences, but all of them contain palindromic sequences at each end. A *palindrome*[11] is a word, phrase, or sentence that has the same

sequence of letters when read backward or forward—for example, "Madam, I'm Adam." In genetics, a palindrome is a region of DNA in which the sequence of nucleotides is

[11]From Greek *palin*, meaning again, and *dramein*, meaning to run.

Table 7.6 Natural Mechanisms of Horizontal Genetic Transfer in Bacteria

Mechanism	Requirements	State of Donor	State of Recipient
Transformation	Free DNA in the environment and a competent recipient	Dead	Living
Transduction	Bacteriophage	Killed by bacteriophage	Living
Conjugation	Cell-to-cell contact and F plasmid (either in cytosol or incorporated into chromosome of donor)	Living	Living

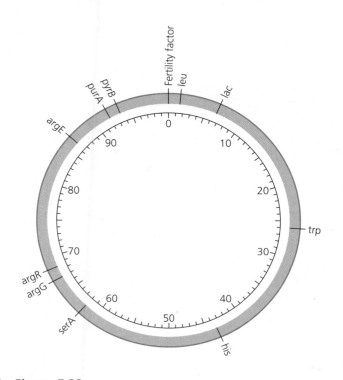

▲ **Figure 7.33**

A genetic map. This map of the circular genome of *E. coli* is divided into 100 minutes because that is the time required to transfer an entire chromosome from an Hfr donor cell to a recipient cell under standard laboratory conditions. The map shows the locations of a few genes, including the *lac* operon (6 minutes) and the *trp* operon (26 minutes). Insertion of the F factor occurs at 0 minutes. *How long does it take for the genes of the leucine biosynthetic pathway (leu) to pass from the donor cell to the recipient cell? How long for the transfer of the histidine (his) operon?*

Figure 7.33 It takes 1.5 minutes for the genes of the leucine biosynthetic pathway (leu) to pass from the donor cell to the recipient, and 43.5 minutes for the transfer of the histidine operon.

▲ **Figure 7.34**

Variation in color of the kernels of corn, which was instrumental in the discovery of transposons. Barbara McClintock discovered that changes in the inherited colors of kernels were caused by the movement of transposons into and out of genes (in this case, the genes involved in the synthesis of kernel pigments), a process called transposition.

identical to an inverted sequence in the complementary strand. For example, GAATTC is the palindrome of CTTAAG. Such a palindromic sequence is also known as an **inverted repeat (IR).**

The simplest transposons, called **insertion sequences (IS),** consist of no more than two inverted repeats and a gene that encodes the enzyme *transposase* (**Figure 7.35a** on page 236). Transposase recognizes its own inverted repeat in a target site, and then it cuts the DNA at that site and inserts the transposon (or a copy of it) into the DNA molecule (**Figure 7.35b**). Such transposition also produces a duplicate copy of the target site.

Complex transposons also contain one or more genes not connected with transposition, such as genes for antibiotic resistance (**Figure 7.35c**). *R factors,* which are plasmids that convey resistance to antibiotics, often contain transposons. R factors are of great clinical concern because they spread antibiotic resistance to innumerable pathogens.

Transposon: Insertion sequence IS1

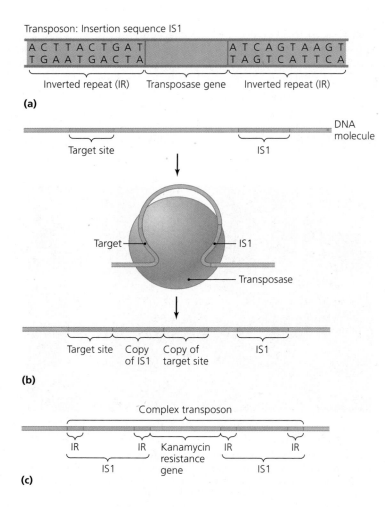

(a)

(b)

(c)

◀ *Figure 7.35*

Transposons. **(a)** A simple transposon, or insertion sequence. Shown here is Insertion Sequence 1 (IS1), which consists of a gene for the enzyme transposase bounded by identical (though inverted) repeats of nucleotides. **(b)** Transposase recognizes its target site (elsewhere in the same DNA molecule, as shown here, or in a different DNA molecule; then it moves the transposon (or, as in the case of IS1, a copy of the transposon) to its new site. The target site is duplicated in the process. **(c)** A complex transposon, which contains genes not related to transposition. Shown here is Tn5, which consists of a gene for kanamycin resistance between two IS1 transposons.

CRITICAL THINKING

Suppose you are a scientist who wants to insert into your dog a gene that encodes a protein that protects dogs from heartworms. A dog's cells are not competent, so they cannot take up the gene from the environment; but you have a plasmid, a competent bacterium, and a related (though incompetent) F^+ bacterium that lives as an intracellular parasite in dogs. Describe a possible scenario by which you could use natural processes to genetically alter your dog to be heartworm resistant.

Chapter 8 discusses how scientists have adapted the natural processes of transformation, transduction, conjugation, and transposition to manipulate the genes of organisms.

CHAPTER SUMMARY

The Structure and Replication of Genomes (pp. 200–208)

1. **Genetics** is the study of inheritance and inheritable traits. Genes are composed of specific sequences of nucleotides that code for polypeptides or RNA molecules. A **genome** is the sum of all the genetic material in a cell or virus. Prokaryotic and eukaryotic cells use DNA as their genetic material; some viruses use DNA, and other viruses use RNA.

2. The two strands of DNA are held together by hydrogen bonds between complementary **base pairs (bp)** of nucleotides. Adenine bonds with thymine, and guanine bonds with cytosine.

3. Prokaryotic genomes consist of one (or rarely, two) **chromosomes,** which are typically circular molecules of DNA associated with protein and RNA molecules, localized in a region of the cytosol called the **nucleoid.** Prokaryotic cells may also contain one or more extrachromosomal DNA molecules called **plasmids,** which contain genes that regulate nonessential life functions, such as bacterial conjugation and resistance to antibiotics.

4. In addition to DNA, eukaryotic chromosomes contain proteins called **histones,** arranged as **nucleosomes** (beads of DNA) that clump to form **chromatin fibers.** Eukaryotic cells also contain extranuclear DNA in mitochondria, chloroplasts, and plasmids.

5. DNA replication is semiconservative; that is, each newly synthesized strand of DNA remains associated with one of the parental strands. After helicase unwinds and unzips the original molecule, synthesis of each of the two daughter strands—called the **leading strand** and the **lagging strand**—occurs from its 5′ end to its 3′ end. Synthesis is mediated by enzymes that prime, join, and proofread the pairing of new nucleotides.

6. After DNA replication, **methylation** occurs. Methylation plays several roles in the control of gene expression, the initiation of DNA replication, recognition of a cell's own DNA, and repair.

Gene Function (pp. 208–221)

1. The **genotype** of an organism is the actual set of genes in its genome, whereas its **phenotype** refers to the physical and functional traits expressed by those genes.

2. RNA has several forms: **messenger RNA (mRNA),** which carries genetic information from DNA to a ribosome; **transfer RNA (tRNA),** which carries amino acids to the ribosome; and **ribosomal RNA (rRNA),** which, together with polypeptides, makes up the structure of ribosomes.

3. The **central dogma** of protein synthesis states that genetic information is transferred from DNA to RNA to polypeptides, which function alone or in conjunction as proteins.

4. The transfer of genetic information begins with **transcription** of the genetic code from DNA to RNA, in which **RNA polymerase** links RNA nucleotides that are complementary to genetic sequences in DNA. Transcription begins at a region of DNA called a **promoter** (recognized by RNA polymerase), and ends with a sequence called a **terminator.** Other proteins may assist in termination, or termination may depend solely on the nucleotide sequence of the transcribed RNA.

5. In **translation,** the sequence of genetic information carried by mRNA is used by ribosomes to construct polypeptides with specific amino acid sequences.

6. The genetic code consists of triplets of mRNA nucleotides, called **codons.** These bind with complementary anticodons on transfer RNAs (tRNA), which are molecules that carry specific amino acids. Ribosomal RNA (rRNA) catalyzes the bonding of one amino acid to another to form a polypeptide. The sequence of nucleotides thus codes for the sequence of amino acids.

7. Eukaryotic mRNA is synthesized as pre-messenger RNA. Before translation can occur, the cell removes noncoding **introns** from pre-mRNA and splices together the **exons,** which are the coding sections.

8. A ribosome contains three tRNA binding sides: an **A site** (associated with incoming amino acids), a **P site** (associated with elongation of the polypeptide), and an **E site** from which tRNA exits the ribosome.

9. An **operon** is a series of genes, a **promoter,** and an **operator** sequence controlled by one regulatory gene. The operon model explains gene regulation in prokaryotes.

10. **Inducible operons** are normally "turned off" and are activated by inducers, whereas **repressible operons** are normally "on" and are deactivated by repressors.

Mutations of Genes (pp. 221–229)

1. A **mutation** is a change in the nucleotide base sequence of a genome. **Point mutations** involve a change in one or a few nucleotide base pairs and include **substitutions** and two types of **frameshift mutations: insertions** and **deletions.** Mutations can be categorized by their effects as **silent, missense,** or **nonsense mutations.**

2. Mutations can be introduced by physical or chemical agents called **mutagens.** Physical mutagens include ionizing radiation such as X-rays and gamma rays, and nonionizing ultraviolet light. Ultraviolet light causes adjacent thymine bases to bond to one another to form **thymine dimers.** Mutagenic chemicals include **nucleotide analogs,** chemicals that are structurally similar to nucleotides and can result in mismatched base pairing, and chemicals that insert or delete nucleotide base pairs, producing frameshift mutations.

3. Cells repair damaged DNA via **light repair** and **dark repair** of thymine dimers, **base-excision repair,** and **mismatch repair.** When damage is so extensive that these mechanisms are overwhelmed, bacterial cells may resort to an **SOS response.**

4. Researchers have developed methods to distinguish **mutants,** which carry mutations, from normal **wild type** cells. These methods include **positive selection, negative (indirect) selection,** and the **Ames test,** which is used to identify potential **carcinogens.**

Genetic Recombination and Transfer (pp. 229–236)

1. Organisms acquire new genes through **genetic recombination,** which is the exchange of homologous pieces of DNA. **Crossing over** occurs during gamete formation during sexual reproduction in eukaryotes.

2. Vertical gene transfer is the transmission of genes from parents to offspring. In **horizontal gene transfer,** DNA from a **donor cell** is transmitted to a **recipient cell.** A **recombinant cell** results from genetic recombination between donated and recipient DNA. Transformation, transduction, and bacterial conjugation are types of horizontal gene transfer.

3. In **transformation,** a **competent** recipient prokaryote takes up DNA from its environment. Competency is found naturally and can be created artificially in some cells.

4. In **transduction,** a virus such as a bacteriophage carries DNA from a donor cell to a recipient cell. Donor DNA is accidentally incorporated in such **transducing phages.**

5. In **conjugation,** an F$^+$ bacterium—that is, one containing an **F (fertility) plasmid (factor)**—forms a **conjugation pilus** that attaches to an F$^-$ recipient bacterium. Plasmid genes are transferred to the recipient, which becomes F$^+$ as a result.

6. **Hfr (high frequency of recombination) cells** result when an F plasmid integrates into a prokaryotic chromosome. Hfr cells form conjugation pili and transfer cellular genes more frequently than normal F$^+$ cells.

7. **Transposons** are DNA segments that code for the enzyme transposase bounded by palindromic sequences known as **inverted repeats (IR)** at each end. Transposons move among locations in chromosomes in eukaryotes and prokaryotes. The simplest transposons, known as **insertion sequences (IS),** consist only of inverted repeats and transposase. **Complex transposons** contain other genes as well.

QUESTIONS FOR REVIEW

(Answers to multiple choice and fill in the blanks are on the web, along with additional review questions. Visit www.microbiologyplace.com.)

Multiple Choice

1. Which of the following is most likely the number of base pairs in a bacterial chromosome?
   a. 4,000,000   c. 400
   b. 4000   d. 40

2. Which of the following is a true statement concerning prokaryotic chromosomes?
   a. They typically have two or three origins of replication.
   b. They contain single-stranded DNA.
   c. They are located in the cytosol.
   d. They are associated in linear pairs.

3. A plasmid is
   a. a molecule of RNA found in bacterial cells.
   b. distinguished from a chromosome by being circular.
   c. a structure in bacterial cells formed from plasma membrane.
   d. extrachromosomal DNA.

4. Which of the following forms ionic bonds with eukaryotic DNA and stabilizes it?
   a. chromatin
   b. bacteriocin
   c. histone
   d. nucleosome

5. Nucleotides used in the replication of DNA
   a. are energy storage molecules.
   b. are found in four forms, each with a deoxyribose sugar, a phosphate, and a base.
   c. are present in the cytosol of cells as triphosphate nucleotides.
   d. all of the above.

6. Which of the following molecules functions as a "proofreader" for a newly replicated strand of DNA?
   a. DNA polymerase
   b. primase
   c. helicase
   d. ligase

7. The addition of —CH_3 to a cytosine nucleotide after DNA replication is called
   a. methylation.
   b. restriction.
   c. transcription.
   d. translation.

8. In translation, the binding site through which discharged tRNA molecules exit is called the
   a. A site.
   b. X site.
   c. P site.
   d. E site.

9. The Ames test
   a. uses auxotrophs and liver extract to reveal potential mutagens.
   b. is time intensive and costly.
   c. involves the isolation of a mutant by eliminating wild type phenotypes with specific media.
   d. proves that suspected chemicals are carcinogenic.

10. Which of the following methods of DNA repair involves enzymes that recognize and correct nucleotide errors in unmethylated strands of DNA?
    a. light repair of thymine dimers
    b. dark repair of thymine dimers
    c. mismatch repair
    d. SOS response

11. Which of the following is *not* a mechanism of natural genetic transfer and recombination?
    a. transduction
    b. transformation
    c. transcription
    d. conjugation

12. Cells that have the ability to take up DNA from their environment are said to be
    a. Hfr cells.
    b. transposing.
    c. genomic.
    d. competent.

13. Which of the following statements is true?
    a. Conjugation requires sex pili extending from the surface of a cell.
    b. Conjugation involves an H^+ factor.
    c. Conjugation is an artificial genetic engineering technique based primarily on the positive or negative charge of the genome involved.
    d. Conjugation primarily involves DNA that has been released into the environment from dead organisms.

14. Which of the following are called "jumping genes"?
    a. Hfr cells
    b. transducing phages
    c. palindromic sequences
    d. transposons

15. Although cells P and Q are totally unrelated, cell Q receives DNA from cell P and incorporates this new DNA into its chromosome. This process is
    a. crossing over of P and Q.
    b. vertical gene transfer.
    c. horizontal gene transfer.
    d. transposition.

16. Which of the following is part of each molecule of mRNA?
    a. palindrome
    b. codon
    c. anticodon
    d. base pair

17. A nucleotide is composed of
    a. a five-carbon sugar.
    b. phosphate.
    c. a nitrogenous base.
    d. all of the above

18. In DNA, adenine forms _____ hydrogen bonds with _____.
    a. three/guanine
    b. two/guanine
    c. two/thymine
    d. three/thymine

19. A sequence of nucleotides formed during replication of the lagging DNA strand is
    a. a palindrome.
    b. an Okazaki fragment.
    c. a template strand.
    d. an operon.

20. Which of the following is *not* part of an operon?
    a. operator
    b. promoter
    c. origin
    d. gene

21. Repressible operons are important in regulating prokaryotic
    a. DNA replication.
    b. RNA transcription.

c. rRNA processing.
d. sugar catabolism.

22. Transcription produces _____.
    a. DNA molecules
    b. RNA molecules
    c. polypeptides
    d. palindromes

23. Ligase plays a major role in
    a. lagging strand replication.
    b. mRNA processing in eukaryotes.
    c. polypeptide synthesis by ribosomes.
    d. RNA transcription.

24. Before mutations can affect a population they must be
    a. permanent.
    b. inheritable.
    c. beneficial.
    d. all of the above

25. The *trp* operon is repressible. This means it is usually _____ and is directly controlled by _____.
    a. active/an inducer
    b. active/a repressor
    c. inactive/an inducer
    d. inactive/a repressor

Fill in the Blanks

1. The three steps in RNA transcription are _____, _____, and _____.

2. A triplet of mRNA nucleotides that specifies a particular amino acid is called a _____.

3. Three effects of point mutations are _____, _____, and _____.

4. Insertions and deletions in the genetic code are also called _____ mutations.

5. An operon consists of _____, _____, and _____ and is associated with a regulatory gene.

6. In general, _____ operons are inactive until the substrate of their genes is present.

7. A daughter DNA molecule is composed of one original strand and one new strand because DNA replication is _____.

8. A gene for antibiotic resistance can move among bacterial cells by _____, _____, and _____.

9. _____ are nucleotide sequences containing palindromes and genes for proteins that cut DNA strands.

10. _____ _____ is a recombination event that occurs during gamete formation in eukaryotes.

11. _____ RNA carries amino acids.

Short Answer

1. How does the genotype of a bacterium determine its phenotype? Use the terms *gene, mRNA, ribosome,* and *polypeptide* in your answer.

2. List several ways in which eukaryotic messenger RNA differs from prokaryotic mRNA.

3. Compare and contrast introns and exons.

4. Polypeptide synthesis requires large amounts of energy. How do cells regulate synthesis to conserve energy? Describe one specific example.

5. Describe the operon model of gene regulation.

6. How can knowledge of nucleotide analogs be useful to a cancer researcher?

7. Compare and contrast the structure and components of DNA and RNA in prokaryotes.

8. Besides the fact that it synthesizes RNA, how does RNA polymerase differ in function from DNA polymerase?

9. Describe the formation and function of mRNA, rRNA, and tRNA in prokaryotes and eukaryotes.

10. Describe how DNA is packaged in both prokaryotes and eukaryotes.

11. Explain the central dogma of genetics.

12. Compare and contrast the processes of transformation, transduction, and conjugation.

13. Fill in the following table:

Process	Purpose	Beginning Point	Ending Point
Replication			Origin or end of molecule
Transcription		Promoter	
Translation	Synthesis of polypeptides		

CRITICAL THINKING

1. A scientist uses a molecule of DNA composed of nucleotides containing radioactive sugar molecules as a template for replication and transcription in a nonradioactive environment. What percentage of DNA strands will be radioactive after three DNA replication cycles? What percentage of RNA molecules will be radioactive?

2. If molecules of mRNA have the following nucleotide base sequences, what will be the sequence of amino acids in polypeptides synthesized by eukaryotic ribosomes? The genetic code is displayed in Figure 7.9.

    a. AUGGGGAUACGCUACCCC

    b. CCGUACAUGCUAAUCCCU

    c. CCGAUGUAACCUCGAUCC

    d. AUGCGGUCAGCCCCGUGA

3. The drugs ddC and AZT are used to treat AIDS.

ddc (2′,3′-dideoxycytidine)   AZT (3′-azido-2′,3′-dideoxythymidine)

Based on their chemical structures, what is their mode of action?

4. Explain why an insertion of three nucleotides is less likely to result in a deleterious effect than an insertion of a single nucleotide.

5. Suppose that *E. coli* sustains a mutation in its gene for the *lac* operon repressor such that the repressor is ineffective. What effect would this have on the bacterium's ability to catabolize lactose? Would the mutant strain have an advantage over wild type cells? Explain your answer.

6. A student claims that nucleotide analogs can be carcinogenic. Another student in the study group insists that nucleotide analogs are sometimes used to treat cancer. Explain why both students are correct.

Recombinant DNA Technology

Recombinant DNA technology affects our lives in many ways. Genetically altered corn, soybeans, or canola are ingredients in over 60% of processed foods in the United States. Scientists now have the ability to take a gene from a bacterium and insert it into a corn plant, enabling the plant to produce a toxin that kills insects but is harmless to humans. Geneticists can insert daffodil and bacterial genes into rice to boost the grain's nutritional content, and they can insert growth-hormone genes into salmon to make the fish grow faster. The possibilities are limitless and intriguing—and frequently controversial, as questions about safety and environmental impact arise. Could crops genetically altered to resist pests spread those resistance genes to weeds? Could genetically modified foods introduce new allergens? What are the effects of these technologies on wildlife populations?

Despite the concerns, recombinant DNA technology holds much promise, not only for the food industry and agriculture, but for medicine as well. This chapter explores the fascinating field of recombinant DNA technology, and the roles of microorganisms within it.

CHAPTER OUTLINE

The Role of Recombinant DNA Technology in Biotechnology

The Tools of Recombinant DNA Technology
Mutagens
The Use of Reverse Transcriptase to Synthesize cDNA
Synthetic Nucleic Acids
Restriction Enzymes
Vectors
Gene Libraries

Techniques of Recombinant DNA Technology
Multiplying DNA *in vitro:* The Polymerase Chain Reaction (PCR)
Selecting a Clone of Recombinant Cells
Separating DNA Molecules: Gel Electrophoresis and the Southern Blot
Inserting DNA into Cells

Applications of Recombinant DNA Technology
Genome Mapping
Pharmaceutical and Therapeutic Applications
Agricultural Applications

The Ethics and Safety of Recombinant DNA Technology

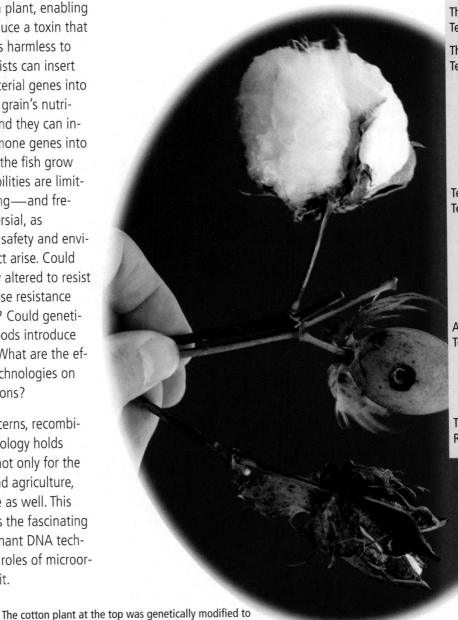

The cotton plant at the top was genetically modified to better resist insects. The plant at the bottom was not.

MicroPrep Pre-Test: Take the pre-test for this chapter on the web.
Visit **www.microbiologyplace.com.**

This chapter examines how genetic researchers have adapted the natural enzymes and processes of DNA recombination, replication, transcription, transformation, transduction, and conjugation to manipulate genes for industrial, medical, and agricultural purposes. Together these techniques are termed *recombinant DNA technology,* commonly called "genetic engineering." We will study the tools and techniques gene researchers use to manipulate DNA and modify genomes for pharmaceutical, therapeutic, and agricultural purposes. We end the chapter with a discussion of the ethics and safety of these techniques.

The Role of Recombinant DNA Technology in Biotechnology

Learning Objectives

✓ Define biotechnology and recombinant DNA technology.

✓ List several examples of useful products made possible by biotechnology.

✓ Identify the three main goals of recombinant DNA technology.

Biotechnology—the use of microorganisms to make practical products—is not a new field. For thousands of years, humans have used microbes to make products such as bread, cheese, soy sauce, and alcohol. During the 20th century, scientists industrialized the natural metabolic reactions of bacteria to make large quantities of acetone, butanol, and antibiotics. More recently, scientists have adapted microorganisms for use in the manufacture of paper, textiles, and vitamins; to assist in cleaning up industrial wastes, oil spills, and radioactive isotopes; and to aid in mining copper, gold, uranium, and other metals. Chapter 26 discusses such applications of biotechnology.

Until recently, microbiologists were limited to working with naturally occurring organisms and their mutants for achieving such industrial and medical purposes. Since the 1990s, however, scientists have become increasingly adept at intentionally modifying the genomes of organisms, by natural and artificial processes, for a variety of practical purposes. This is **recombinant DNA technology,** and it has expanded the possibilities of biotechnology in ways that seemed like science fiction only a few years ago. Today, scientists isolate specific genes from almost any so-called donor organism, such as a human, a plant, or a bacterium, and insert it into the genome of almost any kind of recipient organism.

Scientists who manipulate genomes have three main goals:

- *To eliminate undesirable phenotypic traits in humans, animals, plants, and microbes.* For example, scientists have inserted genes from microbes into plants to make them resistant to pests or freezing, and since 1999 they have cured children born with a fatal and previously untreatable genetic disorder called severe combined immunodeficiency disease (SCID).

- *To combine beneficial traits of two or more organisms to create valuable new organisms,* such as laboratory animals that mimic human susceptibility to HIV.

- *To create organisms that synthesize products that humans need,* such as vaccines, antibiotics, hormones, and enzymes. For instance, gene therapists have successfully inserted the human gene for insulin into bacteria so that the bacteria synthesize human insulin, which is cheaper and safer than insulin derived from animals.

Recombinant DNA technology is not a single procedure or technique, but rather a collection of tools and techniques scientists use to manipulate the genomes of organisms. In general, they isolate a gene from a cell, manipulate it *in vitro,*[1] and insert it into another organism. **Figure 8.1** illustrates the basic processes involved in recombinant DNA technology.

The Tools of Recombinant DNA Technology

Scientists use a variety of physical agents, naturally occurring enzymes, and synthetic molecules to manipulate genes and genomes. These tools of recombinant DNA technology include *mutagens, reverse transcriptase, synthetic nucleic acids, restriction enzymes,* and *vectors.* Scientists use these molecular tools to create *gene libraries,* which are a time-saving tool for genetic researchers.

Mutagens

Learning Objective

✓ Describe how gene researchers use mutagens.

As we saw in Chapter 7, mutagens are physical and chemical agents that produce mutations. Scientists deliberately utilize mutagens to create changes in microbes' genomes so that the microbe's phenotypes are changed. They then select for and culture cells with characteristics considered beneficial for a given biotechnological application. For example, scientists exposed the fungus *Penicillium* to mutagenic agents and then selected strains that produce greater amounts of penicillin. In this manner, they developed a strain of *Penicillium* that secretes over 25 times the amount of penicillin as the strain originally isolated by Alexander Fleming. Today, with recombinant DNA techniques (discussed shortly), researchers can isolate mutated genes rather than dealing with entire organisms.

---

[1]Latin, meaning within glassware.

▲ *Figure 8.2*

Reverse transcription, in which the viral enzyme reverse transcriptase catalyzes the synthesis of cDNA from an RNA template. The flow of information here—from RNA to DNA—is the opposite of what occurs in cells.

The Use of Reverse Transcriptase to Synthesize cDNA

Learning Objective

✓ Explain the function and use of reverse transcriptase in synthesizing cDNA.

As we discussed in Chapter 7, transcription involves the transmission of genetic information from molecules of DNA to molecules of RNA. The discovery of retroviruses, which have genomes consisting of RNA instead of DNA, led to the discovery of an unusual enzyme—**reverse transcriptase.** Reverse transcriptase in effect creates a flow of genetic information in the opposite direction from what occurs in conventional transcription: It uses an RNA template to transcribe a molecule of DNA, which is called **complementary DNA (cDNA)** because it is complementary to an RNA template **(Figure 8.2).** Because hundreds to millions of copies of mRNA exist for every active gene, it is frequently easier to produce a desired gene by first isolating the mRNA molecules that code for a particular polypeptide and then use reverse transcription to synthesize a cDNA gene from the mRNA template. Further, eukaryotic DNA is not normally expressible by prokaryotic cells, which cannot remove the introns (noncoding sequences) present in eukaryotic pre-mRNA. However, since eukaryotic mRNA has already been processed to remove introns, cDNA produced from it lacks noncoding sequences. Therefore, scientists can successfully insert cDNA into prokaryotic cells, making it possible for the prokaryotes to produce eukaryotic proteins such as human growth factor, insulin, and blood clotting factors.

Synthetic Nucleic Acids

Learning Objectives

✓ Explain how gene researchers synthesize nucleic acids.
✓ Describe three uses of synthetic nucleic acids.

Not only do the enzymes of DNA replication and RNA transcription function *in vivo;*[2] they also function *in vitro;*

---

[2]Latin, meaning in life (that is, within a cell).

▲ *Figure 8.1*

Overview of recombinant DNA technology.

therefore, scientists are able to produce molecules of DNA and RNA in cell-free solutions for genetic research. In fact, scientists have so mechanized the processes of nucleic acid replication and transcription that they can produce molecules of DNA and RNA with any nucleotide sequence; all they must do is enter the desired sequence into a synthesis machine's four-letter keyboard. A computer controls the actual synthesis, using a supply of nucleotides and other required reagents. Nucleic acid synthesis machines synthesize molecules over 100 nucleotides long in a few hours, and scientists can join two or more of these molecules end to end with ligase to create even longer synthetic molecules.

Researchers have used synthetic nucleic acids in many ways, including:

- *Elucidating the genetic code.* Using synthetic molecules of varying nucleotide sequences and observing the amino acids in the resulting polypeptides, scientists elucidated the genetic code. For example, synthetic DNA consisting only of adenine nucleotides yields a polypeptide consisting solely of the amino acid phenylalanine. Therefore, the mRNA codon UUU (transcribed from the DNA triplet AAA) must code for phenylalanine.

- *Creating genes for specific proteins.* Once they know the genetic code and the amino acid sequence of a protein, scientists can create a gene for that protein. In this manner, scientists synthesized a gene for human insulin. Of course, such a synthetic gene may well consist of a nucleotide sequence different from that of its cellular counterpart because of the redundancy in the genetic code (see Table 7.2 on p. 213).

- *Synthesizing DNA and RNA probes to locate specific sequences of nucleotides.* **Probes** are nucleic acid molecules with a specific nucleotide sequence that have been labeled with radioactive or fluorescent chemicals so that their locations can be detected. The use of probes to locate specific sequences of nucleotides is based on the fact that any given nucleotide sequence will preferentially bond to its complementary sequence. Thus a probe constructed of a single strand of DNA with the nucleotide sequence ATGCT will bond to a DNA strand with the sequence TACGA, and the probe's label allows researchers to then detect the complementary site. Probes are essential tools for locating specific DNA sequences such as genes for particular polypeptides.

CRITICAL THINKING

Even though some students correctly synthesize a fluorescent cDNA probe complementary to mRNA for a particular yeast protein, they find that the probe does not attach to any portion of the yeast's genome. Explain why the students' probe does not work.

Restriction Enzymes

Learning Objectives

✓ Explain the source and names of restriction enzymes.
✓ Describe the importance and action of restriction enzymes.

An important development in recombinant DNA technology was the discovery of **restriction enzymes** in bacterial cells. Such enzymes cut DNA molecules and are restricted in their action—they cut DNA only at locations with specific and usually palindromic nucleotide sequences called *restriction sites.* In nature, bacterial cells use restriction enzymes to protect themselves from phages by cutting phage DNA into nonfunctional pieces. (Recall from Chapter 7 that bacterial cells protect their own DNA by methylation of some of their nucleotides, which hides the DNA from the restriction enzymes.)

Researchers name restriction enzymes with three letters (denoting the genus and specific epithet of the source bacterium) and Roman numerals (to indicate the order in which enzymes from the same bacterium were discovered). In some cases, a fourth letter denotes the strain of the bacterium. Thus *Escherichia coli* (esh-ĕ-rik′ē-ă kō′lī) strain R produces the restriction enzymes *Eco*RI and *Eco*RII, and *Hind*III is the third restriction enzyme isolated from *Haemophilus influenzae* (hē-mof′i-lŭs in-flu-en′zē) strain Rd.

Scientists have discovered several hundred restriction enzymes, and categorize them in two groups based on the types of cuts they make. The first type, as exemplified by *Eco*RI, makes staggered cuts of the two strands of DNA, producing fragments that terminate in mortise-like *sticky ends,* each of which is composed of up to four nucleotides that form hydrogen bonds with its complementary sticky end **(Figure 8.3a).** Scientists can use these bits of single-stranded DNA to combine pieces of DNA from different organisms into a single recombinant DNA molecule (the enzyme ligase unites the sugar-phosphate backbones of the pieces) **(Figure 8.3b).** Other restriction enzymes, such as *Hind*II and *Sma*I (from *Serratia marcescens,* sĕ-ră′tē-ă măr-sĕs′enz), cut both strands of DNA at the same point, resulting in *blunt ends* **(Figure 8.3c).** It is more difficult to make recombinant DNA from blunt-ended fragments because they are not sticky, but they have a potential advantage—blunt ends are nonspecific. This enables any blunt-ended fragments, even those produced by different restriction enzymes, to be combined **(Figure 8.3d).** In contrast, sticky-ended fragments bind only to complementary, sticky-ended fragments produced by the same restriction enzyme.

Table 8.1 identifies several restriction enzymes and their target DNA sequences.

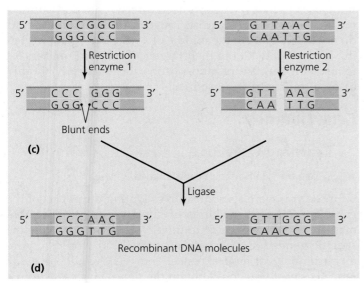

◀ *Figure 8.3*

Actions of restriction enzymes, which recognize and cut both strands of a DNA molecule at a specific (usually palindromic) restriction site. **(a)** Certain restriction enzymes produce staggered cuts with complementary "sticky ends." **(b)** When two complementary sticky-ended fragments produced by a given restriction enzyme come from different organisms, their bonding (catalyzed by ligase) produces recombinant DNA. **(c)** Other restriction enzymes produce blunt-ended fragments. **(d)** A lack of specificity enables blunt-ended fragments produced by different restriction enzymes to be combined into recombinant DNA. *Which restriction enzymes act at the restriction sites shown?*

Figure 8.3 (a) EcoRI (b) HindIII, and (c) SmaI and HpaI.

Table 8.1	Properties of Some Restriction Enzymes	
Enzyme	**Bacterial Source**	**Restriction Site***
*Bam*HI	*Bacillus amyloliquefaciens* H	G↓GATCC CCTAG↑G
*Eco*RI	*Escherichia coli* RY13	G↓AATTC CTTAA↑G
*Eco*RII	*E. coli* R245	CC↓GG GG↑CC
*Hind*II	*Haemophilus influenzae* Rd	GTPy↓PuAC CAPu↑PyTG
*Hind*III	*H. influenzae* Rd	A↓AGCTT TTCGA↑A
*Hinf*I	*H. influenzae* Rf	G↓ANTC CTNA↑G
*Hpa*I	*H. parainfluenzae*	GTT↓AAC CAA↑TTG
*Msp*I	*Moraxella* sp.	CC↓GG GG↑CC
*Sma*I	*Serratia marcescens*	CCC↓GGG GGG↑CCC

*Arrows indicate sites of cleavage; Py = pyrimidine (either T or C); Pu = purine (either A or G); N = any nucleotide (A, T, G, or C).

Vectors

Learning Objective

✓ Define a vector as the term applies to genetic manipulation.

One goal of recombinant DNA technology is to insert a useful gene into a cell so that the cell has a new phenotype—for example, the ability to synthesize a novel protein. To deliver a gene into a cell, researchers use **vectors,** which are nucleic acid molecules such as viral genomes, transposons, and plasmids.

Genetic vectors share several useful properties:

• *Vectors are small enough to manipulate in a laboratory.* Large DNA molecules the size of entire chromosomes are generally too fragile to serve as vectors.

• *Vectors survive inside cells.* Plasmids, which are circular DNA, make good vectors because they are more stable than are linear fragments of DNA, which are typically degraded by cellular enzymes. However, some linear vectors such as transposons and certain viruses insert themselves rapidly into a host's chromosome before they can be degraded.

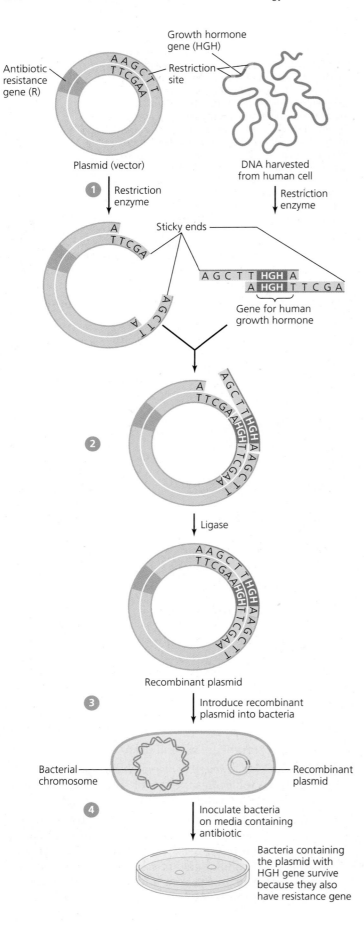

- *Vectors contain a recognizable genetic marker* so that researchers can identify the cells that have received the vector and thereby the specific gene of interest. Genetic markers can either be phenotypic markers, such as those that confer antibiotic resistance or code for enzymes that metabolize a unique nutrient, or *tags* using radioactive or fluorescent labels.

- *Vectors can ensure genetic expression* by providing required genetic elements such as promoters.

An example of the process used to produce a vector containing a specific gene is depicted in **Figure 8.4.** After a given restriction enzyme cuts both the DNA molecule containing the gene of interest (in this example, the human growth hormone gene) and the vector DNA (here a plasmid containing a gene for antibiotic resistance as a marker) into fragments with sticky ends ❶, ligase anneals the fragments to produce a recombinant plasmid ❷. After the recombinant plasmid has been inserted into a bacterial cell ❸, the bacteria are grown on a medium containing the antibiotic ❹; only those cells that contain the recombinant plasmid (and thus the human growth hormone gene as well) can grow on the medium.

Generally, viruses and transposons are able to carry larger genes than can plasmids. Researchers are developing vectors from adenoviruses, poxviruses, and a genetically modified form of the human immunodeficiency virus (HIV). HIV in particular might make an excellent vector for use in humans because HIV inserts itself directly into human chromosomes; however, scientists must ensure that viral vectors do not insert DNA in the middle of a necessary gene, mutating and possibly killing their target cells. **New Frontiers 8.1** focuses on the use of transposons to genetically modify disease-transmitting mosquitos.

Gene Libraries

Learning Objective

✓ Explain the significance of gene libraries.

Suppose you were a scientist investigating the effects of the genes for 24 different kinds of interleukins (proteins that mediate certain aspects of immunity). Having to isolate the specific genes for each type of interleukin would require much time, labor, and expense. Your task would be made much easier if you could obtain the genes you need from a **gene library,** a collection of bacterial or phage clones, each of which contains a portion of the genetic material of interest. In effect, each clone is like one book in a library in that it

◀ *Figure 8.4*

An example of the process for producing a recombinant vector—in this case, a plasmid. *Which restriction enzyme was used in this example?*

Figure 8.4 HindIII

What if scientists could genetically engineer mosquitoes to be incapable of transmitting deadly diseases, such as malaria? Researchers are attempting to do just that. One possible strategy involves using transposons to carry genes into mosquitoes that would boost their production of defensive proteins that destroy pathogens. Mosquitoes naturally produce such proteins, but not in high enough doses to kill off all pathogens. Scientists theorize that if they can genetically engineer and breed mosquitoes that destroy pathogens, these mosquitoes could then be released into the environment to breed with wild mosquitoes, creating subsequent populations that can no longer spread disease.

To date, researchers have successfully used transposons as vectors to insert defense-boosting genes into *Aedes* mosquitoes, which transmit yellow fever. While this is a promising achievement, *Aedes*

is only one of over 100 disease-transmitting mosquito species. Recombinant technologies that work on *Aedes* may not necessarily work on, for instance, *Anopheles gambia* mosquitoes, which are the primary carriers of the most deadly form of malaria. In addition, some worry that unleashing transgenic mosquitoes into the wild may have adverse consequences. Given that malaria alone kills over 2 million people each year, however, the quest for a safer mosquito is one that scientists are compelled to continue.

Aedes mosquito

contains one fragment (typically a single gene) of an organism's entire genome. Alternatively, a gene library may contain clones with all the genes of a single chromosome or of the set of cDNA that is complementary to an organism's mRNA.

As depicted in **Figure 8.5** on page 248, genetic researchers can create each of the clones in a gene library by using restriction enzymes to generate fragments of the DNA of interest and then ligase to synthesize recombinant vectors. They insert the vectors into bacterial cells, which are then grown on culture media. Once a scientist isolates a recombinant clone and places it in a gene library, the gene the clone carries becomes available to other investigators, saving them the time and effort required to isolate that gene. Many gene libraries are now commercially available.

Techniques of Recombinant DNA Technology

Scientists use the tools of recombinant DNA technology in a number of basic techniques to multiply, identify, manipulate, isolate, map, and sequence the nucleotides of genes.

Multiplying DNA *in vitro:* The Polymerase Chain Reaction (PCR)

Learning Objective

✓ Describe the purpose and application of the polymerase chain reaction.

The **polymerase chain reaction (PCR)** is a technique by which scientists produce a large number of identical molecules of DNA *in vitro*. Using PCR, researchers start with a single molecule of DNA and generate billions of exact replicas within hours. Such rapid amplification of DNA is critical in a variety of situations. For example, epidemiologists used PCR to amplify the genome of an unknown pathogen

that killed 90% of the crows and seven people in New York City in 1999. The large number of identical DNA molecules produced by PCR allowed scientists to determine the nucleotide sequence, which was found to be identical to that of West Nile virus—unknown in the Americas until then. Similarly in 2001, scientists used PCR to amplify DNA from *Bacillus anthracis* (ba-sil'ŭs an-thras'is) spores used in the bioterrorism attacks. The DNA molecules were sequenced and the sequence was compared to those of known *Bacillus* strains in an attempt to identify the source of the spores.

PCR is a repetitive process that alternately separates and replicates the two strands of DNA. Each cycle of PCR consists of the following three steps (**Figure 8.6a** on page 249):

1 *Denaturation.* Exposure to heat (about 94°C) separates the two strands of the target DNA by breaking the hydrogen bonds between base pairs but otherwise leaves the two strands unaltered.

2 *Priming.* After a mixture containing an excess of DNA primers (synthesized such that they are complementary to nucleotide sequences near the ends of the target DNA), DNA polymerase, and an abundance of the four deoxyribonucleotide triphosphates (A, T, G, and C) is added to the target DNA; cooling to about 65°C enables double-stranded DNA to re-form. Because there is an excess of primers, the single strands are more likely to bind to a primer than to one another. The primers provide DNA polymerase with the 3' hydroxyl group it requires for DNA synthesis.

3 *Extension.* Raising the temperature to about 72°C increases the rate at which DNA polymerase replicates each strand to produce more DNA.

These steps are repeated over and over, so the number of DNA molecules increases exponentially **(Figure 8.6b)**. After only 30 cycles—which requires only a few hours to complete—PCR produces over 1 billion identical copies of the original DNA molecule.

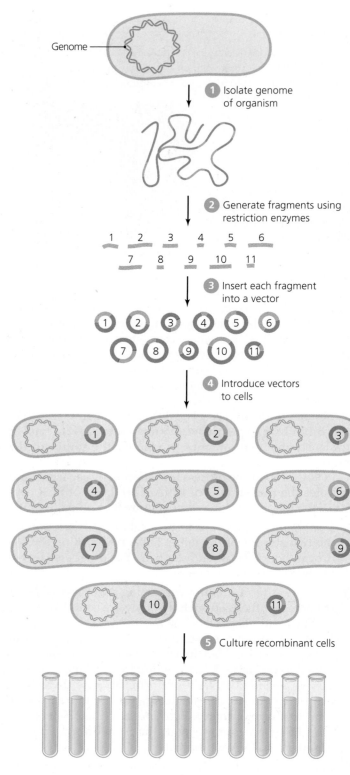

Genome

1 Isolate genome
of organism

2 Generate fragments using
restriction enzymes

1   2   3   4   5   6

7   8   9   10   11

3 Insert each fragment
into a vector

1   2   3   4   5   6

7   8   9   10   11

4 Introduce vectors
to cells

1   2   3

4   5   6

7   8   9

10   11

5 Culture recombinant cells

▲ *Figure 8.5*

Production of a gene library, the population of all cells or phages that together contain all of the genetic material of interest. In this figure, each clone of cells carries a portion of a bacterium's genome.

The process can be automated using a *thermocycler,* a device that automatically performs PCR by continuously cycling all the necessary reagents—DNA polymerase, primers, and triphosphate deoxynucleotides—through the three temperature regimes. A thermocycler uses DNA polymerase derived from hyperthermophilic archaea such as *Thermus aquaticus* (ther'mus a-kwa'ti-cus). This enzyme is not denatured at 94°C, so the machine need not be replenished with DNA polymerase after each cycle.

Selecting a Clone of Recombinant Cells

Learning Objective

✓ Explain how researchers use DNA probes to identify recombinant cells.

Before recombinant DNA technology can have practical application, a scientist must be able to select and isolate recombinant cells that contain particular genes of interest. For example, once researchers have created a gene library, they must find the clone containing the DNA of interest. To do so, scientists use probes—which, you may recall, bind specifically and exclusively to their complementary nucleotide sequences and have either radioactive or fluorescent markers. Researchers then isolate and culture cells that have the radioactive or fluorescent marker, which also aids in identifying the specific location of the genes of interest, as performed in a technique called gel electrophoresis (discussed next).

Separating DNA Molecules: Gel Electrophoresis and the Southern Blot

Learning Objective

✓ Describe the process and use of gel electrophoresis, particularly as it is used in a Southern blot.

Electrophoresis (ē-lek-trō-fōr'ē-sis) is a technique that involves separating molecules based on their electrical charge, size, and shape. In recombinant DNA technology, scientists use **gel electrophoresis** to isolate fragments of DNA molecules that can then be inserted into vectors, multiplied by PCR, or preserved in a gene library.

In gel electrophoresis, DNA molecules, which have an overall negative charge, are drawn through a semisolid gel by an electric current toward the positive electrode within an electrophoresis chamber (**Figure 8.7** on page 250). The gel is typically composed of a purified sugar component of agar, called *agarose,* which in addition to making a more uniform gel than agar, acts as a molecular sieve that retards the movement of DNA fragments down the chamber and separates the fragments by size. Smaller DNA fragments move faster and farther than larger ones. Scientists can determine the size of a fragment by comparing the distance it travels to the distances traveled by standard DNA fragments of known sizes.

▲ Figure 8.6

The use of the polymerase chain reaction (PCR) to replicate DNA. **(a)** Each cycle of PCR consists of three steps: ① During *denaturation* at 94°C, the hydrogen bonds between DNA strands are broken and the strands separate. ② Cooling to 65°C in the presence of deoxyribonucleotide triphosphates, primers, and DNA polymerase allows *priming* to occur. ③ During *extension* at 72°C, warming speeds the action of DNA polymerase in replicating strands to produce more DNA. **(b)** Each cycle of PCR doubles the amount of DNA; over 1 billion copies of the original DNA molecule are produced by 30 cycles of PCR.

As we have seen, DNA probes allow a researcher to find specific DNA sequences such as genes in a cell. Scientists could also use probes to localize specific sequences in electrophoresis gels, but because gels are flimsy, easily broken, and deform as they dry, it is difficult to probe gels.

In 1975, Ed Southern (1938–) devised a method, called the **Southern blot,** to transfer DNA from agarose gels to nitrocellulose membranes, which are less delicate. The Southern blot technique begins with the procedures of gel electrophoresis just described. Once the DNA fragments have been separated by size, the liquid in the electrophoresis gel is blotted out, the DNA it contains is transferred and bonded to a nitrocellulose membrane, and radioactive probes are used to localize DNA sequences of interest in the membrane (**Figure 8.8** on page 250). A *northern blot* is a similar technique used to detect specific RNA molecules.

Researchers use Southern blots for a variety of purposes, including genetic "fingerprinting" (discussed shortly) and diagnosis of infectious diseases. For example, scientists can detect the presence of genetic sequences unique to hepatitis B virus in a blood sample of an infected patient even before the patient shows symptoms or an immune response.

Scientists also use Southern blotting to demonstrate the incidence and prevalence in an environmental sample of archaea, bacteria, and viruses, particularly those that cannot be cultured. Most microorganisms have never been grown in laboratory; indeed, scientists know them only by unique DNA patterns in electrophoresis gels and Southern blot membranes, called their DNA fingerprints or "signatures." For example, based on the many unique DNA signatures

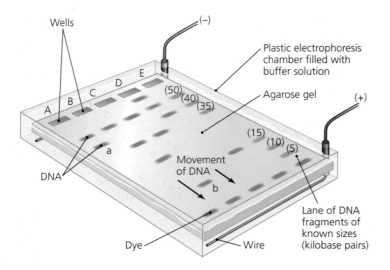

Wells

A B C D E

Plastic electrophoresis chamber filled with buffer solution

Agarose gel

(50) (40) (35)

(15) (10) (5)

Movement of DNA

DNA

a

b

Dye

Wire

Lane of DNA fragments of known sizes (kilobase pairs)

(a)

(b)

▲ *Figure 8.7*

Gel electrophoresis. **(a)** After DNA is cleaved into fragments by restriction enzymes, it is loaded into wells, which are small holes cut into the agarose gel. DNA fragments of known sizes are often loaded into one well (in this case, E) to serve as standards. After the DNA fragments are drawn toward the positive electrode by an electric current, they are stained with a dye. **(b)** Ethidium bromide dye fluoresces under ultraviolet illumination to reveal the locations of DNA within a gel. *Compare the positions of the fragments in lanes A and B to the positions of the fragments of known sizes. What sizes are the fragments labelled a and b?*

Figure 8.7 a–40 kilobase pairs, b–10 kilobase pairs.

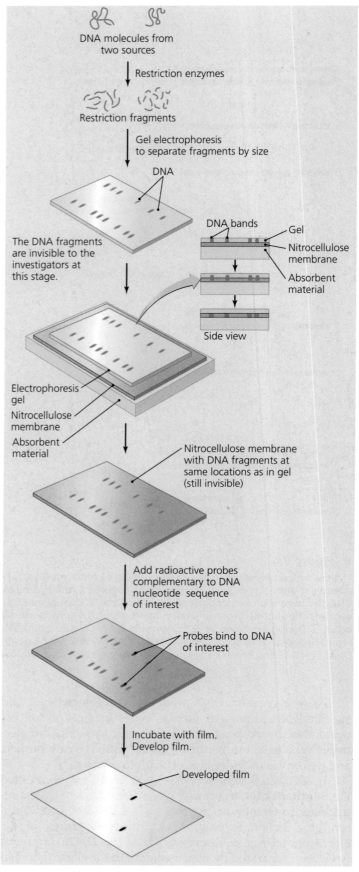

DNA molecules from two sources

Restriction enzymes

Restriction fragments

Gel electrophoresis to separate fragments by size

DNA

The DNA fragments are invisible to the investigators at this stage.

DNA bands — Gel — Nitrocellulose membrane

Absorbent material

Side view

Electrophoresis gel

Nitrocellulose membrane

Absorbent material

Nitrocellulose membrane with DNA fragments at same locations as in gel (still invisible)

Add radioactive probes complementary to DNA nucleotide sequence of interest

Probes bind to DNA of interest

Incubate with film. Develop film.

Developed film

Figure 8.8 ▲

The Southern blot technique, which enables scientists to locate DNA sequences of interest.

(a) Electroporation

(b) Protoplast fusion

(c) Gene gun

(d)

▲ *Figure 8.9*

Artificial methods of inserting DNA into cells. **(a)** Electroporation, in which an electrical current applied to a cell makes it competent to take up DNA. **(b)** Protoplast fusion, in which enzymes digest cell walls to create protoplasts that fuse at a high rate when treated with polyethylene glycol. **(c)** A gene gun, which fires DNA-coated beads into a cell. **(d)** Microinjection, in which a solution of DNA is introduced into a cell through a micropipette.

isolated from soil that do not match signatures from known microorganisms, scientists calculate that fewer than 1% of soil-dwelling microbes have ever been grown in a laboratory.

Inserting DNA into Cells

Learning Objective

✓ List and explain three artificial techniques for introducing DNA into cells.

A goal of recombinant DNA technology is the insertion of a gene into a cell. In addition to using vectors and the natural methods of transformation, transduction, and conjugation,

scientists have developed several artificial methods to introduce DNA into cells, including:

- *Electroporation* **(Figure 8.9a).** Electroporation involves using an electrical current to puncture microscopic holes through a cell's membrane so that DNA can enter the cell from the environment. Electroporation can be used on all types of cells, though the thick-walled cells of fungi and algae must first be converted to *protoplasts*, which are cells whose cell walls have been enzymatically removed. Cells treated by electroporation repair their membranes and cell walls after a time.

- *Protoplast fusion* **(Figure 8.9b).** When protoplasts encounter one another, their cytoplasmic membranes may fuse to form a single cell that contains the genomes of both "parent" cells. Exposure to polyethylene glycol increases the rate of fusion. The DNA from the two fused cells recombines to form a recombinant molecule. Scientists often use protoplast fusion for the genetic modification of plants.

- *Injection* (**Figure 8.9c** and **d**). Two types of injection are commonly used. Researchers use a *gene gun* powered by a blank .22-caliber cartridge or compressed gas to fire tiny tungsten or gold beads coated with DNA into a target cell. The cell eventually eliminates the inert metal beads. In *microinjection,* a geneticist inserts DNA into a target cell with a glass micropipette having a tip diameter smaller than that of the cell or, in the case of eukaryotes, smaller than that of the nucleus. Unlike electroporation and protoplast fusion, injection can be used on intact tissues such as plant seeds.

In every case, foreign DNA that enters a cell will remain in a cell's progeny only if the DNA is self-replicating, as in the case of plasmid and viral vectors, or if the DNA integrates into a cellular chromosome by recombination.

Applications of Recombinant DNA Technology

The importance of recombinant DNA technology does not lie in the novelty, cleverness, or elegance of its procedures, but in its wide range of applications. In this section we consider how recombinant DNA technology is used to solve various problems and produce products in research, medicine, and agriculture.

Genome Mapping

Learning Objecetive

✓ Describe genome mapping and genomics, and explain their usefulness.

One application of these tools and techniques is **genome mapping,** which involves locating genes on a nucleic acid molecule. Genome maps provide scientists with useful facts, including information concerning an organism's metabolism and growth characteristics, as well as its potential relatedness to other microbes. For example, in the late 1990s, scientists discovered a virus with a genome map similar to those of certain hepatitis viruses. As a result, they named the new discovery *hepatitis G virus* because it presumably causes hepatitis, though it has not been demonstrated that the virus actually causes hepatitis.

Locating Genes

Until about 1970, scientists identified the specific location of genes on chromosomes by labor-intensive methods. In eukaryotes, the process involved assessing the frequency with which eukaryotic genes remained linked during meiotic crossing over (see Figure 12.1); in bacteria, the process involved timing the transfer of genes during conjugation (see Figure 7.30). Both these techniques are cumbersome, time consuming, and applicable to only some organisms. Recombinant DNA techniques provide simpler and universal methods for genome mapping.

One technique for locating genes, called *restriction fragmentation,* was one of the earliest applications of restriction enzymes. In this technique, which is used for mapping the relative locations of genes in plasmids and viruses, researchers compare DNA fragments resulting from cleavages by several restriction enzymes to determine each fragment's location relative to the others. Consider the following simplified example of the basic procedure.

Researchers cleave identical circular DNA molecules that are 2400 base pairs (2.4 kilobase pairs or kbp) long using three restriction enzymes: *Eco*RI, *Bam*HI, and *Hin*dIII. After they use the enzymes singly and in combinations, they identify the sizes of the fragments by gel electrophoresis. Their results are as follows:

Enzyme	Fragment Size (kilobase pairs)
*Bam*HI	2.4
*Hin*dIII	2.4
*Eco*RI	1.2
*Eco*RI and *Hin*dIII	0.24, 0.96, 1.2
*Eco*RI and *Bam*HI	0.36, 0.84, 1.2

Since either *Hin*dIII or *Bam*HI produces a single fragment, the original DNA must be a circular molecule—cutting a linear molecule always produces more than one fragment. *Eco*RI produces 1.2 kbp fragments, so there must be two restriction sites located on opposite sides of the circle recognized by this enzyme. The *Hin*dIII and *Bam*HI restriction sites must be located in one or the other of the two halves produced by *Eco*RI. **Figure 8.10** illustrates three possible genome maps consistent with these observations. The researchers can determine which of the three is correct by further analysis with other combinations of restriction enzymes. If the researchers know the locations of specific genes on specific fragments, then elucidation of the correct arrangement of the fragments will reveal the relative locations of the genes on the DNA molecule.

Using this method, scientists first completed the entire gene map of a cellular microbe—the bacterium *H. influenzae*—in 1995. Since then, geneticists have elucidated complete gene maps of numerous viral, prokaryotic, and eukaryotic microbes.

CRITICAL THINKING

If the restriction enzymes *Hin*dIII and *Bam*III together produce restriction fragments 1.08 kbp and 1.32 kbp, then which of the three maps shown in Figure 8.10 is correct?

◀ *Figure 8.10*

Genome mapping using restriction fragmentation. In this technique, a researcher determines the relative locations of restriction sites by an analysis of fragment lengths. Shown are three possible maps corresponding to the actions of the restriction enzymes discussed in the text.

Nucleotide Sequencing

An exciting development in the world of genetics is **genomics,** the sequencing and analysis of the nucleotide bases of genomes. At first, scientists sequenced DNA molecules by selectively cleaving DNA at A, T, G, or C bases, separating the fragments by gel electrophoresis, and mapping the order in which the fragments occur in a complete DNA molecule. Such time-consuming, labor-intensive, and cumbersome sequencing was limited to short DNA molecules such as those of plasmids.

Today, scientists use a faster technique that utilizes cDNA synthesized with nucleotides that have been tagged with four different fluorescent dyes—a different color for each nucleotide base; then an automated DNA sequencer determines the sequence of base colors emitted by the dyes **(Figure 8.11).** Such machines, often running 24 hours a day for months, have sequenced the entire genomes of numerous viruses, bacteria, and eukaryotic organisms. Scientists reached a milestone in 2001 by sequencing the 3 billion nucleotide base pairs that constitute the human genome. **New Frontiers 8.2** on page 254 explains how scientists are even using the nucleotide sequence in the human genome to detect new infectious agents.

Elucidation of the gene sequences of pathogens, particularly those affecting hundreds of millions of people and those with potential bioterrorist uses, is a current priority of researchers. For example, in 2002 scientists finished sequencing the genome of *Plasmodium falciparum* (plaz-mō′dē-ŭm fal-sip′ar-ŭm), one of the protozoa that causes malaria, and hope to use the information to develop novel antimalarial drugs and an effective vaccine. Other researchers are sequencing the genomes of the pathogens that cause anthrax, bubonic plague, and cholera in order to develop more effective therapies and vaccines.

Another use for genomics is to relate DNA sequence data to protein function. For instance, scientists are investigating the genes and proteins of *Deinococcus radiodurans* (dī-nō-kok′ŭs rā-dē-ō-dur′anz), a microorganism that is remarkably resistant to damage of its DNA by radiation.

ATAGTGTCACCTAAATAGCTTGGCGTAATCATGGTCATAGCTGTTTCCTG TGTGAAATTC

▲ *Figure 8.11*

Automated DNA sequencing. Each colored line corresponds to a different nucleotide base; each peak indicates the location of a particular base. The lines are continuous because they are drawn by four colored pens onto a moving strip of paper. *What is the nucleotide sequence for the nucleotides in positions 35–45?*

Figure 8.11 TCATAGCTGTT

New Frontiers 8.2   Using the Human Genome to Detect Disease-Causing Microbes

Could unidentified microorganisms cause chronic autoimmune diseases such as lupus and multiple sclerosis? When new infectious disease agents emerge—as HIV, Ebola virus, hantavirus, and West Nile virus have in recent decades—how can we identify the microbes? Scientists have recently developed an intriguing new approach to investigating these questions: using the human genome.

The Human Genome Project was begun in 1990 with two ambitious goals: identifying all the genes in human DNA, and determining the sequences of the 3 billion base pairs that make up that DNA. In June 2000, in an announcement that made headlines worldwide, scientists proclaimed they had completed a working draft of the human genome. This was followed in February 2001 by reports containing the revelation that humans likely have only 26,000 to 40,000 genes—only two or three times the number possessed by the fruit fly *Drosophila melanogaster,* which has about 13,000 genes. In comparison, pathogenic *E. coli* has 5,416 genes.

Equipped with DNA sequence data compiled by the Human Genome Project, scientists at the Dana-Farber Cancer Institute have developed a technique called "computational subtraction," in which a portion of DNA from a diseased human tissue sample is sequenced and then compared against the sequence of DNA in the entire human genome. The matches are then eliminated, and what is "left over" is DNA that can then be studied to determine if it is that of a novel microorganism.

Scientists tested this technique by using a tissue sample known to contain a type of cancer-causing human papillomavirus. The method worked—after the matches were eliminated, what was "left over" contained DNA known to be that of the papillomavirus. The hope is that using "computational subtraction" scientists might detect the DNA of either an as-yet-unidentified microbial agent responsible for mysterious chronic illnesses such as lupus, or of newly emerging viruses.

Such studies may lead to methods of reversing genetic damage in cancer patients undergoing radiation therapy. Researchers are also investigating the genetic basis of the enzymes of psychrophiles, which are microorganisms that thrive at temperatures below 20°C. Such enzymes have potential applications in food processing and in the manufacture of drugs.

Table 8.2 summarizes the tools and techniques of recombinant DNA technology.

Pharmaceutical and Therapeutic Applications

Learning Objectives

✓ Describe six potential medical applications of recombinant DNA technology.
✓ Describe the steps and uses of genetic fingerprinting.
✓ Define gene therapy.

Researchers now supplement traditional biotechnology with recombinant DNA technology to produce a variety of pharmaceutical and therapeutic substances, and to perform a host of medically important tasks. Next we explore the use of recombinant DNA technology to synthesize selected proteins, produce vaccines, screen for genetic diseases, match DNA specimens to the organisms from which they came, treat genetic illnesses, and aid in organ transplantation.

Protein Synthesis

Scientists have inserted synthetic genes for insulin, for interferon (a natural antiviral chemical), and for other proteins into bacteria and yeast cells so that the cells synthesize these proteins in vast quantities. In the past, such proteins were isolated from donated blood or from animals—labor-intensive processes that carry the risk of inducing allergies or of transferring pathogens such as hepatitis B and HIV.

"Genetically engineered" proteins are safer and less expensive than their naturally occurring counterparts.

Vaccines

Vaccines contain *antigens*—foreign substances such as weakened bacteria, viruses, and toxins that stimulate the body's immune system to respond to and subsequently remember these foreign materials. In effect, a vaccine primes the immune system to respond quickly and effectively when confronted with pathogens and their toxins. However, the use of some vaccines entails a risk—they may cause the disease they are designed to prevent.

Scientists now use recombinant DNA technology to produce safer vaccines. Once they have inserted the gene that codes for a pathogen's antigenic polypeptides into a vector, they can inject the recombinant vector or the proteins it produces into a patient. Thus the patient's immune system is exposed to a so-called subunit of the pathogen—one of the pathogen's antigens—but not to the pathogen itself. Such **subunit vaccines** are especially useful in safely protecting against pathogens that either cannot be cultured or cause incurable fatal diseases. Hepatitis B vaccine is an example of a successful subunit vaccine. Scientists are also pursuing subunit vaccines against HIV.

A promising future approach to vaccination involves introducing genes coding for antigenic proteins of pathogens into common fruits or vegetables such as bananas or beans. The immune systems of people eating such altered produce would be exposed to the pathogen's antigens and theoretically would develop immunologic memory against the pathogen. Such a vaccine would have the advantages of being painless and easy to administer, and vaccination would not require a visit to a health care provider. **New Frontiers 8.3** focuses on such a vaccine that scientists have developed for cattle to protect them against a disease commonly known as "shipping fever."

Table 8.2   Tools and Techniques of Recombinant DNA Technology

Tool or Technique	Description	Potential Application
Mutagen	Chemical or physical agent that creates mutations	Creating novel genotypes and phenotypes
Reverse transcriptase	Enzyme from RNA retrovirus that synthesizes cDNA from an RNA template	Synthesizing a gene using an mRNA template
Synthetic nucleic acid	DNA molecule prepared *in vitro*	Creating DNA probes to localize genes within a genome
Restriction enzyme	Bacterial enzyme that cleaves DNA at specific sites	Creating recombinant DNA by joining fragments
Vector	Transposon, plasmid, or virus that carries DNA into cells	Altering genome of a cell
Gene library	Collection of cells or viruses, each of which carries a portion of a given organism's genome	Providing a ready source of genetic material
Polymerase chain reaction (PCR)	Produces multiple copies of a DNA molecule	Multiplying DNA for various applications
Gel electrophoresis	Uses electrical charge to separate molecules according to their size	Separating DNA fragments for the Southern blot
Electroporation	Uses electrical current to make cells competent	Inserting a novel gene into a cell
Protoplast fusion	Fuses two cells to create recombinants	Inserting a novel gene into a cell
Gene gun	Blasts genes into target cells	Inserting a novel gene into a cell
Microinjection	Uses micropipette to inject genes into cells	Inserting a novel gene into a cell
Nucleic acid probes	RNA or DNA molecules labeled with radioactive or fluorescent tags	Localizing specific genes in a Southern blot
Southern blot	Localizes specific DNA sequences on a stable membrane	Identifying a strain of pathogen
Genome mapping	Uses restriction enzymes to locate relative positions of restriction sites	Locating genes in an organism's genome
DNA sequencing	Determines the sequence of nucleotide bases in DNA	Comparing genomes of organisms

New Frontiers 8.3   Vaccines on the Menu

Wouldn't it be great if instead of receiving painful flu shots we could be vaccinated simply by eating an antigen-laced chocolate bar? We aren't quite there yet, but scientists are making progress in developing genetically modified oral vaccines.

In one case, researchers are currently testing a plant-based oral vaccine to protect cattle against a disease called pneumonic pasteurellosis, commonly known as "shipping fever" because cattle experience this stress-induced disease while being shipped from cow/calf operations to feedlots. During this respiratory illness, caused by the bacterium *Mannheimia* (formerly *Pasteurella*) *haemolytica,* diseased cattle experience decreased appetite, fever, and nasal discharge. An injectable vaccine against the disease is available, but it is costly and labor intensive, and the injection is so stressful for cattle that it may actually contribute to the problem.

Enter Reggie Lo and his colleagues at the University of Guelph in Canada, who have successfully inserted a particular gene from *M. haemolytica* into white clover, so that the plant expresses one of the bacterium's proteins. The idea is that when cattle are fed the geneti-

cally altered clover, the bacterial protein acts as an antigen, triggering a protective immune response against *M. haemolytica.* This vaccine is easy to prepare, requires minimal processing, and remains stable in dried clover hay for at least several days. It also takes advantage of the fact that cattle regularly regurgitate their food and chew it repeatedly—allowing immune cells in their digestive tract to come into maximum contact with the protein. To date, the vaccine has worked successfully in laboratory animals; field trials with cattle began in 2002.

White clover

Table 8.3   Some Products of Recombinant DNA Technology Used in Medicine

Product	Modified Cell	Uses of Product
Interferons	*Escherichia coli, Saccharomyces cerevisiae*	To treat cancer, multiple sclerosis, chronic granulomatous disease, hepatitis, and warts
Interleukins	*E. coli*	To enhance immunity
Tumor necrosis factor	*E. coli*	In cancer therapy
Erythropoietin	Mammalian cell culture	To stimulate red blood cell formation; to treat anemia
Tissue plasminogen activating factor	Mammalian cell culture	To dissolve blood clots
Human insulin	*E. coli*	For diabetes therapy
Taxol	*E. coli*	In ovarian cancer therapy
Factor VIII	Mammalian cell culture	In hemophilia therapy
Macrophage colony stimulating factor	*E. coli, S. cerevisiae*	To stimulate bone marrow to produce more white blood cells; to counteract side effects of cancer treatment
Relaxin	*E. coli*	To ease childbirth
Human growth hormone	*E. coli*	To correct childhood deficiency of growth hormone
Hepatitis B vaccine	Carried on a plasmid of *S. cerevisiae*	To stimulate immunity against hepatitis B virus

Another type of vaccination involves producing a recombinant plasmid carrying a gene from a pathogen and injecting the plasmid into a human, whose body then synthesizes polypeptides characteristic of the pathogen. The polypeptides stimulate immunologic memory within the human body, readying it to mount a vigorous immune response and prevent infection should it subsequently be exposed to the real pathogen. Clinical trials of such a vaccine against malaria have shown some promise.

Table 8.3 lists some products of recombinant DNA technology with medical applications.

Genetic Screening

Genetic mutations cause some diseases such as inherited forms of breast cancer and Huntington's disease. Laboratory technicians use the Southern blot procedure to screen patients, prospective parents, and fetuses for such mutant genes. This procedure, called *genetic screening,* can also identify viral DNA sequences in a patient's blood or other tissues. For instance, genetic screening can identify HIV in a patient's cells even before the patient shows any other sign of infection.

DNA Fingerprinting

Medical laboratory technicians and forensic investigators use gel electrophoresis and Southern blotting for so-called **genetic fingerprinting** or **DNA fingerprinting**—identifying individuals or organisms by their unique DNA sequences.

DNA fingerprinting involves procuring a sample of DNA, making multiple copies of it via PCR, cutting the copies with restriction enzymes, and separating the fragments by gel electrophoresis to produce a unique pattern. The process is analogous to standard fingerprinting in that the pattern resulting from a particular DNA sample is unique, and it must be compared to patterns produced from other DNA molecules **(Figure 8.12)**, much like a standard fingerprint must be compared to known fingerprints. For example, the patterns from DNA collected at a crime scene either matches or does not match a suspect's or victim's DNA; or the pattern from an environmental sample matches or does not match patterns from known organisms. Genetic fingerprinting is used to determine paternity; to connect blood, semen, or even single skin cells to a particular crime suspect; to identify badly damaged human remains; and to identify pathogens.

Gene Therapy

An exciting use of recombinant DNA technology is **gene therapy,** in which missing or defective genes are replaced with normal copies. Theoretically, a physician could remove a few genetically defective cells—for example, cells that produce a defective protein—from a patient, insert normal genes, and replace the cells into the patient—curing the disease. Alternatively, plasmid or viral vectors could deliver genes directly to target cells within a patient.

Unfortunately, gene therapy has proven difficult in practice because of unexpected results. Specifically, some patients' immune systems react out of control to the presence of vectors, resulting in the death of the patient. Nevertheless, doctors have successfully treated patients for severe combined immunodeficiency disease. Other diseases that may respond well to gene therapy are cystic fibrosis, sickle cell anemia, and some types of hemophilia and diabetes.

Xenotransplants

Xenotransplants[3] are animal cells, tissues, or organs introduced into the human body. For years physicians have performed xenotransplants such as using valves from pig hearts to repair severely damaged human hearts; however, recombinant DNA technology may expand the possibilities. It is theoretically feasible to insert functional human genes into animals to direct them to produce organs and tissues for transplantation into humans. For example, scientists could induce pigs to produce human-like cytoplasmic membrane proteins so that the pigs' organs would not be rejected as foreign tissue by a transplant recipient.

Agricultural Applications

Learning Objective

✓ Identify five agricultural applications of recombinant DNA technology.

Recombinant DNA technology has been applied to the realm of agriculture to produce **transgenic** organisms—recombinant plants and animals that have been altered for specific purposes by the addition of genes from other organisms. The purposes for which transgenic organisms have been produced are many and varied and include herbicide resistance, tolerance to salty soils, resistance to freezing and pests, and improvements in nutritional value and yield.

Herbicide Resistance

The biodegradable herbicide *glyphosate* (Roundup™) normally kills all plants—weeds and crops alike—by blocking an enzyme that is essential for the synthesis of several amino acids. After scientists discovered and isolated a *Salmonella* bacterial gene that conveys resistance to glyphosate, they produced transgenic crop plants containing the gene. As a result of this application of recombinant DNA technology, farmers can now apply glyphosate to a field of transgenic plants to kill weeds without damaging the crop. An added benefit is that farmers do not need to till the soil to suppress weeds during the growing season, reducing soil erosion. In 2001, over 50% of the soybeans grown in the United States were genetically modified in this manner to be "Roundup™ ready."

Salt Tolerance

Years of irrigation have resulted in excessive salt buildup in farmland throughout the world, rendering the land useless for farming. Though salt-tolerant plants that can grow under these conditions exist, they are not edible.

Scientists have now successfully removed the gene for salt tolerance and inserted it into tomato and canola plants to create food crops that can grow in soil so salty it would

▲ *Figure 8.12*

DNA fingerprinting. Shown here is a partial X-ray of bands of DNA from four family members: a mother (M), father (F), and two children (C's). Both children share some bands with each parent, proving they are indeed related. A similar process can be used to compare DNA bands from microbial specimens in order to identify a particular specimen.

poison normal crops. Not only do such transgenic plants survive and produce fruit; they also remove salt from the soil, restoring it and making it suitable to grow unmodified crops as well. Researchers are now attempting to insert the gene for salt tolerance into wheat and corn.

Freeze Resistance

Ice crystals more readily form when bacterial proteins (from natural bacteria present in a field) are available as crystallization nuclei. Scientists have modified strains of the bacterium *Pseudomonas* (soo-dō-mō′nas) with a gene for a polypeptide that prevents ice crystals from forming. Crops sprayed with genetically modified bacteria can tolerate mild freezes, so the farmers no longer lose their crops to unseasonable cold snaps.

Pest Resistance

The bacterium *Bacillus thuringiensis* (ba-sil′ŭs thur-in-jē-en′sis) produces a protein that, when modified by enzymes in the intestinal tracts of insects, becomes *Bt-toxin*. Bt-toxin binds to receptors lining the insect's digestive tract and causes the tissue to dissolve. Unlike many insecticides, Bt-toxin is naturally occurring, harmful only to insects, and biodegradable. Farmers, particularly organic growers, have used Bt-toxin for over 30 years to reduce insect damage to their crops.

[3]From Greek *xenos,* meaning stranger.

Highlight 8.1   Does Genetically Modified Corn Harm Monarch Butterflies?

Controversy often surrounds genetically-modified foods. For example, can pollen from genetically modified corn plants harm monarch caterpillars feeding on nearby milkweed leaves? A 1999 Cornell University study found that under laboratory conditions, the dusting of milkweed leaves with pollen from corn modified with the Bt-toxin gene did indeed stunt or kill monarch caterpillars feeding on the leaves. However, a 2001 study found that in and around actual Bt-toxin cornfields, the amount of pollen present on milkweed leaves is unlikely to reach the levels of pollen used in the Cornell study. Moreover, some researchers argue that butterflies may actually be safer in fields of so-called Bt-corn because they are then not subjected to the chemical pesticides otherwise used. Further research is needed to get the complete story.

Monarch caterpillar on milkweed leaf dusted with Bt-pollen.

▲ *Figure 8.13*

The cotton boll on the left was genetically modified to resist boll weevils; the boll on the right was not.

Now, the gene for Bt-toxin has been inserted into a variety of crop plants, including potatoes, cotton, and corn, so that they produce Bt-toxin for themselves. Pests feeding on such plants are killed **(Figure 8.13)** while humans and most animals that eat them are unharmed. See **Highlight 8.1** for a discussion of the effects of Bt-toxin on monarch butterflies.

Improvements in Nutritional Value and Yield

Genetic researchers have increased crop and animal yields in several ways. For example, MacGregor tomatoes remain firm after harvest because the gene for the enzyme that breaks down pectin has been suppressed. This allows farmers to let the tomatoes ripen on the vine, and it increases the tomatoes' shelf life.

Another example involves bovine growth hormone (BGH), which when injected into cattle enables them to more rapidly gain weight, have meat with a reduced fat content, and produce 10% more milk. Though BGH can be derived from animal tissue, it is more economical to insert the BGH gene into bacteria so that they produce the hormone, which is then purified and injected into farm and ranch animals.

In yet another application, scientists have improved the nutritional value of rice by adding a gene for β-carotene, which is a precursor to vitamin A. Vitamin A is required for human embryonic development and for vision in adults, and it is an important antioxidant that plays a role in ameliorating cancer and atherosclerosis (hardening of arteries).

Recombinant DNA technology has progressed to the point that scientists are now considering transplanting genes coding for entire metabolic pathways, rather than merely genes encoding simple proteins. For instance, researchers are attempting to transfer into corn and rice plants all the genes for the enzymes and structures some bacteria use to convert atmospheric nitrogen into nitrogenous fertilizer, in effect allowing the recombinant plants to produce their own fertilizer.

Recombinant DNA tools and techniques allow scientists to examine, compare, and manipulate the genomes of microorganisms, plants, animals, and humans for a variety of purposes including gene mapping and forensic, medical, and agricultural applications. However, as with many scientific advances, concerns arise about the ethics and safety of genetic manipulations. The next section deals with these issues.

The Ethics and Safety of Recombinant DNA Technology

Learning Objective

✓ Discuss the pros and cons concerning the safety and ethics of recombinant DNA technology.

The procedures of recombinant DNA technology provide the opportunity to transfer genes among unrelated organisms, even among organisms in different kingdoms, but how safe and ethical are they? "Super-broccoli," "frankenfood," "biological Russian roulette," and "designer humans" are some of the terms opponents use to denigrate gene therapy and transgenic agricultural products. Some oppenents question the ethics of raising genetically altered animals solely for producing tissues and organs for human use. They contend that this exemplifies a supremacist view—the view that humans are of greater intrinsic value than animals. Other critics of transgenic crops and animals correctly state that the long-term effects of transgenic manipulations are unknown, and that unforeseen problems arise from every new technology and procedure. Recombinant DNA technology may burden society with complex, as yet unforeseen regulatory, administrative, financial, legal, social, and environmental problems.

Critics also argue that natural genetic transfer by processes such as transformation and transduction could deliver genes from transgenic plants and animals into other organisms. For example, if a herbicide-resistant plant cross-pollinates with a related weed species, we might be cursed with a weed that is nearly impossible to kill. Opponents further express concern that transgenic organisms could trigger allergies or cause harmless organisms to become pathogenic. Therefore, some opponents of recombinant DNA technology desire a ban on all genetically modified agricultural products.

The U.S. National Academy of Sciences, the U.S. National Research Council, and 81 research projects conducted between 1985 and 2001 by the European Union have not revealed any risks to human health or the environment from genetically modified agricultural products beyond the usual uncertainties inherent in conventional plant breeding. In fact, the European Union concluded in 2001 that "the use of more precise technology and the greater regulatory scrutiny probably make them [genetically modified foods] even safer than conventional plants and foods."

As the debate continues, governments continue to impose standards on laboratories involved in recombinant DNA technology. These are intended to prevent the accidental release of altered organisms or exposure of laboratory workers to potential dangers. Additionally, genetic researchers often design organisms to lack a vital gene so that they cannot survive for long outside of a laboratory.

Unfortunately, biologists can apply the procedures used to create beneficial crops and animals to create biological weapons that are more infective and more resistant to treatment than their natural counterparts are, though international treaties prohibit the development of biological weapons. Nevertheless, *B. anthracis* spores were used in bioterrorist attacks in the United States in 2001, though, thankfully, the strain utilized was not genetically altered to realize its deadliest potential.

Emergent recombinant DNA technologies raise numerous other ethical issues. Should people be routinely screened for diseases that are untreatable or fatal? Who should pay for these procedures: individuals, employers, prospective employers, insurance companies, HMOs, government agencies? What rights do individuals have to genetic privacy? If entities other than individuals pay the costs involved in genetic screening, should those entities have access to *all* the genetic information that results? Should businesses be allowed to have patents on and make profits from any living organisms they have genetically altered? Should governments be allowed to require genetic screening and then force genetic manipulations on individuals to correct so-called "genetic abnormalities" that some claim are the bases of criminality, manic depression, risk-taking behavior, and alcoholism? Should anyone other than individuals—HMOs, physicians, or the government—demand genetic screening and then refuse to provide services related to the birth or care of supposedly "defective" children?

We as a society will have to confront these and other ethical considerations as the genomic revolution continues to affect people's lives in many unpredictable ways.

CHAPTER SUMMARY

The Role of Recombinant DNA Technology in Biotechnology (p. 242)

1. **Biotechnology** is the use of microorganisms to make useful products. Historically these have included bread, wine, beer, and cheese.

2. **Recombinant DNA technology** is a new type of biotechnology in which scientists change the genotypes and phenotypes of organisms to benefit humans.

The Tools of Recombinant DNA Technology (pp. 242–247)

1. The tools of recombinant DNA technology include mutagens, reverse transcriptase, synthetic nucleic acids, restriction enzymes, vectors, and gene libraries.

2. **Mutagens** are chemical and physical agents used to create changes in a microbe's genome to effect desired changes in the microbe's phenotype.

3. The enzyme **reverse transcriptase** transcribes DNA from an RNA template; genetic researchers use reverse transcriptase to make complementary DNA (**cDNA).**

4. Scientists used synthetic nucleic acids to elucidate the genetic code, and they now use them to create genes for specific proteins and to synthesize DNA and RNA **probes** labeled with radioactive or fluorescent markers.

5. **Restriction enzymes** cut DNA at specific (usually palindromic) nucleotide sequences and are used to produce recombinant DNA molecules.

6. In recombinant DNA technology, a **vector** is a small DNA molecule (such as a viral genome, transposon, or plasmid) that carries a particular gene and a recognizable genetic marker into a cell.

7. A **gene library** is a collection of bacterial or phage clones, each of which carries a fragment (typically a single gene) of an organism's genome.

Techniques of Recombinant DNA Technology (pp. 247–252)

1. The **polymerase chain reaction (PCR)** allows researchers to replicate molecules of DNA rapidly.

2. **Gel electrophoresis** is a technique for separating molecules (including fragments of nucleic acids) by size, shape, and electrical charge.

3. The **Southern blot** technique allows researchers to stabilize specific DNA sequences from an electrophoresis gel and then localize them using DNA dyes or probes.

4. Geneticists artificially insert DNA into cells by electroporation, protoplast fusion, and injection.

Applications of Recombinant DNA Technology (pp. 252–259)

1. **Genomics** is the sequencing **(genetic mapping),** analysis, and comparison of genomes. Genetic sequencing has been speeded up by an automated machine that distinguishes among flourescent dyes attached to each type of nucleotide base.

2. Scientists synthesize **subunit vaccines** by introducing genes for a pathogen's polypeptides into cells or viruses. When the cells, the viruses, or the polypeptides they produce are injected into a human, the body's immune system is exposed to and reacts against relatively harmless antigens instead of the potentially harmful pathogen.

3. **Genetic screening** can detect certain inherited diseases before a patient shows any sign of the disease.

4. **Genetic fingerprinting (DNA fingerprinting),** which identifies unique sequences of DNA, is used in paternity investigations, crime scene forensics, diagnostic microbiology, and epidemiology.

5. **Gene therapy** cures various diseases by replacing defective genes with normal genes.

6. In **xenotransplants** involving recombinant DNA technology, human genes are inserted into animals to produce cells, tissues, or organs that are then introduced into the human body.

7. **Transgenic** plants and animals have been genetically altered by the inclusion of genes from other organisms.

8. Agricultural uses of recombinant DNA technology include advances in herbicide resistance, salt tolerance, freeze resistance, and pest resistance, as well as improvements in nutritional value and yield.

9. Among the ethical and safety issues surrounding recombinant DNA technology are concerns over the accidental release of altered organisms into the environment, the ethics of altering animals for human use, and the potential for creating genetically modified biological weapons.

QUESTIONS FOR REVIEW   *(Answers to multiple choice are on the web, along with additional review questions. Visit www.microbiologyplace.com.)*

Multiple Choice

1. Which of the following statements is true concerning recombinant DNA technology?
   a. It will replace biotechnology in the future.
   b. It is a single technique for genetic manipulation.
   c. It is useful in manipulating genotypes but not phenotypes.
   d. It involves modification of an organism's genome.

2. A gene synthesized from an RNA template is
   a. reverse transcriptase.
   b. complementary DNA.
   c. recombinant DNA.
   d. probe DNA.

3. After scientists exposed cultures of *Penicillium* to agents X, Y, and Z, they examined the type and amount of penicillin produced by the altered fungi to find the one that is most effective. Agents X, Y, and Z were probably
   a. recombinant cells.
   b. competent.
   c. mutagens.
   d. phages.

4. Which of the following is *false* concerning vectors in recombinant DNA technology?
   a. Vectors are small enough to manipulate outside a cell.
   b. Vectors contain a recognizable genetic marker.
   c. Vectors survive inside cells.
   d. Vectors must contain genes for self-replication.

5. Which recombinant DNA technique is used to replicate copies of a DNA molecule?
   a. PCR
   b. gel electrophoresis

c. electroporation
d. Southern blotting

6. Which of the following techniques is used regularly in the study of genomics?
   a. Clones are selected using a vector with two genetic markers.
   b. Genes are inserted to produce an antigenic protein from a pathogen.
   c. Fluorescent nucleotide bases are sequenced.
   d. Defective organs are replaced with those made in animal hosts.

7. Which application of recombinant DNA technology involves the production of a distinct pattern of DNA fragments on a gel?
   a. genetic fingerprinting
   b. gene therapy
   c. genetic screening
   d. protein synthesis

True/False

Indicate which of the following are true and which are false. Rewrite any false statements to make them true.

___ 1. Restriction enzymes inhibit the movement of DNA.

___ 2. Restriction enzymes act at specific nucleotide sequences within a double-stranded DNA molecule.

___ 3. A thermocycler separates molecules based on their size, shape, and electrical charge.

___ 4. Protoplast fusion is often used in the genetic modification of plants.

Short Answer

1. Describe three artificial methods of introducing DNA into cells.

2. Why is cloning a practical technique for medical researchers?

3. Describe three ways scientists use synthetic nucleic acids.

4. Describe a gene library and its usefulness.

CRITICAL THINKING

1. Examine the restriction sites listed in Table 8.1. Which restriction enzymes produce restriction fragments with sticky ends? Which produce fragments with blunt ends?

2. A cancer-inducing virus, HTLV-1, inserts itself into a human chromosome, where it remains. How can a laboratory technician prove that a patient is infected with HTLV-1?

3. A thermocycler uses DNA polymerase from hyperthermophilic prokaryotes, but it cannot use DNA polymerase derived from *Escherichia coli*. Why not?

CHAPTER 9

Controlling Microbial Growth in the Environment

CHAPTER OUTLINE

Basic Principles of Microbial Control
    Terminology of Microbial Control
    Microbial Death Rates
    Action of Antimicrobial Agents

The Selection of Microbial Control Methods
    Factors Affecting the Efficacy of Antimicrobial Methods
    Methods for Evaluating Disinfectants and Antiseptics

Physical Methods of Microbial Control
    Heat-Related Methods
    Refrigeration and Freezing
    Desiccation and Lyophilization
    Filtration
    Osmotic Pressure
    Radiation

Chemical Methods of Microbial Control
    Phenol and Phenolics
    Alcohols
    Halogens
    Oxidizing Agents
    Surfactants
    Heavy Metals
    Aldehydes
    Gaseous Agents
    Antimicrobics
    Development of Resistant Microbes

In response to the anthrax attacks in the mail during the fall of 2001, the United States Postal Service undertook an extraordinary safety measure: It began irradiating mail at selected postal facilities.

Irradiation systems use high-voltage electron beams to kill bacteria—including endospores of *Bacillus anthracis,* the causative agent of anthrax—that may be present in pieces of mail. The beams of electrons work by disrupting bacterial DNA. The same technology is commonly used to sterilize medical equipment and to irradiate fresh produce; it has proven effective against *Bacillus, Salmonella, Escherichia*, and other bacterial contaminants. Contrary to a common misconception, irradiation does not cause mail or food to become radioactive.

In this chapter we will learn not only about irradiation, but also the wide variety of other methods used to control microorganisms in our environment.

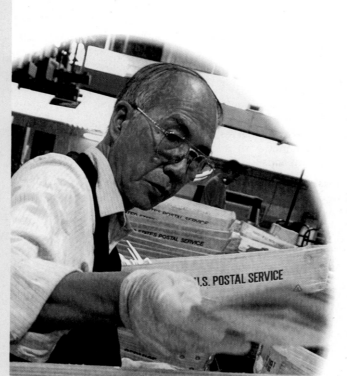

Following the anthrax attacks, postal service employees began wearing protective gloves and mail was irradiated at high-risk facilities.

MicroPrep Pre-Test: *Take the pre-test for this chapter on the web.* Visit ***www.microbiologyplace.com***.

The control of microbes in health care facilities, in laboratories, and at home is a significant and practical aspect of microbiology. In this chapter we will study the terminology and principles of microbial control, consider the factors affecting the efficacy of microbial control, and examine the control of microorganisms and viruses by various chemical and physical means. One important aspect of microbial control—the use of antimicrobial drugs to assist the body's defenses against pathogens—will be considered in Chapter 10.

Basic Principles of Microbial Control

Scientists, health care professionals, researchers, and government workers should use precise terminology in reference to microbial control in the environment. In the following sections we will discuss the terminology of microbial control, examine the concept of microbial death rates, and discuss the action of antimicrobial agents.

Terminology of Microbial Control

Learning Objectives

✓ Contrast sterilization, disinfection, and antisepsis, and describe their practical uses.

✓ Contrast the terms *degerming, sanitization,* and *pasteurization.*

✓ Compare the effects of "-static" versus "-cidal" control agents on microbial growth.

It is important for microbiologists, health care workers, and others to use correct terminology for describing microbial control. While many of these terms are familiar to the general public, they are often misused.

In its strictest sense, **sterilization** refers to the removal or destruction of *all* microbes, including viruses and bacterial endospores, in or on an object. (The term does not apply to *prions,* which are infectious proteins, because standard sterilizing techniques do not destroy them; in fact, scientists do not know the minimum treatment required to destroy prions.)

In practical terms, sterilization indicates only the eradication of harmful microorganisms and viruses; some innocuous microbes may still be present and viable in an environment that is considered sterile. For instance, *commercial sterilization* of canned food does not kill all hyperthermophilic microbes, but they cannot grow and spoil the food at ambient temperatures, so they are of no practical concern. Likewise, some hyperthermophiles may survive sterilization by laboratory methods (discussed shortly), but they are of no practical concern to technicians because they cannot grow or reproduce at normal laboratory temperatures.

The term **aseptic**[1] (ā-sep'tik) describes an environment or procedure that is free of contamination by *pathogens.* For example, vegetables and fruit juices are available in aseptic packaging, and surgeons and laboratory technicians use aseptic techniques to avoid contaminating a surgical field or laboratory equipment.

Disinfection[2] refers to the use of physical or chemical agents known as **disinfectants,** including ultraviolet light, heat, alcohol, and bleach, to inhibit or destroy microorganisms, especially pathogens. Unlike sterilization, disinfection does not guarantee that all pathogens are eliminated; indeed, disinfectants alone cannot inhibit endospores or some viruses. Further, the term *disinfection* is used only when discussing treatment of inanimate objects. When a chemical is used on skin or other tissue, the process is called **antisepsis**[3] (an-tē-sep'sis), and the chemical is called an **antiseptic.** Antiseptics and disinfectants often have the same components, but disinfectants are more concentrated or can be left on a surface for longer periods of time. Of course, some disinfectants, such as steam or concentrated bleach, are not suitable for use as antiseptics.

Degerming is the removal of microbes from a surface by scrubbing, such as when you wash your hands or a nurse prepares an area of skin for an injection. Though chemicals such as soap or alcohol are commonly used during degerming, the action of thoroughly scrubbing the surface is usually more important than the chemical in removing microbes.

Sanitization[4] is the process of disinfecting places and utensils used by the public to reduce the number of pathogenic microbes to meet accepted public health standards. For example, steam, high-pressure hot water, and scrubbing are used to sanitize restaurant utensils and dishes, and chemicals are used to sanitize public toilets. Thus, the only difference between *disinfecting* dishes at home in a dishwasher and *sanitizing* dishes in a restaurant is the arena—private versus public—in which the activity takes place.

Pasteurization[5] is the use of heat to kill pathogens and reduce the number of spoilage microorganisms in food and beverages. Milk, fruit juices, wine, and beer are commonly pasteurized.

So far, we have seen that there are two major types of microbial control—sterilization, which is the elimination of *all* microbes, and disinfection, which is the destruction of vegetative (nonspore) cells and many viruses. Modifications of disinfection include antisepsis, degerming, sanitization, and pasteurization. Some scientists and clinicians apply these terms only to pathogenic microorganisms.

Additionally, scientists and health care professionals use the suffixes *-stasis/-static*[6] to indicate that a chemical or physical agent inhibits microbial metabolism and growth, but doesn't necessarily kill microbes. Thus, refrigeration is bacteriostatic for most bacterial species; it inhibits their growth, but they can resume metabolism when the optimal temperature is restored. By contrast, words ending in *-cide/-cidal*[7] refer to agents that destroy or permanently

---

[1]From Greek *a,* meaning not, and *sepsis,* meaning decay.
[2]From Latin *dis,* meaning reversal, and *inficere,* meaning to corrupt.
[3]From Greek *anti,* meaning against, and *sepsis,* meaning putrefaction.
[4]From Latin *sanitas,* meaning healthy.
[5]Named for Louis Pasteur, inventor of the process.
[6]Greek, meaning to stand—that is, to remain relatively unchanged.
[7]From Latin *cidium,* meaning a slaying.

Table 9.1   Terminology of Microbial Control

Term	Definition	Examples	Comments
Antisepsis	Reduction in the number of microorganisms and viruses, particularly potential pathogens, on living tissue	Iodine; alcohol	Antiseptics are frequently disinfectants whose strength has been reduced to make them safe for living tissues.
Aseptic	Refers to an environment or procedure free of pathogenic contaminants	Preparation of surgical field; hand washing; flame sterilization of laboratory equipment	Scientists, laboratory technicians, and health care workers routinely follow standardized aseptic techniques.
-cide **-cidal**	Suffixes indicating destruction of a type of microbe	Bactericide; fungicide; germicide; virucide	Germicides include ethylene oxide, propylene oxide, and aldehydes.
Degerming	Removal of microbes by mechanical means	Hand washing; alcohol swabbing at site of injection	Chemicals play a secondary role to the mechanical removal of microbes.
Disinfection	Destruction of most microorganisms and viruses on nonliving tissue	Phenolics; alcohols; aldehydes; surfactants	The term is used primarily in relation to pathogens.
Pasteurization	Use of heat to destroy pathogens and reduce the number of spoilage microorganisms in foods and beverages	Pasteurized milk and fruit juices	Heat treatment is brief to reduce alteration of taste and nutrients; microbes still remain and eventually cause spoilage.
Sanitization	Removal of pathogens from objects to meet public health standards	Washing tableware in scalding water in restaurants	Standards of sanitization vary among governmental jurisdictions.
-stasis **-static**	Suffixes indicating inhibition, but not complete destruction, of a type of microbe	Bacteriostatic; fungistatic; virustatic	Germistatic agents include some chemicals, refrigeration, and freezing.
Sterilization	Destruction of all microorganisms and viruses in or on an object	Preparation of microbiological culture media and canned food	Typically achieved by steam under pressure, incineration, or ethylene oxide gas

inactivate a particular type of microbe; *virucides* inactivate viruses, *bactericides* kill bacteria, and *fungicides* kill fungal hyphae, spores, and yeasts. *Germicides* are chemical agents that destroy pathogenic microorganisms in general.

Table 9.1 summarizes the terminology used to describe the control of microbial growth.

CRITICAL THINKING

A student inoculates *Escherichia coli* into two test tubes containing the same sterile liquid medium, except the first tube also contains a drop of a chemical with an antimicrobial effect. After 24 hours of incubation, the first tube remains clear while the second tube becomes cloudy with bacteria. Design an experiment to determine if this amount of the antimicrobial chemical is *bacteriostatic* or *bactericidal* against *E. coli*.

Microbial Death Rates

Learning Objective

✓ Define microbial death rate, and describe its significance in microbial control.

Scientists define **microbial death** as the permanent loss of reproductive ability under ideal environmental conditions. One technique for evaluating the efficacy of an antimicrobial agent is to calculate the **microbial death rate,** which is usually found

to be constant over time for any particular microorganism under a particular set of conditions **(Figure 9.1).** Suppose for example that a scientist treats a broth containing 1 billion (10^9) microbes with an agent that kills 90% of them in 1 minute. The most susceptible cells die first, leaving 100 million (10^8) hardier cells after the first minute. After another minute of treatment, another 90% die, leaving 10 million (10^7) cells that have even greater resistance to and require longer exposure to the agent before they die. Notice that in this case, each full minute decreases the number of living cells tenfold (that is, by one logarithm). The broth will be sterile when all the cells are dead. When these results are plotted on a semilogarithmic graph—in which the *y*-axis is logarithmic, and the *x*-axis is arithmetic—the plot of microbial death rate is a straight line; that is, the microbial death rate is constant.

Action of Antimicrobial Agents

Learning Objective

✓ Describe how antimicrobial agents act against cell walls, cytoplasmic membranes, proteins, and nucleic acids.

There are many types of chemical and physical microbial controls, but their modes of action fall into two basic categories: those that disrupt the integrity of cells by adversely altering their cell walls or cytoplasmic membranes, and

▲ *Figure 9.1*

A plot of microbial death rate. Microbicidal agents do not simultaneously kill all cells. Rather they kill a constant percentage of cells over time—in this case 90% per minute. On this semilogarithmic graph, a constant death rate is indicated by a straight line. *How many minutes are required for sterilization in this case?*

Figure 9.1 For this microbe under these conditions, this microbicidal agent requires 9 minutes to achieve sterilization.

those that interrupt cellular metabolism and reproduction by interfering with the structures of proteins and nucleic acids.

Alteration of Cell Walls and Membranes

As we have seen, a cell wall maintains cellular integrity by counteracting the effects of osmosis when the cell is in a hypotonic solution. If the wall is disrupted by physical or chemical agents, it no longer prevents the cell from bursting as water moves into it by osmosis.

Beneath a cell wall, the cytoplasmic membrane essentially acts as a bag that contains the cytoplasm and controls the passage of chemicals into and out of the cell. Extensive damage to a membrane's proteins or phospholipids by any physical or chemical agent allows the cellular contents to leak out—which, if not immediately repaired, causes death.

In enveloped viruses, the envelope is a membrane composed of proteins and phospholipids that is responsible for the attachment of the virus to its target cell; thus damage to the envelope by physical or chemical agents fatally interrupts viral replication. The lack of an envelope in nonenveloped viruses accounts for their greater tolerance of harsh environmental conditions, including antimicrobial agents.

Damage to Proteins and Nucleic Acids

Proteins in cells act as regulatory compounds in cellular metabolism, function as enzymes in most metabolic reactions, and form structural components in membranes and cytoplasm. As we have seen, a protein's function depends on an exact three-dimensional shape, which is maintained by hydrogen and disulfide bonds between amino acids. When these bonds are broken by extreme heat or certain chemicals, the protein's shape changes (see Figure 5.9). Such *denatured* proteins cease to function, bringing about cellular death.

Chemicals, radiation, and heat can also alter and even destroy nucleic acids. Given that the genes of a cell or virus are composed of nucleic acids, disruption of these molecules can produce fatal mutations. Additionally, that portion of a ribosome that actually catalyses the synthesis of proteins is a *ribozyme*—that is, an enzymatic RNA molecule—so physical or chemical agents that interfere with nucleic acids also stop protein synthesis.

CRITICAL THINKING

Would you expect Gram-negative bacteria or Gram-positive bacteria to be more susceptible to antimicrobial chemicals that act against cell walls? Explain your answer, which you should base solely upon the nature of the cell's walls (see Figure 3.13).

Scientists and health care workers have at their disposal many chemical and physical agents to control microbial growth and activity. In the next section we will consider the factors and conditions that must be considered in choosing a particular control method, as well as some ways for evaluating a method's effectiveness.

The Selection of Microbial Control Methods

Ideally, agents used for the control of microbes should be inexpensive, fast-acting, and stable during storage. Further, a perfect agent would control the growth and reproduction of every type of microbe while being harmless to humans, animals, and objects. Unfortunately, such ideal products and procedures do not exist—every agent and procedure has limitations and disadvantages. In the next two subsections we consider the factors that affect the efficacy of antimicrobial methods and some ways to evaluate disinfectants.

Factors Affecting the Efficacy of Antimicrobial Methods

Learning Objective

✓ List factors to consider in selecting a microbial control method.
✓ Identify the three most-resistant groups of microbes, and explain why they are resistant to many antimicrobial agents.
✓ Discuss environmental conditions that can affect the effectiveness of antimicrobial agents.

In each situation, microbiologists, laboratory personnel, and medical staff must consider at least three factors: the nature of the sites to be treated, the degree of susceptibility of the microbes involved, and the environmental conditions that pertain.

Site to Be Treated

In many cases, the choice of an antimicrobial method depends on the nature of the site to be treated. For example, harsh chemicals and extreme heat cannot be used on humans, animals, and fragile objects such as artificial heart valves and plastic utensils. Moreover, when performing medical procedures, medical personnel must choose a method and level of microbial control based on the site of the procedure, because the site greatly affects the potential for subsequent infection. For example, the use of medical instruments that penetrate the outer defenses of the body such as needles and scalpels carries a greater potential for infection, so they must be sterilized, whereas items that contact only the surface of a mucous membrane or the skin may be disinfected. In the latter case sterilization is required only if the patient is immunocompromised.

Relative Susceptibility of Microorganisms

Though microbial death rate is usually constant for a particular agent acting against a single microbe, death rates do vary—sometimes dramatically—among microorganisms and viruses. Generally these microbes fall along a continuum from most susceptible to most resistant to antimicrobial agents. For example, *enveloped* viruses, such as HIV, are more susceptible to antimicrobial agents and heat than are *nonenveloped viruses*, such as polio virus (see Figures 13.1 and 13.7), because viral envelopes are more easily disrupted than the protein coats of nonenveloped viruses. The relative susceptibility of microbes to antimicrobial agents is illustrated in **Figure 9.2.**

Often, scientists and medical personnel select a method to kill the hardiest microorganisms present, assuming that such a treatment will kill more fragile microbes as well. The most resistant microbes include the following:

- Bacterial endospores. The endospores of *Bacillus* (ba-sil′ŭs) and *Clostridium* (klos-trid′ē-ŭm) are the most resistant forms of life. They can survive environmental extremes of temperature and acidity, and many chemical disinfectants. For example, endospores can survive more than 20 years in 70% alcohol, and scientists have recovered viable endospores that were embalmed with Egyptian mummies thousands of years ago.

- Species of *Mycobacterium* (mī′kō-bak-tēr′ē-ŭm). The cell walls of members of this genus, such as *Mycobacterium tuberculosis*, contain large amounts of waxy lipids. The wax allows these bacteria to survive drying and protects them from most water-based chemicals; therefore, medical personnel must use strong disinfectants or heat to treat whatever comes into contact with tuberculosis patients, including utensils, equipment, and rooms.

- Cysts of protozoa. A protozoan cyst's wall prevents entry of most disinfectants, protects against drying, and shields against radiation and heat.

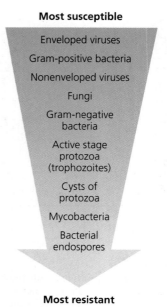

Figure 9.2 ▶
Relative susceptibilities of microbes to antimicrobial agents.

Most susceptible

Enveloped viruses
Gram-positive bacteria
Nonenveloped viruses
Fungi
Gram-negative bacteria
Active stage protozoa (trophozoites)
Cysts of protozoa
Mycobacteria
Bacterial endospores

Most resistant

The effectiveness of germicides can be classified as high, intermediate, or low depending on their effectiveness in inactivating or destroying microorganisms on medical and dental instruments that cannot be sterilized with heat. *High-level germicides* kill all pathogens, including bacterial endospores. Health care professionals use them to sterilize invasive instruments such as catheters, implants, and parts of heart-lung machines. *Intermediate-level germicides* kill fungal spores, protozoan cysts, viruses, and pathogenic bacteria, but not bacterial endospores. They are used to disinfect instruments that come in contact with mucous membranes but are noninvasive, such as respiratory equipment and endoscopes. *Low-level germicides* eliminate vegetative bacteria, fungi, protozoa, and some viruses; they are used to disinfect items that only contact the skin of patients, such as furniture and electrodes.

Environmental Conditions

Temperature and pH affect microbial death rates and the efficacy of antimicrobial methods. Warm disinfectants, for example, generally work better than cool ones because chemicals react faster at higher temperatures **(Figure 9.3)**. Some chemical disinfectants such as sodium hypochlorite ($NaOCl_2$, household chlorine bleach) are more effective at low pH. The antimicrobial effect of heat is also enhanced by acidic conditions.

Organic materials such as fat, feces, vomit, blood, and the intercellular secretions of biofilms interfere with the penetration of heat, chemicals, and some forms of radiation, and in some cases these materials inactivate chemical disinfectants. For this reason, it is important to clean objects be-

▲ **Figure 9.3**

Effect of temperature on the efficacy of an antimicrobial chemical. This semilogarithmic graph shows that the microbial death rate is higher at higher temperatures; to kill the same number of microbes, this disinfectant required only 4 minutes at 45°C, but 12 minutes at 20°C.

fore sterilization or disinfection so that antimicrobial agents can thoroughly contact all the object's surfaces.

Methods for Evaluating Disinfectants and Antiseptics

Learning Objective

✓ Compare and contrast four methods used to measure the effectiveness of disinfectants and antiseptics.

With few exceptions, higher concentrations and fresher solutions of a disinfectant are more effective than more dilute, older solutions. We have also seen that longer exposure times ensure the deaths of more microorganisms. However, anyone using disinfectants must consider whether higher concentrations and longer exposures may damage an object or injure a patient.

Scientists have developed several methods to measure the efficacy of antimicrobial agents. These include the phenol coefficient, the use-dilution test, the disk diffusion method, and the in-use test.

Phenol Coefficient

Recall from Chapter 1 that Lister introduced the use of phenol (also known as carbolic acid) as an antiseptic during surgery in the late 1800s. Since then, researchers have evaluated the efficacy of various disinfectants and antiseptics by calculating a ratio that compares a given agent's ability to control microbes to that of phenol under standardized conditions. This ratio is referred to as the **phenol coefficient.** A phenol

coefficient greater than 1.0 indicates that an agent is more effective than phenol, and the larger the ratio, the greater the effectiveness. For example, *chloramine*, a mixture of chlorine and ammonia, has a phenol coefficient of 133.0 when used against the bacterium *Staphylococcus aureus* (staf'i-lō-kok'us ô'rē-us), and a phenol coefficient of 100.0 when used against *Salmonella typhi* (sal'mŏ-nel'ă tī'fē). This indicates that chloramine is at least 133 times more effective than phenol against *Staphylococcus* but only 100 times more effective against *Salmonella*. Measurement of an agent's phenol coefficient has been replaced by newer methods because scientists have developed disinfectants and antiseptics much more effective than phenol.

CRITICAL THINKING

What is the phenol coefficient of phenol when used against *Staphylococcus*?

Use-Dilution Test

Another method for measuring the efficacy of disinfectants and antiseptics against specific microbes is the **use-dilution test.** In this test, a researcher dips several metal cylinders into broth cultures of bacteria and briefly dries them at 37°C. The bacteria used in the standard test are *Pseudomonas aeruginosa* (soo-dō-mō'nas ā-ru-ji-nō'sa), *Salmonella cholerasuis* (sal'mŏ-nel'ă kol-er-a-su'is), and *S. aureus.* The researcher then immerses each contaminated cylinder into a different dilution of the disinfectants being evaluated. After 10 minutes, each cylinder is removed, rinsed with water to remove excess chemical, and placed into a fresh tube of sterile medium for 48 hours of incubation. The most effective agent is the one that entirely prevents microbial growth at the highest dilution.

The use-dilution test is the current standard test, though it was developed several decades ago, before the appearance of many of today's pathogens, including hepatitis C virus, HIV, and antibiotic-resistant bacteria and protozoa. Moreover, the disinfectants in use at the time were far less powerful than many used today. Some government agencies have expressed concern that the test is neither accurate, reliable, nor relevant; therefore, the American Official Analytical Chemists are developing a new standard procedure.

Disk-Diffusion Method

The **disk-diffusion method,** which is also known as the *Kirby-Bauer method,* is used in teaching laboratories to demonstrate the effectiveness of disinfectants and antiseptics. An investigator soaks sterile paper disks with the agents to be studied and then places them on the surface of a solid medium that has been heavily inoculated with a test microorganism. After incubation overnight, a *zone of inhibition*— an area where bacteria do not grow—is visible around disks containing effective disinfectants (**Figure 9.4** on page 268). A larger zone of inhibition may indicate greater effectiveness

Isopropanol
(rubbing alcohol)

Zone of
inhibition

Tincture of iodine

Hydrogen
peroxide

Mercurochrome

▲ *Figure 9.4*

Results of the disk diffusion method to determine
effectiveness of antimicrobial chemicals. The plate was
inoculated with *Bacillus cereus* before disks soaked
with antiseptics were placed over the inoculum.
Alcohol had no effect, perhaps because it evaporated.
*Which antiseptic was most effective against this bacterium
under these conditions?*

*Figure 9.4 Because it had the largest zone of inhibition,
hydrogen peroxide was the most effective antiseptic.*

against the test microbe, but it may only indicate that a
chemical is more stable or diffuses more rapidly in agar than
another. Of course, a disinfectant that is effective against one
microorganism may have little or no effect against other
microorganisms.

This method has some disadvantages. For example, an
antimicrobial agent's molecular size and solubility in water
influence its ability to diffuse through agar, affecting the size
of the zone of inhibition. Tables relating the sizes of zones of
inhibition to the degree of microbial resistance are available
for many chemicals.

In-Use Test

Though phenol coefficient, use-dilution, and disk-diffusion
tests can be useful for initial screening of disinfectants, they
can also be misleading. All three types of evaluation are measures of effectiveness under controlled conditions against one,
or at most a few, species of microbes, but disinfectants are
generally used in various environments against a diverse

population of organisms that are often associated with one
another in complex *biofilms* affording mutual protection.

A more realistic (though more time-consuming) method
for determining the efficacy of a chemical is called an **in-use
test.** In this procedure, swabs are taken from actual objects,
such as emergency operating room equipment, both before
and after the application of a disinfectant or an antiseptic.
The swabs are then inoculated into appropriate growth media, which after incubation are examined for microbial
growth. The in-use test allows a more accurate determination of the proper strength and application procedure of a
given disinfection agent for each specific situation.

Now that we have studied the terminology and general
principles of microbial control, we turn our attention to the
actual physical and chemical agents available to scientists,
medical personnel, and the general public to control microbial growth.

Physical Methods of Microbial Control

Learning Objective

✓ Describe five types of physical methods of microbial control.

Physical methods of microbial control include exposure of
the microbes to extremes of heat and cold, desiccation, filtration, osmotic pressure, and radiation.

Heat-Related Methods

Learning Objectives

✓ Discuss the advantages and disadvantages of using moist heat in
an autoclave and dry heat in an oven for sterilization.
✓ Explain the use of *Bacillus stearothermophilus* endospores in
sterilization techniques.
✓ Explain the importance of pasteurization, and describe three
different pasteurization methods.

Heat is one of the older and more common means of microbial control. As we have seen, high temperatures denature
proteins, interfere with the integrity of cytoplasmic membranes and cell walls, and disrupt the function and structure
of nucleic acids. Heat can be used for sterilization, in which
case all cells and viruses are deactivated, or for practical
"sterilization," as in the commercial sterilization of canned
goods. In commercial sterilization, endospores of hyperthermophilic prokaryotes remain viable but are harmless because they cannot grow at the normal (room) temperatures
in which canned foods are stored.

Though microorganisms vary in their susceptibility to
heat, it can be an important agent of microbial control. As a
result, scientists have developed concepts and terminology
to convey these differences in susceptibility. **Thermal death
point** is the lowest temperature that kills all cells in a broth
in 10 minutes, while **thermal death time** is the time it takes
to completely sterilize a particular volume of liquid at a set
temperature.

▲ *Figure 9.5*

Decimal reduction time (D) as a measure of microbial death rate. D is defined as the time it takes to kill **90%** of a microbial population. Note that D is a constant that is independent of the initial density of the population. *What is the decimal reduction time of this heat treatment against this organism? What is the thermal death time?*

Figure 9.5 D = 5 minutes; thermal death time = 22.5 minutes.

▲ *Figure 9.6*

The relationship between temperature and pressure. Note that higher temperatures—and in consequence, greater antimicrobial action—are associated with higher pressures. *Ultrahigh-temperature pasteurization of milk requires a temperature of 134°C; what pressure must be applied to the milk to achieve this temperature?*

Figure 9.6 29 psi.

As we have discussed, cell death occurs logarithmically (see Figure 9.1). When measuring the effectiveness of heat sterilization, researchers calculate the **decimal reduction time (D),** which is the time required to destroy 90% of the microbes in a sample **(Figure 9.5).** This concept is especially useful to food processors because they must heat foods to eliminate the endospores of *Clostridium botulinum* (klos-trid'ē-ŭm bo-tū-lī'num), which could germinate and produce botulism toxin inside sealed cans. The standard in food processing is to apply heat such that a population of 10^{12} *C. botulinum* endospores is reduced to 10^0 (1) endospore (a 12-fold reduction), which leaves only a very small chance that any particular can of food contains even a single endospore. Researchers have calculated that the D value for *C. botulinum* endospores at 121°C is 0.204 minute, so it takes 2.5 minutes (0.204 × 12) to reduce 10^{12} endospores to 1 endospore.

Moist Heat

Moist heat, which is commonly used to disinfect, sanitize, sterilize, and pasteurize, kills cells by denaturing proteins and destroying cytoplasmic membranes. Moist heat is more effective in microbial control than dry heat because water is a better conductor of heat than air. This is easily demonstrated in your kitchen—you can safely stick your hand into an oven at 350°F for a few moments, but putting it into boiling water at 212°F would burn you severely. The first method we consider for controlling microbes using moist heat is boiling.

Boiling   Boiling kills the vegetative cells of bacteria and fungi, the trophozoites of protozoa, and most viruses within 10 minutes at sea level. Contrary to popular belief, water at a rapid boil is no hotter than that at a slow boil; boiling water at normal atmospheric pressure cannot exceed boiling temperature (100°C at sea level) because escaping steam carries excess heat away. Therefore, it is impossible to boil something more quickly simply by applying more heat; the added heat is carried away by the escaping steam. Boiling *time* is the critical factor. Further, it is important to realize that water boils at lower temperatures at higher elevations because atmospheric pressure is lower; thus a longer boiling time is required in Denver than in Los Angeles to get the same antimicrobial effect.

Bacterial endospores, protozoan cysts, and some viruses (such as hepatitis viruses) can survive boiling at sea level for many minutes or even hours. In fact, because bacterial endospores can withstand boiling for more than 20 hours, boiling is effective for sanitizing restaurant tableware or disinfecting baby bottles, but it is not recommended when true sterilization is required.

Autoclaving   Achieving true sterilization using heat requires higher temperatures than that of boiling water. To achieve the required higher temperatures, pressure is applied to boiling water to prevent the escape of heat in steam. The reason that applying pressure succeeds in achieving sterilization is that the temperature at which water boils (and steam is formed) increases as pressure increases **(Figure 9.6).** In a kitchen, food can be cooked at a higher

(a)

(b)

▲ *Figure 9.7*

An autoclave. **(a)** A photo of a laboratory autoclave. **(b)** A schematic of an autoclave, showing how it functions.

temperature—and thus much faster—by using a pressure cooker; in a laboratory, scientists and medical personnel routinely use a piece of equipment called an autoclave to sterilize chemicals and objects that can tolerate moist heat. Of course, alternate techniques (discussed shortly) must be used for items such as some plastics and vitamins that are damaged by heat or water.

An **autoclave** consists of a pressure chamber, pipes to introduce and evacuate steam, valves to remove air and control pressure, and pressure and temperature gauges to monitor the procedure **(Figure 9.7).** As steam enters an autoclave chamber, it forces air out, raises the temperature of the contents, and increases the pressure, until a set temperature and pressure are reached.

Scientists have determined that a temperature of 121°C, which requires the addition of 15 pounds per square inch (psi)[8] of pressure above that of normal air pressure (see Figure 9.6), destroys all microbes in a small volume in about 10 minutes. Typically, an autoclave holds the pressure and temperature for 15 minutes to provide a margin of safety. The presence of large volumes to be sterilized slows the process because it requires more time for heat to penetrate. Thus, it requires more time to sterilize 1 liter of fluid in a flask than the same volume of fluid distributed into smaller tubes. Autoclaving requires extra time to sterilize solid substances such as canned meat because it takes longer for heat to penetrate to their centers.

Sterilization in an autoclave requires that steam be able to contact all liquids and surfaces that might be contaminated with microbes; therefore, solid objects must be wrapped in porous cloth or paper, not sealed in plastic or aluminum foil, which are impermeable to steam. Containers

of liquids must be sealed loosely enough to allow steam to escape. Additionally, steam must be able to circulate freely, and all air must be forced out by steam. Since steam is lighter than air, it cannot force air from the bottom of an empty vessel; therefore, empty containers must be tipped so that air can flow out of them.

Scientists use several means to ensure that an autoclave has sterilized its contents. A common one is a chemical that changes color when the proper combination of temperature and time have been reached. Often such a color indicator is impressed in a pattern on tape or paper so that the word *sterile* or a pattern or design appears **(Figure 9.8a).** Another technique uses plastic beads that melt when proper conditions are met.

A biological indicator of sterility uses endospores of the bacterium *Bacillus stearothermophilus* (ba-sil'ŭs ste-rō-ther-ma'fil-ŭs) impregnated into tape. After autoclaving, the tape is aseptically inoculated into sterile broth. If no bacterial growth appears, the original material is considered sterile. In a variation on this technique, the endospores are on a strip in one compartment of a vial that also includes a growth medium containing a pH color indicator. After autoclaving, a barrier between the two compartments is broken, putting the endospores into contact with the medium **(Figure 9.8b).** In this case, the absence of a color change indicates sterility.

CRITICAL THINKING

Where within an autoclave should you place a sterilization indicator? Explain your reasoning.

Pasteurization   In Chapter 1 we learned that Louis Pasteur developed a method of heating beer and wine just enough to destroy the microorganisms that cause spoilage

---

[8]The Standard International (SI) equivalent of 1 psi is 1.01×10^5 pascals.

(a)

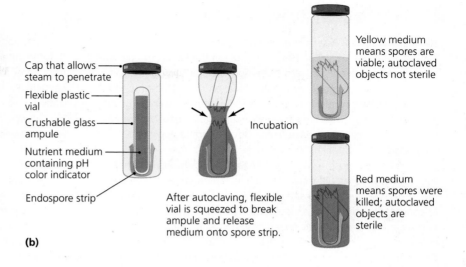

▲ *Figure 9.8*

Sterility indicators. **(a)** A heat-sensitive chemical on this pouch changes color when the conditions for sterility have been met. **(b)** A commercial endospore-test ampule, which is included among objects to be sterilized. After autoclaving is complete, the medium, which contains a pH color indicator, is released onto the endospore strip by breaking the ampule. If the endospores are still alive, their metabolic wastes lower the pH, changing the color of the medium.

without raising the temperature so much that the taste was ruined. Today, pasteurization is also used to kill pathogens in milk, ice cream, yogurt, and fruit juices. *Brucella melitensis* (brū-sel'lă me-li-ten'sis), *Mycobacterium bovis* (mī'kō-bak-tēr'ē-ŭm bō'vis), and *E. coli*, the causative agents of undulant fever, bovine tuberculosis, and one kind of diarrhea respectively, are controlled in this manner.

Pasteurization is not sterilization. *Thermoduric* and *thermophilic*—heat-tolerant and heat-loving—prokaryotes survive pasteurization, but they do not cause spoilage over the relatively short times during which properly refrigerated and pasteurized foods are stored before consumption. In addition, such prokaryotes are generally not pathogenic.

The combination of time and temperature required for effective pasteurization varies with the product. Because milk is the most familiar pasteurized product, we will consider the pasteurization of milk in some detail. Historically, milk was pasteurized by the *batch method* for 30 minutes at 63°C, but most milk processors today use a high-temperature, short-time method known as *flash pasteurization,* in which milk flows through heated tubes that raise its temperature to 72°C for only 15 seconds. This treatment effectively destroys all pathogens. *Ultrahigh-temperature pasteurization* heats the milk to 134°C for only 1 second, but some consumers claim it adversely affects the taste.

Highlight 9.1 on page 272 describes an outbreak of food poisoning on the West Coast that was caused when a popular brand of unpasteurized apple juice became contaminated with *E. coli*.

Ultrahigh-Temperature Sterilization The dairy industry and other food processors also use *ultrahigh-temperature sterilization*, which involves flash heating milk or other liquids to rid them of all living microbes. The process involves passing the liquid through superheated steam at 140°C for

Table 9.2	Moist Heat Treatments of Milk
Process	Treatment
Historical (batch) pasteurization	63°C for 30 minutes
Flash pasteurization	72°C for 15 seconds
Ultrahigh-temperature pasteurization	134°C for 1 second
Ultrahigh-temperature sterilization	140°C for 1 second

1 second, and then cooling it rapidly. Treated liquids can be stored indefinitely at room temperature without microbial spoilage, though after months of storage chemical degradation results in flavor changes. Small packages of dairy creamer served in restaurants are often sterilized by the ultrahigh-temperature method. **Table 9.2** summarizes the dairy industry's use of moist heat for controlling microbes in milk.

Dry Heat

For substances such as powders and oils that cannot be sterilized by boiling or with steam, or for materials that can be damaged by repeated exposure to steam (such as most metal objects), sterilization can be achieved by the use of dry heat, as occurs in an oven. Treatment with dry heat is also a routine part of standard aseptic laboratory procedure; each time laboratory technicians open or close a test tube or flask, they heat the lip of the container in the flame of a Bunsen burner.

Hot air is an effective sterilizing agent because it both denatures proteins and fosters the oxidation of metabolic and structural chemicals; however, in order to sterilize, dry heat requires higher temperatures for longer times than moist heat because dry heat penetrates more slowly. For instance, whereas an autoclave needs only 15 minutes to sterilize an object at 121°C, an oven at the same temperature requires at least 16 hours to achieve sterility. Scientists

---

Highlight 9.1   **Odwalla's Road to Pasteurization**

An apple a day may keep the doctor away, but a batch of bad apples can be deadly. In the fall of 1996, a terrible outbreak of food poisoning occurred in several western states and Canada. Over 60 people fell ill; children were particularly vulnerable, and one 16-month-old girl died. The symptoms included diarrhea, cramping, vomiting, and in the worst cases, a severe kidney disorder called hemolytic uremic syndrome. The culprit: unpasteurized Odwalla brand apple juice contaminated with *Escherichia coli* O157:H7 strain.

At the time, most major juice companies already pasteurized their juice. However, Odwalla had built its reputation on producing the freshest juice possible, and had never pasteurized its products out of concern that the process would make their juice taste bland and strip away nutrients. The company relied on other methods of preventing microbial contamination, such as sorting out rotten fruit and cleaning fruit with sanitizing chemicals. Despite the sorting methods in place, a batch of decayed apples was unintentionally allowed through the production process. It was ultimately linked to the con-

taminated apple juice. Though no one can be certain, it is believed that the initial *E.coli* contamination occurred when the apples fell from a tree and came into contact with animal feces containing the bacteria. Pasteurization would have killed the *E.coli* present in the resulting juice product.

Odwalla accepted responsibility for the outbreak and conducted an extensive review of its quality assurance policies. The company concluded that pasteurization was the only fail-safe method of preventing *E. coli* contamination in juice, and revised its manufacturing processes accordingly. Not all companies have followed suit, however. Many firms in the apple cider industry remain reluctant to pasteurize, contending that fresh, unpasteurized cider is safe for most adults, and that the risk of *E. coli* infection is far greater in hamburgers and seafood. But for young children, the elderly, those with compromised immune systems, and cautious adults everywhere, the only safe apple drink is the one that's been sufficiently heated (to 72°C for at least 15 seconds.)

---

typically use higher temperatures—171°C for 1 hour, or 160°C for 2 hours—to sterilize objects in an oven, but objects made of rubber, paper, and many types of plastic oxidize rapidly (combust) under these conditions.

Complete incineration is the ultimate means of sterilization. As part of standard aseptic technique in microbiological laboratories, inoculating loops are sterilized by heating them in the flame of a Bunsen burner or with an electric heating coil until they glow red (about 1500°C). Health care workers incinerate contaminated dressings, bags, and paper cups; field epidemiologists incinerate the carcasses of animals that have diseases such as anthrax or bovine spongiform encephalopathy (mad cow disease).

Refrigeration and Freezing

Learning Objective

✓ Describe the use and importance of refrigeration and freezing in limiting microbial growth.

In many situations, particularly in food preparation and storage, the most convenient method of microbial control is either refrigeration between 0°C and 7°C or freezing (temperatures below 0°C). These processes decrease microbial metabolism, growth, and reproduction because chemical reactions occur more slowly at low temperatures, and because liquid water is not available at subzero temperatures. Note, however, that psychrophilic microbes can multiply in refrigerated food and spoil its taste and suitability for consumption.

Refrigeration halts the growth of most pathogens, which are predominantly mesophiles. Notable exceptions are the bacteria *Listeria* (lis-tēr-ē-ă), which can reproduce to dangerous levels in refrigerated food, and *Yersinia* (yer-sin'ē-ă), which can multiply in refrigerated blood products

and be passed on to blood recipients. These pathogens are discussed in more detail in Chapters 19 and 20, respectively.

Slow freezing, during which ice crystals have time to form and puncture cell membranes, is more effective than quick freezing in inhibiting microbial metabolism, though microorganisms also vary in their susceptibility to freezing. Whereas the cysts of tapeworms and the roundworm larvae that causes trichinosis perish after several days in frozen meat, many vegetative bacterial cells, bacterial endospores, and viruses can survive subfreezing temperatures for years. In fact, scientists store many bacteria and viruses in low-temperature freezers at −30°C to −80°C and are able to reconstitute the microbes into viable populations by warming them in media containing proper nutrients. Therefore, care must be taken in thawing and cooking frozen food, because it can still contain many pathogenic microbes.

Highlight 9.2 describes how freezing is but one method of microbial control used in the preparation of sushi.

Desiccation and Lyophilization

Learning Objective

✓ Compare and contrast desiccation and lyophilization.

Desiccation or drying has been used for thousands of years to preserve such foods as fruits, peas, beans, grain, nuts, and yeast **(Figure 9.9)**. Desiccation inhibits microbial growth because metabolism requires liquid water. Drying inhibits the spread of most pathogens, including the bacteria that cause syphilis, gonorrhea, and the more common forms of bacterial pneumonia and diarrhea. However, as with freezing, drying is microbistatic; it does not kill all microbes. For example, most molds can grow on dried raisins and apricots, which have as little as 16% water content.

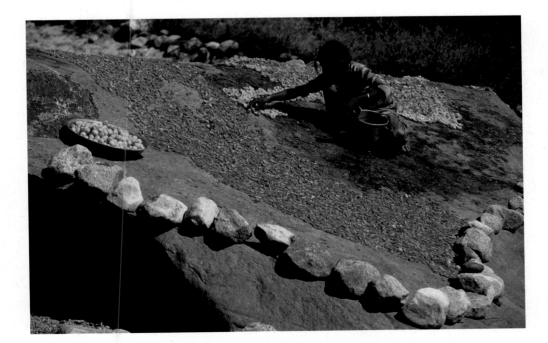

◀ *Figure 9.9*
The use of desiccation as a means of preserving foods. In this time-honored practice, drying inhibits microbial growth in the food by removing the water microbes needed for metabolism.

Highlight 9.2   Sushi: A Case Study in Microbial Control

Sushi—it either makes you squirm or salivate! Although technically the term refers to rice, it's generally understood to refer to bite-sized slices of raw fish served with rice. Long a staple of Japanese cuisine, sushi has become popular throughout the United States. But isn't eating raw fish dangerous? What methods of microbial control are applied in the preparation of sushi?

It's true that raw fish can contain harmful microorganisms. Such parasites as the kind of roundworms called anasakids are commonly found in fish and can cause gastrointestinal symptoms. Accordingly, before fish can be served raw, the Food and Drug Administration requires that it be frozen at −20°C (−4°F) for 7 days, or at −37°C (−35°F) for 15 hours. Unfortunately, freezing does *not* kill all bacterial or viral pathogens. Consumers should be aware that they always assume some risk of food poisoning caused by such bacteria as *S. aureus*, *E. coli*, and species of *Salmonella* and *Vibrio* whenever they eat any kind of raw seafood, including raw oysters, seviche, or carpaccio.

Even though diners tend to fixate on the safety of raw fish, cooked rice left sitting at room temperature is also vulnerable to the growth of pathogens. To counter this potential problem, sushi rice is prepared with vinegar, which acidifies the rice. At pH values below

4.6, rice becomes too acidic to support the growth of pathogens. Sushi bars can also reduce the risk posed by pathogens by keeping restaurant temperatures cool. Additionally, it is believed that wasabi, the fiery horseradish-like green paste commonly eaten with sushi, contains antimicrobial properties, although its mechanism of action is not well understood.

With these antimicrobial precautions in place, the vast majority of diners consume sushi safely meal after meal (although pregnant women and people with compromised immune systems should avoid all raw seafood). When properly prepared, sushi is beautiful, low in calories, and a source of heart-healthy omega 3-fatty acids—and very delicious.

Scientists use **lyophilization** (lī-of′i-li-zā′shŭn), a technique combining freezing and drying, to preserve microbes and other cells for many years. In this process, scientists instantly freeze a culture in liquid nitrogen or frozen carbon dioxide (dry ice); then they subject it to a vacuum that removes frozen water through a process called sublimation, in which the water is transformed directly from a solid to a gas.

Lyophilization prevents the formation of large damaging ice crystals. Although not all cells survive, enough are viable to enable the culture to be reconstituted many years later.

CRITICAL THINKING

Why is liquid water necessary for microbial metabolism?

Nonsterile medium

Filter

To vacuum pump

Sterile medium

(a)

(b)

SEM ⊢—————⊣ 1 µm

◀ *Figure 9.10*

Filtration equipment used for microbial control. **(a)** Assembly for sterilization by vacuum filtration. **(b)** Membrane filters composed of various substances and with pores of various sizes can be used to trap diverse microbes.

Filtration

Learning Objective

✓ Describe the use of filters for disinfection and sterilization.

Filtration is the passage of a fluid (either a liquid or a gas) through a sieve designed to trap particles—in this case, cells or viruses—and separate them from the fluid **(Figure 9.10).** Researchers often use a vacuum to assist the movement of fluid through the filter. Filtration traps microbes larger than the pore size, allowing smaller microbes to pass through. In the late 1800s, filters were able to trap cells, but their pores were too large to trap the pathogens of such diseases as rabies and measles. These pathogens were thus named *filterable viruses*, which today has been shortened to *viruses*[9]. Now, filters with pores small enough to trap even viruses are available, so filtration can be used to sterilize such heat-sensitive materials as ophthalmic solutions, antibiotics, vaccines, liquid vitamins, enzymes, and culture media.

Over the years, filters have been constructed from porcelain, glass, cotton, asbestos, and diatomaceous earth, a substance composed of the innumerable glass-like cell walls of single-celled algae called diatoms. Scientists today typically use thin (only 0.1 mm thick), circular **membrane filters** manufactured of nitrocellulose or plastic and containing specific pore sizes ranging from 25 µm to less than 0.01 µm in diameter (see Figure 9.10b). The pores of the latter filters are small enough to trap small viruses and even some large

protein molecules. Microbiologists also use filtration to estimate the number of microbes in a fluid by counting the number deposited on the filter after passing a given volume through the filter (see Figure 6.21). **Table 9.3** lists some pore sizes of membrane filters and the microbes they allow through.

Health care and laboratory workers routinely use filtration to prevent airborne contamination by microbes. The use of surgical masks prevents contamination of the environment by microbes exhaled by medical personnel, and cotton plugs in culture vessels prevents contamination of cultures by airborne microbes. Additionally, *high-efficiency particulate air (HEPA) filters* are crucial parts of biological safety cabinets **(Figure 9.11),** and HEPA filters are mounted in the air ducts of some operating rooms, rooms occupied by patients with airborne diseases such as tuberculosis, and rooms of immunocompromised patients such as burn victims and AIDS patients.

CRITICAL THINKING

A virologist needs to remove all bacteria from a solution containing viruses without removing the viruses. What size membrane filter should the scientist use?

Osmotic Pressure

Learning Objective

✓ Discuss the use of hypertonic solutions in microbial control.

Another ancient method of microbial control is the use of high concentrations of salt or sugar in foods to inhibit

[9]Latin, meaning poisons.

Table 9.3	Membrane Filters
Pore Size (μm)	**Microbes that Pass Through**
0.01	Prions
0.025	Smallest viruses and prions
0.22	Small viruses, pliable bacteria (mycoplasmas, rickettsias, chlamydias, some spirochetes), and prions
0.45	Viruses, small bacteria, and prions
1.2	Bacteria, viruses, and prions
3	Protozoa, smallest algae, bacteria, viruses, and prions
5	Yeasts, protozoa, algae, bacteria, viruses, and prions

microbial growth by **osmotic pressure.** As we saw in Chapter 3, osmosis is the net movement of water across a semipermeable membrane (such as a cytoplasmic membrane) from an area of higher water concentration to an area of lower water concentration. Cells in a hypertonic solution lose water, whereas cells in a hypotonic solution tend to gain water. When a cell is in an environment with a high salt or sugar concentration, the water in the cell is drawn out by osmosis, and the cell desiccates (see Figure 3.18b). The removal of water inhibits cellular metabolism because enzymes are fully functional only in aqueous environments. Thus, honey, jerky, jams, jellies, salted fish, and some types of pickles are preserved from most microbial attacks.

Fungi have a greater ability than bacteria to tolerate hypertonic environments with little moisture, which explains why jelly in your refrigerator may grow a colony of *Penicillium* (pen-i-sil'ē-ŭm) mold but is not likely to grow the bacterium *Salmonella*.

Radiation

Learning Objective

✓ Differentiate between ionizing radiation and nonionizing radiation as they relate to microbial control.

Another physical method of microbial control is the use of radiation. There are two types of **radiation:** particulate radiation and electromagnetic radiation. Particulate radiation consists of high-speed subatomic particles, such as electrons, that have been freed from their atoms. Electromagnetic radiation can be defined as energy without mass traveling in waves at the speed of light (3×10^5 km/sec). Electromagnetic energy is released from atoms that have undergone internal changes. The *wavelength* of electromagnetic radiation, defined as the distance between two crests of a wave, ranges from very short cosmic and gamma rays, through X rays, ultraviolet light, and visible light, to long

(a)

▲ *Figure 9.11*

The roles of high-efficiency particulate air (HEPA) filters in laminar flow biological safety cabinets. HEPA filters protect workers from exposure to microbes (by maintaining a barrier of filtered air across the opening of the cabinet) and the laboratory environment (by filtering microbes out of exhaust air). HEPA filters are also used in air ducts in operating rooms and in the rooms of highly contagious or immunocompromised patients.

infrared rays, and finally to very long radio waves (see Figure 4.1). Though they are particles, electrons also have a wave nature with wavelengths that are even shorter than gamma rays.

The shorter the wavelength of an electromagnetic wave, the more energy it carries and the farther it can penetrate. Therefore, shorter wavelength radiation is more suitable for microbial control than longer wavelength radiation, which carries less energy and is less penetrating. Scientists describe all types of radiation as either *ionizing* or *nonionizing* according to its effects on the chemicals within cells.

Ionizing Radiation

Electron beams, gamma rays, and X rays, all of which have wavelengths shorter than 1 nm, are **ionizing radiation** because when they strike molecules, they have sufficient energy to create ions by ejecting electrons from atoms. Such ions disrupt hydrogen bonding, oxidize double covalent

(a)   (b)

▲ *Figure 9.12*

Irradiated food. **(a)** A demonstration of the increased shelf life of food achieved by ionizing radiation. **(b)** A radura, the symbol used in the United States to label irradiated foods.

bonds, and create highly reactive hydroxide ions (see Chapter 2). These ions in turn denature other molecules, particularly DNA, causing fatal mutations and cell death.

Electron beams are produced by *cathode ray machines* that operate like an analog television tube. Electron beams are highly energetic and therefore very effective in killing microbes in just a few seconds, but they do not penetrate deep into matter. Thus they cannot sterilize thick objects or objects coated with large amounts of organic matter. They are used to sterilize some sliced meats, microbiological plastic ware, and dental and medical supplies such as gloves, syringes, and suturing material.

Gamma rays, which are emitted by some radioactive elements such as radioactive cobalt, penetrate much farther than electron beams but require hours to kill microbes. The FDA has approved the use of gamma irradiation for microbial control in meats, spices, and fresh fruits and vegetables **(Figure 9.12)**. Irradiation with gamma rays kills not only microbes, but also the larvae and eggs of insects; it also kills the cells of fruits and vegetables, preventing microbial spoilage and overripening.

Consumers have been reluctant to accept irradiated food. A number of reasons have been cited, including fear that radiation makes food radioactive, and claims that it changes the taste and nutritive value of foods or produces potentially carcinogenic (cancer-causing) chemicals. Supporters of irradiation reply that gamma radiation passes through food and cannot make it radioactive any more than a dental X ray produces radioactive teeth, and they cite numerous studies which concluded that irradiated foods are tasty, nutritious, and safe.

X rays travel the farthest through matter, but they have less energy than gamma rays and require a prohibitive amount of time to make them practical for microbial control.

Nonionizing Radiation

Electromagnetic radiation with a wavelength greater than 1 nm does not have enough energy to force electrons out of orbit. However, such radiation does contain enough energy to excite electrons and cause them to make new covalent bonds, which can affect the three-dimensional structure of proteins and nucleic acids.

Ultraviolet (UV) light, visible light, infrared radiation, and radio waves are nonionizing radiation. Of these, only UV light has sufficient energy to be a practical antimicrobial agent. Visible light and microwaves, which are extremely short wavelength radio waves, have little value in microbial control, though microwaves heat food, which can be inhibitory if the food gets hot enough.

UV light with a wavelength of 260 nm is specifically absorbed by adjacent thymine nucleotide bases in DNA, causing them to form covalent bonds with each other rather than forming hydrogen bonds with adenine bases in the complementary DNA strand (see Figure 7.21). As discussed in Chapter 7, such *thymine dimers* distort the shape of DNA, making it impossible for the cell to accurately transcribe or replicate its genetic material. If thymine dimers remain uncorrected, an affected cell may die.

The effectiveness of UV irradiation is tempered by the fact that UV light does not penetrate well. UV light is therefore suitable primarily for disinfecting air, transparent fluids, and the surfaces of objects such as barber's sheers and

Table 9.4   Physical Methods of Microbial Control

Method	Conditions	Action	Representative Use(s)
Moist heat			
Boiling	10 min at 100°C	Denatures proteins and destroys membranes	Disinfection of baby bottles and sanitization of restaurant cookware and tableware
Autoclaving (pressure cooking)	15 min at 121°C		Autoclave: sterilization of medical and laboratory supplies that can tolerate heat and moisture; pressure cooker: sterilization of canned food
Pasteurization	15 sec at 72°C		Destruction of all pathogens and most spoilage microbes in dairy products, fruit juices, beer, and wine
Ultrahigh-temperature sterilization	1–3 sec at 140°C		Sterilization of dairy products
Dry heat			
Hot air	2 h at 160°C or 1 h at 171°C	Denatures proteins, destroys membranes, oxidizes metabolic compounds	Sterilization of water-sensitive materials such as powders, oils, and metals
Incineration	1 sec at more than 1000°C		Sterilization of inoculating loops, flammable contaminated medical waste, and diseased carcasses
Refrigeration	0–7°C	Inhibits metabolism	Preservation of food
Freezing	<0°C	Inhibits metabolism	Long-term preservation of foods, drugs, and cultures
Desiccation (drying)	Varies with amount of water to be removed	Inhibits metabolism	Preservation of food
Lyophilization (freeze drying)	−196°C for a few minutes while drying	Inhibits metabolism	Long-term storage of bacterial cultures
Filtration	Filter retains microbes	Physically separates microbes from air and liquids	Sterilization of air and heat-sensitive ophthalmic and enzymatic solutions, vaccines, and antibiotics
Osmotic pressure	Exposure to hypertonic solutions	Inhibits metabolism	Preservation of food
Ionizing radiation (electron beams, gamma rays, X rays)	Seconds to hours of exposure (depending on wavelength of radiation)	Destroys DNA	Sterilization of medical and laboratory equipment and preservation of food
Nonionizing radiation (ultraviolet light)	Irradiation with 260 nm wavelength radiation	Formation of thymine dimers inhibits DNA transcription and replication	Disinfection and sterilization of surfaces and of transparent fluids and gases

operating tables. Recently, some cities have begun using UV irradiation in sewage treatment. By passing wastewater past banks of UV lights, they reduce the number of bacteria without using chlorine, which can damage the environment.

Table 9.4 summarizes the physical methods of microbial control discussed in the previous pages.

CRITICAL THINKING

During the bioterrorist attack in which anthrax endospores were sent through the mail in the fall of 2001, one news commentator suggested that people should iron all their incoming mail with a regular household iron as a means of destroying endospores. Would you agree that this is a good way to disinfect mail? Explain your answer. Which disinfectant methods would be both more effective and practical?

Chemical Methods of Microbial Control

Learning Objective

✓ Compare and contrast eight major types of antimicrobial chemicals, and discuss the positive and negative aspects of each.

Although physical agents are sometimes used for disinfection, antisepsis, and preservation, more often chemical agents are used for these purposes. As we have seen, chemical agents act to adversely affect microbes' cell walls, cytoplasmic membranes, proteins, or DNA. And as was also the case for physical agents, the effect of a chemical agent varies with temperature, length of exposure, and the amount of contaminating organic matter in the environment; it also varies with pH, concentration, and freshness of the chemical. Chemical agents tend to destroy or inhibit the growth of

▲ *Figure 9.13*

Phenol and phenolics. **(a)** Phenol, a naturally occurring molecule that is also called carbolic acid. **(b)** Phenolics, which are compounds synthesized from phenol, have greater antimicrobial efficacy with fewer side effects. Bisphenols are paired, covalently linked phenolics.

enveloped viruses and the vegetative cells of bacteria, fungi, and protozoa more than fungal spores, protozoan cysts, or bacterial endospores. The latter are particularly resistant to chemical agents, as was demonstrated by numerous failed attempts to decontaminate the United States Senate office building of anthrax endospores sent there by bioterrorists in 2001.

In the following sections we discuss eight major categories of antimicrobial chemicals used as antiseptics and disinfectants: *phenols, alcohols, halogens, oxidizing agents, surfactants, heavy metals, aldehydes,* and *gaseous agents.* Some chemical agents combine one or more of these. Additionally, researchers and food processors sometimes use antimicrobics—substances normally used to treat diseases—as disinfectants.

Phenol and Phenolics

Learning Objective

✓   Distinguish between phenol and the types of phenolics, and discuss their action as antimicrobial agents.

In 1867, Dr. Joseph Lister began using phenol **(Figure 9.13a)** to reduce infection during surgery. As stated previously, the efficacy of phenol remains one standard to which the actions of other antimicrobial agents can be compared.

Phenolics are compounds derived from phenol molecules that have been chemically modified by the addition of halogens or organic functional groups **(Figure 9.13b).** For instance, chlorinated phenolics contain one or more atoms of chlorine and have enhanced antimicrobial action and a less annoying odor than phenol. Natural oils such as pine and clove oils are also phenolics and can be used as antiseptics.

Bisphenolics are composed of two covalently linked phenolics. Two examples of bisphenolics are *orthophenylphenol,* which is the active ingredient in the disinfectant Lysol®, and *triclosan,* which is incorporated into numerous consumer products, including garbage bags, diapers, and cutting boards.

Phenol and phenolics are intermediate- to low-level disinfectants that denature proteins and disrupt cell membranes in a wide variety of pathogens. They are effective even in the presence of contaminating organic material such as vomit, pus, saliva, and feces, and they remain active on surfaces for a prolonged time. For these reasons, phenolics are commonly used in health care settings, laboratories, and households.

Negative aspects of phenolics include their disagreeable odor and possible side effects; for example, phenolics irritate the skin of some individuals. *Hexachlorophene* (see Figure 9.13b), which was once a popular household bisphenolic, was found to cause brain damage in infants. Now it is available only by prescription and is used in nurseries only in response to severe staphylococcal contamination.

Alcohols

Learning Objective

✓   Discuss the action of alcohols as antimicrobial agents, and explain why 70% to 90% alcohols are more effective than pure alcohols.

Alcohols are bactericidal, fungicidal, and virucidal against enveloped viruses; however, they are not effective against fungal spores or bacterial endospores. Alcohols are considered intermediate-level disinfectants. Commonly used alcohols include rubbing alcohol (isopropanol) and drinking alcohol (ethanol):

$$\underset{\text{Isopropanol}}{CH_3-\overset{\displaystyle OH}{\overset{|}{C}H}-CH_3} \qquad \underset{\text{Ethanol}}{CH_3-CH_2OH}$$

Isopropanol is slightly superior to ethanol as a disinfectant and antiseptic. *Tinctures* (tingk′chŭrs), which are solutions of other antimicrobial chemicals in alcohol, are often more effective than the same antimicrobials dissolved in water.

Alcohols denature proteins and disrupt cytoplasmic membranes. Surprisingly, pure alcohol is not an effective antimicrobial agent because the denaturation of proteins requires water; therefore, 70% to 90% alcohol is typically used to control microbes. Alcohols evaporate rapidly, which is advantageous in that they leave no residue, but disadvantageous in that they may not contact microbes long enough to be effective (see Figure 9.4). Swabbing the skin with alcohol prior to an injection removes more microbes by physical action (degerming) than by chemical action.

Halogens

Halogens are the four very reactive, nonmetallic chemical elements: iodine, chlorine, bromine, and fluorine. Halogens are intermediate-level antimicrobial chemicals that are effective against vegetative bacterial and fungal cells, fungal spores, some bacterial endospores and protozoan cysts, and many viruses. Halogens are used both alone and combined with other elements in organic and inorganic compounds. Scientists do not know exactly how halogens exert their antimicrobial effect, but it is thought that they damage enzymes via oxidation or by covalently linking to them, denaturing them.

Iodine is a well-known antiseptic. In the past, backpackers and campers disinfected water with iodine tablets, but experience has shown that protozoan cysts can survive iodine treatment unless the iodine concentration is so great that the water is undrinkable. Knowledgeable campers now filter stream and lake water or carry bottled water, even though standard filtration does not remove toxins from water. Activated charcoal filters remove most toxins.

Medically, iodine is used either as a tincture or as an *iodophor,* which is an iodine-containing organic compound that slowly releases iodine. Iodophors have the advantage of being long-lasting, nonstaining, and nonirritating to the skin. Betadine® **(Figure 9.14)** is an example of an iodophor used in medical institutions to prepare skin for surgery and injections and to treat burns.

Municipalities commonly use *chlorine* in its elemental form (Cl_2) to treat drinking water supplies, swimming pools, and wastewater from sewage treatment plants. Compounds containing chlorine are also effective disinfectants. Examples include *sodium hypochlorite,* which is the active ingredient in chlorine bleach, and *calcium hypochlorite.* The dairy industry and restaurants use these compounds to disinfect utensils and the medical field uses them to disinfect hemodialysis systems. Household bleach diluted by adding two drops to a liter of water can be used in an emergency to make water safer to drink, but it does not kill all protozoan cysts, bacterial endospores, or viruses. *Chlorine dioxide* (ClO_2) is a gas that can be used to disinfect large spaces; for example, it was used in the federal office buildings contaminated with anthrax spores following the bioterrorism attack that occurred in the fall of 2001. Chloramines—chemical

▲ *Figure 9.14*
The use of Betadine®, an iodophor, as a degerming antiseptic in preparation for drawing blood.

combinations of chlorine and ammonia—are used in wound dressings, as skin antiseptics, and in some municipal water supplies. Chloramines are less effective antimicrobial agents than other forms of chlorine, but they release chlorine slowly and are thus longer lasting, and they don't appear to form carcinogenic compounds.

Bromine is an effective disinfectant in hot tubs because it evaporates more slowly than chlorine at high temperatures.

Oxidizing Agents

Peroxides, ozone, and *peracetic acid* kill microbes by oxidizing their enzymes, thereby preventing metabolism. Oxidizing agents are high-level disinfectants and antiseptics that are particularly effective against anaerobic microorganisms contaminating deep wounds; they work by releasing hydroxyl radicals, which kill anaerobes. Therefore, health care workers use oxidizing agents to kill anaerobes in deep puncture wounds.

Hydrogen peroxide is a common household chemical that can disinfect and even sterilize the surfaces of inanimate objects such as contact lenses, but is often mistakenly used to treat open wounds. Hydrogen peroxide does not make a good antiseptic for open wounds because *catalase*—an enzyme released from damaged human cells—quickly neutralizes it by breaking it down into water and oxygen gas, which can be seen as escaping bubbles. Though aerobes and facultative anaerobes on inanimate surfaces also contain catalase, the volume of peroxide used as a disinfectant overwhelms the enzyme, making hydrogen peroxide a useful disinfectant. Food processors use hot hydrogen peroxide to sterilize packages such as juice boxes.

(a) Ammonium ion

(b) Quaternary ammonium ions

▲ *Figure 9.15*

Quaternary ammonium compounds (quats). Quats are surfactants in which the hydrogen atoms of an ammonium ion **(a)** are replaced by other functional groups **(b)**.

Ozone (O_3) is a reactive form of oxygen that is generated when molecular oxygen (O_2) is subjected to electrical discharge. Ozone gives air its "fresh smell" after a thunderstorm. Some Canadian and European municipalities treat their drinking water with ozone rather than chlorine, because ozone does not produce carcinogenic by-products like chlorine can. Ozone is also a more effective antimicrobial agent than chlorine, but it is more expensive, and it is difficult to maintain an effective concentration of ozone in water.

Peracetic acid is an extremely effective sporicide that can be used to sterilize surfaces. Because it is not adversely affected by organic contaminants, and it leaves no toxic residue, food processors and medical personnel use peracetic acid to sterilize equipment.

Surfactants

Learning Objective

✓ Define surfactants and describe their antimicrobial action.

Surfactants are "surface active" chemicals. One of the ways surfactants act is to reduce the surface tension of solvents such as water by decreasing the attraction among solvent molecules. One result of this reduction in surface tension is that the solvent becomes more effective at dissolving solute molecules.

Two common surfactants involved in microbial control are soaps and detergents. One end of a soap molecule is hydrophobic because it is composed of fatty acids, and the other end is hydrophilic and negatively charged. When soap is used to wash skin, for instance, the hydrophobic ends of soap molecules are effective at breaking oily deposits (such as secretions from oil glands) into tiny droplets, and the hydrophilic ends attract water molecules; the result is that the tiny droplets of oily material—and any bacteria they harbor—are more easily dissolved in and washed away by water. Thus soaps by themselves are good degerming

agents, but poor antimicrobial agents; when household soaps are antiseptic, it is largely because they contain antimicrobial chemicals.

Synthetic **detergents** are positively charged organic surfactants that are more soluble in water than soaps. The most popular detergents for microbial control are **quaternary ammonium compounds** or **quats**, which are composed of an ammonium cation (NH_4^+) in which the hydrogen atoms are replaced by other functional groups or hydrocarbon chains **(Figure 9.15).** Quats are not only antimicrobial; they are colorless, tasteless, and harmless to humans (except at high concentrations), making them ideal for many industrial and medical applications. If your mouthwash foams, it probably contains a quaternary ammonium compound. Examples of quats are Zephiran® (benzalkonium chloride) and Cepacol® (cetylpyridinium chloride).

Quats function by disrupting cellular membranes such that affected cells lose essential internal ions such as potassium ions (K^+). Quats are bactericidal (particularly against Gram-positive bacteria), fungicidal, and virucidal against enveloped-viruses; but they are not effective against nonenveloped viruses, mycobacteria, or endospores. The action of quaternary ammonium compounds is retarded by organic contaminants, and they are deactivated by soaps. Some pathogens, such as *P. aeruginosa*, actually thrive in quats; therefore, quats are classified as low-level disinfectants.

Heavy Metals

Learning Objective

✓ Define heavy metals, give several examples, and describe their use in microbial control.

The ions of high-molecular-weight metals such as arsenic, zinc, mercury, silver, and copper are antimicrobial because they combine with sulfur atoms in molecules of cysteine, an amino acid. Such bonding alters the three-dimensional shapes of proteins, inhibiting or eliminating their function. Heavy metal ions, as they are called, are low-level bacteriostatic and fungistatic agents, and with few exceptions their use has been superceded by more effective antimicrobial agents. **Figure 9.16** illustrates the effectiveness of heavy metals in inhibiting bacterial reproduction on a petri plate.

At one time, many states required that the eyes of newborns be treated with a cream containing 1% *silver nitrate* ($AgNO_3$) to prevent blindness caused by *Neisseria gonorrhoeae* (nī-sē′rē-ă go-nor-rē′ē), which can enter a baby's eyes while it passes through an infected birth canal. Today, silver nitrate has largely been displaced by antibiotic ointments, because they are less irritating and are also effective against *Chlamydia trachomatis* (kla-mid′ē-ă tra-kō′ma-tis) and other pathogens. Silver still plays an antimicrobial role in some surgical dressings, burn creams, and catheters.

For several decades, drug companies have used *thimerosal*, a mercury-containing compound, to preserve vaccines. In 1999, the U.S. Public Health Service recommended that alternatives be used because mercury is a metabolic poison.

Copper, which interferes with chlorophyll, is used to control algal growth in reservoirs, fish tanks, swimming pools, and water storage tanks. In the absence of organic contaminants, copper is an effective algicide in concentrations as low as 1 ppm (part per million). In addition to copper, zinc and mercury are used to control mildew in paint.

CRITICAL THINKING

What common household antiseptic contains a heavy metal as its active ingredient?

▲ *Figure 9.16*
The effect of heavy metal ions on bacterial growth. The zones of inhibition formed because metal ions inhibit bacterial reproduction through their effects on protein function.

Aldehydes

Learning Objective

✓ Compare and contrast formaldehyde and glutaraldehyde as antimicrobial agents.

Aldehydes are compounds containing terminal —CHO groups (see Table 2.3 on p. 40). *Glutaraldehyde,* which is a liquid, and *formaldehyde,* which is a gas, are highly reactive chemicals with the following structural formulas:

$$\underset{\text{Glutaraldehyde}}{CH-CH_2-CH_2-CH_2-CH} \qquad \underset{\text{Formaldehyde}}{HCH}$$

Aldehydes function in microbial control by cross-linking amino, hydroxyl, sulfhydryl, and carboxyl organic functional groups thereby denaturing proteins and inactivating nucleic acids.

Hospital personnel and scientists use 2% solutions of glutaraldehyde to kill bacteria, viruses, and fungi; a 10-minute treatment effectively disinfects most objects, including medical and dental equipment. When the time of exposure is increased to 10 hours, glutaraldehyde sterilizes.

Although glutaraldehyde is less irritating and more effective than formaldehyde, it is more expensive as well.

Morticians and health care workers use formaldehyde dissolved in water to make a 37% solution called *formalin*. They use formalin for embalming and to disinfect isolation rooms, exhaust cabinets, surgical instruments, and reusable kidney dialysis machines. Formaldehyde must be handled with care because it irritates mucous membranes and is carcinogenic.

Gaseous Agents

Learning Objective

✓ Describe the advantages and disadvantages of gaseous agents of microbial control.

Many items, such as heart-lung machine components, sutures, plastic laboratory ware, mattresses, pillows, artificial heart valves, catheters, electronic equipment, and dried or powdered foods, cannot be sterilized easily with heat or water-soluble chemicals; nor is irradiation always practical for large or bulky items. However, they can be sterilized within a closed chamber containing highly reactive microbicidal and sporicidal gases such as *ethylene oxide, propylene oxide,* and *beta-propiolactone*:

$$\underset{\text{Ethylene oxide}}{CH_2-CH_2} \qquad \underset{\text{Propylene oxide}}{CH_3-CH-CH_2}$$

$$\underset{\text{Beta-propiolactone}}{CH_2-CH_2 \atop O-C}$$

Table 9.5   Chemical Methods of Microbial Control

Method	Action(s)	Level of Activity	Use
Phenol (carbolic acid)	Denatures proteins and disrupts cell membranes	Intermediate to low	Original surgical antiseptic; now replaced by less odorous and injurious phenolics
Phenolics (chemically altered phenol; bisphenols are composed of a pair of linked phenolics)	Denature proteins and disrupt cell membranes	Intermediate to low	Disinfectants and antiseptics
Alcohols	Denature proteins and disrupt cell membranes	Intermediate	Disinfectants, antiseptics, and as a solvent in tinctures
Halogens (iodine, chlorine, bromine, and fluorine)	Presumably denature proteins	Intermediate	Disinfectants, antiseptics, and water purification
Oxidizing agents (peroxides, ozone, and peracetic acid)	Denature proteins by oxidation	High	Disinfectants, antiseptics for deep wounds, water purification, and sterilization of food-processing and medical equipment
Surfactants (soaps and detergents)	Decrease surface tension of water and disrupt cell membranes	Low	Soaps: degerming; detergents: antiseptic
Heavy metals (arsenic, zinc, mercury, silver, and copper, etc.)	Denature proteins	Low	Fungistats in paints; silver nitrate cream: surgical dressings, burn creams, and catheters; copper: algicide in water reservoirs, swimming pools, and aquariums
Aldehydes (glutaraldehyde and formaldehyde)	Denature proteins	High	Disinfectant and embalming fluid
Gaseous agents (ethylene oxide, propylene oxide, and beta-propiolactone)	Denature proteins	High	Sterilization of heat- and water-sensitive objects
Antimicrobics	Act against cell walls, cell membranes, protein synthesis, and DNA transcription and replication	Intermediate to low	Disinfectants and treatment of infectious diseases

These gases rapidly penetrate paper and plastic wraps and diffuse into every crack. Over time (usually 4–18 hours), they denature proteins and DNA by cross-linking organic functional groups, thereby killing everything they contact without harming inanimate objects.

Ethylene oxide is frequently used as a gaseous sterilizing agent in hospitals and dental offices, and NASA uses the gas to sterilize spacecraft designed to land on other worlds, lest they accidentally export earthly microbes. Large hospitals often use ethylene oxide chambers, which are similar in appearance to autoclaves, to sterilize instruments and equipment sensitive to heat.

Despite their advantages, gaseous agents are far from perfect: They can be extremely hazardous to the people using them. Because they are often highly explosive, they must be administered by combining them with 10% to 20% nitrogen gas or carbon dioxide. Moreover, they are extremely poisonous, so workers must extensively flush sterilized objects with air to remove every trace of the gas (which adds to the time required to use them). Finally, gaseous agents, especially beta-propiolactone, are potentially carcinogenic.

Antimicrobics

Learning Objective

 Describe the types of antimicrobics and their use in environmental control of microorganisms.

Antimicrobics include antibiotics, semisynthetics, and synthetics. *Antibiotics* are antimicrobial chemicals produced naturally by microorganisms. When scientists chemically modify an antibiotic the agent is called a *semisynthetic*. Scientists have also developed wholly synthetic antimicrobial drugs. The main difference between these antimicrobics

Highlight 9.3   Should You Use Antibacterial Soaps?

Walk through the soap aisle at your local supermarket, and it's likely you will find numerous products advertised as "antimicrobial" or "antibacterial." In fact, it's likely you already have such products in your home and use them every day. But do they really work better than plain, regular soaps? And is it possible that such products may actually do more harm than good?

The answers are less clear than you might think. Although soaps containing antimicrobial drugs are indeed more effective than plain soaps in reducing the presence of microorganisms, there is growing concern that overuse of antimicrobials may actually contribute to the evolution of drug-resistant microorganisms.

Antibacterial soaps kill off weaker bacteria, leaving stronger, more-resistant strains to multiply. Dr. Stuart Levy of Tufts University and some other researchers even contend that overuse of antimicrobial products may be contributing to an increase in hard-to-treat infections caused by the stronger bacteria that antimicrobials allegedly leave behind. Levy's research has focused in particular on triclosan, a common bisphenolic found in a wide range of household products.

In tests on *E. coli*, Levy found that triclosan targeted a specific bacterial gene and contends that this specificity raises the potential for the development of resistant bacterial strains. To date, this hypothesis has not been proven.

Levy's critics point to the absence of any scientific evidence that antimicrobial products pose a health threat. They note that triclosan has been used for decades in cleaning products without any sign of encouraging excessive growth of resistant strains. The CDC and the FDA have taken a cautious stance in the controversy, acknowledging the benefits of antimicrobial soaps in certain circumstances while agreeing that further research into the question is needed.

Who knew cleanliness could be so complicated? And in the meantime, what kind of soap should you use? Experts can't seem to agree on a single guideline, but the CDC recommends using mild, regular soap and washing in warm running water for at least 10–15 seconds in most cases, reserving antimicrobial soaps for limited applications: for use in handling food, for the care of newborns, and for the care of high-risk patients by health care workers.

and the chemical agents we have discussed in this chapter is that antimicrobics are typically used for treatment of disease, and not for environmental control of microbes. Nevertheless, some antimicrobics *are* used for antimicrobial control outside the body. For example, the antimicrobics *nisin* and *natamycin* are used respectively to reduce bacterial and fungal growth in cheese. Chapter 10 discusses in more detail the nature and use of antimicrobics to treat infectious diseases.

Table 9.5 summarizes the chemical methods of microbial control discussed in this chapter.

Development of Resistant Microbes

Many scientists are concerned that Americans have become overly preoccupied with antisepsis and disinfection, as evidenced by the proliferation of products containing antiseptic and disinfecting chemicals. For example, one can now buy hand soap, shampoo, toothpaste, hand lotion, foot pads for shoes, deodorants, and bath sponges that contain antiseptics, as well as kitty, litter, cutting boards, scrubbing pads, garbage bags, children's toys, and laundry detergents that contain disinfectants. There is little evidence that the extensive use of such products adds to human or animal health, but it does promote the development of strains of microbes resistant to antimicrobial chemicals because while susceptible cells die, resistant cells remain to proliferate. Scientists have already isolated strains of pathogenic bacteria, including *Mycobacterium tuberculosis, P. aeruginosa, E. coli,* and *S. aureus,* that are less susceptible to common disinfectants and antiseptics.

Highlight 9.3 discusses a controversy regarding the use of antibacterial soap.

CHAPTER SUMMARY

Basic Principles of Microbial Control (pp. 263–265)

1. **Sterilization** is the eradication of microorganisms and viruses; the term is not usually applied to the destruction of prions.

2. **Antisepsis** is the inhibition/killing of microorganisms (particularly pathogens) on skin or tissue by the use of a chemical **antiseptic,** whereas **disinfection** refers to the use of agents to inhibit microbes on inanimate objects.

3. **Degerming** refers to the removal of microbes from a surface by scrubbing.

4. **Sanitization** is the reduction of a prescribed number of pathogens from surfaces and utensils in public settings.

5. **Pasteurization** is a process using heat to kill pathogens and control microbes that cause spoilage of food and beverages.

6. The suffixes -stasis and -static indicate that an antimicrobial agent inhibits microbes, whereas the suffixes -cide and -cidal indicate that the agent kills or permanently inactivates a particular type of microbe.

7. **Microbial** death is the permanent loss of reproductive capacity. **Microbial death rate** measures the efficacy of an antimicrobial agent.

8. Antimicrobial agents destroy microbes either by altering their cell walls and membranes or by interrupting their metabolism and reproduction via interference with proteins and nucleic acids.

The Selection of Microbial Control Methods (pp. 265–268)

1. Factors affecting the efficacy of antimicrobial methods include the site to be treated, the relative susceptibility of microorganisms, and environmental conditions.

2. Four methods for evaluating the effectiveness of a disinfectant or antiseptic are the **phenol coefficient,** which compares the agent's efficacy to that of phenol; the **use-dilution test;** the **disk-diffusion method** (also known as the Kirby-Bauer method); and the **in-use test,** which provides a more accurate determination of efficacy under real-life conditions.

Physical Methods of Microbial Control (pp. 268–277)

1. **Thermal death point** is the lowest temperature that kills all cells in a broth in 10 minutes, whereas **thermal death time** is the time it takes to completely sterilize a particular volume of liquid at a set temperature. **Decimal reduction time (D)** is the time required to destroy 90% of the microbes in a sample.

2. An **autoclave** uses steam heat under pressure to sterilize chemicals and objects that can tolerate moist heat.

3. Pasteurization, a method of heating foods to kill pathogens and control spoilage organisms without altering the quality of the food, can be achieved by several methods: the historical (batch) method, flash pasteurization, and ultrahigh-temperature pasteurization. The methods differ in their combinations of temperature and time of exposure.

4. Under certain circumstances, microbes can be controlled using ultrahigh temperature sterilization, dry heat sterilization, incineration, refrigeration, or freezing.

5. Antimicrobial methods involving drying are **desiccation,** used to preserve food, and **lyophilization** (freeze drying), used for the long-term preservation of cells or microbes.

6. When used as a microbial control method, **filtration** is the passage of air or a liquid through a material that traps and removes microbes. Some **membrane filters** have pores small enough to trap the smallest viruses and some large protein molecules. HEPA (high-efficiency particulate air) filters remove microbes and particles from air.

7. The high **osmotic pressure** exerted by hypertonic solutions of salt or sugar can preserve foods such as jerky and jams by removing the water from microbes that they need to carry out their metabolic functions.

8. **Radiation** includes high-speed subatomic particles and even faster waves of electromagnetic energy released from atoms. **Ionizing radiation** (electromagnetic radiation with wavelengths shorter than 1 nm) creates ions that create effects leading to the denaturation of important molecules and cell death. **Nonionizing radiation,** which has wavelengths longer than 1 nm, is less effective in microbial control; however, UV light causes thymine dimers, which can kill affected cells.

Chemical Methods of Microbial Control (pp. 277–283)

1. **Phenolics,** which are chemically modified phenol molecules, are intermediate- to low-level disinfectants that denature proteins and disrupt cell membranes in a wide variety of pathogens.

2. **Alcohols** are intermediate-level disinfectants that denature proteins and disrupt cell membranes; they are used either as 70% to 90% aqueous solutions or in a tincture, which is a combination of an alcohol and another antimicrobial chemical.

3. **Halogens** (iodine, chlorine, bromine, and fluorine) are used as intermediate-level disinfectants and antiseptics to kill microbes in water or on medical instruments or skin. Although their exact mode of action is unknown, they are believed to denature proteins.

4. **Oxidizing agents** such as hydrogen peroxide, ozone, and peracetic acid are high-level disinfectants and antiseptics that release oxygen radicals, which are toxic to many microbes, especially anaerobes.

5. **Surfactants** include soaps, which act primarily to break up oils during degerming, and detergents such as **quaternary ammonium compounds (quats),** which are low-level disinfectants.

6. **Heavy metal** ions such as arsenic, silver, mercury, copper, and zinc are low-level disinfectants that denature proteins. For most applications they have been superceded by less-toxic alternatives.

7. **Aldehydes** are high-level disinfectants that cross-link organic functional groups in proteins and DNA. A 2% solution of glutaraldehyde and a 37% aqueous solution of formaldehyde (called formalin) are used to disinfect or sterilize medical or dental equipment and in embalming fluid.

8. **Gaseous agents** of microbial control, which include ethylene oxide, propylene oxide, and beta-propiolactone, are high-level disinfecting agents used to sterilize heat-sensitive equipment and large objects. These gases are explosive and potentially carcinogenic.

9. **Antimicrobics,** which include antibiotics, semisynthetics, and synthetics, are compounds that are typically used to treat diseases but can also function as intermediate-level disinfectants.

QUESTIONS FOR REVIEW
(Answers to multiple choice questions are on the web, along with additional review questions. Visit www.microbiologyplace.com.)

Multiple Choice

1. Which of the following statements provides the definition of sterilization in *practical* terms?
   a. Sterilization eliminates all organisms and their spores or endospores.
   b. Sterilization eliminates harmful microorganisms and viruses.
   c. Sterilization eliminates prions.
   d. Sterilization eliminates hyperthermophiles.

2. Which of the following substances or processes kills microorganisms on laboratory surfaces?
   a. antiseptics
   b. disinfectants
   c. degermers
   d. pasteurization

3. Which of the following terms best describes the disinfecting of cafeteria plates?
   a. pasteurization
   b. antisepsis
   c. sterilization
   d. sanitization

4. The microbial death rate is used to measure
   a. the efficiency of a detergent.
   b. the efficiency of an antiseptic.
   c. the efficiency of sanitization techniques.
   d. all of the above

5. Which of the following statements is true concerning the selection of an antimicrobial agent?
   a. An ideal antimicrobial agent is stable during storage.
   b. An ideal antimicrobial agent is fast acting.
   c. Ideal microbial agents do not exist.
   d. all of the above

6. The spores of which organism are used as a biological indicator of sterilization?
   a. *Bacillus stearothermophilus*
   b. *Salmonella typhi*
   c. *Mycobacterium tuberculosis*
   d. *Staphylococcus aureus*

7. A company that manufactures an antimicrobial cleaner for kitchen counters claims that its product is effective when used in a 50% water solution. By what means might scientists have been able to verify this statement?
   a. disk-diffusion test
   b. phenol coefficient
   c. filter paper test
   d. use-dilution test

8. Which of the following items functions most like an autoclave?
   a. a boiling pan
   b. an incinerator
   c. a desiccator
   d. a pressure cooker

9. The preservation of beef jerky from microbial growth relies on which method of microbial control?
   a. filtration
   b. lyophilization
   c. osmotic pressure
   d. radiation

10. Which of the following types of radiation is more widely used as an antimicrobial technique?
    a. electron beams
    b. visible light waves
    c. radio waves
    d. microwaves

11. Which of the following substances would most effectively inhibit anaerobes in a deep wound?
    a. phenol
    b. hexachlorophene
    c. isopropanol
    d. hydrogen peroxide

12. Which of the following adjectives best describes a surgical procedure that is free of microbial contaminants?
    a. disinfected
    b. sanitized
    c. degermed
    d. aseptic

13. Sterile paper disks are soaked in one of three disinfectants: P, Q, and R. The disks are placed on medium inoculated with organism X and incubated. The following chart indicates the results:

Disinfectants	Zone of inhibition
P	13 mm
Q	4.5 mm
R	10 mm

Assuming the disinfectants diffuse at the same rates and are equally stable, which disinfectant was most effective against organism X?
   a. disinfectant P
   b. disinfectant Q
   c. disinfectant R

14. A sample of *E. coli* has been subjected to heat for a specified time, and 90% of the cells have been destroyed. Which of the following terms best describes this event?
    a. thermal death point
    b. thermal death time
    c. decimal reduction time
    d. none of the above

15. Which of the following substances is least toxic to humans?
    a. carbolic acid
    b. glutaraldehyde
    c. hydrogen peroxide
    d. formalin

16. Which of the following chemicals is active against bacterial endospores?
    a. copper ions
    b. ethylene oxide
    c. ethanol
    d. triclosan

17. Which of the following disinfectants acts against cell membranes?
    a. phenol
    b. peracetic acid
    c. silver nitrate
    d. glutaraldehyde

18. Which of the following disinfectants contains alcohol?
    a. an iodophor
    b. a quat
    c. formalin
    d. a tincture of bromine

Short Answer

1. Describe three types of microbes that are extremely resistant to antimicrobial treatment, and explain why they are resistant.

2. Compare and contrast the four tests that have been developed to measure the efficacy of disinfectants.

3. Why is it necessary to use strong disinfectants in areas exposed to tuberculosis patients?

4. Why do warm disinfectant chemicals generally work better than cool ones?

5. Why are Gram-negative bacteria more susceptible to heat than Gram-positive bacteria?

6. Describe five physical methods of microbial control.

7. What is the difference between thermal death point and thermal death time?

8. Defend the following statement: "Pasteurization is not sterilization."

9. What antimicrobial agents would you recommend for making drinking water safe in your community?

10. Compare and contrast the action of alcohols, halogens, and oxidizing agents in controlling microbial growth.

11. Hyperthermophilic prokaryotes may remain viable in canned goods after commercial sterilization. Why is this situation not dangerous to consumers?

12. Why are alcohols more effective in a 70% solution than in a 100% solution?

13. Contrast the structures and actions of soaps and quats.

14. What are some advantages and disadvantages of using ionizing radiation to sterilize food?

15. How can campers effectively treat stream water to remove pathogenic protozoa, bacteria, and viruses?

CRITICAL THINKING

1. In what ways might it be argued that the widespread commercial use of antiseptics and disinfectants has hurt rather than helped American health?

2. Is desiccation the only antimicrobial effect operating when grapes are dried in the sun to make raisins? Explain.

3. How long would it take to reduce a population of 100 trillion (10^{14}) bacteria to 10 viable cells if the D value of the treatment is 3 minutes?

4. Some potentially pathogenic bacteria and fungi, including strains of *Enterococcus, Staphylococcus, Candida,* and *Aspergillus,* can survive for 1–3 months on a variety of materials found in hospitals, including the synthetic material of privacy curtains, the cotton blends of scrub suits, nurses' clothes, lab coats, and plastics from splash aprons and computer keyboards. What steps could hospital personnel take to reduce the spread of these pathogens to susceptible patients?

5. In the spring of 2001, California health officials identified 23 patients infected with *Salmonella* who ate uncooked alfalfa sprouts. Twenty-one patients developed acute diarrhea, and three had to be hospitalized. Investigators discovered that the sprout producer had subjected the contaminated sprouts to a heat treatment followed by a 15-minute soak in calcium hypochlorite. Based on the discussion in this chapter, what types of heat treatment are available to producers of alfalfa sprouts? Why were the antimicrobial methods not effective in preventing salmonellosis? What other precautions could either the producer or consumers have taken?

6. An over-the-counter medicated foot powder contains camphor, eucalyptus oil, lemon oil, and zinc oxide. Only one of the ingredients is antimicrobial. Which one? How does it act against fungi?

CHAPTER 10

Controlling Microbial Growth in the Body: Antimicrobial Drugs

Meet *Staphylococcus aureus,* a common bacterium that is the number one cause of hospital-acquired infections in the United States. The particular strain of *S. aureus* shown here, however, is remarkable in some very important and alarming ways. Not only is it resistant to methicillin, the antimicrobial that is traditionally used to treat staphylococcal infections; it is also resistant to vancomycin, a drug that was long considered the last line of defense against methicillin-resistant *S. aureus.* Although other drugs may be used to fight it, this "superbug" strain of *S. aureus* leaves patients few options for defense—and indeed it can cause death.

How do antimicrobial drugs work? Why do they work on some microorganisms and not on others? What can be done about the increasing problem of superbugs that resist a wide variety of existing drugs? This chapter focuses on the chemical control of pathogens in the body.

CHAPTER OUTLINE

The History of Antimicrobial Agents

Mechanisms of Antimicrobial Action
Inhibition of Cell Wall Synthesis
Inhibition of Protein Synthesis
Disruption of Cytoplasmic Membranes
Inhibition of Metabolic Pathways
Inhibition of Nucleic Acid Synthesis
Prevention of Virus Attachment

Clinical Considerations in Prescribing Antimicrobial Drugs
Spectrum of Action
Efficacy
Routes of Administration
Safety and Side Effects

Resistance to Antimicrobial Drugs
The Development of Resistance in Populations
Mechanisms of Resistance
Multiple Resistance and Cross Resistance
Retarding Resistance

Vancomycin-resistant *Staphylococcus aureus.*

MicroPrep Pre-Test: *Take the pre-test for this chapter on the web.*
*Visit **www.microbiologyplace.com.***

Chemicals that affect physiology in any manner, such as caffeine, alcohol, and tobacco, are called *drugs*. Drugs that act against diseases are called *chemotherapeutic agents*. Examples include insulin, anticancer drugs, and drugs for treating infections—called **antimicrobial agents (antimicrobials),** the subject of this chapter.

In the pages that follow we'll examine the mechanisms by which antimicrobial agents act, the factors that must be considered in the use of antimicrobials, and several issues surrounding resistance to antimicrobial agents among microorganisms. First, however, we begin with a brief history of antimicrobial chemotherapy.

The History of Antimicrobial Agents

Learning Objectives

✓ Describe the contributions of Paul Ehrlich, Alexander Fleming, and Gerhard Domagk in the development of antimicrobials.

✓ Explain how semisynthetic and synthetic antimicrobials differ from antibiotics.

The little girl lay struggling to breathe as her parents stood mutely by, willing the doctor to do something—anything—to relieve the symptoms that had so quickly consumed their four-year-old daughter's vitality. Sadly, there was little the doctor could do. The thick "pseudomembrane" of diphtheria, composed of bacteria, mucus, blood clotting factors, and white blood cells, adhered tenaciously to her pharynx, tonsils, and vocal cords. He knew that trying to remove it could rip open the underlying mucous membrane, resulting in bleeding, possibly additional infections, and death. In 1902, there was little medical science could offer for the treatment of diphtheria; all physicians could do was wait and hope.

At the beginning of the 20th century, much of medicine involved diagnosing illness, describing its expected course, and telling family members either how long a patient might be sick, or when they might expect her to die. Even though physicians and scientists had recently accepted the germ theory of disease and knew the causes of many diseases, very little could be done to inhibit pathogens, including *Corynebacterium diphtheriae* (kŏ-rī′nē-bak-tēr′ē-ŭm dif-thi′rē-ē), and alter the course of infections. In fact, one-third of children born in the early 1900s died from infectious diseases before the age of five.

It was at this time that Paul Ehrlich (1854–1915), a visionary German scientist, proposed the term *chemotherapy* to describe the use of chemicals that would selectively kill pathogens while having little or no effect on a patient. He wrote of "magic bullets" that would bind to receptors on germs to bring about their death while ignoring host cells, which lacked the receptor molecules.

Ehrlich's search for antimicrobial agents resulted in the discovery of one arsenic compound that killed trypanosome parasites, and another that worked against the bacterial agent of syphilis. A few years later, in 1929, the British bacteriologist Alexander Fleming (1881–1955) reported the antibacterial

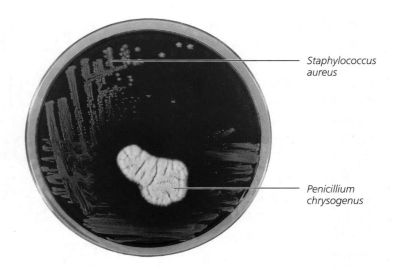

Staphylococcus aureus

Penicillium chrysogenus

▲ *Figure 10.1*

Antibiotic effect of the mold *Penicillium chrysogenum*. Alexander Fleming observed that this mold secretes penicillin, which inhibits the growth of bacteria, as is apparent with *Staphylococcus aureus* growing on this blood agar plate.

action of penicillin released from *Penicillium* (pen-i-sil′ē-ŭm) mold **(Figure 10.1).** Fleming coined the term **antibiotics** to describe antimicrobial agents that are produced naturally by an organism. (See **Highlight 10.1** for a discussion of why microorganisms secrete antibiotics.) In common usage today "antibiotic" means "an antibacterial agent," excluding agents with antiviral and antifungal activity.

Though arsenic compounds and penicillin were discovered first, they were not the first antimicrobials in widespread use because Ehrlich's arsenic compounds are toxic to humans, and because penicillin was not available in large enough quantities to be useful until the late 1940s. Instead, *sulfanilamide,* discovered in 1932 by the German chemist Gerhard Domagk (1895–1964), was the first practical antimicrobial agent efficacious in treating a wide array of bacterial infections.

Other microorganisms besides the mold *Penicillium* are sources of useful antimicrobials, most notably the fungus *Cephalosporium* (sef′ă-lō-spor′ē-um) and species of soil-dwelling bacteria in the genera *Bacillus* (ba-sil′ŭs) and *Streptomyces* (strep-tō-mī′sēz). Additionally, by altering the chemical structure of antibiotics, scientists produce **semisynthetics**—drugs that are more effective, longer lasting, or easier to administer than the naturally occurring antibiotics. Antimicrobials that are completely synthesized in a laboratory are called **synthetics.** More than half of all antibiotics and semisynthetics are derived from species of *Streptomyces*.

Table 10.1 provides a partial list of common antibiotics and semisynthetics and their sources.

CRITICAL THINKING

Why aren't antibiotics effective against the common cold?

Highlight 10.1   Why Do Microbes Make Antibiotics?

We all know that antibiotics benefit humans, but what good are they to the microorganisms that secrete them? From the viewpoint of evolutionary theory, the answer may seem obvious: Antibiotics are weapons that confer an advantage to the secreting organisms in their struggle for survival. In reality, however, the answer is not so simple.

Antibiotics are members of an extremely diverse group of metabolic products known as *secondary metabolites,* which typically are complex organic molecules that are not essential for normal cell growth and reproduction and are produced only *after* an organism has established itself in its environment. The production of secondary metabolites results in a metabolic cost for the cell; that is, antibiotic production consumes energy and raw materials that could be used for growth and reproduction. Tetracycline, for example, is the end result of 72 separate enzymatic steps, and erythromycin requires 25 different chemical reactions—none of which contribute to *Streptomyces*'s normal growth or reproduction. Therefore, the question can be modified: Of what use are metabolically "expensive" antibiotics to organisms that are already secure in their environment?

Adding to the conundrum is the fact that antimicrobials against bacteria have never been discovered in natural soil at high enough concentrations to be inhibitory to *any* neighboring cell. For example, it is almost impossible to detect tetracyclines, aminoglycosides, chloramphenicol, or other antibiotics produced by *Streptomyces*, except when the bacteria are grown in a laboratory. Production of antibiotics in inconsequential quantities hardly seems to represent a significant adaptive edge.

Some scientists have suggested that antibiotics are evolutionary vestiges—leftovers of metabolic pathways that were once useful but no longer have a significant role. Others point out that evolutionary theory suggests that there would be tremendous selective pressure against the slightest continued manufacture of complex antibiotics if they truly have little purpose for the microorganism. Further research is required before we may satisfactorily answer the question, if we ever can.

Table 10.1   Sources of Some Common Antibiotics and Semisynthetics

Microorganism	Antimicrobial
Fungi	
Penicillium chrysogenum	Penicillin
Penicillium griseofulvin	Griseofulvin
Cephalosporium spp.	Cephalothin
Bacteria	
Bacillus licheniformis	Bacitracin
Bacillus polymyxa	Polymyxin
Micromonospora purpurea	Gentamicin
Streptomyces griseus	Streptomycin
Streptomyces fradiae	Neomycin
Streptomyces aureofaciens	Tetracycline
Streptomyces orientalis	Vancomycin
Streptomyces venezuelae	Chloramphenicol
Streptomyces erythraeus	Erythromycin
Streptomyces mediterranei	Rifampin
Streptomyces nodosus	Amphotericin B
Streptomyces avermitilis	Ivermectin

Mechanisms of Antimicrobial Action

Learning Objectives

✓ Explain the principle of selective toxicity.

✓ List six mechanisms by which antimicrobial drugs affect the growth of pathogens.

As Ehrlich foresaw, the key to successful chemotherapy against microbes is **selective toxicity;** that is, an effective antimicrobial agent must be more toxic to a pathogen than to the pathogen's host. Selective toxicity is possible because of differences in structure or metabolism between the pathogen and its host. Typically, the more differences, the easier it is to discover or create an effective antimicrobial agent.

Because there are many differences between the structure and metabolism of pathogenic bacteria and their eukaryotic hosts, antibacterial drugs constitute the greatest number and diversity of antimicrobial agents. Similarly, fewer antifungal, antiprotozoan, and anthelmintic drugs are available because fungi, protozoa, and helminths—like their animal and human hosts—are eukaryotic and thus share many common features. The number of effective antiviral drugs is even more limited, despite major differences in structure, because viruses utilize their host cells' enzymes and ribosomes to metabolize and replicate. Therefore, drugs that are effective against viral replication are likely toxic to the host as well.

Although they can have a variety of effects on pathogens, antimicrobial drugs can be categorized into several general groups according to their mechanisms of action (**Figure 10.2** on page 290):

- Drugs that inhibit cell wall synthesis. These drugs are selectively toxic to certain fungal or bacterial cells, which have cell walls, but not to animals, which lack cell walls.

- Drugs that inhibit protein synthesis (translation) by targeting the differences between prokaryotic and eukaryotic ribosomes.

Figure 10.2 ▶

Mechanisms of action of antibacterial drugs. Also listed are representative drugs for each type of action.

Inhibition of cell wall synthesis
Penicillins
Cephalosporins
Vancomycin
Bacitracin
Isoniazid
Ethambutol

Inhibition of pathogen's attachment to, or recognition of, host
Arildone

Inhibition of protein synthesis
Aminoglycosides
Tetracyclines
Chloramphenicol
Macrolides

Inhibition of DNA or RNA synthesis
Actinomycin
Nucleotide analogs
Quinolones
Rifampin

Disruption of cytoplasmic membrane
Polymyxins
(Polyenes are antifungal)

Inhibition of general metabolic pathway
Sulfonamides
Trimethoprim

- Drugs that disrupt the cytoplasmic membrane.
- Drugs that inhibit general metabolic pathways.
- Drugs that inhibit nucleic acid synthesis.
- Drugs that block a pathogen's recognition of or attachment to its host.

In the following sections we examine these mechanisms in turn.

Inhibition of Cell Wall Synthesis

Learning Objective

✓ Describe the actions and give examples of drugs that affect the cell walls of bacteria.

A cell wall protects a cell from the effects of osmotic pressure. The major structural component of a bacterial cell wall is its peptidoglycan layer. As we discussed in Chapter 3, peptidoglycan is a single, huge macromolecule composed of polysaccharide chains of alternating N-acetylglucosamine

(NAG) and N-acetylmuramic acid (NAM) molecules that are cross-linked by short peptide chains extending between NAM subunits (see Figure 3.12). To enlarge or divide, a cell must synthesize more peptidoglycan by adding new NAG and NAM subunits to existing NAG–NAM chains, and the new NAM subunits must then be bonded to neighboring NAM subunits **(Figure 10.3a and b).**

The most common antibacterial agents act by preventing the cross-linkage of NAM subunits. Most prominent among these drugs are **beta-lactams,** such as penicillins and cephalosporins, which are antimicrobials whose functional portions are called *beta-lactam (β-lactam) rings* **(Figure 10.3c).** Beta-lactams inhibit peptidoglycan formation by irreversibly binding to the enzymes that cross-link NAM subunits **(Figure 10.3d).** In the absence of correctly formed peptidoglycan, growing bacterial cells have weakened cell walls that are less resistant to the effects of osmotic pressure. The underlying cytoplasmic membrane bulges through the weakened portions of cell wall as water moves into the cell, and eventually the cell lyses **(Figure 10.3e).**

A bacterial cell wall is composed of a macro-molecule of peptidoglycan composed of NAG-NAM chains that are cross-linked by peptide bridges between the NAM subunits

New NAG and NAM subunits are inserted into the wall by enzymes, allowing the cell to grow. Normally, other enzymes link new NAM subunits to old NAM subunits with peptide crosslinks.

Growth

Normal bacterial cell wall

NAM
NAG
Crosslink
NAG-NAM chain

(a)

(b)   SEM   0.5 μm

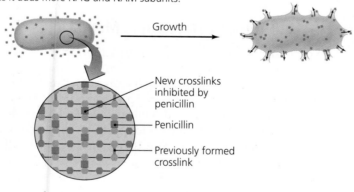

Penicillin G (natural)

Methicillin (semisynthetic)
Penicillins

β-lactam ring

Cephalothin (semisynthetic)
COONa
Cephalosporin

Aztreonam (semisynthetic)
Monobactam

(c)

Penicillin interferes with the linking enzymes, and NAM subunits remain unattached to their neighbors. However, the cell continues to grow as it adds more NAG and NAM subunits.

The cell bursts from osmotic pressure because the integrity of peptidoglycan is not maintained.

Growth

New crosslinks inhibited by penicillin

Penicillin

Previously formed crosslink

(d)

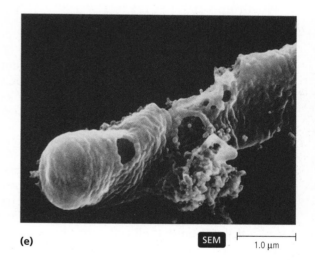

(e)   SEM   1.0 μm

▲ *Figure 10.3*

Bacterial cell wall synthesis, and the inhibitory effects of beta-lactams on it. (a) A schematic depiction of a normal peptidoglycan cell wall showing NAG–NAM chains and cross-linked NAM subunits. **(b)** A scanning electron micrograph of a bacterial cell with a normal cell wall. **(c)** Structural formulas of some beta-lactam drugs. Their functional portion is the β-lactam ring. **(d)** A schematic depiction of the effect of penicillin on peptidoglycan in preventing NAM–NAM cross-links. **(e)** A scanning electron micrograph of bacterial lysis due to the effects of penicillin. *After a beta-lactam weakens a peptidoglycan molecule by preventing NAM–NAM cross-linkages, what force actually kills an affected bacterial cell?*

Figure 10.3 Cells lyse due to osmotic movement of water into the cell.

Chemists have made alterations to natural beta-lactams, such as penicillin G, to create semisynthetic derivatives such as methicillin and cephalothin (see Figure 10.3c), which are more stable in the acidic environment of the stomach, more readily absorbed in the intestinal tract, less susceptible to deactivation by bacterial enzymes, or more active against more types of bacteria. The simplest beta-lactams are *monobactams*, which are seldom used because they are effective only against aerobic Gram-negative bacteria.

Other antimicrobials such as **vancomycin** (van-kō-mī′sin), which is obtained from *Streptomyces orientalis*, and **cycloserine**, a semisynthetic, disrupt cell wall formation in a different manner. They directly interfere with particular alanine–alanine bridges that link the NAM subunits in many Gram-positive bacteria. Those bacteria that lack alanine–alanine crossbridges are naturally resistant to these drugs. Still another drug that prevents cell wall formation, **bacitracin** (bas-i-trā′sin), blocks the secretion of NAG and NAM from the cytoplasm. Like beta-lactams, vancomycin, cycloserine, and bacitracin result in cell lysis due to the effects of osmotic pressure.

Since all these drugs prevent bacteria from *increasing* the amount of cell wall material but have no effect on existing peptidoglycan, they are effective only on bacterial cells that are growing or reproducing; dormant cells are unaffected. Of course, they do not harm plant or animal cells because such cells lack peptidoglycan cell walls.

Bacteria of the genus *Mycobacterium* (mī′kō-bak-tēr′ē-ŭm), notably the agents of leprosy and tuberculosis, are characterized by unique, complex cell walls that have a layer of arabinogalactan-mycolic acid in addition to the usual peptidoglycan of prokaryotic cells. **Isoniazid** (ī-sō-nī′ă-zid, or **INH**[1]) and **ethambutol** (eth-am′boo-tol) disrupt the formation of this extra layer. Mycobacteria typically only reproduce every 12–24 hours, in part because of the complexity of their cell walls, so antimicrobial agents that act against mycobacteria must be administered for months or even years to be effective. It is often difficult to ensure that patients continue such a long regimen of treatment.

Inhibition of Protein Synthesis

As we discussed in Chapter 2, cells use proteins for structure and regulation, as enzymes in metabolism, and as channels and pumps to move materials across cell membranes. Thus, a consistent supply of proteins is vital for the active life of a cell. Given that all cells use ribosomes to translate proteins using information from messenger RNA templates, it is not immediately obvious that drugs could selectively target differences related to protein synthesis. Recall, however, that prokaryotic ribosomes differ from eukaryotic ribosomes in structure and size: Prokaryotic ribosomes are 70S and composed of 30S and 50S subunits, whereas eukaryotic ribosomes are 80S with 60S and 40S subunits.

Many antimicrobial agents take advantage of the differences between ribosomes to selectively target bacterial protein translation without significantly affecting eukaryotes. Note, however, that because some of these drugs affect eukaryotic mitochondria, which also contain 70S ribosomes like those of prokaryotes, such drugs may be harmful to animals and humans.

Understanding the actions of antimicrobials that inhibit protein synthesis requires an understanding of the process of translation, because various parts of ribosomes are the targets of antimicrobial drugs. Recall from the discussion of translation in Chapter 7 (see p. 212) that both the 30S and 50S subunits of a prokaryotic ribosome play a role in the initiation of protein synthesis, in codon recognition, and in the docking of tRNA–amino acid complexes, and that the 50S subunit contains the enzymatic portion that actually forms peptide bonds.

Among the antimicrobials that target the 30S ribosomal subunit are aminoglycosides and tetracyclines. **Aminoglycosides** (am′i-nō-glī′kō-sīds), such as *streptomycin, amikacin, tobramycin,* and *gentamicin,* change the shape of the 30S subunit, making it impossible for the ribosome to read the codons of mRNA correctly **(Figure 10.4a)**. **Tetracyclines** (tet-ră-sī′klēns) block the tRNA docking site, which then prevents the incorporation of additional amino acids into a growing polypeptide **(Figure 10.4b)**.

Other antimicrobials interfere with the function of the 50S subunit. *Chloramphenicol* and similar drugs block the enzymatic site of the 50S subunit **(Figure 10.4c)**, which prevents translation. *Clindamycin* and antimicrobial agents called **macrolides** (mak′rō-līds), including *erythromycin,* bind to a different portion of the 50S subunit, preventing movement of the ribosome from one codon to the next **(Figure 10.4d)**; as a result, translation is frozen and protein synthesis is halted.

Disruption of Cytoplasmic Membranes

Some antimicrobial drugs disrupt the cytoplasmic membrane of targeted cells, often by becoming incorporated into the membrane and damaging its integrity. This is the mechanism of action of a group of drugs called **polyenes** (pol-ē-ēns′). One polyene called *amphotericin B* **(Figure 10.5a)** is fungicidal because it attaches to *ergosterol,* a lipid constituent of fungal membranes **(Figure 10.5b)**, in the process disrupting the membrane and causing lysis of the cell. The cytoplasmic membranes of humans are somewhat susceptible to amphotericin B because they contain cholesterol, which is similar to ergosterol, though cholesterol does not bind amphotericin B as well as does ergosterol.

Most bacterial membranes lack sterols, so these bacteria are naturally resistant to amphotericin B; however, there are other agents that disrupt bacterial membranes. An example of these antibacterial agents is *polymyxin*, produced by *Bacillus polymyxa* (ba-sil′ŭs po-lē-miks′ă). Polymyxin is effective against Gram-negative bacteria, particularly *Pseudomonas* (soo-dō-mō′nas), but because it is toxic to human kidneys it

---

[1]From *isonicotinic acid hydrazide,* the correct chemical name for isoniazid.

(a)

(b)

(c)

(d)

◄ *Figure 10.4*

The mechanisms by which antimicrobials inhibit protein synthesis by targeting prokaryotic ribosomes. **(a)** Aminoglycosides change the shape of the 30S subunit, causing incorrect pairing of tRNA anticodons with mRNA codons. **(b)** Tetracyclines block the tRNA docking site on the 30S subunit, preventing protein elongation. **(c)** Chloramphenicol blocks enzymatic activity of the 50S subunit, preventing the formation of peptide bonds between amino acids. **(d)** Macrolides bind to the 50S subunit, preventing movement of the ribosome along the mRNA. *Which tRNA anticodon should align with the codon CUG?*

Figure 10.4 Anticodon GAC is the correct complement for the codon CUG.

(a)

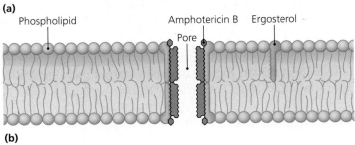

(b)

▲ *Figure 10.5*

Disruption of the cytoplasmic membrane by the antifungal amphotericin B. **(a)** The structure of amphotericin B. **(b)** The proposed action of amphotericin B. The drug binds to molecules of ergosterol, which then congregate, forming a pore.

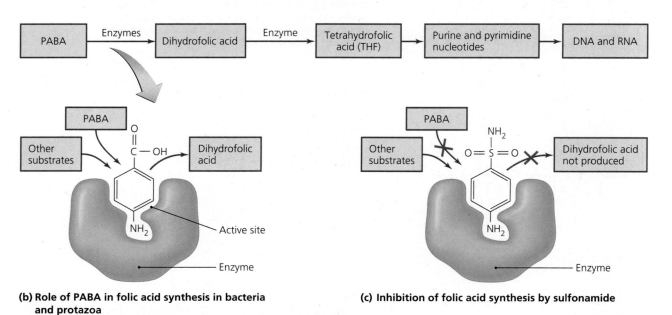

(a) Para-aminobenzoic acid (PABA) and its structural analogs, the sulfonamides

(b) Role of PABA in folic acid synthesis in bacteria and protozoa

(c) Inhibition of folic acid synthesis by sulfonamide

▲ *Figure 10.6*

The antimetabolic action of sulfonamides in inhibiting nucleic acid synthesis. **(a)** Para-aminobenzoic acid (PABA) and representative members of its structural analogs, the sulfonamides. The analogous portions of the compounds are shaded. **(b)** The metabolic pathway in bacteria and protozoa by which folic acid is synthesized from PABA. **(c)** The inhibition of folic acid synthesis by the presence of a sulfonamide, which deactivates the enzyme by binding irreversibly to the enzyme's active site.

is usually reserved for use against external pathogens that are resistant to other antibacterial drugs.

Inhibition of Metabolic Pathways

As we discussed in Chapter 5, metabolism can be defined simply as the sum of all chemical reactions that take place within an organism. Whereas most living things share certain metabolic reactions—for example, glycolysis—other chemical reactions are unique to certain organisms. Whenever differences exist between the metabolic processes of a pathogen and its host, *antimetabolic agents* can be effective.

A variety of kinds of antimetabolic agents are available, including heavy metals (such as arsenic, mercury, and antimony), which inactivate enzymes; agents that rid the body of parasitic worms by paralyzing them; drugs that block the activation of viruses; and metabolic antagonists such as sulfanilamide, the first commercially available antimicrobial agent.

Sulfanilamide and similar compounds, called **sulfonamides,** act as antimetabolic drugs because they are *structural analogs* of—that is, are chemically very similar to—*para-aminobenzoic acid (PABA;* **Figure 10.6a**), a compound that is crucial in the anabolic reactions in the synthesis of nucleotides required for DNA and RNA synthesis. Many organisms, including some pathogens, enzymatically convert PABA into dihydrofolic acid, and dihydrofolic acid into tetrahydrofolic acid (THF), a form of folic acid that is then used as a coenzyme in the synthesis of purine and pyrimidine nucleotides **(Figure 10.6b)**. As analogs of PABA, sulfonamides compete with PABA molecules for the active site of the enzyme involved in the production of dihydrofolic acid **(Figure 10.6c)**, which leads to a decrease in the production of THF, and thus of DNA and RNA, thereby slowing the production of proteins. Thus the end result of sulfonamide competition with PABA is the cessation of cell metabolism, which leads to cell death.

Another antimetabolic agent, *trimethoprim*, also interferes with nucleic acid synthesis. However, instead of binding to the enzyme that converts PABA to dihydrofolic acid, trimethoprim binds to the enzyme involved in the conversion of dihydrofolic acid to THF, the next step in this metabolic pathway.

Note that humans do not synthesize THF from PABA, as some bacteria and protozoa do; instead, we take simple folic acids found in our diets and convert them into THF. As a result, human metabolism is unaffected by sulfonamides.

Antiviral agents can also target the unique aspects of the metabolism of viruses. After attachment to a host cell, many viruses must penetrate the cell's membrane and be uncoated to release viral genetic instructions and assume control of the cell's metabolic machinery. Some viruses of eukaryotes are uncoated as a result of the acidic environment within phagolysosomes. *Amantadine, rimantadine,* and weak organic bases can neutralize the acid of phagolysosomes and thereby prevent viral uncoating; thus, these are antiviral drugs. Amantadine is used exclusively to prevent infections by influenza type A virus.

CRITICAL THINKING

It would be impractical and expensive for every American to take amantadine during the entire flu season to prevent influenza infections. For what group of people might amantadine prophylaxis be cost effective?

Inhibition of Nucleic Acid Synthesis

As we saw in Chapter 7, the nucleic acids DNA and RNA are built from purine and pyrimidine nucleotides and are critical to the survival of cells. Several drugs function by blocking either the replication of DNA or its transcription into RNA.

Because only slight differences exist between the DNA of prokaryotes and eukaryotes, drugs that affect DNA replication often act against both types of cells. For example, *actinomycin* binds to DNA and effectively blocks DNA synthesis and RNA transcription not only in bacterial pathogens, but in their hosts as well. Generally, drugs of this kind are not used to treat infections, though they are used in research of DNA replication and may be used judiciously to slow replication of cancer cells.

Other compounds that can act as antimicrobials by interfering with the function of nucleic acids are called **nucleotide analogs** because of the compounds' structural similarities to the normal nucleotide building blocks of nucleic acids **(Figure 10.7** on page 296). The structures of certain nucleotide analogs enable them to be incorporated into the DNA or RNA of pathogens, where they distort the shapes of the nucleic acid molecules and prevent further replication, transcription, or translation. Nucleotide analogs are most often used against viruses because viral DNA polymerases are tens to hundreds of times more likely to incorporate nonfunctional nucleotides into nucleic acids than are host polymerase enzymes. Additionally, viral nucleic acid synthesis is usually more rapid than nucleic acid synthesis in cells. These characteristics make viruses more susceptible to nucleotide analogs than their hosts are, though nucleotide analogs are also effective against rapidly dividing cancer cells.

The synthetic drugs called *quinolones* and *fluoroquinolones* are unusual because they are active against prokaryotic DNA specifically. These antibacterial agents inhibit *DNA gyrase,* an enzyme necessary for correct coiling and uncoiling of replicating bacterial DNA; they typically have little effect on eukaryotes or viruses. See **Highlight 10.2** on page 297 for a closer look at ciprofloxacin (sip-rō-floks'ă-sin), a quinolone drug that is prescribed for respiratory and urinary tract infections as well as for suspected cases of anthrax.

Other antimicrobial agents function by binding to and inhibiting the action of RNA polymerases during the synthesis of RNA from a DNA template. Several drugs, including *rifampin* (rif'am-pin), bind more readily to prokaryotic RNA polymerase than to eukaryotic RNA polymerase; as a result, rifampin is more toxic to prokaryotes than to their eukaryotic hosts. Rifampin is used primarily against *Mycobacterium tuberculosis* and other pathogens that metabolize slowly and thus are less susceptible to antimicrobials targeting active metabolic processes and protein synthesis.

Clofazimine binds to the DNA of *Mycobacterium leprae,* the causative agent of leprosy, and prevents normal replication and transcription. It is also used to treat tuberculosis and other mycobacterial infections.

Prevention of Virus Attachment

Many pathogens, particularly viruses, must attach to their host's cells via the chemical interaction between attachment proteins on the pathogen and complementary receptor proteins on a host cell. Attachment of viruses can be blocked by peptide and sugar analogs of either attachment or receptor proteins. When these sites are blocked by analogs, viruses can neither attach to nor enter their host cells. The use of such substances, called *attachment antagonists,* is still in the developmental stage. For example, small peptide chains from HIV attachment proteins may prove effective at preventing infection by HIV. *Arildone,* an antagonist of the receptor of polioviruses and some cold viruses, blocks attachment of these viruses and is being evaluated as a deterrent to infections.

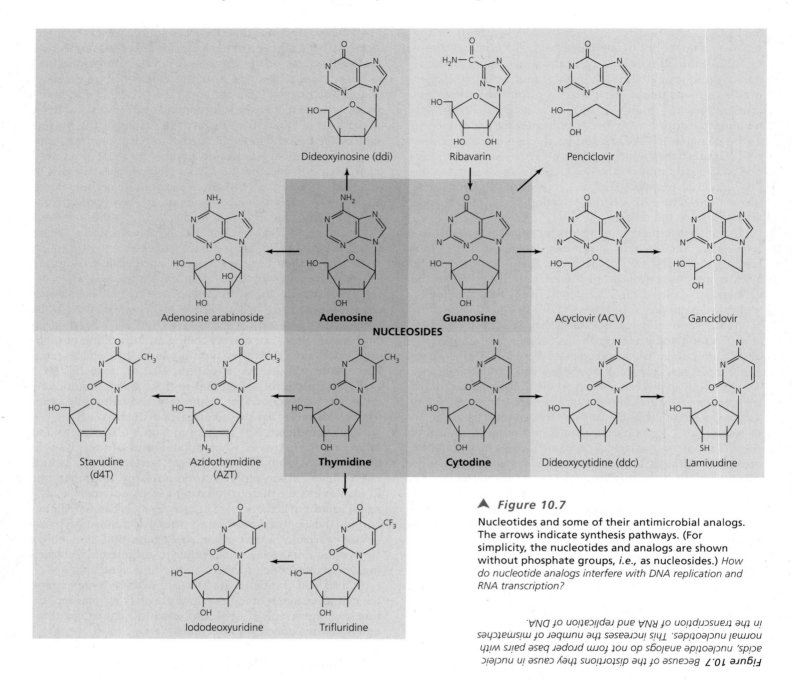

▲ *Figure 10.7*

Nucleotides and some of their antimicrobial analogs. The arrows indicate synthesis pathways. (For simplicity, the nucleotides and analogs are shown without phosphate groups, *i.e.,* as nucleosides.) *How do nucleotide analogs interfere with DNA replication and RNA transcription?*

Figure 10.7 Because of the distortions they cause in nucleic acids, nucleotide analogs do not form proper base pairs with normal nucleotides. This increases the number of mismatches in the transcription of RNA and replication of DNA.

Clinical Considerations in Prescribing Antimicrobial Drugs

Even though antibiotics are produced commonly by many fungi and bacteria, most of these chemicals are not effective for treating diseases because they are toxic to humans and animals, are too expensive, are produced in minute quantities, or lack potency. The ideal antimicrobial agent to treat an infection or disease would be one that is:

- readily available
- inexpensive
- chemically stable (so that it can be transported easily and stored for long periods of time)
- easily administered
- nontoxic and nonallergenic
- selectively toxic against a wide range of pathogens

No antimicrobial agent has all of these qualities, so doctors and medical laboratory technicians must evaluate antimicrobials with respect to several characteristics: the range of pathogens against which they are effective, called their spectrum of action; their efficacy, including the dosages required to be effective; the routes they can be administered; and their overall safety, and the side effects they produce. We consider

Ciprofloxacin hydroxide (commercially known as "Cipro") is a synthetic antibiotic used to treat a variety of bacterial infections. In the fall of 2001, following the anthrax attacks through the U.S. mail, Cipro made headlines for its efficacy against *Bacillus anthracis*, the bacterium that causes anthrax. Thousands of postal workers were given Cipro as a preventative measure against inhalational anthrax, and the U.S. government bought 100 million Cipro tablets from a drug manufacturer for an emergency stockpile. Demand for the drug surged as frightened consumers sought Cipro in doctor's offices, pharmacies, and over the Internet, spurring concerns that the drug was being oversold and misused.

Cipro works by inhibiting the DNA synthesis of bacteria. Specifically, it targets an enzyme called DNA gyrase (topoisomerase II), which is responsible for the coiling and uncoiling of bacterial DNA. Cipro thus interferes with the ability of the bacterial cell to wind and unwind its DNA, which consequently affects the cell's ability to reproduce. It is not the only drug that has proven effective against *B. anthracis*—other antibiotics, including penicillin, can also work—but because it is a relatively new drug, bioterrorists are less likely to have developed strains of *B. anthracis* that can resist it.

Cipro is a powerful drug that should never be taken without the close supervision of a physician. Its side effects can be serious and include vomiting, diarrhea, headaches, dizziness, and central nervous system problems. Indeed, the Centers for Disease Control and Prevention discourages its use by anyone who has not been exposed to anthrax or who has not had the drug prescribed for a medical condition. Random, widespread use of Cipro is not only potentially harmful to individuals; it can also pose public health problems by increasing the chances that a drug-resistant form of *B. anthracis* will develop.

each of these characteristics of microbial agents in the following sections.

Spectrum of Action

Learning Objective

✓ Distinguish between narrow-spectrum drugs and broad-spectrum drugs in terms of their targets and side effects.

The number of different kinds of pathogens a drug acts against is known as its **spectrum of action (Figure 10.8** on page 298); drugs that work against only a few kinds of pathogens are **narrow-spectrum drugs,** whereas those that are effective against many different kinds of pathogens are **broad-spectrum drugs.** For instance, because erythromycin acts against Gram-negative bacteria, Gram-positive bacteria, and chlamydias and rickettsias, it is considered a broad-spectrum antibiotic. In contrast, penicillin cannot easily penetrate the outer membranes of Gram-negative bacteria to reach and prevent the formation of their peptidoglycan cell walls, so its efficacy is largely limited to Gram-positive bacteria; thus penicillin has a narrower spectrum of action than erythromycin, which acts against protein synthesis.

The use of broad-spectrum antimicrobials is not always as desirable as it might seem. Besides killing pathogens, broad-spectrum antimicrobials can also open the door to serious secondary infections by transient pathogens or *superinfections* by members of the normal microbiota unaffected by the antimicrobial. This results because the killing of normal microbiota reduces *microbial antagonism,* the competition between normal microbes and pathogens for nutrients and space that reinforces the body's defense by limiting the

ability of pathogens to colonize the skin and mucous membranes. Thus a woman using erythromycin to treat strep throat (a bacterial disease) could develop vaginitis resulting from the excessive growth of *Candida albicans* (kan'did-ă al'bi-kanz), a yeast that is unaffected by erythromycin and is freed from microbial antagonism when an antibiotic kills normal bacteria in the vagina.

Efficacy

Learning Objective

✓ Compare and contrast *Kirby-Bauer, Etest,®️ MIC, and MBC* tests.

To effectively treat infectious diseases, physicians must know which antimicrobial agent is most effective against a particular pathogen. To ascertain the efficacy of antimicrobials, microbiologists conduct a variety of tests, including diffusion susceptibility tests, the minimum inhibitory concentration test, and the minimum bacteriocidal concentration test.

Diffusion Susceptibility Test

Diffusion susceptibility tests, also known as *Kirby-Bauer tests,* are a variation of the *disk-diffusion method* of determining the effectiveness of disinfectants discussed in Chapter 9 (p. 267). Diffusion susceptibility tests, which are simple, inexpensive, and widely used, involve uniformly inoculating a Petri plate with a standardized amount of the pathogen in question. Then small disks of paper containing standard concentrations of the drugs to be tested are firmly arranged on the surface of the plate. The plate is incubated, and the

The Spectrum of Activity of Antibiotics and Other Antimicrobial Drugs								
Prokaryotes				**Eukaryotes**				
Mycobacteria	Gram-negative bacteria	Gram-positive bacteria	Chlamydias, rickettsias	Protozoa	Fungi	Helminths		Viruses

Drug spectrum bars:
- Arildone (Viruses)
- Ribavirin (Viruses)
- Isoniazid (Mycobacteria)
- Niclosamide (Helminths)
- Polymyxin (Gram-negative bacteria)
- Azoles (Protozoa–Fungi)
- Acyclovir (Viruses)
- Penicillin (Gram-negative–Gram-positive bacteria)
- Praziquantel (Helminths)
- Streptomycin (Mycobacteria–Gram-negative bacteria)
- Erythromycin (Gram-negative–Chlamydias, rickettsias)
- Tetracycline (Gram-negative–Chlamydias, rickettsias)
- Sulfonamides (Gram-positive bacteria–Protozoa)

▲ *Figure 10.8*

Spectrum of action for selected antimicrobial agents. The more kinds of pathogens a drug affects, the broader its spectrum of action.

bacteria grow and reproduce everywhere but the areas where an effective antimicrobial drug diffuses through the agar. After incubation, the plates are examined for the presence of a **zone of inhibition**—that is, a clear area surrounding the disk where the microbe does not grow **(Figure 10.9).** A zone of inhibition is measured as the diameter (to the closest millimeter) of the clear region.

Generally, the larger the zone of inhibition, the more effective that drug is, though the size of the zone also depends on the rate of diffusion of the antimicrobial agent—lower-molecular-weight drugs generally diffuse more quickly than those with higher molecular weights. The size of a zone of inhibition can be compared to a standard table for that particular drug. Diffusion susceptibility tests enable scientists to classify pathogens as *susceptible, intermediate,* or *resistant* to each drug.

CRITICAL THINKING

Sometimes it is not possible to conduct a Kirby-Bauer or other susceptibility test, either because of a lack of time or an inability to access the bacteria (in an inner ear infection, for instance). How could a physician select an appropriate therapeutic agent in such cases?

Minimum Inhibitory Concentration (MIC) Test

Once scientists identify an effective antimicrobial agent, they quantitatively express its potency as a **minimum inhibitory concentration (MIC).** As the name suggests, the MIC is the smallest amount of the drug that will inhibit growth and reproduction of the pathogen. The MIC

▲ *Figure 10.9*

Zones of inhibition in a diffusion susceptibility (Kirby-Bauer) test. In general, the larger the zone of inhibition around disks, which are impregnated with an antimicrobial agent, the more effective that antimicrobial is against the organism growing on the plate. The organism is classified as either susceptible, intermediate, or resistant to the antimicrobials tested based on the sizes of the zones of inhibition. *If all eight of these antimicrobial agents diffuse at the same rate and are equally safe and easily administered, which one would be the "drug of choice" for treating this pathogen?*

Figure 10.9 Drug 6 has the largest zone of inhibition, so all other things being equal, it is the drug of choice.

Turbid tubes   Clear tubes

Increasing concentration of drug

◀ *Figure 10.10*

Minimum inhibitory concentration (MIC) test in test tubes. *What is the MIC for the drug acting against this bacterium?*

Figure 10.10 The MIC is 1.6 mcg/ml.

is often determined via a **broth dilution test,** in which a standardized amount of bacteria is added to serial dilutions of antimicrobial agents in tubes or wells containing broth. After incubation, turbidity (cloudiness) indicates bacterial growth; lack of turbidity indicates that the bacteria were either inhibited or killed by the antimicrobial agent **(Figure 10.10).**

There are many benefits of dilution tests over diffusion tests. For instance, many dilution tests can be conducted simultaneously in wells, and the entire process can be automated, with turbidity measured by special scanners connected to computers.

Another test that determines minimum inhibitory concentration combines aspects of an MIC test and a diffusion susceptibility test. This test, called an **Etest,**[®2] involves placing a plastic strip containing a gradient of the antimicrobial agent being tested on a plate uniformly inoculated with the organism of interest **(Figure 10.11).** After incubation, an elliptical zone of inhibition indicates antimicrobial activity, and the minimum inhibitory concentration can be noted where the zone of inhibition intersects a scale printed on the strip. The advantage of an Etest® over a standard diffusion susceptibility test is that only a single plate need be inoculated to determine both susceptibility and MIC.

Minimum Bactericidal Concentration (MBC) Test

An extension of the MIC test is a **minimum bactericidal concentration (MBC) test.** In an MBC test, samples taken from clear MIC tubes (or alternatively, from zones of inhibition from a series of diffusion susceptibility tests) are transferred to plates containing a drug-free growth medium

▲ *Figure 10.11*

An Etest,® which combines aspects of Kirby-Bauer and MIC tests. The plastic strip contains a gradient of the antimicrobial agent of interest. The MIC is estimated to be the concentration printed on the strip where the zone of inhibition intersects the strip. In this example, the MIC is 1.5 μg/ml.

(Figure 10.12 on page 300). The appearance of bacterial growth in these subcultures after appropriate incubation indicates that at least some bacterial cells survived that concentration of the antimicrobial drug and were able to grow and multiply once placed in a drug-free medium. Any drug concentration at which growth occurs in subculture is *bacteriostatic,* not *bactericidal,* for that bacterium. The lowest

[2]The name Etest® has no specific origin.

Concentration of antibacterial drug (μg/ml)

Clear Mic tube

8   16   25

Bacterial colonies   No colonies   No colonies

Drug-free media

▲ *Figure 10.12*

A minimum bactericidal concentration (MBC) test. In this test, plates containing a drug-free growth medium are inoculated with samples taken from zones of inhibition or from clear MIC tubes. After incubation, growth of bacterial colonies on a plate indicates that that concentration of antimicrobial drug (in this case, 8 μg/ml) is bacteriostatic. The lowest concentration for which no bacterial growth occurs on the plate is the minimum bactericidal concentration; in this case, the MBC is 16 μg/ml.

concentration of drug for which no growth occurs in the subcultures is the minimum bactericidal concentration (MBC).

Routes of Administration

Learning Objective

✓ Discuss the advantages and disadvantages of the different routes of administration of antimicrobial drugs.

An adequate amount of an antimicrobial agent must reach a site of infection if it is to be effective. For external infections such as athlete's foot, drugs can be applied directly. This is known as *topical* or *local* administration. For internal infections, drugs can be administered orally, *intramuscularly (IM)*, or *intravenously (IV)*. Each route has advantages and disadvantages.

Even though the oral route is simplest (it requires no needles and is self-administered), the drug concentrations it achieves are lower than occur via other routes of administration **(Figure 10.13)**. Further, because patients need not rely on a health care provider to administer the drug properly, they do not always follow prescribed timetables for taking oral medications.

IM administration via a hypodermic needle allows a drug to diffuse slowly into the many blood vessels within

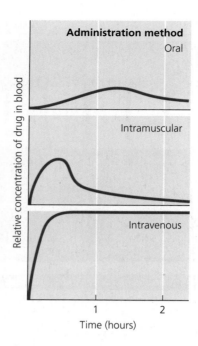

Administration method
Oral

Intramuscular

Intravenous

Time (hours)

▲ *Figure 10.13*

The effect of route of administration on blood levels of a chemotherapeutic agent. Although intravenous (IV) and intramuscular (IM) administration achieve higher drug concentrations in the blood, oral administration has the advantage of simplicity.

muscle tissue, but the concentration of the drug in the blood is never as high as that achieved by IV administration, which delivers the drug directly into the bloodstream through either a needle or a catheter (a plastic or rubber tube). Even though the amount of the drug in the blood is initially very high for the IV route, the concentration rapidly diminishes as the liver and kidneys remove the drug from the circulation, unless the drug is continuously administered.

In addition to route of administration, physicians must consider how antimicrobial agents will be distributed to infected tissues by the blood. For example, an agent removed rapidly from the blood by the kidneys might be the drug of choice for a bladder infection but would not be chosen to treat an infection of the heart. Finally, given that blood vessels in the brain, spinal cord, and eye are almost impermeable to many antimicrobial agents (because of the so-called blood-brain barrier), infections in these organs are often difficult to treat.

Safety and Side Effects

Learning Objective

✓ Identify three main categories of side effects of antimicrobial therapy.

Another aspect of chemotherapy that physicians must consider is the possibility of side effects. These fall into three main categories—toxicity, allergies, and disruption of normal microbiota—though some side effects are benign.

(a)

(b)

▲ *Figure 10.14*

Some side effects resulting from toxicity of antimicrobial agents. **(a)** "Black hairy tongue," caused by the antiprotozoan drug metronidazole (Flagyl®). **(b)** Discoloration and damage of tooth enamel caused by tetracycline. *Who should avoid taking tetracycline?*

Figure 10.14 Pregnant women and children should not use tetracycline.

Toxicity

Though antimicrobial drugs are ideally selectively toxic against microbes and harmless to humans, many in fact have toxic side effects. The exact cause of many adverse reactions is poorly understood, but drugs may be toxic to the kidneys, the liver, or nerves. For example, polymyxin and aminoglycosides can be fatally toxic to kidneys. Not all toxic side effects are so serious. *Metronidazole (Flagyl®)*, an antiprotozoan drug, may cause a harmless temporary condition called "black hairy tongue," which results when the breakdown products of hemoglobin accumulate in the papillae of the tongue **(Figure 10.14a).**

Doctors must be especially careful when prescribing drugs for pregnant women, as many drugs that are safe for adults can have adverse affects when absorbed by a fetus. For instance, tetracyclines form complexes with calcium that can become incorporated into bones and developing teeth, causing malformation of the skull and staining of tooth enamel **(Figure 10.14b).**

Allergies

In addition to toxicity, some drugs trigger allergic immune responses in sensitive patients. Although relatively rare, such reactions may be life threatening, especially in an immediate, violent reaction called *anaphylactic shock*. For example, about 0.1% of Americans have an anaphylactic reaction to penicillin, which results in approximately 300 deaths per year. However, not every allergy to an antimicrobial agent is so serious. Recent studies indicate that patients with mild allergies to penicillin frequently lose their sensitivity to it over time. Thus, an initial mild reaction to penicillin need not preclude its use in treating future infections. Chapter 18 discusses allergies in more detail.

Disruption of Normal Microbiota

As we have seen, drugs that disrupt normal microbiota and their microbial antagonism of opportunistic pathogens may result in secondary infections. In instances when a member of the normal microbiota is not affected by a drug, it can overgrow, causing a superinfection. For example, long-term use of broad-spectrum antimicrobials often results in explosions in the growth rate of *Candida albicans* in the vagina (vaginitis) or mouth (thrush), and the multiplication of *Clostridium difficile* (klos-trid'ē-ŭm dif'fi-sil) in the colon that causes a potentially fatal condition called *pseudomembranous colitis*. Such secondary infections are of greatest concern for hospitalized patients, who are often not only debilitated, but also more likely to be exposed to pathogens with resistance to antimicrobial drugs—the topic of the next section.

Resistance to Antimicrobial Drugs

Among the major challenges facing microbiologists today are the problems presented by pathogens that are resistant to antimicrobial agents. In the sections that follow we will examine the development of resistant populations of pathogens, the mechanisms by which pathogens are resistant to antimicrobials, and some ways that resistance can be retarded.

The Development of Resistance in Populations

Learning Objectives

✓ Describe how populations of resistant microbes can arise.
✓ Describe the relationship between R-plasmids and resistant cells.

Not all pathogens are equally sensitive to a given therapeutic agent; a population may contain a few organisms that are

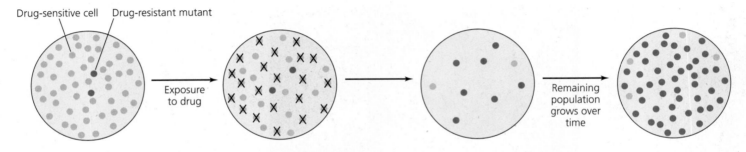

(a) Population of microbial cells

(b) Sensitive cells inhibited by exposure to drug

(c) Most cells are now resistant

▲ *Figure 10.15*

The development of a resistant strain of bacteria. **(a)** A bacterial population contains both drug-sensitive and drug-resistant cells, although sensitive cells constitute the vast majority of the population. **(b)** Exposure to an antimicrobial drug inhibits the sensitive cells; so long as the drug is present, reduced competition from sensitive cells facilitates the multiplication of resistant cells. **(c)** Eventually resistant cells constitute the majority of the population. *Why do resistant strains of bacteria more often develop in hospitals and nursing homes than in college dormitories?*

Figure 10.15 Resistant strains are more likely to develop in hospitals and other health care facilities because the extensive use of antimicrobial agents in those places inhibits the growth of sensitive strains and selects for the growth of resistant strains.

naturally either partially or completely resistant. Among bacteria, individual cells can acquire such resistance in two ways: through new mutations of chromosomal genes, or by acquiring resistance genes on extrachromosomal pieces of DNA called **R-plasmids** (or *R-factors*) via the processes of transformation, transduction, or conjugation (see Chapter 7). We focus here on the development of resistance in populations of bacteria, but resistance to antimicrobials is known to occur among viruses as well—including HIV, as discussed in **New Frontiers 10.1.**

The process by which a resistant strain of bacteria develops is depicted in **Figure 10.15.** In the absence of an antimicrobial drug, resistant cells are usually less efficient than their normal neighbors because they must expend extra energy to maintain resistance genes and proteins. Under these circumstances, resistant cells remain the minority in a population because they reproduce more slowly. However, when an antimicrobial agent is present, the majority of cells (which are sensitive to the antimicrobial) are inhibited or die while the resistant cells continue to grow and multiply, often more rapidly because they then face less competition. The result is that resistant cells soon replace the sensitive cells as the majority in the population. It should be noted that the presence of the chemotherapeutic agent does not *produce* resistance, but instead facilitates the replication of resistant cells that were already present in the population.

CRITICAL THINKING

Enterococcus faecium (en'ter-ō-kok'ŭs fē-sē-um) is frequently resistant to vancomycin. Why might this be of concern in a hospital setting in terms of developing resistant strains of *other* genera of bacteria?

Mechanisms of Resistance

Learning Objective

✓ List five ways by which microorganisms can be resistant to antimicrobial drugs.

Pathogens are resistant to antimicrobial drugs by one of five types of mechanisms:

1. Resistant bacteria may produce an enzyme that destroys or deactivates the drug. This common mode of resistance is exemplified by **β-lactamases** (penicillinases), enzymes that break the β-lactam rings of penicillin and similar molecules, rendering them inactive **(Figure 10.16).** Over 200 different lactamases have been identified. Frequently their genes are located on R-plasmids.

2. Resistant pathogens may slow or prevent the entry of the drug into the cell. This mechanism typically involves changes in the structure or electrical charge of the cytoplasmic membrane proteins that constitute channels or pores. Such proteins in the outer membranes of Gram-negative bacteria are called *porins* (see Figure 3.13b). Altered pore proteins result from mutations in chromosomal genes. Resistance against tetracycline and penicillin are known to occur via this mechanism.

3. Resistant cells may alter the receptor for the drug so that it either cannot attach to or binds less effectively to its target. This form of resistance is often seen against antimetabolites (such as sulfonamides) and against drugs (such as erythromycin) that thwart protein translation.

New Frontiers 10.1   The AIDS "Cocktail" and the Problem of Drug Resistance

One of the most important weapons in the fight against AIDS is a drug regimen known as the "AIDS cocktail." This "cocktail" is a combination of three drugs: a *protease inhibitor* and two kinds of *reverse transcriptase inhibitors*, including zidovudin (commonly known as AZT). Reverse transcriptase inhibitors interfere with enzymes that HIV (the virus that causes AIDS) needs very early in its replication cycle. The protease inhibitor, which revolutionized AIDS treatment upon its introduction in 1995, inhibits an enzyme that HIV needs near the end of its replication cycle. Between 1995 and 2000, this potent "cocktail" of drugs effectively reduced the death rate of HIV patients by more than 50%.

The "cocktail" has its side effects, including anemia, muscle weakness, pneumonia, and bronchitis. It is also an expensive and challenging regimen to maintain; some drug combinations require patients to take up to 20 pills each day. The biggest problem, however, is the emergence of strains of HIV that are resistant to even the powerful drugs that make up the "cocktail." According to a recent study, the number of new patients infected with such drug-resistant strains of HIV is rising—particularly in areas such as San Francisco, where the AIDS "cocktail" has been widely used for many years and the virus has had a greater opportunity to develop resistance.

As a result of the emergence of resistant strains of HIV, researchers are hard at work exploring new anti-HIV agents, including ones that can interfere with other stages of the HIV life cycle. Thus, whereas existing drugs focus on HIV's replication stage, which occurs after the virus has already invaded a cell, new drugs may include "entry inhibitors" that focus on preventing HIV from invading cells in the first place. Early studies of such agents are showing promise, fueling hopes that a new generation of AIDS drugs is about to enter the fray.

References: Stephenson, J. 2002. Researchers Explore New Anti-HIV Agents. *Journal of the American Medical Association.* 287:1635–1637.

Trachtenberg, J. D., M. A. Sande. 2002. Emerging Resistance to Nonnucleoside Reverse Transcriptase Inhibitors: A Warning and a Challenge. *Journal of the American Medical Association.* 288:239.

▲ *Figure 10.16*

How β-lactamase (penicillinase) renders penicillin inactive. The enzyme acts by breaking a bond in the β-lactam ring, the functional portion of the drug.

Lactam ring — β-lactamase (penicillinase) breaks this bond — Penicillin — Inactive penicillin

4. Resistant cells may alter their metabolic chemistry, or they may abandon the sensitive metabolic step altogether. For example, a cell may become resistant to a drug by producing more enzyme molecules for the affected metabolic pathway, effectively reducing the power of the drug. Alternatively, cells become resistant to sulfonamides by abandoning the synthesis of folic acid, absorbing it from the environment instead.

5. Resistant cells may pump the drug out of the cell before the drug can act.

Multiple Resistance and Cross Resistance

Learning Objective

✓ Define cross resistance, and distinguish it from multiple resistance.

A given pathogen can acquire resistance to more than one drug at a time, especially when resistance is conferred by R-plasmids, which are exchanged among bacterial cells. Such multiresistant strains of bacteria frequently develop in hospitals and nursing homes, where the constant use of many kinds of antimicrobial agents eliminates sensitive cells and encourages the development of resistant strains.

Pathogens that are resistant to most antimicrobial agents are sometimes called *superbugs.* Superbug strains of *Staphylococcus, Streptococcus* (strep-tō-kok′ŭs), *Enterococcus* (en′ter-ō-kok′ŭs), *Pseudomonas,* and *Plasmodium* (plaz-mō′dē-ŭm), (the protozoan that causes malaria) pose unique problems for health care professionals, who must treat infected patients without effective antimicrobials while taking extra care to protect themselves and other patients from infection.

Resistance to one antimicrobial agent may confer resistance to similar drugs, a phenomenon called **cross resistance.** Cross resistance typically occurs when drugs are similar in structure. For example, resistance to one aminoglycoside drug, such as streptomycin, may confer resistance to similar aminoglycoside drugs.

Retarding Resistance

Learning Objective

✓ Describe four ways that development of resistance can be retarded.

The development of resistant populations of pathogens can be averted in at least four ways. First, sufficiently high concentrations of the drug can be maintained in a patient's body for a long enough time to kill all sensitive cells and inhibit others long enough for the body's defenses to defeat them. Discontinuing a drug before all of the pathogens have been neutralized promotes the development of resistant strains. For this reason, it is important that patients finish their entire antimicrobial prescription and resist the temptation to "save some for another day."

A second way to avert resistance is to use antimicrobial agents in combination so that pathogens resistant to one drug will be killed by the second, and vice versa. Additionally, one drug sometimes enhances the effect of a second drug in a process called **synergism** (sin'er-jizm) **(Figure 10.17)**. In one example of a synergistic drug combination, the inhibition of cell wall formation by penicillin makes it easier for streptomycin molecules to enter bacteria and interfere with protein synthesis. Synergism can also result from combining an antimicrobial drug and a chemical, as occurs when *clavulanic acid* enhances the effect of penicillin by deactivating β-lactamase.

Not all drugs act synergistically; some combinations of drugs can be *antagonistic*—interfering with each other. For example, drugs that slow bacterial growth are antagonistic to the action of penicillin, which acts only against growing and dividing cells.

A third way to reduce the development of resistance is to limit the use of antimicrobials to necessary cases. Unfortunately, many antimicrobial agents are used indiscriminately, in both developed countries and in less-developed regions where many are available without a physician's prescription. In the United States, an estimated 50% of prescriptions for antibacterial agents to treat sore throats, and 30% of prescriptions for ear infections, are inappropriate because the diseases are viral in nature. Likewise, because antibacterial drugs have no effect on cold and flu viruses, 100% of antibacterial prescriptions for treating these diseases are

Disk with antibiotic amoxicillin-clavulanic acid

Disc with antibiotic aztreonam

▲ *Figure 10.17*

An example of synergism between two antimicrobial agents. The portion of the zone of inhibition outlined in green represents the synergistic enhancement of antimicrobial activity beyond the activities of the individual drugs (outlined in white).

superfluous. As discussed previously, the use of antimicrobial agents encourages the reproduction of resistant bacteria by limiting the growth of sensitive cells; therefore, inappropriate use of such drugs increases the likelihood that resistant strains of bacteria will multiply.

Finally, scientists can combat resistant strains by developing new variations of existing drugs, in some cases by adding novel side chains to the original molecule. In this way, scientists develop semisynthetic *second-generation* drugs. If resistance develops to these drugs, *third-generation* drugs may be developed to replace them. Many health care professionals and research scientists are concerned about how long drug developers can stay ahead of the development of resistance by pathogens.

Selected antimicrobial agents, their modes of action, clinical considerations, and other features are summarized in **Tables 10.2** through **10.6.** Particular use of antimicrobial drugs against specific pathogens is covered in the relevant chapters of this book.

Table 10.2   **Antibacterial Drugs**

Drug	Description and Mode of Action	Clinical Considerations	Resistance
Antibacterial Drugs that Inhibit Cell Wall Synthesis			
Beta-lactams Representative natural penicillins: Penicillin G Penicillin V Representative semisynthetic penicillins: Methicillin Ampicillin Amoxicillin Representative natural cephalosporin: Cephalothin Representative semisynthetic cephalosporins: Cefaclor Cephalexin Monobactam	Large number of natural and semisynthetic derivatives from *Penicillium* (penicillins) and *Cephalosporium* (cephalosporins, *e.g.,* cephalothin); bind to, and deactivate, the enzyme that cross links the NAM subunits of peptidoglycan Monobactams have only a single ring instead of the two rings seen in other beta-lactams	**Spectrum of Action:** Natural drugs have limited action against most G− bacteria because they do not readily cross the outer membrane; semisynthetics have broader spectra of action Monobactams have a limited spectrum of action, affecting only aerobic, G− bacteria **Route of Administration:** Penicillin V, a few cephalosporins (*e.g.,* cephalexin), and monobactams: oral; penicillin G, and many semisynthetics (*e.g.,* methicillin, ampicillin, carbenicillin, cephalothin): IM or IV **Adverse Effects:** Allergic reactions against beta-lactams are seen in some adults; monobactams are least allergenic	Develops in three ways. G− bacteria: • change their outer membrane structure to prevent entrance of the drug • modify the enzyme so that the drug no longer binds to it • synthesize beta-lactamases that cleave the functional lactam ring of the drug; genes for lactamases are often carried on R-plasmids
Vancomycin	Produced by *Streptomyces orientalis*; directly interferes with the formation of alanine–alanine crossbridges between NAM subunits	**Spectrum of Action:** Effective against most G+ bacteria, but generally reserved for use against strains resistant to other drugs such as methicillin-resistant *Staphylococcus aureus* (MRSA) **Route of Administration:** IV **Adverse Effects:** Damage to ears and kidneys, allergic reactions	G− bacteria are naturally resistant because the drug is too large to pass through the outer membrane; some G+ bacteria (*e.g., Ersipelothrix, Lactobacillus*) are naturally resistant because they do not form alanine–alanine bonds between NAM subunits
Cycloserine	Analog of alanine that interferes with the formation of alanine–alanine crossbridges between NAM subunits	**Spectrum of Action:** Some G+ bacteria, mycobacteria **Route of Administration:** Oral **Adverse Effects:** Toxic to nervous system, producing depression, aggression, confusion, and headache	Some G+ bacteria are naturally resistant.
Bacitracin	Isolated from *Bacillus licheniformis* growing on a patient named Tracy; appears to have three modes of action: • interference with the movement of peptidoglycan precursors through the bacterial cell membrane to the cell wall • inhibition of RNA transcription • damage to the bacterial cytoplasmic membrane The latter two modes of action have not been proven definitely	**Spectrum of Action:** G + bacteria **Route of Administration:** Topical **Adverse Effects:** Toxic to kidneys	Resistance most often involves changes in bacterial cell membranes that prevent bacitracin from entering the cell
Isoniazid (Isonicotinic acid hydrazide, INH)	Blocks the gene for an enzyme that forms mycolic acid; analog of the vitamins nicotinamide and pyridoxine	**Spectrum of Action:** Mycobacteria, including *M. tuberculosis* and *M. leprae* **Route of Administration:** Oral **Adverse Effects:** Occasionally toxic to liver	Resistance is due to random mutations of bacterial chromosomes that result in reduced drug uptake or alteration of target sites

Table 10.2   *(continued)*

Drug	Description and Mode of Action	Clinical Considerations	Resistance
Ethambutol	Prevents the formation of mycolic acid; used in combination with other antimycobacterial drugs	**Spectrum of Action:** Mycobacteria, including *M. tuberculosis* and *M. leprae* **Route of Administration:** Oral **Adverse Effects:** None	Resistance is due to random mutations of bacterial chromosomes that result in reduced drug uptake or alteration of target sites
Antibacterial Drugs that Inhibit Protein Synthesis			
Aminoglycosides Representatives: Kanamycin Streptomycin Gentamicin Amikacin Tobramycin	Compounds in which two or more amino sugars are linked with glycosidic bonds; were originally isolated from species of the bacterial genera *Streptomyces* and *Micromonospora*; inhibit protein synthesis by irreversibly binding to the 30S subunit of prokaryotic ribosomes. This either causes the ribosome to mistranslate mRNA, producing aberrant proteins, or causes premature release of the ribosome from mRNA, which stops synthesis	**Spectrum of Action:** Broad: effective against both G+ and G− bacteria **Route of Administration:** IV; do not traverse blood-brain barrier **Adverse Effects:** May be toxic to kidneys or auditory nerves, causing deafness	Because uptake of these drugs is oxygen-dependent, anaerobic bacteria are naturally resistant; aerobic bacteria alter membrane pores to prevent uptake, or synthesize enzymes that alter or degrade the drug once it enters; rarely, bacteria alter the binding site on the ribosome
Tetracyclines Representatives: Tetracycline Doxycycline Trimocycline	Composed of four hexagonal rings with various side groups; prevent tRNA molecules, which carry amino acids, from binding to ribosomes at the 30S subunit's docking site	**Spectrum of Action:** Most are broad: effective against many G+ and G− bacteria as well as against bacteria that lack cell walls, such as *Mycoplasma* **Route of Administration:** Oral, cross poorly into brain **Adverse Effects:** Nausea, diarrhea, sensitivity to light; forms complexes with calcium, which stains developing teeth and adversely affects the strength and shape of bones	Develops in three ways. Bacteria: • alter chromosomes in gene for pores in outer membrane, preventing drug from entering cell • alter binding site on the ribosome to allow tRNA to bind even in presence of drug • actively pump drug from cell
Chloramphenicol	Rarely used drug that prevents prokaryotic ribosomes from moving along mRNA by binding to their 50S subunits	**Spectrum of Action:** Broad, but rarely used except in treatment of typhoid fever **Route of Administration:** Oral; traverses blood-brain barrier **Adverse Effects:** In 1 of 24,000 patients, causes aplastic anemia, a potentially fatal condition in which blood cells fail to form; can also cause neurologic damage	Develops via gene carried on an R-plasmid that codes for an enzyme that deactivates drug
Macrolides Representative: Erythromycin Clarithromycin Azithromycin	Group of antimicrobials typified by a macrocyclic lactone ring; the most prescribed is erythromycin, which is produced by *Streptomyces erythraeus*; act by binding to the 50S subunit of prokaryotic ribosomes and preventing the elongation of the nascent protein	**Spectrum of Action:** Effective against G+ and a few G− bacteria **Route of Administration:** Oral; do not traverse blood-brain barrier **Adverse Effects:** Nausea, mild gastrointestinal pain, vomiting	Develops via changes in ribosomal RNA that prevent drugs from binding, or via R-plasmid genes coding for the production of macrolide-digesting enzymes; resistance genes are same as those of lincosamides
Clindamycin	Binds to 50S ribosomal subunit and stops protein elongation	**Spectrum of Action:** Effective against G+ and anaerobic G− bacteria **Route of Administration:** Oral or IV; does not traverse blood-brain barrier **Adverse Effects:** Gastrointestinal distress, including nausea, diarrhea, vomiting, and pain	Develops via changes in ribosomal structure that prevent drug from binding; resistance genes are same as those of aminoglycosides

Table 10.2   *(continued)*

Drug	Description and Mode of Action	Clinical Considerations	Resistance
Antibacterial Drugs that Alter Cell Membranes			
Polymyxin	Produced by *Bacillus polymyxa*; destroy cytoplasmic membranes of susceptible cells	**Spectrum of Action:** Effective against G− bacteria, particularly *Pseudomonas* **Route of Administration:** Topical **Adverse Effects:** Toxic to kidneys	Results from changes in cell membrane that prohibit entrance of the drug
Antibacterial Drugs that Are Antimetabolites			
Sulfonamides	Synthetic drug; first produced as a dye; is an analog of PABA that binds irreversibly to enzyme that produced folic acid; synergistic with trimethoprim	**Spectrum of Action:** Broad: effective against G+ and G− bacteria and some protozoa and fungi; however, resistance is widespread **Route of Administration:** Oral **Adverse Effects:** Rare: allergic reactions, anemia, jaundice, mental retardation of fetus if administered in last trimester of pregnancy	*Pseudomonas* is naturally resistant due to permeability barriers; cells that require folic acid as a vitamin are also naturally resistant; chromosomal mutations result in lowered affinity for the drug
Trimethoprim	Blocks second metabolic step in the formation of folic acid from PABA; synergistic with sulfonamides	**Spectrum of Action:** Broad: effective against G+ and G− bacteria and some protozoa and fungi; however, resistance is widespread **Route of Administration:** Oral **Adverse Effects:** Allergic reactions in some patients	*Pseudomonas* is naturally resistant due to permeability barriers; cells that require folic acid as a vitamin are also naturally resistant; chromosomal mutations result in lowered affinity for the drug
Antibacterial Drugs that Inhibit Nucleic Acid Synthesis			
Quinolones Nalidixic acid Norfloxacin Ciprofloxacin	Synthetic agents that inhibit DNA gyrase, which is needed to correctly replicate bacterial DNA; penetrate cytoplasm of cells	**Spectrum of Action:** Broad: G+ and G− bacteria are affected **Route of Administration:** Oral **Adverse Effects:** None of major significance	Results from chromosomal mutations that lower affinity of drug or reduce uptake
Rifampin	Semisynthetic derivative of drug produced by *Streptomyces mediterranei* that binds to bacterial RNA polymerase, preventing transcription of RNA; used with other antimicrobial bacterial drugs	**Spectrum of Action:** Bacteriostatic against aerobic G+ bacteria; bactericidal against mycobacteria **Route of Administration:** Oral **Adverse Effects:** None of major significance	Results from chromosomal mutation that alters binding site on enzyme; G− bacteria are naturally resistant due to poor uptake
Clofazimine	Binds to mycobacterial DNA preventing replication and transcription	**Spectrum of Action:** Mycobacteria, especially *M. tuberculosis* and *M. leprae* **Route of Administration:** Oral **Adverse Effects:** Diarrhea, discoloration of skin and eyes	None

Table 10.3   Antiviral Drugs

Drug	Description and Mode of Action	Clinical Considerations	Resistance
Attachment Antagonists			
Arildone	Block attachment molecule on host cell or pathogen	**Spectrum of Action:** Picornaviruses (*e.g.*, poliovirus, some cold viruses) **Route of Administration:** Not approved for antiviral use by FDA **Adverse Effects:** None	None reported
Antiviral Drugs that Inhibit Viral Uncoating			
Amantadine **Rimantadine**	Neutralize acid environment within phagolysosomes that is necessary for viral uncoating	**Spectrum of Action:** Influenza A virus **Route of Administration:** Oral; rimantadine is approved for adults only **Adverse Effects:** Toxic to central nervous system; results in nervousness, irritability, insomnia, and blurred vision	None reported
Antiviral Drugs that Inhibit Nucleic Acid Synthesis			
Acyclovir (ACV) Ganciclovir	Phosphorylation by virally coded kinase enzyme activates the drug; inhibits DNA and RNA synthesis	**Spectrum of Action:** Viruses that code for kinase enzymes: herpes, Epstein-Barr, cytomegalovirus, varicella virus **Route of Administration:** Oral **Adverse Effects:** None	Mutations in genes for kinase enzymes may render them ineffective at drug activation
Ribavirin	Phosphorylation by virally coded kinase enzyme activates the drug; inhibits DNA and RNA synthesis; viral DNA polymerase more likely to incorporate the drugs	**Spectrum of Action:** Respiratory syncytial, hepatitis C, influenza A, measles, some hemorrhagic fever viruses **Route of Administration:** Oral, aerosol, IV **Adverse Effects:** Perhaps harmful to developing fetus	None
Adenosine arabinoside	Phosphorylation by cell-coded kinase enzyme activates the drug: inhibits DNA synthesis; viral DNA polymerase more likely to incorporate the drugs than human DNA polymerase	**Spectrum of Action:** Herpes virus **Route of Administration:** IV **Adverse Effects:** Fatal to host cells that incorporate the drug into cellular DNA	Results from mutation of viral DNA polymerase
Nucleotide analogs Representative (see also Figure 10.7): Azidothymidine (AZT)	Phosphorylation by cell-coded kinase enzyme activates these drugs: inhibits DNA synthesis; viral reverse transcriptase more likely to incorporate these drugs; used in conjunction with protease inhibitor	**Spectrum of Action:** HIV **Route of Administration:** Oral **Adverse Effects:** Nausea, bone marrow toxicity	Results from mutation of viral reverse transcriptase
Antiviral Drugs that Inhibit Viral Proteins			
Protease inhibitors	Computer-assisted modeling of protease enzyme, which is unique to HIV, allowed the creation of drugs that block the active site; used in conjunction with drugs active against nucleic acid synthesis	**Spectrum of Action:** HIV **Route of Administration:** Oral **Adverse Effects:** None	Result from mutation of protease gene

Table 10.4   Antifungal Drugs

Drug	Description and Mode of Action	Clinical Considerations	Resistance
Antifungal Drugs that Inhibit Cell Membrane			
Polyenes Representatives: Amphotericin B Nystatin	Associate with molecules of ergosterol, forming a pore through the fungal membrane, which leads to leakage of essential ions from the cell; amphotericin B is produced by *Streptomyces nodosus*	**Spectrum of Action:** Fungi **Route of Administration:** Amphotericin B: IV; nystatin: topical **Adverse Effects:** Chills, vomiting, fever	None
Azoles Representatives: Miconazole Ketoconazole	Antifungal action due to inhibition of ergosterol synthesis	**Spectrum of Action:** Fungi, G+ bacteria, and parasitic protozoa **Route of Administration:** Topical, IV **Adverse Effects:** Possibly causes cancer in humans	None
Other Antifungal Drugs			
5-fluorocytosine	Fungi, but not mammals, have an enzyme that converts this drug into 5-fluorouracil, an analog of uracil that inhibits RNA function	**Spectrum of Action:** *Candida, Cryptococcus, Aspergillus* **Route of Administration:** Oral **Adverse Effects:** None	Develops from mutations in the genes for enzymes necessary for utilization of uracil
Griseofulvin	Isolated from *Penicillium griseofulvum*; deactivates tubulin, preventing cytokinesis and segregation of chromosomes during mitosis	**Spectrum of Action:** Molds of ringworm (tinea) **Route of Administration:** Topical, oral **Adverse Effects:** None	None

Table 10.5   Antihelminthic Drugs

Drug	Description and Mode of Action	Clinical Considerations	Resistance
Mebendazole **Thiabendazole**	Drugs appear to inhibit some enzymes and inhibit microtubule function, thus preventing cytokinesis and segregation of chromosomes during mitosis	**Spectrum of Action:** Helminths, *Giardia* (protozoan) **Route of Administration:** Oral **Adverse Effects:** None	None
Pyrantel Pamoate **Diethylcarbamazine**	Bind to neurotransmitter receptors, causing complete muscular contraction of helminths	**Spectrum of Action:** Nematodes **Route of Administration:** Oral **Adverse Effects:** None	None
Praziquantel	Attracts calcium ions, which are required for muscular contraction; induces complete muscular contraction in helminths	**Spectrum of Action:** Cestodes, trematodes **Route of Administration:** Oral **Adverse Effects:** None	None
Piperazine **Ivermectin**	Produce flaccid paralysis by blocking neurotransmitters	**Spectrum of Action:** Nematodes **Route of Administration:** Oral **Adverse Effects:** Adverse allergic reactions may result from antigens of dead worms	None
Niclosamide	Prevents production of ATP by oxidative phosphorylation	**Spectrum of Action:** Cestodes **Route of Administration:** Oral **Adverse Effects:** None; humans do not absorb the drug	None
Suramin	Inhibits specific enzymes in some nematodes	**Spectrum of Action:** Nematodes **Route of Administration:** Oral **Adverse Effects:** None	None

Table 10.6   Antiprotozoan Drugs

Drug	Description and Mode of Action	Clinical Considerations	Resistance
Heavy metals (Hg, As, Cr, Sb) Representatives: Salvarsan Antimony sodium gluconic acid	Deactivate enzymes by breaking hydrogen bonds necessary for effective tertiary structure; drugs containing arsenic were the first recognized chemotherapeutic agents that were selectively toxic	**Spectrum of Action:** Metabolically active cells **Route of Administration:** Topical, oral **Adverse Effects:** Toxic to active cells, such as those of the brain, kidney, liver, and bone marrow	None

Quinines
Representatives:
  Natural quinine
Semisynthetic quinines:
  Chloroquine
  Primaquine
	Natural and semisynthetic drugs derived from the bark of cinchona tree: inhibits DNA synthesis in malaria parasites	**Spectrum of Action:** *Plasmodium*
		Route of Administration: Oral
		Adverse Effects: Allergic reactions, visual disturbances
			Results from presence of quinine pumps that remove the drug from the cells

Sulfonamides	Synthetic drug; first produced as a dye; is an analog of PABA that binds irreversibly to enzyme that produced folic acid; synergistic with trimethoprim	**Spectrum of Action:** Broad: effective against G+ and G− bacteria and some protozoa and fungi cells; however, resistance is widespread **Route of Administration:** Oral **Adverse Effects:** Rare: Allergic reactions, anemia, jaundice, mental retardation of fetus if administered in last trimester of pregnancy	Cells that require folic acid as a vitamin are also naturally resistant; chromosomal mutations result in lowered affinity for the drug
Trimethoprim	Blocks second metabolic step in the formation of folic acid from PABA; synergistic with sulfonamides	**Spectrum of Action:** Broad: effective against G+ and G− bacteria and some protozoa and fungi; however, resistance is widespread **Route of Administration:** Oral **Adverse Effects:** Allergic reactions in some patients	Cells that require folic acid as a vitamin are naturally resistant; chromosomal mutations result in lowered affinity for the drug
Nitroimidazoles Representative: Metronidazole	Mode of action unknown	**Spectrum of Action:** *Trichomonas, Giardia,* and *Entamoeba* **Route of Administration:** Oral **Adverse Effects:** Metronidazole suspected to cause cancer	None

CHAPTER SUMMARY

The History of Antimicrobial Agents (pp. 288–289)

1. Chemotherapeutic agents are chemicals used to treat diseases. Among them are **antimicrobial agents (antimicrobials),** which include **antibiotics** (biologically produced agents), **semisynthetics** (chemically modified antibiotics), and **synthetic** agents.

Mechanisms of Antimicrobial Action (pp. 289–295)

1. Successful chemotherapy against microbes is based on **selective toxicity;** that is, using antimicrobial agents that are more toxic to pathogens than to the patient.

2. Antimicrobial drugs affect pathogens by inhibiting cell wall synthesis, inhibiting the translation of proteins, disrupting cytoplasmic membranes, inhibiting general metabolic pathways, inhibiting the replication of DNA, blocking the attachment of viruses to their hosts, and by blocking a pathogen's recognition of its host.

3. Beta-lactams—penicillins, cephalosporins, and monobactams—have a functional **lactam ring.** They prevent bacteria from cross-linking NAM subunits of peptidoglycan in the bacterial cell wall during growth. **Vancomycin** and **cycloserine** also disrupt cell wall formation in many Gram-positive bacteria.

Bacitracin blocks NAG and NAM secretion from the cytoplasm. **Isoniazid (INH)** and **ethambutol** block mycolic acid synthesis in the walls of mycobacteria.

4. Antimicrobial agents that inhibit protein synthesis include **aminoglycosides** and **tetracyclines,** which inhibit functions of the 30S ribosomal subunit, and chloramphenicol and **macrolides,** which inhibit 50S subunits.

5. Amphotericin B and polymyxin disrupt the cytoplasmic membranes of fungi and Gram-negative bacteria respectively.

6. **Sulfonamides** are **structural analogs** of para-aminobenzoic acid (PABA), a chemical needed by some microorganisms but not by humans. The substitution of sulfonamides in the metabolic pathway leading to nucleic acid synthesis kills those organisms. Trimethoprim is another antimetabolite that blocks this pathway.

7. Drugs that inhibit nucleic acid replication in pathogens include actinomycin, **nucleotide analogs,** fluoroquinolones, quinolones, and rifampin.

Clinical Considerations in Prescribing Antimicrobial Drugs (pp. 296–301)

1. Chemotherapeutic agents have a **spectrum of action** and may be classed as either **narrow-spectrum drugs** or **broad-spectrum drugs** depending on the number of kinds of pathogens they affect.

2. **Diffusion susceptibility tests** such as the Kirby-Bauer test reveal which drug is most effective against a particular pathogen;

in general, the larger the **zone of inhibition** around a drug-soaked disk on a Petri plate, the more effective the drug.

3. The **minimum inhibitory concentration (MIC),** usually determined by either a **broth dilution test** or an **Etest,**® is the smallest amount of a drug that will inhibit a pathogen.

4. A **minimum bactericidal concentration (MBC) test** ascertains whether a drug is bacteriostatic and the lowest concentration of a drug that is bactericidal.

5. In choosing antimicrobials, physicians must consider how a drug is best administered, orally, intramuscularly, or intravenously—and possible side effects, including toxicity and allergic responses.

Resistance to Antimicrobial Drugs (pp. 301–304)

1. Some members of a pathogenic population may develop resistance to a drug due to extra DNA pieces called **R-plasmids** or to the mutation of genes. Microorganisms may resist a drug by producing enzymes such as **β-lactamase** that deactivate the drug, by inducing changes in the cell membrane that prevent entry of the drug, by altering the drug's receptor to prevent its binding, by altering the cell's metabolic pathways, or by pumping the drug out of the cell.

2. **Cross resistance** occurs when resistance to one chemotherapeutic agent confers resistance to similar drugs.

3. **Synergism** describes the interplay between drugs that results in efficacy that exceeds the efficacy of either drug alone. Some drug combinations are antagonistic.

QUESTIONS FOR REVIEW
(Answers to multiple choice questions are on the web, along with additional review questions. Visit www.microbiologyplace.com.)

Multiple Choice

1. Diffusion and dilution tests that expose pathogens to antimicrobials are designed to
   a. determine the spectrum of action of a drug.
   b. determine which drug is most effective against a particular pathogen.
   c. determine the amount of a drug to use against a particular pathogen.
   d. both b and c

2. In a Kirby-Bauer susceptibility test, the presence of a zone of inhibition around disks containing antimicrobial agents indicates
   a. that the microbe does not grow in the presence of the agents.
   b. that the microbe grows well in the presence of the agents.
   c. the smallest amount of the agent that will inhibit the growth of the microbe.
   d. the minimum amount of an agent that kills the microbe in question.

3. The key to successful chemotherapy is
   a. selective toxicity.
   b. the diffusion test.
   c. the minimum inhibitory concentration test.
   d. the spectrum of action.

4. Which of the following statements is relevant in explaining why sulfonamides are effective?
   a. Sulfonamides attach to sterol lipids in the pathogen, disrupt the membranes, and lyse the cells.
   b. Sulfonamides prevent the incorporation of amino acids into polypeptide chains.
   c. Humans and microbes use folic acid and PABA differently in their metabolism.
   d. Sulfonamides inhibit DNA replication in both pathogens and human cells.

5. Cross resistance is
   a. the deactivation of an antimicrobial agent by a bacterial enzyme.
   b. alteration of the resistant cells so that an antimicrobial agent cannot attach.
   c. the mutation of genes that affect the cell membrane channels so that antimicrobial agents cannot cross into the cell's interior.
   d. resistance to antimicrobial agent P because of its similarity to antimicrobial agent Q.

6. Superbugs
   a. are resistant to most antimicrobial agents.
   b. can be found in the genus *Plasmodium.*
   c. frequently develop in hospitals.
   d. all of the above

7. Which of the following is most closely associated with a β-lactam ring?
   a. penicillin
   b. vancomycin
   c. bacitracin
   d. isoniazid

8. Drugs that act against protein synthesis include
   a. beta-lactams.
   b. trimethoprim.
   c. polymyxin.
   d. aminoglycosides.

9. Which of the following statements is *false* concerning antiviral drugs?
   a. Macrolide drugs block attachment sites on the host cell wall and prevent viruses from entering.
   b. Drugs that increase the acidity of phagolysosomes prevent viral uncoating and thus prevent viral genetic "takeover."
   c. Nucleotide analogs in antiviral drugs can be used to stop viral replication.
   d. Drugs containing protease inhibitors retard viral growth by blocking the production of essential viral proteins.

10. PABA is
    a. a substrate used in the production of penicillin.
    b. a type of β-lactamase.
    c. molecularly similar to cephalosporins.
    d. a substrate used to synthesize folic acid.

Short Answer

1. What characteristics would an ideal chemotherapeutic agent have? Which drug fits these qualities?

2. Contrast narrow-spectrum and broad-spectrum drugs. Which are more effective?

3. Why is the fact that drug Z destroys the NAM portions of a cell's wall structure an important factor in considering the drug for chemotherapy?

4. Given that both human cells and pathogens synthesize proteins at ribosomal sites, how can antimicrobial agents that target this process be safe to use in humans?

5. Support or refute the following statement: "Antimicrobial agents produce resistant cells."

6. Given that resistant strains of pathogens are a concern to the general health of a population, what can be done to prevent their development?

7. Why are antiviral drugs difficult to develop?

8. A man has been given a broad-spectrum antibiotic for his stomach ulcer. What unintended consequences could arise from this therapy?

CRITICAL THINKING

1. AIDS is often treated with a "cocktail" of several antiviral agents at once. Why is the cocktail more effective than a single agent? What is a physician trying to prevent by prescribing several drugs at once?

2. How does *Penicillium* escape the effects of the penicillin it secretes?

3. How might a colony of *Bacillus licheniformis* escape the effects of its own bacitracin?

4. Fewer than 1% of known antibiotics have any practical value in treatment of disease. Why is this so?

5. In the summer issue of *News of the Lepidopterist's Society* in 2000, a recommendation was made to moth and butterfly collectors to use antibiotics to combat disease in the young of these insects. What are the possible ramifications for human health of such usage of antibiotics?

6. Even though aminoglycosides such as gentamicin can cause deafness, there are still times when they are the best choice for treating some infections. What laboratory test would a clinical scientist use to show that gentamicin is the best choice to treat a particular *Pseudomonas* infection?

7. Your pregnant neighbor has a sore throat and tells you that she is taking some tetracycline she had left over from a previous infection. Give two reasons why her decision is a poor one.

8. Acyclovir has replaced adenosine arabinoside as treatment for herpes infections. Compare the ways these drugs are activated (see Table 10.3). Why is acyclovir a better choice?

9. Why might amphotericin B affect the kidneys more than other human organs?

10. *Clostridium difficile* is an obligate anaerobic opportunistic pathogen. It can cause serious, life-threatening pseudomembranous colitis in patients undergoing long-term antimicrobial therapy. Why might aminoglycosides be used to treat this disease more often than tetracyclines?

In 1980, microbiologist Tony Walsby startled the scientific world by announcing the discovery of something no one had seen before: square-shaped bacteria. These novel prokaryotes were named *Haloarcula*, meaning "salt box," because of their unusual boxy shape and their habitat, the salt-encrusted pools of Egypt's Sinai peninsula. Until this extraordinary find, scientists had only seen spherical, rod-shaped, or spiral prokaryotes. *Haloarcula* was a surprise, one that forced scientists to reconsider some of their assumptions about prokaryotic life forms.

More recently, we have discovered species of archaea that may offer clues to whether or not there is life on other planets. And among the millions of unknown microorganisms that live on Earth, there may very well exist entirely new categories of life. The vast and diverse world of prokaryotes continues to surprise and amaze us.

Characterizing and Classifying Prokaryotes

CHAPTER 11

CHAPTER OUTLINE

General Characteristics of Prokaryotic Organisms
Morphology of Prokaryotic Cells
Reproduction of Prokaryotic Cells
Arrangements of Prokaryotic Cells
Endospores

Modern Prokaryotic Classification

Survey of Archaea
Extremophiles
Methanogens

Survey of Bacteria
Deeply Branching and Phototrophic Bacteria
Low G + C Gram-Positive Bacteria
High G + C Gram-Positive Bacteria
Gram-Negative Proteobacteria
Other Gram-Negative Bacteria

Haloarcula are square- or rectangular-shaped archaea.

MicroPrep Pre-Test: *Take the pre-test for this chapter on the web.*
*Visit **www.microbiologyplace.com.***

Prokaryotes are by far the most diverse group of organisms. They thrive in various habitats: from Antarctic glaciers to thermal hot springs, from the colons of animals to the cytoplasm of other prokaryotes, from distilled water to supersaturated brine, and from disinfectant solutions to basalt rocks thousands of meters below the Earth's surface. In part because of such great diversity, only a very few prokaryotes have enzymes, toxins, or cellular structures that enable them to colonize humans and cause disease. In this chapter we will begin by examining general prokaryotic characteristics and conclude with a survey of specific prokaryotic taxa. We will briefly mention human pathogens throughout the chapter but reserve detailed discussion of these important microbes for Chapters 19–21.

General Characteristics of Prokaryotic Organisms

In previous chapters we considered the general characteristics of prokaryotic cells, including their metabolism, growth, and genetics. Here we will consider prokaryotes not merely as cells, but as distinct organisms, focusing on their cellular shapes, their reproductive processes, their spatial arrangements, and the ability of some to survive unfavorable conditions by forming resistant structures within themselves.

Morphology of Prokaryotic Cells

Learning Objective

✓ Identify six basic shapes of prokaryotic cells.

Prokaryotic cells exist in a variety of shapes, or morphologies **(Figure 11.1)**. The three basic shapes are **cocci** (kok'sī, roughly spherical), **bacilli** (bă-sil'ī, rod-shaped), and **spirals.** Cocci are not all perfectly spherical; for example, there are pointed, kidney-shaped, and oval cocci. Similarly, bacilli vary in shape; for example, some bacilli are pointed, spindle-shaped, or threadlike (filamentous). Spiral-shaped prokaryotes are either **spirilla,** which are stiff, or **spirochetes** (spī'rō-kētz), which are flexible. Slightly curved rods are **vibrios,** and the term **coccobacillus** is used to describe cells that are intermediate in shape between cocci and bacilli; that is, when it is difficult to ascertain if a cell is an elongated coccus or a short bacillus. In addition to these basic shapes, there are star-shaped, triangular, and square prokaryotes, as well as prokaryotes that are **pleomorphic**[1] (plē-ō-mōr' fik); that is, they vary in shape and size (see Figure 11.11).

Reproduction of Prokaryotic Cells

Learning Objectives

✓ List three common types of reproduction in prokaryotes.
✓ Describe snapping division as a type of binary fission.

Coccus

Coccobacillus

Bacillus

Vibrio

Spirillum

Spirochete

Pleomorphic

▲ *Figure 11.1*

Typical prokaryotic morphologies. *What is the difference between a spirillum and a spirochete?*

Figure 11.1 Generally, spirilla are stiff and have fewer than a dozen coils, whereas spirochetes are flexible and have numerous coils.

All prokaryotes reproduce asexually; none reproduce sexually. The most common method of asexual reproduction is **binary fission,** which proceeds as follows **(Figure 11.2):** ❶ The cell replicates its DNA; each DNA molecule is attached to the cytoplasmic membrane. ❷ The cell grows; as the cytoplasmic membrane elongates, it moves the daughter molecules of DNA apart. ❸ The cell forms a cross wall, invaginating the cytoplasmic membrane. ❹ The cross wall completely divides daughter cells. ❺ The daughter cells may or may not separate. The parental cell disappears with the formation of progeny.

A variation of binary fission called **snapping division** occurs in some Gram-positive species **(Figure 11.3).** In snapping division, only the inner portion of the cell wall is deposited across the dividing cell. The thickening of this new transverse wall puts tension on the outer layer of the old cell wall, which still holds the two cells together. Eventually, as the tension increases, the outer wall breaks at its weakest

---

[1]From Greek *pleon,* meaning more and *morphe,* meaning form.

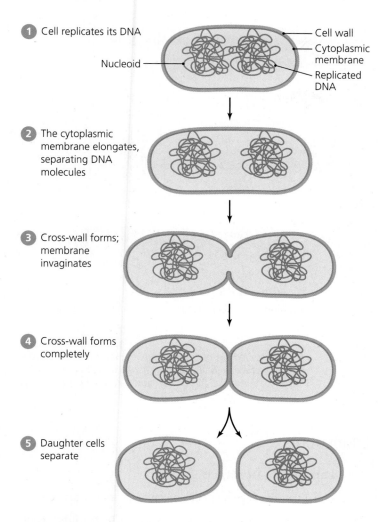

1. Cell replicates its DNA

Nucleoid

Cell wall

Cytoplasmic membrane

Replicated DNA

2. The cytoplasmic membrane elongates, separating DNA molecules

3. Cross-wall forms; membrane invaginates

4. Cross-wall forms completely

5. Daughter cells separate

▲ *Figure 11.2*

Binary fission. The cell replicates its DNA, elongates, forms a cross-wall, and divides into two equal-sized daughter cells.

Outer portion of cell wall

Inner portion of cell wall

(a)

Rupture of outer wall

Hinge

(b)

0.5 μm

▲ *Figure 11.3*

Snapping division, a variation of binary fission. **(a)** Only the inner portion of the cell wall forms a cross wall. **(b)** As the daughter cells grow, tension snaps the outer portion of the cell wall, leaving the daughter cells connected by a hinge of old cell wall material.

Arrangements of Prokaryotic Cells

Learning Objective

✓ Draw and label five arrangements of prokaryotes.

The arrangements of prokaryotic cells result from two aspects of division during binary fission: the planes in which cells divide, and whether or not daughter cells separate completely or remain attached to each other. Thus cocci that remain attached in pairs are **diplococci** (**Figure 11.6a** on page 317), and long chains of cocci are called **streptococci**[2] (**Figure 11.6b**). Some cocci divide in two planes and remain attached to form **tetrads (Figure 11.6c)**; others divide in three planes to form cuboidal packets called **sarcinae**[3] (sar´si-nī) (**Figure 11.6d**). Clusters called **staphylococci**[4] (staf´i-lō-kok-sī), which look like bunches of grapes, form when the planes of cell division are random (**Figure 11.6e**).

Bacilli are less varied in their arrangements than cocci because bacilli divide transversely—that is, perpendicular to the long axis. Daughter bacilli may separate to become single cells or stay attached as either pairs or chains (**Figure 11.7a−c** on page 317). Because the cells of *Corynebacterium diphtheriae* (kŏ-rī´nē-bak-tēr´ē-ŭm dif-thi´rē-ē), the

point with a snapping movement that tears it most of the way around. The daughter cells then remain hanging together, held at an angle by a small remnant of the original outer wall that acts like a hinge.

A few prokaryotes have other methods of reproduction. The parental cell retains its identity during and after these methods. The *actinomycetes* (ak´ti-nō-mī-sētz) produce reproductive cells called **spores** at the ends of their filamentous cells (**Figure 11.4** on page 316). Each spore can develop into a clone of the original organism. Some *cyanobacteria* reproduce by fragmentation into small motile filaments that glide away from the parental strand. Still other prokaryotes reproduce by **budding,** in which an outgrowth of the original cell (a bud) receives a copy of the genetic material and enlarges. Eventually the bud is cut off from the parental cell, typically while it is still quite small (**Figure 11.5** on page 316).

[2]From Greek *streptos*, meaning twisted, because long chains tend to twist.
[3]Latin for bundles.
[4]From Greek *staphyle*, meaning bunch of grapes.

Nucleoid replicates

New nucleoid moves into bud

Young bud

Daughter cell

▲ *Figure 11.5*

Budding. *How does budding differ from binary fission?*

Figure 11.5 In binary fission, the parent cell disappears with the formation of two equal-sized offspring; in contrast, a bud is often much smaller than its parent, and the parent remains to produce more buds.

▲ *Figure 11.4*

Prokaryotic reproductive structures. Spores of actinomycetes grow from filamentous vegetative cells.

causative agent of diphtheria, divide by snapping division, the daughter cells remain attached to form V-shapes and a side-by-side arrangement called a **palisade**[5] **(Figure 11.7d).**

Note that the same word can be used to refer either to a general shape and/or arrangement or to a specific genus. Thus the characteristic shape of the genus *Bacillus* (ba-sil'ŭs) is a bacillus, a rod-shaped bacterium, and the characteristic arrangement of bacteria in the genus *Sarcina* (sar'si-nă) is cuboidal. In such potentially confusing cases the meaning can be distinguished because genus names are always capitalized and italicized. In other cases the genus name uses the singular form while the arrangement uses the plural form; thus streptococci—spherical cells arranged in a chain—are characteristic of the genus *Streptococcus* (strep'tō-kok'ŭs), and staphylococci—spherical cells arranged in grape-like clusters—are characteristic of the genus *Staphylococcus* (staf' i-lō-kok'ŭs).

Endospores

Learning Objective

✓ Describe the formation and function of bacterial endospores.

The Gram-positive bacteria *Bacillus* and *Clostridium* (klos-trid'ē-ŭm) are characterized by the ability to produce unique structures called **endospores,** which are important for several reasons, including their durability and potential pathogenicity (discussed shortly). Though some scientists refer to endospores simply as "spores," endospores should not be confused with the reproductive spores of actinomycetes, algae, and fungi. Each single bacterial cell, called a *vegetative* cell to distinguish it from an endospore, transforms into only one endospore, which then germinates to grow into only one vegetative cell; therefore, endospores are not reproductive structures. Instead, endospores constitute a defensive strategy against hostile or unfavorable conditions.

A vegetative cell normally transforms itself into an endospore only when one or more nutrients (such as carbon or nitrogen) are in limited supply. The process of endospore formation, called *sporulation,* requires 8 to 10 hours and proceeds in seven steps **(Figure 11.8** on page 318). During the process, two membranes, a thick layer of peptidoglycan, and a spore coat form around a copy of the genome and a

---

[5]From Latin *palus,* meaning stake, referring to a fence made of adjoining stakes.

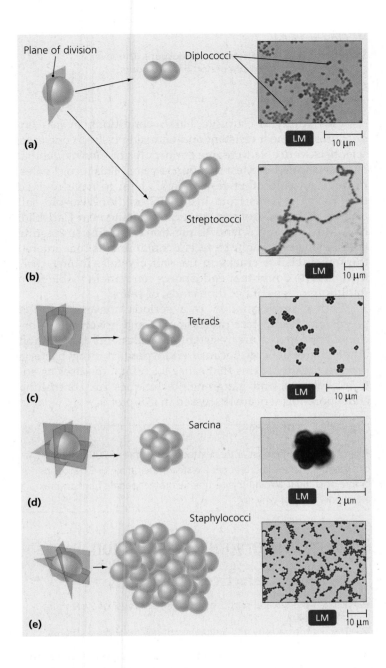

▲ Figure 11.6

Arrangements of cocci. **(a)** The diplococci of *Neisseria gonorrheae*. **(b)** The streptococci of *Streptococcus pyogenes*. **(c)** A tetrad, in this case of *Micrococcus luteus*. **(d)** A sarcina of *Sarcina maxima*. **(e)** The staphylococci of *Staphylococcus aureus*.

small portion of cytoplasm. The cell deposits large quantities of dipicolinic acid, calcium, and DNA-binding proteins within the endospore while removing most of the water. Depending on the species, a cell forms an endospore either *centrally*, *subterminally* (near one end), or *terminally* (at one end) (**Figure 11.9** on page 319). Sometimes an endospore is so large it swells the original cell.

▲ Figure 11.7

Arrangements of bacilli. **(a)** A single bacillus of *Escherichia coli*. **(b)** Diplobacilli in a young culture of *Bacillus cereus*. **(c)** Streptobacilli in an older culture of *Bacillus cereus*. **(d)** V-shapes and a palisade of *Corynebacterium diphtheriae*.

Endospores are extremely resistant to drying, heat, radiation, and lethal chemicals. For example, they remain alive in boiling water for several hours; are unharmed by alcohol, peroxide, bleach, and other toxic chemicals; and can tolerate over 400 rad of radiation, which is more than four times the dose that is lethal to humans. Endospores are stable resting stages that barely metabolize—they are essentially in a state of suspended animation—and they germinate only when conditions improve. Scientists do not know how endospores are able to resist harsh conditions, but it appears that the double membrane, spore coats, dipicolinic acid, calcium, and DNA-binding proteins may serve to stabilize DNA and proteins, protecting them from adverse conditions.

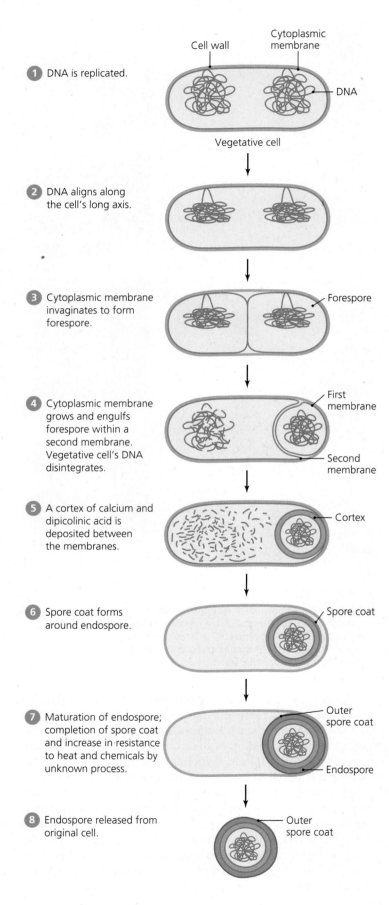

① DNA is replicated.

Cell wall · Cytoplasmic membrane

DNA

Vegetative cell

② DNA aligns along the cell's long axis.

③ Cytoplasmic membrane invaginates to form forespore.

Forespore

④ Cytoplasmic membrane grows and engulfs forespore within a second membrane. Vegetative cell's DNA disintegrates.

First membrane

Second membrane

⑤ A cortex of calcium and dipicolinic acid is deposited between the membranes.

Cortex

⑥ Spore coat forms around endospore.

Spore coat

⑦ Maturation of endospore; completion of spore coat and increase in resistance to heat and chemicals by unknown process.

Outer spore coat

Endospore

⑧ Endospore released from original cell.

Outer spore coat

◀ *Figure 11.8*

Sporulation, the formation of an endospore. The seven steps depicted occur over a period of 8–10 hours.

The ability to survive harsh conditions makes endospores the most resistant and enduring cells. In one case, scientists were able to revive endospores of *Clostridium* that had been sealed in a test tube for 34 years. This record pales, however, beside other researchers' claim to have revived *Bacillus* endospores from inside 250-million-year-old salt crystals retrieved from an underground site near Carlsbad, New Mexico. Some scientists question this claim, suggesting that the bacteria might be recent contaminants that entered through invisible cracks in the salt crystals. In any case, there is little doubt that endospores can remain viable for a minimum of tens, if not thousands, of years.

Endospore formation is a serious concern to food processors, health care professionals, and governments because endospores are resistant to treatments that inhibit other microbes, and because endospore-forming bacteria produce deadly toxins that cause such fatal diseases as anthrax, tetanus, and gangrene. Techniques for controlling endospore-formers are discussed in Chapter 9.

CRITICAL THINKING

Following the bioterrorist anthrax attacks in the fall of 2001, a news commentator suggested that people steam their mail for 30 seconds before opening it. Would this technique protect people from anthrax infections? Why or why not?

Modern Prokaryotic Classification

Learning Objectives

✓ Explain the general purpose of *Bergey's Manual of Systematic Bacteriology*.

✓ Discuss the veracity and limitations of any taxonomic scheme.

As we saw in Chapter 4, scientists called *taxonomists* group similar organisms into categories called *taxa.* At one time the smallest taxa of prokaryotes (that is, genera and species) were based solely on the characteristics we considered in the previous section, especially morphology and arrangement. More recently, the classification of living things has been based more on genetic relatedness. Accordingly, modern taxonomists have placed all organisms into three *domains*— Archaea, Bacteria, and Eukarya—which are the largest, most inclusive taxa and have essentially replaced kingdoms. Each domain is divided into phyla primarily on the basis of similarities in DNA, RNA, and protein sequences.

We previously noted as well that the vast majority of prokaryotes—perhaps as many as 99%, and probably millions of species—have never been isolated or cultured and are known only from their rRNA "fingerprints"; that is, they

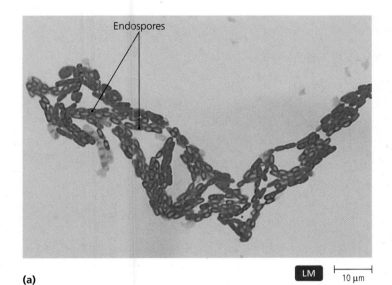

Endospores

(a)

LM 10 μm

(b)

LM 10 μm

▲ *Figure 11.9*

Locations of endospores. **(a)** Central endospores of *Bacillus*. **(b)** Terminal endospores of *Clostridium botulinum*. The enlarged endospore has swollen the cell that produced it.

are known only from sequences of rRNA that do not match any known RNA sequences. In light of this information, taxonomists now construct modern classification schemes of prokaryotes based on the relative similarities of rRNA sequences found in various prokaryotic groups (**Figure 11.10** on page 320).

Perhaps the most authoritative reference in modern prokaryotic systematics is *Bergey's Manual of Systematic Bacteriology,* which classifies prokaryotes into 25 phyla—two in Archaea and 23 in Bacteria. When completed, the five vol-

umes of the second edition of *Bergey's Manual* will discuss the great diversity of prokaryotes based in large part (but not exclusively) on their possible evolutionary relationships as reflected in their rRNA sequences. (As of 2003, only Volume I had been published.)

Our examination of prokaryotic diversity in this text is for the most part organized to reflect the taxonomic scheme that will appear in the second edition of *Bergey's Manual,* but it's important to note that as authoritative as it is, *Bergey's Manual* is not an "official" list of prokaryotic taxa. The reason is that taxonomy is partly a matter of opinion and judgment, and not all taxonomists agree. Because there is room in taxonomy for honest differences of opinion, and because legitimately differing views often change as more information is uncovered and examined, *Bergey's Manual* is merely a consensus of experts at a given time. More information about *Bergey's Manual* can be found in Appendix C.

CRITICAL THINKING

A scientist stated that *Bergey's Manual* would never truly be finished. Support her statement.

In the following sections we examine representative prokaryotes in each major phylum. Human pathogens are discussed in greater detail in Chapters 19–21. We begin our exploration of prokaryotic diversity with a survey of Archaea.

Survey of Archaea

Learning Objective

✓ Identify the common features of microbes in the domain Archaea.

Scientists originally identified archaea as a distinct type of prokaryotes on the basis of unique rRNA sequences. Archaea also share other common features that distinguish them from bacteria:

- Archaea lack peptidoglycan in their cell walls
- Their cell membrane lipids have branched hydrocarbon chains
- The intial amino acid in their polypeptide chains, coded by the AUG start codon, is methionine (as in eukaryotes and in contrast to the N-formylmethionine used by bacteria).

These and other features of archaea, bacteria, and eukaryotes are compared and contrasted in Chapter 4.

Archaea are currently classified in two phyla—Crenarchaeota and Euryarchaeota—based primarily upon rRNA sequences (see Figure 11.10).

Archaea reproduce by binary fission, budding, or fragmentation. Most archaeal cells are cocci, bacilli, or spiral forms, but some unusual shapes also exist (**Figure 11.11** on page 321). Archaeal cell walls vary among taxa and are

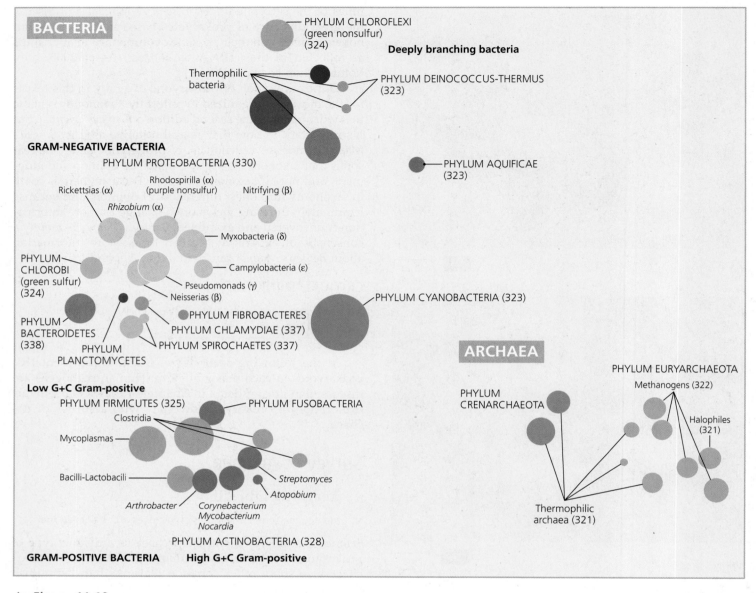

▲ *Figure 11.10*

A taxonomic scheme for prokaryotes based on their relatedness according to rRNA sequences. The closer together the disks, the more similar are the rRNA sequences of the organisms within the group; the sizes of disks are proportional to the number of sequences known for that group. Note that archaea are distinctly separate from bacteria. The discussion in this chapter is largely based on the scheme depicted in this figure; the numbers in parentheses correspond to the page in this book on which the discussion of that group begins. (Adapted from *Road Map to Bergey's.* 2002, Bergey's Manual Trust.)

composed of a variety of compounds, including proteins, glycoproteins, lipoproteins, and polysaccharides; the feature they have in common is a lack of peptidoglycan. Another interesting feature of archaea is that not one of them is known to cause diseases in humans or animals.

Though most archaea live in moderate environmental conditions, the most noted archaea are *methanogens,* organisms that generate methane gas (discussed shortly), and *extremophiles,* which we discuss next.

Extremophiles

Learning Objective

✓ Compare and contrast the two kinds of extremophiles discussed in this section.

Extremophiles are microbes that require extreme conditions of temperature, pH, and/or salinity to survive. Prominent among the extremophiles are *thermophiles* and *halophiles*.

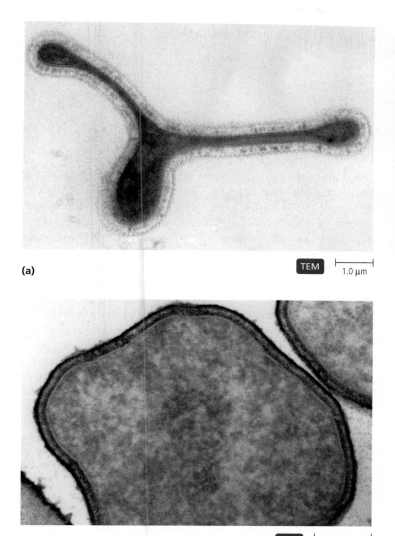

(a)   TEM   1.0 μm

(b)   TEM   0.2 μm

▲ *Figure 11.11*

Thermophiles. **(a)** *Pyrodictium*, which has disc-shaped cells with filamentous extensions.
(b) *Sulfolobus*, which is pleomorphic.

Thermophiles

As discussed in Chapter 6, **thermophiles**[6] are prokaryotes whose DNA, RNA, cytoplasmic membranes, and proteins require them to live at temperatures over 45°C because these cellular components do not function properly at lower temperatures. Prokaryotes that require temperatures over 80°C are called **hyperthermophiles.** Most thermophilic archaea are in the phylum Crenarchaeota, though some thermophilic species are also found in the phylum Euryarchaeota.

Three representative genera of thermophiles are *Acidanus, Pyrodictium,* and *Sulfolobus* (see Figure 11.11). These microorganisms live in sulfur-rich, acidic hot springs such as those found in Yellowstone National Park **(Figure 11.12)** and similar terrestrial volcanic habitats. The cells of *Pyrodic-*

▲ *Figure 11.12*

A typical habitat of hyperthermophilic archaea: a sulfur-rich hot spring in Yellowstone National Park.

tium, which lives in deep-sea hydrothermal vents, are irregular disks with elongated protein tubules that attach them to grains of sulfur, which they use as final electron acceptors in respiration. **Table 11.1** on page 322 summarizes the characteristics of these three genera.

Scientists use thermophiles and their enzymes in recombinant DNA technology applications because thermophiles' cellular structure and enzymes are stable and functional at temperatures that denature most proteins and nucleic acids and kill other cells. As discussed in Chapter 8, DNA polymerase from hyperthermophilic archaea makes possible the automated amplification of DNA in a thermocycler. Heat-stable enzymes are also ideal for many industrial applications, including their use as additives in laundry detergents.

Halophiles

Halophiles[7] are classified in the phylum Euryarchaeota based on rRNA sequences. They inhabit extremely saline habitats such as the Dead Sea, the Great Salt Lake, and solar evaporation ponds used to concentrate salt for use in seasoning and for the production of fertilizer (**Figure 11.13** on page 322). Halophiles can also colonize and spoil such foods as salted fish, sausages, and pork.

The distinctive characteristic of halophiles is their absolute dependence on a concentration of NaCl greater than 9% (that is, a 1.5 molar solution) to maintain the integrity of their cell walls. Most halophiles grow and reproduce within an optimum range of 17–23% NaCl, and many species can survive in a saturated saline solution (35% NaCl). Halophiles contain red to orange pigments that probably

---

[6]From Greek *thermos,* meaning heat, and *philos,* meaning love.
[7]From Greek *halos,* meaning salt.

Table 11.1   Characteristics of Three Representative Genera of Thermophilic Archaea

Genus	Morphology	Optimum Temperature	Optimum pH	Oxygen Requirement
Sulfolobus	Lobed coccus	75°C	2.5	Obligate aerobe
Acidanus	Coccus	88°C	2	Facultative anaerobe
Pyrodictium	Disk-shaped with filaments	105°C	6	Anaerobe

▲ **Figure 11.13**

One habitat of halophiles: the highly saline water in solar evaporation ponds near San Francisco as seen in an aerial photo. Halophiles often contain red to orange pigments to protect them from intense solar energy.

play a role in protecting them from the intense visible and ultraviolet light.

The most studied halophile is *Halobacterium salinarium* (hā′lō-bak-tēr′ē-ŭm sal-ē-nar′ē-um), which is unusual because even though it uses light energy to drive the synthesis of ATP, it does not synthesize organic compounds from CO_2. Further, *Halobacterium* lacks photosynthetic pigments—chlorophylls and bacteriochlorophylls. Instead, it synthesizes purple proteins, called **bacteriorhodopsins** (bak-tēr′ē-ō-rō-dop′sinz), that absorb light energy and use it to pump protons (hydrogen ions) across the cytoplasmic membrane to establish a proton gradient. (Recall from Chapter 5 that cells use the energy of proton gradients to produce ATP via chemiosmosis.) *Halobacterium* also rotates its flagella with energy from the proton gradient to position itself at the proper water depth for maximum light absorption.

Methanogens

Learning Objective

✓ List at least four significant roles played by methanogens in the environment.

Methanogens are obligate anaerobes in the phylum Euryarchaeota that convert CO_2, H_2, and organic acids into methane gas (CH_4). These microbes constitute the largest group of archaea. Most species are mesophilic, though a few thermophilic methanogens are known. For example, *Methanopyrus*[8] has an optimum growth temperature of 98°C and grows in 110°C seawater around submarine hydrothermal vents. Scientists have also discovered halophilic methanogens.

Methanogens play significant roles in the environment by converting organic wastes in pond, lake, and ocean sediments into methane. Other methanogens such as *Methanobacterium* (meth′a-nō-bak-tēr′ē-ŭm) living in the colons of animals are one of the primary sources of environmental methane. Methanogens dwelling in the intestinal tract of a cow, for example, can produce 400 liters of methane a day. Methanogens have produced about 10 trillion tons of methane—twice the known amount of oil, natural gas, and coal combined—that lies buried in the mud on the ocean floor. Sometimes the production of methane in swamps and bogs is so great that bubbles rise to the surface as "swamp gas." Methane is a so-called *greenhouse gas;* that is, methane in the atmosphere traps heat, which adds to global warming. It is about 25 times more potent as a greenhouse gas than carbon dioxide. If all the methane trapped in ocean sediments were released, it would wreak havoc with the world's climate.

Methanogens also have useful industrial applications. An important step in sewage treatment is the digestion of sludge by methanogens, and some sewage treatment plants burn the methane to heat buildings and generate electricity.

Though we have concentrated our discussion on extremophiles and methanogens, many archaea live in more moderate habitats. For example, archaea make up about a third of the prokaryotic biomass in coastal Antarctic water, providing food for marine animals.

CRITICAL THINKING

A scientist who discovers a prokaryote living in a hot spring at 100°C suspects that it belongs to the archaea. Why does he think it might be archaeal? How could he prove that it is not bacterial?

---

[8]From Greek *pyrus,* meaning fire.

Survey of Bacteria

As we noted previously, our survey of prokaryotes in this chapter reflects the classification scheme that will be featured in the as-yet incomplete second edition of *Bergey's Manual*. Whereas the classification scheme for bacteria in the first edition of the *Manual* emphasized morphology, Gram reaction, and biochemical characteristics, the second edition will largely base its classification of bacteria on differences in their 16S rRNA sequences. The result is a scheme that includes 23 bacterial phyla grouped in a way that generally reflects our current understanding of the genetic relationships among bacteria. We begin our survey of bacteria by considering the deeply branching and phototrophic bacteria.

Deeply Branching and Phototrophic Bacteria

Learning Objectives

✓ Provide a rationale for the name "deeply branching bacteria."
✓ Explain the function of heterocysts in terms of both photosynthesis and nitrogen fixation.

Deeply Branching Bacteria

The **deeply branching bacteria** are so named because their rRNA sequences and growth characteristics lead scientists to conclude that these organisms are similar to the earliest bacteria; that is, they branched off the "tree of life" at an early stage. For example, the deeply branching bacteria are autotrophic, and early organisms must have been autotrophs because heterotrophs by definition must derive their carbon from autotrophs. Further, many of the deeply branching bacteria live in habitats similar to those some scientists think existed on the early Earth—hot, acidic, anaerobic, and exposed to intense ultraviolet radiation from the sun.

One representative of these microbes—the Gram-negative, microaerophilic *Aquifex,* a bacterium in the Phylum Aquificae—is considered to represent the earliest branch of bacteria. It is chemoautotrophic, hyperthermophilic, and anaerobic, deriving energy and carbon from inorganic sources in very hot habitats containing little oxygen.

Another representative of deeply branching bacteria is *Deinococcus* (dīn-ō-kok'ŭs), in the proposed Phylum "Deinococcus-Thermus" which grows as tetrads of Gram-positive cocci. Interestingly, the cell wall of *Deinococcus* has an outer membrane similar to that of Gram-negative bacteria, but the cells stain purple like typical Gram-positive microbes. *Deinococcus* is extremely resistant to radiation because of the way it packages its DNA and the presence of radiation-absorbing pigments and unique lipids within its membranes. Even when exposed to 5 million rad of radiation, which is enough energy to shatter its chromosome into hundreds of fragments, its enzymes can repair the damage. Not surprisingly, researchers have isolated *Deinococcus* from sites severely contaminated with radioactive wastes.

Phototrophic Bacteria

Phototrophic bacteria acquire the energy needed for anabolism by absorbing light with pigments located in thylakoids called *photosynthetic lamellae.* Of course, being prokaryotes, they lack the membrane-bound thylakoids seen in eukaryotic chloroplasts.

Most phototrophic bacteria are also autotrophic—they produce organic compounds from carbon dioxide. Phototrophs are a diverse group of microbes that are taxonomically confusing. Based on their pigments and their source of electrons for photosynthesis, phototrophic bacteria can be divided into the following five groups (**Table 11.2** on page 325):

- Blue-green bacteria (cyanobacteria)
- Green sulfur bacteria
- Green nonsulfur bacteria
- Purple sulfur bacteria
- Purple nonsulfur bacteria

Although these organisms are classified into four phyla, we consider all of them together because of their common phototrophic metabolism.

Cyanobacteria   The organisms known as **cyanobacteria** are Gram-negative phototrophs that vary greatly in shape, size, and method of reproduction. They range in size from 1 μm to 10 μm in diameter and are either coccoid or disc shaped. Coccal forms can be single or arranged in pairs, tetrads, chains, or sheets (**Figure 11.14a and b** on page 324); disc-shaped forms are often tightly appressed end to end to form filaments, which can be either straight, branched, or helical and are frequently contained in a gelatinous glycocalyx called a *sheath* (**Figure 11.14c**). Some filamentous cyanobacteria are motile, moving along surfaces by *gliding.* Cyanobacteria generally reproduce by binary fission, with some species also reproducing by motile fragments or by thick-walled spores called akinetes (ā-kin-ēts'; see Figure 11.14a).

Like plants and algae, cyanobacteria utilize chlorophyll *a* and generate oxygen (are oxygenic) during photosynthesis:

$$12H_2O + 6CO_2 \xrightarrow{\text{light}} C_6H_{12}O_6 + 6H_2O + 6O_2 \qquad (1)$$

For this reason, cyanobacteria were formerly called *blue-green algae;* however, the name *cyanobacteria* properly emphasizes their true prokaryotic nature.

Photosynthesis by cyanobacteria is thought to have transformed the anaerobic atmosphere of the early Earth into our oxygen-containing one, and according to the endosymbiotic theory, chloroplasts developed from cyanobacteria. Indeed, chloroplasts and cyanobacteria have similar rRNA and structures such as 70S ribosomes and photosynthetic membranes.

Nitrogen is an essential element in proteins and nucleic acids. Though nitrogen constitutes about 79% of the atmosphere, relatively few organisms can utilize this gas. A few prokaryotes—species of filamentous cyanobacteria and

(a)

LM   10.0 μm

- Akinete
- Vegetative cell
- Heterocyst

(b)

LM   5 μm

(c)

LM   10 μm

- Sheath

◀ *Figure 11.14*

Examples of cyanobacteria with different growth habits. **(a)** *Anabaena*, which grows as a filament of cocci with differentiated cells. Heterocysts fix nitrogen; akinetes are reproductive cells. **(b)** *Merismopedia*, which grows as a flat sheet of cocci surrounded by a gelatinous glycocalyx. **(c)** *Oscillatoria*, which forms a filament of tightly appressed disc-shaped cells.

which produces oxygen. Nitrogen-fixing cyanobacteria solve this problem in one of two ways. Most cyanobacteria isolate the enzymes of nitrogen fixation in specialized, thick-walled, nonphotosynthetic cells called **heterocysts** (see Figure 11.14a). Heterocysts transport reduced nitrogen to neighboring cells in exchange for glucose. Other cyanobacteria photosynthesize during daylight hours and fix nitrogen at night, thereby separating nitrogen fixation from photosynthesis in time rather than in space.

Green and Purple Phototrophic Bacteria   Green and purple bacteria differ from plants, algae, and cyanobacteria in two ways: They use *bacteriochlorophylls* for photosynthesis instead of chlorophyll *a*, and they are *anoxygenic;* that is, they do not generate oxygen during photosynthesis. Green and purple phototrophic bacteria commonly inhabit anaerobic, hydrogen sulfide–rich muds at the bottoms of ponds and lakes. These microbes are not necessarily green and purple in color; rather, the terms refer to pigments in some of the better-known members of the groups.

As previously indicated, the green and purple phototrophic bacteria include both sulfur and nonsulfur forms. Whereas nonsulfur bacteria derive electrons for the reduction of CO_2 from organic compounds such as carbohydrates and organic acids, sulfur bacteria derive electrons from the oxidation of hydrogen sulfide to sulfur, as follows:

$$12H_2S + 6CO_2 \xrightarrow{\text{light}} C_6H_{12}O_6 + 6H_2O + 12S \qquad (2)$$

Green sulfur bacteria deposit the resultant sulfur outside their cells, whereas purple sulfur bacteria deposit sulfur within their cells **(Figure 11.15).**

At the beginning of the 20th century, a prominent question in biology concerned the origin of the oxygen released by photosynthetic plants. It was initially thought that oxygen was derived from carbon dioxide, but a comparison of photosynthesis in cyanobacteria (Equation 1) with that in sulfur bacteria (Equation 2) provided evidence that free oxygen is derived from water.

Whereas green sulfur bacteria are placed in Phylum Chlorobi, green nonsulfur bacteria are members of Phylum Chloroflexi (see Figure 11.10). The purple bacteria (both sulfur and nonsulfur) are placed in three classes of Phylum Proteobacteria (see Table 11.2 and Figure 11.10), which is composed of Gram-negative bacteria and is discussed shortly. First, however, we turn our attention to various groups of Gram-positive bacteria, and to a different characteristic of microbes that is used in the classification of Gram-positive bacteria.

proteobacteria (discussed later in the chapter)—reduce nitrogen gas (N_2) to ammonia (NH_3) via a process called **nitrogen fixation.** Nitrogen fixation is essential for life on Earth because nitrogen-fixers are not only able to enrich their own growth, but also provide nitrogen in a usable form to other organisms.

Because the enzyme responsible for nitrogen fixation is inhibited by oxygen, nitrogen-fixing cyanobacteria are faced with a problem—how to segregate nitrogen fixation, which is inhibited by oxygen, from oxygenic photosynthesis,

Table 11.2   **Characteristics of the Major Groups of Phototrophic Bacteria**

Phylum	Cyanobacteria	Chlorobi	Chloroflexi	Proteobacteria	Proteobacteria
Class	Cyanobacteria	Chlorobia	Chloroflexi	"γ-proteobacteria"[a]	"α-proteobacteria" and one genus in "β-proteobacteria"
Common name(s)	Blue-green bacteria ("blue-green algae")	Green sulfur bacteria	Green nonsulfur bacteria	Purple sulfur bacteria	Purple nonsulfur bacteria
Major photosynthetic pigments	Chlorophyll *a*	Bacteriochlorophyll *a* plus *c, d,* or *e*	Bacteriochlorophylls *a* and *c*	Bacteriochlorophyll *a* or *b*	Bacteriochlorophyll *a* or *b*
Types of photosynthesis	Oxygenic	Anoxygenic	Anoxygenic	Anoxygenic	Anoxygenic
Electron donor in photosynthesis	H_2O	H_2, H_2S, or S	Organic compounds	H_2, H_2S, or S	Organic compounds
Sulfur deposition	None	Outside of cell	None	Inside of cell	None
Nitrogen fixation	Some species	None	None	None	None
Motility	Nonmotile or gliding	Nonmotile	Gliding	Motile with polar or peritrichous flagella	Nonmotile or motile with polar flagella

[a]The names of classes in quotations are not officially recognized.

LM | 4.0 μm

▲ *Figure 11.15*

Deposits of sulfur within purple sulfur bacteria in the genus *Chromatium*. These bacteria oxidize H_2S to produce the granules of elemental sulfur evident in this photomicrograph. *Where do green sulfur bacteria deposit sulfur grains?*

Figure 11.15 Green sulfur bacteria deposit sulfur grains externally.

Taxonomists have come to use the *G + C ratio*—the percentage of all base pairs in a genome that are guanine–cytosine base pairs—as a useful criterion in classifying microbes. Bacteria with G + C ratios below 50% are considered "low G + C bacteria"; the remainder are considered "high G + C bacteria." Because taxonomists have discovered that Gram-positive bacteria with low G + C ratios have similar sequences in their 16S rRNA, and that those with high G + C ratios also have rRNA sequences in common, they have assigned low G + C bacteria and high G + C bacteria to different phyla. We discuss the low G + C bacteria first.

Low G + C Gram-Positive Bacteria

Learning Objectives

✓ Discuss the lack of cell walls in mycoplasmas.

✓ Identify significant beneficial or detrimental effects of the genera *Clostridium, Bacillus, Listeria, Lactobacillus, Sreptococcus,* and *Staphylococcus.*

The low G + C Gram-positive bacteria are classified within Phylum **Firmicutes** (fer-mik′ū-tēz), which includes three groups: clostridia, mycoplasmas, and low G + C Gram-positive bacilli and cocci. Next we consider these three groups in turn.

Clostridia

Clostridia are rod-shaped, obligate anaerobes, many of which from endospores. The group is named for the genus

Clostridium botulinum produces botulinum toxins, some of the deadliest toxins known. When absorbed in the body, botulism toxins interfere with the release of acetylcholine, the neurotransmitter that signals muscles to contract. As a result, muscle cells cannot contract, and a slow but progressive paralysis spreads throughout the body. Death occurs when paralysis of respiratory muscles results in respiratory failure. This disease, called botulism, is discussed in more detail in Chapter 19.

Amazingly, one toxin known as type A botulism toxin is being increasingly used as part of a cosmetic procedure to erase frown lines, crow's feet, and other facial wrinkles. Purified type A botulinum toxin is marketed as Botox, extremely small doses of which are injected into those facial muscles that cause skin wrinkles, paralyzing or weakening them. Such treatments only last approximately 3–6 months and must be repeated in order to maintain the desired effects.

The same type A botulinum toxin is also injected into the appropriate muscles to treat two eye conditions caused by abnormal muscle contractions: *strabismus*, the inability of the eyes to maintain

binocular vision due to an imbalance between eye muscles, and *bleparospasm*, a condition caused by involuntary spasmodic contractions of facial muscles surrounding the eye. The photos in this box show the effects of the toxin on a patient with bleparospasm. The photo at left shows the patient before treatment; she is unable to open her eyes. The middle photo shows the patient, still prior to treatment, forcing her eyes open with her fingers. The photo at right shows the patient after treatment; she is now able to keep her eyes open because the toxin has suppressed the excessive contractions of the treated muscles.

Clostridium[9] (klos-trid′ē-ŭm), which is important both in medicine—in large part because its members produce potent toxins that cause a variety of diseases in humans—and in industry because their endospores enable them to survive harsh conditions, including many types of disinfection and antisepsis. Examples of clostridia include *C. tetani* (which causes tetanus), *C. perfringens* (gangrene), *C. botulinum* (botulism), and *C. difficile* (severe diarrhea). Chapter 19 discusses these pathogens and the diseases they cause in greater detail; **Highlight 11.1** describes how a deadly toxin produced by *C. botulinum* has been put to use for cosmetic purposes.

The giant bacterium *Epulopiscium* (ep′ū-lō-pis-sē-ŭm) is an unusual relative of *Clostridium*. As discussed in Highlight 3.2 on page 59, *Epulopiscium* is large enough to be seen without the use of a microscope and has numerous short flagella, which make it appear ciliated. For these reasons it was first considered eukaryotic. *Epulopiscium* is also unusual in that it does not reproduce by binary fission but instead forms new cells within the parent cell and releases them through a slit in the parent's cell wall.

Other microbes related to *Clostridium* include sulfate-reducing microbes, which produce H_2S from elemental sulfur during anaerobic respiration, and *Veillonella* (vī-lō-nel′ă), a genus of anaerobic cocci that live as part of the biofilm (plaque) that forms on the teeth of warm-blooded animals. *Veillonella* is unusual because even though it has a typical Gram-positive cell wall structure, it has a negative Gram reaction—it stains pink.

Mycoplasmas

A second class of low G + C bacteria are the **mycoplasmas**[10] (mī′kō-plaz′mas). These facultative or obligate anaerobes lack cell walls, which means they stain pink when Gram stained. Indeed, until their nucleic acid sequences proved their similarity to Gram-positive organisms, mycoplasmas were classified in a phylum of Gram-negative microbes, instead of in Phylum Firmicutes with other low G + C Gram-positive bacteria.

Mycoplasmas are able to survive without cell walls because they colonize osmotically protected habitats such as animal and human bodies, and because they have tough cytoplasmic membranes, many of which contain lipids called *sterols* that give the membranes strength and rigidity. Because they lack cell walls, they are pleomorphic. They were named "mycoplasmas" because their filamentous forms resemble the filaments of fungi. Mycoplasmas have diameters ranging from 0.2 μm to 0.8 μm, making them the smallest free-living cells, and many mycoplasmas have a terminal structure that is used to attach them to eukaryotic cells and that gives the bacteria a pear-like shape. They require organic growth factors, such as cholesterol, fatty acids, vitamins, amino acids, and nucleotides, which they either acquire from their host or which must be added to laboratory media. When growing on solid media, most species form a distinctive "fried egg" appearance because cells in the center

---

[9]From Greek *kloster*, meaning spindle.
[10]From Greek *mycos*, meaning fungus, and *plassein*, meaning to mold.

of the colony grow into the agar while those around the perimeter only spread across the surface **(Figure 11.16).**

In animals, mycoplasmas colonize mucous membranes of the respiratory and urinary tracts and are associated with pneumonia and urinary tract infections. Pathogenic mycoplasmas and the diseases they cause are discussed more fully in Chapter 21.

Gram-Positive Bacilli and Cocci

A third group of low G + C Gram-positive organisms is composed of bacilli and cocci that are significant in environmental, industrial, and health care settings. Among the genera in this group are *Bacillus, Listeria* (lis-tēr′ē-ă), *Lactobacillus* (lak′tō-ba-sil′ŭs), *Streptococcus, Enterococcus,* and *Staphylococcus.* Chapter 19 discusses the pathogens in these genera in greater detail.

Bacillus   The genus *Bacillus* includes endospore-forming aerobes and facultative anaerobes that typically move by means of peritrichous flagella. Numerous species of *Bacillus* are common in soil.

A few species of *Bacillus* are directly beneficial to humans. *Bacillus thuringiensis* (thur-in-jē-en′sis) is beneficial to farmers and gardeners. During sporulation, this bacterium produces a crystalline protein that is toxic to caterpillars that ingest it **(Figure 11.17).** Gardeners spray *Bt toxin,* as preparations of the bacterium and toxin are known, on plants to protect them from caterpillars. Scientists have achieved the same effect, without the need of spraying, by introducing the gene for Bt toxin into plants' chromosomes. Other beneficial species of *Bacillus* include *B. polymyxa* and *B. licheniformis,* which synthesize the antibiotics polymyxin and bacitracin, respectively. Chapter 10 discusses the production and effects of antibiotics in more detail.

Bacillus anthracis (an-thrā′sis), which causes anthrax, gained notoriety in 2001 as an agent of bioterrorism. Its endospores are either inhaled or enter the body through breaks in the skin. When they germinate, the vegetative cells produce toxins that kill surrounding tissues (see Figure 19.11). Untreated *cutaneous anthrax* is fatal in 20% of patients; untreated *pulmonary anthrax* is generally 100% fatal.

Endospores of *B. cereus* (se′rē-us) occasionally contaminate rice and cause disease in people who ingest them. These heat-resistant endospores survive cooking, germinate in the intestinal tract, and produce a toxin that induces nausea, vomiting, and abdominal cramping. As discussed in Chapter 9, refrigeration prevents the excessive growth of these and other contaminates.

Listeria   Another pathogenic low G + C Gram-positive rod is *Listeria monocytogenes* (mo-nō-sī-to′je-nēz), which can contaminate milk and meat products. This microbe, which does not produce endospores, is notable because it continues to reproduce under refrigeration, and it can survive inside phagocytic white blood cells. *Listeria* rarely causes disease in adults, but it kills fetuses when the organism crosses

▲ *Figure 11.16*

The distinctive "fried egg" appearance of *Mycoplasma* colonies growing on an agar surface. This visual feature is unique to this group of bacteria.

▲ *Figure 11.17*

Crystals of Bt toxin, produced by the endospore-forming *Bacillus thuringiensis.* The crystalline protein kills caterpillars that ingest it.

the placenta from their infected mothers. It also causes meningitis[11] and bacteremia[12] when it infects immunocompromised patients such as the aged and patients with AIDS, cancer, or diabetes.

Lactobacillus   Organisms in the genus *Lactobacillus* are nonspore-forming rods normally found growing in the human mouth, stomach, intestinal tract, and vagina. These organisms rarely cause disease; instead, they protect the body by inhibiting the growth of pathogens—a situation called *microbial antagonism.* Lactobacilli are used in industry in the

---

[11]Inflammation of the membranes covering the brain and spinal cord; from Greek *meninx,* meaning membrane.
[12]The presence of bacteria, particularly those that produce disease symptoms, in the blood.

Lactobacillus: A Probiotic to Watch

For over a century, people have eaten yogurt and other dairy products containing live bacterial cultures, believing this would benefit their health. As it turns out, they may have been on to something. Recent studies have shown that the ingestion of some microorganisms—in particular, *Lactobacillus,* the same bacterium that converts milk into cheese and yogurt—can reduce symptoms of diarrhea in children, especially in cases caused by rotaviruses. Scientists are also studying the use of *Lactobacillus* in relieving antibiotic-associated diarrhea, milk allergies, and certain respiratory infections. Indeed, *Lactobacillus* is the focus of many studies in the growing field of **probiotics,** the administration of microorganisms to promote health.

What makes *Lactobacillus* so promising as a medical therapy? The mechanisms are not completely understood, but one of the central ideas is that *Lactobacillus* (which lives naturally and harmlessly in our intestines and other parts of the body) competes with harmful microorganisms for resources, keeping pathogenic microbes in check. It may also interact with the immune system in ways that can help alleviate milk allergies. Further evidence suggests that *Lactobacillus* can induce the expression of certain proteins that protect the intestines against harmful bacteria and viruses. In addition, their natural presence in dairy products makes them easy to administer to patients.

Much more research remains to be done, but the preliminary findings are encouraging. In the exciting field of probiotics, *Lactobacillus* is the star on center stage.

Source: Friedrich, M.J. 2000. A bit of culture for children: Probiotics may improve health and fight disease. *Journal of the American Medical Association.* 284(11): 1365–1366.

production of yogurt, buttermilk, pickles, and sauerkraut. **New Frontiers 11.1** discusses the growing role of *Lactobacillus* in promoting human health.

Streptococcus* and *Enterococcus   The genera *Streptococcus* and *Enterococcus* are diverse groups of Gram-positive cocci associated in pairs and chains (see Figure 11.6b). They produce numerous human diseases, including pharyngitis (strep throat), scarlet fever, impetigo, fetal meningitis, wound infections, pneumonia, and diseases of the inner ear, skin, blood, and kidneys. In recent years, health care providers have become concerned over strains of streptococci that have developed resistance to antimicrobial drugs. Of particular concern are so called "flesh-eating," drug-resistant streptococci that produce toxins that destroy muscle and fat tissue.

Staphylococcus   Among the common inhabitants of humans is *Staphylococcus aureus*[13] (aw'rē-ŭs), which is typically found growing harmlessly on the skin in clusters (see Figure 11.6e). A variety of toxins and enzymes allow some strains of *S. aureus* to invade the body and cause such diseases as bacteremia, pneumonia, wound infections, food poisoning, toxic shock syndrome, and diseases of the joints, bones, heart, and blood.

The characteristics of the low G + C Gram-positive bacteria are summarized in **Table 11.3.** To summarize, these bacteria, which are classified into three classes within Phylum Firmicutes, include the anaerobic endospore-forming rod *Clostridium;* the pleomorphic *Mycoplasma;* the aerobic and facultative aerobic endospore-forming rod *Bacillus,* the nonendospore-forming rods *Listeria* and *Lactobacillus,* and the cocci *Streptococcus* and *Staphylococcus.*

Next we consider Gram-positive bacteria that have high G + C ratios.

---

[13]From Latin *aurum, meaning* gold, because it produces yellow pigments.

High G + C Gram-Positive Bacteria

Learning Objectives

✓ Explain the slow growth of *Mycobacterium*.
✓ Identify significant beneficial or detrimental properties of the genera *Corynebacterium, Mycobacterium, Actinomyces, Nocardia,* and *Streptomyces*.

Taxonomists classify Gram-positive bacteria with a G + C ratio greater than 50% in the phylum Actinobacteria, which includes species with rod-shaped cells (many of which are significant human pathogens) and filamentous bacteria, which resemble fungi in their growth habit and in the production of reproductive spores. Here we will discuss briefly some prominent high G + C Gram-positive bacteria; the numerous pathogens in this group are discussed more fully in Chapter 19.

Corynebacterium

Members of the genus *Corynebacterium* (kŏ-rī'nē-bak-tēr'ē-ŭm) are pleomorphic—though generally rod-shaped—aerobes and facultative anaerobes. They reproduce by snapping division, which often causes the cells to form V-shapes and palisades (see Figure 11.7d). Corynebacteria are also characterized by their stores of phosphate within inclusions called **metachromatic granules,** which stain differently from the rest of the cytoplasm when the cells are stained with the blue dyes methylene blue or toluidine blue. The best known species is *Corynebacterium diphtheriae,* which causes diphtheria.

Mycobacterium

The genus *Mycobacterium* (mī'kō-bak-tēr'ē-um) is composed of aerobic species that are slightly curved to straight rods that sometimes form filaments. Mycobacteria grow very slowly, often requiring a month or more to form a visible

Table 11.3		Characteristics of Selected Gram-Positive Bacteria		
Phylum/Class	G + C Ratio	Representative Genera	Special Characteristics	Diseases
Firmicutes				
"Clostridia"[a]	Low (less than 50%)	*Clostridium*	Obligate anaerobic rods; endospore formers	Tetanus Botulism Gangrene Severe diarrhea
		Epulopiscium	Giant rods	
		Veillonella	Part of oral biofilm on human teeth; stain like Gram-negative (pink)	Dental caries
Firmicutes				
Mollicutes	Low (less than 50%)	*Mycoplasma*	Lack cell wall; pleomorphic; smallest free-living cells; stain like Gram-negative (pink)	Pneumonia Urinary tract infections
Firmicutes				
"Bacilli"	Low (less than 50%)	*Bacillus*	Facultative anaerobic rods; endospore formers	Anthrax
		Listeria	Contaminates of dairy products	Listeriosis
		Lactobacillus	Produce yogurt, buttermilk, pickles, sauerkraut	Rare blood infections
		Streptococcus	Cocci in chains	Strep throat, scarlet fever, and others
		Staphylococcus	Cocci in clusters	Bacteremia, food poisoning, and others
Actinobacteria				
Actinobacteria	High (greater than 50%)	*Corynebacterium*	Snapping division; metachromatic granules in cytoplasm	Diphtheria
		Mycobacterium	Waxy cell walls (mycolic acid)	Tuberculosis and meningitis
		Actinomyces	Filaments	Actinomycosis
		Nocardia	Filaments; degrade pollutants	Lesions
		Streptomyces	Produce antibiotics	Rare sinus infections

[a]The names of classes in quotations are not officially recognized.

colony on an agar surface. Their slow growth is partly due to the time and energy required to enrich their cell walls with high concentrations of long carbon-chain waxes called **mycolic acids,** which make the cells resistant to desiccation and to staining with water-based dyes. As discussed in Chapter 4, microbiologists developed the *acid-fast stain* for mycobacteria because they are difficult to stain by standard staining techniques such as the Gram stain. Though some mycobacteria are free-living, the most prominent species are pathogens of animals and humans, including *Mycobacterium tuberculosis* (too-ber'kū-lō-sis) and *Mycobacterium leprae* (lep'rī), which cause tuberculosis and leprosy, respectively.

Mycobacteria should not be confused with the low G + C mycoplasmas discussed earlier.

Actinomycetes

Actinomycetes (ak'ti-nō-mī-sētz) are high G + C Gram-positive bacteria that form branching filaments resembling fungi (**Figure 11.18** on page 330). Of course, in contrast to fungi, the filaments of actinomycetes are composed of prokaryotic cells. As we have seen, some actinomycetes also resemble fungi in the production of chains of reproductive spores at the ends of their filaments (see Figure 11.4). These spores should not be confused with endospores, which are

Branched filaments

LM   20.0 μm

▲ *Figure 11.18*

The branching filaments of actinomycetes. This photograph shows filaments of *Streptomyces* sp. at the edge of a colony growing on agar. *How do the filaments of actinomycetes compare to the filaments of fungi?*

Figure 11.18 Filaments of actinomycetes are thinner than those of fungi, and they are composed of prokaryotic cells.

resting stages and not reproductive cells. Actinomycetes may cause disease, particularly in immunocompromised patients. Among the important actinomycete genera are *Actinomyces* (which gives this group its name), *Nocardia,* and *Streptomyces.*

Actinomyces   Species of *Actinomyces* (ak'ti-nō-mī'sēz) are facultative capneic[14] filaments that are normal inhabitants of the mucous membranes lining the oral cavity and throats of humans. *Actinomyces israelii* (is-rā'el-ē-ē) growing in humans destroys tissue to form abscesses and can spread throughout the abdomen, consuming every vital organ.

Nocardia   Species of *Nocardia* (nō-kar'dē-ă) are soil- and water-dwelling aerobes that typically form aerial and subterranean filaments that make them resemble fungi. *Nocardia* is notable because it can degrade many pollutants of landfills, lakes, and streams, including waxes, petroleum hydrocarbons, detergents, benzene, polychlorinated biphenyls (PCBs), pesticides, and rubber. Some species of *Nocardia* cause lesions in humans.

Streptomyces   Bacteria in the genus *Streptomyces* (strep-tō-mī'sēz) are important in several realms. Ecologically, they recycle nutrients in the soil by degrading a number of carbohydrates, including cellulose, lignin (the woody part of plants), chitin (the skeletal material of insects and crustaceans), latex, aromatic chemicals (organic compounds containing a benzene ring), and keratin (the protein that forms hair, nails, and horns). The metabolic by-products of *Streptomyces* give soil its musty smell. Medically, *Streptomyces* produce most of the important antibiotics, including chloram-

phenicol, erythromycin, and tetracycline, as discussed in Chapter 10.

Table 11.3 on page 329 includes a summary of the characteristics of the genera of high G + C Gram-positive bacteria (Phylum Actinobacteria) discussed in this chapter.

CRITICAL THINKING

In Chapter 4, we discussed the identification of microbes using dichotomous taxonomic keys (see page 121). Design a key for all the genera of Gram-positive bacteria listed in Table 11.3 (p. 329).

To this point we have discussed archaea, deeply branching bacteria, phototrophic bacteria, and Gram-positive bacteria. Now it's time to turn our attention to the Gram-negative bacteria that are grouped together within the phylum Proteobacteria.

Gram-Negative Proteobacteria

The phylum **Proteobacteria**[15] constitutes the largest and most diverse group of bacteria (see Figure 11.10). Though they have a variety of shapes, reproductive strategies, and nutritional types, they are all Gram-negative and share common 16S rRNA nucleotide sequences. The G + C ratio of Gram-negative species is not critical in delineating taxa of most Gram-negative organisms, so we will not consider this characteristic in our discussion of Gram-negative bacteria.

There are five distinct classes of proteobacteria, designated by the first five letters of the Greek alphabet—α, β, γ, δ, and ε (alpha, beta, gamma, delta, and epsilon). These classes are distinguished by minor differences in their rRNA sequences. Here we will focus our attention on species with novel characteristics as well as species with practical importance. Chapters 20 and 21 discuss the medically important proteobacteria in more detail.

"Alphaproteobacteria"

Learning Objective

✓ Describe the appearance and function of prosthecae in alphaproteobacteria.

Alphaproteobacteria are typically aerobes capable of growing at very low nutrient levels. Many have unusual methods of metabolism, as we will see shortly. They may be rods, curved rods, spirals, coccobacilli, or pleomorphic in shape. Many species have unusual extensions called *prosthecae* (pros-thē'kē), which are composed of cytoplasm surrounded by the cytoplasmic membrane and cell wall **(Figure 11.19).** They use prosthecae for attachment and to increase surface area for nutrient absorption. Some prosthecate species produce buds at the ends of the extensions.

---

[14]Meaning they grow best with a relatively high concentration of carbon dioxide.
[15]Named for the Greek god *Proteus,* who could assume many shapes.

TEM   1 μm

▲ *Figure 11.19*

A prostheca, an extension of an α–proteobacterial cell that increases surface area for absorbing nutrients and serves as an organ of attachment. This prosthecate bacterium, *Hyphomicrobium facilis,* also produces buds from its prostheca.

▲ *Figure 11.20*

Nodules on soybean roots stimulated by the growth of *Rhizobium,* a nitrogen-fixing α-proteobacterium.

Nitrogen Fixers   Two genera of nitrogen fixers in the class α-proteobacteria—*Azospirillum* (ā-zō-spī′ril-ŭm) and *Rhizobium* (rī-zō′bē-ŭm)—are important in agriculture. They grow in association with the roots of plants, where they make atmospheric nitrogen (N_2) available to the plants as ammonia (NH_3), so-called fixed nitrogen. *Azospirillum* associates with the outer surfaces of roots of tropical grasses, such as sugar cane. In addition to supplying nitrogen to the grass, this bacterium also releases chemicals that stimulate the plant to produce numerous root hairs, which increase a root's surface area and thus its uptake of nutrients. *Rhizobium* grows within the roots of leguminous plants such as peas, beans, and clover, stimulating the formation of nodules on their roots **(Figure 11.20).** *Rhizobium* cells within the nodules make ammonia available to the plant, encouraging growth. Scientists are actively seeking ways to successfully insert the genes of nitrogen fixation into plants such as corn, which require large amounts of nitrogen.

Nitrifying Bacteria   As discussed in Chapter 6, chemoautotrophic organisms derive their carbon from carbon dioxide, and their energy from inorganic chemicals. Those that derive electrons from the oxidation of nitrogenous compounds are called **nitrifying bacteria.** These microbes are important in the environment and in agriculture because they convert reduced nitrogen compounds, such as ammonia (NH_3), ammonium (NH_4^+), and nitrite (NO_2^-), into nitrate (NO_3^-)—a process called **nitrification.** Nitrate moves more easily through soil than reduced nitrogenous compounds and is thereby more available to plants. Nitrifying α-proteobacteria oxidize ammonia or ammonium and are in the genus *Nitrobacter.*

CRITICAL THINKING

Contrast the processes of nitrogen fixation and nitrification.

Purple Nonsulfur Phototrophs   With one exception (the β-proteobacterium *Rhodocyclus*), purple nonsulfur phototrophs are classified as α-proteobacteria. Purple nonsulfur bacteria grow in the upper layer of mud at the bottoms of lakes and ponds. As we discussed earlier, they harvest light as an energy source by using bacteriochlorophylls, and they do not generate oxygen during photosynthesis. Morphologically, they may be rods, curved rods, or spirals; some species are prosthecate. Refer to Table 11.2 (p. 325) to review the characteristics of all phototrophic bacteria.

Pathogenic Alphaproteobacteria   Notable pathogens among the α-proteobacteria include *Rickettsia* (ri-ket′sē-ă), *Brucella* (broo-sel′lă), and *Ehrlichia* (er-lik′ē-ă).

Rickettsia is a genus of small, Gram-negative, aerobic rods that live and reproduce by binary fission inside mammalian cells. They cause a number of human diseases, including typhus and Rocky Mountain spotted fever. Rickettsias cannot use glucose as a nutrient; instead they oxidize amino acids and Krebs cycle intermediates such as glutamic acid and succinic acid. For this reason they are obliged to live within other cells where these nutrients are produced.

Brucella is a coccobacillus that causes *brucellosis,* a disease of mammals characterized by spontaneous abortions and sterility in animals. In contrast, infected humans suffer chills, sweating, fatigue, and fever. *Brucella* is notable because

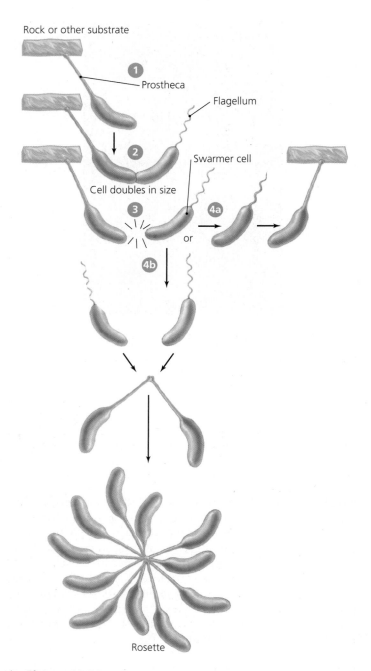

▲ *Figure 11.21*

Growth and reproduction of *Caulobacter.* ① A cell attached to a substrate by its prostheca. ② Growth and production of an apical flagellum. ③ Division by asymmetrical binary fission to produce a flagellated daughter cell called a swarmer cell. ④ⓐ Attachment of a swarmer cell to a substrate by a prostheca that replaces the flagellum. ④ⓑ Attachment of a swarmer cell to other cells to form a rosette.

it survives phagocytosis by white blood cells—normally an important step in the body's defense against disease.

Ehrlichia is a tick-borne α-proteobacterium that causes disease in humans by living within white blood cells. Symptoms of *ehrlichiosis* include high fever, headache, muscle pain, and *leukopenia,* an abnormal decrease in the number of white blood cells.

▲ *Figure 11.22*

A plant gall resulting from infection by the α-proteobacterium *Agrobacterium.* The gall is formed by the proliferation of undifferentiated plant cells caused by the expression of a plant growth hormone gene inserted into a plant chromosome by a plasmid carried by the infecting *Agrobacterium* cells. The gall cells synthesize nutrients for the bacteria.

Other Alphaproteobacteria Other α-proteobacteria are important in industry and the environment. For example, *Acetobacter* (as'ē-tō-bak-ter) and *Gluconobacter* (gloo-kon'ō-bak-ter) are used to synthesize acetic acid (vinegar). *Caulobacter* (kaw-lō-bak-ter) is a common prosthecate rod-shaped microbe that inhabits nutrient-poor seawater and fresh water; it can also be found in laboratory water baths.

Caulobacter has a unique reproductive strategy **(Figure 11.21).** A cell that has attached to a substrate with its prostheca ❶ grows until it has doubled in size, at which time it produces a flagellum at its apex ❷ ; it then divides by asymmetric binary fission ❸. The flagellated daughter cell, which is called a swarmer cell, then swims away. The swarmer cell can either attach to a substrate with a new prostheca that replaces the flagellum ❹ⓐ or to other swarmer cells to form a rosette of cells ❹ⓑ . The process of reproduction repeats about every two hours.

Scientists are very interested in the potential usefulness of another α-proteobacterium, *Agrobacterium* (a'grō-bak-tēr-ē-ŭm), which infects plants to form tumors called *galls* **(Figure 11.22).** The bacterium inserts a plasmid, which carries a gene for a plant growth hormone, into a chromosome of the plant. The growth hormone causes the cells of the plant to proliferate into a gall and to produce nutrients for the bacterium. Scientists have discovered that they can insert almost any DNA sequence into the plasmid, making it an ideal vector for genetic manipulation of plants.

Characteristics of selected members of the alphaproteobacteria are listed in **Table 11.4** on page 334.

"Betaproteobacteria"

Learning Objective

✓ Name three pathogenic and three useful betaproteobacterial species.

Betaproteobacteria are another diverse group of Gram-negative bacteria that thrive in habitats with low levels of nutrients. They differ from α-proteobacteria in their rRNA sequences, though metabolically the two groups overlap. One example of the metabolic overlap of the β-proteobacteria with the α-proteobacteria is *Nitrosomonas* (nī-trō-sō-mō'nas), an important nitrifying bacterium of soils; instead of oxidizing ammonia or ammonium to nitrate, however, it oxidizes nitrite to nitrate. In this section we will discuss a few of the interesting β-proteobacteria.

Pathogenic Betaproteobacteria　Species of *Neisseria* (nī-sē'rē-ă) are Gram-negative diplococci that inhabit the mucous membranes of mammals and cause such diseases as gonorrhea, meningitis, pelvic inflammatory disease, and inflammation of the cervix, pharynx, and external lining of the eye.

Other pathogenic β-proteobacteria include *Bordetella* (bōr-dĕ-tel'ă), which is the cause of pertussis (whooping cough), and *Burkholderia* (burk-hol-der'ē-ă), which recycles numerous organic compounds in nature. *Burkholderia* commonly colonizes moist environmental surfaces (including laboratory and medical equipment) and the respiratory passages of patients with cystic fibrosis.

Chapter 20 discusses these pathogens in more detail.

Other Betaproteobacteria　Members of the genus *Thiobacillus* (thī-ō-ba-sil'ŭs) are colorless sulfur bacteria that are important in recycling sulfur in the environment by oxidizing hydrogen sulfide (H_2S) or elemental sulfur (S^0) to sulfate (SO_4^{2-}). Miners use *Thiobacillus* to leach metals from low-grade ore, though the bacterium does cause extensive pollution when it releases metals and acid from mine wastes. Chapter 26 discusses the sulfur cycle.

Sewage treatment supervisors are interested in *Zoogloea* (zō'ō-glē-ă) and *Sphaerotilus* (sfēr-ōl'til-us), two genera that form *flocs*—slimy, tangled masses of bacteria and organic matter in sewage. *Zoogloea* forms compact flocs that settle to the bottom of treatment tanks and assist in the purification process. *Sphaerotilus,* in contrast, forms loose flocs that do not settle and thus impede the proper flow of waste through a treatment plant.

Spirillum (spī-ril'ŭm) is a Gram-negative, spiral-shaped β-proteobacterium whose stained polar flagella are easily visible with a light microscope (see Figure 3.6c). *Spirillum minus* causes rat-bite fever.

The characteristics of the genera of β-proteobacteria discussed in this section are summarized in Table 11.4 on page 334.

"Gammaproteobacteria"

Learning Objective

✓ Describe the gammaproteobacteria.

The **gammaproteobacteria** make up the largest and most diverse class of proteobacteria; almost every shape, arrangement of cells, metabolic type, and reproductive strategy is represented in this group. Ribosomal studies indicate that γ-proteobacteria can be divided into several subgroups:

- Purple sulfur bacteria
- Intracellular pathogens
- Methane oxidizers
- Facultative anaerobes that utilize Emdem-Meyerhof glycolysis and the pentose phosphate pathway
- Pseudomonads, which are aerobes that catabolize carbohydrates by the Entner-Doudoroff and pentose phosphate pathways

Here we will examine some representatives of these groups. As noted previously, the pathogens are discussed in more detail in Chapters 20 and 21.

Purple Sulfur Bacteria　Whereas the purple *non*sulfur bacteria are distributed among the α- and β-proteobacteria, **purple sulfur bacteria** are γ-proteobacteria. Purple sulfur bacteria are obligate anaerobes that oxidize hydrogen sulfide to sulfur, which they deposit as internal granules. They are found in sulfur-rich zones in lakes, bogs, and oceans. Some species form intimate relationships with marine worms, covering the body of a worm like strands of hair **(Figure 11.23)**.

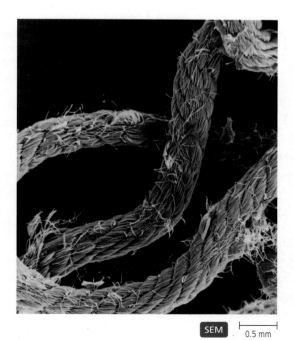

SEM　0.5 mm

▲ *Figure 11.23*

Purple sulfur bacteria covering the surface of a nematode (round worm). The presence of so many bacteria produces the appearance of a woven rope.

Table 11.4   Characteristics of Selected Gram-Negative Bacteria

Phylum/Class	Representative Members	Special Characteristics	Diseases
Proteobacteria			
"α-proteobacteria"[a]	*Azospirillum*	Nitrogen fixers	
	Rhizobium	Nitrogen fixers	
	Nitrobacter	Nitrifying bacteria	
	Purple nonsulfur bacteria	Anoxygenic phototrophs	
	Rickettsia	Intracellular pathogens	Typhus and Rocky Mountain spotted fever
	Brucella	Coccobacilli	Brucellosis
	Ehrlichia	Live inside white blood cells	Leukopenia
	Acetobacter, Glucanobacter	Synthesize acetic acid	
	Caulobacter	Prosthecate bacteria	
	Agrobacterium	Cause galls in plants; vectors for gene transfer in plants	
Proteobacteria			
"β-proteobacteria"	*Nitrosomonas*	Nitrifying bacteria	
	Neisseria	Diplococci	Gonorrhea and meningitis
	Bordetella		Pertussis
	Burkholderia		Lung infection of cystic fibrosis patients
	Thiobacillus	Colorless sulfur bacteria	
	Zoogloea	Used in sewage treatment	
	Sphaerotilus	Block sewage treatment pipes	
	Spirillum	Amphitrichous polar flagella	Rat bite fever
Proteobacteria			
"γ-proteobacteria"	Purple sulfur bacteria		
	Legionella	Intracellular pathogens	
	Coxiella	Intracellular pathogens	
	Methane oxidizers		
	Glycolytic facultative anaerobes	Facultative anaerobes that catabolize carbohydrates via glycolysis and the pentose phosphate pathway	See Table 11.5 on page 336
	Pseudomonas	Aerobes that catabolize carbohydrates via Entner-Doudoroff and pentose phosphate pathways	Urinary tract infections, external otitis
	Azotobacter Azomonas	Nitrogen fixers not associated with plant roots	
Proteobacteria			
"δ-proteobacteria"	*Desulfovibrio*	Sulfate reducers	
	Bdellovibrio	Pathogens of Gram-negative bacteria	
	Myxobacteria	Reproduce by forming differentiated fruiting bodies	
Proteobacteria			
"ε-proteobacteria"	*Campylobacter*	Curved rods	Gastroenteritis
	Helicobacter	Spirals	Gastric ulcers
Chlamydiae			
"Chlamydiae"	*Chlamydia*	Intracellular pathogens; lack peptidoglycan	Neonatal blindness and lymphogranuloma venereum

[a]The names of classes in quotations are not officially recognized.

Table 11.4   *(continued)*

Phylum/Class	Representative Members	Special Characteristics	Diseases
Spirochaetes			
"Spirochaetes"	*Treponema*	Motile by axial filaments	Syphilis
	Borrelia	Motile by axial filaments	Lyme disease
Bacteroidetes			
"Bacteroides"	*Bacteroides*	Anaerobes that live in animal colons	Abdominal infections
	Cytophaga	Digest complex polysaccharides	

Intracellular Pathogens   Organisms in the genera *Legionella* (lē-jŭ-nel′lă) and *Coxiella* (kok-sē-el′ă) are pathogens of humans that avoid digestion by white blood cells, which are normally part of a body's defense; in fact, they thrive inside these defensive cells. *Legionella* derives energy from the metabolism of amino acids, which are more prevalent inside cells than outside. *Coxiella* grows best at low pH, such as is found in the phagolysomes of white blood cells. The bacteria in these genera cause Legionnaire's disease and Q fever, respectively.

Methane Oxidizers   Organisms that are **methane oxidizers** are Gram-negative bacteria that utilize methane both as a carbon and as an energy source. Like methanogens, which are a type of archaea, methane oxidizers inhabit anaerobic environments worldwide, growing just above the anaerobic layers that contain methanogens. Although methane is one of the so-called greenhouse gases that retains heat in the atmosphere, methane oxidizers digest most of the methane in their local environment before it can adversely affect the world's climate.

Glycolytic Facultative Anaerobes   The largest group of γ-proteobacteria is composed of Gram-negative, facultatively anaerobic rods that catabolize carbohydrates by glycolysis and the pentose phosphate pathway. This group, which is divided into three families (**Table 11.5** on page 336), contains numerous human pathogens. Members of the family Enterobacteriaceae, including *Escherichia coli* (esh-ĕ-rik′ē-ă kō′lī), are frequently used for laboratory studies of metabolism, genetics, and recombinant DNA technology. The pathogenic members are discussed in detail in Chapter 20.

Pseudomonads   Bacteria called **pseudomonads** are Gram-negative, aerobic, flagellated, straight to slightly curved rods that catabolize carbohydrates by the Entner-Doudoroff and pentose phosphate pathways. These organisms are noted for their ability to break down numerous organic compounds. Many of them are important pathogens of humans and animals and are involved in the spoilage of refrigerated milk, eggs, and meat because they can grow and catabolize proteins and lipids at 4°C. The pseudomonad

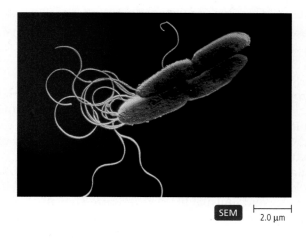

SEM   2.0 µm

▲ *Figure 11.24*
Pseudomonas and its characteristic polar flagella.

group is named for its most important genus: *Pseudomonas* (soo-dō-mō′s′nas; **Figure 11.24**), which causes diseases such as urinary tract infections and external otitis (swimmers' ear). Other pseudomonads such as *Azotobacter* (ā-zō-tō-bak′ter) and *Azomonas* (ā-zō-mō′nas) are soil dwelling, nonpathogenic nitrogen fixers; however, in contrast to nitrogen-fixing α-proteobacteria, these γ-proteobacteria do not associate with the roots of plants.

The characteristics of the members of the γ-proteobacteria discussed in this section are summarized in Table 11.4.

"Deltaproteobacteria"

Learning Objective

✓ List several members of deltaproteobacteria.

The **deltaproteobacteria** are not a large assemblage, but like other proteobacteria they include a wide variety of metabolic types. *Desulfovibrio* (dē′sul-fō-vib-rē-ō) is a sulfate-reducing microbe that is important in the sulfur cycle. It is also an important member of bacterial communities living in the sediments of polluted streams and sewage treatment lagoons, where its presence is often apparent by the odor of

Table 11.5   Representative Glycolytic Facultative Anaerobes of the Class "Gammaproteobacteria"

Family	Special Characteristics	Representative Genera	Typical Human Diseases
Enterobacteriaceae	Straight rods; oxidase negative; peritrichous flagella or nonmotile	*Escherichia*	Gastroenteritis
		Enterobacter	(Rarely pathogenic)
		Serratia	(Rarely pathogenic)
		Salmonella	Enteritis
		Proteus	Urinary tract infection
		Shigella	Shigellosis
		Yersinia	Plague
		Klebsiella	Pneumonia
Vibrionaceae	Vibrios; oxidase positive; polar flagella	*Vibrio*	Cholera
Pasteurellaceae	Cocci or straight rods; oxidase positive; nonmotile	*Haemophilus*	Meningitis in children

hydrogen sulfide that it releases during anaerobic respiration. Hydrogen sulfide reacts with iron to form iron sulfide, so sulfate-reducing bacteria play a primary role in the corrosion of iron pipes in heating systems, sewer lines, and other structures.

Bdellovibrio (del-lō-vib′rē-ō) is not pathogenic to eukaryotes, but instead attacks other Gram-negative bacteria. It has a complex and unusual life cycle **(Figure 11.25a):**

1 A free *Bdellovibrio* swims rapidly through the medium until it collides with a Gram-negative bacterium.

2 It rapidly drills through the cell wall of its prey by secreting hydrolytic enzymes and rotating in excess of 100 revolutions per second **(Figure 11.25b).**

3 Once inside, *Bdellovibrio* lives in the periplasmic space—the space between the cytoplasmic membrane and the cell wall. It kills its host by disrupting the host's cytoplasmic membrane and inhibiting DNA, RNA, and protein synthesis.

4 The invading bacterium uses the nutrients released from its dying prey and grows into a long filament.

5 Eventually, the filament divides into many smaller cells, each of which, when released from the dead cell, produces a flagellum, and swims off to repeat the process. Multiple fission to produce many offspring is a rare form of reproduction.

Myxobacteria are Gram-negative, aerobic, soil-dwelling bacteria with a unique life cycle for prokaryotes in that individuals cooperate to produce differentiated reproductive structures. The life cycle of myxobacteria can be summarized as follows **(Figure 11.26a on page 338):**

1 Vegetative myxobacteria glide on slime trails through their environment, digesting yeasts and other bacteria or scavenging nutrients released from dead cells. When nutrients and cells are plentiful, the myxobacteria divide by binary fission; when nutrients are depleted, however, they aggregate into a mound of cells.

2 Myxobacteria within the mound differentiate to form a macroscopic **fruiting body** ranging in height from 50 μm to 700 μm.

3 Some cells within the fruiting body develop into dormant *myxospores* that are enclosed within walled structures called *sporangia* (singular: *sporangium:* **Figure 11.26b**).

4 The sporangia release the myxospores, which can resist desiccation and nutrient deprivation for a decade or more.

5 When nutrients are again plentiful, the myxospores germinate and become vegetative cells.

Myxobacteria live worldwide in soils that have decaying plant material or animal dung. Though certain species live in the arctic and others in the tropics, most myxobacteria live in temperate regions.

CRITICAL THINKING

What do the names *Desulfovibrio* and *Bdellovibrio* tell you about the morphology of these δ-proteobacteria?

"Epsilonproteobacteria"

Epsilonproteobacteria are Gram-negative rods, vibrios, or spirals. The most important genera are *Campylobacter* (kam′pi-lō-bak′ter), which causes blood poisoning and inflammation of the intestinal tract and *Helicobacter* (hel′ĭ-kō-bak′ter), which causes ulcers. They are discussed more fully in Chapter 20.

The characteristics of the δ- and ε-proteobacteria are summarized in Table 11.4.

CRITICAL THINKING

Design a dichotomous key for the proteobacteria discussed in this chapter.

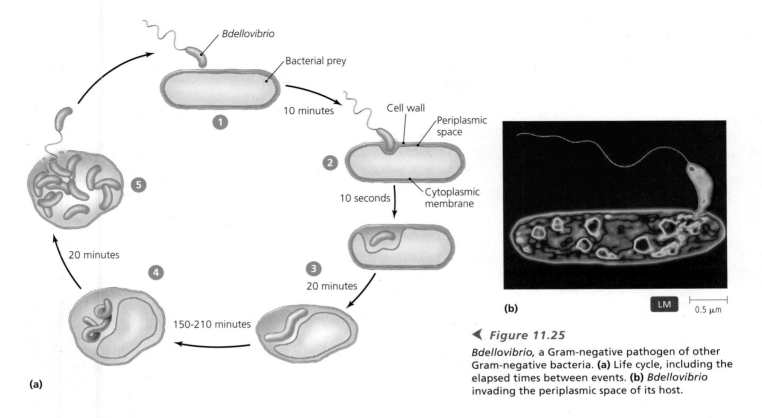

Bdellovibrio

Bacterial prey

10 minutes

Cell wall

Periplasmic space

Cytoplasmic membrane

10 seconds

20 minutes

20 minutes

150-210 minutes

(a)

(b)

LM   0.5 μm

◀ *Figure 11.25*

Bdellovibrio, a Gram-negative pathogen of other Gram-negative bacteria. **(a)** Life cycle, including the elapsed times between events. **(b)** *Bdellovibrio* invading the periplasmic space of its host.

Other Gram-Negative Bacteria

Learning Objectives

✓ Describe the unique features of chlamydias and spirochetes.
✓ Describe the ecological importance of bacteroids.

In the final section of this chapter we consider an assortment of Gram-negative bacteria that are classified in the second edition of *Bergey's Manual* into nine phyla that are grouped together for convenience, not because of genetic relatedness. Species in six of the nine phyla are of relatively minor importance, and here we discuss representatives from the three phyla that are either of particular ecological concern or significantly affect human health: the chlamydias (Phylum Chlamydiae), the spirochetes (Phylum Spirochaetes), and the bacteroids (Phylum Bacteroidetes) (see Figure 11.10).

Chlamydias

Microorganisms called **chlamydias** (kla-mid′ē-ăz) are small, Gram-negative cocci that grow and reproduce only within the cells of mammals, birds, and a few invertebrates. The smallest chlamydias—0.2 μm in diameter—are smaller than the largest viruses (poxviruses); however, in contrast to viruses, chlamydias have both DNA and RNA, cytoplasmic membranes, functioning ribosomes, reproduction by binary fission, and metabolic pathways. Like other Gram-negative prokaryotes, chlamydias have two membranes, but in contrast they lack peptidoglycan in their cell wall.

Because chlamydias and rickettsias share an obvious characteristic—both have a requirement for intracellular

life—these two types of organisms were grouped together in a single taxon in the first edition of *Bergey's Manual*. Now, however, the rickettsias are typically classified with the γ-proteobacteria, and the chlamydias are in their own phylum. Organisms in the genus *Chlamydia* are the most studied of the chlamydias because they cause neonatal blindness, pneumonia, and a sexually transmitted disease called *lymphogranuloma venereum*; in fact, they are the most common sexually transmitted bacteria in the United States.

Chlamydias have a unique method of reproduction. After invading a host cell, a chlamydial cell forms an **initial body** (also called a *reticulate body*). The initial body grows and undergoes repeated binary fissions until the host cell is filled with reticulate bodies. These then change into **elementary bodies,** which contain electron-dense material and have rigid outer boundaries formed when they cross-link their two membranes with disulfide bonds. Each elementary body, which is relatively resistant to drying, is an infective stage. When the host cell dies, the elementary bodies are released to drift until they contact and attach to other host cells, triggering their own phagocytosis by the new host cells. Once inside a host cell, elementary bodies transform back into initial bodies, and the cycle repeats. The chlamydial life cycle is illustrated in Figure 21.8.

Spirochetes

Spirochetes (spī′rō-kētz) are unique helical bacteria that are motile by means of axial filaments—flagellarlike structures that lie in the periplasmic space (see Figure 3.7). When the

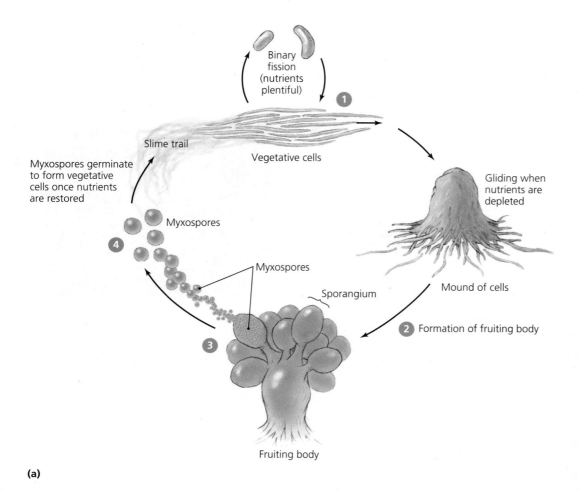

Slime trail

Myxospores germinate to form vegetative cells once nutrients are restored

Myxospores

Myxospores

Sporangium

Fruiting body

Binary fission (nutrients plentiful)

Vegetative cells

Gliding when nutrients are depleted

Mound of cells

Formation of fruiting body

(a)

◀ *Figure 11.26*

(a) Myxobacteria life cycle. ① Vegetative myxobacteria divide by binary fission when nutrients are plentiful. They aggregate in mounds when nutrients are depleted. ② The mound forms fruiting bodies. ③ Some cells in a fruiting body develop into dormant myxospores within a sporangium. ④ The sporangia release the myxospores, which remain dormant until nutrients are again plentiful. **(b)** Fruiting body.

(b) LM 100 mm

axial filaments rotate, the entire cell corkscrews through the medium.

Spirochetes have a variety of types of metabolism and live in diverse habitats. They are frequently isolated from the human mouth, marine environments, moist soil, and the surfaces of protozoa that live in termites' guts. In the latter case, they may coat the protozoan so thickly that they look, and act, like cilia. The spirochetes *Treponema* (trep-ō-nē'mă) and *Borrelia* (bō-rē'lē-ă) cause syphilis and Lyme disease, respectively, in humans.

Bacteroids

Bacteroids are yet another diverse group of Gram-negative microbes that are grouped together on the basis of similarities in their rRNA nucleotide sequences. The group is named for *Bacteroides* (bak-ter-oy'dēz), a genus of obligately anaerobic rods that normally inhabit the digestive tracts of humans and animals. Bacteroids assist in digestion by catabolizing substances such as cellulose and other complex carbohydrates that are indigestible by mammals. About 30% of the bacteria isolated from human feces are *Bacteroides*. As

discussed in Chapter 20, some species of *Bacteroides* cause abdominal, pelvic, blood, and other infections; in fact, they are the most common anaerobic human pathogen.

Bacteroids in the genus *Cytophaga* (sī-tof'ă-gă) are aquatic, gliding, rod-shaped aerobes with pointed ends. These bacteria degrade complex polysaccharides such as agar, pectin, chitin, and even cellulose, so they cause damage to wooden boats and piers; they also play an important role in the degradation of raw sewage. Their ability to glide allows these bacteria to position themselves at sites with optimum nutrients, pH, temperature, and oxygen levels. Organisms in *Cytophaga* differ from other gliding bacteria, such as cyanobacteria and myxobacteria, in that they are nonphotosynthetic and do not form fruiting bodies.

The characteristics of these groups of Gram-negative bacteria are listed in Table 11.4.

CRITICAL THINKING

Design a key for the genera of Gram-negative bacteria listed in Table 11.4.

CHAPTER SUMMARY

General Characteristics of Prokaryotic Organisms (pp. 314–318)

1. Three basic shapes of prokaryotic cells are spherical **cocci,** rod-shaped **bacilli,** and **spirals.** Spirals may be stiff **(spirilla)** or flexible **(spirochetes).**

2. Other variations in shapes include **vibrios** (slighty curved rods), **coccobacilli** (intermediate to cocci and bacilli), and **pleomorphic** (variable shape and size).

3. Cocci may typically be found in groups, including long chains **(streptococci),** pairs **(diplococci),** cuboidal packets **(sarcinae),** and clusters **(staphylococci).**

4. Bacilli are found singly, in pairs, in chains, or in a folded **palisade** arrangement.

5. Prokaryotes reproduce asexually by **binary fission, snapping division** (a type of binary fission), **spore** formation, and **budding.**

6. Environmentally resistant **endospores** are produced within vegetative cells of the Gram-positive genera *Bacillus* and *Clostridium.* Depending on the species in which they are formed, the endospores may be terminal, subterminal, or centrally located.

Modern Prokaryotic Classification (pp. 318–319)

1. Living things are now typically classified into three domains—Archaea, Bacteria, and Eukarya—based largely on genetic relatedness.

2. The most authoritative reference in modern prokaryotic systematics is *Bergey's Manual of Systematic Bacteriology,* which classifies prokaryotes into two phyla in Archaea and 23 phyla in Bacteria. The organization of this text's survey of prokaryotes largely follows the classification scheme proposed in the second edition of *Bergey's Manual.*

Survey of Archaea (pp. 319–322)

1. The domain Archaea includes **extremophiles,** microbes that require extreme conditions of temperature, pH, and/or salinity to survive.

2. **Thermophiles** and **hyperthermophiles** (in the phyla Crenarchaeota and Euryarchaeota) live at temperatures over 45°C and 80°C, respectively, because their DNA, membranes, and proteins do not function properly at lower temperatures.

3. **Halophiles** (Phylum Euryarchaeota) depend on high concentrations of salt to keep their cell walls intact. The halophile *Halobacterium salinarium* synthesizes purple proteins called **bacteriorhodopsins** that harvest light energy to synthesize ATP.

4. **Methanogens** (Phylum Euryarchaeota) are obligate anaerobes that produce methane gas and are useful in sewage treatment.

Survey of Bacteria (pp. 323–338)

1. **Deeply branching bacteria** have rRNA sequences thought to be similar to those of earliest bacteria. They are autotrophic and live in hot, acidic, and anaerobic environments, often with intense exposure to sun.

2. Phototrophic bacteria trap light energy with photosynthetic lamellae. The five groups of phototrophic bacteria are **cyanobacteria,** green sulfur bacteria, green nonsulfur bacteria, purple sulfur bacteria, and purple nonsulfur bacteria.

3. Many cyanobacteria reduce atmospheric N_2 to NH_3 via a process called **nitrogen fixation.** Cyanobacteria must separate (in either time or space) the metabolic pathways of nitrogen fixation from those of oxygenic photosynthesis because nitrogen fixation is inhibited by the oxygen generated during photosynthesis. Many cyanobacteria fix nitrogen in thick-walled cells called **heterocysts.**

4. Green and purple bacteria use bacteriochlorophylls for anoxygenic photosynthesis. Nonsulfur forms derive electrons from organic compounds; sulfur forms derive electrons from H_2S.

5. The phylum **Firmicutes** contains bacteria with a G + C ratio (the proportion of all base pairs that are guanine–cytosine base pairs) of less than 50%. Firmicutes includes clostridia, mycoplasmas, and low G + C cocci and bacilli.

6. Clostridia include the genus *Clostridium* (pathogenic bacteria that cause gangrene, tetanus, botulism, and diarrhea), *Epulopiscium* (which is large enough to be seen without a microscope), and *Veillonella* (often found in dental plaque).

7. **Mycoplasmas** are Gram-negative, pleomorphic, facultative anaerobes and obligate anaerobes that lack cell walls; they are frequently associated with pneumonia and urinary tract infections.

8. Low G + C Gram-positive bacilli and cocci important to human health and industry include *Bacillus* (which contains species that cause anthrax and food poisoning, and includes beneficial Bt-toxin bacteria), *Listeria* (which cause bacteremia and meningitis), *Lactobacillus* (used to produce yogurt and pickles), *Streptococcus* (which causes strep throat and other diseases), and *Staphylococcus* (which causes a number of human diseases).

9. High G + C bacteria (*Corynebacterium, Mycobacterium,* and actinomycetes) are classified in Phylum Actinobacteria.

10. Bacteria in *Corynebacterium* store phosphates in **metachromatic granules;** *C. diphtheriae* causes diphtheria.

11. Members of the genus *Mycobacterium,* including species that cause tuberculosis and leprosy, grow slowly, and have unique, resistant cell walls containing waxy **mycolic acids.**

12. **Actinomycetes** resemble fungi in that they produce spores and form filaments; this group includes *Actinomyces* (normally found in human mouths), *Nocardia* (useful in degradation of pollutants), and *Streptomyces* (produces important antibiotics).

13. Phylum **Proteobacteria** is a very large group of Gram-negative bacteria divided into five classes—the alpha-, beta-, gamma-, delta-, and epsilonproteobacteria.

14. The **alphaproteobacteria** include a variety of aerobes, many of which have unusual attachment extensions of the cell called prosthecae. *Azospirillum* and *Rhizobium* are nitrogen fixers that are important in agriculture.

15. Some members of the alphaproteobacteria are **nitrifying bacteria,** which oxidize NH_3 to NO_3 via a process called **nitrification.** Nitrifying alphaproteobacteria are in the genus *Nitrobacter.*

16. Most purple nonsulfur phototrophs are alphaproteobacteria.

17. Pathogenic α-proteobacteria include *Rickettsia* (some cause typhus and Rocky Mountain spotted fever), *Brucella* (brucellosis), and *Ehrlichia* (ehrlichiosis).

18. There are many beneficial α-proteobacteria, including *Acetobacter* and *Gluconobacter,* both of which are used to synthesize acetic acid. *Caulobacter* is of interest in reproductive studies, and *Agrobacterium* is used in genetic recombination in plants.

19. The **betaproteobacteria** include the nitrifying *Nitrosomonas* and pathogenic species such as *Neisseria* (gonorrhea), *Bordetella* (whooping cough), and *Burkholderia* (which colonizes the lungs of cystic fibrosis patients).

20. Other β-proteobacteria include *Thiobacillus* (ecologically important), *Zoogloea* (useful in sewage treatment), *Sphaerotilus* (hampers sewage treatment), and *Spirillum.*

21. The **gammaproteobacteria** constitute the largest class of proteobacteria; they include purple sulfur bacteria, intracellular pathogens, **methane oxidizers,** facultative anaerobes that utilize glycolysis and the pentose phosphate pathway, and pseudomonads.

22. Both *Legionella* and *Coxiella* are intracellular, pathogenic γ-proteobacteria.

23. Numerous human pathogens are facultatively anaerobic γ-proteobacteria that catabolize carbohydrates by glycolysis.

24. **Pseudomonads,** including pathogenic *Pseudomonas* and nitrogen-fixing *Azotobacter* and *Azomonas,* utilize the Entner-Doudoroff and pentose phosphate pathways for catabolism of glucose.

25. The **deltaproteobacteria** include *Desulfovibrio* (important in the sulfur cycle and in corrosion of pipes), *Bdellovibrio* (pathogenic to bacteria), and **myxobacteria.** The latter form stalked **fruiting bodies** containing resistant, dormant myxospores.

26. The **epsilonproteobacteria** include some important human pathogens, including *Campylobacter* and *Helicobacter.*

27. Chlamydias are Gram-negative cocci typified by the genus *Chlamydia;* they cause neonatal blindness, pneumonia, and a sexually transmitted disease. Within a host cell, chlamydias form **initial bodies,** which change into smaller **elementary bodies** released when the host cell dies.

28. **Spirochetes** are helical bacteria that live in diverse environments. *Treponema* (syphilis) and *Borrelia* (Lyme disease) are important spirochetes.

29. **Bacteroids** include *Bacteroides,* an obligate anaerobic rod that inhabits the digestive tract, and *Cytophaga,* an aerobic rod that degrades wood and raw sewage.

QUESTIONS FOR REVIEW

(Answers to multiple choice and matching questions are on the web, along with additional review questions. Visit www.microbiologyplace.com.)

Modified True-False

For each of the following statements that is true, write "true" in the blank. For each statement that is false, write the word(s) that should be substituted for the italicized word(s) to make the statement correct.

____ 1. All prokaryotes reproduce *sexually.*

____ 2. A *bacillus* is a bacterium with a slightly curved rod shape.

____ 3. If you were to view *staphylococci,* you should expect to see clusters of cells.

____ 4. *Initial* bodies are stable resting stages that do not metabolize but will germinate when conditions improve.

____ 5. Archaea are classified into phyla based primarily on *tRNA sequences.*

____ 6. *Halophiles* inhabit extremely saline habitats such as the Great Salt Lake.

____ 7. Pigments located in thylakoids in phototrophic bacteria trap *light* energy for metabolic processes.

____ 8. Most *cyanobacteria* form heterocysts in which nitrogen fixation occurs.

____ 9. A giant bacterium that is large enough to be seen without a microscope is *Veillonella.*

____10. When environmental nutrients are depleted, myxobacteria aggregate in mounds to form fruiting bodies.

Matching

Match the bacterium on the left with the term with which it is most closely associated.

____ 1. *Bacillus anthracis*   A. wood damage
____ 2. *Veillonella*   B. dental plaque
____ 3. *Clostridium perfringens*   C. gangrene
____ 4. *Clostridium botulinum*   D. botulism poisoning
____ 5. *Bacillus licheniformis*   E. anthrax
____ 6. *Streptococcus*   F. lymphogranuloma venereum
____ 7. *Streptomyces*   G. leprosy
____ 8. *Corynebacterium*   H. tetracycline
____ 9. *Glucanobacter*   I. vinegar
____10. *Bordetella*   J. yogurt
____11. *Zoogloea*   K. impetigo
____12. *Azotobacter*   L. bacitracin
____13. *Desulfovibrio*   M. iron pipe corrosion
____14. *Chlamydia*   N. pertussis
____15. *Cytophaga*   O. nitrogen fixation
                                       P. floc formation
                                       Q. diphtheria

Multiple Choice

1. The type of reproduction in prokaryotes that results in a palisade arrangement of cells is called
   a. pleomorphic division.
   b. endospore formation.
   c. snapping division.
   d. binary fission.

2. The thick-walled reproductive spores produced in the middle of cyanobacterial filaments are called
   a. akinetes.
   b. terminal endospores.
   c. metachromatic granules.
   d. myxospores.

3. Which of the following terms best describes stiff, spiral-shaped prokaryotic cells?
   a. cocci
   b. bacilli
   c. spirilla
   d. spirochetes

4. Endospores
   a. can remain alive for decades.
   b. can remain alive in boiling water.
   c. live in a state of suspended animation.
   d. all of the above

5. *Halobacterium salinarium* is distinctive because
   a. it is absolutely dependent on high salt concentrations to maintain its cell wall.
   b. it is found in terrestrial volcanic habitats.
   c. it photosynthesizes without chlorophyll.
   d. it can survive 5 million rad of radiation.

6. Photosynthetic bacteria that also fix nitrogen are
   a. mycoplasmas.
   b. spirilla.
   c. bacteroids.
   d. cyanobacteria.

7. Which genus is the most common anaerobic human pathogen?
   a. *Bacteroides*
   b. *Spirochetes*
   c. *Chlamydias*
   d. *Methanopyrus*

Short Answer

1. Whereas the first edition of *Bergey's Manual* relied on morphological and biochemical characteristics to classify microbes, the new edition focuses on ribosomal RNA sequences. List several other criteria for grouping and classifying bacteria.

2. What are extremophiles? Describe two kinds and give examples.

3. Name and describe three types of bacteria mentioned in this chapter that "glide."

4. Name three groups of low G + C Gram-positive bacteria.

5. Sketch and label six shapes of bacteria.

6. A student was memorizing the arrangements of bacteria and noticed that there are more arrangements for cocci than for bacilli. Why might this be so?

7. How are bacterial endospores different from the spores of actinomycetes, algae, and fungi?

8. How is *Agrobacterium* used in recombinant DNA technology?

9. Name and describe five distinct classes of Phylum Proteobacteria.

10. Explain why organisms formerly known as blue-green algae are now called cyanobacteria.

11. Louis Pasteur said, "The role of the infinitely small in nature is infinitely large." Explain what he meant by using examples of the roles of microorganisms in health, industry, and the environment.

CRITICAL THINKING

1. A microbiology student described "deeply branching bacteria" as having a branched filamentous growth habit akin to *Streptomyces*. Do you agree with this description? Why or why not?

2. Iron oxide (rust) forms when iron is exposed to oxygen, particularly in the presence of water. Nevertheless, iron pipes typically corrode more quickly when they are buried in moist *anaerobic* soil than when they are buried in soil containing oxygen. Explain why this is the case.

3. Why is it that Gram-positive species don't have axial filaments?

4. Even though *Clostridium* is strictly an anaerobic bacterium, it can be isolated easily from the surface of your skin. Explain how this can be.

CHAPTER 12

Characterizing and Classifying Eukaryotes

CHAPTER OUTLINE

General Characteristics of Eukaryotic Organisms
Reproduction in Eukaryotes
The Classification of Eukaryotic Organisms

Protozoa
Distribution of Protozoa
Morphology of Protozoa
Nutrition of Protozoa
Reproduction in Protozoa
Classification of Protozoa

Fungi
The Significance of Fungi
Morphology of Fungi
Nutrition of Fungi
Reproduction in Fungi
Classification of Fungi
Lichens

Algae
Distribution of Algae
Morphology of Algae
Reproduction in Algae
Classification of Algae

Water Molds and Slime Molds
Water Molds
Slime Molds

Other Eukaryotes of Microbiological Interest: Parasitic Helminths and Vectors

The mushrooms in this photo constitute only the tips of what may be the largest living organism on Earth—a fungus known as *Armillaria ostoyae* that lives 3 feet underground in Oregon's Malheur National Forest. Also known as the honey mushroom, it is 3.5 miles long, covers an area equivalent to 1665 football fields, and is estimated to be at least 2400 years old!

Amazingly, this "humongous fungus" began life as a microscopic spore. During its lifespan of over two millennia, it has used root-like structures called rhizomorphic hyphae to draw water and nutrients from tree roots. Despite its gigantic proportions, what is visible to us are ordinary mushrooms.

Fungi are eukaryotes—a vast category that also includes protozoa, algae, parasitic helminths, as well as all plants and animals. In this chapter we will take a closer look at the eukaryotic organisms of microbiological importance.

Mushrooms of *Armillaria ostoyae*

MicroPrep Pre-Test: Take the pre-test for this chapter on the web.
Visit **www.microbiologyplace.com**.

Eukaryotic microbes include a fascinating and almost bewilderingly diverse assemblage. Eukaryotic microbes include unicellular and multicellular protozoa,[1] fungi,[2] algae,[3] water molds, and slime molds. Additionally, microbiologists study parasitic helminths[4] (worms) because they have microscopic stages and they study arthropod vectors because they are intimately involved in the transmission of microbial pathogens. Eukaryotes include both human pathogens and organisms that are vital for human life. For example, one kind of marine algae called diatoms, and a kind of protozoa called dinoflagellates (dī"nō-flaj'ĕ-lātz), provide the basis for the oceans' food chains and produce most of the world's oxygen. Eukaryotic fungi produce penicillin, and tiny bakers' and brewers' yeasts are essential for making bread and alcoholic beverages.

Among the 20 most frequent microbial causes of death worldwide, six are eukaryotic, including the agents of malaria, African sleeping sickness, and amoebic dysentery. *Pneumocystis* pneumonia, toxoplasmosis, and cryptosporidosis—common afflictions of AIDS patients—are all caused by eukaryotic pathogens. A recently discovered dinoflagellate (*Pfiesteria*) appears to secrete volatile toxins that induce hours of vomiting and fits of rage, and even persistent nerve damage, in humans that inhale them.

In previous chapters we discussed characteristics of *cells*—their metabolism, growth, and genetics. In this chapter we discuss four major groups of eukaryotic *organisms* of importance in microbiology—protozoa (single-celled "animals"), fungi, algae, and the water molds and slime molds—and conclude with a brief discussion of the relationship of parasitic helminths and vectors to microbiology.

We begin by discussing general features of eukaryotic reproduction and classification; the following sections survey some representative members of microbiologically important eukaryotic groups, focusing on beneficial, environmentally significant, and unusual species. Chapters 22 and 23 discuss fungal and parasitic agents and vectors of human disease in more detail.

General Characteristics of Eukaryotic Organisms

Our discussion of the general characteristics of eukaryotes begins with a survey of the events in eukaryotic reproduction; then we consider some aspects of the complex matter of classifying the great variety of eukaryotic organisms.

Reproduction in Eukaryotes

Learning Objectives

✓ State four reasons why eukaryotic reproduction is more complex than prokaryotic reproduction.

✓ Describe the phases of mitosis, mentioning chromosomes, chromatids, centromeres, and spindle.

✓ Contrast meiosis with mitosis, mentioning homologous chromosomes, tetrads, and crossing over.

✓ Distinguish among nuclear division, cytokinesis, and schizogony.

As discussed in Chapter 3, a unique characteristic of living things is the ability to reproduce themselves. Prokaryotic reproduction typically involves replication of DNA and binary fission of the cytoplasm to produce two identical offspring. Reproduction of eukaryotes is more complicated and varied than reproduction in prokaryotes for a number of reasons:

- Most of the DNA in eukaryotes is packaged with histone proteins as *chromosomes* in the form of *chromatin* (krō'ma-tin) *fibers* located within nuclei. The remaining DNA in eukaryotic cells is found in mitochondria and chloroplasts, organelles that reproduce by binary fission in a manner similar to prokaryotic reproduction. In this chapter we will discuss only the nuclear portion of eukaryotic genomes.

- Eukaryotes have a variety of methods of asexual reproduction, including binary fission, budding, fragmentation, spore formation, and *schizogony* (ski-zog'ō-nē) (discussed later).

- Many eukaryotes reproduce sexually—that is, via a process that involves the formation of sexual cells called *gametes*, and the subsequent fusion of two gametes to form a cell called a *zygote*.

- Additionally, algae, fungi, and some protozoa reproduce both sexually and asexually. (Animals generally reproduce only one way or the other.)

Every form of eukaryotic reproduction involves two types of division: nuclear division and cytoplasmic division (also called cytokinesis). After we discuss the various aspects of these two types of cell division, we will consider a special type of eukaryotic reproduction called schizogony.

Nuclear Division

Typically, a eukaryotic nucleus has either one or two complete copies of the chromosomal portion of a cell's genome. A nucleus with a single copy of each chromosome is called a **haploid**[5] or $1n$ nucleus, and one with two sets of chromosomes is a **diploid**[6] or $2n$ nucleus. Generally, each organism has a consistent number of chromosomes. For example, each haploid cell of the brewer's yeast, *Saccharomyces cerevisiae*[7] (sak"ă-rō-mī'sēz se-ri-vis'ē-ē), has 16 chromosomes.

Whereas the cells of most fungi, many algae, and some protozoa are haploid, the cells of most plants and animals and the remaining fungi, algae, and protozoa are diploid. Typically, gametes are haploid, and a zygote (formed from the union of gametes) is diploid.

---

[1]From Greek *protos*, meaning first, and *zoion*, meaning animal.
[2]Plural of Latin *fungus*, meaning mushroom.
[3]Plural of Latin *alga*, meaning seaweed.
[4]From Greek *helmins*, meaning worm.
[5]From Greek *haploos*, meaning single.
[6]From Greek *diploos*, meaning double.
[7]From Greek *sakcharon*, meaning sugar, and *mykes*, meaning fungus, and Latin *cerevisiae*, meaning beer.

A cell divides its nucleus so as to pass a copy of its chromosomal DNA to each of its descendants, so that each new generation has the necessary genetic instructions to carry on life. There are two types of nuclear division—*mitosis* (mī-tō′sis) and *meiosis* (mī-ō′sis).

Mitosis   Cells have two main stages in their life cycle: a stage called *interphase*,[8] during which cells grow and eventually duplicate their DNA, and a stage during which the cell's nucleus divides. In the type of nuclear division called **mitosis**,[9] which begins after the cell has duplicated its DNA such that there are two exact DNA copies (see Figure 7.5), the cell partitions its replicated DNA equally between two nuclei. Thus mitosis maintains the ploidy of the parent nucleus; that is, a haploid nucleus that undergoes mitosis forms two haploid nuclei, and a diploid nucleus that undergoes mitosis produces two diploid nuclei.

Mitosis has four phases: prophase,[10] metaphase,[11] anaphase,[12] and telophase.[13] The events of mitosis proceed as follows **(Figure 12.1a):**

1. **Prophase.** The cell condenses the DNA molecules into visible threads called *chromatids* (krō′mă-tidz). Two identical chromatids, sister DNA molecules, are joined together in a region called a *centromere* to form one chromosome. Also during prophase, a set of microtubules is constructed in the cytosol to form a *spindle*. In most cells, the nuclear envelope disintegrates during prophase so that mitosis occurs freely in the cytosol; however, many fungi and some unicellular microbes (for example, diatoms and dinoflagellates) maintain their nuclear envelopes so that mitosis occurs within a nucleus.

2. **Metaphase.** The chromosomes line up in the middle of the cell and attach to microtubules of the spindle near their centromeres.

3. **Anaphase.** Sister chromatids separate and crawl along the microtubules toward opposite poles of the spindle. Each chromatid is now called a chromosome.

4. **Telophase.** The cell restores its chromosomes to their less compact, nonmitotic state, and nuclear envelopes form around the daughter nuclei. Though a cell may divide during telophase, as shown in Figure 12.1a, mitosis is nuclear division, not cell division.

Though certain specific events distinguish each of the four phases of mitosis, the phases are not discrete steps; that is, mitosis is a continuous process, and there are no clear boundaries between succeeding phases—one phase leads seamlessly to the next. For example, late anaphase and early telophase are indistinguishable.

Students sometimes confuse the terms *chromosome* and *chromatid,* in part because early microscopists used the word *chromosome* for two different things. During prophase and metaphase, a chromosome consists of two chromatids (DNA molecules) joined at a centromere. However, during anaphase and telophase, the chromatids separate, and each chromatid is then called a chromosome. In other words, a "chromosome" is a pair of chromatids during the first two

Figure 12.1 ▶

The two kinds of nuclear division: mitosis and meiosis. The events occurring during each of the numbered phases are described in the text. **(a)** Mitosis, in which the number of chromosomes (ploidy) in the daughter nuclei is the same as in the parent nucleus. Even though cell division (cytokinesis) may occur simultaneously with mitosis (as shown here), mitosis is not cell division. **(b)** Meiosis, which results in four nuclei, each with half the number of chromosomes as the parent nucleus.

phases, while "chromatid" and "chromosome" are synonymous terms during the latter two phases of mitosis.

Meiosis   In contrast to mitosis, **meiosis**[14] is nuclear division that involves the partitioning of chromatids into four nuclei such that each nucleus receives only half the original amount of DNA. Thus, diploid nuclei use meiosis to produce haploid daughter nuclei. Meiosis is a necessary condition for sexual reproduction (in which nuclei from two different cells fuse to form a single nucleus), because if cells lacked meiosis, with each nuclear fusion to form a zygote the number of chromosomes doubles, and their number would soon become unmanageable.

Meiosis occurs in two stages known as *meiosis I* and *meiosis II* **(Figure 12.1b).** As in mitosis, each stage has four phases, named prophase, metaphase, anaphase, and telophase. The events in meiosis as they occur in a diploid nucleus proceed as follows:

1. Early prophase I (prophase of meiosis I). As with mitosis, DNA replication during interphase has resulted in pairs of identical chromatids, forming chromosomes. But now an additional pairing occurs: *homologous chromosomes*—that is, chromosomes carrying similar or identical genetic sequences—line up side by side. Because these are prophase chromosomes, each of them consists of two identical chromatids; therefore, four DNA molecules are involved in this pairing. An aligned pair of homologous chromosomes is known as a *tetrad*.

2. Late prophase I. Once tetrads have formed, the homologous chromosomes exchange sections of DNA in a random fashion via a process called *crossing over*. This results in recombinations of their DNA. It is because of meiotic crossing over that the offspring produced by sexual reproduction have different genetic makeups from their siblings. Prophase I can last for days or longer.

3. Metaphase I. Tetrads align in the center of the cell and attach to spindle microtubules. Metaphase I differs from

---

[8]Latin, meaning between phases.
[9]From Greek *mitos,* meaning thread, after the threadlike appearance of chromosomes during nuclear division.
[10]From Greek *pro,* meaning before, and *phasis,* meaning appearance.
[11]From Greek *meta,* meaning in the middle.
[12]From Greek *ana,* meaning back.
[13]From Greek *telos,* meaning end.
[14]From Greek *meioun,* meaning to make smaller.

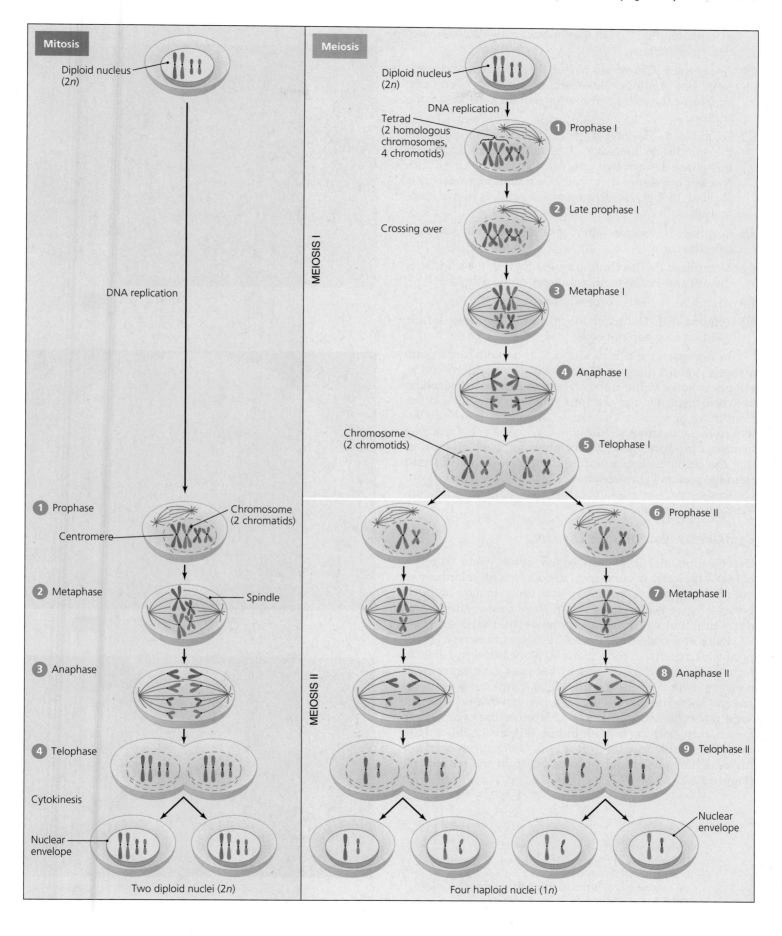

Mitosis

Diploid nucleus (2n)

DNA replication

1 Prophase

Centromere

Chromosome (2 chromatids)

2 Metaphase

Spindle

3 Anaphase

4 Telophase

Cytokinesis

Nuclear envelope

Two diploid nuclei (2n)

Meiosis

Diploid nucleus (2n)

DNA replication

Tetrad (2 homologous chromosomes, 4 chromotids)

1 Prophase I

Crossing over

2 Late prophase I

3 Metaphase I

4 Anaphase I

Chromosome (2 chromotids)

5 Telophase I

MEIOSIS I

6 Prophase II

7 Metaphase II

8 Anaphase II

MEIOSIS II

9 Telophase II

Nuclear envelope

Four haploid nuclei (1n)

metaphase of mitosis in that homologous chromosomes remain as tetrads.

④ Anaphase I. Chromosomes of the tetrads move apart from one another; however, in contrast to mitotic anaphase, the sister chromatids remain attached to one another.

⑤ Telophase I. The first stage of meiosis is completed as the spindle disintegrates. Typically, the cell divides at this phase to form two cells (as shown in Figure 12.1b). Nuclear envelopes may form. Each daughter nucleus is haploid, though each chromosome consists of two chromatids.

⑥ Prophase II. Nuclear envelopes disintegrate, and new spindles form.

⑦ Metaphase II. The chromosomes align in the middle of the cell and attach to microtubules of the spindle.

⑧ Anaphase II. Sister chromatids separate.

⑨ Telophase II. Daughter nuclei form. The cells divide, yielding four haploid cells.

In summary, meiosis produces four haploid nuclei from a single diploid nucleus. Meiosis can be considered back-to-back mitoses without the DNA replication of interphase between them, though the four phases of meiosis I differ from those of mitosis. The phases of meiosis II are equivalent to those in mitosis. Additionally, crossing over during meiosis I produces genetic recombinations, which ensures that the chromosomes resulting from meiosis are different from the parental chromosomes. This provides genetic variety in the next generation. **Table 12.1** compares and contrasts mitosis and meiosis.

Cytokinesis (Cytoplasmic Division)

Cytoplasmic division—also called **cytokinesis** (sī″tō-ki-nē′sis)—typically occurs simultaneously with telophase of mitosis, though in some algae and fungi it may be postponed or may not occur at all. In these cases, mitosis produces multinucleate cells called **coenocytes** (sē′nō-sītz).

In plant and algal cells, cytokinesis occurs as vesicles deposit wall material at the equatorial plane between nuclei to form a *cell plate*, which eventually becomes a transverse wall between daughter cells **(Figure 12.2a).** Cytokinesis of protozoa and some fungal cells occurs when an equatorial ring of actin microfilaments contracts just below the cytoplasmic membrane, pinching the cell in two **(Figure 12.2b).** Single-celled fungi called *yeasts*, form a bud, which receives one of the daughter nuclei and pinches off from the parent cell **(Figure 12.2c).**

Figure 12.2 ▶

Different types of cytoplasmic division. **(a)** Cytokinesis in a plant cell, in which vesicles form a cell plate. **(b)** Cytokinesis as it occurs in animals, protozoa, and some fungi. **(c)** Budding in yeast cells.

Wall of parent cell   Nuclei of daughter cells   Vesicles forming cell plate

(a)   TEM   2 μm

Cleavage furrow

(b)   SEM   100 μm

(c)   SEM   6 μm

Table 12.1   Characteristics of the Two Types of Nuclear Division

	Mitosis	Meiosis
DNA replication	During interphase, before nuclear division	During interphase, before meiosis I begins
Phases	Prophase, metaphase, anaphase, telophase	Meiosis I—prophase I, metaphase I, anaphase I, telophase I
		Meiosis II—prophase II, metaphase II, anaphase II, telophase II
Formation of tetrads (alignment of homologous chromosomes)	Does not occur	Early in prophase I
Crossing over	Does not occur	Following formation of tetrads during prophase I
Number of accompanying cytoplasmic divisions that may occur	One	Two
Resulting nuclei	Two nuclei with same ploidy as the original	Four nuclei with half the ploidy of the original

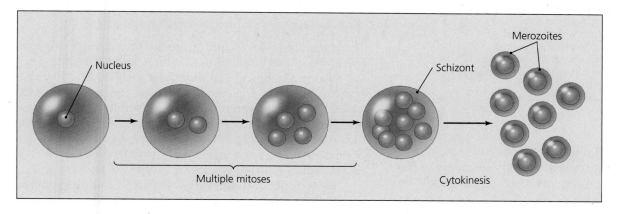

▲ *Figure 12.3*

Schizogony. Sequential mitoses without intervening cytokineses produce a multinucleate schizont, which later undergoes cytokinesis to produce many daughter cells.

Schizogony

The protozoan *Plasmodium* (plaz-mō′dē-ŭm), which causes malaria, reproduces asexually within red blood cells and liver cells via a special type of reproduction called **schizogony** (ski-zog′ō-nē; **Figure 12.3**). In schizogony, multiple mitoses form a multinucleate **schizont** (skiz′ont); only then does cytokinesis occur, simultaneously releasing numerous uninucleate daughter cells called *merozoites* (mer-ō-zō′ītz). The body of infected hosts responds to the release of huge numbers of merozoites with the cyclic fever and chills characteristic of malaria. Chapter 23 discusses *Plasmodium* and malaria in more detail.

The Classification of Eukaryotic Organisms

Learning Objectives

✓ Briefly describe the major groups of eukaryotes as they were first classified in the late 18th century, and as they were classified in the late 20th century.

✓ List some of the problems involved in the classification of protists in particular.

Historically, the classification of many of the eukaryotic microbes has been fraught with difficulty and characterized by change. Since the late 18th century, when Linnaeus first classified living things, until near the end of the 20th century, taxonomists grouped organisms together largely according to readily observable structural traits. Linnaeus classified unicellular algae and fungi as plants, and he classified protozoa—as the name suggests—as animals (**Figure 12.4a** on page 348). By the late 20th century, some taxonomists, including Whittaker, had placed fungi in their own kingdom and grouped protozoa and algae together within the kingdom Protista **(Figure 12.4b);** other taxonomists, however, kept the green algae in the kingdom Plantae.

This scheme is troublesome, in part because the kingdom Protista includes both large multicellular algae (such as kelps) and nonphotosynthetic unicellular protozoa. Adding to the confusion is that taxonomists who classify plants and fungi use the term *divisions* to refer to the same taxonomic level that zoologists call *phyla*.

More recently, many taxonomists have abandoned classification schemes that are so strongly grounded in large-scale

Linnaeus Late 18th century (a)	Animalia						Plantae			Animalia	
Whittaker Late 20th century (b)	Protista							Plantae	Fungi	Protista	Animalia
Modern Early 21st century (c)	Alveolata	Euglenozoa	Diplomonadida	Parabasala	Stramenopila	Rhodophyta	Plantae	Fungi		Mycetozoa	Animalia

▲ *Figure 12.4*

The changing classification of eukaryotes over the centuries. **(a)** In the late 18th century, Linnaeus classified all organisms as either plants or animals. **(b)** In the late 20th century, Whittaker placed fungi in their own group and established a new kingdom Protista. **(c)** Today, microbial eukaryotes are classified into numerous kingdoms based largely on their genetic relatedness. Though not all taxonomists would agree about every detail of this scheme, it forms the basis for the discussion of eukaryotic organisms in this chapter.

structural similarities in favor of schemes based on characteristics that are observable only at the molecular level. More specifically, modern taxonomy attempts to clarify the genetic relatedness of microbes, to place them in groups based on similarities in nucleotide sequences and cellular ultrastructure as revealed by electron microscopy. One of the most evident results of such taxonomic studies is that most schemes no longer include a group of "protists"; instead, the latest studies suggest that such eukaryotic microbes belong in several kingdoms.

Though no one classification scheme has garnered universal support, and more thorough understanding based on new information will almost certainly dictate changes, many taxonomists favor a scheme similar to the one shown in **Figure 12.4c.** In this scheme, on which the discussions of eukaryotic microbes in this chapter are largely based, the organisms we commonly refer to as protozoa are classified in the kingdoms Alveolata, Euglenozoa, Diplomonadida, and Parabasala; fungi are in the kingdom Fungi; algae are distributed among the kingdoms Stramenopila, Rhodophyta, and Plantae; water molds join the algae in Kingdom Stramenopila, and slime molds are in Kingdom Mycetozoa. As we study the eukaryotic microbes discussed in this chapter, bear in mind that because the relationships among eukaryotic microbes are not fully understood, not all taxonomists would completely agree with this scheme, and that new information will shape future alterations in the taxonomy of eukaryotic microbes.

We begin our survey of eukaryotic microbes with the diverse group of organisms commonly known as protozoa.

Protozoa

Learning Objective

✓ List three characteristics shared by all protozoa.

The microorganisms called **protozoa** (prō-tō-zō′ă) are a diverse group that are defined by three characteristics: They are eukaryotic, unicellular, and lack a cell wall. Note that "protozoa" is not a currently accepted taxon. With the exception of one subgroup (called apicomplexans), protozoa are also motile by means of cilia, flagella, and/or pseudopodia. By these criteria, protozoa include a diverse assemblage of microbes. The scientific study of protozoa is *protozoology,* and scientists who study these microbes are *protozoologists.*

In the following sections we discuss the distribution, morphology, nutrition, reproduction, and classification of various groups of protozoa. Chapter 23 discusses the parasitic (and disease-causing) protozoa of humans more fully.

Distribution of Protozoa

Protozoa require moist environments because their lack of a cell wall subjects them to desiccation. Therefore, most species live worldwide in ponds, streams, lakes, and oceans, where they are critical members of the *plankton*—free-living, drifting organisms that form the basis of aquatic food chains. Some protozoa live in moist soil, beach sand, and decaying organic matter, and a very few are pathogens—that is, disease-causing microbes, of animals and humans.

Morphology of Protozoa

Though protozoa have most of the features of eukaryotic cells discussed in Chapter 3 and illustrated in Figure 3.2, this group of eukaryotic microbes is characterized by great morphologic diversity. Indeed, taxonomists once used the variety in a prominent morphological feature—their locomotory structures—to classify the protozoa as flagellates, ciliates, or protozoa having pseudopodia. These features no longer figure prominently in the taxonomic classification of protozoa because the presence of a given locomotory feature may not indicate evolutionary relatedness.

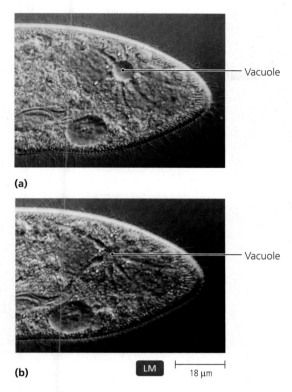

(a)

(b)   LM   18 µm

▲ *Figure 12.5*

A contractile vacuole, a prominent morphological feature of many protozoa such as *Paramecium*. **(a)** A vacuole being filled by water that entered the cell via osmosis. **(b)** A vacuole that is contracting to pump water out of the cell. *Is the environment in this case hypertonic, hypotonic, or isotonic to the cell? Explain.*

Figure 12.5 The environment is hypotonic to the cell; water moves down its concentration gradient—into the cell—by osmosis.

Some ciliates (*e.g., Paramecium*) have two nuclei: often a larger *macronucleus,* which contains many copies of the genome (often more than $50n$) and controls metabolism, growth, and sexual reproduction, and a smaller *micronucleus,* which is involved in genetic recombination, sexual reproduction, and regeneration of macronuclei.

Protozoa also show variety in the number and kind of mitochondria they contain. Several groups completely lack mitochondria, while all the others have mitochondria with discoid or tubular cristae rather than the platelike cristae seen in animals, plants, fungi, and many algae. Additionally, some protozoa have *contractile vacuoles* that actively pump water from the cells, protecting them from osmotic lysis **(Figure 12.5).**

All free-living aquatic and pathogenic protozoa exist as a motile feeding stage called a **trophozoite** (trof-ō-zō′ĭt), and many have a hardy resting stage called a **cyst,** which is characterized by a thick capsule and a low metabolic rate. Cysts of protozoa are not reproductive structures, because one trophozoite forms one cyst, which later becomes one tropho-

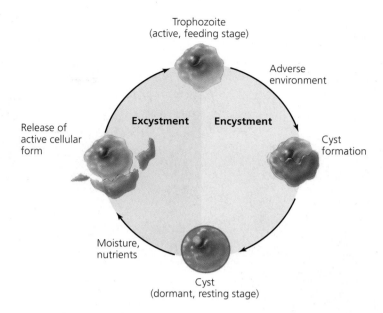

▲ *Figure 12.6*

The cyclical production of trophozoite and cyst stages in protozoa in response to environmental conditions. All protozoa have a trophozoite, but only some protozoa form cysts. *Why is cyst production a common feature of parasitic intestinal protozoa but not of parasitic blood-borne protozoa?*

Figure 12.6 Cysts allow intestinal protozoa to survive in the relatively dry conditions outside the body and infect a new host; blood-borne protozoa are typically transmitted by arthropod vectors and thus need not form resistant cysts.

zoite **(Figure 12.6).** Such cysts allow intestinal protozoa to pass from one host to another and to survive harsh environmental conditions such as desiccation, nutrient deficiency, extremes of pH and temperature, and lack of oxygen.

Nutrition of Protozoa

Most protozoa are chemoheterotrophic, obtaining nutrients by phagocytizing bacteria, decaying organic matter, other protozoa, or the tissues of a host; a few protozoa absorb nutrients from the surrounding water. Because the protozoa called dinoflagellates and euglenoids (discussed shortly) are photoautotrophic, botanists historically classified them with the algae rather than as protozoa.

Reproduction in Protozoa

Most protozoa reproduce asexually only, by binary fission or schizogony; a few protozoa also have sexual reproduction in which two individuals exchange genetic material. Some sexually reproducing protozoa become **gametocytes** (gametes) that fuse to form a diploid zygote. Ciliates, such as *Paramecium,* reproduce sexually via a complex process called *conjugation,* which involves the coupling of two compatible

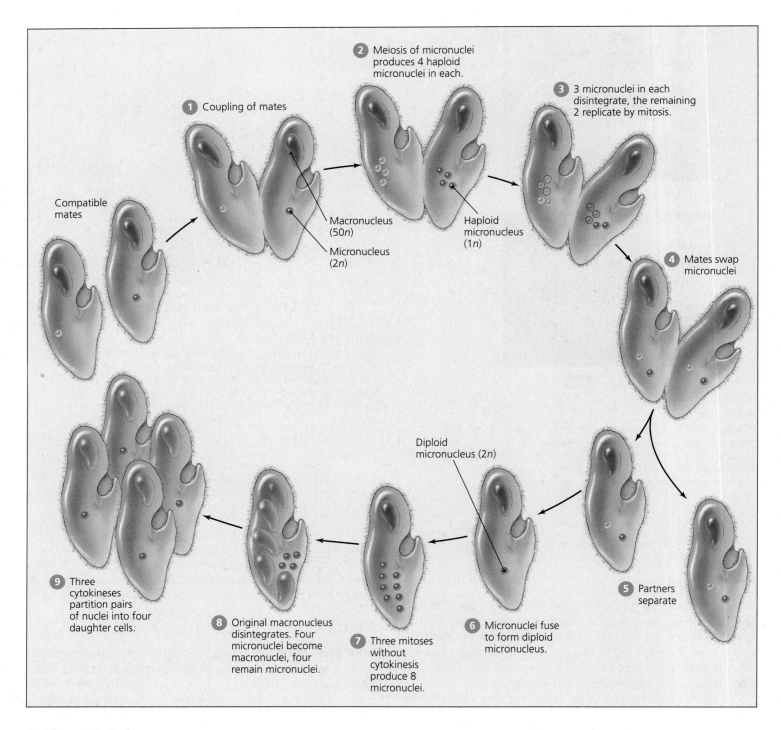

① Coupling of mates

Compatible mates

② Meiosis of micronuclei produces 4 haploid micronuclei in each.

③ 3 micronuclei in each disintegrate, the remaining 2 replicate by mitosis.

Macronucleus (50n)

Micronucleus (2n)

Haploid micronucleus (1n)

④ Mates swap micronuclei

⑤ Partners separate

Diploid micronucleus (2n)

⑥ Micronuclei fuse to form diploid micronucleus.

⑦ Three mitoses without cytokinesis produce 8 micronuclei.

⑧ Original macronucleus disintegrates. Four micronuclei become macronuclei, four remain micronuclei.

⑨ Three cytokineses partition pairs of nuclei into four daughter cells.

▲ *Figure 12.7*

Sexual reproduction via conjugation in the ciliate *Paramecium*.

mating cells, meiosis of diploid micronuclei, exchange of haploid micronuclei between the coupled cells, uncoupling of the cells, fusion of haploid micronuclei to form a diploid micronucleus, three mitoses of the micronucleus to form eight micronuclei, disintegration of the macronucleus and the subsequent formation of a new macronucleus from four micronuclei, and three cytokineses to produce four daughter cells **(Figure 12.7).**

Classification of Protozoa

Learning Objectives

✓ Discuss the reasons for the many different taxonomic schemes for protozoa.

✓ Compare and contrast the three types of alveolates.

✓ Name one distinguishing characteristic of the apicomplexans.

✓ Explain why dinoflagellates and euglenoids were historically classified as both protozoa and algae.

✓ Compare and contrast amoeboid movement and euglenoid movement.

✓ Identify several features of a typical euglenoid.

✓ Describe three types of archaezoa.

As we have seen, over two centuries ago Linnaeus classified protozoa as animals, and in 1959 Whittaker grouped protozoa (and algae as well) into Kingdom Protista in his four-kingdom scheme for eukaryotic classification (see Figure 12.4a and b). Furthermore, some taxonomists divided the protozoa into four groups based on the organisms' mode of locomotion: Sarcodina (which are motile by means of pseudopodia), Mastigophora (flagella), Ciliophora (cilia), and Sporozoa (which are nonmotile). Other taxonomists lumped the first two groups together into a single group called Sarcomastigophora. Grouping of the protozoa according to locomotory features is still in common usage for many practical applications.

Many taxonomists today, however, recognize that these schemes do not reflect genetic relationships, either between protozoa and other organisms or among protozoa. Accordingly, these taxonomists continue to revise and refine the classification of protozoa based on nucleotide sequencing and features made visible by electron microscopy. One such phylogenetic scheme classifies protozoa into the four taxa shown at the left of Figure 12.4c, which different scientists consider kingdoms, subkingdoms, or phyla.

In the following sections we will briefly discuss members of the four groups of protozoa, formed largely according to similarities in nucleotide sequences and ultrastructure, that are shown in Figure 12.4c. We begin with the alveolates.

Alveolates

Alveolates (al-vē'ō-lātz) are protozoa with small membrane-bound cavities called *alveoli*[15] (al-vē'ō-lī) beneath their cell surfaces **(Figure 12.8).** Scientists do not know the purpose of alveoli. Alveolates share at least one other characteristic—tubular mitochondrial cristae. This group is further divided into three subgroups: *ciliates* and *apicomplexans*, which are chemoheterotrophic, and *dinoflagellates*, which are photoautotrophic.

Ciliates As their name indicates, **ciliates** (sil'ē-āts) have cilia by which they either move themselves or move water past their cell surfaces. (The structure and function of cilia were discussed on page 77.) Some ciliates are covered with cilia, whereas others have only a few isolated tufts. All ciliates have two nuclei—one macronucleus and one micronucleus. Some taxonomists consider them the sole members of Phylum Ciliophora.

Notable ciliates include *Vorticella* (vōr-ti-sel'ă), whose apical cilia create a whirlpool-like current to direct food into

---

[15]Latin, meaning small hollows.

Alveolus

TEM | 0.2 μm

▲ *Figure 12.8*

An alveolus, a membrane-bound structure found in some protozoa. Even though the alveolus' function is not yet known, it is present in eukaryotic microbes with similar rRNA sequences, indicating genetic relatedness and forming the basis of the group of eukaryotes called alveolates.

SEM | 25 μm

▲ *Figure 12.9*

A predatory ciliate, *Didinium*, devouring another ciliate, *Paramecium*.

its "mouth"; *Balantidium* (bal-an-tid'ē-ŭm), which is the only ciliate pathogenic to humans (see Figure 3.23b); and the carnivorous *Didinium* (dī-di'nē-ŭm), which phagocytizes other protozoa such as the well-known pond-water ciliate *Paramecium* **(Figure 12.9).** Chapter 23 discusses *Balantidium* in more detail.

Apicomplexans The alveolates called **apicomplexans** (ap-i-kom-plek'sănz) are all pathogens of animals. The name of this group refers to the *complex* of special intracellular organelles, located at the *apices* of the infective stages of

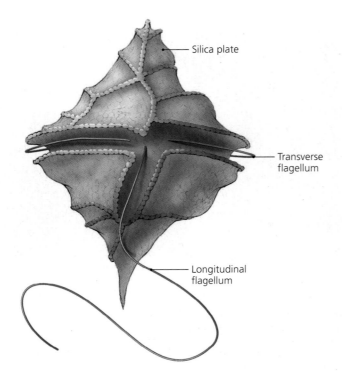

Silica plate

Transverse
flagellum

Longitudinal
flagellum

▲ *Figure 12.10*

Peridinium, a motile armored dinoflagellate with two
flagella. The transverse flagellum spins the cell; the
longitudinal flagellum propels the cell forward.

these microbes, that enables them to penetrate host cells. Examples of apicomplexans are *Plasmodium*, *Babesia* (bă-bē′zē-
ă), and *Toxoplasma* (tok-sō-plaz′mă), which cause malaria,
anemia, and toxoplasmosis, respectively. Apicomplexans
and the diseases they cause are discussed in more detail in
Chapter 23.

Dinoflagellates   The group of alveolates called **dinoflagellates** are unicellular microbes that have photosynthetic
pigments such as carotene and chlorophylls *a*, c_1, and c_2.
Like many plants and algae, their food reserves are starch
and oil, and their cell walls are composed of cellulose. Many
dinoflagellates are armored with protective plates of *silica*
(SiO_2), which is the mineral found in quartz and opal. Even
though botanists have historically classified the dinoflagellates as algae (in the division Pyrrhophyta) because of their
plantlike features, 18S rRNA sequences and the presence of
alveoli indicate that dinoflagellates are more closely related
to ciliates and apicomplexans than they are to either plants
or algae. Interestingly, unlike other eukaryotic chromosomes, dinoflagellate chromosomes lack histone proteins.

Dinoflagellates make up a large proportion of freshwater
and marine plankton. Motile dinoflagellates have two flagella of unequal length **(Figure 12.10)**. The transverse flagellum wraps around the equator of the cell in a groove in the
cell wall, and its beat causes the cell to spin; the second flagellum extends posteriorly and propels the cell forward.

Many dinoflagellates are bioluminescent—that is, able
to produce light via metabolic reactions. When luminescent
dinoflagellates are present in large numbers, the ocean
water lights up with every crashing wave, passing ship, or
jumping fish. Other dinoflagellates produce a red pigment,
and their abundance in marine water is called a **red tide.**

Some dinoflagellates, such as *Gymnodinium* (jīm-nō-
din′ē-um) and *Gonyaulax* (gon-ē-aw′laks), produce *neurotoxins*—poisons that act against the nervous system. In 1991,
over a billion fish died from dinoflagellate poisoning off the
coast of North Carolina. Dinoflagellates can also poison humans who eat shellfish that have concentrated the toxins of
planktonic dinoflagellates they have ingested.

The neurotoxin of another dinoflagellate, *Pfiesteria*[16]
(fes-ter′ē-ă), may be even more potent: It has been claimed
the toxin poisons people who merely handle infected fish or
breathe air laden with the microbes and that *Pfiesteria* toxin
may also cause memory loss, confusion, headache, respiratory difficulties, skin rash, muscle cramps, diarrhea, nausea,
and vomiting. The Centers for Disease Control and Prevention (CDC) calls such poisoning *possible estuary-associated
syndrome (PEAS).*

Amoebae

The evolutionary history of the group of protozoa we consider next is unclear, and its place within the taxonomic
scheme depicted in Figure 12.4c is also unclear. We discuss
these protozoa here, even though some members more
properly belong in the Euglenozoa—they have disc-shaped
mitochondrial cristae and share similar 18S rRNA sequences
with *euglenoids* (discussed next).

The organisms called **amoebae**[17] are protozoa that have
two characteristics in common: They move and feed by
means of pseudopodia (see Figures 3.27 and 3.28), and they
lack mitochondria. Beyond these two characteristics, and
the fact that they all reproduce via binary fission (an unusual
aspect of which is explored in **New Frontiers 12.1**), amoebae
exhibit little uniformity. Though many amoebae lack any
regular shape, others have loose-fitting protein shells with
holes through which they extend pseudopodia to feed and
move around; some species of amoebae form cysts. And
though a few types are pathogenic in animals and humans,
most species live in freshwater, seawater, or soil.

Radiolarians and *foraminifera* are armored, primarily marine amoebae with shells made of silica and calcium carbonate, respectively **(Figure 12.11)**. Over 90% of known
foraminiferans are fossil species, some of which may form
layers of limestone hundreds of meters thick. The great
pyramids of Giza outside Cairo, Egypt, are built of
foraminiferan limestone. Geologists correlate the ages of
sedimentary rocks from different parts of the world by finding identical foraminiferan fossils embedded in them.

---

[16]Named for dinoflagellate biologist Lois Pfiester.
[17]From Greek *ameibein,* meaning to change.

New Frontiers 12.1 Amoeba Midwives

Amoebae divide by binary fission, during which a cell splits into two daughter cells. Occasionally, however, an amoeba fails to divide completely, and the two daughter cells remain attached, as has been observed in

An "assisting" amoeba (yellow) barges into a dividing amoeba (blue).

recent studies of *Entamoeba invadens,* an amoeba found in the guts of snakes. When stalled in the division process, these amoebae appear to signal for the assistance of neighboring amoebae. Like a midwife helping with the birthing process, an assisting amoeba will advance toward a dividing amoeba and barge into it—completing the separation of the two daughter cells in a fascinating example of microbial cooperation.

Source: Biron, D., Libros, P., Sagi, D., Mirelman, D., and Moses, E. 2001. 'Midwives' Assist Dividing Amoebae. *Nature* 410:430.

(a) SEM 0.25 mm

(b) SEM 5 μm

◀ *Figure 12.11*

Armored marine amoebae. The pseudopodia of these protozoa (not visible) extend through holes in the shells. **(a)** Radiolarians have ornate shells of silica. **(b)** Foraminifera have multichambered shells of calcium carbonate.

The normally free-living amoebae *Naegleria* (nā-glē′rē-ă) and *Acanthamoeba* (ă-kan-thă-mē′bă) cause diseases of the brain in humans and animals that swim in water containing these amoebae. Other amoebae, such as *Entamoeba* (ent-ă-mē′bă), always live inside animals, where they produce potentially fatal amoebic dysentery. Chapter 23 discusses these and other pathogenic amoebae in more detail.

Euglenozoa

Part of the reason that taxonomists established the kingdom Protista in the 1960s was to create a "dumping ground" for *euglenoids,* eukaryotic microbes that share certain characteristics of both plants and animals. More recently, based on similar 18S rRNA sequences and the presence of mitochondria with disc-shaped cristae, some taxonomists have created a new taxon: kingdom Euglenozoa (see Figure 12.4c). The euglenozoa include some of the free-living amoebae (*Naegleria* and *Acanthamoeba*), the euglenoids, and some flagellated protozoa called *kinetoplastids.* We discussed amoebae in the previous section, so we now turn our attention to the euglenoids and kinetoplastids.

Euglenoids The group of euglenozoa called **euglenoids,** which are named for the genus *Euglena* (**Figure 12.12a** on page 354), are photoautotrophic, unicellular microbes with chloroplasts containing light-absorbing pigments—chlorophyll *a,* chlorophyll *b,* and carotene. For this reason, botanists historically classified euglenoids in the division Euglenophyta of the kingdom Plantae. However, one reason for not including euglenoids with plants is that euglenoids store food as a unique polysaccharide called *paramylon* instead of as starch. Euglenoids are similar to animals in that they lack cell walls, have flagella, are chemoheterotrophic phagocytes (in the dark), and move by using their flagella as well as by flowing, contracting, and expanding their cytoplasm. Such a squirming movement, which is similar to amoeboid movement but does not involve pseudopodia, is called *euglenoid movement.*

A euglenoid has a semi-rigid, proteinaceous, helical *pellicle* that underlies its cytoplasmic membrane and helps maintain its shape. Typically each euglenoid also has a red "eyespot," which plays a role in positive phototaxis—movement toward light. Euglenoids reproduce by mitosis followed by longitudinal cytokinesis. They form cysts when exposed to harsh conditions.

Kinetoplastids Euglenozoa called **kinetoplastids** (**Figure 12.12b**) each have a single large mitochondrion that

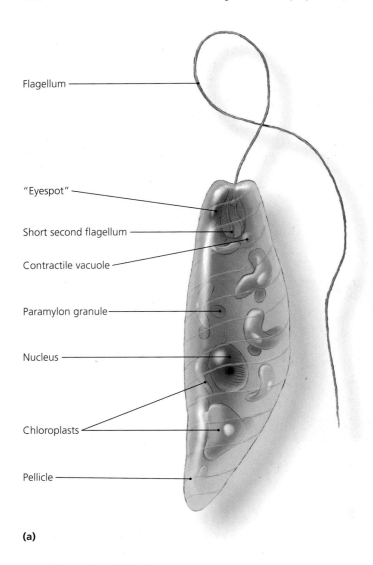

Flagellum

"Eyespot"

Short second flagellum

Contractile vacuole

Paramylon granule

Nucleus

Chloroplasts

Pellicle

(a)

(b)

LM 5 μm

▲ *Figure 12.12*

Two representatives of the kingdom Euglenozoa.
(a) The euglenoid *Euglena.* Euglenoids have
characteristics that are similar to both plants and
animals. **(b)** The kinetoplastid *Trypanosoma cruzi.*
What is the function of the "eyespot" in Euglena?

Figure 12.12 The "eyespot" functions in positive phototaxis.

contains a unique region of mitochondrial DNA called a
kinetoplast. Kinetoplastids live inside animals, and some are
pathogenic. Among the latter is the genus *Trypanosoma,* certain species of which cause African sleeping sickness and
Chagas' disease—potentially fatal diseases of mammals, including humans. Chapter 23 discusses the trypanosomes
and other pathogenic kinetoplastids in more detail.

Archaezoa

Because members of the group of eukaryotic microbes called
archaezoa[18] lack mitochondria, Golgi bodies, and peroxisomes, biologists once thought these organisms were descended from ancient eukaryotes that had not yet
phagocytized the prokaryotic ancestors of mitochondria.
More recently, however, geneticists have discovered mitochondrial genes in the chromosomes of archaezoa, a finding
that suggests that archaezoa might be descended from typical eukaryotes that somehow lost their organelles. In the
following sections we discuss the two taxa of archaezoa—
Diplomonadida and Parabasala.

Diplomonadida The taxon Diplomonadida includes
diplomonads and microsporidia. Diplomonads have two
equal-sized nuclei and multiple flagella. A prominent example is *Giardia* (jē-ar'dē-ă), a diarrhea-causing pathogen of
animals and humans that is spread to new hosts when they
ingest resistant cysts.

Microsporidia are small protozoa with unusual polar filaments. They are obligatory intracellular parasites; that is,
organisms that must live within their host's cells. Microsporidia spread from host to host as small resistant cysts
called "spores." An example is *Nosema* (nō-sē'mă), which is
parasitic on insects such as silkworms and honeybees. The
Environmental Protection Agency has approved one species
of *Nosema* as a biological control agent for grasshoppers.
Seven genera of microsporidia, including *Nosema* and
Microsporidium, are known to cause diseases in immunocompromised patients.

Parabasala Parabasalids also lack mitochondria, but each
has a single nucleus and a *parabasal body,* which is a Golgi-
like structure. *Trychonympha* (trik-ō-nimf'ă), a parabasalid
with prodigious flagella **(Figure 12.13),** inhabits the guts of
termites, where it assists in the digestion of wood. Another
well-known parabasalid is *Trichomonas* (trik-ō-mō'nas),
which lives in the human vagina (see Figure 23.8). When the
normally acidic pH of the vagina is raised, *Trichomonas* proliferates and causes severe inflammation that can lead to
sterility. It is spread by sexual intercourse and is usually
asymptomatic in males.

In summary, protozoa are a heterogeneous collection of single-celled, mostly chemoheterotrophic organisms that lack a
cell wall, and whose relationships with one another and

---

[18]From Greek *arche,* meaning beginning, and *zoion,* meaning animal.

Flagella

LM | 15 μm

▲ *Figure 12.13*

Trychonympha, a parabasalid with prodigious flagella. *What evidence suggests that the progenitors of archaezoa possessed mitochondria?*

Figure 12.13 Scientists have discovered DNA sequences associated with mitochondrial proteins in the chromosomes of archaezoa.

with other eukaryotic organisms is still unclear. **Table 12.2** on page 356 summarizes the incredible diversity of these microbes.

We next turn our attention to another group of chemoheterotrophs: the fungi, which chiefly differ from the protozoa in that they have cell walls.

CRITICAL THINKING

In Chapter 4 (see p. 121), we discussed identifying microbes using dichotomous taxonomic keys. Design a key for the genera of protozoa discussed in this section.

Fungi

Learning Objective

✓ Cite at least three characteristics that distinguish fungi from other groups of eukaryotes.

Organisms in the kingdom **Fungi** (fŭn′jī; see Figure 12.4c), such as molds, mushrooms, and yeasts, are like most protozoa in that they are chemoheterotrophic; however, unlike protozoa they have cell walls, which typically are composed of a strong, flexible, nitrogenous polysaccharide called **chitin.** (The chitin in fungi is chemically identical to that in the exoskeletons of insects and other arthropods such as grasshoppers, lobsters, and crabs.) Fungi differ from plants in that they lack chlorophyll and do not perform photosynthesis; they differ from animals by having cell walls, although genetic sequencing of fungal and animal genomes has shown that fungi and animals are related. The study of

fungi is *mycology,*[19] and scientists who study fungi are *mycologists.*

The Significance of Fungi

Learning Objective

✓ List five ways in which fungi are beneficial.

Fungi are extremely beneficial microorganisms. In nature, they decompose dead organisms (particularly plants) and recycle their nutrients. Additionally, the roots of about 90% of vascular plants form *mycorrhizae,*[20] which are beneficial associations with fungi that assist the plants to absorb water and dissolved minerals.

Humans use fungi for food (mushrooms and truffles), in religious ceremonies (because of their hallucinogenic properties), and in the manufacture of foods and beverages, including bread, alcoholic beverages, citric acid (the basis of the soft drink industry), soy sauce, and some cheeses. Fungi also produce antibiotics such as penicillin and cephalosporin; the immunosuppressive drug *cyclosporine,* which makes organ transplants possible; and *mevinic acids,* which are cholesterol-reducing agents.

Fungi are also important research tools in the study of metabolism, growth, and development, and in genetics and biotechnology. For instance, based on their work with *Neurospora* (noo-ros′pōr-ă) in the 1950s, George Beadle (1903–1999) and Edward Tatum (1909–1975) developed their Nobel Prize–winning theory that one gene codes for one enzyme. Because of similar research, *Saccharomyces* (brewer's yeast) is the best understood eukaryote and the first eukaryote to have its entire genome sequenced. The uses of fungi in agriculture and industry are highlighted in Chapter 26.

But not all fungi are beneficial—about 30% of known fungal species produce **mycoses** (mī-kō′sēz), which are fungal diseases of plants, animals, and humans. For example, Dutch elm disease is a fungal disease of elm trees, and athlete's foot is a fungal disease of humans. We will discuss human and animal mycoses in more detail in Chapter 22. Because fungi tolerate concentrations of salt, acid, and sugar that inhibit bacteria, fungi are responsible for the spoilage of fruit, pickles, jams, and jellies exposed to air.

In the following sections we will consider the basic characteristics of fungal morphology, nutrition, and reproduction before turning to a brief survey of the major groups of fungi.

Morphology of Fungi

Learning Objective

✓ Distinguish among septate hyphae, aseptate hyphae, and mycelia.

---

[19]From Greek *mykes,* meaning mushroom, and *logos,* meaning discourse.
[20]From Greek *rhiza,* meaning root.

Table 12.2 **Characteristics of Protozoa**

Category	Distinguishing Feature(s)	Representative Genera Mentioned in the Text
Alveolates	Alveoli (membrane-bound cavities underlying cytoplasmic membrane); tubular mitochondrial cristae	
Ciliates	Cilia	Balantidium Paramecium Stentor
Apicomplexans	Apical complex of organelles	Babesia Plasmodium Toxoplasma
Dinoflagellates	Photosynthesis; two flagella	Gymnodinium Gonyaulax Pfiesteria
Amoebae	Pseudopodia; lack mitochondria	
Foraminifera		
Radiolarians		
Free-living types		Naegleria Acanthamoeba
Parasitic types		Entamoeba
Euglenozoa	Flagella; disc-shaped mitochondrial cristae	
Euglenoids	Photosynthesis; pellicle	Euglena
Kinetoplastids	Single large mitochondrion with DNA localized in kinetoplast	Trypanosoma
Diplomonadida	Lack mitochondria, Golgi bodies, and peroxisomes	
Diplomonads	Two equal-sized nuclei; multiple flagella	Giardia
Microsporidia	Polar filaments	Nosema
Parabasala		
Parabasalids	Single nucleus	Trichonympha Trichomonas

The vegetative (nonreproductive) body of a fungus is called its **thallus**[21] (plural: *thalli*). The morphology of fungal thalli is variable. The thalli of *yeasts* are small, globular, and composed of a single cell **(Figure 12.14a)**, whereas the thalli of *molds* are large and composed of long, branched, tubular filaments called **hyphae**[22] **(Figure 12.14b)**. Hyphae are either **septate** (divided into cells by crosswalls called *septa*[23]) or **aseptate** (not divided by septa; **Figure 12.14c**). Aseptate hyphae are also coenocytic (multinucleate).

In response to environmental conditions such as temperature or carbon dioxide concentration, some fungi produce both yeastlike thalli and moldlike thalli **(Figure 12.14d)**; fungi that produce two types of thalli are said to be **dimorphic** (which means "two-shaped"). Many medically important fungi are dimorphic, including *Histoplasma capsulatum* (his-tō-plaz′mă kap-soo-lā′tŭm), which causes a respiratory disease called histoplasmosis, and *Coccidioides immitis* (kok-sid-ē-oy′dēz im′mi-tis), which causes a flulike disease called *coccidioidomycosis* (kok-sid-ē-oy′dō-mī-kō′sis).

The thallus of a mold is composed of hyphae intertwined to form a tangled mass called a **mycelium** (plural: *mycelia*; **Figure 12.15** on page 358). Mycelia are typically subterranean and thus usually escape our notice, though they

can be very large. In fact, as mentioned in the chapter opener, the largest organisms on Earth are fungi in the genus *Armillaria*, the mycelia of which can spread through thousands of acres of forest to a depth of several feet and weigh many hundreds of tons. (In contrast, blue whales, the largest living animals, weigh only about 150 tons.) *Fruiting bodies*, such as puffballs and mushrooms, are the reproductive structures of molds and are only small visible extensions of vast underground mycelia.

Nutrition of Fungi

Fungi acquire nutrients by absorption; that is, they secrete catabolic enzymes outside their thalli to break large organic molecules into smaller molecules, which they then transport into their thalli. Most fungi are **saprobes**[24] (sap′rōbz)—they absorb nutrients from the remnants of dead organisms—but some species trap and kill microscopic soil-dwelling

---

[21]From Greek *thallos*, meaning young shoot.
[22]From Greek *hyphe*, meaning weaving or web.
[23]Latin, meaning partitions or fences.
[24]From Greek *sapros*, meaning rotten, and *bios*, meaning life.

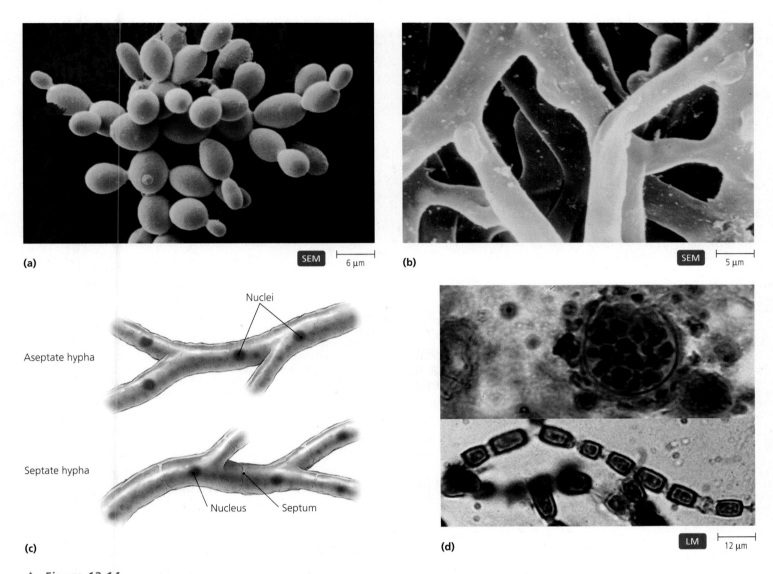

(a) SEM 6 μm

(b) SEM 5 μm

Nuclei

Aseptate hypha

Septate hypha

Nucleus   Septum

(c)

(d) LM 12 μm

▲ *Figure 12.14*

Fungal morphology. **(a)** The thalli of *Saccharomyces* (bakers' or brewers' yeast), which are unicellular and spherical to irregularly oval in shape. **(b)** The thallus of a filamentous fungus, which is composed of cylindrical hyphae. **(c)** Examples of septate and aseptate hyphae. **(d)** The thalli of a dimorphic fungus, *Mucor rouxii*, showing both yeastlike and moldlike growth in response to local environmental conditions.

nematodes (worms) (**Figure 12.16** on page 358). Fungi that derive their nutrients from living plants and animals usually have modified hyphae called **haustoria**[25] (haw-stō′rē-ă), which penetrate the tissue of the host to withdraw nutrients. Absorptive nutrition is important in the role that fungi play as decomposers and recyclers of organic waste. Cytoplasmic streaming frequently transports nutrients and organelles, including nuclei, throughout a mycelium. Streaming between cells of septate mycelia occurs through pores in the septa.

Most fungi are aerobic, though many yeasts (for example, *Saccharomyces*) are facultative anaerobes that obtain energy from fermentation, such as occurs in the reactions that produce alcohol. Anaerobic fungi are found in the digestive systems of many herbivores, such as cattle and deer, where they assist in the catabolism of plant material.

Reproduction in Fungi

Learning Objectives

✓ Describe asexual reproduction in fungi.
✓ List three basic types of asexual spores found in molds.

---

[25]From Latin *haustor,* meaning someone who draws water from a well.

▲ *Figure 12.15*
A fungal mycelium growing on wood.

Whereas all fungi have some means of asexual reproduction involving mitosis followed by cytokinesis, most fungi also reproduce sexually. In the next subsections we briefly examine asexual and sexual reproduction in fungi.

Budding and Asexual Spore Formation

Yeasts typically bud in a manner similar to prokaryotic budding. Following mitosis, one daughter nucleus is sequestered in a small bleb of cytoplasm that is isolated from the parent cell by the formation of a new wall (see Figure 12.14a). In some species, especially *Candida albicans* (kan'did-ă al'bi-kanz), which causes human oral thrush and vaginal yeast infections, a series of buds remain attached to one another and to the parent cell, forming a long filament called a *pseudohypha*. *Candida* invades human tissues by means of such pseudohyphae, which can penetrate intercellular cracks.

Filamentous fungi reproduce asexually by producing lightweight spores, which enable the fungi to disperse vast distances on the wind. Researchers have isolated fungal spores from wind currents many miles above the surface of the earth. Scientists categorize the asexual spores of molds according to their mode of development:

- *Sporangiospores* form inside a sac called a *sporangium*,[26] which is often borne on a spore-bearing stalk, called a *sporangiophore*,[27] at either the tips or sides of hyphae **(Figure 12.17a).**

- *Chlamydospores* form with a thickened cell wall inside hyphae **(Figure 12.17b).**

- *Conidiospores* (also called *conidia*) are produced at the tips or sides of hyphae, but not within a sac. There are

---
[26]From Greek *spora*, meaning seed, and *angeion*, meaning vessel.
[27]From Greek *phoros*, meaning bearing.

Nematode   Hypha

▲ *Figure 12.16*

Predation of a nematode by the fungus *Arthrobotrys*. The fungus produces special looped hyphae that constrict when the worm contacts the inside of the loop. The fungus secretes enzymes that digest the nematode and then absorbs the resulting nutrients. *What is the more typical mode of nutrition found in fungi?*

Figure 12.16 *Most fungi are saprobic.*

many types of conidia, including *arthroconidia*, which develop from hyphae that fragment into individual spores; *blastoconidia*, which form as buds, and others that develop in chains on stalks called *conidiophores* **(Figure 12.17c).**

Medical lab technologists use the presence and type of asexual spores in clinical samples to identify many fungal pathogens.

Sexual Spore Formation

Scientists designate fungal mating types as "+" and "−" rather than as male and female because their thalli are morphologically indistinguishable. The process of sexual reproduction in fungi has four basic steps **(Figure 12.18** on page 360):

1. Haploid (*n*) cells from a + thallus and a − thallus fuse to form a *dikaryon*, a cell containing both + and − nuclei. The dikaryotic stage is neither diploid nor haploid, but instead is designated (*n* + *n*).

2. After a period of time that typically ranges from hours to years but can be centuries, a pair of nuclei within the dikaryon fuse to form one diploid (2*n*) nucleus.

3. Meiosis of diploid nucleus restores the haploid state.

4. The haploid nuclei are partitioned into + and − spores.

Fungi differ in the ways they form dikaryons and in the site at which meiosis occurs.

Sporangium   Sporangiophore

(a)   LM   13 µm

(b)   LM   10 µm

(c)   SEM   15 µm

◀ *Figure 12.17*

Representative asexual spores of molds.
(a) Sporangiospores, which develop within a sac called a sporangium that is borne on a sporangiophore.
(b) Chlamydospores, which are thick-walled spores that form inside hyphae. **(c)** The conidiospores (conidia) of *Penicillium*, which develop on conidiophores at the ends of hyphae. *How do conidia differ from sporangiospores?*

Figure 12.17 Conidia are never enclosed in a sac.

CRITICAL THINKING

Fungi tend to reproduce sexually when nutrients are limited or other conditions are unfavorable, but reproduce asexually when conditions are more ideal. Why is this a successful strategy?

Classification of Fungi

Learning Objectives

✓ Compare and contrast the three divisions of fungi with respect to the formation of sexual spores.

✓ Describe the deuteromycetes, and explain why this group no longer constitutes a formal taxon.

✓ List several beneficial roles or functions of lichens.

In the following sections we will consider the four major subgroups into which taxonomists traditionally divided the kingdom Fungi seen in Figure 12.4c. Three of these subgroups, which are taxons called *divisions* that are equivalent to phyla in kingdom Animalia, are based on the type of sexual spore produced (divisions Zygomycota, Ascomycota, and Basidiomycota); the fourth (the deuteromycetes) was a repository of fungi for which no sexual stage is known. We begin by considering the Zygomycota.

Division Zygomycota

Fungi in the division **Zygomycota** are coenocytic molds called *zygomycetes* (zī′gō-mī-sēts). Of the approximately 1100 species known, most are saprobes; the rest are obligate parasites of insects and other fungi.

Figure 12.19 on page 361 illustrates the life cycle of a typical zygomycete: the black bread mold *Rhizopus* (rī-zō′pŭs) *nigricans*.[28] Zygomycetes reproduce asexually via sporangiospores (steps ① to ④ in Figure 12.19), but the distinctive feature of zygomycetes is the formation of sexual structures called **zygospores.** Zygospores of *R. nigricans* are black, rough-walled structures that develop from a zygote produced by the fusion of two nuclei within a short-lived dikaryon formed by the fusion of sexually compatible hyphae (see steps ⑤ to ⑧ in Figure 12.19). Like other fungal

[28]From Greek *rhiza,* meaning root, and *pous,* meaning foot, and Latin *niger,* meaning black.

Figure 12.18 ➤

The process of sexual reproduction in fungi. The steps in the process are described in the text.

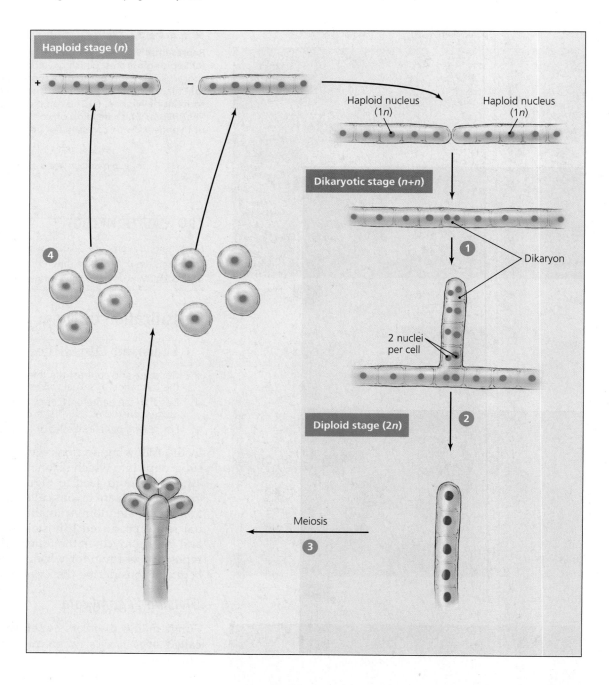

spores, zygospores can withstand desiccation and other harsh environmental conditions.

Each zygospore undergoes meiosis without cytokinesis, but only one of the four meiotic nuclei survives. A zygospore germinates, producing an asexual sporangium that produces numerous haploid spores via mitosis and cell divisions. The sporangium releases its spores, each of which germinates to produce either a + or a − mycelium. This completes the life cycle.

Division Ascomycota

The division **Ascomycota** contains about 32,000 known species of molds and yeasts that are characterized by the for-

mation of haploid **ascospores** within sacs called **asci.**[29] Asci occur in fruiting bodies called *ascocarps*, which have various shapes **(Figure 12.20).** Ascomycetes (as'kō-mī-sēts), as they are called, also reproduce asexually by conidiospores, as illustrated for a representative ascomycete. **(Figure 12.21,** steps ❶ to ❹ on page 362).

Also Figure 12.21 (steps ❺ to ⓫) illustrates the sexual reproduction of an ascomycete, which proceeds as follows:

❺ Multinucleate, hyphal tips of opposite mating types fuse to form a dikaryon.

---

[29]From Greek *askus,* meaning wineskin.

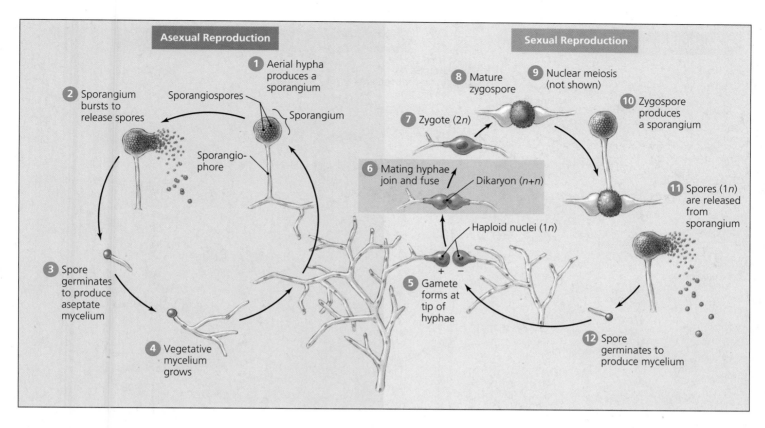

▲ Figure 12.19

Life cycle of the zygomycete *Rhizopus*. During the asexual cycle, the fungus reproduces via sporangiospores that germinate and produce hyphae. In the sexual cycle, the tips of + and − hyphae fuse and form a diploid zygospore that matures, undergoes meiosis, germinates, and produces a sporangium containing haploid sporangiospores, which germinate to form new mycelia.

⑥ The dikaryon reproduces to form hyphae whose cells are all dikaryontic.

⑦ In the dikaryon, opposite mating type nuclei fuse to form a diploid nucleus.

⑧ Each diploid nucleus undergoes meiosis and cytokinesis to form four haploid cells within an ascus.

⑨ Each haploid cell may undergo mitosis and cytokinesis to form two haploid ascospores. Thus, the process produces eight ascospores, which line up inside the ascus.

⑩ The asci open to release their ascospores.

⑪ The ascospores germinate to produce + and − hyphae.

Ascomycetes are familiar and economically important fungi. For example, most of the fungi that spoil food are ascomycetes. This group also includes plant pathogens such as the causative agents of Dutch elm disease and chestnut blight, which have almost eliminated their host trees in many parts of the United States. *Claviceps purpurea* (klav′i-seps poor-poo′rē-ă) growing on grain produces *lysergic acid diethylamide (LSD)*, which causes abortions in cattle and hallucinations in humans.

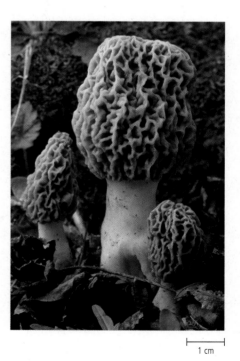

1 cm

▲ Figure 12.20

Ascocarps (fruiting bodies) of the common morel, *Morchella esculenta*, a delectable edible ascomycete. The pits visible in this photograph are lined with asci, sacs that contain numerous ascospores.

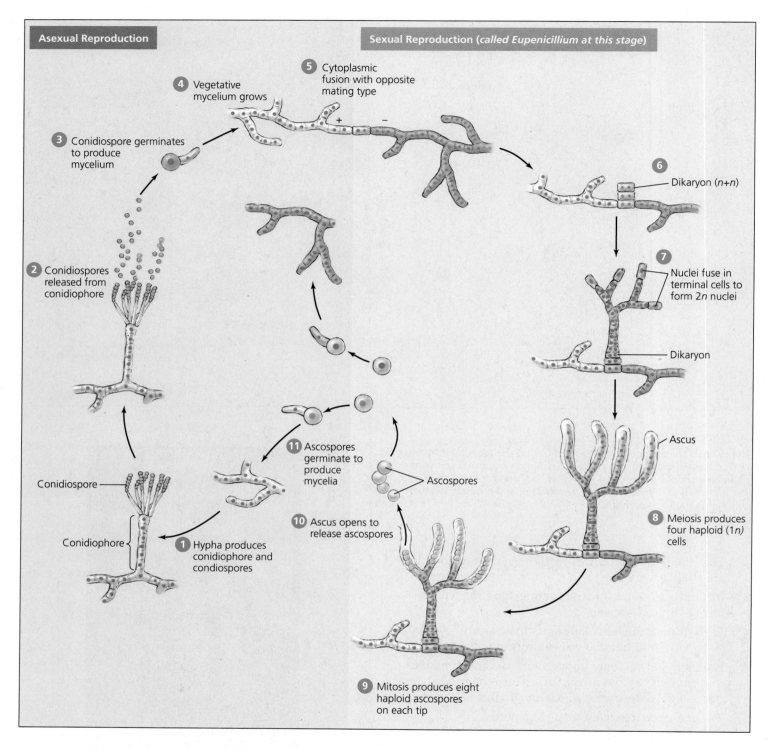

Asexual Reproduction

Sexual Reproduction (*called Eupenicillium at this stage*)

5 Cytoplasmic fusion with opposite mating type

4 Vegetative mycelium grows

3 Conidiospore germinates to produce mycelium

+ −

6 Dikaryon (*n+n*)

2 Conidiospores released from conidiophore

7 Nuclei fuse in terminal cells to form 2*n* nuclei

Dikaryon

Conidiospore

Ascus

11 Ascospores germinate to produce mycelia

Ascospores

Conidiophore

1 Hypha produces conidiophore and condiospores

10 Ascus opens to release ascospores

8 Meiosis produces four haploid (1*n*) cells

9 Mitosis produces eight haploid ascospores on each tip

▲ *Figure 12.21*
Life cycle of an ascomycete.

On the other hand, many ascomycetes are beneficial. For example, *Penicillium* (pen-i-sil'ē-ŭm) mold is the source of penicillin; *Saccharomyces*, which ferments sugar to produce alcohol and carbon dioxide gas, is the basis of the baking and brewing industries; and *truffles* (varieties of *Tuber*) grow as mycorrhizae in association with oak and beech trees to form delectable culinary delights (see **Highlight 12.1**). Another ascomycete, the pink bread mold *Neurospora*, has been an important tool in genetics and biochemistry. Many ascomycetes partner with green algae or cyanobacteria to form *lichens*, which are discussed in more detail shortly.

Truffles—rare, intensely flavored mushrooms that grow underground—are one of the most luxurious foods on Earth, selling at more than $500 per pound. There are many different varieties of truffles. The

Tuber melanosporum

most coveted include *Tuber melanosporum*, a black truffle that is also known as the "black diamond," and *Tuber magnatum pico*, a white truffle that can sell for up to $1800 per pound! Because truffles are very difficult to find, truffle hunters often use pigs and dogs trained to sniff them out. (Dogs are preferred, because pigs are more likely to eat the truffles.) The underground habitat of truffles requires them to form symbiotic relationships with trees (for nutrients) and with animals such as squirrels and chipmunks (for spore dispersal; the animals dig up the truffles and thus help spread the spores to other locations).

Incidentally, chocolate truffles only derive their name from their prized fungal counterparts!

Division *Basidiomycota*

A walk through fields and woods in most parts of the world may reveal mushrooms, puffballs, stinkhorns, jelly fungi, bird's nest fungi, or bracket fungi, all of which are the visible fruiting bodies of the almost 22,000 known species of fungi in the division **Basidiomycota**. Poisonous mushrooms are sometimes called *toadstools*, but there is no sure way to always distinguish between an edible "mushroom" and a poisonous "toadstool" except by eating them—a truly risky practice!

Mushrooms and other fruiting bodies of *basidiomycetes* (ba-sid′ē-ō-mi-sēts) are called **basidiocarps (Figure 12.22)**. The entire structure of a basidiocarp consists of tightly woven hyphae that extend into multiple, often club-shaped projections called *basidia*, the ends of which produce sexual *basidiospores* (typically four on each basidium). **Figure 12.23** on page 364 illustrates the life cycle of *Amanita muscaria* (am-ă-nī′tă mus-ka′rē-ă), a poisonous mushroom.

Besides the edible mushrooms—most notably, the cultivated *Agaricus* and *Cortinellis*—basidiomycetes affect humans in several ways. Most basidiomycetes are important decomposers that digest chemicals such as cellulose and lignin in dead plants and return nutrients to the soil. Many mushrooms produce hallucinatory chemicals or toxins. An example of the latter is the extremely poisonous "Death-cap mushroom," *Amanita* (see Figure 12.23). The basidiomycete yeast *Cryptococcus neoformans* is the leading cause of fungal meningitis. Other basidiomycetes are *rusts* and *smuts*, which cause millions of dollars in crop loss each year.

Deuteromycetes

As noted previously, the divisions Zygomycota, Ascomycota, and Basidiomycota were based on type of sexual spore produced. But because scientists have not observed sexual reproduction in all fungi, taxonomists in the middle of the 20th century created the division **Deuteromycota** to contain the heterogeneous collection of fungi whose sexual stages are unknown—either because they do not produce sexual spores, or because their sexual spores have not been observed. More recently, however, the identification of rRNA sequences has revealed that most deuteromycetes in fact

▲ *Figure 12.22*

Basidiocarps (fruiting bodies) of the bird's nest fungus, *Crucibulum*. The familiar shapes of mushrooms are also basidiocarps of extensive mycelia.

belong in the division Ascomycota, and thus modern taxonomists have abandoned Deuteromycota as a formal taxon. Nevertheless, many medical laboratory technologists, health care practitioners, and scientists continue to refer to "deuteromycetes" because it is a traditional name.

Most deuteromycetes are either terrestrial saprobes, pathogens of plants, or pathogens of other fungi. Several genera are pathogenic to humans, including the filamentous genus *Trichophyton* (tri-kof′i-tŏn), one of the causes of ringworm, and the dimorphic fungus *Histoplasma capsulatum*, which causes a flu-like illness called *histoplasmosis*. Chapter 22 discusses these and other pathogenic fungi in more detail.

Lichens

A discussion of fungi is incomplete without considering **lichens,** which are partnerships between fungi and photosynthetic microbes—either green algae (discussed shortly) or, more commonly, cyanobacteria. The hyphae of the fungus

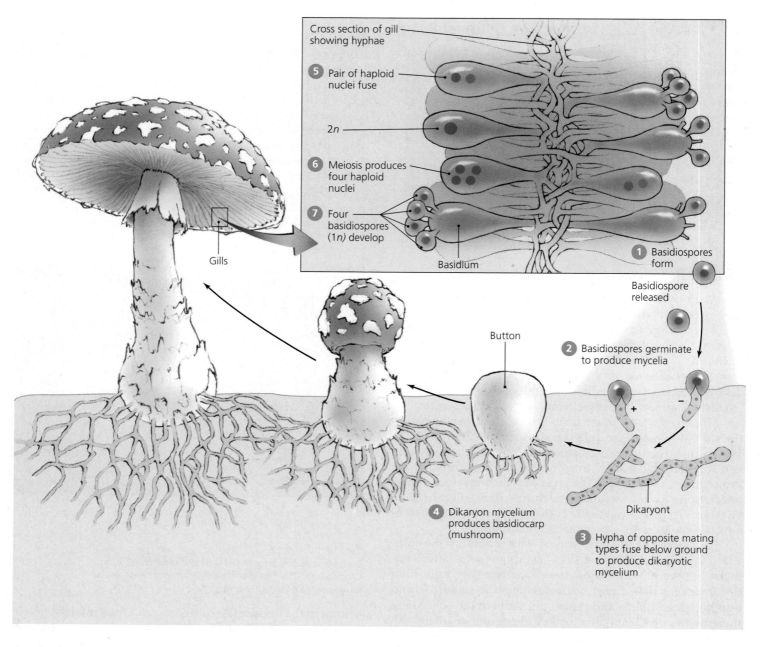

Cross section of gill showing hyphae

5 Pair of haploid nuclei fuse

2n

6 Meiosis produces four haploid nuclei

7 Four basidiospores (1n) develop

Basidium

1 Basidiospores form

Basidiospore released

Gills

Button

2 Basidiospores germinate to produce mycelia

+

−

4 Dikaryon mycelium produces basidiocarp (mushroom)

Dikaryont

3 Hypha of opposite mating types fuse below ground to produce dikaryotic mycelium

▲ *Figure 12.23*

Life cycle of *Amanita muscaria*, **a poisonous basidiomycete.** *How can a novice distinguish between edible and poisonous mushrooms?*

Figure 12.23 There is no sure way to distinguish between edible and poisonous mushrooms except by assuming the risk of eating them.

in a lichen, which is usually an ascomycete, surround the photosynthetic cells **(Figure 12.24)** and provide them nutrients, water, and protection from desiccation and harsh light. In return, each alga or cyanobacterium provides the fungus with products of photosynthesis—carbohydrates and oxygen. In some lichens, the phototroph releases 60% of its carbohydrates to the fungus.

The partnership in a lichen is not always mutually beneficial; in some lichens, the fungus produces haustoria that penetrate and kill the photosynthetic member. Such lichens are maintained only because the phototroph's cells reproduce faster than the fungus can devour them.

The fungus of a lichen reproduces by spores, which must germinate and develop into hyphae that capture an appropriate alga or cyanobacterium. Alternatively, bits of lichen called *soredia*, which contain both phototrophs and fungal hyphae, break away and establish a new lichen when they reach a suitable substrate.

▼ *Figure 12.24*

Makeup of a lichen. The hyphae of a fungus (most commonly an ascomycete) constitute the major portion of the thallus of the lichen; cells of the photosynthetic member of the lichen are concentrated near the lichen's surface. Soredia, which are bits of fungus surrounding phototrophic cells, are the means by which some lichens propagate themselves.

Ascocarp of fungus

Fungal hyphae

Soredium

Phototrophic layer

Substrate

Algal cell

SEM   10 μm

Scientists have identified over 14,000 species of lichens, which are named for their fungal half. Lichens are abundant throughout the world, particularly in pristine unpolluted habitats, growing on soil, rocks, leaves, tree bark, other lichens, and even the backs of tortoises. Indeed, lichens grow in almost every habitat—from high-elevation alpine tundra to submerged rocks on the ocean's shores, from frozen Antarctic soil to hot desert climes. The only unpolluted places where lichens do not consistently grow are in the depths of the oceans and the dark world of caves—after all, lichens require light.

Lichens occur in three basic shapes **(Figure 12.25)**:

- *Crustose* lichens grow appressed to their substrates and may extend into the substrate for several millimeters.

- *Foliose* lichens are leaflike, with margins that grow free from the substrate.

- *Fruticose* lichens are either erect or hanging cylinders.

Lichens are important agents in the creation of soil from weathered rocks, and lichens containing nitrogen-fixing cyanobacteria often provide nitrogen to nutrient-poor environments. Many animals eat lichens; for example, reindeer and caribou subsist primarily on lichens throughout the winter. Birds use lichens for nesting materials, and some insects camouflage themselves with bits of living lichen. Humans also use lichens in the production of foods, dyes,

Fruticose

Crustose

Foliose

10 mm

▲ *Figure 12.25*

Gross morphology of lichens. Crustose forms are flat and tightly joined to the substrate, foliose forms are leaflike with free margins, and fruticose forms are either erect (as shown) or pendulant.

clothing, perfumes, and medicines, and as sensitive living assays for monitoring air pollution because lichens will not grow in polluted environments.

Table 12.3   Characteristics of Fungi

Group	Type of Sexual Spore	Comments	Representative Genera
Division Zygomycota	Zygospores	Coenocytic (aseptate)	*Rhizopus*
Division Ascomycota	Ascospores	Septate; some associated with cyanobacteria or green algae to form lichens	*Claviceps* *Saccharomyces* *Penicillium* *Tuber* *Neurospora*
Division Basidiomycota	Basidiospores	Septate	*Agaricus* *Cortinellis* *Cryptococcus* *Amanita*
Deuteromycetes	None known	Septate	*Histoplasma* *Trichophyton*

In the preceding sections, we have seen that fungi are chemoheterotrophic yeasts and molds that function primarily to decompose and degrade dead organisms. Some fungi are pathogenic, and others associate with algae or cyanobacteria to form lichens. All fungi reproduce asexually either by budding or via asexual spores, and most fungi also produce sexual spores, by which taxonomists classify them. **Table 12.3** summarizes the characteristics of fungi.

CRITICAL THINKING

Design a key for the genera of fungi discussed in this section.

Algae

Learning Objective

✓ Describe the distinguishing characteristic of algae.

The Romans used the word *alga* (al´ga) to refer to any simple aquatic plant, particularly one found in marine habitats. Their usage thus included the organisms we recognize as algae (al´jē) and cyanobacteria, as well as sea grasses and other aquatic plants. Today, the word *algae* properly refers to simple, eukaryotic, phototrophic organisms that, like plants, carry out oxygenic photosynthesis using chlorophyll *a*. Algae differ from plants such as sea grass in having sexual reproductive structures in which every cell becomes a gamete. In plants, by contrast, a portion of the reproductive structure always remains vegetative.

Algae are not a unified group; rather, they differ widely in distribution, morphology, reproduction, and biochemical traits. Moreover, the word *algae* is not synonymous with any taxon; in the taxonomic scheme shown in Figure 12.4c, algae can be found in the kingdoms Stramenopila, Rhodophyta, and Plantae. The study of algae is **phycology**,[30] and the scientists that study them are called *phycologists*.

Distribution of Algae

Even though some algae grow in such diverse habitats as in soil and ice, in intimate association with fungi as lichens, and on plants; most algae are aquatic, living in the *photic zone* (penetrated by sunlight) of fresh, brackish, and salt bodies of water. This watery environment both provides some benefits and presents some difficulties for photosynthetic organisms. Whereas the water contains sufficient dissolved chemicals to provide nutrients for an entire large alga, the water also differentially absorbs longer wavelengths of light (including red light), so only shorter (blue) wavelengths penetrate more than a meter below the surface. This is problematic for algae because their primary photosynthetic pigment, chlorophyll *a*, captures red light. Thus to grow in deeper waters, algae must have *accessory photosynthetic pigments* that trap the energy of penetrating, short-wavelength light and pass that energy to chlorophyll *a*. Members of the group of algae known as red algae, for example, contain *phycoerythrin*, a red pigment that absorbs blue light, enabling red algae to inhabit even the deepest parts of the photic zone.

Morphology of Algae

Algae are either unicellular, colonial, or have simple multicellular bodies called thalli, which are commonly composed of branched filaments or sheets. The thalli of large marine algae, commonly called seaweeds, can be relatively complex, with branched *holdfasts* to anchor them to rocks, stem-like *stipes,* and leaflike *blades*. The thalli of many of the larger marine algae are buoyed in the water by gas-filled bulbs called *pneumocysts* (see Figure 12.29). Though the thalli of some marine algae can surpass land plants in length, they lack the well-developed transport systems common to vascular plants.

---

[30]From Greek *phykos,* meaning seaweed, and *logos,* meaning discourse.

Reproduction in Algae

Learning Objective

✓ Describe the alternation of generations in algae.

In unicellular algae, asexual reproduction involves mitosis followed by cytokinesis. In unicellular algae that reproduce sexually, each algal cell acts as a gamete and fuses with another such gamete to form a zygote, which then undergoes meiosis to return to the haploid state.

Multicellular algae reproduce asexually by fragmentation, in which each piece of a parent alga develops into a new individual, and by motile or nonmotile asexual spores. As noted previously, in multicellular algae that reproduce sexually, every cell in the reproductive structures of the alga becomes a gamete—a feature that distinguishes algae from all other photosynthetic eukaryotes.

Many multicellular algae reproduce sexually with an **alternation of generations** of haploid and diploid individuals (**Figure 12.26** on page 368). In such life cycles, diploid individuals undergo meiosis to produce male and female haploid spores that develop into haploid male and female thalli, which may look identical to the diploid thallus. Each of these thalli produces gametes that fuse to form a zygote, which grows into a new diploid thallus. Both haploid and diploid thalli may reproduce asexually as well.

Classification of Algae

Learning Objectives

✓ List four divisions of algae and describe the distinguishing characteristics of each.
✓ List several economic benefits derived from algae.

The classification of algae is not yet settled. Historically, taxonomists have used differences in photosynthetic pigments, storage products, and cell wall composition to classify algae into several groups that are named for the colors of their photosynthetic pigments: green algae, red algae, golden algae, yellow-green algae, and brown algae. The following sections present algal divisions—less inclusive groups than the kingdoms shown in Figure 12.4c. We begin with the green algae of the division Chlorophyta.

Division Chlorophyta (Green Algae)

Chlorophyta[31] are green algae that share numerous characteristics with plants—they have chlorophylls *a* and *c*, use sugar and starch as food reserves, and have cell walls composed of cellulose. In addition, the 18S rRNA sequences of green algae and plants are comparable. Because of these similarities, the green algae are often considered to be the progenitors of plants, and in some taxonomic schemes the Chlorophyta are placed in the kingdom Plantae.

Most green algae are unicellular or filamentous (see Figure 1.6a) and live in freshwater ponds, lakes, and pools, where they form green to yellow scum. Some multicellular forms grow in the marine intertidal zone—that is, in the region exposed to air during low tide.

Prototheca (prō-tō-thē'kǎ) is an unusual green alga in that it lacks pigments, making it colorless. This chemoheterotrophic alga causes a skin rash in sensitive individuals. *Codium* is a member of a group of marine green algae that do not form crosswalls after mitosis; thus, its entire thallus is a single, large, multinucleate cell. Some Polynesians dry and grind *Codium* for use as seasoning pepper. The green alga *Trebouxia* is the most common alga found associated with fungi in lichens.

CRITICAL THINKING

Since *Prototheca* is colorless, how do scientists know that it is really a green alga?

Division Rhodophyta (Red Algae)

The algae of the division **Rhodophyta**[32], which have been placed in the kingdoms Plantae and Protista, are now in their own taxon—Rhodophyta (see Figure 12.4c). They are characterized by the red accessory pigment **phycoerythrin**, the storage molecule glycogen (also known as *floridean*[33] *starch*), cell walls of **agar** or **carrageenan** (kar-ǎ-gē'nan), and nonmotile male gametes called *spermatia*. As previously noted, phycoerythrin allows red algae to absorb short-wavelength blue light and photosynthesize at depths greater than 100 m. Because the relative proportions of phycoerythrin and chlorophyll *a* vary, red algae range in color from green to black in the intertidal zone to red in deeper water (**Figure 12.27** on page 369). Most red algae are marine, though a few freshwater genera are known.

The gel-like polysaccharides agar and carrageenan, once they have been isolated from red algae such as *Gelidium* (jel-i'dē-ǔm) and *Chondrus* (kon'drǔs), are used as thickening agents for the production of solid microbiological media and numerous consumer products, including ice cream, toothpaste, syrup, salad dressings, and snack foods.

Division Chrysophyta (Golden Algae, Yellow-Green Algae, and Diatoms)

Though the division **Chrysophyta**[34] is a group of algae that are diverse with respect to cell wall composition and pigments, it is unified in using the polysaccharide *chrysolaminarin* as a storage product. (Some also store oils.) Many modern taxonomists group these algae with the brown algae and water molds (discussed shortly) in the kingdom Stramenopila (see Figure 12.4c) based on similarities in nucleotide sequences. Whereas some chrysophytes lack cell walls, others have ornate external coverings such as scales or plates. **Diatoms** (dī'ǎ-tomz) are unique in having cell

---

[31]From Greek *chloros*, meaning green, and *phyton*, meaning plant.
[32]From Greek *rhodon*, meaning rose.
[33]Named for a taxon of red algae, *Florideophycidae*.
[34]From Greek *chrysos*, meaning gold.

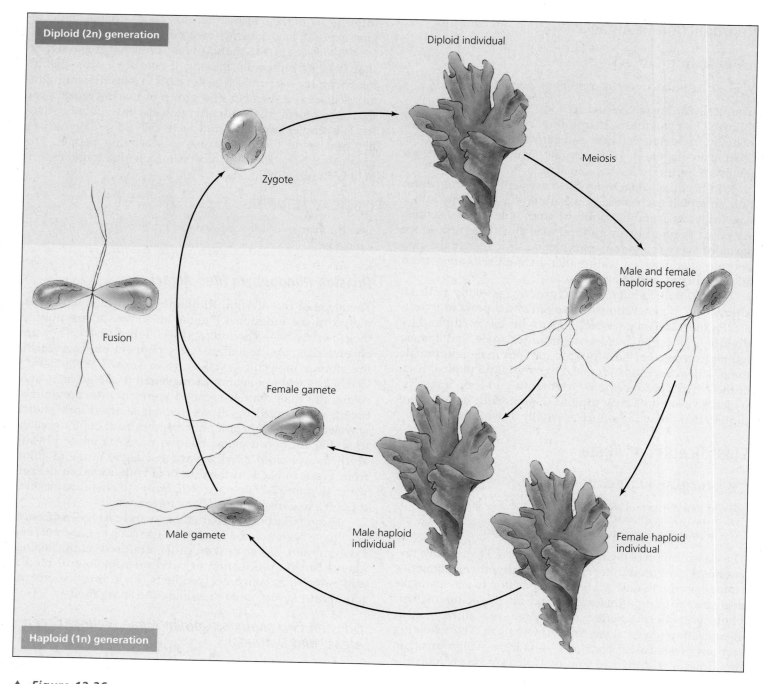

▲ *Figure 12.26*

Alternation of generations in algae, as occurs in *Ulva*. A diploid individual meiotically produces haploid spores that germinate and grow into male and female thalli. These haploid algae produce gametes that fuse to form a diploid zygote, which grows into a new diploid individual. Each generation can also reproduce asexually by fragmentation and by spores.

walls composed of two halves called *frustules* that fit together like a Petri dish **(Figure 12.28).**

Most chrysophytes are unicellular or colonial. All chrysophytes contain more orange-colored *carotene* pigment than they do chlorophyll, which accounts for the common names

of two major classes of chrysophytes—golden algae and yellow-green algae.

Diatoms are a major component of marine *phytoplankton* —free-floating photosynthetic microorganisms that form the basis of food chains in the oceans. Further, because of

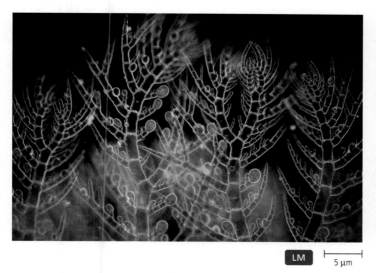

▲ *Figure 12.27*

Red alga.

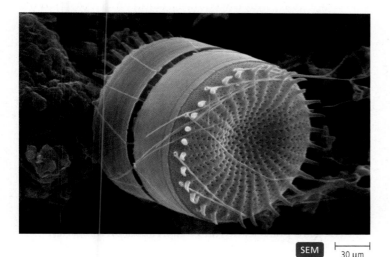

▲ *Figure 12.28*

Diatom. Diatoms have frustules, composed of silica and cellulose, that fit together like a Petri dish.

their enormous number, diatoms are the major source of the world's oxygen. The frustules of diatoms are composed of silica and contain minute holes for the exchange of gases, nutrients, and wastes with the environment. Organic gardeners use *diatomaceous earth*, composed of innumerable frustules of dead diatoms, as a pesticide against harmful insects and worms. Diatomaceous earth is also used in polishing compounds, detergents, paint removers, and as a component of firebrick, soundproofing products, and reflective paints.

Division Phaeophyta (Brown Algae)

The **Phaeophyta**[35] are in the kingdom Stramenopila (see Figure 12.4c) along with chrysophytes and diatoms. They

---

[35]From Greek *phaeo*, meaning brown.

Pneumocyst

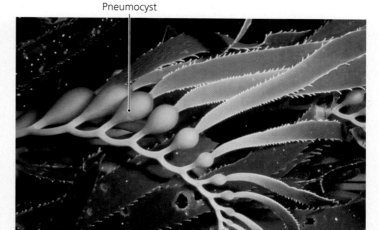

▲ *Figure 12.29*

The giant kelp *Macrocystis*, a brown alga. A kelp's blades are kept afloat by pneumocysts.

have chlorophylls *a* and *c*, carotene, and brown pigments called *xanthophylls* (zan'thō-fils). Depending on the relative amounts of these pigments, brown algae may appear dark brown, tan, yellow-brown, greenish brown, or green. Most brown algae are marine organisms, and some of the giant kelps, such as *Macrocystis* **(Figure 12.29)**, rival the tallest trees in length, though not in girth. Brown algae produce gametes and spores that are motile by means of two flagella—one tinsel-like and one whiplike **(Figure 12.30 on page 370)**.

Brown algae use the polysaccharide *laminarin* and oils as food reserves and have cell walls composed of cellulose and **alginic acid** (alginate). Alginic acid is used in numerous foods as a thickening agent and emulsifier; brewers use it to maintain a foamy head on beer.

In summary, algae are unicellular and multicellular photoautotrophs characterized by sexual reproductive structures in which every cell becomes a gamete. The colors produced by the combination of their primary and accessory photosynthetic pigments give them their common names and provide the basis of at least one classification scheme. **Table 12.4** on page 371 summarizes the characteristics of the major groups of algae; the dinoflagellates and euglenoids are included in it because botanists often classify these phototrophic protozoa as algae.

CRITICAL THINKING

Design a key for the genera of algae discussed in this section.

Water Molds and Slime Molds

The microbes commonly known as *water molds* and *slime molds* were once classified as fungi because they resemble filamentous fungi in having finely branched filaments; however, neither water molds nor slime molds are true fungi.

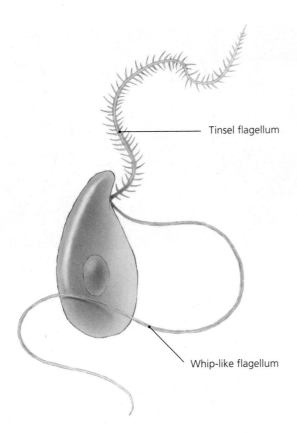

▲ *Figure 12.30*

The two types of flagella on the sperm of the brown alga *Fucus*.

▲ *Figure 12.31*

An example of the important role of water molds in recycling organic nutrients in aquatic habitats.

Water Molds

Learning Objective

✓ List four ways in which water molds differ from true fungi.

Water molds differ from fungi in the following ways:

- They have tubular cristae in their mitochondria.
- They have cell walls of cellulose instead of chitin.
- Their spores have two flagella—one whiplike and one tinsel-like.
- They have true diploid thalli rather than haploid or dikaryon thalli.

As is the case with many eukaryotic microorganisms, taxonomists do not agree on the placement of water molds in classification schemes. Because they often have tinsel-like flagella and certain similarities in rRNA sequence, many scientists think water molds should be grouped together with diatoms, other chrysophytes, and brown algae in the taxon Stramenopila (see Figure 12.4c); other mycologists prefer to classify the water molds in their own kingdom or phylum.

Water molds decompose dead animals and return nutrients to the environment **(Figure 12.31).** Some species are detrimental pathogens of crops such as grapes, tobacco, and potato. In 1845, the water mold *Phytophthora* (fī-tof'tho-ră) was accidentally introduced into Ireland and devastated the potato crop, causing the great famine that killed over 1 million people and forced a greater number to immigrate to the United States and Canada.

Slime Molds

Learning Objectives

✓ List two ways in which slime molds differ from true fungi.
✓ Contrast plasmodial (acellular) and cellular slime molds.

Slime molds differ from true fungi in two main ways:

- They lack cell walls, more closely resembling the amoebae in this regard.
- They are phagocytic rather than absorptive in their nutrition.

Once again, not all taxonomists agree on the classification of slime molds. Some scientists categorize slime molds in a phylum of the kingdom Protista; others classify them in a special division of the kingdom Fungi; still others classify them in a kingdom of their own—kingdom Mycetozoa (see Figure 12.4c).

Species in the two groups of slime molds—plasmodial (acellular) and cellular slime molds—differ based in their morphology, reproduction, and 18S rRNA sequences. Slime molds are important to humans primarily as excellent laboratory systems for the study of developmental and molecular biology.

Plasmodial (Acellular) Slime Molds

Plasmodial slime molds, also known as *acellular* slime molds (*e.g., Physarum*), exist as streaming, coenocytic, colorful filaments of cytoplasm that creep like amoebae through forest litter, feeding by phagocytizing organic debris and bacteria. The thallus, called a *plasmodium* (not to be confused with the agent of malaria—*Plasmodium*), may contain

Table 12.4   Characteristics of Various Algae

Group (Common Name)	Pigments	Storage Product (s)	Cell Wall Component(s)	Habitat	Representative Genera
Chlorophyta (green algae)	Chlorophylls a and b, carotene, xanthophylls	Sugar, starch	Cellulose	Fresh, brackish, and salt water; terrestrial	*Spirogyra* *Prototheca* *Codium* *Trebouxia*
Rhodophyta (red algae)	Chlorophyll a, phycoerythrin, phycocyanin, xanthophylls	Glycogen (floridean starch)	Agar, carrageenan	Mostly salt water	*Chondrus* *Gelidium*
Chrysophyta (golden algae, yellow-green algae, diatoms)	Chlorophylls a, c_1 and c_2; carotene, xanthophylls	Chrysolaminarin	Cellulose, silica, calcium carbonate	Fresh, brackish, and salt water; terrestrial; ice	*Navicula*
Phaeophyta (brown algae)	Chlorophylls a and c, xanthophylls	Laminarin, oils	Cellulose, alginic acid	Brackish and salt water	*Macrocystis*
Pyrrhophyta (dinoflagellates)	Chlorophylls a, c_1, and c_2, carotene	Starch, oils	Cellulose	Fresh, brackish, and salt water	*Gymnodinium* *Gonyaulax* *Pfiesteria*
Euglenophyta (euglenoids)	Chlorophylls a and b, carotene	Paramylon, oils, sugar	Absent	Fresh, brackish, and salt water; terrestrial	*Euglena*

millions of diploid nuclei and cover many square centimeters (**Figure 12.32a** on page 372). Nutrients are distributed throughout the plasmodium by cytoplasmic streaming.

When food or water are in short supply, the plasmodium divides into individual masses of cytoplasm, each of which produces a stalked sporangium. Meiosis occurs within the sporangia to generate haploid spores. These spores germinate to produce *myxamoebae*, which look and act like protozoan amoebae; in the presence of water, however, myxamoebae produce flagella and swim about. (When the water disappears, they become amoeboid again.)

Compatible myxamoebae of opposite mating types fuse to form a diploid zygote. The nucleus of the zygote undergoes numerous mitoses—without cytokineses—to form a new coenocytic plasmodium.

Cellular Slime Molds

Cellular slime molds, such as *Dictyostelium*, exist as individual haploid myxamoeba that phagocytize bacteria, yeasts, dung, and decaying vegetation. **Figure 12.32b** illustrates their life cycle, all of which is haploid—there is no diploid phase. Myxamoebae reproduce by mitosis and cytokinesis when food is abundant; however, in scarcity, some secrete cyclic adenosine monophosphate (cAMP), which acts as a chemotactic attractant for other myxamoebae. The myxamoebae congregate into a sluglike *pseudoplasmodium*, which can migrate for several days. Unlike the true plasmodium of acellular slime molds, the cells of a pseudoplasmodial slug retain their individuality and can be separated mechanically.

Some cells of the pseudoplasmodium form a stalked sporangium; the remaining cells climb the stalk and become spores. In contrast to the spores of plasmodial slime molds, the spores of cellular slime molds do not result from meiosis and are not enclosed in a common wall.

CRITICAL THINKING

Why are cellular slime molds called "cellular"?

In summary, water molds and slime molds are not true molds; they differ from fungi in many aspects of structure, reproduction, and nutrition.

Other Eukaryotes of Microbiological Interest: Parasitic Helminths and Vectors

Learning Objectives

✓ Explain why microbiologists study large organisms such as parasitic worms.

✓ Discuss the inclusion of vectors in a study of microbiology.

Microbiologists are also interested in two other groups of eukaryotes, although they are not microorganisms. The first group are the parasitic *helminths*, commonly called parasitic worms. Microbiologists became interested in parasitic helminths because they observed the microscopic infective and diagnostic stages of the helminths—usually eggs or larvae (immature forms)—in samples of blood, feces, and urine. Thus microbiologists study parasitic helminths in

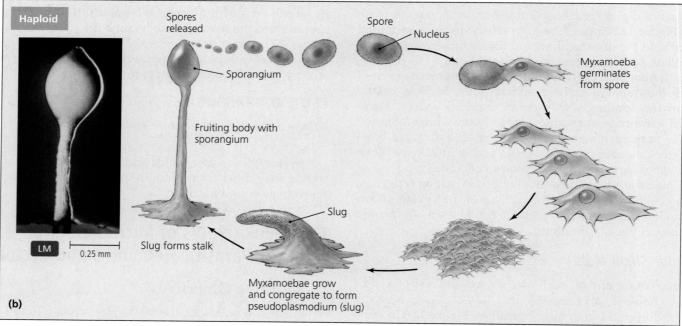

▲ *Figure 12.32*

Life cycles of slime molds. **(a)** A plasmodial (acellular) slime mold. The coenocytic plasmodium feeds on organic debris and bacteria as it creeps through its environment. Under adverse conditions, it produces sporangia that undergo meiosis to produce haploid spores. Compatible spores fuse to form a zygote that undergoes multiple mitoses, but not cytokinesis, to form a new plasmodium. **(b)** A cellular slime mold. Individual myxamoebae congregate during times of starvation to form a multicellular slug, which produces a stalked sporangium. The sporangium releases spores that germinate when conditions are more favorable. All phases are haploid.

part because they must distinguish the parasites' microscopic forms from other microbes. Chapter 14 discusses parasitism and other relationships that exist among microbes and other organisms.

Microbiologists are also interested in another group of macroscopic animals called *vectors*,[36] which are animals such as ticks, lice, and mosquitoes that carry and transmit microscopic pathogens, including bacteria, protozoa, and viruses, to new hosts. Chapter 23 discusses vectors and parasitic helminths in more detail.

---

[36]From Latin *vectus*, meaning carried.

CHAPTER SUMMARY

General Characteristics of Eukaryotic Organisms (pp. 343–348)

1. A typical eukaryotic nucleus may be **haploid** (having a single copy of each chromosome) or **diploid** (having two copies). It divides by **mitosis** in four phases—**prophase, metaphase, anaphase,** and **telophase**—resulting in two nuclei with the same ploidy as the original.

2. **Meiosis** is nuclear division that results in four nuclei with half the ploidy of the original.

3. A cell's cytoplasm divides by **cytokinesis** either during or after nuclear division.

4. **Coenocytes** are multinucleate cells resulting from repeated mitosis but postponed or no cytokinesis.

5. Some microbes undergo multiple mitoses by **schizogony** to form a multinucleate **schizont**, which then undergoes cytokinesis.

6. The classification of eukaryotic microbes is problematic and has changed frequently. Historical schemes based on similarities in morphology and chemistry have been replaced with schemes based on nucleotide sequences and ultrastructural features.

Protozoa (pp. 348–355)

1. **Protozoa** are eukaryotic, unicellular organisms that lack cell walls. Most of them are chemoheterotrophs.

2. A motile **trophozoite** is the feeding stage of a typical protozoan; a **cyst**, a resting stage that is resilient to environmental changes, is formed by some protozoa.

3. A few protozoa undergo sexual reproduction by forming **gametocytes** that fuse to form a **zygote.**

4. Protozoa may be classified into four groups: **alveolates, euglenozoa, diplomonads,** and **parabasalids.** Amoebae, protozoa that move and feed with pseudopodia, are found in several kingdoms.

5. Alveolates, with cavities called alveoli beneath their cell surfaces, include **ciliates** (characterized by cilia), **apicomplexans** (all are pathogenic), and **dinoflagellates** (responsible for **red tides**).

6. Unicellular flagellated **euglenoids** are euglenozoans that store food as paramylon, lack cell walls, and have eyespots used in positive phototaxis. Because they exhibit characteristics of both animals and plants, they are a taxonomic problem.

7. **Kinetoplastids** are euglenozoans with a single, large, apical mitochondrion that contains a kinetoplast, which is a region of DNA.

8. Members of the Diplomonadida lack mitochondria, Golgi bodies, and peroxisomes; they include diplomonads (*e.g., Giardia*) and microsporidia (*e.g., Nosema*).

9. Parabasalids (*e.g., Trichomonas*) are characterized by a Golgi-like structure called a parabasal body.

Fungi (pp. 355–366)

1. **Fungi** (studied by mycologists) are chemoheterotrophic eukaryotes with cell walls that are usually composed of **chitin.**

2. Most fungi are beneficial, but some produce **mycoses** (fungal diseases).

3. The nonreproductive body of a filamentous fungus (mold) or yeast (unicellular fungus) is a **thallus.** Mold thalli are composed of tubular filaments called **hyphae.**

4. Hyphae are described as either **septate** or **aseptate** depending on the presence of crosswalls. A **mycelium** is a tangled mass of hyphae.

5. A **dimorphic** fungus has either type of thallus, depending on environmental conditions.

6. Most fungi are **saprobes**—they absorb nutrients by **absorption** from dead organisms; others get nutrients from living organisms using **haustoria** that penetrate host tissues.

7. Fungi reproduce asexually both by budding and via asexual spores, which are categorized according to their mode of development. Most fungi also reproduce sexually via spores.

8. Fungi in the division **Zygomycota** produce rough-walled **zygospores**.

9. Fungi in the division **Ascomycota,** a group of economically important fungi, produce **ascospores** within sacs called **asci.**

10. Fungi of the division **Basidiomycota,** including mushrooms, puffballs, and bracket fungi, produce **basidiospores** at the ends of **basidia.**

11. **Deuteromycota** is an informal grouping of fungi having no known sexual stage.

12. **Lichens** are economically and environmentally important organisms composed of fungi living in partnership with photosynthetic microbes, either green algae or cyanobacteria.

Algae (pp. 366–369)

1. Algae (studied by phycologists) typically reproduce by an **alternation of generations** in which a haploid thallus alternates with a diploid thallus.

2. Large algae have multicellular thalli with stemlike **stipes,** leaflike **blades,** and **holdfasts** that attach them to substrates.

3. Division **Chlorophyta** contains green algae, which are similar to vascular plants.

4. **Rhodophyta,** red algae, contain the pigment **phycoerythrin,** the storage molecule floridean starch, and cell walls of **agar** or **carrageenan,** substances used as thickening agents.

5. **Chrysophyta**—the golden algae, yellow-green algae, and **diatoms**—contain chrysolaminarin as a storage product. The cell walls of diatoms, which are made of silica, are arranged in nesting halves called **frustules.**

6. **Phaeophyta,** brown algae, contain xanthophylls, laminarin, and oils. They have cell walls composed of cellulose and **alginic acid,** which is another thickening agent. A brown algal spore is motile by means of one tinsel-like flagellum and one whiplike flagellum.

Water Molds and Slime Molds (pp. 369–371)

1. **Water molds** have tubular cristae in their mitochondria, cell walls of cellulose, spores having two different flagella, and diploid thalli. They are placed in the kingdom Stramenopila along with diatoms, other chrysophytes, and brown algae.

2. **Slime molds** lack cell walls and are phagocytic in their nutrition. They are similar to both amoebae and fungi.

3. **Plasmodial** (acellular) **slime molds** are composed of multinucleate cytoplasm. **Cellular slime molds** are composed of myxamoebae that phagocytize bacteria and yeasts.

Other Eukaryotes of Microbiological Interest: Parasitic Helminths and Vectors (pp. 371–373)

1. Parasitic helminths are significant to microbiologists because their infective stages are usually microscopic.

2. Also important to microbiologists are vectors, macroscopic animals that carry and transmit microscopic pathogens.

QUESTIONS FOR REVIEW

(Answers to multiple choice, fill in the blanks, and matching questions are on the web, along with additional review questions. Visit www.microbiologyplace.com.)

Multiple Choice

1. Haploid genomes
   a. contain one set of chromosomes.
   b. contain two sets of chromosomes.
   c. are found in zygotes.
   d. are found in the cytosol of eukaryotic organisms.

2. Which of the following sequences reflects the correct order of events in mitosis?
   a. telophase, anaphase, metaphase, prophase
   b. prophase, anaphase, metaphase, telophase
   c. telophase, prophase, metaphase, anaphase
   d. prophase, metaphase, anaphase, telophase

3. Which of the following statements accurately describes prophase?
   a. The cell appears to have a line of chromosomes across the mid-region.
   b. The nuclear envelope becomes visible.
   c. The cell constructs microtubules to form a spindle.
   d. Chromatids separate and become known as chromosomes.

4. Multiple nuclear divisions without cytoplasmic divisions result in cells called
   a. mycoses.
   b. coenocytes.
   c. haustoria.
   d. pseudohypha.

5. Branched, tubular filaments with crosswalls found in large fungi are
   a. septate hyphae.
   b. aseptate hyphae.
   c. aseptate haustoria.
   d. dimorphic mycelia.

6. The type of asexual fungal spore that buds from vegetative hyphae is called a
   a. sporangiospore.
   b. conidiospore.
   c. blastospore.
   d. chlamydospore.

7. A phycologist studies which of the following?
   a. Classification of eukaryotes
   b. Alternation of generations in algae
   c. Rusts, smuts, and yeasts
   d. Parasitic worms

8. The stemlike portion of a seaweed is called its
   a. thallus.
   b. holdfast.
   c. stipe.
   d. blade.

9. Carrageenan is found in the cell walls of which group of algae?
   a. Red algae
   b. Green algae
   c. Dinoflagellates
   d. Yellow-green algae

10. Chrysolaminarin is a storage product found in which group of microbes?
    a. Dinoflagellates
    b. Euglenoids
    c. Golden algae
    d. Brown algae

11. Which of the following features characterizes dinoflagellates?
    a. Laminarin and oils as food reserves
    b. Protective plates of silica in their cell walls
    c. Chlorophylls *a* and *c*, carotene, and xanthophylls
    d. Paramylon as a food storage molecule

12. Water molds differ from true fungi in which of the following ways?
    a. They have tubular cristae in their mitochondria.
    b. Their cell walls are made of chitin instead of cellulose.
    c. Their spores are encased in silica shells.
    d. Their thalli are haploid rather than diploid.

13. The motile feeding stage of protozoan is called
    a. an apicomplexa.
    b. a gametocyte.
    c. a cyst.
    d. a trophozoite.

Matching

1. ___ Mitosis   A. Cytoplasmic division
2. ___ Meiosis   B. Diploid nuclei producing haploid nuclei
3. ___ Homologous   C. Results in genetic variation
         chromosomes
4. ___ Crossing over   D. Carry similar genes
5. ___ Cytokinesis   E. Diploid nuclei producing diploid nuclei

1. ___ Chitin   A. Fungal cell wall component
2. ___ Basidiospore   B. fungus + alga
3. ___ Zygospore   C. Fungal body
4. ___ Thallus   D. Fungal spore formed in a sac
5. ___ Ascospore   E. Diploid fungal zygote with a thick wall
6. ___ Lichen   F. Fungal spore formed in club-shaped
                                              hyphae

1. ___ Chlorophyta   A. Yellow-green algae
2. ___ Rhodophyta   B. Green algae
3. ___ Chrysophyta   C. Brown algae
4. ___ Phaeophyta   D. Red algae

Short Answer

1. Compare and contrast the following closely related terms:

   Chromatid and chromosome

   Mitosis and meiosis II

   Hypha and mycelium

   Algal thallus and fungal thallus

   Water molds and slime molds

2. How do fungi transport nutrients?

3. How are lichens useful in environmental protection studies?

4. What are the taxonomic challenges in classifying euglenoids?

5. List several economic benefits of algae.

6. Why are relatively large animals such as parasitic worms studied in microbiology?

7. Why are microbiologists interested in macroscopic ticks, fleas, lice, and mosquitoes?

8. Name two ways that slime molds differ from true fungi.

9. What is the role of rRNA sequencing in the classification of eukaryotic microbes?

10. Describe the nuclear divisions that produce eight ascospores in an ascus.

Fill in the Blanks

1. The study of protozoa is called _____.
2. The study of fungi is called _____.
3. The study of algae is called _____.
4. Fungal diseases are called _____.

CRITICAL THINKING

1. How are cysts of protozoa similar to bacterial endospores? How are they different?

2. The host of a home improvement show suggests periodically emptying a package of yeast into a drain leading to a septic tank. Explain why this would be beneficial.

3. Why doesn't penicillin act against any of the pathogens discussed in this chapter?

4. How can one distinguish between a filamentous fungus and a colorless alga?

5. Why do scientists as a group spend more time and money studying protozoa than they do studying algae?

6. Why are there more antibacterial drugs than antifungal drugs?

7. Which type of metabolic pathways are present in protozoa that lack mitochondria (amoebae and archaezoa)? Which metabolic pathways are absent?

8. Mycologists are certain that none of the filamentous deuteromycetes will be shown to make zygospores. How can they be certain when the sexual stages of deuteromycetes are still unknown?

9. Twenty years ago, *Pneumocystis carinii*, a pathogen of immunocompromised patients that causes pneumonia, was classified as a protozoan because it is a chemoheterotroph that lacks a cell wall. However, taxonomists today classify *Pneumocystis* as a fungus. Why do you think this pathogen has been reclassified from a protozoan to a fungus?

They may look like aliens landing on another world, but the structures in this photo are actually bacteriophages, viruses that attach to and infect bacteria. Because they are relatively inexpensive and easy to grow, bacteriophages have been the focus of many studies on virus behavior; indeed, much of our knowledge of viral biology has been attained by studying them. Bacteriophages have also inspired renewed interest in the scientific community for their potential as weapons against drug-resistant bacteria.

This chapter is an introduction to the characteristics and behavior of viruses and other pathogenic particles called viroids and prions—how they infect cells, how they multiply, and how they differ from cellular pathogens. We will discuss specific viral diseases in Chapters 24 and 25.

Characterizing and Classifying Viruses, Viroids, and Prions

CHAPTER 13

CHAPTER OUTLINE

Characteristics of Viruses
    Genetic Material of Viruses
    Hosts of Viruses
    Sizes of Viruses
    Capsid Morphology
    Viral Shapes
    The Viral Envelope

Classification of Viruses

Viral Replication
    Lytic Replication of Bacteriophages
    Lysogeny
    Replication of Animal Viruses

The Role of Viruses in Cancer

Culturing Viruses in the Laboratory
    Culturing Viruses in Whole Organisms
    Culturing Viruses in Embryonated Chicken Eggs
    Culturing Viruses in Cell (Tissue) Culture

Other Parasitic Particles: Viroids and Prions
    Characteristics of Viroids
    Characteristics of Prions

Are Viruses Alive?

Esherichia coli infected by bacteriophages.

MicroPrep Pre-Test: Take the pre-test for this chapter on the web. Visit www.microbiologyplace.com.

Not all pathogens are eukaryotic or prokaryotic cells. Many infections of humans, animals, and plants, and even of bacteria, are caused by **acellular** (noncellular) agents, including viruses and other parasitic particles called viroids and prions. Although these agents are like some eukaryotic and prokaryotic microbes in that they can cause disease when they invade susceptible cells, they are simple compared to a cell, lacking cell membranes and being composed of only a few organic molecules. In addition to lacking a cellular structure, they lack most of the characteristics of life described in Chapter 3: They cannot carry out any metabolic pathway, they can neither grow nor respond to the environment, and they cannot reproduce independently but instead must utilize the chemical and structural components of the cells they infect. Not only are they obligate intracellular parasites—that is, they must invade a living cell—they must also recruit the cell's metabolic chemicals and ribosomes in order to increase their numbers.

In this chapter we will first examine a range of topics concerning viruses: their many characteristics, how they are classified, how they replicate, the role they play in some kinds of cancers, and the techniques used to maintain them in the laboratory. Then we consider the nature of viroids and prions before concluding with further discussion concerning whether or not viruses are alive.

Characteristics of Viruses

Viruses cause most of the diseases that still plague the industrialized world: the common cold, influenza, herpes, and AIDS, to name a few. Although we have immunizations against many viral diseases and are adept at treating the symptoms of others, the characteristics of viruses and the means by which they attack their hosts can make cures for viral diseases elusive. Throughout this section, try to consider the clinical implications of the viral characteristics we discuss.

Let's begin by looking at the characteristics viruses have in common. A **virus** is a miniscule, acellular, infectious agent having one or several pieces of nucleic acid—either DNA or RNA, but never both. The nucleic acid is the genetic material (genome) of the virus. Being acellular, viruses have no cytoplasmic membrane (though, as we will see, some viruses possess a membrane-like *envelope*). Viruses also lack cytosol, and, with one exception, organelles. They are not capable of metabolic activity of their own; instead, once viruses have invaded a cell, they take control of the cell's metabolic machinery to produce more molecules of viral nucleic acid and viral proteins, which are then assembled into new viruses via a process we will examine shortly.

Viruses have an extracellular and an intracellular state. Outside of a cell, in the extracellular state, a virus is called a **virion** (vir′ē-on). Basically, a virion consists of a protein coat, called a **capsid,** surrounding a nucleic acid core **(Figure 13.1a).** Together the nucleic acid and its capsid are also called a *nucleocapsid,* which in many cases can crystallize like

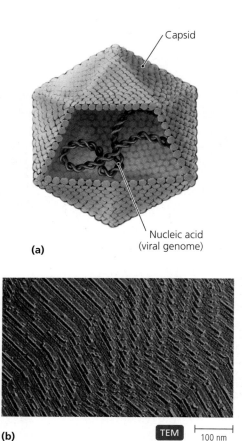

Capsid

Nucleic acid
(viral genome)

(a)

(b)

TEM   100 nm

▲ *Figure 13.1*

Virions, complete virus particles, include a nucleic acid, a capsid, and in some cases an envelope. **(a)** A drawing of a nonenveloped polyhedral virus containing DNA. **(b)** A transmission electron micrograph of crystallized virions. Like many chemicals and unlike cells, some viruses can form crystals.

crystalline chemicals **(Figure 13.1b).** Some virions have a phospholipid membrane called an envelope surrounding the nucleocapsid. The outermost layer of a virion (capsid or envelope) provides a virus both protection and recognition sites that bind to complementary chemicals on the surfaces of cells. These are involved when a virus penetrates a cell. Envelopes typically fuse with the cell's membrane and the virus moves into the cell. Once inside, the intracellular state is initiated; the capsid is removed, and the virus exists solely as nucleic acid.

Now that we have examined ways in which viruses are alike, let's consider the characteristics that are used to distinguish different viral groups. Viruses differ in the type of genetic material they contain, the kinds of cells they attack, their size, the nature of their capsid coat, their shapes, and the presence or absence of an envelope.

Genetic Material of Viruses

Learning Objective

✓ Discuss viral genomes in terms of dsDNA, ssDNA, ssRNA, dsRNA, and number of segments of nucleic acid.

Viruses show more variety in the nature of their genomes than do cells. Whereas the genome of every cell is double-stranded DNA, the genome of a virus may be either DNA or RNA, but never both. The primary way in which scientists categorize and classify viruses is based on the type of genetic material that makes up the viral genome.

Some viral genomes, such as those of herpes virus and chickenpox virus, are double-stranded DNA (dsDNA), like the genomes of cells. Other viruses use either single-stranded RNA (ssRNA), single-stranded DNA (ssDNA), or double-stranded RNA (dsRNA) as their genomes. These molecules never function as the genome of any cell—in fact, ssDNA and dsRNA are almost nonexistent in cells. Further, a viral genome may be either linear and composed of several molecules of nucleic acid, as in eukaryotic cells, or circular and singular, as in most prokaryotic cells. For example, the genome of the influenza (flu) virus is composed of eight linear segments of single-stranded RNA, whereas the genome of poliovirus is one molecule of single-stranded RNA.

Viral genomes are much smaller than the genomes of cells. Whereas the genome of the smallest chlamydial bacterium has almost 1000 genes, the genome of the bacteriophage MS2 has only three genes. **Figure 13.2** compares the genome of bacteriophage T7, which contains 56 genes, with the genome of the bacterium *Escherichia coli* (esh-ĕ-rik′ē-ă kō′lī), which contains over 4000 genes.

CRITICAL THINKING

Some viral genomes, composed of single-stranded RNA, act as mRNA. What advantage might these viruses have over other kinds of viruses?

Hosts of Viruses

Learning Objectives

✓ Explain how viruses are specific for their host cells.
✓ Compare and contrast viruses of fungi, plants, animals, and bacteria.

Most viruses infect only particular host's cells. This specificity is due to the specific affinity of viral surface proteins or glycoproteins for complementary proteins or glycoproteins on the surface of the host cell. Viruses may be so specific they infect not only a particular host, but a particular kind of cell in that host. For example, HIV (human immunodeficiency virus, the agent that causes AIDS) specifically attacks helper T lymphocytes (a type of white blood cell) in humans and has no effect on, say, human muscle cells or bone cells. By contrast, some viruses are *generalists;* they infect many kinds of cells in many different hosts. An example of a gen-

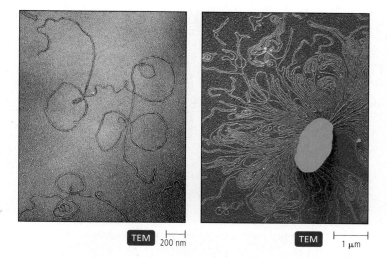

TEM |—| 200 nm   TEM |—| 1 μm

▲ **Figure 13.2**

The relative sizes of the genomes of *E. coli* (on the right) and bacteriophage T7 on the left. The *E. coli* genome was released by rupture of the bacterium.

eralist virus is rabies, which can infect most mammals, from humans to bats.

All types of organisms are susceptible to viral attack. There are viruses that infect plant, bacterial, and animal cells (**Figure 13.3** on page 380), and even fungal cells. Most viral research and scientific study has focused on bacterial and animal viruses. As we have seen, a virus that infects bacteria is referred to as a **bacteriophage** (bak-tēr′ē-ō-fāj), or simply a **phage** (fāj). (**Highlight 13.1** on page 381 discusses the potential use of bacteriophages as an alternative to antibiotics.) We will return our attention to bacteriophages and animal viruses later in this chapter.

Viruses of plants are less well known than bacterial and animal viruses, even though viruses were originally identified and isolated from tobacco plants. Plant viruses infect many food crops, including corn, beans, sugarcane, tobacco, and potatoes, resulting in billions of dollars in losses each year. Viruses of plants are introduced into plant cells either through abrasions of the cell wall or by plant parasites such as nematodes and aphids. Plant viruses follow the replication cycle discussed below for animal viruses.

Fungal viruses have been little studied and are not well known. We do know that fungal viruses are different from animal and bacterial viruses in that fungal viruses exist only within cells; that is, they have no extracellular state. Presumably, fungal viruses cannot penetrate the thick fungal cell wall. However, because fusion of cells is typically a part of the fungal life cycle, viral infections can easily be propagated by the fusion of an infected fungal cell with an uninfected one.

(a)

(b)   SEM   0.5 µm

(c)   TEM   100 nm

(d)   TEM   100 nm

◄ *Figure 13.3*

Some examples of plant, bacterial, and human hosts of viral infections. **(a)** Left, a tobacco leaf infected with tobacco mosaic virus, the first virus discovered. **(b)** A bacterial cell under attack by bacteriophages. **(c)** Influenza viruses budding from an infected cell. **(d)** A human white blood cell's cytoplasmic membrane, to which two enveloped human immunodeficiency viruses (HIV) are attached.

Sizes of Viruses

In the late 1800s, scientists hypothesized that the cause of many diseases, including polio and smallpox, was an agent smaller than a bacterium. They named these tiny agents "viruses," from the Latin word for "poison." Viruses are so small that most cannot be seen by light microscopy. One hundred million polioviruses could fit on the period at the end of this sentence. The smallest viruses have a diameter of 10 nm, whereas the largest are approximately 300 nm in diameter, which is about the size of the smallest bacterial cell. **Figure 13.4** compares the sizes of selected viruses to *E. coli* and a human red blood cell.

In 1892, Russian microbiologist Dmitri Ivanowsky first demonstrated that viruses are acellular with an experiment designed to elucidate the cause of tobacco mosaic disease. He filtered the sap of infected tobacco plants through a porcelain filter fine enough to trap even the smallest of bacterial cells. The viruses, however, were not trapped, but instead passed though the filter with the liquid, which remained infectious to tobacco plants. This experiment proved the existence of an acellular disease-causing entity smaller than a bacterium. Tobacco mosaic virus (TMV) was isolated and characterized by an American chemist, Wendell Stanley, in 1935. Shortly thereafter, the invention of electron microscopy allowed TMV and other viruses to be seen for the first time.

Capsid Morphology

Learning Objective

✓ Discuss the structure and function of the viral capsid.

As we have seen, viruses have capsids—protein coats that provide both protection for the viral nucleic acid and a means by which many viruses attach to a host's cells. The capsid of a virus is composed of proteinaceous subunits called **capsomeres** (or *capsomers*). Some capsomeres (and therefore some capsids) are composed of only a single type of protein, whereas others are composed of several different protein subunits. Recall that a viral nucleic acid surrounded by its capsid is termed a *nucleocapsid*.

CRITICAL THINKING

In some viruses, the capsomeres act enzymatically as well as structurally. What advantage might this provide the virus?

Highlight 13.1   Prescription Bacteriophages?

As more and more bacteria become resistant to antibiotics, interest in the use of bacteriophages to fight bacterial pathogens has increased. One recent study by Rockefeller University researchers, for example, found that enzymes coded by bacteriophages were able to eliminate the presence of *Streptococcus pneumoniae* (a leading cause of illness in young, elderly, and immunocompromised patients) from mucous membranes in mice. The theory is that using a nasal spray containing these enzymes could help prevent certain infections before they even begin. Other scientists propose using intact phages.

There are several potential advantages to the use of phages over antibiotics. Unlike antibiotics, phages target specific bacteria, so phages can be effective without also destroying "good" bacteria in the normally present microbiota. (The destruction of "good" bacteria is linked to the diarrhea suffered by some users of antibiotics.) Furthermore, phages are inexpensive, easy to produce, effective in small doses (because they replicate), have no known deleterious effects on a patient's cells or metabolism, and are unlikely to contribute to antibiotic resistance—all the more reason to further investigate this intriguing alternative to the use of antibiotics.

Reference: Loeffler, J. M., D. Nelson, V. A. Fischetti, 2001. Rapid killing of *Streptococcus pneumoniae* with a bacteriophage cell wall hydrolase. *Science* 294:2170–2172.

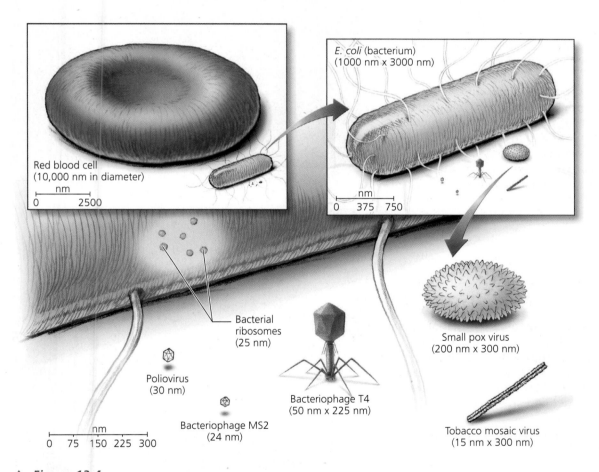

▲ *Figure 13.4*

The sizes of selected viruses as compared to a bacterium, *E. coli,* and a human red blood cell. *How can viruses be so small and yet still be pathogenic?*

Figure 13.4 Viruses utilize a host cell's enzymes, organelles, and membranes to replicate.

(a)   TEM   50 nm

Figure 13.5 ➤

The shapes of virions. **(a)** A helical virus, tobacco mosaic virus. The tubular shape of the capsid results from the tight arrangement of several rows of helical capsomeres. **(b)** Polyhedral virions of a virus that causes the common cold. **(c)** Complex virions of smallpox virus. **(d)** The complex shape of rabies virus, which results from the shapes of the capsid and bullet-shaped envelope surrounding it.

Viral Shapes

The shape of the virion is another characteristic by which viruses are classified. There are three basic types of viral shapes: helical, polyhedral, and complex **(Figure 13.5).** The capsid of a helical virus is composed of capsomeres that bond together in a spiral fashion to form a tube around the nucleic acid. The capsid of a polyhedral virus is roughly spherical, with a shape similar to a geodesic dome. The most common type of polyhedral capsid is an icosahedron, which has 20 sides.

Complex viruses have capsids of many different shapes that do not readily fit into either of the other two categories. An example of a complex virus is smallpox virus, which has several covering layers (including lipid) and no easily identifiable capsid. The complex shapes of many bacteriophages include icosahedral heads, which contain the genome, attached to helical tails with tail fibers. The complex capsids of bacteriophages somewhat resemble NASA's lunar lander **(Figure 13.6).**

(b)   TEM   100 nm

The Viral Envelope

Learning Objective

✓ Discuss the origin, structure, and function of the viral envelope.

All viruses lack cell membranes (after all, they are not cells), but some, particularly animal viruses, have a membrane similar in composition to a cytoplasmic membrane surrounding their capsids. Such a membrane is called an **envelope,** and thus a virus with a membrane is an **enveloped virion (Figure 13.7);** a virion without an envelope is called a *nonenveloped* or *naked virion.*

Enveloped viruses acquire their envelope from the host cell during viral replication or release (discussed shortly). Indeed, the envelope of a virus is a portion of the membrane system of a host cell. Like a cytoplasmic membrane, a viral envelope is composed of a phospholipid bilayer and proteins. Some of the proteins are virally-coded glycoproteins, which appear as spikes protruding outward from the envelope's surface (see Figure 13.5d). Host DNA carries the genetic code required for the assembly of the phospholipids and some of the proteins in the envelope, while the viral genome specifies the other integral proteins.

The envelope's proteins and glycoproteins often play a role in the recognition of host cells. A viral envelope does not perform the other physiological roles of a cytoplasmic membrane such as endocytosis or active transport.

(c)   TEM   200 nm

(d)   TEM   20 nm

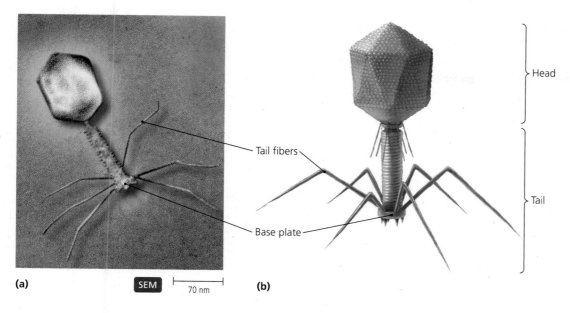

(a) SEM 70 nm (b)

- Head
- Tail fibers
- Base plate
- Tail

◀ *Figure 13.6*
The complex shape of bacteriophage T4, which includes an icosahedral head and an ornate tail that enables viral attachment and penetration.

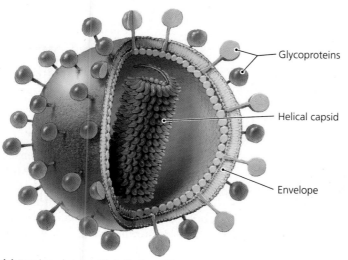

- Glycoproteins
- Helical capsid
- Envelope

(a) Enveloped virus with helical capsid

TEM 50 nm

◀ *Figure 13.7*
Enveloped virions.
(a) Artist's rendition and electron micrograph of a coronavirus, an enveloped virus with a helical capsid. **(b)** Artist's rendition and electron micrograph of a togavirus, an enveloped virus with a polyhedral capsid.

- Glycoproteins
- Polyhedral capsid
- Envelope

(b) Enveloped virus with polyhedral capsid

TEM 50 nm

Table 13.1	The Novel Properties of Viruses
Viruses	**Cells**
Are inert macromolecules outside of a cell, but become active inside a cell	Metabolize on their own
Do not divide or grow	Divide and grow
Acellular	Cellular
Obligate intracellular parasites	Most free-living
Contain either DNA or RNA, never both	Contain both DNA and RNA
Genome can be dsDNA, ssDNA, dsRNA, or ssRNA	Genome is dsDNA
Ultramicroscopic in size, ranging from 10 nm to 300 nm	200 nm to 12 cm in diameter
Have a proteinaceous capsid around genome; some have an envelope around the capsid	Surrounded by a phospholipid membrane and often a cell wall
Replicate in an assembly-line manner using the enzymes and organelles of a host cell	Self-replicating by asexual and/or sexual means

The novel properties of viruses, and how those properties differ from the corresponding characteristics of cells, are summarized in **Table 13.1.** Next we turn our attention to the criteria by which viruses are classified.

Classification of Viruses

Learning Objective

✓ List the characteristics by which viruses are classified.

The International Committee on Taxonomy of Viruses (ICTV) was established in 1966 to provide a single taxonomic scheme for viral classification and identification. Viruses are categorized by their type of nucleic acid, presence of an envelope, shape, and size. For most viruses, families are the highest taxonomic groups established so far by the ICTV; viruses are assigned to certain genera within these families. Kingdoms, divisions, and classes have not been determined for most viruses because the relationships among viruses are not well understood. Only three viral orders have been established.

Family names are typically derived either from special characteristics of viruses within the family or from the name of an important member of the family. For example, family *Picornaviridae* contains very small[1] RNA viruses, and *Hepadnaviridae* contains a DNA virus that causes hepatitis A. Family *Herpesviridae* is named for herpes simplex, a virus that can cause genital herpes. **Table 13.2** lists the major families of human viruses, grouped according to the type of nucleic acid each contains.

At this time, Latinized names are not given to viruses at the species level. Instead, species names for viruses are their common English designations. Accordingly, the taxonomy for two important viral pathogens, HIV and rabies virus, is as follows:

	HIV	*Rabies virus*
Order:	—	Mononegavirales
Family:	*Retroviradae*	*Rhabdoviridae*
Genus:	*Lentivirus* (len'ti-vī-rŭs)	*Lyssavirus* (lis'ă-vī-rŭs)
Species:	Human immuno-deficiency virus	Rabies virus

As you know, viruses cause a variety of human diseases, from mild colds to life-threatening illnesses such as AIDS and Ebola hemorrhagic fever. DNA viruses that are pathogenic to humans are discussed in Chapter 24 and are summarized in Table 24.1; those containing RNA genomes are the subject of Chapter 25 and are summarized in Table 25.1.

CRITICAL THINKING

Why has it been difficult to develop a complete taxonomy for viruses?

Viral Replication

As previously noted, viruses cannot reproduce themselves because they have neither the genes for all the enzymes necessary for replication nor functional ribosomes for protein synthesis. Instead, viruses are dependent on their host's organelles and enzymes to produce new virions. Once a host cell falls under control of a viral genome, it is forced to replicate viral genetic material and translate new proteins, including viral capsomeres and viral enzymes.

The replication cycle of a virus usually results in the death and lysis of the host cell. Because the cell undergoes *lysis* near the end of the cycle, this type of replication is

---

[1]Pico means one trillionth.

Table 13.2   Families of Human Viruses

Family	Strand Type	Representative Genera (Diseases)
DNA Viruses		
Poxviridae	Double	*Orthopoxvirus* (smallpox)
Herpesviridae	Double	*Simplexvirus,* Herpes type 1 (fever blisters, respiratory infections), Herpes type 2 (genital infections); *Varicellovirus* (chicken pox); *Lymphocryptovirus,* Epstein-Barr virus (infectious mononucleosis, Burkitt's lymphoma); *Cytomegalovirus* (birth defects); *Roseolavirus* (roseola)
Papillomaviridae	Double	*Papillomavirus* (benign tumors, warts, cervical and penile cancers)
Polyomaviridae	Double	*Polyomavirus* (progressive multifocal leukoencephalopathy)
Adenoviridae	Double	*Mastadenovirus* (conjunctivitis, respiratory infections)
Hepadnaviridae	Partial single and partial double	*Hepadnavirus* (hepatitis B)
Parvoviridae	Single	*Erythrovirus* (erythema infectiosum)
RNA Viruses		
Picornaviridae	Single, +[a]	*Enterovirus* (polio); *Hepatovirus* (hepatitis A); *Rhinovirus* (common cold)
Caliciviridae	Single, +	*Norovirus* (gastroenteritis)
Astroviridae	Single, +	*Astrovirus* (gastroenteritis)
Hepeviridae	*Single, +*	*Hepevirus* (hepatitis E)
Togaviridae	Single, +	*Alphavirus* (encephalitis); *Rubivirus* (rubella)
Flaviviridae	Single, +	*Flavivirus* (yellow fever); Japanese encephalitis virus; *Hepacivirus* (hepatitis C)
Coronaviridae	Single, +	*Coronavirus* (common cold, severe acute respiratory syndrome)
Retroviridae	Single, +, segmented	Human T cell leukemia virus; *Lentivirus* (HIV)
Orthomyxoviridae	Single, −[b], segmented	*Influenzavirus* (flu)
Paramyxoviridae	Single, −	*Paramyxovirus* (colds, respiratory infections); *Pneumovirus* (pneumonia, common cold); *Morbillivirus* (measles); *Rubulavirus* (mumps)
Rhabdoviridae	Single, −	*Lyssavirus* (rabies)
Bunyaviridae	Single, −, segmented	*Bunyavirus* (California encephalitis virus); *Hantavirus*
Filoviridae	Single, −	*Filovirus* (Ebola); Marburg viruses
Arenaviridae	Single, −, segmented	*Lassavirus* (hemorrhagic fever)
Reoviridae	Double, segmented	*Orbivirus* (encephalitis); *Rotavirus* (diarrhea); *Coltivirus* (Colorado tick fever)

[a]+RNA is equivalent to mRNA, *i.e.,* instructs ribosomes in protein translations.
[b]−RNA is complementary to mRNA; it cannot be directly translated.

termed **lytic replication.** In general, a lytic replication cycle consists of the following five stages:

- **Attachment** of the virion to the host cell
- **Entry** of the virion or its genome into the host cell
- **Synthesis** of new nucleic acids and viral proteins by the host cell's enzymes and ribosomes
- **Assembly** of new virions within the host cell
- **Release** of the new virions from the host cell

In the following sections we will examine the events that occur in the replication of bacteriophages and animal viruses. We will begin with lytic replication in bacteriophages, turn to a modification of replication (called lysogenic replication) in bacteriophages, and then consider the replication of animal viruses.

Lytic Replication of Bacteriophages

Learning Objective

✓ Sketch and describe the five stages of the lytic replication cycle as it typically occurs in bacteriophages.

Bacteriophages make excellent tools for the general study of viruses because they are easier and less expensive to culture than animal or human viruses. Studies of bacteriophages revealed the basics of viral biology.

Here we will examine the replication of a much-studied dsDNA bacteriophage of *E. coli* called *type 4* or *T4*. T4 virions are complex, having the polyhedral heads and helical tails seen in many bacteriophages (see Figure 13.6). We begin with the first stage of replication: attachment.

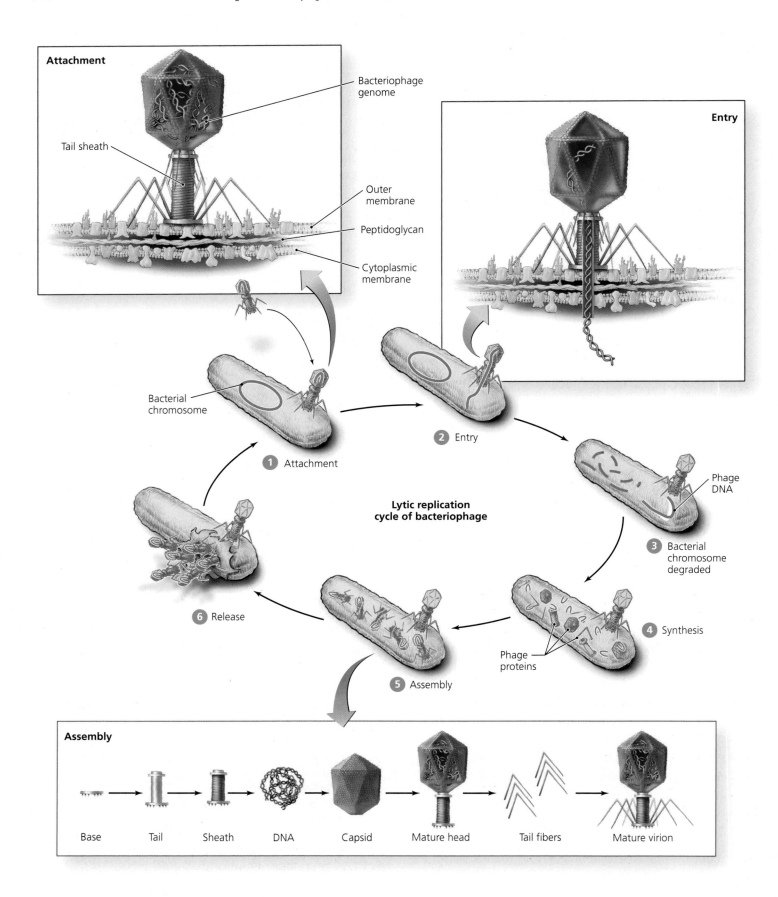

Attachment

Bacteriophage genome

Tail sheath

Outer membrane

Peptidoglycan

Cytoplasmic membrane

Entry

Bacterial chromosome

1 Attachment

2 Entry

Phage DNA

3 Bacterial chromosome degraded

Lytic replication cycle of bacteriophage

6 Release

5 Assembly

4 Synthesis

Phage proteins

Assembly

Base → Tail → Sheath → DNA → Capsid → Mature head → Tail fibers → Mature virion

◀ *Figure 13.8*

The lytic replication cycle in bacteriophages. The phage shown in this illustration is T4, and the bacterium shown is *E. coli*. The circular bacterial chromosome is represented diagrammatically; in reality it would be much longer.

Attachment ❶

Because virions are nonmotile, contact with a bacterium occurs by purely random collision, brought about as Brownian motion and currents move virions through the environment. As is the case for many bacteriophages, the structures responsible for the attachment of T4 to its host bacterium are its tail fibers **(Figure 13.8)**. Attachment is dependent on the chemical attraction and precise fit between attachment proteins on the phage's tail fibers and complementary receptor proteins on the surface of the host's cell wall. The specificity of the attachment proteins for the receptors ensures that the virus will attach only to *E. coli*. Bacteriophages may attach to receptor proteins on bacterial cell walls, flagella, or pili.

Entry ❷

Now that phage T4 has attached to the bacterium's cell wall, it must still overcome the formidable barrier posed by the cell wall and cell membrane if it is to enter the cell. T4 overcomes this obstacle in an elegant way. Upon contact with *E. coli*, T4 releases *lysozyme* (lī'sō-zīm), a protein enzyme carried within the capsid that weakens the peptidoglycan of the cell wall. The phage's tail sheath then contracts, which forces an internal hollow tube within the tail through the cell wall and membrane, much as a hypodermic needle penetrates the skin (see Figure 13.8). The phage genome then moves through the tube and into the bacterium. The empty capsid, having performed its task, is left on the outside of the cell looking like an abandoned spacecraft.

New Frontiers 13.1 describes recent research that reveals the pressure inside the capsid of another phage named phi29 is about 2200 pounds per square inch—comparable to the pressure inside a scuba tank. Such pressure may be used to inject the phage's genome into its host bacterium. In the absence of such high pressure, a phage's genome diffuses into a host cell.

Synthesis ❸ – ❹

After entry, viral enzymes (either carried within the capsid or coded by viral genes and made by the bacterium) degrade the bacterial DNA into its constituent nucleotides. As a result, the bacterium stops synthesizing its own molecules and begins synthesizing new viruses under control of the viral genome.

For dsDNA viruses like T4, protein synthesis is straightforward and similar to cellular transcription and translation, except that mRNA is transcribed from viral DNA instead of cellular DNA. Translation by the host cell's ribosomes results in viral proteins, including head capsomeres, components of the tail, DNA polymerase (which replicates viral DNA), and lysozyme (which weakens the bacterial cell wall from within, enabling the virions to leave the cell once they have been assembled).

Assembly ❺

Scientists do not understand completely how phages are assembled inside a host cell, but it appears that as capsomeres accumulate within the cell, they spontaneously attach to one another to form new capsid heads. Likewise, tails assemble and attach to heads and tail fibers attach to tails to form

---

New Frontiers 13.1 **High-Pressure Viruses**

Imagine the pressure inside an unopened bottle of champagne. Now consider that the DNA inside some viruses can be packed at an internal pressure as high as 10 times that of champagne pressures!

To reach this conclusion, researchers at the University of California at Berkeley and the University of Minnesota used an ingenious method of measuring the force required to package DNA into an empty capsid of bacteriophage phi29. They glued a capsid to a plastic bead held stationary with suction through a micropipette. They then positioned a molecule of DNA glued to a second bead near the capsid. A laser light was focused onto the second bead.

When they added ATP to the system, a capsid enzyme stuffed the DNA into the capsid, which pulled the two beads closer together and deflected the laser light. By measuring the amount of deflection, the researchers could measure the force on the bead and thus on the DNA attached to it. At first, the DNA was packaged smoothly at a rate of 100 base pairs per second, but as the capsid filled with DNA, the pressure inside it increased, and the process slowed. By measur-

ing the force and calculating the volume of the capsid, researchers concluded that the pressure inside the head is 2200 pounds per square inch. This tremendous pressure may help some viruses inject their DNA into cells. Among the viruses suspected of packing their DNA in this way are those that cause herpes, chicken pox, shingles, and other human diseases.

Further research in this area may reveal ways to halt the process—ways that may result in the development of drugs designed to interfere with the ability of these viruses to cause disease.

---

1. Sanders, R. October 18, 2001. Molecular motor powerful enough to pack DNA into viruses at greater than champagne pressures. University of California at Berkeley, press release.

2. Smith, D. E., S. J. Tans, S. B. Smith, S. Grimes, D. L. Anderson, C. Bustamante. 2001. The bacteriophage φ29 portal motor can package DNA against a large internal force. *Nature* 413:748–752.

Figure 13.9 ▶

Virion abundance over time during a single replication cycle in lytic viruses. New virions are not observed in the culture medium until synthesis, assembly, and release (lysis) are complete, at which time (the burst time) the new virions are released all at once. Burst size is the number of new virions released per lysed host cell.

mature virions (see Figure 13.8). Such capsid assembly is a spontaneous process, requiring little or no enzymatic activity. For many years it was assumed that all capsids formed around a genome in just such a spontaneous manner. However, recent research (including that described in New Frontiers 13.1) has shown that for some viruses, enzymes insert the genome into the capsid after it has assembled. This process resembles stuffing a strand of cooked spaghetti into a matchbox through a single hole.

Sometimes a capsid assembles around leftover pieces of host DNA instead of viral DNA. A virion formed in this manner is still able to attach to a new host by means of its tail fibers, but instead of inserting phage DNA it transfers DNA from the first host into a new host. This process, known as *transduction*, was described in Chapter 8.

Release ⑥

Newly assembled virions are released from the cell as the lysozyme completes its work on the cell wall and the bacteria disintegrate. Areas of disintegrating bacterial cells in a lawn of bacteria in a Petri plate looks as if the lawn were being eaten, and it was this appearance that prompted early scientists to give the name *bacteriophage*, "bacterial eater," to these viruses.

For phage T4, the process of lytic replication takes about 25 minutes and can produce as many as about 100–200 new virions for each bacterial cell lysed **(Figure 13.9).** For any phage undergoing lytic replication, the period of time required to complete the entire process, from attachment to release, is called the *burst time*, and the number of new virions released from each lysed bacterial cell is called the *burst size.*

CRITICAL THINKING

If a colony of 1.5 billion *E. coli* cells were infected with a single phage T4, and each lytic replication cycle of the phage produced 200 new phages, how many replication cycles would it take for T4 phages to overwhelm the entire bacterial colony? (Assume for the sake of simplicity that every phage completes its replication cycle in a different cell, and that the bacteria themselves do not reproduce.)

Lysogeny

Learning Objective

✓ Compare and contrast the lysogenic replication cycle of viruses with the lytic cycle and with latency.

Not all viruses follow the lytic pattern of phage T4 we just examined. Some bacteriophages have a modified replication cycle in which infected host cells grow and reproduce normally for many generations before they lyse. The additional portion of such a replication cycle is called a **lysogenic replication cycle** or **lysogeny** (lī-soj′ĕ-nē), and the phages are called **lysogenic phages** or **temperate phages.**

We will examine lysogenic replication as it occurs in much-studied temperate phage, *lambda phage,* which is another parasite of *E. coli.* A lambda phage has a linear molecule of dsDNA in a complex capsid consisting of an icosahedral head attached to a tail that lacks tail fibers **(Figure 13.10).**

Figure 13.11 illustrates lysogeny with lambda phage. First, the virion randomly contacts an *E. coli* cell and attaches to it via its tail ❶. The viral DNA enters the cell, just as occurs with phage T4, but the host cell's DNA is not destroyed, and the phage's genome does not immediately assume control of the cell. Instead, the virus remains inactive. Such an inactive bacteriophage is called a **prophage** (prō′fāj) ❷. A prophage remains inactive by coding for a protein that suppresses prophage genes. A side effect of this repressor protein is that it renders the bacterium resistant to additional infection by other viruses of the same type.

Another difference between the lysogenic cycle and the lytic cycle is that a prophage is inserted into the DNA of the bacterium, becoming a physical part of the bacterial chromosome. For DNA viruses like lambda phage, this is a simple process of fusing two pieces of DNA: One piece of DNA, the virus, is fused to another piece of DNA, the chromosome of the cell. Every time the cell replicates its chromosome, the prophage is also replicated ❸. All daughter cells of a lysogenic cell are thus infected with the quiescent virus. A prophage and its descendents may remain a part of the bacterial chromosome for generations.

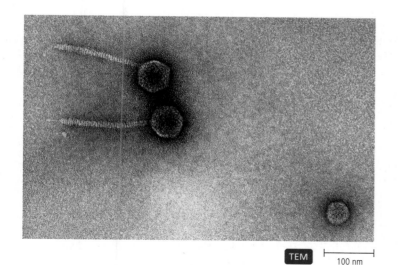

◀ **Figure 13.10**

Bacteriophage lambda. Note the absence of tail fibers.
Phage T4 attaches by means of molecules on its tail fibers.
How does lambda phage, which lacks fibers, attach?

Figure 13.10 Lambda has attachment molecules at the end of its tail rather than on its tail fibers.

TEM | 100 nm

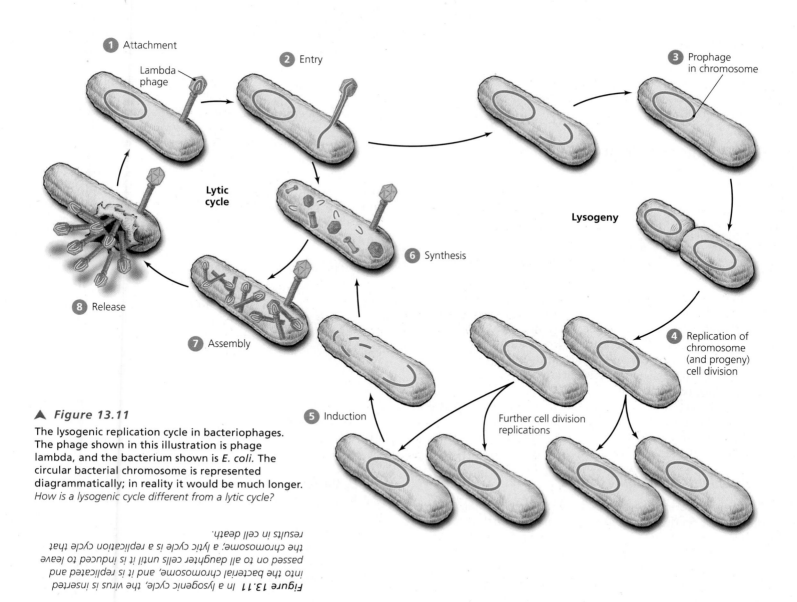

▲ **Figure 13.11**

The lysogenic replication cycle in bacteriophages.
The phage shown in this illustration is phage
lambda, and the bacterium shown is *E. coli*. The
circular bacterial chromosome is represented
diagrammatically; in reality it would be much longer.
How is a lysogenic cycle different from a lytic cycle?

Figure 13.11 In a lysogenic cycle, the virus is inserted into the bacterial chromosome, and it is replicated and passed on to all daughter cells until it is induced to leave the chromosome; a lytic cycle is a replication cycle that results in cell death.

At some later time the prophage may be excised from the chromosome by recombination or some other genetic event and reenters the lytic phase. The process whereby a prophage is excised from the host chromosome is called **induction** ⑤ . Inductive agents are typically the same physical and chemical agents that damage DNA molecules, including ultraviolet light, X-rays, and carcinogenic chemicals.

After induction, the lytic steps of synthesis ⑥ , assembly ⑦ , and release ⑧ resume from the point at which they stopped. The cell becomes filled with virions and breaks open.

Bacteriophages T4 and lambda demonstrate two replication strategies that are typical for many DNA viruses. RNA viruses and enveloped viruses present variations on the lytic and lysogenic cycles we have examined. We will examine some of these variations as they occur with animal viruses.

CRITICAL THINKING

What differences would you expect in the replication cycles of RNA phages from those of DNA phages? (Hints: Think about the processes of transcription, translation, and replication of nucleic acids. Also, note that RNA cannot be inserted into a DNA molecule.)

Replication of Animal Viruses

Learning Objectives

✓ Explain the differences between bacteriophage replication and animal viral replication.

✓ Compare and contrast the replication and synthesis of DNA, −RNA, and +RNA viruses.

✓ Compare and contrast the release of viral particles by lysis and budding.

Animal viruses have the same basic replication pathway as bacteriophages—that is, attachment, entry, synthesis, assembly, and release. However, there are significant differences in the replication of animal viruses that result in part from the presence of envelopes around some of the viruses, and in part from the eukaryotic nature of animal cells as well as their lack of a cell wall.

In this section we will examine the replication processes that are shared by DNA and RNA animal viruses, compare these processes with those of bacteriophages, and discuss how the synthesis of DNA and RNA viruses differ.

Attachment of Animal Viruses

As with bacteriophages, attachment of an animal virus is dependent on the chemical attraction and exact fit between proteins or glycoproteins on the virion and complementary protein or glycoprotein receptors on the animal cell's plasma membrane. Unlike many bacteriophages, animal viruses lack both tails and tail fibers. Instead, animal viruses typically have glycoprotein spikes or other attachment molecules on their capsids or envelopes.

Entry and Uncoating of Animal Viruses

Animal viruses enter a host cell shortly after attachment. Even though entry of animal viruses is not as well understood as for bacteriophages, there appear to be at least three different mechanisms: direct penetration (for naked viruses), and membrane fusion and phagocytosis (for enveloped viruses).

In naked viruses, as with phages, the genome enters the host cell while the capsid remains on the cell's surface **(Figure 13.12a)**. Exactly how the genome enters the host cell remains a mystery. Poliovirus is an example of a virion that infects host cells via direct penetration.

In enveloped viruses, by contrast, the entire capsid and its contents (including the genome) enter the host cell. In a few viruses, including the measles and mumps viruses, the viral envelope and the host cell membrane fuse, releasing the capsid into the cell's cytoplasm and leaving the envelope glycoproteins as part of the cell membrane **(Figure 13.12b)**. Most enveloped viruses, however, enter host cells after attachment of the virus to receptor molecules on the cell's surface stimulates the cell to phagocytize the entire virus **(Figure 13.12c)**. Herpes viruses enter human host cells via phagocytosis.

For those enveloped viruses that penetrate the host cell with their capsids intact, the capsids must be removed to release their genomes before the viruses can continue to replicate. The removal of a viral capsid within a host cell is called **uncoating,** a process that remains poorly understood. It apparently occurs via different means in different viruses; some viruses are uncoated within vesicles by cellular enzymes, whereas others are uncoated by enzymes within the cell's cytosol.

Synthesis of Animal Viruses

Synthesis of animal viruses also differs from synthesis of bacteriophages. Each type of animal virus requires a different strategy for synthesis that depends on the kind of nucleic acid involved—whether it is DNA or RNA, and whether it is double stranded or single stranded. As we discuss the synthesis and assembly of each type of animal virus, consider the following two questions:

- How is mRNA (needed for the translation of viral proteins) synthesized?

- What molecule serves as a template for nucleic acid replication?

dsDNA Viruses   Synthesis of new double-stranded DNA (dsDNA) virions is similar to the normal replication of DNA and translation of proteins in cells. After uncoating, the genomes of most dsDNA viruses enter the nucleus of the cell, where cellular enzymes replicate the viral genome in the same manner as host dsDNA is normally replicated—using each strand of viral DNA as a template for its complement. After messenger RNA is transcribed from viral DNA in the nucleus, and capsomere proteins are made in the

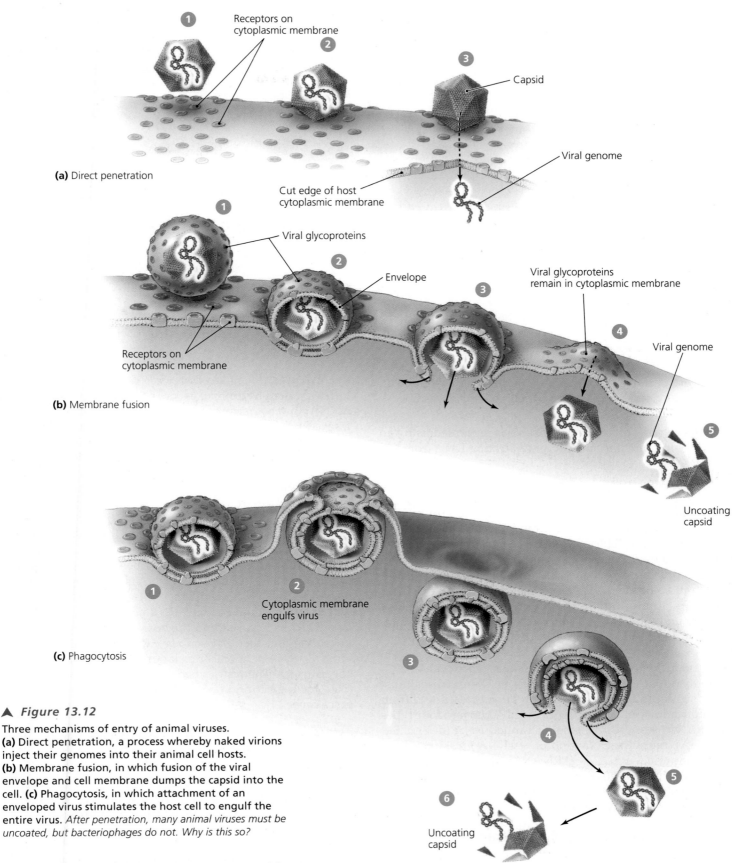

(a) Direct penetration

Receptors on cytoplasmic membrane

Capsid

Viral genome

Cut edge of host cytoplasmic membrane

(b) Membrane fusion

Viral glycoproteins

Envelope

Receptors on cytoplasmic membrane

Viral glycoproteins remain in cytoplasmic membrane

Viral genome

Uncoating capsid

(c) Phagocytosis

Cytoplasmic membrane engulfs virus

Uncoating capsid

▲ *Figure 13.12*
Three mechanisms of entry of animal viruses.
(a) Direct penetration, a process whereby naked virions inject their genomes into their animal cell hosts.
(b) Membrane fusion, in which fusion of the viral envelope and cell membrane dumps the capsid into the cell. **(c)** Phagocytosis, in which attachment of an enveloped virus stimulates the host cell to engulf the entire virus. *After penetration, many animal viruses must be uncoated, but bacteriophages do not. Why is this so?*

Figure 13.12 Generally bacteriophages inject their DNA during penetration, so the capsid does not enter the cell.

Table 13.3   **Synthesis Strategies of Animal Viruses**

Genome	How is mRNA Synthesized?	What Molecule is the Template for Genome Replication?
dsDNA	By RNA polymerase (in nucleus or cytoplasm of cell)	Each strand of DNA serves as template for its complement (except for hepatitis B, which synthesizes RNA to act as the template for new DNA)
ssDNA	By RNA polymerase (in nucleus of cell)	Complementary strand of DNA is synthesized to act as template
+ssRNA	Genome acts as mRNA	−RNA is synthesized to act as template
+ssRNA (*Retroviridae*)	DNA is synthesized from RNA by reverse transcriptase; mRNA is transcribed from DNA by RNA polymerase	DNA
−ssRNA	By RNA-dependent RNA transcriptase	+RNA (mRNA)
dsRNA	Positive strand of genome acts as mRNA	Each strand of genome acts as template for its complement

cytoplasm by host ribosomes, capsomeres enter the cell's nucleus, where new virions spontaneously assemble. This method of replication is seen with herpes and papilloma (wart) viruses.

There are two exceptions to this regimen of dsDNA viruses:

- Every part of a poxvirus is synthesized and assembled in the cytoplasm of the host's cell; the nucleus is not involved.
- The genome of hepatitis B viruses is replicated using an RNA intermediary instead of replicating DNA from DNA.

These two viruses are discussed further in Chapter 24.

ssDNA Viruses   The only significant human viruses with genomes composed of single-stranded DNA (ssDNA) are parvoviruses (see Table 13.2). When a parvovirus enters the nucleus of a host cell, host enzymes produce a new strand of DNA complementary to the viral genome. This complementary strand binds to the ssDNA of the virus to form a dsDNA molecule. Transcription of mRNA, replication of new ssDNA, and viral assembly then follow the DNA virus pattern just described.

As previously noted, RNA is not used as genetic material in cells, so it follows that the synthesis of RNA viruses must differ significantly from typical cellular processes, and from the replication of DNA viruses as well. There are four types of RNA viruses: positive single-stranded RNA (designated +ssRNA), retroviruses (a kind of +ssRNA virus), negative single-stranded RNA (−ssRNA), and double-stranded (dsRNA). The synthesis process for these RNA viruses is varied and sometimes rather complex. We start with the synthesis of +ssRNA viruses.

+ssRNA Viruses   Viral single-stranded RNA that can act directly as mRNA is called **positive strand RNA (+RNA).** An example of a +ssRNA virus is poliovirus. In many +ssRNA viruses, a complementary **negative strand RNA**

(−RNA) is transcribed from the +ssRNA genome by viral RNA polymerase; −RNA then serves as the template for the transcription of +ssRNA genomes. Such transcription of RNA from RNA is unique to viruses; no cell can transcribe RNA from RNA.

Retroviruses   Unlike other +ssRNA viruses, the +ssRNA viruses called **retroviruses** do not use their genome as mRNA. Instead, retroviruses use a DNA intermediary that is transcribed from +RNA by *reverse transcriptase,* an enzyme carried within the capsid. This DNA intermediary then serves as the template for the synthesis of additional +RNA molecules, which act both as mRNA and as genomes for new virions. Human immunodeficiency virus (HIV) is a prominent retrovirus.

−ssRNA Viruses   Other single-stranded RNA virions are −ssRNA viruses, which must overcome a unique problem. In order to synthesize a protein, a ribosome can only utilize mRNA (*i.e.,* +RNA), because −RNA is not recognized by ribosomes. The virus overcomes this problem by carrying within its capsid an enzyme, *RNA-dependent RNA transcriptase,* which is released into the host cell's cytoplasm during uncoating and then transcribes +RNA molecules from the virus's −RNA genome. Translation of proteins can then occur as usual. The transcribed +RNA also serves as a template for transcription of additional copies of −RNA. Diseases caused by −ssRNA viruses include rabies and flu.

dsRNA Viruses   Viruses that have double-stranded RNA use yet another method of synthesis. The positive strand of the molecule serves as mRNA for the translation of proteins, one of which is a different RNA polymerase that transcribes more dsRNA. Each strand of RNA acts as a template for transcription of its opposite, which is reminiscent of DNA replication in cells. Double-stranded RNA viruses cause most cases of diarrhea in infants.

The various strategies by which animal viruses are synthesized are summarized in Table 13.3.

Viral glycoproteins

Viral capsid

Cross section of host
cytoplasmic membrane

Budding of
enveloped virus

Enveloped virion

▲ *Figure 13.13*

The process of budding in enveloped viruses. *What term describes a nonenveloped virus?*

Figure 13.13 Naked.

CRITICAL THINKING

Although many +ssRNA viruses use their genome directly as mRNA, retroviruses do not. Instead, their +RNA is transcribed into DNA by reverse transcriptase. What advantage do retroviruses gain by using reverse transcriptase?

Assembly and Release of Animal Viruses

As with bacteriophages, once the components of animal viruses are synthesized, they are assembled into virions that are then released from the host cell. Most DNA viruses assemble in and are released from the nucleus into the cytosol, whereas most RNA viruses develop solely in the cytoplasm. The number of viruses produced and released depends on both the type of virus and the size and initial health of the host cell.

Replication of animal viruses takes more time than replication of bacteriophages. Herpesviruses, for example, require almost 24 hours to replicate, as compared to 25 minutes for bacteriophage T4.

Enveloped animal viruses are often released via a process called **budding (Figure 13.13).** As virions are assembled, they are extruded through one of the cell's membranes—the nuclear membrane, the endoplasmic reticulum, or the cytoplasmic membrane. Each virion acquires a portion of membrane, which becomes the viral envelope. During synthesis, some viral glycoproteins are inserted into cellular membranes, and these proteins become the glycoprotein spikes on the surface of the viral envelope.

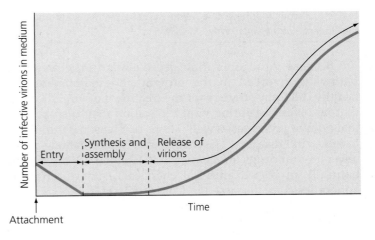

Number of infective virions in medium

Entry | Synthesis and assembly | Release of virions

Attachment

Time

▲ *Figure 13.14*

A generalized curve of virion abundance for persistent infections by budding enveloped viruses. Because the curve does not represent any actual infection, units for the graph's axes are omitted.

Because the host cell is not quickly lysed, as occurs in bacteriophage replication, budding allows an infected cell to remain alive for some time. Infections with enveloped viruses in which host cells shed viruses slowly and relatively steadily are called *persistent infections*; a curve showing virus abundance over time during a persistent infection lacks the burst of new virions seen in lytic replication cycles **(Figure 13.14).**

Table 13.4	A Comparison of Bacteriophage and Animal Virus Replication	
	Bacteriophage	**Animal Virus**
Attachment	Proteins on tails attach to proteins on cell wall	Spikes, capsids, or envelope proteins attach to proteins or glycoproteins on cell membrane
Penetration	Genome is injected into cell or diffuses into cell	Capsid enters cell by direct penetration, fusion, or phagocytosis
Uncoating	None	Removal of capsid by cell enzymes
Site of synthesis	In cytoplasm	RNA viruses in cytoplasm; most DNA viruses in nucleus
Site of assembly	In cytoplasm	RNA viruses in cytoplasm; most DNA viruses in nucleus
Mechanism of release	Lysis	Naked virions: lysis; enveloped virions: budding
Nature of chronic infection	Lysogeny, always incorporated into host chromosome, may leave host chromosome	Latency, with or without incorporation into host DNA, incorporation is permanent

Naked animal viruses may be released in one of two ways: They may either be extruded from the cell by exocytosis, in a manner similar to budding, but without the acquisition of an envelope, or they may cause lysis and death of the cell, reminiscent of bacteriophage release.

CRITICAL THINKING

If an enveloped virus were somehow released from a cell without budding, it would not have an envelope. What effect would this have on the virulence of the virus? Why?

Because viral replication uses cellular structures and pathways involved in the growth and maintenance of healthy cells, any strategy for the treatment of viral diseases that involves disrupting viral replication may disrupt normal cellular processes as well. This is one reason it is difficult to treat viral diseases. The modes of action of the few available antiviral drugs are discussed in Chapter 10; the body's naturally produced antiviral chemicals—interferons and antibodies—are discussed in Chapters 15 and 16.

Latency of Animal Viruses

Some animal viruses, including chickenpox and herpes viruses, may remain dormant in cells in a process known as **latency**; the viruses involved in latency are called **latent viruses** or **proviruses**. Latency may be prolonged for years with no viral activity, signs, or symptoms. Though latency is similar to lysogeny as seen with bacteriophages, there are differences. Some latent viruses do not become incorporated into the chromosomes of their host cells, whereas lysogenic phages always do.

On the other hand, some animal viruses (*e.g.,* HIV) are more like lysogenic phages in that they do become integrated into a host chromosome as a **provirus**. However, when a provirus is incorporated into its host DNA, the condition is permanent; induction never occurs, as it can in lysogeny. Thus an incorporated provirus becomes a perma-

nent, physical part of the host's chromosome, and all descendants of the infected cell will carry the provirus.

An important feature of HIV infections is that proviruses become a permanent part of the DNA of the host's white blood cells. But given that RNA cannot be incorporated directly into a DNA molecule, how does the ssRNA of HIV become a provirus incorporated into the DNA of its host cell? HIV can become a permanent part of a host's DNA because it, like all retroviruses, carries reverse transcriptase, which transcribes the genetic information of the +RNA molecule to a DNA molecule—which *can* become incorporated into the host cell's genome.

CRITICAL THINKING

A latent virus that is incorporated into a host cell's chromosome is never induced; that is, it never emerges from the host cell's chromosome to become a free virus. Given that it cannot emerge from the host cell's chromosome, can such a latent virus be considered "safe"? Why or why not?

A comparison of the features of the replication of bacteriophages and animal viruses is presented in **Table 13.4**. Next we turn our attention to the part viruses can play in cancer, beginning with a brief consideration of the terminology needed to understand the basic nature of cancer.

The Role of Viruses in Cancer

Learning Objectives

✓ Define the terms *neoplasia, tumor, benign, malignant, cancer,* and *metastasis.*

✓ Explain in simple terms how a cell may become cancerous, with special reference to the role of viruses.

Under normal conditions, the division of cells in a mature multicellular animal is under strict genetic control; that is, the animal's genes dictate that some types of cells can no longer divide at all, and that those that can divide are prevented from unlimited division. In this genetic control, either genes for cell division are "turned off," or genes that inhibit division are "turned on," or some combination of both these genetic events occur. However, if something upsets the genetic control, cells begin to divide uncontrollably. This phenomenon of uncontrolled cell division in a multicellular animal is called **neoplasia**[2] (nē-ō-plā′zē-ă). Cells undergoing neoplasia are said to be neoplastic, and a mass of neoplastic cells is a **tumor.**

Some tumors are **benign;** that is, they remain in one place and are not generally harmful, although occasionally such noninvasive tumors are painful and rob adjacent normal cells of space and nutrients. However, some tumors are **malignant;** they invade neighboring tissues and even travel throughout the body and invade other organs and tissues to produce new tumors—a process called **metastasis** (mĕ-tas′tă-sis). Malignant tumors are also called **cancers.** Cancers rob normal cells of space and nutrients and cause pain; in some kinds of cancer, malignant cells derange the function of the affected tissues, until eventually the body can no longer withstand the loss of normal function and dies.

Several theories have been proposed to explain the development of cancers, and the role viruses play. These theories revolve around the presence of *protooncogenes* (prō-tō-ong′kō-jēnz), genes that play a role in cell division. So long as the protooncogenes are repressed, no cancer results. However, activity of oncogenes (their name when they are active) or inactivation of oncogene repressors can cause cancer to develop. In most cases, several genetic changes must occur before cancer develops. Put another way, "multiple hits" to the genome must occur for cancer to result **(Figure 13.15).**

A variety of environmental factors contribute to the inhibition of oncogene repressors and the activation of oncogenes. Ultraviolet light, radiation, certain chemicals called *carcinogens* (kar-si′nō-jenz), and viruses have all been implicated in the development of cancer.

Viruses cause cancers in several ways. Some viruses carry copies of oncogenes as part of their genomes; other viruses stimulate oncogenes already present in the host; still other viruses interfere with normal tumor repression when they insert (as a provirus) into a repressor gene.

That viruses cause some animal cancers is well established. In the first decade of the 1900s, virologist F. Peyton Rous (1879–1970) proved that viruses induce cancer in chickens. Though several DNA and RNA viruses are known to cause about 15% of human cancers, the link between viruses and other human cancers has been more difficult to document. Among the virally induced cancers in humans are Burkitt's lymphoma, Hodgkin's disease, Kaposi's sarcoma, and cervical cancer. DNA viruses in the families *Adenoviridae, Herpesviridae, Hepadnaviridae, Papillomaviridae,* and *Polyomaviridae,* and two RNA viruses in the family *Retroviri-*

---

[2]From Greek *neo,* meaning new, and *plassein,* meaning to mold.

▲ *Figure 13.15*

The oncogene theory of the induction of cancer in humans. The theory suggests that more than one "hit" to the DNA (that is, any change or mutation), whether caused by a virus (as shown here) or various physical or chemical agents, is required to induce cancer.

dae, cause these and other human cancers. The prevention of infection and treatment of viral diseases, and the cancers caused by viruses, are discussed in Chapter 24 (DNA viruses) and Chapter 25 (RNA viruses).

CRITICAL THINKING

Why are DNA viruses more likely to cause neoplasias than are RNA viruses?

Culturing Viruses in the Laboratory

Learning Objectives

✓ Describe some ethical and practical difficulties to overcome in culturing viruses.
✓ Describe three types of media used for culturing viruses.

Viruses must be cultured in order to conduct research and develop vaccines and treatments, but because viruses cannot metabolize or replicate by themselves, they cannot be grown in standard microbiological broths or on agar plates. Instead, they must be cultured inside suitable host cells, a requirement that complicates the detection, identification, and characterization of viruses. Virologists have developed three types of media for culturing viruses: media consisting of whole organisms (bacteria, plants, or animals), embryonated (fertilized) eggs, and cell cultures. We begin by considering the culture of viruses in whole organisms.

Culturing Viruses in Whole Organisms

Learning Objective

✓ Explain the use of a plaque assay in culturing viruses in bacteria.

In the following subsections we consider the use of bacterial cells as a virus culture medium before considering the issues involved in growing viruses in living animals.

Culturing Viruses in Bacteria

Most of our knowledge of viral replication has been derived from research on bacteriophages, which are relatively easy to culture because bacteria are easily grown and maintained. Phages can be grown in bacteria maintained in either liquid cultures or on agar plates. In the latter case, bacteria and phages are mixed with warm (liquid) nutrient agar and poured in a thin layer across the surface of an agar plate. During incubation, bacteria infected by phages lyse and release new phages that infect nearby bacteria, while uninfected bacteria grow and reproduce normally. After incubation, the appearance of the plate includes a uniform bacterial lawn interrupted by clear zones called **plaques,** which are areas where phages have lysed the bacteria **(Figure 13.16).** Such plates enable the estimation of phage numbers via a technique called **plaque assay** (aš'sā), in which each plaque corresponds to a single phage in the original bacterium/virus mixture.

Culturing Viruses in Plants and Animals

Most experiments designed to study disease processes and immune responses, including those involving viruses, require the use of living organisms. Plant and animal viruses can be grown in laboratory plants and animals. Recall that the first discovery and isolation of a virus was the discovery of tobacco mosaic virus in tobacco plants. Rats, mice, guinea pigs, rabbits, pigs, and primates have been used to culture and study animal viruses.

However, maintaining laboratory animals can be difficult and expensive, and this practice raises ethical issues for some. Growing viruses that infect only humans raises additional ethical complications. Therefore, scientists have developed alternative ways of culturing animal and human viruses using fertilized chicken eggs or cell cultures.

Culturing Viruses in Embryonated Chicken Eggs

Chicken eggs are a useful culture medium for viruses because they are inexpensive, among the largest of cells, free of

Viral plaques

Bacterial lawn

▲ *Figure 13.16*

Viral plaques in a lawn of bacterial growth on the surface of an agar plate. *What is the cause of viral plaques?*

Figure 13.16 Each plaque is an area in a bacterial lawn where bacteria have succumbed to phage infections.

contaminating microbes, and contain a nourishing yolk (which makes them self-sufficient). Most suitable for culturing viruses are chicken eggs that have been fertilized and thus contain a developing embryo, the tissues of which (called membranes, which should not be confused with cellular membranes) provide ideal inoculation sites for growing viruses **(Figure 13.17).** Samples of virus can be injected into embryonated eggs at the sites that are best suited for the virus's maintenance and replication.

Vaccines against some viruses can also be prepared in egg cultures. You may have been asked if you are allergic to eggs before you received such a vaccine, because egg protein may remain as a contaminant in the vaccine.

Culturing Viruses in Cell (Tissue) Culture

Learning Objective

✓ Compare and contrast diploid cell culture and continuous cell culture.

Viruses can also be grown in **cell culture,** which consists of cells isolated from an organism and grown on the surface of a medium or in broth **(Figure 13.18).** Such cultures became practical when antibiotics provided a way to limit the growth of contaminating bacteria. Cell culture can be less expensive than maintaining research animals, plants, or eggs, and it avoids some of the moral problems associated with experiments performed on animals and humans. Cell cultures are sometimes called *tissue cultures,* but the term *cell culture* is more accurate because only a single type of cell is

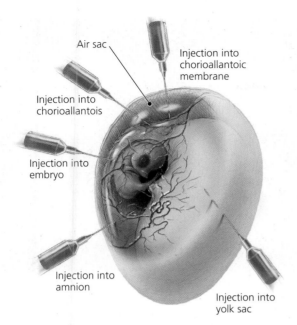

▲ *Figure 13.17*

Inoculation sites for the culture of viruses in embryonated chicken eggs. *Why are eggs often used to grow animal viruses?*

Figure 13.17 Eggs are large, sterile, self-sufficient cells that contain a number of different sites that are suitable for viral replication.

▲ *Figure 13.18*

An example of cell culture. The bag contains a colored nutrient medium for growing cells in which viruses can be cultured.

used in a culture. (By definition, a tissue is composed of at least two kinds of cells.)

Cell cultures are of two types. The first type, **diploid cell cultures,** are created from embryonic animal, plant, or human cells that have been isolated and provided appropriate growth conditions. The cells in diploid cell culture generally last no more than about 100 generations (cell divisions) before they die.

The second type of culture, **continuous cell cultures,** are more long lasting because they are derived from tumor cells. Recall that a characteristic of neoplastic cells is that they divide relentlessly, providing a never-ending supply of new cells. One of the more famous continuous cell cultures is of HeLa cells, derived from a woman named *Henrietta Lacks* who died of cervical cancer in 1951. Though she is dead, Mrs. Lacks's cells live on in laboratories throughout the world.

It is interesting that HeLa cells have lost some of their original characteristics. For example, they are no longer diploid because they have lost many chromosomes. HeLa cells provide a semi-standard[3] human tissue culture medium for studies on cell metabolism, aging, and (of course) viral infection.

CRITICAL THINKING

HIV replicates only in certain types of human cells, and one early problem in AIDS research was culturing those cells. How do you think scientists are now able to culture HIV?

Other Parasitic Particles: Viroids and Prions

Viruses are not the only submicroscopic entities capable of causing disorders within cells. In this section we will consider the characteristics of two molecular particles that infect cells: viroids and prions.

Characteristics of Viroids

Learning Objectives

✓ Define and describe viroids.
✓ Compare and contrast viroids and viruses.

Viroids are extremely small, circular pieces of RNA that are infectious and pathogenic in plants (**Figure 13.19** on page 398). Viroids are similar to an RNA virus except that they lack a capsid surrounding the nucleic acid. Even though they are circular, viroids may appear linear because of hydrogen bonding within the molecule. Several plant diseases, including some of chrysanthemum, potato, cucumber,

---

[3]HeLa cells are "semi-standard" because different strains have lost different chromosomes, and mutations have occurred over the years. Thus, HeLa cells in one laboratory may be slightly different from HeLa cells in another laboratory.

Genome of bacteriophage T7   PSTVd

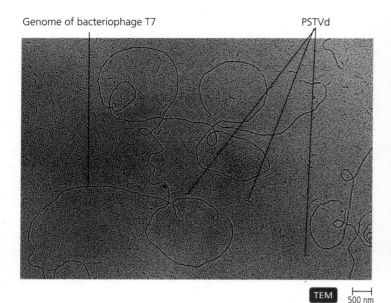

TEM   500 nm

▲ *Figure 13.19*

The RNA strand of the small potato spindle tuber viroid (PSTV). Also shown for comparison is the longer DNA genome of bacteriophage T7. *How are viroids similar to and different from viruses?*

Figure 13.19 Viroids are similar to certain viruses in that they are infectious and contain a single strand of RNA; they are different from viruses in that they lack a proteinaceous capsid.

coconut, and avocado, are caused by viroids, including the stunting shown in **Figure 13.20.**

Viroid-like agents—infectious, pathogenic RNA particles that lack capsids but do not infect plants—affect some fungi. (They are not called viroids because they do not infect plants.) No animal diseases are known to be caused by viroid-like molecules, though the possibility exists that they may be responsible for some diseases in humans.

Characteristics of Prions

Learning Objectives

✓ Define and describe prions, including their replication process.
✓ Compare and contrast prions and viruses.
✓ List four diseases caused by prions.

In 1982, Stanley Prusiner discovered a proteinaceous infectious agent that was different from any other known infectious agent in that it lacked nucleic acid. Prusiner named these new agents of disease **prions** (prē′onz), for *proteinaceous infective* particles. Before his discovery, the diseases now known to be caused by prions were thought to be caused by what were known as "slow viruses," which were so named because of the long delay between infection and the onset of signs and symptoms. Through convincing ex-

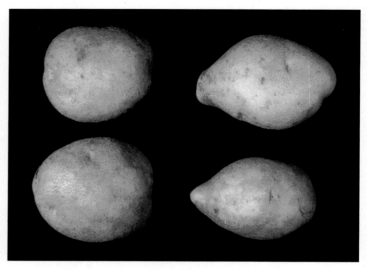

▲ *Figure 13.20*

One effect of viroids on plants. The potatoes at right are stunted as the result of infection with PSTV viroids.

periments, Prusiner and his colleagues showed that prions are not viruses because they lack any nucleic acid, but their work was not without controversy.

For years some scientists resisted the concept of prions because particles that lack any nucleic acid violate the "universal" rule of protein synthesis—that proteins are translated from a molecule of mRNA, which is a copy of a portion of a DNA molecule. So, argued these detractors, if Prusiner's "infectious proteins" lack nucleic acid, how can they carry the information required to replicate themselves?

Prusiner ascertained that prions are composed of a single protein, called PrP, and that all mammals contain a gene that codes for the primary sequence of amino acids in PrP. Moreover, he determined that the sequence of amino acids in PrP is such that the protein can fold into at least two stable tertiary structures: a normal, functional structure that folds with several α-helices, called *cellular PrP*, and the disease-causing form having β-pleated sheets, called *prion PrP* **(Figure 13.21).**

Prusiner determined that prion PrP is able to convert normal cellular PrP into prion PrP by inducing a conformational change in the shape of cellular PrP. In other words, cellular PrP exposed to prion PrP refolds to form prion PrP. As cellular ribosomes synthesize more cellular PrP, the prions already present cause these new molecules to refold, becoming more prions. Thus, the rule of protein synthesis is maintained, because prions do not code for new prions, but instead convert extant cellular PrP into prions. For his work, Prusiner was awarded the Nobel Prize in 1997.

Why don't prions develop in all mammals, given that all mammals have PrP? Under normal circumstances, it appears that other nearby proteins and polysaccharides force PrP into the correct (cellular) shape. Only excess PrP production or mutations in the PrP gene result in the initial

(a)   **(b)**

▲ *Figure 13.21*

The two stable, three-dimensional forms of prion protein (PrP). **(a)** Normal functional PrP, which has a preponderance of alpha helices. **(b)** Prion PrP, which has the same amino acid sequence but is folded to produce a preponderance of beta-pleated sheets.

Vacuole

LM   25 nm

▲ *Figure 13.22*

A brain showing the large vacuoles and spongy appearance typical in prion-induced diseases. Shown here is the brain of a sheep with the prion disease called scrapie.

formation of prion PrP. For instance, it has been determined that human cellular PrP can misfold only if it contains methionine as the 129th amino acid. About 40% of humans have this type of cellular PrP and are thus susceptible to prion disease.

All known prion diseases involve fatal neurological degeneration, the deposition of fibrils in the brain, and the loss of brain matter such that eventually large vacuoles form. The latter process results in a characteristic spongy appearance **(Figure 13.22),** which is why as a group, prion diseases are called *spongiform encephalopathies* (spŭn'ji-fōrm en-sef'ă-lop'ă-thēz). The way in which prion PrP causes these changes in the brain is unknown.

Prions are known to cause several spongiform encephalopathies of animals and of humans, including *bovine spongiform encephalitis* (*BSE,* so-called "mad cow disease"), *scrapie* in sheep, and *Creutzfeldt-Jakob disease (CJD)* in humans. **Highlight 13.2** on page 400 describes *kuru* (koo'roo), a spongiform encephalopathy that has been eliminated. All these diseases are transmitted through either the ingestion of infected tissue, transplants of infected tissue, or contact of mucous membranes or skin abrasions with infected tissue.

Prions are not destroyed by normal cooking or sterilization, though they are destroyed by incineration. There is no treatment for any prion disease, though scientists have determined that the antimalarial drug quinacrine and the an-

tipsychotic drug chlorpromazine forestall spongiform encephalopathy in prion-infected mice. Human trials of these drugs are ongoing.

PrP proteins of different species are slightly different in primary structure, and at one time it was thought unlikely that prions of one species could cross over into another species to cause disease. However, an epidemic of BSE in Great Britain in the late 1980s and in France in the 2000s resulted in the spread of prions not only to other cattle, but also to humans who ingested infected beef. To prevent further infection, the European Union has banned the use of animal-derived protein supplements in animal feed. Unfortunately, this step was too late to prevent the spread of bovine prions to more than 100 people who have developed fatal *variant Creutzfeldt-Jakob disease.*[4]

Prions composed of different proteins may lie behind other muscular and neuronal degenerative diseases, such as Alzheimer's disease, Parkinson's disease, and amyotrophic lateral sclerosis (ALS, or "Lou Gehrig's disease").

CRITICAL THINKING

Why did most scientists initially resist the idea of an infectious, self-replicating protein?

In this chapter we have seen that humans, animals, plants, fungi, and bacteria are susceptible to infection by acellular pathogens: viruses, viroids, and prions. **Table 13.5** on page 400 summarizes the differences and similarities between these pathogenic agents and bacterial pathogens.

---

[4]"Variant" because it is derived from BSE prions in cattle, as opposed to the regular form of CJD, which is a genetic disease of humans.

Highlight 13.2   Kuru and Cannibalism in the Fore Tribe of Papua New Guinea

Between 1957 and 1968, over 1100 members of the Fore tribe in the highlands of Papua New Guinea died from a frightening disease. This condition, called kuru—the Fore word for trembling or shivering—resulted in tremors, loss of the ability to stand or walk, visual impairment, mental deterioration, and eventually death. Initially the cause of kuru was a mystery to researchers, who were intrigued by the observation that kuru affected women and children at significantly higher rates than men. Subsequent research has since attributed kuru to prions acquired when tribe members handled and ate the brains of deceased relatives during mourning rituals. The high rates in women and children afflicted with the disease are consistent with this finding, as they were the primary participants in cannibalistic mourning rituals, in which men rarely participated.

Due to the influence of missionaries and government decree, cannibalism in Papua New Guinea ceased in the late 1950s, and cases of kuru have since vanished as a result. Elsewhere, however, other spongiform encephalopathies remain—including bovine spongiform encephalitis ("mad-cow disease") and Creutzfeldt-Jakob disease in humans.

Table 13.5   Comparison of Viruses, Viroids, and Prions to Bacterial Cells

	Bacteria	Viruses	Viroids	Prions
Width	200–2000 nm	10–300 nm	2 nm	5 nm
Length	200–550,000 nm	20–800 nm	40–130 nm	5 nm
Nucleic acid?	Both DNA and RNA	Either DNA or RNA, never both	RNA only	None
Protein?	Present	Present	Absent	Present (PrP)
Cellular?	Yes	No	No	No
Cytoplasmic membrane?	Present	Absent (though some viruses do have a membranous envelope)	Absent	Absent
Ribosomes?	Present	Absent (with one exception)	Absent	Absent
Growth?	Present	Absent	Absent	Absent
Self-replicating?	Yes	No	No	Yes; transform PrP protein already present in cell
Responsiveness?	Present	Some bacteriophages respond to a host cell by injecting their genomes	Absent	Absent
Metabolism?	Present	Absent	Absent	Absent

Are Viruses Alive?

Now that we have studied the characteristics and replication processes of viruses, let's return to a question first posed in Chapter 3: Are viruses alive?

To be able to wrestle with the answer, we must first recall the characteristics of life: growth, self-reproduction, responsiveness, and the ability to metabolize, all within structures called cells. According to these criteria, viruses seem to lack the qualities of living things, prompting some scientists to consider them nothing more than complex pathogenic chemicals. For other scientists, however, at least three observations—that viruses use sophisticated methods to invade cells, have the means of taking control of their host cells, and possess genomes containing instructions for replicating themselves—indicate that viruses are the ultimate parasites, because they use cells to make more viruses. According to this viewpoint, viruses are the least complex living entities.

In any case, viruses are right on the threshold of life—outside of cells they do not appear to be alive, but within cells they direct the synthesis and assembly required to make copies of themselves.

What do you think? Are viruses alive?

CHAPTER SUMMARY

1. **Acellular** disease-causing agents, which lack cell structure and cannot metabolize, grow, reproduce, or respond to their environment, include viruses, viroids, and prions.

Characteristics of Viruses (pp. 378–384)

1. A **virus** is a tiny infectious agent with nucleic acid surrounded by proteinaceous **capsomeres** that form a coat called a **capsid.**

A virus exists in an extracellular state and an intracellular state. A **virion** is a complete viral particle, including a nucleic acid and a capsid, outside of a cell.

2. The genomes of viruses includes either DNA or RNA, but never both. Viral genomes may be dsDNA, ssDNA, dsRNA, or ssRNA. They may exist as linear or circular and singular or multiple molecules of nucleic acid, depending on the type of virus.

3. A **bacteriophage** (or **phage**) is a virus that infects a bacterial cell.

4. Virions can have a membranous **envelope** or be naked—that is, without an envelope.

Classification of Viruses (p. 384)

1. Viruses are classified based on type of nucleic acid, presence of an envelope, shape, and size.

2. The International Committee on Taxonomy of Viruses (ICTV) has recognized viral family and genus names. With the exception of three orders, higher taxa are not established.

Viral Replication (pp. 384–394)

1. Viruses depend on random contact with a specific host cell type for replication. Typically a virus in a cell proceeds with a **lytic replication cycle** with five stages: **attachment, entry, synthesis, assembly**, and **release.**

2. Once attachment has been made between virion and host cell, the nucleic acid enters the cell. With phages, only the nucleic acid enters the host cell. With animal viruses, the entire virion often enters the cell, where the capsid is then removed in a process called **uncoating.**

3. Within the host cell, the viral nucleic acid directs synthesis of more viruses using metabolic enzymes and ribosomes of the host cell.

4. Assembly of synthesized virions occurs in the host cell, typically as capsomeres surround replicated or transcribed nucleic acids to form new virions.

5. Virions are released from the host cell either by lysis of the host cell (seen with phages and animal viruses) or by the extrusion of enveloped virions through the host's cell membrane (called **budding**), a process seen only with certain animal viruses. If budding continues over time, the infection is a persistent infection. An envelope is derived from a cell membrane.

6. **Temperate (lysogenic)** bacteriophages enter a bacterial cell and remain inactive, in a process called **lysogeny** or a **lysogenic cycle.** Such inactive phages are called **prophages** and are inserted into the chromosome of the cell and passed to its daughter cells. At some point in the generations that follow, a prophage may be excised from the chromosome in a process known as **induction.** At that point the prophage again becomes a lytic virus.

7. In a similar process called **latency,** an animal virus remains inactive in a cell, possibly for years. A **latent** virus is also known as a **provirus.** A provirus that has become incorporated into a host's chromosome remains there.

8. dsDNA viruses act like cellular DNA in transcription and replication.

9. Some ssRNA viruses have **positive strand RNA (+RNA),** which can be directly translated by ribosomes to synthesize protein. From the +RNA, complementary **negative strand RNA (−RNA)** is transcribed to serve as a template for more +RNA.

10. **Retroviruses,** such as HIV, are +ssRNA viruses that carry an enzyme, *reverse transcriptase,* to transcribe DNA from their RNA. This reverse process (DNA transcribed from RNA) is reflected in the name *retrovirus.*

11. −ssRNA viruses carry an RNA-dependent RNA transcriptase for transcribing mRNA from the −RNA genome so that protein can then be translated. Transcription of RNA from RNA is not found in cells.

12. When **double-stranded RNA (dsRNA)** functions as the genome, one strand of the RNA molecule functions as the genome, and the other strand functions as a template for RNA replication.

The Role of Viruses in Cancer (pp. 394–395)

1. **Neoplasia** is uncontrolled cellular reproduction in a multicellular animal. A mass of neoplastic cells, called a **tumor,** may be relatively harmless **(benign)** or invasive **(malignant).** Malignant tumors are also called **cancer.** Metastasis describes the spreading of malignant tumors. Environmental factors or oncogenic viruses may cause neoplasia.

Culturing Viruses in the Laboratory (pp. 395–397)

1. In the laboratory, viruses must be cultured inside whole organisms, in embryonated chicken eggs, or in cell cultures because they cannot metabolize or replicate alone.

2. When a mixture of bacteria and phages is grown on an agar plate, bacteria infected with phages lyse, producing clear areas called **plaques** on the bacterial lawn. A technique called **plaque assay** enables the estimation of phage numbers.

3. Viruses can be grown in two types of **cell cultures.** Whereas **diploid cell cultures** last about 100 generations, **continuous cell cultures,** derived from cancer cells, last longer.

Other Parasitic Particles: Viroids and Prions (pp. 397–400)

1. **Viroids** are small circular pieces of RNA with no capsid that infect and cause disease in plants. Similar pathogenic RNA molecules are known from fungi.

2. **Prions** are infectious protein particles that lack nucleic acids and replicate by converting similar normal proteins into new prions. Diseases caused by prions are spongiform encephalopathies, which involve fatal neurological degeneration.

Are Viruses Alive? (p. 400)

1. Outside of cells, viruses do not appear to be alive, but within cells, they exhibit lifelike qualities such as the ability to replicate themselves.

QUESTIONS FOR REVIEW   (Answers to multiple choice and matching are on the web, along with additional review questions. Visit www.microbiologyplace.com.)

Multiple Choice

1. Which of the following is *not* an acellular agent?
    a. viroid
    b. virus
    c. rickettsia
    d. prion

2. Which of the following statements is true?
    a. Viruses contain both DNA and RNA in specific proportions.
    b. Viruses are capable of metabolism.
    c. Viruses lack a cell membrane.
    d. Viruses grow in response to their environmental conditions.

3. A virus that is specific for a bacterial host is called a
    a. phage.
    b. prion.
    c. virion.
    d. viroid.

4. A naked virus
    a. has no membranous envelope.
    b. has injected its DNA or RNA into the host cell.
    c. is devoid of capsomeres.
    d. is one that is unattached to a host cell.

5. Which of the following statements is *false*?
    a. Viruses may have circular DNA.
    b. ssDNA and dsRNA are found in bacteria more often than in viruses.
    c. Viral DNA may be linear.
    d. Viruses may have DNA or RNA, but not both.

6. When a cell is infected with an enveloped virus and sheds viruses slowly over time, this infection is
    a. called a lytic infection.
    b. a prophage cycle.
    c. called a persistent infection.
    d. caused by a quiescent virus.

7. Another name for a complete virus is
    a. virion.
    b. viroid.
    c. prion.
    d. capsid.

8. Which of the following viruses can be latent?
    a. HIV
    b. chicken pox
    c. herpes
    d. all of the above

9. Which of the following is *not* a criterion for specific family classification of viruses?
    a. the type of nucleic acid present
    b. envelope structure
    c. capsid type
    d. lipid composition

Matching

Place the letter of the appropriate description in the correct blank in the left column.

___ 1. uncoating   A. dormant virus in a eukaryotic cell

___ 2. prophage   B. a virus that infects a bacterium

___ 3. retrovirus   C. DNA transcribed from RNA

___ 4. bacteriophage D. protein coat of virus

___ 5. capsid   E. a membrane outside a viral capsid

___ 6. envelope   F. complete viral particle

___ 7. virion   G. inactive virus within bacterial cell

___ 8. provirus   H. removal of capsomeres from a virion

___ 9. benign tumor   I. invasive neoplastic cells

                     J. harmless neoplastic cells

Short Answer

1. Compare and contrast a bacterium and a virus by writing either "Present" or "Absent" for each of the following structures.

Structure	Bacterium	Virus
Cell membrane	_____	_____
Ribosomes	_____	_____
Cytoplasm	_____	_____
Nucleic acid	_____	_____
Nuclear membrane	_____	_____

2. Describe the five phases of a generalized lytic replication cycle.

3. Why is it difficult to treat viral infections?

4. Describe four different ways that viral nucleic acid can enter a host cell.

5. Contrast lysis and budding as means of release of virions from a host cell.

6. What is the difference between a virion and a virus particle?

7. How is a provirus like a prophage? How is it different?

8. Describe lysogeny.

CRITICAL THINKING

1. Many viruses have a single-stranded genome, but it has been observed that larger viruses usually have a double-stranded genome. What reasonable explanation can you offer for this observation?

2. What are the advantages and disadvantages to bacteriophages of the lytic and lysogenic reproductive strategies?

3. How are computer viruses similar to biological viruses? Are computer viruses alive? Why or why not?

4. Compare and contrast lysogeny by a prophage and latency by a provirus.

CHAPTER 14

Infection, Infectious Diseases, and Epidemiology

CHAPTER OUTLINE

Symbiotic Relationships Between Microbes and Their Hosts
Types of Symbiosis
Normal Microbiota in Hosts
How Normal Microbiota Become Opportunistic Pathogens

The Movement of Microbes into Hosts: Infection
Exposure to Microbes: Contamination and Infection
Portals of Entry
The Role of Adhesion in Infection

The Nature of Infectious Disease
Manifestations of Disease: Symptoms, Signs, and Syndromes
Causation of Disease: Etiology
Virulence Factors of Infectious Agents
The Stages of Infectious Diseases

The Movement of Pathogens Out of Hosts: Portals of Exit

Sources of Infectious Diseases in Humans
Animal Reservoirs
Human Carriers
Nonliving Reservoirs

Modes of Infectious Disease Transmission
Contact Transmission
Vehicle Transmission
Vector Transmission

Classification of Infectious Diseases

Epidemiology of Infectious Diseases
Frequency of Disease
Epidemiological Studies
Hospital Epidemiology: Nosocomial Infections
Epidemiology and Public Health

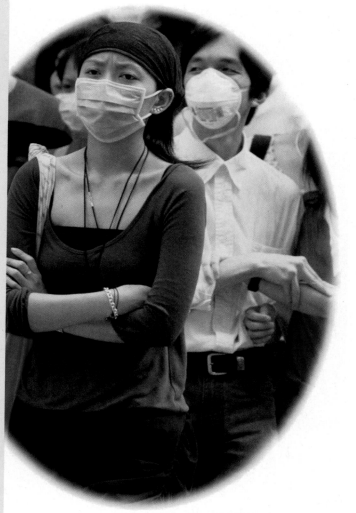

Fearing SARS, residents of Hong Kong don face masks.

During the winter of 2002–2003, a mysterious new viral illness began appearing in China's Guangdong province. Characterized by fever, shortness of breath, and atypical pneumonia, the highly infectious illness spread rapidly, particularly among food handlers and hospital workers tending to the already stricken. Initially confined to mainland China, the disease is believed to have begun its worldwide spread in February 2003, when a physician who had contracted the disease in Guangdong traveled to Hong Kong. During his stay in Hong Kong's Metropole Hotel, the physician passed the disease on to at least twelve other hotel guests, including travelers who continued on to Canada, Singapore, and Vietnam, unknowingly advancing the disease's spread to these countries. By March 2003, the disease had a name—severe acute respiratory syndrome (SARS)—and the World Health Organization (WHO) issued a global alert, triggering quarantines, travel cancellations, and other measures intended to curb the spread of this frightening new, sometimes-fatal disease. (See Chapter 25 for more information on SARS and the virus that causes it.)

How do microorganisms infect and cause disease in people? And how do experts investigate the transmission and spread of infectious diseases in populations? Read on.

MicroPrep Pre-Test: *Take the pre-test for this chapter on the web.*
Visit **www.microbiologyplace.com**.

In previous chapters we have discussed characteristics of microorganisms and their metabolism. We examined ways microbes can be controlled, viewed, identified, and used for our benefit. Now it is time to examine how microbes directly affect our bodies. In this discussion, continually bear in mind that most microorganisms are neither harmful nor expressly beneficial to humans—they live their lives, and we live ours. Relatively few microbes either benefit us or harm us.

In this chapter we will first examine the general types of relationships that microbes can have with the bodies of their hosts. Then we will discuss how microbes enter and attach to their hosts; the nature of microbial diseases, called *infectious diseases;* and how microbes leave hosts to become available to enter new hosts. Next we will explore the sources of infectious diseases for humans, and the ways that infectious diseases are spread among hosts. Finally, we will consider *epidemiology,* the study of the occurrence and spread of diseases within groups of humans, and the methods by which we can limit the spread of pathogens within society.

Symbiotic Relationships Between Microbes and Their Hosts

Symbiosis (sim-bī-o′sis) means "to live together." Each of us has symbiotic relationships with countless microorganisms, although often we are completely unaware of them. The 10 trillion or so cells in your body provide homes for more than 100 trillion bacteria and fungi and a huge number of viruses. We begin this section by considering the types of relationships between microbes and their hosts.

Types of Symbiosis

Learning Objectives

✓ Distinguish among the types of symbiosis, listing them in order from most beneficial to most harmful to the host.

✓ Describe the relationships among the terms parasite, host, and pathogen.

Scientists recognize three primary types of symbiosis: mutualism, commensalism, and parasitism.

Mutualism

One common type of symbiosis is **mutualism** (mū′tū-ăl-izm). In a mutualistic relationship between two organisms, both members benefit from their interaction. For example, bacteria in your colon receive a warm, moist, nutrient-rich environment in which to thrive, while you absorb vitamins released from the bacteria when they die. In this example, the relationship is beneficial to microbe and human alike but is not required by either. The bacteria could live elsewhere, and you could get vitamins from your diet.

Some mutualistic relationships provide such important benefits that one or both of the parties cannot live without

▲ *Figure 14.1*

Wood-eating termites in the genus *Reticulitermes* (shown above) cannot digest the cellulose in wood, but the protozoan *Trichonympha,* which lives in their intestines, can digest cellulose. The two organisms maintain a mutualistic relationship that is crucial for the life of the termite. *What benefits accrue to each of the symbionts in this mutalistic relationship?*

Figure 14.1 The termite gets digested cellulose from the protozoa, which get a constant supply of wood pulp to digest.

the other. For instance, termites, which cannot digest the cellulose in wood by themselves, would die without a mutualistic relationship with colonies of wood-digesting protozoa living in their intestines **(Figure 14.1)**. When the protozoa break down the wood pulp within the termite intestines, the termites share in the nutrients released from the wood. **Highlight 14.1** on p. 406 describes another relationship that benefits two diverse organisms.

Commensalism

In **commensalism** (kŏ-men′săl-izm), a second type of symbiosis, one member of the relationship benefits without significantly affecting the other. For example, **Staphylococcus** (staf′i-lō-kok′us) growing on the skin typically causes no measureable harm to a person, though an absolute example of commensalism is difficult to prove because the host may experience some unobserved benefits.

Parasitism

Of concern to health care professionals is a third type of symbiosis called **parasitism**[1] (par′ă-si-tizm). A **parasite** derives

---

[1] From Greek *parasitos,* meaning one who eats at the table of another.

Highlight 14.1   A Tale of Two Organisms

An unusual relationship exists between the marine unsegmented roundworm (nematode) *Eubostrichus parasitiferus,* and the rod-shaped, aerobic, sulfur-oxidizing bacteria that colonize its surface. The bacteria attach themselves to the nematode and completely cover the outside of the worm, producing a rope-like appearance.

This relationship is mutualistic because both symbionts benefit: The worm's mobility moves the bacteria through the marine sediments that contain their nutrients, while the worm in effect cultivates its own food supply by providing a habitat for the bacteria, its main (perhaps only) source of food.

Eubostrichus parasitiferus covered by bacteria

Table 14.1   The Three Types of Symbiotic Relationships

	Organism 1	Organism 2	Example
Mutualism	Benefits	Benefits	Bacteria in human colon
Commensalism	Benefits	Neither benefits nor is harmed	Tapeworm in human intestine
Parasitism	Benefits	Is harmed	Tuberculosis bacteria in human lung

benefit from its **host** while harming it, though some hosts sustain only slight damage. In the most severe cases, a parasite kills its host, in the process destroying its own home. Any parasite that causes disease is called a **pathogen** (path′ō-jen).

A variety of protozoa, fungi, bacteria, and viruses are microscopic parasites of humans. Larger parasites of humans include parasitic worms and biting arthropods,[2] including mites (chiggers), ticks, mosquitos, fleas, and flies. Biting arthropods are discussed as transmitters of disease later in this chapter; we will discuss parasitic worms in Chapter 23.

The three types of symbiosis are summarized in **Table 14.1.**

CRITICAL THINKING

Corals are small colonial animals that feed by filtering small microbes from seawater in tropical oceans worldwide. Biologists have discovered that the cells of most (if not all) corals are hosts to microscopic algae called zooxanthellae. Design an experiment to ascertain whether corals and zooxanthellae coexist in a mutual, commensal, or parasitic relationship.

Normal Microbiota in Hosts

Learning Objective

✓ Describe the normal microbiota, including resident and transient members.

Even though many parts of your body are *axenic*[3] environments—that is, sites that are free of any microbes—other parts of your body are home to millions of mutualistic and commensal symbionts. Each square centimeter of your skin, for example, contains about 3 million bacteria. The organisms that colonize the surfaces of the body without normally causing disease constitute the body's **normal microbiota (Figure 14.2),** also sometimes called the *normal flora*[4] or the *indigenous microbiota.* The normal microbiota are of two main types: resident microbiota and transient microbiota.

Resident Microbiota

Resident microbiota remain a part of the normal microbiota of a person throughout life. These organisms are found on the skin and on the mucous membranes of the digestive tract, upper respiratory tract, distal[5] portion of the urethra,[6] and vagina. **Table 14.2** on page 408 illustrates these environments and lists many of the resident bacteria, fungi, and protozoa that live there. Most of the resident microbiota are commensal; that is, they feed on excreted cellular wastes and dead cells without causing harm.

Transient Microbiota

Transient microbiota remain in the body for only a few hours, days, or months before disappearing. They are found in the same locations as the resident members of the normal mi-

---

[2]From Greek *arthros,* meaning jointed, and *pod,* meaning foot.
[3]From Greek *a,* meaning no, and *xenos,* meaning foreigner.
[4]This usage of the term *flora,* which literally means plants, derives from the fact that bacteria and fungi were considered plants in early taxonomic schemes. *Microbiota* is preferable because it properly applies to all microbes (whether eukaryotic or prokaryotic) and to viruses.
[5]*Distal* refers to the end of some structure, as opposed to *proximal,* which refers to the beginning of a structure.
[6]The tube that empties the bladder.

(a)

SEM | 4 μm

(b)

▲ *Figure 14.2*

Two examples of the normal microbiota of humans.
(a) Bacteria on the surface of the skin. **(b)** Bacteria in the colon. *What types of symbiotic relationships do the bacteria in parts (a) and (b) have with their host?*

Figure 14.2 Bacteria on the surface of the skin are commensals, though the host may derive some benefit in that the normal microbiota inhibit the growth of pathogens; the bacteria in the colon are mutualistic with the host.

crobiota but cannot persist because of competition from other microorganisms, elimination by the body's defense cells, or chemical and physical changes in the body that dislodge them.

Acquisition of Normal Microbiota

You developed in your mother's womb without a normal microbiota because you had surrounded yourself with an amniotic membrane and fluid (bag of waters), which generally kept microorganisms at bay, and because your mother's uterus was essentially an axenic environment.

Your normal microbiota began to develop when your surrounding amniotic membrane ruptured and microorganisms came in contact with you during birth. Microbes entered your mouth and nose as you passed through the birth canal, and your first breath was loaded with microorganisms that quickly established themselves in your upper respiratory tract. Your first meals provided the progenitors of resident microbiota for your colon, while *Staphylococcus epidermidis* (ep-i-der'mī-dis) and other microbes transferred from the skin of both the medical staff and your parents began to colonize your skin. Although you continue to add to your transient microbiota, most of the resident microbiota was initially established during your first months of life.

How Normal Microbiota Become Opportunistic Pathogens

Learning Objective

✓ Describe three conditions that create opportunities for normal microbiota to cause disease.

Under ordinary circumstances, normal microbiota do not cause disease. However, these same microbes may become harmful if an opportunity to do so arises. In this case, normal microbiota become **opportunistic pathogens** or *opportunists.* Conditions that create opportunities for pathogens include the following:

1. *Immune suppression.* Anything that suppresses the body's immune system—including disease, malnutrition, emotional or physical stress, extremes of age (either very young or very old), the use of radiation or chemotherapy to combat cancer, or the use of immunosuppressive drugs in transplant patients—can enable normal microbiota to become opportunistic pathogens. AIDS patients often die from opportunistic infections that are typically controlled by a healthy immune system because HIV infection suppresses immune system function.

2. *Changes in the normal microbiota.* Normal microbiota use nutrients, take up space, and release toxic waste products, all of which make it less likely that arriving pathogens can compete well enough to become established and produce disease. This situation is known as **microbial antagonism** or **microbial competition.** However, changes in the relative abundance of normal microbiota, for whatever reason, may allow a member of the normal microbiota to become an opportunistic pathogen and thrive. For example, when a woman must undergo long-term antimicrobial treatment for a bacterial blood infection, the antimicrobial may also kill normal bacterial microbiota in the vagina. In the absence of competition from bacteria, *Candida albicans* (kan'did-ă al-bi'kanz), a yeast and also a member of the normal vaginal microbiota, grows prolifically, producing an opportunistic vaginal yeast infection. Other conditions

Table 14.2 Resident Microbiota[a]

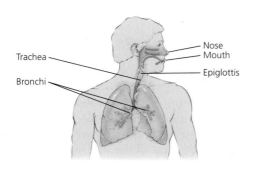

Upper Respiratory Tract

Genera

Staphylococcus, Streptococcus, Moraxella, Haemophilus, Lactobacillus, Veillonella, Fusobacterium, Candida (fungus)

Notes

The nose is cooler than the rest of the respiratory system and has some unique microbiota. The trachea and bronchi have a sparse microbiota compared to the nose and mouth. The alveoli of the lungs, which are too small to see at this magnification, have no natural microbiota.

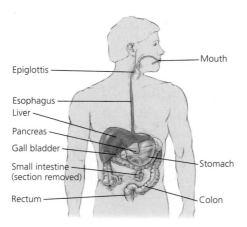

Upper Digestive Tract

Genera

Lactobacillus, Haemophilus, Actinomyces, Bacteroides, Treponema, Neisseria, Corynebacterium, Entamoeba (protozoan), *Trichomonas* (protozoan)

Notes

Microbes colonize surfaces of teeth, gingiva, lining of cheeks, and pharynx, and are found in saliva in large numbers. Dozens of species have never been identified.

Lower Digestive Tract

Genera

Bacteroides, Fusobacterium, Escherichia, Lactobacillus, Clostridium, Bifidobacterium, Enterococcus, Proteus, Shigella, Candida (fungus), *Entamoeba* (protozoan), *Trichomonas* (protozoan)

Notes

Bacteria are mostly strict anaerobes, though some facultative anaerobes are also resident.

Female Urinary and Reproductive Systems

Genera

Lactobacillus, Streptococcus, Staphylococcus, Bacteroides, Clostridium, Candida (fungus), *Trichomonas* (protozoan)

Notes

Microbiota change as acidity in the vagina changes during menstrual cycle. The flow of urine prevents extensive colonization of the urethra.

Male Urinary and Reproductive Systems

Genera

Staphylococcus, Streptococcus, Mycobacterium, Bacteroides, Fusobacterium, Peptostreptococcus

Notes

The flow of urine prevents extensive colonization of the urethra.

Eyes and Skin

Genera

Skin: *Propionibacterium, Staphylococcus, Corynebacterium, Micrococcus, Malassezia* (fungus), *Candida* (fungus)

Conjunctiva: *Staphylococcus*

Notes

Microbiota live on the outer, dead layers of the skin and in hair follicles and pores of glands. The deeper layers (dermis and hypodermis) are axenic.

Tears wash most microbiota from the eyes, so there are few compared to the skin.

[a]Genera are bacteria unless noted.

that can disrupt the normal microbiota are hormonal changes, stress, changes in diet, and exposure to overwhelming numbers of pathogens.

3. *Introduction of a member of the normal microbiota into an unusual site in the body.* As we have seen, normal microbiota are present only in certain body sites, and each site has only certain species of microbiota. If a member of the normal microbiota in one site is introduced into a site it normally does not inhabit, the organism may become an opportunistic pathogen. In the colon, for example, *Escherichia coli* (esh-ĕ-rik′ē-ă kō′lī) is mutualistic, but should it enter the urethra it becomes an opportunist that can produce an infection.

Now that we have considered the relationships of the normal microbiota to their hosts, we turn our attention to the movement of microbes into new hosts.

The Movement of Microbes into Hosts: Infection

In this section we examine the events that occur when hosts are exposed to microbes, the sites at which microbes can gain entry into hosts, and how entering microbes become established in new hosts.

Exposure to Microbes: Contamination and Infection

Learning Objective

✓ Describe the relationship between contamination and infection.

In the context of the interaction between microbes and their hosts, **contamination** refers to the mere presence of microbes in or on the body. Some microbial contaminants reach the body in food, drink, or the air, whereas others are introduced via wounds, biting arthropods, or sexual intercourse. Several outcomes of contamination by microbes are possible. Some microbial contaminants remain where they first contacted the body (such as the skin or mucous membranes) without causing harm and subsequently become part of the resident microbiota; other microbial contaminants remain on the body for only a short time, as part of the transient microbiota. Still others overcome the body's external defenses, multiply, and become established in the body; such a successful invasion of the body by a pathogen is called an **infection.**

Next we consider in greater detail the various sites through which pathogens gain entry into the body.

Portals of Entry

Learning Objective

✓ Identify and describe the portals through which pathogens invade the body.

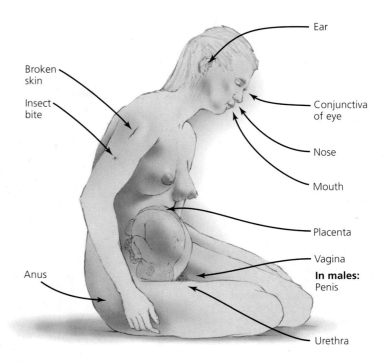

▲ *Figure 14.3*

Portals of entry for invading pathogens. The portals include the skin, placenta, conjunctiva, and mucous membranes of the respiratory, gastrointestinal, urinary, and reproductive tracts.

The sites through which pathogens enter the body can be likened to the great gates or portals of a castle, because those sites constitute the routes by which microbes gain entry. Pathogens thus enter the body at several sites, called **portals of entry (Figure 14.3),** which are of four major types: the skin, the mucous membranes, the placenta, and the so-called *parenteral route* (which is not a portal per se, but a way of circumventing the usual portals). Next we consider each type of portal of entry in turn.

Skin

Because the outer layer of skin is composed of relatively thick layers of tightly packed, dead, dry cells (**Figure 14.4** on p. 410), it forms a formidable barrier to most pathogens so long as it remains intact. Still, some pathogens can enter the body through natural openings in the skin, such as hair follicles and sweat glands; abrasions, cuts, and scrapes open the skin to infection by contaminants. Additionally, the larvae of some parasitic worms are capable of burrowing through the skin to reach underlying tissues, and some fungi can digest the dead outer layers of skin to gain entry to deeper, moister sites.

Mucous Membranes

Other portals of entry for pathogens are the *mucous membranes,* which line all the body cavities that are open to the outside world. They include the linings of the respiratory,

Figure 14.4 ▶

A cross-section of skin, showing the layers of cells that constitute a barrier to most microbes so long as the skin remains intact. Some pathogens can enter the body through hair follicles and the ducts of sweat glands.

gastrointestinal, urinary, and reproductive tracts, as well as the *conjunctiva* (kon-jŭnk-tī′vă), the thin membrane covering the surface of the eyeball and the underside of the eyelids. Like the skin, mucous membranes are composed of tightly packed cells; but unlike the skin, mucous membranes are relatively thin, moist, and warm, and their cells are living. Therefore, pathogens find mucous membranes more hospitable and often easier portals of entry.

The respiratory tract is the most frequently used portal of entry. Pathogens enter the mouth and nose in the air, on dust particles, and in drops of moisture. The bacteria that cause whooping cough, diphtheria, pneumonia, strep throat, and meningitis, as well as some fungi, viruses, and protozoa, enter through the respiratory tract.

Surprisingly, many viruses enter the respiratory tract via the eyes. They are introduced onto the conjunctiva by contaminated fingers and are washed into the nasal cavity with tears. Cold and influenza viruses typically enter the body in this manner.

Some parasitic protozoa, helminths, bacteria, and viruses infect the body through the gastrointestinal mucous membranes. These parasites are able to survive the acidic pH of the stomach and the digestive juices of the intestinal tract.

Placenta

The developing embryo forms an organ, called the *placenta,* through which it obtains nutrients from the mother. The placenta is in such intimate contact with the wall of the mother's uterus that nutrients and wastes can diffuse between the blood vessels of the developing child and of the

mother, but because the two blood supplies do not actually contact each other, the placenta typically forms an effective barrier to most pathogens. However, in about 2% of pregnancies, pathogens cross the placenta and infect the embryo, or fetus, sometimes causing spontaneous abortion, birth defects, or premature birth. Some pathogens that can cross the placenta are listed in **Table 14.3.**

The Parenteral Route

The **parenteral** (pă-ren′ter-ăl) **route** is not technically a portal of entry, but instead a means by which the portals of entry are sometimes circumvented. To enter the body by the parental route, pathogens must be deposited directly into tissues beneath the skin or mucous membranes, such as occurs in punctures (by a nail, thorn, or hypodermic needle), cuts, bites, stab wounds, deep abrasions, or surgery.

The Role of Adhesion in Infection

Learning Objectives

✓ List the types of adhesion factors and the roles they play in infection.

✓ Explain how a biofilm may facilitate contamination and infection.

After entering the body, symbionts must adhere to cells if they are to be successful in establishing colonies. The process by which microorganisms attach themselves to cells is called **adhesion** (ad-hē′zhŭn). To accomplish adhesion, pathogens use **adhesion factors**, which are either specialized structures or attachment proteins. Examples of specialized structures that serve as adhesion factors are adhesion

Table 14.3   Some Pathogens that Cross the Placenta

	Pathogen	Condition in the Adult	Effect on Embryo or Fetus
Protozoan	*Toxoplasma gondii*	Toxoplasmosis	Abortion, epilepsy, encephalitis, microcephaly, mental retardation, blindness, anemia, jaundice, rash, pneumonia, diarrhea, hypothermia, deafness
Bacteria	*Treponema pallidum*	Syphilis	Abortion, multiorgan birth defects, syphilis
	Listeria monocytogenes	Listeriosis	Granulomatosis infantiseptica (nodular inflammatory lesions and infant blood poisoning), death
DNA viruses	*Cytomegalovirus*	Usually asymptomatic	Deafness, microcephaly, mental retardation
	Parvovirus B19	Erythema infectiosum	Abortion
RNA viruses	*Lentivirus* (HIV)	AIDS	Immunosuppression (AIDS)
	Rubivirus rubella	German measles	Severe birth defects or death

disks in some protozoa and suckers and hooks in some helminths (see Chapter 23). In contrast, viruses and many bacteria have surface lipoprotein and glycoprotein molecules called *ligands* that enable them to bind to complementary receptors on host cells **(Figure 14.5).** Ligands are also called *adhesins* on bacteria and *attachment proteins* on viruses. Adhesins are found on fimbriae, flagella, and glycocalyces of many pathogenic bacteria. Receptor molecules on host cells are typically glycoproteins containing mannose and galactose. If ligands or their receptors can be changed or blocked, infection can often be prevented.

The specific interaction of adhesins and receptors with chemicals on host cells often determines the specificity of pathogens for particular hosts. For example, *Neisseria gonorrhoeae* (nī-sē′rē-ă go-nōr-rē′ē) has adhesins on its fimbriae that adhere only to cells lining the urethra and vagina of humans. Thus, this pathogen cannot affect other hosts.

Some bacteria, including *Bordetella* (bōr-dĕ-tel′ă), the cause of whooping cough, have more than one type of adhesin, and other pathogens can change their adhesins over time. This helps the pathogen evade the body's immune system and allows the pathogen to attack more than one kind of cell. Bacterial cells and viruses that have lost the ability to make adhesins or attachment proteins—whether as the result of some genetic change (mutation) or exposure to certain physical or chemical agents (as occurs in the production of some vaccines)—become harmless, or **avirulent** (ā-vir′ū-lent).

Some bacterial pathogens do not attach to host cells directly, but instead interact with each other to form a sticky web of bacteria and polysaccharides called a *biofilm,* which adheres to a surface within a host. A prominent example of a biofilm is dental plaque **(Figure 14.6** on p. 412), which contains the bacteria that cause dental caries (tooth decay).

(a)   (b)   TEM ⊢————⊣ 2μm

▲ *Figure 14.5*

The adhesion of pathogens to host cells. **(a)** An artist's rendition of the attachment of a microbial ligand to a complementary surface receptor on a host cell. **(b)** A photomicrograph of cells of a pathogenic strain of *E. coli* attached to the mucous membrane of the intestine. Although the thick glycocalyces of the bacteria are visible, the ligands are too small to be seen at this magnification.

CRITICAL THINKING

If a mutation occurred in *Escherichia coli* that deleted the gene for an adhesin, what effect might it have on the ability of *E. coli* to cause urinary tract infections?

The Nature of Infectious Disease

Learning Objective

✓ Compare and contrast the terms *infection, disease, morbidity, pathogenicity,* and *virulence.*

Most infections succumb to the body's defenses (discussed in detail in Chapters 15 and 16), but some infectious agents not only evade the defenses and multiply, but also affect

SEM ⊢————⊣ 3μm

▲ *Figure 14.6*
Dental plaque, a biofilm of bacteria that adhere to one another and to teeth by means of sticky capsules, slime layers, and fimbriae.

body function. By definition, all parasites injure their hosts. When the injury is significant enough to interfere with the normal functioning of the body, the result is termed **disease.** Thus, disease, also known as **morbidity,** is any change from a state of health.

Infection and disease are not the same thing. Infection is the invasion of a pathogen; disease results only if the pathogen multiplies sufficiently to adversely affect the body. Some illustrations will help to clarify this point. While caring for a patient, a nurse may become contaminated with the bacterium *Staphylococcus aureus* (staf′i-lō-kok′ŭs o′rē-ŭs). If the pathogen is able to gain access to his body, perhaps through a break in his skin, he will become infected. Then, if the multiplication of *Staphylococcus* results in the production of a boil, the nurse will have a disease. Another example: A drug addict who shares needles may become contaminated and subsequently infected with HIV without experiencing visible change in bodily function for years. Such a person is infected but not yet diseased.

Manifestations of Disease: Symptoms, Signs, and Syndromes

Learning Objective

✓ Contrast symptoms, signs, and syndromes.

Diseases may become manifest in different ways. **Symptoms** are *subjective* characteristics of a disease that can be felt by the patient alone, whereas **signs** are *objective* manifestations of disease that can be observed or measured by others. Symptoms include pain, headache, dizziness, and fatigue; signs include swelling, rash, redness, and fever. Sometimes pairs of symptoms and signs reflect the same underlying cause: Nausea is a symptom, and vomiting a sign; chills are a symptom, whereas shivering is a sign. Note, however, that even though signs and symptoms may have the same un-

Table 14.4	Typical Manifestations of Disease
Symptoms (sensed by the patient)	Signs (detected or measured by an observer)
Pain, nausea, headache, chills, sore throat, fatigue or lethargy (sluggishness, tiredness), malaise (discomfort), itching, abdominal cramps	Swelling, rash or redness, vomiting, diarrhea, fever, pus formation, anemia, leukocytosis/leukopenia (increase/decrease in the number of circulating white blood cells), bubo (swollen lymph node), tachycardia/bradycardia (increase/decrease in heart rate)

derlying cause, they need not occur together. Thus, for example, a viral infection of the brain may result in both a headache (a symptom) and the presence of viruses in the cerebrospinal fluid (a sign), but viruses in the cerebrospinal fluid are not invariably accompanied by headache. Symptoms and signs are used in conjunction with laboratory tests to make diagnoses. Some typical symptoms and signs are listed in **Table 14.4.**

A **syndrome** is a group of symptoms and signs that collectively characterizes a particular disease or abnormal condition. For example, acquired immunodeficiency syndrome (AIDS) is characterized by malaise, loss of certain white blood cells, diarrhea, weight loss, pneumonia, toxoplasmosis, and tuberculosis.

Some infections go unnoticed because they have no symptoms. Such cases are **asymptomatic** or **subclinical** infections. Note that even though asymptomatic infections by definition lack symptoms, in some cases certain signs may still be detected if the proper tests are performed. Thus, for example, leukocytosis (an excess of white blood cells) may be detected in a blood sample from an individual who feels completely healthy.

Table 14.5 lists some prefixes and suffixes used to describe various aspects of diseases and syndromes.

Causation of Disease: Etiology

Learning Objectives

✓ Define etiology.
✓ List Koch's postulates, explain their function, and describe their limitations.

Even though our focus in this chapter is infectious disease, it's obvious that not all diseases result from infections. Some diseases are *hereditary,* which means they are genetically transmitted from parents to offspring. Other diseases, called *congenital diseases,* are diseases that are present at birth, regardless of the cause (whether hereditary, environmental, or infectious). Still other diseases are classified as *degenerative, nutritional, endocrine, mental, immunological,* or *neoplastic.*[7]

---

[7]Neoplasms are tumors, which may either remain in one place (benign tumors) or spread (cancers).

Table 14.5   Terminology of Disease

Term	Meaning	Example
carcino-	cancer	carcinogenic, give rise to cancer
col-, colo-	colon	colitis, inflammation of the colon
dermato-	skin	dermatitis, inflammation of the skin
-emia	pertaining to the blood	viremia, viruses in the blood
endo-	inside	endocarditis
-gen, gen-	give rise to	pathogen, give rise to disease
-genesis	development	pathogenesis, development of disease
hepat-	liver	hepatitis, inflammation of the liver
idio-	unknown	idiopathic, pertaining to a disease of unknown cause
-itis	inflammation of a structure	meningitis, inflammation of the meninges (covering of the brain); endocarditis, inflammation of the lining of the heart
-oma	tumor	melanoma, cancer of melanocytes (pigment cells of the skin)
-osis	condition of	toxoplasmosis, being infected with *Toxoplasma*
-patho	abnormal	pathology, study of disease
septi-	literally *rotting*, refers to presence of pathogens	septicemia, pathogens in the blood
terato-	defects	teratogenic, causing birth defects
tox-	poison	toxin, harmful compound

The various categories of diseases are described in **Table 14.6** on p. 415. Note that some diseases can fall into more than one category. For example, liver cancer, which is usually associated with infection by hepatitis B virus, can be classified as both neoplastic and infectious.

The study of the cause of a disease is called **etiology** (ē-tē-ol′ō-jē). Because our focus here and in subsequent chapters is on infectious diseases, we next examine how microbiologists investigate the causation of infectious diseases, beginning with a reexamination of the work of Robert Koch, previously described in Chapter 1.

Using Koch's Postulates

In our modern world, we take for granted the idea that specific microbes cause specific diseases, but this has not always been the case. In the past, disease was thought to result from a variety of causes, including bad air, imbalances in body fluids, or spiritual forces. In the 19th century, Louis Pasteur, Robert Koch, and other microbiologists proposed the **germ theory of disease,** which states that disease is caused by infections of pathogenic microorganisms (at the time, called *germs*).

But which pathogen causes a specific disease? How can we distinguish between the pathogen that causes a disease and all the other biological agents (fungi, bacteria, protozoa, and viruses) that are part of the normal microbiota and are in effect "innocent bystanders"?

Koch developed a series of essential conditions, or *postulates*, that scientists must demonstrate or satisfy to prove that a particular microbe was pathogenic and caused a particular disease. Using his postulates, Koch proved that *Bacillus* (ba-sil′ŭs an-thra′sis) causes anthrax, and that *Mycobacterium tuberculosis* (mī-kō-bak-tēr′ē-ŭm too-ber-kyu-lō′sis) causes tuberculosis.

In order to prove that a given infectious agent causes a given disease, a scientist must satisfy all of **Koch's postulates** (**Figure 14.7** on p. 414):

1. The suspected pathogen must be present in every case of the disease.

2. That pathogen must be isolated and grown in pure culture.

3. The cultured pathogen must cause the disease when it is inoculated into a healthy, susceptible experimental host.

4. The same pathogen must be reisolated from the diseased experimental host.

It is critical that all the postulates be satisfied in order. The mere presence of a pathogen does not prove that it causes a disease. Although Koch's postulates have been used to prove the cause of many infectious diseases in humans, animals, and plants, inadequate attention to the postulates has resulted in incorrect conclusions concerning disease causation. For instance, in the early 1900s *Haemophilus influenzae* (hē-mof′i-lŭs in-flū-en′zē) was found in the respiratory systems of flu victims and identified as the causative agent of influenza. Later, flu victims that lacked *H. pneumoniae* in their lungs were discovered. This discovery violated the first postulate, so *H. influenzae* cannot be the cause of flu. Today we know that an RNA virus causes flu, and that *H. influenzae* was part of the normal microbiota of

1 The suspected germ must be present in every case of the disease.

Germ not typically found in healthy subjects.

Diseased subjects

Healthy subjects

Microscope

Petri plate

Bacterial colonies

2 The germ must be isolated and grown in pure culture.

Streaked plates

Injection

3 The cultured germ must cause the disease when it is inoculated into a healthy, susceptible experimental host (animal or plant).

4 The same germ must be reisolated from the diseased experimental host.

▲ *Figure 14.7*

Koch's postulates. ① The suspected pathogen must be present in every case of the disease. ② The pathogen must be isolated and grown in pure culture. ③ The cultured pathogen must cause the disease when it is inoculated into a healthy, susceptible experimental host (animal or plant). ④ The same pathogen must be reisolated from the diseased experimental host.

those early flu patients. (Later studies, correctly using Koch's postulates, showed that *H. influenzae* can cause an often-fatal meningitis in children.)

Exceptions to Koch's Postulates

Clearly, Koch's postulates are a cornerstone of the etiology of infectious diseases, but using them is not always feasible, for the following reasons:

- Some pathogens cannot be cultured in the laboratory. For example, pathogenic strains of *Mycobacterium leprae* (mī-kō-bak-tēr′ē-ŭm lep′rē), which causes leprosy, have never been grown on laboratory media.

- Some diseases are caused by a combination of pathogens, or by a combination of a pathogen and physical, environmental, or genetic cofactors. In such cases, the pathogen alone is avirulent, but when accompanied by another pathogen or the appropriate cofactor, disease results. For example, liver cancer can result when liver cells are infected by both the hepatitis B and hepatitis D viruses, but seldom when only one of the viruses infects the cells.

- Ethical considerations prevent applying Koch's postulates to diseases and pathogens that occur in humans only. In such cases the third postulate, which involves

Table 14.6   Categories of Diseases[a]

	Description	Examples
Hereditary	Caused by errors in the genetic code received from parents	Sickle-cell anemia, diabetes mellitus, Down syndrome
Congenital	Anatomical and physiological (structural and functional) defects present at birth; caused by drugs (legal and illegal), X-ray exposure, or infections	Fetal alcohol syndrome, deafness from rubella infection
Degenerative	Result from aging	Renal failure, age-related far-sightedness
Nutritional	Result from lack of some essential nutrients in diet	Kwashiorkor, rickets
Endocrine (hormonal)	Due to excesses or deficiencies of hormones	Dwarfism
Mental	Emotional or psychosomatic	Skin rash, gastrointestinal distress
Immunological	Hyperactive or hypoactive immunity	Allergies, autoimmune diseases, agammaglobulinemia
Neoplastic (tumor)	Abnormal cell growth	Benign tumors, cancers
Infectious	Caused by an infectious agent	Colds, influenza, herpes infections
Iatrogenic[b]	Caused by medical treatment or procedures; are a subgroup of hospital acquired diseases	Surgical error, yeast vaginitis resulting from antimicrobial therapy
Idiopathic[c]	Unknown cause	Alzheimer's disease, multiple sclerosis

[a]Some diseases may fall in more than one category.
[b]From Greek *iatros*, meaning physician.
[c]From Greek *idiotes*, meaning ignorant person, and *pathos*, meaning disease.

inoculation of a healthy susceptible host, cannot be satisfied within ethical boundaries. For this reason, scientists have never attempted to apply Koch's postulates to proving that HIV causes AIDS; however, observations of fetuses that were naturally exposed to HIV by their infected mothers, as well as accidentally infected health care workers, have in effect satisfied the third postulate.

Additionally, some circumstances may make satisfying Koch's postulates difficult:

- It is not possible to establish a single cause for such infectious diseases as pneumonia, meningitis, and hepatitis, because the names of these diseases refer to conditions that can be caused by more than one pathogen. For these diseases, laboratory technicians must identify the etiologic agent involved in any given case.

- Some pathogens have been ignored. For example, gastric ulcers were long thought to be caused by excessive production of stomach acid in response to stress, but the majority of such ulcers are now known to be caused by a long-overlooked bacterium, *Helicobacter pylori* (hel'i-kō-bak-ter pī'lō-rē).

If Koch's postulates cannot be applied to a disease condition for whatever reason, how can we positively know the causative agent of a disease? *Epidemiological* studies, discussed later in this chapter, can give statistical support to causation theories, but not absolute proof. For example, some researchers propose that the bacterium *Chlamydia pneumoniae* (kla-mid'ē-ă noo-mō'nē-ē) causes most cases of arteriosclerosis on the basis of the bacterium's presence in

most patients with the disease. Debate among scientists about such cases fosters a continued drive for knowledge and discovery. For example, we are still searching for the causes of Parkinson's disease, multiple sclerosis, and Alzheimer's disease.

Virulence Factors of Infectious Agents

Learning Objective

✓ Explain how microbial extracellular enzymes, toxins, adhesion factors, and antiphagocytic factors affect virulence.

Microbiologists characterize the disease-related capabilities of microbes by using two related terms. The ability of a microorganism to cause disease is termed **pathogenicity** (path'ō-jě-nis'i-tē), and the *degree* of pathogenicity is termed **virulence.** In other words, virulence is the relative ability of a pathogen to infect a host and cause disease. Organisms can be placed along a virulence continuum (**Figure 14.8** on p. 416); highly virulent organisms almost always cause disease, while less virulent organisms (including opportunistic pathogens) cause disease only in weakened hosts or when present in overwhelming numbers.

Pathogens have a variety of traits that enable them to enter a host, adhere to host cells, gain access to nutrients, and escape detection or removal by the immune system. These traits are collectively called **virulence factors.** Virulent pathogens have one or more virulence factors that nonvirulent microbes lack. We discussed adhesion factors previously; now we examine three other virulence factors: extracellular enzymes, toxins, and antiphagocytic factors.

More virulent

Francisella tularensis
(rabbit fever)

Yersinia pestis (plague)

Bordetella pertussis
(whooping cough)

*Pseudomonas
aeruginosa*
(infections of burns)

Clostridium difficile
(antibiotic-induced
colitis)

Candida albicans
(vaginitis, thrush)

Lactobacilli,
diphtheroids

Less virulent

▲ *Figure 14.8*
Relative virulence of some bacterial pathogens.

Extracellular Enzymes

Many pathogens secrete enzymes that enable them to dissolve structural chemicals in the body and thereby maintain an infection, invade further, and avoid body defenses. **Figure 14.9a** illustrates the action of some extracellular enzymes of bacteria:

- Hyaluronidase and collagenase degrade specific molecules to enable bacteria to invade deeper tissues. Hyaluronidase digests hyaluronic acid, the "glue" that holds animal cells together, and collagenase breaks down collagen, the body's chief structural protein.

- Coagulase coagulates blood proteins, providing a "hiding place" for bacteria within a clot.

- Kinases such as staphylokinase and streptokinase digest blood clots, allowing subsequent invasion of damaged tissues.

Many bacteria with these enzymes are virulent; mutant strains of the same species that have defective genes for these extracellular enzymes are usually avirulent.

Figure 14.9 ▶
Virulence factors. **(a)** Extracellular enzymes. Hyaluronidase and collagenase digest structural materials in the body. Coagulase in effect "camouflages" bacteria inside a blood clot, whereas kinases digest clots to release bacteria. **(b)** Toxins. Exotoxins (including cytotoxin, shown here) are released from living pathogens and harm neighboring cells. Endotoxin is released from many dead Gram-negative bacteria and can trigger widespread disruption of normal bodily functions. **(c)** Antiphagocytic factors. Capsules, one antiphagocytic factor, can prevent phagocytosis or stop digestion by a phagocyte. *How can bacteria prevent a phagocytic cell from digesting them once they have been engulfed by pseudopodia*

Figure 14.9 Some bacteria secrete a chemical that prevents the fusion of lysosomes with the phagosome containing the bacteria.

Toxins

Toxins are chemicals produced by pathogens that either harm tissues or trigger host immune responses that cause damage. The distinction between extracellular enzymes and toxins is not always clear, because many enzymes are toxic, and many toxins have enzymatic action. In a condition called **toxemia** (tok-sē'mē-ă), toxins enter the bloodstream and are carried to other parts of the body, including sites that may be far removed from the site of infection. There are two types of toxin: exotoxins and endotoxin.

Exotoxins Many microorganisms secrete **exotoxins** that are central to their pathogenicity in that they destroy host cells or interfere with host metabolism. Exotoxins are of three principal types: (1) *cytotoxins,* which kill host cells in general or affect their function **(Figure 14.9b);** (2) *neurotoxins,* which specifically interfere with nerve cell function; and (3) *enterotoxins,* which affect cells lining the gastrointestinal tract. Examples of pathogenic bacteria that secrete exotoxins are the clostridia bacteria that cause gangrene, botulism, and tetanus; pathogenic strains of *Staphylococcus aureus* that cause food poisoning and other ailments; and diarrhea-causing *E. coli.* Some fungi and marine dinoflagellates (protozoa) also secrete exotoxins. Specific exotoxins are discussed in the chapters that cover specific pathogens.

The body protects itself with **antitoxins,** which are protective molecules called *antibodies* that bind to specific toxins and neutralize them. Health care workers stimulate the production of antitoxins by administering immunizations composed of *toxoids,* which are toxins that have been treated with heat, formaldehyde, chlorine, or other chemicals to make them nontoxic but still capable of stimulating the production of antibodies. Antibodies, toxoids, and immunizations are discussed further in Chapter 17.

Endotoxin Recall from Chapter 3 that Gram-negative bacteria have an outer (cell wall) membrane composed of lipopolysaccharide, phospholipids, and proteins (see Figure 3.15). **Endotoxin,** also called **lipid A,** is the lipid portion of the membrane's lipopolysaccharide.

Hyaluronidase and collagenase

Invasive bacteria reach epithelial surface

Bacteria produce hyaluronidase and collagenase

Bacteria invade deeper tissues

(a) Extracellular enzymes

Coagulase and kinase

Bacteria produce coagulase

Clot forms

Bacteria later produce kinase, dissolving clot and releasing bacteria

Exotoxin

Bacteria secrete exotoxins, in this case a cytotoxin

Cytotoxin kills host's cells

(b) Toxins

Endotoxin

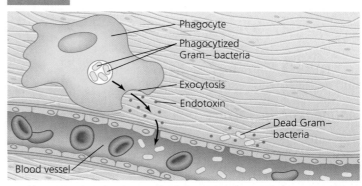

Dead bacteria release endotoxin (lipid A) which induces effects such as fever, inflammation, diarrhea, shock, and blood coagulation

Phagocytosis blocked by capusle

(c) Antiphagocytic factors

Incomplete phagocytosis

Endotoxin is released only when bacterial cells die naturally or are digested by phagocytic cells such as macrophages (see Figure 14.9b). Many types of lipid A stimulate the body to release chemicals that cause fever, inflammation, diarrhea, hemorrhaging, shock, and blood coagulation.

Although infections by bacteria that produce exotoxins are generally more serious than infections with Gram-negative bacteria, most Gram-negative pathogens can be potentially life threatening because the release of endotoxin from dead bacteria can produce serious, systemic effects in the host.

Table 14.7	A Comparison of Bacterial Exotoxins and Endotoxin	
	Exotoxins	Endotoxin
Source	Mainly Gram-positive and Gram-negative bacteria	Gram-negative bacteria
Relation to bacteria	Metabolic product secreted from living cell	Portion of outer (cell wall) membrane released upon cell death
Chemical nature	Protein or short peptide	Lipid (lipid A) portion of lipopolysaccharide of outer (cell wall) membrane
Toxicity	High	Low, but may be fatal in high doses
Heat stability	Unstable at temperatures above 60°C	Stable for up to 1 hour at autoclave temperatures (121°C)
Effect on host	Variable depending on source; may be cytotoxin, neurotoxin, enterotoxin	Fever, lethargy, malaise, shock, blood coagulation
Fever producing?	No	Yes
Antigenicity[a]	Strong: stimulates antitoxin (antibody) production	Weak
Toxoid formation for immunization?	By treatment with heat or fomaldehyde	Not feasible
Representative diseases	Botulism, tetanus, gas gangrene, diphtheria, cholera, plague, staphylococcal food poisoning	Typhoid fever, tularemia, endotoxic shock, urinary tract infections, meningococcal meningitis

[a]Refers to the ability of a chemical to trigger a specific immune response, particularly the formation of antibodies.

Table 14.7 summarizes differences between exotoxins and endotoxin.

Antiphagocytic Factors

Typically, the longer a pathogen remains in a host, the greater the damage and the more severe the disease. To limit the extent and duration of infections, the body's phagocytic cells, such as the white blood cells called macrophages, engulf and remove invading pathogens. Here we consider some virulence factors related to the evasion of phagocytosis, beginning with bacterial capsules.

Capsules   The capsules of many pathogenic bacteria (see Figure 3.4a) are effective virulence factors because many capsules are composed of chemicals that are normally found in the body (including polysaccharides); as a result, they do not stimulate the host's immune response. For example, hyaluronic acid capsules in effect deceive phagocytic cells into treating them and the enclosed bacteria as if they were a normal part of the body. Additionally, capsules are often slippery, making it difficult for phagocytes to surround and phagocytize them—their pseudopodia cannot grip the capsule, much as wet hands have difficulty holding a wet bar of soap **(Figure 14.9c)**. Even when encapsulated bacteria are successfully engulfed, they may retard digestion long enough to reproduce inside the phagocyte.

Antiphagocytic Chemicals   Some bacteria, including those that cause gonorrhea, produce chemicals that prevent the fusion of lysosomes with phagocytic vesicles, which allows the bacteria to survive inside of phagocytes (see Figure 14.9c). *Streptococcus pyogenes* (strep-tō-kok′ŭs pī-aj′en-ēz) produces a protein on its cell wall and fimbriae, called *M protein*, that

resists phagocytosis and thus increases virulence. Other bacteria produce *leukocidins*, which are chemicals capable of destroying phagocytic white blood cells outright.

The Stages of Infectious Diseases

Learning Objective

✓ List and describe the five typical stages of infectious diseases.

Following contamination and infection, a definite sequence of events called the **disease process** occurs. Many infectious diseases have five stages following infection: an incubation period, the prodromal period, illness, decline, and convalescence **(Figure 14.10)**.

Incubation Period

The **incubation period** is the time between infection and occurrence of the first symptoms or signs of disease. The length of the incubation period depends on the virulence (degree of pathogenicity) of the infective agent, the infective dose (initial number of pathogens), the state and health of the patient's immune system, the nature of the pathogen and its generation time, and the site of infection. Some diseases have typical incubation periods, whereas for others the incubation period varies considerably. **Table 14.8** lists incubation periods for selected diseases.

Prodromal Period

The **prodromal**[8] (prō-drō′măl) **period** is a short time of generalized, mild symptoms (such as malaise and muscle aches)

---
[8]From Greek *prodromos,* meaning forerunner.

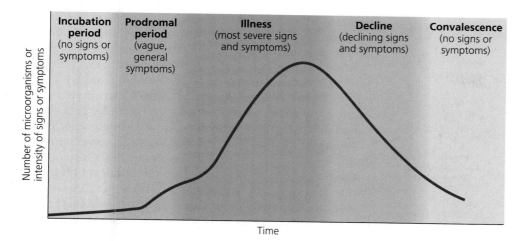

◀ *Figure 14.10*

The stages of infectious diseases. Not every stage occurs in every disease.

Table 14.8	**Incubation Periods of Selected Infectious Diseases**
Disease	**Incubation Period**
Salmonella food-borne infection	<1 day
Influenza	1 day
Cholera	2 days
Genital herpes	5 days
Tetanus	3–21 days
Syphilis	2–4 weeks
Hepatitis B	1–6 months
AIDS	1–>8 years
Leprosy	10–>30 years

that precedes illness. Not all infectious diseases have a prodromal stage.

Illness

Illness is the most severe stage of an infectious disease. Signs and symptoms are most evident during this time. The patient's immune system has not yet fully responded to the pathogens, and their presence is harming the body. This stage is usually when a physician first sees the patient.

Decline

During the period of **decline,** the body gradually returns to normal as the patient's immune response and/or medical treatment vanquish the pathogens. Fever and other signs and symptoms subside. Normally the immune response and its products (such as antibodies in the blood) peak during this stage. If the patient doesn't recover, then the disease is fatal.

Convalescence

During **convalescence** (kon-vă-les'ens), the patient recovers from the illness, and tissues and systems are repaired and returned to normal. The length of a convalescent period

depends on the amount of damage, the nature of the pathogen, the site of infection, and the overall health of the patient. Thus, whereas recovery from staphylococcal food poisoning may take a day, recovery from Lyme disease may take years.

A patient is likely to be infectious during every stage of disease. Even though most of us may realize we are infective during the symptomatic periods, many people are unaware that infections can be spread during incubation and convalescence as well. For example, a patient who no longer has any obvious herpes sores is still capable of transmitting herpesviruses. Good aseptic technique can often limit the spread of pathogens from recovering patients.

The Movement of Pathogens Out of Hosts: Portals of Exit

Just as infections occur through portals of entry, so pathogens must leave infected patients through **portals of exit** in order to infect others (**Figure 14.11** on p. 420). Many portals of exit are essentially identical to portals of entry. However, pathogens often exit hosts in materials that the body secretes or excretes. Thus, pathogens may leave hosts in secretions (ear wax, tears, nasal secretions, saliva, sputum, and respiratory droplets), in blood (via arthropod bites, hypodermic needles, or wounds), in vaginal secretions or semen, in milk produced by the mammary glands, and in excreted bodily wastes (feces and urine). As we will see, health care personnel must consider the portals of entry and exit in their efforts to understand and control the spread of diseases within populations.

Sources of Infectious Diseases in Humans

Learning Objective

✓ Describe three types of reservoirs of infection in humans.

Once pathogens of humans have left the body through the various portals of exit, most cannot survive for long in the

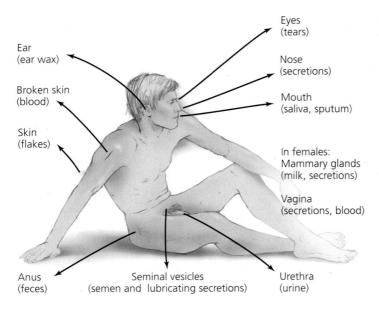

Ear
(ear wax)

Eyes
(tears)

Nose
(secretions)

Broken skin
(blood)

Mouth
(saliva, sputum)

Skin
(flakes)

In females:
Mammary glands
(milk, secretions)

Vagina
(secretions, blood)

Anus
(feces)

Seminal vesicles
(semen and lubricating secretions)

Urethra
(urine)

▲ **Figure 14.11**

Portals of exit. Many portals of exit are the same sites as portals of entry. Pathogens often leave the body via bodily secretions and excretions produced at those sites.

relatively harsh conditions they encounter outside their hosts. Thus if these pathogens are to enter new hosts, they must survive in some site from which they can infect new hosts. These sites where pathogens are maintained as a source of infection are called **reservoirs of infection.** Next we consider three types of reservoirs: animal reservoirs, human carriers, and nonliving reservoirs.

Animal Reservoirs

Many pathogens that normally infect either domesticated or sylvatic[9] (wild) animals can also affect humans. The more similar an animal's physiology is to human physiology, the more likely its pathogens could affect human health. Diseases that are naturally spread from their usual animal hosts to humans are called **zoonoses** (zō-ō-nō′sēz) or **zoonotic diseases**. Over 150 zoonoses have been identified throughout the world. Well-known examples include yellow fever, anthrax, bubonic plague, and rabies.

Humans may acquire zoonoses from animal reservoirs via a number of routes, including various types of direct contact with animals and their wastes, by eating animals, and via bloodsucking arthropods. (Modes of transmission of infectious diseases are discussed in the next section.) Human infections with zoonoses are difficult to eradicate because extensive animal reservoirs are often involved. The larger the animal reservoir (that is, the greater the number of infected animals) and the greater the contact between humans and the animals, the more difficult it is to control the spread of the

disease to humans. This is especially true when the animal reservoir consists of both sylvatic and domesticated animals. In the case of rabies, for example, the disease typically spreads from a sylvatic reservoir (often foxes, bats, and skunks) to domestic pets (especially dogs), from which humans may be infected. The wild animals constitute a reservoir for the rabies virus, but because domestic pets can be vaccinated, transmission to humans can be largely prevented.

Table 14.9 provides a brief view of some common zoonoses. Humans are usually dead-end hosts for zoonotic pathogens—that is, they do not act as significant reservoirs for the reinfection of animal hosts—largely because the circumstances under which zoonoses are transmitted favor movement from animals to humans, and not in the opposite direction. For example, animals do not often eat humans these days, and animals less frequently have contact with human wastes than humans have contact with animal wastes. Zoonotic diseases that are transmitted via the bites of blood-sucking arthropods are most likely to be transmitted back to animal hosts.

Human Carriers

Experience tells you that humans with active diseases are important reservoirs of infection for other humans. What may not be so obvious is that some infected people remain both asymptomatic and infective for years. This is true of syphilis and AIDS, for example. Whereas some of these **carriers** incubate the pathogen in their body and eventually develop the disease, others remain a continued source of infection without ever becoming sick. An example of such a carrier is given in **Case Study 14.1.**

Nonliving Reservoirs

Soil, water, and food can be **nonliving reservoirs** of infection. Soil, especially if fecally contaminated, can harbor *Clostridium* (klos-trid′ē-ŭm) bacteria, which cause botulism, tetanus, and other diseases. Water can be contaminated with

---

Case Study 14.1   A Deadly Carrier

In 1937, a man employed to lay water pipes in Croyden, England, was found to be the source of a severe epidemic of typhoid fever. The man—who was an asymptomatic carrier of *Salmonella typhi*, the bacterium that causes typhoid—habitually urinated at his job site, in the process contaminating the town's water supply with bacteria residing in his bladder. Over 300 cases of typhoid fever developed, and 43 people died, before the man was identified as the carrier.

1. How do you think health officials were able to identify the source of this typhoid epidemic?

2. Given that antibiotics were not generally available in 1937, how could health officials end the epidemic short of removing the man from the job site?

---

[9]From Latin *sylva*, meaning woodland.

Table 14.9   Some Common Zoonoses

Disease	Causative Agent	Animal Reservoir	Mode of Transmission
Helminthic (worms; ch. 23)			
Tapeworm infestation	*Dipylidium caninum*	Dogs	Ingestion of larvae transmitted in dog saliva
Trichinosis	*Trichinella spiralis*	Pigs, bears	Ingestion of undercooked pork or bear meat containing larvae
Protozoan (Ch. 23)			
Malaria	*Plasmodium* spp.	Monkeys	Bite of *Anopheles* mosquito
Toxoplasmosis	*Toxoplasma gondii*	Cats and other animals	Ingestion of contaminated meat, inhalation of pathogen, direct contact with infected tissues
Fungal (Ch. 22)			
Ringworm	*Trichophyton* sp. *Microsporum* sp. *Epidermophyton* sp.	Domestic animals	Direct contact
Bacterial (Chs. 19–21)			
Anthrax	*Bacillus anthracis*	Domestic livestock	Direct contact with infected animals, inhalation
Bubonic plague	*Yersinia pestis*	Rodents	Flea bites
Lyme disease	*Borrelia burgdorferi*	Deer	Tick bites
Salmonellosis	*Salmonella* spp.	Birds, rodents, reptiles	Ingestion of fecally contaminated water or food
Typhus	*Rickettsia prowazeki*	Rodents	Louse bites
Viral (Chs. 24–25)			
Rabies	*Lyssavirus* sp.	Bats, skunks, foxes, dogs	Bite of infected animal
Hantavirus pulmonary syndrome	*Hantavirus.* sp.	Deer mice	Inhalation of viruses in dried feces and urine
Yellow fever	*Flavivirus* sp.	Monkeys	Bite of *Aedes* mosquito

feces and urine containing parasitic worm eggs, pathogenic protozoa, bacteria, and viruses. Meats and vegetables can also harbor pathogens. Milk can contain many pathogens, which is why it is routinely pasteurized in the United States.

Modes of Infectious Disease Transmission

By definition, an infectious disease agent must be transmitted from either a reservoir or a portal of exit to another host's portal of entry. Transmission can occur by numerous modes that are somewhat arbitrarily categorized into three groups: contact transmission, vehicle transmission, and vector transmission (**Table 14.10** on p. 422).

Contact Transmission

Contact transmission is the spread of pathogens from one host to another by direct contact, indirect contact, or respiratory droplets.

Direct contact transmission, including *person-to-person spread,* typically involves body contact between hosts. Touching, kissing, and sexual intercourse between humans are involved in the transmission of such diseases as warts, herpes, and gonorrhea. Touching, biting, or scratching can transmit

zoonoses such as rabies, ringworm, and tularemia from an animal reservoir to a human. The transfer of pathogens from an infected mother to a developing baby across the placenta is another form of direct contact transmission. Direct transmission within a single individual can also occur if the person transfers pathogens from a portal of exit directly to a portal of entry—as occurs, for example, when a person with poor personal hygiene unthinkingly places a finger contaminated with fecal pathogens into the mouth.

Indirect contact transmission occurs when pathogens are spread from one host to another by **fomites** (fōm′i-tēz), which are inanimate objects that are inadvertently used to transfer pathogens to new hosts. Fomites include needles, toothbrushes, paper tissues, toys, money, diapers, drinking glasses, bed sheets, medical equipment, and other objects that can harbor or transmit pathogens. Contaminated needles are a major source of infection of the hepatitis B and AIDS viruses.

Droplet transmission is a third type of contact transmission. Pathogens can be transmitted within *droplet nuclei* (mucus droplets) that exit the body during exhaling, coughing, and sneezing (**Figure 14.12** on p. 422). Pathogens such as cold and flu viruses may be spread in this manner. If pathogens travel more than 1 meter in respiratory droplets, it is considered to be *airborne transmission* (discussed shortly), rather than contact transmission.

Table 14.10	Modes of Disease Transmission
Mode of Transmission	**Diseases Spread Include:**
Contact Transmission	
Direct Contact: *e.g.* handshaking, kissing, sex, bites	Cutaneous anthrax, genital warts, gonorrhea, herpes, rabies, staphylococcus infections, syphilis
Indirect Contact: *e.g.* drinking glasses, toothbrushes, toys, punctures, droplets from sneezing and coughing (within one meter)	Common cold, enterovirus infections, influenza, measles, Q fever, pneumonia, tetanus, whooping cough
Vehicle Transmission	
Airborne: *e.g.* dust particles	Chicken pox, coccidiomycosis, histoplasmosis, influenza, measles, pulmonary anthrax, tuberculosis
Waterborne: *e.g.* streams, swimming pools	*Campylobacter* infections, cholera, *Giardia* diarrhea
Foodborne: *e.g.* poultry, seafood, meat	Food poisoning (botulism, staphylococcal); hepatitis A, listeriosis, tapeworms, toxoplasmosis, typhoid fever
Vector Transmission	
Mechanical: *e.g.* (on insect bodies) flies, roaches	*E. coli* diarrhea, salmonellosis, trachoma
Biological: *e.g.* lice, mites, mosquitoes, ticks	Chagas' disease, Lyme disease, malaria, plague, Rocky Mountain spotted fever, typhus fever, yellow fever

Vehicle Transmission

Vehicle transmission is the spread of pathogens via air, drinking water, and food, as well as bodily fluids being handled outside the body.

 Airborne transmission involves the spread of pathogens to the respiratory mucous membranes of a new host via an **aerosol** (ār'ō-sol), a cloud of small droplets and solid particles suspended in the air. Aerosols may contain pathogens either on dust or inside droplets. (Recall that transmission via droplet nuclei that travel less than 1 meter is considered to be a form of direct contact transmission.) Aerosols can come from sneezing and coughing, or they can be generated by such things as air conditioning systems, sweeping, mopping, changing clothes or bed linens, or even from flaming inoculating loops in microbiological labs. Dust particles can carry *Staphylococcus, Streptococcus,* and hantavirus, whereas measles virus and tuberculosis bacilli can be transmitted in dried, airborne droplets. Fungal spores of *Histoplasma* and *Coccidioides* are typically inhaled.

 Waterborne transmission is important in the spread of many gastrointestinal diseases, including giardiasis, amoebic dysentery, and cholera. Note that water can act as a reservoir as well as a vehicle of infection. **Fecal-oral infection** is a major source of disease in the world. Some waterborne pathogens, such as *Schistosoma* (skis-tō-sō'mă) worms (see Chapter 23) and enteroviruses (see Chapter 25), are shed in the feces, enter through the gastrointestinal mucous membrane or skin, and subsequently cause disease elsewhere in the body.

 Food-borne transmission involves pathogens in and on foods that are poorly processed, undercooked, or poorly refrigerated. Foods may be contaminated with normal microbiota (*e.g.,* *E. coli* and *S. aureus*), with zoonotic pathogens

▲ *Figure 14.12*

Droplet transmission—in this case, droplets being propelled, primarily from the mouth, during a sneeze. By convention, such transmission is considered contact transmission only if droplets transmit pathogens to a new host within 1 meter of their source. *By what portal of entry does airborne transmission most likely occur?*

Figure 14.12 The most likely portal of entry of airborne pathogens is the respiratory mucous membrane.

such as *Mycobacterium bovis* (mī-kō-bak-tēr'ē-ŭm bō'vis) and *Toxoplasma* (tok-sō-plaz'mă), and with parasitic worms that alternate between human and animal hosts. Contamination of food with feces and pathogens such as hepatitis A virus is another kind of fecal-oral transmission. Because milk is

Case Study 14.2   Unusual Transmission of West Nile Virus

Although West Nile virus is primarily transmitted through mosquito bites, it can also be transmitted through organ transplantation. On August 1, 2002, four organs harvested from a single donor were transplanted into four patients. Within 17 days of transplantation, all four recipients became sick with West Nile virus-associated illnesses. Three of the patients developed meningoencephalitis, an inflammation of the brain and of the membranes surrounding the brain and spinal cord; the fourth patient developed West Nile virus fever. The organ donor's plasma tested positive for West Nile virus, but the source of the donor's infection could not be identified. Complicating matters, the organ donor had received blood products from 63 different blood donors. As this book went to press, an investigation into the source of the donor's infection remained ongoing.

1. In addition to patients who had received organs from the infected donor, what other patient populations might be at risk of West Nile virus infection?

2. As a scientist, what questions would you seek to answer about the infected donor?

3. What precautions would you advise instituting in order to prevent transmission of West Nile virus infection via blood transfusion and organ transplantation?

Source: *Morbidity and Mortality Weekly Report* 51(37): 833–836, 2002.

---

particularly rich in nutrients that microorganisms use (protein, lipids, vitamins, and sugars), it would be associated with the transmission of many diseases from infected animals and milk handlers if it were not properly pasteurized.

Because blood, urine, saliva, and other bodily fluids can contain pathogens, everyone—but especially health care workers—must take precautions when handling these fluids in order to prevent **bodily fluid transmission.** Special care must be taken to prevent any bodily fluid, which should be considered potentially contaminated with pathogens, from contacting the conjunctiva or any breaks in the skin or mucous membranes. Examples of diseases that can be transmitted via bodily fluids are AIDS, hepatitis, and herpes, which as we have seen can also be transmitted via direct contact. **Case Study 14.2** examines the transmission of West Nile virus via the blood of an infected organ donor.

Vector Transmission

Vectors are animals (typically arthropods) that transmit diseases from one host to another. Vectors can be either biological vectors or mechanical vectors.

Biological vectors not only transmit pathogens; they also serve as hosts for the multiplication of the pathogen during some stage of the pathogen's life cycle. The biological vectors of diseases affecting humans are typically biting arthropods, including mosquitoes, ticks, lice, fleas, blood-sucking flies, blood-sucking bugs, and mites. After the pathogens replicate within a biological vector, often in its gut or salivary gland, the pathogens enter a new host through a bite; either the bite site becomes contaminated with the vector's feces, or the bite directly introduces the pathogens into the new host.

Mechanical vectors are not required as hosts by the pathogens they transmit; they only passively carry pathogens to new hosts on their feet or other body parts. Mechanical vectors such as houseflies and cockroaches may introduce pathogens such as *Salmonella* (sal'mō-nel'ă) and *Shigella* (shē-gel'lă) into drinking water and food or onto the skin.

Table 14.11 on p. 424 lists some arthropod vectors and the diseases they transmit.

Classification of Infectious Diseases

Learning Objectives

✓ Describe the basis for each of the various classification schemes of infectious diseases.

✓ Distinguish among acute, subacute, chronic, and latent diseases.

✓ Distinguish among communicable, contagious, and noncommunicable infectious diseases.

Infectious diseases can be classified in a number of ways. No one way is "*the* correct way"; each has its own advantages. One scheme groups infectious diseases according to the body systems they affect. The appendix lists diseases organized in this way. A difficulty with this method of classification is that many diseases involve more than one organ system. For example, AIDS may begin as either a sexually transmitted infection (reproductive system) or a parenteral infection of the blood (cardiovascular system). It then becomes an infection of the lymphatic system as viruses invade lymphocytes. Finally, the syndrome involves diseases and degeneration of the respiratory, nervous, digestive, and lymphatic systems.

Infectious diseases may also be classified based on the taxonomic groups of their causative agents. The discussions of diseases in Chapters 19–25 are based on a taxonomic approach.

Every disease (not just infectious diseases) can also be classified according to its longevity and severity. If a disease develops rapidly but lasts only a short time, it is called an **acute disease.** An example is the common cold. In contrast, **chronic diseases** develop slowly (usually with less severe symptoms) and are continual or recurrent. Infectious mononucleosis, hepatitis C, tuberculosis, and leprosy are chronic diseases. **Subacute diseases** have durations and severities that lie somewhere between acute and chronic.

Table 14.11 **Selected Arthropod Vectors**

	Disease	Causative Agent
Biological Vector		
Mosquitoes		
Anopheles, Aedes	Malaria	*Plasmodium* spp.
	Yellow Fever	*Flavivirus* sp.
	Elephantiasis	*Wuchereria bancrofti*
	Dengue	*Flavivirus* sp.
	Viral encephalitis	*Alphavirus* spp.
Ticks		
Ixodes, Dermacentor	Lyme disease	*Borrelia burgdorferi*
	Rocky Mountain spotted fever	*Rickettsia rickettsii*
Fleas		
Xenopsylla	Bubonic plague	*Yersinia pestis*
	Endemic typhus	*Rickettsia prowazekii*
Lice		
Pediculus	Epidemic typhus	*Rickettsia typhi*
Blood-sucking flies		
Simulium, Glossina	African sleeping sickness	*Trypanosoma brucei*
	River blindness	*Onchocerca volvulus*
Blood-sucking bug		
Triatoma	Chagas' disease	*Trypanosoma cruzi*
Mites (chiggers)		
Leptotrombidium	Scrub typhus	*Orienta tsutsugamushi*
Mechanical Vectors		
Houseflies		
Musca	Food-borne infections	*Shigella* spp., *Salmonella* spp., *Escherichia coli*
Cockroaches		
Perplaneta, Blattella, Supella	Food-borne infections	*Shigella* spp., *Salmonella* spp., *Escherichia coli*

Subacute sclerosing panencephalitis, a complication of some measles infections, is an example discussed in Chapter 25. **Latent diseases** are those in which a pathogen remains inactive for a long period of time before becoming active. Herpes is an example of a latent disease.

When an infectious disease comes from another infected host, either directly or indirectly, it is a **communicable** disease. Influenza, herpes, and tuberculosis are examples of communicable diseases. If a communicable disease is easily transmitted between hosts, as is the case for chickenpox or measles, it is also called a **contagious** disease. **Noncommunicable** diseases arise outside of hosts or from normal microbiota. In other words, they are not spread from one host to another, and diseased patients are not a source of contamination for others. Tooth decay, acne, and tetanus are examples of noncommunicable diseases. **Table 14.12** defines these and other terms used to classify infectious diseases.

Yet another way in which all infectious diseases may be classified is by the effects they have on populations, rather than on individuals. Is a certain disease consistently found in a given group of people or geographic area? Under what circumstances is it more prevalent than normal in a given geographic area? How prevalent is "normal"? How is the disease transmitted throughout a population? We next examine issues related to diseases at the population level.

Epidemiology of Infectious Diseases

Learning Objective

✔ Define epidemiology.

Our discussion so far has centered on the negative impact of microorganisms on *individuals*. Now we turn our attention to the effects of pathogens on *populations*. **Epidemiology**[10] (ep-i-dē-mē-ol'ō-jē) is the study of where and when diseases occur, and how they are transmitted within populations. During the 20th century, epidemiologists expanded the scope of their work beyond infectious diseases to also consider injuries and deaths related to automobile and fireworks accidents, cigarette smoking, lead poisoning, and other causes; however, we will limit our discussion primarily to the epidemiology of infectious diseases.

Frequency of Disease

Learning Objectives

✔ Contrast between incidence and prevalence.
✔ Differentiate among the terms *endemic, sporadic, epidemic,* and *pandemic.*

[10]From Greek *epidemios,* meaning among the people, and logos, meaning study of.

Table 14.12 Terms Used to Classify Infectious Diseases

Term	Definition
Acute disease	Disease in which symptoms develop rapidly and that runs its course quickly
Chronic disease	Disease with usually mild symptoms that develop slowly and last a long time
Subacute disease	Disease with time course and symptoms between acute and chronic
Asymptomatic disease	Disease without symptoms
Latent disease	Disease that appears a long time after infection
Communicable disease	Disease transmitted from one host to another
Contagious disease	Communicable disease that is easily spread
Noncommunicable disease	Disease arising from outside of hosts or from opportunistic pathogen
Local infection	Infection confined to a small region of the body
Systemic infection	Widespread infection in many systems of the body; often travels in the blood or lymph
Focal infection	Infection that serves as a source of pathogens for infections at other sites in the body
Primary infection	Initial infection within a given patient
Secondary infection	Infections that follow a primary infection; often by opportunistic pathogens

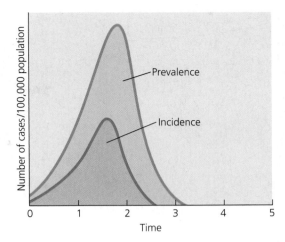

▲ *Figure 14.13*

Curves representing the incidence (orange) and the prevalence (blue) of a given disease over time. Both measures are expressed as a ratio of cases to a selected portion of the population at risk. *Why can the incidence of a disease never exceed the prevalence of that disease?*

Figure 14.13 Because prevalence includes all cases, both old and new, prevalence must always be larger than incidence.

Epidemiologists keep track of the occurrence of diseases by using two measures: incidence and prevalence. **Incidence** is the number of *new* cases of a disease in a given area or population during a given period of time; **prevalence** is the *total number* of cases, both new and already existing, in a given area or population during a given period of time. Both measures are typically expressed as a ratio of the number of cases divided by the number of people at risk:

$$\text{Incidence} = \frac{\text{No. of new cases}}{\text{No. of people at risk}}$$

$$\text{Prevalence} = \frac{\text{No. of old and new cases}}{\text{No. of people at risk}}$$

Thus, for example, the reported number of new cases of tuberculosis in the United States in 2002 was 11,878, and the population at risk was about 289,939,000, so the incidence of tuberculosis was about 4.10 people. However, the prevalence of tuberculosis in 2002 was about 6.87/100,000 people because over 8,000 patients who got the disease prior to 2002 still had the disease. **Figure 14.13** illustrates the relationships between incidence and prevalence.

Epidemiologists report their data in many ways including maps, graphs, charts, and tables (**Figure 14.14** on page 426). Why do they report their data in so many different ways? A variety of formats enables epidemiologists to observe patterns that may give clues about the causes of or ways to prevent diseases. For example, shigellosis (an acute bacterial infection characterized by diarrhea, abdominal pain, and fever) had an overall prevalence of 85 cases per million in the U.S. population in 2000, but when the data are reported by age group, it becomes apparent that the highest prevalence (350 cases per million) occurs in young children (see Figure 14.14c on p. 426).

The occurrence of a disease can also be considered in terms of a combination of frequency and geographic distribution (**Figure 14.15** on p. 427). A disease that normally occurs continually (at moderately regular intervals) at a relatively stable frequency within a given population or geographical area is said to be **endemic**[11] to that population or region. A disease is considered **sporadic** when only a few scattered cases occur within an area or population. Whenever a disease occurs at a greater frequency than is usual for an area or population, the disease is said to be **epidemic** within that area or population.

The commonly held belief that a disease must infect thousands or millions to be considered an epidemic is mistaken. The time period and the number of cases necessary for an outbreak of disease to be classified as an epidemic are not specified; the important fact is that there are more cases than historical statistics indicate are expected. For example,

---

[11]From Greek *endemos,* meaning native.

(a)

(b)

(c)

▲ *Figure 14.14*

Epidemiologists report data in a number of ways. Here incidence of shigellosis in the U.S. in 2000 is presented **(a)** on a map by state, **(b)** by month, and **(c)** as cases per million by age group. The latter graph reveals that shigellosis in young children is much more prevalent (as high as 350 cases per million) than the prevalance in the U.S. population as a whole, a fact not apparent unless the data are sorted by age.

less than 200 cases of hantavirus pulmonary syndrome occurred in the Four Corners area[12] of the United States in 1993, but because they were the first cases of this disease ever reported for the region, the outbreak was considered an epidemic. At the same time and place, thousands of cases of flu occurred, but there was no flu epidemic because the number of cases observed did not exceed the number expected. **Figure 14.16** illustrates how epidemics are defined according to the number of expected cases, and not according to the absolute number of cases.

If an epidemic occurs simultaneously on more than one continent, it is referred to as a **pandemic** (see Figure 14.15d). AIDS is pandemic worldwide.

Obviously, for disease prevalence to be classified as either endemic, sporadic, epidemic, or pandemic, good records must be kept for each region and population. From such records, incidence and prevalence can be calculated, and then changes in these data can be noted. Health departments at the local and state levels require doctors and hospitals to report certain infectious diseases. Some are also nationally notifiable; that is, their occurrence must be reported to the Centers for Disease Control and Prevention (CDC) in Atlanta, Georgia, which is the headquarters and clearinghouse for national epidemiological research. Nationally notifiable diseases are listed in **Table 14.13** on page 429. Each week the CDC reports the number of cases of all nationally notifiable diseases in the *Morbidity and Mortality Weekly Report (MMWR)* (**Figure 14.17** on p. 428).

Epidemiological Studies

Learning Objective

✓ Explain three approaches epidemiologists use to study diseases in populations.

Epidemiologists conduct research to study the dynamics of diseases in populations by taking three different approaches, called descriptive, analytical, and experimental epidemiology.

Descriptive Epidemiology

Descriptive epidemiology involves the careful tabulation of data concerning a disease. Relevant information includes the location and time of cases of the disease, as well as information about the patients, including age, gender, occupation, health history, and socioeconomic status. Because the time course and chains of transmission of a disease are an important part of descriptive epidemiology, epidemiologists strive to identify the **index case** (the first case) of the disease in a given area or population. Sometimes it is difficult or impossible to identify the index case because the patient has recovered, moved, or died.

---

[12]Four Corners is the geographic region surrounding the point at which Arizona, Nevada, New Mexico, and Colorado meet.

Figure 14.15 ➤

Illustrations of the different terms for the occurrence of disease. The colored areas indicate the normal range for the disease, and dots represent new cases of the disease. **(a)** An endemic disease, which is normally present in a region. **(b)** A sporadic disease, which occurs irregularly and infrequently. **(c)** An epidemic disease, which is present in greater frequency than is usual. **(d)** A pandemic disease, which is an epidemic disease occurring on more than one continent at a given time. *Regardless of where you live, name a disease that is endemic, one that is sporadic, and one that is epidemic in your state.*

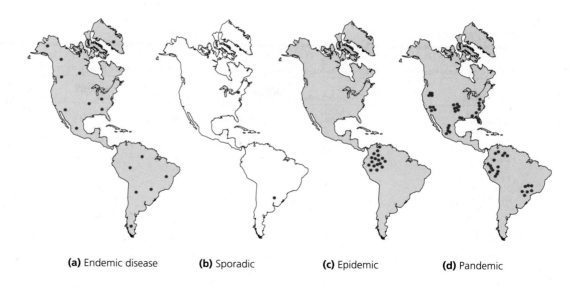

(a) Endemic disease   **(b)** Sporadic   **(c)** Epidemic   **(d)** Pandemic

Figure 14.15 Some possible answers: Flu is endemic in every state; tuberculosis is sporadic in most states; AIDS is epidemic in every state.

◀ *Figure 14.16*

Two graphs that demonstrate the independence between the absolute number of cases and a disease's designation as an epidemic. Even though the number of cases of disease A always exceeds the number of cases of disease B, only disease B is ever considered epidemic—at those times when the number of cases observed exceeds the number of cases expected. *How can scientists know the normal prevalence of a given disease?*

Figure 14.16 Scientists can record every case of the disease so that they will have a "baseline" prevalence, which then becomes the expected prevalence for each disease.

Epidemic 1

Epidemic 2

Number of cases per 100,000 individuals

1999   2000   2001

Time

Expected number of cases
Actual number of cases Disease A
Actual number of cases Disease B

Vol. 51 / No. 32   MMWR   727

TABLE. (*Continued*) Reported cases of notifiable diseases, by geographic division and area — United States, 2001

Reporting area	Legionellosis	Listeriosis	Lyme disease	Malaria	Measles Indigenous	Measles Imported*	Meningococcal disease
United States	**1,168**	**612**	**17,027**	**1,544**	**62**	**54**	**2,326**
New England	**74**	**57**	**5,526**	**107**	**4**	**1**	**113**
Maine	8	2	108	5	—	—	8
N.H.	12	4	129	2	—	—	14
Vt.	5	3	18	1	1	—	7
Mass.	21	30	1,164	53	2	1	57
R.I.	13	3	510	16	1	—	7
Conn.	15	15	3,597	30	—	—	20
Mid. Atlantic	**285**	**119**	**8,907**	**440**	**7**	**13**	**256**
Upstate N.Y.	82	36	4,020	76	—	4	72
N.Y. City	43	26	63	250	3	4	42
N.J.	24	20	2,020	65	—	1	43
Pa.	136	37	2,804	49	4	4	99
E.N. Central	**316**	**88**	**720**	**177**	**—**	**10**	**359**
Ohio	143	17	44	27	—	3	89
Ind.	23	8	26	19	—	4	47
Ill.	24	24	32	71	—	3	88
Mich.	82	25	21	40	—	—	83
Wis.	44	14	597	20	—	—	52
W.N. Central	**55**	**22**	**540**	**77**	**2**	**4**	**173**
Minn.	15	4	461	45	2	2	29
Iowa	8	2	36	9	—	—	31
Mo.	22	10	37	15	—	2	58
N. Dak.	1	—	—	—	—	—	8
S. Dak.	3	—	—	—	—	—	5
Nebr.	5	1	4	2	—	—	27
Kans.	1	5	2	6	—	—	15
S. Atlantic	**223**	**77**	**1,039**	**317**	**3**	**2**	**383**
Del.	12	NN	152	2	—	—	6
Md.	32	16	608	112	2	1	42
D.C.	8	—	17	13	—	—	—
Va.	39	15	156	55	1	—	46
W. Va	NN	6	16	1	—	—	15
N.C.	11	NA	41	19	—	—	63
S.C.	15	5	6	9	—	—	33
Ga.	12	16	—	45	—	—	57
Fla.	94	19	43	61	—	1	121
E.S. Central	**63**	**23**	**72**	**38**	**2**	**—**	**144**
Ky.	14	7	23	14	2	—	27
Tenn.	32	9	31	14	—	—	63
Ala.	13	7	10	6	—	—	35
Miss.	4	—	8	4	—	—	19
W.S. Central	**31**	**34**	**87**	**91**	**—**	**1**	**336**
Ark.	—	1	4	3	—	—	25
La.	7	—	8	6	—	—	78
Okla.	7	2	—	5	—	—	32
Tex.	17	31	75	77	—	1	201
Mountain	**57**	**38**	**15**	**68**	**1**	**1**	**102**
Mont.	—	—	—	3	—	—	4
Idaho	3	1	5	4	—	1	8
Wyo.	3	2	1	1	—	—	5
Colo.	16	9	—	25	—	—	37
N. Mex.	3	7	1	3	—	—	11
Ariz.	21	10	3	19	1	—	21
Utah	7	2	1	4	—	—	8
Nev.	4	7	4	9	—	—	8
Pacific	**64**	**154**	**121**	**229**	**43**	**22**	**460**
Wash.	10	14	9	19	13	2	71
Oreg	NN	12	15	17	3	—	63
Calif.	48	122	95	179	25	15	310
Alaska	1	—	2	1	—	—	3
Hawaii	5	6	—	13	2	5	13
Guam	—	—	—	1	—	—	—
P.R.	2	—	—	6	1	—	9
V.I.	NA	NA	NA	NA	NA	NA	NA
American Samoa	—	—	—	—	—	—	3
C.N.M.I.	—	—	—	—	—	—	—

NA: Not available   NN: Not notifiable   —: No reported cases
* Imported cases include only those resulting from importation from other countries.

◀ *Figure 14.17*

A page from the CDC's *Morbidity and Mortality Weekly Report (MMWR),* which reports epidemiological data state by state for the current week, for the current year to date, and for the previous year to date. The *MMWR* also publishes reports on epidemiological case studies. This page shows the incidence of six diseases in one week in 2001.

An early systematic, epidemiologic study was Dr. John Snow's descriptive study of a cholera outbreak in London in 1854. By carefully mapping the locations of the cholera cases in a particular part of the city, Snow found that the cases were clustered around the Broad Street water pump (**Figure 14.18** on p. 430). This distribution of cases, plus the voluminous watery diarrhea of cholera patients, suggested that the disease was spread via contamination of drinking water by sewage.

Analytical Epidemiology

Analytical epidemiology investigates a disease in detail, including analysis of data acquired in descriptive epidemiological studies, to determine the probable cause, mode of transmission, and possible means of prevention of the disease. Analytical epidemiology may be used in situations where it is not ethical to apply Koch's postulates. Thus, even though Koch's third postulate has never been fulfilled in the case of AIDS (because it is unethical to intentionally inoculate a human with HIV), analytical epidemiological studies indicate that HIV causes AIDS and that it is transmitted primarily sexually.

Often analytical studies are *retrospective;* that is, they attempt to identify causation and mode of transmission after an outbreak has occurred. Epidemiologists compare a group of people who had the disease with a group who did not. The groups are carefully matched by factors such as gender, environment, and/or diet, and then compared to determine which pathogens and factors may play a role in morbidity.

Experimental Epidemiology

Experimental epidemiology involves testing a hypothesis concerning the cause of a disease. The application of Koch's postulates is an example of experimental epidemiology. Experimental epidemiology also involves studies to test a hypothesis resulting from an analytical study such as the efficacy of a preventative measure or certain treatment. For example, analytical epidemiological studies have suggested that the bacterium *Chlamydia pneumoniae* causes arteriosclerosis, resulting in heart attacks. Some scientists have hypothesized that antibacterial drugs could prevent heart attacks by killing the causative agent. To test this hypothesis the scientists administered either the antimicrobial drug azithromycin or a medicinally inactive placebo to 7,700 patients, all of whom had a history of heart disease and were infected with *C. pneumoniae.* The researchers observed no

Table 14.13	Nationally Notifiable Infectious Diseases[a]
Acquired Immunodeficiency Syndrome (AIDS)	Pertussis
Anthrax	Plague
Botulism	Poliomyelitis, paralytic
Brucellosis	Psittacosis
Chancroid	Q fever
Chlamydia trachomatis, genital infections	Rabies, animal
	Rabies, human
Cholera	Rocky Mountain spotted fever
Coccidioidomycosis	
Cryptosporidiosis	Rubella
Cyclosporiasis	Rubella, congenital syndrome
Diphtheria	
Ehrlichiosis	Salmonellosis
Encephalitis/meningitis, arboviral	Shigellosis
Enterohemorrhagic *E. coli*	Streptococcal disease, invasive, Group A
Giardiasis	Streptococcal toxic shock syndrome
Gonorrhea	
Haemophilus influenzae, invasive disease	*Streptococcus pneumoniae,* drug resistant
Hansen disease (leprosy)	*Streptococcus pneumoniae,* invasive in children
Hantavirus pulmonary syndrome	
Hemolytic uremic syndrome, postdiarrheal	Syphilis
	Syphilis, congenital
Hepatitis A	Tetanus
Hepatitis B	Toxic-shock syndrome
Hepatitis C	Trichinosis
HIV infection	Tuberculosis
Legionellosis	Tularemia
Listeriosis	Typhoid fever
Lyme disease	Varicella (deaths only)
Malaria	West Nile encephalitis/ meningitis
Measles	
Meningococcal disease	Yellow fever
Mumps	

[a]Diseases for which hospitals, physicians, and other health care workers are required to report cases to state health departments, who then forward the data to the CDC.

significant difference in the number of heart attacks suffered by patients in the two groups over a two-year period. Thus an experimental epidemiological study disproved a hypothesis suggested by an analytical epidemiological analysis.

Case Study 14.3 on page 431 illustrates the work of epidemiologists.

Figure 14.18 ➤
A map showing cholera deaths in a section of London, 1854. From the map he compiled, Dr. John Snow correctly deduced that the Broad Street pump was the source of this cholera epidemic. Snow's work was a landmark in epidemiologic research.

Hospital Epidemiology: Nosocomial Infections

Learning Objectives

✓ Explain how nosocomial infections differ from other infections.
✓ Describe the factors that influence the development of nosocomal infections.
✓ Describe three types of nosocomial infections and how they may be prevented.

Of special concern to epidemiologists and health care workers are nosocomial[13] (nos-ō-kō′mē-ăl) infections and nosocomial diseases. **Nosocomial infections** are those acquired by patients while they are in health care facilities, including hospitals, dental offices, nursing homes, and doctor's waiting rooms. The CDC estimates that about 10% of American patients—over 2 million people—acquire a nosocomial infection each year. **Nosocomial diseases**—those acquired in a health care setting—increase the duration and cost of medical care and result in some 90,000 deaths annually.

Types of Nosocomial Infections

When most people think of nosocomial infections, what likely comes to mind are **exogenous** (eks-oj′ĕ-nŭs) infections, which are caused by pathogens acquired from the health care environment. After all, hospitals are filled with sick people shedding pathogens from every type of portal of exit. However, we have seen that members of the normal microbiota can become opportunistic pathogens as a result of hospitalization or medical treatments such as chemotherapy. Such opportunists cause **endogenous** (en-doj′ĕ-ñŭs) nosocomial infectious; that is, they arise from normal microbiota within the patient that become pathogenic because of factors within the health care setting.

---

[13]From Greek *nosokomeian,* meaning hospital.

Legionella in the Produce Aisle

On October 31, 1989, the Lousiana state health department received reports of 33 cases of Legionnaire's disease in the town of Bogalusa (population 16,000). Legionnaire's disease, or legionellosis, is a potentially fatal respiratory disease caused by the growth of a bacterium, _Legionella pneumophila,_ in the lungs of patients. The bacterium enters humans via the respiratory portal in aerosols produced by cooling towers, air conditioners, whirlpool baths, showers, humidifiers, and respiratory therapy equipment.

Epidemiologists began trying to ascertain the source of Bogalusa's Legionnaire's disease outbreak by interviewing the victims and their relatives to develop complete histories, and to identify areas of commonality among the victims that were lacking among nonvictims. Victims reported a range of ages, occupations, hobbies, religions, and types and locations of dwellings. No significant differences were identified among the lifestyles, ages, or smoking habits of victims and nonvictims, but one curious fact was discovered: All the victims did their grocery shopping at the same store. However, healthy individuals also shopped at that store.

The air conditioning system of the grocery store proved to be free of _Legionella,_ but the vegetable misting machine did not. The strain of _Legionella_ isolated from the misting machine was identical to the strain recovered from the lungs of the victims.

1. _Would this outbreak be classified as endemic, epidemic, or pandemic?_
2. _Was this a descriptive, analytical, or experimental epidemiological study?_
3. _Knowing the epidemiology and causative agent of Legionnaire's disease, what questions would you have asked of the victims or of their surviving relatives?_
4. _What, as an epidemiologist, would you have examined at the store?_
5. _How did the victims become contaminated? Why didn't everyone who bought vegetables at the store get legionellosis? What could the owners of the store do to limit or prevent future infections?_

Source: _Morbidity and Mortality Weekly Report 39_(7):108–109, 1990.

Iatrogenic (ī-at-rō-jen′ik) **infections** (literally meaning "doctor induced" infections) are a subset of nosocomial infections that ironically are the direct result of modern medical procedures such as the use of catheters, invasive diagnostic procedures, and surgery.

Superinfections may also result from the use of antimicrobial drugs that, by inhibiting some resident microbiota, allow others to thrive in the absence of competition. For instance, long-term antimicrobial therapy to inhibit a bacterial infection may allow _Clostridium difficile_ (klos-trid′ē-ŭm dif′fi-sil), a transient microbe of the colon, to grow excessively and cause a painful condition called pseudomembranous colitis. Such superinfections are not limited to health care settings.

Nosocomial infections most often occur in the urinary, respiratory, cardiovascular, and integumentary (skin) systems, though surgical wounds can become infected and result in nosocomial infections in any part of the body.

Factors Influencing Nosocomial Infections

Nosocomial infections arise from the interaction of several factors in the health care environment, which include the following (**Figure 14.19** on p. 432):

- Exposure to numerous pathogens present in the health care setting, including many that are resistant to antimicrobial agents
- The weakened immune systems of patients who are ill, making them more susceptible to opportunistic pathogens
- Transmission of pathogens among patients, and health care workers—from staff and visitors, to patients, and even from one patient to another via activities of staff members (including invasive procedures and other iatrogenic factors)

Control of Nosocomial Infections

Reduction in the incidence of nosocomial infections involves procedures and precautions designed to reduce the influencing factors, especially the transfer of pathogens within the health care setting. These include disinfection; medical asepsis (including good housekeeping, hand asepsis, bathing, sanitary handling of food, proper hygienic attitudes toward bodily functions, and precautionary measures to avoid the spread of pathogens among patients); surgical asepsis and sterile procedures (including thorough cleansing of the surgical field, use of sterile instruments, and use of sterile gloves, gowns, caps, and masks); isolation of particularly contagious and particularly susceptible patients; and establishment of a nosocomial infection control committee charged with surveillance of nosocomial diseases and review of control measures.

The CDC has instituted the following Universal Precautions to limit occupational exposure to pathogens and reduce the transfer of pathogens to patients:

- Do not sheath used needles, scalpels, or other sharp instruments. Instead, immediately dispose of them in proper "sharps" containers.
- Use protective barriers to prevent exposure to blood, body fluids containing visible blood, and other fluids such as saliva and phlegm.
- Immediately and thoroughly wash hands and other skin surfaces that are contaminated by blood or other fluids.

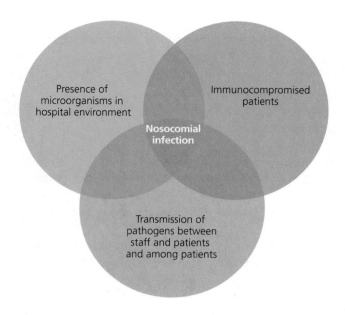

▲ *Figure 14.19*

The interplay of factors that result in nosocomial infections. Although nosocomial infections can result from any one of the factors shown, most nosocomial infections are the product of the interaction of all three factors.

Numerous studies have shown that the single most effective way to reduce nosocomial infections is effective hand antisepsis by all medical and support staff. In one study, deaths from nosocomial infections were reduced by over 50% when hospital personnel followed strict guidelines about washing their hands frequently.

Epidemiology and Public Health

Learning Objective

✓ List three ways public health agencies work to limit the spread of diseases.

As you have likely realized by now, epidemiologists gather information concerning the spread of disease within populations so that they can take steps to reduce the number of cases and improve the health of individuals within a community. In the following sections we will examine how the various public health agencies share epidemiological data, facilitate the interruption of disease transmission, and educate the public about public health issues. Public health agencies also implement immunization programs; immunization is discussed in Chapter 17.

The Sharing of Data Among Public Health Organizations

Numerous agencies at the local, state, national, and global levels work together with the entire spectrum of health care personnel to promote public health. By submitting reports on cases of disease to public officials, practicing physicians can subsequently learn of current disease trends. Additionally, public health agencies often provide physicians with laboratory and diagnostic assistance.

City and county health departments report data on disease incidence to state agencies. Because state laws govern disease reporting, state agencies play a vital role in epidemiological studies. Most states publish a bulletin similar to the MMWR and assist local health departments and medical practitioners with diagnostic testing for diseases such as rabies and Lyme disease.

Data collected by the states is forwarded to the CDC, which is but one branch of the United States Public Health Service, our national public health agency. In addition to epidemiological studies, the CDC and other branches of the Public Health Service conduct research in disease etiology and prevention, make recommendations concerning immunization schedules, and work with public health organizations of other countries.

The World Health Organization (WHO) coordinates efforts to improve public health throughout the world, particularly in poorer countries, and the WHO has undertaken ambitious projects to eradicate such diseases as smallpox, polio, measles, and mumps. Some other current campaigns involve AIDS education, malaria control, and childhood immunization programs in poor countries.

The Role of Public Health Agencies in Interrupting Disease Transmission

As we have seen, pathogens can be transmitted in air, food, and water, as well as by vectors and via fomites. Public health agencies work to limit disease transmission by enforcing standards of cleanliness in water and food supplies, and by working to reduce the numbers of vectors and reservoirs available for transmitting diseases.

A water supply that is **potable** (fit to drink) is vital to good health. Organisms that cause dysentery, cholera, and typhoid fever are just some of the pathogens that can be spread in water contaminated by sewage. Filtration and chlorination processes are used to reduce the number of pathogens in water supplies. Local, state, and national agencies work to ensure that water supplies remain clean and healthful by monitoring both sewage treatment facilities and the water supply.

Food can harbor infective larvae of parasitic worms, infective stages of protozoa, bacteria, and viruses. National and state health officials ensure the safety of the food supply by enforcing standards in the use of canning, pasteurization, irradiation, and chemical preservatives, and by insisting that food preparers and handlers wash their hands and use sanitized utensils. The Department of Agriculture also provides for the inspection of meats for the presence of pathogens such as *E. coli* and *Trichinella* (trik'i-nel'ǎ).

Milk is an especially rich food, and the same nutrients that nourish us also facilitate the growth of many microorganisms. In the past, contaminated milk has been responsible for epidemics of tuberculosis, brucellosis, typhoid fever, scarlet fever, and diphtheria. Today, public health agencies require the pasteurization of milk, and as a result disease transmission via milk has been practically eliminated in the United States.

Individuals should assume responsibility for their own health by washing their hands before and during food preparation, using disinfectants on kitchen surfaces, and by using proper refrigeration and freezing procedures. It is also important to thoroughly cook all meats.

Public health officials also work to control vectors, especially mosquitoes and rodents, by eliminating breeding grounds such as stagnant pools of water and garbage dumps. Insecticides have been used with some success to control insects but fail in the long term because of the development of resistance.

Public Health Education

Diseases that are transmitted through the air or sexually are particularly difficult for public health officials to control. In these cases, individuals must take responsibility for their own health, and health departments can only educate the public to make healthful choices.

Colds and flu remain the most common diseases in the United States because of the ubiquity of the viral pathogens and their mode of transmission in aerosols and via fomites. Health departments can encourage afflicted people to remain at home, use disposable tissues to reduce the spread of viruses, and avoid crowds of coughing, sneezing people. As we have discussed, hand washing is also important in preventing the introduction of cold and flu viruses onto the conjunctiva.

Sexually transmitted diseases such as AIDS, syphilis, gonorrhea, and genital warts are completely preventable if the chain of transmission is interrupted by abstinence or mutually faithful monogamy, and their incidence can be reduced by the use of barrier contraceptives (*e.g.* condoms). Still, we as a society are faced with several epidemics of sexually transmitted diseases. Based on the premise that "an ounce of prevention is worth a pound of cure," public health agencies expend considerable effort in public campaigns to educate individuals to make good choices—those that can result in healthier individuals and greater health for the public at large.

CHAPTER SUMMARY

Symbiotic Relationships Between Microbes and Their Hosts (pp. 405–409)

1. Microbes live with their hosts in **symbiotic** relationships, including **mutualism,** in which both members benefit; **parasitism,** in which a **parasite** benefits while the host is harmed; and **commensalism,** in which one member benefits while the other is relatively unaffected. Any parasite that causes disease is called a **pathogen.**

2. Organisms called **normal microbiota** live in and on the body. Some of these microbes are resident, whereas others are transient.

3. **Opportunistic pathogens** cause disease when the immune system is suppressed, when normal **microbial antagonism (competition)** is affected by certain changes in the body, and when normal microbiota are introduced into an area of the body unusual for that microbe.

The Movement of Microbes into Hosts: Infection (pp. 409–411)

1. Microbial **contamination** refers to the mere presence of microbes in or on the body or object. Microbial contaminants include harmless resident and transient members of the microbiota, as well as pathogens, which after a successful invasion cause an **infection.**

2. **Portals of entry** of pathogens into the body include skin, mucous membranes, the placenta, and the **parenteral route,** by which microbes are directly deposited into deeper tissues.

3. Pathogens can attach to cells—a process called **adhesion**—via a variety of structures or attachment proteins called **adhesion factors.** Some bacteria and viruses lose the ability to make adhesion factors called adhesins and thereby become **avirulent.**

4. Some bacteria interact to produce a sticky web of cells and polysaccharides called a biofilm that adheres to a surface.

The Nature of Infectious Disease (pp. 411–419)

1. **Disease,** also known as **morbidity,** is a sufficiently adverse condition to interfere with normal functioning of the body.

2. **Symptoms** are subjectively felt by a patient, whereas an outside observer can observe **signs.** A **syndrome** is a group of symptoms and signs that collectively characterizes a particular disease or abnormal condition.

3. **Asymptomatic** or **subclinical** infections may go unnoticed due to the absence of symptoms, even though clinical tests might reveal signs of disease.

4. **Etiology** is the study of the cause of a disease.

5. Nineteenth-century microbiologists proposed the **germ theory of disease,** and Robert Koch developed a series of essential conditions called **Koch's postulates** to prove the cause of infectious diseases. Certain circumstances can make the use of these postulates difficult or even impossible.

6. **Pathogenicity** is a microorganism's ability to cause disease; **virulence** is a measure of pathogenicity. **Virulence factors** such as adhesion factors, extracellular enzymes, toxins, and antiphagocytic factors affect the relative ability of a pathogen to infect and cause disease.

7. **Toxemia** is the presence in the blood of poisons called **toxins.** **Exotoxins** are secreted by pathogens into their environment. **Endotoxin,** also known as **lipid A,** is released from the cell wall of dead and dying Gram-negative bacteria and can cause fatal effects.

8. **Antitoxins** are antibodies the host forms against toxins.

9. The stages of infectious diseases are **incubation period, prodromal period, illness, decline,** and **convalescence.**

The Movement of Pathogens Out of Hosts: Portals of Exit (pp. 419–421)

1. **Portals of exit,** such as the nose, mouth, and urethra, allow pathogens to leave the body and are of interest in studying the spread of disease.

2. Living and nonliving continuous sources of infectious disease are called **reservoirs of infection.** Animal reservoirs harbor agents of **zoonotic diseases (zoonoses),** which are diseases of animals that may be spread to humans via direct contact with the animal or its waste products, or via an arthropod **vector.** Humans may be asymptomatic **carriers.**

3. **Nonliving reservoirs** of infection include soil, water, and inanimate objects.

Modes of Infectious Disease Transmission (pp. 421–423)

1. **Direct contact transmission** of infectious diseases involves person-to-person spread by bodily contact. **Indirect contact transmission** occurs when inanimate objects (called **fomites**) inadvertently transmit pathogens.

2. **Droplet transmission** (a third type of contact transmission) occurs when pathogens travel in droplets of mucus less than 1 meter to a new host as a result of speaking, coughing, or sneezing.

3. **Vehicle transmission** involves **airborne, waterborne,** and **food-borne** transmission. **Aerosols** are clouds of water droplets that travel more than 1 meter in airborne transmission. **Fecal-oral infection** can result from sewage-contaminated drinking water or from ingesting fecal contaminents.

4. **Biological vectors** are animals, usually biting arthropods, that serve as both host and vector of pathogens. **Mechanical vectors** are not infected by the pathogens they carry.

Classification of Infectious Diseases (pp. 423–424)

1. There are various ways in which infectious disease may be grouped and studied. When grouped by time course and sever-

ity, disease may be described as **acute, subacute, chronic,** or **latent.**

2. When an infectious disease comes either directly or indirectly from another host, it is considered a **communicable** disease. If a communicable disease is easily transmitted from a reservoir or patient, it is called a **contagious** disease. **Noncommunicable** diseases arise either from outside of hosts or from normal microbiota.

Epidemiology of Infectious Diseases (pp. 424–433)

1. **Epidemiology** is the study of where and when diseases occur, and how they are transmitted within populations.

2. Epidemiologists track the **incidence** and **prevalence** of a disease, and classify disease outbreaks as **endemic** (usually present), **sporadic** (occasional), **epidemic** (more cases than usual), or **pandemic** (epidemic on more than one continent). The Centers for Disease Control and Prevention (CDC) report these findings weekly in the *Morbidity and Mortality Weekly Report* (*MMWR*).

3. **Descriptive epidemiology** is the careful recording of data concerning a disease; it often includes detection of the **index case**—the first case of the disease in a given area or population. **Analytical epidemiology** seeks to determine the probable cause of a disease. **Experimental epidemiology** involves testing a hypothesis resulting from analytical studies.

4. **Nosocomial infections** and **nosocomial diseases** are acquired by patients in health care facilities. They may be **exogenous** (acquired from the health care environment), **endogenous** (derived from normal microbiota that become opportunistic while in the hospital setting), or **iatrogenic** (induced by treatment or medical procedures).

5. Health care workers can protect themselves against occupational exposure to pathogens by following the CDC's Universal Precautions.

6. Public health organizations such as the World Health Organization (WHO) use epidemiological data to promulgate rules and standards for clean, **potable** water and safe food, to prevent disease by controlling vectors and animal reservoirs, and to educate people to make healthful choices concerning the prevention of disease.

QUESTIONS FOR REVIEW

(Answers to multiple choice and fill in the blanks are on the web, along with additional review questions. Visit www.microbiologyplace.com.)

Multiple Choice

1. In which type of symbiosis do both members benefit from their interaction?
   a. mutualism
   b. parasitism
   c. commensalism
   d. pathogenesis

2. An axenic environment is one that
   a. exists in the human mouth.
   b. is free of microbes.
   c. exists in the human colon.
   d. both a and b

3. Which of the following is *false* concerning microbial contaminants?
   a. Contaminants may become opportunistic pathogens.
   b. Most microbial contaminants will eventually cause harm.
   c. Contaminants may be a part of the transient microbiota.
   d. Contaminants may be introduced in a mosquito bite.

4. The most frequent portal of entry for pathogens is
   a. the respiratory tract.
   b. the skin.
   c. the conjunctiva.
   d. a cut or wound.

5. The process by which microorganisms attach themselves to cells is
   a. infection.
   b. contamination.
   c. disease.
   d. adhesion.

6. When pathogenic bacterial cells lose the ability to make adhesins, they
   a. become avirulent.
   b. produce endotoxin.
   c. absorb endotoxin.
   d. increase in virulence.

7. Which of the following are most likely to cause disease?
   a. opportunistic pathogens in a weakened host
   b. pathogens lacking the enzyme kinase
   c. pathogens lacking the enzyme collagenase
   d. highly virulent organisms

8. The nature of bacterial capsules
   a. causes widespread blood clotting.
   b. allows phagocytes to readily engulf these bacteria.
   c. affects the virulence of these bacteria.
   d. has no effect on the virulence of bacteria.

9. Which of the following is the correct sequence of events in infectious diseases?
   a. incubation, prodromal period, illness, decline, convalescence
   b. incubation, decline, prodromal period, illness, convalescence
   c. prodromal period, incubation, illness, decline, convalescence
   d. convalescence, prodromal period, incubation, illness, decline

10. A disease in which a pathogen remains inactive for a long period of time before becoming active is termed a(n)
    a. subacute disease.
    b. acute disease.
    c. chronic disease.
    d. latent disease.

11. Which of the following statements is the best definition of a pandemic disease?
    a. It normally occurs in a given geographic area.
    b. It is a disease that occurs more frequently than usual for a geographical area or group of people.
    c. It occurs infrequently at no predictable time scattered over a large area or population.
    d. It is an epidemic that occurs on more than one continent at the same time.

12. Which of the following types of epidemiologists is most like a detective?
    a. a descriptive epidemiologist
    b. an analytical epidemiologist
    c. an experimental epidemiologist
    d. a reservoir epidemiologist

13. Consider the following case. An animal was infected with a virus. A mosquito bit the animal, was contaminated with the virus, and proceeded to bite and infect a person. Which was the vector?
    a. animal
    b. virus
    c. mosquito
    d. person

14. A patient contracted athlete's foot after long-term use of a medication. His physician explained that the malady was directly related to the medication. Such infections are termed
    a. nosocomial infections.
    b. exogenous infections.
    c. iatrogenic infections.
    d. endogenous infections.

15. Which of the following phrases describes a contagious disease?
    a. a disease arising from a fomite
    b. a disease that is easily passed from host to host in aerosols
    c. a disease that arises from opportunistic, normal microbiota
    d. both a and b

Fill in the Blanks

1. A parasite that causes disease is called a _____.

2. Infections that may go unnoticed due to the absence of symptoms are called _____ infections.

3. The study of the cause of a disease is _____.

4. The study of where and when diseases occur and how they are transmitted within populations is _____.

5. Diseases that are naturally spread from their usual animal hosts to humans are called _____.

6. A nonliving reservoir of disease, such as a toothbrush, drinking glass, or needle, is called a _____.

7. _____ infections are those acquired by patients and staff while in health care facilities.

Short Answer

1. List three types of symbiotic relationships, and give an example of each.

2. List three conditions that create opportunities for pathogens to become harmful in a human.

3. List three portals through which pathogens may enter the body.

4. List Koch's four postulates, and describe situations in which they may not be applicable.

5. List in the correct sequence the five stages of infectious diseases.

6. Describe three modes of disease transmission.

CRITICAL THINKING

1. Explain why patient Q, a menopausal woman, may have developed gingivitis from normal microbiota.

2. Patient W died of *E. coli* infection after an intestinal puncture. Explain why this microbe, which normally lives in the colon, could kill this patient.

3. Examine the following graphs of the RED epidemic and the BLUE epidemic. Neither RED disease nor BLUE disease is treatable. A person with either disease is ill for only 1 day and recovers fully. Both epidemics began at the same time. Which epidemic affected more people during the first 3 days? What could explain the short time course for epidemic RED? Why was epidemic BLUE longer lasting?

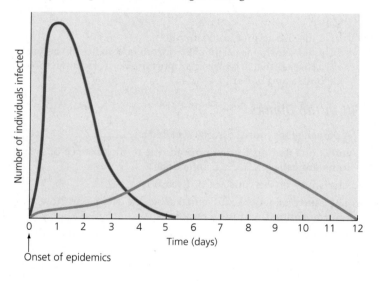

4. A 27-year-old female came to her doctor's office with a widespread rash, fever, malaise, and severe muscle pain. The symptoms had begun with a mild headache 3 days previously. She reported being bitten by a tick 1 week prior to that.

   The doctor correctly diagnosed Rocky Mountain spotted fever (RMSF) and prescribed tetracycline. The rash and other signs and symptoms disappeared in a couple of days, but she continued on her antibiotic therapy for 2 weeks.

   Draw a graph showing the course of disease and the relative numbers of pathogens over time. Label the stages of the disease.

5. Over 30 children younger than 3 years of age developed gastroenteritis after visiting a local water park. These cases represented 44% of the park visitors in this age group on the day in question. No older individuals were affected. The causative agent was determined to be a member of the bacterial genus *Shigella*. The disease resulted from oral transmission to the children.

   Based only on the information given, can you classify this outbreak as an epidemic? Why or why not? If you were an epidemiologist, how would you go about determining which pools in the water park were contaminated? What factors might account for the fact that no older children or adults developed disease? What steps could the park operators take to reduce the chance of future outbreaks of gastroenteritis?

6. Review lichen biology in Chapter 12. Describe the relationship between the photosynthetic member of the lichen and the fungus in terms of the type of symbiosis.

Nonspecific Lines of Defense

CHAPTER 15

Each day the equivalent of a small room full of air enters your respiratory tract through your nose. With that air come dust, smoke, bacteria, viruses, fungi, pollen, soot, fuzz, sand, and more. Acting as a first line of defense, your respiratory mucous membrane uses nose hairs, ciliated epithelium, and mucus to cleanse the inhaled air of pathogens and certain harmful pollutants. Each day you swallow and subsequently digest about a liter of mucus, along with the trapped pathogens and pollutants it contains. Still more mucus clumps around microbes and pollutants to form masses of nasal mucus that may dry out or remain slimy, depending on how rapidly you're breathing and the humidity of the air. Yellowish or greenish mucus contains a large population of trapped bacteria and their waste products.

Mucus is just one example of the body's general, nonspecific defenses against pathogens. The skin, and certain protective cells, chemicals, and processes within the body, are others. In this chapter we will focus on each of these aspects of the body's defenses.

CHAPTER OUTLINE

An Overview of the Body's Defenses

The Body's First Line of Defense
The Role of Skin in Nonspecific Defense
The Role of Mucous Membranes in Nonspecific Defense
The Role of Normal Microbiota in Nonspecific Defense
Other First-Line Defenses

An Overview of the Body's Second Line of Defense
Defense Components of Blood

The Body's Second Line of Defense
Phagocytosis
Extracellular Killing by Leukocytes
Nonspecific Chemical Defenses Against Pathogens
Inflammation
Fever

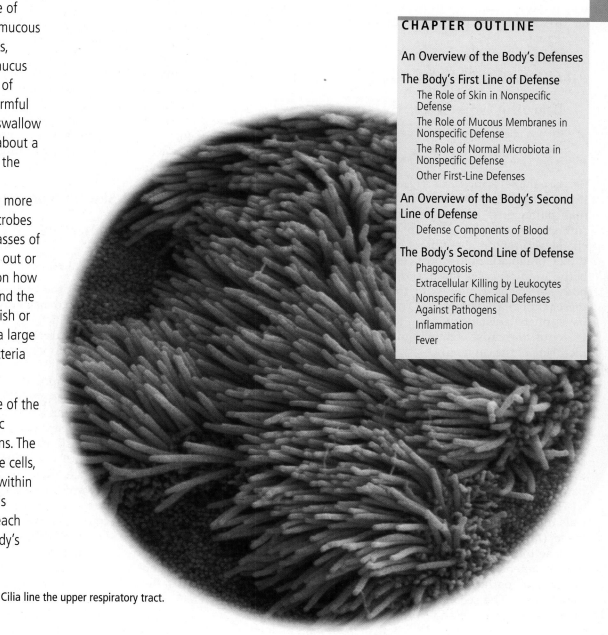

Cilia line the upper respiratory tract.

As we saw in Chapter 14, a pathogen can cause a disease only if it can (1) gain access, either by penetrating the surface of the skin or by entering through some other portal of entry; (2) attach itself to host cells; and (3) evade the body's defense mechanisms long enough to produce harmful changes. In this chapter we will examine the structures, processes, and chemicals that respond in a general way to protect the body from all types of pathogens. Chapter 16 examines the body's specific defense against specific pathogens.

An Overview of the Body's Defenses

Learning Objectives

✓ List and briefly describe the three lines of defense in the human body.

✓ Explain the phrase *nonspecific lines of defense*.

Because the cells and certain basic physiological processes of humans are incompatible with those of most plant and animal pathogens, humans have what is termed **innate (inborn) resistance** to these pathogens. In many cases the chemical receptors these plant or animal pathogens require if they are to attach to a host cell do not exist in the human body; in other cases the pH or temperature of the human body are incompatible with the conditions under which these pathogens can survive. Thus, for example, all humans have innate resistance to both tobacco mosaic virus and to the virus that causes feline immunodeficiency syndrome in members of the cat family.

However, we are confronted every day with a multitude of pathogens that can cause disease in humans. Bacteria, viruses, fungi, protozoa, and parasitic worms come in contact with your body in the air you breathe, the water you drink, the food you eat, and during the contacts you have with other people. Your body must defend itself from these potential pathogens, and in some cases from members of the normal microbiota, which may become opportunistic pathogens.

It is convenient to cluster the structures, cells, and chemicals that act against pathogens into three main lines of defense **(Figure 15.1)**. In actuality, however, each line of defense overlaps and reinforces the other two. The first line of defense is chiefly composed of external physical barriers to pathogens, especially the skin and mucous membranes. The second line is internal and is composed of protective cells, blood-borne chemicals, and processes that inactivate or kill invaders. Both of these lines of defense are **nonspecific;** that is, they respond against a wide variety of pathogens, including parasitic worms, protozoa, fungi, bacteria, and viruses. In other words, nonspecific defenses operate in a generalized way independent of the nature of the intruder.

By contrast, the third line of defense is a *specific immune response* that responds against a specific type (species or strain) of pathogen. We will examine immune responses in

some detail in Chapter 16, but first we will turn our attention to the two lines of nonspecific defense, the topics of this chapter.

The Body's First Line of Defense

The body's initial line of defense is made up of structures, chemicals, and processes that work together to prevent pathogens from entering the body in the first place. Here we discuss the main components of the first line of defense: the skin and the mucous membranes of the respiratory, digestive, urinary, and reproductive systems. These structures provide a formidable barrier to the entrance of microorganisms. As we discussed in Chapter 14, if these barriers are pierced, broken, or otherwise damaged, they become portals of entry for pathogens. In this section we examine aspects of the first line of defense, including the role of the normal microbiota.

The Role of Skin in Nonspecific Defense

Learning Objective

✓ Identify the physical and chemical aspects of skin that enable it to prevent the entrance of pathogens.

The skin—the organ of the body with the greatest surface area—is composed of two major layers: an outer **epidermis,** and a deeper **dermis,** which contains hair follicles, glands, and nerve endings **(Figure 15.2)**. Both the physical structure and the chemical components of skin enable it to act as an effective defense.

The epidermis is composed of multiple layers of tightly packed cells separated by very little extracellular material. It constitutes a physical barrier to most bacteria, fungi, and viruses; very few pathogens can penetrate the layers of epidermal cells unless the skin has been burned, broken, or cut.

The deepest cells of the epidermis continually divide, which pushes their daughter cells toward the surface. As they are pushed toward the surface, the daughter cells flatten and die, and are eventually shed in flakes. Microorganisms that attach to the skin's surface are sloughed off with the flakes of dead cells. **Highlight 15.1** on page 440 describes the fate of lost epidermal cells.

The epidermis also contains phagocytic cells called epidermal **dendritic cells** or *Langerhans* (lahng'ĕr-hahnz) *cells*. The slender, finger–like processes of dendritic cells extend among the surrounding cells, forming an almost continuous network to intercept invaders. These cells both phagocytize pathogens nonspecifically and play a role in specific immunity, as we will see in Chapter 16.

The combination of the barrier function of the epidermis, its continual replacement, and the presence of phagocytic dendritic cells provides significant defense against colonization and infection by pathogens.

The dermis also defends nonspecifically. It contains tough fibers of a protein called collagen. These give the skin

Figure 15.1 ▶

The three lines of defense that protect the body from pathogens.

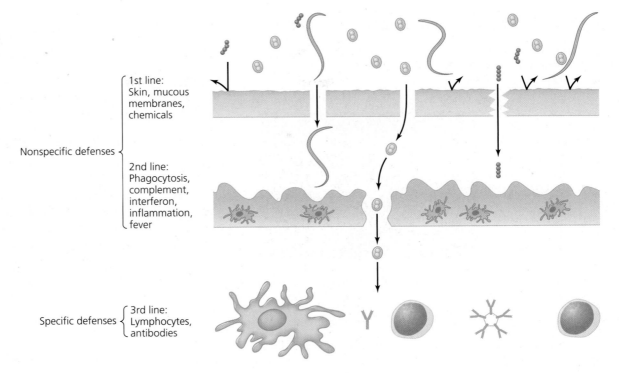

Nonspecific defenses
{
1st line:
Skin, mucous membranes, chemicals

2nd line:
Phagocytosis, complement, interferon, inflammation, fever
}

Specific defenses
{
3rd line:
Lymphocytes, antibodies
}

SEM   40μm

▲ *Figure 15.2*

A scanning electron micrograph of a section of skin. Note that the epidermal cells are tightly packed, which provides an effective barrier to most microorganisms.

strength and pliability to prevent jabs and scrapes from penetrating the dermis and introducing microorganisms. Blood vessels in the dermis deliver defensive cells and chemicals, which will be discussed shortly.

In addition to its physical structure, the skin has a number of chemical substances that nonspecifically defend against pathogens. Sweat glands secrete perspiration (sweat), which contains salt and lysozyme. Salt draws water osmotically from invading cells, which inhibits their growth and kills them. **Lysozyme** (lī′sō-zīm) is an enzyme that destroys the cell walls of bacteria by cleaving the bonds between the sugar subunits of the walls. Bacteria without cell walls are more susceptible to osmotic shock and digestion by other enzymes within phagocytes (discussed with respect to the second line of defense).

The skin also contains sebaceous (oil) glands that secrete **sebum** (sē′bŭm), an oily substance that not only helps keep the skin pliable and less sensitive to breaking or tearing, but also contains fatty acids that lower the pH of the skin's surface to about pH 4, which is inhibitory to many bacteria.

Although salt, lysozyme, and acidity make the surface of the skin an inhospitable environment for most microorganisms, some bacteria, such as *Staphylococcus epidermidis* (staf′i-lō-kok′ŭs ep-ē-der-mi′dis), find the skin a suitable environment for growth and reproduction (see Figure 14.2a). Such bacteria are particularly abundant in crevices around hairs and in the ducts of glands. Usually they are nonpathogenic, but they may become opportunistic pathogens.

The Role of Mucous Membranes in Nonspecific Defense

Learning Objectives

✓ Identify the locations of the body's mucous membranes.
✓ Explain how mucous membranes protect the body both physically and chemically.

Mucus-secreting (mucous) membranes, a second part of the first line of defense, line all body cavities that are open to the

Highlight 15.1   What Happens to All That Skin?

Your body sheds tens of thousands of skin flakes every time you walk or move, and you shed at only a slightly lower rate when you stand still. That comes to about 10 billion skin cells per day, or 250 grams (about half a pound) of skin every year! What happens to all that skin?

Much of household dust is skin that you and your housemates have shed as you go about your lives. The skin flakes fall to the rug and upholstery, where they become food for microscopic mites that live sedentary

Dust mite.

SEM |————| 100 μm

and harmless lives waiting patiently for meals to rain down on them from above. They dwell not only in the rug, but also in your mattress and pillow, and even in the hair follicles of your eyebrows, where they catch skin cells cascading down your forehead.

By the way, house dust also contains mite feces and mite skeletons, which can trigger allergies. So after reading this chapter, you just might want to pull out the vacuum cleaner.

CRITICAL THINKING

Some strains of *Staphylococcus aureus* produce exfoliative toxin, a chemical that causes portions of the entire outer layer of the skin to be sloughed off in a disease called scalded skin syndrome. Given that cells of the outer layers are going to fall off anyway, why is this disease dangerous?

---

outside environment. Thus mucous membranes line the lumens[1] of the respiratory, urinary, digestive, and reproductive tracts. Like the skin, mucous membranes act nonspecifically to limit infection both physically and chemically.

Mucous membranes have two distinct layers: an outer covering of superficial (closest to the surface, in this case the lumen) cells called the *epithelium*, and a deeper connective tissue layer that provides mechanical and nutritive support for the epithelium. Epithelial cells of mucous membranes are packed closely together, like those of the epidermis, but they form only a thin tissue. Indeed, in some mucous membranes, the epithelium is only a single cell thick. Unlike surface epidermal cells, surface cells of mucous membranes are alive and play roles in the diffusion of nutrients and oxygen (in the digestive, respiratory, and female reproductive systems) and in the elimination of wastes (in the urinary, respiratory, and female reproductive systems).

The thin epithelium on the surface of a mucous membrane provides a less efficient barrier to the entrance of pathogens than the multiple layers of dead cells found at the skin's surface. So how are microorganisms kept from invading through these thin mucous membranes? In some cases they are not, which is why some mucous membranes, especially those of the respiratory and reproductive systems, can be common portals of entry for pathogens. However, the epithelial cells of mucous membranes are tightly packed to prevent the entry of pathogens. Moreover, epithelial cells are continually shed and then replaced via the cytokinesis of **stem cells,** which are generative cells capable of dividing to form daughter cells of a variety of types; one effect of such shedding is that it carries attached microorganisms away.

In addition, the epithelia of some mucous membranes have still other means of removing pathogens. In the mucous membrane of the trachea, for example, the stem cells produce both **goblet cells,** which secrete a sticky mucus that traps bacteria and other pathogens, and ciliated columnar cells, whose cilia propel the mucus (and the particles and pathogens trapped within it) up from the lungs **(Figure 15.3).** The effect of the action of the cilia is often likened to that of an escalator. Mucus carried into the throat is coughed up and either swallowed or expelled. Because the poisons and tars in tobacco smoke damage cilia, the lungs of smokers are not properly cleared of mucus, so smokers may develop severe coughs as their respiratory tracts attempt to expel excess mucus from the lungs. Smokers also typically succumb to more respiratory pathogens because they are unable to effectively clear pathogens from their lungs.

In addition to these physical actions, some mucous membranes produce chemicals that defend against pathogens. Nasal mucus, for example, contains lysozyme, which chemically destroys bacterial cell walls. **Table 15.1** on page 442 compares the physical and chemical actions of the skin and mucous membranes in the body's first line of defense.

The Role of Normal Microbiota in Nonspecific Defense

As we saw in Chapter 14, the skin and mucous membranes of the body are normally home to a variety of protozoa, fungi, bacteria, and viruses. This normal microbiota plays a role in protecting the body by competing with potential pathogens in a variety of ways, a situation called **microbial antagonism.**

A variety of activities of the normal microbiota make it less likely that a pathogen can compete with them and produce disease. Some microbiota secrete antimicrobial sub-

---

[1]A lumen is a cavity or channel within any tubular structure or organ.

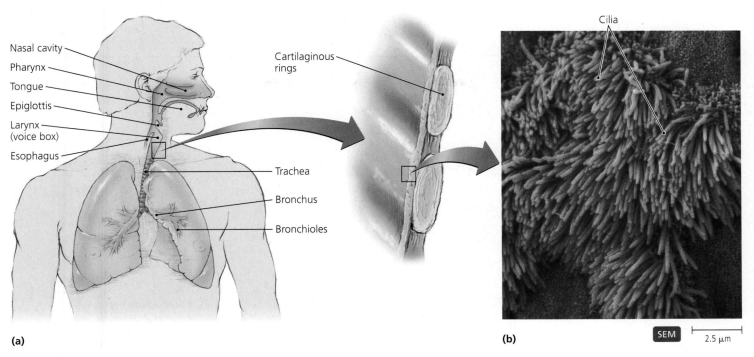

Cilia

Nasal cavity
Pharynx
Tongue
Epiglottis
Larynx
(voice box)
Esophagus

Cartilaginous
rings

Trachea
Bronchus
Bronchioles

(a)

(b) SEM 2.5 μm

▲ *Figure 15.3*

The action of the respiratory mucous membrane in expelling pathogens. **(a)** The structure of the respiratory system, which is lined with a mucous membrane. **(b)** The epithelium of the trachea, which contains mucus-secreting goblet cells, and ciliated cells whose cilia propel the mucus (and the microbes trapped within it) up to the larynx for removal. *What is the function of stem cells within the respiratory epithelium?*

Figure 15.3 Stem cells in the respiratory epithelium undergo cytokinesis to form both ciliated cells and goblet cells to replace those lost during normal shedding.

stances that limit the growth of potential pathogens. Microbiota also consume nutrients, making them unavailable to pathogens. Additionally, the normal microbiota can change the pH, creating an environment that is favorable for themselves but unfavorable to other microorganisms.

Additionally, the presence of the normal microbiota stimulates the body's second line of defense (discussed shortly). Researchers have observed that animals raised in an *axenic* (ā-zēn′ik) environment—that is, one free of all germs—are slower to defend themselves when exposed to a pathogen.

Finally, the normal microbiota of the intestines improve overall health by providing several vitamins, including biotin and pantothenic acid (vitamin B$_5$), which are important in glucose metabolism; folic acid, which is essential for the production of the purine and pyrimidine bases of nucleic acids; and vitamin K, which has an important role in blood clotting.

Other First-Line Defenses

Besides the physical barrier of the skin and mucous membranes, other structures hinder microbial invasion. Notable among these is the lacrimal apparatus of the eye, which is a group of structures that produce and drain away tears (**Figure 15.4** on page 442).

Lacrimal glands, located above and to the sides of the eyes, secrete tears into lacrimal gland ducts and onto the surface of the eyes. The tears either evaporate or drain into small lacrimal canals, which carry them into nasolacrimal ducts that empty into the nose. There, the tears join the nasal mucus and flow into the pharynx, where they are swallowed. The blinking action of the eyelids spreads the tears and washes the surfaces of the eyes. Normally, evaporation and flow into the nose balances the flow of tears onto the eye. However, if the eyes are irritated, increased tear production can flood the eyes, and the overflow can carry the irritant away. In addition to their washing action, tears contain lysozyme, which destroys bacteria.

Many other body organs contribute to the first line of defense by secreting chemicals with antimicrobial properties that are secondary to their prime function. For example, whereas stomach acid is primarily present to aid digestion of proteins, it also prevents the growth of many potential pathogens. Likewise, saliva not only contains digestive

Table 15.1   The First Line of Defense: A Comparison of the Skin and Mucous Membranes

	Skin	Mucous Membrane
Number of layers	Many	One to a few
Cells tightly packed?	Yes	Yes
Dead or alive?	Outer layers: dead; inner layers: alive	Alive
Mucus present?	No	With some
Lysozyme present?	Yes	With some
Sebum present?	Yes	No
Cilia present?	No	With some
Constant shedding and replacement of cells?	Yes	Yes

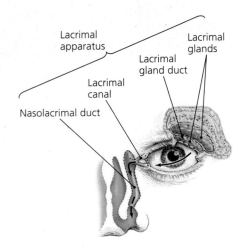

Anterior view

◀ *Figure 15.4*

The lacrimal apparatus, which functions in the body's first line of defense by bathing the eye with tears. Arrows indicate the route tears take across the eye and into the throat. *What antimicrobial protein is found in tears?*

Figure 15.4 Tears contain lysozyme, an antimicrobial protein that acts against the peptidoglycan of bacterial cell walls.

enzymes and provides lubrication for swallowing, it also contains lysozyme and washes microbes from the teeth. The contributions of these and other processes and chemicals to the first line of defense are listed in **Table 15.2.**

An Overview of the Body's Second Line of Defense

Learning Objective

✓   Compare and contrast the body's first and second lines of defense against disease.

When pathogens succeed in penetrating the skin or mucous membranes, the body's second line of defense comes into play. Like the first line of defense, the second line is nonspecific; that is, it operates against a wide variety of pathogens, from parasitic worms to viruses. But unlike the first line of defense, the second line includes no barriers; instead, it is composed of cells (especially phagocytes), antimicrobial chemicals (complement, interferons), and processes (inflammation, fever) (see Figure 15.1). We will consider each of these components of the second line of defense in some detail shortly, but because many of them are either contained in or originate in the blood, we first consider the components of blood.

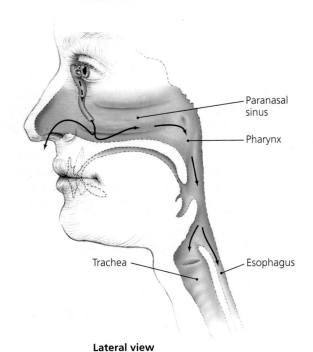

Lateral view

Table 15.2	Secretions and Activities that Contribute to the First Line of Defense
Secretion/Activity	**Function**
Digestive System	
Saliva	Washes microbes from teeth, gums, tongue, and palate; contains lysozyme, an antibacterial enzyme
Stomach acid	Digests and/or inhibits microorganisms
Gastroferritin	Sequesters iron during its absorption, making it unavailable for microbial use
Bile	Inhibitory to most microorganisms
Intestinal secretions	Digests and/or inhibits microorganisms
Peristalsis	Moves gastrointestinal contents through GI tract—constantly eliminating potential pathogens
Defecation	Eliminates microorganisms
Vomiting	Eliminates microorganisms
Nervous System	
Tears	Wash eyes; contain lysozyme
Urinary System	
Urine	Contains lysozyme; its acidity inhibits microorganisms; may wash microbes from ureters and urethra during urination
Reproductive System	
Vaginal secretions	Acidity inhibits microorganisms; contains iron-binding proteins that sequester iron, making it unavailable for microbial use
Menstrual flow	Cleanses uterus and vagina
Prostate secretion	Contains iron-binding proteins that sequester iron, making it unavailable for microbial use
Cardiovascular System	
Blood flow	Removes microorganisms from wounds
Coagulation	Prevents entrance of many pathogens
Transferrin	Binds iron for transport, making it unavailable for microbial use

Defense Components of Blood

Learning Objectives

✓ Discuss the components of blood and their functions in the body's defense.

✓ Explain how macrophages are named.

Blood is a complex liquid tissue composed of cells and portions of cells within a fluid called *plasma*. We begin our discussion of the defense functions of blood by briefly considering plasma.

Plasma

Plasma is mostly water containing electrolytes (ions), dissolved gases, nutrients, and—most relevant to the body's defenses—a variety of proteins. Some plasma proteins are involved in blood clotting, a defense mechanism that both reduces blood loss and the risk of infection. When the clotting factors have been removed from the plasma, as for instance when blood clots, the remaining liquid is called *serum*.

Another group of proteins, called *complement proteins,* is an important part of the second line of defense and is discussed shortly. Still another group of proteins, called *antibodies* or *immunoglobulins,* is a part of specific immunity, the body's third line of defense, which we will explore in Chapter 16.

Defensive Blood Cells: Leukocytes

The cells and cell fragments suspended in the plasma are called **formed elements.** In a process called *hematopoiesis,*[2] blood stem cells located principally in the bone marrow within the hollow cavities of the large bones produce three types of formed elements: **erythrocytes**[3] (ĕ-rith′rō-sītz), **platelets**[4] (plāt′letz), and **leukocytes**[5] (lū′kō-sītz) (**Figure 15.5** on page 444). Erythrocytes, the most numerous of the formed elements, carry oxygen and carbon dioxide in the blood. Platelets, which are pieces of large cells called *megakaryocytes* that have split into small portions of cytoplasm surrounded by cytoplasmic membranes, are involved in blood clotting. Leukocytes, the formed elements that are directly involved in

---

[2]From Greek *haima,* meaning blood, and *poiein,* meaning to make.
[3]From Greek *erythro,* meaning red, and *cytos,* meaning cell.
[4]French for small plates. Platelets are also called thrombocytes, from Greek *thrombos,* meaning lump, and *cytos,* meaning cell, though they are technically not cells, but instead pieces of cells.
[5]From Greek *leuko,* meaning white, and *cytos,* meaning cell.

Figure 15.5 ▶

A schematic representation of hematopoiesis, the process by which the division of stem cells in the bone marrow produces three types of formed (cellular) elements: erythrocytes, platelets, and leukocytes.

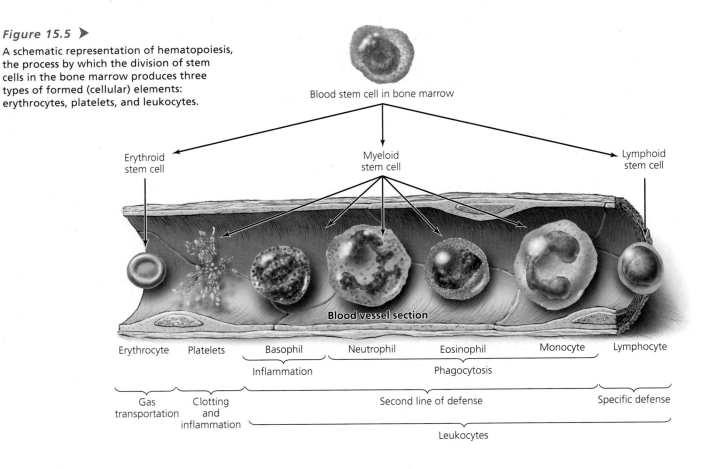

Blood stem cell in bone marrow

Erythroid stem cell

Myeloid stem cell

Lymphoid stem cell

Blood vessel section

Erythrocyte   Platelets   Basophil   Neutrophil   Eosinophil   Monocyte   Lymphocyte

Inflammation   Phagocytosis

Gas transportation | Clotting and inflammation | Second line of defense | Specific defense

Leukocytes

defending the body against invaders, are commonly called white blood cells because they form a whitish layer when the components of blood are separated within a test tube.

Based on their appearance in stained blood smears when viewed under the microscope, leukocytes are divided into two groups: **granulocytes** (gran'ū-lō-sītz) and **agranulocytes** (ă-gran'ū-lō-sītz) **(Figure 15.6).**

Granulocytes have large granules in their cytoplasm that stain different colors depending on the type of granulocyte and the dyes used: **basophils** (bā'sō-fils) stain blue with the basic dye methylene blue; **eosinophils** (ē-ō-sin'ō-fils) stain red to orange with the acidic dye eosin; and **neutrophils** (noo'trō-fils) stain lilac with a mixture of acidic and basic dyes. Both neutrophils (also known as *polymorphonuclear leukocytes*) and eosinophils can phagocytize pathogens, and both can exit the blood to attack invading microbes in the tissues by squeezing between the cells lining capillaries (the smallest blood vessels), a process called **diapedesis**[6] (dī'ă-pĕ-dē'sis) or **emigration (Figure 15.7).** As we will see later in the chapter, eosinophils are also involved in defending the body against parasitic worms and are present in large number during many allergic reactions, though their exact function in allergies is disputed. Basophils can also leave the blood, though they are not phagocytic; instead, they release histamine during inflammation, an aspect of the second line of defense that will be discussed shortly.

The cytoplasm of agranulocytes appears uniform when viewed via light microscopy, though granules do become

visible with an electron microscope. Agranulocytes are of two types: **lymphocytes** (lim'fō-sītz), which are the smallest leukocytes and have nuclei that nearly fill the cells, and **monocytes** (mon'ō-sītz), which are large agranulocytes with slightly lobed nuclei. Although most lymphocytes are involved in specific immunity (see Chapter 16), *natural killer (NK) lymphocytes* function in nonspecific defense and thus are discussed later in this chapter. Monocytes leave the blood and mature into **macrophages** (mak'rō-fāj-ĕz), which are phagocytic cells of the second line of defense. Their initial function is to devour foreign objects, including bacteria, fungi, spores, and dust, as well as dead body cells.

Macrophages are named for their location in the body. **Wandering macrophages** leave the blood via diapedesis and perform their scavenger function while travelling throughout the body, including extracellular spaces. Other macrophages are fixed and do not wander. These include the Langerhans cells of the epidermis, **alveolar** (al-vē'ō-lăr) **macrophages**[7] of the lungs, **microglia** (mī-krog'lē-ă) of the central nervous system, and **Küpffer** (kyūp'fĕr) **cells** of the liver. Fixed macrophages generally phagocytize within a specific organ, where they associate with a *reticulum* (network) of fibers that join the organ's cells together **(Figure 15.8).**

---

[6]From Greek *dia*, meaning through, and *pedan*, meaning to leap.
[7]Alveoli are small pockets at the end of respiratory passages where oxygen and carbon dioxide exchange occurs between the lungs and the blood.

Neutrophil 60–70%   `LM` |—— 7.5 μm ——|

Eosinophil 2–4%   `LM` |—— 7.5 μm ——|

Basophil 0.5–1%   `LM` |—— 7.5 μm ——|

Monocyte 3–8%   `LM` |—— 7.5 μm ——|

Lymphocyte 20–25%   `LM` |—— 7.5 μm ——|

◄ *Figure 15.6*

The appearances of three granulocytes (a neutrophil, an eosinophil, and a basophil) and two agranulocytes (a monocyte and a lymphocyte), as seen in stained blood smears. The numbers are the normal percentages of each cell type among all leukocytes.

`TEM` |—— 10 μm ——|

▲ *Figure 15.7*

Diapedesis, the process whereby leukocytes leave intact blood vessels. Shown here is a neutrophil squeezing between the cells lining a capillary on its way to fight an infection in the tissues.

Küpffer cells

`SEM` |—— 7.5 μm ——|

▲ *Figure 15.8*

A Küpffer cell, a fixed macrophage in the liver. Fixed macrophages remain attached to a reticulum of fibers within specific organs, where they survey the surrounding tissue for the presence of pathogens.

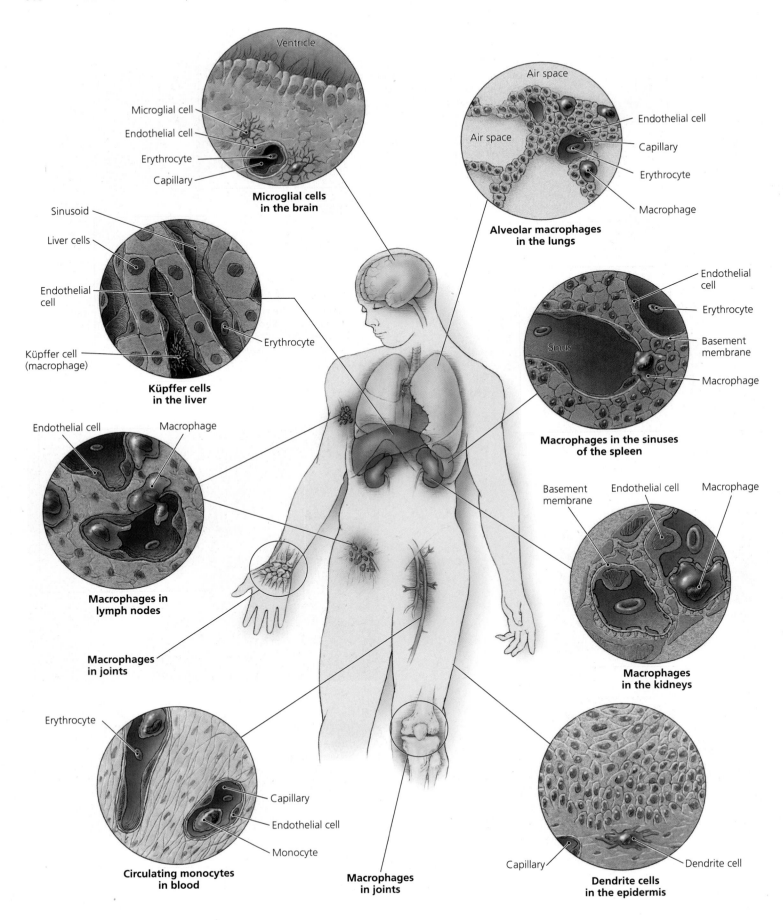

Ventricle

Microglial cell
Endothelial cell
Erythrocyte
Capillary

**Microglial cells
in the brain**

Air space
Air space

Endothelial cell
Capillary
Erythrocyte
Macrophage

**Alveolar macrophages
in the lungs**

Sinusoid
Liver cells
Endothelial
cell
Küpffer cell
(macrophage)

Erythrocyte

**Küpffer cells
in the liver**

Endothelial cell
Erythrocyte
Basement
membrane
Macrophage

Sinus

**Macrophages in the sinuses
of the spleen**

Endothelial cell
Macrophage

**Macrophages in
lymph nodes**

Basement
membrane
Endothelial cell
Macrophage

**Macrophages
in the kidneys**

**Macrophages
in joints**

Erythrocyte
Capillary
Endothelial cell
Monocyte

**Circulating monocytes
in blood**

**Macrophages
in joints**

Capillary
Dendrite cell

**Dendrite cells
in the epidermis**

◀ *Figure 15.9*

The mononuclear phagocytic (reticuloendothelial) system. *Why is this system considered a functional system rather than an anatomical (structural) system?*

Figure 15.9 The mononuclear phagocytic system is considered a functional system because it is composed of cells, located in many body systems, that share a common function: phagocytizing invaders.

As a group, macrophages of all types (fixed or wandering), plus monocytes that are attached to *endothelial cells* (which line the heart chambers, blood vessels, and lymphatic[8] vessels), constitute a system called either the **reticuloendothelial** (rē-tik′yū-lō-en-dō-thē′lē-ăl) **system** or the **mononuclear phagocytic system (Figure 15.9).** Though these terms refer to the same thing, they emphasize different aspects of the nature of this system. The first term emphasizes the association of the cells with reticular fibers, whereas the second term emphasizes their microscopic ap-

[8]The lymphatic system is discussed in Chapter 16.

pearance and primary function. The latter name is preferred because the system is primarily a functional system, not an anatomical one.

Lab Analysis of Leukocytes   Analysis of blood for diagnostic purposes, including white blood cell counts, is one task of medical lab technologists. The proportions of leukocytes, as determined in a **differential white blood cell count,** can serve as a sign of disease. For example, an increase in the percentage of eosinophils can indicate allergies or infection with parasitic worms; bacterial diseases typically exhibit an increase in the number of leukocytes and an increase in the percentage of neutrophils, while viral infections are associated with an increase in the relative number of lymphocytes. The ranges for the normal values for each kind of white blood cell, expressed as a percentage of the total leukocyte population, are shown in Figure 15.6 (on page 445).

CRITICAL THINKING

A medical laboratory technologist argues that granulocytes are a natural group, whereas agranulocytes are an artificial grouping. Based on Figure 15.5, do you agree or disagree with the lab tech? What evidence can you cite to justify your conclusion?

Case Study 15.1   Evaluating a Complete Blood and Differential Leukocyte Count (CBC) Profile

An important diagnostic tool is a count of the number and relative percentages of each type of blood cell in a given volume of blood. Values can increase or decrease as the body responds to infections by bacteria and viruses, as well as to cancer. Consider this portion of a Complete Blood and Differential Leukocyte Count (CBC) profile for a patient named Dorothy. Study the profile and then answer the question concerning her condition on September 7.

1. Which blood cells are at abnormally low levels in Dorothy's blood?

2. Which differential leukocyte counts are outside their normal range?

3. Which of the three lines of defense will be most adversely affected in this patient?

Patient: **Dorothy** _____

CBC Profile

		9/7/00 09:05	7/3/00 14:36	Normal range of values for women
CH				
WBC		6.5	6.9	3.6-10.8 K/µL
RBC	L	3.94	3.62	4.20-5.40 m/Ul
HGB	L	11.7	9.9	12.0-16.0 g/dL
HCT	L	34.8	30.6	37.0-47.0%
MCV		88.4	84.5	81.0-99.0 fL
MCH		29.6	27.3	27.0-31.0 pg
MCHC		33.5	32.3	32.0-36.0 g/dL
RDW	H	15.5	17.0	11.6-14.5%
PLT	H	432	422	130-400 K/µL
MPV	L	6.4	5.7	7.8-11.0 fL
NE%	H	76.3	78.1	37.0-75.0%
LY%	L	10.0	7.1	20.5-51.1%
MO%		10.6	12.7	1.7-15.0%
EO%		2.9	1.6	0.0-10.0%
BA%	L	0.2	0.5	1.0-8.0%
NE#		4.9	5.4	1.4-6.5 K/µL
LY#	L	0.7	0.5	1.2-3.4 K/µL
MO#		0.7	0.9	0.1-0.9 K/µL
EO#		0.2	0.1	0.0-0.7 K/µL
BA#		0.0	0.0	0.0-0.2 K/µL
Plat		Adequate	Adequate	Adequate
Promylo				0-0%

Coulter counter determines the # of each kind of cell (per µL)

Number of platelets

Percentage of leukocytes of each type

Actual number of each type of leukocyte

Values from two dates

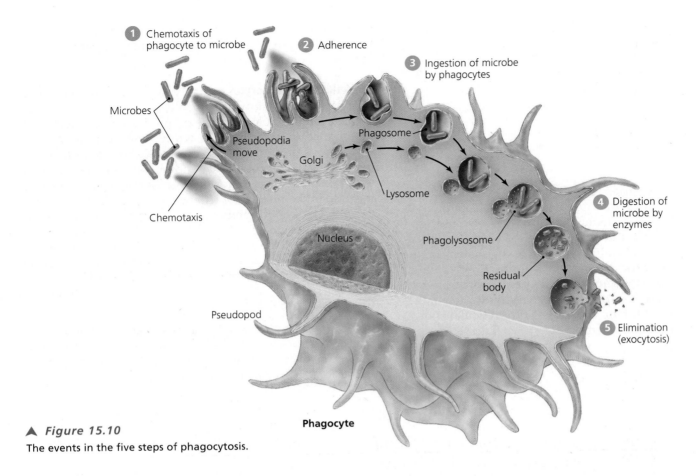

① Chemotaxis of phagocyte to microbe

② Adherence

③ Ingestion of microbe by phagocytes

Microbes

Pseudopodia move

Phagosome

Golgi

Chemotaxis

Lysosome

④ Digestion of microbe by enzymes

Phagolysosome

Nucleus

Residual body

Pseudopod

⑤ Elimination (exocytosis)

Phagocyte

▲ *Figure 15.10*
The events in the five steps of phagocytosis.

The Body's Second Line of Defense

Now that we have some background concerning the defensive properties of plasma components and leukocytes, we turn our attention to the various components of the body's second line of defense: phagocytosis, extracellular killing by leukocytes, nonspecific chemical defenses, inflammation, and fever.

Phagocytosis

Learning Objective

✓ Name and describe the five stages of phagocytosis.

In Chapter 3 we examined phagocytosis as a way by which a cell can transport substances across the cytoplasmic membrane and into the cytoplasm. Here we consider how those cells of the body that are capable of phagocytosis—collectively known as **phagocytes** (fag'ō-sītz), most of which are leukocytes or their derivatives—play a role against pathogens that get past the body's first line of defense.

Although phagocytosis was discovered over a century ago **(see Highlight 15.2),** it is a complex process that is still not completely understood. For the purposes of our discussion, we will divide the continuous process of phagocytosis into five steps: (1) chemotaxis, (2) adherence, (3) ingestion, (4) digestion, and (5) elimination **(Figure 15.10).**

Chemotaxis

Recall from Chapter 3 that *chemotaxis* is movement of a cell either toward a chemical stimulus (positive chemotaxis) or away from a chemical stimulus (negative chemotaxis). In the case of phagocytes, positive chemotaxis involves the use of *pseudopodia* (sū-dō-pō'dē-ǎ) to move toward microorganisms at the site of an infection (step ① in Figure 15.10). Chemicals that attract phagocytic leukocytes include microbial waste products and secretions, components of damaged tissues and white blood cells, and **chemotactic** (kem-ō-tak'tik) **factors.** Chemotactic factors include peptides derived from complement (discussed later in this chapter) and chemicals called **cytokines** (sī'tō-kīnz) (or *chemokines*), which are released by leukocytes already at a site of infection and are considered in Chapter 16.

Adherence

After arriving at the site of an infection, phagocytes attach to microorganisms through the binding of complementary chemicals such as glycoproteins found on the membranes of cells (step ② in Figure 15.10). This process is called **adherence.**

Highlight 15.2   The Discovery of Phagocytes

During the latter part of the 19th century, when microbiology was coming of age in the labs of Louis Pasteur and Robert Koch, a Russian microbiologist Elie Metchnikoff (1845–1916) surprised the scientific world with the announcement that animals were protected from pathogens by the action of wandering cells he called *phagocytes*. Metchnikoff had observed these cells moving in the bodies of starfish larvae by means of pseudopodia, and he hypothesized that because they moved like free-living *Amoeba*, they might also eat like *Amoeba*, phagocytizing invasive pathogens.

He tested his hypothesis by stabbing a starfish larva with a rose thorn from his garden. The next morning he observed that thousands of phagocytes had migrated to the site of infection and were actively devouring germs. He subsequently demonstrated the same phenomenon in other animals, and with these discoveries pointed the way toward understanding both what we now call the second line of defense and cellular immunity, which is part of the body's specific defense.

In 1908, Metchnikoff received a Nobel Prize in Medicine or Physiology for his discoveries.

Some bacteria have virulence factors, such as M protein of *Streptococcus pyogenes* (strep-tō-kok'ŭs pī-aj'en-ēz), or slippery capsules that hinder adherence of phagocytes and thereby increase the virulence of the bacteria. Such bacteria are more readily phagocytized if they are pushed up against a surface such as connective tissue, the wall of a blood vessel, or a blood clot.

All pathogens are more readily phagocytized if they are first covered with either the antimicrobial proteins of complement (discussed later), or the specific antimicrobial proteins called antibodies (discussed in Chapter 16). This coating process is called **opsonization,**[9] and the proteins are called **opsonins.**

Ingestion

After phagocytes adhere to pathogens, they extend *pseudopodia* to surround the microbe (step ③ in Figure 15.10, and **Figure 15.11**). The encompassed microbe is internalized as the pseudopodia fuse to form a sac called a **phagosome.**

Digestion

Digestion begins when lysosomes within the phagocyte fuse with the newly formed phagosomes to form **phagolysosomes** (fag-ŏ-lī'sō-sōmz), or digestive vesicles (step ④ in Figure 15.10). Lysosomes contain digestive enzymes and other antimicrobial substances in an environment with a pH of about 4.0 due to the presence of lactic acid. Among the 30 or so different digestive enzymes within lysosomes are lipases, proteases, nucleases, and a variety of others. Not surprisingly, they are most active at acidic pH.

Lysosomes may also contain toxic oxygen derivatives, including singlet oxygen ($^1O_2$), hydrogen peroxide (H_2O_2), nitric acid (NO), superoxide radicals (O_2^-), hydroxyl radical (OH·), and hypochlorite ions (OCl^-). (Hypochlorite is the ingredient in household bleach that accounts for both its

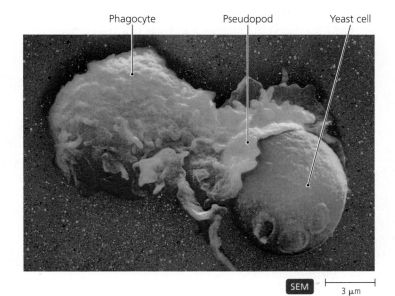

Phagocyte   Pseudopod   Yeast cell

SEM   3 μm

▲ *Figure 15.11*

A phagocyte (in this case, a macrophage) in the process of ingesting a yeast cell by surrounding it with pseudopodia to form a phagosome.

bleaching action and its antimicrobial action.) All of these substances damage proteins and lipids in cell membranes, allowing digestive enzymes access to the contents of microbial cells.

Digestion of most pathogens is complete within 30 minutes, though some bacteria contain virulence factors (such as M protein or waxy cell walls) that resist digestion. After digestion, a phagolysosome is known as a *residual body*.

Elimination

Digestion is not always complete, and phagocytes eliminate undigested remnants of microorganisms via *exocytosis*, a process that is essentially the reverse of ingestion (step ⑤ in Figure 15.10). Some microbial components may remain attached to the cytoplasmic membranes of some phagocytes, a

---

[9]From Greek *opsonein*, meaning to supply food, and *izein*, meaning to cause; thus, loosely, "to prepare for dinner".

Highlight 15.3 **The Fight for Iron**

Most cells require iron for metabolism: It is a component of cytochromes of the electron transport chain and functions as an enzyme cofactor. Because iron is relatively insoluble, in humans it is transported through the blood and to cells that need it by a transport protein called *transferrin*. When transferrin-iron complexes reach cells with receptors for transferrin, the binding of the protein to the receptor stimulates the cell to take up the iron via endocytosis. Excess iron is stored within liver cells bound to another protein called *ferritin*.

Though the main function of iron-binding compounds is transporting and storing iron, they play a secondary, defensive role—sequestering iron so that it is unavailable to microorganisms.

Some bacteria respond to a shortage of iron by secreting their own iron-binding proteins called *siderophores*. Because siderophores have a greater affinity for iron than does transferrin, bacteria that produce them in effect steal iron from the body. In response, the body produces *lactoferrin*, which retakes iron from the bacteria due to its even greater affinity. Thus the body and the pathogens engage in a kind of chemical "tug-of-war" for the possession of iron.

Neisseria meningitidis, a pathogen that causes meningitis, bypasses this contest altogether by producing receptors for transferrin and plucking iron from the bloodstream as it flows by.

phenomenon that plays a role in the specific immune response (discussed in Chapter 16). How is it that phagocytes destroy invading pathogens, and leave the body's own healthy cells unharmed? There are at least two mechanisms responsible for this:

- Some phagocytes have cytoplasmic membrane receptors for various bacterial surface components lacking on the body's cells, such as cell wall components or flagellar proteins.

- Opsonins such as complement and antibody provide a signal to the phagocyte.

Extracellular Killing by Leukocytes

Learning Objective

✓ Describe the role of eosinophils and NK cells in extracellular killing of microorganisms and parasitic helminths.

Phagocytosis involves killing a pathogen once it has been ingested—that is, once it is inside the phagocyte. In contrast, eosinophils and natural killer cells can accomplish *extracellular killing*.

As discussed earlier, eosinophils can phagocytize; however, this is not their usual mode of attack. Instead, they primarily attack parasitic helminths (worms) by attaching to the worm's surface, often binding to antibodies that themselves have bound to chemicals that are specific to the worm. Once bound, eosinophils secrete extracellular protein toxins onto the surface of the parasite. These weaken the helminth and may even kill it. **Eosinophilia** (ē-ō-sin′ō-fil-e-ă), an abnormally high number of eosinophils in the blood, is often indicative of helminth infection.

Natural killer lymphocytes (or **NK cells**) are another type of nonspecific, defensive leukocyte that works by secreting toxins onto the surfaces of virally infected cells and neoplasms (tumors). NK cells identify and spare normal body cells because the latter express membrane proteins similar to those on the NK cells. The ability to distinguish one's own healthy cells from diseased cells and pathogens is discussed more fully in Chapter 16.

Nonspecific Chemical Defenses Against Pathogens

Learning Objectives

✓ Describe the complement system, including its classical and alternate pathways.

✓ Explain the roles of interferons and defensins in nonspecific defense.

Chemical defenses augment phagocytosis in the second line of defense. The chemicals involved assist phagocytic cells either by directly attacking pathogens or by enhancing other features of nonspecific resistance. Defensive chemicals include lysozyme (which we have already discussed), complement, interferons, and defensins. Additionally, **Highlight 15.3** describes the antimicrobial activity of *transferrin*, a chemical whose primary function is transporting iron.

Complement

Complement—or more properly, the **complement system**—is a set of serum proteins designated numerically according to the order of their discovery. These proteins initially act as opsonins and chemotactic factors, and trigger inflammation and fever. The end result of complement activation is lysis of foreign cells.

Complement is activated in several ways. Here we consider two of them **(Figure 15.12)**:

- In the *classical pathway*, the binding of antibodies to foreign antigens activates complement.

- In the *alternate pathway* (also called the *properdin pathway*, after one of the proteins involved), pathogens or pathogenic products (such as bacterial endotoxins and glycoproteins) activate complement.

As Figure 15.12 shows, the two pathways merge, and then complement proteins react with one another in an amplifying sequence of chemical reactions, in which the product of each reaction becomes an enzyme that catalyzes the next reaction many times over. Such reactions are called *cascades* because they progress in a way that can be likened to a

Classical pathway

Antigen

Antibodies

Complement
proteins
1, 2, 4

**Alternate (properdin)
pathway**

C3b

Endotoxin and
glycoproteins

C3b — Properdin B, D and P

Activation
(C3 → C3a + C3b)

C5 convertases
C5 → C5a + C5b

Complement cascade

Opsonization   Inflammation   Membrane attack
complex and cell lysis

▲ *Figure 15.12*

The two pathways by which complement is activated. In
the classical pathway, the binding of antibodies to
antigens activates complement, whereas in the alternate
pathway the binding of properdin to endotoxin or
glycoproteins in the cell walls of bacteria or fungi
activates complement. *How did complement get that name?*

*Figure 15.12 Complement proteins add to—or
complement—the action of antibodies.*

rock avalanche in which one rock dislodges several other
rocks, each of which dislodges many others until a whole
cascade of rocks is tumbling down the mountain. The products of each step in the complement cascade initiate other
reactions, often with wide-ranging effects in the body.

The Classical Pathway   Complement got its name from
events in the originally discovered "classical" pathway, in
which the various complement proteins act nonspecifically
to "complement," or act in conjunction with, the action of
antibodies. As you study the depiction of the classical complement cascade in **Figure 15.13** on page 452, keep the following concepts in mind:

• Complement enzymes cleave other complement molecules to form *fragments*.

• Most fragments have specific and important roles in
  achieving the functions of the complement system.
  Some combine with other complement proteins to form
  new enzymes; some act to increase vascular permeabil-

ity, which increases diapedesis; others enhance inflammation; still others are involved as chemotactic factors
and in opsonization.

• The end products of the entire cascade are **membrane
  attack complexes (MAC),** which form circular holes in
  a pathogen's membrane. The production of numerous
  MACs (**Figure 15.14** on page 453) leads to lysis in a
  wide variety of bacterial and eukaryotic pathogens.
  Gram-negative bacteria are particularly sensitive to the
  production of MACs via the complement cascade because their outer membranes are exposed. In contrast, a
  Gram-positive bacterium, which has a thick layer of
  peptidoglycan overlying its cytoplasmic membrane, is
  typically resistant to the MAC-induced lytic properties
  of complement, though they are susceptible to the
  other effects of the complement cascade.

In addition to its enzymatic roles, fragment C3b acts as
an opsonin, and fragments C3a and C5b function as chemotactic factors, attracting phagocytes to the site of infection.
C3a and C5b are also inflammatory agents that trigger increased vascular permeability and dilation. The inflammatory roles of these fragments are discussed in more detail
shortly.

The Alternate (Properdin) Pathway   As previously mentioned, antibodies bound to antigens are necessary for the
classical activation of complement, whereas activation of the
alternate properdin pathway occurs independent of antibodies. Instead, it is initiated by an interaction between three
proteins (called *properdin factors B, D,* and *P*) and the endotoxins and lipopolysaccharides from bacteria and fungi.
These molecules from microorganisms stabilize molecules of
C3b, which are always found in small quantities in the blood.
Stabilized C3b combines with properdin factors B, P, and D
to form an enzyme that cleaves many copies of C3 to release
many more molecules of C3b. C3b molecules also combine
with properdin B to form a second enzyme that cleaves molecules of C5, and the complement cascade then continues as
in the classical pathway.

Even though the alternate pathway is less efficient than
the classical pathway, it is useful in the early stages of infections by fungi and Gram-negative bacteria—before the specific immune response has created the antibodies needed to
activate the classical pathway.

CRITICAL THINKING

A patient has a genetic disorder that makes it impossible for her to
synthesize complement protein 8 (C8). Is her complement system
nonfunctional? What major effects of complement could still be
produced?

Inactivation of Complement   We have seen that the complement system is nonspecific, and that MACs can form on
any cell's exposed membrane. How do the body's own cells
withstand the action of complement? Membrane-bound

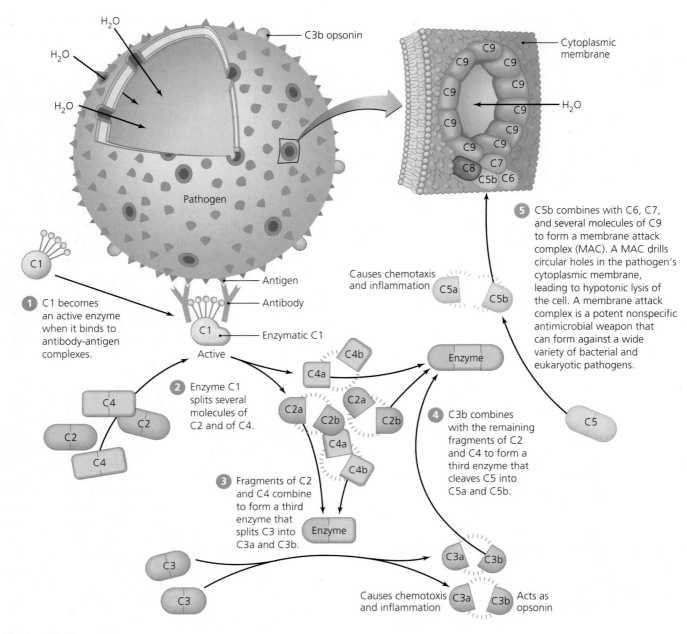

▲ Figure 15.13

The classical complement cascade. *What proteins would be involved if this were the alternate pathway?*

proteins on many cells bind with and break down activated complement proteins, which interrupts the complement cascade before damage can occur. Additionally, because most human cells replace their cell membrane at a very high rate, old membrane and any membrane attack complexes within it are shed or taken in via endocytosis and digested.

Interferons

So far in this chapter we have focused primarily on how the body defends itself against bacteria and eukaryotes. Now we consider how chemicals in the second line of defense act against viral pathogens.

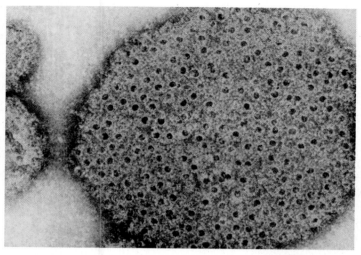

▲ *Figure 15.14*

Transmission electron micrograph of a cell damaged by numerous punctures produced by membrane attack complexes.

As mentioned in Chapter 3, viruses use a host's metabolic machinery to produce new viruses. For this reason, it is often difficult to interfere with virus replication without also producing deleterious effects on the host. **Interferons** (in-ter-fēr′onz) are protein molecules released by host cells to nonspecifically inhibit the spread of viral infections. Its lack of specificity means that interferon produced against one viral invader protects somewhat against infection by other types of viruses as well.

Interferons are particularly effective against viruses with RNA genomes and are often crucial in reducing their effects; however, interferons also cause the malaise, muscle aches, chills, and fever typically associated with viral infections.

Each of the three basic classes of interferon—alpha, beta, and gamma—is produced by different cell types, and interferons within any given class share certain physical and chemical features though they are specific to the species that produces them. In general, alpha and beta interferons are present early in viral infections, whereas gamma interferon appears somewhat later in the course of infection. Because their actions are identical, we examine alpha- and beta-interferons together before discussing gamma-interferon.

Alpha- and Beta-Interferons   Within hours after infection, virally infected monocytes, macrophages, and some lymphocytes secrete small amounts of **alpha-interferons (α-IFN);** similarly, fibroblasts, which are undifferentiated cells in such connective tissues as cartilage, tendon, and bone, secrete small amounts of **beta-interferon (β-IFN)** when infected by viruses. The structures of alpha- and beta-interferons are similar, and their actions are identical.

Interferons do not protect the cells that secrete them—these cells are already infected with viruses. Instead, interferons trigger protective steps in neighboring uninfected cells. Alpha- and beta-interferons bind to interferon receptors on the cytoplasmic membranes of neighboring cells. Such binding triggers the production of **antiviral proteins (AVPs),** which remain inactive within these cells until AVPs bind to double-stranded RNA, a molecule that is common among viruses but generally absent in eukaryotic cells (**Figure 15.15** on page 454).

At least two types of AVPs are produced: *oligoadenylate synthetase,* the action of which results in the degradation of mRNA, and *protein kinase,* which inhibits protein synthesis by ribosomes. Between them, these AVP enzymes essentially destroy the protein production system of a cell, preventing viruses from being replicated. Of course, cellular metabolism is also negatively affected. The antiviral state lasts 3–4 days, which may be long enough for a cell to rid itself of viruses and still a short enough period for the cell to survive without protein production.

Gamma-Interferon   The third type of interferon, **gamma-interferon (γ-IFN),** is produced by activated T lymphocytes and NK lymphocytes. Because T lymphocytes are usually produced as part of the specific immune response (see Chapter 16) days after an infection has occurred, gamma-interferon appears later than either alpha- or beta-interferon. Its action in stimulating the activity of macrophages gives γ-IFN its other name: *macrophage activation factor.*

Although gamma-interferon plays a significant role in protecting the body against viral infections, another of its actions—stimulating the phagocytic activity of neutrophils—is also significant. This is particularly evident when one considers *chronic granulomatous disease (CGD),* a congenital condition that results from the inability to synthesize γ-IFN. Because their neutrophils cannot kill bacteria, children with CGD are susceptible to opportunistic bacterial infections, reducing their life expectancy to about 10 years.

Table 15.3 on page 455 summarizes various properties of the three types of interferon in humans.

Interferon Therapy   As scientists learned more about the effects of interferons, many thought these proteins might be a "magic bullet" against viral infections. Many types of interferons in all three classes have been produced in laboratories using recombinant DNA technology, in the hopes that antiviral therapy can be improved. Recombinant interferons have two significant advantages over interferons collected from blood serum: They are pure and can be manufactured in large quantities.

The initial enthusiasm that greeted both the discovery of interferon's antiviral role and the production of recombinant interferon was not long lived. Interferons have only a slight

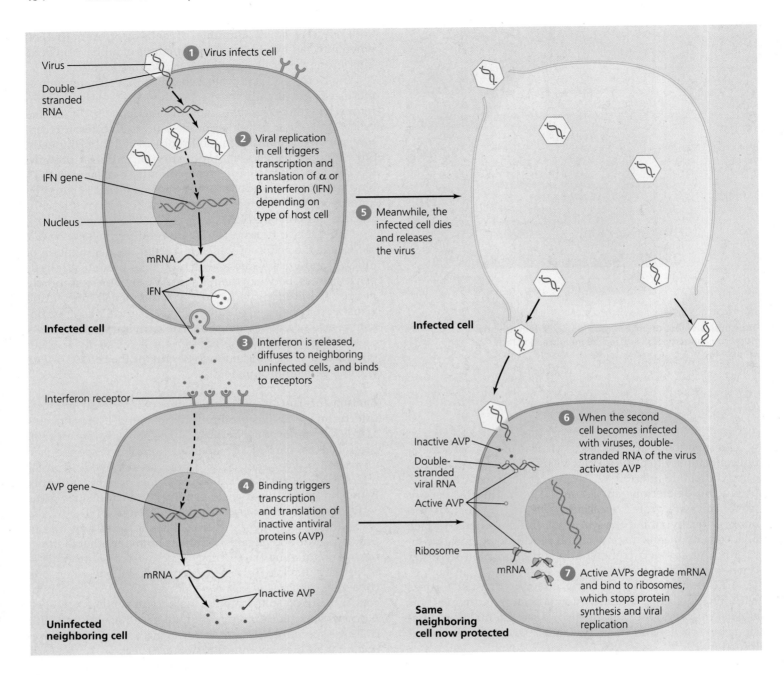

▲ *Figure 15.15*
The actions of alpha- and beta-interferons.

effect against some major virally induced diseases, such as herpes infections, Kaposi's sarcoma (a cancer of blood vessels primarily affecting AIDS patients), and hepatitis B and C infections. Many viral infections do not respond to interferon therapy at all.

One condition that has responded well to recombinant IFN therapy, however, is chronic granulomatous disease. Injections of γ-IFN enable neutrophils in patients with CGD to kill bacteria normally. Gamma-interferon is not a cure for this condition, but it can be an effective lifelong treatment.

Defensins

Defensins (dē-fen´sinz) are small peptides (about 40 amino acids in length) that function nonspecifically to protect against a broad range of pathogens. They occur in a wide variety of animals, including invertebrates such as insects. Different classes of defensins typically act against a specific group of pathogens (for example, Gram-negative bacteria, Gram-positive bacteria, yeasts, and so on).

Although defensins do not harm the cells of the animals that secrete them, they do act against the cells of pathogens

Table 15.3   The Characteristics of Human Interferons

Property	Alpha-interferon	Beta-interferon	Gamma-interferon
Principal source	Epithelium, leukocytes	Fibroblasts	Activated T lymphocytes and NK lymphocytes
Inducing agent	Viruses	Viruses	Specific immune responses
Action	Stimulates production of antiviral proteins	Stimulates production of antiviral proteins	Stimulates phagocytic activity of macrophages and neutrophils
Molecular weight (kdaltons)	16–23	23	20–25
Glycosylated?	No	Yes	Yes
Stable at pH 2?	Yes	Yes	No
Other names	Leukocyte-IFN	Fibroblast-IFN	Immune-IFN, macrophage activation factor

in a variety of ways. Some defensins punch holes in cytoplasmic membranes (in a fashion similar to complement), some interfere with internal signaling and other metabolic processes, and one interferes with the heat shock protein *DnaK,* which normally preserves the three-dimensional shape of bacterial proteins during heat stress. In humans, defensin production increases in response to the chemicals of inflammation, which we discuss next.

Inflammation

Learning Objective

✓ Discuss the process and benefits of inflammation.

Inflammation is a general, nonspecific response to tissue damage resulting from a variety of causes, including heat, chemicals, ultraviolet light (sunburn), abrasions, cuts, and pathogens. There are two types of inflammation: **Acute inflammation,** which develops quickly and is short lived, is typically beneficial and results in the elimination or resolution of whatever condition precipitated it; whereas **chronic inflammation** develops slowly, lasts a long time, and can cause damage (even death) to tissues, resulting in disease. Both acute and chronic inflammation exhibit similar signs and symptoms, including **rubor** (redness), **calor** (localized heat), **tumor** or *edema* (swelling), and **dolor** (pain). Loss of function may also accompany severe inflammation of either type.

It may not be obvious from this list of signs and symptoms that acute inflammation is beneficial; however, acute inflammation is an important part of the second line of defense because it results in (1) dilation and increased permeability of blood vessels, (2) migration of phagocytes, and (3) tissue repair. Although the chemical details of inflammation are beyond the scope of our study, we now consider these three aspects of acute inflammation.

Dilation and Increased Permeability of Blood Vessels

Part of the body's initial response to an injury or invasion of pathogens is localized dilation (increase in diameter) of

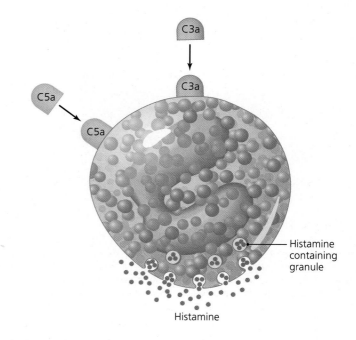

▲ *Figure 15.16*

The stimulation of inflammation by complement. Complement fragments C3a and C5a bind to platelets, basophils, and mast cells, causing them to release histamine, which in turn stimulates the dilation of small blood vessels.

blood vessels in the affected region. Damaged cells release various chemicals, including **histamine** (his′tă-mēn), **prostaglandins** (pros-tă-glan′dinz), and **leukotrienes** (loo-kō-trī′ēnz). In addition, basophils, platelets, and specialized cells located in connective tissue—called **mast cells**—release histamine when they are exposed to complement fragments C3a and C5a **(Figure 15.16).** Recall that these complement peptides were cleaved from larger proteins during the complement cascade.

Histamine causes vasodilation of the body's smallest blood vessels **(Figure 15.17** on page 456). Vasodilation results in more blood being delivered to the site of infection, which in turn delivers more phagocytes, oxygen, and nutrients to

Figure 15.17 ➤

The dilating effect of histamine on small blood vessels. The release of histamine from damaged or infected tissue causes nearby arterioles and capillaries to dilate, which enables more blood to be delivered to the affected site. The increased blood flow causes the reddening and heat associated with inflammation.

Before   **After**

Normal permeability of capillary

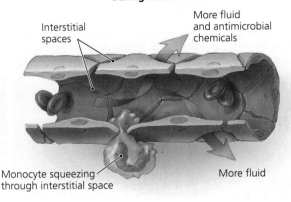

Increased permeability of capillary during inflammation

▲ *Figure 15.18*

The production of increased vascular permeability during inflammation. The presence of prostaglandins, leukotrienes, and histamine causes the cells lining capillaries (shown here, but those lining arterioles and venules as well) to pull apart, allowing phagocytes to leave the bloodstream and more easily reach a site of infection. The leakage of fluid and cells causes the edema and pain associated with inflammation.

the site. Prostaglandins, leukotrienes, and (to some extent) histamine also make blood vessels more permeable—that is, they cause cells lining the vessel to contract and pull apart, leaving gaps in the vessel's wall through which macrophages and neutrophils can move into the damaged tissue and fight invaders **(Figure 15.18).** Increased permeability also allows delivery of more blood-borne antimicrobial chemicals to the site.

Dilation of blood vessels in response to histamine results in the redness and localized heat associated with inflammation. At the same time, prostaglandins and leukotrienes cause fluid to leak from the more permeable blood vessels and accumulate in the surrounding tissue, resulting in edema, which is responsible for much of the pain of inflammation as pressure is exerted on nerve endings.

Vasodilation and increased permeability also deliver fibrinogen, the blood's clotting protein. Clots forming at the site of injury or infection wall off the area and help prevent pathogens and their toxins from spreading. One result is the formation of *pus*, a fluid containing dead tissue cells and dead leukocytes in the walled-off area of tissue destruction.

Pus may push up toward the surface and erupt, or it may remain isolated in the body, where it is slowly absorbed over a period of days. Such an isolated site of infection is called an **abscess.** Pimples, boils, and pustules are examples of abscesses.

The signs and symptoms of inflammation can be treated with antihistamines, which block histamine receptors on blood vessel walls, or with antiprostaglandins. One of the ways aspirin and ibuprofen reduce pain is by acting as antiprostaglandins.

Migration of Phagocytes

Increased blood flow due to vasodilation delivers monocytes and neutrophils to the site of infection. As they arrive, these leukocytes stick to the walls of the blood vessels in a process called **margination.** They then squeeze between the cells of the vessel's wall (diapedesis) and enter the site of infection, usually within an hour of tissue damage. The phagocytes then destroy the pathogens via phagocytosis.

Figure 15.19 ▶

An overview of the events in inflammation. The process, which is characterized by redness, swelling, heat, and pain, ends with tissue repair. *In general, what types of cells are involved in tissue repair?*

Figure 15.19 Tissue repair is effected by cells that are capable of cytokinesis and differentiation. If fibroblasts are among them, scar tissue is laid down.

As mentioned previously, phagocytes are attracted to the site of infection by chemotactic factors, including leukotrienes, microbial components and toxins, and C3a and C5a. The first phagocytes to arrive are often neutrophils, which are then followed by monocytes. Once monocytes leave the blood, they change and become wandering macrophages, which are especially active phagocytic cells that devour pathogens, damaged tissue cells, and dead neutrophils. Wandering macrophages are a major component of pus.

Tissue Repair

The final stage of inflammation is tissue repair, which in part involves the delivery of extra nutrients and oxygen to the site. Although not all sites are repairable, areas of the body where cells regularly undergo cytokinesis, such as the skin and mucous membranes, are repaired rapidly.

If the damaged tissue contains undifferentiated stem cells, tissues can be fully restored. For example, a minor skin cut is repaired to such an extent it is no longer visible. However, if cells called *fibroblasts* are involved to a significant extent, scar tissue is formed, inhibiting normal function. Some tissues, such as cardiac muscle and parts of the brain, do not replicate and thus tissue damage cannot be repaired. As a result, these tissues remain damaged following heart attacks and strokes.

An overview of the entire inflammatory process is illustrated in **Figure 15.19. Table 15.4** on page 458 summarizes the chemicals involved in inflammation.

CRITICAL THINKING

While using a microscope to examine a sample of pus from a pimple, Maria observed a large number of macrophages. Is the pus from an early or a late stage of infection? How do you know?

Fever

Learning Objective

✓ Explain the benefits of fever in fighting infection.

Fever is a body temperature above 37°C. Fever augments the beneficial effects of inflammation, but like inflammation it also has unpleasant side effects, including malaise, body aches, and tiredness.

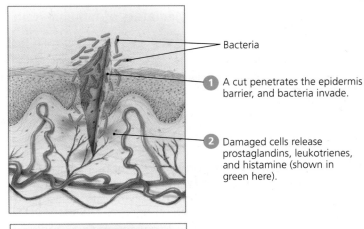

— Bacteria

1 A cut penetrates the epidermis barrier, and bacteria invade.

2 Damaged cells release prostaglandins, leukotrienes, and histamine (shown in green here).

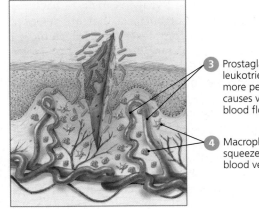

3 Prostaglandins and leukotrienes make vessels more permeable. Histamine causes vasodilation, increasing blood flow to the site.

4 Macrophages and neutrophils squeeze through walls of blood vessels (diapedesis).

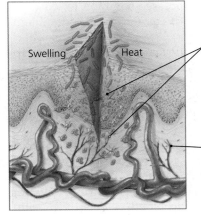

Swelling   Heat

5 Increased permeability allows antimicrobial chemicals and clotting proteins to seep into damaged tissue but also results in swelling, pressure on nerve endings, and pain.

— Nerve ending

6 Blood clot forms.

7 More phagocytes migrate to the site and devour bacteria.

8 Accumulation of damaged tissue and leukocytes forms pus.

9 Undifferentiated stem cells repair the damaged tissue. Blood clot is absorbed or falls off as a scab.

Table 15.4	Chemical Mediators of Inflammation
Vasodilating chemicals	Histamine, serotonin, bradykinin, prostaglandins
Chemotactic factors	Fibrin, collagen, mast cell chemotactic factors, bacterial peptides
Substances with both vasodilating and chemotactic effects	Complement fragments C5a and C3a, interferons, interleukins, leukotrienes, platelet secretions

The hypothalamus, a portion of the brain just above the brain stem, controls the body's internal (core) temperature. Fever results when the presence of chemicals called **pyrogens**[10] (pī′rō-jenz) trigger the hypothalamic "thermostat" to reset at a higher temperature. Pyrogens include bacterial toxins, cytoplasmic contents of bacteria that are released upon lysis, antibody-antigen complexes, and **interleukin 1 (IL-1),** a pyrogen released by phagocytes that have phagocytized bacteria. Although the exact mechanism of fever production is not known, the following discussion and **Figure 15.20** present one possible explanation.

IL-1 produced by phagocytes causes the hypothalamus to secrete prostaglandin, which resets the hypothalamic thermostat by some unknown mechanism. The hypothalamus then communicates the new temperature setting to other parts of the brain, which initiate nerve impulses that produce rapid and repetitive muscle contractions (shivering), an increase in metabolic rate, and constriction of blood vessels of the skin. These processes combine to raise the body's core temperature until it equals the prescribed temperature setting.

Because blood vessels in the skin constrict as fever progresses, one effect of inflammation (vasodilation) is undone. The constricted vessels carry less blood to the skin, causing it to feel cold to the touch, even though the body's core temperature is higher. This symptom is the *chill* associated with fever.

Fever continues as long as IL-1 is present. As an infection comes under control and fewer active phagocytes are involved, the level of IL-1 decreases, the thermostat is reset to 37°C, and the body begins to cool by perspiring, lowering the metabolic rate, and dilating blood vessels in the skin. These processes, collectively called the *crisis* of a fever, are a sign that the infection has been overcome and that body temperature is returning to normal.

The increased temperature of fever enhances the effects of interferons, inhibits the growth of some microorganisms, and is thought to enhance the performance of phagocytes, the activity of cells of specific immunity, and the process of tissue repair. However, if fever is too high, critical proteins are denatured; additionally, nerve impulses are inhibited, resulting in hallucinations, coma, and even death.

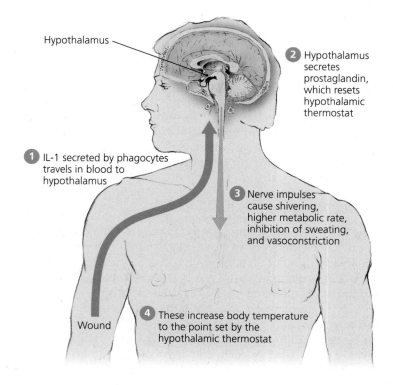

Hypothalamus

2 Hypothalamus secretes prostaglandin, which resets hypothalamic thermostat

1 IL-1 secreted by phagocytes travels in blood to hypothalamus

3 Nerve impulses cause shivering, higher metabolic rate, inhibition of sweating, and vasoconstriction

Wound

4 These increase body temperature to the point set by the hypothalamic thermostat

▲ *Figure 15.20*
One theoretical explanation for the production of fever in response to infection.

Because of the potential benefits of fever, many doctors recommend that patients refrain from taking fever-reducing drugs unless the fever is prolonged and/or extremely high. Other physicians believe that the benefits of fever are too slight to justify enduring the adverse symptoms.

CRITICAL THINKING

How do drugs such as aspirin and ibuprofen act to reduce fever?

The barriers, cells, chemicals, and processes involved in the body's first two, nonspecific lines of defense are summarized in **Table 15.5.**

---

[10]From Greek *pyr,* meaning fire, and *genein,* meaning to produce.

Table 15.5	A Summary of Some Nonspecific Components of the First and Second Lines of Defense						
First Line	**Second Line**						
Barriers and Associated Chemicals	Phagocytes	Extracellular Killing	Complement	Interferons	Defensins	Inflammation	Fever
Skin and mucous membranes prevent the entrance of pathogens; chemicals (e.g. sweat, lysozyme, mucus) enhance the protection	Macrophages, neutrophils, and eosinophils ingest and destroy pathogens	Eosinophils and NK lymphocytes kill pathogens without phagocytizing them	Components attract phagocytes, stimulate inflammation, and attack a pathogen's cytoplasmic membrane	Increase resistance of cells to viral infection, slow the spread of disease	Interfere with membranes, internal signaling, metabolism, and heat shock protein	Increases blood flow, capillary permeability, and migration of leukocytes into infected area; walls off infected region; increases local temperature	Mobilizes defenses, accelerates repairs, inhibits pathogens

CHAPTER SUMMARY

An Overview of the Body's Defenses (p. 438)

1. Humans have **innate (inborn) resistance** to certain pathogens, as well as three overlapping lines of defense. The first two lines of defense are **nonspecific** and protect the body against a wide variety of potential pathogens. The third is a specific immune response to particular pathogens.

The Body's First Line of Defense (pp. 438–442)

1. The first line of defense includes the skin, composed of an outer **epidermis** and a deeper **dermis. Dendritic cells** of the epidermis devour pathogens. Sweat glands of the skin produce salty sweat containing an antibacterial protein called **lysozyme. Sebum** is an oily substance of the skin that lowers pH, which deters the growth of many pathogens.

2. The mucous membranes, part of the body's first line of defense, are composed of tightly packed cells often coated with sticky mucus secreted by **goblet cells,** which are produced by **stem cells.**

3. **Microbial antagonism,** the competition between normal microbiota and potential pathogens, also contributes to the body's first line of defense.

4. Tears contain antibacterial lysozyme and also flush invaders from the eyes. Saliva similarly protects the teeth. The low pH of the stomach inhibits most microbes that are swallowed.

An Overview of the Body's Second Line of Defense (pp. 442–447)

1. The second line of defense includes cells (especially **phagocytes**), antimicrobial chemicals (complement, interferons, defensins), and processes (inflammation and fever).

2. Blood is composed of **formed elements** (cells and parts of cells) within a fluid called **plasma.** Serum is that portion of plasma without clotting factors. The formed elements are **erythrocytes** (red blood cells), **leukocytes** (white blood cells), and **platelets.**

3. Based on their appearance in stained blood smears, leukocytes are grouped as either **granulocytes (basophils, eosinophils,** and **neutrophils)** or **agranulocytes (monocytes, lymphocytes).** When monocytes leave the blood they become macrophages.

4. Basophils function to release histamine during inflammation, whereas eosinophils and neutrophils phagocytize pathogens. They exit the capillaries via **diapedesis (emigration).**

5. Macrophages are the principal phagocytic cells of the second line of defense. They are named for their location in the body: **wandering macrophages, alveolar macrophages** (the lungs), **microglia** (the nervous system), and **Küpffer cells** (the liver). The collection of macrophages—along with monocytes attached to the inner layer of blood and lymphatic vessels—are called either the **reticuloendothelial system** or the **mononuclear phagocytic system.**

6. A **differential white blood cell count** is a lab technique that indicates the relative numbers of leukocytes; it can be helpful in diagnosing disease.

The Body's Second Line of Defense (pp. 448–458)

1. **Chemotactic factors,** such as chemicals called **cytokines,** attract phagocytic leukocytes to the site of damage or invasion. The phagocytes attach to pathogens via a process called **adherence.**

2. **Opsonization,** the coating of pathogens by proteins called **opsonins,** makes those pathogens more vulnerable to phagocytes. A phagocyte's pseudopodia then surround the microbe to form a sac called a **phagosome,** which fuses with a lysosome to form a **phagolysosome,** in which the pathogen is digested.

3. Leukocytes can distinguish between the body's normal cells and foreign cells because they have receptor molecules for foreign cells' components or because the foreign cells are opsonized by complement or antibodies.

4. Eosinophils and **natural killer (NK) lymphocytes** attack extra-cellularly, especially in the case of helminth infections and cancerous cells. Eosinophilia—an abnormally high number of eosinophils in the blood—typically indicates such a helminth infection.

5. The **complement system** is a set of proteins that act as chemo-tactic attractants, trigger inflammation and fever, and ulti-mately effect the destruction of foreign cells via the formation of **membrane attack complexes (MAC),** which result in multi-ple, fatal holes in pathogens' membranes. Complement is acti-vated by a classical pathway involving antibodies and by an alternate pathway triggered by bacterial chemicals.

6. **Interferons** are protein molecules that inhibit the spread of vi-ral infections. **Alpha-interferons** and **beta-interferons,** which are released within hours of infection, trigger **antiviral pro-teins** to prevent viral reproduction in neighboring cells. **Gamma-interferons,** produced days after initial infection, ac-tivate macrophages and neutrophils.

7. **Defensins** are small peptide chains, produced by most ani-mals, that act against a broad range of pathogens.

8. **Acute inflammation** develops quickly and damages pathogens, whereas **chronic inflammation** develops slowly and can cause bodily damage that can lead to disease. Signs and symptoms of inflammation include **rubor** (redness), **calor** (heat), **tumor** (swelling), and **dolor** (pain).

9. Infection causes damaged cells to release chemicals such as **histamine,** which triggers **vasodilation,** and **prostaglandins** and **leukotrienes,** which increase permeability of blood ves-sels. **Mast cells** also release histamine when exposed to pep-tides from the complement system. Blood clots may isolate an infected area to form an **abscess** such as a pimple or boil.

10. When leukocytes arrive near the site of infection, they stick to the walls of blood vessels in a process called **margination,** and they then undergo diapedesis to arrive at the site of tissue damage. The increased blood flow of inflammation also brings extra nutrients and oxygen to the infection site to aid in repair.

11. **Fever** results when chemicals called **pyrogens,** including **in-terleukin 1,** affect the hypothalamus in a way that causes it to reset body temperature at a higher level. The exact process of fever and its control are not fully understood.

QUESTIONS FOR REVIEW
(Answers to multiple choice and matching questions are on the web, along with additional review questions. Visit www.microbiologyplace.com.)

Multiple Choice

1. Fixed phagocytes of the epidermis are called
   a. lysozymes.
   b. Goblet cells.
   c. Küpffer cells.
   d. dendritic cells.

2. Mucus-secreting membranes are found in the
   a. urinary system.
   b. digestive cavity.
   c. respiratory passages.
   d. all of the above

3. The complement system involves
   a. the production of antigens and antibodies.
   b. serum proteins involved in nonspecific defense.
   c. a set of genes that distinguish foreign cells from body cells.
   d. the elimination of undigested remnants of microorganisms.

4. The alternate pathway
   a. is also called the properdin pathway.
   b. involves the cleavage of C5 to form C9.
   c. is more efficient than the classical pathway.
   d. involves recognition of antigens bound to specific antibodies.

5. Complement must be inactivated because if it were not,
   a. viruses could continue to multiply inside host cells using the host's own metabolic machinery.
   b. necessary interferons would not be produced.
   c. protein synthesis would be inhibited, thus halting impor-tant cell processes.
   d. it would make holes in the body's own cells.

6. The type of interferon present late in an infection is
   a. alpha-interferon.
   b. beta-interferon.
   c. gamma-interferon.
   d. delta-interferon.

7. Interferons
   a. do not protect the cell that secretes them.
   b. stimulate the activity of neutrophils.
   c. cause muscle aches, chills, and fever.
   d. all of the above

8. In which of the following choices are events in the correct order?
   a. the formation of a MAC, holes in the pathogen's cell mem-brane, a cascade of reactions, the binding of C1
   b. the binding of C1, a cascade of reactions, the formation of a MAC, holes in the pathogen's cell membrane
   c. a cascade of reactions, the binding of C1, the formation of a MAC, holes in the pathogen's cell membrane
   d. the binding of C1, a cascade of reactions, holes in the pathogen's cell membrane, the formation of a MAC

Matching

In the blank beside each cell, chemical, or process in the left-hand column, write the letter of the line of defense that best applies. Each letter may be used several times.

1. ___ Inflammation
2. ___ Monocytes
3. ___ Cleavage of C3
4. ___ Fever
5. ___ Dendritic cells
6. ___ Alpha-interferon
7. ___ Mucous membrane of the digestive tract
8. ___ Neutrophils
9. ___ Epidermis

A. First line of defense
B. Second line of defense
C. Third line of defense

10. ___ Lysozyme

11. ___ Goblet cells

12. ___ Phagocytes

13. ___ Sebum

14. ___ T lymphocytes

15. ___ Defensins

Write the letter of the description that applies to each of the following.

1. ___ Goblet cell

2. ___ Lysozyme

3. ___ Küpffer cell

4. ___ Dendritic cell

5. ___ Cell from sebaceous gland

6. ___ Bone marrow stem cell

7. ___ Eosinophil

8. ___ Alveolar macrophage

9. ___ Microglia

10. ___ Wandering macrophage

A. Leukocyte that primarily attacks parasitic worms

B. Phagocytic cell in lungs

C. Secrets sebum

D. Devours pathogens in epidermis of skin

E. Breaks bonds in bacterial cell wall

F. Phagocytic cell in central nervous system

G. Phagocytic cell in liver

H. Develops into formed elements of blood

I. Intercellular scavenger

J. Secretes mucus

Modified True/False

Indicate which statements are true. Correct all false statements by writing in the blank a word that, if substituted for the italicized word, would make the statement true.

1. _____ The surface cells of the epidermis of the skin are *alive*.

2. _____ The surface cells of mucous membranes are *alive*.

3. _____ Wandering macrophages experience *diapedesis*.

4. _____ *Monocytes* are immature macrophages.

5. _____ *Lymphocytes* are large agranulocytes.

6. _____ *Phagocytes* exhibit chemotaxis toward a pathogen.

7. _____ Adherence involves the binding between complementary chemicals on a *phagocyte* and on the membrane of a body cell.

8. _____ *Opsonins* result when a phagocyte's pseudopodia surround a microbe and fuse to form a sac.

9. _____ Lysosomes fuse with phagosomes to form *digestive vesicles*.

10. _____ A membrane attack complex drills circular holes in a *macrophage*.

11. _____ Rubor, calor, tumor, and dolor are associated with *fever*.

12. _____ Acute and chronic inflammation exhibit *similar* signs and symptoms.

13. _____ The *hypothalamus* of the brain controls body temperature.

14. _____ Basophils release *leukotrienes* when they are exposed to complement fragments C3a and C5a.

15. _____ Defensins are *phagocytic cells* of the second line of defense.

Short Answer

1. In order for a pathogen to cause disease, what three things must happen?

2. How does a phagocyte "know" it is in contact with a pathogen instead of another body cell?

3. Give three characteristics of the epidermis that make it an intolerable environment for most microorganisms.

4. Other than the different molecules involved, what is the difference between classical and alternate pathways?

5. Describe the classical complement cascade pathway from C1 to the MAC.

6. Why is complement protein C4 "out of order"?

CRITICAL THINKING

1. Neighbor A received a chemical burn on his arm and was instructed by his physician to take an over-the-counter, anti-inflammatory medication for the painful, red, swollen lesions. When Neighbor B suffered pain, redness, and swelling from an infected cut on his foot, he decided to take the same anti-inflammatory drug because his symptoms matched Neighbor A's symptoms. How is Neighbor B's inflammation like that of Neighbor A? How is it different? Is it appropriate for B to medicate his cut with the same medicine A used?

2. What might happen to someone whose body did not produce C3? C5?

3. Duncan, age 65, has had diabetes for 40 years, with resulting damage to the small blood vessels in his feet and toes. His circulation is impaired. How might this condition affect his vulnerability to infection?

4. A patient's chart shows that eosinophils make up 8% of his white blood cells. What does this lead you to suspect? Would your suspicions change if you learned that the patient had spent the previous 3 years as an anthropologist living among an African tribe? What is the normal percentage of eosinophils?

5. There are two kinds of agranulocytes in the blood—monocytes and lymphocytes. Janice noted that monocytes are phagocytic and that lymphocytes are not. She wondered why two agranulocytes would be so different. What facts of hematopoiesis can answer her question?

6. A patient has a genetic disorder that prevents him from synthesizing C8 and C9. What effect would this have on his ability to resist blood-borne Gram-negative and Gram-positive bacteria? What would happen if C3 and C5 fragments were also inactivated?

CHAPTER 16

Specific Defense: The Immune Response

Imagine that your friend has been bitten on the hand by a dog. The skin is broken, and the wound is deep. He washes the wound with soap and water, but does not use iodine or any of the more powerful antimicrobial agents we discussed in Chapter 9. Within 24 hours, his hand and arm are swollen, red, and painful, and he is feeling seriously ill. His body's inflammatory response is in full force, but it isn't strong enough to overcome the invading microorganisms.

Fortunately, infections like this trigger the body's *specific immunity,* which is far more sophisticated than the nonspecific defense responses discussed in Chapter 15. In this chapter we will explore what happens when cells of the specific immune system—especially lymphocytes—enter the battle against microbial invaders.

CHAPTER OUTLINE

Elements of Specific Immunity
Antigens
The Cells, Tissues, and Organs of the Lymphatic System
B Lymphocytes (B Cells) and Antibodies
T Lymphocytes (T Cells)
Immune System Cytokines

The Body's Preparation for a Specific Immune Response
Lymphocyte Editing by Clonal Deletion
The Roles of the Major Histocompatibility Complex
Antigen Processing

The Humoral Immune Response
B Cell Activation and Clonal Selection
Memory B Cells and the Establishment of Immunological Memory

The Cell-Mediated Immune Response
The Perforin-Granzyme Cytotoxic Pathway
The CD95 Cytotoxic Pathway
Memory T Cells
T Cell Regulation

Types of Acquired Immunity
Naturally Acquired Active Immunity
Naturally Acquired Passive Immunity
Artificially Acquired Active Immunity
Artificially Acquired Passive Immunity

White blood cells called lymphocytes are an important component of the specific defense system.

MicroPrep Pre-Test: *Take the pre-test for this chapter on the web.*
Visit ***www.microbiologyplace.com.***

In Chapter 15 we discussed the first two lines of defense against microbial pathogens. These *nonspecific defenses* include intact skin and mucous membranes, phagocytosis, complement, inflammation, and fever. We saw as well that humans have innate (inborn) resistance against pathogens of plants and certain pathogens of animals because human cells lack surface receptors to which these microbes can attach.

As is evident in the chapter opening vignette, nonspecific defenses do not always offer the ideal approach to defending the body. Although the mechanisms of nonspecific defense are readily available and fast-acting, they do not change greatly in either intensity or effectiveness in response to the great variety of pathogens confronting us—a fever is a fever, whether it is triggered by a mild influenza virus or the deadly Ebola virus. Additionally, we have seen that the inflammatory response can have undesirable, even highly damaging effects; imagine how uncomfortable life would be if the body mounted a full-blown inflammatory response every time even a single bacterium entered the body.

The body augments the mechanisms of nonspecific defense with another line of defense that destroys invaders while "learning" in the process. This third line of defense is a "smart" system that has "memory" about specific pathogens, so that when it encounters them a second time, it responds rapidly and effectively. This response is called **specific immunity**—the body's ability to recognize and defend itself against distinct invaders and their products, whether they are protozoa, fungi, bacteria, viruses, or toxins.

Elements of Specific Immunity

In contrast to innate defenses, specific immunity is *acquired;* that is, even though it is present only in a rudimentary form at birth, it develops as the cells of specific immunity encounter and respond to various pathogens and abnormal body cells. *Immunologists*—scientists who study the cells and chemicals involved in specific immunity—are continually refining and revising our knowledge of specific immune processes as well as revealing entirely new aspects of this third line of defense. In the following sections we will survey aspects of specific immunity. Just as the program at a dramatic presentation introduces the actors and their roles, so the following sections introduce the "cast of characters" involved in specific immunity. First we will examine molecules called *antigens* that trigger specific immune responses, followed by a brief description of the cells, tissues, and organs of specific immunity. Then we will examine in some detail two groups of white blood cells—B lymphocytes and T lymphocytes—that function in specific immunity, and conclude by considering special chemical signals or mediators involved in coordinating a specific immune response.

Antigens

Learning Objectives

✓ Identify the characteristics of effective antigens.
✓ Distinguish among exogenous antigens, endogenous antigens, and autoantigens.

The body does not direct specific defense against whole bacteria, fungi, protozoa, or viruses, but instead against portions of cells or viruses, and even single molecules that the body recognizes as foreign and worthy of attack. Immunologists call the molecules that trigger a specific immune response **antigens**[1] (an'ti-jenz). Examples of antigens include components of bacterial cell walls, capsules, pili, and flagella, as well as the external and internal proteins of viruses, fungi, and protozoa. Many toxins are also antigenic. Invading microorganisms are not the only source of antigens; for example, food may contain antigens that provoke allergic reactions, and inhaled dust contains mite feces, pollen grains, dander (flakes of skin), and other antigenic particles.

Pathogens (and their antigens) can enter the body via the various portals of entry: through breaks in the skin and mucous membranes of the respiratory, genitourinary, or digestive tracts; through direct injection, as with a snake's bite or a needle administering a vaccine or blood transfusion; or through organ transplants and skin grafts.

Properties of Antigens

Not every molecule is an effective antigen. Among the properties that make certain molecules "more antigenic"—that is, more effective at provoking an immune response—are a molecule's shape, size, and complexity.

The body recognizes antigens by the three-dimensional shapes of regions called **antigenic determinants,** which are also known as *epitopes* (**Figure 16.1a** on page 464). Therefore, stable molecules—those that assume and maintain a definite shape—are more effective antigens than molecules like the protein gelatin, which have changeable shapes.

In general, larger molecules with molecular masses (often called molecular weights) between 5000 and 100,000 daltons are better antigens than smaller ones. The most effective antigens are large foreign macromolecules such as proteins and glycoproteins, but complex carbohydrates and lipids, as well as some bacterial DNA, can be antigenic. Small molecules, especially those with a molecular mass under 5000 daltons, make poor antigens by themselves because they evade detection. Such small molecules, called *haptens,* can become antigenic when bound to larger, carrier molecules (often proteins). For example, proteins in the blood bind to the fungal product, penicillin, which by itself is too small (molecular mass: 302 daltons) to trigger a specific immune response. Bound to a carrier, penicillin can become antigenic and trigger an allergic response in some patients.

---

[1]From *anti*body *gen*erator; antibodies are specific proteins secreted during a specific immune response.

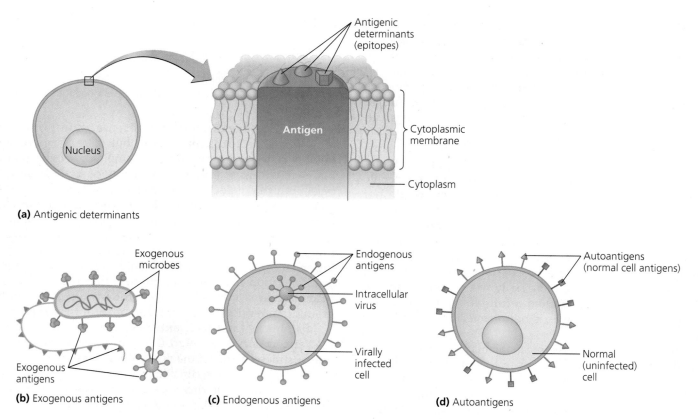

▲ *Figure 16.1*

Antigens, molecules that provoke a specific immune response. **(a)** Antigenic determinants or epitopes, the three-dimensional regions of antigens whose shapes are recognized by cells of the immune system. **(b–d)** Categories of antigens based on their relationship to the body. Exogenous antigens originate from microbes located outside the body's cells; endogenous antigens are produced by intracellular microbes and are typically incorporated into a host cell's cytoplasmic membrane; autoantigens are components of normal, uninfected body cells that are mistakenly recognized as foreign.

Complex molecules make better antigens than simple ones. For example, starch, which is a large polymer of repeating glucose subunits, is not a good antigen despite its very large size because it lacks structural complexity. In contrast, complex molecules, such as glycoproteins and phospholipids, have distinctive shapes and novel combinations of subunits that cells of the immune system recognize as foreign antigenic determinants.

Types of Antigens

Though immunologists categorize antigens in various ways, one especially important way is to group antigens according to their relationship to the body:

- Microbes outside the cells of the body, such as eukaryotic parasites, bacteria, and viruses, are exogenous[2] (eks-oj'en-us) and bear **exogenous antigens (Figure 16.1b).** Exogenous antigens include toxins and other secretions and components of microbial cell walls, membranes, flagella, and pili. Some DNA released from dead bacteria is also antigenic.

- Protozoa, fungi, bacteria, and viruses that reproduce inside a body's cells produce **endogenous**[3] (en-doj'e-nus) **antigens.** The immune system cannot "see" into cells and can respond to endogenous antigens only if the body's cells incorporate such antigens into their cytoplasmic membranes **(Figure 16.1c).**

- **Autoantigens**[4] are antigenic molecules found on an individual's normal (uninfected) cells **(Figure 16.1d).** As we will discuss more fully in a later section, immune cells that recognize autoantigens as foreign are normally eliminated during the development of the immune system. This phenomenon, called *self-tolerance,* prevents the body from mounting an immune response against itself.

In order to understand how the body mounts a specific immune response against antigens, we must first examine

---

[2]From Greek *exo,* meaning without, and *genein,* meaning to produce.
[3]From Greek *endon,* meaning within.
[4]From Greek *autos,* meaning self.

the lymphatic system, which plays an important role in the production, maturation, and housing of the cells that function in specific immunity.

The Cells, Tissues, and Organs of the Lymphatic System

Learning Objectives

✓ Contrast the flow of lymph with the flow of blood.
✓ Describe the tissues and organs of the lymphatic system.
✓ Describe the importance of red bone marrow, the thymus, lymph nodes, and other lymphoid tissues.

The **lymphatic system** is composed of the *lymphatic vessels,* which conduct the flow of a liquid called *lymph;* and lymphatic cells, tissues, and organs, which are directly involved in specific immunity (**Figure 16.2** on page 466). Taken together, the cells, tissues, and organs of the lymphatic system constitute a surveillance system that screens the tissues of the body—particularly possible points of entry such as the throat and intestinal tract—for foreign antigens. We begin by examining the lymphatic vessels.

The Lymphatic Vessels and the Flow of Lymph

The **lymphatic vessels** form a one-way system that conducts **lymph** (limf)—a colorless, watery liquid similar in composition to blood plasma—from local tissues and returns it to the circulatory system. Indeed, lymph arises from fluid that has leaked out of blood vessels into the surrounding tissues. Lymph is first conducted by remarkably permeable *lymphatic capillaries,* which are located in most parts of the body (exceptions are the bone marrow, brain, and spinal cord) and collect fluid from intercellular spaces. From the lymphatic capillaries, lymph passes into increasingly larger lymphatic vessels until it finally flows via two large *lymphatic ducts* into blood vessels near the heart. Unlike the cardiovascular system, the lymphatic system has no unique pump and is not circular; that is, lymph flows in only one direction—toward the heart. One-way valves ensure that lymph flows only toward the heart as muscular activity squeezes the lymphatic vessels. Located at various points within the system of lymphatic vessels are numerous *lymph nodes* (discussed shortly), which house leukocytes that recognize and attack foreign antigens present in the lymph.

Lymphoid Cells, Organs, and Tissues

The site of production of the cells of specific immunity is the *red bone marrow,* located in the ends of long bones and in the centers of flat bones such as the ribs and hip bones. These sites contain *stem cells,* which are cells that give rise to all types of blood cells, including those that have a central role in specific immunity—**lymphocytes** (lim'fō-sītz). Lymphocytes are the smallest leukocytes and each is characterized by a large, round, central nucleus surrounded by a thin rim of cytoplasm (**Figure 16.3a** on page 467).

Although lymphocytes are indistinguishable microscopically, scientists distinguish two main types, called *B lymphocytes* and *T lymphocytes* (discussed shortly), according to the integral surface glycoproteins present in each lymphocyte's cytoplasmic membrane. These glycoproteins are given internationally accepted designations consisting of a number following the prefix *CD* (for *cluster of differentiation*). These numbers—for example, CD4, CD8, CD26, and CD95—reflect the order in which the glycoproteins were discovered, not the order in which they are produced or function. Scientists have identified about 250 CD molecules, only some of which are present on lymphocytes.

After lymphocytes are produced in the bone marrow, they must undergo a maturation process. The type called B lymphocytes mature within the bone marrow, but T lymphocytes travel to and mature in the *thymus,* located in the chest, just above the heart (see Figure 16.2). The thymus is most active early in life and then decreases in size and activity after puberty, after its function has been largely completed.

Once lymphocytes have arisen and have matured in either the bone marrow or thymus, they migrate to what are often considered secondary lymphoid organs and tissues, including the lymph nodes, the spleen, and other, less-organized accumulations of lymphoid tissue.

Even though lymphocytes can circulate throughout the body in blood and lymph, many remain in lymphoid organs and tissues, where they in effect lie in wait for foreign antigens (see Figure 16.3b). Among these sites are hundreds of **lymph nodes** located throughout the body but concentrated in the cervical (neck), inguinal (groin), axillary (armpit), and abdominal regions (see Figure 16.2). Each lymph node receives lymph from numerous afferent lymphatic vessels and drains lymph into just one or two efferent lymphatic vessels (**Figure 16.4** on page 467). A node consists of a medullary (central) maze of passages that filter the lymph passing through it and houses numerous leukocytes, which survey the lymph for foreign antigens and mount a specific immune response against them. The cortex (outer) portion of a lymph node consists of a tough capsule surrounding *germinal centers,* where some cells of specific immunity replicate.

CRITICAL THINKING

The cross-sectional area of the afferent lymphatic vessels arriving at a lymph node is greater than the cross-sectional area of the efferent lymphatics exiting the lymph node. The result is that lymph moves slowly through a lymph node. What advantage does this provide the body?

The lymphatic system contains other lymphoid tissues and organs, including the spleen, the tonsils, and mucosa-associated lymphatic tissue (MALT; see Figure 16.2). The spleen is similar in structure and function to lymph nodes, except that it filters blood instead of lymph. Thus the spleen removes bacteria, viruses, toxins, and other foreign matter from the blood. It also cleanses the blood of old and damaged

Figure 16.2 ➤

The lymphatic system consists of lymphatic vessels containing lymph, plus various lymphoid structures, including lymph nodes, the thymus, tonsils, and mucosa-associated lymphatic tissue (MALT).

blood cells, stores blood platelets (which are required for the proper clotting of blood), and stores blood components such as iron. The tonsils and MALT lack the tough outer capsules of lymph nodes and spleen, but they function in somewhat the same way by physically trapping foreign particles and microbes. MALT includes the appendix, lymphoid tissue of the respiratory tract, and discrete bits of lymphoid tissue called *Peyer's patches* in the wall of the small intestine. MALT contains leukocytes that defend against foreign antigens entering the blood and the respiratory and digestive tracts.

CRITICAL THINKING

As part of the treatment for some cancers, physicians kill the cancer patients' dividing cells, including the stem cells that produce leukocytes, and then give the patients a bone marrow transplant from a healthy donor. Which cell is the most important cell in such transplanted marrow?

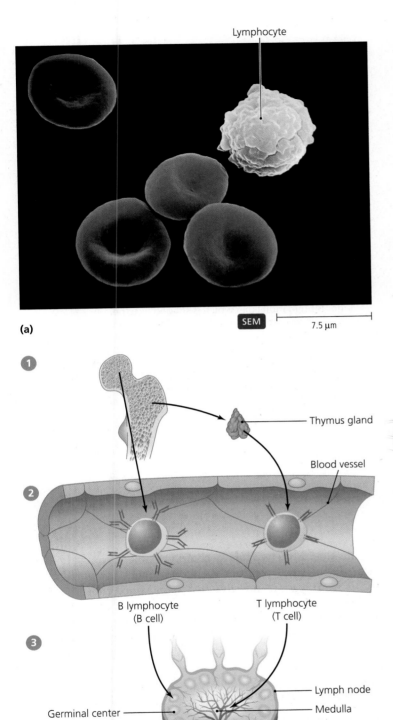

(a)

SEM   7.5 μm

1

Thymus gland

Blood vessel

2

B lymphocyte
(B cell)

T lymphocyte
(T cell)

3

Germinal center
in cortex

Lymph node

Medulla

(b)

◀ *Figure 16.3*

Lymphocytes, the smallest leukocytes, play a central role in specific immunity. **(a)** A lymphocyte is the smallest leukocyte—about the size of a red blood cell. **(b)** The production, maturation, and deployment of lymphocytes. After lymphocytes are produced in the red bone marrow ①, they mature in either the bone marrow (B cells) or the thymus (T cells) before ② circulating in the blood and lymph or ③ waiting in the lymph nodes (or other lymphoid tissues) for foreign antigens.

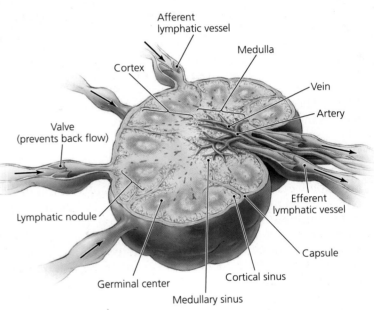

Afferent
lymphatic vessel

Medulla

Cortex

Vein

Artery

Valve
(prevents back flow)

Efferent
lymphatic vessel

Lymphatic nodule

Capsule

Germinal center

Cortical sinus

Medullary sinus

▲ *Figure 16.4*

A lymph node. Lymph enters the node through afferent lymphatic vessels and exits through one or two efferent lymphatic vessels. The outer cortex contains germinal centers, where B lymphocytes proliferate; the inner medulla contains numerous lymphocytes that encounter foreign antigens. A tough capsule surrounds the entire node. *Why do lymph nodes enlarge during an infection?*

Figure 16.4 During an infection, lymphocytes multiply profusely in lymph nodes. This proliferation, and swelling caused by the secretion of inflammatory chemicals, causes lymph nodes to enlarge.

We have seen that lymphocytes can either circulate or remain in lymphoid tissues, where they represent a surveillance system that screens the tissues of the body, particularly the portals of entry for foreign antigens. Next we turn our attention to each of the two types of lymphocytes, beginning with B lymphocytes.

B Lymphocytes (B Cells) and Antibodies

Learning Objectives

✓ Describe the characteristics of B lymphocytes.

✓ Describe the basic structure of an antibody (immunoglobulin) molecule.

✓ Contrast the structure and function of the five classes of immunoglobulins.

✓ Describe five functions of antibodies.

As we have seen, **B lymphocytes,** which are also called **B cells,** arise in the red bone marrow in adults, and they also mature there. However, the designation "B" does not stand for bone marrow, but instead for the *bursa of Fabricius,* which is a unique lymphoid organ located near the terminus of the intestinal tract in birds. Mammals do not have this bursa, which is where B lymphocytes were originally identified, but the name "B cell" has been retained.

B cells, which are identified by surface proteins (for example, CD21, CD22, and CD35), are found primarily in the spleen, lymph nodes, red bone marrow, and Peyer's patches; only a small percentage of B cells circulate in the blood. The major function of B cells is the secretion of protective antibodies, which we discuss next.

Basic Antibody Structure

B cells that are actively fighting against exogenous antigens are called **plasma cells,** which secrete soluble, proteinaceous antigen binding molecules called **antibodies** or **immunoglobulins** (im'yū-nō-glob'yū-linz) **(Ig)** into the blood or lymph. Historically, bodily fluids such as lymph and blood were called *humors* (the Latin word for moisture), and thus the activity of B cells is said to be a **humoral immune response.**

A basic antibody molecule is formed of four peptide chains—two identical longer chains called *heavy chains,* and two identical shorter *light chains* **(Figure 16.5).** The terms *heavy* and *light* refer to their relative molecular weights. Disulfide bonds—covalent bonds between sulfur atoms in amino acids—link the light chains to the heavy chains in such a way that a basic antibody molecule looks like the letter Y with three regions: a *stem* and two *arms* connected by a flexible *hinge region.*

An antibody stem, also called the **F_C region** (because it forms a *fragment* that is *crystallizable*), is formed of the lower portions of the two heavy chains. Among all B cells, there are only five types of heavy chains, designated by the Greek letters *gamma, mu, alpha, epsilon,* and *delta.* These five types of heavy chains give their names to five classes of antibodies (immunoglobulins): IgG, IgM, IgA, IgE, and IgD.

The arms of each heavy and each light chain varies in amino acid sequence among B cells, and thus each is called a *variable region.* Though they vary from B cell to B cell, the variable regions of every antibody molecule formed by a given B cell is identical because for antibodies produced by any particular B cell, the two light chain variable regions are

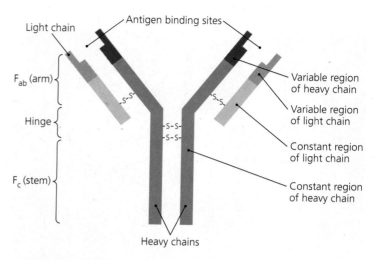

▲ *Figure 16.5*

Basic immunoglobulin (antibody) structure. Each immunoglobulin molecule, which is shaped like the letter Y, consists of two identical heavy chains and two identical light chains held together by disulfide bonds. Five different kinds of heavy chains form the immunoglobulin's stem (F_C region). The arms terminate in variable regions to form two antigen-binding sites. The hinge region is flexible, allowing the arms to bend in almost any direction. *Why are the two antigen-binding sites on a typical antibody molecule identical?*

Figure 16.5 Because the amino acid sequences of the two light chains of an antibody molecule are identical, as are the sequences of the two heavy chains, the binding sites—each composed of the ends of one light chain and one heavy chain—must be identical.

identical and the two heavy chain variable regions are identical. Each B cell randomly chooses genes for variable regions from a set of possible genes and once chosen, makes only those particular variable regions. Together the two variable regions form an **antigen-binding site.** Because the arms of an antibody molecule contain the antigen-binding sites, they are also known as the F_{ab} regions (fragment, antigen-binding).

Antibody Function

Antigen-binding sites are complementary to antigenic determinants (epitopes); in fact, the shapes of the two can match so very closely that most water molecules are excluded from the area of contact, producing a strong, noncovalent, hydrophobic interaction. Additionally, hydrogen bonds and other attractions mediate antibody binding to antigenic determinant. Once antibodies are bound to their target antigen, they function in several ways. These include activation of complement, stimulation of inflammation, agglutination, neutralization, and opsonization. Complement activation and inflammation—nonspecific responses of the second line of defense—are discussed in detail in Chapter 15 (see Figures 15.13 and 15.18); here we examine the other antibody functions.

Agglutination   Because each basic antibody has two antigen-binding sites, each can attach to two antigenic determinants at once. The result of several immunoglobulin molecules binding with two microbial cells each is **agglutination** (ă-glū-ti-nā′shŭn) or clumping **(Figure 16.6a).** Agglutination hinders the activity of pathogenic organisms and increases the chance that they will be phagocytized.

Neutralization   Antibodies can **neutralize** a toxin by binding to a critical portion of the toxin so that it can no longer function against the body. Similarly, antibodies can block attachment molecules on the surfaces of bacteria and viruses such that they cannot adhere to target cells **(Figure 16.6b).**

Opsonization   Antibodies act as **opsonins**[5]—molecules that stimulate phagocytosis. Because neutrophils and macrophages have receptors for the F_c regions of IgG molecules, these leukocytes bind to the stems of antibodies. Once the antibodies are so bound, the leukocytes phagocytize them, along with the antigens they carry. Enhanced phagocytosis is called **opsonization** (op′sŏ-nī-zā′shŭn) **(Figure 16.6c).**

CRITICAL THINKING

Two students are studying for an exam on the body's defensive systems. One of them insists that complement is part of the nonspecific second line of defense, but the partner insists that complement is part of a humoral immune response in the third line of defense. How would you explain to them that they are both correct?

Classes of Antibodies (Immunoglobulins)

The threat confronting the immune system can be extremely variable, so it is not surprising that a single type of antibody is not sufficient. Next we consider the structure and functions of the five classes of antibodies that may participate in a humoral immune response; the class involved in any given humoral immune response depends on the type of invading foreign antigens, the portal of entry involved, and the antibody function required.

The most common class of antibody in the blood, accounting for about 85% of serum antibodies, is **immunoglobulin G (IgG).** IgG molecules play a major role in antibody-mediated defense mechanisms, including complement activation, agglutination, opsonization, and neutralization.

Each molecule of IgG is formed from two identical light chains and two identical gamma heavy chains held together with disulfide bonds **(Figure 16.7a** on page 470). Because of its

---

[5]From Greek *opsonein*, meaning to supply food.

(a) Agglutination

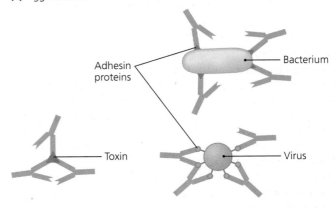

Adhesin proteins

Bacterium

Toxin

Virus

(b) Neutralization

Pseudopod of phagocyte

F_c receptor protein

(c) Opsonization

▲ *Figure 16.6*

Three functions of antibodies. **(a)** Agglutination (clumping). **(b)** Neutralization of toxins and microbes. **(c)** Opsonization. Two other functions of antibodies—activation of complement and participation in inflammation—are discussed in Chapter 15.

small size, IgG can leave blood vessels more easily than can the other immunoglobulins. This is especially important during inflammation, because it enables IgG to bind to invading pathogens in tissues before they get into the circulatory systems. IgG molecules are also the only type of antibody that can cross the placenta to protect a developing child.

The body produces **immunoglobulin M (IgM),** the second most common antibody class in the blood, during the initial stages of an immune response. An IgM molecule is more than five times larger than a molecule of IgG because it

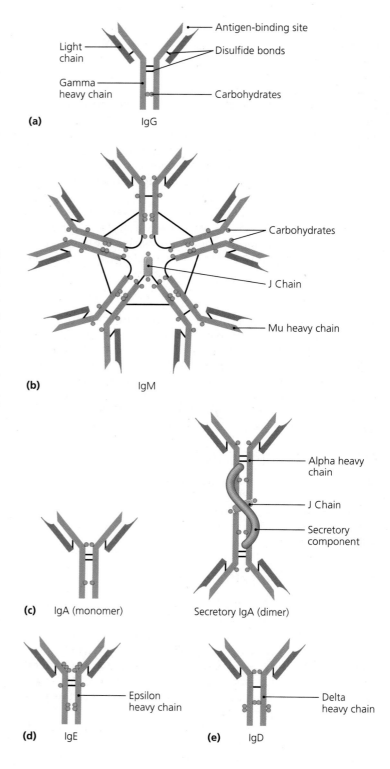

Figure 16.7 ▶

Classes of antibodies. **(a)** IgG, the most common immunoglobulin in blood and lymph, and the only class of antibody that crosses the placenta. **(b)** IgM, the largest antibody and the class first secreted during a humoral immune response. **(c)** IgA. Secretory IgA, a dimer composed of two IgA monomers, can cross membrane surfaces due to the presence of the secretory component. **(d)** IgE, which is present in the serum in low concentrations. Its binding to mast cells stimulates the release of histamine, a chemical of inflammation. **(e)** IgD, the rarest class of antibody and an antigen receptor in the membranes of B cells in humans.

is a pentamer, consisting of five Y-shaped subunits linked together in a circular fashion via disulfide bonds and a short polypeptide *joining (J) chain* **(Figure 16.7b)**. Each IgM subunit has a conventional immunoglobulin structure, consisting of two light chains and two mu heavy chains. IgM is more efficient than IgG at complement activation, neutralization of virions, and agglutination because of its numerous antigen binding sites.

Immunoglobulin A (IgA), which has alpha heavy chains, is the immunoglobulin most closely associated with various body secretions, including tears and milk. Some of the body's IgA is a monomer with a basic Y shape that circulates in the blood, constituting about 5% of total serum antibody. However, plasma cells in various mucous membranes and in the tear ducts and mammary glands take up IgA monomers from the blood and connect them via a J chain and another short polypeptide (called a *secretory component*) to form **secretory IgA (Figure 16.7c).**

Secretory IgA agglutinates and neutralizes antigens and is of critical importance in protecting the body from infections arising in the gastrointestinal, respiratory, urinary, and reproductive tracts. Secretory component both aids in the transport of secretory IgA across mucous membranes and protects secretory IgA from digestion by intestinal enzymes. Because secretory IgA is secreted into the milk in the mammary glands, it can provide newborns some protection against foreign antigens.

Immunoglobulin E (IgE) is a typical Y-shaped immunoglobulin with two epsilon heavy chains **(Figure 16.7d)**. Because it is found in extremely low concentrations in serum (less than 1% of total antibody), it is not important in agglutination, neutralization, or opsonization. Instead, IgE molecules act as signal molecules—they attach to receptors found on certain white blood cells (basophils and mast cells) to trigger the rapid release of histamine, causing inflammation, part of the body's nonspecific defense. As a result, IgE mediates between the specific and nonspecific responses—a process particularly important in allergic diseases. Infections by parasitic worms also stimulate the synthesis of IgE.

Immunoglobulin D (IgD) (Figure 16.7e) is characterized by delta heavy chains. IgD molecules are not secreted,

but are membrane-bound antigen receptors on B cells that are often seen during the initial phases of a humoral immune response. IgD is not found in all mammals, and animals that lack IgD show no observable ill effects; therefore, scientists do not know the importance of IgD.

Table 16.1 compares the different classes of immunoglobulins.

Table 16.1   Characteristics of the Five Classes of Immunoglobulins

Name	Function(s)	Structure (Molecular Weight in Daltons)	Location(s)	Percentage in Serum
IgG	Complement activation, agglutination, opsonization, and neutralization; crosses placenta to protect fetus	Monomer (180,000)	Serum, intercellular fluid	85
IgM	Complement activation, agglutination, and neutralization	Pentamer (900,000)	Serum	5–10
IgA	Agglutination and neutralization	Monomer (150,000); dimer (385,000)	External secretions, including milk	5
IgE	Triggers release of histamines from basophils and mast cells	Monomer (200,000)	Serum, mast cell surfaces	<1
IgD	Unknown	Monomer (180,000)	B cell surface	<1

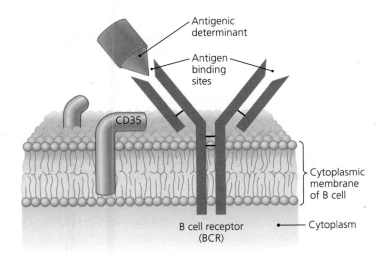

▲ Figure 16.8

Surface structures of B cells. In addition to glycoproteins such as CD35, B cells are characterized by B cell receptors (BCRs), which are symmetrical, Y-shaped antibody molecules with two antigen-binding sites.

B Cell Receptors

The surface of each B lymphocyte is covered with about 250,000 to 500,000 identical copies of **B cell receptor (BCR),** an antibody that remains integral to the cytoplasmic membrane **(Figure 16.8).** Like a secreted antibody molecule, each BCR has two antigen-binding sites identical to the antigen-binding site of the antibodies that particular cell secretes. B lymphocytes do not form BCRs in response to antigens; instead, each cell's randomly generated antibody variable regions determine its specific BCR shape. A BCR is complementary in shape to an antigenic determinant, which the body may or may not encounter. Scientists estimate that each person forms at least 10^{11} different B lymphocytes—each with a distinct set of BCRs! Because there are trillions of lymphocytes, at least one of them fortuitously bears a BCR that is complementary to any given specific antigenic determinant. Thus, for example, each of us has lymphocytes with

BCRs complementary to rattlesnake venom proteins, even though it is unlikely that we have been or ever will be bitten by a rattlesnake.

An antigenic molecule (for example, a bacterial protein) typically has numerous antigenic determinants of various shapes and thus will be recognized by many different BCRs; however, it must be emphasized that each B lymphocyte recognizes only one antigenic determinant. Collectively, the cellular receptors on all of an individual's B cells are capable of recognizing millions of different antigenic determinants; thus, the repertoire of BCRs is also the repertoire of B lymphocytes.

In most cases, B cells do not mount a humoral immune response directly. Instead, they respond to most antigens only with the assistance of certain T lymphocytes. Thus, antigens of this type are known as *T-dependent antigens.* We will discuss the details of humoral immunity against T-dependent antigens shortly—after we consider the various T lymphocytes.

T Lymphocytes (T Cells)

Learning Objectives

✓ Describe the importance of the thymus to the development of T lymphocytes.

✓ Describe the basic characteristics of T lymphocytes.

✓ Compare and contrast three types of T cells.

✓ Describe five types of cytokines.

An adult's red bone marrow produces lymphocytes that leave the bone marrow to mature in the thymus. In the thymus, each of these cells, called **T lymphocytes** or **T cells,** randomly generates multiple copies of a specific **T cell receptor (TCR)** for its cytoplasmic membrane. A TCR is composed of two different polypeptide chains with a groove between them that acts as an antigen binding site **(Figure 16.9** on page 472). As with BCRs of B cells, each specific TCR recognizes and binds to a complementary antigenic determinant, and there are at least 10^{11} different TCRs—at least one for every possible antigenic shape. T lymphocytes act

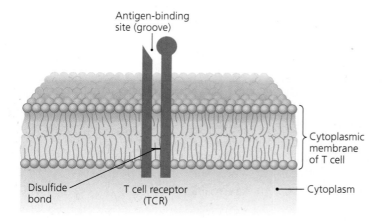

▲ Figure 16.9

A T cell receptor (TCR), an asymmetrical surface molecule composed of two polypeptides containing a single antigen-binding site between them.

directly against endogenous invaders, many of the body's cells that harbor intracellular pathogens, and abnormal body cells such as cancer cells that produce abnormal cell surface proteins. Because T cells act directly against antigens—they do not secrete immunoglobulins—their activity is **cell-mediated immune response.**

After they have matured in the thymus, T cells circulate in the lymph and blood and migrate to the lymph nodes, spleen, and Peyer's patches. They account for about 90% of all lymphocytes in the blood. Based on their surface glycoproteins and their characteristic functions, immunologists recognize three types of T cells: *cytotoxic T cells* and two kinds of *helper T cells.*

Cytotoxic T Cells (T_C Cells)

A **cytotoxic T (T_C) cell** is distinguished by hundreds of copies of its unique TCR as well as the presence of the CD8 cell-surface glycoprotein. The latter gives these lymphocytes

their nickname: *T8 cells.* As the name *cytotoxic* T cell implies, these cells directly kill other cells—those infected with viruses and other intracellular pathogens, as well as abnormal cells, such as cancer cells.

Helper T Lymphocytes (T_H Cells)

Immunologists distinguish **helper T (T_H) cells** by the presence of the CD4 glycoprotein; in fact, many medical personnel call these cells *T4 cells.* These cells are called "helpers" because their function is to "help" in regulating the activity of B cells and cytotoxic T cells during an immune response. There are two subpopulations of helper T cells: type 1 helper T cells (T_H1), which assist cytotoxic T cells, and type 2 helper T cells (T_H2), which function in conjunction with B cells. T helper cells secrete various soluble protein messengers, called *cytokines,* that determine which immune response will be activated. (We will consider the types and effects of cytokines shortly.) **Highlight 16.1** describes some of the effects of the destruction of helper T cells in individuals infected with the human immunodeficiency virus (HIV). Immunologists distinguish between T_H1 and T_H2 cells on the basis of their cytokine secretions and by characteristic cell surface proteins: T_H1 cells express CD26 and a cytokine receptor named CCR5,[6] whereas the cytoplasmic membranes of T_H2 cells have CCR3 and CCR4.

Some scientists distinguish a fourth type of T cell, a *suppressor T cell,* which is thought to repress specific immune responses. Other researchers dispute the existence of a specific category of suppressor T cells, contending instead that other regulatory cells and molecules turn off specific immune responses.

Table 16.2 compares and contrasts the features of the various types of lymphocytes.

---

[6]Immunologists name some cytokine receptors with the prefixes CCR to indicate the presence of two cysteine amino acids in their primary structure.

Highlight 16.1   The Loss of Helper T Cells in AIDS Patients

Neither B cells nor cytotoxic T cells respond effectively to most antigens without the participation of helper T cells, because signals are passed more effectively among leukocytes when the CD4 molecules of helper T cells bind to certain leukocytes.

We also saw in a previous chapter that in order for viruses to enter cells and replicate, they must first attach to a specific protein on their target cells' surface. In the case of human immunodeficiency virus (HIV), the causative agent of AIDS, the surface protein to which it attaches is CD4, and its target cells are helper T cells. Once HIV enters a helper T cell, it, like other viruses, takes over the cell's protein-synthesizing machinery and directs the cell to produce more viruses.

Because they are essential to mounting an effective immune re-

sponse, the body normally has a surplus of helper T cells—usually about three times as many as it needs. As a result, individuals infected with HIV can typically lose up to about two-thirds of their helper T cells before signs of immune deficiency appear. Cell-mediated immunity is usually affected first; even though B cells require the assistance of helper T cells to function optimally, they can still be stimulated directly by large quantities of antigen. Normal human blood typically contains about three CD4 helper T cells for every two CD8 cytotoxic T cells—a CD4 to CD8 ratio of 3:2. During the course of HIV infection, however, CD4 helper T cells are lost, reducing the CD4 to CD8 ratio such that individuals with full-blown AIDS may have a ratio as low as 0.5:1.

Table 16.2   Characteristics of Selected Lymphocytes

Lymphocyte	Site of Maturation	Representative Cell Surface Glycoproteins	Selected Secretions
B cell	Red bone marrow and germinal center of lymph node	Each has distinct BCR and CD21, CD22, CD35	Antibodies
Helper T cell type 1 (T$_H$1)	Thymus	Each has distinct TCR and CD4, CD26, CCR5	IL-2, IFN-γ
Helper T cell type 2 (T$_H$2)	Thymus	Each has distinct TCR and CD4, CCR3, CCR4	IL-4
Cytotoxic T (T$_C$)	Thymus	Each has distinct TCR and CD8, CD95L	Perforin, granzyme

Immune System Cytokines

Cytokines (sī'tō-kīnz) are soluble regulatory proteins that act as intercellular signals when released by certain body cells including kidney cells, skin cells, and immune cells. Here we are concerned with immune system cytokines, which signal among the various leukocytes. For example, B cells and cytotoxic T cells do not respond fully (if at all) to antigens unless they are first signaled by cytokines.

Many different immune system cytokines are secreted by various leukocytes and affect diverse cells. Many cytokines are redundant; that is, they have almost identical effects. Such complexity has given rise to the concept of a *cytokine network*—a complex web of signals among all the cell types of the immune system. The nomenclature of cytokines is not based on a systematic relationship among them; instead, scientists named cytokines after their cells of origin, their function, and/or the order in which they were discovered. Cytokines of the immune system include the following substances:

- **Interleukins**[7] (in-ter-lū'kinz) **(ILs).** As their name suggests, ILs signal among leukocytes. Immunologists named interleukins sequentially as they were discovered. Currently, scientists have identified about 27 interleukins.

- **Interferons** (in-ter-fēr'onz) **(IFNs).** These antiviral proteins, discussed in Chapter 15, may also act as cytokines. The most important interferon with such a dual function is interferon-gamma (IFN-γ), which is secreted by T$_H$1 cells.

- **Growth factors.** These proteins stimulate stem cells to divide, ensuring that the body is supplied with sufficient leukocytes of all types.

- **Tumor necrosis[8] factors (TNFs).** Macrophages and T cells secrete TNFs to kill tumor cells and to regulate immune responses and inflammation.

- **Chemokines** (kē'mō-kīnz). Chemokines signal leukocytes to rush to a site of inflammation or infection and activate other leukocytes.

To this point, we have examined the "cast of characters" involved in specific immunity—antigens, B cells, antibodies, T cells, and cytokines. Before we consider the detailed processes involved in specific immunity—humoral immune responses and cell-mediated immune responses—we need to consider some initial preparations of the body.

The Body's Preparation for a Specific Immune Response

The body prepares for specific immune responses by killing lymphocytes with receptors complementary to autoantigens, making so-called *major histocompatibility complex* proteins, and processing antigens so that they can be recognized by lymphocytes.

Lymphocyte Editing by Clonal Deletion

Learning Objective

✓ Describe apoptosis, and explain its role in lymphocyte editing by clonal deletion.

Given that lymphocytes randomly generate the shapes of their receptors, every population of maturing lymphocytes includes cells with receptors that will bind normal body components—the autoantigens mentioned earlier. Because it is vitally important that specific immune responses not be directed against autoantigens, the body "edits" lymphocytes via a process by which self-reactive cells are eliminated.

In humans, lymphocyte editing occurs in lymphoid organs: the bone marrow for B lymphocytes (**Figure 16.10** on page 474) and the thymus for T lymphocytes (which is why these cells are so named). During their maturation, lymphocytes are exposed to many autoantigens. Under normal circumstances, when a lymphocyte binds to an autoantigen, that lymphocyte responds by undergoing **apoptosis**[9] (cell suicide). Thus, apoptosis is the critical feature in the development of self-tolerance, because lymphocytes with receptors

---

[7]From Latin *inter*, meaning between, and Greek *leukos*, meaning white.
[8]From Latin *necare*, meaning to kill.
[9]Greek meaning falling off.

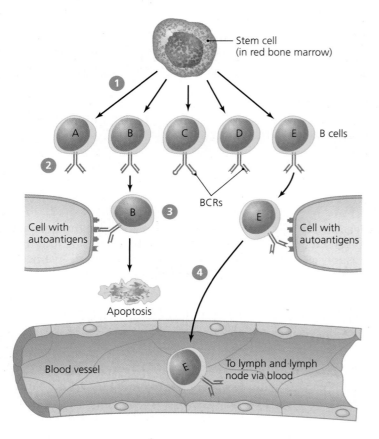

▲ *Figure 16.10*

Lymphocyte editing (clonal deletion) of B cells.
① Stem cells in the red bone marrow generate a host of lymphocytes. ② Each newly formed B cell randomly generates BCRs with a particular shape. ③ Cells whose BCRs are complementary to some autoantigen bind with that autoantigen, stimulating the cell to undergo apoptosis. Subsequently, an entire set of potential daughter B cells (a clone) that are reactive with the body's own cells are eliminated—thus the term *clonal deletion*. ④ B cells with a BCR that is not complementary to any autoantigen are released from the bone marrow and into the blood. Note that T lymphocytes are also edited via the process of clonal deletion—in their case, in the thymus. *Of the B cells shown, which (besides B) is likely to undergo apoptosis?*

Figure 16.10 The B lymphocyte labeled "C" will also undergo apoptosis.

for autoantigens die before they can reproduce, and their potential daughter cells (clones) are deleted from the repertoire of potential lymphocytes. For this reason, lymphocyte editing is also called **clonal deletion.** Surviving lymphocytes and their descendents respond only to foreign antigens. When self-tolerance is impaired, the result is an *autoimmune disease,* which we examine in Chapter 17.

The Roles of the Major Histocompatibility Complex

Learning Objective

✓ Describe the two classes of major histocompatibility complex (MHC) proteins with regard to their location and function.

When scientists first began grafting organs from one animal into another, they discovered that if the animals were not closely related, the recipients swiftly rejected the grafts. When they analyzed the reason for such rapid rejection, they found that graft recipients were mounting a very strong immune response against a specific group of antigens found on the cells of the graft. When the antigens on the graft's cells were sufficiently dissimilar from the antigens on the host's cells, as seen with unrelated animals, the grafts were rejected. This is how scientists came to understand how the body is able to distinguish self from nonself.

Immunologists named this class of antigens *major histocompatibility[10] antigens* to indicate the importance of these molecules in determining the compatibility of tissues in successful grafting. Further research revealed that major histocompatibility antigens are glycoproteins found in the membranes of most cells of vertebrate animals. In humans, these antigens were first identified on leukocytes and were named *human leukocyte antigens (HLA).* Major histocompatibility antigens are coded by a cluster of genes called the **major histocompatibility complex (MHC).** In humans, the MHC is located on each copy of chromosome 6.

Because organ grafting is a modern surgical procedure with no counterpart in nature, scientists reasoned that MHC molecules must have some other "real" function. Indeed, immunologists have determined that MHC proteins in cytoplasmic membranes function to hold and position antigenic determinants for presentation to T cells. Each MHC molecule has an *antigen-binding groove* that lies between two polypeptides; variations in the amino acid sequences of the polypeptides modify the shapes of MHC binding sites and determine which antigenic determinants can be bound and presented.

MHC proteins are of two classes **(Figure 16.11).** Class I MHC molecules are found on the cytoplasmic membranes of all nucleated cells. Thus red blood cells, which do not have nuclei, do not express MHC proteins, while skin and muscles do express MHC class I molecules. In contrast, class II MHC proteins are found only on B lymphocytes and special cells called **antigen presenting cells (APCs).** APCs include monocytes, macrophages, and their functional relatives, such as *dendritic cells,* which are found in lymphoid organs and under the surface of the skin and are characterized by many long, thin cytoplasmic processes called *dendrites.[11]*

The cytoplasmic membrane of an APC has about 100,000 MHC II molecules, which vary among themselves in

---

[10]From Greek *histos,* meaning tissue, and Latin *compatibilis,* meaning agreeable.
[11]From Greek *dendron,* meaning tree.

Antigen-binding groove

Polypeptides

Cytoplasmic membrane of any nucleated cell

Cytoplasm

Class I MHC protein

(a)

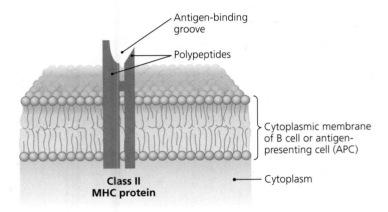

Antigen-binding groove

Polypeptides

Cytoplasmic membrane of B cell or antigen-presenting cell (APC)

Cytoplasm

Class II MHC protein

(b)

▲ *Figure 16.11*

The two classes of major histocompatibility complex (MHC) proteins, which are each composed of two polypeptides that form an antigen-binding groove. **(a)** MHC class I glycoproteins, which are found on all cells except red blood cells. **(b)** MHC class II glycoproteins, which are expressed only by B cells and special antigen presenting cells (APCs).

the antigens they can bind. Their diversity is dependent on an individual's genotype. If an antigen fragment cannot be bound to an MHC molecule, it typically does not trigger an immune response. Thus, MHC molecules determine which antigen fragments trigger immune responses.

Antigen Processing

Learning Objectives

✓ Contrast T-independent and T-dependent antigens in terms of their structures.

✓ Explain the roles of macrophages, dendritic cells, and MHC molecules in antigen processing and antigen presentation.

✓ Contrast endogenous antigen processing with exogenous antigen processing.

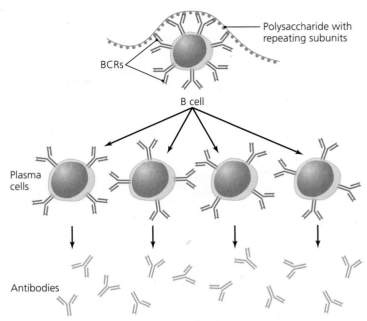

Polysaccharide with repeating subunits

BCRs

B cell

Plasma cells

Antibodies

▲ *Figure 16.12*

The effects of the binding of a T-independent antigen by a B cell. When a molecule with multiple repeating antigenic determinants (such as the polysaccharide shown here) cross-links the BCRs on a B cell, the cell is activated: It proliferates, and its daughter cells become plasma cells that secrete antibodies.

B cells—but not T cells—can bind directly to large antigens that have numerous, readily accessible, and repeating antigenic determinants such as the repeating polysaccharide subunits of a bacterial capsule; they do not require help. The cross-linking of the repeating subunits to numerous BCRs on a B cell stimulates that cell to proliferate, become a plasma cell, and produce antibodies, all without the help of T cells. Therefore, large antigens with repeating antigenic determinants are called **T-independent antigens (Figure 16.12).**

However, most antigens are smaller and have less accessible antigenic determinants, and B cells targeted against these antigens cannot function without the involvement of helper T cells. These antigens are therefore termed **T-dependent antigens.** Additionally the helper T cells require communication with and assistance from other leukocytes that process the antigen in a way that makes antigenic determinants more accessible. Communication among leukocytes is dependent on MHC glycoproteins and cytokines.

Antigen processing occurs via somewhat different processes according to whether the antigen is exogenous or endogenous. For exogenous antigens such as extracellular pathogens, an APC internalizes the invading pathogen and enzymatically catabolizes the pathogen's antigenic molecules, producing fragments, or antigenic determinants which are contained in a phagolysosome (**Figure 16.13** on

Processing of T-dependent exogenous antigens:

Antigenic determinants in phagolysosome

MHC II protein in membrane of vesicle

MHC II-antigenic determinant complex

Vesicles fuse and antigenic determinants bind to complementary MHC II molecules

Vesicle fuses with cytoplasmic membrane

MHC II-antigenic determinants complex on cell surface

Cytoplasmic membrane

(a) MHC II-antigenic determinant complex displayed on cytoplasmic membrane

Processing of endogenous antigens within body cells:

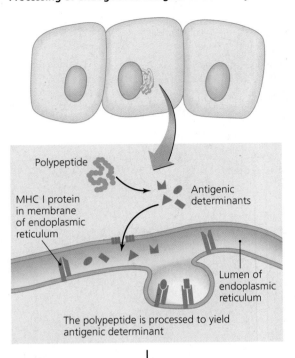

Polypeptide

MHC I protein in membrane of endoplasmic reticulum

Antigenic determinants

Lumen of endoplasmic reticulum

The polypeptide is processed to yield antigenic determinant

MHC I-antigenic determinant complex

Antigenic determinant-MHC I complexes packaged in vesicle

Vesicle fuses with cytoplasmic membrane

MHC I-antigenic determinants complex on cell surface

Cytoplasmic membrane

(b) MHC I-antigenic determinant complex displayed on cytoplasmic membrane

▲ *Figure 16.13*

The processing of exogenous and endogenous antigens. **(a)** Exogenous antigens are phagocytized by an APC, which then digests them to release antigenic determinants that are inserted into the antigen-binding grooves of MHC II molecules. The MHC II–antigenic determinant complexes are then displayed on the outside of the APC's cytoplasmic membrane. **(b)** Polypeptides synthesized within a cell are cleaved to release antigenic determinants, which are imported into the endoplasmic reticulum; there they bind to complementary MHC I proteins. The MHC I-antigenic determinant complexes are then packaged in vesicles and exported to the cytoplasmic membrane.

page 476). A vesicle containing MHC Class II molecules in its membrane fuses with the phagolysosome and each antigenic determinant then binds to an antigen-binding groove of a complementary MHC class II molecule. The fused vesicle then inserts the complexes of antigenic determinants and MHC class II molecules into the cytoplasmic membrane such that the antigen is presented on the outside.

The endogenous antigens of intracellular bacteria and viruses must also be processed. In fact, some of the new proteins produced within nucleated cells—including proteins produced by intracellular bacteria or viral proteins produced by the cell's metabolic machinery—are catabolized into antigenic determinants and attached to antigen-binding grooves of MHC class I molecules located in endoplasmic reticulum membrane. This membrane is packaged into a vesicle by a Golgi body and then inserted into the cytoplasmic membrane to display the MHC class I-antigenic determinant complex on the cell's surface. All nucleated cells express MHC class I molecules.

Now that we have discussed the organs, cells, secretions, and signaling molecules of the immune system, as well as the preparatory steps of editing, antigen processing, and presentation, we will examine the humoral and cell-mediated immune responses in more detail.

The Humoral Immune Response

As we have discussed, the body mounts humoral immune responses against the antigens of exogenous pathogens. Further, such an immune response against T-dependent antigens requires the assistance of helper T (T_H) cells. The following section examines a humoral immune response in more detail.

B Cell Activation and Clonal Selection

Learning Objectives

✓ Describe the formation and functions of plasma cells and memory B cells.

✓ Describe the steps and effect of clonal selection.

Humoral immunity can be viewed as consisting of the following series of cellular interactions (**Figure 16.14** on page 478), which are mediated by membrane-bound proteins as well as cytokines:

1 Antigen presentation. APCs present antigenic determinants to all T_H cells they encounter, but APCs activate only those T_H cells that have binding sites complementary to the presented antigen. Antigen presentation is a random process that depends on chance encounters between APCs and T_H cells. Recall that T_H cells express a unique T cell receptor (TCR) and CD4, and it is CD4 that recognizes and binds to MHC II, which helps stabilize the binding of the antigenic determinant to the TCR.

2 Differentiation of T_H into T_H2 cells. In a humoral immune response, APCs signal the T_H cell with IL-1 to become a T_H2 cell.

3 Clonal selection. Of all the B cells available, only the B cell with BCRs complementary to the antigenic determinant will be recognized—a process called **clonal selection.**

4 Activation of B cell. A complementary T_H2 cell binds to a B cell by recognizing its antigenic determinant–MHC II complex. The T_H cell secretes IL-4, which activates the B cell such that it proliferates rapidly. The offspring of the proliferating B cell differentiate into two types of daughter cells—*memory B cells* (discussed shortly) and *plasma cells.*

The majority of cells produced during B cell proliferation are antibody-secreting **plasma cells,** which are large, activated B cells with an extensive cytoplasm that is rich in rough endoplasmic reticulum. All of the plasma cell offspring of any single activated B cell produce only antibody molecules with active sites identical to the antigen-binding sites of the parent cell's BCR. That is, the antibodies are identical to one another and complementary to the specific antigenic determinant recognized by the parent cell. Each plasma cell produces thousands of molecules of antibody per second. Because of their high activity level, plasma cells are short-lived; they die within a few days of activation, although their antibodies can remain in body fluids for several months, and a clone can persist for years. As discussed previously, antibodies activate complement, trigger inflammation, agglutinate and neutralize antigen, and act as opsonins.

CRITICAL THINKING

What sorts of pathogens would successfully attack a patient with an inability to synthesize B lymphocytes?

Memory B Cells and the Establishment of Immunological Memory

Learning Objective

✓ Contrast primary and memory immune responses.

A small percentage of the cells produced by B cell proliferation do not secrete antibodies but survive as **memory B cells**—that is, long-lived cells with BCRs complementary to the specific antigenic determinant that triggered their production. In contrast to plasma cells, memory cells have BCRs and persist in the lymphoid tissues. They can survive for months or even years, ready to initiate antibody production if the same antigen is encountered again. Let's examine how memory cells provide the basis for immunization to prevent disease, using tetanus immunization as an example.

Because the body produces an enormous variety of B cells (and therefore BCRs), a few T_H cells and B cells bind to and respond to antigens of *tetanus toxoid* (deactivated

Figure 16.14 ➤

A humoral immune response. ① Antigen presentation, in which an APC presents antigen to a complementary T_H cell. ② Differentiation of the T_H cell into a T_H2 cell. ③ Clonal selection, in which the T_H2 cell binds to the B cell clone that recognizes the antigen. ④ Activation of the B cell in response to secretion of IL-4 by the T_H2 cell, which causes the B cell to differentiate into antibody-secreting plasma cells and long-lived memory cells.

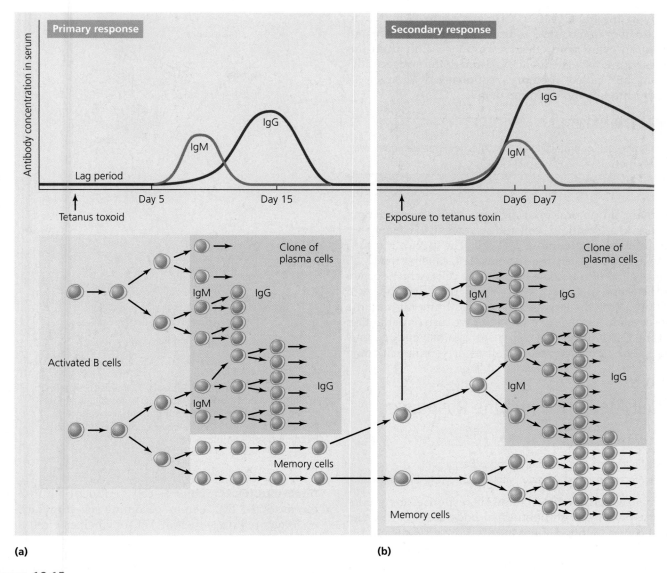

▲ *Figure 16.15*

The production of primary and secondary humoral immune responses. This example depicts the events following the administration of a tetanus toxoid in immunization. **(a)** Primary response. After the tetanus toxoid is introduced into the body, the body slowly removes the toxoid while producing memory B cells. **(b)** Secondary response. Upon exposure to active tetanus toxin during the course of an infection, memory B cells immediately differentiate into plasma cells and proliferate, producing a response that is faster and results in greater antibody production than occurs in the primary response.

tetanus toxin), which is used for tetanus immunization. In a **primary response (Figure 16.15a),** relatively small amounts of antibodies are produced, and it may take days or weeks before sufficient antibodies are made to completely eliminate the toxoid from the body. Though some antibody molecules may persist for 6 months or more, a primary immune response basically ends when the clone of plasma cells has lived out its normal life span.

Memory B cells, surviving in lymphoid tissue, constitute a reserve of antigen-sensitive cells that become active when there is another exposure to tetanus toxin, perhaps

many years later. Thus, tetanus toxin produced during the course of a bacterial infection will encounter a population of memory cells, which proliferate and differentiate rapidly into plasma cells without requiring any interaction with APCs. The newly differentiated plasma cells produce large amounts of antibody within a few days **(Figure 16.15b),** and the tetanus toxin is neutralized before it can cause disease. Because so many more cells are able to recognize and respond to the antigenic determinant, such a **secondary immune response** is much faster and more effective than the initial response.

As you might expect, a third exposure (whether to tetanus toxin or to more toxoid in an immunization booster) results in an even more effective response. Enhanced immune responses to subsequent exposure to the antigenic determinant are called **memory responses.**[12] Chapter 17 discusses immunization in more detail.

CRITICAL THINKING

Plasma cells are vital for protection against infection, but memory B cells are not. Why not?

In summary, a humoral immune response occurs when an APC binds to a specific T_H cell and secretes IL-1, causing the T_H cell to differentiate into a T_H2 cell. The T_H2 cell in turn binds to a specific B cell and secretes IL-4, causing the B cell to multiply and differentiate into plasma cells (which secrete antibodies) and memory cells (which provide long-term protection). Now that we have discussed humoral immune responses against exogenous antigens, we turn our attention to cell-mediated immune responses against endogenous antigens. The two types of responses are similar in some ways.

The Cell-Mediated Immune Response

Learning Objectives

✔ Describe a cell-mediated immune response.
✔ Compare and contrast the two pathways of cytotoxic T-cell action.

The body uses **cell-mediated immune responses** to fight intracellular pathogens and abnormal body cells. Given that the most common intracellular invaders are viruses, our discussion of cell-mediated immunity will focus on these pathogens; however, cell-mediated immune responses are also effective against intracellular bacterial pathogens such as *Mycobacterium tuberculosis,* which causes tuberculosis. **New Frontiers 16.1** on page 482 describes an experimental use of T cells to attack one form of cancer.

As previously discussed, when a cell synthesizes certain proteins, it displays fragments of them on its cytoplasmic membrane in the antigen-binding grooves of MHC class I molecules. Thus, when viruses are replicated inside cells, antigenic determinants of viral proteins are displayed on the host cell's surface.

Certain APCs secrete IL-12, which stimulates nearby T_H cells to become T_H1 cells. T_H1 cells secrete IL-2 and IFN-γ, which activate those cytotoxic T (T_C) cells that have TCRs complementary to an antigenic determinant; in fact, IL-2 drives T cell division. Activated T_C cells reproduce to form memory T cells (discussed shortly) and more T_C progeny

<hr>

[12]Sometimes also called anamnestic responses, from Greek *ana,* meaning again, and *mimneskein,* meaning to call to mind.

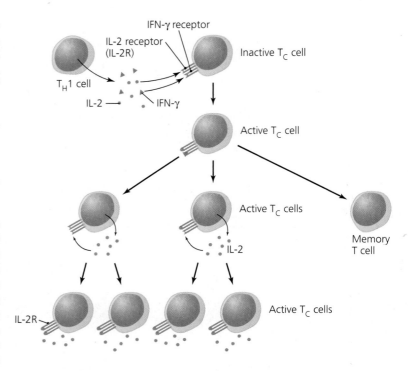

▲ *Figure 16.16*

Activation of cytotoxic T (T_C) cells. T_H1 cells secrete IL-2 and IFN-γ, which together activate T_C cells. IL-2 triggers active T_C cells to divide, forming more active T_C cells as well as memory T cells. Active T_C cells secrete IL-2, becoming self-stimulatory.

(Figure 16.16). Daughter T_C cells produce both IL-2 and IL-2 receptors (IL-2R), thereby becoming self-stimulating—they no longer require either an APC or a helper T cell.

The initial events in cell-mediated immunity are depicted in **Figure 16.17.** An activated T_C cell binds to an infected cell via its TCR, which is complementary to the antigenic determinant, and via its CD8, which is complementary to the MHC class I protein of the infected cell **(Figure 16.17a).** Next, cytotoxic T cells kill their targets through one of two pathways: The *perforin-granzyme pathway,* which involves the synthesis of special killing proteins, or the *CD95 pathway,* which is mediated through CD95, a cell surface glycoprotein.

The Perforin-Granzyme Cytotoxic Pathway

The cytoplasm of cytotoxic T cells has vesicles containing two key protein molecules called *cytotoxins*—**perforins** (per-fōr-inz) and **granzymes** (gran′zīmz). When a cytotoxic T cell first attaches to its target, these vesicles move toward the region of contact, where they release their cytotoxins. Perforin molecules aggregate into a tubular structure in the infected cell's membrane, forming a channel similar to the MAC channels of complement. Granzymes move through the perforin channel and into the target cell, where they activate apoptotic enzymes **(Figure 16.17b).** Having forced its

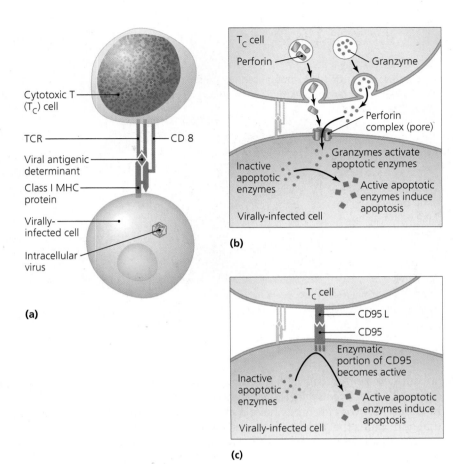

(a)

(b)

(c)

Figure 16.17 ▶

A cell-mediated immune response. **(a)** The binding of a virus-infected cell by a cytotoxic T (T_C) cell. **(b)** The perforin-granzyme cytotoxic pathway. After perforins and granzymes have been released from T_C cell vesicles, granzymes enter the infected cell through the perforin complex pore and activate the enzymes of apoptosis. **(c)** The CD95 cytotoxic pathway. Binding of CD95L on the T_C cell activates the enzymatic portion of the infected cell's CD95 such that apoptosis is induced.

target to commit suicide, the cytotoxic T cell disengages and moves on to another infected cell.

The CD95 Cytotoxic Pathway

The **CD95 pathway** of cell-mediated cytotoxicity involves an integral glycoprotein called *CD95* that is present in the cytoplasmic membranes of most body cells. Its receptor, *CD95L*, is present on activated T_C cells. When an activated T_C cell comes into contact with its target, its CD95L binds to CD95 on the target, which then activates enzymes that trigger apoptosis **(Figure 16.17c)**.

Memory T Cells

Learning Objective

✓ Describe the establishment of memory T cells.

Just as some B cells become memory cells, so too some activated T cells become **memory T cells.** When stimulated by endogenous antigen and cytokines from T_H1 cells, T cells

undergo repeated division and differentiation. Most become cytotoxic T cells and eliminate infected cells, but others simply divide and persist as memory T cells for months or years in lymphoid tissues. If a memory T cell subsequently contacts an antigenic determinant matching its TCR, it responds immediately (without a need for interaction with APCs) and produces cytotoxic T cells.

Because the number of memory T cells is greater than the number of T cells that recognized the antigen during initial exposure, a secondary cell-mediated immune response is much more effective than a primary response. As with humoral immunity, an enhanced cell-mediated immune response upon subsequent exposure to the same antigen is called a memory response.

T Cell Regulation

Learning Objective

✓ Explain the process and significance of the regulation of cell-mediated immunity.

The body carefully regulates cell-mediated immune responses so that T cells do not respond to autoantigens. To prevent this from happening, T cells require additional signals from an antigen presenting cell. If they do not receive these signals, they will not respond. Thus, when a T cell and

Attacking Cancer with Lab-Grown T Cells

The body naturally manufactures T cells that attack cancer cells, but it sometimes does not make enough of them to shrink tumors and effectively halt the disease's progress. A recent study from the National Institutes of Health, however, shows that cancer-fighting T cells can be mass-produced in a laboratory and injected into patients, where they survive, multiply, and effectively reduce tumors. The study focused on 13 patients with malignant melanoma, an aggressive form of skin cancer. Researchers took samples of the pa-

T cells attacking cancer cells.

tients' own cancer-fighting T cells and cloned them in a laboratory until they numbered in the billions. These billions of identical T cells were then injected back into the individuals that first produced them. The results were encouraging. Four of the 13 patients became "virtually cancer free," while tumors were substantially reduced in two other patients.

Such T cell therapy remains highly experimental and is still years away from becoming a generally accepted cancer treatment. Scientists do not yet understand why the therapy works in some patients and not others. In addition, because the T cells attack malignant melanoma tumor antigens that are similar to, or the same as, antigens that are present on certain normal body cells, the therapy can cause autoimmune disease in some patients. To date, the therapy has only been tested against malignant melanoma, but planning to test it against other types of cancer is underway, and some scientists believe that a similar therapy may be effective against viral diseases such as hepatitis.

an antigen presenting cell interact, a specialized contact area, called an *immunological synapse,* forms between them; in effect, the two cell types have a chemical "conversation" that stimulates the T cell to fully respond to the antigen.

If a T cell does not receive the signals required for its activation, it will "shut down" as a precaution against autoimmune responses.

In summary, the body's response to infectious agents seldom relies on one mechanism alone, because this course of action would be far too risky. Therefore, the body typically uses several different mechanisms to combat infections. In the example from the chapter opener, the body's initial response to intruders was inflammation (a nonspecific response), but a specific immune response against the invading microorganisms was also necessary. APCs would phagocytize some of the invaders, process their antigenic determinants, and stimulate both humoral and cell-mediated immune responses, the results of which include the neutralization of toxins and microbial enzymes by antibody; the lysing of bacteria by antibody-activated complement, opsonization by antibodies, and the killing of intracellular bacteria and viruses by cytotoxic T cells. The relative importance of each of these pathways depends on the type of pathogen involved and on the mechanisms by which they cause disease.

CRITICAL THINKING

What sorts of pathogens could successfully attack a patient with an inability to produce T lymphocytes?

Types of Acquired Immunity

Learning Objective

✓ Contrast active versus passive acquired immunity, and naturally acquired versus artificially acquired immunity.

As we have seen, specific immunity is acquired during an individual's life. Immunologists categorize specific immunity as either naturally or artificially acquired. Naturally acquired immunity occurs when the body mounts an immune response against antigens, such as influenza viruses or food antigens, encountered during the course of daily life. Artificial immunity is the body's response to antigens introduced in vaccines, as occurs with immunization against the flu. Immunologists further distinguish acquired immune responses as either *active* or *passive;* that is, the immune system either responds actively to antigens via humoral or cell-mediated responses, or the body passively receives antibodies from another individual. Next we consider each of four types of acquired immunity.

Naturally Acquired Active Immunity

Naturally acquired active immunity occurs when the body responds to exposure to pathogens and environmental antigens by mounting specific immune responses. The body is naturally and actively engaged in its own protection. As we have seen, once an immune response occurs, immunological memory persists—on subsequent exposure to the same antigen, the immune response will be rapid and powerful and often provides the body complete protection.

Table 16.3   A Comparison of the Types of Acquired Immunity

Naturally Acquired		Artificially Acquired	
Active	Passive	Active	Passive
The body responds to antigens that enter naturally, such as during infections	Antibodies are transferred from mother to offspring, either across the placenta (IgG) or in breast milk (secretory IgA)	Health care workers introduce antigens in vaccines; the body responds with humoral or cell-mediated immune responses	Health care workers introduce preformed antibodies (obtained from immune individuals) as antisera and antitoxins into a patient

Naturally Acquired Passive Immunity

Although newborns possess the cells and tissues needed to mount an immune response, they respond slowly to antigens. If required to protect themselves solely via naturally acquired active immunity, they might die of infectious disease before their immune systems were mature enough to respond adequately. However, they are not on their own; in the womb, IgG molecules cross the placenta from the mother's bloodstream to provide protection, and after birth, babies receive secretory IgA in breast milk. Via these two processes, a mother provides her child with antibodies that protect it during its early months. Because the child is not actively producing its own antibodies, this type of protection is known as **naturally acquired passive immunity.**

Artificially Acquired Active Immunity

Physicians induce immunity in their patients by introducing antigens in the form of vaccines. The patients' own immune systems then mount active responses against the foreign antigens, just as if the antigens were part of a naturally acquired pathogen. Such **artificially acquired active immunity** is the basis of immunization, which is discussed in detail in Chapter 17.

Artificially Acquired Passive Immunity

Active immunity usually requires days to weeks to develop fully, and in some cases such a delay can prove detrimental or even fatal. For instance, an active immune response is too slow to protect against rattlesnake venom or infection with hepatitis A virus. Therefore, medical personnel routinely harvest antibodies against toxins and pathogens that are so deadly or so fast-acting that an individual's active immune response is inadequate. They can acquire these antibodies from the blood of immune humans or animals, typically a horse or other large animal. Physicians then inject such *antitoxins* or *antisera* into infected patients to confer **artificially acquired passive immunity.** Chapter 17 discusses this type of treatment in greater detail.

Active immune responses, whether naturally or artificially induced, are advantageous because they result in immunological memory and protection against future infections. However, they are slow-acting. Passive immunity, in which individuals are provided fully formed antibodies, has the advantage of speed but does not confer immunological memory because B and T lymphocytes are not activated. **Table 16.3** summarizes the four types of acquired immunity.

CHAPTER SUMMARY

Elements of Specific Immunity (pp. 463–473)

1. Immunologists study **specific immunity,** which is the ability of a vertebrate to recognize and defend against distinct species or strains of invaders.

2. **Antigens** are substances that trigger specific immune responses. The three-dimensional shape of a region of an antigen that is recognized by the immune system is the **antigenic determinant** (or epitope). Effective antigen molecules are large, complex, stable, degradable, and foreign to their host.

3. **Exogenous** antigens are found on microorganisms that multiply outside the cells of the body; **endogenous** antigens are produced by pathogens multiplying inside the body's cells.

4. Ideally, the body does not attack antigens on the surface of its normal cells, called **autoantigens;** this phenomenon is called self-tolerance.

5. The **lymphatic system** is composed of lymphatic vessels, which conduct the flow of **lymph,** and lymphoid tissues and organs that are directly involved in specific immunity. The latter include lymph nodes, the thymus, the spleen, the tonsils, and mucosa-associated lymphoid tissue (MALT). Cells with a central role in specific immunity, **lymphocytes,** migrate to and persist in various lymphoid organs, where they are available to encounter foreign invaders in the blood and lymph. Lymphocytes originate and mature in the red bone marrow and express characteristic membrane proteins that are designated cluster of differentiation (CD) molecules.

6. **B lymphocytes (B cells),** which mature in the red bone marrow secrete soluble proteins called **antibodies (immunoglobulins, Ig)** that attack exogenous antigens in **humoral immune responses.**

7. Antibodies consist of two **light chains** and two **heavy chains** joined via disulfide bonds to form Y-shaped molecules whose stems, called F_C **regions,** are of five basic classes. Each B cell randomly chooses (once in its life) genes for the upper portion of the "arms" of its antibody molecules; therefore, the upper portions are called variable regions. Together the variable regions of a heavy and light chain form an **antigen-binding site** and the upper portions of antibody molecules are called F_{ab} regions. Each basic antibody molecule has two antigen-binding sites and can potentially bind to two antigenic determinants.

8. Antibodies function in complement activation, inflammation, **agglutination** (clumping), **neutralization** (blocking the action of a toxin or attachment of a pathogen), and **opsonization** (enhanced phagocytosis).

9. The five types of heavy chains in antibody stems distinguish five different classes of antibodies (immunoglobulins). **IgG** is the predominant antibody found in the bloodstream and is largely responsible for defense against invading bacteria. IgG can cross a placenta to protect the fetus. **IgM,** a pentamer with ten antigen-binding sites, is the predominant antibody produced first during a primary humoral response. Two molecules of IgA are attached via J chains and a polypeptide secretory component to produce **secretory IgA,** which is found in milk, tears, and mucous membrane secretions. **IgE** triggers inflammation and allergic reactions. It also functions during helminth infections. Immunoglobulin D **(IgD)** is found in cytoplasmic membranes of some animals.

10. B lymphocytes also express **B-cell receptors (BCRs),** which are antibody-like proteins that are complementary to antigenic determinants but which are inserted into the cytoplasmic membranes of B cells rather than secreted. A BCR has the same shaped binding sites as do the antibodies that can be synthesized by that cell.

11. **T lymphocytes** (T cells) have **T-cell antigen receptors (TCRs),** mature in the thymus, and attack cells that harbor endogenous pathogens during **cell-mediated immune responses.**

12. In cell-mediated immunity, T_H1 cells activate **cytotoxic T cells** (CD8 cells), which then secrete perforins and granzymes that destroy infected or abnormal body cells. Two types of helper T cells—T_H1 and T_H2—are characterized by CD4 and direct cell-mediated and humoral immune responses respectively.

13. **Cytokines** are soluble regulatory proteins that act as intercellular signals to direct activities in immune responses. Cytokines include **interleukin (ILs), interferons (IFNs), growth factors, tumor necrosis factors (TFCs),** and **chemokines.**

The Body's Preparation for a Specific Immune Response (pp. 473–477)

1. Cells with receptors that respond to autoantigens are selectively killed via **apoptosis**—a process known as **clonal deletion.** Only cells that respond to foreign antigens survive to defend the body.

2. The initial step in mounting an immune response is that antigens are captured, ingested, and degraded into antigenic determinants by **antigen presenting cells (APCs)** such as macrophages and dendritic cells. Antigenic determinants are then bound to membrane molecules called **major histocompatibility complex (MHC)** class II molecules, which are inserted to present the antigenic determinant on the outer surface of the APCs' cytoplasmic membrane.

The Humoral Immune Response (pp. 477–480)

1. The combination of antigenic determinant and MHC II proteins activates the helper T cell that has a complementary TCR to become a T_H2 cell. The T_H2 cell secretes the cytokine interleukin-4 (IL-4), which activates the matching B cell to divide and differentiate.

2. **Plasma cells,** which arise from B cells, produce antibodies. Plasma cells live for only a short time but secrete very large amounts of antibodies. B cells that migrate to lymphoid tissues to await a subsequent encounter with the same antigen are called **memory B cells.**

3. The **primary response** to an antigen is slow to develop and of limited effectiveness. When that antigen is encountered a second time, the activation of memory cells ensures that the immune response is rapid and strong. This is a **memory response.**

The Cell-Mediated Immune Response (pp. 480–482)

1. A cell-mediated immune response is directed primarily against infected or abnormal body cells, including virus-infected cells, bacteria-infected cells, some fungal- or protozoan-infected cells, some cancer cells, and foreign cells that enter the body as a result of organ transplantation.

2. Cytotoxic T cells recognize abnormal molecules expressed on the surface of an infected, cancerous, or foreign cell. Cytotoxic T cells function only in the presence of cytokines from T_H1 cells.

3. Cytotoxic T cells destroy their target cells via two pathways: the **perforin-granzyme pathway,** which kills the affected cells by secreting **perforins** and **granzymes,** or the **CD95 pathway,** in which CD95L binds to CD95 on the target cell, which triggers target cell apoptosis. Cytotoxic T cells may also form **memory T cells.**

Types of Acquired Immunity (pp. 482–483)

1. When the body mounts a specific immune response against an infectious agent, the result is called **naturally acquired active immunity.** The passing of maternal IgG to the fetus, or of secretory IgA in milk to a baby, are examples of **naturally acquired passive immunity. Artificially acquired active immunity** is achieved by deliberately injecting someone with antigens in vaccines to provoke an active response, as in the process of immunization. **Artificially acquired passive immunity** involves the administration of preformed antibodies in antitoxins or antisera to a patient.

QUESTIONS FOR REVIEW

(Answers to multiple choice and matching questions are on the web, along with additional review questions. Visit www.microbiologyplace.com.)

Multiple Choice

1. Antibodies function to
   a. directly destroy foreign organ grafts.
   b. mark invading organisms for destruction.
   c. kill intracellular viruses.
   d. promote cytokine synthesis.
   e. stimulate T cell growth.

2. MHC class II molecules bind to _____ and trigger _____.
   a. endogenous antigens, cytotoxic T cells
   b. exogenous antigens, cytotoxic T cells
   c. antibodies, B cells
   d. endogenous antigens, helper T cells
   e. exogenous antigens, helper T cells

3. Rejection of a foreign skin graft is an example of
   a. destruction of virus-infected cells.
   b. tolerance.
   c. antibody-mediated immunity.
   d. a secondary immune response.
   e. a cell-mediated immune response.

4. An autoantigen is
   a. an antigen from bacteria.
   b. a normal body component.
   c. an artificial antigen.
   d. any carbohydrate antigen.
   e. a nucleic acid.

5. Among the key molecules that mediate cell-mediated cytotoxicity are
   a. perforins.
   b. immunoglobulins.
   c. complement.
   d. cytokines.
   e. interferons.

6. Which of the following lymphocytes predominates in blood?
   a. cytotoxic T cells
   b. helper T cells
   c. B cells
   d. memory cells
   e. all are about equally prevalent

7. The major class of immunoglobulin found on the surfaces of the walls of the intestines and airways is
   a. IgG.
   b. IgM.
   c. IgA.
   d. IgE.
   e. IgD.

8. Which cells express MHC class I molecules?
   a. red cells
   b. antigen presenting cells only
   c. neutrophils only
   d. all nucleated cells
   e. dendritic cells only

9. In which of the following sites in the body can B cells be found?
   a. lymph nodes
   b. spleen
   c. red bone marrow
   d. intestinal wall
   e. all of the above

Modified True/False

Mark each statement as either true or false. Rewrite false statements to make them true by changing the italicized words.

1. MHC class II molecules are found on *T cells*.
2. *Apoptosis* is the term used to describe cellular suicide.
3. Lymphocytes with CD8 glycoprotein are *helper* T cells.
4. *Cytotoxic T cells* secrete immunoglobulin.
5. Secretion of antibodies by activated B cells is a form of *cell-mediated* immunity.

Matching

1. Match each cell in the left column with its associated protein from the right column.

   ___ Plasma cell   A. MHC II molecule
   ___ Cytotoxic T cell   B. Interleukin-4
   ___ T$_H$2 cell   C. Perforin and granzyme
   ___ Dendritic cell   D. Immunoglobulin

2. Match each type of immunity in the left column with its associated example from the right column.

   ___ Artificially acquired passive immunity   A. Production of IgE in response to pollen
   ___ Naturally acquired active immunity   B. Acquisition of maternal antibodies in breast milk
   ___ Naturally acquired passive immunity   C. Administration of tetanus toxoid
   ___ Artificially acquired active immunity   D. Administration of antitoxin

Essay Questions

1. When is antigen processing an essential prerequisite for an immune response?

2. Why does the body have both humoral and cell-mediated immune responses?

CRITICAL THINKING

1. Why is it advantageous that the lymphatic system lacks a pump?

2   Contrast nonspecific defenses with specific immunity.

3. What is the benefit to the body of requiring the immune system to process antigen?

4. Scientists can develop genetically deficient strains of mice. Describe the immunological impairments that would result in mice deficient in each of the following: class I MHC, class II MHC, TCR, BCR, IL-2 receptor, or IFN-γ.

5. Human immunodeficiency virus (HIV) preferentially destroys CD4 cells. Specifically, what effect does this have on humoral and cell-mediated immunity?

6. What would happen to a person who failed to make MHC molecules?

7. Why does the body make five different classes of immunoglobulins?

8. Some materials such as metal bone pins and plastic heart valves can be implanted into the body without fear of rejection by the patient's immune system. Why is this? What are the ideal properties of any material that is to be implanted?

A 13-month-old boy is brought into the hospital, coughing and crying. His mother tells the pediatrician that she had noticed a mucous discharge pooling in the corners of her baby's eyes the previous evening. The pediatrician checks the baby's medical record and notes that he has not yet received his first measles/mumps/rubella (MMR) vaccination. He then swabs the baby's throat, draws a blood sample, and tells the mother that he is going to have the lab perform a "special immune test" to confirm his diagnosis: measles. Because he has seen several children with measles in the past few weeks and the baby's symptoms support the diagnosis, the pediatrician administers an intramuscular injection of antimeasles immunoglobulin. He explains that this injection will protect the child against the most severe form of the illness, which can be fatal. He also schedules her son for a routine measles/mumps/rubella vaccination. Mother and son return home to rest and recuperate.

How do immunoglobulins and vaccines work? How do immunologic tests aid in the diagnosis of specific diseases? This chapter is an introduction to the important topics of immunization and immune testing.

Immunization and Immune Testing

CHAPTER 17

CHAPTER OUTLINE

Immunization
Brief History of Immunization
Active Immunization
Passive Immunization

Immune Testing
Precipitation Tests
Agglutination Tests
Neutralization Tests
The Complement Fixation Test
Labeled Antibody Tests
Recent Developments in Immune Testing

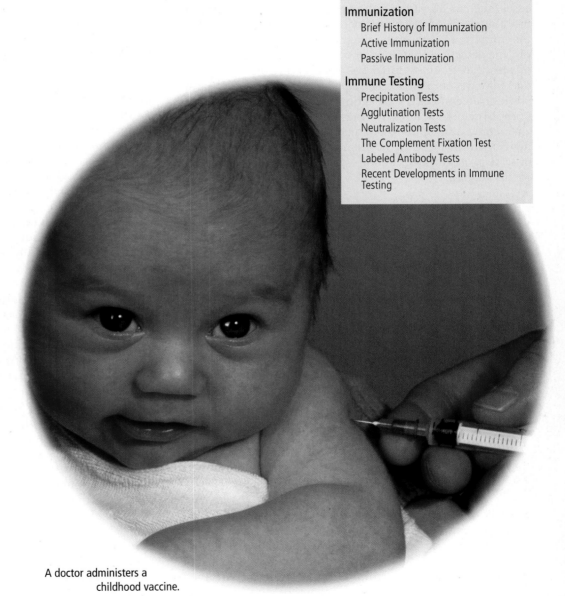

A doctor administers a childhood vaccine.

MicroPrep Pre-Test: Take the pre-test for this chapter on the web.
Visit www.microbiologyplace.com.

In this chapter we will discuss three applications of immunology: active immunization, or vaccination; passive immunization using immunoglobulins; and immune testing. Vaccination has proven the most efficient and cost-effective method of controlling infectious diseases. Without the use of effective vaccines, millions more people worldwide would suffer each year from potentially fatal infectious diseases, including measles, diphtheria, and pertussis (whooping cough). The administration of immunoglobulins has further reduced morbidity and mortality from certain infectious diseases, such as hepatitis A and yellow fever, in unvaccinated individuals. Medical personnel also make practical use of the immune response as a diagnostic procedure. For example, the detection of antibodies to HIV in a person's blood indicates that the individual has been exposed to that virus and may develop AIDS. The remarkable specificity of antibodies also enables the detection of drugs in urine, recognition of pregnancy at early stages, and the identification or characterization of other biological material. The many tests developed for these purposes are the focus of the discipline of *serology* (sĕ-rol′ō-jē) and are discussed in the second half of this chapter.

Immunization

As we saw in Chapter 16, an individual may be made immune to an infectious disease by two artificial methods: *active immunization,* which involves administering a vaccine to a patient so that the patient actively mounts a protective immune response, and *passive immunization,* in which a susceptible individual acquires temporary immunity through the transfer of antibodies formed by other individuals or animals.

In the following sections we will review the history of immunization before examining active and passive immunization in more detail.

Brief History of Immunization

Learning Objective

✓ Discuss the history of vaccination from the 12th century through the present.

As early as the 12th century, the Chinese noticed that children who recovered from smallpox never contracted the disease a second time. They therefore adopted a policy of deliberately infecting young children with particles of ground smallpox scabs from children who had survived mild cases. By doing so, they succeeded in significantly reducing the population's overall morbidity and mortality from the disease. News of this procedure, called *variolation* (var′ē-ō-lā′shŭn), spread westward through central Asia, and the technique was widely adopted.

Lady Mary Montagu, the wife of the English ambassador to the Ottoman Empire, learned of the procedure, had it performed on her own children, and told others about it upon her return to England in 1721. As a result, variolation came into use in England and in the American colonies; however, although effective and usually successful, it still caused death from smallpox in 1–2% of recipients and in people exposed to recipients, so in time variolation was outlawed.

Thus, when the English physician Edward Jenner demonstrated in 1796 that protection against smallpox could be conferred by inoculation with crusts from a person infected with cowpox—a related but very mild disease—the new technique was adopted with enthusiasm. Because cowpox was also called *vaccinia*[1] (vak-sin′ē-ă), Jenner called the new technique **vaccination** (vak′si-nā′shŭn), and the protective inoculum a **vaccine** (vak-sēn′)—terms that are still commonly used to describe all immunizations. For many years thereafter, vaccination against smallpox was widely practiced, even though no one understood how it worked or whether similar techniques could be used for other diseases.

In 1879, Louis Pasteur conducted experiments on the bacterium *Pasteurella multocida* (pas-ter-el′ă mul-tŏ′si-da) and demonstrated that he could make an effective vaccine against this organism (which causes a disease in birds called fowl cholera). Once the basic principle of vaccine manufacture was understood, vaccines against anthrax and rabies rapidly followed. Moreover, once it was discovered that these vaccines provide protection through the actions of antibodies, the technique of transferring protective antibodies to susceptible individuals—that is, passive immunization—was developed soon thereafter.

By the late 1900s, immunologists and health care providers had formulated vaccines that significantly reduced the number of cases of polio, measles, mumps, and rubella in the United States **(Figure 17.1)**. Now we also have successful vaccines against many other diseases, including chickenpox, hepatitis A, hepatitis B, and many strains of microbes that cause diarrhea, influenza, and pneumonia. Health care providers, governments, and international organizations working together have rid the world of naturally occurring smallpox and are well on the way to the worldwide eradication of polio, measles, mumps, and rubella. **Highlight 17.1** on page 490 discusses why a vaccine for the common cold does not yet exist.

Even though immunologists have produced vaccines that protect people against many deadly diseases, a variety of political, social, and economic problems prevent vaccines from reaching all those who need them. In developing nations worldwide, over 3 million children still die each year from vaccine-preventable infectious diseases because of socioeconomic and political obstacles. Additionally, some pathogens such as the protozoa of malaria and the virus of AIDS, still frustrate all attempts to develop effective vaccines against them. Furthermore, the existence of vaccine-associated risks—both medical risks (the low but persistent incidence of vaccine-caused diseases) and financial risks (the high costs of developing and producing

---

[1] From Latin *vacca*, meaning cow.

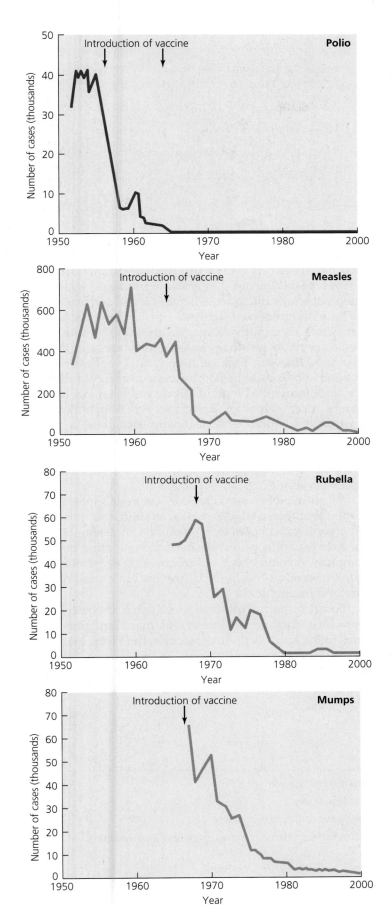

vaccines)—has in recent years discouraged investment in new vaccines. Thus although the history of immunization is marked by stunning advancements in public health, the future of immunization poses immense challenges.

Next we take a closer look at active immunization, commonly known as vaccination.

Active Immunization

Learning Objectives

✓ Describe the advantages and disadvantages of the three types of vaccines.

✓ Describe three methods by which molecular biology techniques can be used to develop improved vaccines.

✓ Delineate the risks and benefits of routine vaccination in healthy populations.

In the following subsections we examine the types of vaccines, the roles of technology in producing modern vaccines, and issues concerning vaccine safety.

Vaccine Types

Scientists are constantly striving to develop vaccines of maximal efficacy and safety, but not all types of vaccines are equally safe or effective. The three general types of vaccines, each of which has its own combination of strengths and weaknesses, are attenuated (live) vaccines, killed (or inactivated) vaccines, and toxoid vaccines.

Attenuated (Live) Vaccines Virulent viruses and organisms are normally not used in vaccines because they cause disease. Instead, immunologists reduce virulence so that, although still living, the pathogens no longer cause disease. The process of reducing virulence is called **attenuation** (ă-ten-ū-ā'shŭn). The most common method for attenuating viruses involves raising them for numerous generations in tissue culture cells, until the viruses lose the ability to produce disease. Thus for rabies viruses, which preferentially attack nerve cells, avirulent viruses for use in vaccines are produced by prolonged tissue culture in rabbit spinal cords, until the viruses' virulence in nerve cells is lost. Bacteria may be made avirulent by culturing them under unusual conditions or by using genetic manipulation.

◀ *Figure 17.1*

The effect of immunization in reducing the incidence of four infectious diseases in the United States. The two arrows in the graph of polio cases represent the introduction of two different types of vaccine. Only polio has been entirely eradicated in the United States.

Attenuated vaccines—those containing attenuated microbes—are also called *modified live vaccines*. Because they contain living but avirulent organisms or viruses, these vaccines cause very mild infections but no serious disease. Attenuated viruses in such a vaccine infect host cells and replicate; the infected cells then process endogenous viral antigens. As a result, modified live viral vaccines trigger a cell-mediated immune response dominated by type 1 helper T cells (T_H1) and cytotoxic T cells (T_C). Because modified live vaccines contain active microbes, a large number of antigen molecules are available to stimulate an immune response. Further, vaccinated individuals can infect those around them, providing so-called *herd immunity*—that is, immunity beyond the individual receiving the vaccine.

While usually very effective, attenuated vaccines can be hazardous because the modified microbes may retain enough residual virulence to cause disease in immunosuppressed people. Additionally, pregnant women should not receive live vaccines because of the danger that the attenuated pathogen will cross the placenta and harm the fetus. Occasionally, modified viruses actually revert to wild type or mutate to a form that causes persistent infection or disease. For example, in 2000 a polio epidemic in the Dominican Republic and Haiti resulted from the reversion of an attenuated live virus in oral polio vaccine to the virulent poliovirus. For this reason, we no longer use live polio vaccine to immunize children in the United States.

Inactivated (Killed) Vaccines   For some diseases live vaccines have been replaced by **inactivated vaccines,** which are of two types: *Whole agent vaccines* are produced with deactivated but whole microbes, whereas *subunit vaccines* are produced with antigenic fragments of microbes. Because neither whole agent nor subunit vaccines can replicate, revert, mutate, or retain residual virulence, they are safer than live vaccines. However, because they cannot replicate, several "booster" doses must be administered to achieve full immunity, and immunized individuals do not stimulate herd immunity. Also, with whole agent vaccines, nonantigenic portions of the microbe occasionally stimulate a painful inflammatory response in some individuals. As a result, pertussis vaccine, formerly available as a whole agent vaccine, is now being replaced with a subunit vaccine called acellular pertussis vaccine.

When microbes are killed for use in vaccines, it is important that their antigens remain as similar to those of living organisms as possible. If chemicals are used, they must not alter the antigens responsible for stimulating protective immunity. A commonly used inactivating agent is *formaldehyde* (fōr-mal'dĕ-hīd), which irreversibly cross-links proteins and nucleic acids.

Because the microbes of inactivated vaccines cannot reproduce, they do not present as many antigenic molecules to the body as do live vaccines; therefore, inactivated vaccines are antigenically weak. To be effective, they must be administered in high doses or in multiple doses, or incorporated with materials called **adjuvants** (ad'joo-

Table 17.1   **Some Common Adjuvants**

Adjuvant	Effects
Aluminum phosphate (alum)	Slows processing and degradation of antigen
Saponin (soap-like plant product)*	Stimulates T cell responses
Mineral oil*	Slows processing and degradation of antigen
Freund's complete adjuvant (mineral oil containing killed mycobacteria*)	Slows processing and degradation of antigen; stimulates T cell responses

*Considered too toxic for humans; used in vaccinating animals only.

văntz), substances that increase the effective antigenicity of the vaccine. Unfortunately, the use of adjuvants to increase antigenicity may stimulate local inflammation, and high individual doses and multiple dosing increase the risk of producing allergies. **Table 17.1** lists some common adjuvants and their effects in enhancing the efficacy of vaccines.

Because all types of killed vaccines are recognized by the immune system as exogenous antigens, they stimulate a T_H2 response, promoting antibody-mediated (rather than cell-mediated) immunity.

Toxoid Vaccines   For some bacterial diseases, notably tetanus and diphtheria, it is more efficacious to induce an immune response against bacterial toxins than against cellular antigens. **Toxoid** (tok'soyd) **vaccines** are chemically or thermally modified toxins that are used in vaccines to stimulate active immunity. As with killed vaccines, toxoids

stimulate antibody-mediated immunity. Because toxoids have few antigenic determinants, effective immunization requires multiple childhood doses as well as reinoculations every 10 years for life.

The 2003 immunization schedule recommended by the CDC is presented in **Figure 17.2.** The types of vaccines available for immunizing against each of the diseases in the vaccination schedule, as well as some other available vaccines, are listed in **Table 17.2** on page 492.

Modern Vaccine Technology

Although live, inactivated, and toxoid vaccines have been highly successful in controlling infectious diseases, researchers are always seeking ways to make vaccines more effective, cheaper, and safer. Scientists can use a variety of

▲ *Figure 17.2*
The CDC's recommended immunization schedule, 2002.

Table 17.2   Principal Vaccines to Prevent Human Diseases

Vaccine	Disease Agent	Disease	Vaccine Type
Recommended by CDC			
Hepatitis B	Hepatitis B virus	Hepatitis	Inactivated subunit
Diphtheria/	Diphtheria toxin	Diphtheria	Toxoid
tetanus/	Tetanus toxin	Tetanus	Toxoid
acellular pertussis (DTaP)	*Bordetella pertussis*	Whooping cough	Inactivated subunit (inactivated whole also available)
Haemophilus influenzae Type B	*Haemophilus influenzae*	Meningitis, pneumonia	Inactivated subunit
Polio	Poliovirus	Poliomyelitis	Inactivated (attenuated also available)
Measles/	Measles virus	Measles	Attenuated
mumps/	Mumps virus	Mumps	Attenuated
rubella	Rubella virus	Rubella (German measles)	Attenuated
Varicella	Chickenpox virus	Chickenpox	Attenuated
Pneumococcal	*Streptococcus pneumoniae*	Pneumonia	Inactivated subunit
Hepatitis A	Hepatitis A virus	Hepatitis	Inactivated whole
Influenza	Influenza viruses	Flu	Inactivated subunit
Not recommended for general population			
Anthrax	*Bacillus anthracis*	Anthrax	Inactivated whole
BCG (bacillus of Calmette and Guerin)	*Mycobacterium tuberculosis, M. leprae*	Tuberculosis, leprosy	Attenuated
Yellow fever	Yellow fever virus	Yellow fever	Attenuated
Rabies	Rabies virus	Rabies	Inactivated whole
Vaccinia (cowpox)	Smallpox virus	Smallpox	Attenuated

recombinant DNA techniques to make improved vaccines. For example, they could selectively delete virulence genes from a pathogen, producing an irreversibly attenuated microbe, one that cannot revert to a virulent pathogen **(Figure 17.3a).**

Additionally, scientists could use recombinant DNA techniques to produce large quantities of very pure viral or bacterial antigens for use in vaccines. In this process, scientists isolate DNA that codes for an antigen and then insert it into a bacterium, yeast, or other cell, which then expresses the antigen **(Figure 17.3b).** However, immunity produced by such pure antigens is often inferior to that produced by cruder vaccines containing whole organisms and numerous antigens; thus such vaccines may require much higher doses to stimulate comparable protection.

Alternatively, a genetically altered microbial cell or virus may itself be used as a live recombinant vaccine **(Figure 17.3c).** Experimental recombinant vaccines of this type have used adenoviruses, herpesviruses, poxviruses, and bacteria such as *Salmonella*. Vaccinia virus (a poxvirus) is widely used because it is easy to administer by dermal scratching or orally, and because its large genome makes inserting a new gene into it relatively easy.

Another innovative method of immunization involves injection not of antigens, but instead of the DNA that codes for the antigen. For example, the DNA coding for a pathogen's antigen can be inserted into a plasmid vector, which is then injected into an individual, where it is taken up by a cell **(Figure 17.3d).** The cell then transcribes and translates the gene to produce antigen, which triggers a cell-mediated immune response. Alternatively, DNA for antigens may be coated onto microscopic gold beads and fired directly into cells via a "gene gun." Once the DNA reaches the nucleus, its genes are expressed. Immunization with viral DNA in this way results in presentation of viral antigens in their native form and synthesized in the same way they are synthesized during an infection.

Vaccine Safety

Health care providers must carefully weigh the risks associated with vaccines against their benefits. The most common vaccine-associated problem is mild toxicity. Some vaccines —especially whole agent vaccines that contain adjuvants— may cause pain at the injection site for several hours or days after injection. In rare cases, toxicity may result in general malaise and possibly a fever high enough to induce seizures. Although not usually life threatening, the potential for these symptoms may be sufficient to discourage people from being vaccinated or having their infants vaccinated.

▲ *Figure 17.3*

Some potential uses of recombinant DNA technology for making improved vaccines. **(a)** Deletion of virulence gene(s) to create an irreversibly attenuated pathogen for use in a vaccine. **(b)** Insertion of an isolated gene that codes for a selected antigen into a cell (in this case, a yeast cell), which then produces large quantities of the antigen for use in a vaccine. **(c)** Insertion of a gene that codes for a selected antigen into a cell or virus (in this case, vaccinia virus), which produces the antigen. The entire recombinant is used in a vaccine. **(d)** Injection of DNA containing a selected gene (in this case, as part of a plasmid) into an individual. Once some of this DNA is incorporated into a cell's genome, that cell synthesizes and processes the antigen, which stimulates a cell-mediated immune response.

Highlight 17.2   Smallpox: To Vaccinate or Not to Vaccinate?

Medical personnel in the United States regularly administered the smallpox vaccine to the general public until 1971, at which time the risk of contracting smallpox was deemed too low to justify required vaccinations. Indeed, in

Vaccinia necrosum

1980 the World Health Assembly declared smallpox successfully eradicated from the natural world. However, recent concerns about the potential use of smallpox virus as an agent of bioterrorism has sparked debate about whether or not citizens should once again be vaccinated against it.

Although safe and effective for most healthy adults, for others the smallpox vaccine, which is attenuated cowpox virus, can result in

serious side effects—even death. Individuals with compromised immune systems (such as AIDS patients or cancer patients undergoing chemotherapy) are considered to be at particularly high risk for developing adverse reactions. Pregnant women, infants, and individuals with a history of the skin condition eczema are also considered poor candidates for the vaccine. Though rare, certain otherwise healthy individuals may also develop adverse reactions to the vaccine. The more serious side effects include vaccinia necrosum (characterized by progressive necrosis in the area of vaccination; see the box photo) and encephalitis (inflammation of the brain.) Approximately 1 in every 1 million individuals receiving the vaccine for the first time develops a fatal reaction to it. Some experts also fear that vaccinated individuals may pose a risk to immunocompromised people whom they contact.

Is the risk of a bioterrorist smallpox attack great enough to warrant the exposure to the known risks of administering smallpox vaccine to the general population? If you were a public health official, what would you decide?

A much more severe problem associated with vaccination is the risk of anaphylactic shock, an allergic reaction, which may develop to some component of the vaccine, such as egg proteins, adjuvants, or preservatives. Because people are rarely aware of such allergies ahead of time, vaccination recipients should remain for several minutes in the physician's office, where epinephrine is readily available to counter any signs of an allergic reaction.

A third major problem associated with vaccination is that of residual virulence, which we previously discussed. Attenuated vaccine viruses occasionally cause disease, not only in fetuses and immunosuppressed patients, but also in healthy children and adults. A good example is the attenuated oral poliovirus vaccine (OPV) commonly used in the United States until the late 1990s. While a very effective vaccine, it causes clinical poliomyelitis in 1 of every 2 million recipients or their close contacts. Medical personnel in the United States eliminated this problem by switching to inactivated polio vaccine (IPV).

Over the past two decades, several lawsuits in the United States and Europe have alleged that certain vaccines against childhood diseases cause or trigger disorders such as autism, diabetes, and asthma. Extensive research has failed to substantiate these allegations, and to date no conclusive evidence for such a link has been found. Indeed, vaccine manufacturing methods have improved tremendously in recent years, ensuring that modern vaccines are much safer than those in use even a decade ago. The U.S. Food and Drug Administration has established a Vaccine Adverse Event Reporting System for monitoring vaccine safety. **Highlight 17.2** discusses the issues surrounding the administration of smallpox vaccinations to the general public.

Passive Immunization

Learning Objectives

✓ Identify three sources of antibodies for use in passive immunization.

✓ Compare the relative advantages and disadvantages of active immunization and passive immunization.

Passive immunization involves the administration of preformed antibodies to a patient, typically to provide immediate protection against a recent infection or a disease in progress. For much of the time since passive immunization was first discovered, immunologists have harvested the desired antibodies from human or animal donors that have either experienced natural causes of a disease or were actively immunized against it. To do so, immunologists remove the cells and clotting factors from the blood to acquire *serum*, which contains a variety of antibodies. When used for passive immunization, such serum is called **antiserum** (an-tē-sē'rŭm) or sometimes *immune serum*. Antisera are typically collected from large animals intentionally exposed to the disease agent of interest because their large blood volume contains more antibodies than can be obtained from smaller animals.

Antisera have the following limitations:

• Because they contain antibodies against many different antigens, they are not limited solely to the antigen of interest.

• Repeated injections of antisera collected from a different species can trigger serious allergic responses in recipients. The result is that the recipient mounts an immune response against the animal antibodies found in the antisera.

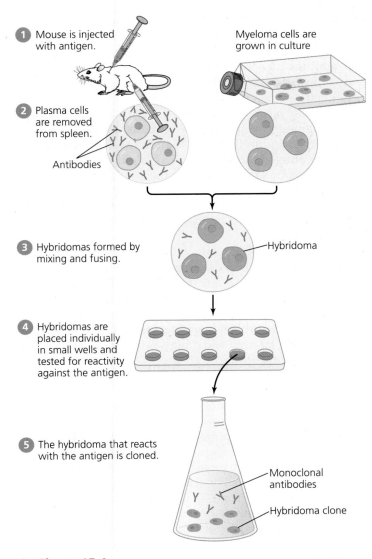

① Mouse is injected with antigen.

Myeloma cells are grown in culture

② Plasma cells are removed from spleen.

Antibodies

③ Hybridomas formed by mixing and fusing.

Hybridoma

④ Hybridomas are placed individually in small wells and tested for reactivity against the antigen.

⑤ The hybridoma that reacts with the antigen is cloned.

Monoclonal antibodies

Hybridoma clone

▲ *Figure 17.4*

The production of hybridomas. After a laboratory animal is injected with the antigen of interest ①, plasma cells, which of course secrete antibodies, are removed from the animal's spleen and isolated ②. When these plasma cells are fused with cultured cancer cells called myelomas, the results are hybridomas—long-lived tumor cells that produce antibodies while grown in culture③. Once the hybridomas are cultured individually and the hybridoma that produces antibodies against the antigen of interest is identified ④, it is cloned to produce a large number of hybridomas, all of which secrete identical antibodies called monoclonal antibodies ⑤.

- Antisera may be contaminated with viral pathogens.
- The antibodies of antisera are degraded relatively quickly by the recipient. The half-life of such antibodies is about 3 weeks; that is, half the molecules are degraded every 3 weeks.

Scientists have overcome most of the limitations of antisera by developing **hybridomas** (hī-brid-ō′măz), which are tumor cells created by fusing antibody-secreting plasma cells with cancerous plasma cells called *myelomas* (mī-ĕ-lō′măz) **(Figure 17.4).** Each hybridoma divides continuously

(because of the cancerous plasma cell component) to produce clones of itself, and each clone secretes large amounts of a single antibody molecule. These identical antibodies are called **monoclonal** (mon-ō-klō′năl) **antibodies** because all of them are secreted by clones originating from a single plasma cell. Once scientists have identified the hybridoma that secretes the antibody complementary to the antigen of interest, they maintain it in tissue culture to produce the antibodies needed for passive immunization.

Physicians use passive immunization when protection against a recent infection or an ongoing disease is needed quickly. Passive immunization provides rapid protection because it does not require the body to mount a response; instead, preformed antibodies are immediately available to bind to antigen, enabling agglutination, neutralization, and opsonization to proceed without delay. For example, if a previously unvaccinated patient has a wound contaminated with *Clostridium tetani* (klos-trid′ē-ŭm te′tan-ē), passive immunization with preformed antibodies against the bacterium and its toxin can provide protection quickly enough to save the patient's life. Antisera directed against toxins are also called *antitoxins* (an-tē-tok′sinz); *antivenin* used to treat snakebites is another antitoxin. In some cases, infections with certain viruses—hepatitis A and B, measles, rabies, rubella, chickenpox, and shingles—are treated with antisera directed against the causative viruses.

Active and passive immunization are used in different circumstances because they provide protection with different characteristics **(Figure 17.5).** As just noted, passive

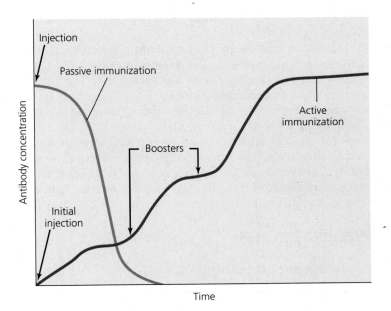

Injection

Passive immunization

Boosters

Active immunization

Antibody concentration

Initial injection

Time

▲ *Figure 17.5*

The characteristics of immunity produced by active and passive immunization. Whereas passive immunity provides strong and immediate protection, it dissipates relatively quickly. Active immunity takes some time and may require additional booster inoculations to reach protective levels, but it is long lasting and capable of restimulation.

immunization with preformed antibodies is used whenever immediate protection is required. However, because the preformed antibodies are gradually removed from the blood, the protection they provide wanes, and thus eventually the recipient becomes susceptible again. Whenever long-term protection is the goal, active immunization provides a slower buildup of protection that is capable of restimulation. Thus, when initiated before any exposure to *C. tetani* has occurred, active immunization using a tetanus toxoid develops long-lasting protection that is readily available upon exposure to the toxin.

Immune Testing

Learning Objectives

✓ In general terms, compare and contrast precipitation, agglutination, neutralization, complement fixation, and labeled antibody tests.

✓ Define serology.

Laboratory scientists use immunological processes in two general diagnostic ways. They may either use known antibodies to detect or identify antigens associated with an infectious agent (such as a particular strain of virus in tissue culture or in body tissues), or they may use selected antigens to detect specific antibodies in a patient's blood, thereby determining whether an individual has been exposed to a specific pathogen. The latter process can be used either to monitor the spread of an infection through a population or to establish a diagnosis in an individual. For example, the presence of antibodies against HIV in an adult's serum is strong evidence that this individual has been infected with HIV.

The study and diagnostic use of antigen-antibody interactions in blood serum is called **serology** (sĕ-rol'ō-jē). A wide variety of serologic tests are available to microbiologists, ranging from simple, automated processes to complex tests requiring skilled technicians. Microbiologists choose a particular test based on the suspected diagnosis, the cost to perform the test, and the speed with which a result can be obtained. In the following subsections we will consider several types of serologic tests: precipitation tests, agglutination tests, neutralization tests, the complement fixation test, and several tagged antibody tests.

Precipitation Tests

Learning Objectives

✓ Describe the general principles of precipitation testing.

✓ Compare and contrast immunodiffusion and immunoelectrophoresis.

✓ Discuss the production of anti-antibodies for immune testing.

One of the simplest of serologic tests relies on the fact that when antigens and antibody are mixed in proper proportions, they form huge lattice-like macromolecular complexes

called precipitates. When, for example, a solution of a soluble antigen is mixed with an antiserum containing antibodies against the antigen, the mixture becomes cloudy within a few minutes due to formation of an insoluble precipitate consisting of antigen-antibody complexes.

When a given amount of antibody is added to each of a series of test tubes containing increasing amounts of antigen, the amount of precipitate increases gradually until it reaches a maximum **(Figure 17.6a)**. In test tubes containing still more antigen molecules, the amount of precipitate declines; in fact, in test tubes containing antigen in great excess over antibody, no precipitate at all develops. Thus, a graph of the amount of precipitate versus the amount of antigen has a maximum in the middle and lower values on either side **(Figure 17.6b)**.

The reasons behind this pattern of precipitation reactions are simple. Complex antigens are generally multivalent—each possesses many antigenic determinants—and antibodies have pairs of active sites and therefore can simultaneously cross-link the same antigenic determinant on two antigen molecules. When there is excess antibody, each antigen molecule is covered with many antibody molecules, preventing cross-linkage and thus precipitation **(Figure 17.6c)**. When the reactants are in optimal proportions, the ratio of antigen to antibody is such that cross-linking and lattice formation are extensive. As this lattice grows, it becomes insoluble and eventually precipitates. In mixtures in which antigen is in excess, each antigen molecule is bound to only two antibody molecules. There is no cross-linkage; because these complexes are small and soluble, no precipitation occurs.

In the body, phagocytic cells do not easily remove small immune-complexes (such as those formed in antigen excess). As a result, small immune complexes may be deposited in the joints and in the tiny blood vessels of the kidneys, where they trigger allergic reactions, as discussed in Chapter 18.

Because precipitation requires that antigen and antibody are mixed in optimal proportions, it's not possible to perform a precipitation test by combining just any two solutions containing these reagents. To ensure that the optimal concentrations of antibody and antigen come together, scientists use two techniques involving movement of the molecules through an agar gel: immunodiffusion and immunoelectrophoresis.

Immunodiffusion

In the technique called *immunodiffusion,* double *immunodiffusion,* or an *Ouchterlony test,* a researcher cuts cylindrical holes called *wells* about 5 mm in diameter in an agar plate. One well is filled with a solution of antigen, and the other with a solution of antibodies against the antigen. The two solutions diffuse in all directions out of the wells and into the surrounding agar, and where they meet in optimal proportions, a line of precipitation appears **(Figure 17.7a)**. If the solutions contain many different antigens and antibod-

Same amount of antibody in each tube

No precipitate   Precipitate   No precipitate

(a)

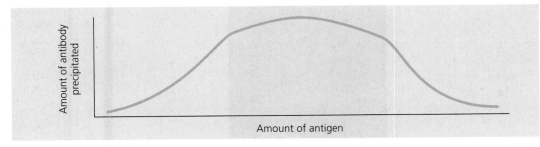

Amount of antibody precipitated

Amount of antigen

(b)

Antibody

Antigen

Antibody excess   **Optimal proportions**   **Antigen excess**

(c)

◀ *Figure 17.6*

Characteristics of precipitation reactions. (a) When increasing amounts of antigen are placed in tubes containing a constant amount of antibody, precipitate forms only in tubes having moderate amounts of antigen. **(b)** A graph of the amount of precipitate formed versus the amount of antigen present. **(c)** In the presence of excess antigen or antibody, immune complexes are small and soluble; only when antigen and antibody are in optimal proportions do large complexes form and precipitate.

▶ *Figure 17.7*

Immunodiffusion, a type of precipitation reaction. (a) Antigen and antibody placed in wells diffuse out and through the agar; where they meet in optimal proportions, a line of precipitation forms. **(b)** When multiple antigens and antibodies are placed in the wells, multiple lines of precipitation mark the sites where different antigen-antibody combinations occurred in optimal proportions. *Why did only three lines of precipitation occur when the antigen well contained four antigens?*

Figure 17.7. None of the antibodies used in the test was complementary to the fourth antigen.

ies, each complementary pair of reactants reaches optimal proportions at different positions, and numerous lines of precipitation are produced—one for each interacting antigen-antibody pair **(Figure 17.7b).**

Another useful variation on precipitation is called **radial immunodiffusion** (rā′dē-ăl im′ū-nō-di-fū′zhŭn) (**Figure 17.8a** on page 498). In this technique, an antigen in solution is allowed to diffuse from a well into agar that contains a

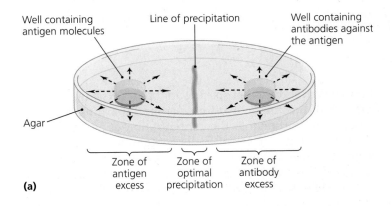

Well containing antigen molecules   Line of precipitation   Well containing antibodies against the antigen

Agar

Zone of antigen excess   Zone of optimal precipitation   Zone of antibody excess

(a)

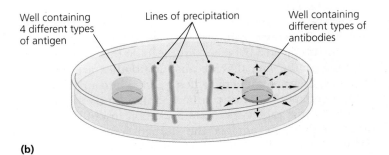

Well containing 4 different types of antigen   Lines of precipitation   Well containing different types of antibodies

(b)

(a)

(b)

▲ *Figure 17.8*

Radial immunodiffusion, a type of precipitation reaction. **(a)** Antigen placed into wells in increasing, known concentrations diffuses into agar containing a known concentration of antibody. The diameter of each ring of precipitation formed is proportional to the concentration of antigen in the well. **(b)** Plotting ring diameter versus known antigen concentration produces a standard curve; subsequently, the concentration of antigen in a solution can be determined by comparing the ring diameter produced by radial diffusion to the standard curve. *What is the antigen concentration of Sample Y?*

Figure 17.8 Based on the graph in (b), the concentration in Sample Y is about 1900 mg/100 ml.

known concentration of specific antibodies. The diameter of the ring of precipitate that forms around the antigen well is directly proportional to the concentration of antigen in the well. By using a series of wells containing increasing but known concentrations of antigen, researchers construct a standard curve that relates ring diameter to antigen concentration **(Figure 17.8b)**. Subsequently, the amount of antigen in a solution can be accurately assayed by comparing the ring diameter produced in other radial diffusion tests to the standard line.

Scientists commonly use radial immunodiffusion to measure the concentrations of specific antibodies or immunoglobulins in a person's serum. To do this, they take advantage of the fact that because antibodies are complex proteins, they are antigenic when injected into an individual of another species. Thus purified human immunoglobulins injected into a rabbit stimulate the rabbit to produce rabbit immunoglobulins against the human immunoglobulins. Such antibodies directed against immunoglobulins are called

anti-antibodies. So, in a radial immunodiffusion test to measure the concentration of a specific human immunoglobulin, the "antigen" in the test is the human immunoglobulin, and the antibody in the test is rabbit anti-antibody.

It is possible to make anti-antibodies that specifically target a particular class of antibody; for example, anti-IgG anti-antibodies target gamma heavy chains, and anti-IgD anti-antibodies target delta heavy chains. By incorporating anti-IgG, anti-IgA, anti-IgM, anti-IgE, and anti-IgD anti-antibodies into different agar gels and filling the wells of each gel with a sample of an individual's serum, scientists can determine the concentrations of the five antibody classes in a serum.

Immunoelectrophoresis

Although conventional gel-diffusion techniques produce a separate precipitation line for each antigen-antibody combination present, it is sometimes difficult to distinguish all the lines if they form within a small area of gel. In **immunoelectrophoresis** (im'ū-nō-ē-lek'trō-fō-rē'sis), which improves the resolution of an immunodiffusion test, the first step is to separate antigens from one another by *electrophoresis*, which as we saw in Chapter 8 separates molecules on the basis of electrical charge and size by running an electrical current through the gel **(Figure 17.9a)**. Positively charged antigens move toward the negative pole, negatively charged antigens move toward the positive pole, and antigens without a charge and very large molecules do not move at all.

After electrophoresis, a trough is cut in the agar parallel to the line of separated antigens. Antiserum placed in the trough diffuses laterally, and where diffusing antibodies encounter their complementary antigens in the correct proportions, curved lines of precipitation form **(Figure 17.9b)**. One line of precipitation forms for each pair of complementary reactants.

Immunoelectrophoresis can resolve more than 30 distinct antigens at once. Diagnosticians use the technique either to demonstrate the absence of a normal antigen or to detect the presence of excessive amounts of an antigen—for example, the excess of so-called *prostate specific antigen* in the serum of a patient with prostate cancer.

Agglutination Tests

Learning Objectives

✓ Contrast agglutination and precipitation tests.
✓ Describe how agglutination is used in immunological testing, including titration.

Not all antigens are soluble proteins that can be precipitated by antibody. Because of their multiple antigen-binding sites, antibodies can also cross-link *particulate* antigens, such as whole bacteria, causing **agglutination** (ă-gloo-ti-nā'shŭn) (clumping). The difference between agglutination and precipitation is that agglutination involves the clumping of insoluble particles, whereas precipitation involves the aggregation of soluble molecules. Agglutination reactions are easy to see and interpret with the unaided eye. IgM antibodies, which have 10 active sites, are more efficient than IgG antibodies (with only two active sites) at causing agglutination. When the particles agglutinated are red blood cells, the reaction is called *hemagglutination* (hē-mă-gloo'ti-nā'shŭn). One use of hemagglutination is to determine blood type in humans. Blood is considered type A if the red blood cells possess surface antigens called A antigen, type B if they possess B antigens, type AB if they possess both antigens, and type O if they have neither antigen. In a hemagglutination reaction to determine blood type **(Figure 17.10)**,

(a)

(b)

▲ *Figure 17.9*

Immunoelectrophoresis, a type of precipitation reaction used to identify antigens in a mixture. **(a)** Antigens in a mixture are placed in a well and separated by electrophoresis. **(b)** After a trough running parallel to the direction of the electrical current is cut and filled with antiserum, arcs of precipitate form where complementary antigen and antibody meet in optimal concentrations.

(a) Negative result: no agglutination of blood cells   Positive result: agglutination of blood cells

(b)

▲ *Figure 17.10*

The use of hemagglutination to determine blood types in humans. **(a)** Antibodies with active sites that bind to either of two red blood cell surface antigens (antigen A or antigen B) are added to portions of a given blood sample. Where the antibodies react with the surface antigens, the blood cells can be seen to agglutinate or clump together. **(b)** Photo of actual test. *What is the blood type of the person whose blood was used in this hemagglutination reaction?*

Figure 17.10 The individual who donated the blood sample has type B blood.

two portions of a given blood sample are placed on a slide, and anti-A antibodies are added to one portion, and anti-B antibodies to the other; the antibodies agglutinate (clump) those blood cells that possess complementary antigens.

CRITICAL THINKING

Draw a picture showing, at *both the molecular and cellular levels*, IgM agglutinating red blood cells.

Another use of agglutination is in a type of test that determines the concentration of antibodies in a clinical sample. Although the simple *detection* of antibodies is sufficient for many purposes, it is usually more desirable to measure the *amount* of antibodies in serum. By doing so, clinicians can determine whether a patient's antibody levels are rising, as occurs in response to the presence of active infectious disease, or falling, as occurs during the successful conclusion of a fight against an infection. One way of measuring antibody levels in blood sera is by **titration** (tī-trā'shŭn). In titration, the serum being tested is serially diluted, and each dilution is then tested for agglutinating activity **(Figure 17.11)**. Eventually, the antibodies in the serum become so dilute that they can no longer cause agglutination. The highest dilution of serum giving a positive reaction is called the **titer** (tī'ter), which provides an estimate of the amount of antibody in that serum. Thus, a serum that must be greatly diluted before agglutination ceases (for example, has a titer of 1000) contains more antibodies than a serum that no longer agglutinates after minimal dilution (has a titer of 10).

Serum added in increasing dilutions

1:1   1:10   1:100   1:1000   1:10,000   Control (no specimen added)

Antigen (identical in each well)

++++   +++   ++   +   —   Control
Very strong agglutination   No agglutination

▲ *Figure 17.11*

Titration, the use of agglutination to quantify the amount of antibody in a serum sample. Serial dilutions of serum are added to wells containing a constant amount of antigen. At lower serum dilutions (higher concentrations of antibody), agglutination occurs; at higher serum dilutions, antibody concentration is too low to produce agglutination. The serum's titer is the highest dilution at which agglutination can be detected—in this case, 1:1000.

Neutralization Tests

Learning Objectives

✓ Explain the purpose of neutralization tests.
✓ Contrast a viral hemagglutination inhibition test with a hemagglutination test.

Neutralization tests work because antibodies can *neutralize* the biological activity of many pathogens and their toxins. For example, combining antibodies against tetanus toxin (which is lethal to mice) with a sample of toxin renders the sample harmless to mice because the antibodies have reacted with and neutralized the toxin. Next we briefly consider two neutralization tests that, although not simple to perform, effectively reveal the biological activity of antibodies.

Viral Neutralization

One neutralization test is **viral neutralization,** which is based on the fact that many viruses introduced into appropriate cell cultures will invade and kill the cells, a phenomenon called a *cytopathic effect*. However, if the viruses are first mixed with specific antibodies against them, their ability to kill culture cells is neutralized. In a viral neutralization test, the lack of cytopathic effects when a mixture containing serum and a known virus is introduced into a cell culture indicates the presence of antibodies against that virus in the serum. For example, if a mixture containing an individual's serum and a sample of hantavirus produces no cytopathic effect in a culture of susceptible cells, then it can be concluded that the individual's serum contains antibodies to hantavirus. Viral neutralization tests are sufficiently sensitive and specific to ascertain whether an individual has been exposed to a particular virus or viral strain.

Viral Hemagglutination Inhibition Test

Because not all viruses are cytopathic, a neutralization test cannot be used to identify all viruses. However, many viruses (including influenza viruses) have surface proteins that naturally clump red blood cells. This natural process, called *viral hemagglutination,* must not be confused with the hemagglutination test we discussed previously—viral hemagglutination is not an antibody-antigen reaction. However, antibodies against influenza virus inhibit viral hemagglutination; therefore, if serum from an individual stops viral hemagglutination, we know that the individual's serum contains antibodies to the influenza virus. Such **viral hemagglutination inhibition tests** are used to detect antibodies against influenza, measles, mumps, and other viruses that naturally agglutinate red blood cells.

The Complement Fixation Test

Learning Objectives

✓ Briefly explain the phenomenon that is the basis for a complement fixation test.

✓ Describe the use of the complement fixation test.

As we discussed in Chapter 15, activation of the classical complement system by antibody leads to the generation of membrane attack complexes (MACs) that disrupt cytoplasmic membranes (see Figure 15.17). This phenomenon is the basis for the **complement fixation** (kom′plĕ-ment fik-sā′shŭn) **test,** which is used to detect the presence of specific antibodies in an individual's serum. A complement fixation test can detect the presence of small amounts of antibody—amounts too small to detect by agglutination. A complement fixation test is performed in three stages:

1 A serum sample from an individual is heated to 56°C to destroy any naturally present complement.

2 Antigen from the disease agent of interest and a standard amount of complement are added to the serum.

3 Sheep red blood cells (RBCs) and antibodies against sheep RBCs (anti-RBC antibodies) are added to the mixture, which is then examined for the lysis of the RBCs.

Figure 17.12 illustrates the steps in a complement fixation test and depicts the events that occur in both positive and negative complement fixation tests. In a positive test—that is, one in which the individual's serum contains antibodies against the disease agent of interest (Figure 17.12a), the antibodies bind the antigen, and the resulting antigen-antibody complexes bind complement, which is thus no longer available to bind to the anti-RBC antibodies. As a result, the sheep RBCs remain intact, and the solution remains cloudy. In a negative test—that is, one in which the individual's serum lacks antibodies against the disease agent of interest (Figure 17.12b)—the lack of antigen-antibody complexes enables complement to bind to the anti-RBC antibodies (which have attached to the sheep RBCs), triggering lysis of the RBCs by MACs and turning the solution a transparent red.

▶ *Figure 17.12*

The complement fixation test. To conduct the test, ① an individual's serum is heated to destroy naturally present complement; ② known amounts of the antigen of interest and complement are added; and ③ sheep red blood cells (RBCs) and antibodies against them (anti-RBC antibodies) are added. **(a)** A positive test. In the presence of antibodies against the antigen, antigen-antibody complexes bind the complement, which then cannot bind with anti-RBC antibodies to lyse the RBCs. **(b)** A negative test. The absence of antibodies (and thus of antigen-antibody complexes) leaves complement free to bind to anti-RBC antibodies attached to the RBCs, which are then lysed. *Why must the serum be heated?*

Figure 17.12 Heat destroys excess complement so that only the standard amount is available.

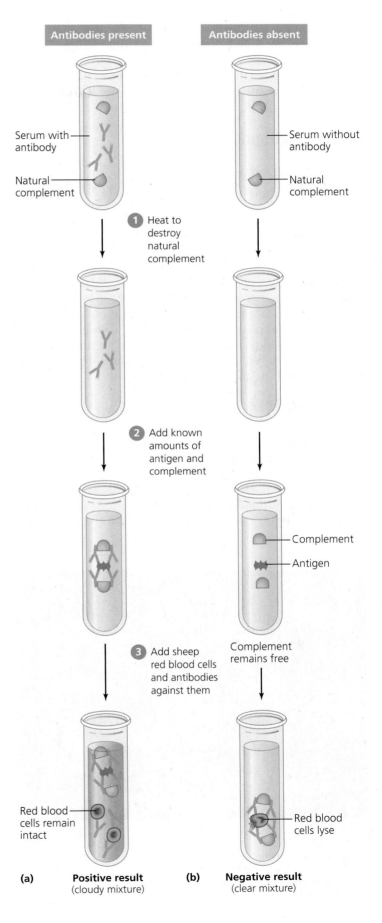

Antibodies present | **Antibodies absent**

Serum with antibody

Serum without antibody

Natural complement

Natural complement

1 Heat to destroy natural complement

2 Add known amounts of antigen and complement

Complement

Antigen

Complement remains free

3 Add sheep red blood cells and antibodies against them

Red blood cells remain intact

Red blood cells lyse

(a) **Positive result** (cloudy mixture)

(b) **Negative result** (clear mixture)

A complement fixation test called the Wasserman test was once the standard diagnostic test for syphilis. Though the Wasserman test has been replaced by labeled antibody tests (discussed next), complement fixation is still used in diagnosis of some infections of fungi, viruses, and rickettsial bacteria.

Labeled Antibody Tests

Learning Objectives

✓ List three tests that use labeled antibodies to detect either antigen or antibodies.

✓ Compare and contrast the direct and indirect fluorescent antibody tests, and identify at least three uses for these tests.

✓ Compare and contrast the methods, purposes, and advantages of ELISA and the western blot test.

A different form of serologic testing involves *labeled* (or *tagged*) *antibody tests,* so-named because these tests use antibody molecules that are linked to some molecular "label" that enables them to be detected easily. Labeled antibody tests can be used to detect either antigens or antibodies. In the following subsections we will consider fluorescent antibody tests, ELISA, and the western blot test.

Fluorescent Antibody Tests

Fluorescent dyes are used as labels in several important serologic tests. The most important of these dyes is *fluorescein* (flōr-es'ē-in), which can be chemically linked to an antibody without affecting the antibody's ability to bind antigen. When exposed to fluorescent light (as in a fluorescent microscope), fluorescein glows bright green. Fluorescein-labeled antibodies are used in direct and indirect fluorescent antibody tests.

▲ *Figure 17.13*

LM | 30 μm

The direct fluorescent antibody test. The fluorescence from fluorescein-labeled antibodies against rabies viruses indicates the presence of the viruses in a section of brain tissue.

Direct fluorescent antibody tests identify the presence of antigen in a tissue. The test is straightforward: A scientist floods a tissue sample suspected of containing the antigen with labeled antibody, waits a short time to allow the antibody to bind to the antigen, washes the preparation to remove any unbound antibody, and examines it with a fluorescent microscope. If the suspected antigen is present, labeled antibody will adhere to it, and the scientist will see fluorescence. This is not a quantitative test—the amount of fluorescence observed is not directly related to the amount of antigen present.

Scientists use direct fluorescent antibody tests to identify small numbers of bacteria in patient tissues. This technique is used to detect *Mycobacterium tuberculosis* (mī-kō-bak-tēr'ē-ŭm tū-ber-kyū-lō'sis) in sputum and viruses infecting tissues. In one use, medical laboratory technologists employ a direct fluorescent antibody test to detect the presence of rabies virus in the brain of an infected animal (**Figure 17.13**).

Like direct fluorescent antibody tests, **indirect fluorescent antibody tests** are used to detect antigens in cells or patient tissues, but they can also be used to detect the presence of specific antibodies in an individual's serum via a two-step process (**Figure 17.14a**):

1 After the antigen of interest is fixed to a microscope slide, the individual's serum is added for long enough to allow serum antibodies to bind to the antigen. The serum is then washed off, leaving any specific antibodies bound to the antigen (but not yet visible).

2 Anti-antibodies (in this example, anti-IgG) labeled with fluorescein is added to the slide and binds to the antibodies already bound to the antigen. After washing to remove unbound anti-antibodies, the slide is examined with a fluorescent microscope.

The presence of fluorescence indicates the presence of the labeled anti-antibodies, which are bound to serum antibodies bound to the fixed antigen; thus fluoresence indicates that the individual has serum antibodies against the antigen of interest.

Indirect fluorescent antibody testing is used to detect antibodies against many viruses and some bacterial pathogens, including *Neisseria gonorrhoeae* (nī-sē'rē ă gon-ō-rē'ē), the causative agent of gonorrhea (**Figure 17.14b**).

ELISAs

In another type of labeled antibody test, called an **enzyme-linked immunosorbent assay** (im'ū-nō-sōr'bent as'sā) (ELISA), the label is not a dye, but instead an enzyme that reacts with its substrate to produce a colored product that indicates a positive test. The most common form of ELISA is used to detect the presence (and quantify the abundance) of antibodies in serum. This test, which often takes place in commercially produced polystyrene-coated plates, has five steps (**Figure 17.15** on page 504):

1 Each of the wells in the plate is coated with antigen molecules in solution.

① Antigen is attached to slide and flooded with patient's serum

Antigen bound to slide

IgG from patient's serum

② Fluorescent labeled anti-Ig antiglobulin is added

Fluorescent label

Anti-IgG (antiglobulin)

(a)

LEM　10 μm

(b)

▲ *Figure 17.14*

The indirect fluorescent antibody test. **(a)** The test procedure. ① Antigen is attached to the slide, which is then flooded with an individual's serum to allow specific antibodies in the serum to bind to the antigen. ② After fluorescein-labeled anti-antibodies are added and then washed off, the slide is examined with a fluorescent microscope. **(b)** A positive indirect antibody test, in which fluorescence indicates the presence of antibodies against *Neisseria gonorrhoeae* (the agent of gonorrhea) in the individual's serum.

② Excess antigen molecules are washed off, and another protein (such as gelatin) is added to the well to completely coat any of the polystyrene surface not coated with antigen.

③ A sample of each of the sera being tested is added to a separate well. Whenever a serum sample contains antibodies against the antigen, they bind to the antigen affixed to the plate.

④ Anti-antibodies labeled with an enzyme are added to each well.

⑤ The enzyme's substrate is added to each well. The enzyme and substrate are chosen because their reaction produces products that cause a visible color change.

A positive reaction in a well, indicated by the development of color, can occur only if the labeled anti-antibody has bound to antibodies attached to the antigen of interest. The intensity of the color, which can be estimated visually or measured accurately using a spectrophotometer, is proportional to the amount of antibody present in the serum.

ELISA has become the test of choice for many diagnostic procedures, such as determination of HIV infection, because of its many advantages:

- Like other labeled antibody tests, ELISA can detect either antibody or antigen.
- Unlike some diffusion and fluorescent tests, ELISA can quantify amounts of antigen or antibody.
- ELISAs are easy to perform, are relatively inexpensive, and can simultaneously test many samples quickly;

moreover, they lend themselves to efficient automation and can be read easily, either by direct observation or by machine.

- Plates coated with antigen and gelatin can be stored for testing, whenever they are needed.

A modification of the ELISA technique, called an *antibody sandwich ELISA*, is commonly used to detect antigen (**Figure 17.16** on page 504). In testing for the presence of HIV in blood serum, for example, the polystyrene plates are first coated with antibody (instead of antigen). Then the sera from individuals being tested for HIV are added to the wells, and any antigen in the sera will bind to the antibody attached to the well. Finally, each well is flooded with enzyme-labeled antibodies specific to the antigen. The name "antibody sandwich ELISA" refers to the fact that the antigen being tested for is "sandwiched" between two antibody molecules. Such tests can also be used to quantify the amount of antigen in a given sample.

CRITICAL THINKING

A diagnostician used an ELISA to show that a newborn had antibodies against HIV in her blood. However, 6 months later the same test was negative. How can this be?

Western Blot Test

A technique for detecting antibodies against multiple antigens in a complex mixture is a **western blot test**. The name "western blot" is a play on words that refers to the similarity

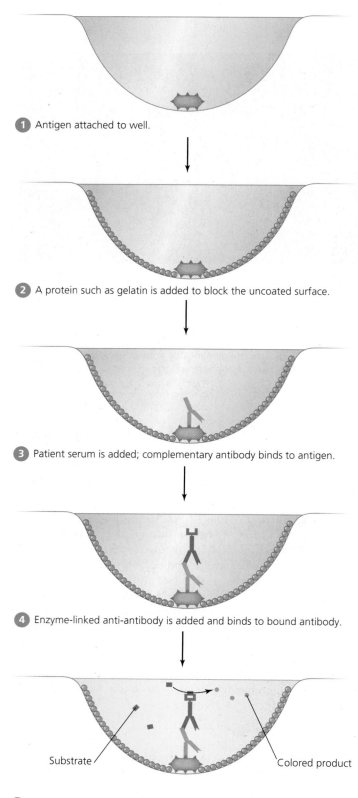

① Antigen attached to well.

② A protein such as gelatin is added to block the uncoated surface.

③ Patient serum is added; complementary antibody binds to antigen.

④ Enzyme-linked anti-antibody is added and binds to bound antibody.

⑤ Enzyme's substrate is added, and reaction produces a visible color change.

◀ *Figure 17.15*

The enzyme-linked immunosorbent assay (ELISA). Shown is one well in a polystyrene plate. ① Antigen added to the well attaches to it irreversibly. ② After excess antigen is removed by washing, gelatin is added to cover any portion of the well not covered by antigen. ③ Test serum is added to the well; any specific antibodies in it bind to the antigen. ④ Enzyme-labeled anti-antibodies are added to the well and bind to any bound antibody. ⑤ The enzyme's substrate is added to the well, and the enzyme converts the substrate into a colored product; the amount of color, which can be measured via spectrophotometry, is directly proportional to the amount of antibody bound to the antigen.

of this technique to the Southern blot test (see Figure 8.8), named for the man who developed it.

Western blot tests are currently used to verify the presence of antibodies against HIV in the serum of individuals who are antibody-positive by ELISA. Compared to other tests, western blot tests can detect more types of antibodies and are less subject to misinterpretation. A western blot test has three steps **(Figure 17.17a):**

① Electrophoresis. Antigens in a solution (in this case, HIV proteins) are placed into wells and separated by gel electrophoresis. Each of the proteins in the solution is resolved into a single band, producing a pattern of protein bands.

② Blotting. Each of the protein band patterns is transferred to an overlying nitrocellulose membrane by absorbing the solution into absorbent paper. The nitrocellulose membrane is then cut into strips.

③ ELISA. Each nitrocellulose strip is incubated with a test solution—in this case, a urine sample and a serum sample from each of three individuals are being tested for antibodies against HIV. After the strips are washed, an enzyme-labeled anti-antibody solution is added for a time; then the strips are washed again and exposed to the enzyme's substrate.

Color develops wherever antibodies against the HIV proteins in the test solutions have bound to the HIV proteins. In this case, the individual tested in strips 1 and 2 is negative for antibodies against HIV, whereas the other two individuals are positive for antibodies against HIV.

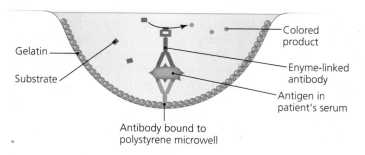

▲ *Figure 17.16*

An antibody sandwich ELISA. Because this variation of ELISA is used to test for the presence of antigen, antibody is attached to the well in the initial step.

Recent Developments in Immune Testing

Learning Objectives

✓ Discuss the benefits of using immunofiltration assays rather than an ELISA.

✓ Contrast immunofiltration and immunochromatographic assays.

Recent years have seen the development of simple immunoassays that give clinicians useful results within minutes. The most commonly used of these are immunofiltration and immunochromatography assays. These tests, which generally are not quantitative but simply give a positive or negative result, are very useful in arriving at a preliminary diagnosis.

Immunofiltration (im′ū-nō-fil-trā′shŭn) **assays** are rapid ELISAs based on the use of antibodies bound to a membrane filter rather than to polystyrene plates. Because of the large surface area of a membrane filter, reactions proceed faster and assay times are significantly reduced as compared to a traditional ELISA.

Immunochromatographic (im′ū-nō-krō′mat-ō-graf′ik) **assays** are still faster and easier to read ELISAs. In these systems, an antigen solution (such as infected blood) flowing through a porous strip encounters antibody labeled with either pink colloidal gold or blue colloidal selenium. Where antigen and antibody bind, colored immune complexes form in the fluid, which then flows through a region where the complexes encounter antibody against them, resulting in a clearly visible pink or blue line. These assays are used for pregnancy testing, which test for *human chorionic growth hormone*—a hormone produced only by an embryo or fetus, and for rapid identification of infectious agents such as HIV.

Table 17.3 on page 506 lists the types of immune tests currently used to diagnose some selected bacterial and viral diseases. **New Frontiers 17.1** on page 506 describes a novel approach to detecting pathogens that might hold promise as a new development in immune testing.

(a)

(b)

◀ *Figure 17.17*

A western blot, a technique for demonstrating the presence of antibodies against multiple antigens in a complex mixture. **(a)** Steps in a western blot. ① Antigens (in this case, HIV proteins) are separated by gel electrophoresis. ② Separated proteins are transferred to a nitrocellulose membrane. ③ Test solutions (in this case, urine and serum that may or may not contain antibodies against HIV), enzyme-labeled anti-antibody, and the enzyme's substrate are added; color changes are detected wherever antibody in the test solutions has bound to HIV proteins. **(b)** The results of this western blot. The individual tested in the top 2 strips is negative for antibodies to HIV; the other two individuals have antibodies against HIV.

Table 17.3   **Immunological Tests and Some of Their Uses**

Test	Use
Immunodiffusion (precipitation)	Diagnosis of syphilis, pneumococcal pneumonia
Immunoelectrophoresis (precipitation)	Assay production of particular classes of antibodies
Agglutination	Blood typing; pregnancy testing; diagnosis of salmonellosis, brucellosis, gonorrhea, rickettsial infection, mycoplasma infection, yeast infection, typhoid fever, meningitis caused by *Haemophilus*
Viral neutralization	Diagnosis of infections by specific strains of viruses
Viral hemagglutination inhibition	Diagnosis of viral infections including influenza, measles, mumps, rubella, mononucleosis
Complement fixation	Diagnosis of measles, influenza A, syphilis, rubella, rickettsial infections, scarlet fever, rheumatic fever, infections of respiratory syncytial virus and *Coxiella*
Direct fluorescent antibody	Diagnosis of rabies, infections of group A *Streptococcus*
Indirect fluorescent antibody	Diagnosis of syphilis, mononucleosis
ELISA	Pregnancy testing; presence of drugs in urine; diagnosis of hepatitis A, hepatitis B, rubella; initial diagnosis of HIV infection
Western blot	Verification of infection with HIV, diagnosis of Lyme disease

New Frontiers 17.1   Listening for Pathogens

Scientists at the University of Cambridge have developed a fascinating new method of detecting the presence of pathogens: by listening for them. They conducted an experiment in which they coated one side of a thin slice of quartz

Quartz resonator

crystal with antibodies against herpes virus, and then introduced herpes viruses, which promptly bound to the antibodies. They then passed an electrical current through the crystal, gradually increasing the voltage until the quartz began to vibrate. They found that the force of the vibration eventually caused the virus and the antibody to break apart, creating a faint popping sound that the crystal registered and converted into electrical signals detectable by researchers.

Why is this important? Because whereas traditional immunological assays detect the presence of antibodies, many of them do not tell us whether a given pathogen is actually present in a patient. The method used by the Cambridge scientists, called *rupture event scanning*, could possibly be adapted to create an instrument that can confirm the presence of a pathogen by its sound. Although herpes virus was used in the Cambridge study, rupture event scanning could work on a wide variety of pathogens, including the causative agents of anthrax, smallpox, foot-and-mouth disease, and Ebola, and might provide quicker, less expensive results than existing serological tests.

CHAPTER SUMMARY

Immunization (pp. 488–496)

1. The first vaccine was developed in the late 1700s by Edward Jenner against smallpox. Vaccines against rabies, anthrax, measles, mumps, rubella, polio and other diseases were developed thereafter.

2. Individuals can be made immune to many infections by either active immunization or passive immunization.

3. Active immunization involves giving antigen in the form of either **attenuated vaccines, inactivated (killed) vaccines,** or **toxoid vaccines.** Pathogens in attenuated vaccines are weakened so that they no longer cause disease, though they are still alive or active and residual virulence may remain a problem. Inactivated vaccines are either whole agent or subunit

vaccines and often contain **adjuvants,** which are chemicals added to increase their ability to stimulate active immunity. Toxoid vaccines use modified toxins to stimulate antibody-mediated immunity.

4. Passive immunization involves administration of an **antiserum** containing preformed antibodies.

5. The fusion of myelomas (cancerous plasma cells) with plasma cells results in **hybridomas,** the source of monoclonal antibodies, which can be used in passive immunization.

Immune Testing (pp. 496–506)

1. **Serology** is the study and use of immunologic tests to diagnose and treat disease or identify antibodies or antigens.

2. The simplest of the serologic tests is a precipitation test, in which antigen and antibody meet in optimal proportions to form an insoluble precipitate. Often this test is performed in clear gels, where it is called immunodiffusion.

3. **Radial immunodiffusion** is a quantitative variant of an immunodiffusion test; another variant used to analyze complex antigen mixtures is **immunoelectrophoresis.**

4. **Agglutination tests** involve the clumping of antigenic particles by antibodies. The amount of these antibodies, called the **titer,** is measured by diluting out the serum in a process called **titration.**

5. Antibodies to viruses or toxins can be measured using a **neutralization test,** such as a viral neutralization test. Infection by viruses that naturally agglutinate red blood cells can be demonstrated using a **viral hemagglutination inhibition test.**

6. The **complement fixation test** is a complex assay used to determine the presence of specific antibodies by the ability of the antibodies to nullify the action of complement on sheep red blood cells.

7. Fluorescently–labeled antibodies—those chemically linked to a fluorescent dye such as fluorescein—can be used in a variety of **direct** and **indirect fluorescent antibody** tests. The presence of labeled antibodies is visible through a fluorescent microscope.

8. **Enzyme-linked immunosorbent assays (ELISAs)** are a family of simple tests that can be readily automated and read by machine. These tests are among the most common serologic tests used. A variation of the ELISA is the **western blot test,** which is used to detect antibodies against multiple antigens in a mixture. Together, ELISA and the western blot are used to detect and confirm infection with HIV.

9. **Immunofiltration assays** and **immunochromatographic assays** are very rapid modifications of ELISA tests that can give rapid diagnostic results.

QUESTIONS FOR REVIEW
(Answers to multiple choice, True/False, and matching questions are on the web, along with additional review questions. Visit www.microbiologyplace.com.)

Multiple Choice

1. In order to obtain immediate immunity against tetanus, a patient should receive
   a. an attenuated vaccine of *Clostridium tetani*.
   b. a modified live vaccine of *C. tetani*.
   c. tetanus toxoid.
   d. immunoglobulin against tetanus toxin (antitoxin).
   e. a subunit vaccine against *C. tetani*.

2. Which of the following vaccine types is commonly given with an adjuvant?
   a. an attenuated vaccine
   b. a modified live vaccine
   c. a chemically killed vaccine
   d. an immunoglobulin
   e. an agglutinating antigen

3. Which of the following viruses is widely used in living recombinant vaccines?
   a. coronavirus
   b. vaccinia virus
   c. influenza virus
   d. retrovirus
   e. myxovirus

4. When antigen and antibodies combine, maximal precipitation occurs when
   a. antigen is in excess.
   b. antibody is in excess.
   c. antigen and antibody are at equivalent concentrations.
   d. antigen is added to the antibody.
   e. antibody is added to the antigen.

5. An anti-antibody is used when
   a. an antigen is not precipitating.
   b. an antibody is not agglutinating.
   c. an antibody does not activate complement.
   d. an antigen is insoluble.
   e. the antigen is an antibody.

6. The many different proteins in serum can be analyzed by
   a. an anti-antibody test.
   b. a complement fixation test.
   c. a precipitation test.
   d. an agglutination test.
   e. immunoelectrophoresis.

7. A complement fixation test requires which of the following?
   a. heat-inactivated serum
   b. normal guinea pig serum
   c. peritoneal fluid
   d. heated plasma
   e. antibodies against complement

8. An ELISA uses which of the following reagents?
   a. an enzyme-labeled anti-antibody
   b. a radioactive anti-antibody
   c. a source of complement
   d. an enzyme-labeled antigen
   e. an enzyme-labeled antibody

9. A direct fluorescent antibody test can be used to detect the presence of
   a. hemagglutination
   b. specific antigens
   c. antibodies
   d. complement
   e. precipitated antigen-antibody complexes

10. Which of the following is the best test to detect rabies virus in the brain of a dog?
    a. agglutination
    b. hemagglutination inhibition
    c. virus neutralization
    d. western blot
    e. direct immunofluorescence

True/False

___ 1. Passive immunization provides more prolonged immunity than active immunization.

___ 2. It is standard to attenuate killed virus vaccines.

___ 3. One single serologic test is inadequate for an accurate diagnosis of HIV infection.

___ 4. ELISA is very easily automated.

___ 5. ELISA has basically replaced the western blot test.

Matching

Match the characteristic in the first column with the therapy it most closely describes in the second column. Some choices may be used more than once.

___ 1. Induces rapid onset of immunity

___ 2. Induces mainly an antibody response

___ 3. Induces good cell-mediated immunity

___ 4. Increases antigenicity

___ 5. Uses antigen fragments

___ 6. Uses attenuated microbes

A. Attenuated viral vaccine

B. Adjuvant

C. Subunit vaccine

D. Immunoglobulin

E. Residual virulence

CRITICAL THINKING

1. Is it ethical to approve the use of a vaccine that causes significant illness in 1% of patients, if it protects immunized survivors against a serious disease?

2. Which is worse: to use a diagnostic test for HIV that may falsely indicate that a patient is not infected (false negative), or to use one that sometimes falsely indicates that a patient is infected (false positive)? Defend your choice.

3. Discuss the importance of costs and technical skill in selecting a practical serologic test. Under what circumstances does automation become important?

4. What body fluids, in addition to blood serum, might be usable for immune testing?

5. Why might a serologic test give a false positive result?

6. Some researchers want to distinguish between B cells and T cells in a mixture of lymphocytes. How could they do this without killing the cells?

It is summer, and you are hiking along a densely wooded trail wearing shorts and a T-shirt. You spot a beautiful wildflower just a few feet away and wander briefly off the trail to get a better look, wading through knee-high foliage. The next morning, an itchy red rash has developed on both of your legs. Within days, the rash has progressed to painful, oozing blisters. You realize, unhappily, that you have contacted poison ivy and will be treating the blisters with calamine lotion until they heal—typically within 14 to 20 days.

A poison ivy rash is caused by an allergic response to urushiol, a chemical that coats poison ivy leaves. The body's immune system essentially overreacts to the urushiol: It identifies the chemical as a foreign substance and initiates an immune response that results in allergic contact dermatitis (inflammation of the skin). In the absence of an immune response, poison ivy would be harmless. About 85% of the population experiences an allergic reaction when exposed to urushiol.

In this chapter we examine a variety of immune system disorders, including allergies and other immune hypersensitivities, autoimmune diseases, and immune deficiencies.

Hypersensitivities, Autoimmune Diseases, and Immune Deficiencies

CHAPTER 18

CHAPTER OUTLINE

Hypersensitivities
 Type I (Immediate) Hypersensitivity
 Type II (Cytotoxic) Hypersensitivity
 Type III (Immune-Complex Mediated) Hypersensitivity
 Type IV (Delayed or Cell-Mediated) Hypersensitivity

Autoimmune Diseases
 Causes of Autoimmune Diseases
 Single Tissue Autoimmune Diseases
 Systemic Autoimmune Diseases

Immunodeficiency Diseases
 Primary Immunodeficiency Diseases
 Acquired Immunodeficiency Diseases

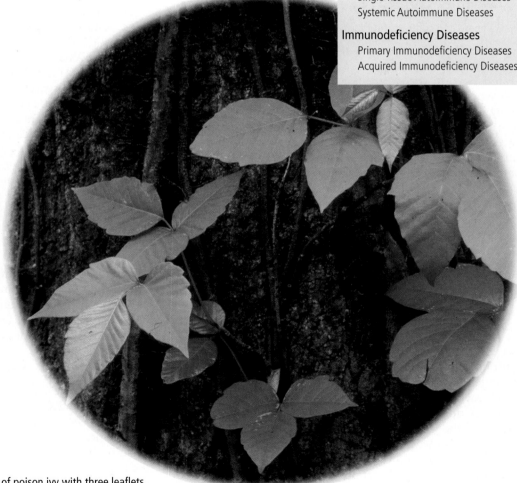

The characteristic foliage of poison ivy with three leaflets.

MicroPrep Pre-Test: Take the pre-test for this chapter on the web.
*Visit **www.microbiologyplace.com.***

As we have seen, the immune system is an absolutely essential component of the body's defenses. However, if the immune system functions abnormally—either by overreacting or underreacting—the malfunction may cause significant disease, even death. If the immune system functions excessively, the body develops any of a variety of *immune hypersensitivities*. An important example of a hypersensitivity is an allergy, which occurs when the immune system reacts to generally harmless substances such as pollen. Another type of hypersensitivity, *graft rejection,* occurs when the immune system attacks transplanted tissues. Further, *autoimmune diseases* develop when the immune system attacks the body's own tissues. Finally, diseases that result from a failure of the immune system are called *immunodeficiency diseases*. In this chapter we discuss each of these three categories of derangements of the immune system.

Hypersensitivities

Learning Objective

✓ Compare and contrast the four types of hypersensitivity.

Hypersensitivity (hī′per-sen-si-tiv′i-tē) may be defined as any immune response against a foreign antigen that is exaggerated beyond the norm. For example, most people can wear wool, smell perfume, or dust furniture without experiencing itching, wheezing, runny nose, or watery eyes. When these symptoms do occur, the person is said to be experiencing a hypersensitivity response. In the following sections we examine each of the four main types of hypersensitivity response, designated as type I through type IV.

Type I (Immediate) Hypersensitivity

Learning Objectives

✓ Describe the two-part mechanism by which type I hypersensitivity occurs.
✓ Explain the roles of three inflammatory chemicals (mediators) released from mast cell granules.
✓ Describe three disease conditions resulting from type I hypersensitivity mechanisms.

Type I hypersensitivities are localized or systemic (whole body) reactions that result from the release of inflammatory molecules (such as histamine) in response to an antigen. These reactions are termed *immediate* hypersensitivity because they develop within seconds or minutes following exposure to the antigen. They are also commonly called **allergies**[1] (al′er-jēz), and the antigens that stimulate them are called **allergens** (al′er-jenz).

In the next two subsections we will examine the two-part mechanism of a type I hypersensitivity reaction: sensitization upon an initial exposure to an allergen, and the degranulation of sensitized cells.

[1]From Greek *allos,* meaning other, and *ergon,* meaning work.

Figure 18.1 ▶

The mechanisms of a type I hypersensitivity reaction. **(a)** Sensitization. After normal processing of an antigen by an antigen-presenting cell ①, T_H2 cells are stimulated to secrete IL-4 ②. This cytokine stimulates a B cell ③, which becomes a plasma cell that secretes IgE ④, which then binds to and sensitizes mast cells, basophils, and eosinophils ⑤. **(b)** Degranulation. When the same antigen (allergen) is subsequently encountered, its binding to the IgE molecules on the surfaces of sensitized cells triggers rapid degranulation and the release of inflammatory chemicals from the cells.

Sensitization upon Initial Exposure to an Allergen

All of us are exposed to antigens in the environment, against which we typically mount immune responses that result in the production of IgG. But when cytokines (especially IL-4) from type 2 helper T (T_H2) cells stimulates B cells in *allergic* individuals, the B cells become plasma cells that produce not IgG, but instead IgE **(Figure 18.1a)**. Typically, IgE is directed against parasitic worms, which rarely infect most Americans, so IgE is not often found in the blood serum—except in allergic individuals.

The precise reason that only some people produce IgE (and thereby suffer from allergies) is a matter of intense research. Some researchers think that environmental factors—including exposure during infancy and childhood to high levels of indoor antigens such as cigarette smoke, dust mites, and molds—sensitize people, making them hypersensitive (allergic). Other researchers believe that childhood exposure to such antigens is protective, that young children who experience multiple bacterial and viral infections as well as exposure to pet dander (flakes of animal skin) are less likely to develop allergies than those who have been relatively sheltered from infectious diseases and unexposed to pets. **New Frontiers 18.1** on page 513 examines the second hypothesis.

In any case, following initial exposure to allergens, the plasma cells of allergic individuals secrete IgE, which binds very strongly via its stem to three types of defense cells—*mast cells, basophils,* and *eosinophils*—sensitizing these cells to respond to subsequent exposures to the allergen.

Degranulation of Sensitized Cells

When the same allergen subsequently enters the body, it binds to the active sites of IgE molecules on the surfaces of sensitized cells. This binding triggers a cascade of internal biochemical reactions that causes the sensitized cells to release the inflammatory chemicals from their granules into the surrounding intercellular space—an event called *degranulation* (dē-gran-ū-lā′shŭn) **(Figure 18.1b)**. It is these inflammatory mediators that generate the characteristic symptoms of type I hypersensitivity reactions: respiratory distress, rhinitis (inflammation of the nasal mucous membranes commonly called "runny nose"), watery eyes, inflammation, and reddening of the skin.

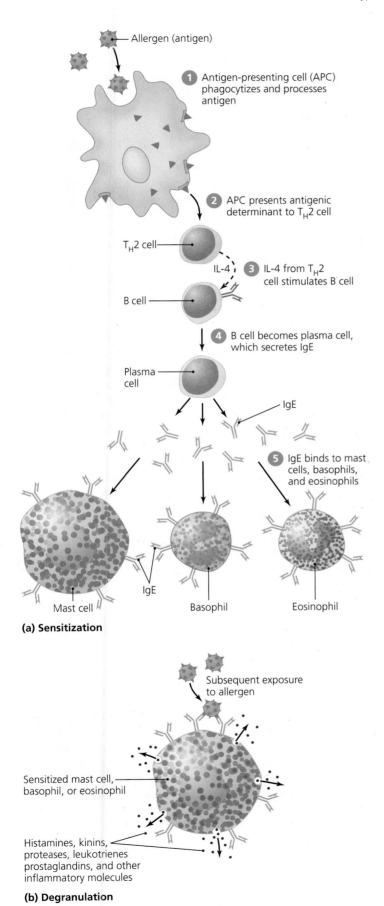

1 Antigen-presenting cell (APC) phagocytizes and processes antigen

2 APC presents antigenic determinant to T$_H$2 cell

T$_H$2 cell

IL-4

3 IL-4 from T$_H$2 cell stimulates B cell

B cell

4 B cell becomes plasma cell, which secretes IgE

Plasma cell

IgE

5 IgE binds to mast cells, basophils, and eosinophils

IgE

Mast cell   Basophil   Eosinophil

(a) Sensitization

Subsequent exposure to allergen

Sensitized mast cell, basophil, or eosinophil

Histamines, kinins, proteases, leukotrienes prostaglandins, and other inflammatory molecules

(b) Degranulation

The Roles of Degranulating Cells in an Allergic Reaction

Unlike most defensive cells, **mast cells** are not white blood cells; indeed, they are not blood cells at all, but rather are derived from other stem cells and are distributed throughout the body in connective tissue other than blood. These large, round cells are most often found in sites close to body surfaces, including the skin and the walls of the intestines and airways. Their characteristic feature is a cytoplasm packed with very large granules (**Figure 18.2a** on page 512) that are loaded with a mixture of very potent inflammatory chemicals.

The most significant of the chemicals released from mast cell granules is **histamine** (his'-tă-mēn), a small molecule related to the amino acid histidine. Histamine stimulates strong contractions in the smooth muscles of the bronchi, gastrointestinal tract, uterus, and bladder, and it also makes blood capillaries dilate (expand) and become leaky. As a result, tissues in which mast cells degranulate become red and swollen. Histamine also stimulates nerve endings, causing itching and pain. Finally, histamine is a potent stimulator of bronchial mucus secretion, tear formation, and salivation.

Other mediators of allergies released by degranulating mast cells include **kinins** (kī'ninz), which are powerful inflammatory chemicals, and **proteases** (prō'tē-ās-ez), enzymes that destroy nearby cells and activate the complement system (see Figure 15.16), which in turn results in the release of still more inflammatory chemicals. Proteases account for more than half the proteins in a mast cell granule. In addition, the binding of allergens to IgE on mast cells activates other enzymes that trigger the production of **leukotrienes** (lū-kō-trī'ēnz) and **prostaglandins** (pros-tă-glan'dinz), lipid molecules that are very potent inflammatory agents. **Table 18.1** on page 514 summarizes the inflammatory molecules released from mast cells.

Basophils, the least numerous type of leukocyte, contain cytoplasmic granules (**Figure 18.2b**) that stain intensely with basophilic dyes and are filled with inflammatory chemicals similar to those found in mast cells. Sensitized basophils bind IgE and degranulate in the same way as mast cells when they encounter allergens.

The blood and tissues of allergic individuals also accumulate many **eosinophils,** leukocytes that contain numerous cytoplasmic granules (**Figure 18.2c**) that stain intensely with the red dye eosin and function primarily to destroy parasitic worms. The process during type I hypersensitivity reactions that results in the accumulation of eosinophils in the blood—a condition termed *eosinophilia*—begins with mast cell degranulation, which releases peptides that stimulate the release of eosinophils from the bone marrow. Once in the bloodstream, eosinophils are attracted to the site of mast cell degranulation, where they themselves degranulate. Eosinophil granules contain unique inflammatory mediators and produce large amounts of leukotrienes, which increase vascular permeability and stimulate smooth muscle contraction, thereby contributing greatly to the severity of a hypersensitivity response.

(a) TEM ⊢ 1 μm

(b) TEM ⊢ 1 μm

(c) TEM ⊢ 1 μm

◀ *Figure 18.2*

Examples of sensitized cells that participate in type I hypersensitivity reactions. **(a)** A mast cell. **(b)** A basophil. **(c)** An eosinophil. Note the many granules, which are membrane-bound vesicles filled with inflammatory chemicals.

Clinical Signs of Localized Allergic Reactions

Type I hypersensitivity reactions are usually mild and localized. The site of the reaction depends on the portal of entry of the antigens. For example, inhaled allergens may provoke a response in the upper respiratory tract commonly known as **hay fever,** a local allergic reaction marked by a profuse watery nasal discharge, sneezing, itchy throat and eyes, and excessive tear production. Mold spores; pollens from grasses, flowering plants, and some trees; and feces and dead bodies of house dust mites, are among the more common allergens **(Figure 18.3).**

If inhaled allergen particles are sufficiently small, they may reach the lungs. The resulting type I hypersensitivity there can cause an episode of severe difficulty in breathing known as **asthma,** characterized by wheezing, coughing, excessive production of a thick, sticky mucus, and constriction of the smooth muscles of the bronchi. Although a localized reaction, asthma can be life threatening: Without medical intervention, the increased mucus and bronchial constriction can quickly cause suffocation.

Some foods may also contain powerful allergens, which may provoke diarrhea and other gastrointestinal signs and symptoms. Common food allergens include high-protein foods such as eggs, soy, dairy products, shellfish, and peanuts.

Other allergens—including latex, wool, certain metals, and the venom or saliva of wasps, bees, fire ants, deer flies, fleas, and other stinging or biting insects—may cause a local dermatitis. As a result of the release of histamine and other mediators into nearby skin tissues, plus the ensuing leakage of serum from local blood vessels, the individual suffers round, raised, red areas called *hives* or **urticaria**[2] (er'ti-kar'i-ă; **Figure 18.4**). These lesions are very itchy because histamine irritates local nerve endings.

Clinical Signs of Systemic Allergic Reactions

Following a sensitized individual's contact with an allergen, many mast cells may degranulate simultaneously, releasing copious amounts of histamine and other inflammatory mediators into the bloodstream. The release of chemicals may exceed the body's ability to adjust, resulting in **acute anaphylaxis**[3] (an'ă-fī-lak'sis) or **anaphylactic shock.**

A common cause of acute anaphylaxis is a bee sting in an allergic and sensitized individual. The first sting produces sensitization and the formation of IgE antibodies, and

[2]From Latin *urtica,* meaning nettle.
[3]From Greek *ana,* meaning away from, and *phylaxis,* meaning protection.

New Frontiers 18.1   Can Pets Help Decrease Children's Allergy Risks?

A recent study headed by Dennis Ownby of the Medical College of Georgia suggests that children who grow up in households with two or more cats or dogs may be less likely to develop

common allergies than children raised without pets. Researchers suspect that exposure to bacteria harbored by cats and dogs can somehow suppress allergic reactions—not only to pets, but also to other

common allergens such as dust mites and grass. Ownby and his team studied 474 children from birth until 6–7 years of age, at which point the children were tested for IgE against common allergens. They found that children exposed to two or more dogs or cats during their first year of life were, on average, 66–77% less likely to have antibodies to common allergens than were children exposed to only one or no pets during the same period. Exactly how the pets and/or their bacteria may suppress the allergic response remains unclear.

Source: Ownby, Dennis R. et al. 2002. Exposure to dogs and cats in the first year of life and risk of allergic sensitization at 6 to 7 years of age. *Journal of the American Medical Association* 288(8): 963–972.

(a) SEM 1 µm

(b) SEM 10 µm

(c) SEM 10 µm

▲ *Figure 18.3*

Some common allergens. **(a)** Spores of the fungus, *Stachybotrys chartarum*. **(b)** Pollen of *Ambrosia trifida* (ragweed), the most common cause of hay fever in the United States. **(c)** A house dust mite. Mites' fecal pellets and dead bodies may become airborne and are a common cause of hay fever.

Figure 18.4 ▶

Urticaria, the presence of round, red, fluid-filled patches on the skin prompted by the release of histamine in response to an allergen. *What causes fluid to accumulate in the patches seen in urticaria?*

Figure 18.4 The fluid accumulates because histamine from degranulated mast cells causes blood capillaries to become more permeable.

Table 18.1   Inflammatory Molecules Released from Mast Cells

Molecules	Role in Hypersensitivity Reactions
Released during degranulation	
Histamine	Causes smooth muscle contraction, increased vascular permeability, and irritation
Kinins	Cause smooth muscle contraction, inflammation, and irritation
Proteases	Damage tissues and activate complement
Synthesized in response to inflammation	
Leukotrienes	Cause slow, prolonged smooth muscle contraction
Prostaglandins	Some contract smooth muscle; others relax it

the second sting—which may occur years later—produces an anaphylactic reaction. Other allergens commonly implicated in acute anaphylaxis are certain foods, vaccines, antibiotics such as penicillin, iodine dyes, local anesthetics, certain narcotics such as morphine, and blood products.

The clinical signs of acute anaphylaxis are those of rapid suffocation. Bronchial smooth muscle, which is highly sensitive to histamine, contracts violently. In addition, increased leakage of fluid from blood vessels causes swelling of the larynx and other tissues. The patient also experiences contraction of the smooth muscle of the intestines and bladder. Without prompt administration of *epinephrine* (ep′i-nef′rin) (discussed shortly), an individual in anaphylactic shock may suffocate, collapse, and die within minutes.

Diagnosis of Type I Hypersensitivity

Type I hypersensitivity is commonly diagnosed by injecting a very small quantity of a dilute solution of the allergens being tested into the skin. In most cases the individual is screened for more than a dozen potential allergens simultaneously, using the forearms as injection sites. Wherever the individual tested is sensitive to an allergen, local histamine release will cause redness and swelling at the injection site within a few minutes **(Figure 18.5)**.

Prevention of Type I Hypersensitivity

Because many different allergens provoke type I hypersensitivity, prevention begins with *identification* and *avoidance* of the allergens responsible. Air conditioning and avoidance of rural areas during pollen season can reduce upper respiratory allergies provoked by some pollens. Encasing bedding in mite-proof covers, frequent vacuuming, and avoidance of home furnishings that trap dust (such as wall-to-wall carpeting and heavy drapes) can reduce the severity of other household allergies. A dehumidifier can reduce mold in a home. The best way for individuals to avoid allergic reactions to pets is to find out if they are allergic before they get a pet. If an allergy develops afterward, they may have to find a new home for the animal.

Food allergens can be identified and then avoided by using a medically supervised elimination diet in which

▲ *Figure 18.5*

Skin tests for diagnosing type I hypersensitivity. The presence of redness and swelling at an injection site indicates sensitivity to the allergen injected at that site.

foods are removed one at a time from the diet to see when signs of allergy cease. Peanuts and shellfish are among the foods more commonly implicated in anaphylactic shock; allergic individuals must scrupulously avoid consuming even minute amounts (see **New Frontiers 18.2**). For example, individuals sensitized against shellfish can suffer anaphylactic shock from consuming meat sliced on an unwashed cutting board previously used to prepare shellfish. Some vaccines may contain egg components and should not be given to individuals allergic to eggs.

Health care workers must assess a patient's medical history carefully before administering penicillin, iodine dyes, and other potential allergens. Therapies such as antidotes and respiratory stimulants must be immediately available if a patient's sensitivity status is unknown.

In addition to avoidance of allergens, type I hypersensitivity reactions can be prevented by *immunotherapy* (im′ū-nō-thār′ă-pē), commonly called "allergy shots," which involves the administration of a series of injections of dilute allergen, usually once a week for many months. It is unclear

New Frontiers 18.2   When Kissing Triggers Allergic Reactions

Individuals with particularly severe food allergies not only need to watch what they eat—they also need to watch who kisses them. In a recent study from the University of California at Davis, researchers found that out of 316 patients with severe allergies to peanuts, tree nuts, and/or seeds, 20 (about 6%) reported developing allergic reactions after being kissed. In nearly all of the cases, the kisser had eaten nuts to which the allergic individual was sensitive. In some of the cases, the allergic individuals were so sensitive that they developed reactions even though the kissers had brushed their teeth or had consumed the nuts several hours earlier. Food allergists believe that the duration and intensity of a kiss may also be a factor—that is, longer kisses involving exchange of saliva may be more likely to elicit allergic reactions than a quick buss on the cheek. The moral of the story: If you have a severe food allergy, you might want to tell your romantic partner about it!

Source: Hallett, R., L. A. D. Haapanen, and S. S. Teuber. 2002. Food allergies and kissing. *New England Journal of Medicine* 346:1833.

just how allergy shots work, but they may alter the helper T cell balance from T_H2 cells to T_H1 cells, reducing the production of antibodies, or they may stimulate the production of IgG, which binds antigen before the antigen can react with IgE on mast cells, basophils, and eosinophils. Immunotherapy reduces the severity of allergy symptoms by about 50% in about two-thirds of patients with upper respiratory allergies; however, the series of injections must be repeated every 2–3 years. Immunotherapy is not effective in treating asthma.

Treatment of Type I Hypersensitivity

One way to treat type I hypersensitivity is to administer drugs that specifically counteract the inflammatory mediators released by degranulating cells. Thus **antihistamines** (an-tē-his'tă-mēnz) are administered to neutralize histamine. However, because histamine is but one of many mediators released in type I hypersensitivity, antihistamines do not completely eliminate all clinical signs. Indeed, because antihistamines are essentially useless in patients with asthma, asthmatics are typically prescribed an inhalant containing *corticosteroid* (kōr'ti-kō-stēr'oyd) and a *bronchodilator* (brong-kō-dī-lā'ter), which counteract the effects of histamine and the other mediators.

The hormone epinephrine neutralizes many of the lethal mechanisms of anaphylaxis by relaxing smooth muscle and reducing vascular permeability. Epinephrine is thus the drug of choice for the emergency treatment of both severe asthma and anaphylactic shock. Patients who suffer from severe type I hypersensitivities may carry a prescription epinephrine kit so that they can inject themselves before their allergic reactions become life threatening.

Type II (Cytotoxic) Hypersensitivity

Learning Objectives

✓ Discuss the mechanisms underlying transfusion reactions.

✓ Construct a table comparing the key features of the four blood types in the ABO blood group system.

✓ Describe the mechanisms and treatment of hemolytic disease of the newborn.

The second major form of hypersensitivity results when cells are destroyed by an immune response—typically by the combined activities of complement and antibodies (see Figure 15.16). So called *cytotoxic* (sī-tō-tok'sik) *hypersensitivity* is part of many autoimmune diseases, which we will discuss later in this chapter, but the most significant examples of type II hypersensitivity are the destruction of donor red blood cells following an incompatible blood transfusion, and the destruction of fetal red blood cells in hemolytic disease of the newborn. We begin our discussion by focusing on these two manifestations of type II hypersensitivity.

The ABO System and Transfusion Reactions

Red blood cells, like nucleated cells, have many different glycoprotein and glycolipid molecules on their surface. The surface molecules of red blood cells, called **blood group antigens,** have various functions, including transportation of glucose and ions across the cytoplasmic membrane. Although most of these surface molecules are cell-membrane components, some are soluble molecules found free in serum, saliva, and other body fluids that have simply stuck to red cell surfaces.

Blood group antigens vary in complexity. The ABO group, for example, consists of just two antigens arbitrarily given the names A antigen and B antigen. Each person's red blood cells have either A antigen, B antigen, both A antigen and B antigen, or neither antigen. Individuals with neither antigen are said to have blood type O. In contrast, the MN blood group is much more complex, with 29 known antigens.

As you probably know, blood can be readily transfused from one person to another. However, if blood is transfused to an individual with a different blood type, then the donor's blood group antigens may stimulate in the recipient the production of antibodies that bind to and eventually destroy the transfused cells. The result can be a potentially life-threatening *transfusion reaction.*

Transfusion reactions, the most problematic of which involve the ABO group, develop as follows:

- If the recipient has preexisting antibodies to foreign blood group antigens, then the donated blood cells will

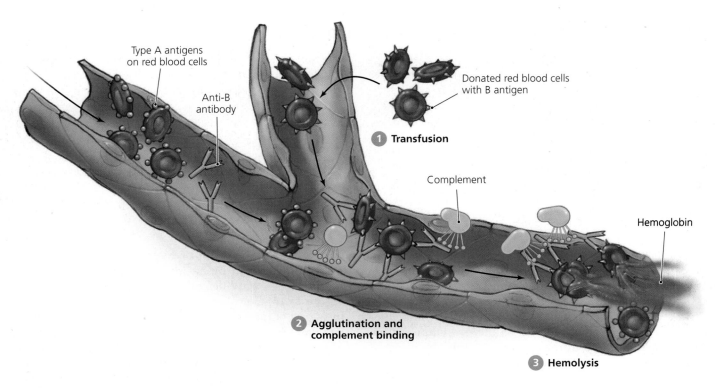

Type A antigens
on red blood cells

Anti-B
antibody

Donated red blood cells
with B antigen

1 **Transfusion**

Complement

Hemoglobin

2 **Agglutination and
complement binding**

3 **Hemolysis**

▲ *Figure 18.6*

Events leading to hemolysis, one of the negative consequences of a transfusion reaction. When antibodies against a foreign ABO antigen combine with the antigen on transfused red blood cells ①, the cells are agglutinated, and the complexes bind complement ②. The resulting hemolysis releases large amounts of hemoglobin into the bloodstream ③, producing one of several additional negative consequences throughout the body.

be destroyed immediately—either the antibody-bound cells will be phagocytized by macrophages and neutrophils, or the antibodies will agglutinate the cells, and complement will rupture them, a process called *hemolysis* **(Figure 18.6).** Hemolysis releases hemoglobin into the bloodstream, which may cause severe kidney damage. At the same time, the membranes of the ruptured blood cells trigger blood clotting within blood vessels, blocking them and causing circulatory failure, fever, difficulty in breathing, coughing, nausea, vomiting, and diarrhea. The liver, which is responsible for breaking down dead red blood cells, becomes overwhelmed. If the patient survives, recovery follows the elimination of all the foreign red blood cells.

- If the recipient has no preexisting antibodies to the foreign blood group antigens, then the transfused cells circulate and function normally, but only for a while—that is, until the recipient's immune system mounts a primary response against the foreign antigens that produces enough antibody to destroy the foreign cells. This happens over a long enough time that the severe symptoms and signs mentioned above do not occur.

To prevent transfusion reactions, laboratory personnel must cross-match blood types between donors and recipients.

Thus, before a recipient receives a blood transfusion, the red cells and serum of the donor are mixed with the serum and red cells of the recipient. If any signs of agglutination are seen, then that blood is not used, and an alternative donor is sought. **Table 18.2** presents the characteristics of the ABO blood group, and the compatible donor/recipient matches that can be made.

In most people, production of antibodies against foreign ABO antigens is not stimulated by exposure to foreign blood cells, but instead by exposure to antigenically similar molecules found on a wide range of plants and bacteria. Individuals encounter and make antibodies against these antigens on a daily basis, but only upon receipt of a mismatched blood transfusion are they problematic.

CRITICAL THINKING

During the Gulf War in 2003, an Army corporal with type AB blood received a life-saving blood transfusion from his sergeant who had type O blood. Later the sergeant was involved in a traumatic accident and needed blood desperately. The corporal wanted to help but was told his blood was incompatible. Explain the immunological reasons the corporal could receive from but could not give to the sergeant.

Table 18.2 **ABO Blood Group Characteristics and Donor/Recipient Matches**

ABO Blood Group	ABO Antigen(s) Present	Antibodies Present	Can Donate To	Can Receive From
A	A	Anti-B	A or AB	A or O
B	B	Anti-A	B or AB	B or O
AB	A and B	None	AB	A, B, AB. or O (universal recipient)
O	None	Both anti-A and anti-B	A, B, AB, or O (universal donor)	O

Figure 18.7 ➤

Events in the development of hemolytic disease of the newborn. **(a)** During an initial pregnancy, Rh+ red blood cells from the fetus enter the Rh− mother's circulation, often during childbirth. As a result, the mother produces anti-Rh antibodies. **(b)** During subsequent pregnancies with another Rh+ child, anti-Rh IgG molecules cross the placenta and destroy the fetus's red blood cells. *How is it possible that the child of an Rh− mother is Rh+?*

Figure 18.7 The father of the child is Rh positive.

(a) First pregnancy

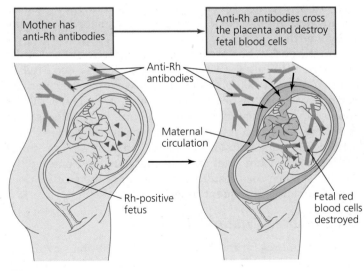

(b) Subsequent pregnancy

The Rh System and Hemolytic Disease of the Newborn

Many decades ago, researchers discovered the existence of an antigen common to the red blood cells of humans and rhesus monkeys. They called this antigen, which is a cytoplasmic membrane protein that transports anions and glucose across the membrane, *rhesus* (rē′sŭs) *antigen* or **Rh antigen.** Laboratory analysis of human blood samples eventually showed that the Rh antigen is present on the red blood cells of about 85% of humans; that is, about 85% of the population is *Rh positive (Rh+),* and about 15% of the population is *Rh negative (Rh−).*

In contrast to the situation with ABO antigens, preexisting antibodies against Rh antigen do not occur. Although a transfusion reaction can occur in an Rh negative patient who receives more than one transfusion of Rh positive blood, such a reaction is usually minor because Rh antigen molecules are less abundant than A or B antigens. Instead, the primary problem posed by the Rh system is the risk of **hemolytic** (hē-mō-lit′ik) **disease of the newborn (Figure 18.7).**

This hypersensitivity reaction develops when an Rh− mother is pregnant with her second or subsequent Rh+ fetus (who inherited an Rh gene from its father). Normally, the placenta keeps fetal red blood cells separate from the mother's blood, so fetal cells do not enter the mother's bloodstream. However, in 20–50% of pregnancies—especially during the later weeks of pregnancy, during a clinical or spontaneous abortion, or during childbirth—fetal red blood cells escape into the mother's blood. The mother's immune system recognizes these Rh+ fetal cells as foreign and

Figure 18.8 ▶

Events in the development of immune thrombocytopenic purpura, a disease resulting from a drug-induced type II (cytotoxic) hypersensitivity reaction. The destruction of platelets through the action of bound antibodies and complement produces the inhibition of blood clotting characteristic of this disease. *What similar disease results from type II (cytotoxic) destruction of red blood cells?*

Figure 18.8 Hemolytic anemia is the type II (cytotoxic) destruction of red blood cells.

Platelet

Drug

① Drug molecules bind to platelets forming drug-platelet complex

Drug-platelet complex

② Complexes are antigenic, triggering a humoral immune response

③ Antibodies bind to drug molecules; complement binds to antibodies

Complement

④ Complement's membrane attack complexes lyse platelet

initiates a humoral immune response by developing antibodies against the Rh antigen. Initially, only IgM antibodies are produced, and because IgM is a very large molecule that cannot cross the placenta, no problems arise during the first pregnancy.

However, IgG antibodies can cross the placenta, so if during subsequent pregnancies the fetus is Rh+, anti-Rh IgG molecules produced by the mother cross the placenta and destroy the fetus's Rh+ red blood cells. Such destruction may be limited, or it may be severe enough—especially in a third or subsequent pregnancy—to cause grave problems. One classic feature of hemolytic disease of the newborn is severe jaundice from excessive *bilirubin* (bil-i-roo'bin), a yellowish blood pigment released during the degradation of hemoglobin from lysed red blood cells. The liver is normally responsible for removing bilirubin, but because of the immaturity of the fetal liver and overload from the hemolytic process, the bilirubin may instead be deposited in the brain, causing severe neurological damage or death during the latter weeks of pregnancy or shortly after the baby's birth.

In the past, when prevention of hemolytic disease of the newborn was impossible, this terrible disease occurred in about 1 of every 300 births. Today, however, physicians can routinely and drastically reduce the prevalence of the disease by administering anti-Rh serum, called *Rhogam*, to Rh-negative pregnant women at 28 weeks gestation and also within 72 hours following abortion, miscarriage, or childbirth to destroy any fetal red cells that may have entered the mother's body. As a result, sensitization of the mother does not occur, and subsequent pregnancies are safer.

Drug-Induced Cytotoxic Reactions

Another kind of type II hypersensitivity involves cytotoxic reactions to drugs. Although the molecules of such drugs as quinine, penicillin, or sulfanilamide are too small to trigger an immune response by themselves, they can be haptens; that is, when they spontaneously bind to blood cells or platelets, they become antigenic and stimulate production of antibodies (a humoral immune response).

When such stimulated antibodies and then complement bind to drug molecules bound to platelets, the platelets are lysed, producing a disease called **immune thrombocytopenic**[4] **purpura** (throm'bō-sī-tō-pē'nik pūr'pū-ră; **Figure 18.8**). The destruction of the platelets inhibits the ability of the blood to clot correctly, which leads to the production of purple hemorrhages, called *purpura*[5] under the skin. Similar destruction of leukocytes is called **agranulocytosis,** and that of red blood cells is called **hemolytic anemia** (ă-nē'mē-ă).

Type III (Immune-Complex Mediated) Hypersensitivity

Learning Objective

✓ Outline the basic mechanism of type III hypersensitivity.

As we have seen, the formation of antigen-antibody complexes, also called **immune-complexes,** initiates several molecular processes, of which the most significant is complement activation. Normally, complement-activating immune-complexes are eliminated from the body via

---

[4]*Thrombocyte* is another name for platelet.
[5]From Latin *porphyra,* meaning purple.

Figure 18.9 ▶

The mechanism of type III (immune-complex mediated) hypersensitivity. After immune complexes form ①, the complexes that are not phagocytized lodge in certain tissues (in this case, the wall of a blood vessel) ②. Trapped complexes then activate complement ③, which attracts and activates neutrophils ④. The release of inflammatory chemicals from the activated neutrophils leads to the destruction of tissue ⑤.

① Antigens combine with antibodies to form antigen-antibody complexes

— Antigen

— Antibody

— Antigen-antibody complex

② Phagocytes remove most of the complexes, but some lodge in the walls of blood vessels

③ There the complexes activate complement

— Inactive complement

— Active complement

④ Antigen-antibody complexes and activated complement attract and activate neutrophils which release inflammatory chemicals

— Neutrophil

— Inflammatory chemicals

⑤ Inflammatory chemicals damage underlying blood vessell wall

phagocytosis. However, in type III (immune-complex mediated) hypersensitivity reactions, the immune complexes are small enough to escape phagocytosis, and so circulate freely in the bloodstream until they become trapped in organs, joints, and tissues (such as the walls of blood vessels). In these sites they trigger mast cells to degranulate, releasing inflammatory chemicals that damage the tissue. **Figure 18.9** illustrates the mechanism of type III reactions.

Later in this chapter we will discuss two systemic autoimmune disorders for which type III hypersensitivity is a factor: systemic lupus erythematosus and rheumatoid arthritis. Here we briefly discuss two more-localized conditions resulting from immune-complex mediated hypersensitivity: hypersensitivity pneumonitis and glomerulonephritis.

Hypersensitivity Pneumonitis

Type III hypersensitivities can affect the lungs, causing a form of pneumonia called **hypersensitivity pneumonitis** (noo-mō-nī′tus). Individuals become sensitized when minute mold spores or other antigens are inhaled deep into the lungs, stimulating the production of antibodies. A hypersensitivity reaction occurs when the subsequent inhalation of the same antigen stimulates the formation of immune complexes that then activate complement.

One form of hypersensitivity pneumonitis, called *farmer's lung,* occurs in farmers chronically exposed to spores from moldy hay. Many other syndromes in humans develop via a similar mechanism and are usually named after the source of the inhaled allergen. Thus, *pigeon breeder's lung* arises following exposure to the dust from pigeon feces, *mushroom grower's lung* is a response to the soil or fungal spores encounted when cultivating mushrooms, and *librarian's lung* results from inhalation of dust from old books.

Glomerulonephritis

Glomerulonephritis (glō-mār′ū-lō-nef-rī′tis) occurs when immune complexes circulating in the bloodstream are deposited in the walls of *glomeruli,* which are networks of minute blood vessels in the kidneys. Immune complexes damage the glomerular cells, leading to enhanced local production of cytokines that stimulate nearby cells to produce more of the intercellular proteins that underlie the cells, impeding blood filtration. Sometimes, the immune complexes are deposited in the center of the glomeruli, where they stimulate local cells to proliferate and compress nearby blood vessels, again interfering with kidney function. The net result is kidney failure; the glomeruli lose their ability to filter wastes from the blood, ultimately resulting in death.

Type IV (Delayed or Cell-Mediated) Hypersensitivity

Learning Objectives

✓ Outline the mechanism of type IV hypersensitivity.
✓ Describe the significance of the tuberculin test.
✓ Identify four types of grafts.
✓ Explain the concept of privileged sites, and explain why a mother does not reject a fetus even though the uterus is not a privileged site.
✓ Compare four types of drugs commonly used to prevent graft rejection.

When certain antigens contact the skin of sensitized individuals, they provoke inflammation that begins to develop at the site only after 12–24 hours. Such **delayed hypersensitivity reactions** result not from the interactions of antigen and antibodies, but rather from interactions among antigen, antigen-presenting cells, and T cells; thus, a type IV reaction is also called *cell-mediated hypersensitivity.* The delay in this cell-mediated response reflects the time it takes for macrophages and T cells to migrate to and proliferate at the site of the antigen. We begin our discussion of type IV reactions by considering two common examples: the tuberculin response and allergic contact dermatitis. Then we discuss two type IV hypersensitivity reactions involving the interactions between the body and tissues grafted to (transplanted into) it—graft rejection, and graft versus host disease—before considering donor-recipient matching and tissue typing.

The Tuberculin Response

The **tuberculin** (too-ber'kū-lin) **response** is an important example of a delayed hypersensitivity reaction in which the skin of an individual exposed to tuberculosis or tuberculosis vaccine reacts to a subcutaneous injection of *tuberculin*—a protein solution obtained from *Mycobacterium tuberculosis* (mī-kō-bak-tēr'ē-ŭm tū-ber-kyū-lō'sis). Health care providers use tuberculin tests to diagnose contact with antigens of *M. tuberculosis.*

When tuberculin is injected into the skin of a healthy, never infected or vaccinated individual, no response occurs. In contrast, when tuberculin is injected into someone currently or previously infected with *M. tuberculosis* or an individual previously immunized with tuberculosis vaccine, a red hard swelling (10 mm or greater in diameter) indicating a positive tuberculin test develops at the site **(Figure 18.10).** Such inflammation reaches its greatest intensity within 24–72 hours and may persist for several weeks before fading. Microscopic examination of the lesion reveals that it is infiltrated with lymphocytes and macrophages.

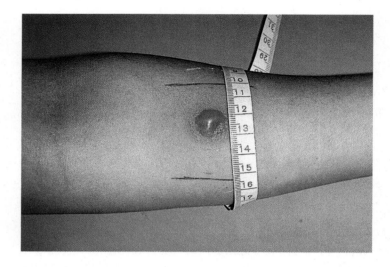

▲ *Figure 18.10*

A positive tuberculin test. The hard, red swelling 10 mm or greater in diameter that is characteristic of the tuberculin response indicates that the individual has been vaccinated against *Mycobacterium tuberculosis* or is now or has been previously infected with the bacterium.

A tuberculin response is mediated by memory T cells. When an individual is first infected by *M. tuberculosis,* the resulting cell-mediated immune response generates memory T cells that persist in the body. When a sensitized individual is subsequently injected with tuberculin, phagocytic cells migrate to the site and attract memory T cells, which secrete a mixture of cytokines that attract still more T cells and macrophages, giving rise to a slowly developing inflammation. The macrophages ingest and destroy the injected tuberculin, allowing the tissues eventually to return to normal.

Allergic Contact Dermatitis

Urushiol, the oil of poison ivy *(Toxicodendron radicans)* and related plants, is a highly reactive hapten that binds to almost any protein it contacts—including proteins in the skin of anyone who rubs against the plant. The body regards these chemically modified skin proteins as foreign, which triggers a cell-mediated immune response resulting in an intensely irritating skin rash called **allergic contact dermatitis** (der-mă-tī'tis). In severe cases, T_C cells destroy so many skin cells that acellular, fluid-filled blisters develop **(Figure 18.11).**

Other haptens that combine chemically with skin proteins can also induce allergic contact dermatitis. Examples include formaldehyde; some cosmetics, dyes, and drugs; metal ions; and chemicals used in the production of latex for hospital gloves and tubing.

Because T cells mediate allergic contact dermatitis, epinephrine and other drugs used for the treatment of immediate hypersensitivity reactions are ineffective. T cell activities and inflammation can, however, be suppressed by

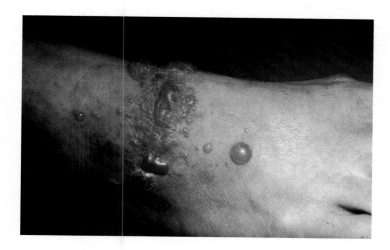

▲ *Figure 18.11*

Allergic contact dermatitis, a type IV hypersensitivity response—in this case, to exposure to poison ivy. Note the large, acellular, fluid-filled blisters that result from the destruction of skin cells.

corticosteroid treatment. Good strategies for dealing with exposure to poison ivy include washing the area thoroughly and immediately with a strong soap, and washing all exposed clothes as soon as possible.

Graft Rejection

A special case of type IV hypersensitivity is the rejection of **grafts,** which are tissues or organs that have been transplanted, whether between sites within an individual or between a donor and a recipient. Even though advances in surgical technique enable surgeons to move grafts freely from site to site, grafts perceived as foreign by the recipient undergo rejection, a highly destructive phenomenon that severely limits the success of organ and tissue transplantation. As discussed in Chapter 16, graft rejection is a normal immune response against foreign major histocompatibility complex (MHC) proteins on the surface of graft cells. The likelihood of graft rejection depends on the degree to which the graft is foreign to the recipient, which in turn is related to the type of graft.

Types of Grafts   Scientists name grafts according to the degree of relatedness between the donor and the recipient **(Figure 18.12).** A graft is called an **autograft** (aw'tō-graft) when tissues are moved to a different location within the same individual. Autografts do not trigger immune responses because they do not express foreign antigens. Examples of autografts include the grafting of skin from one area of the body to another to cover a burn area, or the use of a leg vein to bypass blocked coronary arteries.

Isografts (ī'sō-graftz) are grafts transplanted between two genetically identical individuals; that is, between identical siblings or clones. Because these individuals have

identical MHC proteins, the immune system of the recipient cannot differentiate between the grafted cells and its own normal body cells. As a result, isografts are not rejected.

Allografts (al'ō-graftz) are grafts transplanted between genetically distinct members of the same species. Most grafts performed in humans are allografts. Because the MHC proteins of the allograft are different from those of the recipient, allografts induce a strong type IV hypersensitivity, resulting in graft rejection. Rejection must be suppressed with immunosuppressive drugs if the graft is to survive.

Xenografts (zen'ō-graftz) are grafts transplanted between individuals of different species. Thus, the transplant of a baboon's heart into a human is a xenograft. Because xenografts are usually very different from the tissues of the recipient, both biochemically and immunologically, they usually provoke a rapid, intense rejection that is very difficult to suppress. Therefore, xenografts from mature animals are not commonly used therapeutically.

Privileged Sites   Certain sites in the body, including certain parts of the eye such as the cornea, the thymus, the testes, and the brain, are said to be **privileged sites** because grafts involving them are not likely to be rejected. Different sites are privileged for different reasons. The brain lacks lymphatic vessels, and its blood vessel walls are impermeable to lymphocytes such as T cells. The cornea also lacks extensive blood vessels, while the eyes and the testes contain naturally high levels of immunosuppressive molecules. Other sites either lack dendritic cells or express low levels of MHC molecules, so antigen processing does not occur.

Why Fetuses Are Not Rejected   The uterus is not a privileged site; allographs placed in the uterine wall experimentally are readily rejected. Given that a fetus has different

Autograft

Isograft
Genetically identical sibling or clone

Allograft
Genetically different member of same species

Xenograft

▲ *Figure 18.12*

Types of grafts, which are based on the degree of relatedness between donor and recipient. Autografts are grafts moved from one location to another within a single individual. In isografts, the donor and recipient are either genetically identical siblings or clones. In allografts, the donor and recipient are genetically distinct individuals of the same species. In xenografts, the donor and recipient are of different species.

MHC genes than those of its mother, why is a fetus in the uterus not rejected like an allograft?

Immunological rejection of a fetus is prevented by the combined activities of many different immunosuppressive mechanisms:

- Early embryos do not express MHC class I or II molecules on the placental layer that is in contact with maternal tissues.

- Cytokines that enhance MHC expression have no effect on placental cells.

- T cells are prevented from functioning in the placenta to reject the fetus. The reason is that certain fetal placental cells produce *indoleamine 2,3-dioxygenase*, an enzyme, which destroys tryptophan, an amino acid that T cells cannot synthesize for themselves. The lack of available tryptophan in their environment prevents T cell division and promotes their destruction via apoptosis.

Graft-versus-Host Disease

Bone marrow allografts are often used as a component of the treatment for cancers of leukocytes (leukemias and lymphomas). In this procedure, physicians use total body irradiation combined with cytotoxic drugs to kill tumor cells in the patient's bone marrow. In the process, the patient's existing leukocytes and all stem cells are also destroyed, which completely eliminates the body's ability to mount any kind of immune response. Physicians then inject the patient with donated stem cells, which migrate to the bones and multiply to produce a new set of leukocytes. Ideally, within a few months, a fully functioning bone marrow is restored.

Unfortunately, when they are transplanted, the donated bone marrow cells may regard the patient's cells as foreign, mounting an immune response against them and giving rise to a condition called **graft-versus-host disease.** If the donor and recipient mainly differ in MHC class I molecules, the grafted T cells attack all of the recipient's tissues, producing especially destructive lesions in the skin and intestine. If the donor and recipient differ mainly in MHC class II molecules, then the grafted T cells attack the antigen-presenting cells of the host, leading to immunosuppression, leaving the recipient vulnerable to infections. The same immunosuppressive drugs used to prevent graft rejection (discussed shortly) can limit graft-versus-host disease.

Donor-Recipient Matching and Tissue Typing

Although it is usually not difficult to ensure that donor and recipient have identical blood groups, MHC compatibility is much harder to achieve. The reason is the very high degree of MHC variability, which ensures that unrelated individuals differ widely in their MHCs. In general, the more closely donor and recipient are related, the smaller will be their MHC difference. Given that in most cases an identical sibling is not available, it is usually preferable that grafts are donated by a parent or sibling possessing MHC antigens similar to those of the recipient.

In practice, of course, closely related donors are not always available. And even though a paired organ (such as a kidney) or a portion of an organ (such as liver tissue) may be available from a living donor, unpaired organs (such as the heart) almost always come from an unrelated cadaver. In such cases, attempts are made to match donor and recipient as closely as possible by means of *tissue typing*. Physicians examine the white cells of potential recipients waiting for an organ graft to determine what MHC proteins they have. When a donor organ becomes available, it too is typed. Then the individual whose MHC proteins most closely match those of the donor is chosen to receive the graft. Though a perfect match is rarely achieved, the closer the match, the less intense will be the rejection process, and the greater the chance of successful grafting. A match of 50% or less of the MHC proteins is usually acceptable for most organs, but near absolute matches are required for successful bone marrow transplants.

The major characteristics of the four types of hypersensitivity are summarized in **Table 18.3.** Next we turn our attention to the actions of various types of immunosuppressive drugs used in situations in which the immune system is overactive.

CRITICAL THINKING

A patient arrives at the doctor's office with a rash covering her legs. How could you determine if the rash were a type I or a type IV hypersensitivity?

The Actions of Immunosuppressive Drugs

The development of highly potent immunosuppressive drugs has played a large role in the dramatic success of modern transplantation procedures. Note that these drugs can also be effective in combating certain autoimmune diseases (discussed shortly). In the following discussion we consider four classes of immunosuppressive drugs: corticosteroids, cytotoxic drugs, immunophilins, and lymphocyte-depleting therapies.

Corticosteroids, commonly called *steroids,* have been used as immunosuppressive agents for many years. Corticosteroids such as prednisone (pred'ni-sōn) and methylprednisone suppress the response of T cells to antigen and inhibit such mechanisms as T cell cytotoxicity and cytokine production. These drugs have a much smaller effect on B cell function. Synthetic glucocorticoids also suppress inflammation by inhibiting increases in vascular permeability and vasodilatation, and preventing the production of leukotrienes and prostaglandins.

Cytotoxic (sī-tō-tok'sik) **drugs** inhibit mitosis and cytokinesis (cell division). Given that lymphocyte proliferation is a key feature of specific immunity, blocking cellular reproduction is a potent, although very nonspecific, method of

Table 18.3   **The Characteristics of the Four Types of Hypersensitivity Reactions**

Descriptive	Name	Cause	Time Course	Characteristic Cells Involved
Type 1	Immediate hypersensitivity	IgE on sensitized cells' membranes binds antigen, causing degranulation	Seconds to minutes	Mast cells, basophils, and eosinophils
Type II	Cytotoxic hypersensitivity	Antibodies and complement lyse target cells	Minutes to hours	Red blood cells
Type III	Immune-complex mediated hypersensitivity	Nonphagocytized immune complexes trigger mast cell degranulation	Several hours	Neutrophils
Type IV	Delayed hypersensitivity	T_C cells attack the body's cells	Several days	Activated T cells

immunosuppression. Among the cytotoxic drugs currently in common use, the following three drugs are noteworthy:

- *Cyclophosphamide* (sī-klō-fos′fă-mīd) cross-links daughter DNA molecules in mitotic cells, preventing their separation and blocking mitosis. It impairs both B cell and T cell responses.

- *Azathioprine* (ā-ză-thī′ō-prēn) is a purine analog; it competes with purines during the synthesis of nucleic acids, thus blocking DNA replication and suppressing both primary and secondary antibody responses.

- *Methotrexate* (meth-ō-trek′sāt) is a folic acid antagonist that inhibits the production of thymidine and purine nucleotides, thereby suppressing DNA replication, RNA transcription, and (indirectly) antibody formation.

Three newly developed cytotoxic drugs used for immunosuppression include *mycophenolate mofetil*, which inhibits purine synthesis, and *brequinar sodium* and *leflunomide*, which inhibit pyrimidine synthesis.

Drugs called **immunophilins** (im′u-nō-fil′inz) constitute a family of potent inhibitors of T cell function. The most important of these drugs is *cyclosporine* (sī-klō-spōr′ēn), a polypeptide derived from fungi. It prevents T cell production of interleukins and interferons, thereby blocking T_H1 responses. Because cyclosporine acts only on activated T cells and has no effect on resting T cells, it is far less toxic than the nonspecific drugs just described. When given to prevent allograft rejection, only activated T cells attacking the graft are suppressed. Because steroids have a similar effect, the combination of corticosteroids and cyclosporine is especially potent and can enhance survival of allografts.

Scientists have developed techniques involving relatively specific *lymphocyte-depleting therapies* in an attempt to reduce the many adverse side effects associated with the use of less-specific immunosuppressive drugs. One technique involves administering an antiserum called *antilymphocyte globulin*, which is specific for lymphocytes. Another, even more specific antilymphocyte technique uses monoclonal antibodies against CD3, which is found only on T cells. An even more specific monoclonal antibody is directed against

the interleukin 2 receptor (IL-2R), which is expressed only on activated lymphocytes. Such immunosuppressive therapies are effective in reversing graft rejection, and because they target a narrow range of cells, they produce fewer undesirable side effects than less-specific drugs.

Table 18.4 on page 524 lists some drugs, and their actions, for each of the four classes of the immunosuppressive drugs.

Autoimmune Diseases

Just as today's military must control its arsenal of sophisticated weapons to avert losses of its own soldiers from "friendly fire," the immune system must be carefully regulated so that it does not damage the body's own tissues. However, the immune system does occasionally produce antibodies and cytotoxic T cells that target normal body cells—a phenomenon called *autoimmunity*. Although such responses are not always damaging, they can give rise to **autoimmune diseases,** some of which are life threatening.

Causes of Autoimmune Diseases

Learning Objective

✓ Briefly discuss seven hypotheses concerning the causes of autoimmunity and autoimmune diseases.

Most autoimmune diseases appear to develop spontaneously and at random, with no obvious predisposing cause. Nevertheless, scientists have noted some common features of autoimmune disease. For example, they occur more often in older individuals, and they are also much more common in women than in men, although the reasons for this gender difference are unclear.

Hypotheses to explain the etiology of autoimmunity abound. They include the following:

- Estrogen may stimulate the destruction of tissues by cytotoxic T cells.

- During pregnancy, some maternal cells may cross the placenta and colonize the fetus; because these cells are

Table 18.4	The Four Classes of Immunosuppressive Drugs	
Class	**Examples**	**Action**
Corticosteroids	Prednisone, methylprednisone	Anti-inflammatory; kills T cells
Cytotoxic drugs	Cyclophosphamide, azathioprine, methotrexate, mycophenolate mofetil, brequinar sodium, leflunomide	Blocks cell division nonspecifically
Immunophilins	Cyclosporine	Blocks T cell responses
Lymphocyte-depleting therapies	Antilymphocyte globulin, monoclonal antibodies	Kills T cells nonspecifically, kills activated T cells

more likely to survive in a daughter than in a son (because they respond to female hormones), they might trigger an autoimmune disease in a daughter later in her life. Fetal cells may also cross the placenta and trigger autoimmunity in the mother.

- Environmental factors may contribute to the development of autoimmune disorders. Some autoimmune diseases—type I diabetes mellitus and rheumatoid arthritis, for example—develop in a few patients following their recovery from viral infections, though many individuals who develop autoimmune diseases have no history of such viral infections.

- Genetic factors may play a role in autoimmune diseases. MHC genes that in some way promote autoimmunity are found in individuals with autoimmune diseases, whereas MHC genes that somehow protect against autoimmunity may dominate in other individuals. MHC genes may also determine susceptibility to autoimmune disease by regulating the elimination of self-reactive T cells in the thymus.

- Some autoimmune diseases may develop when T cells encounter self-antigens that are normally "hidden" in sites where T cells rarely go. For example, because sperm develop within the testes during puberty, long after the body has selected its T cell population, men may have T cells that recognize their own sperm as foreign. This is normally of little consequence because sperm are sequestered from the blood, but should the testes be injured, T cells may enter the site of damage and mount an autoimmune response against the sperm, resulting in male infertility.

- Infections with a variety of microorganisms may trigger autoimmunity as a result of **molecular mimcry,** which occurs when an infectious agent has an antigenic determinant that is very similar or identical to a self-antigen. In responding to the invader, the body produces *autoantibodies* (antibodies against self-antigens) that damage body tissues. For example, children infected with some strains of group A *Streptococcus pyogenes* (strep-tō-kok′ŭs pī-aj′en-ēz) may produce antibodies to heart muscle and so develop heart disease. Other strains of streptococci trigger the produc-

tion of antibodies that cross-react with glomerular basement membranes and so cause kidney disease. It is possible that some virally triggered autoimmune diseases may also result from molecular mimicry.

- Other autoimmune responses may result from failure of the normal control mechanisms of the immune system. Thus even though harmful self-reactive lymphocytes are normally destroyed in the thymus via apoptosis triggered through the cell surface receptor CD95 (see Chapter 16), defects in CD95 can cause autoimmunity by permitting abnormal T cells to survive and cause disease.

Regardless of the specific mechanism that causes an autoimmune disease, immunologists categorize them into two major categories: single tissue autoimmune diseases, which affect a single organ or tissue, and systemic autoimmune diseases.

Single Tissue Autoimmune Diseases

Learning Objective

✓ Describe a serious autoimmune disease associated with each of the following: blood cells, endocrine glands, and nervous tissue.

Among the many single tissue autoimmune diseases recognized, common examples affect blood cells, endocrine glands, or nervous tissue.

Autoimmunity Affecting Blood Cells

Some patients make autoantibodies to their leukocytes. As a result, they have difficulty combating infections. Other patients make autoantibodies to blood platelets, and as a result their blood will not clot.

As we have seen, individuals with hemolytic anemia produce antibodies against their own red blood cells. These autoantibodies hasten the destruction of the red blood cells, and the patient becomes severely anemic. Different hemolytic anemia patients make antibodies of different classes. Some patients make IgM autoantibodies, which bind to red blood cells and activate the classical complement pathway; the red blood cells are lysed, and degration products from hemoglobin released into the bloodstream are excreted in the urine. Other hemolytic anemia patients make IgG autoantibodies, which are less effective in activating complement and instead serve as opsonins that promote phagocytosis. In this case, red blood cells are removed by macrophages in the liver, spleen, and bone marrow. Even though these patients have no hemoglobin in their urine, they are still severely anemic.

The precise causes of all cases of autoimmune hemolytic anemia are unknown, but some cases follow infections with viruses or treatment with certain drugs, both of which alter the surface of red blood cells such that they are recognized as foreign and trigger an immune response.

Autoimmunity Affecting Endocrine Organs

Other common targets of autoimmune attack are the endocrine (hormone-producing) organs. Patients can develop autoantibodies or produce T cells against cells of the islets of Langerhans within the pancreas, against cells of the thyroid gland, or against cells of other endocrine glands. In most cases, the ensuing autoimmune reaction results in damage to or destruction of the gland, and in hormone deficiencies as endocrine cells are killed.

Immunological attack on the islets of Langerhans results in a loss of the ability to produce the hormone *insulin,* which leads to the development of **type I diabetes mellitus** (dī-ă-bē′tēz mě-lī′tĕs) (also known as juvenile-onset diabetes). As with other autoimmune diseases, the exact cause of type I diabetes is unknown, but many patients endured a severe viral infection some months before the onset of diabetes. Additionally, some patients are known to have a genetic predisposition to developing type I diabetes that is associated with the possession of certain class I MHC molecules. Some physicians have been successful in delaying the onset of type I diabetes by treating at-risk patients with immunosuppressive drugs before damage to the islets of Langerhans becomes apparent.

An autoimmune response can lead to stimulation rather than to inhibition or destruction of glandular tissue. An example of this is **Graves' disease,** which involves the thyroid gland—a major endocrine gland located in the neck region. Affected patients make autoantibodies that bind to and stimulate receptors on the cytoplasmic membranes of cells in the anterior pituitary gland. The stimulated pituitary cells produce *thyroid-stimulating hormone,* which elicits excessive production of *thyroid hormone* and growth of the thyroid gland. Such patients develop a *goiter* (enlarged thyroid gland), protruding eyes, rapid heartbeat, fatigue, and weight loss despite increased appetite. Physicians treat Graves' disease by surgically removing excess thyroid tissue or by administering immunosuppressive drugs.

Autoimmunity Affecting Nervous Tissue

Of the group of autoimmune diseases affecting nervous tissue, the most important is **multiple sclerosis** (sklĕ-rō′sis) **(MS).** In this disease, cytotoxic T cells attack and destroy the *myelin sheaths* that normally insulate brain and spinal cord neurons and increase the speed of nerve impulses along the length of the neurons. Consequently, MS patients experience deficits in vision, speech, and neuromuscular function that may be quite mild and intermittent, or may ultimately lead to death. Like other autoimmune diseases, MS may be triggered by a viral infection in individuals with certain genetic backgrounds.

Systemic Autoimmune Diseases

Learning Objective

✓ Discuss two important systemic autoimmune diseases.

Two classic examples of autoimmune diseases that affect multiple organs or the entire body are systemic lupus erythematosus and rheumatoid arthritis.

Systemic Lupus Erythematosus

Systemic lupus erythematosus (loo′pŭs er-ĭ-thē′mă-tō-sus) **(SLE),** often shortened to *lupus,* is a generalized immunologic disorder that results from a loss of control of both humoral and cell-mediated immune responses. Affected individuals make autoantibodies against numerous antigens found in normal organs and tissues, giving rise to many different pathological lesions and clinical manifestations. One consistent feature of SLE is the development of autoantibodies against nucleic acids, especially DNA. These autoantibodies combine with free DNA released from dead cells to form immune complexes that are deposited in glomeruli, causing glomerulonephritis and kidney failure. Thus, SLE results from a type III (immune-complex mediated) hypersensitivity reaction. Immune complexes may also be deposited in joints, where they provoke arthritis.

The disease's curious name—systemic lupus erythematosis—stems from two features of the disease: *Systemic* simply reflects the fact that it affects different organs throughout the body; *lupus* is Latin for wolf, and *erythematosus* refers to a redness of the skin, so these latter two words describe the characteristic red, butterfly-shaped rash that develops on the face of many patients, giving them what is sometimes described as a wolf-like appearance (**Figure 18.13** on page 526). This rash is caused by deposition of nucleic acid–antibody complexes in the skin, and is worse in skin areas exposed to sunlight.

Although autoantibodies to nucleic acids are characteristic of SLE, many other autoantibodies are also produced. Autoantibodies to red blood cells cause hemolytic anemia; autoantibodies to platelets give rise to bleeding disorders; antilymphocyte antibodies alter immune reactivity; and antibodies against muscle cells provoke muscle inflammation and, in some cases, damage to the heart. Because of its variety of symptoms, SLE is often misdiagnosed.

The cause or causes of lupus are unknown. In experimental animal models, a lupus-like disease can be induced by some drugs, viral infections, or genetic mutations. It is likely that SLE in humans also has many causes. Some

▲ *Figure 18.13*

The characteristic facial rash of systemic lupus erythematosus. Its shape corresponds to areas most exposed to sunlight, which worsens the condition.

▲ *Figure 18.14*

The crippling distortion of joints characteristic of rheumatoid arthritis.

patients are known to develop lupus in response to a complement deficiency; the absence of certain complement components prevents the removal of immune complexes from the circulation, leading to their deposition in the kidneys and joints.

Physicians treat lupus with immunosuppressive drugs that reduce autoantibody formation, and with corticosteroids that reduce the inflammation associated with the deposition of immune-complexes.

Rheumatoid Arthritis

Rheumatoid arthritis (roo'mă-toyd ar-thrī'tis) is another crippling, systemic, autoimmune disease resulting from a type III hypersensitivity reaction. This disease commences when B cells in the joints produce autoantibodies against collagen, a protein in cartilage, which commonly covers joint surfaces. The resulting immune-complexes and complement bind to mast cells, which release the inflammatory chemicals in their granules. The resulting inflammation causes the tissues to swell, thicken, and proliferate into the joint. As the altered tissue extends into the joint, the inflammation further erodes and destroys joint cartilage and the neighboring bony structure, until the joints begin to break down and fuse; as a result, affected joints become distorted and lose their range of motion **(Figure 18.14)**. The course of rheumatoid arthritis is often intermittent; however, with each recurrence, the lesions and damage get progressively more severe.

As with other autoimmune diseases, the cause or causes of rheumatoid arthritis is not well understood. The fact that there are no animal models (because the disease appears to affect humans only) significantly hinders research on its cause. Many cases demonstrate that rheumatoid arthritis commonly follows an infectious disease in a genetically susceptible individual. Possession of certain MHC genes appears to promote susceptibility.

Physicians treat rheumatoid arthritis by administering anti-inflammatory drugs such as corticosteroids to prevent additional joint damage, and methotrexate to inhibit the humoral immune response.

Immunodeficiency Diseases

Learning Objective

✓ Differentiate between primary and acquired immunodeficiencies, and cite two diseases caused by each form of immunodeficiency.

You may have noticed that during periods of increased emotional or physical stress, you are more likely to succumb to a cold or a bout of the flu. You are not imagining this phenomenon; stress has long been known to decrease the efficiency of the immune system. Similarly, chronic defects in the immune system typically first become apparent when affected individuals show increased susceptibility to infections of opportunistic pathogens (or even of organisms not normally considered pathogens). Such opportunistic infections, which were discussed in Chapter 14, are the hallmarks of *immunodeficiency* (im'ū-nō-dē-fish'en-sē) diseases, which are conditions resulting from defective immune mechanisms.

Researchers have characterized a large number of immunodeficiency diseases in humans, which are of two general types:

- **Primary immunodeficiency diseases,** which are detectable near birth and develop in infants and young children, result from some genetic or developmental defect.

Highlight 18.1   SCID: "Bubble Boy" Disease

As a result of the widely publicized case of David Vetter—a Houston Texas resident commonly known as the "bubble boy" because he lived from birth until age 12 in a sterile, plastic-enclosed environment—*severe combined immunodeficiency disease (SCID)* has been known as "bubble boy" disease since the 1970s. Like all SCID patients, David had no immune system—his body produced neither B cells nor T cells—so the slightest infection could be lethal. As a result, any contact with him could occur only through a pair of antiseptic rubber gloves built into one of the walls of his plastic enclosure. At age 12, David underwent an experimental

David Vetter

bone marrow transplant in the hope that marrow donated by his older sister would enable him to build up an immune system. Unfortunately, the donated cells turned out to be infected with Epstein-Barr virus (a common virus that causes mononucleosis), ultimately killing David.

Today, there is much more hope for patients with SCID. Virus-free bone marrow transplants are now possible. More recently, some children have been treated via gene therapy, in which the gene for an enzyme (adenine deaminase) missing in SCID patients is inserted, via retroviral vectors, into clones of the patient's bone marrow cells. These genetically altered cells are then returned to the child, where they grow and synthesize sufficient enzyme to normalize the immune response. While successful in several children studied in a French trial, the therapy was halted in September 2002 after one child developed a leukemia-like illness following treatment. Investigations into whether or not the illness was caused by the gene therapy are underway; however, scientists remain hopeful that gene therapy can achieve a cure for SCID—without side effects—in the near future.

- **Acquired (secondary) immunodeficiency diseases** develop in later life (in older children, adults, and the elderly) as a direct consequence of some other recognized cause, such as malnutrition, severe stress, or infectious disease.

Next we consider each general type of immunodeficiency disease in turn.

CRITICAL THINKING

A 43-year-old woman has been diagnosed with rheumatoid arthritis. Unable to find relief from her symptoms, she seeks treatment from a doctor in another country who injects a daily regimen of antibodies and complement into her afflicted joints. Do you expect the treatment to improve her condition? Explain your reasoning.

Primary Immunodeficiency Diseases

Many different inherited defects have been identified in all of the body's lines of defense, affecting first and second lines of defense as well as humoral and cell-mediated immune responses.

One of the more important inherited defects in the second line of defense is **chronic granulomatous disease,** in which children have recurrent infections characterized by the development of large masses of inflammatory cells in lymph nodes, lungs, bones, and skin. The reason is that these children have neutrophils that are unable to kill ingested bacteria due to an inherited biochemical defect in one of the polypeptides of *NADPH oxidase,* an enzyme that is crucial in the increased metabolic activity that occurs during phagocytosis.

Most primary immunodeficiencies are associated with defects in the components of the third line of defense—specific immunity. For example, some children fail to develop any lymphoid stem cells whatsoever, and as a result they produce neither B cells nor T cells and cannot mount any sort of immune response. The resulting defects in the immune system cause **severe combined immunodeficiency disease (SCID),** which is discussed in **Highlight 18.1.**

Other children suffer from T cell deficiencies alone. For example, **DiGeorge anomaly** results from a failure of the thymus to develop. In consequence, there are no T cells. The importance of T cells in protecting against viruses is emphasized by the observation that individuals with DiGeorge anomaly generally die of viral infections while remaining resistant to most bacteria. Physicians cure DiGeorge anomaly with a thymic stem cell transplant.

B cell deficiencies also occur in children. The most severe of the B cell deficiencies, called **Bruton-type agammaglobulinemia**[6] (ā-gam'ă-glob'ū-li-nē'mē-ă), is an inherited disease in which affected babies, usually boys, cannot make immunoglobulins. These children experience recurrent bacterial infections but are usually resistant to viral, fungal, and protozoan infections.

Inherited deficiencies of individual immunoglobulin classes are more common than a deficiency of all classes. Among these, IgA deficiency is the most common. Because affected children cannot produce secretory IgA, they experience recurrent infections in the respiratory and gastrointestinal tracts.

---

[6]Agammaglobulinemia means an absence of gamma globulin (IgG).

Table 18.5	Primary Immunodeficiency Diseases	
Disease	**Defect**	**Manifestation**
Chronic granulomatous disease	Ineffective neutrophils	Uncontrolled infections
Severe combined immunodeficiency disease	A lack of T cells and B cells	No resistance to any type of infection, leading to rapid death
Bruton-type agammaglobulinemia	A lack of B cells and thus a lack of immunoglobulins	Death from overwhelming bacterial infections
DiGeorge anomaly	A lack of T cells and thus no cell-mediated immunity	Death from overwhelming viral infections

CRITICAL THINKING

Two boys have autoimmune diseases: One has Bruton-type agammaglobulinemia, and the other has DiGeorge anomaly. On a camping trip, each boy is stung by a bee and each falls into poison ivy. What hypersensitivity reactions might each boy experience as a result of his camping mishaps?

Table 18.5 summarizes the major primary immunodeficiency diseases.

Acquired Immunodeficiency Diseases

Unlike inherited primary immunodeficiency diseases, acquired immunodeficiency diseases affect older individuals who had a previously healthy immune system.

Acquired immunodeficiencies result from a number of causes. In all humans the immune system (but especially T cell production) deteriorates with increasing age; as a result, older individuals normally have less effective immunity, especially cell-mediated immunity, than younger individuals, leading to an increased incidence of both viral diseases and certain types of cancer. Severe stress can also lead to immunodeficiencies by prompting the secretion of increased quantities of corticosteroids, which are toxic to T cells and thus suppress cell-mediated immunity. This is why, for ex-

ample, cold sores commonly "break out" on the faces of students during final exams: The causative herpes simplex viruses, which are normally controlled by a fully functioning cell-mediated immune response, escape immune control in stressed individuals. Malnutrition and certain environmental toxins can also cause acquired immunodeficiency diseases by inhibiting the normal production of B cells and T cells.

Of course, the most significant example of the result of an acquired immunodeficiency is **acquired immunodeficiency syndrome (AIDS),** caused by human immunodeficiency virus (HIV), which infects and kills helper T (CD4) cells. Because the body's cytotoxic T (CD8) cells mount a powerful response to HIV infection, the rate of viral replication is slowed, and thus HIV infection has a prolonged latent period (up to several years) before helper T cell counts drop below critical levels. When this occurs, infected individuals gradually lose the ability to fight off infections by viruses and other intracellular pathogens. Once immunodeficiency is well established, affected individuals suffer from a wide variety of problems, including fungal pneumonia, virus-mediated skin cancer, opportunistic yeast infections, persistent diarrhea, shingles, and frequent outbreaks of herpes simplex. AIDS is discussed in detail in Chapter 25.

CHAPTER SUMMARY

1. Immunological responses may give rise to inflammatory reactions called **hypersensitivities.** An immunological attack on normal tissues gives rise to autoimmune disorders. A failure of the immune system to function normally may give rise to **immunodeficiency diseases.**

Hypersensitivities (pp. 510–523)

1. Type I hypersensitivity gives rise to **allergies.** Antigens that trigger this response are called **allergens.**

2. Allergies result when allergens bind to IgE that were previously bound to **mast cells, basophils,** and **eosinophils.** This causes the sensitized cells to degranulate and release **histamine, kinins, proteases, leukotrienes,** and **prostaglandins.**

3. Depending on the amount of and site at which these molecules are released, the result can produce various clinical syndromes, including **hay fever, asthma, urticaria** (hives), or various other allergies. When the inflammatory mediators exceed the body's coping mechanisms, **acute anaphylaxis** or **anaphylactic shock** may occur. The specific treatment for anaphylaxis is epinephrine.

4. Type I hypersensitivity can be diagnosed by skin testing and can be partially prevented by avoidance of allergens and immunotherapy; some type I hypersensitivities are treated with **antihistimines.**

5. Type II cytotoxic hypersensitivities, such as incompatible blood transfusions and hemolytic disease of the newborn, result when cells are destroyed by an immune response.

6. Red blood cells have **blood group antigens** on their surface. If incompatible blood is transfused into a recipient, a severe transfusion reaction can result.

7. The most important of the blood group systems is the **ABO system,** which is largely responsible for transfusion reactions.

8. Approximately 85% of the human population carries **Rh antigens,** which are also found in rhesus monkeys. If an Rh negative pregnant woman is carrying an Rh positive fetus, the fetus may be at risk of **hemolytic disease of the newborn,** in which antibodies made by the mother against the Rh antigens may cross the placenta and destroy the fetus's red blood cells. Rhogam administered to pregnant Rh− women may prevent this disease.

9. Drugs bound to blood platelets may subsequently bind antibodies and complement, causing **immune thrombocytopenic purpura,** in which the platelets lyse.

10. If excessive amounts of **immune-complexes** are deposited in tissues, they can cause significant tissue damage. Immune-complexes deposited in the lung cause a **hypersensitivity pneumonitis,** of which the most common example is farmer's lung. If large amounts of immune-complexes form in the bloodstream, they may be filtered out by the glomeruli of the kidney, causing **glomerulonephritis,** which can result in kidney failure.

11. Type IV hypersensitivity, also known as **delayed hypersensitivity,** is a T cell-mediated inflammatory reaction that is so called because it takes 24−72 hours to reach maximal intensity.

12. A good example of a delayed hypersensitivity reaction is the **tuberculin response,** generated when tuberculin, a protein extract of *Mycobacterium tuberculosis,* is injected into the skin of an individual who has been infected with or vaccinated against *M. tuberculosis.*

13. Another example of a type IV hypersensitivity reaction is **allergic contact dermatitis,** which is T cell-mediated damage to chemically modified skin cells. The best known example is a reaction to poison ivy.

14. An organ or tissue **graft** can be made between different sites within a single individual (an **autograft**), between genetically identical individuals (an **isograft**), between genetically dissimilar individuals (an **allograft**), or between individuals of different species (a **xenograft**). Most surgical organ grafting involves allografts, which, if not treated with immunosuppressive drugs, will lead to graft rejection.

15. Some tissues, such as portions of the eye and the testes, are regarded as **privileged sites** because allografts transplanted into these areas are not rejected. Despite being similar to an allograft, a fetus is not rejected by a pregnant mother because a variety of immunosuppressive mechanisms prevent rejection.

16. Commonly used immunosuppressive drugs include **corticosteroids, cytotoxic drugs, immunophilins,** and lymphocyte-depleting therapies, which involve treatment with antibodies against cells of specific immunity.

Autoimmune Diseases (pp. 523–526)

1. **Autoimmune diseases** may result when an individual begins to make autoantibodies or cytotoxic T cells against normal body components.

2. There are many hypotheses concerning the cause of autoimmune disease. One involves **molecular mimicry,** in which microorganisms with antigenic determinants similar to self-antigens trigger autoimmune tissue damage. Others examine estrogen, pregnancy, and environmental and genetic factors.

3. One group of autoimmune diseases involve only a single organ or cell type. Examples of such diseases include **autoimmune hemolytic anemia,** or **type I diabetes mellitus, Graves' disease,** and **multiple sclerosis.**

4. A second group of autoimmune diseases involves multiple organs or body systems. Examples include **systemic lupus erythematosus** and **rheumatoid arthritis,** in which antibodies made against many different autoantigens are part of immune-complexes deposited in tissues. In rheumatoid arthritis, the immune-complexes result in the growth of inflammatory tissue within joints.

Immunodeficiency Diseases (pp. 526–528)

1. Immunodeficiency diseases may be classified into **primary immunodeficiency diseases,** which result from mutations or developmental anomalies and occur in young children, and **acquired** or **secondary immunodeficiency diseases,** which result from other known causes such as viral infections.

2. Examples of primary immunodeficiency diseases include **chronic granulomatous disease,** in which a child's neutrophils are incapable of killing ingested bacteria. Inability to produce both T cells and B cells is called **severe combined immunodeficiency disease.** In **DiGeorge anomaly,** the thymus fails to develop. In **Bruton-type agammaglobulinemia,** B cells fail to function.

3. Previously healthy immune systems may become damaged in very old age and by severe stress, malnutrition, or environmental toxins. Insufficient T cell production may enable viruses to invade successfully.

4. A significant example of virus-mediated immunosuppression occurs following invasion by the human immunodeficiency virus (HIV). After several years of latency, infection with this virus results in **acquired immunodeficiency disease (AIDS),** a syndrome characterized by fungal pneumonia, opportunistic infections, and viral outbreaks.

QUESTIONS FOR REVIEW

(Answers to multiple choice, True/False, and matching questions are on the web, along with additional review questions. Visit www.microbiologyplace.com.)

Multiple Choice

1. The immunoglobulin class that mediates type I hypersensitivity is
   a. IgG.
   b. IgM.
   c. IgD.
   d. IgE.
   e. IgA.

2. The major inflammatory mediator released by degranulating mast cells in type I hypersensitivity is
   a. immunoglobulin.
   b. complement.
   c. histamine.
   d. interleukin.
   e. prostaglandin.

3. The major blood group antigens responsible for hemolytic disease of the newborn are
   a. MHC proteins.
   b. MN antigens.
   c. ABO antigens.
   d. rhesus antigen.
   e. type II proteins.

4. Farmer's lung is a hypersensitivity pneumonitis resulting from
   a. a type I hypersensitivity reaction to grass pollen.
   b. a type II hypersensitivity to red cells in the lung.
   c. a type III hypersensitivity to mold spores.
   d. a type IV hypersensitivity to bacterial antigens.
   e. none of the above.

5. A positive tuberculin skin test indicates that a patient not immunized against tuberculosis
   a. is free of tuberculosis.
   b. is shedding *Mycobacterium*.
   c. has been exposed to tuberculosis antigens.
   d. is susceptible to tuberculosis.
   e. is resistant to tuberculosis.

6. Which of the following is an autoimmune disease?
   a. a heart attack
   b. acute anaphylaxis
   c. farmer's lung
   d. graft-versus-host disease
   e. systemic lupus erythematosus

7. When a surgeon conducts a cardiac bypass operation by transplanting a piece of vein from a patient's leg to the same patient's heart, this is
   a. a xenograft.
   b. an autograft.
   c. an allograft.
   d. an isograft.
   e. a coreograft.

8. A deficiency of both B cells and T cells is most likely
   a. a secondary immunodeficiency.
   b. a complex immunodeficiency.
   c. an acquired immunodeficiency.
   d. a primary immunodeficiency.
   e. an induced immunodeficiency.

9. Infection with HIV causes
   a. primary immunodeficiency disease.
   b. acquired hypersensitivity syndrome.
   c. acquired immunodeficiency syndrome.
   d. anaphylactic immunodeficiency diseases.
   e. combined immunodeficiency diseases.

True/False

Indicate whether each statement is true or false. If the statement is false, change the italicized word or phrase to make the statement true.

____ 1. *Cyclosporine* is released by degranulating mast cells.

____ 2. *Type III* hypersensitivity reactions may lead to the development of glomerulonephritis.

____ 3. ABO blood group antigens are found on *nucleated* cells.

____ 4. The tuberculin reaction is a *type I* hypersensitivity.

____ 5. Graft-versus-host disease can follow a bone marrow *isograft*.

Matching

____ 1. Acute anaphylaxis

____ 2. Allergic contact dermatitis

____ 3. Systemic lupus erythematosus

____ 4. Allograft rejection

____ 5. AIDS

____ 6. Graft-versus-host disease

____ 7. Milk allergy

____ 8. Farmer's lung

A. Type I hypersensitivity
B. Type II hypersensitivity
C. Type III hypersensitivity
D. Type IV hypersensitivity
E. Not a hypersensitivity

CRITICAL THINKING

1. What possible advantages might an individual gain from making class E antibodies (IgE)?

2. Why can't physicians use skin tests similar to the tuberculin reaction to diagnose other bacterial diseases?

3. In both Graves' disease and juvenile onset diabetes mellitus, autoantibodies are directed against cytoplasmic membrane receptors. Speculate on the clinical consequences of an autoimmune response to estrogen receptors.

4. In general, people with B cell defects acquire numerous bacterial infections, whereas those with T cell defects get viral diseases. Explain why this is so.

5. What types of illnesses cause death in patients with combined immunodeficiencies or AIDS?

6. Because of the severe shortage of organ donors for transplants, many scientists are examining the possibility of using organs from nonhuman species such as pigs. What special clinical problems might be encountered when using these xenografts?

CHAPTER 19

Pathogenic Gram-Positive Cocci and Bacilli

CHAPTER OUTLINE

Staphylococcus
Structure and Physiology
Pathogenicity
Epidemiology
Staphylococcal Diseases
Diagnosis, Treatment, and Prevention

Streptococcus
Group A Streptococcus: *S. pyogenes*
Group B Streptococcus: *S. agalactiae*
Other Beta-Hemolytic Streptococci
Alpha-Hemolytic Streptococci: The Viridans Group
Streptococcus pneumoniae

Enterococcus
Structure and Physiology
Pathogenesis, Epidemiology, and Diseases
Diagnosis, Treatment, and Prevention

Bacillus
Structure, Physiology, and Pathogenicity
Epidemiology
Disease
Diagnosis, Treatment, and Prevention

Clostridium
Clostridium perfringens
Clostridium difficile
Clostridium botulinum
Clostridium tetani

Listeria
Pathogenesis , Epidemiology, and Disease
Diagnosis, Treatment, and Prevention

Corynebacterium
Pathogenesis, Epidemiology, and Disease
Diagnosis, Treatment, and Prevention

Mycobacterium
Tuberculosis
Leprosy
Other Mycobacterial Infections in AIDS Patients

Propionibacterium

Nocardia and **Actinomyces**
Nocardia asteroides
Actinomyces

We are all familiar with acne, the common skin disorder characterized by pimples and other blemishes. But did you know that acne is often caused by a Gram-positive bacterium called *Propionibacterium acnes*? *P. acnes* grows in the sebaceous (oil) glands of the skin. Teenagers are especially prone to developing acne because the hormonal changes of adolescence can trigger the production of excess oil, stimulating bacterial growth. The bacteria secrete chemicals that attract the body's leukocytes, which phagocytize the bacteria and trigger inflammation. Pus-filled pimples result when lymphocytes and dead bacteria clog the skin's pores.

In this chapter we will learn more about *P. acnes,* and about other medically important Gram-positive pathogens, including *Staphylococcus, Streptococcus,* and *Bacillus.* A general characteristic of Gram-positive pathogens is that the disease symptoms they cause often result from the toxins they secrete. In Chapter 20 we will turn to Gram-negative pathogens, which not only secrete toxins but also possess a toxic outer membrane that can produce disease symptoms in humans.

Propionibacterium acnes

MicroPrep Pre-Test: *Take the pre-test for this chapter on the web.* Visit **www.microbiologyplace.com**.

In Chapter 11 we examined the general features of prokaryotes, including some pathogens. In this chapter we explore the major Gram-positive pathogens in more detail. Recall that Gram-positive pathogens stain purple when Gram-stained and generally fall within the phylum *Firmicutes*. We will discuss the major Gram-negative pathogens, as well as pathogens with less clear-cut classifications, in Chapters 20 and 21, respectively.

Taxonomists currently group Gram-positive pathogens into two major groups: low G + C Gram-positive bacteria and high G + C Gram-positive bacteria. As the names indicate, these classifications are based on the ratio of guanine-cytosine nucleotide base pairs to adenine-thymine nucleotide base pairs in these organisms' DNA. Low G + C Gram-positive bacteria include three genera of pathogenic spherical cells (cocci)—*Staphylococcus, Streptococcus,* and *Enterococcus*—and three genera of pathogenic rod-shaped cells (bacilli)—*Bacillus, Clostridium,* and *Listeria.* The high G + C Gram-positive pathogens include rod-shaped genera *Corynebacterium, Mycobacterium,* and *Propionibacterium,* and the filamentous, fungus-like *Nocardia* and *Actinomyces.*

It is worth mentioning that the *mycoplasmas,* a group of bacteria that lack cell walls, have historically been classified as Gram-negative bacteria because they stain pink when Gram-stained. However, studies of their DNA nucleotide sequences have revealed that they are genetically more similar to low G + C Gram-positive bacteria. We will discuss the mycoplasmas in detail in Chapter 21, but keep in mind that at a molecular level, these pathogens have much in common with the low G + C Gram-positive pathogens that we will discuss in this chapter.

We begin our discussion of Gram-positive bacteria with *Staphylococcus.*

Staphylococcus

Bacteria in the genus *Staphylococcus* (staf'i-lō-kok'ŭs) are living and reproducing on almost every square inch of your skin right now. They are normal members of every human's microbiota that usually go unnoticed. However, they can be opportunistic pathogens, causing minor to life-threatening diseases.

Structure and Physiology

Learning Objectives

✓ Characterize the cells of *Staphylococcus,* particularly those features that make it virulent.

✓ Contrast the virulence of *S. aureus* with that of *S. epidermis* in humans.

Staphylococcus[1] is a genus of Gram-positive, facultatively anaerobic prokaryotes whose spherical cells are typically clustered in grapelike arrangements **(Figure 19.1).** As we saw in Chapter 11, this arrangement results from two characteristics of cell division: Cell divisions occur in succes-

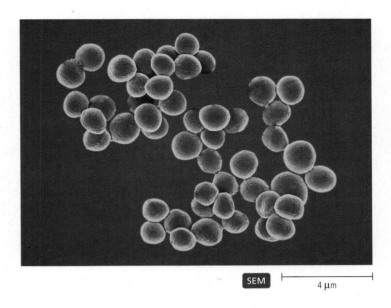

SEM | 4 μm

▲ *Figure 19.1*

A scanning electron micrograph of *Staphylococcus,* showing the grapelike clusters of cells characteristic of *S. aureus* (shown) and *S. epidermidis.*

sively different planes, and daughter cells remain attached to one another. Staphylococcal cells are 0.5–1.0 μm in diameter, nonmotile, and salt tolerant—they are capable of growing in media that are up to 10% NaCl, which explains how they tolerate the salt deposited on human skin by sweat glands. Additionally, they are tolerant of desiccation, radiation, and heat (up to 60°C for 30 min), allowing them to survive on environmental surfaces in addition to skin. Further, *Staphylococcus* synthesizes catalase—a characteristic that distinguishes this genus from other low G + C Gram-positive cocci.

Two species are commonly associated with staphylococcal diseases in humans:

- *Staphylococcus aureus*[2] (o'rē-ŭs) is the more virulent, producing a variety of disease conditions and symptoms depending on the site of infection.

- *Staphylococcus epidermidis* (e-pi-der'mi-dis), as its name suggests, is a part of the normal microbiota of human skin, but it is an opportunistic pathogen in immuno-compromised patients or when introduced into the body via intravenous catheters or on prosthetic devices such as artificial heart valves.

---

[1]From Greek *staphle,* meaning small bunch of grapes, and *kokkos,* meaning a berry.
[2]Latin for golden, for the golden color of its colonies growing on the surface of a solid medium.

Pathogenicity

Learning Objective

✓ Discuss the structural and enzymatic features and toxins of *Staphylococcus* that enable it to be pathogenic.

What are commonly called "staph" infections result when staphylococci breach the body's physical barriers (skin or mucous membranes); entry of only a few hundred bacteria can ultimately result in disease. The pathogenicity of *Staphylococcus* results from three features: structures that enable it to evade phagocytosis, the production of enzymes, and the production of toxins.

Structural Defenses Against Phagocytosis

The cells of *S. aureus* are uniformly coated with a protein, called *protein A,* that interferes with humoral immune responses by binding to the stems of class G antibodies (IgG). Recall from Chapter 16 that antibodies are opsonins—they enhance phagocytosis—precisely because phagocytic cells have receptors for antibody stems; therefore, protein A, by binding stems, effectively inhibits opsonization. Protein A also inhibits the complement cascade, which is triggered by antibody molecules bound to antigen (see Figure 15.13).

The outer surfaces of most pathogenic strains of *S. aureus* also contain *bound coagulase,* an enzyme that converts the soluble blood protein fibrinogen into long, insoluble fibrin molecules, which are threads that form blood clots around the bacteria. Fibrin clots in effect hide the bacteria from phagocytic cells.

Both *S. aureus* and *S. epidermidis* also evade the body's defenses by synthesizing loosely organized polysaccharide slime layers (sometimes called capsules) that inhibit chemotaxis of and phagocytosis by leukocytes, particularly neutrophils. The slime layer also facilitates attachment of *Staphylococcus* to artificial surfaces such as catheters, shunts, artificial heart valves, and synthetic joints.

In summary, protein A, bound coagulase, and a slime layer allow *S. aureus* to evade the body's defenses, whereas *S. epidermidis* relies almost exclusively on its slime layer.

Enzymes

Staphylococci produce a number of enzymes that contribute to their survival and pathogenicity:

- *Cell-free coagulase,* like bound coagulase, triggers blood clotting. Cell-free coagulase does not act on fibrin directly, but instead combines with a blood protein before becoming enzymatic and converting fibrinogen to fibrin threads. Only *S. aureus* synthesizes coagulase; *S. epidermis* and other species of *Staphylococcus* are coagulase negative.
- *Hyaluronidase* breaks down hyaluronic acid, which is a major component of the matrix between cells. Hyaluronidase, found in 90% of *S. aureus* strains, enables the bacteria to spread between cells throughout the body.

- *Staphylokinase* (produced by *S. aureus*) dissolves fibrin threads in blood clots, allowing *S. aureus* to free itself from clots. Thus, *Staphylococcus* can escape the immune system by enclosing itself in a fibrin clot (via coagulase), and then, when space and nutrients become limiting, it can digest its way out of the clot with staphylokinase and spread to new locations.
- *Lipases* digest lipids, allowing staphylococci to grow on the surface of the skin and in cutaneous oil glands. All staphylococci produce lipases.
- *β-lactamase (penicillinase),* now present in over 90% of *S. aureus* strains, breaks down penicillin. Though β-lactamase plays no role in inhibiting the natural defenses of the body, it does allow the bacteria to survive treatment with beta-lactam antimicrobial drugs such as penicillin and cephalosporin.

Toxins

Staphylococcus aureus and (less frequently) *S. epidermidis* also possess several toxins that contribute to their pathogenicity, including the following:

- *Cytolytic toxins.* So called alpha, beta, gamma, and delta toxins are proteins, coded by chromosomal genes, that disrupt the cytoplasmic membranes of a variety of cells including leukocytes. Leukocidin is a fifth cytolytic toxin that lyses leukocytes specifically, providing *Staphylococcus* with some protection against phagocytosis.
- *Exfoliative toxins.* Each of two distinct proteins causes the dissolution of epidermal *desmosomes* (intercellular bridge proteins that hold adjoining cytoplasmic membranes together), causing the patient's skin cells to separate from each another and slough off the body.
- *Toxic-shock-syndrome (TSS) toxin.* This protein causes toxic shock syndrome (discussed shortly).
- *Enterotoxins.* These five proteins (designated A through E) stimulate the intestinal muscle contractions, nausea, and intense vomiting associated with staphylococcal food poisoning. Enterotoxins are heat stable, remaining active at 100°C for up to 30 minutes.

Table 19.1 compares and contrasts the virulence factors possessed by *S. aureus* and *S. epidermidis.*

Epidemiology

Staphylococcus epidermidis is ubiquitous on human skin, whereas *S. aureus* is commonly found only on moist skin folds. Both species also grow in the upper respiratory, gastrointestinal, and urogenital tracts of humans. Both bacteria are transmitted through direct contact between individuals, as well as via fomites such as contaminated clothing, bed sheets, and medical instruments; as a result, proper hand antisepsis and aseptic techniques are essential in preventing their transfer in health care settings.

Table 19.1	A Comparison of the Virulence Factors of Staphylococcal Species	
	S. aureus	*S. epidermidis*
Protein A	+	−
Coagulase	+	−
Catalase	+	+
Hyaluronidase	+	−
Staphylokinase	+	−
Lipase	+	+
β-lactamase (penicillinase)	+	−

▲ *Figure 19.2*

Scalded skin syndrome, in which an exfoliative toxin produced by some strains of *Staphylococcus* causes reddened patches of the epidermis to slough off.

Staphylococcal Diseases

Learning Objectives

✓ Describe the symptoms and prevention of staphylococcal food poisoning.

✓ List and describe six pyogenic lesions caused by *Staphylococcus aureus*.

✓ Discuss five systemic and potentially fatal diseases caused by *Staphylococcus*.

Staphylococcus causes a variety of medical problems, depending on the site of infection, the immune state of its host, and the toxins and enzymes a particular species or strain secretes. Staphylococcal syndromes and diseases can be categorized as noninvasive, cutaneous, and systemic diseases.

Noninvasive Disease

Staphylococcus aureus is one of the more common causes of food poisoning (strictly speaking, food intoxication, because disease is caused by enterotoxin-contaminated food rather than by invasion of the bacteria). Commonly affected foods include processed meats, custard pastries, potato salad, and ice cream that have been contaminated with bacteria from human skin. (Unlike many other forms of food poisoning, animals are not involved.) The food must remain at room temperature or warmer for several hours for the bacteria to grow and secrete toxin. Warming or reheating inoculated food does not inactivate enterotoxins, which are heat stable, although heating does kill the bacteria. Food contaminated with staphylococci does not appear or taste unusual.

Symptoms, which include nausea, severe vomiting, diarrhea, headache, sweating, and abdominal pain, usually appear within 4 hours following ingestion. Consumed staphylococci do not continue to produce toxins, so the course of the disease is rapid, usually lasting 24 hours or less.

Cutaneous Diseases

Staphylococcus aureus causes localized *pyogenic*[3] (pī-ō-jen′ik) lesions. **Scalded skin syndrome** is a reddening of the skin that typically begins near the mouth, spreads over the entire body, and is followed by large blisters that contain clear fluid lacking bacteria or white blood cells—consistent with the fact that the disease is caused by a toxin released by bacteria growing on the skin. Within 2 days the affected outer layer of skin (epidermis) peels off in sheets, as if it had been dipped into boiling water **(Figure 19.2)**. The seriousness of scalded skin syndrome results from secondary bacterial infections in denuded areas.

Small, flattened, red patches on the face and limbs, particularly of children whose immune systems are not fully developed, characterize **impetigo** (im-pe-tī′gō) **(Figure 19.3 on page 536)**. The patches develop into pus-filled vesicles that eventually crust over. The pus is filled with bacteria and white blood cells, which distinguishes impetigo from scalded skin syndrome. *S. aureus* acting alone causes about 80% of impetigo cases; about 20% of cases also involve streptococci.

Folliculitis (fo-lik′ū-lī′tis) is an infection of a hair follicle in which the base of the follicle becomes red, swollen, and pus filled. When this condition occurs at the base of an eyelid, it is called a **sty.** A **furuncle** (fū′rŭng-kl) or boil is a large, painful, raised nodular extension of folliculitis into surrounding tissue. When several furuncles coalesce, they form a **carbuncle** (kar′bŭng-kl), which extends deeper into the tissues, triggering the fever and chills that are characteristic of action by the body's second line of defense. As staphylococci spread into underlying tissues, multiple organs and systems can become involved.

Systemic Diseases

Staphylococcus aureus, and to a lesser extent *S. epidermidis,* cause a wide variety of potentially fatal systemic infections

---

[3]From Greek *pyon,* meaning pus, and *genein,* meaning to produce.

▲ *Figure 19.3*

Impetigo, a staphylococcal disease in which reddened patches of skin become pus-filled vesicles that eventually crust over.

▲ *Figure 19.4*

Toxic shock syndrome (TSS). Fatal infections not only involve cutaneous tissues (as shown here) but internal organs as well.

when they are introduced into deeper tissues of the body, including the blood, heart, lungs, and bones.

Toxic Shock Syndrome When strains of *Staphylococcus* that produce TSS toxin grow in a wound or in the vagina, the toxin can be absorbed into the blood and cause **toxic shock syndrome (TSS),** characterized by fever, vomiting, red rash, extremely low blood pressure, and loss of sheets of skin **(Figure 19.4).** TSS is fatal to 5% of patients when their blood pressure falls so low that the brain, heart, and other vital organs have an inadequate supply of oxygen—a condition known as *shock.*

Though toxic shock syndrome occurs in males and females, in 1980 epidemiologists noted an epidemic of TSS among menstruating women. Researchers subsequently discovered that *S. aureus* grows exceedingly well in super-absorbent tampons, especially when the tampon remains in place for a prolonged period. As a result of the withdrawal of this type of tampon from the market, plus government-mandated reduction in the absorbency of all tampons and mandatory inclusion of educational information concerning the risks of TSS on every package of tampons, the number of cases of TSS declined rapidly **(Figure 19.5).**

Bacteremia *S. aureus* is a common cause of **bacteremia** (bak-tēr-ē′mē-ă), the presence of bacteria in the blood. After staphylococci enter the blood from a site of infection, they travel to other organs of the body, which may become infected. The initial site of infection and point of entry may be a furuncle, a vaginal infection, a surgical wound, or a contaminated medical device such as an intravascular catheter. Nosocomial (hospital-acquired) infections account for about half of all cases of staphylococcal bacteremia. The initial site of infection is never identified in about a third of patients, but is presumed to be an innocuous skin infection.

Endocarditis *S. aureus* may attack the lining of the heart (including its valves), producing a condition called **endocarditis**[4] (en′dō-kar-dī′tis). Typically, patients with endocarditis have nonspecific, flulike symptoms, but their condition quickly deteriorates as the amount of blood pumped from the heart drops precipitously. About 50% of patients with endocarditis do not survive.

Pneumonia and Empyema *Staphylococcus* in the blood can invade the lungs, causing **pneumonia** (noo-mō′nē-ă)—an inflammation of the lungs in which the alveoli (air sacs) and bronchioles (smallest airways) become filled with fluid. In 10% of patients with staphylococcal pneumonia this fluid is pus—a condition known as **empyema**[5] (em-pī-ē′mă).

Osteomyelitis When *Staphylococcus* invades a bone, either through a traumatic wound or via the blood during bacteremia, it causes **osteomyelitis**[6] (os′tē-ō-mī-e-lī′tis)—inflammation of the bone marrow and the surrounding bone. Osteomyelitis is characterized by pain in the infected bone accompanied by high fever. In children the disease typically occurs in the growing regions of long bones, which are areas with well-developed blood supplies. In adults, osteomyelitis is more commonly seen in vertebrae.

---

[4]From Greek *endon*, meaning within; *kardia*, meaning heart; and *itis*, meaning inflammation.
[5]From Greek *en*, meaning in, and *pyon*, meaning pus.
[6]From Greek *osteon*, meaning bone; *myelos*, meaning marrow; and *itis*, meaning inflammation.

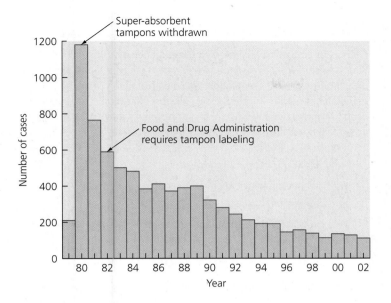

▲ *Figure 19.5*

The incidence of toxic shock syndrome in the United States since 1979. *Use of what product puts women at increased risk of TSS?*

Figure 19.5 Use of tampons is a risk factor for TSS.

Diagnosis, Treatment, and Prevention

Learning Objectives

✓ Describe how staphylococcal species are distinguished from one another during diagnosis.

✓ Discuss briefly the history of staphylococcal resistance to antimicrobial drugs.

Diagnosis of staphylococcal infection involves the detection of Gram-positive bacteria in grapelike arrangements isolated from pus, blood, or other fluids. If staphylococci isolated from an infection are able to clot blood, then they are coagulase-positive *S. aureus.* Coagulase-negative staphylococci are usually *S. epidermidis,* which is a normal part of the microbiota of the skin; their presence in a clinical sample is usually not indicative of a staphylococcal infection.

During the latter half of the 20th century, genes for β-lactamase, which convey resistance to natural penicillin, spread among *Staphylococcus* species. Thus, 90% of staphylococci were susceptible to penicillin in 1945, but only 5% are susceptible today. For this reason, the semisynthetic form of penicillin—*methicillin,* which is not inactivated by β-lactamase—became the drug of choice for staphylococcal infections. Unfortunately, **methicillin-resistant *Staphylococcus aureus* (MRSA)** has emerged as a major nosocomial problem. MRSA is also resistant to many other common antimicrobial drugs, including penicillin, macrolides, aminoglycosides, and cephalosporin; as a result, *vancomycin* has been used to treat MRSA infections. Physicians are very concerned about the increasing prevalence of vancomycin-resistant strains of *S. aureus* (see page 538).

Clinical practice has shown how crucial it is that abscesses be drained of pus and cleansed if subsequent antibiotic therapy is to be effective. Systemic infections such as endocarditis and osteomyelitis require long-term therapy with antimicrobial drugs.

Because strains of *Staphylococcus* resistant to antimicrobial agents have become more common, especially in hospitals, it is imperative that health care workers take precautions against introducing the bacterium into patients. Of course, given that *Staphylococcus* is ubiquitous on human skin, this source of staphylococcal infections cannot be eliminated. Fortunately, because a large inoculum is required to establish an infection, proper cleansing of wounds and surgical openings, attention to aseptic use of catheters and indwelling needles, and the appropriate use of antiseptics will prevent infections in most healthy patients. Health care workers with staphylococcal infections may be barred from delivery rooms, nurseries, and operating rooms. The most important measure for protecting against nosocomial infection is frequent hand antisepsis.

Scientists are currently testing the efficacy and safety of a vaccine that has proven effective in protecting dialysis patients from *S. aureus* infections. Such immunization, if it proves safe and effective in preventing other infections, may have a significant effect on nosocomial morbidity and mortality. It would be especially good news for health care providers who are battling resistant strains of *Staphylococcus.*

Streptococcus

Learning Objectives

✓ Describe the classification of streptococcal strains.

✓ Describe seven diseases caused by *S. pyogenes,* and the treatments available.

The genus *Streptococcus* (strep-tō-kok'ŭs) is a diverse assemblage of Gram-positive cocci 0.5–1.2 μm in diameter and arranged in pairs or chains. They are catalase negative (unlike *Staphylococcus*), although they do synthesize peroxidase and thus are facultatively anaerobic.

Researchers differentiate species of *Streptococcus* using several different, overlapping schemes, including serological classification based on the reactions of antibodies to specific bacterial antigens, type of hemolysis (alpha, beta, or gamma; see Figure 6.12), cell arrangement, and physiological properties as revealed by biochemical tests. In this chapter we will largely use a serological classification scheme, developed in 1938 by Rebecca Lancefield (1895–1981), that divides streptococci into serotype groups based on the bacteria's antigens (known, appropriately, as Lancefield antigens). The serotypes in this scheme include Lancefield groups A through H and K through V. Whereas the more significant streptococcal pathogens of humans are in groups A and B, two other significant streptococcal pathogens of humans lack Lancefield antigens. We begin our survey of the streptococci with group A streptococcus.

Case Study 19.1 A Fatal Case of Methicillin-Resistant *S. aureus*

A 5-year-old girl was admitted to the hospital with a temperature of 103°C and pain in her right hip. After pus was surgically drained from the hip joint, she was treated with a semisynthetic cephalosporin. Her physicians changed the antibiotic regimen after 24-hour

Methicillin-resistant *S. aureus*

TEM   0.5 μm

cultures of blood and pus revealed the presence of MRSA. On the third day, she suffered respiratory failure and empyema and was placed on mechanical ventilation. She died from pulmonary hemorrhage and pneumonia after 5 weeks of hospitalization. The girl had been previously healthy with no recent hospitalizations, though she had skinned her knee while learning to ride a bicycle 2 days before admittance to the hospital.

1. How might the girl have been infected?
2. How did her hip joint become infected?
3. Describe the series of diseases she suffered.
4. What was likely the second antibiotic she received?

Group A Streptococcus: *Streptococcus pyogenes*

Learning Objectives

✓ Describe two structures in *Streptococcus pyogenes* that enable this organism to survive against the body's defenses.

✓ Identify four enzymes and three toxins that facilitate the spread of *S. pyogenes* in the body.

✓ Identify the conditions under which group A streptococci cause disease.

Group A streptococcus, which is synonymously known as *S. pyogenes* (pī-oj'en-ēz), is a coccus that forms 1–2 mm white colonies surrounded by a large zone of beta-hemolysis after 24 hours on blood agar plates. Pathogenic strains of this species often form capsules. The following sections discuss the pathogenesis and epidemiology of this species, as well as the diagnosis, treatment, and prevention of the diseases it causes.

Pathogenicity

Strains of *Streptococcus pyogenes* have a number of structures, enzymes, and toxins that enable them to survive as pathogens in the body.

Two main structural features enable cells of *S. pyogenes* to evade phagocytosis:

- Protein M. Virulent group A streptococci have a membrane protein—called *protein M*—that destabilizes complement, thereby interfering with opsonization and lysis.

- Hyaluronic acid capsule. Because hyaluronic acid is normally found in the body, white blood cells may ignore bacteria "camouflaged" by this type of capsule.

Researchers have identified two *streptokinases* that break down blood clots, presumably enabling group A *streptococci* to rapidly spread through infected and damaged tissues. Similarly, four distinct *deoxyribonucleases* depolymerize DNA that has been released from dead cells in abscesses, reducing the firmness of the pus surrounding the bacteria and facili-

tating bacterial spread. *C5a peptidase* breaks down the complement protein C5a, which, as we discussed in Chapter 15, acts as a chemotactic factor. Thus, *S. pyogenes* decreases the movement of white blood cells into a site of infection. Finally, hyaluronidase facilitates the spread of streptococci through tissues by breaking down hyaluronic acid.

Group A streptococci secrete three distinct *pyrogenic*[7] (pī-rō-jen'ik) *toxins* that stimulate macrophages and helper T lymphocytes to release cytokines that in turn stimulate fever, a widespread rash, and shock. Because these toxins cause blood capillaries near the surface to dilate, producing a red rash, some scientists call them *erythrogenic*[8] (e-rith-rō-jen'ik) *toxins*. As the genes for these toxins are carried on temperate bacteriophages, only lysogenized bacteria secrete the toxins. Fever-stimulating *pyrogenic* toxins should not be confused with pus-producing *pyogenic* toxins.

S. pyogenes also produces two different, membrane-bound proteins, called *streptolysins,* which lyse red blood cells, white blood cells, and platelets; thus, these proteins interfere with the oxygen-carrying capacity of the blood, immunity, and blood clotting. After they have been phagocytized, group A streptococci release streptolysins into the cytoplasm of the phagocyte, causing lysosomes to release their contents, which lyses the phagocyte and releases the bacteria.

Epidemiology

Group A streptococci frequently infect the pharynx or skin, but the resulting abscesses are usually temporary, lasting only until specific immune responses against bacterial antigens (particularly M protein and streptolysins) clear the pathogens. Typically, group A streptococci cause disease only when normal competing microbiota are depleted; when a large inoculum enables the streptococci to gain a rapid foothold before antibodies are formed against them; or

---

[7]From Greek *pyr*, meaning fire, and *genein*, meaning to produce.
[8]From Greek *erythros*, meaning red, and *genein*, meaning to produce.

Highlight 19.1   An Outbreak of Group A Streptococci in the Marine Corps

In December 2002, over 100 recruits at a Marine Corps training depot in San Diego fell ill with pneumonia-like symptoms. Tests confirmed that an outbreak of group A streptococcal infection was responsible for many of the illnesses. This diagnosis, coupled with the sudden, unexplained deaths of two recruits in the weeks preceding the confirmed outbreak, caused alarm to spread

throughout the base, and officials decided to administer penicillin shots to more than 5000 recruits as a precautionary measure. On the same day the shots were administered, an 18-year-old private died within 4 hours after seeing a doctor for a mysterious rash that had appeared on his ankle. For days, the Marine private's case prompted fears that he had contracted the dreaded "flesh-eating" variety of streptococcus. However, an autopsy later revealed that he had actually died of a severe meningococcal bacteria infection. (Meningococcus, a cause of fatal meningitis, is discussed in Chapter 20.) Streptococcus was also ultimately ruled out as the cause of death of the other two recruits, but the outbreak was no less frightening for all involved.

Source: *MMWR* 52(6):100–109. 2003.

---

when specific immunity is impaired. Following colonization of the skin or a mucous membrane, *S. pyogenes* can invade deeper tissues and organs through a break in such barriers. People spread *S. pyogenes* among themselves via respiratory droplets, especially under crowded conditions such as those in classrooms and day care centers.

Before the discovery of antimicrobials, a major cause of human disease was group A streptococcal infection, which claimed the lives of millions. Because it is sensitive to antimicrobial drugs, its significance as a pathogen has declined, but group A streptococci still sicken thousands of Americans annually. (**Highlight 19.1** focuses on an outbreak of group A streptococcus among Marine Corps recruits.)

Group A Streptococcal Diseases

Group A streptococci cause a number of diseases, depending on the site of infection, the strain of bacteria, and the immune responses of the patient.

Pharyngitis   A sore throat caused by streptococci, commonly known as "strep throat," is a kind of **pharyngitis** (far-in-jī′tis)—inflammation of the pharynx (**Figure 19.6**)—that is accompanied by fever, malaise,[9] and headache. The back of the pharynx typically appears red, with swollen lymph nodes and *purulent* (pus-containing) abscesses covering the tonsils. Some microbiologists estimate that only 50% of patients diagnosed with strep throat actually have it; the rest have viral pharyngitis. Given that the symptoms and signs for the two diseases are identical, a sure diagnosis requires bacteriological or serological tests. Correct diagnosis is essential because bacterial pharyngitis is treatable with antibacterial drugs, which of course have no effect on viral pharyngitis.

▲ *Figure 19.6*

The reddened appearance of a throat with pharyngitis.

Scarlet Fever   The disease known as **scarlet fever,** also known as *scarlatina,* often accompanies streptococcal pharyngitis when the infection involves a lysogenized strain of *S. pyogenes*. After 1–2 days of pharyngitis, pyrogenic toxins released by the streptococci trigger a diffuse rash that typically begins on the chest and spreads across the body. The tongue usually becomes strawberry red. The rash disappears after about a week and is followed by sloughing of the skin.

Pyoderma and Erysipelas   A **pyoderma** (pī-ō-der′mă) is any confined, pus-producing lesion that usually occurs on the exposed skin of the face, arms, or legs. One cause is group A streptococcal infection following direct contact with an infected person or a contaminated fomite. This condition is also known as *impetigo* because of its similar appearance

---

[9]French, meaning discomfort.

▲ *Figure 19.7*

The localized, pus-filled lesions of erysipelas, a condition caused by group A streptococcus *(Streptococcus pyogenes.). What is the meaning of the word* pyogenes?

Figure 19.7 Pyogenes means pus-producing.

to the staphylococcal disease of the same name. After a pus-filled lesion breaks open, it forms a yellowish crust. This stage is highly contagious, and scratching may spread the bacteria to the surrounding skin, spreading the lesions.

When a streptococcal infection also involves surrounding lymph nodes and triggers pain and inflammation, the condition is called **erysipelas**[10] (er-i-sip'ĕ-las) **(Figure 19.7).** Erysipelas occurs most commonly on the faces of children.

Streptococcal Toxic Shock Syndrome   Group A streptococci can spread, albeit rarely, from an initial site of infection, particularly in patients infected with HIV or suffering with cancer, heart disease, pulmonary disease, or diabetes mellitus. Such spread leads to bacteremia and severe multisystem infections producing **streptococcal toxic shock syndrome.** Patients experience inflammation at sites of infection, as well as pain, fever, chills, malaise, nausea, vomiting, and diarrhea. These signs and symptoms are followed by increased pain, organ failure, shock—and, in over 40% of patients, death.

Necrotizing Fasciitis   Another serious disease caused by *S. pyogenes* is **necrotizing**[11] **fasciitis** (ne'kro-tī-zing fas-ē-ī'tis), sensationalized by the news media as "flesh-eating strep." In this disease, streptococci enter the body through

breaks in the skin, secrete enzymes and toxins that destroy tissues, and eventually destroy muscle and fat tissue **(Figure 19.8).** The bacteria spread deep within the body along the *fascia,* which are fibrous sheets of connective tissue surrounding muscles and binding them to one another—hence the word *fasciitis* in the disease's name. Necrotizing fasciitis also involves *toxemia* (toxins in the blood), failure of many organs, and death in more than 50% of cases.

Rheumatic Fever   A complication of untreated group A streptococcal pharyngitis is **rheumatic** (rū-mat'ik) **fever,** in which inflammation leads to damage of heart valves and muscle. Though the exact cause of the damage is unknown, it appears that rheumatic fever is not caused directly by *Streptococcus,* but instead is an autoimmune response in which antibodies directed against streptococcal antigens cross-react with heart antigens. Damage to the heart valves may be so extensive that they must be replaced when the patient reaches middle age. Rheumatic fever was much more prevalent before the advent of antimicrobial drugs.

Glomerulonephritis   For an undetermined reason, antibodies bound to the antigens of some strains of group A streptococcus are not removed from circulation, but instead accumulate in the *glomeruli* (small blood vessels) of the kidneys' *nephrons* (filtering units). The result is **glomerulonephritis** (glō-mār'ū-lō-nef-rī'tis)—inflammation of the glomeruli and nephrons, which obstructs blood flow through the kidneys and leads to hypertension (high blood pressure) and low urine output. Blood and proteins are often secreted in the urine. Young patients usually recover fully from glomerulonephritis, but progressive and irreversible kidney damage may occur in adults.

Diagnosis, Treatment, and Prevention

Because *Streptococcus* is not a normal member of the microbiota of the skin, the observation of Gram-positive bacteria in short chains or pairs in cutaneous specimens can provide a rapid preliminary diagnosis of pyoderma, erysipelas, and necrotizing fasciitis. In contrast, streptococci are normally in the pharynx, so their presence in a respiratory sample is of little diagnostic value; instead, physicians must use immunological tests that identify the presence of group A streptococcal antigens.

Penicillin is very effective against *S. pyogenes.* Erythromycin or cephalosporin is used to treat penicillin-sensitive patients. *S. pyogenes* is also susceptible to the topical antimicrobial bacitracin—a characteristic that distinguishes it from group B streptococcus (discussed shortly). Necrotizing fasciitis must be treated with aggressive surgical removal of nonviable tissue and infecting bacteria. Because rheumatic fever and glomerulonephritis are the result of an immune response against group A streptococci, they cannot

---

[10]From Greek *erythros,* meaning red, and *pella,* meaning skin.
[11]From Greek *nekros,* meaning corpse.

▲ *Figure 19.8*

Necrotizing fasciitis produced by so-called "flesh-eating" group A streptococci.

be treated directly; instead, the underlying infection must be arrested.

Antibodies against M protein provide long-term protection against *S. pyogenes*. However, antibodies directed against the M protein of one strain provide no protection against other strains; this explains why a person can have strep throat more than once.

Group B Streptococcus: *Streptococcus agalactiae*

Learning Objectives

✓ Contrast group B streptococcus with group A streptococcus in terms of structure.

✓ Discuss the epidemiology, diagnosis, treatment, and prevention of infections with *streptococcus agalactiae*.

Group B streptococcus, or *S. agalactiae* (a-ga-lak′tē-ē), is a Gram-positive coccus, 0.6–1.2 μm in diameter, that divides to form chains. Like group A streptococcus, *S. agalactiae* is beta-hemolytic, but it can be distinguished from the former by three qualities: It has group-specific, polysaccharide cell wall antigens; it forms buttery colonies that are 2–3 mm in diameter and have a small zone of beta-hemolysis after 24 hours of growth on blood agar; and it is bacitracin resistant.

Pathogenicity

Even though *S. agalactiae* forms capsules, antibodies target its capsular antigens, so the capsules are not protective. For this reason *S. agalactiae* has a predilection for newborns who have not yet formed type-specific antibodies and whose mothers are uninfected (and so do not provide passive immunity across the placenta or in milk).

Group B streptococci produce enzymes—*proteases* (that catabolize protein), *hemolysins* (that lyse red blood cells), deoxyribonuclease, and hyaluronidase—that probably play a role in causing disease, but such a role has not been proven.

Epidemiology

Group B streptococci normally colonize the lower gastrointestinal (GI), genital, and urinary tracts. Diseases in adults primarily follow wound infections and childbirth.

Sixty percent of newborns are inoculated with group B streptococcal strains either during passage through the birth canal or by health care personnel. Such infections do not cause disease when maternal antibodies have crossed the placenta; but mortality rates can exceed 50% in children of uninfected mothers. Infections in newborns less than 1 week old cause *early-onset disease*; infections occurring in infants 1 week to 3 months of age cause *late-onset disease*.

Diseases

Even though microbiologists initially described *Streptococcus agalactiae* as the cause of *puerperal*[12] *(childbirth) fever* in women, today the bacterium is most often associated with neonatal bacteremia, meningitis, and pneumonia, at least one of which occurs in approximately 3 of every 1000 newborns. Mortality has been reduced to about 5% as a result of rapid diagnosis and supportive care, but about 25% of infants surviving group B streptococcal meningitis have permanent neurological damage, including blindness, deafness, or severe mental retardation. Immunocompromised older patients are also at risk from group B streptococcal infections, and about 25% of them die from streptococcal diseases.

Diagnosis, Treatment, and Prevention

Medical laboratory technologists identify group B streptococcal infections by means of agglutination and ELISA tests utilizing antibodies directed against the bacteria's distinctive cell wall polysaccharides. Samples of clinical specimens can also be incubated in blood media containing the antimicrobial drug bacitracin, which inhibits the growth of other beta-hemolytic bacteria.

Penicillin G is the drug of choice for use against group B streptococcus, though some strains tolerate concentrations of the drug more than 10 times greater than that needed to inhibit group A streptococcus. For this reason, physicians

---

[12]From Latin *puer,* meaning child, and *pario,* meaning to bring forth.

often prescribe an aminoglycoside such as streptomycin in addition to penicillin.

The CDC recommends prophylactic administration of penicillin at birth to children whose mothers' urinary tracts are colonized with group B streptococci. Implementation of this guideline in 1996 reduced early-onset disease morbidity and mortality by 70% as of 2001. Additionally, physicians can immunize women against group B streptococci, preventing infection of future children.

Other Beta-Hemolytic Streptococci

Learning Objective

✓ Contrast *Streptococcus equisimilis* and *S. anginosus* in terms of polysaccharide composition, diseases caused, and treatment.

Streptococcus equisimilis and *S. anginosus* are the only other pathogenic beta-hemolytic streptococci. Although members of both species typically have group C polysaccharides, some strains of *S. anginosus* have group F or group G antigens instead—an example of the confusing status of the classification of the streptococci.

S. equisimilis causes pharyngitis (and occasionally glomerulonephritis) but, unlike group A streptococci, these cases of pharyngitis never lead to rheumatic fever. *S. anginosus* produces purulent abscesses. Penicillin is effective against both species.

Alpha-Hemolytic Streptococci: The Viridans Group

Learning Objective

✓ Identify the normal sites of viridans streptococci in the human body, and list three serious diseases they cause.

Many alpha-hemolytic streptococci lack group-specific carbohydrates, and thus they are not part of any Lancefield group. Instead, microbiologists classify them as the **viridans**[13] **streptococci** (vir'i-danz strep'tō-kok'sī) because many of them produce a green pigment when grown on blood media. The taxonomic relationships of these microorganisms is poorly understood; European and American microbiologists do not always agree about the names assigned to species, and some microbiologists place these microbes in a separate genus: *Abiotrophia*. Some of the names given by Americans to viridans streptococci are *S. mitis,* and *S. sanguis.* Members of the group are alpha-hemolytic and susceptible to penicillin.

Viridans streptococci normally inhabit the mouth, pharynx, GI tract, genital tract, and urinary tract of humans. They are opportunists that produce purulent abdominal infections, and are one cause of dental **caries**[14] (kār'ēz) or cavities. They stick to dental surfaces via an insoluble polysaccharide, *dextran*, which they produce from glucose. Large quantities of dextran allow viridans streptococci and other bacteria to colonize the enamel of teeth, forming

LM   10 μm

▲ *Figure 19.9*

Streptococcus pneumoniae is typically arranged in pairs, as shown above. *What is pneumonia?*

Figure 19.9 Pneumonia is inflammation of the lungs.

biofilms known as *dental plaque* (see Figure 14.6). Viridans streptococci are not highly invasive, entering the blood only through surgical wounds and lacerations of the gums, including undetectable cuts produced by chewing hard candy, brushing the teeth, and dental procedures. Once in the blood they can cause meningitis and endocarditis.

Streptococcus pneumoniae

Learning Objectives

✓ Describe how the structure of *Streptococcus pneumoniae* affects its pathogenicity.

✓ Describe the route of *Streptococcus pneumoniae* through the body, describing the chemical and physical properties that allow it to cause pneumonia.

✓ Discuss the diagnosis, treatment, and prevention of pneumococcal diseases.

Louis Pasteur discovered *Streptococcus pneumoniae* (strep-tō-kok'ŭs nū-mō'nē-ī) in pneumonia patients about 120 years ago. The bacterium is a Gram-positive coccus, 0.5–1.2 μm in diameter, that forms short chains or, more commonly, pairs **(Figure 19.9)**. In fact, it was once classified in its own genus, "Diplococcus." Ninety two different strains of *S. pneumoniae,* collectively called *pneumococci,* are known to infect humans.

---

[13]From Latin *viridis,* meaning green.
[14]Latin, meaning dry rot.

Colonies of *S. pneumoniae* grown for 24 hours are 1–3 mm in diameter, round, mucoid, unpigmented, and dimpled in the middle because of the death of older cells. Colonies are alpha-hemolytic on blood agar when grown aerobically, and beta-hemolytic when grown anaerobically. This bacterium lacks Lancefield antigens but does incorporate a species-specific teichoic acid into its cell wall.

Pathogenicity

Streptococcus pneumoniae is a normal member of the pharyngeal microbiota that can colonize the lungs, sinuses, and middle ear. Even though microbiologists have studied the pneumococci extensively—the entire genomes of more than 10 strains have been sequenced—they still do not fully understand their pathogenicity; nevertheless, certain structural and chemical properties are known to be required.

The cells of virulent strains of *S. pneumoniae* are surrounded by a polysaccharide capsule, which protects them from digestion after phagocytosis. The capsule is required for virulence; unencapsulated strains are avirulent. Microbiologists distinguish 90 unique serotypes based on differences in the antigenic properties of the capsules among various strains.

In addition, *S. pneumoniae* insert into their cell walls a chemical called *phosphorylcholine,* which by binding to receptors on cells in the lungs, in the meninges, and blood vessel walls, stimulates the cells to phagocytize the bacteria. Thus together, the polysaccharide capsule and phosphorylcholine enable pneumococci to "hide" inside body cells. *S. pneumoniae* can then pass across these cells into the blood and brain.

Pathogenic pneumococci secrete *protein adhesin,* a poorly defined protein that mediates binding of the cells to epithelial cells of the pharynx. From there the bacteria enter the lungs.

The body acts to limit migration of bacteria into the lungs by binding the microbes with the active sites of secretory IgA. The rest of the antibody molecule then binds to mucus, enabling mucus-enveloped bacteria to be swept from the airways by the action of ciliated epithelium. The bacteria counteract this defense by secreting *secretory IgA protease,* which destroys IgA, and *pneumolysin,* which binds to cholesterol in the cytoplasmic membranes of ciliated epithelial cells, producing transmembrane pores that result in the lysis of the cells. Pneumolysin also suppresses the digestion of phagocytized bacteria by interfering with the action of lysosomes.

Epidemiology

Streptococcus pneumoniae grows in the mouths and pharynges of 75% of humans, without causing harm; however, when pneumococci travel to the lungs, they cause disease. Typically, the incidence of pneumococcal disease is highest in children and the elderly, groups whose immune responses are not fully active.

Pneumococcal Diseases

Streptococcus pneumoniae causes a variety of diseases, which we explore next.

Pneumococcal Pneumonia   The most prevalent disease caused by *S. pneumoniae* is **pneumococcal pneumonia** (nū-mō′nē-ă), which constitutes about 85% of all cases of pneumonia. The disease results when pneumococci are inhaled from the pharynx into lungs damaged either by a previous viral disease, such as influenza or measles, or by other conditions, such as alcoholism, congestive heart failure, or diabetes mellitus. As the bacteria multiply in the alveoli (air sacs) of the lower lobes of the lungs, they damage the lining of the alveoli, allowing red blood cells and leukocytes to enter the lungs. The leukocytes attack the bacteria, in the process secreting inflammatory and pyrogenic chemicals. The onset of clinical symptoms is abrupt and includes a fever of 39–41°C and severe shaking chills. Most patients have a productive cough, slightly bloody sputum, and chest pain.

Sinusitis and Otitis Media   Following viral infections of the upper respiratory tract, *S. pneumoniae* can also invade the sinuses and middle ear, where it causes **sinusitis** (sī-nŭ-sī′tis; inflammation of the nasal sinuses) and **otitis media** (ō-tī′tis mē′dē-ă; inflammation of the middle ear). Pus production and inflammation in these cavities create pressure and pain. Whereas sinusitis occurs in patients of all ages, otitis media is more prevalent in children because in younger individuals the narrow auditory tubes connecting the pharynx with the middle ears are nearly horizontal, which facilitates the flow of infected fluid from the pharynx into the middle ears. The tubes become wider and more vertical when the shape of the head changes as children grow into adults.

Bacteremia and Endocarditis   *Streptococcus pneumoniae* can enter the blood either through lacerations (as might result from vigorous tooth brushing, chewing hard foods, or dental procedures) or as a result of tissue damage in the lungs during pneumonia. Bacteria generally do not enter the blood during sinusitis or otitis media. As with *Staphylococcus, S. pneumoniae* can colonize the lining of the heart, causing endocarditis. The heart valves, once involved, are typically destroyed.

Pneumococcal Meningitis   Pneumococci can spread to the meninges via bacteremia, during sinusitis or otitis media, or following head or neck surgery or trauma that opens a passage between the subarachnoid space of the meninges and the pharynx. The mortality rate of pneumococcal meningitis, which is primarily a disease of children, is up to 20 times that of meningitis caused by other microorganisms.

Diagnosis, Treatment, and Prevention

Medical laboratory technologists can quickly identify pneumococci in Gram stains of sputum smears (see Figure 19.9), and confirm their presence with the *Quellung*[15] *reaction* in which anti-capsular antibodies cause the capsule to swell. Antibodies against particular strains trigger Quellung reactions only against those strains. Antibody agglutination tests are also used to identify specific strains. Culture of sputum samples is often difficult because pneumococci have fastidious nutritional requirements and cultures are often overgrown by normal oral microbiota. Because the bacteria are sensitive to most antimicrobial drugs, samples for culture must be obtained before antibacterial therapy has begun. Laboratory technologists can distinguish pneumococcal colonies from other colonies of alpha-hemolytic strains by adding a drop of bile to a colony. Bile triggers chemicals present in pneumococci to lyse the cells, dissolving the colony in just a few minutes.

Penicillin has long been the drug of choice against *S. pneumoniae*, though penicillin-resistant strains have emerged in the last decade; about a third of peumococcal isolates are now resistant. Cephalosporin, erythromycin, and chloramphenicol are also effective treatments.

Prevention of pneumococcal diseases is focused on a vaccine made from purified capsular material from the 23 most common pathogenic strains. The vaccine is immunogenic and long lasting in normal adults, but unfortunately it is not as efficacious in patients at greatest risk, such as the elderly, young children, and AIDS patients.

Enterococcus

Learning Objective

✓ Identify two species of *Enterococcus*, and describe their pathogenicity and the diagnosis, treatment, and prevention of their diseases.

We have discussed two Gram-positive cocci that are pathogenic in humans: catalase-positive *Staphylococcus* and catalase-negative *Streptococcus*. Now we turn our attention to another genus of Gram-positive, catalase-negative cocci—*Enterococcus* (en'ter-ō-kok'ŭs), so named because all enterococci are spherical and live in the intestinal tracts of animals. Lancefield classified enterococci as group D streptococci, but they differ significantly from other members of group D in that *Enterococcus* is unencapsulated, produces gas during fermentation of sugars, and is typically nonhemolytic (*i.e.* gamma-hemolytic). Because of these and other differences, microbiologists now classify *Enterococcus* as a separate genus.

Structure and Physiology

Enterococci form short chains and pairs; they do not form capsules **(Figure 19.10)**. Enterococci grow at temperatures up to 45°C, at pH as high as 9.6, and in 6.5% NaCl or 40% bile salt broths—all conditions that severely inhibit the growth of *Streptococcus*. The two species that are signifi-

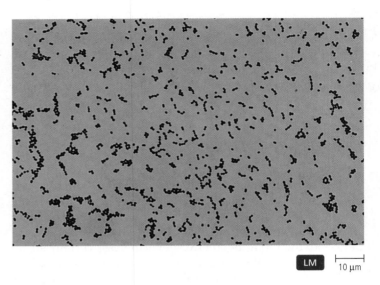

LM   10 μm

▲ *Figure 19.10*

A Gram stain of *Enterococcus faecalis*, which (like streptococci) are arranged in pairs and short chains, but (unlike *S. pneumoniae*) lack capsules. *Why are these bacteria called "enterococci"?*

Figure 19.10 They are called "enterococci" because they are a spherical bacteria that live in the intestinal tracts of animals.

cant pathogens of humans are *E. faecalis*[16] (fē-kā'lis) and *E. faecium*[17] (fē'sē-ŭm).

Pathogenesis, Epidemiology, and Diseases

E. faecalis is ubiquitous in the human colon. *E. faecium* is found less often. Because both species lack structural and chemical elements that make them virulent, they are rarely pathogenic in the intestinal tract. Nevertheless, they do have the ability to adhere to human epithelial cells, and they secrete *bacteriocins*, which are chemicals that inhibit the growth of other bacteria.

E. faecalis and *E. faecium* can cause serious disease if they are introduced into other parts of the body, such as the urinary tract and bloodstream, via poor personal hygiene or intestinal laceration, respectively. Enterococci account for about 10% of nosocomial infections and cause bacteremia, endocarditis, and wound infections.

Diagnosis, Treatment, and Prevention

Even though a Gram-stain of *Enterococcus* looks like one of *S. pneumoniae*, the two genera can be distinguished readily in that *Enterococcus* is not sensitive to bile.

Enterococcal infections are very difficult to treat because strains resistant to frequently used antimicrobials—lactams,

---

[15]German, meaning swelling.
[16]Latin, meaning pertaining to feces.
[17]Latin, meaning of feces.

aminoglycosides, and vancomycin—are relatively common. This is particularly troublesome in that the genes for antimicrobial resistance occur on plasmids that can be transmitted to other bacteria.

It is difficult to prevent enterococcal infections, particularly in health care settings, where patients' immune systems are taxed and weakened. Health care workers should always use good hygiene and aseptic techniques to minimize the transmission of these microorganisms.

Table 19.2 on page 546 summarizes the features of the major pathogenic, Gram-positive streptococci—*Streptococcus* and *Enterococcus*. Now we turn our attention to the Gram-positive pathogenic bacilli, beginning with the low G + C, endospore-forming genera *Bacillus* and *Clostridium*.

Bacillus

Scientists divide Gram-positive bacilli (rod-shaped cells) into endospore-forming and nonendospore-forming genera. The endospore-forming (sometimes shortened to merely "spore-forming") genera are *Bacillus* (ba-sil'ŭs) and *Clostridium* (klos-trid'ē-ŭm). In this section we examine *Bacillus anthracis*—the only one of the 51 species of *Bacillus* that is a strict pathogen of animals and humans. The following section discusses *Clostridium*.

Structure, Physiology, and Pathogenicity

Learning Objective

✓ Identify the structural features of *Bacillus* that contribute to its pathogenicity.

Bacillus anthracis[18] (ba-sil'ŭs an-thrā'sis) is a large (1 μm × 3–5 μm) rod-shaped, facultatively anaerobic bacterium that normally dwells in soil. Its cells are arranged singly, in pairs, or in chains **(Figure 19.11)**. In laboratory culture, prominent endospores form in the middle of the cells (see Figure 4.22), which lack capsules. In contrast, bacteria in clinical specimens typically lack endospores and possess capsules. As we discussed in Chapter 11, the tough external coat and the internal chemicals of endospores make these structures resistant to harsh environmental conditions, enabling *Bacillus* to survive in the environment for centuries or perhaps even longer. *Bacillus* can survive in the body because it has a capsule, which is composed solely of glutamic acid. The capsule prevents effective phagocytosis by white blood cells.

Pathogenic strains of *B. anthracis* cause disease because they secrete *anthrax toxin,* composed of three distinct polypeptide components that in combination are lethal. Scientists are investigating the precise mechanisms by which the three components of anthrax toxin work in hopes of developing techniques for neutralizing them. Anthrax can be

SEM |——| 3 μm

▲ *Figure 19.11*
Bacillus anthracis growing in tissue.

deadly even after treatment, because antimicrobial drugs do not inactivate accumulated anthrax toxin.

Epidemiology

Learning Objective

✓ List three methods of transmission of anthrax.

Anthrax is primarily a disease of herbivores; humans contract the disease from infected animals. Anthrax is not normally transmitted from individual to individual, but can invade via one of three routes: inhalation of spores, inoculation of spores into the body through a break in the skin, or ingestion of spores. Ingestion anthrax is the normal means of transmission among animals but is rare in humans. In the 25 years between January 1976 and September 2001, only 15 cases of anthrax were reported in the United States; thus epidemiologists suspected bioterrorism when over a dozen cases of anthrax were reported in New York, Florida, and Washington, D.C. in the fall of 2001. (See New Frontiers 1.1 on page 22, as well as **Case Study 19.2** on page 547.)

Disease

Learning Objectives

✓ List and describe three clinical manifestations of *Bacillus anthracis* infections.
✓ Identify the diagnosis, treatment, and prevention of anthrax.

---

[18]Greek, meaning charcoal.

Table 19.2   Characteristics of Pathogenic Streptococci

Lancefield Group	Scientific Name	Hemolytic Pattern	Significant Characteristics	Characteristic Diseases
A	S. pyogenes	Large zone of beta-hemolysis	1- to 2-mm white colonies on blood agar; bacitracin sensitive	Pharyngitis, scarlet fever, pyoderma, erysipelas, streptococcal toxic shock syndrome, necrotizing fasciitis, rheumatic fever, glomerulonephritis
B	S. agalactiae	Small zone of beta-hemolysis	2- to 3-mm buttery colonies on blood agar; bacitracin resistant	Puerperal fever, neonatal bacteremia, meningitis, pneumonia
C	S. equisimilis	Large zone of beta-hemolysis	1- to 2-mm white colonies on blood agar	Pharyngitis, glomerulonephritis
C, F, or G	S. anginosus	Small zone of beta-hemolysis	1- to 2-mm white colonies on blood agar	Purulent abscess
-	S. mutans	Alpha-hemolysis	Viridans group (produce green pigment when grown on blood agar)	Dental caries; rarely bacteremia, meningitis, endocarditis
-	S. pneumoniae	Alpha-hemolysis (aerobic); beta-hemolytic (anaerobic)	Diplococci; capsule required for pathogenicity	Pneumonia, sinusitis, otitis media, bacteremia, endocarditis, meningitis
D	Enterococcus faecalis, E. faecium	None (gamma-hemolysis)	Diplococci; no capsule	Urinary tract infections, bacteremia, endocarditis, wound infections

Bacillus anthracis causes only one disease—anthrax—but it can have three clinical manifestations. The first, *gastrointestinal anthrax,* is very rare in humans but is common in animals; it results in intestinal hemorrhaging and eventually death.

Cutaneous anthrax begins when a painless, solid, raised nodule forms on the skin at the site of infection. The cells in the affected area die, and the nodule spreads to form a painless, swollen, black, crusty ulcer called an *eschar* (**Figure 19.12**). It is from the black color of eschars that anthrax, which means "charcoal" in Greek, gets its name. *B. anthracis* growing in the eschar releases anthrax toxin into the blood, producing toxemia. Untreated cutaneous anthrax is fatal for 20% of patients.

Inhalation anthrax is also rare in humans, as it requires inhalation of airborne endospores. After the spores germinate in the lungs, they secrete toxins that are absorbed into the bloodstream, producing toxemia. Early signs and symptoms include fatigue, malaise, fever, aches, and cough—all of which are common to many pulmonary diseases (see Table 25.1 on page 705). In a later phase, victims of inhalation anthrax have a high fever and labored breathing due to localized swelling, and they go into shock. Historically, mortality rates have been high—approaching 100% even when treated—perhaps in part because the disease was not suspected until it was irreversible, and because antimicrobial drugs do not neutralize toxins released during the course of

the disease. During the bioterrorism attack of 2001, physicians learned that early and aggressive treatment of inhalation anthrax with antimicrobial drugs accompanied by persistent drainage of fluid from around the lungs increased the survival rate to greater than 50%.

Diagnosis, Treatment, and Prevention

Large, nonmotile, Gram-positive bacilli in clinical samples from the lungs or skin are diagnostic. As mentioned previously, endospores are not typically seen in clinical samples but are produced after a few days in culture.

Penicillin, erythromycin, chloramphenicol, and many other antimicrobial agents are effective against *B. anthracis.* Ciproflaxacin (Cipro®) gained fame in 2001 as physicians were encouraged to use it in the fear that the bioterrorists had genetically altered the microorganism to be penicillin resistant—a fear that proved unfounded.

Prevention of naturally occurring disease in humans requires control of the disease in animals. Farmers in areas where anthrax is endemic must vaccinate their stock and bury or burn the carcasses of dead animals. Anthrax vaccine has proven effective and safe for humans, but requires six doses over 18 months plus annual boosters. The National Academy of Sciences has called for the development of an alternative vaccine.

Case Study 19.2   Anthrax in the Mail: The Case of Tom Brokaw's Assistant

One of the first confirmed cases of anthrax resulting from the bioterrorist attacks in the fall of 2001 was that of 38-year-old Erin O'Connor, an assistant to NBC "Nightly News" anchor Tom Brokaw. Between September 18 and September 25, O'Connor handled two letters addressed to Brokaw that contained powdery or sandlike substances—one postmarked St. Petersburg, Florida, and the other postmarked Trenton, New Jersey. On September 25, O'Connor noticed a rash beneath her left collarbone. Over the following week, the rash developed into an oval-shaped lesion, approximately 5 cm long, that eventually resembled the black eschar characteristic of cutaneous anthrax (see Figure 19.12). Subsequent tests confirmed the presence of

Bacillus anthracis in samples of O'Connor's tissues. A link to one of the letters to Brokaw was established when the Trenton-postmarked letter tested positive for traces of anthrax. (The powder in the Florida letter tested negative.) O'Connor was prescribed ciprofloxacin and recovered fully.

1. What are two major virulence factors of *B. anthracis*?
2. What feature of *B. anthracis* allowed it to survive?
3. In this case, anthrax was acquired via inoculation of spores into a break in the skin. What other routes of infection are possible?

▲ **Figure 19.12**

A black eschar characteristic of cutaneous anthrax, produced when *Bacillus anthracis* in the lesion release anthrax toxin.

Clostridium

Learning Objective

✓ Characterize the four major species of *Clostridium*.

Clostridium is an anaerobic, Gram-positive, endospore-forming bacillus that is ubiquitous in soil, water, sewage, and the gastrointestinal tracts of animals and humans. Several species are significant human pathogens. Pathogenicity is due in great part to the ability of endospores to survive harsh conditions, and to the secretion by vegetative cells of potent *histolytic toxins, enterotoxins* (toxins affecting the GI tract), and *neurotoxins*. The most common pathogenic clostridia are *C. perfringens* (per-frin'jens), *C. difficile* (di-fi'sēl), *C. botulinum* (bo-tū-lī'nŭm), and *C. tetani* (tet'ă-nē).

Clostridium perfringens

Learning Objectives

✓ Identify the mechanisms accounting for the pathogenesis of *Clostridium perfringens* infections.

✓ Describe the diagnosis, treatment, and prevention of *Clostridium perfringens* infections.

Clostridium perfringens, the clostridium most frequently isolated from clinical specimens, is a large, almost rectangular, Gram-positive bacillus. Although it is nonmotile, its rapid growth enables it to spread across the surface of laboratory media, resembling the growth of motile bacteria. Spores are rarely observed either in clinical samples or in culture. Type A *C. perfringens,* known by its specific antigens, is the most virulent serotype.

Pathogenesis, Epidemiology, and Disease

C. perfringens produces 11 toxins that lyse erythrocytes and leukocytes, increase vascular permeability, reduce blood pressure, and kill cells, resulting in irreversible damage. Because *C. perfringens* commonly grows in the digestive tracts of animals and humans, it is nearly ubiquitous in fecally contaminated soil and water.

The severity of diseases caused by *C. perfringens* ranges from mild food poisoning to life-threatening illness. Clostridial food poisoning is a relatively benign disease characterized by abdominal cramps and watery diarrhea, but not fever, nausea, or vomiting. It lasts for less than 24 hours. Clostridial food poisoning typically results from the ingestion of large numbers (10^8 or more) of type A *C. perfringens* in contaminated meat.

C. perfringens is not invasive, but when some traumatic event (such as a surgical incision, a puncture, a gunshot wound, crushing trauma, or a compound fracture) introduces endospores into the body, they can germinate in the anaerobic environment of deep tissues. The immediate result is intense pain at the initial site of infection as clostridial toxins induce swelling and tissue death. The rapidly growing bacteria can then spread into the surrounding tissue,

▲ **Figure 19.13**

Gas gangrene, a life-threatening disease caused by *Clostridium perfringens*. The blackening is the result of the death of muscle tissue (myonecrosis), whereas the "bubbling" appearance results from the production of gaseous waste products by the bacteria.

causing myonecrosis (the death of muscle and connective tissue) that is typically accompanied by the production of abundant, foul-smelling gaseous bacterial waste products—hence the common name for the disease: **gas gangrene**[19] **(Figure 19.13).** Shock, kidney failure, and death can follow, often within a week of infection.

Diagnosis, Treatment, and Prevention

Medical laboratory technologists show that *Clostridium* is involved in food poisoning by demonstrating more than 10^5 bacteria in a gram of food or 10^6 cells per gram of feces. The appearance of gas gangrene is usually diagnostic by itself; though the detection of large Gram-positive bacilli is confirmatory.

Clostridial food poisoning is typically self-limiting—the pathogens and their toxins are eliminated in the resulting watery stool. In contrast, physicians must quickly and aggressively intervene to stop the spread of necrosis in gas gangrene by surgically removing dead tissue and administering large doses of penicillin. Despite all therapeutic care, mortality of gas gangrene exceeds 40%.

It is difficult to prevent infections of *C. perfringens* because the organism is so common; however, refrigeration of food prevents toxin formation and reduces the chance of clostridial food poisoning. Alternatively, reheating contaminated food destroys any toxin that has formed. Given that myonecrosis occurs when endospores are introduced deep in the tissues, proper cleaning of wounds can prevent many cases of the disease.

---

[19]From Greek *gangraina*, meaning an eating sore.

Clostridium difficile

Learning Objectives

✓ Discuss the role of antimicrobial drugs in the development of gastrointestinal diseases caused by *Clostridium difficile*.

✓ Discuss the diagnosis, treatment, and prevention of *C. difficile* infections.

Clostridium difficile is a motile, anaerobic intestinal bacterium with cells about 1.5 μm in width and 3–6.5 μm in length that form oval, subterminal endospores. The bacterium produces two toxins (called toxin A and toxin B) and the enzyme hyaluronidase.

Pathogenesis, Epidemiology, and Disease

Although so-called "*C. diff.*" is a common member of the intestinal microbiota, it can be an opportunistic pathogen in patients treated with broad-spectrum antimicrobial drugs, such as penicillin and cephalosporin. In such patients, the normal proportions of different bacteria in the colon can be significantly altered. In many cases the hardy endospores of *C. difficile* germinate, enabling it to become the predominant intestinal bacterium, such that the toxins and enzymes it produces cause hemorrhagic necrosis of the intestinal wall. In minor infections these lesions result in a self-limiting explosive diarrhea; however, in more serious cases, *C. difficile* produces life-threatening **pseudomembranous colitis,** in which large sections of the colon wall slough off, potentially perforating the colon, and leading to massive internal infection by fecal bacteria and eventual death.

Diagnosis, Treatment, and Prevention

Whereas diarrhea in patients undergoing antimicrobial therapy is suggestive of *C. difficile* infection, laboratory microbiologists confirm the diagnosis either by isolating the organism from feces using selective media, or by demonstrating the presence of the toxins via immunoassays.

Discontinuation of the implicated antimicrobial drug, which allows the microbiota to return to normal, usually resolves minor infections with *C. difficile*. More serious cases are treated with vancomycin or metronidazole, though endospores often survive such therapy in about a third of patients, causing a relapse. Treatment of relapses with the same antimicrobial drug is often successful.

C. difficile is frequently found in hospitals, and hospital personnel can easily transmit it between patients. Proper hygiene—particularly frequent hand antisepsis—is critical for limiting nosocomial infections.

Clostridium botulinum

Learning Objectives

✓ Contrast the three manifestations of botulism.

✓ Describe the use of mice in the diagnosis of botulism.

✓ Describe three treatments of botulism, and explain how to prevent it.

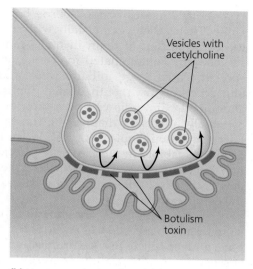

(a) Normal neuromuscular junction

(b) Neuromuscular junction with botulism toxin present

▲ *Figure 19.14*

How botulism toxin acts at a neuromuscular junction. **(a)** Normal function at a neuromuscular junction. ① A nerve impulse from the central nervous system causes vesicles filled with acetylcholine to fuse with the neuron's cytoplasmic membrane, ② releasing acetylcholine into the synaptic cleft. The binding of acetylcholine to receptors on the muscle cell's cytoplasmic membrane stimulates a series of events that result in contraction of the muscle cell (not shown). **(b)** Botulism toxin blocks the fusion of the vesicles with the neuron's cytoplasmic membrane, thereby preventing secretion of the neurotransmitter into the synaptic cleft; as a result, the muscle cell does not contract.

Clostridium botulinum[20] is an anaerobic, endospore-forming, Gram-positive bacillus that is common in soil and water worldwide. Its endospores survive improper canning of food, germinating to produce vegetative cells that grow and release into the jar or can a powerful neurotoxin that causes **botulism** (bot'ū-lizm).

Pathogenesis

Different strains of *C. botulinum* produce one of seven antigenically distinct **botulism toxins** (A through G). Scientists consider botulism toxins the deadliest toxins known. Thirty grams of pure toxin would be enough to kill every person in the United States. Each of the seven toxins is a quaternary protein composed of a single neurologically active polypeptide associated with one or more nontoxic polypeptides that prevent the inactivation of the toxin by stomach acid. To understand the action of botulism toxins, we must consider the way the nervous system controls muscle contractions.

Each of the many ends of a motor neuron (nerve cell) forms an intimate connection with a muscle cell at a *neuromuscular junction*; however, the two cells do not actually touch—a small gap, called a *synaptic cleft*, remains between them **(Figure 19.14a)**. The neuron stores a chemical, called *acetylcholine*, in vesicles near its terminal cytoplasmic membrane. Acetylcholine is one of a family of chemicals called neurotransmitters that mediate communication among neurons and between neurons and muscle cells.

When a signal arrives at the terminus of a motor neuron, the vesicles containing acetylcholine fuse with the neuron's cytoplasmic membrane, releasing acetylcholine into the synaptic cleft. Molecules of acetylcholine then diffuse across the cleft and bind to receptors on the cytoplasmic membrane of the muscle cell. The binding of acetylcholine to its receptor then triggers a series of events inside the muscle cell that results in muscle contraction (not shown).

Botulism toxins act by binding irreversibly to neuronal cytoplasmic membranes, thereby preventing the fusion of vesicles and secretion of acetylcholine into the synaptic cleft **(Figure 19.14b)**. Thus, these neurotoxins prevent muscular contraction, resulting in a *flaccid paralysis*.

Botulism toxins are extremely potent; even a small taste of contaminated food can cause full-blown illness or death. To learn about a positive use for botulism toxins, see Highlight 11.1 on page 326.

Epidemiology and Diseases

Botulism is not an infection, but instead an *intoxication* (poisoning) caused by botulism toxin. Clinicians recognize three manifestations of botulism: food-borne botulism, infant botulism, and wound botulism. Fortunately, all three are rare.

---

[20]Latin, meaning sausage.

About 25 cases of food-borne botulism occur in the United States each year, usually within 1–2 days following the consumption of toxin in home-canned foods or preserved fish. Contaminated food may not appear or smell spoiled. Patients are initially weak and dizzy and have blurred vision, dry mouth, dilated pupils, constipation, and abdominal pain, followed by a progressive paralysis that eventually affects the diaphragm. The patient remains mentally alert throughout the ordeal. When fatal, death results from the inability of muscles of respiration to effect inhalation. Survivors recover very slowly as their nerve cells grow new endings over the course of months or years, replacing the debilitated termini.

In contrast to food-borne botulism, infant botulism results from the ingestion of endospores, which then germinate and colonize the gastrointestinal tract. Infants are susceptible to colonization because their GI tracts do not have a sufficient number of benign bacteria to compete with *C. botulinum* for nutrients and space. Botulism toxin is absorbed from an infected infant's GI tract, causing nonspecific symptoms: crying, constipation, and "failure to thrive." Fortunately, paralysis and death are rare. About 100 cases of infant botulism are reported in the United States each year, though some cases reported as sudden infant death syndrome may be fatal infant botulism instead.

Wound botulism usually begins 4 or more days following the contamination of a wound by endospores. With the exception that the gastrointestinal system is not typically involved, the symptoms are the same as those of food-borne botulism.

Diagnosis, Treatment, and Prevention

The symptoms of botulism are diagnostic; culturing the organism from contaminated food, from feces, or from the patient's wounds confirms the diagnosis. Further, toxin activity is detected by using a mouse bioassay. In this laboratory procedure, specimens of food, feces, and/or serum are divided into two portions. Botulism antitoxin is mixed with one of the portions, and the portions are then inoculated into two sets of mice. If the mice receiving the antitoxin survive while the other mice die, botulism is confirmed.

Treatment of botulism entails three approaches:

- Repeated washing of the intestinal tract to remove *Clostridium.*

- Administration of antibodies against botulism toxin to neutralize toxin in the blood, before it can bind to neurons.

- Administration of antimicrobial drugs to kill clostridia in infant botulism cases. (This approach is not effective in treating food-borne botulism because this disease results from the ingestion of toxin, not of the bacteria themselves.)

Note that the damage resulting from any prior binding of toxin to any particular nerve ending is irreversible, so botulism may progress despite aggressive medical care.

LM   10 μm

▲ *Figure 19.15*

Cells of *Clostridium tetani*, whose terminal endospores give the cells a distinctive "lollipop" shape. *How does the location of* C. tetani *endospores distinguish this bacterium from* Bacillus anthracis?

Figure 19.15 The endospores of C. tetani are terminal, whereas those of Bacillus anthracis are centrally located.

Food-borne botulism is prevented by destroying all endospores in contaminated food by proper canning techniques, by preventing endospores from germinating by using refrigeration or establishing an acidic environment (< pH 4.5), or by destroying the toxin by heating to at least 80°C for 20 minutes or more. Because infant botulism is often associated with the consumption of honey, parents are advised not to feed honey to infants under age 1 because their intestinal microbiota are not sufficiently developed to inhibit the germination of *C. botulinum* endospores and their growth as toxin-producing vegetative cells.

Clostridium tetani

Learning Objectives

✓ Describe the epidemiology of tetanus.

✓ List treatments for and preventative measures against *Clostridium tetani* infections.

Clostridium tetani[21] is a small, motile, obligate anaerobe that produces a terminal endospore, giving the cell a distinctive "lollipop" appearance (**Figure 19.15**). *C. tetani* is ubiquitous in soil, dust, and the GI tracts of animals and humans. Its vegetative cells are extremely sensitive to oxygen and live

---

[21]From Greek *tetanos,* meaning to stretch.

only in anaerobic environments, but its endospores survive for years. Its toxin causes the disease **tetanus.**

Pathogenesis

Contrary to popular belief, deep puncture wounds by rusty nails are not the only (or even primary) source of tetanus. Even a tiny break in the skin or mucous membranes can allow endospores of *C. tetani* access to an anaerobic environment in which they germinate, grow, and produce a fatal disease caused by *tetanus toxin*—a potent neurotoxin released by *C. tetani* cells when they die. To understand the action of tetanus toxin, we must further consider the control of muscles by the central nervous system.

As we have seen, the secretion of acetylcholine at neuromuscular junctions stimulates muscles to contract. Skeletal muscles are typically arranged in antagonistic pairs on opposite sides of a joint: When a motor nerve stimulates one muscle of the pair to contract, the muscle on the other side of the joint relaxes. Relaxation occurs because the muscle is *not* stimulated by nerve impulses; rather, inhibitory neurons of the central nervous system release inhibitory neurotransmitters that prevent motor neurons from stimulating the muscle (**Figure 19.16a** on page 552).

Tetanus toxin is composed of two polypeptides held together by a disulfide bond. The heavier of the two polypeptides binds to a receptor on the neuron's cytoplasmic membrane. The neuron then phagocytizes the toxin, cleaves off the lighter of the two polypeptides, and transports it to the central nervous system. There the smaller polypeptide enters an inhibitory neuron and blocks the release of inhibitory neurotransmitter. With inhibition blocked, excitatory activity is unregulated, and both muscles in an antagonistic pair are signaled to contract simultaneously **(Figure 19.16b).** The result is that muscles on both sides of the joint contract and do not relax. Contractions can be so severe they break bones.

Disease and Epidemiology

The incubation period of tetanus ranges from a few days to a week depending on the distance of the site of infection from the central nervous system. Typically the initial and diagnostic sign of tetanus is tightening of the jaw and neck muscles—which is why tetanus is also called *lockjaw.* Other early symptoms include sweating, drooling, grouchiness, and constant back spasms. If the toxin spreads to autonomic neurons, then heartbeat irregularities, fluctuations in blood pressure, and extensive sweating result. Spasms and contractions may spread to other muscles, becoming so severe that the arms and fists curl tightly, the feet curl down, and the body assumes a stiff backward arch as the heels and back of the head bend toward one another **(Figure 19.16c).** Complete, unrelenting contraction of the diaphragm results in a final inhalation—patients die because they cannot exhale.

Over a million cases of tetanus occur annually worldwide, mostly in underdeveloped countries where vaccina-

tion is unavailable or medical care is inadequate. The effect of tetanus toxin is irreversible at any synapse, so recovery depends on the growth of new neuronal terminals to replace those affected. The mortality rate of tetanus is about 50% among all patients, but the mortality of neonatal tetanus, resulting most commonly from infection of the umbilical stump, exceeds 90%.

Diagnosis, Treatment, and Prevention

The diagnostic feature of tetanus is the characteristic muscular contraction, which is often noted too late to save the patient. The bacterium itself is rarely isolated from clinical samples because it grows slowly in culture and is extremely sensitive to oxygen.

Treatment involves thorough cleaning of wounds to remove all endospores, immediate passive immunization with immunoglobulin directed against the toxin, the administration of antimicrobials such as penicillin, and active immunization with tetanus toxoid. Cleansing and antimicrobials eliminate the bacteria, whereas the immunoglobulin binds to and neutralizes toxin before it can attach to neurons. Active immunization stimulates the formation of antibodies that neutralize the toxin. Once the toxin binds to a neuron, treatment is limited to supportive care.

The number of cases of tetanus in the United States has steadily declined as a result of effective immunization with tetanus toxoid. The CDC currently recommends three doses beginning at 6 months of age, followed by a booster every 10 years for life.

In the previous two sections we examined two pathogenic, Gram-positive bacilli that form endospores: *Bacillus,* which is facultatively anaerobic, and *Clostridium,* which is anaerobic. Next we consider some pathogenic, Gram-positive bacilli that do not form endospores: the low G + C genus *Listeria,* and the high G + C genera *Corynebacterium, Mycobacterium, Propionibacterium, Nocardia,* and *Actinomyces.*

Listeria

Learning Objectives

✓ Describe the structures of *Listeria monocytogenes* that account for its pathogenicity.
✓ Discuss the signs and symptoms, diagnosis, treatment, and prevention of listeriosis.

Listeria monocytogenes (lis-tēr′ē-ă mo-nō-sī-to′je-nēz) is a Gram-positive, nonspore-forming coccobacillus found in soil, water, mammals, birds, fish, and insects. It enters the body in contaminated food and drink.

Pathogenesis, Epidemiology, and Disease

L. monocytogenes is a facultative intracellular parasite that binds to the surfaces of a macrophage or epithelial cell, triggering its own phagocytosis. Once inside a human cell's

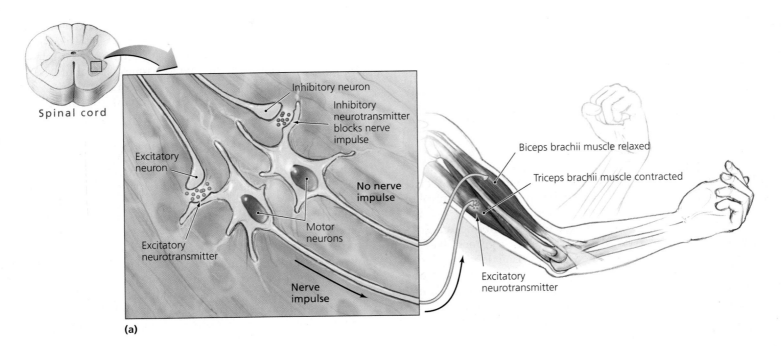

Spinal cord

Inhibitory neuron

Inhibitory neurotransmitter blocks nerve impulse

Excitatory neuron

Excitatory neurotransmitter

No nerve impulse

Motor neurons

Nerve impulse

(a)

Biceps brachii muscle relaxed

Triceps brachii muscle contracted

Excitatory neurotransmitter

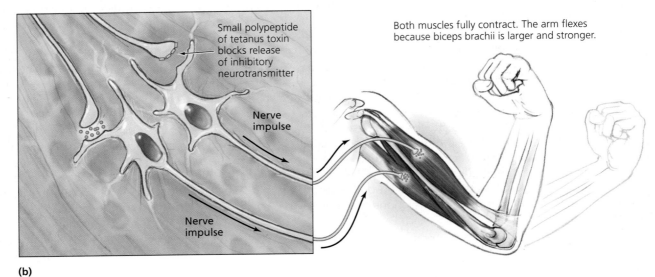

Small polypeptide of tetanus toxin blocks release of inhibitory neurotransmitter

Nerve impulse

Nerve impulse

(b)

Both muscles fully contract. The arm flexes because biceps brachii is larger and stronger.

(c)

▲ *Figure 19.16*
The action of tetanus toxin on a pair of antagonistic muscles. **(a)** Normally, when the muscle on one side of a joint is stimulated to contract, the neuron controlling the antagonistic muscle is inhibited, so the antagonistic muscle relaxes. **(b)** Tetanus toxin blocks the inhibitory neurotransmitter. As a result, motor neurons controlling the antagonistic muscle are not inhibited, but instead can generate nerve impulses that stimulate muscle contraction. **(c)** A patient with tetanus—prolonged muscular contractions—as a result of the action of tetanus toxin.

(a)

(b)

TEM | 2 μm

▲ *Figure 19.17*

How *Listeria* avoids the host's immune system while infecting new cells. **(a)** After
Listeria is initially phagocytized ①, the intracellular bacterium escapes the phagosome
② and reproduces within the phagocyte ③. The bacterium then polymerizes the host
cell's actin filaments into a "tail" ④ that pushes the bacterium into a pseudopod ⑤ that
is subsequently phagocytized by a new host cell ⑥. **(b)** A photomicrograph showing the
phagocytosis of a pseudopod containing a *Listeria* cell.

phagosome, *Listeria* synthesizes a pore-forming protein,
called *listeriolysin O*, that forms a hole in phagosome membranes, releasing the bacterium into the host cell's cytosol
before a lysosome can fuse with the phagosome. *Listeria* then
grows and reproduces in the cytosol, sheltered from the humoral immune system.

Listeria continues to avoid exposure to the immune system via a unique method of transferring itself to neighboring cells without having to leave host cells **(Figure 19.17)**.
The pathogen polymerizes the host cell's actin molecules to
form stiff actin filaments that lengthen and push the bacterium through the cytosol to the cell's surface, where it
forms a pseudopod. A neighboring macrophage or epithelial
cell then phagocytizes the pseudopod, and *Listeria* once
again "tunnels" its way out of the phagosome to continue its
intracellular parasitic existence within a new host cell.

Listeria's virulence is directly related to its ability to live
within cells, which is conferred by listeriocin O and the
membrane protein that triggers phagocytosis; it produces no
toxins or enzymes that make it virulent.

Listeria is rarely pathogenic in healthy adults, who experience either no symptoms or only a mild flulike illness.
In contrast, infection in pregnant women, fetuses, newborns, the elderly, and immunocompromised patients (par-

ticularly those with suppressed cell-mediated immunity)
can be quite severe. In these patients, *Listeria* travels via the
bloodstream to the brain, causing meningitis and possibly
death. Human-to-human transmission is limited to the
transfer of *Listeria* from pregnant women to fetuses, leading
to premature delivery, miscarriage, stillbirth, or meningitis
in the newborn.

Diagnosis, Treatment, and Prevention

Because *Listeria* most commonly causes meningitis, clinicians look for it in the cerebrospinal fluid (CSF) of suspected
patients. Unfortunately, only a few *Listeria* cells are required
to produce disease, so the bacterium is rarely seen in Gram-
stained preparations of phagocytes.

Listeria can be cultured from blood and CSF specimens
on many laboratory media, especially when a technique
called *cold enrichment* is used. In this procedure, the specimen is held at 4°C—a temperature that inhibits most bacteria, but not *Listeria*—and samples are periodically
inoculated onto media. Even with cold enrichment, colonies
of *Listeria* may take 4 weeks or longer to become visible.

The bacterium exhibits a characteristic end-over-end
"tumbling" motility that occurs at room temperature but not

V shapes   Palisade arrangement

LM   ⊢──────┤
          10 μm

▲ *Figure 19.18*

The characteristic arrangements of *Corynebacterium diphtheria* cells, which result from the type of binary fission called snapping division.

at 37°C. Tumbling motility is so distinctive that it can provide the basis for an initial diagnosis, although serological testing is required for positive identification.

Most antimicrobial drugs, including penicillin and erythromycin, inhibit *Listeria*; however, the bacterium is resistant to tetracycline and trimethropin. Prevention of infections is difficult because the organism is ubiquitous. Individuals at greatest risk should avoid undercooked vegetables, unpasteurized milk, undercooked meat, and all soft cheeses such as feta, Brie, and Camembert. Aged cheeses made from raw milk should be particularly avoided because *Listeria* can grow during refrigeration. Further, people at a high risk of infection should thoroughly cook raw meat and heat prepared meats—including cold cuts, luncheon meats, and hot dogs—before consuming them.

Corynebacterium

Learning Objectives

✓ Characterize the arrangements of *Corynebacterium cells*.
✓ Describe the transmission of *Corynebacterium diphtheriae* and the effect of diphtheria toxin.
✓ Discuss the diagnosis, treatment, and prevention of diphtheria.

Corynebacterium (kŏ-rī′nē-bak-tēr′ē-ŭm) is a genus of high G + C, pleomorphic, nonendospore-forming bacteria that are ubiquitous on plants and in animals and humans, where they colonize the skin and the respiratory, gastrointestinal, urinary, and genital tracts. The bacteria divide via a type of binary fission called *snapping division,* in which daughter cells remain attached to form characteristic V shapes and

side-by-side *palisade* arrangements **(Figure 19.18).** Although all species of corynebacteria can be pathogenic, the agent of diphtheria is most widely known.

Pathogenesis, Epidemiology, and Disease

Corynebacterium diphtheriae[22] (dif-thi′rē-ē) is transmitted from person to person via respiratory droplets or skin contact. Diphtheria is endemic in poorer parts of the world that lack adequate immunization. The bacterium normally contains a lysogenic bacteriophage that codes for *diphtheria toxin,* which is directly responsible for the signs and symptoms of diphtheria.

Like many bacterial toxins, diphtheria toxin consists of two polypeptides. One polypeptide binds to a growth factor receptor found on many types of human cells, triggering endocytosis of the toxin by these cells. Once inside a cell, proteolytic enzymes cleave the toxin, releasing the other polypeptide into the cytosol. This portion of the toxin enzymatically destroys *elongation factor,* a protein required for synthesis of polypeptides in eukaryotes. Because the action of the toxin is enzymatic, a single molecule of toxin sequentially destroys every molecule of elongation factor in the cell, completely blocking all polypeptide synthesis and resulting in cell death. Diphtheria toxin is thus one of the more potent toxins known.

Infections with *C. diphtheriae* have different effects depending on the host's immune status and the site of infection. Infections in immune individuals are asymptomatic, whereas infections in partially immune individuals result in a mild respiratory disease. Respiratory infections of nonimmune patients are most severe, resulting in the sudden and rapid signs and symptoms of **diphtheria** (dif-thē-rē′ă), including sore throat, localized pain, fever, pharyngitis, and the oozing of a fluid composed of intracellular fluid, blood clotting factors, leukocytes, bacteria, and the remains of dead cells lining the larynx and pharynx. The fluid thickens into a *pseudomembrane* **(Figure 19.19)** that can adhere so tightly to the tonsils, uvula, palate, pharynx, and larynx that it cannot be dislodged without causing bleeding in the underlying tissue. In severe cases a pseudomembrane can completely occlude the respiratory passages, resulting in death by suffocation.

In a similar way, cutaneous diphtheria also causes cell death and the formation of a pseudomembrane on the skin. With both cutaneous and respiratory infections, toxin absorbed into the blood can kill heart cells and nerve cells throughout the body, resulting in cardiac arrhythmia and coma, and possibly death.

Diagnosis, Treatment, and Prevention

Initial diagnosis is based on the presence of a pseudomembrane. Laboratory examination of the membrane or of tissue collected from the site of infection does not always reveal

---

[22]From Greek *koryne,* meaning club; *bakterion,* meaning small rod; and *diphthera,* meaning leather membrane.

Figure 19.19
A pseudomembrane characteristic of diphtheria.

bacterial cells because the effects are due largely to the action of diphtheria toxin and not the cells directly. Culture of specimens on *Loffler's medium*, which was developed especially for the culture of *C. diphtheriae*, produces three colonial morphologies. Some colonies are large, irregular, and gray; others are small, flat, and gray; and still others are small, round, convex, and black. Even though observations of these distinctive colonies are useful, absolute certainty of diagnosis results only from an immunodiffusion assay, called an **Elek test,** in which antibodies against the toxin react with toxin in a sample of fluid from the patient.

The most important aspect of treatment is the administration of antitoxin (immunoglobulins against the toxin) to neutralize toxin before it binds to cells; once the toxin binds to a cell, it enters via phagocytosis and kills the cell. Penicillin or erythromycin kills the bacteria, preventing the synthesis of more toxin. In severe cases, a blocked airway must be opened surgically or bypassed with a tracheostomy tube.

Because humans are the only known host for *C. diphtheriae*, the most effective way to prevent diphtheria is immunization. Before immunization, hundreds of thousands of cases occurred in the United States each year; in contrast, only 10 cases total were reported between 1997 and 2002. Toxoid (deactivated toxin) is administered as part of the DPT vaccine, which combines diphtheria and tetanus toxoids with antigens of the pertussis bacterium, at 2, 4, and 6 months of age, followed by booster immunizations without the pertussis antigens (Td) every 10 years.

CRITICAL THINKING

Why must diphtheria and tetanus vaccines be boosted every 10 years?

Mycobacterium

Learning Objective

✓ Characterize mycobacteria in terms of endospore formation, cell wall composition, growth rate, and resistance to antimicrobial drugs.

Another devastating high G + C, nonendospore-forming pathogen is *Mycobacterium* (mī-kō-bak-tēr′ē-ŭm). Species in this genus have cell walls containing an abundance of a waxy lipid, called **mycolic** (mī-kol′ik) **acid,** that is composed of chains of 60–90 carbon atoms. This unusual cell wall is directly responsible for the unique characteristics of this pathogen. Specifically, mycobacteria

- grow slowly (due to the time required to synthesize numerous molecules of mycolic acid). The generation time varies from hours to several days.
- are protected from lysis once they are phagocytized.
- are capable of intracellular growth.
- are resistant to Gram-staining, detergents, many common antimicrobial drugs, and desiccation. Because mycobacteria stain only weakly with the Gram procedure (if at all), the acid-fast staining procedure was developed to differentially stain mycobacteria (see Figure 4.21).

Even though almost 75 species of mycobacteria are known, most mycobacterial diseases in humans are caused by three species: *M. tuberculosis* (tū-ber-kyū-lō′sis), *M. leprae* (lep′rē), and *M. avium-intracellulare* (ā′vē-ŭm in′tra-sel-ū-lă-rē), which cause tuberculosis, leprosy, and opportunistic infections in AIDS patients, respectively.

Tuberculosis

Learning Objectives

✓ Identify two effects of cord factor of *Mycobacterium tuberculosis*.
✓ Describe the transmission of *M. tuberculosis*, and its subsequent action within the human body.
✓ Discuss the diagnosis, treatment, and prevention of tuberculosis.

Tuberculosis (TB), the primary mycobacterial disease, is fundamentally a respiratory disease caused by *Mycobacterium tuberculosis*. This bacterium forms dull-yellow raised colonies after growth for weeks on a special differential medium called Lowenstein-Jensen agar (**Figure 19.20a** on page 556). Virulent strains of *M. tuberculosis* have a cell wall component, called **cord factor,** that produces strands of daughter cells that remain attached to one another in parallel alignments **(Figure 19.20b).** Cord factor also inhibits migration of neutrophils and is toxic to mammalian cells. Mutant mycobacteria that are unable to synthesize cord factor do not cause disease.

(a)

(b)

LM   5 µm

▲ *Figure 19.20*

Mycobacterium tuberculosis. **(a)** The raised, granular, dull yellow colonies the bacteria produce on Lowenstein-Jensen agar. **(b)** The pink appearance of the bacteria when prepared with an acid-fast stain. Note the corded growth (parallel alignments) of daughter cells, a result of the presence of cord factor in the cell walls of virulent strains.

Pathogenesis and Disease

The waxy wall protects the pathogen from desiccation, and it can remain viable in dried aerosol droplets for 8 months. *Mycobacterium* is not particularly virulent, as only about 5% of people infected with the bacterium develop disease; however, it kills about 50% of untreated patients. Clinicians divide tuberculosis into three types; *primary TB, secondary* (or *reactivated*) *TB*, and *disseminated TB*.

Primary Tuberculosis   Although *M. tuberculosis* can infect any organ, 85% of infections remain in the lungs. Primary infection, which typically occurs in children, involves the formation of small, hard nodules in the lungs called **tubercles** (too'ber-klz), which are characteristic of TB and give it its name. The five stages of a primary infection are as follows **(Figure 19.21a):**

1 *Mycobacterium* typically infects the respiratory tract via inhalation of respiratory droplets formed when infected individuals talk, sing, cough, or sneeze. A respiratory droplet is about 5 µm in diameter and carries 1–3 bacilli. The minimum infectious dose is about 10 cells.

2 Macrophages in the alveoli (air sacs) of the lungs phagocytize the pathogens but are unable to digest them in part because of the waxy cell walls of mycobacteria.

3 Instead, the bacteria replicate freely within macrophages, gradually killing them. Bacteria released from dead macrophages are phagocytized by other macrophages, beginning the cycle anew. This stage of infection, which lasts for a few weeks, is typically asymptomatic or associated with a mild fever.

4 Infected macrophages present antigen to T lymphocytes, which produce lymphokines that attract and activate more macrophages and trigger inflammation. Tightly appressed macrophages surround the site of infection, forming a tubercle.

5 Other cells of the body deposit collagen fibers, enclosing infected macrophages and lung cells within the tubercle. Infected cells in the center of the tubercle die, releasing *M. tuberculosis* and producing *caseous*[23] *necrosis*—the death of tissue that takes on a cheeselike consistency due to the presence of protein and fat released from dying cells. Sometimes, for an unknown reason, the center liquefies and subsequently becomes filled with air. Such a tubercle is called a *tuberculous cavity.*

In most patients the immune system reaches a stalemate with the bacterium at this juncture: The immune system is able to prevent further spread of the pathogen and stop the progression of the disease, but it is not able to rid the body of all mycobacteria. *M. tuberculosis* may remain dormant for decades within macrophages and in the centers of tubercles. If the immune system breaks the stalemate by killing all the mycobacteria, the body deposits calcium around the tubercles, which are then called *Ghon complexes.*

Secondary or Reactivated Tuberculosis   This condition results when *M. tuberculosis* breaks the stalemate, ruptures the tubercule, and reestablishes an active infection in which the bacteria spread through the lungs via the bronchioles **(Figure 19.21b).** Reactivated TB is a common occurrence in TB-infected individuals with suppressed immune systems.

[23]From Latin *caseus,* meaning cheese.

(a) Primary tuberculosis infection

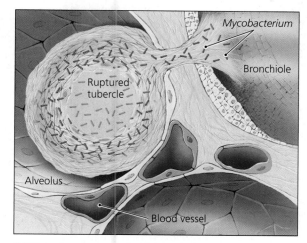

(b) Secondary or reactivated tuberculosis

▲ *Figure 19.21*

The processes involved in the development of tuberculosis in the lungs. **(a)** The formation of tubercles in primary tuberculosis. **(b)** Events in secondary (reactivated) tuberculosis.

Disseminated Tuberculosis   This type of TB results when some macrophages carry the pathogens via the blood and lymph to a variety of sites, including the bone marrow, spleen, kidneys, spinal cord, and brain. The signs and symptoms observed in disseminated TB correspond to complications arising at the various sites involved. The common name for TB in the early 1900s—*consumption*—reflects the wasting away of the body resulting from the involvement of multiple sites in disseminated TB.

Epidemiology

Even though tuberculosis is on the decline in the United States, it is pandemic in other parts of the world. Some scientists estimate that over one-third of the world's population is infected with *M. tuberculosis*. Patients with lowered immunity are at the greatest risk of infection; other risk factors include diabetes, poor nutrition, stress, crowded living conditions, alcohol and drug abuse, and smoking.

TB is prevalent in the countries of the former Soviet Union, where the breakdown of essential medical services has allowed a resurgence. Prison populations are especially hard hit because of cramped living conditions. In some prisons in the former USSR, 10% or more of prisoners have active cases of tuberculosis. When these prisoners are released without being effectively treated, they spread tuberculosis among the rest of the population. One study in the late 1990s indicated that 80% of the tuberculosis cases in Russia could be traced to released prisoners. The incidence of tuberculosis is also on the rise in Africa, where the immunosuppression associated with the AIDS epidemic allows infection and reactivation.

Diagnosis, Treatment, and Prevention

A **tuberculin skin test** is used to screen patients for possible exposure to tuberculosis. In this test, a health care worker injects about 0.1 ml of cell wall antigens of *M. tuberculosis* into a patient's skin. The appearance of a hard, red swelling at the test site within 24 to 72 hours is a positive test **(Figure 19.22a).** The reaction is a type IV (cell-mediated) hypersensitivity response, indicating past infection or vaccination but not necessarily current disease.

A positive tuberculin skin test is not sufficient to distinguish among patients who have been exposed to antigen but are currently uninfected, chronic carriers, and patients with active disease. Chest X-rays can reveal the presence of tubercles in the lungs **(Figure 19.22b);** primary tuberculosis appears as tubercles in the lower and central areas of the lungs, whereas secondary tuberculosis more commonly appears higher in the lungs. The presence of acid-fast cells and cords in sputum (see Figure 19.20b) confirms an active case of tuberculosis.

Common antimicrobials such as penicillin and erythromycin have little effect on *Mycobacterium tuberculosis,* because it grows so slowly that the drugs are cleared from the body by the liver and kidneys before they can have any significant effect. Further, antimicrobial drugs have little impact on *M. tuberculosis* living within macrophages. The currently recommended treatment is a combination of isoniazid (INH), rifampin, and one of a number of other drugs for 2 months, followed by INH and rifampin alone for 4 months.

Epidemiologists and physicians are concerned about strains of *M. tuberculosis* that are resistant to more than one of the standard antituberculosis drugs. Such *multidrug resistant (MDR)* strains have developed in many countries. For example, in a tuberculosis epidemic in New York City in the early 1990s, 9% of the cases were caused by strains resistant to two or more drugs; MDR strains account for over 10% of tuberculosis cases in Germany, Denmark, and New Zealand. No one knows the extent of MDR cases in developing countries, but it is suspected that the situation there is even worse.

For patients in an area where MDR strains are endemic, either ethambutol or streptomycin is added to the standard regimen. Costly, newer drugs such as cycloserine are also used. Further, the World Health Organization and the CDC recommend a strategy of drug delivery called *Directly Observed Treatment, Shortcourse (DOTS),* in which health care workers observe patients to ensure they take their medica-

(a)

(b)

▲ *Figure 19.22*

Diagnosis of tuberculosis. **(a)** The enlarged, reddened, and raised inoculation site of a positive tuberculin skin test. **(b)** An X-ray of a tuberculosis patient showing white patches indicative of tubercules. *What does a positive tuberculin test indicate?*

Figure 19.22 A positive tuberculin skin test indicates that a patient has cell-mediated memory against antigens of M. tuberculosis, as a result of either infection or immunization.

tions on schedule. Obviously, effective treatment of cases of MDR tuberculosis is quite expensive. For instance, the 1990s New York City outbreak cost over $1 billion to quell. Researchers are developing a subcutaneous implant that would gradually deliver drugs directly into the body over a period of months.

Physicians use antibacterial drugs prophylactically to treat patients who have either shown a recent conversion from a negative to a positive tuberculin skin test or undergone significant exposure to active cases of tuberculosis. In countries where tuberculosis is common, health care workers immunize patients with *BCG*[24] vaccine, which is composed of attenuated *M. bovis* (bō'vis), a species that causes tuberculosis in cattle and is only rarely transmitted to humans in contaminated milk. The vaccine is not used for immunocompromised patients because it can cause disease.

Even though immunization produces immunity in less than half of the patients, its use significantly reduces the spread of disease. In the United States and other developed countries, the cost of mass immunization is not warranted because of the relatively low prevalence of tuberculosis. Further, immunized patients will have a positive skin reaction for the rest of their lives, even if they have not been infected with *M. tuberculosis*, and such "false positive" results would hinder the work of epidemiologists trying to track the spread of the disease.

The prediction that tuberculosis would be eliminated in the United States by 2000 was not realized because of a resurgence of tuberculosis among AIDS patients, and its subsequent spread to drug abusers and the homeless. Nevertheless, both renewed efforts to detect and stop infection and the implementation of DOTS reduced the number of reported U.S. cases in 2002 to 11,878.

CRITICAL THINKING

Why don't physicians try to prevent the spread of TB by simply administering prophylactic antimicrobial drugs to everyone living in endemic areas?

Leprosy

Learning Objectives

✓ Compare and contrast tuberculoid leprosy with lepromatous leprosy.

✓ Discuss the diagnosis, treatment, and prevention of leprosy.

Mycobacterium leprae causes **leprosy** (lep'rō-sē)—which is also called by the less dreaded name *Hansen's disease*, after Gerhard Hansen (1841–1912), a Norwegian bacteriologist who discovered its cause in 1874. *M. leprae* is a high G + C, Gram-positive bacillus. Because of the abundance of mycolic acid in the cell wall, bacilli do not stain purple and must instead be stained with an acid-fast stain. *M. leprae* grows best at 30°C, showing a preference for cooler regions

[24]For Bacillus of Calmette-Guerin, named after the two French developers of the vaccine.

▲ *Figure 19.23*
A person with lepromatous leprosy.

of the human body, particularly peripheral nerve endings and skin cells in the fingers, toes, lips, and earlobes. The bacterium has never been grown in cell-free laboratory culture, a fact that has hindered research and diagnostic studies. Armadillos, which have a normal body temperature of 30°C, are its only other host and have proven valuable in studies on leprosy and on the efficacy of leprosy treatments.

Pathogenesis, Epidemiology, and Disease

Leprosy has two different manifestations depending on the immune response of the patient: Patients with a strong cell-mediated immune response are able to kill cells infected with the bacterium, resulting in a nonprogressive form of the disease called *tuberculoid leprosy*. Regions of the skin that have lost sensation as a result of nerve damage are characteristic of this form of leprosy (see **Case Study 19.3** on page 560).

By contrast, patients with a weak cell-mediated immune response develop *lepromatous leprosy* **(Figure 19.23),** in which bacteria multiply in skin and nerve cells, gradually destroying tissue and leading to the progressive loss of facial features, digits (fingers and toes), and other body structures. Development of signs and symptoms is very slow; incubation may take years before any signs or symptoms are evident. Death from leprosy is rare and usually results from the infection of leprous lesions by other pathogens.

Lepromatous leprosy is the more virulent form of the disease, but fortunately it is becoming relatively rare: In the 1990s about 12 million cases were diagnosed annually worldwide, but in 2002 this number had decreased to about 635,000, including 77 cases in the United States. Ninety percent of the U.S. patients were immigrants, even though leprosy is endemic in California, Hawaii, Texas, and Louisiana.

Leprosy is transmitted via person-to-person contact. Given that the nasal secretions of patients with lepromatous leprosy are loaded with mycobacteria, infection presumably

Case Study 19.3   The Man with No Feelings

A 41-year-old man arrives in a hospital emergency room with third-degree burns on the second and third fingers of his left hand. Through an interpreter, the man tells the admitting nurse that he had moved to the United States from Southeast Asia barely 6 months before. This evening he had been sitting in his living room, cigarette in hand, watching the nightly news, when he became aware of an odd and pungent odor. At first he thought the building was on fire, but the smell was not that of wood burning. Instead, it reminded him of roasting meat—which it was. As he rose to investigate, he saw that it was his hand. His blackened fingers were cracked, curling, and smoking where the cigarette had burned down. With a jerk, he discarded his cigarette and marveled that he felt nothing—it was as though his hand were not his own.

1. The physicians suspect the patient has leprosy. How might they verify this?
2. What type of leprosy is this patient more likely to have?
3. What treatment should be prescribed?
4. Are the doctors, nurses, and other emergency room personnel at risk of contracting the disease?

occurs via inhalation of respiratory droplets. Alternatively, infection may occur through breaks in the skin. In any case, leprosy is not particularly virulent; individuals are typically infected only after years of intimate social contact with a victim.

Why, then, does a diagnosis of leprosy elicit such dread? Much of the public's aversion to leprosy stems from Biblical references calling for the quarantine of lepers, and from the horrible disfigurement lepers can suffer. During the Middle Ages lepers were often required to wear bells to warn passers-by of their presence and forced to live in special villages or in isolated areas of the forests.

The 12th-century English king Henry II took lepraphobia (fear of leprosy and lepers) a step further; he ordered that those with the disease be burned at the stake. Not to be outdone, his great-grandson Edward I ordained that lepers be accorded a complete funeral service—before they were led to the cemetery and buried alive. Patients with leprosy are no longer quarantined because the disease is so rarely transmitted and is fully treatable.

Diagnosis, Treatment, and Prevention

Diagnosis of leprosy is based on signs and symptoms of disease—a loss of sensation in skin lesions in the case of tuberculoid leprosy, and disfigurement in the case of lepromatous leprosy. Diagnosis is confirmed by a positive skin test with leprosy antigen (similar to the tuberculin skin test) or through direct observation of acid-fast rods (AFRs) in tissue samples or nasal secretions (in the case of lepromatous leprosy).

As with *M. tuberculosis*, *M. leprae* quickly develops resistance to single antimicrobial agents, so therapy consists of administering multiple drugs, such as clofazimine (a fat-soluble dye), rifampin, or dapsone, for a minimum of 2 years. Treatment can be lifelong for some patients.

BCG vaccine provides some protection against leprosy (as well as against tuberculosis), but prevention is primarily achieved by limiting exposure to the pathogen, and by the prophylactic use of antimicrobial agents when exposure occurs.

Other Mycobacterial Infections in AIDS Patients

Learning Objective

✓ Explain why the incidence of infections with *Mycobacterium avium-intracellulare*, long thought to be harmless to humans, has been increasing.

Mycobacterium avium-intracellulare, which is commonly found in soil, water, and food, was long thought to be a harmless occasional member of the respiratory microbiota. However, the advent of the AIDS epidemic has unmasked it as an important opportunistic pathogen—and the cause of the most common mycobacterial infection among AIDS patients in the United States. (*M. tuberculosis* is more common in countries where tuberculosis is epidemic.) Infections are believed to result from the ingestion of contaminated food or water; direct person-to-person transmission does not occur. The bacteria spread throughout the body via the lymph.

In contrast to infections with other mycobacteria, *M. avium-intracellulare* simultaneously affects almost every organ of the body. In some tissues, every cell is packed with mycobacteria, and during the terminal stages of AIDS the blood is often filled with thousands of bacteria per milliliter. The disease remains asymptomatic until organ failure occurs on a massive scale.

Treatment consists of trial-and-error administration of antimicrobial agents; the disseminated nature of the infection often makes treatment ineffectual.

Propionibacterium

Learning Objectives

✓ Identify the species of *Propionibacterium* most commonly involved in infections of humans.
✓ Explain the role of *P. acnes* in the formation of acne.

Propionibacteria are small, Gram-positive rods commonly found growing on the skin. They are so named because they

① Normal skin

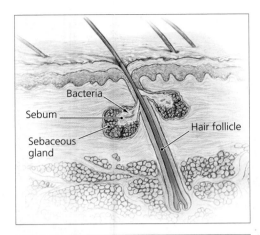

Bacteria

Sebum

Sebaceous
gland

Hair follicle

② Whitehead

◀ *Figure 19.24*

The development of acne. Normally, oily sebum
produced by glands reaches the hair follicle and is
discharged from the pore and onto the skin surface ①.
When inflammation resulting from bacteria infecting
the hair follicle causes the skin to swell over the pore,
sebum and the colonizing bacteria accumulate to form
a whitehead ②. The blockage of the pore by a plug of
dead and dying bacteria and sebum produces a
blackhead ③. When the inflammation of the follicle
becomes severe enough, pustules form ④ and rupture,
producing cystic acne, which is often resolved by the
formation of scar tissue. *Which bacterial pathogen is a
common cause of cystic acne?*

Figure 19.24 Propionibacterium acnes is a common cause
of acne.

③ Blackhead

Blocked pore

④ Pustule formation
(cystic acne)

Pus

produce propionic acid as a by-product of the fermentation
of carbohydrates. The species most commonly involved in
infections of humans is *Propionibacterium acnes* (prō-pē-on-i-
bak-tēr′ē-ŭm ak′nēz), which causes acne in 85% of adolescents and young adults. The bacterium can also be an
opportunistic pathogen in patients with intrusive medical
devices such as catheters, artificial heart valves, artificial
joints, and cerebrospinal fluid shunts.

In its role in the development of acne **(Figure 19.24)**, *Propionibacterium* typically grows in the sebaceous (oil) glands
of the skin ①. Excessive oil production triggered by the
hormones of adolescence, particularly testosterone in males,
stimulates the growth of the bacteria, which secrete chemicals that attract leukocytes. The leukocytes phagocytize the
bacteria and release chemicals that stimulate local inflammation. The combination of dead bacteria and dead and living leukocytes makes up the white pus associated with the
pimples of acne ②. A blackhead is formed when a plug of
dead and dying bacteria blocks the pore ③. In *cystic acne*, a
particularly severe form of the disease, bacteria form inflamed pustules (cysts) ④ that rupture, triggering the formation of scar tissue.

There are many common misconceptions about acne.
Cleaning the surface of the skin is not particularly effective
in preventing the development of acne because the bacteria
live deep in the sebaceous glands; though, frequent cleansing may dry the skin, which helps loosen plugs from hair
follicles. Scientists have not shown any connection between
acne and diet, including chocolate and oily foods.

In most cases the immune system is able to control *Propionibacterium,* and no treatment is required. Dermatologists
may prescribe antimicrobial drugs such as erythromycin
and clindamycin, which have proven effective in controlling
the bacterium, though long-term antibiotic use can destroy
susceptible normal bacterial microbiota, making the individual more vulnerable to opportunistic fungal and bacterial infections that are resistant to antimicrobial drugs.
Accutane,® a derivative of vitamin A, inhibits the formation

of sebum. Because this drug can cause intestinal bleeding, it is prescribed only for severe cases of acne. Further, it causes birth defects and thus should not be used by pregnant women.

In the previous few sections we have discussed pathogenic, nonendospore-forming, high G + C rods in the genera *Corynebacterium*, *Mycobacterium*, and *Propionibacterium*. Now we will consider filamentous, high G + C pathogens in the genera *Nocardia* and *Actinomyces*.

Nocardia and Actinomyces

Nocardia (nō-kar'dē-ă) and *Actinomyces* (ak'ti-nō-mī'sēz) have elongated filamentous cells that resemble fungal hyphae. The cell walls of *Nocardia* contain mycolic acid, so the cells are difficult to Gram stain, but they are acid-fast. *Actinomyces* stains purple in a Gram stain.

Nocardia asteroides

Learning Objective

✓ Characterize *Nocardia asteroides* in terms of appearance, cell wall composition, and role in producing disease.

Nocardia asteroides (as-ter-oy'dēz), a common inhabitant of soils rich in organic matter, is the most common pathogen in this genus, accounting for about 90% of infections.

Pathogenesis, Epidemiology, and Disease

N. asteroides is an opportunistic bacterial pathogen that infects numerous sites, including the lungs, skin, and central nervous system. Pulmonary infections develop following inhalation; cutaneous infections result from introduction of the bacteria into wounds; and infections of the central nervous system and other internal organs follow the spread of the bacterium in the blood.

In the lungs, *Nocardia* causes pneumonia accompanied by signs and symptoms typical of pulmonary infections—cough, shortness of breath, and fever. Cutaneous infections may produce **mycetoma**, a painless, long-lasting infection characterized by swelling, pus production, and draining sores **(Figure 19.25)**.

Diagnosis, Treatment, and Prevention

Microscopic examination of samples of skin, sputum, pus, or cerebrospinal fluid is usually sufficient to establish infection with *Nocardia*; the presence of long, acid-fast, hyphal-like cells is diagnostic. Treatment usually involves 6 weeks of an appropriate antimicrobial drug; sulfanilamide is the drug of choice. Although such treatment is usually sufficient for localized infections in patients with normal immune systems, the prognosis for immunocompromised patients with widespread infections is poor. Prevention of nocardial dis-

▲ *Figure 19.25*
A *Nocardia* infection.

eases involves avoiding exposure to the bacterium in soil, especially by immunocompromised patients.

Actinomyces

Learning Objective

✓ Characterize *Actinomyces* in terms of appearance, cell wall composition, and role in producing disease.

Actinomyces is another genus characterized by hyphal-like cells **(Figure 19.26a)**, as reflected in the name of the genus, which means "ray fungus" in Greek. Despite the name, *Actinomyces* is not a fungus but a bacterium. Unlike *Nocardia*, *Actinomyces* is not acid-fast and stains purple when Gram-stained.

Colonies of *Actinomyces* form visible concretions resembling grains of sand. Even though the yellow of these multicellular structures has prompted the name "sulfur granules," they are in fact held together by calcium phosphate.

Pathogenesis, Epidemiology, and Disease

Actinomyces is a normal member of the surface microbiota of human mucous membranes. It can become an opportunistic pathogen of the respiratory, gastrointestinal, urinary, and female genital tracts, and it sometimes causes dental caries (cavities). It causes a disease called **actinomycosis** when it enters breaks in the mucous membranes resulting from trauma, surgery, or infection by other pathogens. Actinomycosis is characterized by the formation of many abscesses connected by channels in the skin or mucous membranes **(Figure 19.26b)**.

(b)

LM ┣———┫ 30 μm

(a)

▲ *Figure 19.26*

Actinomyces. **(a)** A Gram stain of pus containing the bacteria, which resemble fungal hyphae (compare with Figure 12.14b) and may form visible yellow aggregations that look like grains of sand. **(b)** Lesions of actinomycosis.

Diagnosis, Treatment, and Prevention

Diagnosis of actinomycosis is difficult because other organisms cause similar symptoms, and it is difficult to show that the presence of *Actinomyces* is not merely contamination from its normal site on the mucous membranes. Often a health care worker collects and crushes a "sulfur granule" to reveal the presence of filamentous cells.

Treatment involves surgical removal of the infected tissue and the administration of penicillin, tetracycline, or erythromycin for 4–12 months. Prevention of *Actinomyces* infections involves good oral hygiene, and the prolonged use of prophylactic antimicrobials should trauma or surgery broach the mucous membranes.

CHAPTER SUMMARY

Staphylococcus *(pp. 533–537)*

1. *Staphylococcus aureus* and *Staphylococcus epidermidis* are found on the skin, and in the upper respiratory, gastrointestinal, and urogenital tracts. *S. aureus* is the more virulent, due partly to its production of various enzymes (including coagulase and staphylokinase) and several toxins.

2. Food poisoning is one noninvasive disease caused by *Staphylococcus*. Cutaneous diseases caused by *Staphylococcus* include **scalded skin syndrome, impetigo, folliculitis, sties, furuncles,** and **carbuncles.**

3. Potentially fatal systemic infections caused by *Staphylococcus* include **toxic shock syndrome (TSS), bacteremia, endocarditis, pneumonia, empyema,** and **osteomyelitis.**

Streptococcus *(pp. 537–544)*

1. Cells of **group A streptococcus** *(Streptococcus pyogenes)* produce protein M and a hyaluronic acid capsule, each of which contributes to the pathogenicity of the species. Enzymes and toxins produced by *S. pyogenes* dissolve blood clots, stimulate fever, and lyse blood cells.

2. The diseases caused by group A streptococci include **pharyngitis** ("strep throat"), **scarlet fever** (scarlatina), **pyoderma** (impetigo), **erysipelas, streptococcal toxic shock syndrome, necrotizing fasciitis** ("flesh-eating strep"), **rheumatic fever,** and **glomerulonephritis.**

3. Cells of **group B streptococcus** *(S. agalactiae)* occur normally in the lower GI, genital, and urinary tracts. They are associated with neonatal diseases and are treated with penicillin.

4. The **viridans group** of alpha-hemolytic streptococci include members that inhabit the mouth, pharynx, GI tract, and genitourinary tracts, cause dental **caries (cavities),** and can enter the blood to cause bacteremia, meningitis, and endocarditis.

5. Virulent strains of *Streptococcus pneumoniae* are protected by polysaccharide capsules and phosphorylcholine in their cell walls. These bacteria secrete protein adhesin, secretory IgA protease, and pneumolysin to lyse cells in the lungs.

6. **Pneumonia** is the most prevalent disease caused by *Streptococcus pneumoniae* infection. Other diseases include **sinusitis** (inflammation of the nasal sinuses), **otitis media** (inflammation of the middle ear), bacteremia, endocarditis, and meningitis.

Enterococcus (pp. 544–545)

1. *Enterococci* are normal members of the intestinal microbiota that can cause nosocomial bacteremia, endocarditis, and wound infections.

Bacillus (pp. 545–547)

1. *Bacillus anthracis* secretes anthrax toxin, which can cause three forms of **anthrax:** gastrointestinal anthrax (rare, and fatal, with intestinal hemorrhaging), cutaneous anthrax (can be fatal if untreated), and inhalation anthrax (often fatal).

Clostridium (pp. 547–551)

1. *Clostridium perfringens* produces 11 toxins that lyse blood cells and cause diseases such as food poisoning and gas gangrene.

2. *Clostridium difficile* is an intestinal bacterium that can cause a self-limiting explosive diarrhea or a life-threatening **pseudomembranous colitis.**

3. *Clostridium botulinum* is an anaerobic, endospore-forming bacterium that can release an extremely poisonous toxin in improperly canned food. Infant botulism occurs when the pathogen grows in the gastrointestinal tract of an infant. Wound botulism results when the bacterial endospores germinate in surgical or traumatic wounds.

4. *Clostridium tetani* is a ubiquitous bacterium that enters the body via a break in the skin and causes **tetanus.** Immunization is effective in preventing this disease.

Listeria (pp. 551–554)

1. *Listeria monocytogenes* is rarely pathogenic but can cause severe disease in pregnant women, newborns, the elderly, and immunocompromised patients.

Corynebacterium (pp. 554–555)

1. *Corynebacterium diphtheriae*, transmitted via respiratory droplets, contains a bacteriophage that codes for diphtheria toxin, which causes the symptoms of the potentially fatal disease **diphtheria.** Diagnosis results from an immunodiffusion assay called an **Elek test.**

Mycobacterium (pp. 555–560)

1. Cell walls of *Mycobacterium* contain **mycolic acid.** *Mycobacterium tuberculosis* with a cell wall component called **cord factor** causes **tuberculosis,** a lung disease in which the bacteria replicate within macrophages.

2. *Mycobacterium leprae* causes **leprosy.** Tuberculoid leprosy is a nonprogressive form of the disease, whereas the lepromatous leprosy results in progressive destruction of body structures.

3. *Mycobacterium avium-intracellulare* is an opportunistic pathogen that can become pathogenic in AIDS patients. It adversely affects almost every organ of the body.

Propionibacterium (pp. 560–562)

1. *Propionibacterium acnes* living in the sebaceous glands of the skin causes adolescent **acne.** The bacterium may also infect patients who have intrusive medical devices. Antimicrobial drugs are more effective than topical products in the treatment of acne.

Nocardia and Actinomyces (pp. 562–563)

1. The opportunistic pathogen *Nocardia asteroides* has elongated cells resembling fungal hyphae; it is acid-fast due to the presence of mycolic acid in its cell wall.

2. *Actinomyces* is another opportunistic pathogen with elongated cells, but it is not acid-fast. It causes **actinomycosis.**

QUESTIONS FOR REVIEW
(Answers to multiple choice and matching questions are on the web, along with additional review questions. Visit www.microbiologyplace.com.)

Matching

For each of the following diseases or conditions, indicate the genus (or genera) of bacterium that causes it.

____ 1. Scalded skin syndrome   A. *Staphylococcus*

____ 2. Osteomyelitis   B. *Streptococcus*

____ 3. Pharyngitis   C. *Mycobacterium*

____ 4. Scarlet fever   D. *Listeria*

____ 5. Pyoderma   E. *Propionibacterium*

____ 6. Rheumatic fever   F. *Corynebacterium*

____ 7. Glomerulonephritis   G. *Bacillus*

____ 8. Sinusitis   H. *Clostridium*

____ 9. Otitis media   I. *Actinomyces*

____ 10. Anthrax

____ 11. Myonecrosis

____ 12. Diphtheria

____ 13. Leprosy

____ 14. Dental caries

____ 15. Acne

Multiple Choice

1. Which of the following bacteria causes a type of food poisoning?
   a. *Streptococcus sanguis*
   b. *Clostridium perfringens*
   c. *Staphylococcus aureus*
   d. *Streptococcus pyogenes*

2. How does *Staphylococcus aureus* affect the matrix between cells in the human body?
   a. *S. aureus* triggers blood clotting, which coats the matrix and inhibits cellular communication.
   b. *S. aureus* produces an enzyme that dissolves hyaluronic acid and thus enables it to pass between the cells.

c. *S. aureus* contains a hyaluronic acid capsule that causes leukocytes to ignore the bacteria as if they were camouflaged.

d. *S. aureus* does not affect the matrix but instead it produces a necrotizing agent that dissolves body cells.

3. Which of the following conditions is a "systemic disease" caused by *Staphylococcus*?
   a. impetigo
   b. folliculitis
   c. carbuncle
   d. toxic shock syndrome

4. A bacterium associated with bacteremia, meningitis, and pneumonia in newborns is
   a. *Staphylococcus aureus*.
   b. *Staphylococcus epidermidis*.
   c. *Streptococci pyogenes*.
   d. *Streptococcus agalactiae*.

5. Which type of anthrax is more common in animals than in humans?
   a. cutaneous anthrax
   b. inhalation anthrax
   c. gastrointestinal anthrax
   d. mucoid anthrax

6. Of the following genera, which can survive the harshest conditions?
   a. *Staphylococcus*
   b. *Clostridium*
   c. *Mycobacterium*
   d. *Actinomyces*

7. Pathogenic strains that have become resistant to antimicrobial drugs are found in which of the following genera?
   a. *Staphylococcus*
   b. *Mycobacterium*
   c. *Enterococcus*
   d. all of the above

8. The bacterium causing pseudomembranous colitis is

   a. *Clostridium difficile*.
   b. *Streptococcus pyogenes*.
   c. *Mycobacterium avium-intracellulare*.
   d. *Corynebacterium diptheriae*.

9. In which of the following diseases would a patient experience a pseudomembrane covering the tonsils, pharynx, and larynx?
   a. tuberculoid leprosy
   b. diphtheria
   c. arrhythmia
   d. tetanus

10. The most effective treatment of adolescent acne is
    a. cleaning the skin.
    b. abstaining from eating oily foods.
    c. exposure to ultraviolet light.
    d. the administration of antimicrobial drugs.

Short Answer

1. *Mycobacterium avium-intracellulare* was considered relatively harmless until the late 20th century, when it became common in certain infections. Explain how this bacterium's pathogenicity changed.

2. Contrast tuberculoid leprosy with lepromatous leprosy in terms of pathogenesis. How does the cellular immune response of the patient affect the form of the disease?

3. On a trans-Atlantic flight, a Latvian passenger with tuberculosis sits next to a Canadian businessman, who later develops tuberculosis and dies. Trace the path of *Mycobacterium tuberculosis* from a droplet of mucus leaving the European passenger to within the businessman, explaining how the disease became fatal.

4. Explain how mice are used in the diagnosis of botulism poisoning.

5. Why do pediatricians recommend that children under 1 year never be fed honey?

6. Explain the different actions of pyogenic and pyrogenic toxins.

CRITICAL THINKING

1. In Chapter 4 (p. 121) we discussed the identification of microbes using dichotomous taxonomic keys. Design a key for the genera of Gram-positive cocci discussed in this chapter. Design a second key for identifying the Gram-positive bacilli discussed in the chapter.

2. A few days after the death of a hospitalized child from an MRSA infection, another child, who had been admitted to the hospital with viral pneumonia, worsened and died. An autopsy revealed that the second child also died from complications of MRSA. By what route was the second child likely infected? What should hospital personnel do to limit the transfer of MRSA and other bacteria among patients?

3. An elderly man is admitted to the hospital with severe pneumonia, from which he eventually dies. What species is the most likely cause of his infection? What antimicrobial drug is effective against this species? How could the man have been protected from infection? Is the hospital staff at significant risk of infection from the man? Which groups of patients would be at risk if the man had visited their rooms before he died?

4. Botulism toxin can be used as an antidote for tetanus. Can tetanus toxin be used as an antidote for botulism? Why or why not?

5. A blood bank refused to accept blood from a potential donor who had just had his teeth cleaned by a dental hygienist. Why did they refuse the blood?

CHAPTER 20

Pathogenic Gram-Negative Cocci and Bacilli

CHAPTER OUTLINE

Pathogenic Gram-Negative Cocci: *Neisseria*

Structure and Physiology of *Neisseria*

The Gonococcus: *Neisseria gonorrhoeae*

The Meningococcus: *Neisseria meningitidis*

Pathogenic, Gram-Negative, Facultatively Anaerobic Bacilli

The Enterobacteriaceae: An Overview

Coliform Opportunistic Enterobacteriaceae

Noncoliform Opportunistic Enterobacteriaceae

Truly Pathogenic Enterobacteriaceae

The Pasteurellaceae

Pathogenic, Gram-Negative, Aerobic Bacilli

Bartonella

Brucella

Bordetella

Pseudomonads

Francisella

Legionella

Coxiella

Pathogenic, Gram-Negative, Anaerobic Bacilli

Bacteroides

Prevotella

Late in 2002, a New Mexico couple vacationing in New York City fell ill with flulike symptoms. A few days later they were admitted to a hospital, suffering from chills and a high fever. The wife had developed a "bubo" (a severely swollen lymph node) on her thigh, leading doctors to suspect bubonic plague—the dreaded "black death" that killed one-third of all Europeans during the 14th century but had not been seen in New York City in over 100 years. Subsequent tests confirmed that they were indeed infected with *Yersinia pestis*, the causative bacterium of plague. The wife was administered antibiotics and recovered quickly, but her husband worsened, requiring amputation of both feet and remaining hospitalized for 3 months. Authorities later determined that the couple had most likely contracted the disease from infected fleas encountered near their property in rural New Mexico.

Y. pestis is one example of a Gram-negative pathogen—a group that also includes the genera *Escherichia*, *Salmonella*, and *Legionella*. Gram-negative pathogens cause a wide variety of human diseases. In this chapter we study important Gram-negative cocci and bacilli, and major human diseases they cause.

Yersinia pestis, the causative agent of plague.

MicroPrep Pre-Test: Take the pre-test for this chapter on the web. Visit **www.microbiologyplace.com**.

Numerous Gram-negative, cocci, bacilli, and coc-cobacilli are bacterial pathogens of humans. In fact, Gram-negative bacteria constitute the largest group of human pathogens, in part because the outer membrane of a Gram-negative bacterial cell wall contains lipid A (see Figure 3.13). Lipid A triggers fever, vasodilation, inflammation, shock, and **disseminated intravascular coagulation (DIC)**—the formation of blood clots within blood vessels throughout the body. Almost every Gram-negative bacterium that can breach the skin or mucous membranes, grow at 37°C, and evade the immune system can cause disease and death in humans; however, most of the Gram-negative bacteria that cause disease fall into fewer than 30 genera.

We have seen that it is possible to group these organisms according to a variety of criteria. In Chapter 11 we considered the Gram-negative bacteria according to their placement in *Bergey's Manual*, which is based largely on genomic similarities; here we primarily consider them according to a different organizational scheme that traditionally has been used for many clinical purposes—according to their shapes, oxygen requirements, and other biochemical properties.

First we survey the primary genus of Gram-negative cocci that are pathogenic to humans *(Neisseria);* then we consider 13 genera of pathogenic facultatively anaerobic bacilli in the families Enterobacteriaceae and Pasteurellaceae, which account for almost half of all Gram-negative pathogens. Finally, we examine a heterogenous collection of nine genera of pathogenic aerobic bacilli before concluding by considering two genera of strictly anaerobic bacilli.

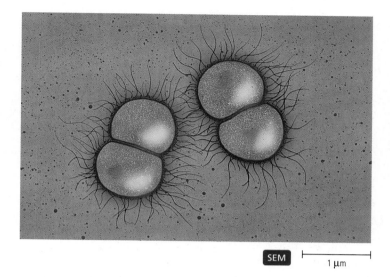

SEM |———| 1 μm

▲ *Figure 20.1*

Diplococci of *Neisseria*. *What function of gonococcal fimbriae contributes to the bacteria's pathogenicity?*

Figure 20.1 The fimbriae act as adhesins, enabling gonococci to adhere to cells lining the cervix and urethra and to sperm, which allows Neisseria to travel far into the female reproductive tract.

Pathogenic Gram-Negative Cocci: *Neisseria*

Neisseria (nī-se′rē-ă) is the only genus of Gram-negative cocci that regularly causes disease in humans. *Neisseria* is placed in the class Betaproteobacteria of the phylum Proteobacteria. Morphologically similar pathogenic bacteria in the genera *Moraxella* (mōr′ak-sel′ă) and *Acinetobacter* (as-i-nē′tō-bak′ter), considered coccobacilli, are now known to be more closely related to aerobic bacilli in class Gammaproteobacteria and thus are discussed in a later section.

Structure and Physiology of *Neisseria*

Learning Objective

✓ List three structural features of *Neisseria* that contribute to its pathogenicity.

The cells of all strains of *Neisseria* are nonmotile and typically arranged as diplococci (pairs) with their common sides flattened in a manner reminiscent of coffee beans **(Figure 20.1).** In addition to their shape, these aerobic bacteria are distinguished from many other Gram-negative pathogens

by being **oxidase positive** (see Figure 20.4); that is, their electron transport chains contain the enzyme *cytochrome oxidase*. Pathogenic strains of *Neisseria* also have fimbriae and polysaccharide capsules, as well as a major cell wall antigen called *lipooligosaccharide* (*LOS*, lip′ō-ol-i-gō-sak′a-rīd), composed of lipid A (endotoxin) and sugar molecules—all of which enable the bacteria to attach to human cells. Cells of *Neisseria* that lack any of these three structural features are typically avirulent.

In the laboratory, *Neisseria* is fastidious in its growth requirements. Microbiologists culture these bacteria on either autoclaved blood agar (called *chocolate agar* because of its appearance) or a selective medium called *modified Thayer-Martin medium. Neisseria* is particularly susceptible to drying and extremes of temperature; a moist culture atmosphere containing 5% carbon dioxide enhances their growth. Laboratory personnel typically use sugar fermentation tests to distinguish among strains of *Neisseria*; pathogenic strains, for example, are unable to ferment either sucrose or lactose.

Two species of *Neisseria* are pathogenic to humans: the so-called gonococcus, *N. gonorrhoeae* (go-nor-rē′ē), which causes gonorrhea, and the meningococcus, *N. meningitidis* (me-nin-ji′ti-dis), which causes a type of meningitis.

The Gonococcus: *Neisseria gonorrhoeae*

Learning Objectives

✓ Compare and contrast the symptoms of gonorrhea in men and women.

✓ Discuss the difficulties researchers face in developing an effective vaccine against *Neisseria gonorrhoeae*.

Pathogenesis, Epidemiology, and Disease

Neisseria gonorrhoeae causes **gonorrhea** (gon-ō-rē'ǎ), which has been known as a sexually transmitted disease for centuries, although the disease was often confused with syphilis until the 19th century. In the 2nd century A.D., the Roman physician Claudius Galen named gonorrhea for what he thought was its cause—an excess of semen (gonorrhea means "flow of seed" in Greek). The disease is sometimes called "clap" from an archaic French word *clapoir,* meaning brothel.

Gonorrhea occurs in humans only. Even though it is one of the more common sexually transmitted diseases in the United States, the number of cases among civilians has been declining over the past few decades **(Figure 20.2a).** Most U.S. cases of gonorrhea in civilians occur in adolescents, particularly among those who engage in a promiscuous sexual lifestyle in several southeastern states **(Figure 20.2b).** The disease is slightly more common among males than females, and four times more common among blacks than nonblacks. Whereas women have a 50% chance of becoming infected during a single sexual encounter with an infected man, men have only a 20% chance of infection during a single encounter with an infected woman. An individual's risk of infection increases with increasing frequency of sexual encounters.

Gonococci adhere, via their fimbriae and capsules, to epithelial cells of the mucous membranes lining the genital, urinary, and digestive tracts of humans. As few as 100 pairs of cells can cause disease. The cocci protect themselves from the immune system by secreting a protease enzyme that cleaves secretory IgA in mucus. Further, phagocytized bacteria survive and multiply within neutrophils that have phagocytized them. As the bacteria multiply, they invade deeper connective tissues, often while being transported intracellularly by phagocytes.

In men, gonorrhea is usually symptomatic—acute inflammation typically occurs 2–5 days after infection in the urethra of the penis, causing extremely painful urination and a purulent (pus-filled) discharge. Rarely, the gonococci invade the prostate or epididymis, where the formation of scar tissue can render the man infertile.

In contrast, gonorrhea in women is often asymptomatic; about 50% of infected women have no symptoms or obvious signs of infection. *Neisseria* cannot attach to cells lining the vagina; instead, the bacteria most commonly infect the cervix of the uterus. Gonococci can infect deep in the uterus and even the uterine (Fallopian) tubes by "hitchhiking" to these locations on sperm to which they attach by means of fimbriae or LOS. In the uterine tubes they can trigger in-

(a)

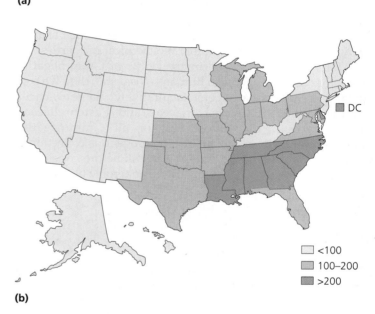

(b)

▲ *Figure 20.2*

The incidence of gonorrhea in the United States. **(a)** The number of new cases annually, 1977–2002. **(b)** The geographic distribution of cases reported in 2000. The incidences shown are the number of new cases per 100,000 individuals in each state. *What organism causes gonorrhea?*

Figure 20.2 Gonorrhea is caused by Neisseria gonorrhoeae.

flammation, fever, and abdominal pain—a condition known as **pelvic inflammatory disease (PID).** Chronic infections can lead to scarring of the tubes, resulting in *ectopic*[1] *pregnancies* or sterility.

---

[1]From Greek *ektopos,* meaning out of place.

Gonococcal infection of organs outside the reproductive tracts can also occur. Because a woman's urethral opening is close to her vaginal opening, the urethra can become infected with gonococci during sexual intercourse. Anal intercourse can lead to *proctitis* (inflammation of the rectum), and oral sexual intercourse can infect the pharynx or gums, resulting in *pharyngitis* and *gingivitis*, respectively. These conditions are most commonly seen in men who have sex with men. In very rare cases, gonococci enter the blood and travel to the joints, meninges, or heart, causing arthritis, meningitis, and endocarditis, respectively.

Gonococcal infection of a child during birth can result in inflammation of the cornea, *ophthalmia neonatorum* (inflammation of the conjunctiva in newborns), and even blindness. Gonococci can also infect the respiratory tracts of newborns during their passage through the birth canal. Infection of older children with *N. gonorrhoeae* is strong evidence of sexual abuse by an infected adult.

Diagnosis, Treatment, and Prevention

The presence of Gram-negative diplococci in pus from an inflamed penis is sufficient for a diagnosis of symptomatic gonorrhea in men. For diagnosis of asymptomatic cases of gonorrhea in men and women, commercially available genetic probes provide direct, accurate, rapid detection of *N. gonorrhoeae* in clinical specimens. Alternatively, the bacteria can be cultured on selective media, such as Thayer-Martin agar.

The treatment of gonorrhea has been complicated in recent years by the worldwide spread of gonococcal strains that are resistant to penicillin, tetracycline, erythromycin, and aminoglycosides. Currently, the CDC recommends broad-spectrum oral cephalosporin or oral quinolones to treat gonorrhea.

There is no long-term specific immunity against *N. gonorrhoeae*, and thus promiscuous people can be infected multiple times. This lack of immunity is explained in part by the highly variable surface antigens in this bacterium; immunity against one strain may protect against homologous strains but provides no protection against other strains. Further, the existence of many different strains has prevented the development of an effective vaccine.

Except for the routine administration of antimicrobial agents to newborns' eyes, which successfully prevents ophthalmic disease, chemical prophylaxis is ineffective in preventing disease. In fact, the use of antimicrobials to prevent genital disease may select for hardier resistant strains, worsening the situation. The most effective prevention is sexual abstinence, followed by monogamy with a monogamous partner, followed by the use of condoms. Efforts to stem the spread of gonorrhea focus on education, aggressive detection, and the screening of all sexual contacts of carriers.

CRITICAL THINKING

Penicillin-resistant strains of *N. gonorrhoeae* produce a plasmid-coded beta-lactamase, which degrades penicillin. What changes in structure or metabolism could explain resistance to other antimicrobial drugs in this bacterium?

The Meningococcus: *Neisseria meningitidis*

Learning Objectives

✓ Describe how meningococci survive and thrive in humans.
✓ Discuss the epidemiology of meningococcal diseases.

Neisseria meningitidis (nī-se′rē-ă me-nin-ji′ti-dis) causes life-threatening diseases when it invades the blood or cerebrospinal fluid (CSF).

Pathogenicity, Epidemiology, and Disease

Of the 13 known antigenic strains of meningococci, strains A, B, C, and W 135 cause most cases of disease in humans. The polysaccharide capsules of *N. meningitidis* resist lytic enzymes of the body's phagocytes, allowing phagocytized meningococci to survive, reproduce, and be carried throughout the body within neutrophils and macrophages. In the United States, the bacterium causes most cases of meningitis in children and adults under 20. Most of the damage caused by *N. meningitidis* results when sufficient amounts of the lipid A component of LOS are released into extracellular spaces, where it triggers fever, vasodilation, inflammation, shock, and DIC.

Humans are the only natural carriers of *N. meningitidis*; in fact, the bacterium is a common member of the normal microbiota in the upper respiratory tracts of up to 40% of healthy people. As a result of crowded living conditions, meningococci are more prevalent in children and young adults from lower economic groups. Even though meningococcal diseases occur worldwide, cases are relatively rare in countries with advanced health care systems and access to antimicrobial drugs.

Respiratory droplets transmit the bacteria among people living in close contact, especially families, soldiers living in barracks, prisoners, and college students living in dormitories. In fact, meningococcal disease is 9–23 times more prevalent in students living in dormitories than in the general population. Individuals whose lungs have been irritated by dust are more susceptible to airborne *Neisseria*. For example, annual outbreaks of meningococcal meningitis occur in sub-Saharan Africa when dry winter winds create irritating dust storms.

The incidence of meningococcal disease in the United States is about 1 case per 100,000 population. Most of these cases are meningitis, usually involving abrupt sore throat, fever, headache, stiff neck, vomiting, and convulsions. (Very young children may have only fever and vomiting.) Arthritis and partial loss of hearing occasionally result. Meningococcal

(a)

(b)

▲ *Figure 20.3*

Petechiae in meningococcal septicemia. These lesions can either be small and relatively diffuse **(a)** or may coalesce in areas containing dead cells **(b)**.

meningitis can progress so rapidly that death results within 6 hours of the initial symptoms.

Meningococcal septicemia (blood poisoning) can also be life threatening. LOS may trigger shock, devastating blood coagulation in many organs, and *petechiae* (pe-tē′kē-ē)—minute hemorrhagic skin lesions—on the trunk and lower extremities **(Figure 20.3a).** Petechiae may coalesce and combine with regions of cell death to form large black lesions **(Figure 20.3b).** In some patients, however, septicemia produces only mild fever, arthritis, and petechiae.

Diagnosis, Treatment, and Prevention

Because meningococcal meningitis constitutes a medical emergency, rapid diagnosis is critical. The demonstration of Gram-negative diplococci within phagocytes of the central nervous system is diagnostic for meningococcal disease. *N. meningitidis* can ferment maltose in laboratory culture, which distinguishes it from *N. gonorrhoeae.* Further, serological tests can demonstrate the presence of antibodies against *N. meningitidis,* though strain B is relatively nonimmunogenic and therefore does not react well in such tests.

Despite the fact that meningococcal meningitis proceeds rapidly and is ordinarily 100% fatal when left untreated, immediate treatment with antimicrobial drugs has reduced mortality to less than 10%. Penicillin, administered intravenously, is the drug of choice, although penicillin resistance is becoming more common. Chloramphenicol and cephalosporin are also effective.

Because healthy asymptomatic carriers are common, eradication of meningococcal disease is unlikely. Prevention involves prophylactic treatment with sulfonamides, tetracycline, or rifampin of anyone exposed to individuals with meningococcal disease, including family, classmates, day care center contacts, roommates, and health care workers.

Researchers have developed a vaccine against strains A, C, and Y, but it cannot be used with children under age 2. As previously noted, strain B is weakly immunogenic, so immunity to it cannot be induced by a vaccine and must instead develop following a natural infection.

CRITICAL THINKING

Investigations of *Mycobacterium tuberculosis* (mī-kō-bak-tēr′ē-ŭm tŭ-ber-kyū-lō′sis) have clearly demonstrated that this pathogen can be transmitted among airline passengers; however, as of 2002, no cases of in-flight transmission of meningococci had been reported to the CDC. What are some possible explanations for the lack of documented cases of in-flight transmission of *N. meningitidis*?

Pathogenic, Gram-Negative, Facultatively Anaerobic Bacilli

Learning Objective

✓ Describe how members of the family Enterobacteriaceae are distinguished from members of the family Pasteurellaceae.

Two families of facultatively anaerobic bacilli and coccobacilli in the class Gammaproteobacteria—Enterobacteriaceae and Pasteurellaceae—contain most of the Gram-negative pathogens of humans. Scientists distinguish between members of the two families by performing an oxidase test: Members of the Enterobacteriaceae are oxidase negative, whereas members of the Pasteurellaceae are oxidase positive **(Figure 20.4).**

We begin our survey of these microorganisms by examining the family Enterobacteriaceae, which contains about 150 species in 41 genera, including some of the more important nosocomial pathogens of humans **(Figure 20.5).**

▲ *Figure 20.4*

Use of the oxidase test to distinguish members of the family Enterobacteriaceae (left), which are oxidase negative, from members of the family Pasteurellaceae, which are oxidase positive, as indicated by a purple color.

The Enterobacteriaceae: An Overview

Learning Objectives

✓ Discuss how members of the family Enterobacteriaceae are distinguished in the laboratory.

✓ List six virulence factors found in members of the family Enterobacteriaceae.

✓ Describe the diagnosis, treatment, and prevention of diseases of enteric bacteria.

As their name suggests, prokaryotes in the family **Entero-bacteriaceae** (en'ter-ō-bak-tēr-ē-ā'sē-ē)—also called **enteric bacteria**—are members of the intestinal microbiota of most animals and humans. They are also ubiquitous in water, soil, and decaying vegetation. Some species are always pathogenic in humans, whereas others are opportunists normally found in the intestinal microbiota that become pathogenic when introduced into other body sites.

As a group, enteric bacteria are the most common Gram-negative pathogens of humans. Though most enteric bacteria can be pathogenic in humans, over 95% of the medically important species are in the 13 genera we discuss shortly.

Structure and Physiology of the Enterobacteriaceae

Enteric bacteria are coccobacilli or bacilli that measure about 1 μm × 1.2–3 μm **(Figure 20.6)**. Those members that are motile have peritrichous flagella. Some have prominent capsules, whereas others have only a loose slime layer. All members of the family reduce nitrate to nitrite and ferment glucose anaerobically, though most of them grow better in aerobic environments. As previously noted, all of them are also oxidase negative.

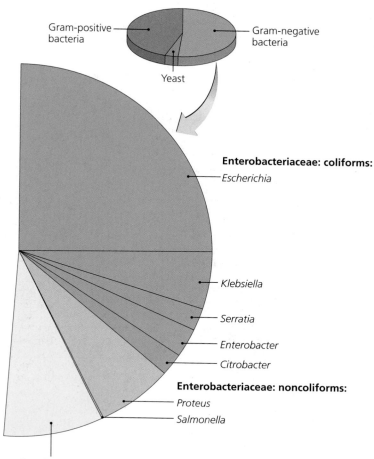

Gram-positive bacteria

Gram-negative bacteria

Yeast

Enterobacteriaceae: coliforms:
  Escherichia

  Klebsiella

  Serratia

  Enterobacter

  Citrobacter

Enterobacteriaceae: noncoliforms:
  Proteus

  Salmonella

Pseudomonas
(not in family Enterobacteriaceae)

▲ *Figure 20.5*

Relative causes of nosocomial infections in the United States. Gram-negative bacteria in the family Enterobacteriaceae account for most nosocomial infections.

(a) LM   25 μm   **(b)** LM   25 μm

▲ *Figure 20.6*

Gram stains of bacteria in the family Enterobacteriaceae. **(a)** *Enterobacter aerogenes*, a coccobacillus. **(b)** *Escherichia coli*, a bacillus. Note the similarity in staining properties.

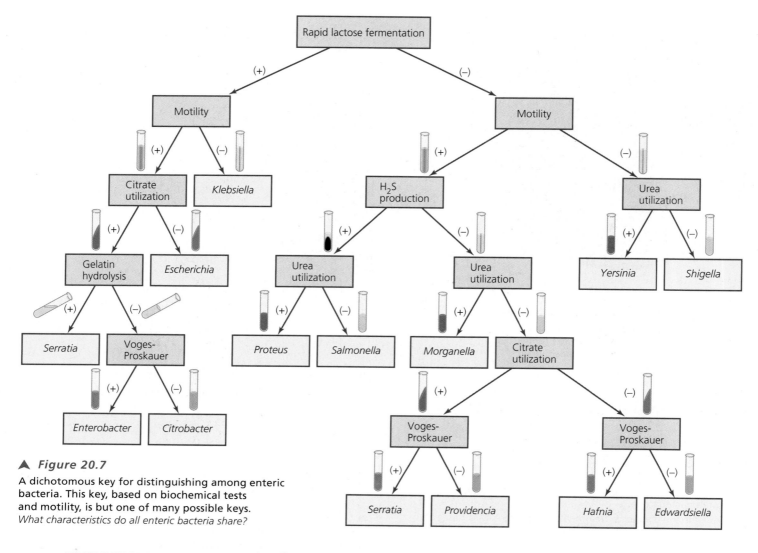

▲ **Figure 20.7**

A dichotomous key for distinguishing among enteric bacteria. This key, based on biochemical tests and motility, is but one of many possible keys. *What characteristics do all enteric bacteria share?*

Figure 20.7 All enteric bacteria are Gram-negative, oxidase negative, and able to metabolize nitrate to nitrite.

Given that all enteric bacteria are similar in microscopic appearance and staining properties, scientists traditionally distinguish among them by using biochemical tests, motility, and colonial characteristics on a variety of selective and nonselective media (for example, MacConkey agar and blood agar). **Figure 20.7** presents one dichotomous key, based on motility and biochemical tests, for distinguishing among the 13 genera of enteric bacteria discussed in this chapter.

Pathogenicity of the Enterobacteriaceae

Members of the Enterobacteriaceae have an outer membrane that contains lipopolysaccharide composed of three antigenic components:

- a core polysaccharide, shared by all of the enteric bacteria and called *common antigen*

- so-called *O polysaccharide,* which has various antigenic varieties among strains and species

- lipid A, which (as noted in our examination of *Neisseria*) can trigger fever, vasodilation, inflammation, shock, and DIC. Because of the presence of lipid A, most of the enteric bacteria can cause serious disease and death.

Other antigens of certain enteric bacteria include protein and polysaccharide capsular antigens called K antigens (Vi antigens in *Salmonella*) and flagellar proteins called H antigens. By controlling the genetic expression of K and H antigens, alternately producing and not producing the antigens, the bacteria survive by evading their host's immune system. Scientists use serological identification of antigens to distinguish among strains and species of enteric bacteria. For example, *Escherichia coli* O157:H7, a potentially deadly

Case Study 20.1   *Neisseria* Nightmare

A 17-year-old male reports to his physician that he has experienced painful urination, as if he were urinating molten solder. He has also noticed a puslike discharge from his penis. A Gram-stained smear of the discharge is shown in the photo.

The patient reports having been sexually active with two or three women in the previous 6 months. Because his partners

LM   10 µm

reported being "absolutely sure" that they carried no sexually transmitted diseases, the patient had not used a condom.

1. What disease caused by *Neisseria* does this patient most likely have? What are the medical and common names for this disease?

2. How did the patient acquire the disease?

3. What could explain the lack of any history of sexually transmitted disease in his sexual partners?

4. What is the likely treatment?

5. Is the patient immune to future infections of *Neisseria*?

bacterium, is so designated in part because of its antigenic components. The specificity of serological tests can also be used in diagnosis and in epidemiological studies to identify the source of a particular infecting agent.

Among the variety of virulence factors possessed by pathogenic enteric bacteria, some (for example, lipid A) are common to all genera, whereas others are unique to certain strains. Virulence factors seen in the Enterobacteriaceae include the following:

- *Capsules* that protect the bacteria from phagocytosis and provide a poorly immunogenic surface and protection from antibodies.

- *Fimbriae* and proteins called *adhesins*, which enable the bacteria to attach tightly to human cells.

- *Exotoxins* that cause a variety of symptoms such as diarrhea. The genes for exotoxins, fimbriae, and adhesins are frequently located on plasmids, which increases the likelihood that they will be transferred among bacteria.

- *Iron-binding compounds* that capture iron and store it in the bacteria.

- *Hemolysins,* which release nutrients such as iron by lysing red blood cells.

- A so-called *type III secretion system,* a complex structure composed of 20 different polypeptides that is synthesized by several pathogenic enteric species. Once assembled, the system spans the two membranes and peptidoglycan of the bacterial cell and inserts into the host cell's cytoplasmic membrane, forming a channel through which bacterial proteins are introduced into the host cell. Because the bacteria synthesize type III systems only after they have contacted a host cell, the proteins of a type III system are never exposed to leukocytes, which helps enteric bacteria evade immune surveillance.

Figure 20.8 on page 574 illustrates the locations of the major types of antigens and virulence factors of enteric bacteria.

Diagnosis, Treatment, and Prevention of Diseases of the Enterobacteriaceae

The presence of enteric bacteria in urine, blood, or cerebrospinal fluid is always diagnostic of infection or disease. To culture members of the Enterobactericeae from clinical specimens, clinicians use selective and differential media such as eosin methylene blue (EMB) agar and MacConkey agar. In their role as selective media, these media inhibit the growth of many members of the normal microbiota and allow the growth of enteric bacteria (**Figure 20.9a** on page 575); in their role as differential media, they enable laboratory technicians to distinguish between enteric bacteria that can and cannot ferment lactose (**Figure 20.9b**). Researchers have developed such sophisticated biochemical tests to distinguish among enteric bacteria that the bacteria can be accurately identified by commercially available systems in 4–24 hours.

Treatment of diarrhea involves treating the symptoms with fluid and electrolyte replacement. Often, antimicrobial drugs are not used to treat diarrhea because diarrhea is typically self-limiting—the expulsion of the organisms from the body is often a more effective therapy than the action of antimicrobials. Also, antimicrobial therapy can worsen the prognosis by killing many bacteria at once, which releases a large amount of lipid A.

Treatment of internal infections involves selection of an appropriate antimicrobial drug, which often necessitates a susceptibility test such as a Kirby-Bauer test or an Etest.® Susceptible enteric bacteria may develop resistance to antimicrobial drugs, presumably because enteric bacteria exchange plasmids readily; many enterics can exchange DNA via conjugation.

Given that enteric bacteria are a major component of the normal microbiota, preventing infections with enterics is almost impossible. Health care practitioners should give prophylactic antimicrobial drugs to patients whenever mucosal barriers are breached by trauma or surgery, while avoiding prolonged unrestricted use of antimicrobials, a practice that

Antigens

Virulence Factors

Outer membrane:
(common antigen,
O antigen,
Lipid A)

Type III
secretion
system

Capsular antigens
(K; Vi in *Salmonella*)

Flagellar
antigens (H)

Nucleoid

Pilus

Exotoxin

Adhesin

Plasmid
(virulence genes)

Iron-binding protein

Hemolysin

▲ *Figure 20.8*

Antigens and virulence factors of typical enteric bacteria. *What is the function of a type III system?*

Figure 20.8 A type III secretion system is a virulence factor that enables enteric bacteria to insert chemicals directly into the cytosol of a eukaryotic cell.

can select for resistant strains of bacteria. Good personal hygiene (particularly hand antisepsis) and proper sewage control are important in limiting the risk of infection. Specific methods of control and prevention associated with particular genera of enteric bacteria are discussed in the following sections.

CRITICAL THINKING

Resistance to antimicrobial agents is more commonly seen in hospital-acquired infections with enteric bacteria than in community-based infections with the same species. Explain why this is so.

Scientists, researchers, and health care practitioners find it useful to categorize the pathogenic members of the Enterobacteriaceae into three groups:

- coliforms, which rapidly ferment lactose, are part of the normal microbiota, and may be opportunistic pathogens
- noncoliform opportunists, which do not ferment lactose
- true pathogens

In the next three sections we will examine the important members of the Enterobacteriaceae according to this scheme, beginning with the opportunistic coliforms.

Coliform Opportunistic Enterobacteriaceae

Learning Objectives

✓ Compare and contrast *Escherichia, Klebsiella, Serratia, Enterobacter, Hafnia,* and *Citrobacter.*
✓ Describe the pathogenesis and diseases of *Escherichia coli* O157:H7.

Coliforms (kol′-i-formz) are defined as aerobic or facultatively anaerobic, Gram-negative, rod-shaped bacteria that ferment lactose to form gas within 48 hours of being placed in a lactose broth at 35°C. Coliforms are found in the intestinal tracts of animals and humans, in soil, and on plants and decaying vegetation. The presence of fecal coliforms in water is indicative of impure water and of poor sewage treatment. The most common of the opportunistic coliform pathogens are bacteria in the genera *Escherichia, Klebsiella, Serratia, Enterobacter, Hafnia,* and *Citrobacter.*

Staphylococcus aureus

Escherichia coli

(a)

Escherichia coli

(b)

Shigella flexneri

▲ *Figure 20.9*

The use of a selective and differential medium (here, MacConkey agar) to distinguish among bacteria in clinical specimens. **(a)** MacConkey agar inhibits Gram-positive bacteria (here, *Staphylococcus aureus*, at left) and allows growth of Gram-negative bacteria (here, *Escherichia coli*, at right). **(b)** Additionally, colonies of enteric bacteria that ferment lactose (here, *E. coli*) are pink, differentiating them from other enteric bacteria that do not ferment lactose (here, *Shigella flexneri*).

Escherichia coli

Escherichia coli (esh-ĕ-rik′ē-ă kō′ lī)—the most common and important of the coliforms—is well known as a laboratory research organism.

Scientists have described numerous O, H, and K antigens of *E. coli*, which epidemiologists use to identify particular strains. Some antigens, such as O157, O111, H8, and H7, are associated with virulence. Virulent strains have genes (located on virulence plasmids) for fimbriae, adhesins, and a variety of exotoxins, enabling these strains to colonize human tissue and cause disease. Such plasmids can be transferred among bacteria.

E. coli is responsible for a number of diseases, including septicemia, urinary tract infections (UTIs), neonatal meningitis, and gastroenteritis. Gastroenteritis, the most common disease associated with *E. coli*, is often mediated by exotoxins, called *enterotoxins*, that bind to proteins on cells lining the intestinal tract. A portion of the toxin then enters the cell and triggers a series of chemical reactions that cause the loss of sodium, chloride, potassium, bicarbonate, and water from the cells, producing watery diarrhea, cramps, nausea, and vomiting. Enterotoxin-producing strains are common in developing countries and are important causes of pediatric diarrhea, accounting for up to a third of the cases of this life-threatening disease in some countries.

E. coli is the most common cause of non-nosocomial urinary tract infections. Women, with their shorter urethras, are 15 times more likely to acquire UTIs than are men. Surprisingly, girls are less likely to acquire UTIs than are women, perhaps because of the hormonal influences of adulthood or

an increased likelihood of UTIs following sexual intercourse. About 32% of adult women will suffer uncomplicated mild bladder infections by *E. coli* sometime during their lifetimes. In some cases, the bacteria may ascend the ureters from the bladder, infecting the kidneys and causing a more serious disease called *acute pyelonephritis,* which involves fever, flank pain, bacteriuria (bacteria in the urine), profuse perspiration, nausea, and vomiting.

E. coli O157:H7, first described as a cause of human illness in 1982, is now the most prevalent strain of pathogenic *E. coli* in developed countries; in the United States it causes an estimated 70,000 cases and 60 deaths annually. The ingestion of as few as 10 organisms can result in disease, which ranges from bloody diarrhea, to severe or fatal hemorrhagic colitis, to a severe kidney disorder called *hemolytic uremic syndrome,* in which renal function ceases. Urinary tract and blood infections with this organism can be fatal, especially in children and immunocompromised adults. Most epidemics of *E. coli* O157:H7 have been associated with the consumption of undercooked ground beef or unpasteurized milk or fruit juice contaminated with feces (see Highlight 9.1 on page 272). Despite its notoriety, only one in 10 million deaths in the United States is caused by a U.S. citizen experiencing death or injury from stain O157:7 in ground beef.

E. coli O157:H7 produces a type III secretion system through which it injects two types of proteins into intestinal cells. Proteins of one type disrupt the cell's metabolism; proteins of the other type become lodged in the cell's cytoplasmic membrane, where they act as a receptors for the attachment of additional *E. coli* O157:H7 bacteria. Such attachment apparently enables this strain of *E. coli* to displace normal harmless strains.

E. coli O157:H7 also produces *Shiga-like toxin,* which inhibits protein synthesis in host cells. Shiga-like toxin, which was first identified in *Shigella* (discussed shortly), attaches to the surfaces of neutrophils and is spread by them throughout the body, causing widespread death of host cells and tissues. Antimicrobial drugs induce *E. coli* O157:H7 to increase its production of Shiga-like toxin, exacerbating the disease.

Klebsiella

Species of the genus *Klebsiella* (kleb-sē-el′ă) grow in the digestive and respiratory systems of humans and animals. They are nonmotile, and their prominent capsules (**Figure 20.10** on page 576) give *Klebsiella* colonies a mucoid appearance and protect the bacteria from phagocytosis, enabling them to become opportunistic pathogens.

The most commonly isolated pathogenic species is *K. pneumoniae,* which causes pneumonia and may be involved in bacteremia, meningitis, wound infections, and UTIs. Pneumonia caused by *K. pneumoniae* often involves destruction of alveoli and the production of bloody sputum. Alcoholics and patients with compromised immunity are at greater risk of pulmonary disease because of their poor ability to clear aspirated oral secretions from their lower respiratory tracts.

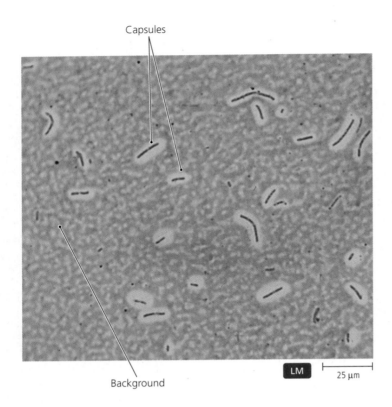

Capsules

Background

LM 25 μm

▲ *Figure 20.10*

The prominent capsule of *Klebsiella pneumoniae*.
How does the capsule function as a virulence factor?

Figure 20.10 Capsules inhibit phagocytosis and intracellular digestion by phagocytic cells.

Serratia

The motile coliform *Serratia marcescens* (se-răt'ē-ă mar-se'senz) produces a red pigment when grown at room temperature **(Figure 20.11)**. For many years, scientists considered *Serratia* to be totally benign; in fact, researchers intentionally introduced this bacterium into various environments to track the movement of bacteria in general. Using this method, scientists have followed the movement of *Serratia* from the mouth into the blood of dental patients, and throughout hospitals via air ducts. In a series of tests performed between 1949 and 1968, the U.S. government released large quantities of *S. marcescens* into the air over New York City and San Francisco to mimic the movement of a discharged biological weapon.

Now, however, we know that *Serratia* can grow on catheters, in saline solutions, and on other hospital supplies, and that it is a potentially life-threatening opportunistic pathogen in the urinary and respiratory tracts of immunocompromised patients. Antibacterial agents are often ineffective in treating *Serratia* infections because the bacteria are frequently resistant to such drugs.

▲ *Figure 20.11*

The red colonies of *Serratia marcescens* grown at room temperature on agar.

Enterobacter, Hafnia, and *Citrobacter*

Other motile coliforms that can be opportunistic pathogens are *Enterobacter* (en'ter-ō-bak'ter), *Hafnia* (haf'nē-ă), and *Citrobacter* (sit'rō-bak-ter). These bacteria, like other coliforms, ferment lactose and reside in the digestive tracts of animals and humans as well as in soil, water, decaying vegetation, and sewage. *Enterobacter* can be a contaminant of dairy products. All three genera are frequently involved in nosocomial infections of the blood, wounds, surgical incisions, and urinary tracts of immunocompromised patients. Treatment of such infections can be difficult because these prokaryotes, especially *Enterobacter*, are often resistant to most antibacterial drugs.

Noncoliform Opportunistic Enterobacteriaceae

Learning Objectives

✓ Differentiate between coliform and noncoliform opportunists.
✓ Describe the diseases caused by noncoliform opportunistic enteric bacteria.

The family Enterobacteriaceae also contains genera that cannot ferment lactose; several of these genera are opportunistic pathogens. These noncoliform opportunists include *Proteus, Morganella, Providencia,* and *Edwardsiella*.

Proteus

Like the mythical Greek god Proteus, who could change shape, the Gram-negative, facultative anaerobe *Proteus* (prō'tē-ŭs) is a typical rod-shaped bacterium with a few polar flagella when cultured in broth, but differentiates into an elongated cell that swarms by means of numerous peritrichous flagella when cultured on agar. Swarming cells produce concentric wavelike patterns on agar surfaces **(Figure 20.12)**. The roles of the two morphologies in human infections are not clear.

Proteus mirabilis (mi-ra'bi-lis) is the most common species of *Proteus* associated with disease in humans, particularly with urinary tract infections in patients with long-term urinary catheters. One study showed that 44% of catheterized patients had *P. mirabilis* growing in their urine.

In the presence of urea, this microorganism releases a large amount of the enzyme *urease*, which breaks down urea into carbon dioxide and ammonia. The ammonia raises the pH of urine in the bladder and kidneys such that ions that are normally soluble at the normal acidic pH of urine (around pH 6.0) precipitate, often around bacterial cells, to form infection-induced kidney stones composed of magnesium ammonium phosphate or calcium phosphate. Kidney stones produced in this manner can contain hundreds of *P. mirabilis* cells.

Proteus is resistant to many antimicrobial drugs so health care workers must determine proper treatment following a susceptibility test. Researchers have been able to immunize mice against urinary tract infections with *Proteus*. This success is encouraging to those seeking protection for patients who must have long-term urinary catheters.

Morganella, Providencia, and Edwardsiella

Nosocomial infections of immunocompromised patients by *Morganella* (mōr'gan-el'-ă), *Providencia* (prov'i-den'sē-ă), and *Edwardsiella* (ed'ward-sē-el'lă) are becoming more frequent. These organisms are primarily involved in urinary tract infections. *Providencia* contains a plasmid that codes for urease, and thus triggers the formation of kidney stones.

Truly Pathogenic Enterobacteriaceae

Learning Objectives

✓ Describe the diseases caused by truly pathogenic enteric bacteria.
✓ Contrast salmonellosis and shigellosis.
✓ Describe the life cycle of *Yersinia pestis*, and contrast bubonic and pneumonic plague.

Three important nonlactose-fermenting bacteria in the family Enterobacteriaceae are *Salmonella*, *Shigella*, and *Yersinia*. These genera are not considered members of the normal microbiota of humans, because they are almost always pathogenic due to their numerous virulence factors. All three synthesize type III secretion systems through which they introduce proteins that inhibit phagocytosis, rearrange the cytoskeletons of eukaryotic cells, or induce apoptosis. We begin our discussion with *Salmonella*, which is the most common of the three pathogens.

▲ *Figure 20.12*
The wavelike concentric rings of the swarming bacterium *Proteus microbilis* on an agar surface. Although this characteristic is beautiful, it makes isolation of *Proteus* difficult.

Salmonella

Salmonella (sal'mŏ-nel'ă) is a genus of motile, Gram-negative, peritrichous bacilli that live in the intestines of virtually all birds, reptiles, and mammals and are eliminated in their feces. These bacteria do not ferment lactose, but they do ferment glucose, usually with gas production. They are urease negative and oxidase negative, and most produce hydrogen sulfide (H_2S). Scientists have identified more than 2000 unique serotypes (strains) of *Salmonella*. Analysis of the DNA sequences of their genes indicates that all of them properly belong to a single species—*S. enterica*. However, because this taxonomically correct approach has not been well received by researchers or medical professionals, we will consider the various strains of this single species according to their traditional specific epithets, such as *S. typhi* (tī'fē), *S. paratyphi* (par'a-ti'fē), and *S. typhimurium* (tī-fē-mur-ē-ŭm).

Most infections of humans with salmonellae result from the consumption of food contaminated with animal feces. The pathogen is particularly common in foods containing poultry or eggs—about one-third of chicken eggs are contaminated. *Salmonella* released during the cracking of an egg on a kitchen counter and inoculated into other foods can reproduce into millions of cells in just a few hours. Some strains, such as *S. enterica* (en-ter'i-ka) Dublin, are acquired through the consumption of inadequately pasteurized contaminated milk. Infections with about 1 million cells of most strains of *Salmonella* are usually asymptomatic in people with healthy immune systems.

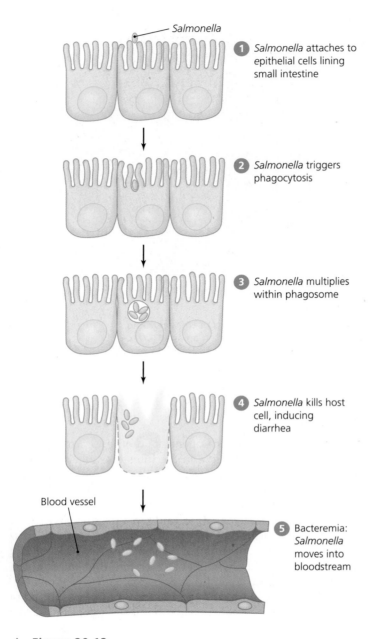

1 *Salmonella* attaches to epithelial cells lining small intestine

2 *Salmonella* triggers phagocytosis

3 *Salmonella* multiplies within phagosome

4 *Salmonella* kills host cell, inducing diarrhea

Blood vessel

5 Bacteremia: *Salmonella* moves into bloodstream

▲ *Figure 20.13*
The events in salmonellosis.

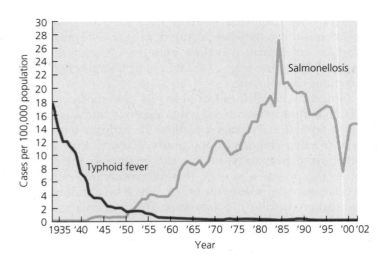

▲ *Figure 20.14*

The incidences of diseases caused by *Salmonella*. Whereas the incidence of typhoid fever has declined due to improved sewage treatment and personal hygiene, the incidence of reported cases of salmonellosis has increased. *What factors might explain the increase in the observed incidence of salmonellosis?*

Figure 20.14 Since salmonellosis and typhoid fever are caused by the same species and transmitted the same way, and the incidence of typhoid fever has decreased significantly, it is likely that the actual number of cases of salmonellosis has declined. The observed increase may be due to better diagnosis and reporting, rather than to an actual increase in the number of cases.

Larger infective doses can result in **salmonellosis** (sal′mō-nel-ō′sis), which is characterized by nonbloody diarrhea, nausea, and vomiting; fever, myalgia, headache, and abdominal cramps are also common. The events in the course of salmonellosis are depicted in **Figure 20.13.** After the salmonellae pass through the stomach, they attach to the cells lining the small intestine **1** . The bacteria then insert proteins into the host cells, inducing the normally nonphagocytic cells to phagocytize the bacteria **2** . The salmonellae then reproduce within phagosome **3** , eventually killing the host cells **4** and inducing the signs and symptoms of salmonellosis. Cells of some strains can subse-

quently enter the blood, causing bacteremia **5** and localized infections throughout the body, including in the lining of the heart, the bones, and the joints.

Humans are the sole hosts of *Salmonella typhi,* which causes **typhoid fever.** (A mild form of typhoid fever is also produced by *S. paratyphi.*) Infection occurs via the ingestion of food or water contaminated with sewage containing bacteria from carriers, who are often asymptomatic. (For the story of perhaps the most notorious asymptomatic carrier ever, see **Highlight 20.1.**) An infective dose of *S. typhi* is only about 1000–10,000 cells. The bacteria pass through the intestinal cells into the bloodstream; there they are phagocytized but not killed by phagocytic cells, which carry the bacteria to the liver, spleen, bone marrow, and gall bladder. Patients typically experience gradually increasing fever, headache, muscle pains, malaise, and loss of appetite that may persist for a week or more. Bacteria may be released from the gall bladder to reinfect the intestines, producing gastroenteritis and abdominal pain, and then a recurrence of bacteremia. In some patients the bacteria ulcerate and perforate the intestinal wall, allowing bacteria from the intestinal tract to enter the abdominal cavity and cause *peritonitis.* **Figure 20.14** contrasts the incidences of salmonellosis and typhoid fever in the United States since 1935.

Highlight 20.1   The Tale of "Typhoid Mary"

Mary Mallon is one of the most famous cooks in history—but not for the tastiness of her dishes. An asymptomatic carrier of typhoid, Mary is believed to have caused outbreaks of typhoid fever in at least seven New York families for whom she worked as a private cook during the early 1900s. She first came to attention in 1906, when Dr. George Soper, a physician hired to investigate an outbreak of typhoid fever in a wealthy Long Island household, became suspicious upon learning that the family's cook—

Mary—had mysteriously disappeared 3 weeks after the illnesses began. Subsequent investigations revealed that Mary had a history of working in households whose members fell ill with typhoid. Public health authorities ultimately tracked Mary down and confirmed the presence of *S. typhi* in her stool and gallbladder.

Uneducated and headstrong, Mary never quite believed that she could be the cause of so much sickness—perhaps because she never fell ill herself—and she refused to have her gallbladder removed, as authorities suggested. Deemed a public health hazard, Mary was quarantined in 1907 and not released until 1910, with the provision that she was never to work as a cook again. Within a few years, however, she was discovered in Manhattan, working as a food preparer, and still spreading typhoid. She was again detained and remained in quarantine until her death, from a stroke, in 1938.

Chloramphenicol and sulfonamides are effective against *S. typhi*. The drugs are not only effective for sick patients, but also can eliminate the bacteria from asymptomatic carriers. Nevertheless, physicians sometimes remove the gallbladder of a carrier to ensure that they do not infect others. Researchers have developed two vaccines that provide temporary protection for travelers to areas where typhoid fever is still endemic.

Shigella

The genus *Shigella* (shē-gel'lă) contains Gram-negative, oxidase negative, nonmotile pathogens of the family Enterobacteriaceae that are primarily parasites of the digestive tracts of humans. These bacteria produce a diarrhea-inducing enterotoxin, are urease negative, and do not produce hydrogen sulfide gas. Some scientists suggest that *Shigella* may actually be a strain of *E. coli* that has become oxidase negative, and some researchers even go so far as to consider *Shigella* an invasive, toxin-producing *E. coli* that is cloaked in *Shigella* antigens. Other scientists contend the converse—that *E. coli* is really a disguised species of *Shigella*!

In any case, taxonomists have historically identified four well-defined species of *Shigella*: *S. dysenteriae* (dis-en-te'rē-ī), *S. flexneri* (fleks'ner-ē), *S. boydii* (boy'dē-ē), and *S. sonnei* (sōn'ne-ē). *Shigella sonnei* is most commonly isolated in industrialized nations, whereas *S. flexneri* is the predominant species in developing countries. All four species cause a severe form of dysentery called **shigellosis** (shig-ĕ-lō'sis), which is characterized by abdominal cramps, fever, diarrhea, and purulent bloody stools.

Shigellosis is primarily associated with poor personal hygiene and ineffective sewage treatment. People become infected primarily by ingesting bacteria on their own contaminated hands, and secondarily by consuming contaminated food. *Shigella* is little affected by stomach acid, so an infective dose may be as few as 200 cells; therefore, person-

to-person spread is also possible, particularly among children and among homosexual men.

Initially, *Shigella* colonizes cells of the small intestine, causing an enterotoxin-mediated diarrhea; the main events in shigellosis, however, begin once the bacteria invade cells of the large intestine (**Figure 20.15** on page 580). When *Shigella* attaches to a large intestine epithelial cell ❶, the cell is stimulated to phagocytize the bacteria ❷, which then multiply within the cell's cytosol ❸. (Note that this differs from salmonellosis, in which bacteria multiply within phagocytic vesicles in cells of the small intestine.) *Shigella* then polymerizes the host's actin fibers, propelling itself out of the host cell and into adjacent cells ❹, in the process evading the host's immune system. As the bacteria kill host cells, abscesses form in the mucosa ❺; any bacteria that enter the blood from a ruptured abscess are quickly phagocytized ❻, so bacteremia is rarely a part of shigellosis. About 20,000 cases of shigellosis are reported each year in the United States, but because mild cases are rarely reported, this figure may represent only 10% of the actual total.

Shigella dysenteriae secretes an exotoxin called **Shiga toxin,** which stops protein synthesis in its host's cells. Shiga toxin is similar to the Shiga-like toxin of *Escherichia coli* O157:H7. Shigellosis caused by *S. dysenteriae* is more serious than that caused by other species of *Shigella*, with a mortality rate as high as 20%.

Treatment of shigellosis involves replacement of fluids and electrolytes. The disease is usually self-limiting, but oral antimicrobial drugs such as ciproflaxacin, sulfonamides, penicillin, or cephalosporin can reduce the spread of shigellae to close contacts of the patient.

A recent live, attenuated vaccine against *S. flexneri* has been successful in preventing dysentery caused by this species, although the participants in the study still experienced mild diarrhea and fever. Researchers are working to perfect a vaccine that will not cause symptoms.

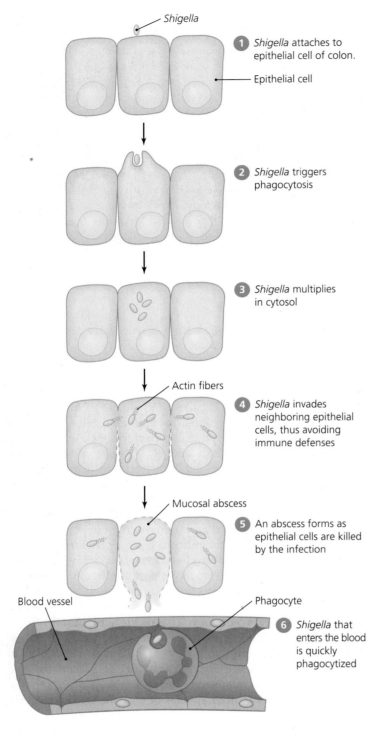

Shigella

1 *Shigella* attaches to epithelial cell of colon.

Epithelial cell

2 *Shigella* triggers phagocytosis

3 *Shigella* multiplies in cytosol

Actin fibers

4 *Shigella* invades neighboring epithelial cells, thus avoiding immune defenses

Mucosal abscess

5 An abscess forms as epithelial cells are killed by the infection

Blood vessel

Phagocyte

6 *Shigella* that enters the blood is quickly phagocytized

▲ *Figure 20.15*
The events in shigellosis.

Yersinia

The genus *Yersinia* (yer-sin′ē-ă) contains three notable species—*Y. enterocolitica* (en′ter-ō-kō-lit′ī-kă), *Y. pseudotuberculosis* (soo-dō tū-ber-kyū-lō′sis), and *Y. pestis* (pes′tis)— that are normally pathogens of animals. All three species

contain virulence plasmids that code for adhesins and type III secretion systems. The adhesins allow *Yersinia* to attach to human cells, after which the type III system is used to inject proteins that trigger apoptosis in macrophages and neutrophils.

Yersinia enterocolitica and *Y. pseudotuberculosis* are enteric pathogens acquired via the consumption of food or water contaminated with animal feces. *Y. enterocolitica* is a common cause of painful inflammation of the intestinal tract; such inflammation is accompanied by diarrhea and fever that can last for weeks or months. Involvement of the terminal portion of the small intestine can result in painful inflammation of the mesenteric lymph nodes, which mimics appendicitis. *Y. pseudotuberculosis* produces a less severe inflammation of the intestines.

Yersinia pestis is an extremely virulent, nonenteric pathogen that has two clinical manifestations: **bubonic** (boo-bon′ik) **plague** and **pneumonic** (noo-mō′nik) **plague.** A major pandemic of plague that lasted from the mid-500s A. D. to the late 700s is estimated to have claimed the lives of more than 40 million people. This devastation was surpassed during a second pandemic that killed about a third of the population of Europe (25 million people) in just 5 years during the 14th century. A third major pandemic spread from China across Asia, Africa, Europe, and the Americas in the 1860s. So great was the devastation of these three great pandemics that the word "plague" still provokes a sense of dread today.

Rats, mice, and voles are the hosts for the natural endemic cycle of *Yersinia pestis* **(Figure 20.16a);** they harbor the bacteria but do not develop disease. In this cycle, fleas are the vectors for the spread of the bacteria among rodents. As the bacteria multiply within a flea, they block the esophagus such that the flea can no longer ingest blood from a host. As the starving flea jumps from host to host seeking a blood meal, it can infect a new host with each bite. When other animals—including prairie dogs, rabbits, deer, camels, dogs, and cats—become infected via flea bites, they act as *amplifying hosts* **(Figure 20.16b);** that is, they support increases in the numbers of bacteria and infected fleas, even though they die from bubonic plague. Humans become infected either when bitten by infected fleas that have left their normal animal hosts, or through direct contact with infected animals **(Figures 20.16c).** Bubonic plague is not spread from person to person.

Bubonic plague is characterized by high fever and swollen, painful lymph nodes called *buboes*[2] **(Figure 20.17),** which develop within a week of a flea bite. Untreated cases progress to bacteremia, which results in disseminated intravascular coagulation, subcutaneous hemorrhaging, and death of tissues, which may become infected with *Clostridium* (klos-trid′ē-ŭm) and develop gangrene. Because of the extensive darkening of dead skin, plague has been called the "Black Death." Untreated bubonic plague is fatal in 75% of

---

[2]From Greek *boubon*, meaning groin, referring to a swelling in the groin.

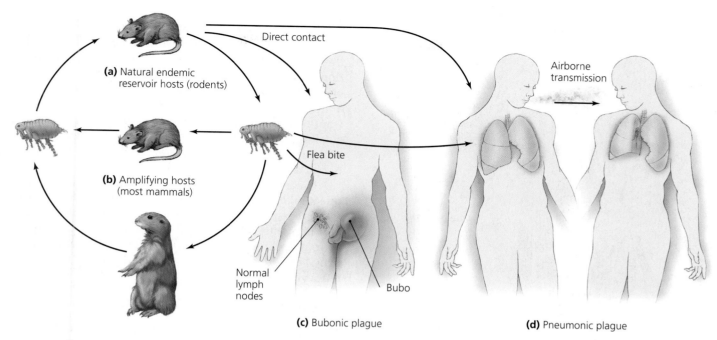

(a) Natural endemic reservoir hosts (rodents)

Direct contact

(b) Amplifying hosts (most mammals)

Flea bite

Normal lymph nodes

Bubo

(c) Bubonic plague

Airborne transmission

(d) Pneumonic plague

▲ *Figure 20.16*

The natural history and transmission of *Yersinia pestis*, the bacterium that causes plague. **(a)** The natural endemic cycle of *Yersinia* among rodents. **(b)** The cycle involving amplifying mammalian hosts. **(c, d)** Humans develop one of two forms of plague as the result of the bite of infected fleas or direct contact with infected hosts. *Yersinia* can move from the bloodstream into the lungs to cause pneumonic plague, which can be spread between humans via the airborne transmission of the bacteria in aerosols. *Why is the plague so much less devastating today than it was in the Middle Ages?*

Figure 20.16 Urban living has reduced contact between people and flea vectors living on wildlife and farm animals; improved hygiene and the use of insecticides have reduced contact with flea vectors living on pets; and antimicrobials are effective against *Y. pestis*.

▲ *Figure 20.17*

A bubo—a lymph node swollen by infection and a characteristic feature of bubonic plague.

cases. Even with treatment, more than 10% of patients die. In fatal infections, death usually occurs within a week of the onset of symptoms.

Pneumonic plague occurs when *Yersinia* in the bloodstream infects the lungs. Disease develops very rapidly—patients develop fever, malaise, and pulmonary distress within a day of infection. Pneumonic plague can be spread from person to person through aerosols and sputum **(see Figure 20.15d),** and if left untreated is fatal in nearly 100% of cases.

Because plague is so deadly and can be so rapid, diagnosis and treatment must also be rapid. The characteristic symptoms, especially in patients that have traveled in areas where plague is endemic, are usually sufficient for diagnosis. Rodent control and better personal hygiene have almost eliminated plague in industrialized countries, although wild animals remain reservoirs. Many antibacterial drugs, including tetracycline, streptomycin, and chloramphenicol, are effective against *Yersinia*.

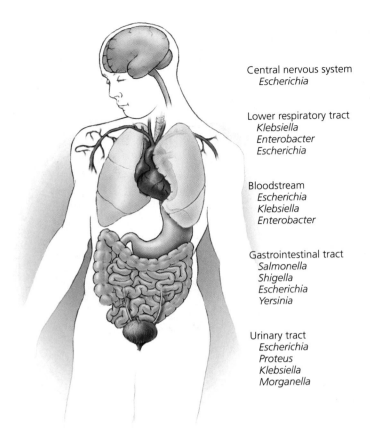

Central nervous system
Escherichia

Lower respiratory tract
Klebsiella
Enterobacter
Escherichia

Bloodstream
Escherichia
Klebsiella
Enterobacter

Gastrointestinal tract
Salmonella
Shigella
Escherichia
Yersinia

Urinary tract
Escherichia
Proteus
Klebsiella
Morganella

▲ *Figure 20.18*
Sites of infection of some common members of the Enterobacteriaceae. For each site, genera are listed in their relative order of prevalence.

Figure 20.18 illustrates the general sites of infections by the more common members of the Enterobacteriaceae.

The Pasteurellaceae

Learning Objectives

✓ Describe two pathogenic genera in the family Pasteurellaceae.
✓ Identify and describe three diseases caused by species of *Haemophilus*.

In the previous sections we examined opportunistic and pathogenic Gram-negative, facultative anaerobes of the family Enterobacteriaceae, all of which are oxidase negative. Now we consider another family of gammaproteobacteria—the **Pasteurellaceae** (pas-ter-el-ă′sē-ē), which contains species that are oxidase positive (see Figure 20.4). These microorganisms are mostly facultative anaerobes that are small, nonmotile, and fastidious in their growth requirements—they require heme or cytochromes. Two genera, *Pasteurella* and *Haemophilus*, contain most of this family's pathogens of humans.

Pasteurella

Pasteurella (pas-ter-el′ă) is part of the normal microbiota in the oral and nasopharyngeal cavities of animals, including domestic cats and dogs. Humans are typically infected via animal bites and scratches or via inhalation of aerosols from animals. Most cases in humans are caused by *P. multocida* (mul-tō′si-dă), which produces localized inflammation and swelling of lymph nodes at the site of infection. Patients with suppressed immunity are at risk of widespread infection and bacteremia.

Diagnosis of infection with *Pasteurella* depends on identification of the bacterium in cultures of specimens collected from patients. As noted, *Pasteurella* is fastidious in its growth requirements, so it must be cultured on blood or chocolate agar (autoclaved blood agar). A wide variety of antibacterial drugs are effective against *Pasteurella*, including penicillin, tetracycline, and cephalosporin.

Haemophilus

Haemophilus[3] (hē-mof′i-lŭs) species are generally small, sometimes pleomorphic bacilli **(Figure 20.19)** that require heme and NAD^+ for growth; as a result, they are obligate parasites, colonizing mucous membranes of humans and some animals. *Haemophilus influenzae* (in-flū-en′zē) is the most notable pathogen in the genus, though *H. ducreyi* (doo-krā-ē) and *H. aphrophilus* (a-frō-fil′ŭs) are agents of a sexually transmitted disease and a rare type of endocarditis, respectively. Other species in the genus primarily cause opportunistic infections.

Haemophilus influenzae   Most strains of *Haemophilus influenzae* have polysaccharide capsules that resist phagocytosis. Researchers distinguish among six strains by differences in the K antigens of the capsules. Before the introduction of an effective vaccine in the 1980s, 95% of *H. influenzae* diseases were caused by strain b.

Belying its name, *H. influenzae* does not cause the flu; in fact, it is rarely found in the upper respiratory tract. Instead, *H. influenzae* b was the most common cause of meningitis in children 3–18 months of age before immunization brought it under control. The pathogen also causes inflammation of subcutaneous tissue, infantile arthritis, and life-threatening epiglottitis in children under 4 years old. In the latter disease, swelling of the epiglottis and surrounding tissues can completely block the pharynx.

Prompt treatment with cephalosporin is required for meningitis and epiglottitis. Over the past decade, the use of an effective vaccine—Hib, which is composed of capsular antigens conjugated to protein molecules—has virtually eliminated all the diseases caused by *H. influenzae* in the United States.

The incidence of infections with other strains of *H. influenzae* has remained fairly constant in the United States.

---

[3]From Greek *haima*, meaning blood, and *philos*, meaning love.

Pleomorphism

LM |⊢⎯⎯⎯|
      25 µm

▲ *Figure 20.19*

The pleomorphic bacilli of *Haemophilus influenzae* in a Gram stain. *What simple lab test can be used to distinguish* Haemophilus *from* Escherichia?

Figure 20.19 An oxidase test can distinguish Haemophilus, which is oxidase positive, from Escherichia, which is oxidase negative.

These strains cause a variety of diseases, including conjunctivitis, sinusitis, dental abscesses, middle ear infections, meningitis, bronchitis, and pneumonia. Strain *aegyptius* causes Brazilian purpuric fever, an extremely rapid pediatric disease characterized by conjunctivitis that within a few days is followed by fever, vomiting, abdominal pain, shock, and death. The pathogenesis of this disease is not understood, but fortunately it is rare in the United States.

CRITICAL THINKING

Haemophilus influenzae was so named because researchers isolated the organism from flu patients. Specifically, how could a proper application of Koch's postulates have prevented this misnomer?

Other Species of *Haemophilus* Sexually transmitted *Haemophilus ducreyi* causes a genital ulcer, called *chancroid* or *soft chancre* (**Figure 20.20**), by producing a toxin that kills human epithelial cells. Chancroid is most often diagnosed in men because the ulceration is more visible and often painful in men; the disease is often asymptomatic in women. Although prevalent in Africa and Asia, chancroid is rare in Europe and the Americas.

Thus far we have examined pathogenic Gram-negative cocci and the more common facultatively anaerobic Gram-negative

▲ *Figure 20.20*

Chancroid (soft chancre), a genital ulcer caused by *Haemophilus ducreyi*.

pathogenic bacilli. Now we turn our attention to the pathogenic Gram-negative aerobic bacilli.

Pathogenic, Gram-Negative, Aerobic Bacilli

Pathogenic, Gram-negative, aerobic bacilli are a diverse group of bacteria in several classes of the phylum Proteobacteria. The following sections examine some of the more common species in the order they are presented in *Bergey's Manual: Bartonella* and *Brucella* in the class Alphaproteobacteria; *Bordetella* in the class Betaproteobacteria; and several members of the class Gammaproteobacteria, including the pseudomonads (*Pseudomonas, Moraxella,* and *Acinetobacter*), *Francisella, Legionella,* and *Coxiella.*

Bartonella

Learning Objective

✓ Distinguish among bartonellosis, trench fever, and cat-scratch disease.

Members of *Bartonella* (bar-tō-nel'ă), a genus of Gram-negative aerobic bacilli in the class Alphaproteobacteria, are found in animals, but they are known to cause disease only in humans. They have fastidious growth requirements and must be cultured on special media such as blood agar. Many require prolonged incubation in a humid, 37°C atmosphere supplemented with carbon dioxide before they form visible colonies. Three species (formerly classified in the genus *Rochalimaea*) are pathogenic in humans.

Case Study 20.2   When "Health Food" Isn't Healthy

In a single day, two 19-year-old women and one 20-year-old man sought treatment at a university health clinic, complaining of acute diarrhea, nausea, and vomiting. No blood was found in their stools. One of the women was found to have a urinary tract infection. All three had eaten lunch at a nearby health food store the previous day. The man had a turkey sandwich with tomato, sprouts, pickles, and sunflower seeds. One woman had a pocket sandwich with turkey, sprouts, and mandarin oranges; the other woman had the lunch special, described in the menu as a "delightful garden salad of fresh organic lettuces, sprouts, tomatoes, and cucumbers with zesty raspberry vinaigrette dressing." All had bottled water to drink.

1. Which of the foods is the most likely source of the infections?
2. What media would you use to culture and isolate enteric contaminants in the food?
3. Which enteric bacteria could cause these symptoms?
4. How did the woman likely acquire the urinary tract infection?
5. What is the likely treatment of choice?
6. What steps can the food store's manager and the students take to reduce the chance of subsequent infections?

▲ *Figure 20.21*
The localized swelling at the site of cat scratches or bites characteristic of cat-scratch disease.

Bartonella bacilliformis (ba-sil′li-for′mis) invades and weakens erythrocytes, causing **bartonellosis**—an often fatal disease characterized by fever, severe anemia, headache, muscle and joint pain, and chronic skin infections. Bloodsucking sand flies of the genus *Phlebotomus* transmit the bacterium, so the disease is endemic only in Peru, Ecuador, and Colombia, where such flies live.

Bartonella quintana (kwin′ta-nǎ) causes **trench fever** (*5-day fever*), which was prevalent among soldiers during World War I. Human body lice transmit the bacteria from person to person. Many infections are asymptomatic, whereas in other patients severe headaches, fever, and pain in the long bones characterize the disease. The fever can be recurrent, returning every 5 days, giving the disease its alternate name. *B. quintana* also causes two newly described diseases in immunocompromised patients: *Bacillary angiomatosis* is characterized by fever, inflamed nodular skin lesions, and proliferation of blood vessels; *Bacillary peliosis*

hepatitis is a disease in which patients develop blood-filled cavities in their livers.

Bartonella henselae (hen′sel-ē) causes **cat-scratch disease** when the bacteria are introduced into humans through cat scratches and bites. Fleas may also transmit the bacterium from cats to people. Cat-scratch disease has emerged as a relatively common and occasionally serious infection of children, affecting an estimated 22,000 children annually in the United States. This zoonotic disease involves fever and malaise for a few days, plus localized swelling at the site of infection **(Figure 20.21)** and of local lymph nodes for several months. Serological testing of individuals who exhibit these characteristic signs and symptoms following exposure to cats confirms a diagnosis of cat-scratch disease.

Gentamicin, erythromycin, and cephalosporin are used to treat most *Bartonella* infections, though there is a high rate of relapse. Azithromycin is used to treat cat-scratch disease. The efficacy of other antimicrobial drugs against this species has not been assessed.

Brucella

Learning Objective

✓ Describe brucellosis.

Brucella (broo-sel′lǎ) is a genus of small, nonmotile, aerobic coccobacilli that lack capsules. Analysis of rRNA nucleotide base sequences reveals that *Brucella* is closely related to *Bartonella* in the class Alphaproteobacteria. Although *Brucella*-caused diseases in humans have been ascribed to four species, rRNA analysis has revealed the existence of only a single true species: *Brucella melitensis;* the other variants are strains of this one species. Nevertheless, because the classical species names are so well known, they are retained here.

In animal hosts, the bacteria live as intracellular parasites in organs such as the uterus, placenta, and epididymis, but these organs are not infected in humans. Typically, infections in animals are either asymptomatic or cause a mild disease—**brucellosis** (broo-sel-ō′sis)—though they can cause sterility or abortion. *Brucella melitensis* infects goats and

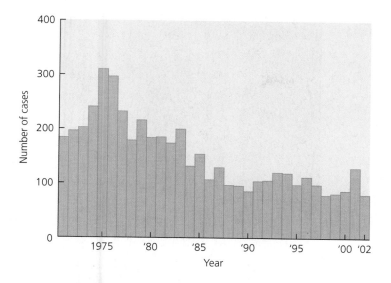

▲ *Figure 20.22*

The incidence of brucellosis in humans in the United States, 1971–2002. The decline in cases is largely the result of improvements in livestock management. *Given that* Brucella abortus *has been practically eliminated from U.S. cattle herds, how is it that any human cases of brucellosis are detected in the United States?*

Figure 20.22 Most cases of brucellosis in the United States involve immigrants and patients who acquired the disease during foreign travel.

sheep, *B. abortus* cattle, *B. suis* swine, and *B. canis* dogs, foxes, and coyotes.

Humans become infected either by consuming unpasteurized contaminated dairy products or through contact with animal blood, urine, and placentas in workplaces such as slaughterhouses, veterinary clinics, and feedlots. The bacteria enter the body through breaks in mucous membranes of the digestive and respiratory tracts.

Brucellosis in humans is characterized by a fluctuating fever—which gives the disease one of its common names: *undulant fever*—and chills, sweating, headache, myalgia, and weight loss. The disease in humans has been given a variety of other names, including *Bang's disease,* after microbiologist Bernhard Bang (1848–1932), who investigated the disease, and *Malta fever, rock fever of Gibraltar,* and *fever of Crete,* after localized epidemics in those locales.

Physicians treat brucellosis with combinations of antibacterial drugs, including tetracycline, rifampin, and streptomycin. An attenuated vaccine for animals exists, but humans are not immunized because the vaccine can cause disease. Through the vaccination of uninfected domesticated animals and the slaughter of infected ones, the threat of brucellosis has been vastly reduced for U.S. residents **(Figure 20.22)**; the few remaining cases usually occur in immigrants or people who have traveled outside the country.

CRITICAL THINKING

Why do physicians substitute trimethropin and sulfanilamide for tetracycline when treating children and pregnant women infected with *Brucella*?

Bordetella

Learning Objectives

✓ Describe five virulence factors of *Bordetella pertussis*.
✓ Identify the four phases of pertussis.

Bordetella pertussis (bōr-dĕ-tel'ǎ per-tus'is) is a small, aerobic, nonmotile, Gram-negative, coccobacillus in the family Betaproteobacteria that is responsible for the disease **pertussis**,[4] commonly called *whooping cough. B. parapertussis* causes a milder form of pertussis.

Pathogenesis, Epidemiology, and Disease

Bordetella pertussis causes disease by interfering with the action of ciliated epithelial cells of the trachea. Various adhesins and toxins mediate the disease.

The bacterium attaches to certain lipids in the cytoplasmic membranes of tracheal cells via two adhesins: *filamentous hemagglutinin* and *pertussis toxin*. Filamentous hemagglutinin also binds to certain glycoproteins on the cytoplasmic membranes of neutrophils, initiating phagocytosis of the bacteria. *B. pertussis* survives within phagocytes, in the process evading the immune system. Pertussis toxin causes infected cells to produce more receptors for filamentous hemagglutinin, leading to further bacterial attachment and phagocytosis.

Four toxins *B. pertussis* produces are:

- *Pertussis toxin,* a portion of which interferes with the ciliated epithelial cell's metabolism, resulting in increased mucus production. (Note that pertussis toxin is both an adhesin and a toxin.)

- *Adenylate cyclase toxin,* which triggers increased mucus production, and inhibits leukocyte movement, phagocytosis, and killing. This function may provide protection for the bacterium early in an infection.

- *Dermonecrotic toxin,* which causes localized constriction and hemorrhage of blood vessels, resulting in cell death and tissue destruction.

- *Tracheal cytotoxin,* which at low concentrations inhibits the movement of cilia on ciliated respiratory cells, and at high concentrations causes the expulsion of the cells from the lining of the trachea.

Pertussis is considered a pediatric disease, as most cases are reported in children younger than 5 years old. More than 60 million children worldwide suffer from pertussis each year, including over 7000 reported cases in the United States

[4]From Latin *per,* meaning intensive, and *tussis,* meaning cough.

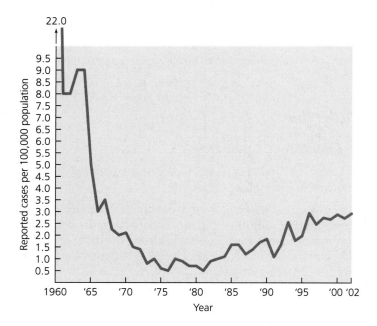

▲ *Figure 20.23*

Reported cases of pertussis in the United States, 1960–2002.

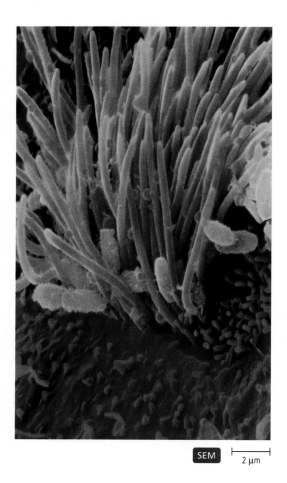

SEM   ⊢——⊣ 2 μm

▲ *Figure 20.24*

Bordetella pertussis among the cilia of a tracheal epithelial cell. The bacteria attach to and infect the cells, eventually causing the loss of the cells.

(Figure 20.23). However, these figures considerably underestimate the actual number of cases, because patients with chronic coughs are not routinely tested for infection with *Bordetella,* and because the disease in older children and adults is typically less severe and is frequently misdiagnosed as a cold or influenza.

Pertussis begins when bacteria, inhaled in aerosols, attach to and multiply in ciliated epithelial cells **(Figure 20.24).** Pertussis then progresses through four stages **(Figure 20.25):**

❶ During 7–10 days of **incubation** the bacteria multiply, but no symptoms are apparent.

❷ The **catarrhal**[5] (kă-tă′răl) **phase** is characterized by signs and symptoms that resemble a common cold. During this phase, which lasts 1–2 weeks, the bacteria are most abundant and the patient is most infectious.

❸ The **paroxysmal**[6] (par-ok-siz′mal) **phase** begins as the ciliary action of the tracheal cells is impaired, even as copious mucus is secreted. The condition worsens as ciliated cells are expelled. To clear the accumulating mucus, the body initiates a series of deep coughs, each of which is followed by a characteristic "whoop" caused by the intake of air through the congested trachea. Each day a patient may experience 40–50 coughing spells that often end with vomiting and exhaustion. During this phase, which can last 2–4 weeks, coughing may be so severe that oxygen exchange is limited, such that the patient may turn blue or even die.

❹ During the **convalescent phase,** the number of bacteria present becomes quite small, the ciliated lining of the trachea grows back, and the severity and frequency of coughing diminishes. During this phase, which typically lasts 3–4 weeks or longer, secondary bacterial infections (such as with *Staphylococcus* or *Streptococcus*) in the damaged epithelium may lead to bacteremia, pneumonia, seizures, and encephalopathy.

Diagnosis, Treatment, and Prevention

The symptoms of pertussis are usually diagnostic, particularly when a patient is known to have been exposed to *Bordetella.* Even though health care workers can isolate *B. pertussis* from respiratory specimens, the bacterium is extremely sensitive to desiccation, so specimens must be inoculated at the patient's bedside onto *Bordet-Genou* medium, which is specially designed to support the growth of the bacterium. If this is not practical, clinicians must use special transport media to get a specimen to a laboratory.

Treatment for pertussis is primarily supportive. By the time the disease is recognized, the immune system has often already "won the battle." Recovery depends on regenera-

---

[5]From Greek *katarrheo*, meaning to flow down.
[6]From Greek *paroxysmos*, meaning to irritate.

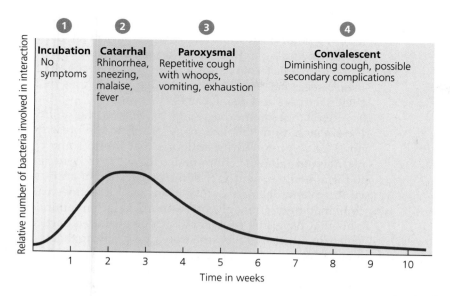

① Incubation
No symptoms

② Catarrhal
Rhinorrhea, sneezing, malaise, fever

③ Paroxysmal
Repetitive cough with whoops, vomiting, exhaustion

④ Convalescent
Diminishing cough, possible secondary complications

Relative number of bacteria involved in interaction

Time in weeks 1 2 3 4 5 6 7 8 9 10

▲ **Figure 20.25**

The approximate time course for the progression of pertussis, from incubation through the catarrhal, paroxysmal, and convalescent phases.

tion of the tracheal epithelium, not on the number of bacteria; therefore, although antibacterial drugs reduce the number of bacteria and the patient's infectivity to others, they have little effect on the course of the disease.

Given that *B. pertussis* has no animal or environmental reservoir, and that an effective vaccine (the "P" of the DPT vaccine) has been available since 1949, whooping cough could be eradicated. Despite this possibility, the number of cases of pertussis in the United States has increased from a low of 0.5 cases per 100,000 in 1976 to 2.92 per 100,000—a number not seen since the 1960s (see Figure 20.23). This increase results in part from the refusal of parents to immunize their children because of publicity concerning adverse reactions to the original attenuated vaccine, and to the fact that immunity, whether acquired artificially or naturally, is not lifelong. The CDC now recommends the use of an acellular pertussis vaccine (aP), which has fewer side effects than the attenuated vaccine but is equally effective.

Pseudomonads

Learning Objectives

✓ Reconcile the apparent discrepancy between the ubiquitous distribution of pseudomonads and their numerous virulence factors and the fact they cause so few diseases.

✓ Identify three genera of opportunistic pathogenic pseudomonads.

✓ Describe *Pseudomonas aeruginosa* as an opportunistic pathogen of burn victims and cystic fibrosis patients.

Pseudomonads (soo-dō-mō′nadz) are Gram-negative, aerobic bacilli in the class Gammaproteobacteria. Unlike the fastidious members of *Bartonella, Brucella,* and *Bordetella,* pseudomonads are not particular about their growth re-

quirements. They are ubiquitous in soil, decaying organic matter, and almost every moist environment, including swimming pools, hot tubs, sponges, wash cloths, and contact lens solutions. In hospitals they are frequently found growing in moist food, vases of cut flowers, sinks, sponges, toilets, floor mops, dialysis machines, respirators, humidifiers, and in disinfectant solutions. Some species can even grow using the trace nutrients in distilled water. Pseudomonads utilize a wide range of organic carbon and nitrogen sources. The Entner-Doudoroff pathway is their major means of glucose catabolism rather than the more common Embden-Mayorof pathway of glycolysis.

Here we examine three genera that are commonly isolated opportunistic pathogens—*Pseudomonas, Moraxella,* and *Acinetobacter*—beginning with the species of greatest medical importance: *Pseudomonas aeruginosa.*

Pseudomonas aeruginosa

Pseudomonas aeruginosa (soo-dō-mō′nas ā-rū-ji-nō′sā) is somewhat of a medical puzzle in that even though it expresses a wide range of virulence factors, it rarely causes disease. This is fortunate because its ubiquity and inherent resistance to a wide range of antimicrobial agents would make a more pathogenic *Pseudomonas* a formidable challenge to health care professionals. That *P. aeruginosa* is *only* an opportunistic pathogen is testimony to the vital importance of the body's protective tissues, cells, and chemical defenses.

P. aeruginosa has fimbriae and other adhesins that enable its attachment to host cells. The fimbriae are similar in structure to those of *N. gonorrhoeae.* The bacterial enzyme *neuraminidase* modifies the fimbriae receptors on the host cell in such a way that attachment of fimbriae is enhanced. A mucoid polysaccharide capsule also plays a role in attachment, particularly in the respiratory system of cystic fibrosis patients (as discussed shortly). The capsule also shields the bacterium from phagocytosis.

P. aeruginosa synthesizes other virulence factors, including toxins and enzymes. Lipid A (endotoxin) is prevalent in the cell wall of *Pseudomonas* and mediates fever, vasodilation, inflammation, shock, and other symptoms. Two toxins—*exotoxin A* and *exoenzyme S*—inhibit protein synthesis in eukaryotic cells, contributing to cell death and interfering with phagocytic killing. The enzyme *elastase* breaks down elastic fibers, degrades complement components, and cleaves IgA and IgG. The bacterium also produces a blue-green pigment, called *pyocyanin,* that triggers the formation of superoxide radical (O_2^-) and peroxide anion (O_2^{2-}), two reactive forms of oxygen that contribute to tissue damage in *Pseudomonas* infections.

Although a natural inhabitant of bodies of water and moist soil, *P. aeruginosa* is rarely part of the normal human microbiota. Nevertheless, because of its ubiquity and its

▲ *Figure 20.26*

A *Pseudomonas aeruginosa* infection. *What is responsible for the blue-green appearance of this infection?*

Figure 20.26 P. aeruginosa secretes pyocyanin, a blue-green pigment.

virulence factors, this opportunistic pathogen colonizes immunocompromised patients and is involved in about 10% of nosocomial infections. Once it breaches the skin or mucous membranes, *P. aeruginosa* can successfully colonize almost every organ and system. It can be involved in bacteremia, endocarditis, and urinary, ear, eye, central nervous system, gastrointestinal, muscle, and skeletal infections **(Figure 20.26)**. Infections in burn victims and cystic fibrosis patients are so common they deserve special mention.

Pseudomonas infections of severe burns are pervasive. The surface of a burned area provides a warm moist environment that is quickly colonized by this ubiquitous opportunist; almost two-thirds of burn victims develop environmental or nosocomial *Pseudomonas* infections.

P. aeruginosa also typically infects the lungs of cystic fibrosis patients, forming a biofilm that protects the bacteria from phagocytes. Such infections exacerbate the decline in pulmonary function in these patients by causing certain lung cells to synthesize large amounts of mucus. As the bacteria feed on the mucus, they signal host cells to secrete more of it, creating a positive feedback loop; the result of such *P. aeruginosa* infections is that cystic fibrosis patients are more likely to require hospitalization, and more likely to die.

Diagnosis of *Pseudomonas* infection is not always easy because its presence in a culture may represent contamination acquired during collection, transport, or inoculation. Certainly, pyocyanin discoloration of tissues is indicative of massive infection.

Treatment of *P. aeruginosa* is also frustrating. The bacterium is notoriously resistant to a wide range of antibacterial agents, including antimicrobial drugs, soaps, antibacterial dyes, and quarternary ammonium disinfectants. In fact, *Pseudomonas* has been reported to live in solutions of antibacterial drugs and disinfectants.

Resistance of the bacterium is due to the ability of *Pseudomonas* to metabolize many drugs, to the presence of nonspecific proton/drug antiports that pump some types of drugs out of the bacterium, and to the ability of *Pseudomonas* to form biofilms, which resist the penetration of antibacterial drugs and detergents. Physicians treat infections with a combination of aminoglycoside and beta-lactam antimicrobials that have first proven efficacious against a particular isolate in a susceptibility test. Polymyxin, a drug that is also toxic to humans, is used as a last resort.

In summary, though *Pseudomonas aeruginosa* is ubiquitous and possesses a number of virulence factors, the bacterium rarely causes diseases because it cannot normally penetrate the skin and mucous membranes, nor ultimately evade the body's other defenses. Only in debilitated patients does this opportunist thrive.

Moraxella and *Acinetobacter*

Moraxella (mōr′ak-sel′ă) and *Acinetobacter* (as-i-nē′tōbak′ter) are aerobic, short, plump bacilli that formerly were classified in the same family as *Neisseria*. Analysis of rRNA has shown that they are more properly classified as pseudomonads.

Moraxella catarrhalis (formerly *Branhamella catarrhalis*) is rarely pathogenic but can cause opportunistic infections of the sinuses, bronchi, ears, and lungs. The bacterium is susceptible to cephalosporins, erythromycin, tetracycline, and most other antibacterial drugs (with the exception of penicillin).

Acinetobacter grows in soil, water, and sewage and is only rarely associated with disease in humans, though it is often isolated from clinical specimens. It is an opportunistic pathogen that causes infections of the respiratory, urinary, and central nervous systems. Endocarditis and septicemia have also been reported in infections with *Acinetobacter*. It is often resistant to antimicrobial drugs, so susceptibility tests must guide the choice of an effective treatment.

Francisella

Learning Objectives

✓ Describe the modes of transmission of *Francisella tularensis*.
✓ List practical measures that can be taken to prevent infection by *F. tularensis*.

Francisella tularensis[7] (fran′si-sel′lă too-lă-ren′sis) is a very small (0.2 μm × 0.2–0.7 μm), nonmotile, strictly aerobic,

---

[7]Named for Tulare County, California.

Gram-negative coccobacillus of the family Gammaproteobacteria that causes a zoonotic disease called **tularemia** (too-lă-rē′mē-ă). *F. tularensis* has a lipid capsule that discourages phagocytosis, but even if phagocytized it survives by inhibiting the fusion of lysosomes with phagocytic vesicles.

F. tularensis is found in temperate regions of the northern hemisphere living in water and as an intracellular parasite of animals. *Francisella* has an amazingly diverse assortment of hosts. It lives in mammals, birds, fish, and blood-sucking ticks and insects. Indeed, one is hard pressed to find an animal that cannot be its host. The most common reservoirs in the United States are rabbits, muskrats, and ticks, which give the disease two of its common names—*rabbit fever* and *tick fever*.

Francisella is also incredibly varied in the ways it can be spread among hosts. Tularemia in humans is most often acquired either through the bite of an infected tick or via contact with an infected animal; the bacterium, by virtue of its small size, can pass through apparently unbroken skin or mucous membranes. Blood-sucking flies, mosquitoes, and mites also transmit *Francisella*, and humans can be infected by consuming infected meat, drinking contaminated water, or inhaling bacteria in aerosols produced in the laboratory or during slaughter. Fortunately, human-to-human spread does not seem to occur.

Francisella is one of the more infectious of all bacteria: Infection requires as few as 10 organisms when transmitted by a biting insect or tick or through unbroken skin or mucous membranes. Only about 50 cells must be inhaled to cause disease, but consumption of 10^8 cells in food or drink is required to contract the disease via the digestive tract.

Tularemia was removed from the list of nationally notifiable diseases in 1994, but concern about the possible use of *Francisella* by bioterrorists led officials in 2000 to again include this disease among those reportable. Only 60 cases were reported in 2002, but the actual number of infections was probably much higher considering its virulence, prevalence in animals, and multiple modes of transmission. Tularemia frequently remains unsuspected because its symptoms—fever, chills, malaise, and fatigue—are not notably different from many other bacterial and viral diseases, and because it is difficult to confirm tularemia by using laboratory tests. Though it is typically innocuous, tularemia is fatal to 10% of untreated patients.

The bacterium produces beta-lactamase, so penicillins and cephalosporins are ineffective, but other antimicrobial drugs have been used successfully. Currently, doxycycline or ciproflaxacin are the recommended antimicrobial drugs for use against *Francisella*. An attenuated vaccine provides some protection for people at risk of exposure. The vaccine is not entirely effective but can lessen the severity of the disease.

To prevent infection, people should avoid the major reservoirs of *Francisella* (rabbits, muskrats, and ticks), especially in endemic areas. They can protect themselves from tick bites by wearing long clothing with tight-fitting sleeves, cuffs, and collars, and by using repellant chemicals. Because *Francisella* is not present in tick saliva, but instead in its feces only, prompt removal of ticks can mitigate infection. Hikers and hunters should never handle ill-appearing wild animals or their carcasses, and should wear gloves and masks when field dressing game.

CRITICAL THINKING

Most U.S. cases of tularemia occur in the late spring and summer months, and few cases occur in January. Why might this be so?

Legionella

Learning Objectives

✓ Explain why *Legionella pneumophila* was unknown before 1976.
✓ Describe the symptoms and treatment of Legionnaires' disease.

In 1976, celebration of the 200th anniversary of the Declaration of Independence was curtailed in Philadelphia when over 200 American Legion members attending a convention were stricken with severe pneumonia; 29 died. After extensive epidemiological research, this disease—dubbed **Legionnaires' disease** or *legionellosis*—was found to be caused by a previously unknown pathogen, which was subsequently named *Legionella* (lē-jŭ-nel′lă).

Pathogenesis, Epidemiology, and Disease

To date, scientists have identified over 40 species of *Legionella*. These aerobic, slender, pleomorphic bacteria in the family Gammaproteobacteria are extremely fastidious in their nutrient requirements, and laboratory media must be enhanced with iron salts and the amino acid cysteine. **Figure 20.27** on page 590 shows colonies growing on one commonly used medium: buffered charcoal yeast extract agar. *Legionella* species are almost universal inhabitants of water, but they had not been isolated previously because they stain poorly and cannot grow on common laboratory media. Nineteen species are known to cause disease in humans, but 85% of all infections in humans are caused by *L. pneumophila*[8] (noo-mō′fi-lă).

Legionella pneumophila presented a conundrum for early investigators: How can such a fastidious bacterium be nearly ubiquitous in moist environmental samples? In the original epidemic, for example, *Legionella* was cultured from condensation in hotel air conditioning ducts, an environment that seems unsuitable for such a fastidious microorganism. Investigations revealed that *Legionella* living in the environment invade freshwater protozoa, typically amoebae, and reproduce inside phagocytic vesicles. Thus the bacteria survive in the environment much as they survive in humans—as intracellular parasites.

Protozoa release bacteria-filled vesicles into the environment; alternatively, *Legionella* forms exit pores in host cells,

[8]From Greek *pneuma*, meaning breath, and *philos*, meaning love.

▲ *Figure 20.27*

Legionella pneumophila growing on buffered charcoal yeast extract agar. This bacterium cannot be grown in culture without such special media.

TEM | 1 µm

▲ *Figure 20.28*

Coxiella burnetii, an intracellular pathogen that grows and reproduces within phagolysosomes, here in a placental cell. An infective body of *Coxiella* can persist in the environment for years.

first in the vesicular membrane and then in the cytoplasmic membrane. Humans acquire the disease by inhaling *Legionella* in aerosols produced by showers, vaporizers, spa whirlpools, hot tubs, air conditioning systems, cooling towers, and grocery store misters. *Legionella* was not a notable pathogen until such devices provided a suitable means of transmitting the bacteria to humans.

Legionnaires' disease is characterized by fever, chills, a dry nonproductive cough, headache, and pneumonia. Complications involving the gastrointestinal tract, central nervous system, liver, and kidneys are common. If not promptly treated, pulmonary function rapidly decreases, resulting in death in 20% of patients with normal immunity. Mortality is much higher in immunocompromised individuals, particularly kidney and heart transplant recipients.

L. pneumophila also causes a flulike illness called *Pontiac fever* (after the Michigan city where it was first described). This disease has symptoms similar to those of Legionnaires' disease, but it does not involve pneumonia and is not fatal.

Diagnosis, Treatment, and Prevention

Diagnosis is made by fluorescent antibody staining or other serological tests that reveal the presence of *Legionella* in clinical samples. The bacterium can also be cultured on suitable commercially available selective media such as charcoal buffered yeast extract.

Erythromycin is the antimicrobial drug of choice for treating Legionnaires' disease. Other antimicrobials, with the exception of beta-lactams, are also effective. Pontiac fever is self-limiting and requires no treatment.

Completely eliminating *Legionella* from water supplies is not feasible; chlorination and heating are only moderately successful. However, the bacterium is not highly virulent, so reducing their number is typically a successful control measure.

Coxiella

Learning Objectives

✓ Explain how *Coxiella burnetii* survives desiccation.
✓ Describe the mode of transmission of Q fever.

Coxiella burnetii (kok-sē-el'ă ber-ne'tē-ē) is an extremely small, aerobic, obligate intracellular parasite (**Figure 20.28**) that grows and reproduces in the acidic environment within phagolysosomes. Its small size and dependence on the cytoplasm of eukaryotes for growth led early investigators to think it was a virus. However, its bacterial nature is unquestionable: It has a typical Gram-negative cell wall (albeit with minimal peptidoglycan), RNA and DNA (no virus has both), functional ribosomes, and Krebs cycle enzymes. *Coxiella* was originally classified with other obligate intracellular bacteria called *rickettsias* (discussed in Chapter 21), but rRNA analysis reveals that it is more properly classified with *Legionella* in the family Gammaproteobacteria.

Coxiella forms an internal, stable, resistant *infective body* (sometimes called a spore) that is similar in structure and function to the endospores of some Gram-positive species. The infective body enables the bacterium to survive harsh environmental conditions (such as desiccation and heat) for years.

C. burnetii infects a wide range of mammalian and avian hosts and is transmitted among them by feeding ticks. Farm animals and pets are the reservoirs most often associated with disease in humans. Transmission to humans does not occur via tick vectors; instead, humans are most frequently infected by inhaling infective bodies that become airborne when released from dried tick feces or from the dried urine, feces, or placentas of host animals. Humans can also become infected by consuming contaminated unpasteurized milk.

Most *Coxiella* infections are asymptomatic, though this bacterium can cause **Q fever,** so named because its cause was *questionable* (unknown) for many years. Q fever occurs worldwide, particularly among ranchers, veterinarians, and food handlers, and it may be either an acute or a chronic condition. Acute Q fever follows an incubation period of 20 days or more; besides a high fever, it involves severe headache, chills, muscle pain, and mild pneumonia. In chronic Q fever, months to years may pass from the time of infection until life-threatening endocarditis develops. Inflammation of the lungs and liver may occur simultaneously.

Q fever is diagnosed via serological testing. Oral tetracycline is the preferred antimicrobial. The biggest problem in treating Q fever is getting antimicrobial drugs to work in the acidic environment of phagolysosomes, where the bacteria live. Researchers have developed an effective vaccine for Q fever, but it is not available in the United States. Prevention involves avoiding the inhalation of dust contaminated with barnyard and pet wastes.

Pathogenic, Gram-Negative, Anaerobic Bacilli

Learning Objective

✓ Compare and contrast *Bacteroides* and *Prevotella* with one another, and with other Gram-negative opportunists.

Over 50 species of strictly anaerobic Gram-negative bacteria are known to colonize the human body. Indeed, such anaerobic bacteria are the predominant microbiota of the gastrointestinal, urinary, reproductive, and lower respiratory tracts—they outnumber aerobic and facultatively anaerobic bacteria by more than 100 times. The abundance of anaerobic bacteria in these locations is important for human health: They inhibit the growth of most pathogens, and in the intestinal tract they synthesize necessary vitamins and assist in the digestion of food.

Relatively few of the normal anaerobic microbiota cause disease, and then only when they are introduced into other parts of the body by trauma or surgery. Almost all of these opportunistic species are aerotolerant anaerobes; they produce catalase or superoxide dismutase, which are enzymes that inactivate hydrogen peroxide and superoxide radicals. The more important anaerobic opportunists are *Bacteroides* and *Prevotella*.

Bacteroides

Bacteria in the genus *Bacteroides* (bak-ter-oy'dēz) are part of the normal microbiota of the intestinal tract and the upper respiratory tract. They are Gram-negative, anaerobic, and have an outer wall membrane containing lipopolysaccharide. However, in contrast to the lipid A of other Gram-negative

▲ *Figure 20.29*

Bacteroides fragilis produces a black color on bile-esculin agar. Bile suppresses the growth of most aerobes and facultative anaerobes, including *Proteus vulgaris* (right).

bacteria discussed in this chapter, the lipid A formed by members of this genus has little endotoxin activity.

The most important pathogen in this genus is *Bacteroides fragilis* (fra'ji-lis). It is associated with about 85% of gastrointestinal diseases, even though it accounts for only about 1% of the bacteria in the colon. *B. fragilis* is a pleomorphic bacillus that produces a number of virulence factors. It attaches to host cells via fimbriae and a polysaccharide capsule, the latter of which also inhibits phagocytosis. Should the bacteria become phagocytized, short-chain fatty acids produced during anaerobic metabolism inhibit the activity of lysosomes, enabling the bacteria to survive within phagocytes.

B. fragilis is frequently involved in a variety of conditions, including abdominal infections (for example, those following a ruptured appendix), genital infections in females (for example, pelvic abscesses), and wound infections of the skin (which can result in life-threatening necrosis of muscle tissue). It is also involved in about 5% of all cases of bacteremia.

If anaerobic bacteria are to be isolated from patients, clinicians must maintain anaerobic conditions while collecting specimens, transporting them to the laboratory, and culturing them. Selective media are often used to reveal the presence of these bacteria. For example, because *B. fragilis* grows well in the presence of bile, it can be cultured on bile-esculin agar in an anaerobic environment—conditions which inhibit aerobic and most anaerobic bacteria **(Figure 20.29)**. Metronidazole is the antimicrobial of choice for treating *Bacteroides* infections.

Prevotella

Members of the genus ***Prevotella*** (prev'ō-tel'ă) are also anaerobic, Gram-negative, pleomorphic bacilli. They differ from *Bacteroides* in that they are sensitive to bile; as a result, they do not grow in the intestinal tract, but are instead found in the urinary, genital, and respiratory tracts as part of the normal microbiota. Their virulence factors, including adhesive and antiphagocytic capsules and proteases that destroy antibodies, make them potential pathogens.

Prevotella is involved with other Gram-negative bacteria in about half of all sinus and ear infections, and in almost all periodontal infections. Additionally, these bacteria cause gynecological infections such as pelvic inflammatory disease, pelvic abscesses, and endometriosis; brain abscesses; and abdominal infections.

Treatment of *Prevotella* infections involves surgical removal of infected tissue and the use of antimicrobial drugs.

CHAPTER SUMMARY

1. Lipid A in the outer membranes of Gram-negative bacteria can stimulate symptoms of fever, vasodilation, and shock, as well as blood clots throughout the body, a condition known as **disseminated intravascular coagulation (DIC).**

Pathogenic Gram-Negative Cocci: Neisseria: (pp. 567–570)

1. *Neisseria* is a pathogenic Gram-negative coccus; its virulence results from the presence of fimbriae, polysaccharide capsules, and lipooligosaccharide in its cell walls.

2. *Neisseria gonorrhoeae* causes **gonorrhea,** a sexually transmitted disease in humans. In men it results in acute inflammation of the urethra, whereas in women it is generally asymtomatic. It can move to the uterus and uterine tubes to cause **pelvic inflammatory disease (PID).**

3. *Neisseria meningitidis* causes a type of meningitis; the bacterium is transmitted on respiratory droplets and is life threatening when it enters the bloodstream or central nervous system.

Pathogenic, Gram-Negative, Facultatively Anaerobic Bacilli (pp. 570–583)

1. Members of the **Enterobacteriaceae,** the **enteric bacteria,** can be pathogenic, are oxidase negative, reduce nitrate to nitrite, and ferment glucose anaerobically. Their outer membranes contain a lipopolysaccharide, contributing to their virulence.

2. Pathogenic enteric bacteria are grouped as coliform opportunists, noncoliform opportunists, and true pathogens. **Coliforms** are found in the intestinal tracts of animals and humans.

3. *Escherichia coli* is the most common and most widely studied coliform. It causes gastroenteritis, non-nosocomial urinary tract infections, fatal hemorrhagic colitis, and hemolytic uremic syndrome.

4. *Klebsiella, Serratia, Enterobacter, Hafnia,* and *Citrobacter* are genera of coliform bacteria. *K. pneumoniae* causes a type of pneumonia. All are involved in nosocomial infections.

5. Noncoliform opportunistic pathogens of the family Enterobacteriaceae include *Proteus* (urinary tract infections), and *Morganella, Providencia,* and *Edwardsiella,* which can cause nosocomial infections in immunocompromised patients.

6. Truly pathogenic enteric bacteria include *Salmonella,* which causes **salmonellosis,** a serious form of diarrhea. *S. typhi* and *S. paratyphi* cause **typhoid fever.**

7. Members of the genus *Shigella* cause **shigellosis,** a severe form of diarrhea. **Shiga toxin,** secreted by some *Shigella,* arrests protein synthesis in host cells.

8. Two enteric species, *Yersinia enterocolitica* and *Y. pseudotuberculosis,* cause different degrees of intestinal distress; the virulent, nonenteric *Y. pestis* causes **bubonic plague** and **pneumonic plague** and has had a major historical impact.

9. Two genera in the family *Pasteurellaceae*—*Pasteurella* and *Haemophilus*—are significant human pathogens. They differ from enteric bacteria by being oxidase positive.

10. *Haemophilus influenzae* causes infantile meningitis, epiglottitis, arthritis, and inflammation of the skin. Widespread immunization with Hib vaccine has almost eliminated disease caused by this pathogen in the United States. *H. ducreyi* is responsible for a sexually transmitted lesion called chancroid.

Pathogenic, Gram-Negative, Aerobic Bacilli (pp. 583–591)

1. *Bartonella* includes pathogenic aerobic bacilli. *B. bacilliformis,* transmitted by sand flies, causes **bartonellosis.** *B. quintanna,* transmitted by lice, causes 5-day fever or **trench fever.** *B. henselae,* transmitted through cat scratches, bites, and fleas, causes **cat-scratch disease.**

2. *Brucella,* transmitted in unpasteurized contaminated milk, causes **brucellosis,** also known as Bang's disease, undulant fever, Malta fever, and other names.

3. *Bordetella pertussis* is responsible for **pertussis** (whooping cough), a pediatric disease in which the ciliated epithelial cells of the trachea are damaged. After incubation, the disease progresses through three additional phases: The **catarrhal phase** resembles a cold; in the **paroxysmal stage** the patient coughs deeply to expel copious mucus from the trachea; and in the **convalescent phase** the disease subsides, but secondary bacterial infections may ensue.

4. Pseudomonads are ubiquitous opportunistic pathogens. *Pseudomonas aeruginosa* is involved in many nosocomial infections and is common in burn victims and cystic fibrosis patients. *Moraxella* and *Acinetobacter* are opportunistic pathogens that rarely cause disease.

5. **Tularemia** (rabbit fever or tick fever) is a zoonotic disease caused by *Francisella tularensis*, which is so small that it can pass through apparently unbroken skin.

6. **Legionnaires' disease** (legionellosis) and Pontiac fever are caused by *Legionella*, a bacterium transmitted in aerosols such as those produced by air conditioning systems.

7. *Coxiella burnetii* live in phagolysosomes of mammal and bird cells; its infective bodies are transmitted via aerosolized dried feces and urine to cause **Q fever.**

Pathogenic, Gram-Negative, Anaerobic Bacilli (pp. 591–592)

1. Gram-negative anaerobes—*Bacteroides* and *Prevotella*—are part of the normal microbiota of the intestinal, urinary, genital, and respiratory tracts.

2. Gram-negative anaerobes may cause disease as a result of virulence factors such as capsules, fimbriae, and proteases that degrade antibodies, particularly when the bacteria are introduced into novel sites in the body.

QUESTIONS FOR REVIEW

(Answers to multiple choice and matching questions are on the web, along with additional review questions. Visit www.microbiologyplace.com.)

Multiple Choice

1. The presence of lipid A in the outer membranes of Gram-negative bacteria
   a. affects the formation of blood clots in the host.
   b. causes these bacteria to be oxidase positive.
   c. triggers the secretion of a protease enzyme to cleave IgA in mucus.
   d. enables enteric bacteria to ferment glucose anaerobically.

2. The only genus of Gram-negative cocci that causes significant disease in humans is
   a. *Pasteurella.*
   b. *Salmonella.*
   c. *Klebsiella.*
   d. *Neisseria.*

3. Which of the following bacterial cells is most likely to be virulent?
   a. a cell with fimbriae and LOS
   b. a cell with a polysaccharide capsule and lipooligosaccharide
   c. a cell with fimbriae, lipooligosaccharide, and a polysaccharide capsule
   d. a cell with fimbriae but no capsule

4. Which of the following statements is true?
   a. PID is a severe type of diarrhea in which infection spreads from the intestines to the bloodstream.
   b. PID can result from *Neisseria* infection.
   c. PID is more common in men than women.
   d. Members of the family Enterobacteriaceae usually cause PID.

5. Capsules on pathogenic enteric bacteria are virulence factors because they
   a. capture iron from hemoglobin and store it in the bacteria.
   b. release hemolysins that destroy red blood cells.
   c. produce fimbriae that enable the bacteria to attach to human cells.
   d. protect the bacteria from phagocytosis and from antibodies.

6. A coliform bacterium that contaminates dairy products is
   a. *Bartonella.*
   b. *Serratia.*
   c. *Enterobacter.*
   d. *Proteus.*

7. Which of the following bacteria might be responsible for the formation of petechiae in a host?
   a. *Neisseria meningitidis*
   b. *Escherichia coli* O157:H7
   c. *Klebsiella*
   d. *Proteus mirabilis*

8. The pathogen *Haemophilus influenzae* b causes
   a. meningitis in children.
   b. upper respiratory flu.
   c. endocarditis.
   d. genital chancroid.

9. Which of the following diseases is typically mild or asymptomatic?
   a. Brazilian purpuric fever   c. pediatric meningitis
   b. bartonellosis   d. cat-scratch disease

10. Which bacterium causes infections in many burn victims?
    a. *Moraxella catarrhalis*   c. *Pseudomonas aeruginosa*
    b. *Escherichia coli*   d. *Bartonella bacilliformis*

11. Which of the following statements is true of Q fever?
    a. For many years its cause was questionable.
    b. It was first described in 1976 during an outbreak in Quincy, Massachusetts.
    c. Researchers found it could be effectively treated with quinine.
    d. The sharp spikes of fever on patients' temperature charts resemble porcupine quills.

12. Which of the following is a bile-tolerant anaerobe?
    a. *Bacteroides*   c. *Shigella*
    b. *Escherichia*   d. *Prevotella*

Matching

Match the bacteria with the disease (or symptoms) it causes.

____ 1. *Escherichia coli*	A. Bubonic plague
____ 2. *Klebsiella pneumoniae*	B. Typhoid fever
____ 3. *Proteus mirabilis*	C. Gastroenteritis
____ 4. *Salmonella typhi*	D. Kidney stones
____ 5. *Shigella flexneri*	E. Purulent bloody stools, cramps, fever, and diarrhea
____ 6. *Yersinia pestis*	F. Pneumonia

Short Answer

1. A physician treating a patient with diarrhea prescribed fluid replacement. Although tests showed that a pathogenic enteric bacterium was the cause of the intestinal distress, why was an antimicrobial drug not prescribed?

2. Distinguish among the pathogenicity of coliforms, noncoliforms, and truly pathogenic enteric bacteria.

3. Why do physicians place antimicrobial agents in babies' eyes at birth?

4. Statistics show that meningococcal diseases are more frequent in college dormitories and military barracks than in the population at large. Suggest an explanation of this observation.

5. Why might an alcoholic be susceptible to pulmonary disease?

6. Name six factors that facilitate the production of disease by *Bordetella pertussis*.

7. Given that pseudomonads are present in almost every moist environment, why do they cause less disease than other, less prevalent Gram-negative bacteria?

8. Although an effective vaccine is available to eradicate pertussis in the United States, why has the number of reported cases increased since the 1970s?

9. What two illnesses can be caused by *Legionella pneumophila*?

10. What attribute of *Coxiella burnetii* enables it to survive desiccation and heat for an extended time?

11. What virulence factors allow Gram-negative anaerobes to cause disease?

CRITICAL THINKING

1. A 3-year old boy complains to his day care worker that his head hurts. The worker calls the child's mother, who arrives 30 minutes later to pick up her son. She is concerned that he is now listless and unresponsive, so she drives straight to the hospital emergency room. Though the medical staff immediately treats the boy with penicillin, he dies—only 4 hours after initially complaining of a headache. What caused the boy's death? Were the day care workers or the hospital staff to blame for his death? What steps should be taken to protect the other children at the day care facility?

2. An epidemiologist notices a statistical difference in the fatality rates between cases of Gram-positive bacteremia treated with antimicrobial drugs and treated cases of Gram-negative bacteremia—that is, patients with Gram-positive bacteremia are much more likely to respond to treatment and survive. Explain one reason why this might be so.

3. In June 2001, local physicians reported 11 cases of severe diarrhea in infants less than 3 years of age. All the children lived in the same neighborhood and played in the same park, which has a wading pool. Cultures of stool specimens from the children revealed a Gram-negative bacillus that gave negative results for the following biochemical tests: oxidase, lactose, urease, and hydrogen sulfide production. The organism was also nonmotile. What is the organism? How was it transmitted among the children? What could officials do to limit infections of other children?

4. Several years ago, epidemiologists noted that the number of cases of salmonellosis increased dramatically in the summer and was lower the rest of the year. In contrast, the number of cases of typhoid fever, caused by the same bacterium, was constant throughout the year. Explain these observations.

5. Most physicians maintain that the viral "stomach flu" is really a bacterial infection. Which genera of bacteria discussed in this chapter are good candidates for causing this disease?

6. A hunter reports to his physician that he has been suffering with fever, chills, malaise, and fatigue. What Gram-negative bacteria may be causing his symptoms? How can a laboratory scientist distinguish among these species?

CHAPTER 21

Mycoplasmas, Rickettsias, Chlamydias, Spirochetes, and Slightly Curved Bacilli

Two days after consuming raw oysters at a cocktail party, a 40-year-old man is admitted to the hospital with a 105° fever, nausea, myalgia (muscle pain), and circular lesions on his right leg. As part of his medical history he reveals that he consumes more than six bottles of beer a day. The hospital's diagnosis is an infection with *Vibrio vulnificus,* a bacterium that can cause septicemia (blood poisoning), particularly in individuals who are immunocompromised or have liver disease. Within 24 hours of the diagnosis, the man goes into septic shock and dies. Investigations reveal that he had a preexisting alcohol-related liver disease, and that the source of his infection was contaminated oysters.

V. vulnificus is a Gram-negative bacterium, but in some ways an atypical one: It has a slightly curved shape and differs metabolically from other flagellated enteric bacteria. In this chapter we focus on pathogenic Gram-negative bacteria that differ in some way from most other Gram-negative bacteria: the mycoplasmas, rickettsias, chlamydias, spirochetes, and curved bacilli (including *Vibrio*).

CHAPTER OUTLINE

Mycoplasmas
  Mycoplasma pneumoniae
  Other Mycoplasmas

Rickettsias
  Rickettsia
  Orienta
  Ehrlichia

Chlamydias
  Chlamydia trachomatis
  Chlamydia pneumoniae
  Chlamydia psittaci

Spirochetes
  Treponema
  Borrelia
  Leptospira

Pathogenic Gram-Negative Curved Bacilli
  Vibrio
  Campylobacter jejuni
  Helicobacter pylori

Vibrio vulnificus

In Chapter 20 we explored the more important pathogenic Gram-negative cocci and bacilli. But a number of bacteria that stain pink in a Gram stain differ from typical Gram-negative organisms in morphology, growth habit, or reproductive strategy—including the wall-less mycoplasmas, the obligate intracellular rickettsias and chlamydias, the spiral-shaped spirochetes, and the slightly curved vibrios. Because of their unique features, these bacteria have traditionally been discussed separately—a tradition we continue here, beginning with the mycoplasmas.

Mycoplasmas

Learning Objectives

✓ List at least four characteristics of all mycoplasmas.

✓ Explain why mycoplasmas have been classified with both Gram-negative and Gram-positive organisms.

✓ Compare and contrast mycoplasmas and viruses.

Mycoplasmas (mī'kō-plaz-măz) are unique bacteria named for their most common representative, the genus *Mycoplasma*. Mycoplasmas also include bacteria in the genus *Ureaplasma* (yū-rē'ă-plaz'mă). These bacteria lack cytochromes (which are present in other organisms' electron transport chains), enzymes of the Krebs cycle, and cell walls—indeed, they are incapable of synthesizing peptidoglycan and its precursors. Moreover, most mycoplasmas have sterols in their cytoplasmic membranes, a feature lacking in other prokaryotes. They are further distinct from all other organisms in that the codon UGA neither serves as a stop codon nor codes for selenocysteine, but instead codes for the amino acid tryptophan.

Before analysis of the nucleic acid sequences of mycoplasmal rRNA revealed that they are similar to Gram-positive organisms, mycoplasmas were classified in a separate phylum of Gram-negative bacteria—Phylum Mollicutes. Modern taxonomists and the second edition of *Bergey's Manual of Systematic Bacteriology* now categorize organisms in Mollicutes as a class of low G + C, Gram-positive bacteria in the phylum Firmicutes. Nevertheless, they appear pink when stained with the Gram-stain (because they lack a cell wall), and thus traditionally they have been discussed with the Gram-negative bacteria examined in this chapter.

Because they lack cell walls, mycoplasmas[1] are pleomorphic, taking on a variety of shapes, including cocci and thin, unicellular filaments up to 150 μm long that resemble the hyphae of fungi **(Figure 21.1)**. Mycoplasmas are able to withstand osmotic stress despite lacking walls because they colonize osmotically protected habitats such as animal and human bodies, and the *sterols* in their membranes convey sufficient strength and rigidity.

Mycoplasmas have diameters ranging from 0.1 μm to 0.8 μm, making them the smallest *free-living* microbes—that is, those that can grow and reproduce independently of

SEM   20 μm

▲ *Figure 21.1*

Pleomorphic forms of *Mycoplasma*. These bacteria can be either coccoid or filamentous because they lack a cell wall.

other cells. Originally, many mycoplasmas were thought to be viruses because their small, flexible cells enabled them to squeeze through the 0.45 μm pores of filters that were then commonly used to remove bacteria from solutions; however, mycoplasmas contain both RNA and DNA, and they divide by binary fission—traits that viruses lack.

Mycoplasmas require organic growth factors, including cholesterol, fatty acids, vitamins, amino acids, and nucleotides; these factors must either be acquired from a host or supplied by laboratory media. With exception of the strictly aerobic *Mycoplasma pneumoniae* (nū-mō'nē-ē), mycoplasmas are facultative anaerobes.

Most mycoplasmas form distinctive colonies on solid media. The colonies are so small that they must be viewed through a microscope, and the colonies of most species resemble fried eggs (because cells at the center of colonies tend to grow into the agar, whereas those at the periphery remain on the surface of the medium; see Figure 11.16). Note that two important pathogenic mycoplasmas, *M. pneumoniae* and *Ureaplasma*, are exceptions in that their colonies do not resemble fried eggs.

Mycoplasmas can colonize mucous membranes of the respiratory and urinary tracts and are associated with pneumonia and urinary tract infections. Of the two genera that cause diseases in humans, *Mycoplasma* is unable to utilize urea (that is, it is urease negative), whereas *Ureaplasma* can hydrolyze urea to form ammonia.

Forty-six species of *Mycoplasma* have been identified, but only a few cause significant diseases in humans. We begin our examination of these species with *M. pneumoniae*.

---

[1]From Greek *mycos*, meaning fungus, and *plasma*, meaning anything formed or molded.

Mycoplasma pneumoniae

Learning Objective

✓ Describe the damage done to respiratory epithelial cells by *Mycoplasma pneumoniae*.

Pathogenesis and Epidemiology

Mycoplasma pneumoniae **(Figure 21.2)** has an adhesive protein that attaches specifically to receptors located at the bases of cilia on epithelial cells lining the respiratory tracts of humans. Attachment causes the cilia to stop beating, and mycoplasmal colonization eventually kills the epithelial cells. This interrupts the normal removal of mucus from the respiratory tract, allowing colonization by other bacteria and causing a buildup of mucus that irritates the upper respiratory tract. Early symptoms of *M. pneumonia* infections—including fever, malaise, headache, and sore throat—are not typical of other types of pneumonia; thus the disease is called **primary atypical pneumonia.** The body subsequently responds with a persistent, unproductive cough in an attempt to clear the lungs.

Primary atypical pneumonia may last for several weeks, but it is usually not severe enough to require hospitalization or to cause death. Because symptoms can be mild, the disease is also sometimes called *walking pneumonia.*

Nasal secretions spread *M. pneumoniae* among people in close contact, such as classmates and family members. The disease is uncommon in children under age 5 or in adults older than 20, and it is probably the most common form of pneumonia seen in children 5–15 years old. However, because primary atypical pneumonia is not a reportable disease and is difficult to diagnose, the actual incidence of infection is unknown.

Primary atypical pneumonia occurs throughout the year. This lack of seasonality is in contrast to pneumococcal pneumonia (caused by *Streptococcus pneumoniae*), which is more commonly seen in the fall and winter.

Diagnosis, Treatment, and Prevention

Diagnosis is difficult because mycoplasmas are small and difficult to detect in clinical specimens or tissue samples. Further, mycoplasmas grow slowly in culture, requiring 2–6 weeks before colonies can be seen. As previously noted, the colonies of *M. pneumoniae* differ from the colonies of other mycoplasmas in lacking a fried egg appearance; instead, they have a uniform granular appearance. Complement fixation, hemagglutination, and immunofluorescent tests are sometimes used to confirm a diagnosis, but such tests are nonspecific and are not by themselves positively diagnostic.

Physicians treat primary atypical pneumonia with erythromycin or tetracycline. Prevention is difficult because patients are often infective for long periods of time without signs or symptoms, and they remain infective even while undergoing antimicrobial treatment. Nevertheless, frequent hand antisepsis, avoidance of contaminated fomites, and reducing aerosol dispersion can limit the spread of the pathogen and the number of cases of disease. No vaccine against *M. pneumonia* is available.

CRITICAL THINKING

Mycoplasma pneumoniae, like many pathogenic bacteria, is resistant to penicillin; however, unlike most resistant species, *Mycoplasma* does not synthesize beta-lactamase. Explain why *Mycoplasma* is resistant to penicillin.

Other Mycoplasmas

Learning Objectives

✓ List three mycoplasmas associated with urinary and genital tract infections.
✓ Describe pelvic inflammatory disease.

Three other mycoplasmas are associated with diseases of humans. *M. hominis* (ho′mi-nis), *M. genitalium* (jen-ē-tal′ē-ŭm), and *Ureaplasma urealyticum* (ū-rē′ă-li′ti-kŭm) often colonize the urinary and genital tracts of newborn girls, but such infections rarely persist through childhood. However, the incidence of genital reinfections increases after puberty as a result of sexual activity; approximately 15% of sexually active adult Americans (males and females) are currently infected with *M. hominis,* and 70% are infected with *Ureaplasma.*

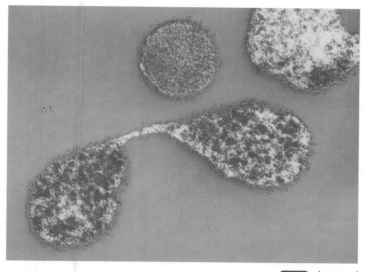

TEM ⊢————⊣ 0.3 μm

▲ *Figure 21.2*

Colored transmission electron micrograph of *Mycoplasma pneumoniae* cells.

M. genitalium and *U. urealyticum* cause *nongonococcal urethritis (NGU)*—that is, inflammation of the urethra not caused by *N. gonorrhoeae*. By contrast, *M. hominis* causes *pyelonephritis* (inflammation of the kidneys) and *postpartum fever*, and can cause **pelvic inflammatory disease (PID)** in women. Pelvic inflammatory disease is characterized by inflammation of the organs in the pelvic cavity, fever, and abdominal pain. Recall from Chapter 20 that *N. gonorrhoeae* can also cause this condition.

Infections of either *M. genitalium* or *U. urealyticum* are treated with erythromycin or tetracycline, whereas clindamycin is used to treat *M. hominis*, which is commonly resistant to tetracycline and erythromycin. Abstinence, mutually faithful monogamy, and proper use of condoms prevent the spread of these sexually transmitted organisms.

CRITICAL THINKING

Why do pediatricians refrain from using tetracycline to treat mycoplasmal infections in children?

Rickettsias

Rickettsias (ri-ket′sē-ăz) are Gram-negative, obligate intracellular parasites that synthesize only a small amount of peptidoglycan and thus appear almost wall-less. These organisms, and the most common genus of them (*Rickettsia*), are named after Dr. Howard Ricketts (1871–1910), who first identified them and described the transmission of one rickettsial species via its tick vector. Rickettsias are extremely small (0.3 μm × 1.0 μm); in fact, they are not much bigger than a smallpox virus. Because of these characteristics, rickettsias were originally considered viruses, but closer examination has revealed that they contain DNA and RNA, functional ribosomes, and Krebs cycle enzymes, and that they reproduce via binary fission—all characteristics of cells, not viruses.

Researchers have proposed several hypotheses to explain why rickettsias are obligate parasites, even though they have functional genes for protein synthesis, ATP production, and reproduction. Primary among these hypotheses is that rickettsias have very "leaky" cytoplasmic membranes and lose small cofactors (such as NAD$^+$) unless they are in an environment that contains an equivalent amount of these cofactors—such as the milieu within a host cell.

Scientists classify rickettsias in the class Alphaproteobacteria of the bacterial phylum Proteobacteria based on the sequence of nucleotides in their rRNA molecules. Three genera of rickettsias—*Rickettsia, Orienta,* and *Ehrlichia*—cause diseases in humans. We begin our discussion with *Rickettsia*.

Rickettsia

Learning Objectives

✓ List three species of *Rickettsia* that are responsible for human infections.

✓ Identify the vectors of *R. rickettsii, R. prowazekii,* and *R. typhi.*

✓ Describe the rash and petechiae of Rocky Mountain spotted fever (RMSF).

✓ Explain the relationship between epidemic typhus and Brill-Zinsser disease.

Rickettsia is a genus of nonmotile, aerobic, Gram-negative, intracellular parasites that live in the cytosol of their host cells. They possess cell walls of peptidoglycan and an outer membrane of lipopolysaccharide, which has little endotoxin activity. A loosely organized slime layer surrounds each cell. Rickettsias do not react well to the Gram stain, coloring only lightly pink, so scientists use special staining procedures such as *Giemsa* or *Gimenez* stains to visualize them **(Figure 21.3).** Most human infections are by three species of *Rickettsia*: *R. rickettsii* (ri-ket′sē-ē), *R. typhi* (tī′fē), and *R. prowazekii* (prō-wă-ze′kē-ē).

All three species of *Rickettsia* are transmitted by arthropod vectors and enter host cells by stimulating phagocytosis. Once inside a host cell, the microbes secrete an enzyme that digests the phagosome membrane, releasing the bacteria into the cytosol. As a result, the rickettsias avoid the digestion that would ensue if a lysosome had merged with the phagosome.

The pathogen grows and reproduces slowly, producing daughter cells every 8–12 hours. *R. rickettsii* and *R. typhi* are continually released via exocytosis from long cytoplasmic extensions of host cells. In contrast, *R. prowazekii* gradually fills the host cell until it lyses, releasing the parasites.

Outside of its host's cells, *Rickettsia* is unstable and dies quickly; therefore, rickettsias require vectors for transmission from host to host. Each rickettsial species is transmitted by different vectors, which also act as hosts and reservoirs.

▶ *Figure 21.3*

A Giemsa stain for visualizing *Rickettsia rickettsii* within infected tick cells.

LM   10 μm

Rickettsia rickettsii

R. rickettsii causes **Rocky Mountain spotted fever (RMSF)**, the most severe and most reported rickettsial illness. Even though the earliest reports of RMSF came from the Rocky Mountain states, the disease is actually more prevalent in the Appalachian Mountains, Oklahoma, and the southeastern states **(Figure 21.4)**. *R. rickettsii* is transmitted among humans and rodents (which act as reservoirs) by hard ticks of the genus *Dermacentor* (see Figure 23.24a). Male ticks infect female ticks during mating. Female ticks transmit the bacteria to the eggs forming in their ovaries. This process is called *transovarian transmission.*

R. rickettsii is typically dormant in the salivary glands of the ticks, and only when the arachnids feed for several hours are the bacteria activated. Active bacteria are released from the tick's salivary glands into the mammalian host's circulatory system, where they infect endothelial cells lining small blood vessels. In rare cases, humans become infected following exposure to tissues and fluids from crushed ticks or to tick feces.

R. rickettsii secretes no toxins, and disease is not the product of the host's immune response. Apparently, damage to the blood vessels leads to leakage of plasma into the tissues, which may result in low blood pressure and insufficient nutrient and oxygen delivery to the body's organs.

About a week after infection, patients experience fever, headache, chills, muscle pain, nausea, and vomiting. In most cases (90%), a spotted, nonitchy rash develops on the trunk and appendages **(Figure 21.5)**—including the palms and soles, sites not involved in rashes caused by the chickenpox or measles viruses. In about 50% of patients, the rash develops into subcutaneous hemorrhages called *petechiae.* In severe cases, the respiratory, central nervous, gastrointestinal, and renal systems fail. Encephalitis may also occur, produc-

▲ *Figure 21.5*

The rash in a case of Rocky Mountain spotted fever.

ing language disorders, delirium, convulsions, coma, and death. Even with treatment, almost 5% of patients die. Patients recovering from life-threatening acute Rocky Mountain spotted fever may experience paralysis of the legs, hearing loss, and gangrenous secondary infections with *Clostridium* (klos-trid′ē-ŭm) that necessitate the amputation of fingers, toes, arms, or legs.

Serological tests such as latex agglutination and fluorescent antibody stains are used to confirm an initial diagnosis based on sudden fever and headache following exposure to hard ticks, plus a rash on the soles or palms. Nucleic acid probes of specimens from rash lesions provide specific and accurate diagnosis, but such tests are expensive and typically are performed only by trained technicians in special laboratories. Early diagnosis is crucial because prompt treatment often makes the difference between recovery and death.

Physicians treat Rocky Mountain spotted fever with doxycycline, tetracycline, or chloramphenicol. An effective vaccine is not available. Prevention of infection involves wearing tight-fitting clothing, using tick repellants, promptly removing attached ticks, and avoiding tick-infested areas, especially in spring and summer, when ticks are most voracious. It is impossible to eliminate the ticks in the wild because they can survive without feeding for more than 4 years.

CRITICAL THINKING

Why do most cases of Rocky Mountain spotted fever occur in May, June, and July?

Rickettsia prowazekii

Rickettsia prowazekii causes **epidemic typhus,**[2] which is also called *louse-borne typhus* because it is vectored by the human body louse, *Pediculis humanus* (see Figure 23.24c).

[2]From Greek *typhos,* meaning stupor.

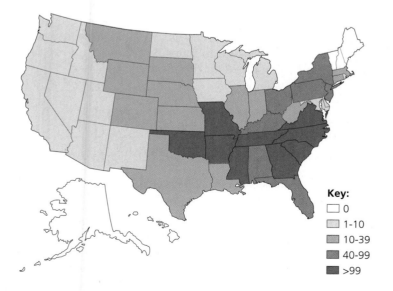

Key:
☐ 0
☐ 1-10
◻ 10-39
▨ 40-99
■ >99

▲ *Figure 21.4*

Incidence of Rocky Mountain spotted fever in the United States, 1994–1998.

R. prowazekii is fatal to its vector within 2–3 weeks, which prevents transovarian transmission but allows sufficient time for the lice to transmit the bacterium between humans.

In contrast to other rickettsias, humans are the primary hosts of *R. prowazekii,* though the bacterium can live in squirrels and their fleas. Epidemiological and laboratory evidence indicates that such fleas can transmit the bacterium from squirrels to humans, but such transmission has not actually been demonstrated in nature. Epidemic typhus can recur many years (even decades) following an initial episode. The recurrent disease (called *Brill-Zinsser*[3] *disease*) is mild, brief, and resembles murine typhus (discussed shortly).

Epidemic typhus occurs in crowded, unsanitary living conditions that favor the spread of body lice; it is endemic in Central and South America, in Africa, and in localized areas in the eastern United States where squirrels are common. The disease is not nationally reportable in the United States, so the actual incidence of epidemic typhus is unknown.

Diagnosis is based on the observation of signs and symptoms—high fever, mental and physical depression, and a rash that lasts for about 2 weeks—following exposure to lice. Diagnosis must be confirmed by the demonstration of the bacteria in tissue samples using fluorescent antibody tests.

Epidemic typhus is treated with doxycycline, tetracycline, or chloramphenicol. Prevention involves controlling lice populations and maintaining good personal hygiene. An attenuated vaccine against epidemic typhus is available for use in high-risk populations.

Rickettsia typhi

Rickettsia typhi causes **murine**[4] (mū'rēn) **typhus,** which is so named because the major reservoir for the bacteria is rodents. The vectors for this disease, which is also known as *endemic typhus,* are fleas. The rat flea *Xenopsylla cheopis* (see Figure 23.24b) and the cat flea *Ctenocephalides felis* (which also feeds on opossums, raccoons, and skunks) transmit the bacteria among the animal hosts and to humans. About 10 days following the bite of an infected flea, an abrupt fever, severe headache, chills, muscle pain, and nausea occur. A rash typically restricted to the chest and abdomen occurs in less than 50% of cases. The disease usually lasts about 3 weeks if left untreated and is not fatal.

Murine typhus is most often seen in the southern United States from Florida to California, and it is estimated that 50 cases occur annually, though the extent and incidence of the disease is not known because national notification requirements for murine typhus were discontinued in 1994. The disease is still endemic to Australia, Africa, Asia, Europe, and South America.

Diagnosis is initially based on signs and symptoms following exposure to fleas. An immunofluorescent antibody stain of a blood smear provides specific confirmation. Treatment is with tetracycline, doxycycline, or chloramphenicol. Prevention, as with other rickettsial diseases, involves avoiding bites by the arthropod vectors, wearing protective clothing, and using chemical repellants. No vaccine is available for murine typhus.

Orienta

Learning Objective

✓ Identify the causative agent, vector, and reservoir of scrub typhus.

The rickettsial organism *Orienta tsutsugamushi* (ōr-ē-en'tă soo-soo-gă-mū'shē) was classified formerly as *Rickettsia tsutsugamushi,* but taxonomists now assign it to a new genus. It differs from *Rickettsia* by having significantly different rRNA nucleotide sequences, a thicker cell wall, and a minimal slime layer. Mites of the genus *Leptotrombidium* (lep'tō-trom-bid'ē-ŭm), also known as red mites or chiggers, transmit *Orienta* among rodents and humans. Infected female mites transmit the rickettsia to their offspring via transovarian transmission, and the offspring transmit the bacteria to a rodent or human while feeding. Because these arachnids feed on a host only once in their lives, an individual mite cannot transmit rickettsia directly from a rodent to a human. In addition to being the vector, mites are the only known reservoir of *Orienta.*

O. tsutsugamushi causes **scrub typhus,** a disease endemic to eastern Asia, Australia, and western Pacific islands, including Japan. It occurs in the United States among immigrants who arrived from endemic areas. Scrub typhus is characterized by fever, headache, and muscle pain, all of which develop about 11 days after a mite bite. Less than half of patients with scrub typhus also develop a spreading rash on their trunks and appendages. In a few cases, death results from failure of the heart and the central nervous system.

Scrub typhus is treated with tetracycline, doxycycline, or chloramphenicol. No vaccine is available. Prevention involves avoiding exposure to mites by wearing appropriate clothing and using repellant chemicals.

Ehrlichia

Learning Objectives

✓ Identify two pathogenic species of *Ehrlichia* and the diseases they cause.
✓ Identify the three developmental stages of *Ehrlichia.*
✓ Discuss the difficulties in diagnosing ehrlichiosis.

Ehrlichia chaffeensis (er-lik'ē-ă chaf-ē-en'sis) and *E. equi* (ē'kwi) cause **human monocytic ehrlichiosis (HME)** and **human granulocytic ehrlichiosis (HGE),** respectively. These two diseases are considered *emerging diseases* in the United States because they were unknown before 1986, and the number of reported cases has increased from a few dozen per year in the 1980s to over 400 cases annually since 2000—the first year they became nationally reportable.

---

[3]After physician Nathan Brill and bacteriologist Hans Zinsser, who studied the condition.
[4]From Latin *muris,* meaning mouse.

Ticks, including the Lone Star tick (*Ambylomma*), the deer tick (*Ixodes*), and the dog tick (*Dermacentor*), transmit *Ehrlichia* to humans. Once in the blood, each bacterium triggers its own phagocytosis by a white blood cell (either a monocyte or a neutrophil). Unlike *Rickettsia*, *Ehrlichia* grows and reproduces within a host cell's phagosomes. Because *Ehrlichia* is killed if a phagosome fuses with a lysosome, the bacteria must somehow prevent fusion, but the mechanism is unknown. Inside a leukocyte the bacterium grows and reproduces through three developmental stages: an *elementary body*, an *initial body*, and a *morula* **(Figure 21.6).** The release of the cells from the morula and into the blood makes them available to feeding ticks.

The diseases of *Ehrlichia* resemble Rocky Mountain spotted fever without a rash, which only rarely occurs in ehrlichiosis. *Leukopenia* (loo-kō-pē′nē-ǎ), which is an abnormally low leukocyte count, is also seen. Mortality of untreated HME is about 5%, and that of untreated HGE approaches 10%, especially in elderly patients.

Diagnosis of ehrlichiosis is difficult because the symptoms resemble other diseases. Physicians consider ehrlichiosis in any case of otherwise unexplained acute fever in patients exposed to ticks in endemic regions **(Figure 21.7).** Immunofluorescent antibodies against *Ehrlichia* can demonstrate the bacterium within blood cells, confirming the diagnosis.

Doxycycline and tetracycline are effective against *Ehrlichia*, but chloramphenicol is not. Treatment should start

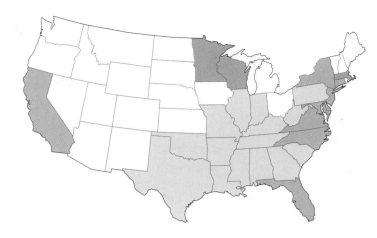

▲ *Figure 21.7*

The geographical distribution of ehrlichiosis in the contiguous United States. Red indicates states where HGE is endemic, blue where HME is endemic, and purple where both diseases are endemic.

immediately, even before the diagnosis is confirmed by serological testing, because the complications of infection and mortality rates increase when treatment is delayed. Prevention involves avoiding tick-infested areas, promptly removing ticks, wearing tight-fitting clothing, and using repellant chemicals. Vaccines against the two species are not available.

Case Study 21.1 describes an interesting human rickettsial infection, and **Table 21.1** on page 602 summarizes the rickettsial species, vectors, reservoirs, and diseases.

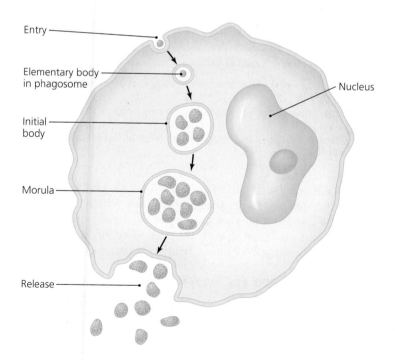

Entry

Elementary body in phagosome

Nucleus

Initial body

Morula

Release

▲ *Figure 21.6*

The growth and reproduction cycle of *Ehrlichia* in an infected leukocyte. *If the rickettsia shown here is* Ehrlichia chaffeensis, *then what kind of leukocyte is it infecting?*

Figure 21.6 Ehrlichia chaffeensis *infects human monocytes.*

Case Study 21.1 Blame It on the Chiggers?

A 75-year-old man arrives one summer morning at the hospital with severe headache, chills, muscle pain, and nausea. He reports that the symptoms developed quite suddenly. He has had no rash. During the interview the man states that he is an immigrant from Poland, a survivor of a Nazi concentration camp, and an avid bird watcher who spends many hours in the woods near his home. He had not noticed any fleas, ticks, or lice on his body, but he had been "almost eaten alive by chiggers." He had not been out of the United States in 40 years.

Chiggers (mites embedded in feet)

1. What rickettsial diseases might the man have?

2. What are the vectors for these diseases?

3. What diagnostic test(s) might verify the initial diagnosis?

4. Is the man's life in danger?

5. What drug will the physician likely prescribe?

Table 21.1   Characteristics of Rickettsias

Organism	Vectors	Reservoirs	Diseases
R. rickettsii	Hard ticks: wood tick (*Dermacentor andersoni*) and dog tick (*D. variabolis*)	Ticks, rodents	Rocky Mountain spotted fever (RMSF)
R. prowazekii	Human body louse (*Pediculus humanus*)	Humans, squirrels, squirrel fleas	Epidemic typhus
R. typhi	Rat flea (*Xenopsylla cheopsis*) and cat flea (*Ctenocephalides felis*)	Rodents	Murine (endemic) typhus
Orienta tsutsugamushi	Mite, chigger (*Leptotrombidium* spp.)	Mites	Scrub typhus
Ehrlichia spp.	Hard ticks: Lone Star tick (*Ambyloma*), deer tick (*Ixodes*), and dog tick (*D. variabolis*)	Ticks	Human monocytic ehrlichiosis (HME) and human granulocytic ehrlichiosis (HGE)

CRITICAL THINKING

Explain the etymology of the names of the diseases caused by *Ehrlichia*.

Chlamydias

Learning Objectives

✓ Describe the life cycle of *Chlamydia*, including the two developmental forms.

✓ Discuss the term *energy parasite* as it relates to *Chlamydia*.

Bacteria of the genus *Chlamydia* (kla-mid′ē-ă) are nonmotile and grow and multiply only within vesicles in host cells. Scientists once considered chlamydias to be viruses because of their small size, obligate intracellular lifestyle, and ability to pass through 0.45 μm pores in filters, which were thought to trap all cells. However, chlamydias possess DNA, RNA, and functional 70S ribosomes, and they are cellular. Each cell is surrounded by two membranes, similar to a typical Gram-negative bacterium, but there is no peptidoglycan between the membranes—thus chlamydias do not have cell walls. In contrast with rickettsias, arthropods do not serve as vectors or hosts for chlamydias. Because of their unusual features and unique rRNA nucleotide sequences, taxonomists now classify chlamydias in their own phylum: Chlamydiae.

Chlamydias have a unique developmental cycle involving two forms, both of which can occur within the phagosome of a host cell (**Figure 21.8a**): tiny (0.2–0.4 μm) cocci called **elementary bodies (EBs),** and larger (0.6–1.5 μm) pleomorphic **reticulate bodies (RBs).** EBs, which are relatively dormant, resistant to environmental extremes, and can survive outside cells, are the infective forms. RBs are noninfective obligate intracellular forms that replicate via binary fission within phagosomes, where they survive by inhibiting the fusion of a lysosome with the phagosome. RBs lack the metabolic enzymes needed to synthesize ATP, so they must depend on their host cells for the high-energy phosphate compounds they require; thus chlamydias have been called *energy parasites.*

In the life cycle of chlamydias (**Figure 21.8b**), once an EB attaches to a host cell ❶, it enters by triggering its own phagocytosis ❷. Once inside the phagosome, the EB converts into an RB ❸, which then divides rapidly into multiple RBs ❹. Once an infected phagosome becomes filled with RBs, it is called an **inclusion body.** About 21 hours after infection, RBs within an inclusion body begin converting back to EBs ❺, and about 19 hours after that, the EBs are released from the host cell via exocytosis ❻, becoming available to infect new cells and completing the life cycle.

Three species of *Chlamydia* cause disease in humans. In order of the prevalence with which they infect humans, they are *C. trachomatis, C. pneumoniae,* and *C. psittaci.* We begin our discussion with the most common species—*C. trachomatis.*

Chlamydia trachomatis

Learning Objectives

✓ List the types of cells in the human body that are most often infected by *Chlamydia*.

✓ Explain how sexually inactive children may become infected with *C. trachomatis.*

✓ Describe the cause and symptoms of lymphogranuloma venereum.

✓ Discuss the prevention of chlamydial infections.

Chlamydia trachomatis (kla-mid′ē-ă tra-kō′ma-tis) has a very limited host range. One exceptional strain, which may eventually be classified as a separate species, causes pneumonia in mice, but all other strains are pathogens of humans, infecting the conjunctiva, lungs, urinary tract, or genital tract.

Pathogenesis and Epidemiology

C. trachomatis enters the body through abrasions and lacerations and infects a limited array of human cells—those that have receptors for elementary bodies, including cells of the conjunctiva and cells lining the mucous membranes of the trachea, bronchi, urethra, uterus, uterine (Fallopian) tubes, anus, and rectum. The clinical manifestations of chlamydial infection result from the destruction of infected cells at the site of infection, and from the inflammatory response this destruction stimulates. Reinfection in the same site by the same or a similar strain triggers a vigorous hypersensitive

(a)

TEM |—— 1 µm ——|

① EB attaches to receptor on host cell (0 hour)

⑥ EBs are released from host cell (40 hours)

② EB enters host cell via phagocytosis

⑤ Most RBs convert back into EBs (21 hours)

EB inside phagosome

③ EB converts into RB in phagosome (10 hours)

Inclusion body

④ RB divides rapidly resulting in multiple RBs in an inclusion body

(b)

immune response that can result in blindness, sterility, or sexual dysfunction.

Infection with *C. trachomatis* is the most common of the reportable sexually transmitted diseases in the United States; 770,894 cases were reported in 2002, but epidemiologists estimate that about 3.5 million asymptomatic cases go unreported annually. Sexually transmitted chlamydial infections are most prevalent among women under the age of 20 because they are physiologically more susceptible to infection.

Researchers further estimate that over 500 million people worldwide, particularly children, contract ocular infections with *C. trachomatis*. Children can be infected as they pass through the birth canal, and the pathogen can be transmitted from eye to eye via droplets, hands, contaminated fomites, or flies. Infected children also harbor the bacteria in their digestive and respiratory tracts, so bacteria can also be transmitted to other children or adults via fecal contamination and respiratory droplets. Chlamydial eye infections are endemic in crowded, poor communities where people have inadequate personal hygiene, inadequate sanitation, and inferior medical care, particularly in the Middle East, North Africa, and India.

Diseases

The diseases caused by the various strains of *Chlamydia trachomatis* are of two main types: sexually transmitted diseases and the ocular disease called trachoma.

Sexually Transmitted Diseases   A transient genital lesion and swollen, painfully inflamed, inguinal lymph nodes characterize **lymphogranuloma venereum**[5] (lim′fō-gran-ū-lō′mă ve-ne′rē-ŭm) **(Figure 21.9** on page 604), which is caused by the so-called *LGV strain* of *C. trachomatis*. The initial lesion of lymphogranuloma venereum occurs at the site of infection on the penis, urethra, scrotum, vulva, vagina, or cervix. This lesion is often overlooked because it is small, painless, and heals rapidly. Headache, muscle pain, and fever may also occur at this stage of the disease.

The second stage of the disease involves the development of buboes (swollen lymph nodes) associated with lymphatic vessels draining the site of infection. The buboes, which are accompanied by fever, chills, anorexia, and muscle pain, may enlarge to the point that they rupture, producing draining sores. In a few cases, lymphogranuloma venereum proceeds to a third stage characterized by genital sores, constriction of the urethra, and genital elephantiasis. Arthritis may also occur during this third stage, particularly in young white males.

◀ *Figure 21.8*

The developmental forms and life cycle of *Chlamydia*. **(a)** Elementary bodies and reticulate bodies within an inclusion body. **(b)** The life cycle of *Chlamydia*. Times in parentheses refer to hours since infection.

_____

[5]From Latin *lympha,* meaning clear water; Latin *granulum,* meaning a small grain; Greek *oma,* meaning tumor (swelling); and Latin *Venus,* the goddess of sexual love.

▲ *Figure 21.9*

An advanced case of lymphogranuloma venereum in a man. *Which microorganism causes lymphogranuloma venereum?*

Figure 21.9 Chlamydia trachomatis causes lymphogranuloma venereum.

Although about 85% of genital tract infections in women are asymptomatic, more than 75% of infections in men have symptoms. Infected men also often have urethral inflammation, which cannot be distinguished from gonorrhea based on symptoms alone. Such chlamydial urethritis accounts for about 50% of cases of *nongonococcal urethritis.*

Proctitis[6] may occur in men and women as a result of lymphatic spread of the bacteria from the vagina, vulvae, cervix, or urethra to the rectum. About 15% of the cases of proctitis in homosexual men result from the spread of *C. trachomatis* via anal intercourse.

An immune response against reinfections of *C. trachomatis* in women can have serious consequences, causing **pelvic inflammatory disease.** PID involves chronic pelvic pain; irreversible damage to the uterine tubes, uterus, and ovaries; and sterility.

Trachoma   The so-called *trachoma strains* of *C. trachomatis* cause a disease of the eye called **trachoma** (tră-kō′mă; **Figure 21.10**), which is the leading cause of nontraumatic blindness in humans. The pathogen multiplies in cells of the conjunctiva and kills them, triggering a copious, purulent[7] (pus-filled) discharge that causes the conjunctiva to become scarred. Such scarring in turn causes the patient's eyelids to

▲ *Figure 21.10*

An eye afflicted with trachoma.

turn inward, such that the eyelashes abrade, irritate, and scar the cornea, which triggers an invasion of blood vessels into this normally clear surface of the eye. A scarred cornea that is filled with blood vessels is no longer transparent, and the eventual result is blindness.

Trachoma is typically a disease of children who have been infected during birth. However, LGV strains of *C. trachomatis* may produce blindness by a similar process in adults when bacteria from the genitalia are introduced into the eyes via fomites or fingers.

Diagnosis, Treatment, and Prevention

Diagnosis of chlamydial infection involves demonstration of the bacteria inside cells from the site of infection. Specimens can be obtained from the urethra or vagina by inserting, rotating, and then removing a sterile swab. Giemsa- or Gimenez-stained specimens may reveal bacteria or inclusion bodies within cells, but the most specific method of diagnosis involves amplifying the number of chlamydia by inoculating the specimen into a culture of susceptible cells. Laboratory technicians then demonstrate the presence of *Chlamydia* in the cell culture by means of specific fluorescent antibodies **(Figure 21.11)** or nucleic acid probes.

Physicians prescribe doxycycline for 21 days to eliminate genital infections of LGV strains of *C. trachomatis* in adults. Erythromycin or sulfonamides are recommended for treatment of pregnant women. Trachoma strains of *C. trachomatis* infecting the eyes of newborns are treated with erythromycin cream for 10–14 days, whereas ocular infections with LGV strains in adults are treated with doxycycline cream for 7 days. Surgical correction of eyelid deformities may prevent the abrasion, scarring, and blindness that typically result from ocular infections.

Prevention of sexually transmitted chlamydial infections is best achieved by abstinence or faithful mutual

---

[6]From Greek *proktos,* meaning rectum, and *itis,* meaning inflammation.
[7]From Latin *pur,* meaning pus.

▲ *Figure 21.11*

A direct fluorescent antibody test for *C. trachomatis.* The bright red color reveals the presence of chlamydia within cells; this result is diagnostic for a *C. trachomatis* infection.

monogamy. Condoms may provide some protection, though some researchers warn that irritation by condoms and their lubricants actually increases the likelihood of infection. Blindness can be prevented only by prompt treatment with antibacterial agents and prevention of reinfection. Unfortunately, chlamydial infections are often asymptomatic and frequently occur among populations that have limited access to medical care.

CRITICAL THINKING

Why is either erythromycin or sulfonamide substituted for doxycycline in treatment of chlamydial infections in children? Why are penicillins and cephalosporins useless against *Chlamydia?*

Chlamydia pneumoniae

Learning Objective

✓ Describe three diseases associated with *Chlamydia pneumoniae.*

Chlamydia pneumoniae (kla-mid′ē-ă nū-mō′nē-ē) causes bronchitis, pneumonia, and sinusitis. Based primarily on circumstantial and epidemiological data, this chlamydia has been implicated also as a cause of some cases of atherosclerosis—lipid deposits on the walls of arteries, and the first stage of arteriosclerosis, or hardening of the arteries. (For more on the possible role of chlamydias in cardiovascular disease, see **New Frontiers 21.1.**) Most infections with *C. pneumoniae* are mild, producing only malaise and a chronic cough, and do not require hospitalization. Some cases, however, are characterized by the development of a severe pneumonia that cannot be distinguished from primary atypical pneumonia, which is caused by *Mycoplasma pneumoniae.*

Fluorescent antibodies are used to demonstrate the intercellular presence of *C. pneumoniae*, which is diagnostic. Such fluorescent antibody testing is not available in most medical laboratories, so physicians must rely on specialized research labs. Tetracycline or erythromycin for 14 days is used to treat infections of *C. pneumoniae*, but the drugs are not always effective, and infections may persist. Prevention of infection is difficult because the bacterium is ubiquitous and spreads via respiratory droplets.

Chlamydia psittaci

Learning Objective

✓ Describe the diseases caused by *Chlamydia psittaci.*

Chlamydia psittaci (kla-mid′ē-ă sit′ă-sē) causes **ornithosis**[8] (ōr-ni-thō′sis), a disease of birds that can be transmitted to humans, in whom it typically causes flulike symptoms. In

[8]From Greek *ornith,* meaning bird.

▲ *Figure 21.12*
The age-related prevalence of ornithosis in the United States, 1995.

some cases severe pneumonia can occur, and rarely nonrespiratory conditions such as endocarditis, hepatitis, arthritis, conjuncitivitis, and encephalitis are observed. The disease is sometimes called *psittacosis* or *parrot fever* because it was first identified in parrots (*psitakos* means parrot in Greek). Several dozen cases of ornithosis are reported in the United States annually, most often in adults **(Figure 21.12)**. Most likely this disease is underreported because symptoms are

often mild and diagnosis is difficult. Zookeepers, veterinarians, poultry farmers, pet shop workers, and owners of pet birds are at greatest risk of infection.

Elementary bodies may be inhaled in aerosolized bird feces or respiratory secretions, or ingested from fingers or fomites that have contacted infected birds. Pet birds may transmit the disease to humans via beak-to-mouth contact. Because very brief exposure to birds can be sufficient for the transmission of elementary bodies, some patients cannot recall having any contact with birds. Symptoms usually occur within 10 days of exposure to the chlamydias. Without treatment, the mortality rate of ornithosis is about 20%, but with treatment death is rare.

Because the symptoms of ornithosis are the same as those of many respiratory infections, diagnosis requires demonstration of *C. psittaci* by serological testing. Doxycycline for 2 weeks is the preferred treatment. Prevention involves wearing protective clothing when handling infected birds, 30-day quarantine of and tetracycline treatment for all imported birds, and preventative husbandry, in which bird cages are regularly cleaned and are positioned so as to prevent the transfer of feces, feathers, food, and other materials from cage to cage. No vaccine for either birds or humans is available.

CRITICAL THINKING

Why do most cases of ornithosis occur in adults 25–64 years of age, as indicated in Figure 21.12?

Table 21.2 compares and contrasts the features of the smallest microbes—mycoplasmas, rickettsias, chlamydias, and viruses. Next we turn our attention to the spirochetes.

Table 21.2   Characteristics of the Smallest Microbes

Feature	Mycoplasmas	Rickettsias	Chlamydias	Viruses
Cellular structure	Small, wall-less, pleomorphic cells with sterol in cytoplasmic membranes	Small cells with little peptidoglycan in cell walls	Small, wall-less cells with two membranes	Acellular
Diameter	0.1–0.3 μm	0.3 μm	EBs: 0.2–0.4 μm RBs: 0.6–1.5 μm	0.01–0.3 μm
Lifestyle	Free-living	Obligate intercellular parasites within cytosol	Obligate intercellular parasites within phagosomes	Obligate intercellular parasites within cytosol or nuclei
Replication	Binary fission	Binary fission	Binary fission of reticulate bodies	Chemical assembly
Nucleic acid(s)	Both DNA and RNA	Both DNA and RNA	Both DNA and RNA	DNA or RNA
Functional ribosomes	Present	Present	Present	Absent
Metabolism	Limited—lack Krebs cycle, electron transport chains, and enzymes for cell wall synthesis	Present	Present	Completely dependent on metabolic enzymes of host cell
ATP-generating system	Present	Present	Absent	Absent
Phylum	Firmicutes	Proteobacteria	Chlamydiae	None officially recognized

Spirochetes

Learning Objective

✓ Describe the morphology and locomotion of spirochetes.

Spirochetes (spī'rō-kētz), which means *coiled hairs* in Greek, are thin (0.1–0.5 μm in diameter), tightly coiled, helically shaped, Gram-negative bacteria that share certain unique features—most notably axial filaments, which are composed of endoflagella located in the periplasmic space between the cytoplasmic membrane and the outer (wall) membrane (see Figure 3.7). As its axial filaments rotate, a spirochete corkscrews through its environment—a type of locomotion thought to enable pathogenic spirochetes to burrow through their hosts' tissues. Mutants lacking axial filaments are rod-shaped rather than helical, indicating that axial filaments play a role in maintaining cell shape as well as in motility.

Taxonomists place spirochetes in their own phylum—Spirochaetes. Three genera, *Treponema, Borrelia,* and *Leptospira,* cause diseases in humans; we begin by discussing *Treponema.*

Treponema

Learning Objectives

✓ Describe the disease caused by *Treponema pallidum pallidum.*
✓ Describe the four phases of untreated syphilis and the treatment for each.
✓ Describe the diseases caused by nonvenereal strains of *Treponema.*

Treponema (trep-ō-nē'mă) is a pathogen in humans only. Four types cause diseases, the most widespread of which is *Treponema pallidum pallidum* (i.e. subspecies or strain *pallidum*). We will first examine this sexually transmitted pathogen and the venereal disease it causes, and then we will briefly consider the nonvenereal diseases caused by the other three strains.—*T. pallidum endemicum* (i.e. subspecies or strain *endemicum*), *T. carateum,* and *T. pallidum pertenue.*

Treponema pallidum pallidum

Treponema pallidum pallidum (pal'li-dŭm), which is usually simply called *T. pallidum,* causes **syphilis.** Its flattened helical cells are so narrow (0.1 μm) that they are difficult to see by regular light microscopy in Gram-stained specimens. Therefore, scientists either use phase contrast microscopy or dark field microscopy, or they intensify the contrast of the cells with special stains composed of dye linked to antitreponemal antibodies **(Figure 21.13).**

The bacterium lives naturally in humans only, and because it is destroyed by exposure to heat, disinfectants, soaps, drying, the concentration of oxygen in air, and pH changes, it cannot survive in the environment. Scientists have not successfully cultured *T. pallidum* in cell-free media, but they have coaxed it to multiply in rabbits, monkeys, and

▲ *Figure 21.13*
Treponema pallidum pallidum stained with a dye linked to antibodies against the bacteria.

rabbit epithelial cell cultures. Outside of humans it multiplies slowly (binary fission occurs once every 30 hours), and only for a few generations.

Pathogenicity and Epidemiology   Scientists have had difficulty identifying the virulence factors of *T. pallidum* because the pathogen cannot be cultured. Researchers have used recombinant DNA techniques to insert genes from *Treponema* into *Escherichia coli* (esh-ĕ-rik'ē-ă kō'lī) and have then isolated the proteins the genes coded. Apparently, some of these proteins enable *Treponema* to adhere to human cells. Virulent strains also produce hyaluronidase, which may enable *Treponema* to infiltrate intracellular spaces. The bacterium also has a glycocalyx, which may protect it from phagocytosis by leukocytes.

Syphilis occurs worldwide. Europeans first recognized the disease in 1495, leading some epidemiologists to hypothesize that syphilis was brought to Europe from the Western Hemisphere by returning Spanish explorers. Another hypothesis is that *T. pallidum* evolved from a less-pathogenic strain of *Treponema* endemic to North Africa, and that its spread throughout Europe was coincidental to Europeans' explorations of the New World. In any case, the discovery and development of antimicrobial drugs over four centuries later has greatly reduced the number of syphilis cases in the United States **(Figure 21.14a** on page 608). The disease remains endemic among sex workers, men who have sex with men, and users of illegal drugs. It occurs throughout the United States, particularly in the Southeast **(Figure 21.14b).**

Because of its fastidiousness and sensitivity, *T. pallidum* is an obligate parasite of humans and is transmitted almost solely via sexual contact, usually during the early stage of infection, when the spirochetes are most numerous. The risk of infection from a single, unprotected sexual contact with an infected partner is 10–30%. *T. pallidum* can also be spread from mother to fetus, and rarely through blood transfusion; it cannot be spread by fomites such as toilet seats, eating utensils, or clothing.

(a)

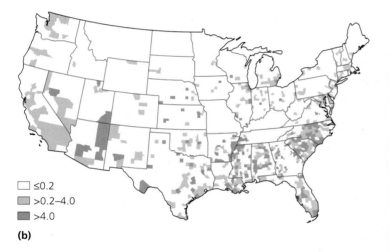

(b)

▲ *Figure 21.14*

The incidence of syphilis in the United States. **(a)** Nationwide incidence of syphilis, 1941–2002. **(b)** Reported cases of syphilis per 100,000 population, by county, 2001.

(a)

(b)

(c)

Disease   Untreated syphilis has four phases: primary, secondary, latent, and tertiary syphilis. In *primary syphilis*, a small, painless, reddened lesion called a **chancre** (shan'ker) forms at the site of infection 10–21 days following exposure **(Figure 21.15a)**. Although chancres typically form on the external genitalia, about 20% form in the mouth, around the anus, or on the fingers, lips, or nipples. Chancres are often unobserved, especially in women, in whom these lesions frequently form on the cervix. The center of a chancre fills with serum that is extremely infectious due to the presence

▶ *Figure 21.15*

The lesions of syphilis. **(a)** A chancre, a hardened and painless lesion of primary syphilis that forms at the site of the infection. **(b)** A widespread rash characteristic of secondary syphilis. **(c)** A gumma, a painful rubbery lesion that often occurs on the skin or bones during tertiary syphilis.

of millions of spirochetes. Chancres remain for 3–6 weeks and then disappear without scarring.

In about a third of cases, the disappearance of the chancre is the end of the disease. However, in most infections *Treponema* has invaded the bloodstream and spread throughout the body to cause the symptoms and signs of *secondary syphilis:* sore throat, headache, mild fever, malaise, myalgia (muscle pain), lymphadenopathy (diseased lymph nodes), and a widespread rash **(Figure 21.15b)** that can include the palms and the soles of the feet. Although this rash does not itch or hurt, it can persist for months, and like the primary chancre, rash lesions are filled with spirochetes and are extremely contagious. People, including health care workers, can become infected when fluid from the lesions enters breaks in the skin, though such nonsexual transmission of syphilis is rare.

After several weeks or months the rash gradually disappears, and the patient enters a *latent (clinically inactive) phase* of the disease. The majority of cases do not advance beyond this point, especially in developed countries where antimicrobial drugs are in use.

Latency may last 30 or more years, after which perhaps a third of the originally infected patients proceed to *tertiary syphilis*. This phase is not associated with the direct effects of *Treponema*, but rather with severe complications resulting from inflammation and a hyperimmune response against the pathogen. Tertiary syphilis may affect virtually any tissue or organ and can cause dementia, blindness, paralysis, heart failure, and syphilitic lesions called **gummas** (gŭm'ăz), which are rubbery, painfully swollen lesions that can occur in bones, in nervous tissue, or on the skin **(Figure 21.15c)**.

Congenital syphilis results when *Treponema* crosses the placenta from an infected mother to her fetus. Transmission to the fetus from a mother experiencing primary or secondary syphilis often results in the death of the fetus. If transmission occurs while the mother is in the latent phase of the disease, the result can be a latent infection in the fetus that causes mental retardation and malformation of many fetal organs. After birth, newborns with latent infections usually exhibit a widespread rash at some time during their first 2 years of life.

Diagnosis, Treatment, and Prevention The diagnosis of primary, secondary, and congenital syphilis is relatively easy and rapid using specific antibody tests against antigens of *Treponema pallidum pallidum*. One such test is **MHA-TP** (microhemagglutination assay against *T. pallidum*), which uses red blood cells that have been artificially coated with *Treponema* antigens. In patients infected with *T. pallidum*, antibodies in their serum will agglutinate (clump) the red blood cells—a positive test for exposure to *Treponema pallidum*. Spirochetes can be observed (see Figure 21.13) in fresh discharge from lesions, but only when microscopic observations of clinical samples are made immediately—*Treponema* does not survive transport to a laboratory. Nonpathogenic spirochetes, which are a normal part of the oral microbiota,

can yield false positive results, so clinical specimens from the mouth cannot be tested for syphilis. Tertiary syphilis is extremely difficult to diagnose because it mimics many other diseases, because few (if any) spirochetes are present, and because the signs and symptoms may occur years apart and seem unrelated to one another.

Penicillin is the drug of choice for treating primary, secondary, latent, and congenital syphilis, but it is not efficacious for tertiary syphilis, because this phase is caused by a hyperimmune response, not an active infection.

The Centers for Disease Control and Prevention has established some national goals for control of syphilis: maintaining an incidence of fewer than 0.2 cases of syphilis per 100,000 population in every county in the United States, and having 90% of counties syphilis free by 2005. A vaccine against syphilis is not available, so abstinence, faithful mutual monogamy, and consistent and proper condom usage are the primary ways to avoid contracting syphilis. All sexual partners of syphilis patients must be treated with prophylactic penicillin to prevent spread of the disease.

Nonvenereal Treponemal Diseases

Other members of *Treponema* cause three nonvenereal diseases in humans: bejel, pinta, and yaws. These diseases are primarily seen in impoverished children in Africa, Asia, and South America who live in unsanitary conditions. The spirochetes that cause these diseases look like *Treponema pallidum pallidum*.

T. pallidum endemicum (en-de'mi-kŭm) causes **bejel** (be'jel), a disease seen in children in Africa, Asia, and Australia. In bejel, the spirochetes are spread by contaminated eating utensils, so it is not surprising that the initial lesion is an oral lesion, which typically is so small that it is rarely observed. As the disease progresses, larger and more numerous secondary lesions form around the lips and inside the mouth. In the later stages of the disease, gummas form on the skin, bones, or nasopharyngeal mucous membranes.

T. carateum[9] (kar-a'tē-ŭm) causes **pinta** (pēn'tă), a skin disease seen in children in Central and South America. The spirochetes are spread among the children by skin-to-skin contact. After 1–3 weeks of incubation, hard, pus-filled lesions called papules form at the site of infection; the papules enlarge and persist for months or years, resulting in scarring and disfigurement.

T. pallidum pertenue[10] (per-ten'ū-ē) causes **yaws** (yăz), a disease of tropical South America, central Africa, and southeast Asia that is characterized initially by granular skin lesions that, although unsightly, are painless. Over time the lesions develop into large, destructive, draining lesions of the skin, bones, and lymph nodes **(Figure 21.16** on page 610). The disease is spread via contact with spirochetes in the fluid draining from the lesions.

---

[9]Some taxonomists consider this organism to be a subspecies: *T. pallidum carateum*.
[10]Some taxonomists consider this organism to be a separate species: *T. pertenue*.

▲ *Figure 21.16*
The draining lesions characteristic of the later stages of yaws.

LM   30 μm

▲ *Figure 21.17*
Borrelia burgdorferi, the Gram-negative spirochete that causes Lyme disease.

Physicians diagnose bejel, pinta, and yaws by their distinctive appearance in children from endemic areas. Clinicians can detect spirochetes in specimens from the lesions of pinta and yaws, but spirochetes are not seen in bejel.

Penicillin, tetracycline, and chloramphenicol are the most favored antimicrobial agents for treating these three tropical diseases. Prevention involves limiting the spread of the bacteria by preventing contact with the lesions.

Borrelia

Learning Objectives

✓ Describe Lyme disease, its vector, and its causative agent.
✓ Discuss the life cycle of the *Ixodes* tick as it relates to Lyme disease.
✓ Describe the two types of relapsing fever, including their causes and vectors.

Members of the genus *Borrelia* (bō-rĕ′lē-ă) are lightly staining, Gram-negative spirochetes that are larger than species of *Treponema*. These spirochetes cause two diseases in humans: Lyme disease and relapsing fever.

Lyme Disease

In 1975, epidemiologists noted that the incidence of childhood rheumatoid arthritis in Lyme, Connecticut, was over

100 times higher than expected. Upon investigation they discovered that a previously unknown disease dubbed Lyme disease—characterized by dermatological, cardiac, and neurological abnormalities in addition to the observed arthritis, resulted when ticks transmitted the spirochete *Borrelia burgdorferi*[11] (bŭrg-dor′fer-ē; **Figure 21.17**) to human hosts. Even though its discovery was relatively recent, retrospective epidemiological studies have shown that Lyme disease had been present in the United States for decades before its discovery.

B. burgdorferi is an unusual bacterium in that it lacks iron-containing enzymes and iron-containing proteins in its electron transport chains. By utilizing manganese rather than iron, the spirochete circumvents one of the body's natural defense mechanisms: the lack of free iron in human tissues and fluids.

Hard ticks of the genus *Ixodes* (ik-sō′dēz)—*I. scapularis* (scap-ū-lar′is) in the northeast and central United States, *I. pacificus* (pas-i′fi-kŭs) on the Pacific coast, *I. ricinus* (ri-ki′nŭs) in Europe, and *I. persulcatus* (per-sool-ka′tŭs) in eastern Europe and Asia—are the vectors of Lyme disease. An understanding of the life cycle of the tick is essential to an understanding of Lyme disease.

An *Ixodes* tick lives for 2 years, during which it passes through three stages of development: a six-legged larva, an eight-legged nymph, and an eight-legged adult. During each stage it attaches to an animal host for a single blood meal. After each of its three feedings, the tick drops off its host and lives in leaf litter or on brush.

The different stages of each species of *Ixodes* typically feed on different hosts. For example, the larvae and nymphs

---

[11]Named for the French bacteriologist Amedee Borrel and a specialist in tick-borne diseases, Willy Burgdorfer.

of *I. scapularis* tend to feed on deer mice, whereas the adults most frequently feed on deer. In contrast, *I. pacificus* adults feed frequently on lizards, a host that does not support the growth of *Borrelia*. All stages of all species of *Ixodes* may feed on humans.

Transovarian transmission of *Borrelia* is rare, so ticks that hatch in the spring are usually uninfected. Larvae become infected during their first blood meal ❶. Over the winter, larvae digest their blood meals while *Borrelia* replicate in the ticks' guts.

In the spring of their second year, the ticks molt into nymphs and feed a second time, infecting the new hosts with *Borrelia* via saliva ❷. (Uninfected nymphs can become infected if they feed on an infected host.) Laboratory studies have shown that infected ticks must remain on a host for 48 hours in order to transmit enough spirochetes to establish a *Borrelia* infection in that host.

Nymphs drop off and undergo further development into adults ❸. In the fall, adult ticks feed a final time ❹, mate, lay eggs, and die. Adults infected with *Borrelia* infect their hosts as they feed. Adult ticks are much larger than nymphs, so humans usually see and remove adults before they can transmit *Borrelia;* thus, nymphs most often infect humans.

Figure 21.18 summarizes the life cycle of *Ixodes* as it relates to the transmission of *Borrelia*.

CRITICAL THINKING

For what other pathogen discussed in this chapter is *Ixodes* also the vector?

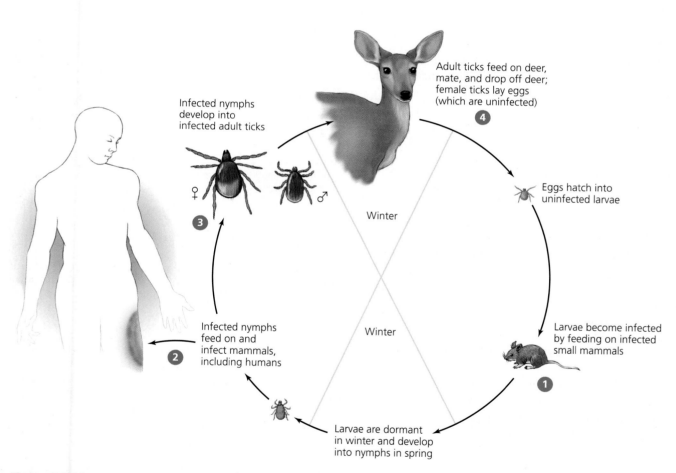

▲ *Figure 21.18*

The life cycle of the deer tick *Ixodes* as it relates to its role as the primary vector of Lyme disease. *Besides the nymph stage, what other stage in the tick's life cycle can infect a human with* B. burgdorferi?

Figure 21.18 Adult ticks infected with B. burgdorferi can also infect humans during a blood meal.

▲ *Figure 21.19*

The distinctive "bull's eye" rash often seen in the initial phase of Lyme disease.

Lyme disease mimics many other diseases, and its range of signs and symptoms is vast. The disease typically has three phases in untreated patients:

1. An expanding red rash, which often resembles a "bull's eye," occurs at the site of infection within 30 days **(Figure 21.19).** About 75% of patients have such a rash, which lasts for several weeks. Other early signs and symptoms include malaise, headaches, dizziness, stiff neck, severe fatigue, fever, chills, muscle and joint pain, and lymphadenopathy.

2. Neurological symptoms (for example, meningitis, encephalitis, and peripheral nerve neuropathy) and cardiac dysfunction typify the second phase, which is seen in only 10% of patients.

3. The final phase is characterized by severe arthritis that can last for years. The pathological conditions of the latter phases of Lyme disease are due in large part to the body's immunological response; rarely is *Borrelia* seen in the involved tissue or isolated in cultures of specimens from these sites.

Two major events have contributed to the increase in cases of Lyme disease in the United States over the past 20 years **(Figure 21.20a):** The human population has encroached on woodland areas, and the deer population has been protected and even encouraged to feed in suburban yards. Thus, humans have been brought into closer association with ticks infected with *Borrelia*.

A diagnosis of Lyme disease, which is based typically on observations of its usual signs and symptoms, is rarely confirmed by detecting *Borrelia* in blood smears; instead, the diagnosis is confirmed through the use of serological tests.

Treatment with antimicrobial drugs such as doxycycline or penicillin effectively cures most cases of Lyme disease in the first phase, though prolonged treatment with large doses may be required. Treatment of later phases is more difficult

(a)

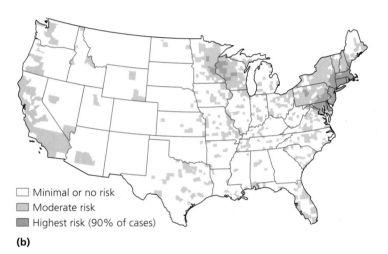

☐ Minimal or no risk
▨ Moderate risk
▣ Highest risk (90% of cases)

(b)

▲ *Figure 21.20*

The occurrence of Lyme disease in the United States. **(a)** Incidence of cases, 1982–2002. **(b)** The geographic distribution of the disease in 2001.

because later symptoms frequently result from immune responses rather than the presence of the spirochetes.

People hiking, picnicking, and working outdoors in areas where Lyme disease is prevalent **(Figure 21.20b)** should take precautions to reduce the chances of infection, particularly during May to July, when nymphs are feeding. People who must be in the woods should wear long-sleeved shirts and long tight-fitting pants, and should tuck the cuffs of their pants into their socks to deny ticks access to skin. Repellants containing *DEET* (N,N diethyl-m-toluamide) that are noxious to ticks should be used. As soon as possible after leaving a tick-infected area, people should thoroughly examine their bodies for ticks or their bites. Though researchers have developed a vaccine against *B. burgdorferi*, it is not widely used because of the ready availability of other preventative measures, and because the vaccine produces the symptoms of Lyme disease in some patients.

CRITICAL THINKING

Health departments recommend that outdoor enthusiasts wear light colored pants while hiking in areas where *Ixodes* is endemic. Why?

Relapsing Fever

Other species of spirochetes in the genus *Borrelia* can cause two types of *relapsing fever*. A disease called **epidemic relapsing fever** (also known as *louse-borne fever*) results when *Borrelia recurrentis* (re-kur-ren'tis) is transmitted among humans by the human body louse *Pediculus humanus*. A disease known as **endemic relapsing fever** occurs when several different species of *Borrelia* are transmitted among humans by soft ticks in the genus *Ornithodoros*.

Lice become infected with *B. recurrentis* when they feed on infected humans, the only reservoir for this spirochete. When infected lice are crushed, spirochetes can infect the bite wounds. Because the lice live only a few months, maintenance of the disease in a population requires crowded, unsanitary conditions such as occur in poor neighborhoods and during wars and natural disasters. Epidemic relapsing fever occurs in central and eastern Africa and in South America.

In contrast to hard ticks, the soft ticks of the genus *Ornithodoros* that transmit the many species of *Borrelia* that cause endemic relapsing fever can live for years between feedings and pass spirochetes to their offspring via transovarian transmission. Soft ticks can also transmit *Borrelia* to animals, particularly mice and rats. Ticks and rodents are the reservoirs of infection. Endemic relapsing fever has a worldwide distribution; in the United States it is found primarily in the West.

Both types of relapsing fever are characterized by recurrent episodes of fever and septicemia separated by symptom-free intervals—a pattern that results from the body's repeated efforts to remove the spirochetes, which continually change their antigenic surface components **(Figure 21.21)**. Within the first week of infection, the spirochetes induce a fever ❶, which is associated with chills, myalgia, and headache. Additionally, the body produces antibodies that are directed against the spirochetes' surface molecules; once the antibodies have agglutinated the spirochetes, phagocytes clear them from the blood, and the body's temperature returns to normal ❷. But some *Borrelia* evade the immune response by changing their antigenic surface molecules; when these spirochetes multiply, they induce a relapse of fever ❸. The body mounts a second immune response, this time against the new antigens, eventually clearing the blood of spirochetes and again returning body temperature to normal ❹. But a few spirochetes yet again change their antigenic components, leading to still another recurrence of fever, continuing the pattern ❺.

Because of their relatively large size and abundance in the blood, observation of spirochetes in blood smears taken during periods of fever is the primary method of diagnosis.

Physicians successfully treat the relapsing fevers with tetracycline or, in pregnant women and children, with eryth-

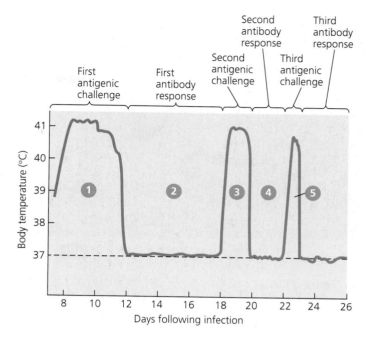

▲ *Figure 21.21*

The time course of recurring episodes of fever in relapsing fever. The events represented by the step numbers are described in the text.

romycin. Prevention involves avoidance of ticks and lice, good personal hygiene, and the use of repellant chemicals. Effective rodent control is crucial to the control of endemic relapsing fever. Vaccines are not available.

Leptospira

Learning Objective

✓ Describe *Leptospira interrogans* and zoonotic leptospirosis.

The last of the three groups of spirochetes that infect humans is *Leptospira interrogans* (lep'tō-spī'rǎ in-ter'rǎ-ganz). The specific epithet *interrogans* alludes to the fact that one end of the spirochete is hooked in a manner reminiscent of a question mark **(Figure 21.22** on page 614). This thin, pathogenic spirochete is an obligate aerobe that is highly motile by means of two axial filaments, each of which is anchored at one end. Clinicians and researchers grow it on special media enriched with bovine serum albumin or rabbit serum.

L. interrogans normally occurs in many wild and domestic animals—in particular, rats, raccoons, foxes, dogs, horses, cattle, and pigs—in which it grows asymptomatically in the kidney tubules. Humans contract the zoonotic disease **leptospirosis** (lep'tō-spī-rō'sis) through direct contact with the urine of infected animals or indirectly via contact with the spirochetes in contaminated streams, lakes, or moist soil, environments in which the organisms can remain viable for 6 weeks or more. Person-to-person spread has not been observed.

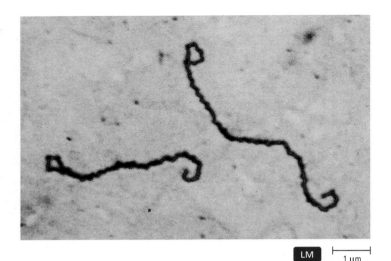

LM | 1 μm

▲ *Figure 21.22*

Leptospira interrogans, a spirochete with one end hooked like a question mark.

After *Leptospira* gains initial access to the body through invisible cuts and abrasions in the skin and mucous membranes, it corkscrews its way through these tissues. It then travels via the bloodstream throughout the body, including the central nervous system, damaging cells lining the small blood vessels and triggering fever and intense pain. Infection may lead to hemorrhaging and to liver and kidney dysfunction. Eventually, the bacteremia resolves, and the spirochetes are found only in the kidneys. As the disease progresses the spirochetes are excreted in urine. Leptospirosis is usually not fatal.

Leptospirosis occurs throughout the world. National reporting in the United States has ceased in part because the disease is rare; a total of only 89 cases were reported in 1993 and 1994, the last two years that leptospirosis was listed as a nationally reportable disease.

Because *Leptospira* is very thin and does not stain well with Gram stain, silver stain, or Giemsa stain, specific antibody tests revealing the presence of the spirochete in clinical specimens is the preferred method of diagnosis. Intravenous penicillin is used to treat severe infections; oral doxycycline is the drug of choice for less severe cases. Rodent control is the most effective way to limit the spread of *Leptospira*, but eradication is impractical because of the spirochete's many animal reservoirs. An effective vaccine is available for livestock and pets.

CRITICAL THINKING

In Chapter 4 we discussed the identification of microbes using dichotomous taxonomic keys. Design such a key for the spirochetes discussed in this chapter.

Pathogenic Gram-Negative Curved Bacilli

To this point, we have discussed a number of unrelated bacteria that appear pink in the Gram stain: the Gram-positive and wall-less mycoplasmas, the obligate intracellular rickettsias and chlamydias, and the spirochetes. We will now turn our attention to a final group—the slightly curved bacteria called *vibrios*. The more important pathogenic vibrios of humans are in the genera *Vibrio*, *Campylobacter*, and *Helicobacter*.

Vibrio

Learning Objectives

✓ Contrast *Vibrio* with enteric bacteria in terms of their flagella and biochemical properties.

✓ Describe the action of cholera toxin in causing cholera.

✓ Name two species of *Vibrio* and describe the resulting diseases.

Based on rRNA nucleotide sequences, biochemical properties, size, and morphology, taxonomists classify slightly curved, Gram-negative bacilli in the genus *Vibrio* (vib'rē-ō; **Figure 21.23**). Members of this genus (which is in class Gammaproteobacteria and phylum Proteobacteria) share many characteristics with enteric bacteria such as *Escherichia* and *Salmonella*, including O polysaccharide antigens that enable epidemiologists to distinguish among strains. Unlike enteric bacteria, which are oxidase negative and have peritrichous flagella, *Vibrio* is oxidase positive and has polar flagella.

Vibrio lives naturally in estuarine and marine environments worldwide. It prefers warm, salty, and alkaline water, and most pathogenic species can multiply inside shellfish. Eleven of the 34 species of *Vibrio* cause diseases in humans; of these, only *V. cholerae* (kol'er-ī) can survive in fresh water, making it the most likely species to infect humans (via contaminated drinking water). In the following sections we examine *V. cholerae*, and the serious disease it causes.

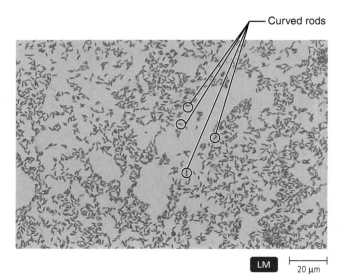

— Curved rods

LM | 20 μm

▲ *Figure 21.23*

Vibrio *cholerae*, the causative agent of cholera.

Pathogenesis and Epidemiology of Vibrio cholerae

Vibrio cholerae (kol′er-ē), particularly strain O1 *el Tor*, causes **cholera** (kol′er-ă, sometimes called *epidemic cholera*), one of the world's more pernicious diseases. As discussed in Chapter 14, the science of epidemiology began with Dr. John Snow's efforts to understand and limit the 1854 cholera epidemic in London.

Ships have introduced the pathogen into the harbors of every continent through their practice of taking on ballast water in one port and dumping it in another; two world wars have only exacerbated the situation. In the seven major pandemics that have occurred since 1800, millions of people have been sickened and thousands have died, causing significant socioeconomic upheaval.

The seventh pandemic of O1 *el Tor* began in 1961 in Asia, spread to Africa and Europe in the 1970s and 1980s, and reached Peru in January 1991. **Figure 21.24** illustrates the progression of this cholera pandemic throughout Latin America in the first half of the 1990s. A new pandemic strain, *V. cholerae* O139 Bengal, arose in India in 1992 and is spreading across Asia; this strain is the first non-O1 strain capable of causing epidemic disease. Other strains of *V. cholerae* do not produce epidemic cholera, only milder gastroenteritis.

Humans become infected with *Vibrio cholerae* by ingesting contaminated food and water. Cholera is most frequent in communities with poor sewage and water treatment. After entering the digestive system, *Vibrio* is confronted with the inhospitable acidic environment of the stomach. Most cells die, which is why a high inoculum—at least 10^8 cells— is typically required for the disease to develop. Fewer cells can achieve infection in people who use antacids.

Recent research indicates that the environment within the human body activates some *Vibrio* genes that make the bacteria more virulent than their counterparts in the environment. Scientists hypothesize that such gene activation may explain the rapid, almost explosive, nature of cholera epidemics.

Although cholera infections may be asymptomatic or cause mild diarrhea, some result in rapid, severe, and fatal fluid and electrolyte loss. Symptoms usually begin 2–3 days following infection, with abrupt watery diarrhea and vomiting. As the disease progresses, the colon is emptied and the stool becomes increasingly watery, colorless, and odorless. The stool, which is typically flecked with mucus, is called *rice-water stool* because it resembles water poured off a pan of boiled rice. Some patients lose 1 liter of fluid an hour, resulting in a 50% loss of body weight over the course of infection.

The most important virulence factor of *V. cholerae* is a potent exotoxin called **cholera toxin,** which is composed of five identical B subunits and a single A subunit. **Figure 21.25** illustrates the action of cholera toxin in producing the severe diarrhea that characterizes cholera. The process proceeds as follows:

① One of the B subunits binds to a glycolipid receptor in the cytoplasmic membrane of an intestinal epithelial cell.

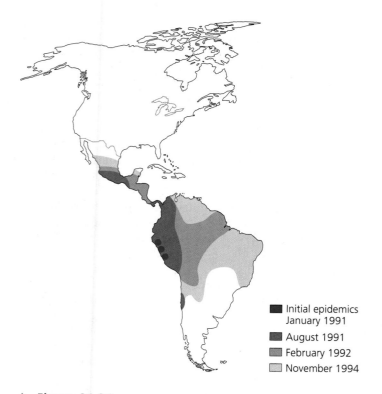

▲ *Figure 21.24*

The spread of cholera caused by *Vibrio cholerae* O1 *el Tor* throughout Latin America in the first half of the 1990s.

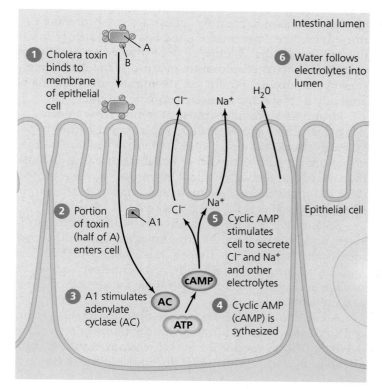

▲ *Figure 21.25*

The action of cholera toxin in intestinal epithelial cells.

2 The A subunit is cleaved, and a portion (called A1) enters the cell's cytosol.

3 A1 acts as an enzyme that activates cellular adenylate cyclase.

4 Activated adenylate cyclase enzymatically converts ATP into cyclic AMP (cAMP).

5 cAMP stimulates the active secretion of excess amounts of electrolytes (sodium, chlorine, potassium, and bicarbonate ions) from the cell.

6 Water follows the movement of electrolytes from the cell and into the intestinal lumen via osmosis.

The severe fluid and electrolyte losses result in dehydration, metabolic acidosis (decreased pH of body fluids) due to loss of bicarbonate ions, hypokalemia,[12] and hypovolemic shock caused by reduced blood volume in the body. These conditions can produce muscle cramping, irregularities in heartbeat, kidney failure, coma, and death.

Cholera mortality is 60% in untreated patients infected by strain O1; death can occur less than 48 hours after infection. Disease produced by strain O139 is even more severe. In nonfatal cases, cholera ends spontaneously after a few days as the pathogens and toxins are flushed from the system by the severe diarrhea.

Diagnosis, Treatment, and Prevention of Cholera

Rarely, an experienced observer may find the quick darting cells of *Vibrio* in the watery stool of a patient, but diagnosis is usually based on the characteristic diarrhea. *Vibrio* can be cultured on many laboratory media designed for stool cultures, but clinical specimens must be collected early in the disease (before the volume of stool dilutes the number of cells) and inoculated promptly because *Vibrio* is extremely sensitive to drying. Strains are distinguished by means of antibody tests specific for each antigenic variant.

Health care providers must promptly treat cholera patients with fluid and electrolyte replacement before hypovolemic shock ensues. Antimicrobial drugs are not as important as with many other bacterial diseases, because they are lost in the watery stool; nevertheless, they may reduce the production of exotoxin and ameliorate the symptoms. Doxycycline is the drug of choice.

Because *Vibrio cholerae* can grow in estuarine and marine water, and because up to 20% of former patients still asymptomatically harbor the pathogen, it is unlikely that cholera will be eradicated worldwide. Nevertheless, adequate sewage and water treatment can limit the spread of *Vibrio* and prevent epidemics.

A killed vaccine has been developed for the O1 strains of *V. cholerae*, but its protective value is unfortunately short lived; there is no vaccine for the O139 strain. Antibiotic prophylaxis of those who travel to endemic areas has not proven effective. Fortunately, because the infective dose for *V. cholerae* is high, proper hygiene generally makes vaccination and prophylaxis unnecessary.

Other Diseases of Vibrio

Vibrio parahaemolyticus (pa-ră-hē-mō-li'ti-kŭs) causes cholera-like gastroenteritis following ingestion of shellfish harvested from contaminated estuaries; fortunately, only rarely is it severe enough to be fatal. Typically the disease is characterized by a self-limiting, explosive diarrhea accompanied by headache, nausea, vomiting, and cramping for 72 hours.

V. vulnificus (vul-nif'i-kŭs) is responsible for septicemia (blood poisoning) following consumption of contaminated shellfish, and for infections resulting from the washing of wounds with contaminated seawater. Wound infections are characterized by swelling and reddening at the site of infection and are accompanied by fever and chills. Infections with *V. vulnificus* are fatal for 50% of untreated patients; prompt treatment with doxycycline is effective.

Campylobacter jejuni

Learning Objectives

✓ List several possible reservoirs of *Campylobacter jejuni*.
✓ Describe gastroenteritis caused by *Campylobacter jejuni*.

Campylobacter jejuni[13] (kam'pi-lō-bak'ter jē-jū'nē) is likely the most common cause of bacterial gastroenteritis in the United States. Like *Vibrio*, it is Gram-negative, slightly curved, oxidase positive, and motile by means of polar flagella; however, genetic analysis has shown that this comma-shaped pathogen is properly classified in the class Deltaproteobacteria (and not in Gammaproteobacteria). *Campylobacter* also differs from *Vibrio* in being microaerophilic and capneic.

Campylobacter infections are zoonotic; many domesticated animals, including poultry, dogs, cats, rabbits, pigs, cattle, and minks, serve as reservoirs. Humans acquire the bacterium by consuming food, milk, or water contaminated with infected animal feces. The most common source of infection is contaminated poultry. One study found that of dozens of Thanksgiving turkeys purchased in stores throughout Chicago, all were contaminated with *Campylobacter*.

The pathogenicity of *C. jejuni* is not well understood. The bacterium possesses adhesins, cytotoxins, and endotoxins that appear to enable colonization and invasion of the jejunum, ileum, and colon, producing bleeding lesions and triggering inflammation. Interestingly, nonmotile mutants are avirulent. *C. jejuni* infections commonly produce malaise, fever, abdominal pain, and bloody and frequent di-

---

[12]From Greek *hypo*, meaning under; Latin *kalium*, meaning potassium; and Greek *haima*, meaning blood.

[13]From Greek *kampylos*, meaning curved, and *jejunum*, the middle portion of the small intestine.

Case Study 21.2   Campylobacter and the Lactovegetarians

Two patients—a woman and her husband, ages 23 and 22, respectively—arrive at the health clinic one morning. They report having had severe abdominal cramps, grossly bloody diarrhea, nausea, and fever for 48 hours. Cultures of stool samples grown under microaerophilic, capneic conditions contain comma-shaped, Gram-negative bacilli. Both the patients are lactovegetarians and report being part of a "cow leasing" program at a local dairy in which patrons lease part of a cow's milk production so they can drink natural, whole, raw milk. The couple devised the program so that they and several neighbors could circumvent state regulations prohibiting the sale of unpasteurized milk. Investigators obtained and cultured a milk sample from the dairy's bulk milk tank; and the cultures contained *Campylobacter jejuni*.

1. How did the couple become infected with *Campylobacter*?

2. Are the couple's colleagues at work at risk of acquiring an infection from the couple?

3. What other foods that are common sources of *Campylobacter infections* can be ruled out in this case?

Source: adapted from *MMWR* 51(25):548–549. 2002.

---

arrhea—10 or more bowel movements per day are not uncommon. The disease is self limiting; as bacteria are expelled from the intestinal tract, the symptoms abate, although a typical infection may last 7–10 days.

Scientists estimate that over 2 million cases of *Campylobacter* gastroenteritis occur each year in the United States—more than those caused by *Salmonella* and *Shigella* combined. However, *Campylobacter* gastroenteritis is not a reportable disease, so its actual incidence is unknown.

The number of infections can be reduced by proper food preparation, including thoroughly washing poultry carcasses, hands, and utensils that have contacted carcasses; cooking food sufficiently to kill bacteria; and pasteurizing cow's milk. Steps must also be taken to prevent contamination of the water supply with feces from stockyards, feedlots, and slaughterhouses.

Helicobacter pylori

Learning Objectives

✓ Discuss the major change in medical opinion concerning the cause of peptic ulcers.

✓ Describe the effect of *Helicobacter pylori* on the lining of the human stomach.

Helicobacter pylori (hel'ĭ-kō-bak'ter pī'lō-rē) is a slightly helical, highly motile bacterium that colonizes the stomachs of its hosts **(Figure 21.26)**. Based on rRNA nucleotide sequences, *Helicobacter* is classified in the class Deltaproteobacteria. It was originally placed in the genus *Campylobacter;* however, unlike *Campylobacter*, *Helicobacter* cannot reduce nitrate, has sheathed flagella, and is urease positive, a characteristic that is essential for colonizing the stomach.

Today, an idea that only a few years ago was very controversial—that *H. pylori* causes most (if not all) **peptic ulcers,** which are erosions of the mucous membrane of the stomach or duodenum (the initial portion of the small intestine)—is accepted as fact. What was once the prevailing view—that stress, alcohol consumption, spicy food, or excess stomach acid production caused ulcers—was discredited beginning in 1982, when Australian gastroenterologists

SEM   ⊢———⊣ 1 μm

▲ *Figure 21.26*

Helicobacter pylori, the cause of peptic ulcers.

Robin Warren and Barry Marshall detected *Helicobacter* colonizing the majority of their patients' stomachs. Following treatment with antibiotics, the bacteria and the ulcers disappeared. It is reported that Dr. Marshall provided final proof of *H. pylori's* involvement by fulfilling Koch's postulates himself—he drank one of his cultures of *Helicobacter*! As he expected, he soon developed painful ulcers, from which he was able to isolate *H. pylori*.

H. pylori possesses numerous virulence factors that enable it to colonize the human stomach: a protein that inhibits acid production by stomach cells; flagella that enable the pathogen to burrow through the mucus lining the stomach; adhesins that facilitate binding to gastric cells;

enzymes that inhibit phagocytic killing; and urease, an enzyme that degrades urea, which is present in gastric juice, to produce highly alkaline ammonia, which neutralizes stomach acid.

The portal of entry for *H. pylori* is not known positively, but its colonization of the stomach suggests a fecal-oral path of infection. Studies have shown that *H. pylori* in feces on the hands, in well water, or on fomites may be a source of human infection. In addition to humans, cats may be an initial source of infection.

The formation of a peptic ulcer occurs as follows **(Figure 21.27):** The process begins when *H. pylori* (protected by urease) burrows through the stomach's protective mucus layer to reach the underlying epithelial cells ❶, where the bacteria attach to the cells' cytoplasmic membranes and multiply. A variety of factors—the triggering of inflammation by bacterial exotoxin and perhaps the destruction of mucus-producing cells by the bacteria—causes the mucus layer to become thin ❷, allowing acidic gastric juice to digest the

stomach lining. Once the epithelial layer has been ulcerated by gastric juice, *H. pylori* gains access to the underlying muscle tissue and blood vessels ❸. Those bacteria that are phagocytized by leukocytes survive through the actions of catalase and superoxide dismutase, enzymes that neutralize part of the phagocytes' killing mechanism.

The presence of *H. pylori* in specimens from the stomach can be demonstrated by a positive urease test within 1–2 hours of culturing; the bacterium can also be seen in Gram-stained specimens. Definitive identification is based on a series of biochemical tests.

Physicians treat ulcers with one or more antimicrobial drugs given in combination with drugs that inhibit acid production and thus allow the stomach lining time to regenerate. Because the exact mode of transmission of *H. pylori* is not yet known, prevention of infection is difficult. Presumably, good personal hygiene, adequate sewage treatment, water purification, and proper food handling can limit the incidence of infections with *Helicobacter*.

❶ Bacteria invade mucus and attach to gastric epithelial cells.

❷ *Helicobacter*, its exotoxin, and inflammation, cause the mucus layer to become thin.

❸ Gastric acid destroys epithelial cells and underlying tissue.

▲ *Figure 21.27*
The role of *Helicobacter pylori* in the formation of peptic ulcers.

CHAPTER SUMMARY

Mycoplasmas (pp. 596–598)

1. **Mycoplasmas,** the smallest free-living microbes, lack cell walls. They are mostly facultative anaerobes. The colonies of most mycoplasmas have a "fried egg" appearance.

2. *Mycoplasma pneumoniae* is a human pathogen that affects the epithelial cells of the respiratory tract and causes **primary atypical pneumonia** (walking pneumonia).

3. *Mycoplasma hominis* causes pyelonephritis, postpartum fever, and **pelvic inflammatory disease** in women.

4. *Mycoplasma genitalium* and *Ureaplasma urealyticum* cause nongonococcal urethritis.

5. Venereal mycoplasmal diseases can be prevented by abstinence, mutually faithful monogamy, and the proper use of condoms.

Rickettsias (pp. 598–602)

1. **Rickettsias** are extremely small, Gram-negative, obligate intracellular parasites.

2. *Rickettsia rickettsii* causes **Rocky Mountain spotted fever (RMSF),** a serious illness transmitted by ticks. Rash, malaise, petechiae, encephalitis, and death (in 5% of cases) characterize the disease.

3. *Rickettsia prowazekii* causes louse-borne or **epidemic typhus,** characterized by fever, depression, and rash. The disease may recur as Brill-Zinsser disease.

4. *Rickettsia typhi* causes endemic typhus, also called **murine typhus** because the major reservoir is rodents. Various fleas act as vectors to transmit the bacterium to humans.

5. Mites (chiggers) transmit *Orienta tsutsugamushi* to cause **scrub typhus,** which is characterized by a spreading rash.

6. *Ehrlichia chaffeensis* causes **human monocytic ehrlichiosis (HME),** and *Ehrlichia equi* causes **human granulocytic ehrlichiosis (HGE).** These bacteria are transmitted to humans via the Lone Star tick, the deer tick, and the dog tick. The bacteria enter leukocytes and grow through three developmental stages: an elementary body, an initial body, and a morula.

Chlamydias (pp. 602–606)

1. Chlamydias are small, nonmotile obligate intracellular parasites; their developmental cycle includes infectious **elementary bodies (EBs)** and noninfectious **reticulate bodies (RBs).** A phagosome full of RBs is called an **inclusion body.**

2. *Chlamydia trachomatis* enters the body through abrasions in mucous membranes of the genitalia, eyes, and respiratory structures. *Chlamydia* causes the most reported sexually transmitted disease in the United States.

3. **Lymphogranuloma venereum** is caused by *Chlamydia trachomatis.* The majority of infections in women are asymptomatic, whereas the majority of infections in men result in buboes, fever, chills, anorexia, and muscle pain. Other symptoms may include proctitis (in men and women), and **pelvic inflammatory disease** and sterility (in women).

4. **Trachoma** is a serious eye disease that is caused by *C. trachomatis* and may result in blindness.

5. *Chlamydia* infections can be prevented by abstinence and faithful mutual monogamy.

6. *Chlamydia pneumoniae* causes bronchitis, pneumonia, and sinusitis.

7. *Chlamydia psittaci* causes **ornithosis** (also called psittacosis or parrot fever), a respiratory disease of birds that can be transmitted to humans.

Spirochetes (pp. 607–614)

1. **Spirochetes** are helical bacteria with axial filaments that cause the organism to corkscrew, enabling it to burrow into a host's tissues.

2. **Syphilis** is caused by *Treponema pallidum pallidum,* a sexually transmitted obligate parasite of humans. The disease is observed worldwide.

3. Primary syphilis is characterized by a **chancre,** a red lesion at the infection site. If *Treponema* moves from the chancre to the

bloodstream, secondary syphilis results, causing rash, aches, and pains. After a period of latency, progression may occur to tertiary syphilis, characterized by swollen **gummas,** dementia, blindness, paralysis, and heart failure.

4. **Congenital syphilis** results when an infected mother infects her fetus.

5. Penicillin is used to treat all phases except tertiary syphilis; no vaccine is available.

6. *Treponema pallidum endemicum* causes **bejel,** an oral disease observed in children in Africa, Asia, and Australia.

7. *Treponema carateum* causes **pinta,** a disfiguring skin disease observed in children in Central and South America.

8. *Treponema pallidum pertenue* causes **yaws,** a skin disease observed in South America, central Africa, and southeast Asia.

9. *Borrelia burgdorferi* causes **Lyme disease,** a disease transmitted by ticks of the genus *Ixodes* and characterized by a "bull's eye" rash, neurological and cardiac dysfunction, and finally severe arthritis.

10. Human body lice transmit *Borrelia recurrentis,* which causes **epidemic relapsing fever** in Africa and South America. In the western United States, soft ticks transmit several species of *Borrelia* causing **endemic relapsing fever.**

11. *Leptospira interrogans* causes **leptospirosis,** a zoonotic disease in humans transmitted via animal urine and characterized by pain, headache, and liver and kidney dysfunction.

Pathogenic Gram-Negative Curved Bacilli (pp. 614–618)

1. *Vibrio* is a genus of Gram-negative curved bacteria with polar flagella that naturally live in marine environments.

2. *Vibrio cholerae* causes **cholera,** a disease contracted via the ingestion of contaminated food and water. Cholera has been pandemic through the centuries.

3. Via a series of biochemical steps, **cholera toxin**—an exotoxin produced by *Vibrio cholerae*—causes the movement of water out of the intestinal epithelium, resulting in potentially fatal diarrhea.

4. *Vibrio parahaemolyticus* enters the body via ingestion of shellfish from contaminated waters; it causes a milder form of cholera-like gastroenteritis.

5. *Vibrio vulnificus,* contracted either by ingestion of contaminated shellfish or by contamination of wounds with seawater, causes a potentially fatal blood poisoning.

6. *Campylobacter jejuni,* which is found in domestic animal reservoirs, commonly causes gastroenteritis when ingested in contaminated food, water, or milk.

7. Once *Helicobacter pylori* reduces the amount of mucus produced in the stomach, acidic gastric juice eats away the stomach lining, causing **peptic ulcers.**

QUESTIONS FOR REVIEW

(Answers to multiple choice and matching questions are on the web, along with additional review questions. Visit www.microbiologyplace.com.)

Multiple Choice

1. Which of the following is *not* characteristic of mycoplasmas?
   a. cytochromes
   b. sterols in cytoplasmic membranes
   c. use of UGA codon for tryptophan
   d. rRNA nucleotide sequences similar to Gram-positive bacteria

2. Most human infections caused by species of *Rickettsia*
   a. are acquired from fomites.
   b. could be prevented by handwashing.
   c. are transmitted via vectors.
   d. are sexually transmitted.

3. The bacterium that causes RMSF is more likely to infect a human
   a. if an infected tick feeds for several hours.
   b. when an infected tick initially penetrates the skin.
   c. when tick feces dries and becomes airborne.
   d. if the human is exposed to rodent feces containing the bacterium.

4. The most severe rickettsial illness is caused by
   a. *Rickettsia typhi.*   c. *Orienta tsutsugamushi.*
   b. *Rickettsia rickettsii.*   d. *Ehrlichia chaffeensis.*

5. The smallest free-living microbes are
   a. rickettsias.   c. chlamydias.
   b. mycoplasmas.   d. vibrios.

6. The most common reportable sexually transmitted disease in the United States is caused by the bacterium
   a. *Mycoplasma genitalium.*
   b. *Chlamydia trachomatis.*
   c. *Chlamydia proctitis.*
   d. *Ureaplasma urealyticum.*

7. Which of the following diseases would be *least* likely in rural areas of the United States?
   a. epidemic typhus
   b. Rocky Mountain spotted fever
   c. murine typhus
   d. lymphogranuloma venereum

8. Treatment of chlamydial infections involves
   a. erythromycin cream.
   b. doxycycline creams.
   c. surgical correction of eyelid deformities.
   d. all of the above

9. *Chlamydia pneumoniae* causes
   a. pneumonia.
   b. sinusitis.
   c. bronchitis.
   d. all of the above

10. Which of the following organisms is transmitted via sexual contact?
    a. *Treponema pallidum endemicum*
    b. *Treponema pallidum pertenue*
    c. *Treponema pallidum pallidum*
    d. *Treponema carateum*

11. Historical journals have described gummas on patients. What disease most likely caused these lesions?
    a. ornithosis
    b. syphilis
    c. trachoma
    d. pneumonia

12. In which stage of syphilis is penicillin ineffective?
    a. primary syphilis
    b. secondary syphilis
    c. tertiary syphilis
    d. all of the above

13. Two weeks after a backpacking trip in Tennessee, a hiker experienced flulike symptoms and noticed a red rash on his thigh. What is the likely cause of his illness?
    a. *Treponema pallidum pertenue*
    b. *Borrelia burgdorferi*
    c. *Borrelia recurrentis*
    d. *Leptospira interrogans*

14. The most common cause of bacterial gastroenteritis in the United States is
    a. *Vibrio parahaemolyticus.*
    b. *Campylobacter jejuni.*
    c. *Helicobacter pylori.*
    d. *Vibrio cholerae.*

15. Which of the following is *not* true of cholera?
    a. The causative agent lives naturally in marine water.
    b. There is an effective vaccine for cholera.
    c. Strain O1 *el Tor* has been responsible for several pandemics.
    d. Rice-water stool is a symptom.

Short Answer

1. Why are mycoplasmas able to survive a relatively wide range of osmotic conditions, even though these bacteria lack cell walls?

2. Describe a hypothesis to explain why rickettsias are obligate parasites.

3. What biochemical test distinguishes between *Mycoplasma* and *Ureaplasma*?

4. Describe the three developmental stages of the bacterium *Ehrlichia*.

5. Explain why the Gram-positive mycoplasmas are pink in a Gram-stained smear.

6. Compare and contrast mycoplasmas and viruses.

7. Why have scientists had problems identifying the virulence factors of *Treponema pallidum pallidum*?

8. Describe the phases of untreated syphilis.

9. Discuss the prospects for the eradication of leptospirosis.

10. Beginning with the ingestion of water contaminated with *V. cholerae* O1 *el Tor*, describe the course of the disease it causes.

Matching

1. Match the disease with the causative pathogen.

___ Rocky Mountain spotted fever	A. *Rickettsia typhi*
___ Endemic typhus	B. *Rickettsia prowazekii*
___ Epidemic typhus	C. *Rickettsia rickettsii*
___ Scrub typhus	D. *Orienta tsutsugamushi*
___ HME	E. *Ehrlichia chaffeensis*

2. Match the pathogen with the vector responsible for transmitting it to humans.

___ *Rickettsia typhi*	A. Rat flea
___ *Rickettsia prowazekii*	B. Body louse
___ *Rickettsia rickettsii*	C. Tick
___ *Orienta tsutsugamushi*	D. Mite
___ *Ehrlichia chaffeensis*	E. Bat
___ *Borrelia burgdorferi*	
___ *Borrelia recurrentis*	

3. Match the pathogen with the disease it causes.

___ *Chlamydia psittaci*	A. Syphilis
___ *Chlamydia pneumoniae*	B. Trachoma
___ *Chlamydia trachomatis*	C. Sinusitis
___ *Treponema pallidum pallidum*	D. Lymphogranuloma venereum
___ *Treponema pallidum pertenue*	E. Proctitis
___ *Treponema pallidum endemicum*	F. Pelvic inflammatory disease
___ *Treponema carateum*	G. Ornithosis
___ *Borrelia burgdorferi*	H. Yaws
	I. Bejel
	J. Pinta
	K. Lyme disease

4. Match the following diseases with the causative bacterium.

___ Peptic ulcers	A. *Vibrio cholerae* O1 *el Tor*
___ Gastroenteritis (various forms)	B. *Vibrio parahaemolyticus*
___ Blood poisoning	C. *Vibrio vulnificus*
___ Cholera	D. *Campylobacter jejuni*
	E. *Helicobacter pylori*

CRITICAL THINKING QUESTIONS

1. Why is it more difficult to rid a community of a disease transmitted by arthropods (for example, Lyme disease or RMSF) than a disease transmitted via contaminated drinking water (for example, cholera)?

2. Some scientists think that syphilis is a New World disease brought to Europe by returning Spanish explorers; others think that syphilis traveled the opposite way. Design an experiment to test the two hypotheses.

3. A patient arrives at a medical emergency room in Austin, Texas, complaining of 48 hours of severe abdominal cramping and persistent diarrhea. While waiting, he falls into a coma. He is admitted into the hospital but never regains consciousness; a week later he dies. No one else in his family is ill. They report that the man has not been out of the country, but that they all dined at a seafood restaurant the night before the man became sick. He alone ate sea urchins. Culture of the man's stool reveals the presence of curved bacilli. What bacterium might account for the man's condition?

4. Scientists discover a mutant strain of *Helicobacter pylori* that is urease negative. The strain is found to cause ulcers only in patients that either consume large quantities of antacids and/or take drugs to block acid production. Explain why only these patients develop ulcers when infected by the mutant strain.

5. Describe how a proper application of Koch's postulates could determine whether either *Mycoplasma* or *Chlamydia* causes atherosclerosis.

Pathogenic Fungi   CHAPTER 22

The fungus *Epidermophyton floccosum* is beautiful when observed on a microscope slide, but it is much less welcome when found between the toes and on the soles of the feet, where it can cause the condition called *tinea pedis* —commonly known as "athlete's foot." Despite its name, this surface fungal disease, which can cause itching, peeling, and other irritation, affects athletes and nonathletes alike; having warm, moist feet is a much more important factor than being physically active. Because fungal spores are constantly being shed from the feet and onto bedding and furniture—and onto moist bathroom floors—athlete's foot is common among college students living in dorms. The fungi are hardy; most survive for a long time in the environment, resisting desiccation, heat, cold, salt, acid, and even some antimicrobial cleansers. The best way to deal with athlete's foot is to keep your feet dry, change your socks frequently, use an anti-fungal cream, and wear a pair of flip-flops in public bathrooms.

In this chapter we examine this and other fungi that cause human diseases.

Epidermophyton floccosum

CHAPTER OUTLINE

An Overview of Medical Mycology
The Epidemiology of Mycoses
Categories of Fungal Agents: True Fungal Pathogens and Opportunistic Fungi
Clinical Manifestations of Fungal Diseases
The Diagnosis of Fungal Infections
Antifungal Therapies

Systemic Mycoses Caused by Pathogenic Fungi
Blastomycosis
Coccidioidomycosis
Histoplasmosis
Paracoccidioidomycosis

Systemic Mycoses Caused by Opportunistic Fungi
Aspergillosis
Candidiasis
Cryptococcosis
Pneumocystis Pneumonia
Zygomycoses
The Emergence of Fungal Opportunists in Immunosuppressed Individuals

Superficial, Cutaneous, and Subcutaneous Mycoses
Superficial Mycoses
Cutaneous and Subcutaneous Mycoses

Fungal Intoxications and Allergies
Mycotoxicoses
Mycetismus: Mushroom Poisoning
Allergies to Fungi

Medical mycology is the field of medicine that is concerned with the diagnosis, management, and prevention of fungal diseases, or **mycoses** (mī-kō′sēz). As discussed in Chapter 12, most fungi exist as saprobes and function as the major decomposers of organic matter in the environment. More than 100,000 species of fungi have been classified in four divisions—Zygomycota, Ascomycota, Basidiomycota, and Deuteromycota—and of these, fewer than 200 have been demonstrated to cause diseases in humans.

Unlike the other disease chapters in this text, this chapter is not arranged according to any taxonomic scheme. Because the classification of fungi is difficult and the subject of considerable debate among mycologists, mycoses are generally grouped according to site of general disease. Even this scheme is imperfect. Under certain conditions, fungi often invade tissues where they are normally not found, where they may present symptoms that are not characteristic of the organism in its usual site. But some framework for discussion is required, so in this chapter we will examine the medically important fungal infections and diseases according to their type of cause—true pathogens or opportunists—and category of clinical manifestation—systemic, superficial, cutaneous, subcutaneous, poisoning, or allergy. First, however, we will briefly consider some of the basic topics in medical mycology.

An Overview of Medical Mycology

Learning Objective

✓ Summarize some of the complexities of identifying and treating fungal infections.

Mycoses are among the more difficult human diseases to diagnose and treat properly. Signs of mycoses are often missed or misinterpreted, and once identified, fungi often prove to be remarkably resistant to antimicrobial agents. Before we begin our discussion of fungal diseases, therefore, it will be helpful to consider some of the characteristics that set fungal infections apart from other microbial infections.

The Epidemiology of Mycoses

Learning Objectives

✓ State the most significant mode of transmission for mycoses.
✓ Explain why the actual prevalence of fungal infections is unknown.

Fungi and the spores they produce are almost everywhere in the environment—in soil, in water, and on or in most multicellular organisms, including plants and animals. They coat the surfaces of almost every object, whether made of wood, glass, metal, or plastic. Thus it is not surprising that most of us will experience a mycosis at some time.

Mycoses are typically acquired via inhalation, trauma, or ingestion; only very infrequently are fungi spread from person to person. Therefore, most mycoses are *not* contagious. Epidemics of mycoses can and do occur, but not as a result of person-to-person contact. Instead, they result from mass exposure to some environmental source of fungi. For example, the cleanup of bird droppings near a building's air conditioning intake vents could stir up fungal spores in the droppings; the airborne spores could then be drawn inside the building and distributed throughout its duct system, potentially infecting numerous people.

One group of fungi that *are* contagious are *dermatophytes* (der′mă-tō-fītz), which live in the dead layers of skin and are found in the environment on fomites that may be involved in transmission. Species of the genera *Candida* (kan′did-a) and *Pneumocystis* (noo-mō-sis′tis) also appear to be transmitted at least some of the time by contact among humans.

Because most mycoses are not contagious, they typically are not reportable; that is, neither local public health agencies nor the CDC must be notified when they are diagnosed. Again, there are a few exceptions: Pathogenic fungi are usually reported where they are endemic, and certain opportunistic fungi that frequently infect AIDS patients are also tracked.

The fact that most mycoses are not reportable creates a problem for epidemiologists. Without reliable data and statistics, it is impossible to know the effects of mycoses on the human population—how often they occur, which segments of the population are most affected, and where research and treatment efforts should be directed. Without such information, it is difficult to ascertain the risks posed by mycoses, and thus it is difficult to identify actions that can prevent them.

CRITICAL THINKING

Discuss the relationship between the following two facts: (1) A huge number of fungal spores are present in the environment, and (2) fungal diseases are not typically acquired via contact with infected individuals. How does not relying on a host for transmission benefit a fungus?

Categories of Fungal Agents: True Fungal Pathogens and Opportunistic Fungi

Learning Objectives

✓ Compare and contrast true fungal pathogens with opportunistic fungi.
✓ Identify factors that predispose people to opportunistic fungal infections.

Of all the fungi known to cause disease in humans, only four—*Blastomyces dermatitidis, Coccidioides immitis, Histoplasma capsulatum,* and *Paracoccidioides brasiliensis*—are usually considered true pathogens; the remaining fungi that cause disease in humans are considered *opportunistic fungi.*

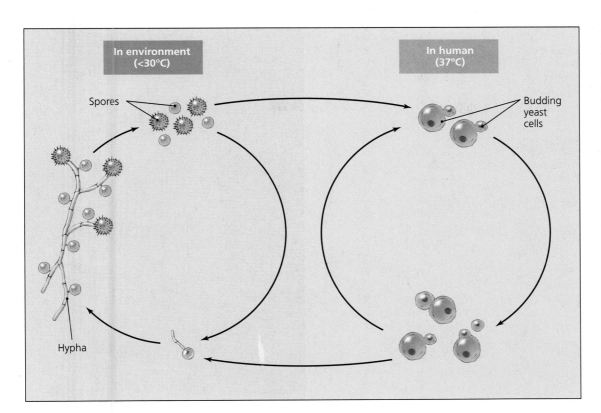

◀ *Figure 22.1*

The dimorphic nature of true fungal pathogens, in which the mycelial form grows in the environment, and the yeast form grows within the host. Shown here is *Histoplasma capsulatum*. *What are the other three genera of pathogenic, dimorphic fungi?*

Figure 22.1 In addition to Histoplasma, members of the genera Blastomyces, Coccidioides, and Paracoccidioides are pathogenic fungi.

Like all pathogens, the pathogenic fungi of humans have the ability to actively attack and invade tissues. All four true fungal pathogens are *dimorphic* (dī-mōr'fik); that is, in the environment, where temperatures are frequently below 30°C, they have mycelial thalli composed of hyphae, whereas within the body (at 37°C) they exist as yeast thalli **(Figure 22.1).** The two forms differ not only structurally, but also physiologically. Yeast forms are invasive because they express a variety of enzymes and other proteins that aid their survival and reproduction within the body by, for example, making them tolerant of higher temperatures and relatively resistant to phagocytic killing.

In contrast, opportunistic fungi lack genetic characteristics associated with true pathogenesis. Because they lack genes for proteins that aid in colonizing the body, they do not cause disease in healthy individuals. Instead, they are often commensals that can take advantage of some weakness in a host's defenses to become established and cause disease. Thus whereas true fungal pathogens can infect any appropriate host, regardless of its immune status, opportunists infect only susceptible individuals.

Four factors increase an individual's risk of experiencing opportunistic mycoses: invasive medical procedures, medical therapies, certain preexisting conditions, and specific lifestyle factors **(Table 22.1).** Surgical insertion of devices such as heart-valve implants can introduce fungal spores and provide a site for fungal colonization, and a vari-

Table 22.1	Factors that Predispose Individuals to Opportunistic Mycoses
Factors	**Examples**
Medical procedures	Surgery; insertion of medical implants (heart valves, artificial joints); catheterization
Medical therapies	Immunosuppressive therapies accompanying transplantation; radiation and other cancer therapies; steroid treatments; long-term use of antibacterial agents
Preexisting conditions	Inherited immune defects; leukemia and lymphomas; AIDS; diabetes and other metabolic disorders; severe burns; preexisting chronic illnesses
Lifestyle factors	Poor diet; poor hygiene; IV drug abuse

ety of medical therapies leave patients with a weakened or dysfunctional immune system. AIDS, diabetes, other serious illnesses, and malnutrition also weaken immunity. Poor hygiene results in reduced skin defenses, and IV drug use can directly introduce fungi into the blood and, in the long term, lead to the erosion of overall health. Note that many of the factors that contribute to opportunistic fungal infections involve the actions of medical personnel in health care

Figure 22.2 ▶

The geographic distributions of the four true fungal pathogens. Shaded areas indicate areas of primary endemicity. *Can a traveler with coccidiomycosis establish an endemic region in northern Russia?*

Figure 22.2 No, because the environmental conditions in northern Russia are too different from those in the desert regions where the fungi are endemic.

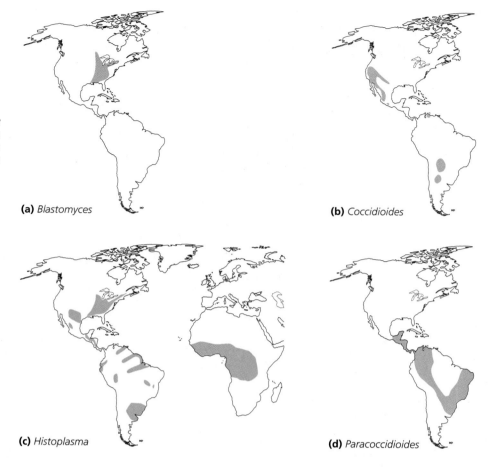

(a) *Blastomyces*

(b) *Coccidioides*

(c) *Histoplasma*

(d) *Paracoccidioides*

settings, and thus health care providers can play a crucial, active role in reducing the incidence of mycoses.

In addition to differences in pathogenesis and matters related to host susceptibility, fungal pathogens and opportunists differ with respect to geographical distribution. Whereas the four pathogenic fungi are endemic to certain regions, primarily in the Americas **(Figure 22.2)**, opportunistic fungi are distributed throughout the world. Even though a few opportunistic fungi display some regional preferences, their distributions are widely dispersed because their typical roles as members of the normal microbiota mean that people carry them wherever they go.

Dermatophytes—fungi that normally live on the skin—are the only fungi that do not fall comfortably into either the pathogenic or opportunistic groupings. They are considered by some researchers to be "emerging" pathogens. Dermatophytes can infect all individuals, not just the immunocompromised, which makes them similar to the pathogens. They are not, however, intrinsically invasive, being limited to body surfaces. They also have a tendency to occur in people with the same predisposing factors that allow access by opportunistic fungi. For these reasons, dermatophytes will be discussed as opportunists, rather than as pathogens.

CRITICAL THINKING

Suggest a way in which each of the risk factors listed in Table 22.1 can be reduced or eliminated to limit opportunistic infections. What can the patient do? What can health care providers do?

Clinical Manifestations of Fungal Diseases

Learning Objective

✓ Describe three primary clinical manifestations of mycoses.

Fungal diseases are grouped into the following three categories of clinical manifestations:

- *Fungal infections,* which are the most common mycoses, are caused by the presence in the body of either true pathogens or opportunists. As previously noted, this chapter discusses mycoses according to general location within the body: Systemic mycoses are discussed first, followed by superficial, cutaneous, and subcutaneous mycoses.

- *Toxicoses* (poisonings) are acquired through ingestion, as occurs when poisonous mushrooms are eaten. Although relatively rare, toxicoses are discussed briefly near the end of the chapter.

- *Allergies* (hypersensitivity reactions) most commonly result from the inhalation of fungal spores and are the subject of the chapter's final section.

The Diagnosis of Fungal Infections

Learning Objective

✓ Discuss why the diagnosis of opportunistic fungal infections can be difficult.

Most patients reporting to a hospital emergency room with a severe respiratory illness are routinely tested for influenza, bacterial pneumonia, and perhaps tuberculosis; without some concrete reason to suspect a fungal infection, physicians may not look for them. However, if an emergency-room patient with respiratory distress reveals that she breeds exotic birds, she is much more likely to also be tested for mycoses. Thus a patient's history—including occupation, hobbies, travel history, and the presence of preexisting medical conditions—is critical for diagnosis of most mycoses. Even when fungal infections are relatively obvious, as when distinctive mycelial growth is observed, definitive diagnosis requires isolation, laboratory culture, and morphological analysis of the fungus involved.

Microbiologists culture fungi collected from patients on Sabouraud dextrose agar, a medium that favors fungal growth over bacterial growth (see Figure 6.11). If a fungal pathogen is suspected, cultures are grown at both 25°C and 37°C to reveal dimorphism. The appearance of colonies, and the microscopic appearance of yeast cells, mycelia, or mold spores, are usually diagnostic.

Several techniques are commonly used in identifying fungi. *Potassium hydroxide (KOH) preparations* dissolve keratin in skin scrapings or biopsy specimens, leaving only fungal cells for examination. *Gomori methenamine silver (GMS) stain* is used on tissue sections to stain fungal cells black (other cells remain unstained). *Direct immunofluorescence* can also be used to detect fungal cells in tissues; however, immunological tests, though very useful for other microbial infectious agents, are not always useful for fungi. Because fungi are so prevalent in the environment and many are commensal, it is often impossible to distinguish between actual infection and simple exposure. Nucleic acid probes can also be used to diagnose the four fungal pathogens.

Diagnosis of opportunistic fungal infections is especially challenging, because fungal types can be confused with one another as well as with other diseases. When a fungal opportunist infects tissues in which it is normally not found, it may display abnormal morphology that confuses identification. In addition, the fungi that produce pulmonary infections—the true pathogens and a few opportunists—produce symptoms and imaging profiles (X-rays, CT scans) that strongly resemble those of tuberculosis. Fungal masses may also resemble endogenous tumors.

Antifungal Therapies

Learning Objective

✓ Discuss the advantages and disadvantages of fungicidal and fungistatic medications.

Mycoses are among the most difficult diseases to heal for two reasons. First, fungi generally possess the biochemical ability to resist the oxidative damage perpetuated by T cells during cell-mediated immune responses. Second, fungi are eukaryotic and thus biochemically similar to human cells, which means that most fungicides are toxic to human tissues. The majority of antifungal agents exploit one of the few differences between human and fungal cells—instead of cholesterol, the membranes of fungal cells contain a related molecule, *ergosterol*. Antifungal drugs target either ergosterol synthesis or its insertion into fungal membranes. However, cholesterol and ergosterol are not sufficiently different to prevent damage to human tissues by such antifungal agents. Serious side effects associated with long-term use of almost all antifungal agents include anemia, headache, rashes, gastrointestinal upset, and serious liver and kidney damage.

The "gold standard" of antifungal agents is the fungicidal drug **amphotericin B,** considered the best drug for treating systemic mycoses and other fungal infections that do not respond to other drugs. Unfortunately, it is also one of the more toxic antifungal agents to humans. The major alternatives to amphotericin B are the azole drugs—*ketoconazole, itraconazole,* and *fluconazole*—which are fungistatic (inhibitory) rather than fungicidal, but also much less toxic to humans.

Two antifungal drugs that do not target ergosterol are *5-fluorocytosine,* which inhibits RNA and DNA synthesis and is given in conjunction with amphotericin B in the treatment of cryptococcal infections, and *griseofulvin,* which interferes with microtubule formation and chromosomal separation in cell division. Griseofulvin accumulates in the outer epidermal layers of the skin, preventing fungal penetration and growth, and it is generally well tolerated by a patient for the duration of treatment. For example, griseofulvin is usually administered orally for up to a year to clear fungal infections of the nails without harming the patient.

Treatment of opportunistic fungal infections in immunocompromised patients involves two steps: a high-dose treatment to eliminate or reduce the number of fungal pathogens, followed by long-term (usually lifelong) maintenance therapy involving the administration of antifungal agents to control ongoing infections and prevent new infections.

Now that we have considered some of the basics of medical mycology, we turn our attention to the first category of important mycoses: systemic mycoses caused by pathogenic fungi.

Systemic Mycoses Caused by Pathogenic Fungi

Learning Objectives

✓ Compare and contrast the endemic areas for the four genera of pathogenic fungi that cause systemic mycoses.

✓ Compare the clinical appearances of the diseases resulting from each of the four pathogenic fungi.

✓ Identify the laboratory techniques used to distinguish among the four pathogenic fungi.

Systemic mycoses—those fungal infections that spread throughout the body—are the result of infections by one of the four pathogenic, dimorphic fungi of the division Ascomycota: *Blastomyces, Coccidioides, Histoplasma,* and *Paracoccidioides*. These pathogenic fungi are uniformly acquired through inhalation, and all begin as a generalized pulmonary infection that then disseminates to the rest of the body.

Whereas all four true fungal pathogens are endemic primarily in the Americas, *Blastomyces, Coccidioides,* and *Histoplasma* are endemic to (but not limited to) the United States (see Figure 22.2). Note, however, that cases have occurred outside the Western Hemisphere.

Dimorphic fungi pose extremely dangerous hazards in the laboratory. Laboratory personnel must take specific precautions to avoid exposure to spores, particularly when culturing the organisms. Biological containment (HEPA) hoods, protective clothing, and masks are thus required when working with these pathogens.

In the following sections we consider each of the systemic conditions caused by true fungal pathogens, beginning with blastomycosis.

Blastomycosis

Blastomycoses (blas′tō-mī-kō′sēz) are caused by an ascomycete *Blastomyces dermatitidis* (blas-tō-mī′sēz der-mă-tit′i-dis), which is endemic across the southeastern United States north to Canada (see Figure 22.2a). Outbreaks have also been reported in Latin America, Africa, Asia, and Europe. *B. dermatitidis* normally grows in soil rich in organic material such as decaying vegetation and animal wastes, where cool, damp conditions favor growth and sporulation. The fungus has been recovered from barns, paddocks, and old wooden buildings. Inhalation of dust disturbed from environmental sources carries fungal spores and pieces of hyphae into the lungs. In humans, both recreational and occupational exposure occurs, and a relatively small inoculum can produce disease. The incidence of human infection is increasing due to increases in the number of immunocompromised individuals in the population. Cats, dogs, and horses can also become infected.

Pulmonary blastomycosis is the most common manifestation of *Blastomyces* infection in humans. After spores enter the lungs, they convert to yeast forms and multiply. Initial pulmonary lesions are asymptomatic in most individuals. If symptoms do develop, they are vague and include cough,

▲ *Figure 22.3*

Cutaneous blastomycosis, which results from the dissemination of *Blastomyces dermatitidis* from the lungs to the skin.

fever, malaise, and weight loss. Purulent (pus-filled) lesions develop and expand as the yeast multiply, resulting in necrosis (death of tissues) and cavity formation. Rarely, a deep productive cough and chest pain accompany the generalized symptoms. In otherwise healthy people, pulmonary blastomycosis typically resolves, but it may become chronic. Respiratory failure and death occur at a high frequency among immunocompromised patients.

The fungus can disseminate, that is, spread, beyond the lungs. *Cutaneous blastomycosis* following dissemination occurs in 60–70% of cases and consists of generally painless lesions on the face and upper body **(Figure 22.3).** The lesions can be raised and wartlike, or they may be craterlike due to the death of tissue. In roughly 30% of cases, the fungus spreads to the spine, pelvis, cranium, ribs, long bones, or subcutaneous tissues surrounding joints, a condition called *osteoarticular*[1] (os′tē-ō-ar-tik′ū-lar) *blastomycosis*. About 40% of AIDS patients experience meningitis resulting from dissemination of *Blastomyces* to the central nervous system. Abscesses may form in other tissue systems as well.

Diagnosis relies on identification of *B. dermatitidis* following culture, or direct examination of various samples such as sputum, bronchial washings, biopsies, cerebrospinal fluid, or skin scrapings. Observation of dimorphism in laboratory cultures coupled with microscopic examination is diagnostic.

Physicians treat blastomycoses with amphotericin B for 10 weeks, though longer treatment may be necessary. Oral itraconazole may be used as an alternative but must be administered for a minimum of 3–6 months. Relapse is common in AIDS patients, and suppressive maintenance therapy with itraconazole is recommended.

---

[1]From Greek *osteon,* meaning bone, and *arthron,* meaning joint.

Outbreaks of blastomycoses have occurred in Latin America even though the organism itself is not normally found there. Based on what you have read, explain why a few cases of blastomycosis might appear outside of endemic areas.

Coccidioidomycosis

Coccidioidomycosis (kok-sid-ē-oy′dō-mī-kō-sis), caused by an ascomycete *Coccidioides immitis* (kok-sid-ē-oy′dēz im′i-tis), is found almost exclusively in the southwestern United States (Arizona, California, Nevada, New Mexico, and Texas) and northern Mexico. Small, focally endemic areas also exist in semiarid parts of Central and South America (see Figure 22.2b). Fungi can be recovered from desert soil, rodent burrows, archaeological remains, and mines. Dust coating materials from endemic areas, including Native American pots and blankets sold to tourists, can serve as a vehicle of infection. Infection rates in endemic areas have risen in recent years due to population expansion and increased recreational activities, such as the use of off-road vehicles in the desert. More ominous threats to human health, however, are windstorms and earthquakes, which can disturb large tracts of contaminated soil, spreading the fungal elements for miles downwind and exposing hundreds or thousands of people.

In the warm and dry summer and fall months, particularly in drought cycles, *C. immitis* grows as a mycelium and produces sturdy chains of asexual spores called *arthroconidia* or *arthrospores*. When mature, arthrospores germinate into new mycelia.

Once inhaled in dust, arthrospores become trapped in the alveolar spaces of the lungs, where they germinate into a parasitic form called a *spherule* (sfer′ool). As each spherule matures, it enlarges and generates a large number of spores via multiple cleavages, until it ruptures and releases the spores into the surrounding tissue. Each spore then forms a new spherule to continue the cycle of division and release. This type of growth accounts for the seriousness of *Coccidioides* infection.

The major manifestation of coccidioidomycosis is pulmonary. About 60% of patients experience either no overt symptoms or mild, unremarkable symptoms that go unnoticed and typically resolve on their own. Other patients develop more severe infections characterized by fever, cough, chest pain, difficulty breathing, coughing up or spitting blood, headache, night sweats, weight loss, and pneumonia; in some individuals a diffuse rash may appear on the trunk. Occasionally the mild forms of the pulmonary disease become more chronic, in which cases the continued multiplication of spherules results in large, permanent cavities in the lungs similar to those seen in tuberculosis patients.

In a very small percentage of cases, generally in those who are severely immunocompromised, *C. immitis* disseminates from the lungs to various other sites. Invasion of the CNS may result in meningitis, headache, nausea,

▲ *Figure 22.4*

Coccidioidomycosis lesions in subcutaneous tissue, the result of dissemination of *Coccidioides immitis* from the lungs.

Spherule   Spores

TEM   25 μm

▲ *Figure 22.5*

Spherules of *Coccidioides immitis.* Note the numerous spores within a spherule.

and emotional disturbance. Dissemination can also occur to the bones, joints, and subcutaneous tissues; subcutaneous lesions are inflamed masses of granular material **(Figure 22.4)**.

The diagnosis of coccidioidomycosis is based on the identification of spherules **(Figure 22.5)** in KOH- or GMS-treated samples collected from patients; diagnosis is confirmed by the isolation of the mycelial form of *C. immitis* in laboratory culture. Serological tests designed to detect fungal antigens or antibodies against the fungus in healthy individuals are useful, but may prove difficult to interpret in immunocompromised individuals.

Although infections in otherwise healthy patients generally resolve on their own without damage, amphotericin B is the drug of choice. CNS involvement is fatal without treatment. In AIDS patients, maintenance therapy with itraconazole or fluconazole is recommended to prevent relapse or reinfection. The wearing of protective masks in endemic areas can prevent exposure to spores, but this practice may be deemed impractical for daily use for all but those whose occupations put them at clear risk of infection.

Histoplasmosis

Histoplasma capsulatum (his-tō-plaz′mă kap-soo-lā′tŭm), the causative agent of **histoplasmosis** (his′tō-plaz-mō′sis), is an ascomycete and the most common fungal pathogen affecting humans. *H. capsulatum* is particularly prevalent in the eastern United States along the Ohio River Valley (see Figure 22.2c), but endemic areas also exist in Africa and South America. *H. capsulatum* is found in moist soils containing high levels of nitrogen such as from the droppings of bats and birds, especially chickens, starlings, and blackbirds.

Two strains of *H. capsulatum* are recognized: strain *capsulatum*, which is the most widely distributed, and strain *duboisii*, which is limited to Africa. Spores of both may become airborne and inhaled when soil containing the fungus is disturbed by wind or by human activities. Cutaneous inoculation can also lead to disease, but such infections are extremely rare.

CRITICAL THINKING

Statistically, men are more likely than women to contract histoplasmosis. What might explain this fact?

H. capsulatum is an intracellular parasite that, upon inhalation, first attacks alveolar macrophages in the lungs. These macrophages then disperse the fungus beyond the lungs via the blood and lymph. Cell-mediated immunity eventually develops, clearing the organism from healthy patients.

In almost 95% of individuals infected with *H. capsulatum*, pulmonary infections are asymptomatic and subclinical, and resolve without damage. Some individuals may experience mild or nonspecific respiratory symptoms (for example, aches and pains or mild coughing) that also disappear on their own. Slightly more pronounced symptoms—fever, night sweats, and weight loss—may occur in a few individuals. About 5% develop clinical histoplasmosis, which manifests as one of four diseases:

- *Chronic pulmonary histoplasmosis* is characterized by severe coughing, blood-tinged sputum, night sweats, loss of appetite, and weight loss. It is often seen in individuals with preexisting lung disease. As with blastomycosis

LM   50 μm

▲ *Figure 22.6*

The characteristic spiny spores of *Histoplasma capsulatum*.

and coccidiodomycosis, pulmonary histoplasmosis can be mistaken for tuberculosis.

- *Chronic cutaneous histoplasmosis*, characterized by ulcerative skin lesions, can follow dissemination from the lungs.

- *Systemic histoplasmosis* can also follow dissemination, but is usually seen only in severely immunocompromised individuals. This syndrome, characterized by enlargement of the spleen and liver, can be rapid, severe, and fatal.

- *Ocular histoplasmosis* is a type I hypersensitivity reaction against *Histoplasma* in the eye characterized by inflammation and redness.

Diagnosis of histoplasmosis is based on the identification of the distinctive budding yeast in KOH- or GMS-prepared samples of skin scrapings, sputum, cerebrospinal fluid, or various tissues; the diagnosis is confirmed by the observation of dimorphism in cultures grown from such samples. Cultured *H. capsulatum* produces distinctively spiny spores that are also diagnostic **(Figure 22.6)**. Antibody tests are not useful indicators of *Histoplasma* infection because many people have been exposed without contracting disease. In the endemic regions of the United States, close to 90% of the population tests positive for *H. capsulatum* exposure.

Infections in immunocompetent individuals typically resolve without treatment. When symptoms do not resolve, amphotericin B is prescribed. Ketoconazole can be used to treat mild infections. Maintenance therapy for AIDS patients is recommended.

Case Study 22.1 presents some of the problems that are encountered when trying to properly diagnose systemic mycoses.

Case Study 22.1 What's Ailing the Bird Enthusiast?

A 24-year-old man arrives at a hospital emergency room with signs and symptoms of serious pulmonary infection. He complains of a deep cough, blood-tinged sputum, night sweats, and weight loss. His liver and spleen feel enlarged upon physical examination. His chest X-ray is suggestive of tuberculosis, but a tuberculin skin test is negative, as is an HIV test. The man, an Oregon native, has recently traveled to Africa and Asia. He is also an avid bird enthusiast who likes to watch birds from his balcony. He keeps many bird feeders on the balcony, attracts numerous birds, and spends a fair amount of time cleaning up bird droppings. Skin tests are positive for histoplasmosis but inconclusive for coccidiodomycosis and blastomycosis.

1. What is the most likely infecting agent? How do you suppose this individual acquired the disease?

2. Is this disease unusual in people who work with birds?

3. Given that some tests were inconclusive, what other tests or lab work would aid in arriving at a specific diagnosis?

4. What treatment would most likely be prescribed?

Paracoccidioidomycosis

Another ascomycete, *Paracoccidioides brasiliensis* (par'ă-kok-sid-ē-oy'dēz bră-sil-ē-en'sis) causes **paracoccidioidomycosis** (par'ă-kok-sid-ē-oy'dō-mī-kō'sis), a chronic fungal disease similar to blastomycosis and coccidiodomycosis. *P. brasiliensis* is found in cool, damp soil from southern Mexico to regions of South America, particularly in Brazil (see Figure 22.2d). Because overall this fungus is far more geographically limited than the other true fungal pathogens, paracoccidioidomycosis is not a common disease. Those most at risk include farm workers.

Infections range from asymptomatic to systemic and follow an incubation period that can last for years. Disease first becomes apparent as a pulmonary form that is slow to develop but manifests as chronic cough, fever, night sweats, malaise, and weight loss. Dissemination almost always follows, producing a chronic inflammatory disease of mucous membranes. Painful ulcerated lesions of the gums, tongue, lips, and palate progressively worsen over the course of weeks to month. Cutaneous lesions spread to the surrounding areas of the mouth and nose **(Figure 22.7)** and can, in some cases, become widespread. Lesions are permanently destructive. Lymphatic involvement can occur in younger patients, and extensive dissemination throughout the lymphatic system leads to lesions in the intestines, liver, brain, and other tissues.

KOH or GMS preparations of tissue samples reveal yeast cells with multiple buds in a "steering wheel" formation that is diagnostic for this organism **(Figure 22.8** on page 632). Laboratory culture at 25°C and at 37°C demonstrating dimorphism is confirmatory. Serological identification of antibodies also is useful. Treatment is with amphotericin B or ketoconazole.

CRITICAL THINKING

All of the true fungal pathogens manifest initially as a pulmonary disease. Explain how you could ascertain which of the four pathogens a patient has.

▲ *Figure 22.7*

Ulcerative lesion of cutaneous paracoccidioidomycosis, resulting from the dissemination of *Paracoccidioides brasiliensis* from the lungs.

Systemic Mycoses Caused by Opportunistic Fungi

Learning Objective

✓ Discuss the difficulties in diagnosing opportunistic fungal infections.

Opportunistic mycoses do not typically affect healthy humans, because the fungi involved lack genes that code for factors that make them actively invasive. Instead, opportunistic mycoses are essentially limited to people with poorly functioning immune systems. Because of the growing number of immunocompromised individuals, opportunistic fungal infections have become one of the more significant causes of human disease and death. Even though technically any fungus can become an opportunist, five genera of fungi are considered "classical" opportunists because

LM | 5 μm

▲ *Figure 22.8*
The "steering wheel" formation of buds characteristic
of *Paracoccidioides brasiliensis*.

▲ *Figure 22.9*
A noninvasive aspergilloma in the eye. *In what group of
individuals is aspergillosis increasingly becoming a problem?*

*Figure 22.9 Aspergillosis is a growing problem among
AIDS patients.*

they are routinely encountered: *Aspergillus, Candida, Crypto-coccus, Pneumocystis,* and *Mucor*.

Opportunistic infections present a formidable challenge to clinicians. Because they appear only when their hosts are weakened, they often display "odd" clinical signatures; that is, their symptoms are often atypical, or they occur in individuals not residing in endemic areas for a particular fungus, and of course they cause more severe signs and symptoms than they would in a healthier person. A fungus that normally manifests as a subcutaneous infection, for example, presents radically different symptoms when growing in the lungs. Increased severity of symptoms is often a reflection of the poor immune status of the patient and frequently results in higher fatality rates, even if the fungus is normally nonthreatening.

In the following subsections we consider some of the mycoses caused by the "classical" fungal opportunists, beginning with aspergillosis.

Aspergillosis

Learning Objective

✓ Describe the clinical manifestations of aspergillosis.

Aspergillosis (as'per-ji-lō-sis) is not a single disease, but instead a term for several diseases caused by the inhalation of spores of fungi in the genus *Aspergillus* (as-per-jil'ŭs) in the division Ascomycota. Although hundreds of species of *Aspergillus* are found in soil, food, compost, agricultural buildings, and the air vents of homes and offices worldwide,

aspergillosis is caused principally by three species: *A. fumigatus, A. niger,* and *A. flavus*. Although exposure to *Aspergillus* most commonly causes only allergies, more serious diseases can occur, and aspergillosis is a growing problem for AIDS patients. Because the fungi are so common in the environment, little can be done to prevent exposure to the spores.

Even though *Aspergillus* species are capable of being opportunistic pathogens of almost all body tissues, these fungi are chiefly responsible for causing three clinical pulmonary diseases:

- *Hypersensitivity aspergillosis* manifests as asthma or other allergic symptoms and results most commonly from inhalation of *Aspergillus* spores. Symptoms may be mild and result in no damage, or they may become chronic, with recurrent episodes leading to permanent damage.

- *Noninvasive aspergillomas*—ball-like masses of fungal hyphae—can form in the cavities resulting from a previous case of pulmonary tuberculosis. Most cases are asymptomatic, though coughing of blood-tinged sputum may occur.

- *Acute invasive pulmonary aspergillosis* is more serious. Signs and symptoms, which include fever, cough, and pain, may present as mild pneumonia. Necrosis of lung tissue can lead to significant respiratory impairment.

Aspergillus also causes nonpulmonary disease when aspergillomas form in paranasal sinuses, the ear canals, eyelids **(Figure 22.9)**, the conjunctiva, the eye socket, or brain.

Figure 22.10 ➤
Four of the many manifestations of candidiasis.
(a) Oropharyngeal candidiasis, or thrush. **(b)** Diaper rash.
(c) Onychomycosis. **(d)** Ocular candidiasis.

(a)

Although it is rare, *cutaneous aspergillosis* results when the fungus is either introduced into the skin by trauma or disseminates from the lungs to the skin in immunocompromised individuals. Lesions begin as raised, red papules, but progressively become necrotic. *Systemic aspergillosis,* which involves invasion of the major organ systems, occurs primarily in AIDS patients, but it has also been seen in IV drug abusers. The fungus produces abscesses in the brain, kidneys, heart, bones, and gastrointestinal tract. Systemic aspergillosis is often fatal, especially when the brain is involved.

Clinical history and radiographs demonstrating abnormal lung structure are suggestive of aspergillosis, but diagnosis must be confirmed via laboratory techniques. The presence of septate hyphae and distinctive conidia in KOH- and GMS-prepared samples taken from the patient are diagnostic. Immunological tests to detect antigen in the blood are confirmatory.

(b)

Treatment of hypersensitivity involves either the use of various allergy medications or desensitization to the allergen. Invasive disease is treated by surgical removal of aspergillomas and surrounding tissues, plus high-dose, intravenous administration of amphotericin B. Systemic infections must be treated with amphotericin B in conjunction with other antifungal agents. Maintenance therapy with itraconazole is suggested for preventing relapse in immunocompromised patients.

Candidiasis

Learning Objective

✓ Explain how candidiasis can develop from a localized infection to a systemic mycosis.

(c)

Candidiasis (kan-di-dī′ă-sis) is any of a variety of opportunistic fungal infections and diseases caused by various species of the genus *Candida* (kan′did-ă)—most commonly *Candida albicans.* Fungi in this ascomycete genus are common members of the microbiota of the skin and mucous membranes; for example, the digestive tracts of 40–80% of all healthy individuals harbor *Candida* species. *Candida* is also one of the very few fungi that can be transmitted between individuals. From its site as a normal inhabitant of the female reproductive tract, for example, it can be passed to babies during childbirth and to adults during sexual contact.

Although *Candida* species can infect tissues in essentially every body system, producing a wide range of disease manifestations in humans **(Figure 22.10),** in all cases the fungus is an opportunist. Localized, superficial infections

(d)

Table 22.2   The Clinical Manifestations of Candidiasis

Type	Clinical Signs and Symptoms	Predisposing Factors
Oropharyngeal (thrush)	White plaques on the mucosa of the mouth, tongue, gums, palate, and/or pharynx	Diabetes, AIDS, cancers, various chemotherapeutic drug regimens
Cutaneous	Moist, macular red rash between skin folds	Moisture, heat, and friction of skin in skin folds and between digits, particularly in the obese
	Diaper rash: raised red rash, pustules in the gluteal region	Infrequently changed, soiled diapers
	Onychomycosis: painful red swelling around the nails, destruction of nail tissue, and loss of the nail	Immunocompromised individuals
Vulvovaginal	Creamy white, curdlike discharge, burning, redness, painful intercourse	Use of broad-spectrum antibiotics, pregnancy, diabetes, changes to the vaginal microbiota
Chronic mucocutaneous	Lesions associated with the skin, nails, and mucous membranes; severe forms may occur	Children with various metabolic problems relating to cell-mediated immunity (e.g., diabetes)
Neonatal and congenital	Meningitis, renal disorders, or generalized, body-wide, red, vesicular rash	Young age, low birth weight, use of antibiotics by the mother
Esophageal	White plaques along the esophagus, burning pain, nausea, and vomiting	AIDS, immunocompromised patients
Gastrointestinal	Ulceration of the stomach and intestinal mucosa	Hematological cancers
Pulmonary	Generalized, nonspecific symptoms; often remains undiagnosed until autopsy	Dissemination from other types of candidiasis
Peritoneal	Fever, pain, tenderness	Indwelling catheters for dialysis or gastrointestinal perforation
Urinary tract	Urinary tract infection: painful urination, possibly discharge	Chemotherapeutic drug regimens, catheterization, diabetes, preexisting bladder problems
	Renal: fever, pain, rigors, fungus ball	
Meningeal	Swelling of the meninges, fever, headache, stiffness of the neck	Dissemination of *Candida* during systemic infection
Hepatic and splenic	Fever, swelling of the liver and spleen, liver dysfunction, lesions and abscesses	Leukemia
Endocardial, myocardial, pericardial	Fever, heart murmur, congestive heart failure, anemia	Preexisting heart valve disease, catheterization plus use of antibiotics, IV drug abuse, valve prosthetics
Ocular	Cloudy vision, lesions within the eye	Disseminated candidiasis, indwelling catheters, IV drug abuse, trauma
Osteoarticular	Pain when weight is placed on joint	Disseminated candidiasis, prosthetic implants
Candidemic	Antibiotic-resistant fevers, tachycardia, hypotension, skin lesions, small abscesses in various organ systems	IV or urinary catheters, use of antibacterial drugs, surgery, severe burns, antibacterial therapy

are seen in individuals with impairment of the barrier function of the epithelial layers (generally newborns and the elderly), or changes in the normal vaginal environment in women. Systemic disease is seen in immunocompromised individuals only, especially AIDS patients. Candidiasis can be acute, subacute, chronic, or episodic, and different segments of the population experience a given manifestation of candidiasis at different rates. For example, whereas oral candidiasis (thrush) is rare in healthy individuals and occurs in only about 5% of newborns, it affects 10% of the elderly and nearly all those infected with HIV once they develop clinical AIDS. Predisposing factors for the development of candidiasis include cancer, invasive hospital proce-

dures, antibacterial treatments, diabetes, severe burns, IV drug abuse, and severe immunosupression. **Table 22.2** summarizes the wide variety of clinical manifestations of candidiasis.

KOH and GMS preparations of skin samples, nail scrapings, cerebrospinal fluid, serum, urine, sputum, or tissue biopsies can reveal clusters of budding yeast and branching pseudohyphae, which are diagnostic when correlated with symptoms.

In immunocompetent patients, for whom excessive friction and body moisture in skin folds or preexisting diseases are the reasons for fungal colonization, resolution involves treating these underlying problems in addition to administering topical antifungal agents. Oral candidiasis in infants and children are treated with nystatin. Azole suppositories, creams, and/or oral administration of fluconazole are used for vaginal candidiasis.

The same infections in immunocompromised individuals are much more difficult to treat because most superficial or mucous membrane infections in these patients do not respond well to topical antifungal treatments. Orally administered fluconazole is used for primary and maintenance therapy for candidiasis in AIDS patients. Invasive candidiasis requires treatment with amphotericin B, and often with 5-fluorocytosine; fluconazole may be substituted for either drug. Prophylactic treatment with fluconazole of AIDS patients, in addition to patients undergoing induced immunosuppressive regimens, is recommended.

CRITICAL THINKING

Even though *Candida* species are not as virulent as some microbial pathogens, the fungus can still invade every human tissue. Propose a possible explanation for this observation.

Cryptococcosis

Learning Objective

✓ Discuss the characteristics of *Cryptococcus* that contribute to the severity of cryptococcoses in immunocompromised patients.

A basidomycete, *Cryptococcus neoformans* (krip-tō-kok'ŭs nē-ō-for'manz), is the main etiological agent of **cryptococcoses** (krip'tō-kok-ō'sēz). Two variants are known. Whereas *C. neoformans neoformans* is found worldwide and predominantly infects immunocompromised hosts, *C. neoformans gattii* is found in Australia, Papua New Guinea, parts of Africa, the Mediterranean, India, southeast Asia, parts of Central and South America, and southern California. It primarily infects immunocompetent individuals. Approximately 50% of all cryptococcal infections reported each year are due to variant *neoformans*.

Human infections result from the inhalation of spores and/or dried yeast forms made airborne when the droppings of pigeons and chickens containing the fungus are disturbed by the activities of the birds or humans. People who work around buildings where birds roost are at increased risk of infection. Because the fungus is such a threat to sick individuals, hospitals, convalescent homes, and other such facilities often place devices that deter the roosting of birds near outside air-intake vents in an effort to prevent *Cryptococcus*-contaminated air from entering the building.

The pathogenesis of *Cryptococcus* is enhanced by several characteristics of this fungus: the presence of a phagocytosis-resistant capsule surrounding the yeast forms; the ability of the yeast forms to produce melanin, which appears to further inhibit phagocytic killing mechanisms; and the organism's predilection for the central nervous system, which is isolated from the immune system by the so-called blood-brain barrier. Cryptococcal infections also tend to appear in terminal AIDS patients when little immune function remains.

Primary pulmonary cryptococcosis is generally asymptomatic in most individuals. Some individuals experience a mild, self-limiting pneumonia consisting of a low-grade fever, cough, and mild chest pain. In a few cases, *invasive pulmonary cryptococcosis* occurs, resulting in chronic pneumonia in which cough and pulmonary lesions progressively worsen over a period of year.

Cryptococcal meningitis, the most common clinical form of cryptococcal infection, follows dissemination of the fungus to the CNS. Symptoms are slow to develop and include headache, dizziness, drowsiness, irritability, confusion, nausea, vomiting, and neck stiffness. In late stages of the disease, loss of vision and coma occur. Acute onset of rapidly fatal cryptococcal meningitis occurs in individuals with widespread disease.

Cryptococcoma (krip'tō-kok-ō'mǎ) is a very rare condition in which solid fungal masses form in the cerebral hemispheres, cerebellum, or (rarely) in the spinal cord. The symptoms of this condition, which can be mistaken for cerebral tumors, are similar to those of cryptococcal meningitis but also include motor and neurological impairment.

Cutaneous cryptococcosis may also develop. Primary infections manifest as ulcerated skin lesions or as inflammation of subcutaneous tissues. Infection may resolve on its own, but patients should be monitored for CNS dissemination. Secondary cutaneous lesions occur following dissemination of *Cryptococcus* to other areas of the body. In AIDS patients, cutaneous cryptococcosis is the second most common manifestation of *Cryptococcus* infection (after meningitis). Lesions are common on the head and neck.

Diagnosis involves collecting specimen samples that correlate with the symptoms. These may include cerebral spinal fluid (CSF), tissue biopsy, or bronchial washings. GMS stains revealing the presence of encapsulated yeast in CSF are highly suggestive of cryptococcal meningitis, even if no overt symptoms are present. Recovery of *Cryptococcus* from respiratory secretions is generally not diagnostic due to general environmental exposure, though any patient testing positive for *C. neoformans* should be monitored for disseminated disease. The preferred method of confirming cryptococcal meningitis is detection of fungal antigen in CSF. In AIDS patients, antigen can be detected in serum as well.

Treatment for cryptococcoses is with amphotericin B and 5-fluorocytosine administered together for 6–10 weeks. The synergistic action of the two drugs allows lower doses of amphotericin B to be used, though toxicity is not completely eliminated. In AIDS patients, primary treatment must be followed by life-long maintenance therapy to prevent relapses. Though AIDS patients may appear well following primary treatment, the fungus typically remains and must be actively suppressed by oral fluconazole.

Pneumocystis Pneumonia

Learning Objective

✓ List the characteristics of *Pneumocystis* that distinguish it from other fungal opportunists.

Pneumocystis jiroveci (noo-mō-sis'tis yē-rō-vĕt'zē), an ascomycete formerly known as *P. carinii,* is one of the more common opportunistic fungal infections seen in AIDS patients. Prior to the AIDS epidemic, disease caused by *Pneumocystis,* which is commonly found as a normal member of the human respiratory microbiota, was extremely rare. *P. jiroveci* is distributed worldwide in humans; based on serological confirmation of antibodies, the majority of healthy children (75%) have been exposed to the fungus by the age of 5.

Even though the organism was first identified in 1909, little is known about it. Originally considered a protozoan, it has been reclassified as a fungus based on rRNA nucleotide sequences and biochemistry. Its morphological and developmental characteristics, however, still resemble those of protozoa more than those of fungi. Because *P. jiroveci* is an obligate parasite and cannot survive on its own in the environment, transmission most likely occurs through inhalation of droplet nuclei containing the fungus.

In immunocompetent people, infection is asymptomatic, and generally clearance of the fungus from the body is followed by lasting immunity. However, some individuals may remain infected indefinitely; in such carriers the organism remains in the alveoli and can be passed in sputum and presumably in respiratory droplets.

Before the AIDS epidemic, **Pneumocystis** **pneumonia** was observed only in malnourished, premature infants and debilitated elderly patients, now the disease is almost diagnostic for AIDS. *Pneumocystis* pneumonia is abbreviated **PCP,** which originally stood for *Pneumocystis carinii* pneumonia but now can be considered to stand for *Pneumocystis* **p**neumonia.

Once the fungus enters the lungs of an AIDS patient, it multiplies rapidly because the patient's defenses are impaired, extensively colonizing the lungs. Widespread inflammation results in fever, difficulty in breathing, and a nonproductive cough. Mild anemia and hypoxia (low tissue oxygen) occurs in some cases. Rarely, extrapulmonary lesions develop in the lymph nodes, spleen, liver, and bone marrow. If left untreated, PCP involves more and more lung tissue until death occurs.

Diagnosis relies on clinical and microscopic findings. Chest X-rays usually reveal abnormal lung features. GMS or other stains of fluid from the lungs or from biopsies can reveal distinctive morphological forms of the fungus **(Figure 22.11).** The use of fluorescent antibody on samples taken from patients is more sensitive and provides a more specific diagnosis. *Pneumocystis* has not been cultured in the laboratory.

LM   10 µm

▲ *Figure 22.11*

Cysts of *pneumocystis jiroveci,* in lung tissue. Such microscopic findings are diagnostic.

Due to the organism's similarities to protozoa, both primary treatment and maintenance therapy are with the antiprotozoan drugs trimethoprim and sulfanilamide. Pentamidine, another antiprotozoan drug, is an alternative. The drugs may be aerosolized to allow for direct inhalation into the lungs.

CRITICAL THINKING

Compare and contrast the clinical features of cryptococcosis and PCP. Why are both of these diseases so dangerous to AIDS patients?

Zygomycoses

Learning Objective

✓ Describe the clinical forms of zygomycoses.

Zygomycoses (zī'gō-mī-kō'sez) are opportunistic fungal infections caused by various genera of fungi classified in the division Zygomycota, especially *Mucor,* and sometimes *Rhizopus* (rī-zō'pŭs), and *Absidia* (ab-sid'ē-ă), all of which have a worldwide distribution and are extremely common in the environment. They can be found in soil, on decaying organic matter, or as contaminants that cause food spoilage (*Rhizopus* is the classic bread mold).

Zygomycoses are commonly seen in patients with uncontrolled diabetes, in people who inject illegal drugs, in some cancer patients, and in some patients receiving antimicrobial agents. Infections generally develop in the face and head area but can also develop elsewhere, including the lungs, gastrointestinal tract, or the skin. Any of these infections may disseminate throughout the body in severely immunocompromised individuals and result in the following conditions:

- *Rhinocerebral zygomycosis* begins with infection of the paranasal sinuses following inhalation of spores. The fungus spreads to the roof of the mouth and to the nose, producing macroscopic cotton-like growths. Deeper invasion enables the fungus to invade the blood vessels, where it produces fibrous clots and causes tissue necrosis. The ensuing invasion of the brain is fatal within days, even with treatment. IV drug use can also lead to infection of the central nervous system that results in brain abscesses.

- *Pulmonary zygomycosis* follows deep inhalation of spores (as from moldy foods) into the bronchioles and alveoli of the lungs. Necrosis leads to the formation of cavities in the lungs.

- *Gastrointestinal zygomycosis* produces necrotic ulcers in the intestinal tract following the ingestion of fungal elements.

- *Cutaneous zygomycosis* results from the introduction of fungal elements through the skin after trauma (such as burns or needles punctures). Lesions range from pustules and ulcers to abscesses and necrotic patches of skin.

Diagnosis is usually made clinically based on patient history and microscopic findings of fungus. Samples may be obtained from scrapings of lesions, lung aspirates, or biopsies. KOH mounts and direct microscopy of GMS-stained tissue sections reveal hyphae with irregular branching and few septate divisions. For example, diagnosis of rhinocerebral zygomycosis is made through clinical observation of fungus in the mouth or nose and confirmed by using laboratory techniques.

Treatment, which should begin as soon as infection is suspected, involves physical removal of infected tissues and management of predisposing factors. Amphotericin B administered intravenously for 8–10 weeks is the drug of choice for all zygomycoses.

The Emergence of Fungal Opportunists in Immunosuppressed Individuals

Learning Objectives

✓ Identify several emerging fungal opportunists seen among AIDS patients.

✓ Identify the factors that contribute to the emergence of new fungal opportunists.

Immunosuppression is not unique to AIDS patients, but because HIV systematically destroys functional immunity, AIDS patients are extremely vulnerable to opportunistic infections. In otherwise healthy people, immunosuppressive episodes are transitory and result from surgery, chemotherapy, or an acute illness. With AIDS, by contrast, immune dysfunction is permanent. For this reason, once an opportunistic infection is established in an AIDS patient, it is not likely to ever be fully cured. In fact, mycoses account for most deaths associated with AIDS. *Candida albicans, Aspergillus fumigatus,* and *Cryptococcus neoformans* are so common in HIV-positive individuals that their mycoses are part of what defines end-stage AIDS.

The emergence of new fungal opportunists results from a combination of factors. First, the number of immunocompromised individuals is increasing as a result of the AIDS epidemic. Then, the widespread use of antifungal drugs in this enlarging group of immunocompromised individuals selects fungi resistant to the drugs. Thus as the immunocompromised population grows, fungi will become increasingly responsible for illness and death. An AIDS patient's lack of immunity allows a normally superficial fungus to gain access to internal systems, with significant, even fatal, results. Over the course of the past decade, several new fungal opportunists have been identified, including species of non-*albicans Candida, Fusarium* spp., and *Penicillium marneffei,* which are ascomycetes, and *Trichosporon beigelii,* which is a basidiomycete.

Fusarium (fū-zā'rē-ŭm) species cause respiratory distress, disseminated infections, and *fungemia* (fungi in bloodstream). These fungi produce toxins that can accumulate to dangerous levels when ingested in food. *Fusarium* spp. are resistant to all antifungal agents, including amphotericin B.

Penicillium marneffei (pen-i-sil'ē-ŭm mar-nef-ē'ī) is a dimorphic, invasive fungus that causes pulmonary disease upon inhalation. It is the third most frequent illness seen in AIDS patients in southeast Asia (behind tuberculosis and cryptococcis).

Trichosporon beigelii (trik-ō-spōr'on be-gĕl-ē'ī) can cause a superficial, easily treated disease of the hair. However, when *T. beigelii* enters an AIDS patient through the lungs, the gastrointestinal tract, or catheters, it causes a disseminated and drug-resistant systemic disease that is typically fatal.

Among the approaches to combating the threat posed by these opportunists are redoubling efforts to maintain rigorous hygiene during medical procedures and the development of new therapies that limit immunosuppression in all patients, including those with AIDS.

Superficial, Cutaneous, and Subcutaneous Mycoses

Learning Objective

✓ Describe the general manifestations of superficial, cutaneous, and subcutaneous mycoses.

All the mycoses we discuss in this section are localized at sites at or near the surface of the body, and they are the most commonly reported fungal diseases. All are opportunistic infections, but unlike those we have just discussed, they can be acquired by healthy individuals via person-to-person contact or through environmental exposure. Most of these fungi are not life threatening, but they often cause chronic, recurring infections and diseases.

Superficial Mycoses

Learning Objective

✓ Compare and contrast the clinical and diagnostic features of dermatophytosis and *Malassezia* infections.

Superficial mycoses are usually acquired by direct contact with the fungus and are the most common fungal infections. They are confined to the outer, dead layers of the skin, nails, or hair, all of which are composed of dead cells filled with a protein called *keratin*. In severely immunocompromised patients, superficial infections can spread to cover significant areas of skin or become systemic.

Black Piedra and White Piedra

Piedra[2] (pē-ā′dră) is a superficial infection that forms nodules on hair shafts. Infections are normally associated with the scalp, but facial and pubic hair can also be infected. Transmission is usually mediated via shared hair brushes and combs, and several members of a family are typically infected at the same time.

Piedraia hortae, an ascomycete which causes **black piedra,** is distributed throughout Central and South America and southeast Asia. Hard, black nodules encircling the hair shafts characterize this disease **(Figure 22.12a). White piedra,** which is found worldwide in the tropics and subtropics, is caused by *Trichosporon beigelii,* which produces soft, gray to white nodules on the hair shafts **(Figure 22.12b).** In both diseases, the clinical appearance of the hair is diagnostic for the disease.

Treatment for both types of piedra typically involves shaving the infected area to remove the fungi with the hair; one shave is usually sufficient. To treat difficult-to-cure cases or women in cultures where cutting the hair is socially devastating, physicians prescribe antifungal agents such as terbinafine.

Dermatophytoses

Dermatophytoses (der′mă-tō-fī-tō′sēz) are infections caused by *dermatophytes,* which are fungi that grow only on skin, nails, and hair. In the past, such infections were called *ringworms,* because dermatophytes produce circular, scaly patches that resemble a worm lying just below the surface of the skin. Even though we know worms are not involved, the medical names for many dermatophytoses use the term *tinea* (tin′ē-a), which is Latin for worm. Dermatophytes use keratin as a nutrient source and thus colonize only dead layers of skin, nails, and hair. No living tissues are infected, but the fungi can provoke cell-mediated immune responses, which can damage living tissue.

2Spanish, meaning a stone.

(a) LM 100 μm

(b) LM 80 μm

▲ *Figure 22.12*

Two forms of piedra, a superficial fungal infection of the hair. **(a)** The hard nodules of black piedra, caused by *Piedraia hortae.* **(b)** The looser, light-colored nodules of white piedra, caused by *Trichosporon beigelii.*

Based on their natural habitats, dermatophytes infecting humans are classified as follows:

- *Anthropophilic* dermatophytes are associated with humans only and are transmitted either by close human contact or through the sharing of contaminated fomites.

- *Zoophilic* dermatophytes are associated with animals; the fungi are transmitted to humans either by close contact with animals such as a pet or through contaminated animal products such as wool.

- *Geophilic* dermatophytes are soil fungi that are transmitted to humans via direct exposure to soil or to dusty animals.

Three genera of ascomycetes are responsible for most dermatophytoses: (1) *Trichophyton* (tri-kof′i-ton) species, (2) *Microsporum* (mī-kros′po-rŭm) species, and (3) *Epidermophyton floccosum*, (ep′i-der-mof′i-ton flŏk′ŏ-sŭm), which we discussed in the chapter-opening vignette. All three cause skin and nail infections, but *Trichophyton* species also infect hair. Rare subcutaneous infections also occur in AIDS patients. Twenty-one species of *Trichophyton* infect humans; *Trichophyton rubrum* is the most common dermatophyte in the United States. Worldwide, *Microsporum canis* is the most common dermatophyte seen in humans.

Dermatophytes are among the few infectious fungi. Spores and bits of hyphae along with dead skin cells and hair, are constantly shed from infected individuals, making recurrent infections common.

Dermatophytoses can have a variety of clinical manifestations, some of which are summarized in **Table 22.3** on page 640. Most such diseases are clinically distinctive and so common that they are readily recognized.

Clinical observation is generally sufficient to diagnose infections with dermatophytes. KOH preparations of skin or nail scrapings or hair samples can reveal hyphae and/or arthroconidia, which confirm the diagnosis. When desired, the determination of the specific identity of a dermatophyte requires laboratory culture, which may take weeks because the fungi grow slowly.

Limited infections can be treated effectively with topical antifungal agents, but more widespread infections of the scalp or skin, as well as nail infections, must be treated with oral agents. Terbinafine, administered for 6–12 weeks, is effective in most cases. Chronic or stubborn cases are treated with griseofulvin until cured. **Case Study 22.2** on page 641 explores some of the difficulties in dealing with dermatophyte infections.

Malassezia *Infections*

Malassezia furfur (mal-ă-sā′zē-ă fur′fur) is basidiomycete that is a normal inhabitant of the skin surfaces of humans worldwide. It feeds on oil produced by the skin and causes various common superficial infections that tend to be chronic.

Chronic infections with *M. furfur* can produce the condition known as **pityriasis** (pit-i-rī′ă-sis), characterized by depigmented or hyperpigmented patches of scaly skin resulting from fungal interference with melanin production **(Figure 22.13)**. This condition, which typically occurs on the trunk, shoulders, and arms, and rarely on the face and neck, is also called *tinea versicolor* in reference to the variable pigmentation seen in affected individuals. The skin takes months to regain its normal pigmentation following successful treatment of pityriasis.

M. furfur causes several other diseases. *Folliculitis* (fŏ-lik-ŭ-lī′tis) involves infection of the hair follicles on the back, chest, and upper arms, producing follicular papules or pustules. *Seborrheic dermatitis* (seb-ō-rē′ik der-mă-tī′tis) and **dandruff** primarily affect the scalp, face, eyebrows, ears, and upper trunk. Red, scaly, itchy patches form, and the skin

▲ *Figure 22.13*

The variably pigmented skin patches of pityriasis, caused by *Malassezia furfur*. *What other, more common condition is caused by M. furfur?*

Figure 22.13 Dandruff is a very common condition caused by M. furfur infection.

begins to flake off continuously. Small lesions may develop in the lungs or other organs as a result of fungal dissemination. *M. furfur* rarely causes fungemia when introduced into the bloodstream via catheters used for liquid replacement therapies in newborns and adults.

Because patches of skin infected with *M. furfur* fluoresce pale green when exposed to ultraviolet light with a wavelength of 365 nm, the quickest way to diagnose *M. furfur* infections is to examine the patient under ultraviolet illumination **(Figure 22.14** on page 640), though some other fungi also floresce. KOH preparations of skin samples reveal masses of budding yeast and short hyphal forms that are also diagnostic for *M. furfur*. Laboratory cultures started from blood or a contaminating catheter are required to diagnose cases of fungemia.

Superficial *Malassezia* infections are treated topically with solutions of antifungal imidazoles, such as ketoconazole shampoos. Alternatively, topical applications of zinc pyrithione, selenium sulfide lotions, or propylene glycol solutions can be used. Oral therapy with ketoconazole may

Table 22.3 Common Dermatophytoses

Disease	Agents	Common Signs	Source
Tinea pedis ("athlete's foot")	*Trichophyton rubrum;* *T. mentagrophytes* var. *interdigitale;* *Epidermophyton floccosum*	Red, raised lesions on and around the toes and soles of the feet; webbing between the toes is heavily infected	Human reservoirs in toe-webbing; carpeting holding infected skin cells
Tinea cruris ("jock itch")	*T. rubrum; T. mentagrophytes* var. *interdigitale; E. floccosum*	Red, raised lesions on and around the groin and buttocks	Usually spreads from the feet
Tinea unguium (onychomycosis)	*T. rubrum; T. mentagrophytes* var. *interdigitale*	*Superficial white onychomycosis:* patches or pits on the nail surface; *Invasive onychomycosis:* yellowing and thickening of the distal nail plate, often leading to loss of the nail	Humans
Tinea corporis	*T. rubrum; Microsporum gypseum; M. canis*	Red, raised, ringlike lesions occurring on various skin surfaces (tinea corporis on the trunk, tinea capitis on the scalp, tinea barbae of the beard)	Can spread from other body sites; can be acquired following contact with contaminated soil or animals
Tinea capitis	*M. canis; M. gypseum; T. equinum; T. verrucosum; T. tonsurans; T. violaceum; T. schoenleinii*	*Ectothrix invasion:* fungus develops arthroconidia on the outside of the hair shafts, destroying the cuticle; *Entothrix invasion:* fungus develops arthroconidia inside the hair shaft without destruction; *Favus:* crusts or scutula form on the scalp, with associated hair loss	Humans; can be acquired following contact with contaminated soil or animals

Fluorescent infected hair — Uninfected hair

LM ⊢──────┤ 5 mm

▲ *Figure 22.14*

An example of the use of ultraviolet illumination for diagnosing infections with *Malassezia furfur*.

be required to treat extensive infections or infections that do not respond to topical treatments. Relapses are common, and prophylactic topical treatment may be necessary.

CRITICAL THINKING

Explain why many superficial fungal infections are chronic, recurring problems.

Cutaneous and Subcutaneous Mycoses

Learning Objectives

✓ Explain why cutaneous and subcutaneous mycoses are not as common as superficial mycoses.
✓ State the specific difference between chromoblastomycosis and phaeohyphomycosis.
✓ Describe how the lesions of lymphocutaneous sporotrichosis correlate with its spread through the lymphatic system.

The fungi involved in cutaneous and subcutaneous mycoses are common soil saprobes, but the diseases they produce are not as common as superficial mycoses because infection requires traumatic introduction of fungal elements beneath the outer, dead layers of skin. Thus, exposure is common, but infection is rare. Most lesions remain localized to the subepidermal tissues in the skin, though infections may rarely become systemic.

Chromoblastomycosis and Phaeohyphomycosis

Chromoblastomycosis (krō′mō-blas′tō-mī-kō′sis) and **phaeohyphomycosis** (fē′ō-hī′fō-mī-kō′sis) are similar cutaneous and subcutaneous mycoses caused by *dematiaceous* (dē-mat-ē-ā′shŭs) (dark-pigmented) fungi. Four ascomycetes cause chromoblastomycosis: *Fonsecaea pedrosoi* (fon-sē-sē′ă pe-drō′sō-ē), *F. compacta* (kom-pak′-ta), *Phialophora verrucosa* (fī-ă-lof′ŏ-ră ver-ū-kō′să), and *Cladophialophora carrionii* (klă-dŏf′ē-ă-lŏf-ō-ră kar-rē-ŏn′ē-ī). Of the over 30 genera of fungi that cause phaeohyphomycosis, the most common are the ascomycetes *Alternaria* (al-ter-nā′rē-ă), *Exophiala* (ek-sō-fī′ă-lă), *Wangiella* (wang-gē-el′ă), and *Cladophiolaphora*.

Initially, chromoblastomycosis uniformly presents as small, itchy but painless, scaly lesions on the skin surface

Case Study 22.2   Is it Athlete's Foot?

A 30-year-old man comes into a community clinic complaining of persistent redness, itching, and peeling skin on the soles of his feet and between his toes. He states that he works out daily and plays on several sports teams. He concluded he has athlete's foot, but every over-the-counter antifungal agent he has tried has failed to cure his condition. He has finally come in because his toenails have begun to thicken, yellow, and detach from the nail bed. His condition is pictured.

1. Given the information here, can athlete's foot be diagnosed definitively? If not, what laboratory tests should be performed to identify the infecting organism?

2. What treatment is likely? Would the recommended treatment be different if two fungal species, and not just one fungal species, were present?

resulting from fungal growth in subcutaneous tissues near the site of inoculation. Over the course of several years, the lesions progressively worsen becoming large, flat to thick, tough, and wartlike. They become tumorlike and extensive if not treated **(Figure 22.15).** Inflammation, fibrosis, and abscess formation occur in surrounding tissues. The fungus can spread throughout the body.

Phaeohyphomycoses are more variable in presentation. *Paranasal sinus phaeohyphomycosis* involves colonization of the nasal passages and sinuses; it occurs in allergy sufferers and AIDS patients. *Cerebral phaeohyphomycosis* is the most serious of these diseases and is caused by *Cladophiolophora bantiana* (ban-tē-ā'nǎ), which has a preference for neural cells and actively invades the brain. Fortunately, this is the rarest form of phaeohyphomycosis and only occurs in the severely immunocompromised.

KOH preparations and GMS staining of skin scrapings, biopsy material, or cerebral spinal fluid (CSF) reveals the key distinguishing feature between the two diseases: the microscopic morphology of the fungal cells within tissues. Tissue sections from chromoblastomycosis cases contain golden brown *sclerotic bodies* that are distinctive and distinguishable from budding yeast forms **(Figure 22.16** on page 642), whereas tissues from phaeohyphomycosis cases contain brown-pigmented hyphae. Determining the species of fungus causing these two diseases relies on laboratory culture, macroscopic examination of colonies on agar, and microscopic examination of spores.

Both diseases are difficult to treat, especially in advanced cases. Some cases of phaeohyphomycosis can be treated with itraconazole, but the disease is permanently destructive to tissues. Chromoblastomycosis requires surgical removal of infected and surrounding tissues followed by antifungal therapy. Extensive lesions may require amputation. Thiabendazole and 5-fluorocytosine, given for 3–12 months, have been effective in some cases, but often the disease cannot be cured. The earlier treatment begins, the more likely it will be successful.

Despite the worldwide occurrence of dematiaceous fungi, the overall incidence of infection is relatively low.

▲ *Figure 22.15*

A leg with extensive lesions of chromoblastomycosis, in this case caused by *Fonsecaea pedrosi.*

People who work daily in the soil with bare feet are at risk due to accidental foot wounds. The simple act of wearing shoes would greatly reduce the number of infections. Phaeohyphomycosis is acquired when spores, which are prevalent within building environments, including hospitals, enter open surgical or traumatic wounds.

Mycetomas

Fungal **mycetomas** (mī-sē-tō'mǎs) are tumorlike infections of the skin, fascia (lining of muscles), and bones of the hands or feet caused by mycelial fungi of several genera in the division Ascomycota: *Madurella* (mad'ū-rel'ǎ), *Pseudallescheria* (sood'al-es-kē'rē-ǎ), *Exophiala* (ek-sō-fī'ǎ-lǎ), and *Acremonium* (ak-rě-mō'nē-ŭm). These fungi are distributed worldwide,

(a)

LM | 50 μm

(b)

LM | 10 μm

▲ *Figure 22.16*

Microscopic differences between chromoblastomycosis and phaeohyphomycosis. **(a)** Sclerotic bodies seen in chromoblastomycoses, here caused by *Fonsecaea pedrosi*. **(b)** Hyphal forms found in tissues during phaeohyphomycosis, here caused by *Cladophiolophora bantiana*.

▲ *Figure 22.17*

A mycetoma of the ankle resulting from the invasion and destruction of bone and other tissues by a fungus (in this case, *Madurella mycetomatis*).

A combination of the symptoms and microscopic demonstration of fungi in samples from the infected area is diagnostic for mycetomas. Laboratory culture of specimens produces macroscopic colonies that can be identified to species.

Treatment involves surgical removal of the mycetoma that may, in severe cases, involve amputation of an infected limb. Surgery is followed by 1–3 years of antifungal therapy with ketoconazole. Though slow, the infection is invasive; thus even combinations of surgery and antifungal agents do not always cure the disease.

Sporotrichosis

Sporothrix schenckii (spōr′ō-thriks shen′kē-ī) is a dimorphic ascomycete that causes **sporotrichosis** (spōr′ō-tri-kō′sis), or *rose-gardener's disease,* a subcutaneous infection usually limited to the arms and legs. *S. schenckii* resides in the soil and is most commonly introduced by thorn pricks or wood splinters. Avid gardeners, farmers, and artisans who work with natural plant materials have the highest incidence of sporotrichosis. Though distributed throughout the tropics and subtropics, most cases occur in Latin America, Mexico, and Africa. The disease is also common in warm, moist areas of the United States.

Fixed cutaneous sporotrichosis initially appears as painless, nodular lesions that form around the site of inoculation. With time, these lesions produce a pus-filled discharge, but they remain localized and do not spread. In *lymphocutaneous sporotrichosis,* however, the fungus enters the lymphatic system near the site of a primary lesion, giving rise to secondary lesions on the skin surface along the course of lymphatic vessels **(Figure 22.18).** The fungus remains restricted to the subcutaneous tissues and does not enter the blood.

but infection is most prevalent in countries near the equator. The cases that occur in the United States are almost always caused by *Pseudallescheria* or *Exophiala*.

These fungi live in the soil and are introduced into humans via the prick wounds and scrapes caused by twigs, thorns, or leaves contaminated with fungi. As with chromoblastomycosis and phaeohyphomycosis, those who work barefoot in soil are most at risk, and wearing protective shoes or clothing can greatly reduce infection.

Infection begins near the site of inoculation with the formation of small, hard, subsurface nodules that slowly worsen and spread as time passes. Local swelling occurs, and ulcerated lesions begin to produce pus. Infected areas release an oily fluid containing fungal "granules" (spores and fungal elements). The fungi spread to more tissues, destroying bone and causing permanent deformity **(Figure 22.17).**

Microscopic observation of pus or biopsy tissue stained with GMS can reveal budding yeast forms in severe infections, but often the fungus is present at a low density, making direct examination of clinical samples a difficult method of diagnosis. The patient's history and clinical signs, plus the observation of the dimorphic nature of *S. schenckii* in laboratory culture, are considered diagnostic for sporotrichosis.

Cutaneous lesions can be treated successfully with topical applications of saturated potassium iodide for several months. Itraconazole and terbinafine are also useful treatments. Prevention requires the wearing of gloves, long clothing, and shoes to prevent inoculation.

CRITICAL THINKING

Make a dichotomous key to distinguish among all the fungal mycoses covered in the previous sections.

Fungal Intoxications and Allergies

Learning Objective

✓ Compare and contrast mycotoxicosis, mycetismus, and fungal allergies.

Some, but not all, fungi produce toxins or cause allergies. Fungal toxins are low-molecular-weight metabolites that can harm humans and animals that ingest them, causing *toxicosis* (poisoning), of which there are two kinds: **Mycotoxicosis** (mī′kō-tok-si-kō-sis) is caused by eating food contaminated with fungal toxins (the fungus itself is not present); whereas **mycetismus** (mī′sē-tiz′mŭs) is mushroom poisoning resulting from eating the fungus. Fungal *allergens* are products of a fungus (usually proteins or glycoproteins) that elicit hypersensitivity reactions in sensitive people who inhale, ingest, or otherwise contact them.

Mycotoxicoses

Learning Objective

✓ Define mycotoxin, and identify the most common group of mycotoxins.

During their normal metabolic activities, fungi produce chemicals called **mycotoxins** (mī′kō-tok-sinz) that are poisonous to animals and humans. People most commonly consume mycotoxins in food crops (grains or vegetables) that have become contaminated with fungal growth while the crop was growing or after it was harvested. Some toxins can also be ingested in milk from a cow that has ingested toxin-contaminated feed. Up to 25% of the world's food supply is contaminated with mycotoxins, but only 20 of the 300 or so known toxins are ever present at dangerous levels. Long-term ingestion of mycotoxins can cause liver and kidney damage, gastrointestinal or gynecological distur-

▲ *Figure 22.18*

Lymphocutaneous sporotrichosis on the leg. The locations of these secondary, subcutaneous lesions correspond to the course of lymphatic vessels leading from sites of primary, surface lesions. *How is sporotrichosis contracted?*

Figure 22.18 Sporotrichosis is contracted when this soil-borne fungus is inoculated into the skin by thorn pricks and scratches.

bances, or cancers; each toxin produces a specific clinical manifestation.

Aflatoxins (af′lă-tok′sinz), produced by *Aspergillus flavus, A. parasiticus,* and *A. nomius,* are the most well-known mycotoxins. Aflatoxins are fatal to many vertebrates and are carcinogenic at low levels when consumed continually. Aflatoxins cause liver damage and liver cancer throughout the world, but *aflatoxicosis* is most prevalent in the tropics, where mycotoxins are more common because of subsistence farming, poor food-storage conditions, and warm, moist conditions that foster the growth of fungi in harvested foods.

Some mycotoxins are considered useful. Among them are ergot alkaloids, produced by some strains of an ascomycete, *Claviceps purpurea* (klav′i-seps poor-poo′rē-ă), which are used in drugs. For example, *ergometrine* is used to stimulate labor contractions and is used after birth to constrict the mother's blood vessels (when she is at risk of bleeding excessively). Another mycotoxin, *ergotamine,* is used to treat migraine headaches.

Mycetismus: Mushroom Poisoning

Learning Objective

✓ Describe at least two types of mushroom poisoning.

▲ *Figure 22.19*
Amanita phalloides, the "death cap" mushroom.

Most mushrooms (the spore-bearing structures of certain basidiomycetes) are not toxic, but those that are produce extremely dangerous poisons capable of causing neurological dysfunction or hallucinations, organ damage, or even death. The deadliest mushroom toxin is produced by the "death cap" mushroom, *Amanita phalloides* (am-ă-nī'tă fal-ōy'dēz) **(Figure 22.19).** The death cap contains two related polypeptide toxins: *phalloidin,* which irreversibly binds actin within cells, disrupting cell structure; and *alpha-amanitin,* which inhibits mRNA synthesis. Both toxins cause liver damage.

Other deadly mushrooms include the false morel, *Gyromitra esculenta* (gī-rō-mē'tră es-kū-len'tă), which can cause bloody diarrhea, convulsions, and death within 2 days after ingestion, and *Cortinarius gentilis* (kōr'ti-nar-ē-us jen'til-is), which can cause excessive thirst, nausea, and kidney failure between 3 days and 3 weeks after ingestion.

Psilocybe cubensis (sī-lō-sī'bē kū-benz'is) produces hallucinogenic *psilocybin,* and *Amanita muscaria* produces two hallucinogenic toxins—*ibotenic acid* and *muscimol.* Such toxins may also cause painful and serious illness in adults, and convulsions in children.

Mushroom poisoning typically occurs when untrained individuals pick and eat wild mushrooms. Young children are especially attracted to colorful mushrooms, which can be among the most poisonous. Treatment involves inducing vomiting followed by oral administration of activated charcoal to absorb toxins. Severe mycetismus may require replacement by transplant of the damaged liver.

Allergies to Fungi

Learning Objective

✓ Identify the types of hypersensitivity reactions fungal allergens produce.

Fungal allergens are common and can be found both indoors and out. Over 80 genera of fungi have been demonstrated to trigger fungal allergies. Epidemiologists estimate that 3–10% of humans worldwide are affected. However, determining the specific cause of allergies is often difficult—the presence of spores in an environment does not automatically mean that they are responsible for the allergies suffered by a group of exposed individuals. Because skin sensitivity testing (discussed in Chapter 18) is not performed on most people, the true impact of fungal allergens remains undetermined, even though they continue to be blamed for a variety of illnesses **(see Highlight 22.1).**

Fungal allergens typically cause type I hypersensitivities in which immunoglobulin E (IgE) binds the allergen, triggering responses such as asthma, eczema, hay fever, and watery eyes and nose. Fungal spores dispersed in the air are common allergens. The concentrations of such spores may peak in autumn or spring, although they are present year-round.

Less frequently, type III hypersensitivities result from chronic inhalation of particular fungal allergens. In these cases, allergens that have penetrated deep into the lungs, encounter complementary antibodies, forming immune complexes in the alveoli that lead to inflammation, fibrosis, and, in some cases, death. Type III fungal hypersensitivities are associated with certain occupations, such as farming, in which workers are constantly exposed to fungal spores in moldy vegetation.

CRITICAL THINKING

Despite the fact that fungi are everywhere, fungal allergies are still not as common as allergies to pollen or dust mites. Propose an explanation to explain this observation.

Highlight 22.1   Does "Killer Mold" Exist?

In 1994, eight infants in Cleveland were hospitalized with life-threatening pulmonary hemorrhage. One child died. Each of the infants had lived in water-damaged buildings, and initial epidemiological investigations pointed at the black mold *Stachybotrys chartarum* as the agent responsible. The cases were widely reported in the media, and *Stachybotrys* became known as "killer mold." Since then, the mold has been blamed for everything from "sick building syndrome" to asthma to other occurrences of pulmonary hemorrhage in infants.

SEM   30 µm

Stachybotrys chartarum

A subsequent review of the Cleveland cases by the CDC, however, found serious shortcomings in the initial investigations. Significantly, *Stachybotrys* could not be recovered from any of the infants, and in most of the cases the level of mold in the water-damaged homes was below detection limits or statistically no different from levels in homes without water damage. In the numerous studies that have been conducted on *Stachybotrys* since 1994, none has established a direct causal relationship between the mold and infant deaths. Nor have researchers found any compelling relationship between *Stachybotrys* and any increased prevalence of asthma or other respiratory illnesses. Within the scientific community, "killer mold" remains an unproven moniker.

In approaching mold contamination of any kind, individuals should exercise their common sense. Some people are naturally sensitive to mold—*Stachybotrys* and others. Immunocompromised or allergy-prone individuals in particular may be at increased risk of experiencing adverse reactions to mold. Decisions regarding the more drastic methods of mold removal, such as burning down a house, should thus be made on a case-by-case basis.

CHAPTER SUMMARY

An Overview of Medical Mycology (pp. 624–627)

1. Medical **mycology** is the study of the diagnosis, management, and prevention of **mycoses** (fungal diseases). Only a few fungi cause serious diseases in humans. The actual incidence of mycoses is unknown because they are not reportable.

2. Most fungi are not considered infectious agents because they are almost never spread by human-to-human contact. The few exceptions include the dermatophytes, species of *Candida*, and *Pneumocystis*.

3. True fungal pathogens are capable of actively invading the tissues of immunocompetent people. Opportunists can infect only weakened or immunocompromised patients.

4. Clinically, mycoses include infections, toxicoses, and hypersensitivity reactions. Infections can be superficial, cutaneous, subcutaneous, or systemic.

5. Diagnosis of mycoses usually correlates signs and symptoms with microscopic examination of tissues or laboratory cultures for identifying unique morphological features. Biochemical and immunological tests exist for some fungi.

6. **Amphotericin B** kills most fungi and can be used to treat most mycoses.

Systemic Mycoses Caused by Pathogenic Fungi (pp. 628–631)

1. Systemic mycoses are traditionally caused by one of four pathogenic, dimorphic fungi: *Blastomyces, Coccidioides, Histoplasma,* or *Paracoccidioides*. Whereas each is geographically limited, they all cause similar pulmonary diseases and can disseminate beyond the lungs.

2. **Blastomycosis** is found in the eastern United States and is caused by *Blastomyces dermatitidis*, which normally lives in soil rich in organic material. *B. dermatitidis* almost always spreads beyond the lungs to produce cutaneous lesions.

3. *Coccidioides immitis*, which is localized to deserts, causes **coccidioidomycosis**, which is common in AIDS patients. Contaminated dust is a major source of transmission.

4. *Histoplasma capsulatum*, which causes **histoplasmosis,** is associated primarily with bat and bird droppings in soil in the Ohio River Valley. Generally, histoplasmosis is an occupationally acquired disease, but recreational exposure does occur. *H. capsulatum* can be carried from the lungs inside macrophages.

5. *Paracoccidioides brasiliensis* is the most geographically limited of the true fungal pathogens, occurring only south of the United States. It causes **paracoccidioidomycosis.** Following a pulmonary phase, the fungus spreads and creates permanently disfiguring lesions on the face and neck.

6. Pathogenic fungi are diagnosed by morphological analysis and by demonstration of dimorphism. Serological tests are also available. If treated promptly, they can be controlled.

7. Opportunistic infections are difficult to diagnose and treat successfully. AIDS patients require both primary treatment to curb infections and maintenance therapy to keep the fungi under control.

Systemic Mycoses Caused by Opportunistic Fungi (pp. 631–637)

1. **Aspergillosis** is a group of diseases caused by *Aspergillus* species, including noninvasive fungus balls in the lungs and invasive diseases that can be fatal. *Aspergillus* is very common in the environment, from which it is inhaled.

2. **Candidiasis,** caused by various species of *Candida,* is one of the most important pathogens of AIDS patients, but it can also causes infections in relatively healthy individuals. Disease ranges from superficial infections to fungemia.

3. *Cryptococcus neoformans,* common in bird droppings and soil, causes **cryptococcoses,** which most frequently manifest as cryptococcal meningitis in AIDS patients. Other clinical manifestations are possible.

4. *Pneumocystis jiroveci* causes **Pneumocystis pneumonia (PCP),** a leading cause of death in AIDS patients in the United States. The organism shows a blend of characteristics similar to those of both protozoans and fungi. PCP is debilitating because the fungi multiply rapidly.

5. **Zygomycoses,** caused by *Mucor,* and various other genera of zygomycetes, can involve the brain resulting in death. Less severe infections also occur.

6. Diseases caused by *Candida albicans, Aspergillus fumigatus,* and *Cryptococcus neoformans* are defining illnesses of AIDS patients. *Fusarium, Penicillium,* and *Trichosporon* are emerging opportunists. Factors contributing to their emergence include an increase in the number of immunocompromised patients and the selective pressure of antifungal agents, which increases the number of resistant strains of fungi.

Superficial, Cutaneous, and Subcutaneous Mycoses (pp. 637–643)

1. Superficial, cutaneous, and subcutaneous mycoses are caused by opportunistic fungi.

2. **Black piedra** and **white piedra** are benign, superficial infections of the hair shafts that can be transmitted among family members through shared combs and brushes.

3. **Dermatophytoses** (tineas, ringworms) encompass superficial skin, nail, and hair infections caused by a variety of fungi transmitted from individual to individual. Three genera predominate: *Trichophyton, Microsporon,* and *Epidermophyton.*

4. *Malassezia furfur* is a fungus that infects the skin. Clinical manifestations include **pityriasis,** in which fungal growth disrupts melanin production to produce discolored patches, and the most common manifestation of *M. furfur* infection—**dandruff.**

5. Most superficial infections can be diagnosed by simple clinical observation of symptoms, and all can be treated successfully. Rarely are such infections severe or invasive, but they have a tendency to recur.

6. **Chromoblastomycosis** and **phaeohyphomycosis** are similar diseases resulting from infection with dematiaceous (dark-pigmented) fungi. Both are acquired by traumatic introduction of fungi into the skin. Lesions may be extensive and can disseminate internally. The two diseases are distinguished by differences in fungal morphology in tissue sections.

7. **Mycetomas** are invasive and destructive infections following introduction of soil fungi through scrapes or pricks from vegetation. Cutaneous lesions can spread to adjacent bone and can be permanently damaging. Surgery or amputation is required to remove infected tissues.

8. **Sporotrichosis** also is acquired from inoculation of soil fungi by thorn pricks. Lymphatic dispersal of the organisms results in multiple lesions along the course of lymphatic vessels.

Fungal Intoxications and Allergies (pp. 643–644)

1. **Mycotoxicosis** results from the ingestion of foods contaminated with **mycotoxins,** secondary metabolites produced by fungi. Some fungal toxins contaminate much of our food supply and are deadly.

2. **Mycetismus** is mushroom poisoning. Mushroom toxins can cause neurological and physiological damage, which can lead to death.

3. Allergens are spores or other fungal elements that cause type I hypersensitivities or, more rarely, type III hypersensitivities. Because fungi are everywhere, everyone is exposed to fungal allergens every day.

QUESTIONS FOR REVIEW

(Answers to multiple choice, true/false, fill in the blanks, and matching questions are on the web, along with additional review questions. Visit www.microbiologyplace.com.)

Multiple Choice

1. A fungus that can infect both healthy and immunocompromised patients is called
   a. a true pathogen.
   b. an opportunistic pathogen.
   c. a commensal organism.
   d. a symbiotic organism.

2. Of the following fungi, which is usually transmitted from person to person?
   a. *Blastomyces dermatitidis*
   b. *Coccidioides immitis*
   c. *Epidermophyton floccosum*
   d. *Aspergillus fumigatus*

3. Which of the following is not used to identify true fungal pathogens?
   a. growth at 25°C and 37°C to show dimorphism
   b. GMS staining of infected tissues
   c. serological testing
   d. clinical symptoms alone

4. Because amphotericin B is extremely toxic to humans, most clinicians only prescribe it for
   a. dermatophyte infections.
   b. *Malassezia* infections.
   c. systemic infections.
   d. black piedra.

5. Ringworm is caused by a
   a. helmith.
   b. dermatophyte.
   c. dimorphic fungus.
   d. commensal fungus.

6. Which of the following is considered a classical opportunistic fungus?
   a. *Piedraia hortae*
   b. *Histoplasma capsulatum*
   c. *Fonsecaea pedrosi*
   d. *Aspergillus niger*

7. Subcutaneous infections tend to
   a. be acquired through inhalation and remain localized.
   b. be acquired through inhalation and become systemic.
   c. be acquired through trauma and remain localized.
   d. be acquired through trauma and become systemic.

8. The term *dematiaceous* refers to
   a. pathogenicity.
   b. pigmentation.
   c. method of spread.
   d. method of antifungal treatment.

9. Which of the following subcutaneous mycoses may exhibit respiratory and cerebral forms?
   a. chromoblastomycosis
   b. mycetoma
   c. phaeohyphomycosis
   d. sporotrichosis

10. Which of the following systemic mycoses is endemic to the deserts of the southwest United States?
    a. blastomycosis
    b. coccidioidomycosis
    c. histoplasmosis
    d. paracoccidioidomycosis

11. A spherule stage is seen in humans infected with what organism?
    a. *Blastomyces dermatitidis*
    b. *Coccidioides immitis*
    c. *Histoplasma capsulatum*
    d. *Paracoccidioides brasiliensis*

12. Of the following fungal diseases, which is found in almost all terminal AIDS patients?
    a. chromoblastomycosis
    b. blastomycosis
    c. candidiasis
    d. piedra

13. Which of the following predisposing factors would leave a patient with the greatest long-term risk of acquiring a fungal infection?
    a. invasive medical procedures
    b. HIV/AIDS
    c. chronic illness such as diabetes
    d. short-term treatment with antibacterial agents

14. Fungal allergens generally stimulate what type of reaction?
    a. type I hypersensitivity
    b. type II hypersensitivity
    c. type III hypersensitivity
    d. type IV hypersensitivity

15. A pathogenic feature of *Cryptococcus neoformans* is
    a. production of destructive enzymes.
    b. production of a capsule.
    c. infection of immune cells.
    d. variation of surface antigens to avoid immune system recognition.

16. The most common manifestation of *Cryptococcus* infection in AIDS patients is
    a. blindness.
    b. cutaneous infection.
    c. meningitis.
    d. pneumonia.

17. Bread mold can cause which disease?
    a. aspergillosis
    b. dermatophytosis
    c. mycetoma
    d. zygomycosis

18. Mycetismus is caused by
    a. inhalation of fungal allergens.
    b. ingestion of mushrooms.
    c. traumatic inoculation of fungi beneath the skin.
    d. close contact with infected individuals.

19. One of the more poisonous mycotoxins is produced by
    a. *Amanita phalloides*.
    b. *Amanita muscaria*.
    c. *Psilocybe cubensis*.
    d. *Claviceps purpurea*.

20. The number of mycoses worldwide is rising because
    a. the number of fungi in the environment is rising.
    b. the number of immunocompromised individuals in the population is rising.
    c. fungi have become more pathogenic.
    d. fungi are developing a tendency to be spread from person to person.

True or False

Indicate which of the following are true and which are false. Rewrite false statements to make them true by changing the italicized word(s).

____ 1. Fungi are generally not transmitted *from person to person*.

____ 2. On the whole, fungal infections are relatively *easy* to treat.

____ 3. *Dermatophytes* are always contracted from the environment.

____ 4. Chromoblastomycosis and phaeohyphomycosis are both caused by *dematiaceous* fungi.

____ 5. *Sporotrichosis* is always caused by traumatic introduction of fungi beneath the skin.

____ 6. Coccidiomycosis does not occur normally outside the *United States*.

____ 7. Treatment of individuals with broad-spectrum antibacterial agents is a predisposing factor for *opportunistic* fungal infections.

____ 8. *Candida albicans* generally causes localized opportunistic infections but can become systemic, particularly in the *immunocompetent*.

____ 9. Relapse of fungal diseases is common in *AIDS patients*.

____10. *Almost everyone has* allergies to fungal elements.

Fill in the Blanks

1. Dimorphic fungi exist as _____ forms in the environment and as _____ forms in their hosts.

2. The true fungal pathogens are _____, _____, _____, and _____.

3. Most antifungal agents target the compound _____ in fungal cytoplasmic membranes.

4. _____ piedra is characterized by hard nodules on the hair shaft, whereas _____ piedra is characterized by soft nodules.

5. Sporotrichosis is caused by the traumatic introduction of _____ _____ into the skin (give genus and species).

6. Which systemic mycosis is associated with bird droppings? _____ _____ (give genus and species).

7. The five most common agents of opportunistic infections are _____, _____, _____, _____, and _____.

8. Thrush is caused by _____ (genus name).

9. *Pneumocystis* was once classified as a _____, but now it is classified as a _____.

10. Ergot alkaloids are produced by some strains of the genus _____.

Matching

Match each of the following diseases with the manner in which fungi enter the body. Answers can be used more than once.

___ Aspergillosis   A. Inhalation

___ Candidiasis   B. Contact

___ Chromoblastomycosis   C. Trauma

___ Coccidioidomycosis   D. Ingestion

___ Cryptococcosis

___ Dermatophytosis

___ Histoplasmosis

___ Hypersensitivity reactions

___ Mushroom poisoning

___ Mycetoma

___ Sporotrichosis

Short Answer

1. Amphotericin B is considered the "gold standard" of antifungal agents and is the drug of choice for most fungal infections. Technically, its mode of action works against most fungal infections. Why, then, isn't it prescribed for most fungal infections?

2. Discuss why it is difficult in many cases to determine the source of superficial fungal infections (that is, from other humans, animals, or the environment).

3. Given that superficial fungal infections are only on the surface, why is it necessary to even try to identify the source of infection?

4. AIDS patients usually die of bacterial or fungal infections. Why do so many fungal infections appear in these individuals, and why are the mycoses so severe given the fact that fungi, for the most part, are benign residents of the environment?

CRITICAL THINKING

1. Correlate the observation that the majority of fungal infections are caused by opportunists with the fact that most infections are acquired from the environment. How does this make the control and diagnosis of fungal infections difficult?

2. The four pathogenic fungi (*Blastomyces, Coccidioides, Histoplasma,* and *Paracoccidioides*) are all dimorphic. Antifungal agents, like all antimicrobials, are generally designed to target something specific about the organism to avoid harming human tissues. Propose how you could use the idea of dimorphism to produce new antifungal agents.

3. What actions could be taken to limit the contamination of foods with fungal toxins? Would these actions differ depending on where in the world the food is grown? Explain.

4. Onychomycoses are nail infections that can be caused by several species of fungi. Explain why these mycoses are so difficult to treat, and why it is generally necessary to treat patients with oral antifungal agents for long periods of time.

5. What factors contribute to the pathogenicity of *Cryptococcus* infections?

6. Clinically, all fungi that cause subcutaneous mycoses produce lesions on the skin around the site of inoculation. What are some of the things you would look for when attempting to distinguish among them?

7. Given the various predisposing factors that make humans susceptible to opportunistic infections, how can health care providers curtail the rising incidence of such infections?

The next time you sit down to a nice meal containing pork or beef, it's probably a good idea to make sure the meat is properly cooked—if not well done. One reason is that the muscles of pigs can be infected with cysts that are the immature stage of *Taenia solium,* a parasitic tapeworm commonly called the pork tapeworm. The cysts remain viable in undercooked pork, and in the human small intestine they develop into mature tapeworms, which can grow as long as 800 cm (more than 26 feet). Cysts of a similar tapeworm, *Taenia saginata* (the beef tapeworm) also grow in the intestines of humans who eat insufficiently cooked infected beef.

This chapter provides a brief survey of parasitology, the study of eukaryotic parasites—in particular, various protozoa and helminths (tapeworms, other flatworms called flukes, and roundworms). Because parasitologists also study arthropod vectors (including mites, ticks, flies, mosquitoes, and fleas), which are important in the transmission of a variety of infectious organisms, we consider vectors in the final section of the chapter.

Parasitic Protozoa, Helminths, and Arthropod Vectors

CHAPTER 23

CHAPTER OUTLINE

Protozoan Parasites of Humans
Ciliates
Amoebae
Flagellates
Apicomplexans

Helminthic Parasites of Humans
Cestodes
Trematodes
Nematodes

Arthropod Vectors
Arachnids
Insects

Attachment organ of a tapeworm

MicroPrep Pre-Test: Take the pre-test for this chapter on the web.
Visit **www.microbiologyplace.com.**

As we saw in Chapter 14, parasites live, feed and grow in or on another organism at the expense of their host's metabolism. Sometimes no obvious damage is done to the host, but in other cases parasitism can lead to a host's death. Many protozoan and helminthic parasites exist worldwide, especially in the tropics and subtropics, and particularly among people living in rural, undeveloped, or overcrowded places. Parasitic diseases are also emerging as serious threats among developed nations in nontropical regions. This rise is due to many factors, but can most generally be attributed to human migration (travel, forced relocation), poor regulation/inspection of food and water supplies, and a changing world climate that alters the parasites' transmission patterns, particularly those transmitted by vectors. As a whole, parasites infect billions of humans each year, making parasitic infections medically, socially, and economically important.

Parasitic infections often involve several hosts—a **definitive host** in which mature (sometimes sexual) forms of the parasite are present and usually reproducing, and with many parasites, one or more **intermediate hosts** in which immature parasites undergo various stages of maturation. In general, parasites infect human hosts in one of three major ways: by being ingested (often by the fecal/oral route), through vector-borne transmission, or via direct contact and penetration of the skin or mucous membranes **(Figure 23.1)**.

We begin our consideration of parasitic infections by discussing the protozoan parasites of humans.

Protozoan Parasites of Humans

Learning Objective

✓ Describe the epidemiology of the more important protozoan parasites.

As discussed in Chapter 12, protozoa are unicellular eukaryotes found in a wide range of habitats. Although the majority of protozoa are free-living and do not adversely affect humans or animals, some are parasites that can cause debilitating and deadly diseases.

Among the many protozoa that enter the body via ingestion, most have two morphological forms: a feeding and reproducing stage called a *trophozoite*, which lives within the host, and a dormant *cyst* stage, which can survive in the environment and is infective to new hosts. Once ingested by a host, cysts undergo **excystment** and develop into new trophozoites, which resume feeding and reproducing. In most cases, trophozoites undergo **encystment** before leaving the host in the feces and becoming available to infect other hosts.

This chapter presents the protozoa in four groups, an approach that is different from the taxonomic presentation of Chapter 12. The scheme presented in this chapter reflects the largely obsolete, but clinically useful, classical taxonomic groupings for these parasites, which were based primarily on mode of locomotion: the ciliates, the amoebae, the flagellates, and the typically nonmotile apicomplexans.

Ciliates

Learning Objective

✓ Describe the characteristics of *Balantidium coli*.

Ciliates (sil'ē-āts), are protozoa that in their trophozoite stages, use cilia for locomotion, to acquire food, or both. *Balantidium coli* (bal-an-tid'ē-ŭm kō'lī), a relatively large (50 μm × 100 μm) ciliate commonly found in animal intestinal tracts, is the only ciliate known to cause disease in humans. Pigs are its most common host, but it is also found in rodents and nonhuman primates. Humans become infected by consuming food or water contaminated with feces containing cysts.

Once ingested, excystment occurs in the small intestine, releasing trophozoites that use their cilia to attach to (and then burrow through) the mucosal epithelium lining the intestine. Eventually, some trophozoites undergo encystment, and both cysts and trophozoites are shed in feces. Trophozoites rapidly die outside the body, but cysts are hardy and infective.

In healthy adults, *B. coli* infection is generally asymptomatic. For those in poor health, however, **balantidiasis** (bal'an-ti-dī'ă-sis) occurs. Persistent diarrhea, abdominal pain, and weight loss characterize the disease. Severe infections produce *dysentery* (frequent and painful diarrhea, often containing blood and mucus) and possibly ulceration and bleeding from the intestinal mucosa. *Balantidium* infection is rarely fatal.

Cysts are few and usually cannot be recovered from stool (fecal) samples, but trophozoites can be detected, and their presence is diagnostic for the disease. Fresh stool samples must be used for diagnostic purposes because the trophozoites do not survive long outside the intestinal tract. The treatment of choice for balantidiasis is the antibacterial drug tetracycline, which does not kill *B. coli* directly but instead modifies the normal microbiota of the digestive tract such that it is unsuitable for infection by the ciliate.

Prevention of balantidiasis relies on good personal hygiene, especially for those who live around or work with pigs. Additionally, efficient water sanitation is necessary to kill or remove cysts from drinking water.

CRITICAL THINKING

Why are diseases such as balantidiasis generally more severe in those who are already in poor health?

Amoebae

Learning Objective

✓ Compare and contrast three amoebae that cause disease in humans.

Amoebae (ă-mē'bē, singular: ă-mē'bă) are protozoa that have no truly defined shape and that move and acquire food

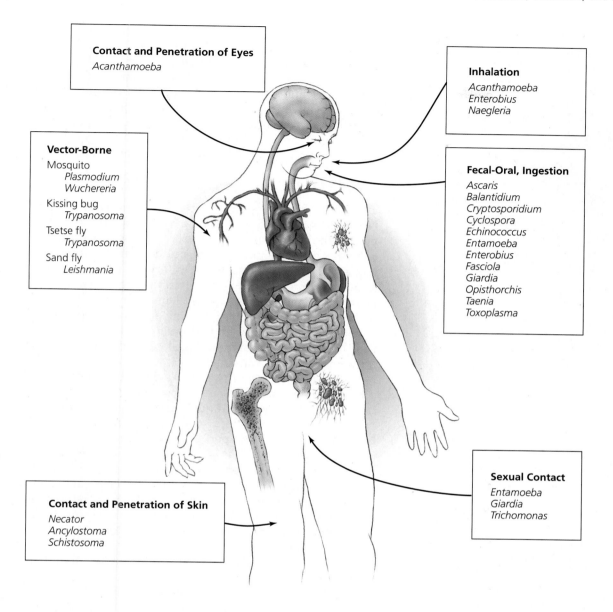

Contact and Penetration of Eyes
Acanthamoeba

Inhalation
Acanthamoeba
Enterobius
Naegleria

Vector-Borne
Mosquito
  Plasmodium
  Wuchereria
Kissing bug
  Trypanosoma
Tsetse fly
  Trypanosoma
Sand fly
  Leishmania

Fecal-Oral, Ingestion
Ascaris
Balantidium
Cryptosporidium
Cyclospora
Echinococcus
Entamoeba
Enterobius
Fasciola
Giardia
Opisthorchis
Taenia
Toxoplasma

Contact and Penetration of Skin
Necator
Ancylostoma
Schistosoma

Sexual Contact
Entamoeba
Giardia
Trichomonas

◀ **Figure 23.1**
The major routes by which humans acquire parasitic infections: the fecal-oral route, the vector-borne route, and direct contact. Occasionally parasites are inhaled. Listed genera are discussed in the chapter.

through the use of pseudopodia. Although amoebae are abundant throughout the world in freshwater, seawater, and moist soil, few cause disease. The most important amoebic pathogen is *Entamoeba*.

Entamoeba

Entamoeba histolytica (ent-ă-mē′bă his-tō-li′ti-kă) is carried asymptomatically in the digestive tracts of roughly 10% of the world's human population. When disease develops, it can be fatal, causing a worldwide mortality of over 100,000 people annually. Carriers predominate in third-world countries, especially in rural areas, where human feces are used to fertilize food crops and where water sanitation is deficient. Travelers, immigrants, and institutionalized populations are at greatest risk within industrialized nations. No animal reservoirs exist, but human carriers are sufficiently numerous to ensure continued transmission of the parasite.

Infection most commonly occurs through the drinking of water contaminated with feces that contain cysts. The parasite can also be ingested following fecal contamination of hands or food, or during oral-anal intercourse.

Excystment in the small intestine releases trophozoites that migrate to the large intestine and multiply. The organism uses pseudopodia to attach to the intestinal mucosa, where it feeds and reproduces via binary fission. Both trophozoites and cysts are shed into the environment in feces, but the trophozoites die quickly, leaving infective cysts.

Depending on the health of the host and the virulence of the particular infecting strain, three types of **amoebiasis** (ă-mē-bī′ă-sis) can result from infections with *E. histolytica*. The least severe form, *luminal amoebiasis,* occurs in otherwise healthy individuals. Infections are asymptomatic; trophozoites remain in the lumen of the intestine, where they do little tissue damage. Invasive *amoebic dysentery,* is a more serious and more common form of infection characterized

LM | 50 μm

▲ *Figure 23.2*

A trophozoite (amoeboid form)
of *Entamoeba histolytica*.

by severe diarrhea, colitis (inflammation of the colon), appendicitis, and ulceration of the intestinal mucosa. Bloody, mucus-containing stools and pain are characteristic of amoebic dysentery. In the most serious disease, *invasive extraintestinal amoebiasis,* trophozoites invade the peritoneal cavity and the bloodstream, which carries them throughout the body. Necrotic lesions formed by the trophozoites occur most commonly in the liver but can also be found in the lungs, spleen, kidneys, and brain. Amoebic dysentery and invasive extraintestinal amoebiasis can be fatal, especially without adequate treatment.

Diagnosis is based on the identification of microscopic cysts or trophozoites **(Figure 23.2)** recovered from either fresh stool specimens or intestinal biopsies. Microscopic analysis is necessary to distinguish amoebic dysentery from dysentery caused by bacteria. Serological identification using antibodies to detect antigens can aid in distinguishing *E. histolytica* from nonpathogenic amoebae.

For asymptomatic infections, the drugs iodoquinol or paromomycin are effective. Physicians prescribe metronidazole followed by iodoquinol, paromomycin, or diloxanide furoate for symptomatic amoebiasis. They may also prescribe antibacterial agents at the same time to prevent secondary bacterial infections. The coupling of oral rehydration therapy with drug treatment is recommended and vital in severe cases.

Several preventative measures interrupt the transmission of *Entamoeba*. Discontinuing the use of human wastes as fertilizer reduces the transmission of amoebiasis, and of many other diseases as well. Normal methods of treating wastewater and drinking water are helpful but not completely effective in removing all cysts. Effective processing of water requires chemical treatment, filtration, or extensive boiling to eliminate all infectious cysts. Good personal hygiene can eliminate transmission via intimate contact.

Acanthamoeba and *Naegleria*

Two other amoebae—*Acanthamoeba* (ă-kan-thă-mē′bă) and *Naegleria* (nā-glē′rē-ă)—cause rare and usually fatal infections of the brain. These amoebae are common free-living inhabitants of warm lakes, ponds, puddles, ditches, mud, and moist soil; they are also found in artificial water systems such as swimming pools, air conditioning units, humidifiers, and dialysis units. Contact lens wearers who use tap water (as opposed to sterile saline solution) to wash and store their lenses create an additional focal point for infection. *Naegleria* may assume a flagellated form in addition to the amoeboid and cyst stages.

Acanthamoeba usually enters a host through cuts or scrapes on the skin, through the conjunctiva via abrasions from contact lenses or trauma, or through inhalation of contaminated water while swimming. *Keratitis,* in which trophozoites invade and perforate the eye, results from conjunctival inoculation. Damage can be extensive enough to require corneal replacement. The more common disease caused by infection with *Acanthamoeba* is **amoebic encephalitis** (inflammation of the brain), characterized by headache, altered mental state, and neurological deficit. Symptoms progressively worsen over a period of weeks until the patient dies.

Infection with *Naegleria* occurs when swimmers inhale water contaminated with trophozoites, which then invade the nasal mucosa. After replicating, the trophozoites migrate to the brain, where they cause an **amoebic meningoencephalitis** (inflammation of the brain and its membranes). Severe headache, fever, vomiting, and neurological tissue destruction lead to hemorrhage, coma, and usually death within 4–5 days after the onset of symptoms.

Acanthamoeba and *Naegleria* infections are diagnosed by the detection of trophozoites in corneal scrapings, cerebrospinal fluid, or biopsy material. Keratitis can be treated topically with propamidine isethionate plus neomycin, polymyxin B, and/or gramicidin. Miconazole, sulfadiazine, or tetracycline used in combination have limited success in treating neurological infections, but *only* if administered very early in the course of the infection.

As both these amoebae are environmentally hardy, control and prevention of infection can be difficult. Natural waterways in which the organisms are known to be endemic should be avoided for recreational purposes. Nonsterile solutions should never be used to clean or store contact lenses. Swimming pools should be properly chlorinated and tested periodically to ensure their safety. Air conditioning systems, dialysis units, and other devices that routinely use water should also be cleaned thoroughly and regularly to prevent these amoebae from becoming resident.

Flagellates

Learning Objectives

✓ Contrast the life cycles of *Trypanosoma brucei* and *Trypanosoma cruzi*.

✓ Describe the life cycle of *Leishmania* and the clinical forms of leishmaniasis.

✓ Compare *Giardia* infections to those caused by *Entamoeba* and *Balantidium*.

✓ Identify the risk factors and preventative measures for *Trichomonas vaginalis* infection.

In 1832 Charles Darwin explored South America as one of the naturalists sailing aboard HMS *Beagle*. His explorations were to impact him profoundly—and medically. During the voyage he began to formulate his ideas concerning the evolution of species, and he was bitten by a blood-sucking bug that infected Darwin with *Trypanosoma cruzi*, a flagellated protozoan that afflicted him for the rest of his life. The following sections explore this and other parasitic flagellates that infect humans.

Flagellates (flaj′e-lātz) are protozoa that possess at least one long flagellum, which is generally used for movement. The number and arrangement of the flagella are important diagnostic features for determining the species of flagellate present within a host.

Because the kinetoplastids and the archaezoans described in Chapter 12 were once classified together as flagellates, we will examine some parasitic kinetoplastids (*Trypanosoma* and *Leishmania*) and some parasitic archaezoans (*Giardia* and *Trichomonas*) here. We begin our discussion of the more important flagellate parasites with *Trypanosoma brucei*.

Trypanosoma brucei

Trypanosoma brucei (tri-pan′ō-sō-mă brūs′ī) causes **African sleeping sickness,** which occurs in equatorial and subequatorial savanna, agricultural, and riverine areas of Africa. Its geographical range depends on the range of its insect vector: the tsetse (tset′se) fly *(Glossina)*. There are two variants of *Trypanosoma brucei*—*T. brucei gambiense* (gam′bē-en′sē), which occurs primarily in western Africa, and *T. brucei rhodesiense* (rō-dē′zē-en′sē) in eastern and southern Africa.

Whereas cattle and sheep are the reservoirs for *T. brucei gambiense,* several species of wild animals apparently serve as reservoirs for *T. brucei rhodesiense*. Although humans usually become infected when bitten by tsetse flies infected while feeding on infected animals, flies infected by feeding on infected humans frequently transmit *T. brucei gambiense* to other humans, and occasionally transmit *T. brucei rhodesiense* to other humans.

The life cycle of *T. brucei* proceeds as follows (**Figure 23.3a** on page 654): Within the salivary gland of a male or female tsetse fly, a form of the parasite called an *epimastigote*[1] matures into an infective *trypomastigote* ❶. During the course of taking a blood meal from an animal or human, tsetse flies inject trypomastigotes into the wounds feeding creates ❷. Trypomastigotes then travel throughout the lymphatic and circulatory systems to other sites ❸, where they reproduce by binary fission ❹. Eventually, some trypomastigotes enter the central nervous system and the spinal fluid while others continue to circulate in the blood, where they can be picked up by feeding tsetse flies ❺. In the midgut of the fly, trypomastigotes multiply by binary fission, producing immature epimastigotes that migrate to the salivary glands ❻, where they mature and become trypomastigotes that are infective to new hosts when the fly feeds once again.

Untreated African sleeping sickness progresses through three clinical stages. First, the wound created at the site of each fly bite becomes a chancre containing necrotic tissue and rapidly dividing parasites. Next, the presence of parasites in the blood generates fever, swelling of lymph nodes, and headaches. Finally, invasion of the central nervous system results in *meningoencephalitis,* characterized by headache, extreme drowsiness, abnormal neurological function, coma, and death. Symptoms may take years to develop with *T. brucei rhodesiense,* but occur within 3–6 months with *T. brucei gambiense.*

T. brucei infections are characterized by cyclical waves of *parasitemia* (parasites in the blood) that occur roughly every 7–10 days. Although the presence of parasites in the blood is in itself serious, these cycles are particularly dangerous because with each wave of replication, *T. brucei* changes its surface glycoproteins, and thus its surface antigens. The result is that by the time the host's immune system has produced antibodies against a given set of glycoproteins, the parasite has already produced a new set, continually leaving the host's immune system one step behind the parasite. Once infected, a patient is incapable of ever clearing the infection and becoming immune.

Microscopic observation of trypomastigotes in blood, lymph, spinal fluid, or a tissue biopsy is diagnostic for both strains of *T. brucei* (**Figure 23.4** on page 655). Trypomastigotes are long and thin with a single long flagellum running along the cell and extending past the posterior end.

To be successful, treatment of African sleeping sickness must begin as soon as possible after infection begins. However, immediate treatment is rare because the poor health care infrastructure in endemic areas often prevents identification of the disease at its earliest stages. Pentamidine isethionate and suramin are used to treat the early stages of African sleeping sickness. Melarsoprol is used when the disease progresses to the central nervous system because this drug crosses the blood-brain barrier. A newer, and expensive,

---

[1]From Greek *epi*, meaning on, and *mastix*, meaning whip. The flagellum of an epimastigote originates on the middle of the cell.

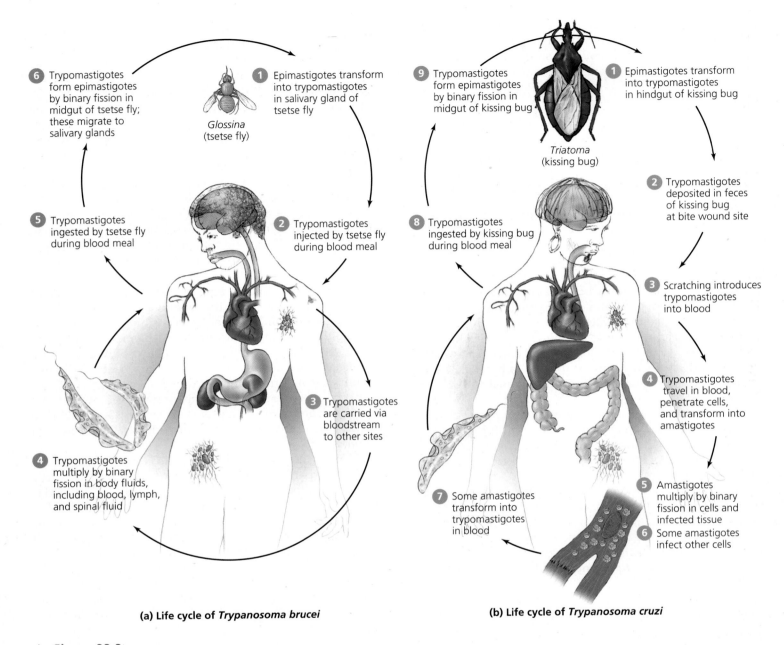

(a) Life cycle of *Trypanosoma brucei*

(b) Life cycle of *Trypanosoma cruzi*

▲ *Figure 23.3*

The life cycles of two trypanosomes that infect humans. **(a)** *Trypanosoma brucei.*
(b) *Trypanosoma cruzi.* Note that whereas *T. brucei* replicates in the blood,
T. cruzi replicates within cells. *By which mode of transmission do trypanosomes infect humans?*

Figure 23.3 Trypanosomes are transmitted to humans by arthropod vectors.

drug called eflornithine is remarkably effective against
T. brucei gambiense (but not against *T. brucei rhodesiense*). In
2001, private pharmaceutical companies guaranteed availability of this drug through the World Health Organization to
those who need it.

Clearing of tsetse fly habitats and broad application of
insecticides have reduced the occurrence of African sleeping

sickness in local areas. However, large-scale spraying of insecticides is impractical, expensive, and may have disas-
trous long-term environmental consequences. Personal
insecticide use is preferable, as is the use of insecticide-
impregnated netting, and long, loose-fitting clothing, which
can prevent insect feeding. No vaccine currently exists for
African sleeping sickness.

Trypanosoma cruzi

Trypanosoma cruzi (tri-pan'ō-sō-mă kroo'zī) causes **Chagas' disease,** named for Brazilian doctor Carlos Chagas (1879–1934), which is endemic throughout Central and South America. Localized outbreaks have also occurred in California and Texas. Opossums and armadillos are the primary reservoirs for *T. cruzi*, but most mammals, including humans, can harbor the organism. Transmission occurs through the bite of insects, which are true bugs in the genus *Triatoma.* These blood-sucking bugs feed preferentially from blood vessels in the lips, which gives the bugs their common name—kissing bugs.

The basic life cycle of *T. cruzi* is similar to that of *T. brucei* (see **Figure 23.3b**). Within the hindgut of the kissing bug, epimastigotes develop into infective trypomastigotes ❶, which are shed in the bug's feces while the bug feeds on a mammalian host ❷. When the host scratches the itchy wounds created by the bug bites, the trypomastigotes in the feces enter the nearby wounds ❸. The trypomastigotes are then carried in the bloodstream throughout the body and penetrate certain cells, especially macrophages and heart muscle cells, where they transform into small, nonflagellated forms called *amastigotes* ❹. The multiplication of these intracellular amastigotes via binary fission eventually causes the cells to rupture ❺, releasing amastigotes that either infect other cells ❻ or revert to trypomastigotes that circulate in the bloodstream ❼. The circulating trypomastigotes are subsequently ingested by a kissing bug when it takes a blood meal ❽. Within the midgut of the kissing bug, binary fission of the trypomastigotes produces epimastigotes ❾, completing the cycle.

The life cycle of *T. cruzi* differs from that of *T. brucei* in several ways. Whereas *T. brucei* matures in the salivary glands of a tsetse fly, *T. cruzi* matures in the hindgut of a kissing bug; additionally, whereas *T. brucei* is injected directly through the vector's bite, *T. cruzi* enters the host when the vector's feces are rubbed into a bite wound. Finally, *T. cruzi* has an intracellular stage, whereas *T. brucei* does not. Chagas' disease progresses over the course of several months through the following four stages:

1. An acute stage characterized by *chagomas,* which are swellings at the sites of each of the bites

2. A generalized stage characterized by fever, swollen lymph nodes, myocarditis (inflammation of the heart muscle), and enlargement of the spleen, esophagus, and colon

3. A chronic stage, which is asymptomatic and can last for years

4. A symptomatic stage characterized primarily by congestive heart failure following the formation of *pseudocysts,* which are clusters of amastigotes in heart muscle tissue

Parasite-induced heart disease is one of the leading causes of death in Latin America.

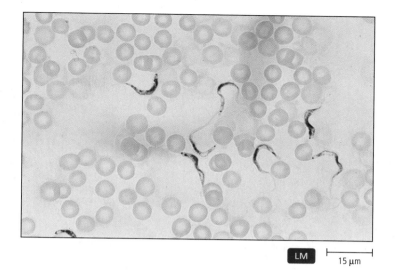

LM | 15 µm

▲ *Figure 23.4*

Mature typomastigotes of *T. brucei* among erthyrocytes in a blood smear. The trypomastigotes of *T. cruzi* are more curled than those of *T. brucei*.

Microscopic identification of trypanosomes or their antigens in blood, lymph, spinal fluid, or a tissue biopsy is diagnostic for *T. cruzi*. Trypomastigotes of *T. cruzi* are generally more curled than those of *T. brucei*. Another simple and practical diagnostic method is *xenodiagnosis.* In this procedure, uninfected kissing bugs are allowed to feed on a person suspected of having a *T. cruzi* infection. Four weeks later, the bugs are dissected. The presence of parasites within the hindgut of the bug indicates the presence of the infection.

In its earliest stages, Chagas' disease, can be treated with benzonidazole or nifurtimox; late stages cannot be treated. Prevention of Chagas' disease involves replacing thatch and mud building materials, which provide homes for the bugs, with concrete and brick. The use of insecticides, both personally and in the home, and sleeping under insecticide-impregnated netting can prevent insect feeding. No vaccines exist for Chagas' disease.

Leishmania

Leishmania (lēsh-man'ē-ă) is a genus of protozoa commonly hosted by wild and domestic dogs and small rodents; that is, it is zoonotic. *Leishmania* is endemic in parts of the tropics and subtropics, including Central and South America, central and southern Asia, Africa, Europe, and the Middle East. Twenty-one of the 30 known species of *Leishmania* can infect humans, causing **leishmaniasis** (lēsh'mă-nī'ă-sis). Of the more than 1.5 million cases of leishmaniasis reported each year, over 45,000 are fatal. The parasites are transmitted through the bite of infected female sand flies of the genera *Phlebotomus* and *Lutzomyia*. *Leishmania* are obligate parasites, and they are so dependent on sand flies for transmission that the parasite's life cycle has become closely adapted to the feeding rhythms of the flies.

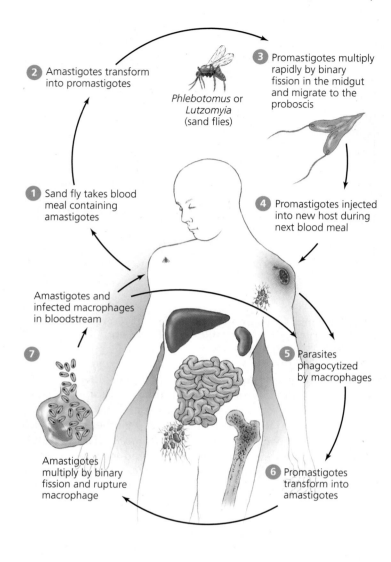

▲ *Figure 23.5*

The life cycle of *Leishmania*. *What is the effect of amastigote replication on the host's immune system?*

Figure 23.5 The loss of macrophages severely inhibits the host's immune response.

Leishmania has two developmental stages: *amastigotes*, which lack flagella and multiply within the mammalian host's macrophages and monocytes, and *promastigotes*, each of which has a single anterior flagellum and develops extracellularly within the vector's gut. In the life cycle of *Leishmania* **(Figure 23.5),** a sand fly feeding on a host ingests phagocytes containing amastigotes during a blood meal **1**. The amastigotes are released from the phagocytes in the fly's midgut, where they transform into promastigotes **2**. The rapidly dividing promastigotes fill the fly's digestive tract and migrate to the proboscis **3**, so that each time the fly feeds, it injects promastigotes into a new host **4**. Macrophages near the bite site phagocytize the promastigotes **5**, which then transform into amastigotes **6**. The

amastigotes reproduce via binary fission until the macrophage ruptures, releasing amastigotes that infect other macrophages and/or circulate in the bloodstream **7**, where they can be ingested by a sand fly during a blood meal.

The clinical manifestations of leishmaniasis depend on the species of *Leishmania*, the geographical location of hosts and flies, and the immune response of the infected host. Upon initial infection, macrophages become activated, but not sufficiently to kill the intracellular parasites. Chemicals released by the macrophages stimulate inflammatory responses that continue to be propagated by the infection of new macrophages. Depletion of macrophage numbers due to the reproduction of amastigotes decreases the efficiency of the immune response. The severity of the resulting immune dysfunction depends on the overall number of macrophages infected.

Three clinical forms of leishmaniasis are commonly observed. *Cutaneous leishmaniasis* involves large painless skin ulcers that form around the bite wounds. Such lesions often become secondarily infected with bacteria. Scars remain when the lesions heal. *Mucocutaneous leishmaniasis* results when skin lesions enlarge to encompass the mucous membranes of the mouth, nose, or soft palate. Damage is severe and permanently disfiguring **(Figure 23.6).** Neither of these forms of leishmaniasis is fatal. However, *visceral leishmaniasis* (also known as *kala-azar*) is fatal in 95% of untreated cases. In this disease, macrophages spread the parasite to the liver, spleen, bone marrow, and lymph nodes. Inflammation, fever, weight loss, and anemia increase in severity as the disease progresses. Visceral leishmaniasis is becoming increasingly problematic as an opportunistic infection among AIDS patients.

Microscopic identification of amastigotes in samples from cutaneous lesions, the spleen, or bone marrow is diagnostic of *Leishmania* infection. Immunoassays using antibodies to detect antigen can be used to confirm the diagnosis and to identify the strain. Molecular techniques such as PCR are needed for definitive speciation.

Most cases of leishmaniasis heal without treatment (though scars may remain) and confer immunity upon the recovered patient. At one time, lesions were purposefully encouraged on the buttocks of small children both to induce immunity and to prevent lesions from forming elsewhere and leaving more-visible scars. Treatment, which is required for more serious infections and for visceral leishmaniasis, generally involves administering sodium stibogluconate or meglumine antimonate. For resistant strains of *Leishmania*, pentamidine has been used with some success.

Prevention is essentially limited to reducing exposure by controlling reservoir host and sand fly populations. For example, rodent nesting sites and burrows can be destroyed around human habitations to reduce contact with potentially infected reservoir populations. In some areas, infected dogs are also destroyed to eliminate this reservoir. Spraying insecticide around homes can reduce the number of sand flies. Personal use of insect repellents, protective clothing, and netting further limits exposure. A vaccine for leishmaniasis does not currently exist.

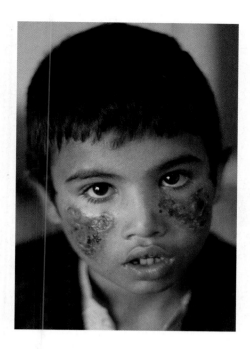

▲ *Figure 23.6*

The large, permanently disfiguring skin lesions
of mucocutaneous leishmaniasis.

LM | 2.5 μm

▲ *Figure 23.7*

A trophozoite of *Giardia intestinalis*, showing
its distinctive resemblance to a face.

CRITICAL THINKING

Given the regions of the world where *Leishmania* and HIV are endemic,
would you expect the incidence of *Leishmania* to increase or decrease in
the next decade? Explain your answer.

Giardia

Giardia intestinalis (jē-ar′dē-ă; formerly called *G. lamblia*)
is the causative agent of **giardiasis** (jē-ar-dī′ă-sis), one of the
more common water-borne gastrointestinal diseases in the
United States. *Giardia* is found in the intestinal tracts of animals and humans worldwide, but as the organism is very
hardy, it can also be found in water, in soil, on food, and on
surfaces that have been contaminated with feces. The organism can survive for months in the environment due to the
protective outer shell of its cyst.

Infection usually results from the ingestion of cysts in
contaminated drinking water. Accidental ingestion during
swimming or other water activities can also lead to infection. Hikers, campers, and their pets are at particular risk because wild animals can shed *Giardia* into mountain streams.
Because beavers are common zoonotic sources of *Giardia*, giardiasis is sometimes referred to as "beaver fever." Even if
humans avoid drinking stream water, they usually don't
think twice about letting their dogs drink it. Unfortunately,
it is often only a matter of time before the dog passes the
protozoa on to its owners. Alternatively, eating unwashed
raw fruits or vegetables that have been contaminated by feces can lead to infection, as can contact with feces during

sex. Giardiasis outbreaks in day care facilities are usually the
result of children putting contaminated toys or eating utensils into their mouths.

Giardia has a life cycle similar to that of *Entamoeba*. Following ingestion, each cyst releases a trophozoite into the
small intestine that begins multiplying via binary fission.
Trophozoites can either remain free in the lumen of the small
intestine or attach to the intestinal mucosa via a ventral
sucking disk; they do not, however, invade the intestinal
wall. As trophozoites pass into the colon, encystment occurs. Cysts are immediately infective upon release, leading
to a significant incidence of person-to-person transmission.

Giardiasis can range from an asymptomatic infection to
significant gastrointestinal disease. Signs and symptoms,
when they occur, include severe watery diarrhea, abdominal
pain, bloating, nausea, vomiting, ineffective absorption of
nutrients, and low-grade fever. Stools are foul smelling, usually with the "rotten-egg" smell of hydrogen sulfide. Incubation lasts roughly 1–2 weeks, and symptoms resolve after
1–4 weeks in normal, healthy adults. In extreme cases, the
attachment of parasites to the intestinal mucosa can cause
superficial tissue damage, and fluid loss can become life
threatening. Chronic giardiasis can occur, often among animals.

Diagnosis of *Giardia*, which involves direct microscopic examination of stool specimens, is based on the observation of flat, pear-shaped trophozoites that resemble a
face when viewed ventrally **(Figure 23.7)**. In addition to
the "face," which results from two nuclei that resemble

eyes, four pairs of flagella extending from the ventral surface may also be visible. Several stool samples may need to be examined, as the parasites are shed only intermittently from the intestinal tract.

Metronidazole is the drug of choice for treatment of giardiasis. If diarrhea is not present, treatment is often waived. Oral rehydration therapy may be required for severe cases of giardiasis, or to treat very young children regardless of the severity of infection.

To prevent infection in regions where *Giardia* is endemic, filtering water is necessary. When hiking, neither humans nor their pets should drink unfiltered stream or river water. Most stores that sell camping and hiking equipment sell portable water filtration kits, making it unnecessary to carry bottled water. Eating utensils should be cleaned with filtered water. In day care facilities, scrupulous hygiene practices and the separation of feeding and diaper-changing areas are essential to preventing transmission. Individuals recovering from infection should also be extremely vigilant with their own personal hygiene to avoid transmitting *Giardia* to family members, and should avoid swimming in public or natural waterways for several weeks after recovery to ensure they do not shed parasites into the water.

CRITICAL THINKING

How does the visually distinctive appearance of *Giardia* trophozoites improve the success of medical treatment of giardiasis as compared to that for amoebic infections?

Trichomonas

Trichomonas vaginalis (trik-ō-mō′nas va-jin-al′is) is globally distributed and is the most common protozoan disease of humans in industrialized nations. The parasite lives on the vulvas and in the vaginas of women, and in the urethras and prostates of men. This obligate parasite, which is incapable of surviving long outside a human host, is transmitted almost exclusively via sex. *T. vaginalis* occurs most frequently in people with a preexisting sexually transmitted disease, such as chlamydial infection, and in people with multiple sex partners. No animal reservoirs exist, but men serve as reservoirs for their female companions.

In women, infection results in **vaginitis** (vaj-i-nī′tis), which is accompanied by a purulent (pus-filled) odorous discharge, vaginal and cervical lesions, abdominal pain, painful urination, and painful intercourse. Trophozoites feed on vaginal tissue, leading to erosion of the epithelium. In men, *T. vaginalis* infection may cause inflammation of the urethra or bladder, but more typically it is asymptomatic.

Microscopic observation of actively motile trophozoites in vaginal and urethral secretions is diagnostic. The trophozoites are flat and possess several flagella and a sail-like undulating membrane **(Figure 23.8).** An immunofluorescent assay can be performed when infection is suspected and microscopy is too insensitive. Patients and all their sexual

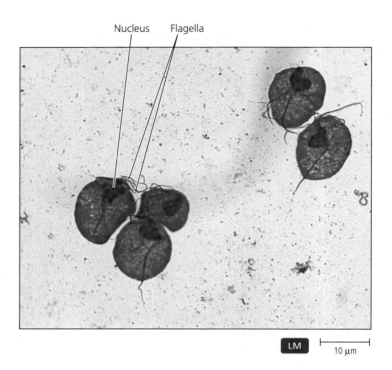

Nucleus   Flagella

LM ⊢———⊣ 10 μm

▲ *Figure 23.8*

Trophozoites of *Trichomonas vaginalis.*

partners must be treated to prevent reinfection. Metronidazole is the drug of choice for treatment, though some resistant strains of *T. vaginalis* exist. Prevention involves abstinence, mutual monogamy, and condom usage.

Apicomplexans

Learning Objectives

✓ Describe the life cycle of *Plasmodium,* and relate malarial symptoms to stages in the life cycle.

✓ Describe the clinical manifestations of *Toxoplasma gondii* infections.

✓ Compare and contrast the intestinal diseases caused by *Cryptosporidium* and *Cyclospora.*

Apicomplexans (ap-i-kom-plek′sănz) are protozoa whose infective forms are characterized by an ornate complex of organelles at their apical ends, which gives the group its name. They have also been called *sporozoa* because during their life cycles they assume nonmotile, sporelike shapes, but this term is not accurate because the term *spores* correctly refers to reproductive structures of multicellular plants and fungi. All apicomplexans are parasites of animals, and all have complicated life cycles involving at least two types of hosts. A major feature of apicomplexan life cycles is *schizogony,* a form of asexual reproduction in which multinucleate *schizonts* form before the cells divide (see Figure 12.3).

Plasmodium, Toxoplasma, Cryptosporidium, and *Cyclospora* are four important apicomplexan parasites. We begin our

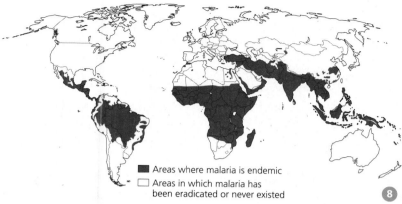

◀ *Figure 23.9*

The geographical distribution of malaria. The disease appears to be returning to areas from which it was once eradicated, including Europe and the United States.

■ Areas where malaria is endemic
☐ Areas in which malaria has been eradicated or never existed

discussion with *Plasmodium*, the causative agent of malaria, and the most infamous protozoan parasite in the world.

Plasmodium

Four species of *Plasmodium* (plaz-mō'dē-ŭm) cause **malaria** in humans: *P. falciparum* (fal-sip'ă-rŭm), *P. vivax* (vī'vaks), *P. ovale* (ō-vă'lē), and *P. malariae* (mă-lār'ē-ī). Malaria is endemic in over 100 countries throughout the tropics and subtropics, where the parasites' mosquito vector breeds **(Figure 23.9).** The movement of infected individuals and mosquitoes continues to expand beyond these endemic areas, threatening to reintroduce *Plasmodium* into countries, including the United States, where mosquito eradication programs eliminated the disease decades ago. Over 400 million people are infected with *Plasmodium*, and 2–3 million (usually children) die annually. Females of sixty different species of the mosquito genus *Anopheles* serve as vectors of *Plasmodium*. (Adult male mosquitoes do not feed.)

The life cycle of *Plasmodium* has three prominent stages **(Figure 23.10):**

1. The **exoerythrocytic cycle,** as its name indicates, occurs outside of red blood cells. During a blood meal on a human, an infected female mosquito injects infective *sporozoites* along with saliva containing an anticoagulant into the blood **①**. The sporozoites reach the liver via the bloodstream, and over the course of 1–2 weeks undergo schizogony within liver cells **②**. The numerous *merozoites* produced via schizogony rupture the liver cells and enter the blood **③**. *P. ovale* and *P. vivax* are further capable of forming *hypnozoites*, which can remain dormant in liver cells for years and can be reactivated at any time to cause relapses of malaria.

2. The **erythrocytic cycle ④** begins when free merozoites penetrate erythrocytes and become *trophozoites* that phagocytize the erythrocytes' hemoglobin. The trophozoites undergo schizogony to produce numerous merozoites that are released from the lysed erythrocytes. Merozoites typically infect new erythrocytes, but some develop into male and female *gametocytes* within

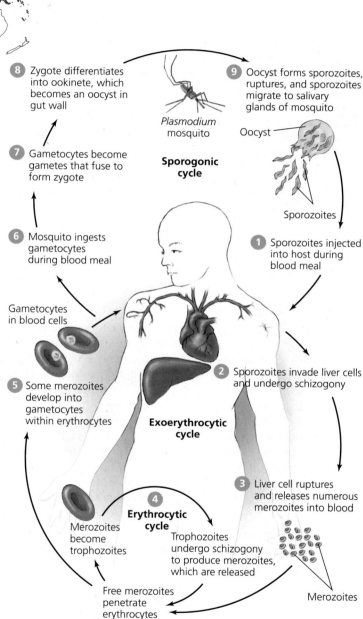

⑧ Zygote differentiates into ookinete, which becomes an oocyst in gut wall

⑨ Oocyst forms sporozoites, ruptures, and sporozoites migrate to salivary glands of mosquito

Plasmodium mosquito

Oocyst

Sporogonic cycle

⑦ Gametocytes become gametes that fuse to form zygote

⑥ Mosquito ingests gametocytes during blood meal

Sporozoites

① Sporozoites injected into host during blood meal

Gametocytes in blood cells

② Sporozoites invade liver cells and undergo schizogony

⑤ Some merozoites develop into gametocytes within erythrocytes

Exoerythrocytic cycle

③ Liver cell ruptures and releases numerous merozoites into blood

④ **Erythrocytic cycle**

Merozoites become trophozoites

Trophozoites undergo schizogony to produce merozoites, which are released

Free merozoites penetrate erythrocytes

Merozoites

▲ *Figure 23.10*

The life cycle of *Plasmodium*, which includes the exoerythrocytic and the erythrocytic cycles in humans, and the sporogonic cycle in mosquitoes. *During which cycle can malaria be diagnosed in humans?*

Figure 23.10 Malaria in humans can be diagnosed during the parasite's erythrocytic cycle.

Case Study 23.1   A Protozoan Mystery

A 20-year-old student was admitted to his college's student health center with headaches and fever shortly after beginning the fall semester. He had spent his summer working with an international aid organization in Nigeria and had returned to the United States only a week earlier. Gross examination revealed a diffuse rash and numerous insect bites. The young man had spent most of his summer outdoors in rural areas, had taken no prophylactic anti-malarial medication, and had spent some time on African game reserves working with the families of local guides. The patient could not specifically remember receiving any of

LM   15 µm

the bite wounds on his body, and he did not always use insect repellent in the field. The young man was admitted to the local hospital, where intermittent fever, nausea, and headache continued. Initial blood smears proved negative for malaria.

1. What are some possible protozoan diseases the patient could have contracted in Africa?

2. Can the disease be identified from the symptoms alone?

3. Pictured is a blood smear. Based on this, what would you conclude as to the cause of the disease?

4. What would the treatment be if the patient had tested positive for malaria?

5. What treatment would you now recommend?

6. What prevention would you have suggested to this individual?

7. Is there a local threat to anyone else for contracting this disease from the young man?

---

circulating blood cells ⑤. Lysis of erythrocytes occurs cyclically—every 48–72 hours, depending on the species of *Plasmodium* involved.

3. The **sporogonic cycle** begins when a female mosquito feeding on an infected human ingests gametocytes within erythrocytes ⑥. Freed from erythrocytes within the mosquito's digestive tract, gametocytes develop into *gametes*, and male gametes fertilize female gametes to produce *zygotes* ⑦. A zygote differentiates into an *ookinete*, which penetrates the mosquito's gut wall and becomes an *oocyst* ⑧. Some 10–20 days later, the oocyst ruptures, releasing thousands of sporozoites that migrate to the mosquito's salivary glands ⑨. The mosquito then transmits the sporozoites to a new host during a blood meal, completing the parasite's life cycle.

People living in endemic areas, and their descendants throughout the world, commonly have one or more of the following genetic traits that increase their resistance to malaria:

- *Sickle-cell trait*. Individuals with this gene produce an abnormal type of hemoglobin called hemoglobin S (hemoglobin A is normal). Hemoglobin S causes erythrocytes to become sickle-shaped, and somehow it also makes erythrocytes resist penetration by *Plasmodium*.

- *Hemoglobin C*. Humans with two genes for hemoglobin C are invulnerable to malaria. The mechanism by which this mutation provides protection is unknown.

- Genetic deficiency of *glucose-6-phosphate dehydrogenase*, an enzyme that trophozoites must acquire from their host before the trophozoites can synthesize DNA.

- Lack of so-called *Duffy antigens* on erythrocytes. Because *P. vivax* requires Duffy antigens to attach to and infect erythrocytes, Duffy-negative individuals are resistant to this pathogen.

The severity of malaria caused by each species varies. *P. ovale* generally causes mild disease; *P. vivax* usually results in chronic malaria, with periodic recurrence of symptoms; *P. malariae* and *P. falciparum* cause more serious malaria, with *P. falciparum* infection most often fatal.

The general symptoms of malaria are associated with synchronous cycles of erythrocyte lysis. Two weeks after the erythrocytic cycle begins, sufficient numbers of parasites exist to cause symptoms of fever, chills, diarrhea, headache, and (occasionally) pulmonary or cardiac dysfunction. Fever correlates with erythrocyte lysis and most likely results from efforts of the immune system to remove cellular debris, toxins, and merozoites. Loss of erythrocytes leads to anemia, weakness, and fatigue. The inability of the liver to process the inordinate amount of hemoglobin released from dying erythrocytes results in *jaundice*.

P. falciparum causes a form of malaria called *blackwater fever*, characterized by extreme fever, large-scale erythrocyte lysis, renal failure, and dark urine discolored by excreted hemoglobin. Protozoan proteins on the surfaces of erythrocytes cause adherence of erythrocytes to the capillary endothelium, blocking blood flow and causing small hemorrhages in various tissues (and ultimately tissue death). *Cerebral malaria* results when tissue death occurs in the brain. Falciparum malaria can be fatal within 24 hours of the onset of symptoms.

New Frontiers 23.1   **In Search of a Malaria Vaccine**

Malaria infects and kills millions worldwide each year. No vaccines currently exist, but among the many approaches being investigated is the development of a vaccine that does not kill the parasite itself, but instead attacks a toxin produced by the parasite.

Even though the classic symptoms of malaria—fever, chills, diarrhea, and headache—are caused by erythrocytic lysis by malarial para-

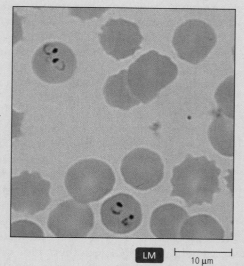

LM   10 μm

Malarial trophozoites inside red blood cells

sites, much of the severity of the disease is believed to result from a malarial toxin. One suspected toxin is glycosylphosphatidylinositol, or GPI. Although GPI has not been conclusively shown to promote pathogenesis and death in humans, studies in mice have shown that vaccinating against malarial-GPI can mitigate some of the most dangerous aspects of malaria, including host immune responses that compound malarial pathogenesis. An anti-GPI vaccine stimulates antibody production that is protective in mice. The results suggest that similar vaccines may be successful in humans.

Antitoxin vaccines work successfully against bacterial toxins such as diphtheria and tetanus toxins, so the potential for an effective anti-GPI vaccine appears good. While it would neither kill the parasite nor protect people from infection, it could alleviate symptoms enough to allow recovery and immunological memory. For people at risk for malaria, an anti-GPI vaccine offers the hope of living long and productive lives.

Source: Schofield, L., et al. 2002. Synthetic GPI as a candidate anti-toxic vaccine in a model of malaria. *Nature* 418:785–789.

If a victim survives the acute stages of malaria, immunity gradually develops. Periodic episodes become less severe over time—unless the victim becomes immunocompromised, in which case episodes can return to original levels of severity.

Microscopy is most commonly used for diagnosing malaria because *Plasmodium* species can be readily identified and distinguished in blood smears. Serology can also be used for differential diagnosis. However, because the symptoms might be confused with those of less severe infections, diagnosis may be missed in nonendemic areas, unless a good case history, including travel history, is obtained.

Treatment varies by species and severity of symptoms. Drug choices are constantly revised based on effectiveness and the existence of resistant parasites. Standard antimalarial drugs include chloroquine, used for uncomplicated malarial infections in areas where drug resistance by *Plasmodium* does not exist, and quinine, which is used in severe infections and when chloroquine fails. Fansidar is an alternative to chloroquine. Primaquine is used to eliminate hypnozoites of *P. vivax* and *P. ovale*. Antifever medication and blood transfusions may be required as supportive measures.

Control of malaria involves limiting contact with mosquitoes carrying *Plasmodium*. Widespread use of insecticides, drainage of wetlands, and removal of standing water can reduce mosquito breeding rates. Personal use of insect repellents, netting, and protective clothing reduces mosquito bites. When traveling from nonendemic to endemic areas, prophylactic medication can prevent infection. *Proguanil*, the prophylactic drug of choice, is usually taken several weeks before and after travel. Chloroquine is some-

times given prophylactically. Several malaria vaccines are currently under development, but scientists have not yet developed an effective vaccine (see **New Frontiers 23.1**).

Toxoplasma

Toxoplasma gondii (tok-sō-plaz′mă gon′dē-ē) is one of the world's most widely distributed protozoan parasites of humans. The CDC estimates that up to 40% of the U.S. population has been exposed to this obligate intracellular parasite. Wild and domestic mammals and birds are the major reservoirs for the protozoan, and cats are the definitive host, in which the protozoan reproduces sexually.

Humans typically become infected by ingesting undercooked meat containing the parasite. People at greatest risk include butchers, hunters, and anyone who tastes food while preparing it. Ingestion or inhalation with contaminated soil can also be a source of infection. The protozoa can also cross the human placenta to infect the fetus. Historically, contact with infected cats and their feces was proposed as a major risk for infection, but recent studies have shown that cats are not the major source of infection for humans because they shed *Toxoplasma* only briefly.

In the life cycle of *Toxoplasma* (**Figure 23.11** on page 662), male and female gametes in the cat's digestive tract fuse to form zygotes, which develop into immature *oocysts* that are shed in the feces **1**. Each day a cat can excrete up to 10 million oocysts, which can survive in moist soil for several months. Vegetation growing in soil containing the parasites becomes contaminated with oocysts. As each oocyst matures, it produces *sporozoites*. **2** When rodents, livestock, or

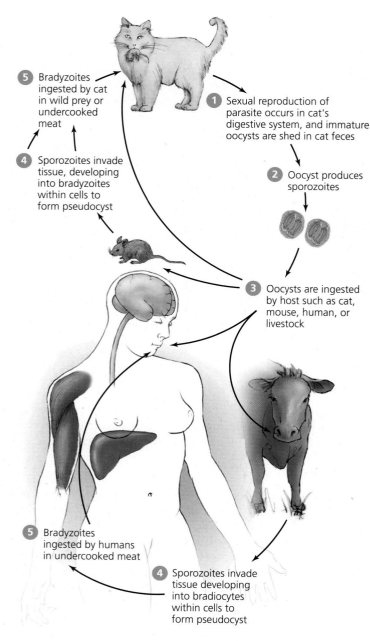

⑤ Bradyzoites ingested by cat in wild prey or undercooked meat

① Sexual reproduction of parasite occurs in cat's digestive system, and immature oocysts are shed in cat feces

④ Sporozoites invade tissue, developing into bradyzoites within cells to form pseudocyst

② Oocyst produces sporozoites

③ Oocysts are ingested by host such as cat, mouse, human, or livestock

⑤ Bradyzoites ingested by humans in undercooked meat

④ Sporozoites invade tissue developing into bradiocytes within cells to form pseudocyst

▲ *Figure 23.11*
The life cycle of *Toxoplasma gondii*.

humans ingest mature oocysts on vegetation ③, digestion releases the sporozoites, which invade the host's heart, tongue, and diaphragm cells and proliferate asexually to produce *pseudocysts* containing numerous *bradyzoites* ④. Pseudocysts remain dormant in the tissues of livestock or rodents until humans or cats, respectively, consume these hosts ⑤.

Although the majority of people infected with *T. gondii* have no symptoms, a small percentage develop **toxoplasmosis** (tok'sō-plaz-mō-sis), a febrile (fever-producing) illness with other symptoms resembling pneumonia, mononucleo-sis, hepatitis, or myocarditis. Toxoplasmosis generally results in no permanent damage and is self-limiting, resolving spontaneously within a few months to a year.

Toxoplasmosis is more severe in two populations: AIDS patients and fetuses. In AIDS patients, symptoms are thought to result as tissue pseudocysts reactivate when the immune system fails. Spastic paralysis, blindness, myocarditis, encephalitis, and death result. Transplacental transfer from mother to fetus is most dangerous in the first trimester of pregnancy; it can result in spontaneous abortion, stillbirth, epilepsy, mental retardation, *microcephaly* (abnormally small head), inflammation of the retina, blindness, anemia, jaundice, and other conditions. Ocular infections may remain dormant for years, at which time blindness develops.

Physicians diagnose *T. gondii* infection via serological detection of organisms in tissue samples, via microscopic identification of parasites in tissue biopsies, or via molecular identification of *T. gondii* genetic material or products in specimens using PCR, Southern blot, or DNA probes. Serology is the most common diagnostic method. Molecular techniques are most useful for congenital toxoplasmosis.

In healthy adults, toxoplasmosis normally resolves without treatment. When treatment is recommended, pyrimethamine, sulfadiazine, and spiramycin are administered together for 3–4 weeks. Treatment of pregnant women prevents most transplacental infections. More aggressive treatment may be needed in AIDS patients, including the addition of steroids to reduce tissue inflammation.

Controlling the incidence of *T. gondii* infection is difficult because so many hosts harbor the parasites. A vaccine for cats is currently under development to reduce the chance of pet-to-owner transmission. The best prevention is to thoroughly cook or deep-freeze meats, and to avoid contact with contaminated soil.

Cryptosporidium

Cryptosporidiosis (krip'tō-spō-rid-ē-ō'sis) is a zoonotic disease caused by *Cryptosporidium parvum* (krip'tō-spō-rid'ē-ŭm par'vŭm). Once thought to only infect livestock and poultry, *Cryptosporidium* is carried asymptomatically by about 30% of people living in developing nations. It is estimated that most natural waterways in the United States are contaminated with the oocysts of *Cryptosporidium* due to the presence of livestock wastes.

Infection most commonly results from drinking water contaminated with oocysts, but direct fecal-oral transmission resulting from poor hygienic practices also occurs, particularly in day care facilities. In the intestine, oocysts release sporozoites, which invade the intestinal mucosa and become intracellular parasites. Eventually, new oocysts are produced, released into the lumen of the intestine, and shed in the feces.

Signs and symptoms of cryptosporidiosis include severe diarrhea that lasts from 1–2 weeks accompanied by headache, muscular pain, cramping, and severe fluid and weight loss. In HIV-positive individuals, chronic cryptosporidiosis is one of the indicator diseases that reveals when a person has reached the clinical stage of AIDS. In AIDS patients, the disease is life threatening.

Most diagnostic techniques are not very sensitive for *Cryptosporidium,* but microscopic examination of concentrated fecal samples or biopsy material can reveal oocysts **(Figure 23.12).** Fluorescent-labeled antibodies can detect low concentrations of oocysts in appropriate specimens.

Treatment essentially consists of oral rehydration therapy, because no drugs are effective against *Cryptosporidium.* Drinking from rivers and streams should be avoided in areas where *Cryptosporidium* is found; filtration is required to remove oocysts from drinking water because they cannot be killed by chlorination. Good personal hygiene can eliminate fecal-oral transmission of the parasite.

Cyclospora

Cyclospora cayetanensis (sī-klō-spōr'ă kī'ē-tan-en'sis) is a water-borne apicomplexan that has been responsible for hundreds of disease cases in the United States since the early 1980s. Outbreaks have been linked in particular to raspberries imported from Central and South America. Infection occurs when humans ingest oocysts, whether on contaminated raw berries, fruits, or vegetables or in contaminated drinking water. The disease is not transmitted between individuals. The environmental reservoir for *Cyclospora* remains unknown.

Once oocysts enter the intestine, they release invasive parasites. Symptoms, which develop after about a week, include cramps, watery diarrhea, myalgia (muscle pain), and fever. The disease usually resolves in days or weeks. Symptoms are more severe in AIDS patients, leading to severe dehydration and weight loss.

Microscopic examination of stool samples can reveal the presence of oocysts in heavier infections, but fluorescent DNA probes more reliably detect the presence of *Cyclospora* in human infections. *Cyclospora* infections are treated with trimethoprim and sulfamethoxazole given in combination for 7 days.

Reliable food testing does not currently exist but would be an important tool in preventing contaminated food from reaching U.S. consumers. The only reliable methods for reducing risk of infection in the United States are thoroughly washing fruits and vegetables prior to eating them raw, or cooking them; freezing also kills many oocysts.

Table 23.1 on page 664 lists the general characteristics of the protozoan parasites discussed in this chapter.

Helminthic Parasites of Humans

Learning Objective

✓ Describe the general morphological and physiological features of parasitic worms.

Helminths are macroscopic, multicellular, eukaryotic worms found throughout the natural world, some in para-

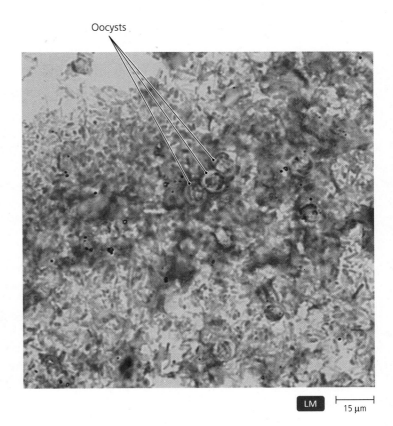

Oocysts

LM   15 μm

▲ *Figure 23.12*
Oocysts of *Cryptosporidium parvum* embedded in the wall of the intestine.

sitic associations with other animals. They are not microorganisms, though microbiologists generally study them.

Taxonomists divide helminths into three groups: **Cestodes** (ses'tōdz), commonly called *tapeworms,* are long, flat, and segmented. **Trematodes** (trem'a-tōdz), also called *flukes,* are also flat, but are leaf-shaped and are unsegmented. **Nematodes** (nēm'a-tōdz), or *roundworms,* are long, thin, unsegmented, cylindrical worms that taper to points at each end.

The reproductive systems and life cycles of parasitic helminths are quite complex. For many helminths, intermediate hosts are required to support larval (immature) stages needed for the worms to reach maturity. Adult worms are either *dioecious,*[2] meaning that the male and female sex organs are in separate worms, or *monoecious,*[3] meaning that each worm has both sex organs. Monoecious organisms either fertilize each other during copulation or are capable of self-fertilization, depending on the species. Most parasitic helminths release large numbers of fertilized eggs into the environment, so that transmission to new hosts is likely to occur.

---

[2]From Greek *di,* meaning two, and *oikos,* meaning house.
[3]From Greek *mono,* meaning one, *and oikos,* meaning house.

Table 23.1 Key Features of Protozoan Parasites of Humans

Organism	Primary Diseases	Geographical Distribution	Mode of Transmission	Host Organisms
Ciliates				
Balantidium coli	Balantidiasis, dysentery	Worldwide	Fecal-oral	Pigs, rodents, primates, humans
Amoebae				
Entamoeba histolytica	Luminal amoebiasis, amoebic dysentery, invasive extraintestinal amoebiasis	Worldwide	Fecal-oral	Humans
Acanthamoeba spp.	Ulcerative keratitis, amoebic encephalitis	Worldwide	Contact	Humans
Naegleria	Primary amoebic meningoencephalitis	Worldwide	Inhalation	Humans
Flagellates				
Trypanosoma brucei	African sleeping sickness	African subcontinent	Tsetse fly (*Glossina*)	Wild game, pigs, humans
Trypanosoma cruzi	Chagas' disease	Central and South America	Kissing bugs (*Triatoma*)	Opossums, armadillos, humans
Leishmania spp.	Cutaneous, mucocutaneous, or visceral leishmaniasis	Tropics, subtropics	Sand flies (*Phlebotomus, Lutzomyia*)	Canines, rodents, humans
Giardia intestinalis (lamblia)	Giardiasis	Developed nations, tropics	Fecal-oral	Humans, wild animals
Trichomonas vaginalis	Vaginitis	Developed nations	Sexual contact	Humans
Apicomplexans				
Plasmodium spp.	Malaria	Tropics, subtropics	Mosquitoes (*Anopheles*)	Humans
Toxoplasma gondii	Toxoplasmosis	Worldwide	Fecal-oral	Cats, livestock, humans
Cryptosporidium parvum	Cryptosporidiosis	Worldwide	Fecal-oral	Livestock, poultry, humans
Cyclospora cayetanensis	Gastrointestinal disorders	North, Central, and South America	Fecal-oral	Humans

Cestodes

Learning Objectives

✓ Describe the common features of the life cycles of tapeworms that infect humans.

✓ List the predominant modes of infection for *Taenia* and *Echinococcus*, and suggest measures to prevent infection.

All tapeworms are intestinal parasites that completely lack digestive systems. Though they differ in size when mature, all possess the same general body plan **(Figure 23.13).** A **scolex** (skō′leks), is a small attachment organ that possesses suckers and/or hooks used to attach the worm to host tissues. There is no mouth—anchorage is the only role of a scolex. Cestodes acquire nutrients by absorption through the worm's *cuticle* (outer "skin").

Behind the scolex is the neck region. Body segments, called **proglottids** (prō-glot′idz), grow from the neck continuously, so long as the worm remains attached to its host. New proglottids displace older ones moving them farther from the neck. Proglottids mature, producing both male and female reproductive organs. Thus a tapeworm's chain of proglottids, called a *strobila*, reflects a sequence of development: Proglottids near the neck are immature, those in the

middle are mature, and those near the end are *gravid*[4]—full of fertilized eggs. Each proglottid is monoecious and may fertilize other proglottids of the same or different tapeworm; a proglottid is not capable of fertilizing itself, but a tapeworm may be.

After fertilization occurs and the proglottids fill with eggs, gravid proglottids break off the chain and pass out of the intestine along with feces. In a few cases, the proglottids actually rupture within the intestine, releasing eggs directly into the feces. Some tapeworms produce proglottids large enough to be obviously visible in stools.

Generalized Tapeworm Life Cycle

The generalized life cycle of most tapeworms of humans is depicted in **Figure 23.14.** Gravid proglottids and/or eggs enter the environment in feces from infected humans (definitive hosts) that harbor adult tapeworms in their intestines ❶. Intermediate hosts, which vary with the species of tapeworm, become infected by ingesting vegetation contaminated with gravid proglottids or eggs ❷. Once the eggs reach the intermediate host's intestines, they hatch into

---

[4]From Latin *gravidus*, meaning heavy—that is, pregnant.

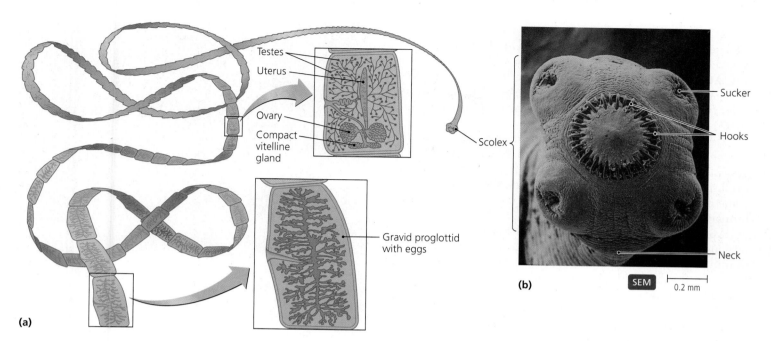

(a)

(b)

SEM   0.2 mm

▲ *Figure 23.13*

Features of tapeworm morphology. Each tapeworm consists of an organ of attachment called a scolex (which in this case bears hooks in addition to suckers), a neck, and a long chain of segments called proglottids.

larvae that penetrate through the intestinal wall and migrate to other tissues, often muscle ③, where the larvae develop into immature forms called **cysticerci** (sis-ti-ser′sī) ④. Cysticerci are generally harmless to the intermediate hosts. Humans become infected by consuming meat containing cysticerci ⑤. In the human intestine the cysticerci excyst, attach to the intestinal wall, and mature ⑥, until they develop into new adult tapeworms ⑦, which eventually shed gravid proglottids, thus completing the cycle.

CRITICAL THINKING

How might humans become intermediate hosts instead of definitive hosts?

Taenia

Taenia saginata (te′ne-a sa-ji-na′ta), the beef tapeworm, and *Taenia solium* (te′ne-a so′lī-um), the pork tapeworm, are so named because cattle and swine, respectively, serve as intermediate hosts. Both tapeworm species are distributed worldwide in areas where beef and pork are eaten. Poor, rural areas with inadequate sewage treatment and where humans and livestock live in close proximity have the highest incidence of human infection. Overall *Taenia* infections in humans are rare in the United States.

Proglottids are deposited in pasture or grazing land, where they can remain viable for months; cattle and swine become infected by eating contaminated vegetation. Humans

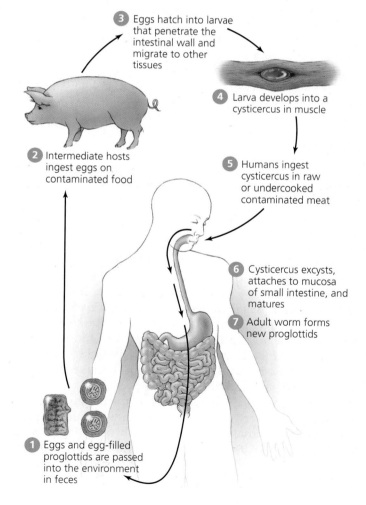

③ Eggs hatch into larvae that penetrate the intestinal wall and migrate to other tissues

④ Larva develops into a cysticercus in muscle

② Intermediate hosts ingest eggs on contaminated food

⑤ Humans ingest cysticercus in raw or undercooked contaminated meat

⑥ Cysticercus excysts, attaches to mucosa of small intestine, and matures

⑦ Adult worm forms new proglottids

① Eggs and egg-filled proglottids are passed into the environment in feces

▲ *Figure 23.14*

A generalized life cycle of some tapeworms of humans.

ingest cysticerci in raw or undercooked meat. Tapeworms attached to the intestinal epithelium can grow quite long. Infected human hosts pass approximately six proglottids per day. At maturity, *T. saginata* worms have 1000–2000 proglottids, each of which can contain 100,000 eggs. *T. solium* adults average 1000 proglottids, each of which contains about 50,000 eggs.

Rarely do humans become intermediate hosts of *T. solium* instead of definitive hosts when they ingest eggs or gravid proglottids rather than cysticerci. Larvae released from the eggs can then invade human muscle tissue and become cysticerci. For the parasite this is a dead end because humans are not normally cannibals.

Most people often actively shed proglottids without experiencing symptoms, though in some cases, nausea, abdominal pain, weight loss, and diarrhea accompany infection. If the tapeworm is particularly large, blockage of the intestine is possible.

Diagnosis is achieved by microscopic identification of proglottids (or sometimes eggs) in fecal samples at least 3 months after infection—after the worms have sufficiently matured. Examination of the scolex is required to differentiate between *T. saginata* and *T. solium*.

Treatment is with praziquantel. Rarely, surgery may be necessary to remove the tapeworms from the intestinal tract. Thoroughly cooking or freezing meat is the easiest method of prevention. Because cysticerci are readily visible in the meat, giving it a "mealy" look, inspecting meat either before shipment to market or before purchase can also reduce the chances that infected meat will be consumed.

Echinococcus

Echinococcus granulosus (ĕ-kī′nō-kok′ŭs gra-nū-lō′sŭs) is an unusual tapeworm of canines in that its body consists of only three proglottids—one immature, one sexually mature, and one gravid. A gravid proglottid is released each time the neck forms a new immature proglottid. Canines are infected by eating cysticerci in various herbivore hosts such as cattle, sheep, and deer. Incidence is highest in areas wherever livestock are kept, including the western United States.

Humans become accidental intermediate hosts by consuming food or water contaminated with *Echinococcus* eggs shed in dogs' feces. The eggs release larvae into the intestine, the larvae invade the circulatory system and are carried throughout the body where they form **hydatid**[5] (hī′da-tid) cysts and cause **hydatid disease.** Hydatid cysts form in any organ, but occur primarily in the liver.

Over several years, hydatid cysts can enlarge to the size of grapefruits (20 cm in diameter), filling with fluid and several million granular *protoscoleces*. Each protoscolex is the potential scolex of a new worm, but since dogs do not normally eat humans, the protoscoleces of hydatid cysts in humans are a dead end for *Echinococcus*.

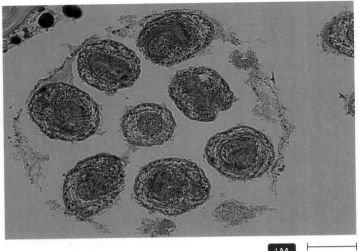

LM   30 µm

▲ *Figure 23.15*
Micrograph of a cross section of a hydatid cyst. Note the many protoscoleces, each of which can become a tapeworm's scolex if ingested by a dog.

Symptoms of hydatid disease follow the enlargement of cysts in infected tissue and result from tissue dysfunction. For example, hydatid cysts in the liver produce abdominal pain and obstruction of the bile ducts. If a hydatid cyst ruptures, the protoscoleces spread throughout the body via the bloodstream, and each protoscolex grows into a new hydatid cyst. Hydatid cysts in the brain, lungs, bones, heart, or kidneys can be fatal.

Diagnosis involves visualization of cyst masses using ultrasound, CT scan, or radiography, followed by serological confirmation via immunoassays. Microscopy reveals the presence of hydatid cysts **(Figure 23.15)**. Serology can be performed on aspirates or biopsies from liver, lungs, sputum, or urine, although biopsies should be avoided if possible as rupture of the cysts can lead to dissemination.

Treatment involves surgery to remove cysts, followed by the antihelminthic drug albendazole. Human infection is prevented by good hygiene practices to avoid fecal-oral transmission of eggs from infected pets. It is also a good idea to avoid drinking untreated water from streams or other waterways in areas where foxes, coyotes, or wolves roam.

Trematodes

Learning Objectives

✓ Explain how the life cycles of blood flukes differ from those of other flukes.

✓ Discuss the life cycles and disease manifestations of *Schistosoma* and *Fasciola*.

Trematodes, commonly known as **flukes** (flūks), are flat, leaf-shaped worms **(Figure 23.16)**. They lack complete digestive systems as they have no anus. Oral and ventral suckers

---

[5]From Greek *hydatis*, meaning a drop of water, presumably in reference to the fluid that accumulates within hydatid cysts.

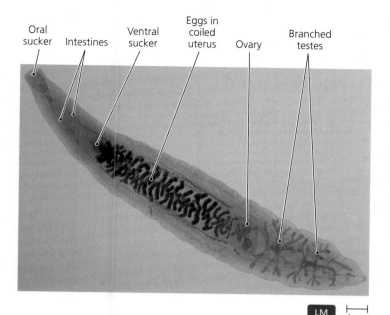

Oral sucker | Intestines | Ventral sucker | Eggs in coiled uterus | Ovary | Branched testes

LM | 1 mm

▲ *Figure 23.16*

Some features of fluke morphology. Note the two suckers, which keep the leaf-shaped flatworm attached to its host's tissues.

enable the parasites to attach to host tissues and obtain nutrients. The geographical distribution of most flukes is extremely limited, because the geographical distribution of the specific species of snails they require as intermediate hosts is also limited.

Researchers group flukes together according to the sites in the body they parasitize: blood, liver, lungs, or intestine. The following sections examine some representative blood flukes, and a representative liver fluke, but first we consider the common features of fluke life cycles.

Common Features of the Fluke Life Cycle

Flukes that cause disease in humans share similar complex life cycles (**Figure 23.17** on page 668). After fluke eggs reach the environment ❶, those deposited in freshwater hatch to release free-swimming larvae called *miracidia* ❷. Within 24 hours, miracidia actively seek out and burrow into a snail of a specific species ❸, where two or three cycles of asexual reproduction produce larvae called *cercariae* ❹, which leave the snail as free-swimming forms. In blood flukes, the cercariae aggressively seek out and penetrate the skin of humans ❺; in other flukes, cercariae encyst to become *metacercariae* on vegetation ❻ or within a second intermediate host ❼. For liver, lung, and intestinal flukes, humans become infected by ingesting metacercariae ❽. Once inside the definitive host, the parasites migrate to appropriate sites and develop into adults ❾, which then produce eggs that pass into the environment in feces or urine.

With the exception of one dioecious genus (*Schistosoma*), all flukes are monoecious. They are not self-fertile.

CRITICAL THINKING

Compare the general tapeworm life cycle to the general fluke life cycle. What is similar? What is different?

Blood Flukes: Schistosoma

The dioecious blood flukes in the genus *Schistosoma* (shis-tō-sō′mă) cause **schistosomiasis** (shis′tō-sō-mī′ă-sis)—a potentially fatal disease and one of the major public health problems in the world. The World Health Organization estimates that over 200 million people are infected worldwide, of which 20 million suffer serious disease and about 20,000 die annually. Most cases of schistosomiasis occur in sub-Saharan Africa, but *Schistosoma* is endemic in 76 countries.

Three geographically limited species of *Schistosoma* infect humans:

- *S. mansoni* (man-sō′nī) is common to the Caribbean, Venezuela, Brazil, Arabia, and large areas of Africa.
- *S. haematobium* (hē′mă-tō′bē-ŭm) is found only in Africa and India.
- *S. japonicum* (jă-pon′i-kŭm) occurs in China, Taiwan, the Philippines, and Japan, although infections in Japan are relatively rare.

Humans are the principal definitive host for most species of *Schistosoma* (*S. japonicum* can infect other mammals). Cercariae burrow through the skin of humans who contact contaminated water while washing clothes and utensils, bathing, or swimming. The larvae enter the circulatory system where they mature and mate. Females lay eggs, which have distinctive spines (**Figure 23.18** on page 668), in the walls of the blood vessels. The eggs work their way to the lumen of the intestine (*S. mansoni* and *S. japonicum*) or lumens of the urinary bladder and ureters (*S. haematobium*) to be eliminated into the environment.

The number of cases of schistosomiasis has increased over the past few decades as economic stability in endemic regions has allowed improved irrigation systems and water reservoirs which provide more habitat for intermediate host snails.

A transient dermatitis called *swimmer's itch* may occur in the area where cercariae burrow in. Migration through the circulatory system is generally asymptomatic.

S. mansoni and *S. japonicum* infections become chronic when eggs lodge in the liver, lungs, brain, and other organs. Trapped eggs die and calcify in the liver, leading to tissue damage that is generally fatal. In *S. haematobium* infections, movement of eggs into the bladder and ureters results in blood in the urine, blockage due to long-term fibrosis and calcification and, in some geographical regions, possibly fatal bladder cancer.

Diagnosis is most effectively made by microscopic identification of spiny eggs in either stool (*S. mansoni*, *S. japonicum*) or urine (*S. haematobium*) samples. The species of blood fluke causing a given infection can be ascertained based on the shape of the egg and the location of its spine. If

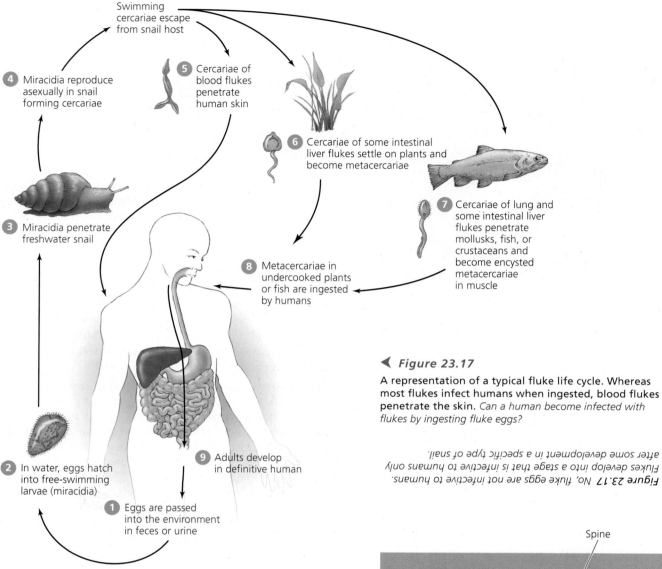

Swimming cercariae escape from snail host

4 Miracidia reproduce asexually in snail forming cercariae

5 Cercariae of blood flukes penetrate human skin

6 Cercariae of some intestinal liver flukes settle on plants and become metacercariae

7 Cercariae of lung and some intestinal liver flukes penetrate mollusks, fish, or crustaceans and become encysted metacercariae in muscle

3 Miracidia penetrate freshwater snail

8 Metacercariae in undercooked plants or fish are ingested by humans

9 Adults develop in definitive human

2 In water, eggs hatch into free-swimming larvae (miracidia)

1 Eggs are passed into the environment in feces or urine

◄ *Figure 23.17*

A representation of a typical fluke life cycle. Whereas most flukes infect humans when ingested, blood flukes penetrate the skin. *Can a human become infected with flukes by ingesting fluke eggs?*

Figure 23.17 No, fluke eggs are not infective to humans. Flukes develop into a stage that is infective to humans only after some development in a specific type of snail.

eggs are not seen but infection is suspected, immunological assays can be used to identify antigen.

The drug of choice for treatment of schistosomiasis is praziquantel. Metrifonate, niridazole, oxamniquine, and oltipraz are alternative treatments, but they do not work equally well against all species of *Schistosoma*.

Prevention of infection depends on improved sanitation, particularly sewage treatment, and avoiding contact with contaminated water. A recombinant vaccine for *S. mansoni* is currently in clinical trials.

CRITICAL THINKING

Propose some additional methods that could be used to prevent transmission of *Schistosoma* infections to humans.

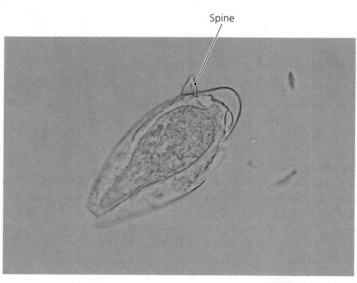

Spine

LM

0.15 mm

▲ *Figure 23.18*

A *Schistosoma mansoni* egg in a stool sample showing the distinctive spine.

Case Study 23.2   A Fluke Disease?

A 40-year-old female of Cambodian descent was admitted to the hospital with fatigue, constipation, weight loss, abdominal pain, and abdominal swelling. Before the onset of symptoms, the woman was healthy, as are the majority of her close relatives. Laboratory testing indicated signs of liver dysfunction, prompting a CT scan. Multiple hepatic lesions and inflammation of the bile ducts were evident. Analysis of a tissue biopsy revealed carcinoma in the ducts. Stool samples examined over several successive days revealed the presence of numerous fluke eggs. The woman had visited Cambodia recently, but her symptoms had begun before she traveled there.

1. What type of patient history would you take? Why is this necessary and important?
2. What would you try to do to ascertain which parasite was causing the woman's illness?
3. What is the treatment for this condition?
4. What type of preventative measures could have kept this woman disease free?
5. Are this woman's relatives at risk of contracting the disease?

Representative Liver Fluke: Fasciola

Two species of liver fluke—*Fasciola hepatica* (fa-sē'ō-lă hepa'ti-kǎ) and *F. gigantica* (ji-gan'ti-kǎ)—infect sheep and cattle worldwide. *Fasciola* also can infect humans, who become accidental definitive hosts when they ingest metacercariae encysted on watercress or other vegetation. Following excystment in the intestine, the larvae burrow through the intestinal wall, the peritoneal cavity, or the liver to reach the bile ducts, where they mature within 3–4 months.

Acute disease characterized by tissue necrosis, abdominal pain, fever, nausea, vomiting, and diarrhea accompanies the migration of the parasite from the intestine to the liver; chronic infection begins when flukes take up residence in the bile ducts. Symptoms coincide with episodes of bile duct obstruction and inflammation, and heavy infections cause liver atrophy and cirrhosis. Lesions can occur in other tissues as well.

Diagnosis involves the correlation of symptoms and patient history with microscopic identification of eggs in fecal matter. *Fasciola* eggs are not always readily distinguishable from those of other intestinal and liver flukes, and symptoms alone mimic other liver disorders; however, liver dysfunction following the consumption of watercress or other raw vegetables grown in endemic areas is diagnostic.

Treatment is dependent on accurate diagnosis, as *F. hepatica* does not respond to praziquantel as other flukes do. Triclobendazole or bithionol is effective.

CRITICAL THINKING

Thoroughly cooking raw vegetables prior to consumption would kill the metacercariae of *Fasciola* and prevent infection. Why is this method of prevention not practical?

Nematodes

Learning Objectives

✓ Describe the common characteristics of nematodes.
✓ Compare and contrast the three most common nematode infections of humans.

LM   1.5 mm

▲ *Figure 23.19*

A roundworm. Shown here is an adult female pinworm, *Enterobius vermicularis*.

✓ Discuss the life cycle of filarial nematodes, and contrast it with the life cycles of intestinal nematodes.

Nematodes or roundworms are long, cylindrical worms that taper at each end **(Figure 23.19)**, possess complete digestive tracts, and have a protective cuticle. All nematodes are dioecious, and females are generally larger than males.

Features of the Life Cycle of Roundworms

Nematodes are highly successful parasites of almost all vertebrate animals, including humans. They have a number of reproductive strategies:

- Most intestinal nematodes shed their eggs into the lumen of the intestine, where they are eliminated with the feces. The eggs are then consumed by the definitive host either in contaminated food (particularly raw fruits and vegetables that grow close to the soil) or in contaminated drinking water.

- For a few intestinal nematodes, infection occurs via a different method. The eggs release larvae into the soil, and these larvae then actively penetrate the skin of new hosts. Once in the body, they travel a circuitous route (discussed in a later section) to the intestine.

- Other nematodes encyst in muscle tissue and are consumed in raw or undercooked meat.

- Mosquitoes transmit a few species of nematodes, called *filarial nematodes*, among hosts.

All nematodes develop through four larval stages either within eggs, in intermediate hosts, or in the environment before becoming adults. Adult, sexually mature stages are found only in definitive hosts. Copulation of male and female worms leads to the production of fertilized eggs and perpetuation of the cycle.

We will examine representative intestinal and filarial nematodes in the following sections. We begin with the largest nematode parasite of humans, *Ascaris lumbricoides*.

Ascaris

Ascaris lumbricoides (as'kă-ris lŭm'bri-koy'dez), the causitive agent of **ascariasis** (as-kă-rī'ă-sis), is the most common nematode infection of humans worldwide, infecting approximately 1 billion people, primarily in tropical and subtropical regions. *Ascaris* is also the largest nematode to infect humans, growing as large as 30 cm in length. *A. lumbricoides* is endemic in rural areas of the southeastern United States.

Adult worms grow and reproduce in the small intestine, where each day females produce about 200,000 eggs that are released with feces into the environment. Eggs can remain viable in shaded, moist, warm soil for years. Within 3 weeks of entering the environment, an embryo develops within each egg. After eggs containing such infective larvae are ingested in water or on vegetables, the larvae are released, invade the intestinal wall, and enter the circulatory or lymphatic systems. From there the larvae reach the lungs, where they develop through two more larval stages in approximately 2 weeks. The worms are then coughed up into the trachea and pharynx and swallowed. The parasites subsequently develop into adults in the intestine.

Most infections are asymptomatic, though if the worm burden (number of worms) is high, intestinal symptoms and signs can include abdominal pain, nausea, vomiting, and intestinal obstruction. Transitory pulmonary symptoms, called *Loeffler's syndrome,* may occur during larval migration and include fever, dry cough, difficulty breathing, and bloody sputum. In rare cases, intestinal obstruction may lead to death.

Diagnosis is made by the microscopic identification of eggs in the stool, larvae in sputum, or (rarely) by identification of adult worms passed in the stool or exiting via the nose or mouth, for larger worms a distinctly unpleasant experience! The treatment of *Ascaris* with albendazole, mebendazole, or pyrantel pamoate over 1–3 days is 90% effective. Surgery may be required to alleviate intestinal obstruction. Proper sanitation and hygiene, including sewage and drinking water treatment, are important for prevention. Good

▲ *Figure 23.20*

Mouth of *Ancylostoma duodenale.* The sharp teeth allow the parasite to attach to the intestinal wall and feed on blood.

personal hygiene and cooking practices are also important methods of preventing infection in endemic areas.

Ancylostoma and Necator

Hookworms—so called because adults resemble shepherd's hooks—are the second most common nematode infecting humans, responsible for more than 800 million infections worldwide. Two hookworms infect humans. *Ancylostoma duodenale* (an-si-los'tō-mă doo'ō-de-nā'lē) is distributed throughout Africa, Asia, the Americas, the Middle East, North Africa, and Southern Europe. *Necator americanus* (nē-kā'tor ă-mer-i-ka'nŭs) predominates in the Americas and Australia but can be found in Asia and Africa as well. *N. americanus* is endemic in the southeastern United States.

Eggs passed in the stools of infected humans are deposited in soil, where they hatch in a couple of days if the soil is warm and moist. The initial larvae develop over a span of 5–10 days into infective larvae that can survive up to 5 weeks in soil. Once they come in contact with a human host, the larvae burrow through the skin and are carried by the circulatory system to the heart and the lungs. In the lungs, larvae burrow through the pulmonary mucosa, migrate up the trachea to the esophagus, where they are swallowed. Once larvae reach the small intestine, they attach to the intestinal mucosa, mature into adults, and mate. Most adults can survive in the intestine for 1–2 years or longer.

Adult worms attached to the intestinal mucosa with ghastly-equipped mouths **(Figure 23.20)** feed on the blood of their hosts, causing anemia. *A. duodenale* consumes 150 μl of blood per day per worm, whereas *N. americanus* consumes 40 μl of blood per day per worm. Worms often move to new sites within the intestines, increasing blood loss and creating chronic anemia. Cardiac complications, iron deficiency, and gastrointestinal and metabolic dysfunction can also occur.

Itching, rash, and inflammation that can persist for several weeks characterize *ground itch,* which occurs at the site of skin penetration by larvae. Ground itch is generally worse with successive infections. Pulmonary symptoms (Loeffler's syndrome) can also occur.

Microscopic identification of eggs in the stool is diagnostic for hookworm infection when present with anemia and blood in the feces. Stool samples generally must be processed to concentrate eggs to ensure either early diagnosis or diagnosis in cases in which worm burdens are low. Specific identification of genus and species cannot be based on the eggs alone; they must first be hatched and reared to the infective larval stage, which is different in the two genera.

Treatment of *A. duodenale* is with albendazole, mebendazole, or pyrantel pamoate for several days. *N. americanus* is more difficult to treat effectively. Proper sanitation (sewage treatment) is essential for prevention. In areas where hookworm is endemic, going barefoot should be avoided to prevent exposure.

Enterobius

Enterobius vermicularis (en-ter-ō'bī-ŭs ver-mi-kū-lar'is) infects about 500 million people worldwide, particularly in temperate climates, in school-age children, and in conditions of overcrowding. *Enterobius* is the most common parasitic worm found in the United States, affecting 40 million Americans. It is commonly known as the pinworm, after the shape of the female worm's tail (see Figure 23.19). Humans are the only host for *Enterobius.*

After mating in the colon, female pinworms migrate at night to the anus, where they deposit eggs perianally. Scratching dislodges eggs onto clothes or bedding, where they dry out, become aerosolized, and settle in water or on food that is then ingested. Alternatively, scratching leaves eggs on the skin and under the fingernails, such that infected individuals can continually reinfect themselves by ingesting the eggs on their hands or in food. Adult worms mature in several weeks and live in the intestinal tract for approximately 2 months.

One-third of all *Enterobius* infections are asymptomatic. When symptoms do occur, intense perianal itching is the chief complaint. Scratching can lead to secondary bacterial infections. Rarely, worms enter the female genital tract, where they can cause inflammation of the vulva.

Microscopy is used for diagnosis. In the morning, before bathing or defecation, transparent sticky tape is applied to the perianal area to collect the readily identifiable microscopic eggs **(Figure 23.21)**. Adult worms, if recovered, are also diagnostic.

Treatment is with an initial dose of pyrantel pamoate or mebendazole followed by a second treatment 2 weeks later to kill any newly acquired worms. Prevention of reinfection and spread to family members requires the thorough laun-

LM   50 µm

▲ *Figure 23.21*

Eggs of *Enterobius vermicularis.* Note the characteristic flattening of one side of each egg.

dering of all clothes and bedding of infected individuals. Further, infected individuals should not handle food consumed by others. Thorough hand washing is vital to prevent self-inoculation, and if at all possible, children should be prevented from scratching.

Wuchereria

Wuchereria bancrofti (voo-ker-e'rē-ă ban-krof'tē) is a *filarial nematode*—a type of nematode that does not infect the intestinal tract of vertebrate hosts but rather the lymphatic system or subcutaneous tissues, causing **filariasis** (fil-ă-rī'ă-sis). The word *filarial* refers to the fact that these nematodes are very slender and threadlike. *W. bancrofti* infects the lymphatic system of humans and is found throughout the tropics: Africa, India, Southeast Asia, Indonesia, the Pacific Islands, South America, and the Caribbean. Female mosquitoes in the genera *Culex, Aedes,* and *Anopheles* transmit *W. bancrofti.*

Mosquitoes ingest circulating immature forms called *microfilariae* while taking blood meals from infected human hosts. The microfilariae penetrate the mosquito's stomach wall, burrowing to its thoracic flight muscles, where they further develop as larvae. Ten days later, these larvae migrate to the mosquito's salivary glands. When the mosquito next feeds, parasites are injected into a new human host. Larvae migrate via the circulatory system to deeper tissues, reaching adulthood within 10 days. After 4 months to 1 year, adults penetrate the lymphatic system, where they can live and reproduce for up to 17 years. Adult female worms continue to produce live microfilariae during this time.

Lymphatic filariasis remains asymptomatic for years. Acute symptoms, when they develop, are due to lymphatic dysfunction. Early in the disease course, symptoms may include mild inflammation, headache, and nausea. As the disease progresses, true lymphatic damage occurs—subcutaneous tissues swell grotesquely because blocked

▲ *Figure 23.22*

Elephantiasis in a leg, the result of infection with
Wuchereria bancrofti. Adult Wuchereria *are found in
what human body system?*

*Figure 23.22 Adults of Wuchereria live in the lymphatic
system of humans.*

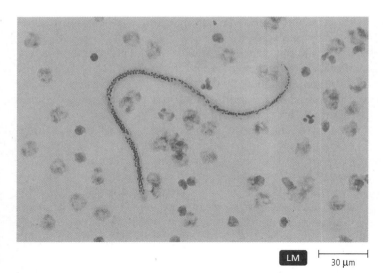

LM   30 μm

▲ *Figure 23.23*

A microfilaria of *Wuchereria bancrofti* in blood.

lymphatic vessels cannot properly drain. The end result is
elephantiasis (el-ĕ-fan-tī′ă-sis), generally in the lower extremities, in which cutaneous and subcutaneous tissues enlarge and harden in areas where lymph has accumulated
(Figure 23.22). Elephantiasis can be further associated with
secondary bacterial infections in affected portions of the
body.

Diagnosis is made by the microscopic identification of
microfilariae in the blood **(Figure 23.23)**. Thick blood smears
or examination of blood that has undergone concentration
procedures (centrifugation or filtration to collect microfilariae) provide best results. Blood samples are collected at
night, when microfilariae circulate; during the day, worm
larvae sequester in capillaries feeding internal organs; they
swim freely in the bloodstream only at night, coinciding
with the feeding habits of most mosquitoes. In addition to
microscopy, immunoassays using antibodies to detect
W. bancrofti antigens in the blood can be used for diagnosis.

Treatment with diethylcarbamazine citrate (DEC) effectively kills microfilariae and adult worms; ivermectin also
kills microfilariae. Antibiotic soaps or creams can be used to
stop bacterial infections in the ulcerated skin of elephantiasis patients.

Prevention relies on avoiding infected mosquitoes. Using insect repellants or mosquito netting, wearing loose,
light-colored clothing to protect the skin, or remaining in-

doors when mosquitoes are most active are all effective personal prevention methods. In endemic areas, particularly
urban areas, eliminating standing water (a requirement of
the mosquito's life cycle) around human dwellings decreases mosquito numbers. Widespread spraying with insecticides to kill mosquitoes has reduced the number of
cases of filariasis.

Table 23.2 summarizes the key features of the common
helminthic parasites of humans.

Protozoan and helminthic diseases are found worldwide,
and much of the world's population is at risk of infection.
Many of the drugs used to treat parasitic diseases have been
around for decades, and whereas most remain effective, resistant parasites are occurring with greater frequency. For
none of the diseases discussed in this chapter have effective
vaccines been produced. The only way to protect oneself,
therefore, is to avoid exposure. For a great many of these diseases, prevention seems simple: provide access to clean
water and safe food. Until both are available to all people
worldwide, however, parasitic diseases will continue to
claim the lives of millions.

Arthropod Vectors

Learning Objective

✓ List the principal arthropod vectors of human pathogens, and give
   one example of a pathogen transmitted by each.

As mentioned in Chapter 12, vectors are animals—usually
arthropods (ar′thrō-podz)—that carry microbial pathogens.
An **arthropod** is an animal with a segmented body, a hard
exoskeleton (external skeleton), and jointed legs. Arthropods are extremely diverse and abundant—insects alone account for more than half of all the known species on Earth.

Table 23.2   Key Features of Representative Helminthic Parasites of Humans

Organism	Primary Infection or Disease	Geographical Distribution	Mode of Transmission	Length of Adult Worms
Cestodes (Tapeworms)				
Taenia saginata; *Taenia solium*	Beef tapeworm infection; pork tapeworm infection	Worldwide with local endemic areas	Consumption of undercooked meat	*T. saginata:* 5–25 m; *T. solium:* 2–7 m
Echinococcus granulosus	Hydatid disease	Worldwide with local endemic areas	Consumption of eggs shed in dog feces.	3–6 mm
Trematodes (Flukes)				
Fasciola hepatica; *F. gigantica*	Fascioliasis	*F. hepatica:* Europe, Middle East, Asia; *F. gigantica:* Asia, Africa, Hawaii	Consumption of watercress or lettuce	up to 30 mm; up to 75 mm
Schistosoma spp.	Schistosomiasis	*S. mansoni:* Caribbean, S. America, Arabia, Africa; *S. haematobium:* Africa, India; *S. japonicum:* East Asia	Direct penetration of the skin	7–20 mm
Nematodes (Roundworms)				
Ascaris lumbricoides	Ascariasis	Tropics and subtropics worldwide	Fecal-oral	Females: 20–35 cm; males: 15–30 cm
Ancylostoma duodenale, *Necator americanus*	Hookworm disease	*Ancylostoma:* Africa, Asia, the Americas, Middle East, North Africa, southern Europe; *Necator:* the Americas, Australia, Asia, Africa	Direct penetration of the skin	*Ancylostoma* females: 10–13 mm, males: 8–11 mm; *Necator* females: 9–11 mm, males: 7–9 mm
Enterobius vermicularis	Pinworm	Worldwide	Anal-oral and fecal- oral, inhalation	Females: 8–13 mm; males: 2–5 mm
Wuchereria bancrofti	Filariasis, elephantiasis	Worldwide, tropics	Mosquitoes	Females: 80–100 mm; males: 40 mm

We also saw in Chapter 14 that some arthropods are *biological vectors,* meaning that they also serve as hosts for the pathogens they transmit. Given that arthropods are small organisms (so small that we generally don't notice them until they bite us), and given that they produce large numbers of offspring, controlling arthropod vectors to eliminate their role in the transmission of important human diseases is an almost insurmountable task.

Disease vectors belong to two classes of arthropods: *Arachnida* and *Insecta.* Ticks and mites are arachnid vectors. (Though spiders are also arachnids, they do not transmit diseases.) Insects account for the greatest number of vectors, and within this group are fleas, lice, true flies (such as tsetse flies and mosquitoes), and true bugs (such as kissing bugs). **Figure 23.24** on page 674 shows representatives of the major types of arthropod vectors. Most arthropod vectors are found on a host only when they are actively feeding. Lice are the only arthropods that may spend their entire lives in association with a single host individual.

Arachnids

Learning Objective

✓ Describe the distinctive features of arachnids.

All adult **arachnids** (ă-rak′nidz) have four pairs of legs. Ticks and mites (commonly known as chiggers) both go through a six-leg stage when they are juveniles, but display the characteristic eight legs as adults. Ticks and mites resemble each other morphologically, having disc-shaped bodies without heads. Ticks are roughly pea-sized, whereas mites are usually the size of sand grains.

Ticks are the most important arachnid vectors. They are distributed worldwide and serve as vectors for bacterial, viral, and protozoan diseases. Ticks are second only to mosquitoes in the number of diseases that they transmit. Hard ticks—those with a hard plate on their dorsal surfaces—are the most prominent disease vectors. They wait on stalks of grasses and brush for animals to walk by. When a human, for example, brushes past them, the ticks leap onto the potential host and begin searching for exposed skin. They use their mouthparts to cut holes in the skin and attach themselves very firmly to the wound, sometimes with a gluelike compound to prevent themselves from being dislodged. As they feed on blood, their bodies swell to several times normal size. Some tick-borne diseases are Lyme disease, Rocky Mountain spotted fever, tularemia, relapsing fever, and tick-borne encephalitis.

Parasitic mites of humans live around the world, wherever humans and animals coexist. A few mite species transmit

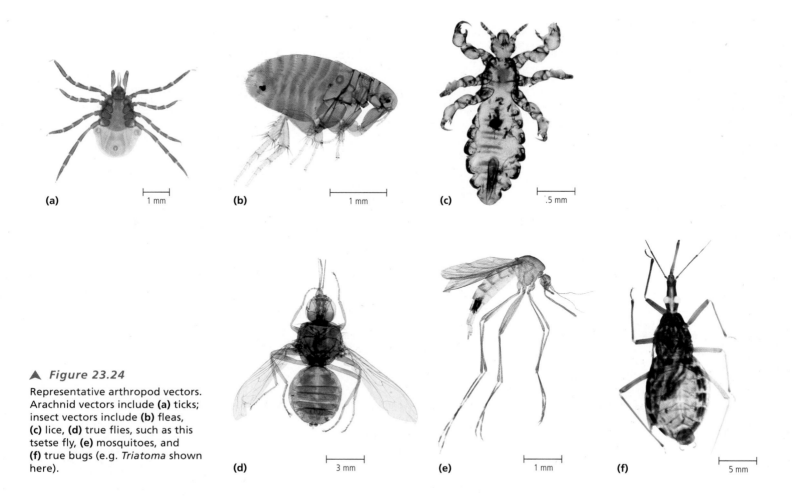

(a) · 1 mm (b) · 1 mm (c) · .5 mm (d) · 3 mm (e) · 1 mm (f) · 5 mm

▲ *Figure 23.24*
Representative arthropod vectors. Arachnid vectors include **(a)** ticks; insect vectors include **(b)** fleas, **(c)** lice, **(d)** true flies, such as this tsetse fly, **(e)** mosquitoes, and **(f)** true bugs (e.g. *Triatoma* shown here).

rickettsial diseases (rickettsial pox and scrub typhus) among animals and humans.

Insects

Learning Objectives

✓ Describe the general physical features of insects, and the features specific to the various groups of insect vectors.
✓ Discuss methods of controlling insect vectors.

As adults, all **insects** have three pairs of legs and three body regions—a head, a thorax (chest), and an abdomen (see Figure 23.24g). Adult insects are, however, far from uniform in appearance. Some have either two wings or four wings, whereas others are wingless; some have long legs, some short; and some have biting mouthparts, whereas others have sucking mouthparts. Many insects also have larval stages that look very different from the adult, complicating the identification of insects. The fact that many insects can fly also has epidemiological implications. Flying insects have broader home ranges than nonflying insects, and some migrate, making control more difficult.

Fleas are small, vertically flattened, wingless arthropods that are found worldwide, though some species do have geographically limited ranges. Most are found in association with wild rodents, bats, and birds, and are not encountered by humans. A few species, however, do feed on humans. Cat and dog fleas are usually just pests (to the animal and its owner), but they can also serve as the intermediate host for a dog tapeworm, *Dipylidium.* The most significant microbial disease transmitted by fleas is plague, carried by the rat flea.

Lice are parasites that can also transmit disease. They are horizontally flat, soft-bodied, wingless insects found worldwide among humans and their habitations. They live in clothing and bedding, and move onto humans to feed. Lice are most common among the poor and those living in severely overcrowded communities. Lice are the vectors involved in epidemic outbreaks of typhus in developing countries.

Flies are among the most common insects, and many different species are found around the world. Flies differ greatly in size, but all have at least two wings and fairly well defined body segments. Not all flies transmit disease, but those that do are generally bloodsuckers. As we have seen, female sand flies *(Phlebotomus)* transit *Leishmania,* the cause of leishmaniasis. Geographically, sand flies are found in North Africa, the Middle East, Europe, and parts of Asia. Tsetse flies *(Glossina)* are limited geographically to tropical Africa, where they are found in brushy areas and transmit

Trypanosoma brucei, the causative agent of African sleeping sickness. Both male and female tsetse flies can transmit *Trypanosoma.*

Mosquitoes are also a type of fly, though they are morphologically distinct from other fly species. Female mosquitoes are thin and have elongated bodies, long antennae, wings, long legs, and a long proboscis for feeding on blood.

Mosquitoes are found throughout the world, but particular species are geographically limited. Mosquitoes are the most important arthropod vector of disease, and they carry some of the world's most devastating diseases, including malaria, yellow fever, dengue fever, filariasis, viral encephalitis, and Rift Valley fever.

Kissing bugs are relatively large, winged true bugs with cone-shaped heads and wide abdomens. As we have discussed, they are called kissing bugs because of their tendency to take blood meals near the mouth of a host. Both sexes feed on blood nocturnally, while their victims sleep. Kissing bugs that transmit disease are limited to Central and South America. The most important disease they transmit is Chagas' disease.

CHAPTER SUMMARY

1. **Parasitology** is the study of protozoan and helminthic parasites of animals and humans. All parasites have a **definitive host** in which adult or sexual stages reproduce, and many parasites also have one or more **intermediate hosts,** in which immature stages live.

Protozoan Parasites of Humans (pp. 650–663)

1. Many protozoa have two life stages: a trophozoite (feeding) stage and a dormant cyst stage. **Encystment** is the process of cyst formation in tissues; **excystment** refers to the process by which cysts become trophozoites.

2. *Balantidium coli* is the only **ciliate** to cause disease in humans. It is responsible for **balantidiasis,** usually a mild gastrointestinal illness.

3. **Amoebae** are protozoa that move and feed using pseudopodia. *Entamoeba histolytica* is a relatively common protozoan parasite that causes **amoebiasis,** which can be mild luminal amoebiasis, more severe amoebic dysentery, or very severe invasive extraintestinal amoebiasis, in which the parasite stimulates lesions in the liver, lungs, brain, and other organs.

4. *Acanthamoeba* and *Naegleria* are free-living amoebae that can cause rare infections in humans. Both are acquired from water. *Acanthamoeba* causes keratitis and **amoebic encephalitis,** which progresses slowly and is usually fatal. *Naegleria* causes a more rapid **amoebic meningoencephalitis,** which is also fatal.

5. Parasitic flagellates include *Trypanosoma, Leishmania, Giardia,* and *Trichomonas.*

6. *Trypanosoma brucei,* transmitted by the tsetse fly (*Glossina*), causes **African sleeping sickness. T. cruzi,** transmitted by the kissing bug (*Triatoma*), causes **Chagas' disease** in the Americas, where it is a frequent cause of heart disease.

7. Both trypanosomes have similar life cycles. Epimastigotes in the insect gut develop into infective trypomastigotes introduced to the human host by the insect vector. *T. brucei* trypomastigotes remain in the blood, are antigenically variable, can spread to the CNS, and can be fatal. *T. cruzi* trypomastigotes develop intracellularly in macrophages and heart muscle as amastigotes.

8. Many species of *Leishmania* infect humans. This parasite has two developmental stages: an intracellular amastigote in humans, and a promastigote form in sandflies, which transmit the disease. Three clinical forms of **leishmaniasis** are seen, each with increasing severity: cutaneous, mucocutaneous, and visceral.

9. *Giardia intestinalis* is a frequent cause of gastrointestinal disease **(giardiasis)** and is endemic worldwide. Giardiasis is rarely fatal.

10. *Trichomonas vaginalis* is the most common protozoan disease of humans. This sexually transmitted protozoan causes **vaginitis,** and prevention relies on abstinence, monogamy, and condom usage.

11. *Plasmodium* species, *Toxoplasma, Cryptosporidium,* and *Cyclospora* are all **apicomplexans,** in which sexual reproduction produces oocysts that undergo schizogony to produce various forms.

12. Four species of *Plasmodium* cause **malaria** and are transmitted by female *Anopheles* mosquitoes. The complex life cycle of *Plasmodium* has three stages: Sporozoites travel to the liver to initiate the **exoerythrocytic cycle.** Rupture of liver cells releases merozoites from intracellular clusters called schizonts. Merozoites infect erythrocytes to initiate the **erythrocytic cycle.** The **sporogonic cycle** in the mosquito produces an ookinete that divides to form sporozoites, which are injected into humans when the mosquito feeds. Hypnozoites, dormant forms, are produced in *P. ovale* and *P. vivax* infections. *P. falciparum* produces the deadliest form of malaria. Several host traits influence whether or not an individual is susceptible to malaria.

13. *Toxoplasma gondii* causes **toxoplasmosis,** a disease of cats that can be acquired by other animals and humans, primarily via the consumption of meat containing the parasites. **Toxoplasmosis** is generally mild in humans, but *Toxoplasma* can cross the placenta to cause serious harm to fetuses.

14. *Cryptosporidium parvum* and *Cyclospora cayetanensis* are both important causes of water-borne gastrointestinal disease in the United States. *Cryptosporidum,* which causes **cryptosporidiosis,** is resistant to standard water treatment methods.

Helminthic Parasites of Humans (pp. 663–672)

1. **Helminths** are multicellular eukaryotic worms, some of which are parasitic.

2. **Cestodes** (tapeworms) all share similar morphologies and general life cycles. A **scolex** attaches the tapeworm to the intestine of its definitive host. From this extends a neck region from which **proglottids** grow. When fertilized eggs (sometimes within gravid proglottids) are eaten by an intermediate host, they hatch to release larval stages that invade tissues of the intermediate host and form cysticerci. When the cysticerci are eaten by a definitive host, they undergo development into adult tapeworms.

3. Humans are the definitive hosts of *Taenia saginata,* the beef tapeworm, and *Taenia solium,* the pork tapeworm. Humans become infected by eating cysticerci in undercooked beef or pork, respectively.

4. *Echinococcus granulosus* forms **hydatid** cysts to cause **hydatid disease** in humans who are accidental intermediate hosts. The definitive host is a canine. Humans may become accidental intermediate hosts by consuming the tapeworm's eggs.

5. **Trematodes,** or flukes, are often divided into four groups: blood, intestinal, liver, and lung flukes. Eggs deposited in water hatch to release miracidia, which burrow into freshwater snails. Following asexual reproduction cercariae are released into the water. In intestinal, liver, and lung flukes, cercariae encyst on plants or in a second intermediate host as metacercariae, which are ingested by human hosts. Cercariae of blood flukes directly penetrate the skin of humans. After maturation occurs in the human, flukes release eggs back into the environment via the feces or urine.

6. *Fasciola* is a liver fluke. Humans acquire *Fasciola* by ingesting metacercariae on watercress, lettuce, water chestnut, or red caltrop.

7. Species of *Schistosoma* (blood flukes) affect millions of people worldwide, causing **schistosomiasis.** Three species infect humans: *S. mansoni, S. japonicum,* and *S. haematobium,* each of which is geographically limited. Heavy infections can significantly damage the liver.

8. **Nematodes** are round, unsegmented worms with pointed ends.

9. *Ascaris lumbricoides,* the largest nematode to infect humans, causes **ascariasis,** which is the most common nematode disease of humans in the world. Eggs containing an infective larvae are consumed with contaminated vegetables. The eggs hatch in the intestine, and larvae enter the lymphatic system. Worm migration through the body may be associated with various symptoms, but infections are usually not fatal.

10. *Ancylostoma duodenale* and *Necator americanus* are both hookworms, the second most common nematodes to infect humans worldwide. Eggs deposited in soil mature to form infective larvae that leave the egg and actively burrow through human skin to initiate infection. The circulatory system carries the worms through the lungs and heart. Eventually worms migrate up the respiratory tract to the esophagus, where they are swallowed. They become adults in the intestine, where they attach to the intestinal mucosa and feed on blood. Hookworm infection can lead to anemia and other blood-related dysfunction.

11. The pinworm *Enterobius vermicularis* is the most common parasitic worm infection of humans in the United States, and the third most common in the world. Eggs deposited in the perianal area are introduced into the mouth following scratching to reinfect the human host. Transmission among family members is also common.

12. *Wuchereria bancrofti* is a filarial (threadlike) nematode of humans. The nematode lives in the lymphatic system, and adult worms block lymph flow, causing **elephantiasis,** in which lymph accumulates, especially in the lower extremities. Bacterial infections may ensue, and subcutaneous tissues harden, producing permanent disfigurement. The parasite is transmitted when microfilariae, which circulate in the blood, are picked up by feeding mosquitoes.

Arthropod Vectors (pp. 672–675)

1. **Arthropods** are organisms with segmented bodies, jointed legs, and a hard exoskeleton. They are found worldwide in nearly all habitats, and some may serve as biological vectors for many organisms. Arthropod vectors can be classified as either **arachnids** (adults having eight legs), or **insects** (adults having six legs).

2. Arachnid vectors include ticks and mites. Ticks are the second most important arthropod vectors, transmitting a variety of diseases to humans. Mites transmit rickettsial disease to humans.

3. Insect vectors include fleas, lice, flies (such as sand flies, tsetse flies, and mosquitoes), and kissing bugs. Mosquitoes are the most important arthropod vectors of diseases of humans.

QUESTIONS FOR REVIEW

(Answers to multiple choice, true/false, fill in the blanks, and matching questions are on the web, along with additional review questions. Visit www.microbiologyplace.com.)

Multiple Choice

1. Parasitology is the study of
   a. parasitic viruses.
   b. parasitic prokaryotes.
   c. parasitic fungi.
   d. parasitic protozoa and helminths.

2. The only ciliate to cause disease in humans is
   a. *Naegleria.*
   b. *Balantidium.*
   c. *Fasciola.*
   d. *Trypanosoma.*

3. Which of the following organisms is regularly transmitted sexually?
   a. *Trichomonas*
   b. *Entamoeba*
   c. *Trypanosoma*
   d. *Enterobius*

4. *Leishmania* species are transmitted by
   a. sand flies.
   b. tsetse flies.
   c. kissing bugs.
   d. mosquitoes.

5. Which of the following is the name of the intracellular infection stage of *Leishmania*?
   a. miracidia
   b. metacercariae
   c. bradyzoites
   d. amastigotes

6. In malaria, which cycle occurs in the mosquito?
   a. exoerythrocytic cycle
   b. erythrocytic cycle
   c. sporogonic cycle
   d. amastigote cycle

7. The definitive host for *Toxoplasma gondii* is
   a. humans.
   b. cats.
   c. birds.
   d. mosquitoes.

8. Tapeworms are generally transmitted via
   a. consumption of an intermediate host.
   b. consumption of the definitive host.
   c. vectors such as mosquitoes.
   d. consumption of adult tapeworms.

9. *Cryptosporidium* cannot be killed by routine boiling. Another parasite that cannot be killed by boiling is
   a. *Giardia.*
   b. *Trypanosoma.*
   c. *Toxoplasma.*
   d. *Plasmodium.*

10. The immature fluke stages that infect snail intermediate hosts are called
    a. metacercariae.
    b. cercariae.
    c. cysticerci.
    d. miracidia.

11. The beef tapeworm is known by the scientific name
    a. *Taenia solium.*
    b. *Taenia saginata.*
    c. *Ancylostoma duodenale.*
    d. *Echinococcus granulosus.*

12. *Enterobius vermicularis* is commonly called
    a. hookworm.
    b. pinworm.
    c. whipworm.
    d. tapeworm.

13. The infective larvae of *Necator americanus* must pass through which human organ to mature?
    a. bladder
    b. brain
    c. lung
    d. liver

14. Both *Plasmodium* species and *Wuchereria bancrofti* can be carried by mosquitoes in the genus
    a. *Aedes.*
    b. *Anopheles.*
    c. *Culex.*
    d. *Ctenocephalides.*

15. Which of the following arthropods is responsible for transmitting the most parasitic diseases?
    a. fleas
    b. ticks
    c. mosquitoes
    d. true bugs

16. The majority of cestodes are transmitted via
    a. ingestion.
    b. vectors.
    c. direct contact.
    d. inhalation.

True or False

Mark each statement as true or false. Rewrite the false statements to make them true by changing the italicized word(s).

____ 1. *Sexual contact* is the most common method of transmitting parasites.

____ 2. Examination of *stool samples* can reveal the presence of *Naegleria* parasites.

____ 3. *Trichomonas vaginalis* is the most common parasitic protozoan of humans in the industrialized world.

____ 4. *Trypanosoma brucei* is transmitted by tsetse flies and *Trypanosoma cruzi* is transmitted by kissing bugs.

___ 5. *Plasmodium falciparum* causes the most serious form of malaria.

___ 6. *Toxoplasma gondii* has been shown to be capable of crossing the *placenta*.

___ 7. Humans can become intermediate hosts for *Taenia solium*.

___ 8. *Fasciola hepatica* can be acquired by eating infected *sheep*.

___ 9. The number of cases of schistosomiasis has *increased* worldwide due in part to improved technology and economic stability in endemic areas.

___10. *Wuchereria bancrofti* is a filarial nematode that infects the *lymphatic system*.

Fill in the Blanks

1. *Balantidium coli* can be distinguished from *Entamoeba histolytica* microscopically because *B. coli* has _____.

2. _____ may be transmitted to humans from cat litter boxes.

3. African sleeping sickness is caused by *Trypanosoma* _____ but not by *Trypanosoma* _____.

4. The parasitic amoeba _____ can be acquired by ingestion.

5. *Plasmodium vivax* and *P. ovale* can both form dormant _____.

6. Of the parasitic helminths discussed in this chapter, the only one transmitted by mosquitoes is _____.

7. The following helminths can directly penetrate the skin of humans to establish infection: _____ and _____ (give genera only).

8. A trematode that can be acquired by eating raw or undercooked vegetables is _____.

9. Hookworm disease is caused by _____ in the Middle East.

10. Both of the following parasites demonstrate nocturnal movement, which is important during diagnosis: _____ and _____.

Matching

Match the terms with the correct organisms.

___ 1. *Balantidium coli*	A.	Miracidia
___ 2. *Echinococcus granulosus*	B.	Bradyzoites
___ 3. *Fasciolopsis buski*	C.	Schizogony
___ 4. *Leishmania* spp.	D.	Microfilaria
___ 5. *Plasmodium falciparum*	E.	Hydatid cyst
___ 6. *Plasmodium vivax*	F.	Cysticerci
___ 7. *Taenia* spp.	G.	Trophozoites
___ 8. *Toxoplasma gondii*	H.	Amastigotes
___ 9. *Trypanosoma*	I.	Hypnozoites
___10. *Wuchereria bancrofti*	J.	Trypomastigotes

Short Answer

1. Why do insect vectors and animal reservoirs increase the difficulty of preventing and controlling parasitic infections in humans?

2. How is the transmission of *Entamoeba* different from the transmission of *Acanthamoeba* and *Naegleria*?

CRITICAL THINKING

1. Compare *Entamoeba, Acanthamoeba,* and *Naegleri*. Based on incidence, which of these parasites is a threat to more people? Clinically, which is the most serious? Explain why your answer is not the same for both questions.

2. Explain how each of the following could lead to the reemergence of malaria in the United States: (a) global warming, (b) increased travel of individuals from endemic regions to the United States, (c) increased immigration of individuals from endemic regions to the United States, (d) regulations against the use of insecticides, and (e) laws protecting wetlands.

3. People who suffer from AIDS are severely immunocompromised because HIV destroys immune cells. Thus all the diseases in this chapter could be a threat to an AIDS patient. Which ones, however, would specifically exacerbate the immune dysfunction of AIDS. Why?

4. Discuss why sickle-cell trait is advantageous to people living in malaria endemic areas, but is not advantageous in malaria-free areas.

5. Present a logical argument to explain the differences between the clinical manifestations of *Trypanosoma brucei* and *T. cruzi*. Relate your argument to the respective life cycles.

6. What feature of the life cycle of *T. brucei* makes it difficult to create a successful vaccine?

Pathogenic DNA Viruses

Everyone reading this textbook has been infected by DNA viruses at one time or another. Fever blisters (cold sores), warts, chickenpox, infectious mononucleosis, and many types of colds are all caused by viruses that have DNA genomes. DNA viruses are also the cause of smallpox, hepatitis B, shingles, and genital herpes, and they are associated with some cancers. Fortunately, many of the more severe diseases caused by DNA viruses, including hepatitis B and hepatic cancer, are relatively uncommon. However, smallpox—a disease that has been successfully eradicated in nature—is once again a public health concern because of its potential as an agent of bioterrorism.

The general characteristics of DNA viruses were covered in Chapter 13. In this chapter we focus on a selection of medically important DNA viruses—the differences among them, the diseases they cause, and the ways they may be transmitted, treated, and prevented.

CHAPTER OUTLINE

Poxviridae
  Smallpox
  Molluscum Contagiosum
  Other Poxvirus Diseases

Herpesviridae
  Herpes Simplex Infections
  Varicella-Zoster Virus Infections
  Epstein-Barr Virus Infections
  Cytomegalovirus Infections
  Other Herpesvirus Infections

Papillomaviridae and **Polyomaviridae**
  Papillomavirus infections
  Polyomavirus infections

Adenoviridae

Hepadnaviridae
  Hepatitis B Infections
  The Role of Hepatitis B Virus in Hepatic Cancer

Parvoviridae

The herpes simplex virus is a DNA virus.

MicroPrep Pre-Test: *Take the pre-test for this chapter on the web.*
Visit **www.microbiologyplace.com.**

DNA viruses that cause human diseases are grouped into seven families based on several factors, including the type of DNA they contain, the presence or absence of an envelope, size, and the host cells they attack (**Table 24.1,** page 682). Recall from Chapter 13 that these DNA viruses have either double-stranded DNA (dsDNA) or single-stranded DNA (ssDNA) for their genomes, and that none of them contains RNA. Double-stranded DNA viruses belong to the families *Poxviridae, Herpesviridae, Papillomaviridae, Polyomaviridae,* and *Adenoviridae.* The family *Parvoviridae* contains viruses with ssDNA genomes. Viruses in the family *Hepadnaviridae* are unusual because they have a genome that is partly dsDNA and partly ssDNA. In the following sections we will examine the infections and diseases caused by the viruses in each of these families.

Poxviridae

Learning Objectives

✓ Name two diseases caused by poxviruses, and discuss their signs and symptoms.

✓ Describe the progression of disease in poxvirus infections.

✓ Discuss the historical importance of poxviruses in immunization and their use in disease eradication.

Poxviruses are double-stranded DNA viruses with complex capsids and envelopes (**Figure 24.1**). They are the largest of viruses, being up to 300 nm in diameter, which is just visible with high-quality, optical, ultraviolet light microscopes. Recall that their large size allows them to be potential vectors for the introduction of genetic material in vaccinations and gene therapy.

Specific poxviruses infect many mammals, including mice, camels, elephants, cattle, and monkeys. (Note that chickenpox is neither caused by a poxvirus nor occurs in chickens; instead, it is caused by a herpesvirus and is discussed later in this chapter.) Most animal poxviruses are species specific; they are unable to infect humans because they cannot attach to human cells, though a few poxviruses of monkeys, sheep, goats, and cattle can infect humans because some humans cells have surface proteins similar to those of animals.

All poxviruses produce lesions that progress through a series of stages (**Figure 24.2**). The lesions begin as flat, reddened **macules**[1] (mak'ūlz), which then become raised sores called **papules**[2] (pap'ūlz); when the lesions fill with clear fluid, they are called **vesicles,** which then progress to pus-filled **pustules**[3] (pŭs'choolz), which are also known as **pocks** or **pox.** (Note that these terms for the stages of poxvirus lesions are also used to describe the lesions of other skin infections.) Poxvirus pustules then dry up to form a crust, and because these lesions penetrate the dermis, they may result in characteristic scars.

[1]From Latin *maculatus,* meaning spotted.
[2]From Latin *papula,* meaning pimple.
[3]From Latin *pustula,* meaning blister.

(a)

200 nm

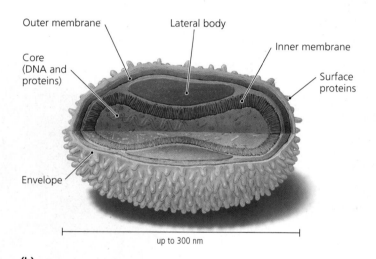

Outer membrane · Lateral body

Core (DNA and proteins)

Inner membrane

Surface proteins

Envelope

up to 300 nm

(b)

▲ *Figure 24.1*

A poxvirus: variola, the causative agent of smallpox. **(a)** An electron micrograph of variola. **(b)** An artist's rendition of variola, showing its complex capsid that includes an outer envelope.

Infection with poxviruses occurs primarily through inhalation of viruses in droplets or dried crusts. Because the envelopes of the viruses are relatively unstable outside the host's body, close contact is necessary for infection by poxviruses. The two main poxvirus diseases of humans are *smallpox* and *molluscum contagiosum.* Three diseases of animals—*orf* (sheep and goat pox), *cowpox,* and *monkeypox*—may be transmitted to humans as well.

Smallpox

Smallpox has played an important role in the history of microbiology, immunization, epidemiology, and medicine. During the Middle Ages, 80% of the European population

1 Macule

2 Papule

Fluid

← 1 cm →

3 Vesicle

Pus

4 Pustule

5 Crust

6 Scar

Epidermis

Dermis

◀ *Figure 24.2*

The stages of the lesions in poxvirus infections. Red macules become hard papules, then fluid-filled vesicles that become pus-filled pustules (pox), which then crust over and leave characteristic scars. *Are the dried viruses in the crust of smallpox and cowpox lesions infective?*

Figure 24.2 Yes, the dried crusts of poxvirus lesions may contain active viruses. The ancient Chinese used the viruses in crusts for variolation, an early attempt at vaccination.

contracted smallpox. Later, European colonists introduced smallpox into susceptible, Native Americans, resulting in the death of as many as 3.5 million people (**Highlight 24.1** on page 683). Recall from Chapter 1 that Edward Jenner first demonstrated immunization using the relatively mild cowpox virus *(Vaccinia)* to protect against smallpox. He was successful because the antigens of cowpox virus are chemically similar to those of smallpox virus, so that exposure to cowpox results in immunological memory and subsequent resistance against both cowpox and smallpox.

Smallpox is the first human disease to be globally eradicated in nature. In 1967, the World Health Organization (WHO) began an extensive campaign of identification and isolation of smallpox victims, as well as vaccination of their contacts, with the goal of completely eliminating the disease. The efforts of WHO were successful, and in 1980 smallpox was declared to be eradicated—one of the great accomplishments of 20th-century medicine. The factors that made it possible for smallpox to be eliminated included the following:

- the existence of an inexpensive, stable, and effective vaccine (vaccinia)
- the specificity of the infection to humans (there are no animal reservoirs)
- the severe, obvious signs and symptoms of smallpox, which enabled quick and accurate diagnosis and quarantine
- the lack of asymptomatic cases
- the fact that the virus is spread only via close contact

Stocks of the virus are still maintained in laboratories in the United States and Russia as research tools for investigations concerning virulence, pathogenicity, protection against bioterrorism, vaccination, and recombinant DNA technology. Now that smallpox vaccination has been discontinued, experts are concerned that the world's population is once again susceptible to smallpox epidemics should the virus be accidentally released from storage or used as a biological weapon. For this reason, many scientists advocate the destruction of all smallpox virus stocks; other scientists insist that we must maintain the virus in secure laboratories so that it is available for the development of more effective vaccines and treatments should the virus be disseminated in a bioterrorism attack.

Table 24.1 **Taxonomy and Characteristics of DNA Viruses of Humans**

Family	Strand Type	Enveloped or Naked	Capsid Symmetry	Size (diameter, nm)	Representative Genera (disease)
Poxviridae	Double	Enveloped	Complex	200–300	*Orthopoxvirus* (smallpox)
Herpesviridae	Double	Enveloped	Icosahedral	150–200	*Simplexvirus*—herpes, type 1 (fever blisters, respiratory infections, encephalitis), type 2 (genital infections), *Varicellovirus* (chickenpox), *Lymphocryptovirus* Epstein-Barr virus (infectious mononucleosis, Burkitt's lymphoma), *Cytomegalovirus* (birth defect), *Roseolovirus* (roseola)
Papillomaviridae	Double	Naked	Icosahedral	45–55	*Papillomavirus* (benign tumors, warts, cervical and penile cancers)
Polyomaviridae	Double	Naked	Icosahedral	45–55	*Polyomavirus* (progressive multifocal leukoencepholopathy)
Adenoviridae	Double	Naked	Icosahedral	60–90	*Mastadenovirus* (conjunctivitis, respiratory infections)
Hepadnaviridae	Partial single and partial double	Enveloped	Icosahedral	42	*Orthohepadnavirus* (hepatitis B)
Parvoviridae	Single	Naked	Icosahedral	18–26	*Erythrovirus* (fifth disease)

Smallpox virus is in the genus *Orthopoxvirus.* Commonly known as **variola** (vă-rī′ō-lă), it exists in two forms or strains. **Variola major** causes severe disease with a mortality rate of 20% or higher, depending upon the age and general health of the host; **variola minor** causes a less severe disease with a mortality rate of less than 1%. Both types of variola infect internal organs and produce fever—42°C (107°F) has been recorded—malaise, delirium, and prostration before moving via the bloodstream to the skin, where they produce the recognizable pox **(Figure 24.3).** Both strains also leave severe scars on the skin of survivors, especially on the face.

CRITICAL THINKING

The United States and Russia have repeatedly agreed to destroy their stocks of the smallpox virus, but the deadline for destruction has been postponed numerous times. In the meantime, the entire genome of variola major has been sequenced. What reasons can governments cite for maintaining smallpox viruses? Should all laboratory stores of smallpox viruses be destroyed? Given that the genome of the virus has been sequenced, and that DNA can be reconstructed if the sequence of nucleotides is known, would elimination of all laboratory stocks really be the extinction of the smallpox virus?

▲ *Figure 24.3*
Smallpox lesions. Facial scarring was often especially severe in smallpox.

Molluscum Contagiosum

Another poxvirus, *Molluscipoxvirus,* causes **molluscum contagiosum** (mo-lŭs′kŭm kon-taj-ē-ō′sum), a skin disease characterized by smooth, waxy, tumorlike nodules on the face, trunk, and limbs **(Figure 24.4).** Molluscum contagiosum occurs worldwide but is seen primarily in Pacific Islanders. Although the disease is typically spread by contact among infected children, sexually active individuals (particularly adolescents) can contract a genital form of the disease. Treatment involves removing the infected nodules by chemicals or freezing.

The virus is unusual in that it produces only a weak immune response. It also causes neighboring uninfected cells to divide rapidly, which means it acts in a manner that is similar to tumor-causing viruses (discussed later).

Highlight 24.1   The "Gift" of Smallpox: Bioterrorism During the French and Indian War

Lord Jeffery Amherst, who was the commanding general of British troops in North America during the final years of what is now called the French and Indian War (1754–1763), was one of the more prominent figures in the Colonies. Now, however, Lord Amherst is more infamous than famous, as he is widely believed to have authorized a plan to infect Native Americans by giving them blankets contaminated with smallpox virus. This is corroborated not only by letters between Amherst and his troops, but also by an entry in the journal of William Trent, commander of the militia at Pittsburgh, who wrote that ". . . two Blankets and an [sic] Handkerchief out of the Small Pox Hospital . . ." were given to the Indians. Trent then added, "I hope it will have the desired effect."

How could smallpox have been transmitted through contaminated blankets? Although transmission usually occurs via respiratory droplets or contact with the lesions, smallpox can be spread by fomites. Variola major, the more virulent of the two smallpox virus strains, can survive for some time in a dried state. Thus it is very possible that at least some cases of smallpox—which did indeed ravage the Native American population during the 18th century—resulted from viruses on the blankets. Native Americans were especially vulnerable to smallpox because they had no immunological memory of previous exposure to this "Old World" disease. Millions perished as a result of naturally acquired smallpox epidemics following the colonization of the Americas by Europeans.

▲ *Figure 24.4*

Pearly-white to light pink, tumorlike lesions of molluscum contagiosum, which is caused by a poxvirus, *Molluscipoxvirus*.

Other Poxvirus Infections

Poxvirus infections also occur in animals. Infections of humans with these viruses are usually relatively mild, resulting in pox and scars but little other damage. In the case of cowpox, the advent of milking machines has reduced the number of cases to near zero in the industrialized world.

Other poxvirus infections, including those in sheep, goats, and monkeys, are less easily contracted by humans. Transmission of these poxviruses to humans requires close contact with infected animals, similar to the contact Edward Jenner observed for milkmaids and cowpox. Still, an upsurge in the number of monkeypox cases in humans has been seen in the past decade. This may be the result of human encroachment into monkey habitats or changes in viral antigens. Some researchers speculate that the cessation of smallpox vaccination may also be involved, because evidence suggests that smallpox vaccination may protect

against monkeypox. Thus, the elimination of smallpox and the cessation of smallpox vaccination may have led to increased human susceptibility to monkeypox. Some scientists have recommended that smallpox vaccination be reinstated in areas where monkeypox is endemic.

Herpesviridae

The family *Herpesviridae* contains a large group of viruses with enveloped polyhedral capsids containing linear dsDNA. A herpesvirus attaches to a host cell's receptor and enters the cell through the fusion of its envelope with the cytoplasmic membrane. After the viral genome is replicated and assembled in the cell's nucleus, the virion acquires its envelope from the nuclear membrane and exits the cell via exocytosis or cell lysis.

Many types of herpesviruses infect humans, and their high infection rates make them the most prevalent of DNA viruses. Indeed, three types of herpesvirus—herpes simplex, varicella-zoster, and Epstein-Barr (discussed shortly)—infect 90% of the adult population in the United States, and several other herpesviruses have infection rates over 50%.

Herpesviruses are often *latent*; that is, they may remain inactive inside infected cells, often for years. Such latent viruses may reactivate as a result of aging, chemotherapy, immunosuppression, or physical and emotional stress, causing a recurrence of the manifestations of the diseases they cause. Additionally, some latent herpesviruses become inserted into a host's chromosomes, where they may cause genetic changes and potentially induce cancer.

Herpesviruses cause a variety of human diseases, including fever blisters (cold sores), genital herpes, chickenpox, shingles, mononucleosis, and cancer. Herpesviruses include the genera *Simplexvirus, Varicellovirus, Lymphocryptovirus, Cytomegalovirus,* and *Roseolovirus*. Additionally, the human herpesviruses have been assigned designations combining "HHV" (for "human herpesvirus") with numbers corresponding to the order in which they were discovered. Thus *Simplexvirus* contains two species (as discussed in the next section) known as HHV-1 and HHV-2. *Varicellovirus* (which causes chickenpox and shingles) is HHV-3, Epstein-Barr virus is HHV-4, *Cytomegalovirus* is HHV-5, and *Roseolovirus* is HHV-6. The newly described HHV-8 is associated with Kaposi's sarcoma in AIDS patients. All these viruses share similar capsids and genes but can be distinguished by genetic variations among them.

HHV-7, which was not mentioned in the previous paragraph, was isolated in the 1980s from the T cells of an AIDS patient and identified as a herpesvirus on the basis of its genome, envelope, capsid, and antigens. At this time, however, no disease has been attributed to HHV-7. It is an *orphan*—a virus without a disease.

Herpes Simplex Infections

Learning Objectives

✓ Describe the diseases caused by herpes simplex types 1 and 2, including their signs and symptoms, treatment, and prevention.

✓ List conditions that may reactivate latent herpesviruses.

The name "herpes," which is derived from a Greek word meaning "to creep," is descriptive of the slowly spreading skin lesions that are often seen with some types of herpesviruses, particularly those of the genus *Simplexvirus*. Viruses in this genus are commonly known as herpes simplex virus or HSV.

HSV Latency and Recurrent Lesions

One of the more distressing aspects of herpes infections is the recurrence of lesions. About two-thirds of patients with HSV will experience recurrences during their lives as a result of the activation of latent viruses. After a primary infection, herpes simplex viruses often enter sensory nerve cells and are carried by cytoplasmic flow to the base of the nerve, where they remain latent in the trigeminal, sacral, or other ganglia[4] **(Figure 24.5).** These latent viruses may reactivate later in life as a result of immune suppression caused by stress, fever, trauma, sunlight, menstruation, or disease, and travel down the nerve to produce recurrent lesions as often as every 2 weeks. Fortunately, the lesions of recurrent infections are rarely as severe as those of initial infections because of the existence of immunologic memory.

Types of HSV Infections

There are two types of herpes simplex: **herpes simplex virus type 1 (HSV-1)** and **herpes simplex virus type 2 (HSV-2).** Historically, HSV-1 has also been known as "above the waist herpes," whereas HSV-2 has been known as "below the waist herpes." In the following subsections we consider the variety of herpes simplex infections caused by these two species of virus.

[4]A ganglion is a collection of nerve cell bodies containing the nuclei of the cells.

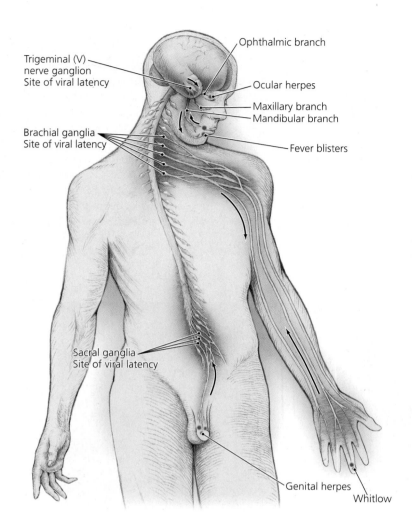

▲ *Figure 24.5*

Sites of events in herpesvirus infections. Infections typically occur when viruses invade the mucous membranes of either the lips (for the fever blisters of oral herpes) or the genitalia (for genital herpes), or through broken skin of a finger (for a whitlow). Viruses may remain latent for years in the trigeminal, brachial, or sacral ganglia before traveling down nerve cells to cause recurrent symptoms in the lips, genitalia, fingers, or eyes (for ocular herpes). *What factors may trigger the reactivation of latent herpesviruses and the recurrence of symptoms?*

Figure 24.5 Stress, fever, trauma, sunlight, menstruation, or diseases such as AIDS may cause recurrent symptoms as a result of immune suppression.

Oral Herpes   Herpes simplex virus type 1 (sometimes called *herpes 1*) was the first human herpes virus to be discovered. About a week after infection, the virions of HSV-1 produce painful, itchy skin lesions on the lips, called **fever blisters** or **cold sores (Figure 24.6);** initial infections may also be accompanied by flu-like signs and symptoms, including malaise, fever, and muscle pain. Severe infections in

▲ *Figure 24.6*

Oral herpes lesions. *Which is the more likely cause of these lesions, simplex 1 or simplex 2?*

Figure 24.6 Approximately 90% of oral herpes lesions are caused by herpes simplex type I.

which the lesions extend into the oral cavity, called *herpetic gingivostomatitis,*[5] are most often seen in young patients and in patients with lowered immune function due to disease, chemotherapy, or radiation treatment. Young adults with sore throats resulting from other viral infections may also develop *herpetic pharyngitis.*

Genital Herpes The second species of herpes simplex that infects humans, called *herpes 2,* differs slightly from herpes 1 in its surface antigens, primary location of infection, and mode of transmission. Type 2 herpes is usually associated with lesions on the genitalia **(Figure 24.7)** because the virions are typically transmitted sexually. However, HSV-2 can cause oral lesions when it infects the mouth. Thus, about 10% of oral cold sores result from HSV-2, and about 15% of genital herpes lesions are caused by herpes type 1. The presence of HSV-2 in the oral region and of HSV-1 in the genital area is presumed to be the result of oral sex.

Ocular Herpes Ocular (ŏk′yū-lăr) or **ophthalmic herpes** is an infection caused by latent herpesviruses. In ocular herpes, latent HSV-1 in the trigeminal ganglion travel down the ophthalmic branch of the trigeminal nerve instead of the mandibular or maxillary branches (see Figure 24.5). Symptoms and signs usually occur in only one eye and include a gritty feeling, conjunctivitis, pain, and sensitivity to light. The viruses may also cause corneal lesions **(Figure 24.8a** on page 686). Recurrent ocular infections can lead to blindness as a result of corneal damage.

Whitlow An inflammed blister called a **whitlow**[6] (hwit′lō) may result if either HSV-1 or HSV-2 enters a cut or break in the skin of a finger **(Figure 24.8b).** A whitlow usually occurs on only one finger. Whitlow is a hazard for children who suck their thumbs, for patients with genital herpes, and especially for health care workers in the fields of obstetrics, respiratory care, gynecology, and dentistry who come into contact with lesions. Such workers should always wear gloves when treating patients with herpes lesions. Since a whitlow is also a potential source of infection, symptomatic personnel should not work with patients until the whitlow has healed. The viruses involved in whitlow may become latent in brachial ganglia and cause recurrent lesions.

Neonatal Herpes Although most herpes simplex infections in adults are painful and unpleasant, they are not life threatening. This is not the case of herpes infections in newborns. A fetus can be infected *in utero* if the virus crosses the placental barrier, but it is more likely that a baby is infected at birth through contact with lesions in the mother's reproductive tract. Also, mothers with oral lesions can infect their babies if they kiss them on the mouth.

HSV-2 is usually the species involved in neonatal herpes. Neonatal herpes infections can be very severe, with a mortality rate of 30% if the infection is cutaneous or oral, and of 80% if the central nervous system is infected. Pregnant women with herpes infections, even if asymptomatic, should inform their doctors. To protect the baby, delivery should be by caesarean section if genital lesions are present at the time of birth.

▲ *Figure 24.7*

Genital herpes lesions. *Are these lesions caused by herpes simplex type 1, or type 2?*

Figure 24.7 Either type of herpes can cause genital lesions, but "below the waste" lesions are usually caused by type 2.

---

[5]From Latin *gingival,* meaning gum, Greek *stoma,* meaning mouth, and Greek *itis,* meaning inflammation.
[6]From Scandanavian *whick,* meaning nail, and *flaw,* meaning crack.

(a)

(b)

▲ *Figure 24.8*

Two manifestations of herpesvirus infections. **(a)** Ophthalmic or ocular herpes, the result of the reactivation of latent viruses in the trigeminal ganglion. **(b)** A whitlow on a health provider's finger, resulting from the entry of herpesviruses through a break in the skin of the finger. *How can health care workers protect themselves from a whitlow?*

Figure 24.8 Health providers should always wear gloves when treating patients who have herpes lesions on any part of their bodies.

Other Herpes Simplex Infections   Athletes may develop herpes lesions almost anywhere on their skin as a result of contact with herpes lesions during wrestling. *Herpes gladiatorum*, as this condition is called, is most often caused by HSV-1.

Herpes simplex viruses may also cause other diseases, including encephalitis, meningitis, and pneumonia. Patients with severe immune suppression, such as those with AIDS, are most likely to develop these rare disorders.

Table 24.2 compares infections caused by herpes simplex types 1 and 2.

Epidemiology and Pathogenesis of HSV Infections

Herpes simplex viruses, which occur worldwide, are transmitted through close bodily contact. Active lesions are the usual source of infection, though asymptomatic carriers can shed HSV-2 genitally. After entering the body through cracks or cuts in mucous membranes, viruses reproduce in epithelial cells near the site of infection and produce inflammation and cell death, resulting in painful, localized lesions on the skin. By causing infected cells to fuse with uninfected neighboring cells to form a structure called a *syncytium*, herpes virions can spread from cell to cell, in the process avoiding the host's immune system.

Initial infections with herpes simplex viruses are usually age specific. Primary HSV-1 infections typically occur via casual contact during childhood, and usually produce no signs or symptoms; in fact, by age 2, about 80% of children have been asymptomatically infected with HSV-1. Most HSV-2 infections, by contrast, are acquired between the ages of 15 and 29 as a result of sexual activity.

Diagnosis, Treatment, and Prevention of HSV Infections

Physicians often diagnose herpes infections by the presence of their characteristic, recurring lesions, especially in the genital region and on the lips. Microscopic examination of infected tissue reveals syncytia. Positive diagnosis is achieved by immunoassay that demonstrates the presence of viral antigen.

Herpes simplex infections are among the few viral diseases that can be controlled with chemotherapeutic agents, notably acyclovir for cold sores and genital herpes, and other nucleoside analogs such as iododeoxyuridine or trifluridine for ocular herpes. Topical applications of the drugs limit the duration of the lesions and reduce viral shedding, though none of these drugs cures the diseases or frees nerve cells of latent viral infections. Evidence suggests that a daily oral dose of acyclovir over a 6–12 month period can be effective in preventing recurrent genital herpes and in reducing the spread of herpes infections to sexual partners. Many obstetricians prescribe acyclovir during the final weeks of pregnancy for women with a history of genital herpes. The actions of these drugs are discussed in Chapter 10.

Health care workers can reduce their exposure to infection with herpes simplex by wearing latex gloves. Health providers with whitlows should not have contact with patients, especially newborns.

Sexual abstinence and faithful monogamy between uninfected partners are the only sure ways of preventing HSV-2 infections. Herpes lesions in women are usually on the external genitalia, so use of a condom during sexual intercourse provides little protection in these cases. Further, if

Table 24.2   Comparative Epidemiology and Pathology of Herpes Simplex Type 1 and Type 2 Infections

	HSV-1	HSV-2
Usual diseases	90% of cold sores/fever blisters; whitlow	85% of genital herpes cases
Mode of transmission	Close contact	Sexual intercourse
Site of latency	Trigeminal and brachial ganglia	Sacral ganglia
Locations of lesions	Face, mouth, and rarely trunk	External genitalia, and less commonly thighs, buttocks, and anus
Other complications	15% of genital herpes cases; pharyngitis; gingivostomatitis; ocular/ophthalmic herpes; herpes gladiatorum; 30% of neonatal herpes cases	10% of oral herpes cases; 70% of neonatal herpes cases

condoms are used, they must be used even in the absence of lesions because asymptomatic individuals can shed viruses genitally. People with active lesions should abstain from sexual activity until the crusts have disappeared.

CRITICAL THINKING

After a patient complains that his eyes are extremely sensitive to light and feel gritty, his doctor informs him that he has ocular herpes. What causes ocular herpes? Which herpes simplex virus, type I or type 2, is more likely to cause ocular herpes? Why?

Varicella-Zoster Virus Infections

Learning Objective

✓ Compare and contrast chickenpox in children with shingles in adults.

A third herpesvirus, called **varicella-zoster virus (VZV),** takes its name from the two diseases it causes: *varicella,* which is typically a disease of children, and *herpes zoster,* which usually occurs in adults. Taxonomically, VZV is in the genus *Varicellovirus.*

Epidemiology and Pathogenesis of VZV Infections

Varicella (var-i-sel′ă), commonly known as **chickenpox,** is a highly infectious disease most often seen in children. Viruses enter the body through the respiratory tract and eyes, replicate in cells at the site of infection, and then travel via the blood throughout the body, triggering fever, malaise, and skin lesions. Viruses are shed before and during the symptoms. Usually the disease is mild, but it may rarely be fatal, especially if associated with secondary bacterial infections. Chickenpox in adults is typically more severe than the childhood illness, presumably because much of the tissue damage results from the patient's immune response, and adults have a more developed immune system than children.

▲ *Figure 24.9*
Characteristic chickenpox lesions.

About 2–3 weeks after infection, characteristic skin lesions first appear on the back and trunk and then spread to the face, neck, and limbs **(Figure 24.9).** In severe cases, the rash may spread into the mouth, pharynx, and vagina. Chickenpox lesions begin as macules, progress in 1–2 days to papules, and finally become thin-walled, fluid-filled vesicles on red bases, which have been called "teardrops on rose petals." The vesicles turn cloudy, dry up, and crust over in a few days. Successive crops of lesions appear over a 3–5-day

Figure 24.10 ▶

Latency and reactivation of varicella-zoster virus. After initial infection, VZV can move up the spinal nerve and become latent in nerve cell bodies in spinal root ganglia. After subsequent reactivation (often years later), the viruses may move down the spinal nerve to the skin, causing a rash known as shingles.

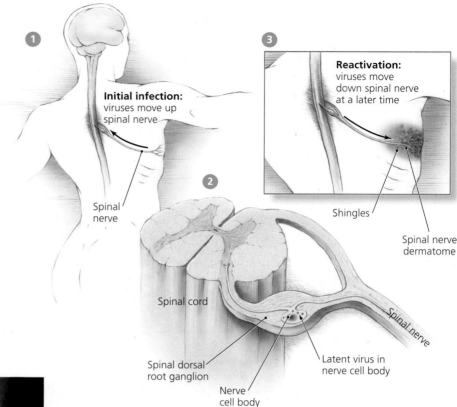

Initial infection: viruses move up spinal nerve

Spinal nerve

Reactivation: viruses move down spinal nerve at a later time

Shingles

Spinal nerve dermatome

Spinal cord

Spinal nerve

Spinal dorsal root ganglion

Nerve cell body

Latent virus in nerve cell body

▲ **Figure 24.11**

Shingles, a rash caused by varicella-zoster virus.

period, so that at any given time all stages can be seen. Viruses are shed through respiratory droplets and the fluid in lesions; dry crusts are not infective.

Like many herpesviruses, VZV can become latent within sensory nerves and may remain dormant for years. In about 15% of individuals who have had chickenpox, stress, aging, or immune suppression cause the viruses to reactivate, travel down the nerve they inhabit, and produce an extremely painful skin rash near the distal end of the nerve **(Figure 24.10)**. This rash is known as **shingles** or **herpes zoster (Figure 24.11)**.

Zoster, a Greek word that means belt, refers to the characteristic localization of shingle lesions along a band of skin, called a *dermatome,* which is innervated by a single sensory

nerve **(Figure 24.12)**. Shingles lesions form only in the dermatome associated with an infected nerve, and thus the lesions are localized and on the same side of the body as the infected nerve cell. Occasionally, lesions from latent infections of cranial nerves form in the eye, ear, or other part of the head. In some patients, the pain associated with shingles remains for months or years after the lesions have healed.

Unlike the multiple recurrences of herpes simplex lesions, shingles only occurs once or twice in an individual's life, usually after age 45. Before vaccination became routine, the majority of children developed chickenpox, but only a small percentage of adults developed shingles, presumably because most individuals develop complete immunity to VZV following chickenpox and therefore never acquire shingles. The observation that VZV from the lesions of shingles can cause chickenpox in individuals who have not had chickenpox revealed that shingles and chickenpox have a common cause: VZV. **Case Study 24.1** explores some of the dynamics of VZV infections within families.

Diagnosis, Treatment, and Prevention of VZV Infections

Chickenpox is usually diagnosed from the characteristic appearance of the lesions. It is sometimes difficult to distinguish shingles from other herpesvirus lesions, though the localization within a dermatome is characteristic. Antibody tests are available to verify the diagnosis of VZV infections.

(a)

(b)

▲ *Figure 24.12*

Dermatomes, the bands of skin innervated by a single sensory nerve. The letters and numbers refer to the spinal nerve that innervates a given dermatome.

Case Study 24.1 **Grandfather's Shingles**

The Davises were excited about their newborn twin boys and couldn't wait to take them to see Mr. Davis's father. Grandfather Davis was excited to see his first grandsons as well, and thought their visit might help take his mind off the pain of his shingles, which had suddenly appeared only days before.

1. What virus is responsible for Grandfather Davis's shingles?

2. Are the new parents at risk of catching shingles from Grandfather Davis?

3. Are the newborns at risk of contracting shingles from their grandfather? If Mrs. Davis asked your advice about visiting her father-in-law with the twins, what would you recommend?

4. Could the twins be vaccinated against shingles before the visit?

Uncomplicated chickenpox is typically self-limiting and requires no treatment other than relief of the symptoms with acetaminophen and antihistamines. Aspirin should not be given to children or adolescents with symptoms of chickenpox because of the risk of contracting Reye's syndrome, a condition associated with several viral diseases and aspirin usage (**Highlight 24.2** on page 691).

Treatment of shingles involves management of the symptoms and bed rest. Nonadherent dressings and loose-fitting clothing may help prevent irritation of the lesions. Acyclovir has proven to provide relief from the painful rash for some patients, but it is not a cure.

It is difficult to prevent exposure to VZV because viruses are shed from patients before obvious signs appear. The FDA has approved an attenuated vaccine, though the vaccine has caused some cases of shingles in Japan, and immunocompromised individuals are at risk of developing diseases from the vaccine. Nevertheless, the benefits of vaccination appear to outweigh the risks, and the Centers for Disease Control and Prevention (CDC) recommends immunization against chickenpox for all children who are least 1 year old.

A more potent form of the live chickenpox vaccine is being tested for efficacy in preventing shingles in adults who developed chickenpox as children.

Epstein-Barr Virus Infections

Learning Objective

✓ Describe the diseases associated with Epstein-Barr virus.

A fourth human herpesvirus is **Epstein-Barr virus (EBV),** which is placed in the genus *Lymphocryptovirus.* EBV is named after its discoverers. The official name for the species is human herpesvirus 4, or HHV-4.

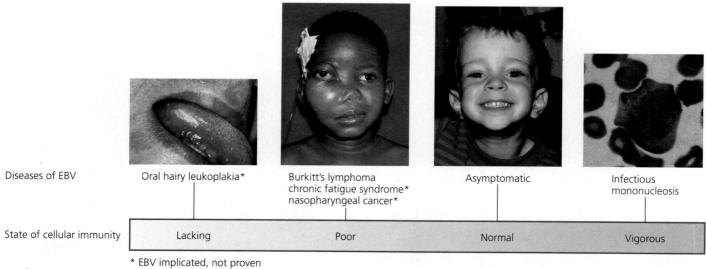

Diseases of EBV

Oral hairy leukoplakia*

Burkitt's lymphoma
chronic fatigue syndrome*
nasopharyngeal cancer*

Asymptomatic

Infectious
mononucleosis

State of cellular immunity

| Lacking | Poor | Normal | Vigorous |

* EBV implicated, not proven

▲ *Figure 24.13*

Diseases associated with Epstein-Barr virus. The disease that results from infection with EBV appears to depend on the relative vigor of the host's cellular immune response, which itself can be related to age. Oral hairy leukoplakia, a precancerous change in the mucous membrane of the tongue, occurs in EBV-infected hosts with severely depressed cellular immunity (for example, AIDS patients). Burkitt's lymphoma, a cancer of the jaw, occurs primarily in African boys whose immune systems have been suppressed by the malaria parasite. EBV is implicated in nasopharyngeal cancer and chronic fatigue syndrome, which are associated with impaired immune function. Asymptomatic EBV infections are typical for children exposed at a young age, before the immune response has become vigorous. Infectious mononucleosis, which is characterized by enlarged B lymphocytes with lobed nuclei, results from infection with EBV during adulthood, when the cellular immune response is vigorous.

Diseases Associated with EBV Infection

Like other herpesviruses, EBV is associated with a variety of diseases. First we consider two diseases EBV is known to cause—Burkitt's lymphoma and infectious mononucleosis—before discussing several other diseases for which EBV is implicated as a cause. As you read about the diseases with which EBV is associated, bear in mind that they are in part the result of the patient's immune status, which itself can be related to the patient's age.

In the 1950s, Dr. Denis Burkitt described a malignant neoplasm of the jaw, primarily in African boys **(Figure 24.13)**, that appeared to be infectious. This cancer is now known as **Burkitt's lymphoma.** Intrigued by the idea of an infectious cancer, Michael Epstein and Yvonne Barr cultured samples of the tumors and isolated Epstein-Barr virus, the first virus shown to be responsible for a human cancer. The way in which a latent virus such as EBV can trigger cancer was discussed in Chapter 13.

Epstein-Barr virus also causes **infectious mononucleosis** (mon'ō-noo-klē-ō'sis) **(mono).** Discovery of the connection between EBV and mono was accidental. A laboratory technician, who worked with EBV and Burkitt's lymphoma, contracted mononucleosis while on vacation. Upon her return to work, her serum was found to contain antibodies against EBV antigens. Serological studies of large numbers of college students have provided statistical verification that EBV causes mono.

Although conclusive proof is lacking, the presence of EBV and antibodies against EBV have implicated this herpesvirus as an etiological agent of *nasopharyngeal cancer* in patients from Southern China, *chronic fatigue syndrome* in the U.S. population, *oral hairy leukoplakia* in AIDS patients, and *Hodgkin's lymphoma.* Chronic fatigue syndrome is associated with debilitating fatigue, memory loss, and enlarged lymph nodes. Hairy leukoplakia is a precancerous change in the mucous membrane of the tongue.

CRITICAL THINKING

Whereas many doctors are convinced that Epstein-Barr virus causes chronic fatigue syndrome, others deny the association between EBV and the syndrome. Why is the etiology of chronic fatigue syndrome debated when Epstein-Barr virus is present in most cases?

Epidemiology and Pathogenesis of EBV Infections

Transmission of Epstein-Barr viruses usually occurs via saliva, often during the sharing of drinking glasses or while kissing. ("Kissing disease" is a common name for infectious mononucleosis.) After initially infecting the epithelial cells of the pharynx and parotid salivary glands, EBV virions enter the blood—a condition called **viremia** (vī-rē′mē-ă)—where they invade B lymphocytes.[7] The viruses become latent in B cells and immortalize them by suppressing *apoptosis* (programmed cell death). These infected B cells are one source of lymphomas, including Burkitt's lymphoma.

Infectious mononucleosis results from a "civil war" between the humoral and cellular branches of the specific immune system, in which cytotoxic T lymphocytes of the cell-mediated branch of specific immunity kill infected B lymphocytes of the humoral branch. This "civil war" is responsible for the symptoms and signs of mono, including sore throat, fever, enlargement of the spleen and liver, and fatigue.

Seventy percent of Americans over the age of 30 have antibodies against EBV. As with other herpesviruses, the age of the host at the time of infection is a determining factor in the seriousness of the disease, in this case because the competence of cytotoxic T cells is in part related to age (see Figure 24.13). Infection during childhood, which is more likely to occur in countries with poor sanitation and inadequate standards of hygiene, is usually asymptomatic because a child's cellular immune system is immature and cannot cause severe tissue damage. Where living standards are higher, childhood infection is less likely, and the postponement of infection until adolescence or later results in a more vigorous cellular immune response that produces the signs and symptoms of mononucleosis.

Cofactors appear to play a role in the development of cancers by EBV. For example, Burkitt's lymphoma is almost exclusively limited to young males exposed to malaria parasites, and the restriction of EBV-induced nasopharyngeal cancer to certain geographical regions is suggestive of cofactors in food, environment, or the genes of regional populations.

Extreme disease, such as oral hairy leukoplakia, arises in individuals with a T cell deficiency, as occurs in malnourished children, the elderly, AIDS patients, and transplant recipients. These individuals are more susceptible to EBV because infected cells are not removed by cytotoxic T lymphocytes and therefore remain a site of virus proliferation.

Diagnosis, Treatment, and Prevention of EBV Infections

Large, lobed, B lymphocytes with atypical nuclei and neutropenia are characteristic of EBV infection. Some diseases associated with EBV, such as hairy leukoplakia and Burkitt's lymphoma, are easily diagnosed by their characteristic signs. Other EBV infections, such as infectious mononucleosis, have symptoms common to many pathogens. Fluorescent antibodies directed against anti-EBV immunoglobulins or ELISA tests provide specific diagnosis.

Burkitt's lymphoma responds well to chemotherapy, and tumors can be removed from affected jaws in geographic areas where surgery is available. Care of mono patients involves relief of the symptoms. Though acyclovir, interferon, and monoclonal antibodies have shown some promise in reducing the viral load (number of viruses), infection is a permanent condition because EBV becomes latent in B cells. There is no effective treatment for other EBV-induced conditions.

Prevention of EBV infection is almost impossible because the viruses are widespread and transmitted readily in saliva. However, only a small proportion of EBV infections result in disease because the immune system is either too immature to cause cellular damage (in children) or cellular immunity is efficient enough to kill infected B lymphocytes.

Cytomegalovirus Infections

Learning Objective

✓ Describe the epidemiology of cytomegalovirus infection.

A fifth herpesvirus that affects humans is **cytomegalovirus** (sī-tō-meg′ă-lō-vī′rŭs) **(CMV),** so called because cells infected with this virus become enlarged. Cytomegalovirus infection is one of the more common infections of humans. Studies have shown that 50% of the adult population of the United States is infected with CMV; in some other countries, 100% of the population tests positive for antibodies against CMV. Like other herpesviruses we have studied, CMV becomes latent in various cells, and infection by CMV lasts for life.

---

[7]These lymphocytes are also known as mononuclear leukocytes, from which the name mononucleosis is derived.

Epidemiology and Pathogenesis of CMV Infections

CMV is transmitted in bodily secretions, including saliva, mucus, milk, urine, feces, semen, and cervical secretions. Individual viruses are not highly contagious, so transmission requires intimate contact involving a large exchange of secretion. Transmission usually occurs via sexual intercourse, but can result from *in utero* exposure, vaginal birth, blood transfusions, and organ transplants. CMV infects 7.5% of all neonates, making it the most prevalent viral infection in this age group.

Most people infected with CMV are asymptomatic, but fetuses, newborns, and immunodeficient patients are susceptible to severe complications of CMV infection. About 10% of congenitally infected newborns develop signs of infection, including enlarged liver and spleen, jaundice, microcephaly, and anemia. CMV may also be **teratogenic**[8] (ter'ă-tō-jen'ik)—that is, cause birth defects—when the virus infects *stem cells*[9] in an embryo or fetus. In the worst cases, mental retardation, hearing and visual damage, or death may result.

AIDS patients and other immunosuppressed adults may develop pneumonia, blindness (if the virus targets the retina), or *cytomegalovirus mononucleosis,* which is similar to infectious mononucleosis caused by Epstein-Barr virus. CMV diseases may result from initial infections or from latent viruses.

Diagnosis, Treatment, and Prevention of CMV Infections

Diagnosis of CMV-induced diseases is dependent on laboratory procedures that reveal the presence of abnormally enlarged cells and inclusions within the nuclei of infected cells **(Figure 24.14).** Viruses and antibodies against them can be detected by ELISA tests and DNA probes (see Chapter 8). Treatment for fetuses and newborns with complications of CMV infection is difficult because in most cases the damage is done before the infection is discovered.

Treatment of adults can also be frustrating. Interferon and gammaglobulin treatment can slow the release of CMV from adults but does not affect the course of disease. None of the drugs effective against other herpesviruses (for example, acyclovir, iododeoxyuridine, or trifluridine) are effective against CMV, and there is no vaccine for preventing infection.

Abstinence, mutual monogamy, and use of condoms can reduce the chances of infection. Organs harvested for transplantation can be made safe by treatment with monoclonal antibodies against CMV, which passively reduces the CMV load before the organs are transplanted.

[8]From Greek *teratos,* meaning monster.
[9]Formative cells, the daughter cells of which develop into several different types of mature cells.

LM   100 μm

▲ *Figure 24.14*

An abnormally enlarged cell, referred to as an owl's eye cell, with an enlarged nucleus containing inclusion bodies, features that are diagnostic for cytomegalovirus infection. The inclusion bodies are sites of viral assembly.

Other Herpesvirus Infections

Learning Objective

✓ Describe the diseases that are associated with HHV-6 and HHV-8.

Another herpesvirus, known simply as **human herpesvirus 6 (HHV-6),** is in the genus *Roseolovirus* and causes **roseola** (rō-zē'ō'lă). Roseola is an endemic illness of children characterized by an abrupt fever, sore throat, enlarged lymph nodes, and a faint pink rash on the face, neck, trunk, and thighs **(Figure 24.15).** The name of the disease is derived from the rose-colored rash.

HHV-6 has recently been linked to multiple sclerosis (MS) by some researchers. These investigators point to the fact that over 75% of MS patients have antibodies against the virus, that patients with higher antibody levels are more likely to develop MS later in life, and that the brain lesions of MS patients typically contain HHV-6. Further, MS patients respond well to treatment with interferon, a natural antiviral protein. This evidence is not proof of causation, but it provides tantalizing clues and ideas for further research.

HHV-6 may also cause mononucleosis-like symptoms and signs, including enlargement of the lymph nodes. There is some evidence that human herpesvirus 6 may also be a cofactor in the pathogenesis of AIDS; that is, infection with HHV-6 may make individuals more susceptible to AIDS.

▲ *Figure 24.15*
Roseola, a rose-colored rash on the body that results from infection with human herpesvirus-6.

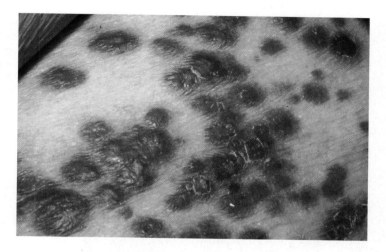

▲ *Figure 24.16*
Kaposi's sarcoma, a rare and malignant neoplasia of blood and blood vessels. This cancer is most often seen in AIDS patients. *What observations support the hypothesis that Kaposi's sarcoma is caused by HHV-8?*

Figure 24.16 HHV-8 is not found in healthy patients, but it is found in Kaposi tumor cells. The virus also causes some cells to become cancerous.

Human herpesvirus 8 (HHV-8) causes Kaposi's sarcoma, a cancer often seen in AIDS patients **(Figure 24.16)**. The virus is not found in cancer-free patients or in normal tissues of victims and has been shown to make cells lining blood vessels cancerous.

Papillomaviridae and *Polyomaviridae*

Papillomaviruses and polyomaviruses each have a single molecule of double-stranded DNA contained in a small, naked, icosahedral capsid. They were formerly classified in a single family, *Papovaviridae,* but recently have been separated into two families, based in part on the different diseases they cause.

Papillomavirus Infections

Learning Objectives

✓ Describe four kinds of warts associated with papillomavirus infections.

✓ Describe the pathogenesis, treatment, and prevention of genital warts.

Papillomaviruses cause **papillomas** (pap-i-lō'măz), which are benign growths of the epithelium of the skin or mucous membranes commonly known as **warts.** Papillomas form on many body surfaces, but are most often found on the fingers or toes (*seed warts*); deep in the soles of the feet (*plantar warts*); on the trunk, face, elbow, or knees (*flat warts*); or on the external genitalia (*genital warts*) **(Figure 24.17** on page 694). Genital warts range in size from almost undetectable small bumps to giant, cauliflower-like growths called **condylomata acuminata**[10] (kon-di-lō'mah-tă ă-kyū-mi-nah'tă) **(Figure 24.18** on page 695). Over 40 varieties of papillomaviruses cause warts in humans.

Although warts are often painful and unsightly, genital warts are even more distressing because of their association with an increased risk of cancer. Compelling evidence suggests that papillomaviruses of genital warts precipitate anal, vaginal, penile, and oral cancers, as well as nearly all cervical cancer.

Epidemiology and Pathogenesis of Papillomavirus Infections

Papillomaviruses are transmitted via direct contact and, because they are stable outside the body, via fomites. They can also be spread from one location to another on a given person by a process called *autoinoculation.*

Viruses that cause genital warts invade the skin and mucous membranes of the penis (particularly when it is

---

[10]From Greek *kondyloma,* meaning knob, and Latin *acuminatus,* meaning pointed.

(a)

(b)

(c)

(d)

◄ *Figure 24.17*

The various kinds of warts, the lesions caused by infections with papillomaviruses. **(a)** Seed warts of the fingers. **(b)** Plantar warts on the sole of a foot. **(c)** Flat warts, which occur on the trunk, face, elbows, or knees. **(d)** Genital warts on the female external genitalia. Genital warts also occur on the anal perineum in both sexes and on the lining of the vagina in females.

uncircumcised), vagina, and anus during sexual intercourse. The incubation time from infection to the development of the wart usually is 3–4 months. Genital warts are now the most common sexually transmitted disease. Over 500,000 new cases are reported each year in the United States, most among young adults.

Diagnosis, Treatment, and Prevention of Papillomavirus Infections

Diagnosis of warts is usually a simple matter of observation, though only DNA probes can elucidate the exact strain of papillomavirus involved. Warts usually regress over time as the cellular immune system recognizes and attacks virally infected cells. Cosmetic concerns and the pain associated with some warts may necessitate removing the infected tissue via surgery, freezing, cauterization (burning), laser, or the use of caustic chemicals. These techniques are not entirely satisfactory because viruses may remain latent in neighboring tissue and produce new warts at a later time. Laser surgery has the added risk of causing viruses to become airborne; some physicians have developed warts in their noses after inhaling airborne viruses during laser treatment.

Effective treatment of cancers caused by papillomaviruses often depends on early diagnosis resulting from thorough inspection of the genitalia in both sexes, and in women by a Papanicolaou (Pap) smear to screen for cervical cancer. Treatment involves radiation or chemical therapy directed against reproducing tumor cells. Advanced cases of genital cancer necessitate removal of the entire diseased organ.

Prevention of most types of warts is difficult, but prevention of genital warts is possible by abstinence or mutual monogamy. Using recombinant DNA techniques, scientists have created a vaccine that successfully prevents infection of women by the four most common strains of sexually transmitted papillomaviruses, including the strains that cause most cervical cancers. Researchers are in the last phase of safety and efficacy testing for the vaccine.

▲ *Figure 24.18*
Condylomata acuminata on the vulva. Such genital warts are considered precancerous.

Case Study 24.2 A Child with Warts

Ten-year-old Rudy has several large warts on the fingers of his right hand. They do not hurt, but their unsightly appearance causes him to shy away from people. He is afraid to shake hands, or to play with other children, out of fear that he may transfer the warts to them. Initially, his mother tells him not to worry about them, but Rudy cannot help feeling self-conscious. Furthermore, Rudy fears that the warts may somehow spread on his own body. After consulting with a physician, Rudy and his mother decide to have the warts surgically removed.

1. Is it possible that Rudy's warts will spread to other parts of his body?

2. Is it possible for someone else to "catch" Rudy's warts by shaking his hand?

3. If Rudy's warts disappear without treatment, are they likely to return to the same sites on his hands?

4. Are Rudy's warts likely to become cancerous?

5. Following surgical removal of the warts, is there a chance that Rudy will develop warts again?

The evidence for the effectiveness of condoms in preventing the spread of genital wart viruses is equivocal—one study suggests that condoms reduce the risk of acquiring genital warts by 25%–50%, whereas other studies have revealed no decrease in the acquisition of warts by condom users or their partners. In 2002, researchers found that women whose sexual partners are circumcised have a smaller chance of getting cervical cancer. This research suggests that if all men were circumcised, the worldwide incidence of cervical cancer would be reduced at least 43 percent.

Case Study 24.2 concerns a young patient with papillomas.

Polyomavirus Infections

Learning Objective

✓ Describe two diseases caused by polyomavirus infection in humans.

The search for cancer-causing viruses led scientists to the discovery of a group of viruses capable of causing several different tumors in animals and humans. Researchers named these viruses **polyomaviruses** by combining *poly* (meaning "many") and *oma* (meaning "tumor").

Polyomaviruses cause other diseases as well. Two human polyomaviruses, the **BK** and **JC** viruses, are endemic worldwide. The initials are derived from the names of the patients from which the viruses were first isolated. Everyone appears to be infected with BK and JC by age 15, probably via the respiratory route. The viruses initially infect lymphocytes, and what happens subsequently depends at least in part on the state of the individual's immune system. If the immune response is normal, latent infections are typically prevented; if the immune system is compromised, however, then latent infections can become established in the kidneys, where reactivation of the viruses leads to the release of new virions. In the case of BK virus, the new virions can cause potentially severe urinary tract infections.

In individuals with latent infections of JC virus, reactivated virions are shed from kidney cells into the blood, disseminating the virions throughout the body. Virions that reach the brain in immunosuppressed individuals can cause a rare disease called **progressive multifocal leukoencephalopathy (PML),** in which the viruses infect and kill cells in the white matter of the central nervous system called oligodendrocytes. Oligodendrocytes produce myelin, a white-colored lipid that surrounds and insulates some nerve cells, so PML patients, as the name of the disease implies, lose various brain functions as the loss of myelin in multiple sites progresses, and the conduction of nerve impulses becomes increasingly impaired. Vision, speech, coordination, and cognitive skills may be affected. Eventually paralysis and death result.

Interferon treatments can prevent kidney damage by BK. The prospect of treatment for PML, however, is bleak. By the time a diagnosis can be made, damage to the brain is severe and irreversible.

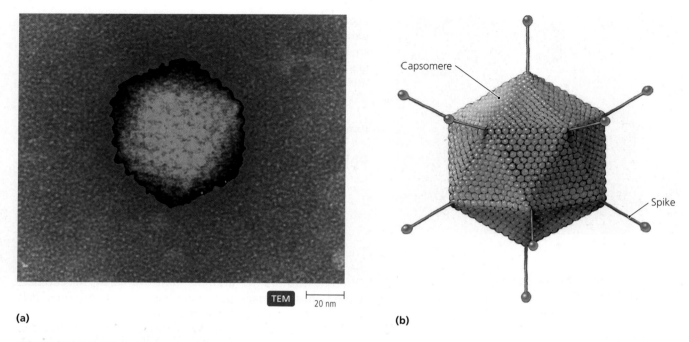

(a)

TEM | 20 nm

Capsomere

Spike

(b)

▲ *Figure 24.19*

Adenovirus. **(a)** This naked icosahedral virus usually has distinctive spikes. **(b)** An artist's rendition.

Adenoviridae

Learning Objective

✓ Discuss the epidemiology and pathogenicity of diseases caused by adenoviruses.

The last family of double-stranded DNA viruses we will discuss is *Adenoviridae.* Adenoviruses have a single, linear dsDNA genome contained in a naked polyhedral capsid with spikes **(Figure 24.19).** Adenoviruses are so-named because they were first discovered infecting cultures of human adenoid[11] cells.

Adenoviruses are but one of the many causative agents of the "common cold." Even though colds typically have a common set of signs and symptoms, their causes are quite varied. At least 30 different respiratory adenoviruses, and over 100 RNA viruses, cause colds.

Adenoviruses, which are spread via respiratory droplets, are very stable outside the body and can survive on fomites and in improperly chlorinated drinking water. Respiratory adenovirus infections become established when adenoviruses are taken into cells lining the respiratory tract via endocytosis. The resulting infections have a variety of manifestations, including sneezing, sore throat, cough, headache, and malaise. (Table 25.1 on page 705 compares the signs and symptoms of colds with those of other respiratory infections.) For some reason, epidemics of respiratory adenoviruses occur on military bases but rarely under the similar conditions found in college dormitories.

Adenoviruses can also cause infections in other systems of the body. After entering the cells lining the intestinal tract,

intestinal adenoviruses can cause mild diarrhea in children. Still other adenoviruses infect the conjunctiva, resulting in a form of conjunctivitis called *pinkeye* **(Figure 24.20).**

Adenoviruses often produce semi-crystallized viral masses within the nuclei of infected cells **(Figure 24.21),** a feature that is diagnostic for adenoviral infections. Although this feature is similar in appearance to the nuclear inclusions seen in cytomegalovirus infections, adenoviruses do not cause infected cells to enlarge.

Adenovirus infection can be treated during the early stages with interferon, and a live, attenuated vaccine is available. The vaccine is used only for military recruits, because some adenoviruses of animals are known to be oncogenic, and there is concern about widespread use of a potentially oncogenic vaccine. **New Frontiers 24.1** on page 698 discusses the use of adenoviruses in research.

Hepadnaviridae

Viruses of the family *Hepadnaviridae*[12] are enveloped DNA viruses with icosahedral capsids that invade and replicate in liver cells. From these characteristics, human **hepatitis B virus (HBV)** acquires the name of its genus, *Orthohepadnavirus.* Its scientific species name, hepatitis B virus, is also its common name.

---

[11]Adenoids are swollen pharyngeal tonsils, which are masses of lymphatic tissue located in the pharynx, posterior to the nose.
[12]From Greek *hepar,* meaning liver, and DNA.

▲ *Figure 24.20*

Adenoviral conjunctivitis. *What signs and symptoms can result from infections with various adenoviruses?*

Figure 24.21 Respiratory adenoviruses cause sore throat, cough, fever, headache, and malaise; intestinal adenoviruses cause mild diarrhea in children; still other adenoviruses cause conjunctivitis (pinkeye).

TEM   ⊢——⊣ 400 mm

▲ *Figure 24.21*

A semi-crystalline mass of adenoviruses replicating and assembling within the nucleus of an infected cell, a feature that is diagnostic of adenovirus infections.

The genome of a hepadnavirus is unique in that it is partly double-stranded DNA and partly single-stranded DNA **(Figure 24.22).** The amount of ssDNA in the genome varies. Some virions are almost entirely dsDNA, whereas others have a significant amount of ssDNA. How does this variation in genomic structure occur?

HBV replicates through an RNA intermediary, a phenomenon that is unique among DNA viruses. One strand of the DNA genome of hepatitis B virus is transcribed into RNA, which then serves as a template for a new DNA strand. The result is an RNA-DNA hybrid molecule. Next, the RNA is removed and the complement of the DNA strand is synthesized. Incomplete dsDNA results when virions are assembled before all their DNA can be completely copied. Thus the earlier in the replication process assembly occurs, the more ssDNA will be in the genome; the later assembly occurs, the more complete replication will be, and the more dsDNA will be in the genome.

Hepatitis B Infections

Learning Objective

✓ Describe the epidemiology, treatment, and prevention of hepatitis B infections.

Hepatitis (hep-ă-tī′tis) is an inflammatory condition of the liver. The liver has many functions, including making blood clotting factors, storing sugar and other nutrients, assisting in the digestion of lipids, and removing wastes from the blood. When a patient's liver is severely damaged by viral

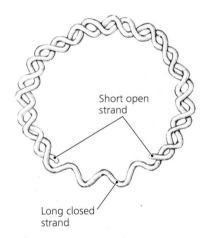

Figure 24.22 ▶

The genome of a hepadnavirus, which is composed of incompletely double-stranded DNA resulting from incomplete DNA replication.

Short open strand

Long closed strand

infection, all these functions are disturbed. **Jaundice** (jawn′dis), a yellowing of the skin and eyes, occurs when a greenish-yellow waste product called bilirubin accumulates in the blood. Hepatitis is also characterized by enlargement of the liver, abdominal distress, and bleeding into the skin and internal organs. The patient becomes stuporous and eventually goes into a coma as a result of the accumulation of nitrogenous wastes in the blood.

Hepatitis B virus is the only DNA virus that causes hepatitis; all other cases of viral hepatitis result from RNA viruses (see Table 25.3 on page 709). HBV infection results in serious liver damage in less than 10% of cases, but co-infection with hepatitis D virus (an RNA virus also called *delta agent*) increases the risk of permanent liver damage.

New Frontiers 24.1   The Importance of Adenoviruses in Research

Early studies of adenovirus replication revealed that a gene in adenoviruses was much longer than the mRNA used to translate the polypeptide product of the gene. How could this be? According to what was known at the time, the nucleotides of mRNA always corresponded in a 1:1 ratio with the nucleotides of DNA, yet in this case the mRNA was much shorter.

Further studies revealed that after the mRNA molecule was transcribed from DNA, portions called *introns* were deleted, and the remaining segments (called *exons*) were then spliced together (see Figure 7.16 on page 218). Ribosomes then translated the polypeptide from the spliced exons only. At first this phenomenon was thought to be a unique feature of adenoviruses, but it was soon discovered that host enzymes carried out the editing function. Further research revealed that eukaryotic genes contain introns, and that RNA editing occurs in almost all eukaryotic cells. Thus, viral research led to the discovery of an important aspect of human biology.

Adenoviruses have also been used to selectively infect and kill cancer cells. In clinical trials to evaluate its efficacy in destroying cancer cells, Onyx-015—an adenovirus with a mutation that enables it to replicate in and lyse only rapidly dividing cells, especially cancer cells—caused head and neck tumors to shrink significantly or disappear altogether in 63% of patients.

In the future, adenoviruses may also play a role in gene therapy, the addition of useful genes into cells with dysfunctional ones. For example, to cure cystic fibrosis (CF), which is caused by mutant genes, normal genes could be incorporated into adenoviral capsids, which could then deliver those genes into the cells of the CF patient. The normal genes could integrate into the patient's chromosomes and be transcribed into mRNA, from which normal polypeptides would then be translated. Past attempts to use adenoviruses for gene therapy were frustrated because the adenovirus capsids were too small to carry large human genes. However, recent experiments have successfully demonstrated that once a large gene has been cut into pieces—each of which is small enough to fit within adenovirus capsids—the entire gene can be inserted into mouse cells piece by piece by several adenovirus capsids. Functional polypeptides have then been translated from the pieces of the gene within the mouse cells.

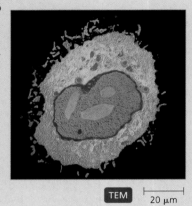

TEM   20 μm

Cell infected with adenoviruses.

Epidemiology and Pathogenesis of HBV Infections

Hepatitis B viruses replicate in liver cells and are released by exocytosis rather than cell lysis. Infected liver cells thus serve as a source for the continual release of virions into the blood, resulting in billions of virions per milliliter of blood. Virions in the blood are shed into saliva, semen, and vaginal secretions.

The viruses are transmitted when infected body fluids, particularly blood, come into contact with breaks in the skin or mucous membranes. The viral load is typically so high, and the infective dose so low, that infection can result from sharing a razor or toothbrush. It is often difficult to determine the source of infection. Improperly sterilized needles used in acupuncture, ear piercing, tattooing, intravenous drug usage, and blood transfusion may spread the virions. They may also be spread by sexual intercourse, especially anal intercourse, which can cause tearing of the rectal lining. Babies can become infected during childbirth.

Studies by the U.S. Centers for Disease Control and Prevention indicate that 12.5 million Americans have been infected with HBV. Many of these individuals are asymptomatic, and most manifest only mild jaundice, low-grade fever, and malaise. Severe liver damage is most likely to result from asymptomatic, chronic infection after 20–40 years. Approximately 10% of infected individuals retain the virus for years and are chronic carriers. The carrier state is age related: Newborns are much more likely to remain chronically infected than are individuals infected as adults.

During the 1970s and 1980s, an estimated 300,000 Americans were infected with HBV annually, including 25,000 children each year. The infection rates are even higher in Asia and Africa, where 50% of the population has been reported to be infected. Today, fewer cases of hepatitis B are seen in the United States as a result of childhood vaccination programs and safer sexual practices, but 150–200 people still die each year from hepatitis B infection.

Diagnosis, Treatment, and Prevention of HBV Infections

Laboratory diagnosis of hepatitis B infection involves the use of labeled antibodies to detect the presence of viral surface antigens released from HBV-infected cells. The body fluids of infected individuals contain copious amounts of viral protein of three forms: so-called *Dane particles*, spherical particles, and filamentous particles **(Figure 24.23)**. Dane particles are complete and infective virions, whereas spherical and filamentous particles are merely capsids composed of surface antigens without genomes. Within infected individuals, the release of so much viral surface antigen in the form of spherical and filamentous particles in effect serves as a decoy: The binding of antibody to empty capsids reduces the antibody response against the infective Dane particles. The large amount of viral antigen released provides one benefit to the patient: It ensures a plentiful substrate for binding labeled antibodies in diagnostic tests.

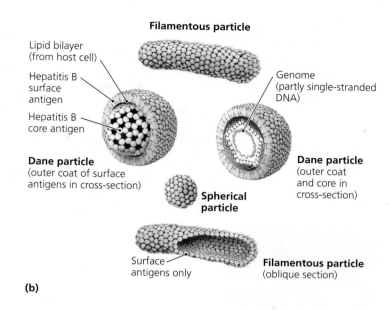

▲ *Figure 24.23*

The three types of viral protein particles produced by hepatitis B viruses. Dane bodies are complete virions, whereas filamentous particles and spherical particles are capsomeres that have assembled without genomes. **(a)** Micrograph. **(b)** Artist's rendition.

There is no universally effective treatment for hepatitis B, though interferon or lamivudine help in 40% of cases. The most effective "treatment" is prevention, which is possible because an effective vaccine is available. Three doses of HBV vaccine given over a 6-month period results in protection against the virus in 95% of individuals. Several studies indicate that immunity against HBV lasts for at least 15 years, and probably for life. Vaccination is recommended for high-risk groups, including health care workers, drug addicts, people receiving blood or blood products, and the sexually promiscuous, especially those who engage in anal intercourse. The CDC recommends HBV immunization for all children because children are at a greater risk of chronic infection, life-threatening liver damage, and (as we will see shortly) liver cancer.

To prevent the spread of HBV, special care should be taken with needles and other sharp instruments, especially in health care settings. Unlike most enveloped viruses, HBV is resistant to detergents, ether, freezing, acid, and moderate heat, but contaminated materials can be disinfected with 10% bleach solutions. Abstinence is the only sure way to prevent sexually transmitted infection.

The Role of Hepatitis B Virus in Hepatic Cancer

Learning Objective

✓ Describe the epidemiology, treatment, and prevention of hepatic cancer.

HBV has been shown to be associated with liver cancer, or *hepatic cancer,* based on the following evidence:

- Epidemiological studies reveal large numbers of hepatic cancer cases in geographic areas that also have a high prevalence of HBV (**Figure 24.24** on page 700).
- The HBV genome has been found integrated into hepatic cancer cells.
- Hepatic cancer cells typically express HBV antigen.
- Chronic carriers of HBV are 200 times more likely to develop hepatic cancer than are noncarriers.

How does HBV produce cancer? Perhaps the gap in the double strand of DNA allows HBV to easily integrate into the DNA of a liver cell, in the process possibly activating oncogenes or suppressing oncogene repressor genes. Another theory is that repair and cell growth in response to liver damage proceeds out of control, resulting in cancer. In any case, because of its association with HBV, hepatic cancer may become the first cancer eliminated as a result of vaccination.

Parvoviridae

Learning Objective

✓ Describe the pathogenesis of erythema infectiosum.

Parvoviruses are the only pathogens of humans with a single-stranded DNA (ssDNA) genome. They are also the smallest of the DNA viruses and have an icosahedral capsid. They cause a number of endemic diseases of animals, including a potentially fatal disease of puppies.

Figure 24.24 ▶

The worldwide distributions of hepatitis B carriers (as evidenced by the prevalence of HBV antigen). Areas with the greatest prevalence of HBV also have the highest incidence of hepatic cancer.

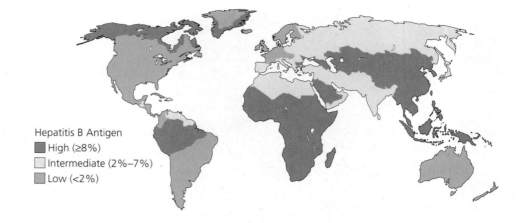

Hepatitis B Antigen
■ High (≥8%)
□ Intermediate (2%–7%)
■ Low (<2%)

◀ **Figure 24.25**

A case of erythema infectiosum (fifth disease). Distinct reddening of facial skin that resembles the result of a slap is characteristic of this parvovirus-caused disease.

The primary parvovirus of humans is **B19** in the genus *Erythrovirus,* which is the cause of **erythema infectiosum** (er-ĭ-thē′mă in-fek-shē-ō′sŭm), also known as **fifth disease** because it was the fifth childhood rash[13] on a 1905 list. Erythema infectiosum, as the name implies, is an infectious reddening of the skin, beginning on the cheeks **(Figure 24.25)** and appearing later on the arms, thighs, buttocks, and trunk. As the rash progresses, earlier lesions fade, though they can reappear if the patient is exposed to sunlight. Sunlight aggravates the eruptions at all stages. No treatment is available.

---

[13]The other four diseases are varicella (chickenpox), rubella (German measles), roseola, and measles.

CHAPTER SUMMARY

Poxviridae (pp. 680–683)

1. Poxviruses, the largest of viruses, cause skin lesions that begin as flat, red **macules,** enlarge into **papules,** fill with clear fluid to become **vesicles,** and then become pus-filled **pustules,** also called **pocks** or **pox.** These crust over and may scar.

2. **Smallpox,** eradicated in nature by 1980, was caused by a poxvirus **(variola).** The smallpox virus strain commonly known as **variola minor** caused less severe cases of smallpox than **variola major.** Governments maintain smallpox viruses in secure laboratories as research tools for studies in pathogenicity, vaccination, and recombinant DNA technology, and as material for developing protection against bioterrorism attacks.

3. The poxvirus *Molluscipoxvirus* causes **molluscum contagiosum,** which results in tumorlike growths on the skin.

4. Although monkeypox and cowpox (both poxviruses) can infect humans, such infections are rare.

Herpesviridae (pp. 683–693)

1. Herpesviruses include viruses that cause fever blisters, genital herpes, chickenpox, shingles, mononucleosis, and cancer.

2. **Herpes simplex type 1 (HSV-1)** and **herpes simplex type 2 (HSV-2)** typically produce painful skin lesions on lips and genitalia, respectively. Oral herpes lesions are called **fever blisters** or **cold sores.**

3. Herpes simplex viruses enter sensory nerve cells, where they may remain latent for years. Lesions recur in about half of all patients in response to emotional stress, sunlight, immune suppression, or other factors that stimulate the latent viruses to reactivate.

4. **Ocular herpes** usually affects one eye to various degrees, from recurring discomfort and light sensitivity to blindness.

5. A **whitlow** is an inflamed blister that may result from infection of herpes simplex viruses via a cut or break in the skin.

6. Herpes simplex infections are usually treatable and are often preventable.

7. **Varicella (chickenpox)** in children and **herpes zoster (shingles)** in adults are both caused by **varicella-zoster virus (VZV)** in the genus *Varicellovirus.* Skin lesions progress from macules to papules to vesicles to crusts. Latent viruses in the sensory nerves can be activated in adults by stress, age, or immune suppression.

8. **Burkitt's lymphoma** is an infectious cancer caused by **Epstein-Barr virus (EBV),** a herpes virus that also causes **infectious mononucleosis (mono).** EBV is also implicated in chronic fatigue syndrome, nasopharyngeal cancer, and hairy leukoplakia in AIDS patients.

9. The herpesvirus **cytomegalovirus (CMV)** causes infected cells to enlarge. This virus is widespread, transmitted by bodily secretions, and often asymptomatic; though infections in immunosuppressed adults and congenitally infected newborns can exhibit severe signs and may be fatal.

10. **Human herpesvirus 6 (HHV-6)** causes **roseola**, a rose-colored rash on the face of children.

11. **Human herpesvirus 8 (HHV-8)** causes Karposi's sarcoma, a cancer often seen in AIDS patients.

Papillomaviridae and *Polyomaviridae* (pp. 693–695)

1. A **wart,** also known as a papilloma, is a benign growth of epithelium or mucous membrane caused by a papillomavirus. Seed warts are most often found on fingers or toes; plantar warts on the soles of the feet; flat warts on the trunk, face, elbow, or knees; and genital warts on the external genitalia. **Condylomata acuminata** are large cauliflower-like genital warts.

2. **Polyomaviruses,** such as **BK** and **JC,** infect the kidneys of most people, but cause disease only in immunosuppressed patients. In these cases, BK viruses compromise kidney function, and JC viruses cause **progressive multifocal leukoen-** **cephalopathy (PML),** which results in severe, irreversible brain damage.

Adenoviridae (p. 696)

1. **Adenoviruses** are one cause of the "common" cold and a cause of one form of conjunctivitis called pinkeye.

Hepadnaviridae (pp. 696–699)

1. Hepadnaviruses are unique in being partially dsDNA and partially ssDNA—a condition resulting from incomplete replication involving an RNA intermediary.

2. **Hepatitis B virus (HBV)** infects the liver, is transmitted in blood and other body fluids, and is the only DNA virus that causes **hepatitis.** A vaccine against HBV is available, but no treatment is universally acceptable.

3. HBV is implicated as a cause of liver cancer.

Parvoviridae (pp. 699–700)

1. **Parvoviruses,** the only human pathogens with ssDNA genomes, include a virus that causes a fatal disease in puppies, and **B19** virus, which causes **erythema infectiosum (fifth disease),** a harmless red rash occurring in children.

QUESTIONS FOR REVIEW   *(Answers to multiple choice and matching questions are on the web, along with additional review questions. Visit www.microbiologyplace.com.)*

Multiple Choice

1. DNA viruses in which of the following families are relatively large, and thus potentially well suited for vaccinations and the introduction of genetic material in gene therapy?
   a. *Herpesviridae*
   b. *Poxviridae*
   c. *Papillomaviridae*
   d. *Hepadnaviridae*

2. For which of the following reasons are most animal poxviruses unable to infect humans?
   a. The animal is not in frequent contact with humans.
   b. The human immune system makes it impossible for the foreign viral particles to reproduce effectively.
   c. Attachment to human cells is unlikely.
   d. Human cells lack the necessary enzymes.

3. The initial flat, red skin lesions of poxviruses are called
   a. macules.
   b. papules.
   c. pustules.
   d. pocks.

4. Which of the following statements is true concerning variola major?
   a. It carries a mortality rate of less than 1%.
   b. It affects internal organs before appearing on the skin.
   c. It has been totally eradicated from the Earth.
   d. The skin lesions it causes are smooth, waxy, tumorlike nodules on the face.

5. Which of the following herpesvirus infections would be the most serious?
   a. a whitlow   c. shingles
   b. ocular herpes   d. cytomegalovirus in fetuses

6. A man experienced a laboratory accident in which he was infected with adenovirus. What signs or symptoms might he exhibit?
   a. fever
   b. headache
   c. cough
   d. pinkeye
   e. all of the above

7. Which of the following conditions is an inflammatory condition of the liver?
   a. fifth disease
   b. PML
   c. seed warts
   d. hepatitis

8. Which of the following statements is *false*?
   a. B19 is the primary parvovirus of humans.
   b. Erythema infectiosum is caused by a parvovirus.
   c. In children, parvovirus infections are accompanied by a high mortality rate.
   d. Parvovirus infection in humans results in infectious reddening of the skin.

9. A distinguishing feature of poxvirus is
   a. its size.
   b. a polyhedral capsid.
   c. the type of RNA it contains.
   d. the production of several types of warts.

10. Monkeypox has been diagnosed in several humans in Zaire. What might be recommended to prevent further risk of infection?
    a. Catch all the male monkeys for inoculation with monkeypox vaccine, and then release them.
    b. Remove infected tissues from humans by chemicals or freezing.
    c. Reinstate smallpox vaccinations for the population of Zaire.
    d. The diagnosis must have been incorrect, because humans are unaffected by monkeypox.

11. Which of the following viral families is most likely to contain viruses that exist in a latent state in humans?
    a. *Herpesviridae*
    b. *Poxviridae*
    c. *Adenoviridae*
    d. *Parvoviridae*

12. If the envelope of a particular virus were unstable outside the host's body, which of the following statements would you expect to be true concerning this virus?
    a. It would be a dsRNA virus.
    b. It would be transmitted by body contact.
    c. Touching a doorknob would easily transmit it.
    d. That virus would eventually cease to be a threat to the population.

13. Being habitually careful not to touch or rub your eyes with unwashed hands would reduce your risk of contracting
    a. chickenpox.
    b. infectious mononucleosis.
    c. seed warts.
    d. a cold.

Short Answer

1. The Smith family seems to get fever blisters regularly. Suggest an explanation of why this problem occurs.

2. You have been given a large grant to do postgraduate research on live smallpox viruses. Where in the world would you find samples?

3. What was the difference in the effects of variola major and variola minor?

4. What observation led scientists to understand the relationship between shingles and chickenpox?

5. Most of the world's population has been infected with EBV by age 1 and show no ill effects, even where medical care is poor. In contrast, individuals in industrialized countries are ordinarily infected after puberty, and these older patients tend to have more severe reactions to infection despite better health and medical care. Explain this apparent paradox.

Matching

In each of the blanks beside the diseases in the left-hand column, write the letter of the viral family that contains that disease's causative agent. Answers may be used more than once.

___ Chickenpox	A. *Poxviridae*
___ Smallpox	B. *Herpesviridae*
___ Cowpox	C. *Papillomaviridae*
___ Molluscum contagiosum	D. *Adenoviridae*
___ HSV-1 infection	E. *Hepadnaviridae*
___ Whitlow	F. *Parvoviridae*
___ Shingles	G. *Polyoma virus*
___ Burkitt's lymphoma	
___ Infectious mononucleosis	
___ Chronic fatigue syndrome	
___ Cytomegalovirus infection	
___ Genital warts	
___ Roseola	
___ Plantar warts	
___ Progressive multifocal leukoencephalopathy	
___ Common cold	
___ Hepatitis B	
___ Fifth disease	

CRITICAL THINKING

1. Most DNA viruses replicate within the nucleus of a host cell, using host enzymes to replicate their DNA. In contrast, poxviruses replicate in the cytoplasm of host cells. What problem does this create for poxvirus replication? How could the virus overcome this problem?

2. Mrs. Rathbone called the pediatrician concerning her young daughter Rene, who had a rosy facial rash and cold-like sniffles for 2 weeks. What is the most likely cause of Rene's problem?

3. Certain features of smallpox viruses allowed them to be eradicated in nature. Which other DNA viruses are suitable candidates for eradication, and what features of their biology make them suitable candidates?

4. A 10-year-old girl at summer camp complains of fever, runny nose, cough, sore throat, and tiredness. Within several hours, 36 other girls report to the infirmary with the same signs and symptoms. Although the girls were assigned to several different cabins and ate in two different dining halls, all of them had participated in outdoor archery, had gone horseback riding, and had been swimming in the camp pool. Infection with what DNA virus could account for their symptoms? The facts point to a common source of infection; what is it? What could the camp management do to limit such an outbreak in the future?

5. A week after spending their vacation rafting down the Colorado River, all five members of the Chen family developed cold sores on their lips. Their doctor told them that the lesions were caused by a herpesvirus. Mr. and Mrs. Chen were stunned: Isn't herpes a sexually transmitted disease? How could it have affected their young children?

CHAPTER 25

Pathogenic RNA Viruses

A patient lies delirious and near death in an African hospital, bleeding from his eyes and gums and into his digestive tract as a result of Ebola hemorrhagic fever. An AIDS patient in Europe continues his expensive daily "cocktail" of antiviral drugs. A hospital in Hong Kong is quarantined to contain an outbreak of severe acute respiratory syndrome (SARS). Thousands of people around the world go to bed with fever, sore throat, cough, and malaise, wishing the flu season was over.

What do these diseases have in common? They are all caused by viruses that use RNA as their genetic material. The manifestations of RNA viral infection are widely variable and pose great challenges in diagnosis, treatment, and prevention. Although RNA viruses can also cause diseases in animals and plants, in this chapter we will focus on RNA viruses that cause human diseases.

Filamentous Ebola viruses

CHAPTER OUTLINE

Naked, Positive ssRNA Viruses: *Picornaviridae, Caliciviridae, Astroviridae,* and *Hepeviridae*
Common Colds Caused by Rhinoviruses
Diseases of Enteroviruses
Acute Gastroenteritis and Hepatitis E

Enveloped, Positive ssRNA Viruses: *Togaviridae, Flaviviridae,* and *Coronaviridae*
Diseases of +RNA Arboviruses
Rubella
Hepatitis C
Diseases of Coronaviruses

Enveloped, Positive ssRNA Viruses with Reverse Transcriptase: *Retroviridae*
Oncogenic Retroviruses
Acquired Immunodeficiency Syndrome

Enveloped, Unsegmented Negative ssRNA Viruses: *Paramyxoviridae, Rhabdoviridae,* and *Filoviridae*
Measles
Diseases of Parainfluenza Virus
Mumps
Disease of Respiratory Syncytial Virus
Rabies
Hemorrhagic Fevers

Enveloped, Segmented Negative ssRNA Viruses: *Orthomyxoviridae, Bunyaviridae,* and *Arenaviridae*
Influenza
Diseases of Bunyaviruses
Diseases of Arenaviruses

Naked, Segmented dsRNA Viruses: *Reoviridae*
Rotaviruses
Coltiviruses

MicroPrep Pre-Test: *Take the pre-test for this chapter on the web.*
*Visit **www.microbiologyplace.com.***

Recall from previous chapters that all cells and DNA viruses use DNA as their genetic material. In contrast, RNA viruses and viroids are the only infective agents that use RNA molecules to store their genetic information.

As we saw in Chapter 13, there are four basic types of RNA viruses:

- positive single-stranded RNA (+ssRNA) viruses
- retroviruses (which are +ssRNA viruses that convert their genome to DNA inside a cell)
- negative single-stranded RNA (−ssRNA) viruses
- double-stranded RNA (dsRNA) viruses

Recall that positive RNA is RNA that can be used by a ribosome to translate protein; it is essentially mRNA. Negative RNA cannot be processed by a ribosome; it must first be transcribed as mRNA. (We discussed the replication strategies of these four types of RNA viruses in Chapter 13, and the information is summarized in Table 13.3 on page 392.) In addition to grouping RNA viruses by the kind of RNA they contain, scientists also note that many RNA viruses have more than one molecule of RNA—in other words, their genomes are *segmented*.

RNA viruses of humans are categorized into 15 families on the basis of (1) their genomic structure, (2) the presence of an envelope, and (3) the size and shape of their capsid. Positive single-stranded RNA viruses are classified into seven families: *Picornaviridae, Caliciviridae, Astroviridae, Hepeviridae, Togaviridae, Flaviviridae, Coronaviridae,* and *Retroviridae.* Negative single-stranded RNA viruses are in six families: *Paramyxoviridae, Rhabdoviridae, Filoviridae, Orthomyxoviridae, Bunyaviridae,* and *Arenaviridae. Reoviridae* is the only family containing double-stranded RNA viruses. We begin our discussion of pathogenic RNA viruses by considering three families that contain positive ssRNA viruses.

Naked, Positive ssRNA Viruses: *Picornaviridae, Caliciviridae, Astroviridae,* and *Hepeviridae*

The families *Picornaviridae, Caliciviridae, Astroviridae,* and *Hepeviridae* contain positive single-stranded RNA viruses with naked polyhedral capsids. The family *Picornaviridae* is a large viral family containing many human pathogens, including viruses that cause common colds, poliomyelitis, and hepatitis A. Picornaviruses range in size from 22–30 nm in diameter, making them the smallest of animal viruses (see Figure 13.4). Five hundred million picornaviruses could sit side by side on the head of a pin. Their small size and positive ssRNA genome is reflected in the name, **picornavirus**[1] (pi-kōr-nă-vī′rus). Representative picornaviruses that cause human diseases are in the genera *Rhinovirus, Enterovirus,* and *Hepatovirus.*

Caliciviruses, astroviruses, and hepeviruses are generally larger than picornaviruses (27–40 nm in diameter) and have a six-pointed star shape; they cause gastrointestinal diseases.

TEM   ⊢——⊣ 100 nm

▲ *Figure 25.1*

Rhinoviruses, the most common cause of colds. *What other types of viruses cause common colds?*

Figure 25.1 Besides rhinoviruses (in the family Picornaviridae), adenoviruses, coronaviruses, reoviruses, and paramyxoviruses cause colds.

In the following sections we will consider in turn three diseases caused by picornaviruses—the common colds, polio, and hepatitis—before examining diseases caused by caliciviruses, astroviruses, and hepeviruses.

Common Colds Caused by Rhinoviruses

Learning Objective

✓ Discuss treatment of the common cold.

Most Americans will sniffle and sneeze their way through at least two colds each year. Though the symptoms of all colds are similar—sneezing, *rhinorrhea* (runny nose), congestion, mild sore throat, headache, malaise, and cough for 7–10 days—there is no single common cause of the "common cold." Rather, many different viruses, including adenoviruses, coronaviruses, reoviruses, and paramyxoviruses, cause colds. However, the over 100 *serotypes* (varieties) of the picornaviruses known as **rhinoviruses**[2] (rī′nō-vī′rŭ-ĕz) cause most colds **(Figure 25.1)**. The symptoms of colds and other respiratory infections are compared in **Table 25.1**.

Epidemiology of Rhinovirus Infections

Rhinoviruses are limited to infecting the upper respiratory tract. They replicate best at a temperature of 33°C, which is the internal temperature of the nasal cavity.

---

[1] From Latin *pico*, meaning small, and RNA virus.
[2] From Greek *rhinos*, meaning nose.

Table 25.1	Manifestations of Respiratory Infections
Ailment	**Manifestations**
Common cold (viral)	Sneezing, rhinorrhea, congestion, sore throat, headache, malaise, cough
Influenza (viral)	Fever, chills, rhinorrhea, headache, body aches, fatigue, cough
"Strep" throat (bacterial)	Fever, red and sore throat, swollen lymphatics in neck
Viral pneumonia	Fever, chills, mucus-producing cough, headache, body aches, fatigue
Bacterial pneumonia	Fever, chills, congestion, cough, chest pain, rapid breathing, possible nausea and vomiting
Bronchitis (viral or bacterial)	Mucus-producing cough, wheezing
Inhalation anthrax (bacterial)	Fever, malaise, cough, chest discomfort, vomiting

Rhinoviruses are extremely infective—a single virus is sufficient to cause a cold in 50% of infected individuals. Symptomatic or not, an infected person can spread viruses by releasing them into the surrounding environment, where they can be transmitted in aerosols produced by coughing or sneezing, via fomites (nonliving carriers of pathogens), or via hand-to-hand contact. Direct person-to-person contact is the most common means of transmitting infections.

When cold symptoms are most severe, over 100,000 virions may be present in each milliliter of nasal mucus. They remain viable for hours outside the body. Infection often results from inoculation by hand into the mucous membranes of the eyes where the viruses are washed by tears into the nasal cavity and readily infect nasal cells.

Although people of all ages are susceptible to rhinoviruses, they can acquire some immunity against serotypes that have infected them in the past. For this reason, children typically have six to eight colds per year, younger adults have two to four, and adults over age 60 have one or fewer. An isolated population may acquire a certain amount of herd immunity (see Chapter 17) to specific varieties of rhinoviruses; however, new serotypes introduced into the population by outsiders and by mutations in the RNA of the viruses ensure that no group is free of all colds.

Diagnosis, Treatment, and Prevention of Rhinovirus Infections

The manifestations of rhinoviruses are usually diagnostic (see Table 25.1), and laboratory tests are required only if the serotype is to be identified.

While many home remedies and over-the-counter medicines exist to treat the common cold, none prevents colds or provides a cure. Studies of one "cure"—large amounts of vitamin C—have yielded conflicting results: Some studies indicate no benefit, whereas others suggest a slight benefit for patients who are treated with vitamin C. A prescription remedy, pleconaril, taken at the onset of symptoms, reduces the seriousness and duration of disease. Antihistamines, decongestants, and pain relievers relieve the symptoms but do not reduce the duration of the disease. Rest and fluids allow the body to mount an effective immune response.

Because rhinoviruses do not share any common antigens that are accessible to the immune system, an effective vaccine for the common cold would have to immunize against over a hundred serotypes, which is impractical.

Antisepsis is probably the most important preventative measure, especially if you have touched the hands of an infected person. Disinfection of fomites is also effective in limiting the spread of rhinoviruses.

CRITICAL THINKING

As you have probably noticed, colds occur more frequently in the fall and winter. One explanation for this observation is that more people are crowded together in buildings when school starts and the weather cools. Design an experiment or epidemiological survey to test the hypothesis that crowded conditions explain the prevalence of colds in the fall and winter.

Diseases of Enteroviruses

Learning Objectives

✓ Describe the effects of polioviruses on humans.
✓ Describe the contributions of Jonas Salk and Albert Sabin toward eliminating polio.
✓ Describe the signs and symptoms of hepatitis A infections.
✓ Compare and contrast the diseases caused by coxsackieviruses.

A second genus of picornaviruses is *Enterovirus*. Contrary to their name, *enteroviruses* do not usually cause diseases of the digestive system, though they can survive the low pH environment of the stomach. Instead, enteroviruses are so named because they are transmitted via the fecal-oral route. They infect the pharynx and intestine, where they multiply in the mucosa and lymphatic tissue without causing symptoms. Enteroviruses spread from these initial sites of infection through the blood (viremia) and infect different target cells, depending on the particular virus. Enteroviruses are cytolytic, killing their host cells.[3] The clinical manifestations of enteroviruses depend on the viral serotype, the size of the infecting dose, the target organ, and the patient's health, gender, and age. There are three main types of enteroviruses: polioviruses, coxsackieviruses, and echoviruses.

---
[3]From Greek *cytos*, meaning cell, and *lysis*, meaning dissolution.

▲ *Figure 25.2*

A hospital ward full of mechanical respirators called "iron lungs," which assisted the paralyzed respiratory muscles of polio patients. Such wards were common before 1955.

Poliomyelitis

Many Americans still remember the dreaded **poliomyelitis** (pō′lē-ō-mī′ĕ-lī′tis) or **polio** epidemics of the past, when the floors of hospitals were filled with iron lungs **(Figure 25.2)** and schoolchildren donated their change to the March of Dimes to develop a polio vaccine. Because polioviruses are stable for prolonged periods in swimming pools and lakes and can be acquired by swallowing contaminated water, parents in the 1930s and 1940s often feared to let their children swim. Thankfully, those days appear to be over. The last case of wild-type poliomyelitis in the Americas occurred in 1979, (though vaccine-induced polio occurred as recently as 2001.) The World Health Organization (WHO) has worked for years to eradicate polio worldwide by 2005 **(Figure 25.3).** Then, like smallpox virus, it is hoped that poliovirus will exist only in laboratories.

There are three serotypes of poliovirus. After being ingested and infecting pharyngeal and intestinal cells, any of these serotypes could cause one of the following four conditions:

1. *Asymptomatic* infections. Most infections (almost 90%) are asymptomatic.

2. *Minor polio,* which includes nonspecific symptoms such as temporary fever, headache, malaise, and sore throat. Approximately 5% of cases are minor polio.

3. *Nonparalytic polio,* resulting from polioviruses invading the meninges and central nervous system, producing muscle spasms and back pain in addition to the general symptoms of minor polio. Nonparalytic polio occurs in about 2% of cases.

4. *Paralytic polio,* in which the viruses invade cells of the spinal cord and motor cortex of the brain, producing paralysis by limiting nerve impulse conduction. The degree of paralysis varies with the type of poliovirus involved, the infective dose, and the health and age of the patient. In **bulbar poliomyelitis,** the brain stem and medulla are infected, resulting in paralysis of muscles in the limbs or of respiratory muscles, in which case iron lungs were used to assist in respiration. In most paralytic cases, complete recovery resulted after 6–24 months, but in some cases paralysis was lifelong. Paralytic polio occurs in less than 2% of infections. Famous victims of paralytic polio include President Franklin Delano Roosevelt and violinist Itzhak Perlman **(Figure 25.4),** as well as novelist Sir Walter Scott.

Postpolio syndrome is a crippling deterioration in the function of polio-affected muscles that occurs in up to 80% of recovered polio patients some 30–40 years after their original bout with poliomyelitis. This condition is not caused by a reemergence of polioviruses, as viruses are not present. Instead, the effects appear to stem from an aging-related aggravation of nerve damage that occurred during the original infection.

Figure 25.3 ▶

The worldwide distribution of cases of naturally occurring polio in 2002. The goal of global eradication of polio by 2005 is within reach.

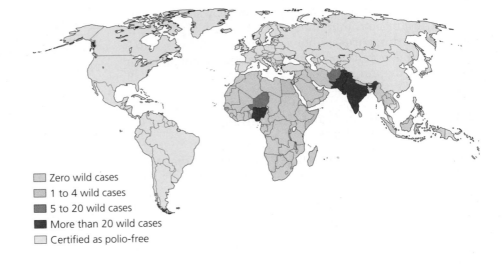

☐ Zero wild cases
☐ 1 to 4 wild cases
☐ 5 to 20 wild cases
■ More than 20 wild cases
☐ Certified as polio-free

The near elimination of polio stands as one of the great achievements of 20th-century medicine. It was made possible by the development of two effective vaccines. Jonas Salk developed an **inactivated polio vaccine (IPV)** in 1955. Six years later, it was replaced in the United States by a live, attenuated, **oral polio vaccine (OPV)** developed by Albert Sabin. Both vaccines are effective in providing immunity against all three strains of poliovirus. Recently, pediatricians have returned to using the IPV for initial vaccination of babies because viruses in the OPV have occasionally mutated into a virulent form and caused polio. **Table 25.2** compares the advantages and disadvantages of the two vaccines.

CRITICAL THINKING

Smallpox is the only disease that has been eradicated worldwide, though scientists are close to eradicating polio. What features do the smallpox and polio viruses share that has allowed medical science to rid the world of these diseases? Compare these features to those of rhinoviruses in discussing the likelihood of eliminating colds caused by rhinoviruses.

Other Diseases of Enteroviruses

The other enteroviruses that cause human disease are **coxsackieviruses**[4] and **echoviruses.** As with all enteroviruses, infection with these viruses is by the fecal-oral route. Most infections are subclinical or result in mild fever, muscle aches, and malaise. There are two types of coxsackieviruses (known as type A and type B), each with several serotypes, as well as numerous serotypes of echoviruses, which cause a variety of human diseases.

Coxsackieviruses Coxsackie A viruses are associated with lesions and fever that last for a few days to weeks and are self-limiting. Several serotypes cause lesions of the mouth and pharynx called *herpangina* because of their resemblance to herpes lesions. Sore throat, pain in swallowing, and vomiting accompany herpangina.

Another serotype of coxsackie A virus causes *hand-foot-and-mouth disease*, a name that is aptly descriptive because it

▲ *Figure 25.4*

President Franklin Delano Roosevelt and violinist Itzhak Perlman, two people who did not let paralytic polio stop them from achieving in their fields of endeavor.

involves lesions on the extremities and in the mouth (**Figure 25.5** on page 708). Yet another coxsackie A virus causes an extremely contagious *acute hemorrhagic conjunctivitis.* Coxsackie A viruses also cause some colds.

Coxsackie B viruses are associated with *myocarditis* (inflammation of the heart muscle) and *pericardial*[5] *infections.* The symptoms in young adults may resemble myocardial infarction (heart attack), but fever is associated with a coxsackie infection. Newborns are particularly susceptible to coxsackie B myocarditis resulting in a rapid onset of heart failure with a high mortality rate.

---

[4]From Coxsackie, New York, where the virus was first isolated.
[5]The pericardium is the membrane that surrounds and protects the heart.

Table 25.2   **Comparison of Polio Vaccines**

	Advantages	Disadvantages
Salk vaccine		
Inactivated polio vaccine (IPV)	Is effective and inexpensive; is stable during transport and storage; poses no risk of vaccine-related disease	Requires booster to achieve life-long immunity; is injected and can be painful; requires higher community vaccination rate than does OPV
Sabin vaccine		
Oral polio vaccine (OPV)	Provides life-long immunity without boosters; induces secretory antibody response similar to natural infection; is easy to administer; can result in herd immunity	Is more expensive than IPV; is less stable than IPV; can mutate to disease-causing form; poses risk of polio developing in immunocompromised patients

(a)

(b)

▲ *Figure 25.5*

Lesions characteristic of hand-foot-and-mouth disease, which is caused by a coxsackie A virus. **(a)** Lesions on the hard palate (roof of the mouth). **(b)** Lesions on a hand. *Why are coxsackieviruses considered tract enteroviruses when they do not cause infections of the gastrointestinal tract?*

Figure 25.5 All enteroviruses, including coxsackievirus, are infective through the gastrointestinal tract, though they may affect other organ systems.

One type of coxsackie B virus causes *pleurodynia* (ploor-ō-din′ē-ă), also known as *devil's grip*. This disease involves the sudden onset of fever and unilateral, severe, low thoracic pain, which may be excruciating. Coxsackie B virus can be transmitted across the placenta to produce severe and sometimes fatal disseminated disease of the brain, pancreas, and liver in the fetus. Infections of the pancreas by coxsackie B is suspected to be a cause of *diabetes mellitus* because the viruses destroy cells in the islets of Langerhans that produce insulin.

Both coxsackie A and B viruses can cause *viral meningitis*, an acute disease accompanied by headache and, with type A infections, a skin rash. Coxsackievirus meningitis is usually self-limiting and uneventful unless the patient is less than 1 year old.

Echoviruses The name *echovirus* is derived from *enteric cytopathic human orphan* virus, because the virus is acquired

intestinally and was not initially associated with any disease. Echoviruses are another cause of viral meningitis and colds.

Epidemiology of Enterovirus Infections

Enteroviruses have a worldwide distribution and occur particularly in areas with inadequate sewage treatment. Transmission is by the fecal-oral route involving ingestion of contaminated food or water or oral contact with infected hands or fomites. For some reason, enterovirus diseases are more common in summer.

As we have seen, fetuses and newborns are at greatest risk from enteroviruses, with the exception of poliovirus, which results in more serious symptoms in older children and young adults.

Diagnosis, Treatment, and Prevention of Enterovirus Infections

Enterovirus infections are usually benign and have mild symptoms, so they are not often diagnosed except in severe cases such as paralytic polio and viral meningitis. The cerebrospinal fluid (CSF) of viral meningitis patients has a normal glucose level, in contrast to that seen with bacterial meningitis. Viruses are rarely found in the CSF, so their presence there is diagnostic for viral meningitis; serological testing can confirm enterovirus infection.

No antiviral therapy is effective against enterovirus infection. Treatment involves support and limitation of pain and fever. Good hygiene and adequate sewage treatment can prevent infection with enteroviruses. No vaccines exist for coxsackievirus and echovirus infections, but effective vaccines do exist for polio. Current vaccination recommendations are given in Chapter 17.

Hepatitis A

Like enteroviruses, **hepatitis A virus,** (genus *Hepatovirus*), is transmitted through the fecal-oral route, but unlike enteroviruses it is not cytolytic. Hepatitis A virus can survive on surfaces such as countertops and cutting boards for days and resists common household disinfectants such as chlorine bleach.

Hepatitis A has an incubation period of about 1 month before fever, fatigue, nausea, anorexia, and jaundice abruptly start. As with hepatitis B (caused by a DNA virus), the patient's own cellular immune system kills infected liver cells, resulting in the signs and symptoms. Children are less likely than adults to develop symptoms, because their cellular immune responses are still immature. Hepatitis A virus does not cause chronic liver disease, and complete recovery occurs 99% of the time. Patients release virions in their feces and are infective even without developing symptoms.

Table 25.3 on page 709 compares five viruses known to cause viral hepatitis. Four of them (hepatitis A, C, D, and E) are RNA viruses discussed in this chapter. The fifth is hepatitis B virus, which was discussed in Chapter 24.

Table 25.3   **Comparison of Hepatitis Viruses**

Feature	Hepatitis A	Hepatitis B	Hepatitis C	Hepatitis D	Hepatitis E
Common names of disease	Infectious hepatitis	Serum hepatitis	Non-A, non-B hepatitis; chronic hepatitis	Delta agent hepatitis	Hepatitis E, enteric hepatitis
Virus family	*Picornaviridae*	*Hepadnaviridae*	*Flaviviridae*	*Arenaviridae*	*Hepeviridae*
Genome	RNA	DNA	RNA	RNA	RNA
Transmission	Fecal-oral	Needles; sex	Needles; sex	Needles; sex	Fecal-oral
Severity (mortality rate)	Mild (<0.5%)	Occasionally severe (1–2%)	Usually subclinical (0.5%–4%)	Requires coinfection with hepatitis B virus; may be severe (high)	Mild (1–2%) except in pregnant women (20%)
Chronic carrier state?	No	Yes	Yes	No	No
Other disease associations	—	Hepatic cancer	Hepatic cancer	Cirrhosis	—

Acute Gastroenteritis and Hepatitis E

Learning Objective

✓ Compare and contrast caliciviruses, astroviruses, and hepatitis E virus.

Two groups of small (about 27 nm diameter), round, enteric viruses that cause acute gastroenteritis are **caliciviruses** (kă-lis'i-vī'rŭs-ĕz) and **astroviruses** (as'trō-vī'rŭs-ĕz), which are slightly larger than picornaviruses and have naked, star-shaped, polyhedral capsids **(Figure 25.6).** Like enteroviruses, caliciviruses and astroviruses enter the body through the digestive system, but in contrast to enteroviruses they cause gastrointestinal disease. The capsids of caliciviruses have indentations, whereas those of astroviruses do not. Antigens of viruses in the two groups can be distinguished through serological tests. The incubation period for these two types of viruses is about 24 hours. Both caliciviruses and astroviruses have caused outbreaks of gastroenteritis in daycare centers, schools, hospitals, nursing homes, restaurants, and on cruise ships. These viruses account for 60% of nonbacterial infections of the gastrointestinal tract, and thus constitute about 10% of all gastroenteritis cases. The symptoms resolve within 12–60 hours.

Caliciviruses cause diarrhea, nausea, and vomiting, though the symptoms of any given serotype vary from patient to patient. The best studied of the caliciviruses are **noroviruses,** first, discovered in the stools of victims during an epidemic of diarrhea in Norwalk, Ohio.

Astroviruses also cause diarrhea, but no vomiting, and they are less likely to infect adults. In the United States, 70% of children have antibodies against astroviruses by age 7.

There is no specific treatment for calicivirus and astrovirus gastroenteritis, except support and replacement of lost fluid and electrolytes. Prevention of infection involves adequate sewage treatment, purification of water supplies, frequent handwashing, and disinfection of contaminated surfaces and fomites.

Hepatitis E virus (genus *Hepevirus*) was formerly considered a calicivirus based on structure and size (27–34 nm),

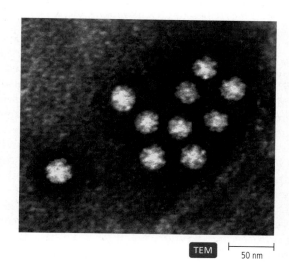

▲ *Figure 25.6*
Viruses of the families *Caliciviridae* and *Astroviridae* have naked, star-shaped capsids.

but now it is placed in its own family, *Hepeviridae*. Hepatitis E virus infects the liver and causes **hepatitis E** which is also known as enteric hepatitis. Hepatitis E is fatal to 20% of infected pregnant women (see Table 25.3). There is no treatment for hepatitis E. Prevention involves interrupting the fecal-oral route of infection with good personal hygiene, water purification, and sewage treatment.

Enveloped, Positive ssRNA Viruses: *Togaviridae, Flaviviridae,* and *Coronaviridae*

Members of the *Togaviridae* and *Flaviviridae* are enveloped, icosahedral, positive, single-stranded RNA viruses. They differ from the picornaviruses in being enveloped ("*toga*" is Latin for "cloak"); though the envelopes of these viruses are

TEM | 50 nm

▲ *Figure 25.7*

Togaviruses. Each virus has a closely appressed envelope around its capsid.

tightly appressed to the capsids (like shrink wrapping), which allows individual capsomeres to be visualized **(Figure 25.7)**. Because most togaviruses and flaviviruses are transmitted by arthropods (including mosquitoes, ticks, flies, mites, and lice), these viruses are designated **arboviruses**[6] (ar′bō-vī′rŭs-ĕz), and this shared characteristic is the reason taxonomists originally included the flaviviruses in the *Togaviridae*. However, differences in antigens, replication strategy, and RNA sequence indicate that flaviviruses belong in their own family.

Viruses of the family *Coronaviridae* are also enveloped, positive, single-stranded RNA viruses. However, in contrast to togaviruses and flaviviruses, they have helical capsids **(Figure 25.8)**.

In this section we will consider diseases of the arboviruses from the families *Togaviridae* and *Flaviviridae*, discuss the togaviruses and flaviviruses that are not transmitted via arthropods, and finally discuss the coronaviruses.

Diseases of +RNA Arboviruses

Learning Objectives

✓ Define zoonoses.
✓ Compare and contrast EEE, WEE, and VEE.
✓ Contrast the two types of dengue fever.
✓ Discuss the signs, symptoms, and prevention of yellow fever.

Specific genera of mosquitoes and ticks transmit arboviruses among animal hosts, usually small mammals or birds, and into humans. Diseases that spread from animal reservoirs to humans are called **zoonoses**[7] or *zoonotic diseases*. Arthropod vectors remain infected with arboviruses and are a continual source of new infections when they bite vertebrate hosts. In animal hosts, virions are released from infected cells into the blood (viremia), but persistent viremia does not occur as readily in humans. Thus, humans are "dead-end" hosts for these zoonoses.

Arboviruses enter target cells through endocytosis and replicate within them. Most cause mild, flu-like symptoms in humans within 3–7 days of infection. Arbovirus disease does not usually proceed beyond this initial manifestation. Occasionally, however, arboviruses in the blood infect the brain, liver, skin, and blood vessels. Diseases associated

[6]From *arthropod borne*.
[7]Plural of Greek *zoos*, meaning animal, and *nosis*, meaning disease.

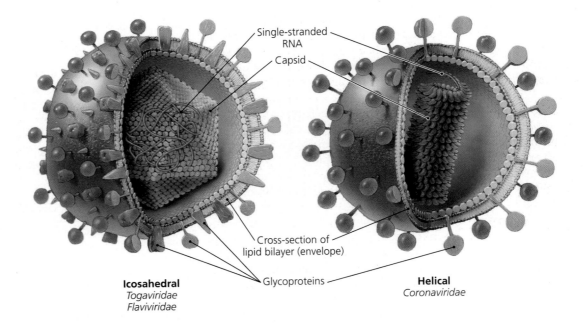

Figure 25.8 ▶

The two types of capsids of enveloped +ssRNA viruses. Whereas viruses in the families *Togaviridae* and *Flaviviridae* have icosahedral capsids, viruses in the family *Coronaviridae* have helical capsids.

Single-stranded RNA

Capsid

Cross-section of lipid bilayer (envelope)

Glycoproteins

Icosahedral *Togaviridae Flaviviridae*

Helical *Coronaviridae*

Table 25.4 Diseases Caused by Arboviruses

Disease	Vector	Natural Host(s)	Distribution	Symptoms
Togaviridae (enveloped, +ssRNA)				
Eastern equine encephalitis (EEE)	*Aedes* and *Culiseta* mosquitoes	Birds	Americas	Flu-like symptoms and encephalitis
Western equine encephalitis (WEE)	*Culex* and *Culiseta* mosquitoes	Birds	Americas	Flu-like symptoms and encephalitis
Venezuelan equine encephalitis (VEE)	*Aedes* and *Culex* mosquitoes	Rodents, horses	Americas	Flu-like symptoms and encephalitis
Flaviviridae (enveloped, +ssRNA)				
Japanese encephalitis	*Culex* mosquitoes	Birds, pigs	Asia	Flu-like symptoms and encephalitis
West Nile encephalitis	*Culex* mosquitoes	Birds	Africa, Europe, Asia, North America	Flu-like symptoms and encephalitis
St. Louis encephalitis	*Culex* mosquitoes	Birds	North America	Flu-like symptoms and encephalitis
Russian spring–summer encephalitis	*Ixodes* and *Dermacentor* ticks	Birds	Russia	Flu-like symptoms and encephalitis
Dengue and dengue hemorrhagic fever	*Aedes* mosquitoes	Monkeys, humans	Worldwide, especially tropics	Severe pain, hemorrhaging, hepatitis, shock
Yellow fever	*Aedes* mosquitoes	Monkeys, humans	Africa, South America	Hepatitis, hemorrhagic fever, shock
Bunyaviridae (enveloped, segmented, −ssRNA)				
LaCrosse encephalitis	Mosquitoes	Rodents, small mammals, birds	North America	Fever, rash, encephalitis
Rift Valley fever	Mosquitoes	Sheep, goats, cattle	Africa, Asia	Hemorrhagic fever, encephalitis
Sand fly fever	*Phlebotomus* flies	Sheep, cattle	Africa	Hemorrhagic fever, encephalitis
Crimean–Congo hemorrhagic fever	Ticks	Horses, cattle, goats, seabirds	Africa, Crimea	Hemorrhagic fever
Reoviridae (naked, dsRNA)				
Colorado tick fever	*Dermacentor* ticks	Small mammals, deer	Western North America	Fever, chills, headache, photophobia, rash, and in children, hemorrhage

with such second-stage infections include several kinds of encephalitis, dengue fever, and yellow fever **(Table 25.4).**

Encephalitis

Different togaviruses cause **Eastern equine encephalitis (EEE), Western equine encephalitis (WEE),** and **Venezuelan equine encephalitis (VEE).** As the names indicate, viral replication can occur in the brains of horses as well as in humans. The normal host for these viruses is either a bird (EEE, WEE, VEE) or a rodent (VEE) **(Figure 25.9** on page 712). Of the three, EEE causes the most severe disease in humans, though any of them may produce fatal brain infections.

Flaviviruses can also cause viral encephalitis that is usually relatively benign. The flaviviruses involved include *St. Louis, Japanese, West Nile,* and *Russian spring–summer* viruses (see Table 25.4). West Nile virus, which was identified for the first time in the United States in 1999, causes a form of encephalitis that can be fatal **(New Frontiers 25.1** on page 715). The incidence of all types of arboviral encephalitis in the United States is seasonal, with peaks that correspond to

the times of the year during which the vectors—adult mosquitoes—are active **(Figure 25.10** on page 712).

Dengue Fever

Dengue (den'gā) **fever** is caused by a flavivirus transmitted by *Aedes* mosquitoes. It afflicts approximately 100 million people in tropical and subtropical areas of the world **(Figure 25.11** on page 713). The disease usually occurs in two phases separated by 24 hours of remission. First, the patient suffers from fever, weakness, edema (swelling) of the extremities, and severe pain in the head, back, and muscles. The severity of the pain is indicated by the common name for the disease, *breakbone fever.* The second phase involves a return of the fever and a bright red rash. Dengue fever is self-limiting, and lasts 6 or 7 days.

Dengue hemorrhagic fever is a more serious disease caused by reinfection with the dengue virus, and it involves a hyperimmune response **(Figure 25.12** on page 714). Inflammatory lymphokines released by activated memory T cells cause rupture of blood vessels, internal bleeding, shock, and possibly death.

Figure 25.9 ▶

Hosts and transmission of equine encephalitis. Mosquitoes carry Western equine encephalitis (WEE), Eastern equine encephalitis (EEE), and Venezuelan equine encephalitis (VEE) viruses from wild birds (the primary reservoir) to wild and domesticated birds, to horses, to other mammals, and to humans. Except for rodents carrying VEE, mammals do not normally develop viremia, making them dead-end hosts for the virus.

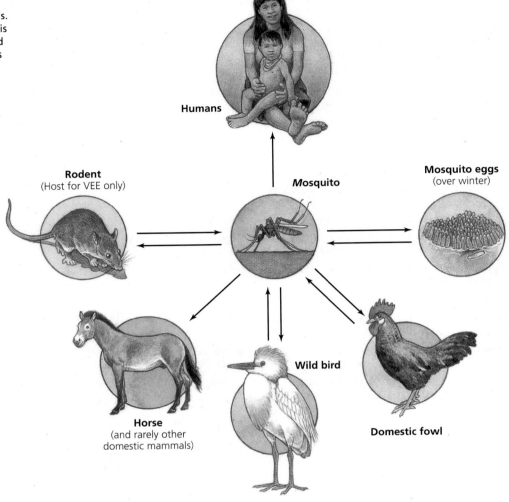

Humans

Rodent
(Host for VEE only)

Mosquito

Mosquito eggs
(over winter)

Horse
(and rarely other
domestic mammals)

Wild bird

Domestic fowl

Figure 25.10 ▶

Togavirus encephalitis in the United States. Note the seasonal nature of the diseases. *Why are there more cases of encephalitis in summer than in winter of every year?*

Figure 25.10 Arboviral encephalitis is dependent on mosquitoes as a vector. In winter it is too cold for adult mosquitoes to be active.

Togavirus encephalitis

Dengue and dengue hemorrhagic fever are epidemic in South America. There is concern that the diseases could be introduced into the United States by immigrants or returning travelers. Given that *Aedes* mosquitoes already live in parts of the United States, dengue could potentially become established here.

Yellow Fever

Another flavivirus causes **yellow fever,** a disease involving degeneration of the liver, kidneys, and heart, as well as massive hemorrhaging. Hemorrhaging in the intestines may

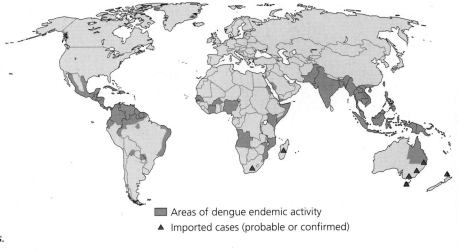

■ Areas of dengue endemic activity
▲ Imported cases (probable or confirmed)

Figure 25.11 ➤

Global distribution of dengue fever in 2000. The U.S. population is susceptible to this disease, which afflicts millions of people worldwide, because of the presence of the disease's vectors, mosquitoes in the genus *Aedes.*

result in "black vomit." Liver damage causes jaundice, from which the disease acquires its name and its nickname, "Yellow Jack." The mortality rate of yellow fever can approach 50%. Before the 1900s, yellow fever epidemics ravaged the Americas, killing thousands. In 1793, it killed over 4,000 people in Philadelphia (then the capital of the United States), and during the Spanish American War of 1898, it killed more American soldiers than bullets did. With mosquito control and the development of a vaccine in the 1900s, however, yellow fever has since been eliminated from the United States.

Diagnosis, Treatment, and Prevention of Arbovirus Infections

Hemagglutination inhibition, ELISA, and agglutination of latex beads to which arbovirus antigens are afixed are used for the diagnosis of arboviral infections. The only treatment for arboviral diseases is to provide supportive care. Only acetaminophen is recommended for managing pain and fever associated with dengue infections, as the anticoagulant properties of aspirin could aggravate the hemorrhagic properties of dengue virus infections.

Prevention of infection involves control of the vectors, which has resulted in elimination of many arboviral diseases in certain geographic areas. Insect repellants and netting should be used to prevent infections with arboviruses.

Vaccines for humans are available against yellow fever, EEE, WEE, Japanese encephalitis, and Russian spring–summer encephalitis viruses. Animal vaccines against VEE and West Nile virus are also available. No vaccine is available for dengue. Vaccination is recommended for people traveling to areas where these arboviral diseases are prevalent.

CRITICAL THINKING

Suppose a vaccine for dengue that induced the production of memory T cells could be developed. After reviewing the characteristics infection and reinfection, would you argue for or against the use of such a vaccine? Why or why not?

Rubella

Rubella (rū-bel'ǎ) virus (*Rubivirus*) shares the structure and reproductive strategy of other togaviruses, but unlike them it is transmitted by the respiratory route and not by arthropod vectors. **Rubella** is one of five childhood diseases that produce skin lesions. (The other four are measles, caused by a −ssRNA virus and discussed later in this chapter; and chickenpox, roseola, and fifth disease, discussed in Chapter 24.) Rubella, first distinguished as a separate disease by German physicians, is commonly known as "*German measles*" or "*three-day measles.*"

Rubella infects humans only, entering the respiratory system and infecting cells of the upper respiratory tract. It spreads from there to lymph nodes and into the blood, and then throughout the body. Then, the characteristic rash of flat, pink to red spots (macules) develops (**Figure 25.13** on page 714) and lasts about three days. Rubella in children is usually not serious, but adults may develop arthritis or encephalitis. Patients shed virions in respiratory droplets for approximately 2 weeks before and 2 weeks after the rash.

Rubella was not seen as a serious disease until 1941, when Dr. Norman Gregg, an Australian ophthalmologist, recognized that rubella infections of pregnant women resulted in severe congenital defects in their babies. These effects include cardiac abnormalities, deafness, blindness, mental retardation, microcephaly,[8] and growth retardation. Death of the fetus is also common. It is now known that a mother is able to transmit the virus across the placenta even if she is asymptomatic.

Diagnosis of rubella is usually made by observation and by serological testing for IgM against rubella. No treatment is available, but immunization has proven effective at reducing the incidence of rubella in industrialized countries (**Figure 25.14** on page 714). The rubella vaccine is made from a live, attenuated virus, and therefore should never be given to pregnant women or immunocompromised patients. Immunization is aimed at reducing the number of rubella cases that might serve to introduce rubella to pregnant women.

---

[8]From Greek *micro,* meaning small, and *cephalos,* meaning head.

Initial Infection

Dengue infects monocyte

↓

Mild febrile illness

↓

Recovery, with antibody to dengue in blood

Reinfection →

Virus-antibody complexes forms

↓ Virus-antibody complexes phagocytized by APC

Activation of memory T cells

↓

Increased production of inflammatory lymphokines

↓

Dengue hemorrhagic fever, severe illness, shock, hemorrhage, death (10–50%)

◄ *Figure 25.12*

Pathogenesis of dengue hemorrhagic fever, a severe hyperimmune response to a second infection with dengue virus. Complexes of virus and antibodies (formed during the initial infection) are phagocytized by antigen presenting cells, which activate T memory cells. These release an abundance of lymphokines that induce hemorrhaging and shock.

▲ *Figure 25.13*

The rash of flattened red spots (macules) characteristic of rubella, also known as German measles or three-day measles.

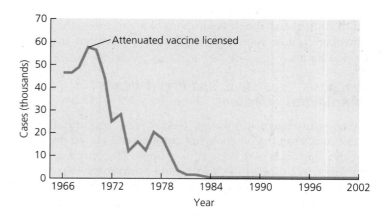

▲ *Figure 25.14*

The efficacy of vaccination against rubella. The use of a live, attenuated virus vaccine has practically eliminated rubella in the United States.

Hepatitis C

Learning Objective

✓ Compare and contrast hepatitis C with the other types of viral hepatitis.

A flavivirus called **hepatitis C virus (HCV)** accounts for 90% of the cases of so-called non-A, non-B hepatitis in the United States (see Table 25.3 on page 709). About 4 million Americans are chronically infected by HCV, which is spread via needles, organ transplants, and sexual activity, and not by arthropod vectors. In the past, blood transfusions accounted for many cases of hepatitis C, but testing blood for HCV has considerably reduced the risk of infection via this means.

Hepatitis C is a chronic disease, often lasting for 20 years or longer, with few if any symptoms. Severe liver damage (20% of cases) and liver failure (5% of cases) occur over time. Hepatic cancer can also result from HCV infection.

Alpha interferon and ribavirin provide some relief for hepatitis C but are not cures. There is no other treatment or vaccine for hepatitis C.

Diseases of Coronaviruses

Learning Objective

✓ Describe the structure of and diseases caused by coronavirus.

Coronaviruses (kō-rō′nǎ-vī′rǔs-ěz) are enveloped, positive, single-stranded RNA viruses with a helical capsid. Their envelopes form corona-like halos around the capsids, giving these viruses their name **(Figure 25.15).**

Coronaviruses, the second most common cause of colds after rhinoviruses, are transmitted from epithelial cells of the upper respiratory tract in large droplets (sneezes and coughs). Like rhinoviruses, coronaviruses replicate best at 33°C, but in contrast, some strains can live in the GI tract and cause gastroenteritis in children. Generally, diseases of coro-

TEM | 50 nm

▲ *Figure 25.15*

Coronaviruses, which have distinctive envelopes with glycoprotein spikes that resemble flares from the corona of the sun.

naviruses are mild, and no antiviral treatment or vaccine is available. Symptoms of mild diseases can be alleviated with antihistamines and analgesics.

| **New Frontiers 25.1** | **West Nile Virus and Severe Acute Respiratory Syndrome (SARS)** |

In 1999, hundreds of birds—mostly crows—suddenly began dying in the New York City area. At the same time, reports surfaced of people feeling ill with flu-like symptoms after being bitten by mosquitoes. Seven patients contracted viral encephalitis and died. As it turned out, these events were related. The culprit was West Nile virus, a pathogen endemic to Africa and Israel that had never before been reported in North America. It was being transmitted by mosquito bites, thereby infecting birds, humans, and other animals. After the initial outbreak in New York, West Nile virus spread throughout the rest of the United States, following migratory bird patterns. Fatal to many bird species, the virus rarely causes serious illness in humans, but in some cases can cause meningitis or encephalitis. While the typical person's chances of contracting West Nile virus-associated illness is slim, one can take preventive measures by applying insect repellent and wearing long-sleeved clothing in order to reduce the chances of being bitten by an infected mosquito.

In the winter of 2002–2003, another newly emerging disease made headlines worldwide. Severe acute respiratory syndrome (SARS)

Inspecting a dead crow for evidence of West Nile virus infection

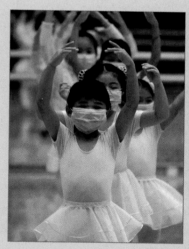

Ballet students in Hong Kong during a SARS outbreak

was first identified in China's Guangdong province, characterized by symptoms of fever above 100.4°F, headache, general discomfort, and respiratory distress. By March of 2003, the disease had spread to Singapore, Vietnam, Taiwan, Canada, and other parts of the world. Amazingly, epidemiologists were able to trace the disease's global spread back to a single individual: an infected physician who had travelled from China to Hong Kong in order to attend his nephew's wedding and unwittingly spread the disease to other travellers (see page 404).

SARS created widespread alarm, due to its seemingly sudden emergence, highly infectious nature (it is spread by close person-to-person contact), initially unknown origin (it has since been attributed to a coronavirus), and five to eight percent mortality rate. As this book went to press, SARS remained a serious problem in countries such as China and Taiwan, but appeared to be remarkably contained in most other countries. At the end of May 2003, over 7000 probable cases of SARS had been reported in China, Hong Kong, and Taiwan, but fewer than 70 cases (with no deaths) had been reported in the United States. In contrast, 36,000 Americans die of influenza during an average winter.

Figure 25.16 ▶

Artist's conception of HIV, a retrovirus. Retroviruses contain 10–50 molecules of reverse transcriptase, integrase, and protease. All retroviruses have envelopes with glycoprotein spikes, but not necessarily gp41 and gp120.

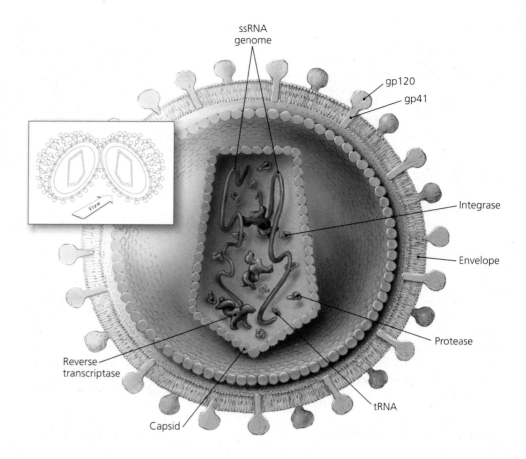

Because diseases of coronaviruses are usually mild, the 2003 pandemic of sometimes fatal **severe acute respiratory syndrome (SARS),** which is caused by a coronavirus, was particularly alarming. This emerging disease initially taxed health care systems in several Asian countries, but unprecedented worldwide cooperation allowed epidemiologists to identify a new virus, track its spread, sequence its genome, and investigate effective quarantine procedures faster than with any previous disease **(New Frontiers 25.1).**

Enveloped, Positive ssRNA Viruses with Reverse Transcriptase: *Retroviridae*

Learning Objectives

✓ Explain how retroviruses do not conform to the central dogma of molecular biology.

✓ Describe the steps of reverse transcription.

✓ List two types of retroviruses.

Retroviruses (re′trō-vī′rŭs-ĕz) have been studied more than any other group of viruses because of their unique features and the diseases they cause. These viruses have polyhedral capsids with spiked envelopes 80–120 nm in diameter, and genomes composed of two identical molecules of positive, single-stranded RNA. Each virion also contains two tRNA molecules and 10–50 copies of the enzymes *reverse transcriptase, protease,* and *integrase,* whose function will be described

shortly. **Figure 25.16** illustrates these features as exemplified by HIV, the most studied of the retroviruses.

The unique characteristic of retroviruses is that they do not conform to what is considered the *central dogma* of molecular biology; that is, that genetic information flows from DNA to RNA to translated proteins. Retroviruses[9] reverse this flow of information; that is, they transcribe dsDNA from ssRNA.

The retroviruses accomplish this amazing feat by means of the complex enzyme, **reverse transcriptase,** which transcribes double-stranded DNA from RNA. Transcription occurs as the enzyme moves along an RNA molecule and can best be understood as occurring in three steps, though in reality the events occur simultaneously **(Figure 25.17):**

1 An RNA-DNA hybrid is made from the +RNA genome using tRNA carried by the virion as a primer for DNA synthesis. The DNA portion of the hybrid molecule is negative DNA.

2 The RNA portion of the hybrid molecule is degraded by reverse transcriptase, leaving −ssDNA.

3 The reverse transcriptase transcribes a complementary +DNA strand to form dsDNA, and the tRNA primer is removed.

The discovery of reverse transcriptase was a momentous event in biology, and it made possible an ongoing revolution

---

[9]From Latin *retro,* meaning reverse.

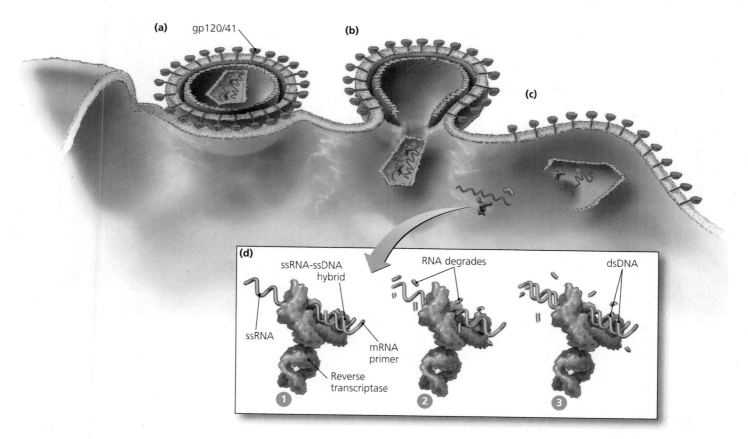

(a) gp120/41 **(b)** **(c)**

(d) ssRNA-ssDNA hybrid | RNA degrades | dsDNA

ssRNA | mRNA primer | Reverse transcriptase

① ② ③

▲ *Figure 25.17*

The process by which reverse transcriptase transcribes dsDNA from ssRNA. After the virus has **(a)** attached to a host cell, **(b)** inserted its capsid into the host cell, and **(c)** uncoated to release its RNA and enzymes, the enzyme catalyzes three nearly simultaneous reactions **(d)**, depicted here as three distinct steps: ① A complementary −ssDNA molecule is transcribed to form a DNA-RNA hybrid. Virion tRNA is used as a primer for this synthesis. ② The RNA portion of the hybrid is degraded leaving −ssDNA. ③ A complementary +DNA strand is synthesized to form dsDNA.

in recombinant DNA technology and biotechnology. For example, it is much easier to isolate mRNA for a particular protein (e.g., human insulin) and then use reverse transcriptase to make a gene, than it is to find the gene itself amidst the DNA of 23 pairs of human chromosomes.

In the following sections we consider the two types of retroviruses, both of which infect humans: those that are primarily oncogenic (genus: *Deltaretrovirus*), and those that are primarily immunosuppressive (genus *Lentivirus*).

Oncogenic Retroviruses

Learning Objective

✓ Describe how oncogenic viruses may induce cancer.

Retroviruses have been linked with cancer ever since they were first isolated from chicken neoplasias. In 1981, Robert Gallo and his associates isolated the first human retrovirus, **human T-lymphotrophic virus 1 (HTLV-1),** from a patient with *adult acute T-cell lymphocytic leukemia*, a cancer in hu-

mans. Since then, two other human retroviruses of this type have been discovered: HTLV-2, which causes a rare cancer called *hairy-cell leukemia* because of distinctive extensions of the cytoplasmic membrane of infected cells (**Figure 25.18** on page 718), and HTLV-5, which has not been linked to cancer.

HTLV-1 and HTLV-2 infect lymphocytes and are transmitted within these cells to other people via sexual intercourse, blood transfusion, and contaminated needles. HTLV-1 is primarily found in populations in Japan and the Caribbean, and among African Americans. HTLV-2 is not as well studied as HTLV-1. An ELISA can reveal infection with HTLV. As with many viral diseases, there is no specific antiviral treatment. Infections are chronic, and the long-term prognosis of patients is poor. Prevention involves the same changes in behavior needed to prevent HIV infection (see pages 723–724).

The exact way oncogenic retroviruses cause cancer is unknown. It is suspected that either an oncogene from the virus is integrated into host chromosomes, or that regulatory proteins (transcribed from viral genes) free the cell from normal control of cytokinesis.

Acquired Immunodeficiency Syndrome

Learning Objectives

✓ Differentiate between a disease and a syndrome, using AIDS as an example.

✓ Describe the virions that cause AIDS.

✓ Explain the roles of gp41 and gp120 in HIV infection.

✓ Describe the relationship of T-helper cells to the course of AIDS.

✓ List four measures that can be effective in preventing AIDS.

From the time of its discovery in 1981 among homosexual males in the United States to its emergence as a worldwide pandemic, no affliction has affected modern life as much as **acquired immunodeficiency syndrome (AIDS).** AIDS is not a disease but a **syndrome;** that is, a complex of signs and symptoms associated with a common cause. At the present time, AID syndrome is defined as several opportunistic or rare infections, listed in **Table 25.5,** in the presence of antibodies against the **human immunodeficiency virus (HIV).**

HIV is restricted to replicating in humans and, as its name indicates, this pathogen destroys the human immune system. Its many characteristics that make it particularly difficult to combat are summarized in **Table 25.6.**

There are two major types of HIV. **HIV-1,** which is more prevalent in the United States and Europe, was discovered by Luc Montagnier and his colleagues at the Pasteur Institute in 1983. **HIV-2,** the prevalent strain in West Africa, appears to reproduce more slowly than HIV-1; therefore, HIV-1 will be the major focus of our discussion.

Before you read on, preview the terms associated with HIV infections and treatment listed in **Table 25.7.**

SEM   10 μm

▲ *Figure 25.18*

The characteristic extensions of the cytoplasmic membrane in hairy-cell leukemia.

Structure of HIV

HIV is a typical retrovirus in size and shape. Its envelope is characterized by two antigenic glycoproteins (see Figure 25.16). The larger of these, named **gp120,**[10] is the primary at-

[10]gp120 is short for glycoprotein with a molecular weight of 120,000.

Table 25.5   Opportunistic Infections and Tumors of AIDS Patients

Type	Manifestations
Viral infections (24)[a]	Cytomegalovirus disseminated in brain, lungs, retina, etc. Herpes simplex virus disseminated in lungs, GI tract, etc. Progressive multifocal leukoencephalopathy (PML) (JC virus) Oral hairy leukoplakia (Epstein-Barr virus)
Bacterial infections (19–21)	Tuberculosis, especially extrapulmonary (*Mycobacterium* spp.) Rectal gonorrhea Recurrent fever and septicemia due to *Salmonella, Haemophilus, Streptococcus*
Protozoal infections (23)	Toxoplasmosis *(Toxoplasma gondii)* Chronic diarrhea *(Cryptosporidium)* Diarrhea, persisting for more than 1 month, due to *Isospora* sp.
Fungal infections (22)	*Pneumocystis jiroveci* pneumonia Thrush, disseminated in trachea, lungs, esophagus *(Candida albicans)* Histoplasmosis *(Histoplasma capsulatum)*
Viral-induced tumors (24)	Kaposi's sarcoma (especially when associated with HHV-8) Lymphoma (induced by Epstein-Barr virus)
Others	Wasting disease; called *slim* in Africa (cause unknown) Dementia

---

[a]Numbers in parentheses refer to chapters where relevant material is discussed.

Table 25.6   Characteristics of HIV That Challenge the Immune System

Characteristic	Effect(s)
Genome consists of two copies of +ssRNA	Reassortment of viral genes possible; virus is highly mutable, reverse transcription produces much genetic variation; genome can integrate into host's chromosomes
Targets T-helper lymphocytes especially, but also macrophages, dendritic cells	Inactivates key aspects of host's immune response
Antigenic variability	Large amount of antigenic drift due to mutations helps virus evade host's immune response
Induces formation of syncytia	Increases routes of infection; intracellular site helps virus evade immune detection.

tachment molecule of HIV. Its antigenicity can change during the course of prolonged infection, making an effective antibody response against it very difficult. The smaller glycoprotein, **gp41,** promotes fusion of the viral envelope to a target cell. The effects of these structural characteristics—antigenic variability and the ability to fuse with host cells—impede immune clearance of HIV from a patient.

Origin of HIV

Evidence suggests that HIV arose from mutation of a similar virus, *simian immunodeficiency virus (SIV),* found in African monkeys. The nucleotide sequence of SIV is similar to that of HIV. Antibodies against HIV have been found in human blood stored since 1959, though the first cases of AIDS were not documented until 1981. Based on mutation rates and the rate of antigenic change in HIV, it is estimated that HIV emerged in the human population in Africa about 1930.

The relationship between the two types of HIV is not clear. HIV-1 and HIV-2 may be derived from different strains of SIV, or one type of HIV may be derived from the other. We may never know the exact temporal relationships among the immunodeficiency viruses.

Replication of HIV

The replication cycle of HIV is shown in **Figure 25.19** on page 720. The cycle is a typical cycle for an animal retrovirus, as discussed in Chapter 13.

Table 25.7   The Terminology of HIV

Term		Definition
AIDS	**Acquired immunodeficiency syndrome**	A syndrome is a complex of signs, symptoms, and in this case diseases, associated with a common cause. This syndrome results from an infection that suppresses the immune system.
HIV	**Human immunodeficiency virus**	Retrovirus that causes AIDS. Two major types of HIV exist. HIV-1 is seen primarily in Europe and North America, HIV-2 occurs more commonly in Africa.
gp120	**Glycoprotein 120**	Antigenic spike of HIV envelope; molecule for attachment to target cell; highly variable and mutable, antigen for possible vaccine.
gp41	**Glycoprotein 41**	Envelope protein required for HIV docking to target cell and syncytia formation; antigen for possible vaccine, but highly mutable
Reverse transcriptase		Enzyme that transcribes dsDNA from +ssRNA; each HIV carries at least 10 copies of the enzyme
Integrase		Enzyme that inserts HIV genome into host's DNA; integration into host's chromosome is permanent
Protease		Enzyme that cleaves a polypeptide to release reverse transcriptase; required for HIV activation and virulence
Protease inhibitor		Therapeutic drug that blocks protease enzymatic function, which renders HIV ineffectual; one of several drugs in a therapeutic "cocktail"
AZT	**Azidothymidine**	Nucleoside analog that inhibits reverse transcriptase; reduces frequency of opportunistic infections, prevents transplacental transmission of HIV; part of AIDS therapeutic "cocktail"
ddI	**Dideoxyinosine**	Nucleoside analog that inhibits reverse transcriptase; part of AIDS therapeutic "cocktail"
ddC	**Dideoxycytidine**	Nucleoside analog that inhibits reverse transcriptase; part of AIDS therapeutic "cocktail"

Figure 25.19 ➤

The replication cycle of HIV. The artist's rendition depicts three steps involved in the replication of the virus within a T-helper cell: ① attachment, penetration, and uncoating, ② latency and synthesis, and ③ assembly and release. Assembly is completed outside the host cell when the action of protease produces maturation of the virion. Photos **(a)**, **(b)**, and **(c)** show attachment and penetration; photos **(d)**, **(e)**, and **(f)** show the budding and release of a virion.

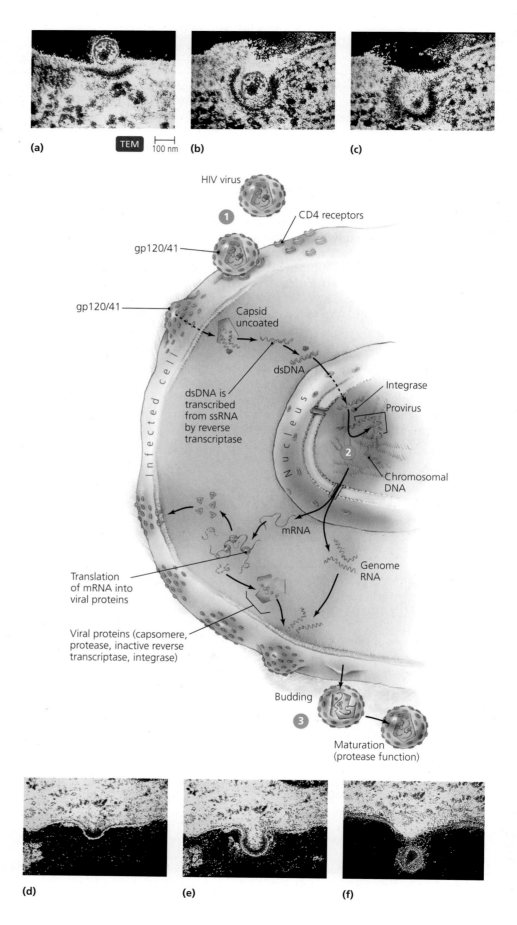

(a) TEM ├──100 nm──┤ **(b)** **(c)**

HIV virus

1

CD4 receptors

gp120/41

gp120/41

Capsid uncoated

dsDNA

dsDNA is transcribed from ssRNA by reverse transcriptase

Infected cell

Nucleus

Integrase

Provirus

2

Chromosomal DNA

mRNA

Genome RNA

Translation of mRNA into viral proteins

Viral proteins (capsomere, protease, inactive reverse transcriptase, integrase)

Budding

3

Maturation (protease function)

(d) **(e)** **(f)**

Figure 25.20 ➤

The process by which HIV attaches to and enters a host cell. **(a)** Binding of gp120 to a CD4 receptor on the cell membrane. **(b)** Removal of the CD4-gp120 complex, allowing gp41 to attach to the cytoplasmic membrane. **(c)** Fusion of the lipid bilayers, introducing the capsid into the cytoplasm. *Which cells have CXCR4 receptors (as shown here)? Which cells have CCR5 receptors?*

Figure 25.20 T-helper cells have CXCR4 receptors, whereas cells of macrophage lineage and smooth muscle cells have CCR5 receptors.

(a)

(b)

(c)

Attachment, Penetration, and Uncoating

HIV infects four kinds of cells: (1) T-helper cells; (2) cells of the macrophage lineage, including monocytes, macrophages, and *microglia* (special phagocytic cells of the central nervous system); (3) smooth muscle cells, such as those in arterial walls; and (4) *dendritic cells,* which are antigen-presenting cells of the skin, mucous membranes, and lymph nodes. B lymphocytes adhere to HIV that has been covered with complement proteins (see Chapter 15). B cells do not themselves become infected with HIV, but they deliver HIV to the lymphoid tissues, where T cells mature.

Attachment of HIV and entry into T cells and macrophages involves three steps **(Figure 25.20).** First, gp120 binds to CD4 receptor molecules on a host cell. Second, the gp120-CD4 complex binds to another receptor, called a *chemokine receptor,* which removes gp120 from the virion. Different chemokine receptors exist on T cells and on macrophage derivatives—*CXCR4*[11] on most T cells, and *CCR5* on macrophages and smooth muscle cells. Finally, once a virion binds to a cell, its envelope contacts the cytoplasmic membrane via gp41 and fuses with it. The HIV capsid is introduced into the cytoplasm while its envelope remains behind as part of the cell's cytoplasmic membrane. Glycoprotein 41 remaining on the cytoplasmic membrane induces the cell to fuse to neighboring cells to form a **syncytium** (sin-sish'ē-ŭm), a giant multinucleate cell.

Dendritic cells are infected in a single step. The gp120 of HIV attaches to a receptor (DC-SIGN) found on dendritic cell cytoplasmic membranes. HIV fuses with the target cell and introduces its capsid into the cytoplasm. Infected dendritic cells can deliver HIV into T cells during antigen presentation.

Latency and Synthesis

In the cytoplasm, reverse transcriptase uses cellular tRNA (packaged in the virion when it was assembled) as a primer to transcribe dsDNA. Reverse transcriptase is very error prone, making about five errors per genome, and this generates many new antigenic variations of HIV during the course of infection. As many as 100 variants may develop in a single patient over the course of the syndrome.

---

[11]Membrane proteins are often designated by number-letter combinations.

HIV is a latent virus. The dsDNA made by reverse transcriptase, which is known as a **provirus,** enters the nucleus and is inserted into a human chromosome by the virion-carried enzyme **integrase.** Once integrated, the provirus remains a part of the cellular DNA for life. Integrated HIV is transcribed and replicated in the same way as any cellular gene. It may remain dormant for years or be activated immediately, depending on its location in the human genome and the availability of enhancer and promoter sequences (see Chapter 7). Macrophages and monocytes are major reservoirs and means of distribution of HIV.

Assembly and Release   RNA transcribed from the provirus may be used in either of two ways:

1. A molecule of RNA may be used as mRNA by ribosomes for the translation of viral proteins, including capsomeres, gp120, and reverse transcriptase.

2. Two molecules of RNA may be incorporated into a new virion along with copies of reverse transcriptase, integrase, and two molecules of tRNA, (the latter were coded by the host's genome).

HIV virions subsequently bud from the cytoplasmic membrane (see Figure 25.19). At this point in its replication cycle, a virion is nonvirulent because its capsid is not assembled and reverse transcriptase is inactive. **Protease** (prō′tē-ās), a viral protein packaged in the virion, releases reverse transcriptase and capsomeres after the virion buds from the cell. This action of protease renders HIV virulent.

Pathogenesis of AIDS

The course of AIDS is directly related to the destruction of T-helper cells in a patient **(Figure 25.21).**

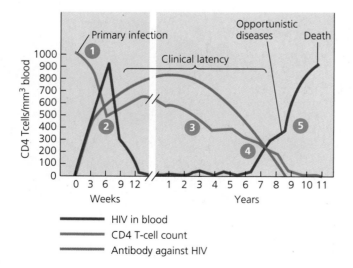

▲ *Figure 25.21*

The course of AIDS, which follows the course of T-helper cell destruction. The circled numbers correspond to the steps described in the text.

① Initially, there is a burst of virion production and release from infected cells. This primary infection is accompanied by fever, fatigue, weight loss, diarrhea, and body aches.

② The immune system responds by producing antibodies, and the number of free virions plummets. The body and HIV are waging an invisible war with few signs or symptoms. During this period, almost a billion virions are destroyed each day, but about 100 million CD4 helper T cells are killed by viruses and by cytotoxic T cells. No specific symptoms accompany this stage, and the patient is often unaware of the infection.

③ As latent viruses continue to replicate and virions are released into the blood, the body cannot adequately replenish helper T cells. Over the course of 5–10 years, the number of T-helper cells declines to a level that severely impairs the immune response.

④ The rate of antibody formation falls precipitously as helper T function is lost.

⑤ HIV production climbs, and the patient dies from infections the body can no longer resist.

Many of the diseases associated with the loss of immune function in AIDS (see Table 25.5 on page 718) are nonlethal infections in other patients, but AIDS patients cannot effectively resist them. Other diseases, such as Kaposi's sarcoma (a cancer of blood vessels) and disseminated herpes **(Figure 25.22),** as well as toxoplasmosis and *Pneumocystis* pneumonia, are rarely seen except in AIDS patients. AIDS often results in dementia during the final stages.

Epidemiology of HIV

AIDS was first recognized in young male homosexuals in the United States, but it is now found throughout the world's population. Current estimates of the number of HIV-infected individuals vary from 27 million to 60 million people worldwide. Approximately 16,000 new infections occur every day. It is disturbing that although the rate of new infections declined throughout most of the 1990s, it is now rising again, presumably because of a false sense of security in richer nations provided by the availability of drugs that slow the progress of the disease, and because of a relaxation of precautions to prevent infection.

About a third of those infected have already developed AIDS; it is presumed that most of the rest will eventually succumb to the syndrome. The spread of AIDS in subsaharan Africa is particularly horrific **(Figure 25.23).**

HIV is found in all secretions of infected individuals; however, it typically exists in sufficient concentration to cause infection only in blood, semen, saliva, vaginal secretions, and breast milk. Virions are found both free and inside infected white blood cells. Infected blood contains 1000–100,000 virions per milliliter, and semen has 10–50 virions per milliliter. Other secretions have lower concentrations and are less infective than blood or semen.

(a)

(a)

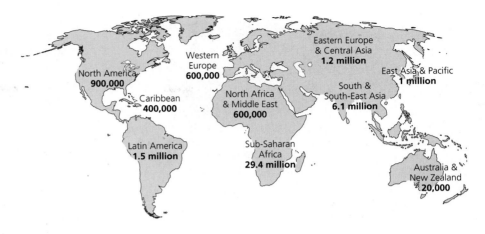

▲ *Figure 25.23*

The global distribution of HIV/AIDS. Figures indicate the estimated number of children and adults living with HIV/AIDS as of the end of 2002. Note the alarmingly high number of people infected in sub-Saharan Africa.

▲ *Figure 25.22*

Diseases associated with AIDS. **(a)** Kaposi's sarcoma. **(b)** Disseminated herpes. *Which viruses cause herpes?*

Figure 25.22 Herpes is caused by herpes simplex virus-1 and herpes simplex virus-2.

Infected fluid must come in contact with a tear or lesion in the skin or mucous membranes, or it must be injected into the body. Because HIV is only about 90 nm in diameter, a break in the protective membranes of the body large enough to allow entry of HIV may be too small to see or feel. Sufficient numbers of virions must be transmitted to target cells to establish an infection. The exact number of virions needed to establish an infection is unknown and probably varies with the strain of HIV and the overall health of the host's immune system.

HIV is transmitted primarily via sexual contact (including vaginal, anal, and oral sex, either homosexual or heterosexual) and intravenous drug abuse (**Figure 25.24** on page 724). HIV is also rarely acquired from blood transfusions, organ transplants, hemophiliac therapy, tattooing, and accidental medical needle-sticks. The risk of infection from a needle-stick injury involving an HIV-infected patient is less than 1%. HIV is also transmitted from mother to baby across the placenta and in breast milk. Approximately one-third of babies born to HIV-positive women are infected with HIV at birth. Behaviors that increase the risk of infection include the following:

- Sexual promiscuity; that is, sex with more than one partner
- Anal intercourse, especially receptive anal intercourse
- Intravenous drug use
- Sexual intercourse with anyone in the previous three categories

The mode of infection is unknown for about 9% of AIDS patients. These patients are often unable or unwilling to answer questions about their sexual and drug abuse history. The CDC has documented a few cases of casual spread of HIV, including infections from sharing razors and toothbrushes and from mouth-to-mouth kissing. In all cases of casual spread, it is suspected that small amounts of the donor's blood may have entered abrasions in the recipient's mouth.

Diagnosis, Treatment, and Prevention of HIV/AIDS

The signs and symptoms of AIDS vary according to the diseases present. By definition, the syndrome indicates that patients have one or more rare diseases and antibodies against HIV. Patients also typically have unexplained weight loss, fatigue, fever, and fewer than 200 CD4 lymphocytes per microliter (µl) of blood.[12]

HIV itself is difficult to locate in patient's secretions and blood because it becomes a provirus that is inserted into a patient's chromosomes, and because for years it is kept in check by the immune system. Therefore, serological diagnosis involves detecting antibodies against HIV using ELISA, agglutination, or western blot testing. Most individuals develop antibodies within 6 months of infection, though some remain without detectable antibodies for up to 3 years. A positive test for antibodies does not mean that HIV is

---

[12]The normal value for CD4 cells is 500–700 cells/µl.

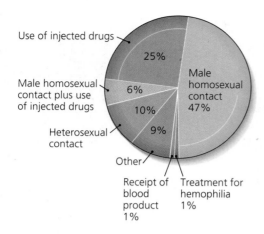

▲ *Figure 25.24*

Modes of HIV transmission in Americans over 12 years of age during 2002. Percentages represent the estimated proportions of HIV infections that result from each type of transmission. (Estimates are rounded to the nearest 1% and thus do not add up to 100%.)

currently present or that the patient has or will develop AIDS; it merely indicates that the patient has been exposed to HIV.

A small percentage of infected individuals, called *long-term nonprogressors,* have not yet developed AIDS. It appears that these individuals are either infected with defective virions, or lack effective coreceptors (CXCR4 and CCR5) needed to maintain an infection, or have unusually well-developed immune systems.

Discovering a treatment for AIDS is an area of intense research and development. Currently physicians prescribe a "cocktail" of several antiviral drugs, including replication and protease inhibitors to reduce viral replication. The action of antiviral drugs was discussed in Chapter 10. Such "cocktails" are expensive and must be taken on a fairly strict schedule. Some studies indicate that such a regimen of drugs can stop the replication of HIV without adversely affecting the patient; however, treatment is not a cure, because the infection remains. As long as treatment continues, the patient can live a relatively normal life until mutations enable the viruses to resist the drugs in the cocktail. Individual diseases associated with AIDS must be managed and treated on a case-by-case basis.

Progress in developing a vaccine against HIV has been disappointing. Among the problems that must be overcome in developing an effective vaccine are the following:

- A vaccine must generate both secretory IgA to prevent sexual transmission and infection, and cytotoxic T lymphocytes to eliminate infected cells.

- IgG induction, while a necessary part of a vaccine, can actually be detrimental to a patient. Because IgG-viral complexes remain infective inside phagocytic cells, a vaccine must stimulate cellular immunity more than humoral immunity so that infected cells can be killed.

- HIV is highly mutable, generating antigenic variants that enable it to evade the immune response. Indeed, a patient with untreated AIDS is likely to possess viruses with every possible single-base error in their RNA.

- HIV can spread through syncytia (formed by gp41) and thus can evade immune surveillance.

- HIV infects and inactivates macrophages, dendritic cells, and helper T cells—cells that combat infections and the very cells a vaccine would stimulate.

- Testing a vaccine presents ethical and medical problems because HIV is a pathogen of humans only. For example, a study in 1994 had to be stopped when five volunteers were infected with HIV despite having received an experimental vaccine.

Notwithstanding the challenges, scientists have developed a vaccine that protects monkeys from SIV disease, and are hopeful that an effective vaccine for humans will become available. In the meantime, the principal way the AIDS epidemic can be slowed and eventually stopped is by effecting changes in behavior:

- Abstinence and faithful monogamy between unifected individuals are the only truly safe forms of sex. Studies have shown that whereas condoms reduce a heterosexual individual's risk of acquiring HIV by about 69%, the benefit of condom usage to a population can be undone by an overall increase in sexual activity. This finding presumably results from a false sense of security that condoms provide.

- Use of new, clean needles and syringes for all injections, as well as caution in dealing with sharp, potentially contaminated objects, can reduce HIV infection rates. If clean supplies are not available, undiluted bleach or other antiviral agents can deactivate HIV.

- AZT given to pregnant women has reduced transplacental transfer of HIV, but infants can still be infected via breast milk. HIV-infected mothers should not breast feed their infants.

- Screening blood, blood products, and organ transplants for HIV and anti-HIV antibodies has virtually eliminated the risk of HIV infection from these sources.

Enveloped, Unsegmented Negative ssRNA Viruses: *Paramyxoviridae, Rhabdoviridae,* and *Filoviridae*

Viruses of the families *Paramyxoviridae, Rhabdoviridae,* and *Filoviridae* are enveloped, helical, negative, single-stranded RNA viruses. Recall that a negative genome cannot be directly translated to a ribosome because its information is nonsensical; however, it is the complement of a readable sequence of nucleotides. Therefore, a −ssRNA virus must make a positive copy of its genome to serve as mRNA for protein translation. The positive copy also serves as a

A 25-year-old male was admitted to the hospital with thrush, diarrhea, unexplained weight loss, and difficulty in breathing. Cultures of pulmonary fluid revealed the presence of *Pneumocystis jiroveci*. The man

Thrush

admitted to being a heroin addict and to sharing needles in a "shooting gallery."

1. What laboratory tests could confirm a diagnosis of AIDS?
2. How did the man most likely acquire an HIV infection?
3. What changes to the man's immune system allowed the opportunistic infections of *Candida* (thrush) and *Pneumocystis* to arise?
4. What precautions should be taken with the patient's blood?

template for transcription of more −ssRNA to be incorporated into new virions.

Viruses in the family *Paramyxoviridae* have similar morphologies and antigens. They also share the ability to cause infected cells to fuse with their neighbors forming giant, multinucleate syncytia **(Figure 25.25),** which enable virions to pass from an infected cell into neighboring cells and to evade immune surveillance and antibodies. The family *Paramyxoviridae* contains four genera that infect humans: *Morbillivirus* (measles virus), *Paramyxovirus* (parainfluenza), *Rubulavirus* (mumps viruses), and *Pneumovirus* (respiratory syncytial virus). **New Frontiers 25.2** on page 726 describes an emerging fatal disease caused by Nipah virus, which appears to belong to a new genus in the *Paramyxoviridae*.

Rhabdoviruses have bullet-shaped envelopes and include a variety of plant and animal pathogens. Rabies is the most significant pathogen among the rhabdoviruses.

Filoviruses are particularly significant because they are pathogens that cause a number of emerging and quite frightening diseases, including Ebola hemorrhagic fever.

Measles

Learning Objective

✓ Describe the signs and symptoms of rubeola and SSPE, and some ways of preventing them.

Measles virus causes one of five classical childhood diseases. **Measles,** also known in the United States as *rubeola*[13] or *red measles,* is one of the more contagious and serious childhood diseases, and it should not be confused with the generally milder rubella (German measles). **Table 25.8** on page 727 compares and contrasts measles and rubella.

Epidemiology of Measles Infections

Measles viruses are spread in the air via respiratory droplets from coughing, sneezing, and talking. The viruses infect cells of the respiratory tract before spreading via lymph and

[13]Unfortunately, rubeola is the name given to German measles (rubella) in some other countries.

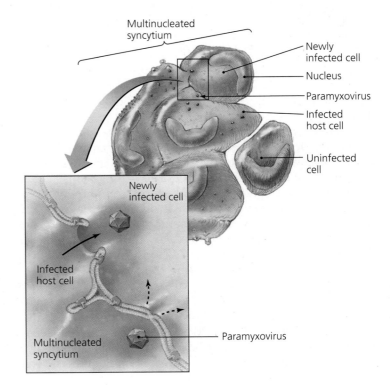

▲ *Figure 25.25*

A syncytium formed when paramyxoviruses trigger infected cells to fuse with uninfected neighboring cells. Virions moving through these large multinucleate cells can infect new cells, all the while evading the host's immune system.

blood throughout the body. Besides the respiratory tract, the conjunctiva, urinary tract, small blood vessels, lymphatics, and central nervous system become infected.

Humans are the only host for measles virus, and a large, dense population of susceptible individuals must be present for the virus to spread. In the United States before vaccination was instituted, measles outbreaks occurred in 1–3 year epidemic cycles as the critical number of susceptible individuals increased in the population. More than 80% of

Nipah Virus: From Pigs to Humans

Many diseases become known to science either when people encroach on a pathogen's home territory and come into contact with natural hosts, or when pathogens are introduced into geographical areas outside their historical range. Diseases resulting under such conditions are called *emerging diseases*. One example of an emerging disease is West Nile Virus encephalitis, which first began to appear in the United States in 1999 (see p. 715). That same year, another new form of encephalitis appeared in Malaysia. The culprit: a never-before-identified virus that became known as "Nipah" virus, after the region in Malaysia in which it was discovered.

The Malaysian victims of Nipah encephalitis experienced high fever, severe headache, muscle pain, drowsiness, disorientation, convulsions, and coma. Over one-third of them died. Serological investigations and RNA sequencing of the Nipah genome revealed that Nipah virus likely belongs to a new genus in the family *Paramyxoviri-*

dae. Epidemiological studies suggest that the victims contracted Nipah virus from infected pigs: 93% of the Malaysian victims reported occupational exposure to pigs, and antibodies against Nipah virus have been detected in the animals.

How, then, did pigs become infected with Nipah virus? The prevailing theory involves human encroachment on bat habitats. Certain species of Malaysian fruit bats are believed to be the natural hosts of Nipah virus. Highway workers pushing into these bats' rainforest habitats could have disturbed their roosts, driving the bats into proximity with pig farms. The virus could then have "jumped" species and infected the pigs. Individuals working on pig farms or in slaughterhouses could subsequently have been infected either via the respiratory route or through breaks in their skin and mucous membranes. However, to date, the exact mode of transmission remains a mystery.

---

susceptible patients exposed to the virus developed symptoms 7–13 days after exposure.

Signs and symptoms include fever, sore throat, headache, dry cough, and conjunctivitis. After two days of illness, lesions called **Koplik's spots** appear on the mucous membrane of the mouth **(Figure 25.26)**. These lesions, which have been described as grains of salt surrounded by a red halo, last 1–2 days and provide a definitive diagnosis of measles. Red, raised (maculopapular) lesions then appear on the head and spread over the body **(Figure 25.27)**. These lesions are extensive and often fuse to form red patches, which gradually turn brown as the disease progresses. Measles is self-limiting in most cases.

Rare complications of measles include pneumonia, encephalitis, and the extremely serious **subacute sclerosing panencephalitis (SSPE)**. SSPE is a slow progressive disease of the central nervous system that involves personality changes, loss of memory, muscle spasms, and blindness. The disease begins 1–10 years after the initial infection with measles virus and lasts a few years before resulting in death. A defective measles virus, which cannot make a capsid, causes SSPE. The virus replicates and moves from brain cell to brain cell via cell fusion, limiting the functioning of infected cells and resulting in the symptoms. SSPE afflicts less than 7 measles patients in 1 million and is becoming much rarer as a result of vaccination of children.

Diagnosis, Treatment, and Prevention of Measles

The signs of measles, particularly Koplik's spots, are sufficient for diagnosis, but serological testing can confirm the presence of measles antigen in respiratory and blood specimens. No antiviral treatment is available for measles.

An effective, live, attenuated vaccine for measles introduced in 1963 has eliminated measles as an endemic disease in the United States **(Figure 25.28)**; however, measles remains a frequent cause of death in other countries. Measles

Koplik's spots

Figure 25.26 ▶
White Koplik's spots on the oral mucous membrane, a diagnostic sign of measles.

Figure 25.27 ▶
The raised lesions of measles, which begin on the face and spread across the body. They may grow and fuse into large reddish patches.

vaccine, in combination with vaccines against mumps and rubella (MMR vaccine), is given to children in the United States who are less than 15 months old. The measles portion of the vaccine is effective 95% of the time but should be

Table 25.8   A Comparison of Measles and Rubella

	Causative Agent	Primary Patient(s)	Complications	Skin Rash	Koplik's Spots
Measles (also known as rubeola and red measles)	*Paramyxoviridae: Morbillivirus*	Child	Pneumonia, encephalitis, subacute sclerosing panencephalitis	Extensive	Present
Rubella (also known as German measles, rubeola, and three-day measles)	*Togaviridae: Rubivirus*	Child, fetus	Birth defects	Mild	Absent

boosted before grade school to ensure protection against measles introduced from outside the country.

Diseases of Parainfluenza Virus

Learning Objective

✓ Identify the specific cause of croup.

Four strains of **parainfluenza** viruses cause respiratory tract disease, particularly in young children. Respiratory droplets and person-to-person contact transmit the virions. Strains 1, 2, and 3 are associated with lower respiratory infections, whereas strain 4 is limited to mild, upper respiratory tract infections. **Croup** (kroop), a severe condition characterized by inflammation and swelling of the larynx, trachea, and bronchi and a "seal bark" cough, is a childhood disease of parainfluenza viruses 1 and 2.

Most patients recover from parainfluenza infections within 2 days. There is no specific antiviral treatment beyond support and careful monitoring to ensure that airways do not become completely occluded. In the latter case, *intubation,* insertion of a tube into the airways, is necessary. Vaccination with deactivated virus is not effective, possibly because it fails to produce significant secretory antibody. A live vaccine has not been developed against parainfluenza.

Mumps

The **mumps** virus is another −ssRNA virus. Mumps virions are spread in respiratory secretions and infect cells of the upper respiratory system. As virions are released from these areas, viremia develops, and a number of other organs are infected, resulting in fever and pain in swallowing. The parotid glands, located just in front of the ears under the skin covering the lower jaw, are particularly susceptible to the virus and become painfully enlarged, a condition known as *parotitis* **(Figure 25.29).** In some patients, the mumps virus causes orchitis,[14] meningitis, pancreatitis, or permanent deafness in one ear. In many other patients, infection is asymptomatic. Recovery from mumps is typically complete, and the patient has effective, lifelong immunity.

---

[14]From Greek *orchis,* meaning testis, and *itis,* meaning inflammation.

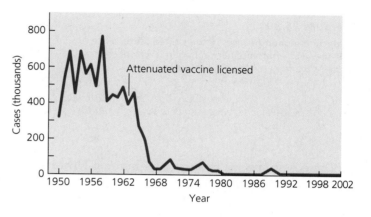

▲ *Figure 25.28*

The number of cases of measles in the United States since 1950. Cases have declined dramatically since vaccination began in 1963. *Why does the measles vaccine pose risks for immunocompromised contacts of vaccinated children?*

Figure 25.28 Measles vaccine is a live attenuated vaccine and can cause disease in the immunocompromised.

▲ *Figure 25.29*

Parotitis—inflammation of the parotid salivary glands—a common sign of mumps.

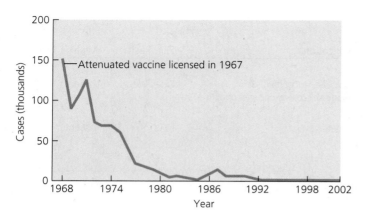

▲ *Figure 25.30*

The decline in mumps cases in the United States since 1968, the year after vaccination began. *With which other vaccines is mumps vaccine administered?*

Figure 25.30 The mumps vaccine is administered with the measles and rubella vaccines.

There is no specific treatment for mumps. However, because an effective attenuated vaccine (MMR) is available and humans are the only natural host for the virus, mumps has almost been eradicated in the industrialized world **(Figure 25.30)**. Epidemics of mumps in the late winter and early spring still occur in countries without vaccination programs, especially in densely populated areas.

Disease of Respiratory Syncytial Virus

Learning Objective

✓ Compare the effects of RSV infection in infants and adults.

Respiratory syncytial virus (RSV) in the genus *Pneumovirus* and the family *Paramyxoviridae* causes a disease limited to the respiratory tract. As its name indicates, the virus causes syncytia formation in the lungs. Additionally, plugs of mucus, fibrin, and dead cells fill the smaller air passages resulting in *dyspnea*—difficulty in breathing. RSV is the leading cause of fatal respiratory disease in infants and young children worldwide, especially those who are premature, immune impaired, or otherwise weakened. The disease in older children and adults is manifested as a cold or is asymptomatic.

Damage to the lungs is increased by the action of cytotoxic T cells and other specific immune responses to infection. Attempts at developing a vaccine with deactivated RSV have proven difficult because the vaccine enhances the severity of the cellular immune response and lung damage.

Epidemiology of RSV Infection

RSV is prevalent in the United States; epidemiological studies reveal that 65–98% of children in day-care centers are infected by age 3. Approximately 95,000 (1%) of these children require hospitalization each year. The virus is transmitted on fomites, hands, and less frequently via respiratory droplets. Because the virus is also extremely contagious, its introduction into a nursery can be devastating; therefore, good aseptic technique by healthcare workers, and good hygiene by day-care center employees, is essential. Hand asepsis and the use of gowns, goggles, masks, and gloves are important measures to reduce nosocomial infections.

Diagnosis, Treatment, and Prevention of RSV Infection

Prompt diagnosis is essential if infected infants are to get the care they need. The signs of respiratory distress provide some diagnostic clues, but verification of RSV infection is made by immunoassay. Specimens of respiratory fluid may be tested by immunofluorescence, ELISA, or complementary DNA probes.

Supportive treatment includes administration of oxygen, intravenous fluids, drugs to reduce fever, and antibiotics to reduce secondary bacterial infections. Immunoglobulins against RSV, derived from blood donations, have proven effective for treating the more severe cases. *Ribavirin* is used to treat extreme cases (for example, in premature and immunocompromised infants). The drug is administered by inhalation in mist form.

Control of RSV is limited to attempts to prevent infection of the most susceptible infants. As mentioned, an experimental deactivated vaccine caused patients to develop more complications upon infection than developed in the unvaccinated.

Rabies

Learning Objective

✓ Describe the effects, treatment, and prevention of rabies.

Although the cry "Mad dog! Mad dog!" may make you apprehensive, in the 19th century—before Pasteur developed a rabies vaccine and treatment—a bite from a rabid dog was a death sentence, and the cry of "Mad dog!" often spawned panic. In all of recorded history, only three people survived rabies before Pasteur's pivotal work.

Rabies[15] virus is a negative, single-stranded RNA (−ssRNA) virus in the genus *Lyssavirus*, family *Rhabdoviridae*. Like viruses in the family *Paramyxoviridae*, all of the rhabdoviruses have helical capsids, but those of rhabdoviruses are supercoiled into cylinders, which gives them a striated appearance, and are surrounded by bullet-shaped envelopes **(Figure 25.31)**.

Glycoprotein spikes on the surface of the envelope serve as attachment proteins. Attachment to skeletal muscle cells triggers endocytosis, and rabies viruses then replicate in the

---

[15]From Latin *rabere*, meaning rage or madness.

Attachment glycoprotein

TEM |—| 20 nm

▲ *Figure 25.31*

A rabies virus. Note the bullet-shaped envelope.

cytoplasm of these cells. Subsequently, virions move into neurons at neuromuscular junctions and then travel to the central nervous system via cytoplasmic flow within the neurons. The spinal cord and brain degenerate as a result of infection, though infected cells show little damage when examined microscopically. Viruses travel back to the periphery, including the salivary glands, through nerve cells. Viruses are secreted in the saliva of infected mammals.

Initial signs and symptoms of rabies include pain or itching at the site of infection, fever, headache, malaise, and anorexia. Once the viruses reach the central nervous system, neurological manifestations characteristic of rabies develop; **hydrophobia**[16] (triggered by the pain involved in attempts to swallow water), seizures, disorientation, hallucinations, and paralysis. Death results from respiratory paralysis and other neurological complications.

Epidemiology of Rabies

Rabies is a classical zoonotic disease of mammals, though not all mammals are reservoirs. Rodents, for instance, rarely get rabies. The distribution of rabies in humans follows the distribution in animals. The animals involved differ from locale to locale and change over time as a result of dynamics of animal populations and interactions among animals and humans. The primary reservoir of rabies in urban areas is the dog. In the wild, rabies can be found in foxes, badgers, raccoons, skunks, cats, bats, coyotes, and feral dogs (**Figure 25.32**). Bats are the source of most cases of rabies in humans, causing about 74% of cases between 1990 and 2002.

Transmission of rabies viruses in the saliva of infected animals usually occurs via a bite but can occur through the introduction of viruses into breaks in the skin or mucous membranes, or rarely through inhalation.

[16]From Greek *hydro*, meaning water, and *phobos*, meaning fear.

Diagnosis, Treatment, and Prevention of Rabies

The neurological symptoms of rabies are unique and generally sufficient for diagnosis. Serological tests for antibodies can confirm the diagnosis. Postmortem laboratory tests are often conducted to determine whether a suspected animal in fact carries rabies virus. These tests include antigen detection by immunofluorescence and the identification of aggregates of virions (called **Negri bodies**) in the brain (**Figure 25.33** on page 730). Unfortunately, by the time symptoms and antibody production occur, it is too late to intervene, and the disease will follow its natural course (**Highlight 25.1** on page 730).

The rabies vaccine, *human diploid cell vaccine (HDCV)*, is prepared from deactivated rabies viruses cultured in human diploid cells. It is administered intramuscularly on days 0, 3, 7, 14, and 28 after exposure. The vaccine can also be administered prophylactically to workers who regularly come into contact with animals (veterinarians, zookeepers, and animal control workers) and people traveling to areas of the world where rabies is prevalent.

Treatment of rabies begins with treatment of the site of infection. The wound should be thoroughly cleansed with water and soap or another substance that deactivates viruses. The World Health Organization recommends anointing the wound with antirabies serum. Initial treatment also involves injection of *human rabies immune globulin (HRIG)*. Subsequent treatment involves five vaccine injections. Rabies is one of the few infections that can be treated with active immunization, because the progress of viral replication and movement

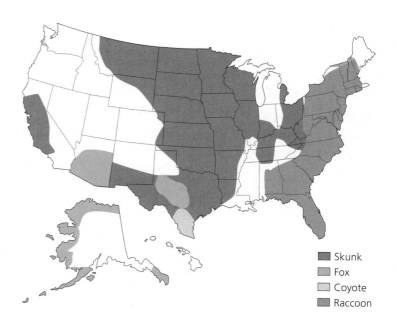

■ Skunk
■ Fox
□ Coyote
■ Raccoon

▲ *Figure 25.32*

Portions of the United States in which skunks, foxes, coyotes, or raccoons are the predominant wildlife reservoirs for rabies. Absence of a color does not indicate the absence of rabies in that region, only that no wildlife reservoir is predominant. Bats, for instance, harbor rabies throughout the country but are rarely the predominant reservoir.

Rabies: A Threat from the Wild

Vaccination of cats and dogs has sharply reduced the number of cases of human rabies in the United States; however, infections from wild animals still result in a few cases of rabies each year.

In one case, a man entered the hospital with neurological disorders that produced uncontrolled facial twitching, anxiety, and feelings of fear. He also had a sore throat, difficulty in swallowing, and complained of itching all over his body. Though the symptoms were typical of rabies, it was not diagnosed because there was no history of an animal bite.

The patient remained alive for several days, but became increasingly agitated and refused to drink because of the pain involved in swallowing. He vomited repeatedly and his temperature rose to 106° F. He died 1 week after being admitted to the hospital.

An autopsy revealed circulating antibodies against rabies virus as well as Negri bodies in the victim's brain. Epidemiologists determined from interviews with the man's neighbors that a bat had landed on his face about a month before the symptoms began, though the bat had not bitten him. Possibly the bat scratched the man during the encounter and transmitted the virus. This story reinforces the CDC recommendation that all contacts with wild animals (particularly bats, skunks, foxes, and feral dogs) be taken seriously and rabies prophylaxis administered.

Not all cases of rabies exposure are so tragic. In 1885, a rabid dog savagely mauled a young boy, Joseph Meister. His distraught mother brought the boy to Louis Pasteur, who was in the process of perfecting the rabies vaccine. The vaccine had never been tested on humans, but with assurance from doctors that the boy would surely die without treatment, Pasteur injected the 14 doses of attenuated virus that made up the vaccine at that time. Joseph survived and Pasteur's fame and fortune were further secured.

to the brain is slow enough to allow effective immunity to develop before disease develops.

Ultimately, control of rabies involves immunization of domestic dogs and cats and the removal of unwanted strays from urban areas. Little can be done to completely eliminate rabies in wild animals because rabies virus can infect so many species, though in the late 1990s an epidemic of rabies was stopped in southern Texas by the successful vaccination of the wild coyote population. This was accomplished by lacing meat with an oral vaccine and dropping it from airplanes into the coyotes' range.

Hemorrhagic Fevers

Learning Objective

✓ Describe the effects of filoviruses on the human body.

Filamentous viruses that cause **hemorrhagic fever** were originally placed in the *Rhabdoviridae* but have now been assigned to their own family, *Filoviridae*, primarily on the basis of the symptoms they cause. These enveloped viruses form long filaments, which sometimes curve back upon themselves (**Figure 25.34**).

There are two types of filoviruses: Marburg virus and Ebola virus. **Marburg virus** was first isolated from laboratory workers in Marburg, Germany, and it has also been found in Zimbabwe, Kenya, Uganda and in the Democratic Republic of the Congo (DRC). Outbreaks of **Ebola virus** were first recorded in the DRC in 1976 near the Ebola River. Ebola epidemics have also occurred in humans in Sudan, Ivory Coast, Gabon, and Uganda, and in monkeys in a research facility in the United States (**Figure 25.35**).

The natural reservoirs and method of transmission to humans of filoviruses are unknown. The virions attack many cells of the body, especially macrophages and liver cells. One predominant manifestation is uncontrolled bleed-

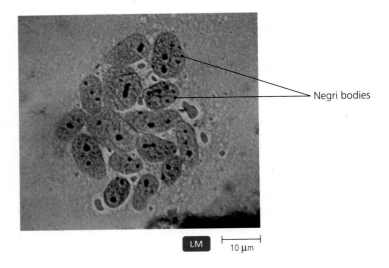

Negri bodies

▲ *Figure 25.33*

Negri bodies, a characteristic of rabies infection, in human cerebellum. *What are Negri bodies?*

Figure 25.33 Negri bodies are stained aggregates of rabies virions.

ing under the skin and from every body opening as the internal organs are reduced to a jelly-like consistency. Hemorrhaging appears to be due to the synthesis of a virally coded glycoprotein that is expressed on the surface of infected cells lining blood vessels. The glycoprotein prevents neighboring cells from adhering to one another, allowing blood to leak out of the vessels. Headache and severe fever accompany this "meltdown." Up to 90% of human victims die, and epidemics sometimes cease only when there are not enough live hosts left to sustain viral transmission.

The Marburg and Ebola viruses are spread from person to person via contaminated bodily fluids, primarily blood,

▲ *Figure 25.34*
Filamentous Ebola viruses.

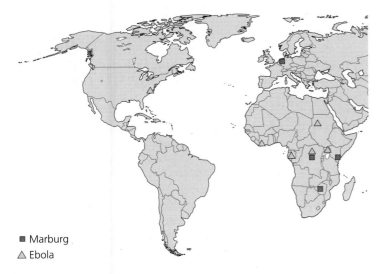

■ Marburg
△ Ebola

▲ *Figure 25.35*
Sites in which localized epidemics of Marburg and Ebola viruses are known to have occurred. Whereas most of the epidemics involved humans, the Ebola outbreak in Virginia only involved disease in monkeys in a research facility.

and through the use of contaminated syringes. Even a small amount of infected blood contacting a mucous membrane or conjunctiva can induce disease. Ebola virus can spread among monkeys through the air, but aerosol transmission from monkeys to humans or among humans has not been observed.

The symptoms of filovirus infection are diagnostic. Additionally, the typical curved virus filaments are found in the

victim's blood. Immunofluorescence, ELISA, or radioactive immunoassay can also detect antigens and antibodies. Extreme caution must be exercised in dealing with Ebola and Marburg patients and their bodily fluids. In fact, level four biocontainment—which involves the use of complete "space suits" to protect laboratory workers from contact with blood, vomit, or any other bodily fluids that might be contaminated with virus (**Figure 25.36** on page 732)—is mandatory. The only treatment for Ebola and Marburg fevers involves fluid replacement.

In 2000, scientists developed a vaccine that protects monkeys from Ebola. Immunization involves injecting pieces of viral nucleic acid (both by themselves and inserted into adenovirus capsids) into the monkeys. It is not known whether the vaccine will protect humans.

CRITICAL THINKING

Ebola hemorrhagic fever belongs to a group of diseases called "emerging diseases"—diseases that were previously unidentified or had never been identified in human populations. Emerging diseases are often first seen in Third World countries such as the DRC. What factors may explain the emergence of new diseases in these countries?

Enveloped, Segmented Negative ssRNA Viruses: *Orthomyxoviridae*, *Bunyaviridae*, and *Arenaviridae*

A second group of negative, enveloped, single-stranded RNA viruses differ from those in the previous section in having a **segmented** genome; that is, their capsids contain more than one molecule of RNA. These viruses include the flu viruses of the family *Orthomyxoviridae*, and hundreds of viruses in the families *Bunyaviridae* and *Arenaviridae* that normally infect animals but can be transmitted to humans.

Influenza

Learning Objective

✓ Describe the roles of hemagglutinin and neuraminidase in the replication cycle of influenzaviruses.

Imagine being the only elementary schoolchild of your sex in your mid-sized Swedish town because all of your peers died six winters ago. Or imagine returning to college after a break, only to learn that half of your fraternity brothers had died during the previous two months. These vignettes are not fictional; they happened to relatives of this author during the great flu pandemic in the winter of 1918–19 (**Figure 25.37** on page 732). During that winter, half the world's population was infected with a new, extremely virulent strain of influenza virus, and approximately 40 million died during that one flu season. In some U.S. cities, 10,000 people died each week for several months. Could it happen again? In this section we will learn about the characteristics

▲ *Figure 25.36*

Laboratory personnel working in level four biocontainment, designed to completely prevent contact between humans and pathogens. The "space suit" constitutes more than a physical barrier; should the material of the suit be breached, positive air pressure within the suit ensures that any pathogens present will be blown back from the opening.

▲ *Figure 25.37*

A scene from the flu pandemic of 1918–19, which afflicted so many people in the United States that gymnasiums were used as hospital wards.

of influenzaviruses that enable flu—a common infection, second only to common colds—to produce such devastating epidemics, and some ways to protect ourselves from the flu.

The causes of of **influenza**[17] (in-flū-en′ză) or **flu** are two species of orthomyxoviruses, designated types A and B. An influenza virion contains a genome consisting of eight different −ssRNA molecules surrounded by a pleomorphic (many-shaped) lipid envelope studded with prominent

glycoprotein spikes composed of either **hemagglutinin**[18] (hě′mă-glū′ti-nin) **(HA)** or **neuraminidase** (nūr-ă-min′i-dāz) **(NA) (Figure 25.38)**. Both HA and NA play roles in the attachment (and thus the virulence) of the viruses: NA spikes provide the virus access to cell surfaces by hydrolyzing mucus in the lungs, whereas HA spikes attach to pulmonary epithelia cells and trigger endocytosis. Because influenzaviruses rarely attack cells outside the lungs, so-called "stomach flu" is probably caused by enteroviruses, and not by orthomyxoviruses.

The genomes of flu viruses are extremely variable, especially with respect to the genes that code for HA and NA. Mutations in the genes coding for these glycoprotein spikes are responsible for the production of new strains of influenzavirus, via processes known as antigenic drift and antigenic shift. **Antigenic** (an-ti-gen′ik) **drift (Figure 25.39a** on page 734) refers to the accumulation of HA and NA mutations within a single strain of virus in a given geographic area; as a result of such relatively minor changes in the virus's antigenicity, localized increases in the number of flu infections occur about every 2 years. **Antigenic shift (Figure 25.39b)** is a major antigenic change resulting from the reassortment of genomes from different influenzavirus strains that occurs within host cells (either human hosts or animal hosts, including ducks and pigs). On average, antigenic shift occurs in influenza A every 10 years. Because influenza B occurs primarily in humans, it does not undergo antigenic shift. Asia is a major site of antigenic shift, and the source of most pandemic strains, because of the very high population densities of humans, ducks, and pigs there.

Once virions are taken into epithelial cells lining the lung, the viral envelopes fuse with the membranes of phagocytic vesicles. Viruses multiply using positive sense RNA molecules both for translation of viral proteins and as templates for transcription of −ssRNA genomes.

During the process by which virions bud from the cytoplasmic membrane of infected cells (**Figure 25.40** on page 735), NA keeps viral proteins from clumping. During budding, genomic segments are enveloped in a random manner, such that each virion ends up with about 11 RNA molecules. However, to be functional, a virion must have at least one copy of each of the eight genomic segments. Numerous defective particles are released for every functional virion formed.

The deaths of the epithelial cells infected with influenzaviruses eliminate the lungs' first line of defense against infection, its epithelial lining. As a result, flu patients are more susceptible to secondary bacterial infection. One common infecting bacterium is *Haemophilus influenzae*, which was mistakenly named because it was found in so many flu victims.

The signs and symptoms of flu—fever, malaise, headache, and myalgia (body aches)—are induced by

---

[17]*Influenza*, which is Italian for "influence," derives from the mistaken idea that the alignment of celestial objects caused or "influenced" the disease.
[18]The word *hemagglutinin* refers to these spikes' ability to attach to and clump (agglutinate) red blood cells.

(a)

TEM   100 nm

◀ *Figure 25.38*

Influenzaviruses. (a) A colorized electron micrograph showing the pleomorphic shape of the envelope and the presence of glycoprotein spikes. **(b)** Artist's rendition of a cross-sectioned virion, revealing the viral genome consisting of eight segments of ssRNA, each of which is enclosed in a helical capsid. *What are the types of glycoprotein spikes, and what is their relationship to viral strain?*

Figure 25.38 Glycoprotein spikes on influenza virus are neuraminidase (NA) and hemagglutinin (HA); variations in these glycoproteins determine the strain of the virus.

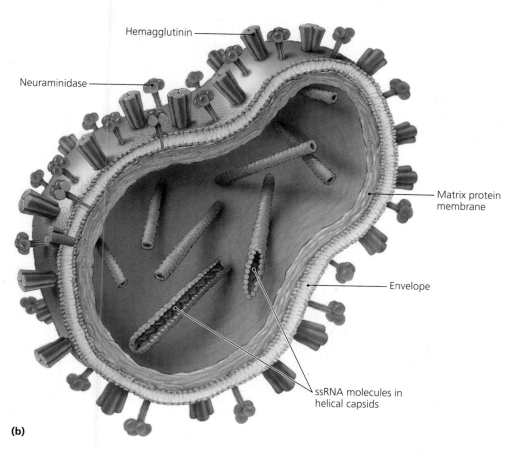

Hemagglutinin

Neuraminidase

Matrix protein membrane

Envelope

ssRNA molecules in helical capsids

(b)

cytokines released as part of the immune response. The symptoms of influenza are compared to those of other respiratory infections in Table 25.1 on page 705.

Epidemiology of Influenza

The changing antigens of orthomyxoviruses guarantee that there will be susceptible people, especially children, each flu season. Infection occurs primarily through inhalation of airborne viruses but can also occur via fomites.

Strains of influenza are named by type (A or B), location and date of original identification, and antigen (HA and NA). For example, A/Singapore/1/80 (H1N2) is influenza type A, isolated in Singapore in January 1980, that contains HA and NA antigens of type 1 and type 2, respectively. If the virus is isolated from an animal, the animal name is appended to the location. From these names, common names such as "Hong Kong flu" or "swine flu" arise. The number of different flu strains is almost infinite.

Every year millions of Americans are infected, but the mortality rate seldom exceeds 0.1%. Patients with weak immune systems, including the very young, the elderly, and those immunocompromised as a result of cancer therapy or disease, are particularly susceptible.

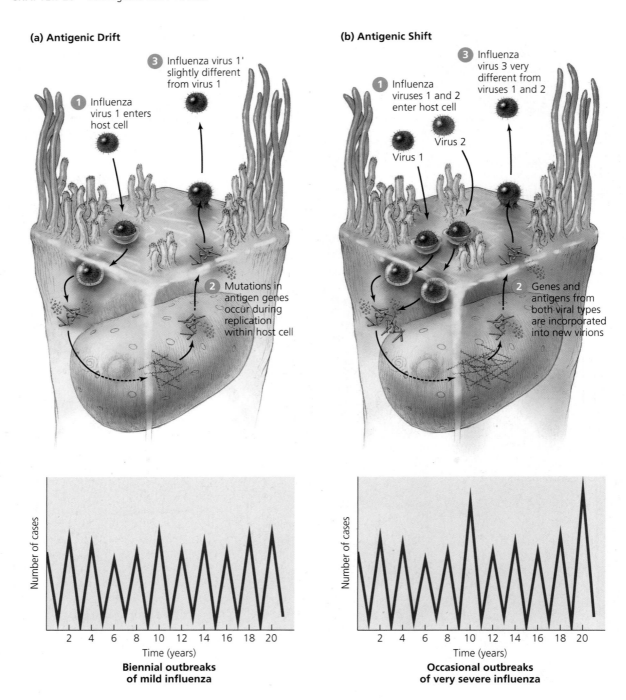

▲ *Figure 25.39*

The development of new strains of flu viruses. **(a)** Antigenic drift, which results from variation in the NA and HA spikes of a single strain of influenzavirus. **(b)** Antigenic shift, which occurs when RNA molecules from two or more influenzaviruses infecting a single cell are incorporated into a single virion. Because antigenic shift produces significantly greater antigenic variability than occurs in antigenic drift, antigenic shifts can result in major epidemics.

▲ *Figure 25.40*

Influenza viruses budding from an infected cell.

Diagnosis, Treatment, and Prevention of Influenza

The manifestation of flu signs and symptoms during a community-wide outbreak of the flu is often sufficient for diagnosis of influenza. Laboratory tests such as immunofluorescence or ELISA can distinguish strains of flu virus.

Hundreds of millions of dollars are spent each year in the United States on antihistamines and pain relievers to alleviate the symptoms of the flu. As discussed in Chapter 24, aspirin and aspirin-like products should not be used to treat the symptoms of flu in children and teenagers because of increased risk of *Reye's syndrome.*

Amantadine pills prevent the uncoating of influenza type A, and *oseltamivir* pills or inhaled *zanamivir* mist are neuraminidase inhibitors that block the release of virions from infected cells. These prescription drugs must be taken during the first 48 hours of infection because they cannot prevent the later immunopathogenic manifestations of the disease.

The greatest success in controlling flu epidemics has come from immunization with multivalent vaccines; that is, vaccines that contain several antigens at once. The CDC has personnel in Asia who detect changes in the HA and NA antigens of emerging flu viruses. The newly discovered antigens are then used to create each season's flu vaccine in advance of the next flu season in the United States. The vaccines are at least 70% effective, but only against the viral antigens they contain.

Flu vaccines are made of deactivated viruses. Contrary to many people's perception, you cannot get flu from the vaccine; however, you may get flu from strains of influenza virus not included in the vaccine. Flu vaccine usually provides protection for no more than 3 years. Natural active immunity as a result of infection lasts much longer but is probably not lifelong.

Diseases of Bunyaviruses

Learning Objective

✔ Contrast hantaviruses with the other bunyaviruses.

Most **bunyaviruses** (bun'ya-vī'rŭs-ĕz) are zoonotic pathogens that are usually transmitted to humans by biting arthropods (mosquitoes, ticks, and flies); thus, most are *arboviruses.* Bunyaviruses have a segmented genome of three −ssRNA molecules (**Figure 25.41** on page 736), which distinguishes them from the unsegmented, +ssRNA arboviruses we have already discussed in the families *Togaviridae* and *Flaviviridae.*

Epidemiology of Bunyavirus Infections

Four genera of *Bunyaviridae* contain human pathogens. Most are arboviruses, including the viruses of *Rift Valley fever, California encephalitis,* and *hemorrhagic fevers,* and thus enter the blood through the bite of an arthropod, which establishes an initial viremia. The blood disseminates viruses to target organs, such as the brain, kidney, liver, or blood vessels. Each bunyavirus has a specific target, and different viruses cause encephalitis, fever, or rash. Most patients have mild symptoms, but symptomatic cases can be severe and fatal.

Hantaviruses are an exception: They are not arboviruses, but are transmitted to humans via inhalation of virions in dried deer-mouse urine or feces. Two American strains[19] of hantaviruses infect the lungs only, causing a rapid, severe, and often fatal pneumonia called **hantavirus pulmonary syndrome (HPS).** This syndrome was first recognized during an epidemic in the Four Corners[20] area of the United States in May 1993.

Bunyaviruses cause human diseases around the world. Each virus is typically found only in certain geographic areas. However, with modern travel patterns, viruses can be introduced into new areas relatively easily.

Diagnosis, Treatment, and Prevention of Bunyavirus Infections

Diseases caused by bunyavirus infections are indistinguishable from illnesses caused by several other viruses, so laboratory tests must be employed in diagnosis. ELISAs are used to detect viral antigens associated with Rift Valley fever, the LaCrosse strain of California encephalitis, and hemorrhagic fevers. Hantaviruses in the United States were first identified in patient's tissue using the *reverse transcriptase polymerase chain reaction:* Viral RNA was transcribed to DNA using retroviral reverse transcriptase, a number of copies were generated through polymerase chain reaction, and the resulting strands were compared to known hantavirus sequences.

---

[19]Sin Nombre and Convict Creek strains.
[20]Four Corners is the geographic area where Arizona, Nevada, New Mexico, and Colorado meet.

Figure 25.41 ▶

A bunyavirus, an enveloped helical virion with a genome consisting of three −ssRNA molecules.

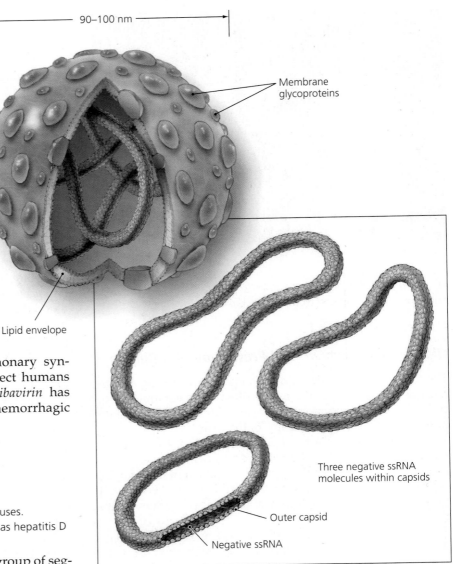

← 90–100 nm →

Membrane glycoproteins

Lipid envelope

Three negative ssRNA molecules within capsids

Outer capsid

Negative ssRNA

No specific treatment exists for most bunyavirus infections. Human disease is prevented by limiting contact with the vectors (the use of netting and insect repellents) and by reducing arthropod abundance (elimination of breeding sites and the use of insecticides). Rodent control is necessary to prevent hantavirus pulmonary syndrome. A vaccine has been developed to protect humans and animals against Rift Valley fever, and *ribavirin* has proved effective in treating Crimean-Congo hemorrhagic fever.

Diseases of Arenaviruses

Learning Objectives

✓ Identify the distinguishing characteristics of arenaviruses.

✓ Describe the unique feature of satellite viruses such as hepatitis D virus.

Arenaviruses[21] (ă-rē′nă-vī′rŭs-ĕz) are the final group of segmented, negative ssRNA viruses we will consider. The segmented genome is composed of two RNA molecules. Most arenaviruses contain ribosomes, which gives them a sandy appearance in electron micrographs **(Figure 25.42).** The ribosomes are functional *in vitro,*[22] but there is no evidence that the viruses utilize them in nature.

Zoonoses

Arenaviruses cause zoonotic diseases that include hemorrhagic fevers (**Lassa** in Africa, and the rarer *Junin* in Argentina and *Machupo* in Bolivia) and **lymphocytic choriomeningitis** (lim-fō-sit′ik kō-rē-ō-men-in-jī′tis) **(LCM).** Hemorrhagic fever viruses in the family *Arenaviridae* are endemic in rodent populations in Africa and South America. They cause severe bleeding under the skin and into internal organs. LCM virus has been found infecting pet hamsters in the United States. Contrary to its name, LCM is rarely associated with meningitis but more typically causes flu-like symptoms.

Humans become infected through the inhalation of aerosols or consumption of contaminated food, or from fomites. Lassa fever virus can be spread from human to human through contact with body fluids.

Diagnosis of arenavirus diseases is based on a combination of symptoms and immunoassay. Fluid samples from hemorrhagic patients are too dangerous for normal laboratories and instead must be processed in maximum biocontainment facilities. LCM samples can be handled with level three procedures (gown, mask, gloves, and labs with laminar flow hoods).

Supportive care is usually all that can be done for patients with arenavirus zoonoses, though ribavirin has shown some benefit in treating Lassa fever. Prevention of arenavirus disease involves rodent control and limiting contact with rodents, their droppings, and their dried urine.

---

[21]From Greek *arenosa,* meaning sandy.
[22]From Latin *in,* meaning within, and *vitreus,* meaning glassware; in other words "in the laboratory."

TEM | 50 nm

▲ *Figure 25.42*

The sandy appearance of arenaviruses, which results from the presence of numerous ribosomes within the capsid. Although the ribosomes are functional, the virions apparently do not use them.

Hepatitis D

Hepatitis D virus *(Deltavirus),* which is also known as *delta agent,* causes hepatitis D and appears to play a role along with hepatitis B virus in triggering liver cancer. Delta agent is unusual in that it does not possess genes for the glycoproteins it requires to attach to liver cells. To become infective, it "steals" the necessary glycoprotein molecules from hepatitis B viruses that are simultaneously infecting the same liver cell. For this reason, hepatitis D virus is called a **satellite virus** because its replication cycle "revolves around" a helper virus.

Like hepatitis B virus, hepatitis D virus is spread in bodily fluids via sexual activity and the use of contaminated needles. Prevention involves the same precautions used to prevent infection with hepatitis B virus—abstinence, monogamy, condom usage, the use of sterile needles, and avoidance of accidental needle sticks. Since hepatitis D virus requires hepatitis B virus to become virulent, hepatitis B vaccine limits the spread of hepatitis D viruses.

Naked, Segmented dsRNA Viruses: *Reoviridae*

Learning Objective

✓ Discuss the meaning of the name "reovirus."

Reoviruses (rē′-ō-vī′rŭs-ĕz) are respiratory and enteric viruses with naked icosahedral capsids. They are unique in being the only microbes, viral or cellular, with genomes composed of double-stranded RNA (dsRNA). Each virus contains 10–12 segments of dsRNA.

Because they were not initially associated with any diseases, they were called *orphan viruses*. The acronym *reo* is derived from *respiratory, enteric, orphan*. Reoviruses include *Rotavirus* and *Coltivirus*.

Rotaviruses

Learning Objective

✓ Describe the effects of rotavirus infections.

Rotaviruses[23] (rō′tă-vī′rŭs-ĕz) are almost spherical in appearance and have prominent glycoprotein spikes (**Figure 25.43** on page 738). During replication, rotaviruses acquire and then lose an envelope. Even without an envelope, they function like enveloped viruses because glycoprotein spikes on their capsids act as attachment molecules that trigger endocytosis, a process common to enveloped viruses.

Rotaviruses cause infantile gastroenteritis and account for approximately 50% of all cases of diarrhea in children requiring hospitalization as a result of fluid and electrolyte loss (55,000 cases per year in the United States). In developing countries, rotaviruses account for more than a million deaths from uncontrolled diarrhea. Transmission is via the fecal-oral route. Infected children may pass as many as 100 trillion virions per gram of stool.

Rotavirus infection is self-limiting, and treatment involves supportive care such as replacement of lost water and electrolytes. Prevention of infection involves good personal hygiene, proper sewage treatment, and vaccination. The newly developed rotavirus vaccine, known as RotaShield, provides protection about 50% of the time, but it has been linked with a rare bowel blockage condition in a very few vaccinated children (0.005%). Its use in the United States has been discontinued.

Coltiviruses

Learning Objective

✓ Discuss the most common disease caused by a coltivirus.

A **coltivirus** (kol′tē-vī′rŭs) is an arbovirus that causes a zoonotic disease, **Colorado tick fever** (see Table 25.4 on page 711) for which these viruses are named. Colorado tick fever is usually a mild disease involving fever and chills, though severe cases can manifest as headache, photophobia, myalgia, conjunctivitis, rash, and, in children, hemorrhaging. Colorado tick fever should not be confused with Rocky Mountain spotted fever, a tick-borne bacterial disease discussed in Chapter 21. The disease is self-limiting, but red blood cells in all stages of development retain the virus for some time. Therefore, patients should not donate blood for several months after recovery.

---

[23]From Latin *rota,* meaning wheel.

Colorado tick fever is diagnosed by detecting antigens in the blood using ELISA or other immunoassays. As with many viral diseases, there is no specific treatment. Prevention involves limiting contact with infected ticks.

RNA viruses are a heterogeneous group of virions capable of causing diseases of varying severity, including several life-threatening ones. **Table 25.9** summarizes the physical characteristics of most RNA viruses of humans, as well as the diseases they cause. As we have seen, RNA viruses pose significant problems for health care personnel with respect to diagnosis, treatment, and prevention, and some of these viruses also pose threats to health care workers who must handle them. Emerging RNA viruses are of particular concern as people move deeper into rainforests and come into contact with new viral agents, but they have also provided the impetus for expanded research into the fields of immunization and treatment of viral diseases.

TEM ⊢ 50 nm

▲ *Figure 25.43*
The wheel-like appearance of rotaviruses, from which they get their name.

Table 25.9   Taxonomy and Characteristics of Human RNA Viruses

Family	Strand Type	Enveloped or Naked	Capsid Symmetry	Size (nm)	Representative Genera (Diseases)
Picornaviridae	Single, positive	Naked	Icosahedral	22–30	*Enterovirus* (polio), *Rhinovirus* (common cold) *Hepatovirus* (hepatitis A)
Caliciviridae	Single, positive	Naked	Icosahedral	35–40	*Norovirus* (acute gastroenteritis),
Heperviridae	Single, positive	Naked	Icosahedral	27–34	*Hepevirus* (hepatitis E)
Astroviridae	Single, positive	Naked	Icosahedral	30	*Astrovirus* (gastroenteritis)
Togaviridae	Single, positive	Enveloped	Icosahedral	40–75	*Alphavirus* (encephalitis), *Rubivirus* (rubella)
Flaviviridae	Single, positive	Enveloped	Icosahedral	37–50	*Flavivirus* (yellow fever), Japanese encephalitis virus, *Hepacivirus* (hepatitis C)
Coronaviridae	Single, positive	Enveloped	Helical	80–160	*Coronavirus* (common cold, severe acute respiratory syndrome)
Retroviridae	Single, positive, segmented	Enveloped	Icosahedral	80–100	*Deltaretrovirus* (leukemia) *Lentivirus* (AIDS)
Paramyxoviridae	Single, negative	Enveloped	Helical	125–250	*Paramyxovirus* (colds, respiratory infections), *Pneumovirus* (pneumonia, common cold), *Morbillivirus* (measles), *Rubulavirus* (mumps)
Rhabdoviridae	Single, negative	Enveloped	Helical	75 × 130–240	*Lyssavirus* (rabies)
Filoviridae	Single, negative	Enveloped	Helical	790–970	*Filovirus* (Ebola, Marburg viruses)
Orthomyxoviridae	Single, negative, segmented	Enveloped	Helical	80–120	*Influenzavirus* (flu)
Bunyaviridae	Single, negative, segmented	Enveloped	Helical	90–100	*Bunyavirus* (encephalitis), *Hantavirus* (pneumonia)
Arenaviridae	Single, negative, segmented	Enveloped	Helical	50–300	*Lassavirus* (hemorrhagic fever) *Deltavirus* (hepatitis D)
Reoviridae	Double, segmented	Naked	Icosahedral	78–80	*Rotavirus* (diarrhea), *Coltivirus* (Colorado tick fever)

CHAPTER SUMMARY

Naked, Positive ssRNA Viruses (pp. 704–709)

1. RNA viruses and viroids are the only infective agents that use RNA molecules as their genetic material.

2. **Picornaviruses** are the smallest of animal viruses and include the **rhinoviruses** and **enteroviruses.** Rhinoviruses are extremely infective in the upper respiratory tract and cause most "common" colds. Enteroviruses, which are mostly cytolytic picornaviruses that infect via the fecal-oral route, include poliovirus, coxsackievirus, echovirus, and hepatitis A virus.

3. **Poliomyelitis** or **polio** is a disease of varied degrees of severity from asymptomatic infections to **bulbar poliomyelitis** (requiring an iron lung to assist in breathing), and postpolio syndrome occurring 30–40 years after infection. Eradication of polio, due to effective vaccination, is imminent.

4. **Hepatitis A virus** is a noncytolytic enterovirus that infects liver cells but is rarely fatal.

5. **Coxsackie A virus** causes lesions in the mouth and throat, respiratory infections, hand-foot-and-mouth disease, viral meningitis, and acute hemorrhagic conjunctivitis. **Coxsackie B virus** causes myocarditis and pericardial infections, which may be fatal in newborns. In addition, coxsackie B viruses cause pleurodynia, viral meningitis, and may be a cause of diabetes mellitus.

6. **Echoviruses** (enteric cytopathic human orphan viruses) cause viral meningitis and colds.

7. Jonas Salk developed **inactivated polio vaccine (IPV)** in 1955. It was replaced by live **oral polio vaccine (OPV),** which was developed by Albert Sabin in 1961. Now, health professionals have switched back to IPV because OPV, which is a live, attenuated vaccine, can revert to a disease-causing form.

8. **Caliciviruses** and **astroviruses** are small viruses that enter through the digestive system, multiply in the cells of the intestinal tract, and cause gastroenteritis and diarrhea. **Noroviruses** are representative calciviruses.

9. **Hepatitis E virus** causes **hepatitis E,** which is fatal to about 20% of infected, pregnant women.

Enveloped, Positive ssRNA Viruses (pp. 709–716)

1. Flaviviruses and most togaviruses are **arboviruses** (viruses transmitted by arthropods) and are enveloped, icosahedral, positive, ssRNA viruses that cause flu-like symptoms, encephalitis, dengue, and yellow fever.

2. Different togaviruses, most transmitted via mosquito vectors, cause zoonoses—diseases of animals that are spread to humans. These are **Eastern equine encephalitis (EEE), Western equine encephalitis (WEE),** and **Venezuelan equine encephalitis (VEE),** all of which may be fatal in humans.

3. **Dengue** and **dengue hemorrhagic fever** are endemic diseases in South America caused by flaviviruses and carried by the *Aedes* mosquito.

4. **Yellow fever,** which is contracted through a mosquito bite carrying a flavivirus, results in internal hemorrhaging and damage to liver, kidneys, and heart.

5. Rubivirus, a togavirus that is not transmitted by an anthropod vector, causes **rubella,** commonly known as "German measles" or "three-day measles." It can cause serious birth defects if the virus crosses the placenta into a fetus.

6. **Hepatitis C Virus (HCV),** caused by a flavivirus and transmitted via the fecal-oral route, is a disease that often results in serious liver damage.

7. **Coronaviruses** have a corona-like envelope and cause colds as well as gastroenteritis in children. A newly discovered coronavirus causes **severe acute respiratory syndrome (SARS).**

Enveloped, Positive ssRNA Viruses with Reverse Transcriptase (pp. 716–724)

1. **Retroviruses** have **reverse transcriptase** and **integrase** enzymes. The first allows retroviruses to make dsDNA from RNA templates; the latter allows them to integrate into a host cell's chromosomes.

2. **Human T-lymphotropic viruses** are oncogenic retroviruses associated with cancer of lymphocytes. HTLV-1 causes adult acute T-cell lymphocytic leukemia; HTLV-2 causes hairy-cell leukemia. Sexual intercourse, blood transfusion, and contaminated needles transmit them.

3. **Acquired immunodeficiency syndrome (AIDS)** is a condition marked by the presence of antibodies against **HIV** in conjunction with certain opportunistic infections. A **syndrome** is a complex of signs and symptoms with a common cause. **Human immunodeficiency viruses (HIV-1** or **HIV-2)** destroy the immune system. HIV is characterized by glycoproteins such as **gp120** and **gp41,** which enable attachment, antigenic variability, and **syncytia** (fused cells) formation. The destruction of the immune system results in vulnerability to any infection.

4. HIV affects T-helper cells, macrophages, and dendritic cells. Using gp120, HIV fuses to a cell, uncoats, enters the cell, and transcribes dsDNA to become a **provirus,** which inserts into a cellular chromosome. After HIV is replicated and released, an internal viral enzyme, **protease,** makes HIV virulent. HIV is spread primarily through contact with an infected individual's secretions or blood.

Enveloped, Unsegmented Negative ssRNA Viruses (pp. 724–731)

1. Paramyxoviruses can cause infected cells to fuse together into giant, multinucleate syncytia. One virus causes **measles** ("rubeola" or "red measles"), recognized by **Koplik's spots** in the mouth followed by lesions on the head and trunk. Complications of measles include **subacute sclerosing panencephalitis (SSPE),** a progressive fatal disease of the central nervous system. Vaccination is effective.

2. **Parainfluenza** viruses usually cause mild coldlike diseases, including **croup.** There is no effective vaccine.

3. **Mumps** virus is a paramyxovirus that causes painfully enlarged parotid glands. Mumps vaccine is very effective, and eradication is possible.

4. **Respiratory syncytial virus (RSV)** is a paramyxovirus that affects the lungs and can be fatal in infants and young children, while in older children and adults the symptoms are similar to those of a cold.

5. **Rabies** is caused by a rhabdovirus—a bullet-shaped virus with glycoprotein spikes—that attaches to muscle cells, replicates, moves into nerve cells and eventually causes brain degeneration. Infection usually results from bites by infected mammals. Symptoms include pain in swallowing (which results in **hydrophobia**), seizures, paralysis, and death. Rabies may be identified by aggregates of virions called **Negri bodies** in the brain of the affected mammal. Human and animal vaccines are effective.

6. **Marburg virus** and **Ebola virus** are filamentous viruses that cause hemorrhagic fever. There is no known treatment or prevention for hemorrhagic fever, though a vaccine is being tested on monkeys.

Enveloped, Segmented Negative ssRNA Viruses (pp. 731–737)

1. **Influenza (flu)** is caused by strain A or B orthomyxoviruses, which enter lung cells by means of **hemagglutinin (HA)** and **neuraminidase (NA)** glycoprotein spikes. Gene mutations for HA and NA account for continual formation of new variants of flu viruses. **Antigenic drift** occurs every 2 or 3 years when a single strain mutates within a local population. **Antigenic shift** of influenza A virus occurs about every 10 years and results in reassortment of genomes from different strains of animal and human viruses infecting a common cell. Vaccination for the flu in any given flu season is prepared against newly identified variants.

2. **Bunyaviruses,** with three −ssRNA molecules as a genome, are zoonotic pathogens. They are usually arboviruses. Each bunyavirus has a specific target and may cause encephalitis, fever, rash, or death.

3. **Hantaviruses** are bunyaviruses that are transmitted to humans via inhalation of virions in dried deer-mouse urine. They cause **hantavirus pulmonary syndrome.**

4. **Arenaviruses** are segmented −ssRNA viruses with apparently nonfunctional ribosomes. They cause **Lassa** hemorrhagic fevers and **lymphocytic choriomeningitis (LCM)**, and hepatitis D.

5. **Hepatitis D virus,** also known as delta agent, is a **satellite virus;** that is, it requires proteins coded by another virus (in this case hepatitis B virus) to complete its replication cycle.

Naked, Segmented dsRNA Viruses (pp. 737–738)

1. The **reoviruses** are unique in having double-stranded RNA as a genome. They include **rotaviruses,** which have glycoprotein spikes for attachment and cause a potentially fatal infantile gastroenteritis, and **coltiviruses,** which cause Colorado tick fever.

QUESTIONS FOR REVIEW
(Answers to multiple choice and matching questions are on the web, along with additional review questions. Visit www.microbiologyplace.com.)

Multiple Choice

1. A segmented genome is one that
   a. has more than one strand of nucleic acid.
   b. has double-stranded RNA.
   c. has both RNA and DNA strands.
   d. has +ssRNA and −ssRNA molecules.

2. What do viruses in the families *Picornaviridae, Caliciviridae, Astroviridae, Coronaviridae, Togaviridae, Flaviviridae,* and *Retroviridae* have in common?
   a. They are arboviruses.
   b. They are nonpathogenic.
   c. They have positive single-stranded RNA genomes.
   d. They have negative single-stranded RNA genomes.

3. The smallest animal viruses are in the family
   a. *Caliciviridae.*   c. *Togaviridae.*
   b. *Astroviridae.*   d. *Picornaviridae.*

4. Which of the following viruses cause most colds?
   a. rhinoviruses   c. pneumoviruses
   b. parainfluenza viruses d. bunyaviruses

5. Arboviruses
   a. are zoonotic pathogens.
   b. are deactivated viruses used in vaccines.
   c. are viruses that are transmitted to humans via the bite of an arthropod.
   d. are found in arbors.

6. Negri bodies are associated with which of the following?
   a. Marburg virus
   b. hantavirus
   c. coltivirus
   d. rabies virus

7. If mosquitoes were eradicated from an area, which of the following diseases would be most affected?
   a. mumps
   b. hantavirus pulmonary syndrome
   c. hepatitis E
   d. break-bone fever

8. Koplik's spots are oral lesions associated with
   a. mumps.
   b. measles.
   c. flu.
   d. colds.

9. Which of the following is an accurate statement concerning zoonoses?
   a. They are diseases that are spread from animals to humans.
   b. They are diseases specifically transmitted by mosquitoes and ticks.
   c. They are mucus-borne viruses, which are transmitted in the droplets of moisture in a sneeze or cough.
   d. They are diseases that can be transmitted from humans to an animal population.

10. A horror movie portrays victims of biological warfare with uncontrolled bleeding from the eyes, mouth, nose, ears, and anus. What actual virus causes these symptoms?
    a. Ebola virus
    b. bunyavirus
    c. hantavirus
    d. human immunodeficiency virus

True/False

___ 1. A single virion is sufficient to cause a cold.

___ 2. All infections of polio are crippling.

___ 3. Postpolio syndrome is due to latent polioviruses that become active 30–40 years after the initial infection.

___ 4. Because the oral polio vaccine contains live attenuated viruses, mutations of these viruses can cause polio.

___ 5. A typical host for a togavirus is a horse.

Matching

Match the diseases on the left with the infecting virus on the right.

___ Myocarditis   A. Rhabdovirus

___ Colorado tick fever   B. Paramyxovirus

___ Rabies   C. Coltivirus

___ Influenza   D. Rubivirus

___ Dengue fever   E. Togavirus

___ German measles   F. Flavivirus

___ Acute gastroenteritis   G. Orthomyxovirus

___ Ebola virus   H. Calicivirus

___ RSV   I. Filovirus

___ Western Equine Encephalitis   J. Coxsackie B virus

Short Answer

1. Why are humans considered "dead end" hosts for many arboviruses?

2. Young Luis has skin lesions. His mother knows from microbiology class that five childhood diseases can produce spots. Name those five diseases, and the viruses that cause each. List some questions to ask to determine which of these viruses Luis has.

3. The patient in room 519 exhibits yellowing skin and eyes, and it is suspected among the nursing staff that the diagnosis will be some kind of viral hepatitis. Make a chart of five kinds of hepatitis mentioned in this chapter, the infecting pathogen, how the patient might have become infected, and the relative degree of seriousness.

4. Why is AIDS more accurately termed a "syndrome" instead of a "disease"?

5. Consider the viruses you studied in this chapter. Which three would you rank as the deadliest?

6. Support or refute the following statement: "Rubeola is common and of little concern as a childhood disease."

7. Translate the following identification label on a vial of influenza virus: B/Kuwait/6/97 (H1N3).

8. Polio and smallpox have been eliminated as natural threats to human health in the United States. (Some risk from bioterrorism remains.) You have considered the features of these diseases that allowed them to be eliminated. From your studies of other viruses, what other viral diseases are candidates for elimination? Why hasn't AIDS been eliminated?

9. Several laboratory tests are used to identify viruses. From your study of this chapter alone, which tests would you surmise are the most common?

CRITICAL THINKING

1. Chapter 4 presented the use of dichotomous keys to identify microorganisms. Design a key for the RNA viruses discussed in this chapter.

2. A 20-year-old man is brought to a South Carolina hospital's emergency room suffering from seizures, disorientation, hallucinations, and an inordinate fear of water. His family members report that he had suffered with fever and headache for several days. The patient had been bitten by a dog approximately 12 days before admission. What disease does the patient have? Will a vaccine be an effective treatment? If the dog is found to be rabies free, what wild animals are the likely source of a rabies infection in South Carolina? What treatment should the patient's family, coworkers, friends, and caregivers receive?

of preserving or flavoring foods with brine (saturated salt water) or acid. The acid may result from microbial fermentation—as is done with dill pickles. Various spices can be added to the pickling solution to enhance taste. Because pickled foods are acidic, few pathogenic microorganisms survive, making pickling an excellent preservation method.

Fermented foods aren't made for human consumption alone. *Silage,* a product used as animal feed on many farms, is made from the natural fermentation of corn, grass, grain stalks, potatoes, and other types of green foliage. The vegetation is cut up and stored in silos that are closed to air. Under such moist, anaerobic conditions, the vegetation in the silos ferment, producing many organic compounds that make the silage both aromatic and tasty to livestock. Some farmers produce silage by baling hay and grasses and then wrapping the bale in plastic to allow fermentation to occur.

Fermented Meat Products

Unless dried, cured, smoked, or fermented in some way, meat has a tendency to spoil rapidly. The combination of fermentation and drying or smoking has been used for centuries to preserve meats, particularly pork and beef. Dry sausages such as salami and pepperoni are made by grinding certain meats, mixing in various spices and starter cultures, allowing the mixture to ferment in the cold, and then stuffing it into casings. Headcheese isn't cheese at all, but rather pork pickled in brine and spices and stuffed into casings. Fermented fish—common in Asian countries—are prepared by grinding fish in brine. Naturally occurring microbes ferment the mixture. The solids are removed and pressed to form a fish paste, while the liquid portion is drained off and mixed with flavoring agents to form fish sauce.

Fermented Dairy Products

Milk fermentation relies, for the most part, on the activities of lactic acid bacteria. Milk in an udder is sterile, but because milking introduces many microorganisms, not all of which are desirable, "raw" milk is pasteurized before it is packaged or used in the production of fermented dairy products.

Buttermilk is made from skim milk (milk from which fat has been removed) by adding a starter culture of *Lactococcus lactis* (lak'tis) subspecies *cremoris* (kre-mōr'is) and *Leuconostoc citrovorum* (sit-rō-vō'rum). The metabolic products of fermentation give buttermilk and other fermented milk products their characteristic textures and aromas.

Yogurt production utilizes starter cultures of *Streptococcus thermophilus* (ther-mo'fil-us) and *Lactobacillus bulgaricus* (bul-gā'ri-kŭs). Pasteurized milk, milk solids, sweeteners, and other ingredients are mixed to a uniform consistency, and then the starter culture is added. The mix is fermented at 43°C and then cooled to stop microbial activity. Flavors are added prior to packaging.

Cheeses are classified as unripened or ripened; furthermore, ripened cheeses can be hard or soft, and mild or sharp. Regardless of the end product, however, the cheese-making

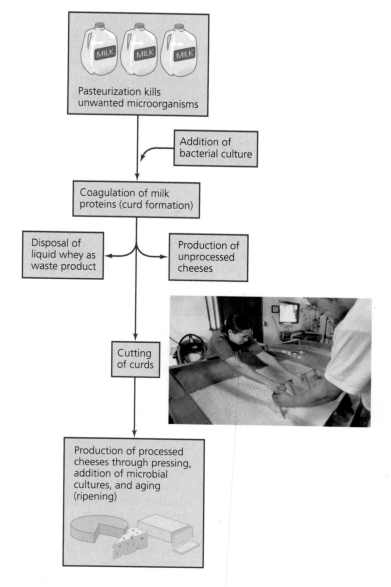

▲ *Figure 26.1*

The cheese-making process. The production of all cheeses begins with the pasteurization of milk; the wide variety of cheeses available on the market results from the nature of subsequent processing steps. *Why do cheeses that are aged longer have "sharper" tastes?*

Figure 26.1 Fermentation during aging results in continued production of acid; the longer the cheese is aged, the more acid produced and the "sharper" the taste.

process always begins with pasteurized milk **(Figure 26.1).** Proteins in milk are coagulated by cultures of *Lactococcus lactis* to form *curds* (solids) and *whey* (liquids). For unripened cheeses such as cottage cheese, the curd is removed, cut into small pieces, and packaged. Alternatively, the curdling process can be accelerated by the addition of the enzyme *rennin.*

Ripened cheeses are made from milk curds that have been pressed to remove water. Hard cheeses, such as Parmesan and cheddar, are pressed such that little water remains, while some soft cheeses, such as Camembert, retain enough moisture to make them spreadable. Once pressed, new microbial cultures are added and the cheese is ripened (aged) for months to years while continued microbial activity imparts to the cheeses their characteristic smells and tastes. The longer a cheese is ripened, the more lactic acid is produced, and the sharper the taste. For "mold-ripened" cheeses such as Roquefort or "blue cheese," fungi are added instead of bacteria.

CRITICAL THINKING

Even though some of the naturally present microorganisms in raw milk are the same as those added with the starter culture, raw milk is not generally used commercially for the production of cheeses. Why not?

Products of Alcoholic Fermentation

Alcoholic fermentation is the process by which various species of microorganisms convert simple sugars such as glucose into alcohol (ethanol) and carbon dioxide (CO_2). These two products of fermentation take on different importance for different foods and beverages. As is the case for the production of fermented dairy products, specific starter cultures are used in the large-scale commercial applications of alcohol fermentation. In the next subsections we consider the role of fermentation in the production of wine and spirits, beer, vinegar, and bread.

Wine and Spirits   In the production of wine, *Saccharomyces cerevisiae* (sak-ah-ro-mī'sēs se-ri-vis'ē-ē) is commonly used to ferment sugar in fruit juice to alcohol. Weather conditions and soil composition affect the sugar content of fruit, which in turn greatly affects the quality of the wine. Generally, wine production proceeds as follows **(Figure 26.2)**:

1. Preparation of must. Wine makers crush the fruit and remove the stems to form *must* (fruit solids and juices). They make red wines from the entire must of dark grapes, whereas white wines are made using only the juice of either dark or light grapes.

2. Fermentation. A thin film of bacteria and yeast naturally covers grapes and other fruit, and in some wineries these microbes produce the wine. However, in most commercial operations, sulfur dioxide (SO_2) is added to retard bacterial growth, and then starter cultures of yeast are added to ensure that wine of a consistent quality is generated from year to year.

3. Clarification. Solids are removed by either filtration or settling.

4. Aging. The wine is aged in wooden barrels. Continued yeast metabolism and the leaching of compounds from the wood add taste and aroma to the wine.

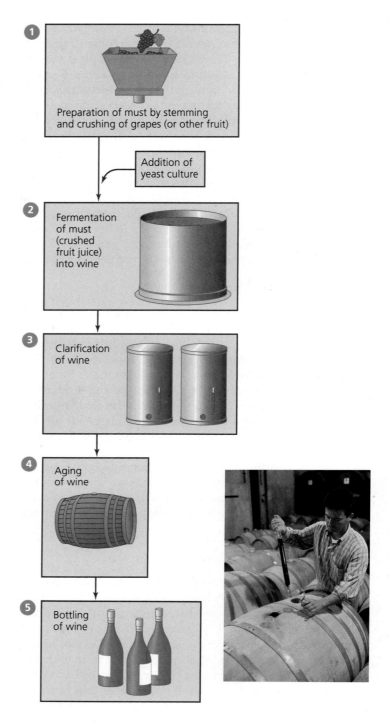

▲ *Figure 26.2*

The wine-making process. Red wine is produced by fermenting must (fruit solids and juice) from dark grape varieties, whereas white wine results from the fermentation of juice from either dark or light grape varieties.

5. Bottling. The wine is then bottled and distributed.

Dry wines are those in which all the sugar has been allowed to ferment, whereas sweet or dessert wines retain some sugar. Most table wines have a final alcohol content of 10–12%, but distilled spirits (discussed next) are added to increase the alcohol content of so-called fortified wines to about 20%. Champagne is produced by the addition of sugar and a second fermentation that produces CO_2 inside the bottle.

Distilled spirits are made in a process similar to wine-making in that *Saccharomyces* ferments fruits, grains, or vegetables. The difference is that during the aging process, the alcohol is concentrated by distillation. In this process, the liquid is heated under a vacuum to drive off water. The net result is an increase in the alcohol content (100 proof spirits have the equivalent content of 50% alcohol). Brandies are made from fruit juices, whiskeys are made from cereal grains, and vodka comes from potatoes.

Beer and Sake　Beer is made from barley in the following steps **(Figure 26.3):**

①　Malting. Barley is moistened and germinated, a process that produces catabolic enzymes, which convert starch into sugars, primarily *maltose*. Drying halts germination, and the barley is crushed to produce *malt*.

②　Mashing. Malt and *adjuncts* (such as starch, sugar, rice, sorghum, or corn) are combined with warm water, allowing the enzymes released during malting to generate more sugars. Adjuncts serve as additional sources of carbohydrate for the enzymes to catabolize.

③　Preparation for fermentation. The spent grain is removed from the sugary liquid *(wort)*, and *hops*—the dried flower-bearing parts of the vine–like hops plant— are added. The mixture is boiled to halt enzymatic activity, extract flavors from the hops, kill most microorganisms, and concentrate the wort.

④　Fermentation. A starter culture of *Saccharomyces* is added to the wort. Most beers are made with the so-called bottom-fermenting yeast *S. carlsbergensis* (karls-berg-en′sis). Ales are produced with the top-fermenting yeast *S. cerevisiae* (so-called because the yeast floats in the vat).

⑤　Aging. The beer is aged, pasteurized or filtered, and bottled or canned. Beer has a considerably lower alcohol content (about 4%) than either wine or distilled spirits.

Sake (so-called rice wine) is more correctly "rice beer" because the starch in cooked rice is first converted to sugar (by the fungus *Aspergillus oryzae*) and then the sugar is fermented by *Saccharomyces*. Sake has an alcohol content of about 14%.

Vinegar　Vinegar[2] is produced when the ethanol resulting from the fermentation of a fruit, grain, or vegetable is further oxidized to acetic acid. This secondary metabolic conversion is performed by *acetic acid bacteria* such as *Acetobacter* or

---

[2]From French *vinaigre,* meaning sour wine.

① Barley is moistened and germinated, producing enzymes that convert starch into sugars. Barley is then dried to halt germination, and crushed to produce malt.

② Mashing malt and adjuncts with warm water allows enzymatic activity to generate more sugars; solids are removed to produce wort.　Mashing kettle

Addition of hops for flavoring

③ Cooking of wort halts enzymatic activity, extracts flavor from hops, and kills microorganisms present.

Remove hops　　Addition of yeast culture

④ Fermentation of wort into beer.

⑤ Aging, filtering or pasturization, and bottling of beer

▲ *Figure 26.3*

The beer-brewing process. The type of beer produced (ale, lager) is largely determined by the type of yeast used (top-fermenting or bottom-fermenting).

Gluconobacter, and generates roughly 4% acetic acid. Different flavors of vinegar are derived from different starting materials. Cider vinegar, for example, is made from apples.

Table 26.1 Some Fermented Foods and the Microorganisms Used in Their Production

Food	Starting Material	Representative Culture Microorganisms
Fermented Vegetables		
Sauerkraut/kimchi	Cabbage	Various lactic acid bacteria
Pickles	Cucumbers, peppers, beets	Various lactic acid bacteria
Soy sauce	Soybeans and wheat	*Aspergillus oryzae* or *Aspergillus soyae* and *Lactobacillus* spp.
Miso	Rice and soybeans or rice and other grains	*Aspergillus oryzae* and *Lactobacillus* spp. and *Torulopsis etchellsii*
Fermented Meat Products		
Dry salami	Pork, beef, chicken	Various lactic acid bacteria
Fish sauces/pastes	Ground fish	Various naturally occurring bacteria
Fermented Dairy Products		
Milks		
Buttermilk	Pasteurized skim milk	*Lactococcus lactis cremoris* and *Leuconostoc citrovorum*.
Yogurt	Pasteurized skim milk	*Streptococcus thermophilus* and *Lactobacillus bulgaricus*
Cheeses		
Cottage cheese	Pasteurized milk	*Streptococcus* or *Lactococcus cremoris* and *Lactococcus lactis*
Hard cheeses (cheddar)	Milk curd	Starter culture as in cottage cheese without further additions
Soft cheeses (Camembert)*	Milk curd	Starter culture as in cottage cheese and *Penicillium camemberti*
Mold-ripened (Roquefort)	Milk curd	Starter culture as in cottage cheese and *Penicillium roqueforti*
Animal Feed		
Silage	Corn, grains, vegetation	Various naturally occurring bacteria
Alcoholic Fermentations		
Wine	Grapes	*Saccharomyces cerevisiae*
Distilled spirits	Fruits, vegetables, grains	*Saccharomyces cerevisiae*
Beer	Barley	*Saccharomyces cerevisiae* or *S. carlsbergensis*
Sake	Cooked rice	*Aspergillus oryzae* and *Saccharomyces* spp.
Vinegar	Fruits, vegetables, grains	*Saccharomyces cerevisiae* and *Acetobacter* or *Gluconobacter*
Bread	Flour, salt, etc.	*Saccharomyces cerevisiae*

*In addition to being a soft cheese, Camembert is also a mold-ripened cheese.

Table 26.1 summarizes types of fermented foods along with the organisms involved in their production.

The Causes and Prevention of Food Spoilage

Learning Objectives

✓ Describe how food characteristics and the presence of microorganisms in food can lead to food spoilage.

✓ List several methods for preventing food spoilage.

Whereas many chemical reactions involving food are desirable because they enhance taste, aroma, or preservation, other reactions are not. Spoilage of food leads not only to economic loss, but potentially to illness and even death. In

this section we examine the causes of food spoilage before considering ways to prevent it.

Causes of Food Spoilage

Foods may spoil—that is, be adversely affected in nutrition or appearance—because of inherent properties of the food itself. Such *intrinsic factors* include qualities such as nutrient content, water activity, pH, and physical structure of the food, as well as the competitive activities of microbial populations. Alternatively, food spoilage may be the result of *extrinsic factors* that have nothing to do with the food itself, but instead with the way it is processed or handled.

Intrinsic Factors in Food Spoilage The nutritional composition of food determines both the types of microbes

Table 26.2 Factors Affecting Food Spoilage

	Foods at Greatest Risk of Spoilage	Food at Least Risk of Spoilage
Intrinsic Factors		
Nutritional composition	Chemically rich or fortified foods (steak, bread, yogurt)	Chemically limited foods (flour, cereals, grains)
Water activity	Moist foods (meat, milk)	Dry foods or those with low water activity (pasta, jam)
pH	Foods with neutral pH (bread)	Foods with low pH (orange juice, pickles)
Physical structure	Foods without rinds, skins, or shells; ground meat	Foods with rinds, skins, or shells; intact foods
Microbial competition	Foods that lack natural microbe populations (ground beef)	Foods with resident microbial populations (pickles)
Extrinsic Factors		
Degree of processing	Unprocessed foods (raw milk, fruit)	Processed foods (pasteurized foods)
Amount of preservatives	Foods without preservatives (meats, natural foods)	Foods with either naturally occurring or added preservatives (garlic, spices, sulfur dioxide)
Storage temperature	Foods left in warm conditions	Foods kept cold
Storage packaging	Foods stored exposed	Foods wrapped or sealed

present and whether they will grow; for example, high-starch foods tend to support a different group of microbes than do high-protein foods. Some foods contain natural antimicrobial agents, such as the benzoic acid in cranberries, and thus are not prone to spoilage. In contrast, *fortified foods*—those enriched in vitamins and minerals to improve the health of humans—may also, inadvertently, facilitate growth of microorganisms by providing more nutrients.

Water activity refers to water that is not bound by solutes or to surfaces and is thus available to microbes. The water activity of pure water is set at 1.0, and most microbes require environments with a water activity of at least 0.90. Moist foods with water activities near 0.90, such as fresh meat, support microbial growth, whereas dry foods with water activities near zero (for example, uncooked pasta) do not support microbial growth. The water activity of foods can be reduced either by drying or by adding salts or sugars. Thus even though jam looks moist, its sugar content is very high, and its water activity is too low to support much microbial growth.

The pH of food can either be an intrinsic chemical property of the food, as in the case of citrus fruit, or the result of fermentation, as with pickles. In either case, a pH below 5.0 typically supports little microbial growth except by a few molds or lactic acid bacteria. A pH closer to neutral (7.0) supports a wider range of microbial growth. Since the pH of most foods is close to neutral, pH is generally not a deterrent to spoilage microorganisms.

Physical structure is a visible intrinsic factor. Rinds or thick skins naturally protect many fruits and vegetables, and eggs have shells. These coverings are dry and nutritionally poor, and thus they support little microbial growth. If the outer covering is broken or cut, however, microbes can reach the moist interior of the food and cause it to rot. Additionally, moisture left on the rind or skin can create conditions that support mold and bacterial growth. Ground hamburger supports microbial growth, and thus spoils faster, than whole steaks or roasts, because grinding creates air pockets and more surfaces that can harbor microbes. In uncut meats, the largest volume of the meat is anaerobic and not exposed to microorganisms.

To some extent, microbial competition can also be an intrinsic factor in food spoilage, especially in fermented foods. Fermented foods are populated with large numbers of fermentative bacteria, but few pathogens, because the latter do not readily grow in the environment produced by fermentation. Furthermore, pathogens that require the rich nutrient levels found in the human body are not typically capable of growing on "lower nutrient" foods.

Table 26.2 summarizes the effects of intrinsic factors on food spoilage.

Extrinsic Factors in Food Spoilage Extrinsic factors, including how food is processed, handled, and stored, also govern food spoilage. Microorganisms enter food in a variety of ways; some are introduced accidentally during harvesting from soil or contamined water, and others are introduced during commercial food processing. Mechanical vectors such as flies can also deposit microbes from their feet onto food. Most microbial contaminants leading to spoilage, however, can be traced to improper handling by consumers. Methods to prevent the introduction of microbes into food, or to limit their growth, will be presented shortly.

Describe the intrinsic and extrinsic factors that would contribute to the spoilage of butter, an apple, and a steak.

Classifying Foods in Terms of Potential for Spoilage

Based on the previous considerations, foods can be grouped into three broad categories based on their likelihood of spoilage. Highly *perishable* foods, such as milk, spoil relatively quickly (within days). Perishable foods tend to be nutrient rich, moist, and unprotected by rinds or coverings, and need to be kept cold. *Semi-perishable* foods, such as tomato sauce, can be stored in sealed containers for months without spoiling so long as they are not opened. Once opened, however, they may spoil within several weeks. Many fermented foods are semi-perishable. *Nonperishable* foods are usually dry foods, such as pasta or uncooked rice, or canned goods that can be stored almost indefinitely without spoiling. Nonperishable foods are typically either nutritionally poor, dried, fermented, or preserved (discussed shortly).

Prevention of Food Spoilage

Food spoilage can occur during production, processing, packaging, or handling. Spoilage can result in significant economic losses, especially to producers—in the form of lost productivity, food recalls, and even the loss of jobs resulting from the shutdown of unsafe food processing facilities—but also to consumers in the form of medical expenses and time lost from work. In the next subsections we examine some of the many methods used to prevent food spoilage: food processing methods, the use of preservatives, and attention to temperature and other storage conditions.

Food Processing Methods   *Industrial canning* is a major food packaging methodology for preserving foods **(Figure 26.4)**. After food is prepared by washing, sorting, and processing, it is then packaged into cans or jars and subjected to high heat (115°C) under pressure for a given amount of time, depending on the food. This is followed by rapid cooling. The heat kills vegetative mesophilic bacteria and destroys endospores formed by *Bacillus* (ba-sil′lŭs) and *Clostridium* (klos-trid′ē-ŭm). Although canning results in the elimination of most contaminating microorganisms, it does not sterilize food; hyperthermophic microbes remain, but because they cannot grow at room temperature, they do not pose a threat.

Canning is usually a very effective means of preservation. Spoilage, however, can occur if the food is underprocessed, or if mesophilic microbes contaminate the can during cooling or thereafter. The two most frequent contaminants found in canned goods are *Clostridium* spp. and coliforms, anaerobic organisms that grow well in low-acid foods. Contamination with the resistant endospores of *Clostridium* is especially problematic because if the cans or

▲ *Figure 26.4*

Industrial canning, in which food is processed, put into cans or jars, and heated under pressure to a temperature sufficient to kill most microbes, including endospore-forming bacteria.

jars are not heated sufficiently, the endospores germinate, grow and release toxins inside the container.

Pasteurization is less rigorous than canning and is typically used primarily with beer, wine, and dairy products because it does not degrade the flavor of these more delicate foods. By heating foods only enough to kill mesophilic, nonendospore-forming microbes (including most pathogens), pasteurization lowers the overall number of microbes, but because some microbes survive, pasteurized foods tend to spoil without refrigeration.

Because moisture is essential for microbial growth, desiccation (drying) is an excellent food preservation technique. Drying greatly reduces, if not eliminates, microbial growth, but it does not kill bacterial endospores. Although drying by leaving foods in the sun is still used for some fruits, today most commercially dried foods are processed in ovens or heated drums that evaporate the water from the foods.

Lyophilization (lī-of′i-li-zā′shŭn), or *freeze-drying*, involves freezing foods and then using a vacuum to draw off the ice crystals. Freeze-dried foods such as soups and sauces **(Figure 26.5)** can be reconstituted by mixing with water.

Gamma radiation, produced primarily by the isotope cobalt-60, can penetrate fruits, vegetables, fish, and poultry to cause irreparable (and thus fatal) damage to the DNA of microbes in these foods. Such ionizing *irradiation* is controversial because some consumers believe that irradiated foods become radioactive, are less nutritious, or contain toxins. Nevertheless, because it can achieve complete sterilization, it is used routinely to preserve some foods such as spices.

UV light is not ionizing and does not penetrate very far. It is used to treat packing and cooling water used in industrial canning and to treat work surfaces and utensils in industrial meat processing plants.

In *aseptic packaging,* paper or plastic containers that cannot withstand the rigors of canning or pasteurization are sterilized with hot peroxide solutions, UV light, or superheated steam. Then a conventionally processed food is added to the aseptic package, which is sealed without the need for additional processing.

The Use of Preservatives   Salt and sugar are two chemical preservatives known even to ancient humans. They draw water out of foods and microbes alike; high concentrations of salt or sugar both kill microbes present during processing and retard the growth of any subsequent microbial contaminants. Ham and bacon are examples of high-salt foods; jams and jellies are examples of high-sugar foods.

Whereas salt and sugar act by removing water, other natural preservatives such as onions, garlic, many herbs, spices, and oils actively inhibit microbial enzymes or disrupt cytoplasmic membranes. Garlic contains *allicin,* an antimicrobial compound that inhibits enzyme function. Benzoic acid, produced naturally by cranberries, also interferes with enzymatic function. Cloves, cinnamon, oregano, and thyme (and to some degree, sage and rosemary) produce oils that interfere with the membrane functions of microorganisms.

As we have seen, fermentation (whether natural or commercial) preserves some foods by producing an acidic environment that is inhospitable to most microbes. For other foods such as meats, the use of wood smoke during the drying process introduces growth inhibitors that help preserve the food. Other naturally occurring chemicals and several synthetic ones can be purposely added to foods as chemical preservatives. Acceptable food additives should be harmless to the consumer and should not alter the taste or appearance of the food to any great extent. Organic acids such as benzoic acid, sorbic acid, and propionic acid are commonly used in beverages, dressings, baked goods, and a variety of other foods. Gases such as sulfur dioxide and ethylene oxide are used to preserve dried fruit, spices, and nuts. All such chemical preservatives inhibit some function of the microbial cell to prevent growth, but many do not actually kill microbes; in other words, they are *germistatic* rather than *germicidal.* Some chemicals work better against bacteria than against fungi, and vice versa. Benzoic acid, for example, has a largely antifungal function and does not affect the growth of many bacteria.

Attention to Temperature During Processing and Storage   In general, higher temperatures are desirable during food processing and preparation to prevent food spoilage, whereas lower temperatures are desirable for food

▲ *Figure 26.5*
Examples of food preserved by freeze-drying.

storage. High temperatures such as those during pasteurization, canning, or cooking kill potential pathogens because proteins and enzymes become irreversibly denatured. (Heat does not inactivate many toxins; for example, botulism toxin may remain in cooked foods even when the bacteria that produced it are dead.) Unlike heat, cold rarely kills microbes, but instead only retards their growth by slowing their metabolism. Even freezing fails to kill all microorganisms and may only lower the level of microbial contamination enough to reduce the likelihood of food poisoning after frozen foods have been thawed.

To prevent food spoilage, foods should be prepared at sufficiently high temperatures and then stored under conditions that do not facilitate microbial growth. Wrapping leftovers or putting them in containers enables foods to be stored away from exposure to air or contact that could result in contamination. Storing foods in the cold of a refrigerator or freezer combines the protection of packaging with growth-inhibiting temperatures.

One microbe for which cold storage does not suppress growth is *Listeria monocytogenes* (lis-tēr'ē-ă mo-nō-sī-tah'je-nēz), the causative agent of listeriosis and a common environmental bacterium, which is prevalent in certain dairy products (such as soft cheeses). Because this bacterium grows quite well under refrigeration, it is best to prevent its entry into and presence in foods during production and packaging.

Food-Borne Illnesses

Learning Objective

✓ Discuss the basic types of illnesses caused by food spoilage or food contamination, and describe how they can be avoided.

Spoiled food, if consumed, can result in illness, but not all food-borne illnesses result from actual food spoilage.

Table 26.3 Most Common Agents of Food-Borne Illnesses

Organism	Affected Food Products	Comments
Campylobacter jejuni	Raw and undercooked meats; raw milk; untreated water	Most common cause of diarrhea of all food-borne agents
Clostridium botulinum	Home-prepared foods	Produces a neurotoxin
Escherichia coli O157:H7	Meat; raw milk	Produces an enterotoxin
Listeria monocytogenes	Dairy products; raw and undercooked meats; seafood; produce	Common in soils and water; contamination from these sources occurs easily
Salmonella spp.	Raw and undercooked eggs; meat; dairy products; fruits and vegetables	Second most common cause of food-borne illness in the United States
Shigella spp.	Salads; milk and other dairy products; water	Third most common cause of food-borne illness in the United States
Staphylococcus aureus	Cooked high-protein foods	Produces a potent toxin that is not destroyed by cooking
Toxoplasma gondii	Meat (pork in particular)	Parasitic protozoan
Vibrio vulnificus	Raw and undercooked seafood	Causes primary septicemia (bacteria in the blood)
Yersinia enterocolitica	Pork; dairy products; produce	Causes generalized diarrhea

Food-borne illnesses may also result from the consumption of harmful microbes or their products in food.

Food-borne illnesses, or *food poisoning*, can be divided into two types: **food infections**, caused by the consumption of living microorganisms, and **food intoxications**, caused by the consumption of microbial toxins instead of the microbes themselves. Typical signs and symptoms of food poisoning are generally the same regardless of the cause and include nausea, vomiting, diarrhea, fever, fatigue, and muscle cramps. Symptoms occur within 2 to 48 hours after ingestion, and the effects of the illness can linger for days. Most outbreaks of food poisoning are *common-source epidemics*, meaning that a single food source is responsible for many individual cases of illness.

According to the Centers for Disease Control and Prevention, roughly 76 million cases of food poisoning occur each year. Of these, about 325,000 people require hospitalization and 5200 die. Known pathogens account for only 14 million of the total cases. The United States Department of Agriculture estimates the economic cost of food poisoning—due to loss of productivity, medical expenses, and death—at roughly $5–10 billion per year.

More than 250 different food-borne diseases have been described. **Table 26.3** lists 10 of the more common causes of food-borne illness and their sources. Except for the protozoan *Toxoplasma gondii* (tok-sō-plaz′mă gon′dē-ī), all are bacterial agents. A few fungi (*Aspergillus* and *Penicillium*) and viruses (hepatitis A, noroviruses) also cause food poisoning.

CRITICAL THINKING

Suggest why neither the original food involved nor the organism responsible can ever be identified in more than 80% of food poisoning cases.

Industrial Microbiology

The potential uses of microorganisms for producing valuable compounds, as environmental sensors, and in the genetic modification of plants and animals makes industrial microbiology one of the more important fields of study within the microbiological sciences. In the subsections that follow we will examine the use of microbes in industrial fermentation, in the production of several industrial products, in the treatment of water and wastewater, in the disposal and cleanup of biological wastes, and in the treatment of mine drainage.

The Roles of Microbes in Industrial Fermentations

Learning Objective

✓ Describe the role of genetically manipulated microorganisms in industrial and agricultural processes and the basics of industrial-scale fermentation.

In industry, the word *fermentation* is used a bit differently than it is used in food microbiology. *Industrial fermentations* involve the large-scale growth of particular microbes for producing some type of beneficial compound. Examples include amino acids and vitamins. Because of its scale, industrial fermentation is performed under highly controlled conditions in huge vats that can be readily filled, emptied, and sterilized **(Figure 26.6)**. Vats are typically made from stainless steel to enable them to be cleaned and sterilized using heat, steam, or chemicals before being used again. Temperature, aeration, and pH are all regulated to maintain optimal microbial growth conditions. Generally, industrial fermentations start with the cheapest growth medium available, often the waste product from another industrial or food preparation process (such as whey from cheese production).

Industrial fermentation equipment is designed to achieve one of two types of production, depending on the nature of the product being produced. In *batch production*, organisms are allowed to ferment their substrate until it is exhausted, and then the end product is harvested all at once. In *continuous flow production*, the vat is continuously fed new medium while wastes and product are continuously removed. For this setup to work, the organism must secrete its product into the surrounding medium.

Industrial products are produced as either primary or secondary metabolites by the microorganisms. *Primary metabolites* such as ethanol are produced during active growth and metabolism because they are either required for the production of new organisms or are by-products of active metabolism. *Secondary metabolites* such as penicillin are produced after cells have left the log phase of growth and have entered the stationary phase, during which time the substances produced are not immediately needed for growth.

In Chapter 8 we examined many of the recombinant DNA techniques used for the production of *recombinant microorganisms*, and some of their uses. Most genetically modified organisms used in industry have been specifically designed either to produce a stable, high-yield output of desirable chemicals and complex molecules or to perform certain novel functions. Still others are designed to deliver pesticides or growth-enhancing genes to plants, hormones to livestock, or disease-curing genes to humans.

CRITICAL THINKING

Which process would you expect to yield more product from the same sized vats over the same amount of time, batch production or continuous flow production? Why?

Industrial Products of Microorganisms

Learning Objective

✓ List some of the various commercial products produced by microorganisms.

Microorganisms, particularly bacteria, produce an incredible array of industrially useful chemicals during their metabolic activities, including enzymes, food additives and supplements, dyes, plastics, and fuels. Recombinant organisms add to this diversity by producing pharmaceuticals such as human insulin that are not normally manufactured by microbial cells, and genetically modified crops. In the subsections that follow we will consider a variety of industrial products produced by microbes, including enzymes and other industrial products, alternative fuels, pharmaceuticals, pesticides and agricultural products, and biosensors and bioreporters.

▲ *Figure 26.6*

Fermentation vats used for growing microorganisms in the vast quantities needed for the large-scale production of many industrial, agricultural, and medical products. *Why are industrial fermentation vats made of stainless steel instead of wood?*

Figure 26.6 Fermentation vats are made of stainless steel so that they can be cleaned more easily to avoid contamination of the product being produced.

Enzymes and Other Industrial Products

Enzymes are among the more important products made by microbes. Most are naturally occurring enzymes for which industrial uses have been devised. Amylase, produced by *Aspergillus oryzae*, is used as a spot remover. Pectinase, obtained from species of *Clostridium*, enzymatically releases cellulose fibers from flax, which are then made into linen. Proteases from a variety of microbes are used as glues, meat tenderizers, and spot removers. Streptokinases and hyaluronidase are used in medicine to dissolve blood clots and enhance the absorption of injected fluids, respectively. The enzymatic tools of recombinant DNA technology are made by microbes and include restriction enzymes, ligases, and polymerases.

Many of the products made naturally by microorganisms for their own growth can be useful to humans in the form of food additives and food supplements. Food additives generally enhance a food in some way, such as color or taste, whereas food supplements make up for nutritional deficiencies, either in a food or in the consumer. Amino acids and vitamins are two important microbial products used as additives or supplements. Vitamins are added directly to foods or sold as single or multivitamin tablets. Amino acids are either used directly or combined to make new compounds, such as the sweetener aspartame (made from the amino acids phenylalanine and aspartic acid). Many organic acids, such as citric acid, gluconic acid, and acetic acid (vinegar), are also used in food manufacturing. Citric acid is used

New Frontiers 26.1   *Making Blue Jeans "Green"*

Blue denim

Soon we may have *Escherichia coli* to thank for making the blue in our blue jeans environmentally friendly.

Most blue denim today is colored with an indigo dye that has been synthetically created via a petroleum-based process, because extracting the dye from indigo plants is labor intensive and expensive. Environmentalists are concerned that the potentially toxic by-products of the coal- and oil-dependent process pose risks to the environment.

To address such concerns, scientists have been investigating an alternative method of producing indigo using *E. coli*, which naturally produces the amino acid tryptophan, with a chemical structure very similar to that of indigo. Since 1983, scientists have known how to take advantage of this similarity and genetically alter *E. coli* so that the bacterium produces indigo instead. The problem? Bacterial synthesis of indigo was both slow and resulted in an unwanted red pigment that remained visible when the dye was used on denim.

These problems may soon be resolved. In a study published in the March 2002 issue of *The Journal of Industrial Microbiology and Biotechnology*, scientists reported that they were successful in modifying the *E. coli* genome such that the red pigment was no longer produced. They also reported that they were able to boost *E. coli's* production of indigo by 60%. Though more work is needed to increase the efficiency of this process, these results provide hope for environmentally sensitive blue jeans in the future.

---

as an antioxidant in foods, while gluconic acid is used medically to facilitate calcium uptake.

Other industrial products made by microbes include dyes, such as indigo, which makes denim jeans blue (see **New Frontiers 26.1**), and cellulose fibers used in woven fabrics. Additionally, microbially produced biodegradable plastics may one day replace nonbiodegradable, petroleum-based plastics, because many bacteria produce different types of a carbon-based storage molecule called poly-β-hydroxyalcanoate (PHA), a polymer with a structure similar to petroleum-based plastics. Because the type of PHA produced depends on the source of carbon available to the bacteria, microbes could be used to produce a host of new plastics—all of which are degradable by other microbes. Even though such biodegradable plastics are expensive to manufacture, some have been commercially available since the early 1990s.

Alternative Fuels

Photosynthetic microorganisms use the energy in sunlight to convert CO_2 in air into carbohydrates that can be used as fuels. This process is one type of **bioconversion,** in which organisms produce organic materials for fuels. Other microorganisms convert biomass (plants, crops, or wastes) into renewable alternative fuels, including ethanol, methane, and hydrogen.

Ethanol, which is made during alcoholic fermentation and is the simplest alternative fuel to synthesize, can be mixed with gasoline to make *gasohol*, which is used in existing cars. Currently, crops such as corn are fermented to ethanol, but in the future new types of crops may be grown specifically for the production of ethanol. The most efficient production of alternative fuel would utilize crop wastes—for example, corn stalks.

▲ *Figure 26.7*

The release of methane gas from a landfill. Such microbially produced methane may one day be an important alternative fuel source. *What are some potential benefits of renewable energy produced by microbes?*

Figure 26.7 Microbial fuels may be easier to produce and less harmful to the environment.

Methane gas, another renewable energy source, can be piped through natural gas lines to be used for cooking and heating. The largest potential source of methane is from landfills **(Figure 26.7)** where methanogens anaerobically convert wastes into methane via fermentation. Some communities already use methane from landfills to produce energy. Methane can also be used as a fuel in cars.

New Frontiers 26.2 Bacterial Batteries

Some bacteria in the class Deltaproteobacteria have an astonishing ability: They can generate electricity from underwater mud.

Researchers have observed that *Desulfuromonas acetoxidans* is able to oxidize organic compounds in underwater sediments and transfer a current of electrons to graphite wires placed in the water by scientists. Essentially, these bacteria act as batteries that use mud as a fuel to initiate chemical reactions that produce enough electricity to power a small calculator.

The current methods of using bacteria to generate electricity are neither practical nor efficient enough for large-scale energy produc-

tion. Intriguing as the idea is, we will likely never have bacteria-powered cars! However, scientists believe that bacteria could be used as inexpensive batteries for underwater sensors. In addition, because these bacteria use organic sediments as fuel, they may be useful in cleaning up organic pollution on the sea floor by converting pollutants to electricity. Sound too good to be true? Microbiologists are working to turn this possibility into reality.

Source: Bond, et al. 2002. "Electrode-Reducing Microorganisms That Harvest Energy from Marine Sediments." *Science* 295: 483–485.

Although hydrogen is produced by a variety of microbes as part of their normal metabolism and can also be produced by photosynthetic microorganisms through the photolysis of water, neither production route is especially efficient. Theoretically, the production of hydrogen fuel using sunlight would have widespread applications and enormous appeal, but much research is needed to develop technologies to produce and efficiently use hydrogen as a viable energy source.

All microorganisms synthesize hydrocarbons as part of their normal metabolism, but only some microbes produce hydrocarbons that could be useful fuels. One colonial alga, *Botryococcus braunii* (bot'rē-ō-kok'ŭs brow'nē-ē) for example, produces hydrocarbons that account for 30% of its dry weight. The technology to harvest such hydrocarbons does not yet exist, but one day this alga may be an important source of fuel.

New Frontiers 26.2 explores the potential use of bacteria in generating electricity directly, without having to burn fuel.

Pharmaceuticals

Foremost among the pharmaceutical substances microorganisms produce are antimicrobial drugs. Between 5000 and 6000 antimicrobial substances have been described since penicillin was first produced during World War II. Of these, 100 or so have current medical applications. Even though *Streptomyces* (strep-tō-mī'sēz), *Bacillus,* and *Penicillium* (pen-i-sil'ē-ŭm) synthesize the majority of useful antimicrobials, recombinant DNA techniques enable the modification of microbes to produce new versions of old drugs.

In addition to antimicrobials, genetically modified microbes are producing hormones and other cell regulators. For example, *Escherichia coli* (esh-ě-rik'ē-ă kō'lī) and *S. cerevisiae* have been manipulated genetically to produce a variety of such molecules. Some of the compounds made by *E. coli* include human insulin, human growth factor, interleukin-2, α-interferon, taxol (used to treat cancer), and epidermal growth factor (used to treat burns and wounds). Genetically altered *S. cerevisiae* produces hepatitis B proteins for use in the hepatitis B vaccine.

Pesticides and Agricultural Products

Microbes and their products are also being used in a variety of agricultural applications, particularly with regard to crop management. Though some people are leery of the prospect of transgenic foods derived from these plants while others contend that the genetic manipulation of crops is crucial if we are to continue to feed the world's population, one thing is clear: The genetic manipulation of plants is big business.

Pseudomonas syringae (sū-dō-mō'nas sēr'in-jē) is one example of an organism with many industrial applications. Whereas this bacterium produces a protein that serves in the nucleation of water to form ice crystals, the protein is not essential to the survival of the organism, and can be removed. When sprayed on crops such as strawberries, the strain lacking this trait, known as *Pseudomonas syringae* (ice⁻, "ice-minus"), inhibits the formation of ice, protecting the plants from freeze damage.

Bacillus thuringiensis (ba-sil'ŭs thur-in-jē-en'sis) is one of the most widely used organisms in agriculture because during sporulation it produces *Bt-toxin* that, when digested by a caterpillar of such pests as gypsy moths, diamondback moths, and tomato hornworms, destroys the lining of the insect's gut wall, eventually killing it. Bt-toxin can be spread as a dust on plants, but through recombinant DNA technology many plants (for example, corn, soy beans, and cotton) can protect themselves by manufacturing their own Bt-toxin.

There is evidence, however, of insect resistance to Bt-toxin. Large tracts of U.S. farmland are devoted to raising Bt-toxin–expressing plants, which places selective pressure on the insect population to spread resistance to the toxin to survive. Although no crops have yet been lost to Bt-toxin-resistant insects, the situation must be closely monitored. Other species of *Bacillus* used as pesticides include *B. sphaericus* (sfe'ri-kŭs), used to control mosquito populations, and *B. popilliae* (pop-pil'ē-ē), used to control Japanese beetle infestations.

Table 26.4 on page 756 lists selected industrial products produced by microorganisms and some of their uses.

Table 26.4	Selected Industrial Products Produced by Microorganisms
Products Made	**Use of Product**
Enzymes	
Amylase, proteases	Spot removers
Streptokinase	Breakdown of blood clots
Restriction enzymes, ligases, polymerases	Molecular biology/research
Food Additives/Supplements	
Amino acids, vitamins	Health supplements
Citric acid	Antioxidant
Sorbic acid	Food preservative
Other Industrial Products	
Indigo	Dye used in manufacturing clothes
Plastics	Biodegradable substitutes for petroleum-based plastics
Alternative Fuels	
Ethanol	Used in gasahol
Methane	Burned to generate heat and electricity
Hydrogen, hydrocarbons	Potential fuels
Pharmaceuticals	
Antimicrobial drugs	Treatment of bacterial infections
Insulin, human growth hormone	Replacement hormones
Taxol	Cancer treatment
Pesticides and Agricultural Products	
Bt-toxin	Insecticide

Biosensors and Bioreporters

Among the relatively new applications of microorganisms to solve environmental problems are biosensors and bioreporters. **Biosensors** are devices that combine bacteria or microbial products (such as enzymes) with electronic measuring devices to detect other bacteria, bacterial products, or chemical compounds in the environment. **Bioreporters** are somewhat simpler sensors that are composed of microbes (again usually bacteria) with innate signaling capabilities, such as the ability to glow in the presence of biological or chemical compounds.

Currently, biosensors and bioreporters are used to detect the presence of environmental pollutants (for example, petroleum) and to monitor efforts to remove harmful substances; they may also be useful in detecting bioterrorist attacks. Because bacteria are so sensitive to their environments, they can detect compounds in very small amounts. Biosensors and bioreporters could serve as early warning systems that would give public health officials more time to respond by detecting the metabolic waste products of weaponized biological agents.

Water and Wastewater Treatment

Learning Objective

✓ Explain how water and wastewater are treated to make them safe and usable.

Clean water is vital for people and their activities. In this section we will consider the treatment of drinking water and of wastewater (sewage). In both cases, treatment is designed to remove microorganisms, chemicals, and other pollutants to prevent human illness.

Treatment of Drinking Water

Potable water is water that is considered safe to drink, but the term *potable* (pō'tăbl) does not imply that the water is devoid of all microorganisms and chemicals. Rather, it implies that the levels of microorganisms or chemicals in the water are low enough that they are not a health concern. Water that is not potable is **polluted;** that is, it contains organisms or chemicals in excess of acceptable values.

The permissible levels of microbes and chemicals in water varies from state to state, and on whether or not the water is used for drinking. The U.S. Environmental Protection Agency (EPA) requires that drinking water have a coliform count of 0 (zero) coliforms per 100 ml of water and that recreational waters have no more than 200 coliforms per 100 ml of water. Recall that coliforms are intestinal bacteria such as *E. coli*. The presence of coliforms in water indicates fecal contamination and thus an increased likelihood that disease-causing microbes are present.

The treatment of drinking water can be divided into three stages (**Figure 26.8**):

1. **Sedimentation.** During sedimentation, water is pumped into holding tanks where particulate materials (sand, silt, and organic material) settle. The partially clarified water is then pumped into a secondary tank for *flocculation,* in which alum (aluminum potassium sulfate) added to the water complexes with suspended particles and microorganisms to form large aggregates, called *flocs,* which settle to the bottom of the tank. The water above the sediment is then pumped into a different tank for filtration.

2. **Filtration.** In this stage, the number of microbes is reduced by about 90% in one of several ways. One method uses sand and other materials to which microbes absorb and form biofilms that trap and remove other microbes. *Slow sand filters* are composed of a 1-meter layer of fine sand or diatomaceous earth and are used in smaller cities or towns to process 3 million gallons per acre of filter per day. Large cities use *rapid sand filters* that contain larger particles and gravel and can process 200 million gallons per acre per day. Both types of filters are cleaned by back-flushing with water. Two other filtration methods are *membrane filtration,* which uses a filter with a pore size of 0.2 μm, and the use of *activated charcoal*

(a)

◀ *Figure 26.8*

The treatment of drinking water. **(a)** A water
treatment facility. **(b)** The stages of water treatment:
sedimentation, filtration, and disinfection. *Why is it
that chemical treatment cannot destroy most viruses?*

*Figure 26.8 Most chemicals are designed either
to inhibit some aspect of active metabolism or damage
a cell structurally; therefore, they don't damage acellular and nonmetabolizing viruses.*

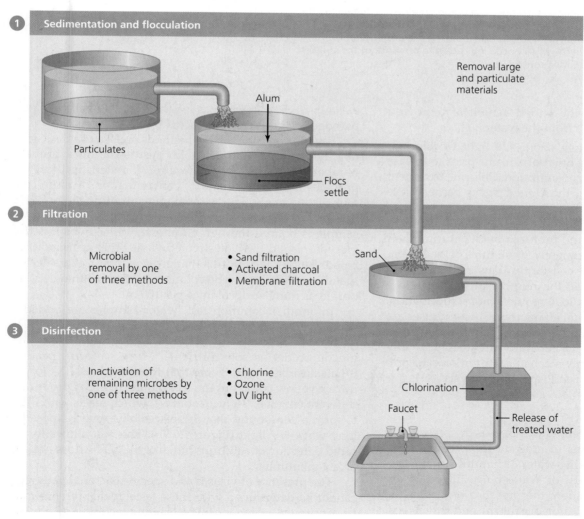

1 **Sedimentation and flocculation**

Removal large
and particulate
materials

Alum

Particulates

Flocs
settle

2 **Filtration**

Microbial
removal by one
of three methods

• Sand filtration
• Activated charcoal
• Membrane filtration

Sand

3 **Disinfection**

Inactivation of
remaining microbes by
one of three methods

• Chlorine
• Ozone
• UV light

Chlorination

Faucet

Release of
treated water

(b)

(a)   (b)

▲ *Figure 26.9*

Two water quality tests. **(a)** Membrane filtration. The grid on the membrane makes it easier to count the typical metallic green colonies of fecal coliforms. **(b)** The ONPG and MUG tests. The yellow color of the ONPG bottle indicates the presence of coliforms, whereas the blue fluorescence in the MUG bottle indicates the presence of the coliform *E. coli*; the clear bottle is the negative control.

filters, which provide the added benefit of removing some organic chemicals from the water.

③ **Disinfection.** In this stage, ozone, UV light, or chlorination is used to kill most microorganisms prior to release of the water for public consumption. Chlorine treatment is most widely used in the United States because it is least expensive.

Chlorine gas, an oxidizing agent, is thought to kill bacteria, algae, fungi, and protozoa by denaturing their proteins within approximately 30 minutes of treatment. Chlorine levels must be constantly adjusted to reflect estimates of *microbial load,* the number of microbes in a unit of water—a higher load requires more chlorination. Chlorination does not kill all microbes: Most viruses are not inactivated by chlorine, and bacterial endospores and protozoan cysts are generally unaffected by any chemical treatment. Only mechanical filtration can completely remove viruses, endospores, and cysts.

Water Quality Testing

Water quality testing is a technique that uses the presence of certain **indicator organisms** to indicate the possible presence of pathogens in drinking water or natural aquatic systems. Because the majority of water-borne illnesses are caused by fecal contamination, the presence of *E. coli* (the most commonly used indicator organism) and other fecal coliforms in water indicates a high probability that pathogens are present as well. *E. coli* clearly meets the criteria for a good indicator: It is consistently prevalent in human

wastes, survives in water as long as (if not longer than) most pathogens, and is easily detected by simple tests.

Several common testing methods are use to assess water quality. An older and unwieldy test is the most probable number (MPN) test (see Figure 6.23), a statistical test that provides only an estimate of contamination. A more common testing method is the membrane filtration method **(Figure 26.9a),** which is simple to perform: A 100-ml water sample is poured through a fine membrane, which is then placed on solid EMB (eosin methylene blue) agar and incubated. Any fecal coliforms that grow exhibit a characteristic metallic green sheen. The colonies are then counted and reported as number of colonies per 100 ml.

In another common test, water samples are added to small bottles containing both *ONPG* (*o*-nitrophenyl-β-D-galactopyranoside) and *MUG* (4-methylumbilliferyl-β-D-glucouronide) as sole nutrients. Most coliforms produce β-galactosidase, an enzyme that reacts with ONPG to produce a yellow color, but the fecal coliform *E. coli* produces a different enzyme, β-glucuronidase, which reacts with MUG to form a compound that fluoresces blue when exposed to long-wave UV light **(Figure 26.9b).** This test allows for the rapid detection of coliforms but, like the MPN, does not give an actual number.

The presence of viruses and specific bacterial pathogens cannot be determined with these tests; their presence must be confirmed by genetic fingerprinting techniques, in which water samples are collected and enriched to cultivate any organisms present. The DNA content of the enriched sample is then genetically screened for the identification of potential pathogens. Tests such as these are generally available only in large cities or through government agencies that have the laboratory capability to perform them.

CRITICAL THINKING

Explain how biosensors might be used for water quality testing. Would biosensors give more accurate identifications of the microorganisms present in water than an ONPG or MUG test? Is such specificity truly necessary?

Treatment of Wastewater

Sewage, or **wastewater**, is typically defined as any water that leaves homes or businesses after being used for washing or flushed from toilets. (Some municipalities also include industrial water or even rainwater as wastewater.) Wastewater contains a variety of contaminants, including suspended solids, biodegradable organic and inorganic compounds, toxic metals, and pathogens. The objective of wastewater treatment is to remove or reduce these contaminants to acceptable levels.

Because sewage is mostly water (less than 1% solids), most of the sewage treatment process involves the removal of microorganisms. At one time, "raw" (unprocessed) sewage was simply dumped into the nearest river or ocean; the idea was that wastes would be diluted to the point at which they would be harmless. Burgeoning populations, and the realization that waterways were becoming increasingly polluted, led to the development of effective wastewater treatment processes.

A key concept in the processing of wastewater is reducing **biochemical oxygen demand (BOD),** which is a measure of the amount of oxygen aerobic bacteria required to fully metabolize organic wastes in water. This amount is proportional to the amount of waste in the water; the higher the concentration of degradable chemicals, the more oxygen is required to catabolize them and the higher the BOD. Effective wastewater treatment reduces the BOD to levels too low to support microbial growth, thus reducing the likelihood that pathogens will survive.

In the following subsections we will consider wastewater treatments of various types: the traditional sewage treatment used in municipal systems, treatments used in nonmunicipal (rural) areas, a treatment used for agricultural wastes, and the use of artificial wetlands to treat wastewater.

Municipal Wastewater Treatment   Today, people living in larger U.S. towns and cities are usually connected to *municipal sewer systems*, pipes that collect wastewater and deliver it to sewage treatment plants for processing. Traditional sewage treatment consists of four phases (**Figure 26.10** on page 760):

➊ **Primary treatment.** Wastewater is pumped into settling tanks, where lightweight solids, grease, and floating particles are skimmed off and heavier materials settle onto the bottom as **sludge.** After alum is added as a floc-

culating agent, the sludge is removed, and the partially clarified water is further treated. Primary treatment removes 25–35% of the BOD in the water.

➋ **Secondary treatment.** The biological activity in this phase reduces the BOD to 5–25% of the original. Most of the pathogenic microorganisms are also removed. The water is aerated to facilitate the growth of aerobic microbes that oxidize dissolved organic chemicals to CO_2 and H_2O. In an *activated sludge system*, aerated water is seeded with primary sludge containing a high concentration of metabolizing bacteria; flocculation also occurs during this step. Any remaining solid material settles and is added to the sludge from primary treatment. The combined sludge is pumped into anaerobic holding tanks. Some smaller communities accomplish secondary treatment using a *trickle filter system,* which is similar to the slow sand filters used in treating drinking water but less effective in removing BOD than activated sludge systems.

➌ **Chemical treatment.** The water from secondary treatment is disinfected, usually by chlorination, after which the water is either released into rivers or the ocean or, in some states, used for spray irrigation of crops and highway vegetation. Some communities remove nitrates, phosphates, and any remaining BOD or microorganisms from the water by passing it over fine sand filters and/or activated charcoal filters. Nitrogen is converted to ammonia and discharged into the air (removes roughly 50% of the nitrogen content), whereas phosphorus is precipitated using lime or alum (removes 70% of the phosphorus content). Such tertiary treatment is generally used in environmentally sensitive areas or in areas where the only outlet for the water is a closed-lake system.

➍ **Sludge treatment.** Sludge is digested anaerobically in three steps: First, anaerobic microbes ferment organic materials to produce CO_2 and organic acids. Second, microbes metabolize these organic acids to H_2, more CO_2, and simpler organic acids such as acetic acid. Finally, the simpler organic acids, H_2, and CO_2 are converted to methane gas. Any leftover sludge is then dried for use as landfill or fertilizer.

Nonmunicipal Wastewater Treatment   Houses in rural areas, which typically are not connected to sewer lines, often use **septic tanks,** essentially the home equivalent of primary treatment (**Figure 26.11** on page 761). Sewage from the house enters a sealed concrete holding tank buried in the yard; solids settle to the bottom and the effluent flows from the top of the tank into an underground *leach field* that acts as a filter. Sludge in the tank and organic chemicals in the water are digested by microorganisms. However, because the tank is sealed, it must occasionally be pumped out to remove sludge buildup. **Cesspools** are similar to septic tanks except that they are not sealed. As wastes enter a system of porous concrete rings buried underground, the water is released

(a)

Traditional sewage treatment. **(a)** A municipal wastewater treatment facility. **(b)** The traditional sewage treatment process. Microbial digestion during secondary treatment removes most of the biochemical oxygen demand (BOD) before the water is chemically treated and released; dried sludge is recycled as landfill.

1 Primary Treatment: Sedimentation

Sewer line

Water effluent

Removal 25–35% BOD

Primary sludge (particulates + flocs)

Primary sludge

2 Secondary Treatment

Activated sludge or trickle filter system promotes microbial degradation organic material

Sludge

Secondary sludge

Removal 75–95% BOD

3 Chemical Treatment

Cl

Chlorination and/or filtration prior to release

4 Sludge Treatment

Anaerobic digestion of sludge

Dried sludge for landfill or agriculture

Methane burned off or trapped for fuel

(b)

into the surrounding soil; solid wastes accumulate at the bottom and are digested by anaerobic microbes.

Treatment of Agricultural Wastes Farmers and ranchers often use **oxidation lagoons** to treat animal waste from livestock raised in feedlots—a penned area where animals are fattened for market. Oxidation lagoons accomplish the equivalent of primary and secondary sewage treatment. Wastes are pumped into deep lagoons and left to sit for up to

3 months; sludge settles to the bottom of the lagoon; and anaerobic microorganisms break down the sediment. The remaining liquid is pumped into shallow, secondary lagoons, where wave-action aerates the water. Aerobic microorganisms, particularly algae, break down the suspended organic chemicals in the water. Eventually, the microbes die, and the clarified water is released into rivers or streams. One problem with oxidation lagoons is that they are open, which can be dangerous if floodwaters inundate

Sedimentation in septic tank | Filtration in leach field

House

Sludge
(must be pumped out eventually)

Pipes beneath ground distribute
water through leach field

◄ *Figure 26.11*
A home septic system.
After wastewater from
the house enters the
septic tank, solids
settle out as sludge,
and the effluent liquid
is filtered by the soil in
the leach fields.

the lagoons and spread largely untreated animal wastes over a wide area.

Artificial Wetlands Since the 1970s, small planned communities and some factories have constructed **artificial wetlands** to treat wastewater. Wetlands use natural processes to break down wastes and to remove microorganisms and chemicals from water before its final release. Individual septic tanks are not needed; instead, wastewater flows into successive ponds where microbial digestion occurs (**Figure 26.12** on page 762). The first pond in the series is aerated to allow aerobic digestion of wastes in the water; anaerobic digestion occurs in the sludge at the bottom. The water then flows through marshland, where organic chemicals are further digested by soil microbes. A second pond, which is still and contains algae, removes additional organic material, and the water then passes through open meadowland, where grasses and plants trap pollutants. By the time the water reaches a final pond, most of the BOD and microorganisms have been removed, and the water can be released for recreational purposes or irrigation. One drawback of an artificial wetland is that it requires considerable space—an artificial wetland to serve a small community can cover 50 acres of land or more.

Bioremediation

Learning Objective

✓ Describe the process of bioremediation.

Each year Americans produce more than 150 million tons of solid wastes, accumulated from household, industrial, medical, and agricultural sources; most of it ends up in landfills. A landfill is essentially a large, open pit into which wastes are dumped, compacted, and buried. Soil microbes anaerobically break down biodegradable wastes; methanogens degrade organic molecules to methane. When a landfill is full, it is covered with soil and revegetated.

To prevent leaching of potentially hazardous materials into the soil and groundwater, a landfill pit is lined with clay or plastic, and sand and drainage pipes lining the bottom

filter out small particulates and some microorganisms—but not chemicals. Unfortunately, chemicals and hazardous compounds might leak from landfills. Some of these substances are carcinogenic (benzene, phenol, petroleum products), whereas others are toxic (gasoline, lead), and still others are *recalcitrant* (resistant to decay, degradation, or reclamation by natural means).

The reason that one synthetic molecule is biodegradable while another is recalcitrant relates to chemical structure; sometimes a subtle variation—perhaps in a single atom or bond—can be enough to make the difference. Most recalcitrant molecules resist microbial degradation because the microbes lack enzymes capable of degrading them; after all, until very recently these synthetic compounds didn't even exist.

Bioremediation is the use of organisms, particularly microorganisms, to clean up toxic, hazardous, and recalcitrant compounds by degrading them to less harmful forms. Although most naturally occurring organic compounds are eventually degraded by microorganisms, synthetic compounds are not so easily removed from the environment. Petroleum, pesticides, herbicides, and industrial chemicals that accumulate in soil and water are degraded only slowly by naturally existing microbes.

Bioremediation uses either natural or artificially enhanced microbes. In *natural bioremediation,* environmentally extant microbes are "encouraged" to degrade toxic substances in soil or water by the addition of nutrients (carbon, nitrogen, or phosphorus) that stimulate growth, or of chemicals that facilitate microbial access and activity by increasing the solubility of toxic compounds. In *artificial bioremediation,* microbes are genetically modified by recombinant DNA technology to specifically degrade certain chemicals. Recombinant microbes also benefit from nutrient enrichment.

The most widely known application of bioremediation is the use of bacteria to clean up oil spills. Species of *Pseudomonas* have proven particularly amenable to the degradation of crude oil, and these microbes played a role in the cleanup of the 1989 Exxon *Valdez* spill in Alaska (see **Highlight 26.1** on page 763).

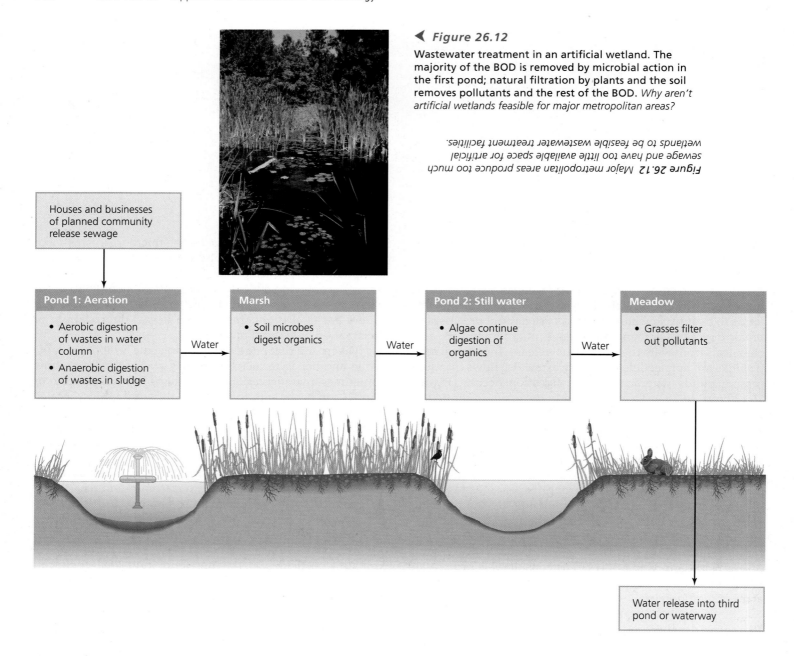

◄ *Figure 26.12*

Wastewater treatment in an artificial wetland. The majority of the BOD is removed by microbial action in the first pond; natural filtration by plants and the soil removes pollutants and the rest of the BOD. *Why aren't artificial wetlands feasible for major metropolitan areas?*

Figure 26.12 Major metropolitan areas produce too much sewage and have too little available space for artificial wetlands to be feasible wastewater treatment facilities.

Houses and businesses of planned community release sewage

Pond 1: Aeration
- Aerobic digestion of wastes in water column
- Anaerobic digestion of wastes in sludge

Water →

Marsh
- Soil microbes digest organics

Water →

Pond 2: Still water
- Algae continue digestion of organics

Water →

Meadow
- Grasses filter out pollutants

Water release into third pond or waterway

The Problem of Acid Mine Drainage

Learning Objective

✓ Discuss the problem of acid mine drainage.

Acid mine drainage is a serious environmental problem resulting from the exposure of certain metal ores to oxygen and microbial action **(Figure 26.13)**. Strip-mining for coal in Indiana, Kentucky, Ohio, Pennsylvania, and Virginia poses a threat to natural waterways. Coal deposits are often found associated with reduced metal compounds such as pyrite (FeS_2). Mining exposes pyrite to oxygen, which oxidizes the iron, and bacteria such as *Thiobacillus* (thī-ō-bă-sil'ŭs) oxidize the sulfur. Rainwater then leaches the oxidized compounds from the soil to form sulfuric acid (H_2SO_4) and iron

hydroxide [$Fe(OH)_3$], which are carried into streams and rivers, reducing the pH low enough (pH 2.5-4.5) to kill fish, plants, and other organisms. Such acidic water is also unfit for human consumption or recreational use. The EPA requires that strip mines be reburied as soon as possible after mining is complete.

Subsurface mining operations pose similar (but somewhat less severe) problems resulting from runoff from mine tailings, the low-grade ores remaining after the extraction of higher-grade ores. Typically, subsurface mines are backfilled as richer veins of minerals are depleted, thus limiting the exposure of iron sulfides to oxygen.

While acid mine drainage is generally devastating to the environment, some microbes—mostly archaea—actually flourish in the acidic conditions. One unique archaeal

Highlight 26.1   Oil-Eating Bacteria to the Rescue

In March 1989, the Exxon *Valdez* ran aground on Alaska's Bligh Reef, spilling 11 million gallons of crude oil into the waters of Prince William Sound. It was the worst oil spill in U.S. history, affecting hundreds of miles of shoreline and killing untold numbers of birds, sea otters, and other wildlife. The subsequent massive cleanup involved the efforts of over 10,000 workers; $2 billion in costs; collaboration among the

Cleaning up after the Exxon *Valdez* oil spill.

Exxon Corp., the EPA, and the state of Alaska; and legions of oil-eating bacteria.

Consisting largely of species native to the contaminated areas (including *Pseudomonas*), these bacteria metabolized the oil, converting it into harmless end products such as carbon dioxide and water. To boost the activity of these bacteria, scientists added such nutrients as nitrogen, phosphorus, and various trace elements to affected shoreline. The addition of nutrients not only speeded up bacterial growth rates; it also increased the rate at which the bacteria degraded the oil and cleaned up the shore.

The long-term effects of the oil spill on Prince William Sound continue to be debated. Even more than 10 years later, some plant and animal populations had not yet fully recovered. However, the Sound's ecosystem has also proven to be surprisingly resilient. Most of the area now looks much as it did before the spill, thanks in part to the bioremediation efforts of oil-eating bacteria.

---

species, found in mine drainage in California, is *Ferroplasma acidarmarnus* (fe′rō-plaz′ma a-sid′ar-mar-nŭs) (**Figure 26.14** on page 764), which lives in a pH near zero and obtains its energy from oxidizing pyrite in mine sediments. Such an organism is just one example of the incredible diversity of microorganisms that colonize every conceivable habitat on Earth.

Environmental Microbiology

Environmental microbiology is the study of microorganisms as they occur in their natural **habitats,** the physical localities in which organisms are found. Because of their vast metabolic capacities and adaptability, the habitats available to microbes include nearly every place on Earth, from Antarctic ice to boiling hot springs to bedrock. Without the activities of microbes, the functioning of the Earth's **ecosystems**—all the organisms living in all the habitats—would cease.

In the following sections we will explore the roles of microbes in the cycling of chemical elements, in soil, and in aquatic habitats. First, however, we turn our attention to the relationships of microbes to other microbes, and to their environment.

Microbial Ecology

Learning Objectives

✓ Define the terms used to describe microbial relationships within the environment.

✓ Explain the influences of competition, antagonism, and cooperation on microbial survival.

As examined in Chapters 5 and 6, microorganisms use a variety of energy sources and grow under a variety of

▲ *Figure 26.13*

The effects of acid mine drainage. Upon exposure to air, iron in water leached from mine tailings is oxidized to Fe^{3+}; the activity of iron-sulfur bacteria reduces the pH of the water to a level that is destructive to plants and animals.

conditions. They adapt to changing conditions, compete with other organisms for scarce resources, and change their environment in many ways. In some cases, their effects on the environment are undesirable from a human perspective, but in other cases they are beneficial or even essential. The study of the interrelationships among microorganisms and their environment is called **microbial ecology.** The first aspects of microbial ecology we will examine are the levels of microbial associations in the environment.

▲ *Figure 26.14*
The filamentous archaea *Ferroplasma acidarmarnus* growing in acid mine runoff from an abandoned mine in California. The ore from this mine is rich in pyrite.

Levels of Microbial Associations in the Environment

In understanding the roles of microorganisms in the environment, relationships are crucial. A variety of terms are used to describe the different levels of microbial associations:

- Individual organisms give rise to *populations* (all the members of a single species) through the processes of growth and reproduction.

- Populations of microorganisms performing metabolically related processes make up groups called *guilds*.

- Sets of guilds constitute *communities* that are typically quite heterogeneous; a variety of relationships exist among the various populations and guilds. Only rarely—and usually only in extreme environments—does a community consist of a single population.

- Populations and guilds within a community typically reside in their own distinct *microhabitats*, specific small spaces where conditions are optimal for survival.

- Groups of microhabitats form larger *habitats* in which microorganisms interact with larger organisms and the environment. Together, the organisms, the environment, and the relationships between the two constitute an *ecosystem*.

- All of the ecosystems taken together constitute the *biosphere*, that region of Earth inhabited by living organisms.

To illustrate these concepts, consider a plot of garden soil **(Figure 26.15)**. Soil is composed of tiny particles of rock and organic material, each of which forms a microhabitat for populations of microorganisms. Populations are distributed among the soil particles according to the available light, moisture, and nutrients. Populations of photosynthetic microbes, for example, live at the very top of the soil and form a photosynthetic guild, which in turn provides nutrients to other guilds living deeper within the soil. A soil community is composed of all of these populations and guilds, whose activities are a major factor affecting soil quality. Garden soil is just one type of soil occupying the soil ecosystem, which in turn is just one ecosystem within the biosphere.

Ecologists use the term **biodiversity** to refer to the number of species living within a given ecosystem, whereas the term **biomass** refers to the quantity (mass) of all organisms in an ecosystem. By either measure, microorganisms are the most abundant of all living things: The number of species of microbes exceeds the number of larger organisms, and the mass of microorganisms living throughout the biosphere—including places that are uninhabitable by plants, animals, or humans—is astronomical.

The Role of Adaptation in Microbial Survival

Lush ecosystems support great biodiversity, but the vast majority of microorganisms live in harsh environments, where nutrient levels are low enough to limit growth. The harsher the environment, the more specially adapted a microbe must be in order to survive. Some environments cycle between periods of excess and depletion, and the microbes that live in them must be capable of adapting to such constantly varying conditions. Some organisms attach to surfaces where nutrients accumulate to form biofilms. Extremophiles—microbes adapted to extremely harsh temperatures, pHs, and salt concentrations—are so adapted to their habitats that they cannot survive anywhere else.

Although a population of microorganisms could potentially exist in a variety of habitats, in reality biodiversity is held in check by *competition*. The best-adapted microorganisms have traits that provide them advantages in nutrient uptake, reproduction, response to environmental changes, or some other factor. *Antagonism*—in which a microbe makes some product that actively inhibits the growth of another also occurs. The production of antibacterial compounds by fungi is but one example.

While competition is a major factor in limiting biodiversity, rarely does a microorganism outcompete all other rivals; communities are the norm, not the exception. Even though *cooperation* (at least that characterized by a specific and close-knit relationship) is rare among organisms in the environment, microbes readily use the waste products of other microbes for their own metabolism. Moreover, one

◀ *Figure 26.15*

The basic relationships between microorganisms, and between microorganisms and the environment.

Biosphere

Soil ecosystem — Heterogeneous microbial communities (all guilds together)

Soil habitats — Guilds of related populations (e.g., photosynthetic microorganisms at soil surface)

Soil particle microhabitats — Populations of microbes inhabit microhabitats (e.g., cyanobacteria, algae)

microbe's metabolic activities sometimes make the environment more favorable for other microbes. Biofilms are examples of such "cooperation."

The relationships among microorganisms and their environment constantly change, with each microbe adapting to subtle changes engendered by its environment and other microbes. The fact that microbes continue to survive is a testament to their versatility. Most habitats support successions of microbes as populations come and go within the community. By studying how these successions occur, we learn much about the microbes, and their effects on the environment.

The Roles of Microorganisms in Biogeochemical Cycles

Learning Objective

✓ Compare and contrast the processes by which microorganisms cycle carbon, nitrogen, sulfur, phosphorus, and trace metals.

Most of the atoms of the chemical elements on Earth are tied up in essentially nonbiodegradable forms within the environment and are thus unavailable to organisms. The release of the elements in rock, for example, requires thousands of years of degradation by rain, wind, and microbes, and thus these processes contribute little to the day-to-day availability of nutrients for living things. As a consequence, the actions of organisms in converting the elements in nonliving organic materials into biomass are the major component of **biogeochemical cycles**—the processes by which organisms cycle nutrients by converting elements from one form to another, typically between oxidized and reduced forms.

Biogeochemical cycling essentially entails three processes: (1) *production,* in which organisms called producers convert inorganic compounds into the organic compounds of biomass; (2) *consumption,* in which organisms called consumers feed on other organisms, in the process converting organic molecules to other organic molecules, and (3) *decomposition,* in which organisms (called decomposers) convert the organic molecules in dead organisms back into inorganic compounds.

As with all cycles, balance is critical. Elements must be cycled continuously among organisms. When one part of a cycle becomes skewed relative to other parts, the cycle becomes inefficient, often with detrimental effects.

In the following subsections we will examine the biogeochemical cycles for carbon, nitrogen, sulfur, phosphorus, and some trace metals.

Figure 26.16 ▶

Simplified carbon cycle. The most mobile form of carbon in the cycle is CO_2; this inorganic molecule is fixed by autotrophs, which incorporate its carbon into organic molecules.

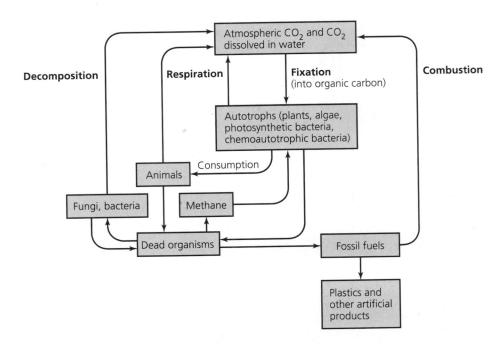

The Carbon Cycle

Carbon is the fundamental element of all organic chemicals. The continual cycling of carbon in the form of organic molecules constitutes the majority of the **carbon cycle (Figure 26.16),** though CO_2 has the highest *turnover rate* (the highest rate of conversion to other forms) within the system. Carbon in rocks and sediments has a very low turnover rate: This form of carbon is incorporated into organic chemicals only slowly over long periods of time.

The core of the carbon cycle is photosynthesis. Photoautotrophic *primary producers*—cyanobacteria, green and purple sulfur bacteria, green and purple nonsulfur bacteria, algae, and plants—convert CO_2 to organic molecules via carbon fixation. Because they require light, photoautotrophs are restricted to the surfaces of soil and water systems. Chemoautotrophs also fix carbon but use H_2S or other inorganic molecules as energy sources. Chemoautotrophs are found in a greater variety of habitats because they do not depend on light energy, but they are not as important in the carbon cycle as photosynthetic organisms.

Heterotrophs catabolize some organic molecules for energy during aerobic respiration, resulting in the release of CO_2. Other organic molecules made by autotrophs are incorporated into the tissues of heterotrophs. There they remain until the organism dies. Then other microorganisms decompose the dead, metabolizing the organic materials and releasing CO_2—the reverse of photosynthesis. Waste products are also broken down by decomposers.

The release of CO_2 starts the cycle over again as primary producers fix CO_2 once more into organic material. A rough balance exists between CO_2 fixation and CO_2 release.

Many scientists are concerned by a growing imbalance in the carbon cycle due to the overabundance of CO_2 in the atmosphere. The burning of fossil fuels and wood sends tons of carbon into the atmosphere each year. Furthermore, in waterlogged soils, sewage treatment plants, landfills, and the digestive systems of ruminants, methanogens actively release methane gas (CH_4), which can be photochemically transformed into CO (carbon monoxide) and CO_2. Methane-oxidizing bacteria, living in conjunction with the methanogens, can also directly convert methane to CO_2. Overall, the rate of CO_2 production exceeds the rate at which it is being incorporated into organic material.

Further, carbon dioxide and methane are called "greenhouse gases" because their abundance in the atmosphere prevents the escape of infrared radiation into space and redirects it back toward the Earth, trapping heat within the atmosphere, much as the glass panes in a greenhouse trap heat within it. The retention of infrared radiation due to these gases adds to global warming, which can cause climatic changes.

The Nitrogen Cycle

Nitrogen is an important nutritional element required by organisms as a component of proteins, nucleic acids, and other nitrogen-containing compounds. Most nitrogen in the environment is in the atmosphere as nitrogen gas. Most organisms acquire nitrogen as part of organic molecules from other organisms or from soluble inorganic nitrogen compounds (nitrate and ammonia) found in limited quantities in soil and water. Microbes cycle nitrogen atoms from dead organic materials and animal wastes to these soluble forms for absorption by other organisms, and they cycle nitrogen between the biosphere and the atmosphere. **Figure 26.17** examines the basic processes involved in the **nitrogen cycle**—nitrogen fixation, ammonification, nitrification, and denitrification. It should be noted that this presentation is a simplified version of the complete cycle.

▲ *Figure 26.17*

Simplified nitrogen cycle. Even though nitrogen gas (N_2) is the most common form of nitrogen in the environment, most organisms cannot use it; to become available, nitrogen from the atmosphere must be incorporated into organic compounds by a very few species of nitrogen-fixing organisms.

Nitrogen fixation, a process whereby gaseous nitrogen (N_2) is reduced to ammonia (NH_3), is an energy-expensive process catalyzed by the enzyme *nitrogenase*. A limited number of prokaryotes, and no eukaryotes, fix nitrogen. Nitrogenase functions only in the complete absence of oxygen, a condition that presents no problem for anaerobes.

Aerobic nitrogen fixers, on the other hand, must protect nitrogenase from oxygen. Some aerobes use oxygen at such a high rate that it does not diffuse into the interior of the cell where nitrogenase is sequestered. Some nitrogen-fixing cyanobacteria form thick-walled, non-photosynthetic cells called *heterocysts* to protect nitrogenase from oxygen in the environment as well as from the oxygen generated during photosynthesis. Other cyanobacteria photosynthesize (and produce oxygen) during daylight hours and fix nitrogen only at night, thereby separating nitrogen fixation from photosynthesis in time rather than in space.

Nitrogen fixers may be free-living or symbiotic. Among the free-living nitrogen fixers are aerobic species of *Azotobacter* and anaerobic species of *Bacillus* and *Clostridium*. When they die, these soil-dwelling microbes release fixed nitrogen into the upper layers of soil where it then becomes available to plants and to animals that eat the plants. None of the free-living genera are directly associated with the plants they fertilize.

Symbiotic nitrogen fixers, in contrast, live in direct association with plants, forming *root nodules* on legumes (for example, peas and peanuts). The predominant genus of nitrogen-fixing, symbiotic bacteria is *Rhizobium* (rī-zō′bē-um). In the symbiosis, the plants provide nutrients and an anaerobic environment in the nodules, while the bacteria provide usable nitrogen to the plant. Farmers do not have to apply nitrogen fertilizers to legume crops because the nitrogen fixers provide enough nitrogen to facilitate plant growth.

Bacteria and fungi in the soil decompose wastes and dead organisms, disassembling proteins into their constituent amino acids, which then undergo *deamination* (removal of their amino groups). The amino groups are then converted to ammonia (NH_3), a process called **ammonification.** In dry or alkaline soils, NH_3 escapes as a gas into the atmosphere, but in moist soils, NH_3 is converted to ammonium ion (NH_4^+), which can be absorbed or oxidized to other forms.

In **nitrification,** NH_4^+ is oxidized to nitrate (NO_3^-) via a two-step process requiring two specific genera of autotrophic bacteria. *Nitrosomonas* spp. convert NH_4^+ to nitrite (NO_2^-), which is toxic to plants. Fortunately, *Nitrosomonas* spp. are usually found in association with species of *Nitrobacter* that rapidly convert nitrite to nitrate (NO_3^-), which is soluble and can be used by plants. The soluble nature of nitrate, however, means that it can be leached from the soil by water to accumulate in groundwater, lakes, and rivers. If converted back to nitrite, it can be toxic to plants. Loss of NO_3^- from soils also decreases soil fertility. Certain microorganisms in waterlogged soils perform **denitrification,** in which NO_3^- is oxidized to N_2 by anaerobic respiration. N_2 gas escapes into the atmosphere.

The Sulfur Cycle

The **sulfur cycle** involves moving sulfur between several oxidation states (**Figure 26.18**). Bacteria decompose dead

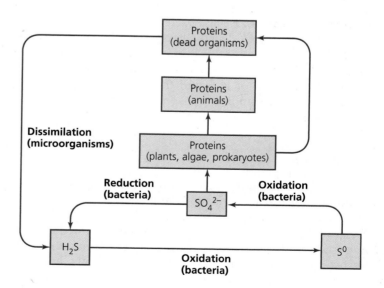

▲ *Figure 26.18*

Simplified sulfur cycle. The two main constituents of this cycle are hydrogen sulfide (H_2S) and sulfate (SO_4^{2-}), the fully reduced and oxidized forms of sulfur, respectively.

organisms, which releases sulfur-containing amino acids into the environment. Sulfur released from amino acids is converted to its most reduced form, hydrogen sulfide (H_2S), by microorganisms via a process called *sulfur dissimulation*. H_2S can be oxidized to elemental sulfur (S^0) and then to sulfate (SO_4^{2-}) under various conditions and by various organisms, including nonphotosynthetic autotrophs such as *Thiobacillus* and *Beggiatoa* (bej′-jē-a-tō-a) and photoautotrophic green and purple sulfur bacteria. Sulfate is the most readily usable form of sulfur for plants and algae, which animals then eat. Anaerobic respiration by the bacterium *Desulfovibrio* (dē′sul-fō-vib-rē-ō), reduces SO_4^{2-} back to H_2S. Thus, the two major inorganic constituents of the sulfur cycle are H_2S and SO_4^{2-}.

The Phosphorus Cycle

Unlike nitrogen and sulfur, phosphorus undergoes little change in oxidation state in the environment; it usually exists in the environment and is utilized by organisms as phosphate ion (PO_4^{3-}). The **phosphorus cycle** involves the movement of phosphorus from insoluble to soluble forms available for uptake by organisms and the conversion of phosphorus from organic to inorganic forms by pH dependent processes. No gaseous form exists to be lost to the atmosphere, but dissolved phosphates do accumulate in water, particularly the oceans, and organic forms of phosphorus are deposited in surface soils following the decomposition of dead animals and plants.

Too much phosphorus can be a problem in the environment; for example, agricultural fertilizers rich in phosphate (and nitrogen) are easily leached from fields by rain. The resulting runoff into rivers and lakes can result in **eutrophication** (ū′trof′ik-ā-shŭn), the overgrowth of microorganisms (particularly algae and cyanobacteria) in nutrient rich waters. Such overgrowth, called a *bloom,* depletes oxygen from the water killing aerobic organisms such as fish. Anaerobic organisms take over the water system, leading to an increased production of H_2S and the release of foul odors. When excess nitrogen and phosphate are removed, such a water system recovers over time.

The Cycling of Metals

Metal ions, including Fe^{2+}, Zn^{2+}, Cu^{2+}, Cd^{2+}, and Mg^{2+}, are important microbial nutrients. Though they are only needed in trace amounts, they can nonetheless be limiting factors in the growth of organisms. Many metals—including the most important trace metal, iron—are present in the environment in insoluble forms in rocks, soils, and sediments, and are thus unavailable for uptake by organisms. The cycling of these metals primarily involves a transition from an insoluble to a soluble form, allowing them to be used by organisms and thus to move through the environment.

▲ *Figure 26.19*

The soil layers, and the distributions of nutrients and microorganisms within them. Whereas topsoils in general are richer in nutrients and microbes than are subsoils, the nutrient and microbial content of topsoils is highly variable.

Soil Microbiology

Learning Objective

✓ Identify five factors affecting microbial abundance in soils.

Soil microbiology (or terrestrial microbiology) examines the roles played by organisms living in soil. Soil microorganisms are responsible for biogeochemical processes and are essential decomposers. They rarely cause human disease, though plant pathogens are prevalent in soils and are agriculturally and economically important.

The Nature of Soils

Soil arises both from the weathering of rocks and through the actions of microorganisms, which produce wastes and organic materials needed to support more complex life forms such as plants. Soil is composed of two major layers **(Figure 26.19):** *topsoil,* which is rich in *humus* (organic chemicals); and *subsoil* or *mineral soil,* which is composed primarily of inorganic materials. Soil overlies *bedrock,* which is solid rock and contains little organic material. The majority of microorganisms are found in topsoil, where the richness of the organic deposits sustain a large biomass. Topsoil itself, however, is highly heterogeneous, and therefore different kinds and amounts of microbes are found in various soils around the world.

Factors Affecting Microbial Abundance in Soils

Several environmental factors influence the density and the composition of the microbial population found within a soil, including water activity, oxygen content, acidity, temperature, and the availability of nutrients.

In soils, moisture and oxygen content are closely linked. Moisture is essential for microbial survival; microbes exhibit lower metabolic activity, are present in lower numbers, and are less diverse in dry soils than in moist soils.

Because oxygen dissolves poorly in water, moist soils have a lower oxygen content than dryer soils. When soil is waterlogged, microbial diversity declines and anaerobes predominate, even at the surface. Weather patterns also affect oxygen content, as the presence or absence of rainwater determines moisture, and thus dissolved oxygen.

The pH of a soil also determines in part whether it is rich in bacteria or rich in fungi. Highly acidic and highly basic soils favor fungi over bacteria, though typically fungi prefer acidic conditions.

Most soil organisms are mesophiles and prefer temperatures between 20°C and 50°C. Thus most soil microbes live quite well in areas where winters and summers are not too extreme. Whereas psychrophiles grow only in consistently cold environments and cannot survive in soils that experience spring thawing, the opposite is true for thermophiles, which cannot survive where winters are harsh.

Nutrient availability also affects microbial diversity in soil habitats. Most soil microbes are heterotrophic, utilizing organic matter in the soil. The size of a microbial community is determined more by the amount of organic material than by the kind of organic material: Any soil that has a relatively constant input of organic material, such as agricultural land, supports a wider array of microorganisms than soil that is more barren.

Microbial Populations in Soils

Because of the variety of soils, microbial populations differ tremendously from soil to soil, and even within the same soil over the course of a season. Wherever present, however, microbes perform a variety of necessary functions. As previously mentioned, they help cycle nitrogen, sulfur, phosphorus, and other elements, converting them into forms that other organisms can use. They degrade dead organisms and their wastes. Soil microbes can also clean up industrial pollutants, and they produce an incredible variety of compounds that have potential human uses. **Bio-mining** involves surveying natural populations of microbes for useful chemicals or biological functions that can be manipulated for such processes as bioremediation.

Bacteria are the most numerous and most diverse inhabitants of soil, and are found in all soil layers, where they often form biofilms. One reason bacteria are so numerous in soils is their greater ability to adapt to changing environmental conditions than other microorganisms. Actinomycetes (ak'ti-nō-mī-sēts) such as *Streptomyces,* which produces numerous antibiotics, are common in soils.

The fungi make up the next most populous group of microorganisms in soil, though free-living and symbiotic fungi are found only in topsoil. Within topsoil they can form large mycelia that cover several acres.

Soil algae live on or near the surface of topsoil because as photoautotrophs they require light. Protozoa are mobile and move through the soil, grazing on other microbes. For the most part, protozoa require oxygen and thus remain in the topsoil. Neither algae nor protozoa can withstand dramatic environmental changes or the introduction of pollutants.

Although viruses also are active within microorganisms living in the soil, they are rarely found free in the soil.

CRITICAL THINKING

Microbes are found mostly in topsoil, but some are found miles deep in bedrock. Nutritionally, how do these deeply buried microbes survive?

Soil-Borne Diseases in Humans and Plants

Although the vast majority of soil microorganisms are not human pathogens, there are some exceptions. Soil-borne infections in humans generally result from either direct contact with, ingestion of, or inhalation of microorganisms deposited in the soil in animal or human feces or urine. In some cases the microbes live and replicate in the soil, but in most cases soil is simply a vehicle for moving the pathogen from one host to another. The majority of soil-borne disease agents are fungal or bacterial. Few soil protozoa or viruses cause disease.

Human pathogens acquired from soil include the bacterium *Bacillus anthracis* (ba-sil'ŭs an-thrā'sis), the causative agent of anthrax, which produces endospores that are shed from the skins of infected livestock. Endospores may remain dormant in soil for decades or centuries. Disturbing the soil can lead to infection if endospores enter cuts or abrasions on the skin (cutaneous anthrax) or are inhaled into the lungs (inhalation anthrax).

Histoplasma capsulatum (his-tō-plaz'ma kap-soo-la'tum), is a fungus that causes the disease histoplasmosis, a serious respiratory tract infection. *Histoplasma* both grows in soil and is deposited there as spores in the droppings of infected birds and bats. The spores can be inhaled by humans when contaminated soil is disturbed.

Hantavirus pulmonary syndrome is a life-threatening, viral respiratory disease acquired through the inhalation of soil contaminated by mouse droppings and urine containing hantavirus. Hantaviruses have been found throughout North America.

Plant diseases are also important to human welfare, and soil contains far more plant pathogens than human pathogens. Microbial plant infections are generally characterized by one or more of the following signs: necrosis (rot), cankers/lesions, wilt (droopiness), blight (loss of foliage), galls (tumors), stunted growth or excessive growth, or bleaching (loss of chlorophyll). Bacteria, fungi, and viruses all cause diseases in plants, and can be spread either as airborne spores, through the roots or wounds, or by insects.

Table 26.5	Selected Soil-Borne Diseases of Humans and Plants	
Microorganism	Host	Disease
Bacteria		
Bacillus anthracis	Humans	Anthrax
Clostridium tetani	Humans	Tetanus
Agrobacterium tumefaciens	Plants	Crown gall disease
Ralstonia solanacearum	Plants	Potato wilt
Streptomyces scabies	Plants	Potato scab
Fungi		
Histoplasma capsulatum	Humans	Histoplasmosis
Blastomyces dermatididis	Humans	Blastomycosis
Coccidioides immitis	Humans	Coccidioidomycosis
Polymyxa spp.	Plants	Root rot in cereals
Fusarium oxysporum	Plants	Root rot in many plants
Phytophthora cinnamomi	Plants	Potato blight; root rot in many plants
Viruses		
Hantavirus	Humans	Hantavirus pulmonary syndrome
Tobacco mosaic virus	Plants	Necrotic spots in various plants
Wheat soil-borne mosaic virus	Plants	Mosaic disease in winter wheat and barley

Table 26.5 lists selected bacterial, fungal, and viral soil-borne diseases of humans and plants.

Aquatic Microbiology

Learning Objectives

✓ Compare the characteristics and microbial populations of freshwater and marine ecosystems.

✓ Describe several microbial illnesses that can be acquired from terrestrial and aquatic environments.

Aquatic microbiologists study the microorganisms living in freshwater and marine environments. Compared to soil habitats, water ecosystems support fewer microbes overall because nutrients are diluted. Most organisms that live in aquatic systems exist in biofilms attached to surfaces. Biofilms allow organisms in aquatic habitats to concentrate enough nutrients to sustain growth; without forming biofilms, they would starve.

Types of Aquatic Habitats

Aquatic habitats are divided into freshwater and marine systems. *Freshwater* systems, which are characterized by low salt content (about 0.05%), include groundwater, water from deep wells and springs, and surface water in the form of lakes, streams, rivers, shallow wells, and springs. *Marine* environments, characterized by a salt content of about 3.5%, encompass the open ocean and coastal waters such as bays, estuaries, and lagoons.

Natural aquatic systems are greatly affected by the release of so-called *domestic water*, which is water resulting from the treatment of sewage and industrial waste. Domestic water released into the environment affects water chemistry and the microorganism content of the water. Changing levels of chemicals can cause increases or decreases in microbial numbers. Furthermore, faulty treatment of sewage can lead to contamination of natural water systems with pathogenic microorganisms.

Freshwater Ecosystems Microorganisms become distributed vertically within lake systems according to oxygen availability, light intensity, and temperature. Surface waters are high in oxygen, well lit, and warmer than deeper waters. In large lakes, wave action continually mixes nutrients, oxygen, and organisms, which allows efficient utilization of resources. In stagnant waters, oxygen is readily depleted, resulting in more anaerobic metabolism and poorer water quality.

Deep lakes can be divided into four zones **(Figure 26.20a).** The **littoral** (li'ter-al) **zone** is the area along the shoreline, where nutrients from runoff and plant material enter the lake. The zone is shallow, and light penetrates deeply. The **limnetic** (lim-net'ik) **zone** is the upper layer of water away from the shore. Photoautotrophs reside here and in the littoral zone. The **profundal** (prō-fun'dal) **zone** is the deeper water beneath the limnetic zone. It has a lower oxygen content and more diffuse light. Some photosynthetic organisms, such as purple and green sulfur bacteria,

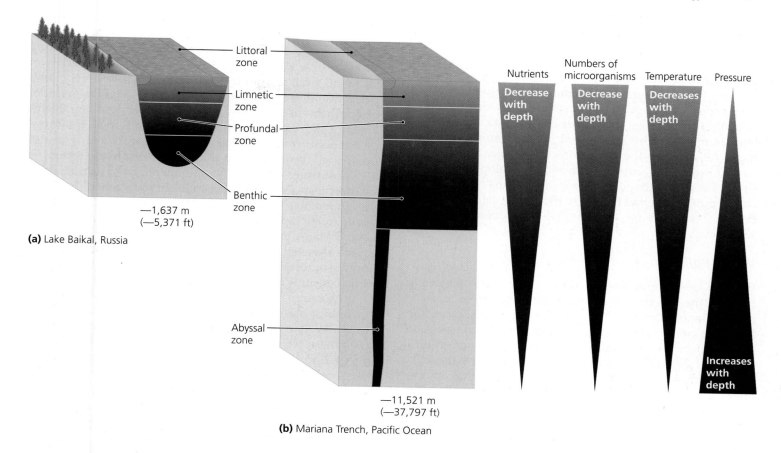

▲ *Figure 26.20*

Vertical zonation in deep bodies of water: **(a)** Lake Baikal, Russia, the world's deepest freshwater lake, and **(b)** the Mariana Trench in the Pacific Ocean, the deepest place in the ocean. Both deep lakes and oceans can be divided into zones that vary with respect to light penetration, concentration of nutrients, temperature, and pressure—and thus the types and abundance of microorganisms. *Why would a bacterium from the bottom of Lake Baikal die in the Mariana Trench?*

Figure 26.20 A bacterium from the bottom of Lake Baikal could not survive the even greater pressure and the salinity of the seawater deep in the Mariana Trench.

perform anaerobic photosynthesis here. Beneath this is the **benthic** (bĕn'thik) **zone,** which encompasses deeper lake waters and sediments. Anaerobic bacteria in the sediments produce H_2S, which is used by organisms nearer the surface.

Unlike lakes, streams and rivers usually lack stratification because organisms and nutrients are swept along and mixed. Biofilms are particularly important in moving waterways, and the majority of organisms live toward the edges of the waterway, where currents are less severe and organic materials enter.

Marine Ecosystems   Marine ecosystems are largely nutrient poor, dark, cold, and subject to great pressures. Microorganisms, mostly photoautotrophic diatoms, dinoflagellates, and algae, are found only near the surface of the ocean. Most marine waters, in fact, are extreme environments inhabited only by highly specialized microorganisms. All microbes in marine systems must be salt tolerant and possess highly efficient nutrient-uptake mechanisms to compensate for the scarcity of nutrients.

Like freshwater lake systems, oceans can be delineated into zones **(Figure 26.20b).** The majority of microorganisms in marine environments are found in the littoral zone along the shore, where nutrient levels are high and light is available for photosynthesis. The benthic zone makes up the majority of the marine environment. The **abyssal** (a-bis'sal) **zone** encompasses deep ocean trenches. Even though the latter two zones are largely devoid of nutrients, they still support some microbial growth, particularly around **hydrothermal vents** located in the abyssal zone. Such vents spew super-heated, nutrient-rich water, providing nutrients and an energy source for thermophilic chemoautotrophic anaerobes, which support in turn a variety of invertebrate and vertebrate animals.

Specialized Novel Aquatic Ecosystems   In addition to the two broad categories of water systems just described, many specialized aquatic ecosystems also exist, including salt lakes, iron springs, and sulfur springs. Each of these systems is inhabited by specialized microorganisms that are highly adapted to the conditions. The Great Salt Lake in Utah, for example, contains the extreme halophile *Halobacterium salinarium* (ha-lō-bak-tēr-ē-ŭm sa-li-nā′rēŭm), an archaeal prokaryote that thrives in the highly saline waters of the lake, which range from 5–27% salt depending on the water level.

Water Pollution

Water becomes polluted in three basic ways: *physically,* through the presence of oil, particulate matter, or an overabundance of microbial or vegetative organisms; *chemically,* from the presence of inorganic and organic compounds, usually derived from industrial activities or agricultural runoff; and *biologically,* through the presence of non-native microorganisms. *Polluted waters* are obvious in that one can either see, taste, or smell the pollutant. *Contaminated waters,* by contrast, contain physical, chemical, or biological pollutants that are not readily visible. Most people wouldn't use polluted water, but contaminated water is more difficult to avoid.

"Clean" water is generally low in nutrients (that is, it contains few organic molecules) and thus does not readily support microbial growth. Microbes that are present come from soil runoff or are naturally present primary producers. Polluted waters, in contrast, are high in organic compounds (particularly nitrogen and phosphorus) and support a greater than normal microbial load. Pollution is generally more problematic in closed lake systems (those that lack a constant influx and efflux of water), but slow-moving rivers and coastal waters can also experience problems with microbial overgrowth due to the dumping of industrial and sewage wastes.

CRITICAL THINKING

Following eutrophication of a lake, microbes reproduce prolifically and then die and sink to the bottom to "feed" anaerobes in the sediments. Even though the surface water may then look clear, why is it still unsafe to drink?

Although physical and chemical pollution are important, of greater concern is biological contamination of water with human pathogens, which may be present without visible sign. They nevertheless can cause significant human diseases that can be prevented only through vigilant treatment of water before consumption.

Water-Borne Illnesses

Consumption of contaminated water, either as drinking water or in water added to foods, can result in a variety of bacterial, viral, and protozoan diseases. Each year contaminated drinking water results in roughly 3–5 billion episodes of diarrheal diseases worldwide, and over 3 million deaths among children ages 5 years or younger. Fungal infections do not occur from water consumption. Intoxication can also occur from the presence of microbial toxins in the water.

Most water-borne pathogens are removed during water treatment, and so water-borne illnesses are generally rare in the United States as compared to countries with inadequate water treatment facilities. Outbreaks that do occur in the United States are *point-source infections,* in which a single source of water has become contaminated leading to a self-limiting outbreak of illness only in those individuals that directly consume the water.

In some marine environments, toxins are produced by eutrophic blooms of marine dinoflagellates such as *Gonyaulax* (gon′ē-o′laks) and *Gambierdiscus* (gam′bē-er-dis-kŭs). These blooms are called *red tides* because of the color imparted to the water by the huge numbers of dinoflagellates. In these cases the toxins are absorbed by shellfish, and human intoxication results from food consumption rather than water consumption.

Table 26.6 lists some of the common human water-borne pathogens and toxins.

Biological Warfare and Bioterrorism

The many properties of microorganisms that allow scientists to manipulate organisms to make advancements in medicine, food production, and industry also enables humans to fashion microbes into *biological weapons* that can be directed at people, livestock, and crops. **Bioterrorism,** the use of microbes or their toxins to terrorize human populations, is a topic of major concern in today's world. Another topic of growing concern is **agroterrorism,** the use of microbes to terrorize humans by destroying the livestock and crops that constitute their food supply.

Assessing Microorganisms as Potential Agents of Warfare or Terror

Learning Objectives

✓ Identify the criteria used to assess microorganisms for potential use as biological weapons or agents of bioterrorism.

✓ List the characteristics of microbes that make them greater threats as agents of biological warfare and bioterrorism.

Many microorganisms cause disease, but not all disease-causing organisms have potential as biological weapons. Recently, the federal government has established criteria for evaluating the potential of microorganisms to be

Table 26.6 **Selected Water-Borne Agents and the Diseases They Cause**

Organism	Disease
Bacteria	
Vibrio cholerae	Cholera
Campylobacter jejuni	Acute gastroenteritis
Salmonella spp.	Salmonellosis, typhoid fever
Shigella spp.	Shigellosis (bacterial dysentery)
Escherichia coli	Acute gastroenteritis
Viruses	
Hepatitis A virus	Infectious hepatitis
Norwalk virus	Acute gastroenteritis
Poliovirus	Poliomyelitis
Protozoa	
Giardia lamblia	Giardiasis
Schistosoma spp.	Schistosomiasis
Entamoeba histolytica	Amoebic dysentery
Toxin-Producers	
Gonyaulax (saxitoxin)	Paralytic shellfish poisoning
Gambierdiscus (ciguatoxin)	Fish poisoning

"weaponized"—to be used to cause significant loss of life or economic devastation. Establishing such criteria helps focus research and defense efforts where they are needed most, and facilitates efforts to develop better response and deterrence capabilities.

Criteria for Assessing Biological Threats to Humans

The assessment of a potential biological threat to humans is based on the following four criteria:

1. *Public health impact:* This criterion relates to the ability of hospitals and clinics to deal effectively with numerous casualties. The more casualties, the more difficult it is for hospitals and clinics to effectively respond to the needs of all patients. If an agent causes serious disease that could not be treated outside a hospital, emergency response systems could be overwhelmed and even cease to function. If an agent is highly lethal, proper disposal of bodies would become difficult, which would contribute to the spread of disease.

2. *Delivery potential:* This criterion evaluates how easily an agent can be introduced into a population. The more people that can be infected at one time, the more devastating a primary attack. If an agent spreads on its own once introduced, secondary and tertiary waves of illness will augment the number of casualties. Also assessed as part of delivery potential are ease of mass production (the easier an agent is to produce in quantity, the greater its potential threat), availability (the more prevalent it is

in the environment, the easier it is to obtain and weaponize), and environmental stability (the longer it remains infective once released, the greater the threat).

3. *Public perception:* This criterion evaluates the effect of public fear on the ability of response personnel to control a disease outbreak following an attack. Agents with high mortality, few treatment options, and no vaccine instill greater fear into a populace, making such control measures as quarantine difficult to enforce. The resulting chaos could dramatically decrease the ability of response personnel to control disease transmission and treat patients.

4. *Public health preparedness:* This criterion assesses existing response measures and attempts to identify improvements needed in the health care infrastructure to prepare for a biological attack. This criterion can be divided into two parts. Diagnosis/recognition involves proper surveillance and training medical personnel to ascertain whether an attack has occurred. Once an attack has been confirmed, predetermined, well-planned responses are required to reduce confusion and allow rapid control of the situation. Public health preparedness also involves funding for research and development of new vaccines, treatments, and diagnostic capabilities.

When assessing a threat, each agent is given a score for each of the criteria. Agents with the highest total scores are considered to be the most serious threats for use as weapons.

Criteria for Assessing Biological Threats to Livestock and Poultry

The criteria used to evaluate biological threats to livestock and poultry are very similar to those used to evaluate potential threats to humans and include agricultural impact, delivery potential, and plausible deniability.

Highly infectious agents prove the most devastating for agricultural livestock kept in large herds or flocks. Because many animal pathogens exist naturally in the soil or are already endemic in livestock and poultry, agents can be easily obtained and readily grown in large quantities. Because they are environmentally stable, such agents can persist even after a herd or flock is destroyed. Although vaccines exist for many potential pathogens, the highly infectious nature of some means that by the time the disease is recognized, much of a herd is already infected, and thus all must be destroyed in an attempt to control the outbreak. Many animal diseases are spread either by contact or inhalation, thus making attack easy for a terrorist. Though highly contagious among animals, most of the agents are not infectious to humans, making them "safe" for terrorists to handle.

Criteria for Assessing Biological Threats to Agricultural Crops

Plant diseases are generally not as contagious as animal or human diseases. Threats to crops are evaluated on predicted

Table 26.7	Bioterrorist Threats to Humans in Order of Concern	
Disease	Agent	Natural Source
Category A Threats: Highest Priority		
Smallpox	Variola major (virus)	None
Anthrax	*Bacillus anthracis* (bacterium)	Soil
Plague	*Yersinia pestis* (bacterium)	Small rodents
Botulism	*Clostridium botulinum* toxin (bacterial)	Soil
Tularemia	*Francisella tularensis* (bacterium)	Wild animals
Viral hemorrhagic fevers	Filoviruses and Arenaviruses	Unknown in most cases
Category B Threats: Moderate		
Q fever (fever + flu-like syndrome)	*Coxiella burnetii* (bacterium)	Sheep, goats, cattle
Brucellosis (severe flu-like syndrome)	*Brucella* spp. (bacteria)	Livestock
Glanders (pulmonary syndrome)	*Burkholderia mallei* (bacterium)	Horses
Melioidosis (severe pulmonary syndrome)	*Burkholderia pseudomallei* (bacterium)	Horses, livestock, rodents, soil
Encephalitis	Alphaviruses	Rodents, birds
Typhus fever	*Rickettsia prowazekii* (bacterium)	Humans
Toxins	Various bacteria	Various
Psittacosis (pneumonia-like syndrome)	*Chlamydia psittaci* (bacterium)	Birds
Food-borne diseases	Various bacteria and viruses	Soil or animals
Water-borne diseases	Various bacteria and viruses	Water
Category C Threats: Low Risk		
Nipah virus (encephalitis)	Paramyxovirus	Pigs
Hantavirus Pulmonary Syndrome	Hantavirus	Rodents, soil

extent of crop loss, delivery and dissemination potential, and containment potential.

Plant pathogens that either cause severe crop loss or produce toxins are considered the greatest threats. Most agents already exist in the environment and can be readily obtained; however, plant pathogens are not as easily mass-produced as animal agents. Infectious agents that can be spread systematically through fields by natural means, such as through contaminated soil or by insects, could remain in the environment even after destruction of the target crop. Successive plantings in contaminated fields would result in continued crop loss for as long as the agent persists. Because the causes of many plant diseases are not easily diagnosed, widespread dissemination of such a pathogen could occur before response measures were instituted. Economic losses would be staggering, particularly given that embargos on these crops would likely remain in place for years after the attack.

Known Microbial Threats

In the following subsections we briefly discuss some of the microorganisms currently considered threats as agents of bioterrorism. Note that as threat assessments become more refined and technology advances, the content of the lists of known bioterrorist agents is likely to change.

Human Pathogens

The government categorizes biological agents that could be used against humans into three categories (**Table 26.7**). *Category A agents* are those with the greatest potentials as weapons. *Category B agents* have some potential as weapons, but for various reasons are not as dangerous as category A agents (most lack the potential to cause mass casualties). *Category C agents* are potential threats; not enough is currently known about them to determine their true weapons-potential.

Smallpox currently tops the list of bioterrorist threats. Fortunately, it is difficult for would-be terrorists to acquire viral samples for propagation; it also takes a high degree of skill, in addition to specific containment facilities, to work with it. An effective vaccine is available, and the vaccine is effective when administered soon after infection.

Animal Pathogens

Biological agents against animals are also divided into categories, with category A agents being the most dangerous. Whereas many of the potential agents are spread via inhalation, others are spread by insect vectors, making them more difficult to use as weapons. Furthermore, some agents infect wild animal populations in addition to livestock, potentially amplifying any outbreak that might occur.

Foot-and-mouth disease virus, the most dangerous of these agents, affects all wild and domestic cloven-hoofed animals. The virus is spread by aerosols and by direct or indirect contact. Humans can transport it from herd to herd on their person, on farm equipment, or through the movement of animals between auctions and farms. Any appearance of foot and mouth disease on a farm requires destruction of entire herds, complete disinfection of all areas occupied by the herd, and disposal of all animals by burning and/or burial. A vaccine exists, but not in sufficient quantities to protect all animals.

Plant Pathogens

Many plant pathogens exist, but the categorization of plant pathogens as warfare or terrorist agents lags behind similar efforts for humans and animals. Most potential agents are fungi whose dissemination could easily result in contamination of soils, which would be difficult to neutralize. Attacks against staple crops (grains, corn, rice, and potatoes) are considered the most dangerous, as they would have significant negative impacts on national economies and food supplies. All of the agents are natural problems, so detecting the difference between a natural outbreak and an intentional attack would be difficult.

Defense Against Bioterrorism

No defense can completely prevent a carefully planned biological attack against any group or nation. However, much can be done either to decrease the chances that such an attack will be successful or to limit its scope once it occurs. The key is coupling *surveillance*—the active diagnosis and tracking of human, animal, and plant diseases—with *effective response protocols*.

Because Category A human biological agents are not common, the appearance of more than a few scattered cases is highly suggestive that an attack of some kind has occurred. This is one reason the diseases of category A are *reportable,* that is, they must be reported to state departments of health whenever they occur. Active monitoring of reportable diseases allows epidemiologists to quickly determine when unusual outbreaks are occurring. Once an unexpected pattern is seen, diagnostic confirmation can be obtained and, if an attack is deemed to have occurred, appropriate responses implemented **(Figure 26.21).** Such responses may entail forced quarantine (isolating those who are sick from the general population), distribution of antimicrobial drugs, or vaccination of those who have contacted the sick individuals.

Agroterrorism has become more of a concern with the realization that very little security protects the nation's agricultural enterprises. Livestock and poultry are routinely

▲ *Figure 26.21*

One aspect of the response to a bioterrorist attack. Shown here are biohazard-suited personnel investigating at the scene of anthrax-containing mail delivered in Washington, D.C., in October 2001.

moved around the country without being tested for disease and without being quarantined prior to introduction into new herds or flocks. Infected animals could therefore spread disease as they pass from facility to facility. Compounding the problem is the fact that farms, ranches, auction houses, livestock shows, and irrigation facilities are all open to the public and impose few security measures to prevent purposeful infection of animals. It has been suggested that a step in defending against agroterrorism would be to restrict public access to such facilities. Additionally, effective screening of imported animals and plants would help prevent the introduction of foreign pathogens into the United States. Further, better diagnostic techniques, vaccines, and treatments need to be developed for animal pathogens, and recombinant DNA technology could be used to create pathogen-resistant crops.

The Roles of Recombinant Genetic Technology in Bioterrorism

Recombinant genetic technology has a wide array of applications, most of them beneficial, but the techniques of gene manipulation could also be used to create new biological threats or modify existing ones, so that, for example, vaccines against them no longer work. Traits of various agents could be combined to create novel agents for which no immunity exists in the population.

In addition to the manipulation of existing threat agents, the techniques of recombinant DNA technology enable the synthesis of agents from scratch. Terrorists could, in theory, make their own microbes; to learn about the production of a completely manufactured poliovirus, see **New Frontiers 26.3.** While poliovirus is a relatively simple virus, and there

New Frontiers 26.3   Could Bioterrorists Manufacture Viruses from Scratch?

Researchers at the State University of New York at Stony Brook recently managed an alarming achievement: They synthesized a fully functional poliovirus from materials that can be readily obtained from any of a number of molecular biology supply companies.

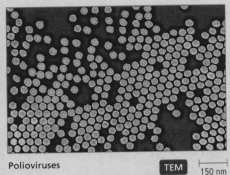

Polioviruses   TEM   150 nm

After they pieced together sequences of RNA to form a full-length poliovirus genome, they successfully replicated and translated this material in cell-free extracts in test tubes. The resulting nucleic acids and proteins were then able to assemble spontaneously into fully infectious viral agents. The scientists began their work from genetic blueprints that exist in the public domain—in published journal articles and on Internet databases.

The ability to manufacture an infectious agent from scratch using preexisting, published knowledge is an unsettling development. If scientists can independently manufacture an infectious poliovirus by purchasing materials and accessing information openly available to any determined individual, then terrorists may be able to manufacture their own agents in similar fashion—rather than needing to steal agents from research facilities or isolate them from natural sources.

As a result of studies like that at Stony Brook, an ethical debate has arisen over whether such research should be pursued—and if so, whether the details of such research should be published. Some argue that the pursuit and publication of such research unwittingly aids would-be terrorists; others argue that the dissemination of this information is necessary for the effective sharing of research within the scientific community, and for science to progress. What do you think?

---

is no guarantee that such a process would work for more complex agents such as smallpox, the threat nevertheless remains.

The techniques of recombinant genetic technology may also be used to thwart bioterrorism. Scientists can identify unique genetic sequences—"fingerprints" or signatures—which may aid in tracking biological agents and determining their source. Further genetic techniques may help in developing vaccines and treatments.

CRITICAL THINKING

Compare human, animal, and plant pathogens that could be used as biological agents in terms of environmental survivability. Why are animal and plant pathogens more common in environmental reservoirs than are human pathogens?

CHAPTER SUMMARY

Food Microbiology (pp. 744–752)

1. The commercial use of microorganisms is referred to as **applied microbiology** and includes two distinct fields: food microbiology and industrial microbiology.

2. **Food microbiology** involves the use of microorganisms in food production and the prevention of food-borne illnesses. **Fermentations** involve desirable changes to food; **spoilage** involves undesirable changes to food.

3. Food fermentations involve the use of **starter cultures**—known organisms that carry out specific and reproducible fermentation reactions. For most of the great variety of fermented vegetables, meats, and dairy products, starter cultures of lactic acid bacteria are used. The acid produced results in "sour" flavors.

4. **Alcoholic fermentations,** usually performed by yeasts, convert sugars to alcohols. Alcoholic fermentation is used in the production of wine, distilled spirits, beer, vinegar, and bread.

5. Intrinsic factors of food spoilage are properties of the food itself, such as moisture content and physical structure, that de-

termine how susceptible a food is to spoilage. Extrinsic factors of food spoilage include ways in which the food is handled.

6. Industrial processes preserve food via canning, pasteurization, drying, freeze-drying **(lyophilization),** irradiation, and aseptic packaging techniques. Natural and artificial preservatives are added to some foods to inhibit microbial growth. In stores and at home, foods should be properly stored in appropriate containers, and cold foods should be kept cold. Consumers should cook foods thoroughly and then refrigerate leftovers to reduce spoilage.

7. Food poisoning is a general term that describes instances of **food infections** (illnesses due to the consumption of living microbes) or **food intoxications** (illnesses due to the consumption of microbial toxins). Food poisoning frequently follows poor food handling.

Industrial Microbiology (pp. 752–763)

1. **Industrial microbiology** is concerned with the use of microorganisms for the production of commercially valuable materials. It relies on industrial-scale (large-scale) fermentations of

desired products and the use of genetically modified microbes designed to perform specific functions.

2. Batch production is the growth of organisms followed by harvesting of the entire culture and its products. Continuous flow production involves the constant addition of nutrients to a culture and the removal of the products formed. Products may either be primary metabolites (produced during active growth) or secondary metabolites (produced during the stationary phase).

3. Microorganisms produce a variety of useful products, including enzymes, dyes, alternative fuels, plastics, pharmaceuticals, pesticides, biosensors, and bioreporters. Alternative fuels are produced via the process of **bioconversion,** the transformation of biomass into fuel. **Biosensors** and **bioreporters** use microbes and electronics to detect microbial activity in the environment.

4. **Potable** drinking water is derived via water treatment, which involves the removal of organic and inorganic contaminants and microorganisms from the water. Polluted water contains organisms or chemicals at unsafe levels.

5. Water treatment involves three steps: sedimentation and flocculation, filtration, and disinfection. Sedimentation removes large materials through flocculation, in which alum complexes with materials to make them precipitate. Filtration, using either slow or rapid sand filters, removes microorganisms and chemicals. Disinfection, usually chlorination, kills most microbes that remain after filtration. Water is potable following treatment if it has zero coliforms per 100 ml of water, as determined by one of several testing methods (MPN, membrane filtration, ONPG/MUG test).

6. **Wastewater** refers to water used for washing or flushed from toilets. Wastewater treatment involves the removal of solids, organic chemicals, and microorganisms. **BOD (biochemical oxygen demand)** is a measure of the amount of oxygen required to fully metabolize organic wastes.

7. Municipal wastewater treatment involves four phases. Primary treatment sediments large materials and involves flocculation. Secondary treatment involves some sedimentation as well as the removal of microorganisms and organic material using activated sludge systems or trickle filters. In the third phase, effluent water is chemically treated (chlorinated) and released. In the fourth phase, primary and secondary sludge is digested and dried to produce landfill. Methane gas can also be recovered during the processing of sludge.

8. **Septic tanks** and **cesspools** are home equivalents of municipal wastewater treatment. After wastewater leaves the home and is deposited in underground tanks, sludge settles and the water is released into the soil, where natural processes remove organic chemicals and microorganisms.

9. **Oxidation lagoons** are used by farmers and ranchers to process animal wastes. Waste is pumped into successive lagoons, where wastes are digested by soil microorganisms and algae prior to the release of water into natural water systems.

10. In **artificial wetlands,** found in some planned communities and industrial sites, successive ponds, marshes, and meadowland remove organic compounds, chemicals, and microorganisms from sewage as the water moves through the wetlands.

11. **Bioremediation** is the use of microorganisms to metabolize toxins in the environment to reclaim soils and waterways. Industrial products are either biodegradable or recalcitrant (resistant to degradation by natural means). Recombinant DNA technology, however, enables some microbes to degrade some recalcitrant chemicals.

12. Acid mine drainage is an environmental problem in places where ores are heavy in iron. Microbial action on leached iron in water effluent from mines results in the production of acid and ferric iron deposits that reduce the pH of waterways to levels that are destructive to plant and animal life.

Environmental Microbiology (pp. 763–772)

1. Microorganisms live in microhabitats (small places) within larger habitats, or physical localities, in the environment. All organisms and habitats together form ecosystems. Single cells give rise to populations; populations performing similar functions form guilds; and many guilds together form a community of organisms living in the habitat. The study of the interactions of microorganisms among themselves and their environment is **microbial ecology.**

2. All the ecosystems on Earth form the biosphere. **Biodiversity** describes the number of species living in a given ecosystem, whereas **biomass** refers to the quantity of all these species.

3. Microbes compete for the scarce resources that characterize the majority of habitats. Some microbes actively oppose the growth of other microbes (antagonism), but many microbes cooperate, forming complex biofilms.

4. **Biogeochemical cycling** involves the movement of elements and nutrients from unusable forms to useable forms due to the activities of microorganisms. This process involves production of new biomass, consumption of existing biomass, and decomposition of dead biomass for reuse in the cycle. The four major biogeochemical cycles are the **carbon, nitrogen, sulfur,** and **phosphorus cycles.** The cycling of trace metals is also of importance.

5. CO_2 is fixed by photoautotrophs and chemoautotrophs into organic molecules, which are used by other organisms. Organic carbon is converted back to CO_2 in aerobic respiration, by decomposition, and by combustion.

6. In the **nitrogen cycle,** nitrogen gas in the atmosphere is converted to ammonia via a process called **nitrogen fixation.** Ammonia may be converted to nitrate via a two-step process called **nitrification.** Organisms also use ammonia and nitrate to make nitrogenous compounds. Such compounds in wastes and dead cells are converted back to ammonia via **ammonification.** Nitrate can be converted to nitrogen gas by **denitrification.**

7. Sulphur is cycled among several inorganic oxidation states (primarily H_2S, SO_4^{2-}, and S^0) and proteins.

8. The phosphorus cycle involves the conversion of PO_4^{2-} among organic and inorganic forms.

9. **Eutrophication**—the overgrowth of microorganisms in aquatic systems—can result from the presence of excess nitrogen and phosphorus, which act as fertilizers. The overgrowth of microbes depletes the oxygen in the water, resulting in the death of fish and other animals.

10. **Soil microbiology** is the study of the roles of microbes in the soil. Soils are fairly diverse and differ greatly in nutrients, water content, pH, oxygen content, and temperature. Microbes inhabit topsoil in high numbers and are less abundant in deeper rock and sediments. **Bio-mining** is the science involved in searching natural populations of soil microbes for useful organisms. The pathogenic microorganisms found in soil can be acquired through contact with contaminated soil, but often disease follows the consumption of food contaminated by soil or soil organisms.

11. Aquatic habitats include freshwater and marine water systems. Scientists recognize four zones in freshwater based on temperature, light, and nutrient levels: the **littoral zone** along the shore, the sunlit **limnetic zone** at and near the surface, the **profundal zone** just below the limnetic zone, and the **benthic zone** on the bottom, which is devoid of light and nutrients. Marine environments have the same four zones plus an **abyssal zone** (below the benthic zone), which is virtually devoid of life except around **hydrothermal vents.**

12. Microorganisms living in aquatic environments typically form biofilms to better accumulate nutrients that are limiting in most nonpolluted water systems. Water pollution can entail both microbial contamination and chemical contamination. Some illnesses are associated with consumption of microbes in water.

Biological Warfare and Bioterrorism (pp. 772–776)

1. Of the relatively few microorganisms that can cause disease in humans, animals, and plants, some can be used to purposely infect individuals and are thus potential agents of biological warfare and **bioterrorism. Agroterrorism** is the deliberate infection of livestock or crops.

2. The degree to which an organism is considered a biological threat depends on several criteria concerning public health impact, dissemination or delivery potential, public perception, and public health preparedness.

3. Human, animal, and plant pathogens are categorized by threat level. Table 26.7 lists microorganisms considered to be significant threats for use as biological agents.

4. Defense against bioterrorism begins with surveillance, the reporting and monitoring required for effective response to biological attacks. Diagnoses must be reliable, and control measures of some kind must exist and be able to be efficiently administered to limit the impact of any attack that occurs.

5. Recombinant genetic technology could potentially lead to the development of novel agents or the modification of existing agents to make them more difficult to control in the event of an attack. Such technology could also lead to potential vaccines and cures.

QUESTIONS FOR REVIEW

(Answers to multiple choice, True/False, matching questions, and fill in the blanks are on the web, along with additional review questions. Visit www.microbiologyplace.com.)

Multiple Choice

1. Food fermentations do all of the following *except*
   a. give foods a characteristic taste.
   b. lower the risk of food spoilage.
   c. sterilize foods.
   d. increase the shelf life of the food.

2. Commercially produced beers and wines are usually fermented with the aid of
   a. naturally occurring bacteria.
   b. naturally occurring yeast.
   c. specific cultured bacteria.
   d. specific cultured yeast.

3. Which of the following lists foods in order, from perishable to nonperishable?
   a. pasta, cheese, fruit, uncooked ground beef
   b. pasta, fruit, uncooked ground beef, cheese
   c. uncooked ground beef, fruit, cheese, pasta
   d. uncooked ground beef, fruit, pasta, cheese

4. Which of the following would be the best growth medium to use for industrial fermentations?
   a. corn that could be sold commercially as food
   b. synthetic medium made by hand
   c. whey from cheese production
   d. brewing mash

5. Biodegradable plastics are made from which of the following microbial metabolites?
   a. sludge
   b. PHA
   c. BOD
   d. alum

6. Strains of the bacterium *Pseudomonas syringae* have been identified as being capable of
   a. producing plastics.
   b. producing alternative fuels.
   c. fermenting foods.
   d. preventing ice formation.

7. Which of the following is added during water or sewage treatment to promote flocculation?
   a. sludge
   b. BOD
   c. alum
   d. PHA

8. During chemical treatment of drinking water and wastewater, which of the following microbes is least likely to be inactivated or killed?
   a. algae
   b. bacterial endospores
   c. fungal spores
   d. viruses

9. In which step is most of the organic content of sewage removed?
   a. primary treatment
   b. secondary treatment
   c. tertiary treatment
   d. sludge treatment

10. Microbial communities are composed of
    a. single, pure populations.
    b. mixed populations of competing organisms.
    c. mixed populations of cooperating organisms.
    d. a biosphere.

11. In the environment, nutrients are generally
    a. limiting.
    b. present in excess.
    c. stable.
    d. artificially induced.

12. Most of the material cycled by microbes exists in the environment as
    a. usable forms in soil and rock.
    b. usable forms in water.
    c. unusable forms in soil and rock.
    d. unusable forms in water.

13. In the carbon cycle, microbes
    a. convert CO_2 into organic material for consumption.
    b. convert CO_2 into inorganic material for storage.
    c. convert fossil fuels into usable organic compounds.
    d. convert oxygen into water as a by-product of photosynthesis.

14. Nitrification
    a. converts organic nitrogen to NH_3.
    b. converts NH_3 to NH_4^+.
    c. converts NH_4^+ to NO_3^-.
    d. converts NO_3^- to N_2.

15. In deep aquatic environments, most microbial life is found in the
    a. littoral zone.
    b. limnetic zone.
    c. profundal zone.
    d. benthic zone.
    e. abyssal zone.

16. Which of the following diseases is *not* caused by category A biological weapons agents?
    a. Smallpox
    b. Plague
    c. Q Fever
    d. Tularemia

17. Of the following characteristics, which would contribute most to making a microorganism an effective biological warfare agent?
    a. Is readily available in the environment
    b. Can be spread by contact after original dissemination
    c. Cannot be treated well outside of a hospital
    d. Is easily identified by symptoms

True/False

Indicate whether each of the following statements is true or false. Rewrite the phrase in italics to make any false statements true.

1. The fermentation of dairy products relies on *mixed acid* fermentation.

2. Sauerkraut production involves the *alcoholic* fermentation of cabbage.

3. Pasteurization kills *mesophilic* microorganisms except endospore formers.

4. Methane is a gas produced by microbial metabolism that can be used directly as a *fuel source*.

5. The treatment of drinking water and sewage are *similar* processes.

6. *Recalcitrant* molecules can be degraded by naturally occurring microorganisms.

7. Biofilms of microorganisms form in *aquatic* environments only.

8. *Cooperation* is common among microorganisms living in microhabitats.

9. Aquatic microorganisms are *more* prevalent near the surface than at the bottom of waterways.

Matching

Match each term with its correct definition.

1. ___ Organisms whose presence in water indicates contamination from feces or soil

2. ___ Compound produced by a bacterium that kills insects; the gene for it is inserted into plants to protect them

3. ___ Refers to water that is fit to drink; has been treated to reduce levels of microbes to undetectable levels

4. ___ Community of organisms surrounded by polysaccharides and attached to surfaces

5. ___ Used in the processing of animal wastes; mimics primary and secondary wastewater treatment

6. ___ Refers to compounds that are resistant to microbial degradation and are typically man-made

7. ___ Water that is not bound by solutes, which can be used by microorganisms

8. ___ Refers to quantity of all organisms present in an environment

9. ___ Process whereby pollutants accumulate to high amounts in waterways, causing overgrowth and anaerobic conditions

10. ___ Process of reducing nitrogen from the atmosphere

11. ___ The process whereby organisms actively inhibit the growth of other organisms

12. ___ Undesirable fermentation reactions in food leading to poor taste, smell, or appearance

13. ___ Brief heating of foods during processing, kills pathogenic organisms

14. ___ Descriptor of the level of organic material present in wastewater

15. ___ Fermentative products produced by microorganisms during stationary phase

A. Spoilage
B. Water activity
C. Coliforms
D. Pasteurization
E. Secondary metabolites
F. Bt-toxin
G. Potable
H. BOD
I. Oxidation lagoon
J. Recalcitrant
K. Biomass
L. Antagonism
M. Nitrogen fixation
N. Eutrophication
O. Biofilm

Fill in the Blank

1. Intrinsic factors affecting food spoilage are properties of _____ rather than _____ .

2. Leaving foods out at room temperature _____ the likelihood of food spoilage.

3. The two types of industrial fermentation equipment are designed for _____ production or _____ production.

4. Potable water is allowed to have _____ coliforms per 100 ml of water tested.

5. Leaching of compounds from mine tailings often results in the oxidation of two elements: _____ and _____ .

6. Biogeochemical cycling involves three primary steps: _____ , _____ , and _____ .

7. Nitrogen exists primarily as _____ in the environment; phosphorus exists as _____ .

CRITICAL THINKING

1. Why does the application of recombinant DNA technology to food production have the potential not only to enhance food quality, but also to enhance food output? Given that it has the potential to feed more people, why are some people opposed to genetically modified foods?

2. Given what you know about microbial nutrition and metabolism, explain why it is technically more difficult to achieve high yields of a secondary metabolite than of a primary metabolite.

3. Compare the types of alternative fuels that could be produced by microbes. Based on starting materials, which would provide the most renewable energy?

4. One way that farmers are attempting to prevent the development of widespread insect resistance to Bt-toxin is by planting non–Bt-producing crops in fields adjacent to Bt-containing crops that will be harvested. Insects can infest both fields, but the insects eating the non–Bt plants should survive at a higher rate than those in the Bt-producing field. Why would this prevent the dissemination of resistance to Bt-toxin among insects?

5. Even though water and wastewater undergo essentially the same forms of treatment, treated wastewater is not put into the water system from which drinking water is derived. Why not?

6. Take a critical look at the garbage in all of your wastebaskets. How much of the material present could possibly be degraded by microbes? How much could be recycled either at a recycling center or in compost? What is left if the degradable or recyclable materials were removed?

7. Given the amount of pollutants and disease-causing microbes that are in soil and water, why don't we see higher incidences of soil- and water-borne illnesses?

8. Explain why influenzaviruses could be potentially devastating biological weapons.

9. Explain why sake (rice wine) would be more accurately described as "rice beer."

10. "Cheap" bulk wines and wines that have been left exposed to air for too long often acquire a "vinegary" smell or taste. Explain why this is so.

APPENDIX A: Metabolic Pathways

Glycolysis (Embden-Meyerhof Pathway) A-2

Pentose-Phosphate Pathway A-4

Entner-Doudoroff Pathway A-5

Krebs Cycle A-6

Calvin-Benson Cycle A-7

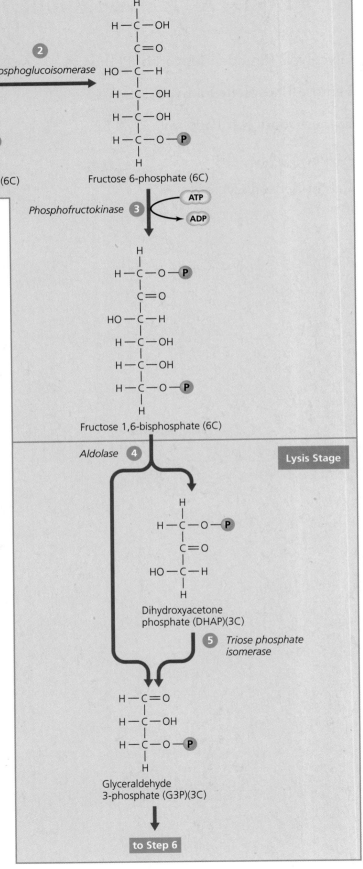

Energy Investment Stage

Glucose (6C)

Glucose 6-phosphate (6C)

Fructose 6-phosphate (6C)

Fructose 1,6-bisphosphate (6C)

Dihydroxyacetone phosphate (DHAP)(3C)

Glyceraldehyde 3-phosphate (G3P)(3C)

Lysis Stage

to Step 6

GLYCOLYSIS (Embden-Meyerhof Pathway):

Energy-Investment Stage

Step ①. *Hexokinase* transfers phosphate from ATP to carbon 6 of glucose, forming glucose 6-phosphate—a charged molecule. Since the cytoplasmic membrane is impermeable to ions, phosphorylated glucose cannot diffuse out of a cell.

Step ②. *Phosphoglucoisomerase* rearranges the atoms of glucose to form an isomer—fructose 6-phosphate.

Step ③. Phosphofructokinase invests more energy by adding another phosphate group from ATP to form fructose 1,6-bisphosphate.

Lysis Stage

Step ④. *Aldolase* splits six-carbon fructose 1,6-bisphosphate into three-carbon glyceraldehyde 3-phosphate (G3P) and three-carbon dihydroxyacetone phosphate (DHAP). It is this cleavage that gives glycolysis its name.

Step ⑤. *Triose phosphate isomerase* rearranges the atoms of DHAP to form another molecule of glyceraldehyde 3-phosphate. From this point, every step of glycolysis occurs twice—once for each of the three-carbon molecules.

Energy-Conserving Stage

Step 6. *Glyceraldehyde 3-phosphate dehydrogenase* catalyzes a reaction with two parts: (a) It oxidizes G3P, transferring electrons (and hydrogen) to NAD$^+$ to form NADH and (b) it adds inorganic phosphate from the cytosol to G3P with a high-energy bond to form 1,3-bisphosphoglyceric acid. This two-part reaction is among the more important in glycolysis because it generates a molecule of NADH for each molecule of G3P and because it creates the first high-energy intermediate.

Step 7. *Phosphoglycerate kinase* conserves the energy in the high-energy bonds in two molecules of ATP, yielding 3-phosphoglyceric acid molecules.

Step 8. *Phosphoglycerate mutase* rearranges the atoms to form 2-phosphoglyceric acid.

Step 9. *Enolase* removes a molecule of water from each substrate, forming a double bond and a high-energy bond with phosphate.

Step 10. *Pyruvate kinase* ends glycolysis by transferring energy to ATP, forming pyruvic acid. In the final analysis, two molecules of ATP are invested to yield four molecules of ATP—a net gain of two molecules of ATP—and two molecules of NADH. Pyruvic acid then undergoes respiration (when conditions are aerobic) or fermentation (when conditions are anaerobic).

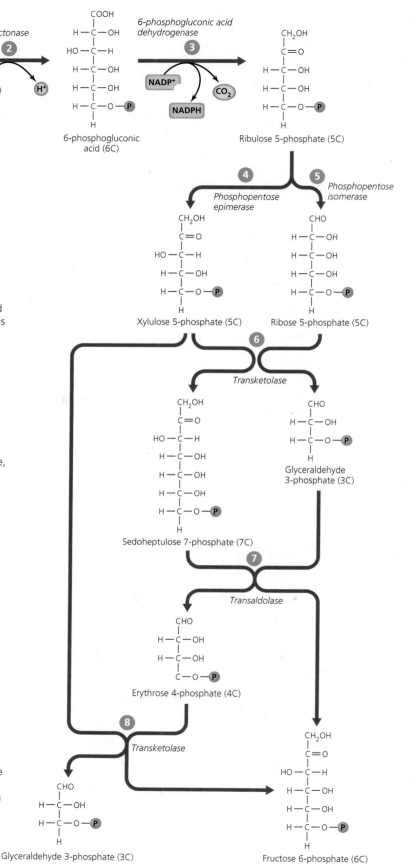

PENTOSE-PHOSPHATE PATHWAY

Step 1. *Glucose 6-phosphate dehydrogenase* oxidizes glucose 6-phosphate to 6-phosphogluconolactone by transferring hydrogen to NADP⁺, forming NADPH.

Step 2. *Lactonase* removes a water molecule, forming 6-phosphogluconic acid.

Step 3. *6-phosphogluconic acid dehydrogenase* oxidizes this intermediate (transferring hydrogen to another molecule of NADPH) and removes a molecule of carbon dioxide to form ribulose 5-phosphate. This five-carbon phosphorylated sugar is used in the synthesis of nucleotides, certain amino acids, and glucose (via photosynthesis).

Step 4. *Phosphopentose epimerase* converts some molecules of ribulose 5-phosphate to xylulose 5-phosphate.

Step 5. Simultaneously, *phosphopentose isomerase* converts other molecules of ribulose 5-phosphate to ribose 5-phosphate.

Step 6. *Transketolase* catalyzes a reaction in which a two-carbon fragment from xylulose 5-phosphate is transferred to ribose 5-phosphate, forming seven-carbon sedoheptulose 7-phosphate and three-carbon glyceraldehyde 3-phosphate (G3P).

Step 7. *Transaldolase* then transfers a three-carbon fragment from sedoheptulose 7-phosphate to G3P, forming four-carbon erythrose 4-phosphate and six-carbon fructose 6-phosphate.

Step 8. *Transketolase* transfers another two-carbon fragment from another molecule of xylulose 6-phosphate to erythrose 4-phosphate, forming another molecule of fructose 6-phosphate and another molecule of G3P. G3P enters glycolysis at step 6; fructose 6-phosphate can enter glycolysis at step 1 or may be converted into glucose 6-phosphate, which reenters the pentose-phosphate pathway.

Glucose 6-phosphate (6C)

Glucose 6-phosphate dehydrogenase **1** — NADP⁺ → NADPH

6-phosphogluconolactone (6C)

Lactonase **2** H_2O

6-phosphogluconic acid (6C)

3 — H_2O — 6-phosphogluconic acid dehydrase

2-keto-3-deoxy-6-phosphogluconic acid (6C)

4 KDPG aldolase

Glyceraldehyde 3-phosphate (3C)

5 via glycolysis steps 6–10

Pyruvic acid (3C) (2 molecules)

ENTNER-DOUDOROFF PATHWAY

The first two steps of the Entner-Doudoroff pathway are the same as the first two steps of the pentose-phosphate pathway:

Step 1. *Glucose 6-phosphate dehydrogenase* oxidizes glucose 6-phosphate to 6-phosphogluconolactone by transferring hydrogen to NADP⁺, forming NADPH.

Step 2. As in step two of the pentose phosphate pathway, *lactonase* removes a water molecule, forming 6-phosphogluconic acid.

Step 3. *6-phosphogluconic acid dehydrase* removes a molecule of water to form a six-carbon molecule. (The enzyme in this step should not be confused with the similarly named 6-phosphogluconic acid dehydrogenase of the pentose phosphate pathway.)

Step 4. *Aldolase* splits 2-keto-3-deoxy-6-phosphogluconic acid into two three-carbon compounds: pyruvic acid and glyceraldehyde 3-phosphate (G3P).

Step 5. G3P is converted to pyruvic acid by steps 6 through 10 of Embden-Meyerhof glycolysis.

KREBS CYCLE

In a complex reaction with several parts, pyruvate dehydrogenase decarboxylates and oxidizes pyruvic acid (from the Embden–Meyerhof or Entner–Doudoroff pathways) and adds the remaining two-carbon acetic acid to coenzyme-A, forming acetyl-CoA. It is this molecule that enters the Krebs cycle. Recall that for every molecule of glucose that enters glycolysis, two molecules of pyruvic acid are produced, necessitating two sets of Krebs cycle reactions.

Step 1. *Citrate synthase* adds the two carbons of acetic acid from acetyl-CoA to oxaloacetic acid, forming citric acid. This step gives the Krebs cycle its alternate name—the citric acid cycle.

Step 2. *Aconitase* forms an isomer, isocitric acid, by removing a molecule of water and then adding another back.

Step 3. *Isocitric acid dehydrogenase* oxidizes and decarboxylates six-carbon isocitric acid to five-carbon α-ketoglutaric acid; a molecule of NADH is formed in the process.

Step 4. *α-ketoglutaric acid dehydrogenase* oxidizes and decarboxylates α-ketoglutaric acid to form a four-carbon fragment that is attached to coenzyme-A, forming succinyl-CoA.

Step 5. *Succinyl-CoA synthetase* forms four-carbon succinic acid and simultaneously phosphorylates GDP to form GTP. The latter molecule subsequently phosphorylates ADP to ATP.

Step 6. *Succinic acid dehydrogenase* oxidizes succinic acid to four-carbon fumaric acid, forming a molecule of FADH₂.

Step 7. *Fumarase* rearranges the four carbons to form malic acid.

Step 8. *Malic acid dehydrogenase* oxidizes malic acid to reform oxaloacetic acid, completing the cycle and forming another molecule of NADH.

NADH and FADH₂ carry electrons to the electron transport chain.

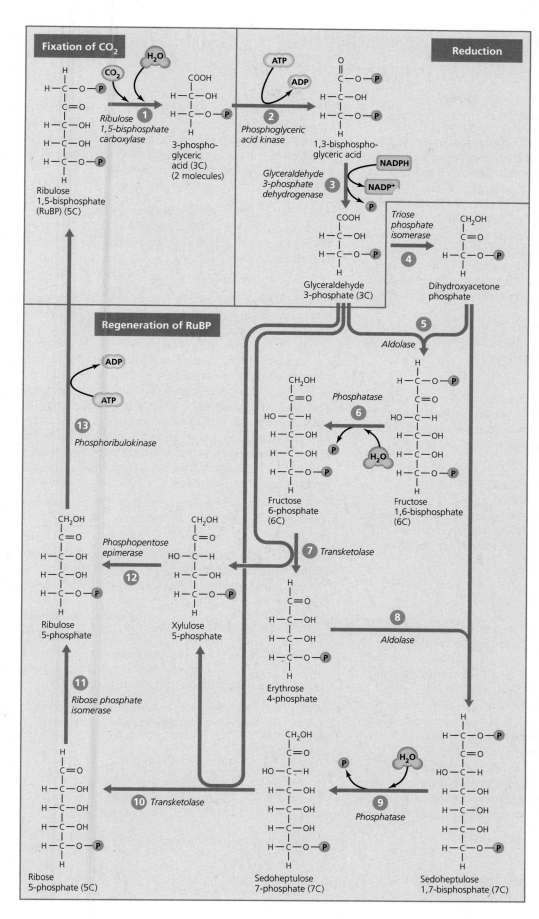

CALVIN-BENSON CYCLE

Fixation of CO₂

Step ❶. *Ribulose 1,5-bisphosphate carboxylase* adds CO_2 and H_2O to five-carbon ribulose 1,5-bisphosphate (RuBP), forming a six-carbon intermediate (not shown) that is immediately split into two, three-carbon molecules of 3-phosphoglyceric acid.

Reduction

Step ❷. *Phosphoglyceric acid kinase* phosphorylates phosphoglyceric acid at the expense of ATP to form 1,3-bisphosphoglyceric acid.

Step ❸. *Glyceraldehyde 3-phosphate dehydrogenase* reduces (using NADPH) and removes one phosphate from 1,3-bisphosphoglyceric acid, forming glyceraldehyde 3-phosphate (G3P).

Regeneration of RuBP

Step ❹. *Triose phosphate isomerase* changes some molecules of G3P to dihydroxyacetone phosphate (DHAP).

Step ❺. *Aldolase* combines a molecule of G3P and a molecule of DHAP to form fructose 1,6-bisphosphate.

Step ❻. *Phosphatase* adds H_2O and removes a phosphate group, forming fructose 6-phosphate.

Step ❼. *Transketolase* combines fructose 6-phosphate with a molecule of G3P (from step 3) to form five-carbon xylulose 5-phosphate and four-carbon erythrose 4-phosphate.

Step ❽. *Aldolase* rearranges the carbons of erythrose 4-phosphate and of a molecule of DHAP (from step 4), forming sedoheptulose 1,7-bisphosphate.

Step ❾. *Phosphatase* adds water and removes a phosphate group to form sedoheptulose 7-phosphate.

Step ❿. *Transketolase* combines this molecule with a molecule of G3P (from step 3) to form five-carbon ribose 5-phosphate and five carbon xylulose 5-phosphate.

Step ⓫. *Ribose 5-phosphate isomerase* rearranges the atoms of ribose 5-phosphate to form its isomer—ribulose 5-phosphate.

Step ⓬. *Phosphatase epimerase* similarly rearranges the atoms of xylulose 5-phosphate, forming another molecule of ribulose 5-phosphate.

Step ⓭. *Phosphoribulokinase* phosphorylates ribulose 5-phosphate to regenerate ribulose bisphosphate (RuBP).

For every three molecules of CO_2 that enter the Calvin-Benson cycle (at step 1), one molecule of G3P leaves the cycle to be used for the synthesis of glucose via the reversal of the early reactions of glycolysis.

APPENDIX B: Some Mathematical Considerations in Microbiology

Scientific Notation

Scientific notation (also called exponential notation) is a mathematical device developed to express numbers, particularly very small and very large numbers, in a manner that is convenient and readable. For example, 1.234×10^{-8} is much less cumbersome than 0.00000001234. Similarly, 5.67×10^{17} is more easily read than 567,000,000,000,000,000.

A number expressed in scientific notation is composed of a *coefficient*, which is always a number with only one whole number digit to the left of the decimal place, times ten to some *exponential power*. In the examples above, the coefficients are 1.234 and 5.67 respectively. The exponents are -8 and 17 respectively.

To write a number in scientific notation, move the decimal point either right or left so that there is only one nonzero number to the left of it. For example,

454 becomes 4.54

In this case we moved the decimal point (which is understood to be at the end of a whole number) two places to the left. Since we moved two places left, the exponent will be positive 2, and the scientific notation is

$$4.54 \times 10^2$$

Similarly, 4,500,264.75 becomes 4.50026475×10^6 since we moved the decimal six places to the left.

As you would expect, when we have to move the decimal point to the right, as in 0.00562 (which becomes 5.62), then the exponent will be a negative number, in this case -3; therefore, $0.00562 = 5.62 \times 10^{-3}$.

Logarithms

A **logarithm** is the number of times a number, called the *base number*, must be multiplied by itself to get a certain number. For example, the number 10 must be multiplied by itself three times to get 1000 ($10 \times 10 \times 10$); therefore, the logarithm of 1000 in base 10 is 3. Similarly, the logarithm of 100,000 is 5. Though any number can be used as the base number, in microbiology, base 10 is commonly used. Microbiologists use base 10 logarithms to express pH values and the size of bacterial populations in culture.

When a number is expressed in scientific notation, the logarithm is the exponent when the coefficient is exactly 1. When the coefficient is a number other than 1, the logarithm must be calculated using the logarithm function on a calculator.

Generation Time

When bacteria and other microbes reproduce by binary fission, the number in the population doubles with each division cycle (generation). Such growth is called *logarithmic* (or *exponential*) *growth*.

The number in a population can be calculated as 2 (because each cell produces two offspring) multiplied by itself as many times as there are generations. In other words, the number of cells in a population arising from a single individual is expressed mathematically as

$$2^{\text{ number of generations}}$$

A cell dividing for five generations would produce a population of 32 ($2^5 = 2 \times 2 \times 2 \times 2 \times 2 = 32$). When the population begins with more than one individual, then the final population equals

Original number of cells $\times\ 2^{\text{ number of generations}}$

For example, seven cells reproducing for five generations would produce a population of 224 cells ($7 \times 2^5 = 7 \times 2 \times 2 \times 2 \times 2 \times 2 = 224$).

In most cases, microbiologists are not concerned with the number of generations required to produce a given population, but they do want to know the **generation time;** that is, the time required for a bacterial cell to grow and divide. This is also the time required for a population of cells to double in number. To calculate generation time, scientists must first calculate how many generations have been produced, knowing the beginning and ending population sizes. When population sizes are converted to logarithms, the number of generations is calculated as

$$\frac{\text{Number of}}{\text{generations}} = \frac{\begin{array}{c}\text{logarithm (log) number of cells} - \text{log number of cells}\\ \text{at the end of reproduction} \qquad \text{initially}\end{array}}{\log 2}$$

Log 2 is used in the formula because each cell produces two offspring each time it divides. Log $2 = 0.301$.

The number of generations is used to calculate generation time:

$$\text{Generation time (min/generation)} = \frac{60 \text{ minutes} \times \text{number of hours}}{\text{number of generations}}$$

For example, if 100 bacteria multiply to produce a population of 3.28×10^6 cells in seven hours, then the generation time for this bacterium is calculated as follows:

$$\frac{\text{Number of}}{\text{generations}} = \frac{\begin{array}{c}\text{logarithm (log) number of cells} - \text{log number of cells}\\ \text{at the end of reproduction} \qquad \text{initially}\end{array}}{\log 2}$$

$$= \frac{\log (3.28 \times 10^6) - \log (100)}{\log 2}$$

$$= 15 \text{ generations}$$

$$\text{Generation time (min/generation)} = \frac{60 \text{ minutes} \times \text{hours}}{\text{number of generations}}$$

$$= \frac{60 \text{ min} \times 7 \text{ h}}{15}$$

$$= 28 \text{ minutes/generation}$$

APPENDIX C: Classification of Bacteria According to *Bergey's Manual**

Domain: Archaea
Phylum I: Crenarchaeota
Class I: Thermoprotei
Order I: Thermoproteales
Family I: Thermoproteaceae
Caldivirga
Pyrobaculum
Thermocladium
Thermoproteus
Family II: Thermofilaceae
Thermofilum
Order II: Desulfurococcales
Family I: Desulfurococcaceae
Acidolobus
Aeropyrum
Desulfurococcus
Igniococcus
Staphylothermus
Stetteria
Sulfophobococcus
Thermodiscus
Thermosphaera
Family II: Pyrodictiaceae
Hyperthermus
Pyrodictium
Pyrolobus
Order III: Sulfolobales
Family I: Sulfolobaceae
Acidianus
Metallosphaera
Stygiolobus
Sulfolobus
Sulfurisphaera
Sulfurococcus
Phylum II: Euryarchaeota
Class I: Methanobacteria
Order I: Methanobacteriales
Family I: Methanobacteriaceae
Methanobacterium
Methanobrevibacter
Methanosphaera
Methanothermobacter
Family II: Methanothermaceae
Methanothermus

Class II: Methanococci
Order I: Methanococcales
Family I: Methanococcaceae
Methanococcus
Methanothermococcus
Family II:
Methanocaldococcaceae
Methanocaldococcus
Methanotorris
Order II: Methanomicrobiales
Family I: Methanomicrobiaceae
Methanoculleus
Methanogenium
Methanolacinia
Methanomicrobium
Methanoplanus
Methanofollis
Family II:
Methanocorpusculaceae
Methanocorpusculum
Family III: Methanospirillaceae
Methanospirillum
Genera incertae sedis
Methanocalculus
Order III: Methanosarcinales
Family I: Methanosarcinaceae
Methanococcoides
Methanohalobium
Methanohalophilus
Methanolobus
Methanomicrococcus
Methanosarcina
Methanosalsum
Family II: Methanosaetaceae
Methanosaeta
Class III: Halobacteria
Order I: Halobacteriales
Family I: Halobacteriaceae
Haloarcula
Halobacterium
Halobaculum
Halococcus
Haloferax
Halogeometricum
Halorhobdus
Halorubrum
Haloterrigena
Natrialba
Natrinema
Natronobacterium
Natronococcus

Natronomonas
Natronorubrum
Class IV: Thermoplasmata
Order I: Thermoplasmatales
Family I:
Thermoplasmataceae
Thermoplasma
Family II: Picrophilaceae
Picrophilus
Family III: Ferroplasmatacea
Ferroplasma
Class V: Thermococci
Order II: Thermococcales
Family I: Thermococcaceae
Palaeococcus
Pyrococcus
Thermococcus
Class VI: Archaeoglobi
Order I: Archaeoglobales
Family I: Archaeoglobaceae
Archaeoglobus
Ferroglobus
Class VII: Methanopyri
Order I: Methanopyrales
Family I: Methanopyraceae
Methanopyrus
Domain: Bacteria
Phylum I: Aquificae
Class I: Aquificae
Order I: Aquificales
Family I: Aquificaceae
Aquifex
Calderobacterium
Hydrogenobacter
Thermocrinis
Genera incertae sedis
Desulfurobacterium
Phylum II: Thermotogae
Class I: Thermotogae
Order I: Thermotogales
Family I: Thermotogaceae
Fervidobacterium
Geotoga
Petrotoga
Thermosipho
Thermotoga
Phylum III: Thermodesulfobacteria
Order I: Thermodesulfobacteriales
Family I:
Thermodesulfobacteriaceae
Thermodesulfobacterium

*Bergey's Manual of Systematic Bacteriology, 2nd ed., 5 vols. (2002), is the reference for classification. *Bergey's Manual of Determinative Bacteriology*, 9th ed. (1994), should be used for identification of culturable bacteria and archaea. For updates to *Bergey's Manual*, visit: http://www.springer-ny.com/bergeysoutline/main.htm

Phylum IV: Deinococcus-Thermus
Class I: Deinococci
Order I: Deinococcales
Family I: Deinococcaceae
Deinococcus
Order II: Thermales
Family I: Thermaceae
Meiothermus
Thermus
Phylum V: Chrysiogenetes
Class I: Chrysiogenetes
Order I: Chrysiogenales
Family I: Chrysiogenaceae
Chrysiogenes
Phylum VI: Chloroflexi
Class I: Chloroflexi
Order I: Chloroflexales
Family I: Chloroflexaceae
Chloroflexus
Chloronema
Heliothrix
Family II: Oscillochloridaceae
Oscillochloris
Order II: Herpetosiphonales
Family I: Herpetosiphonaceae
Herpetosiphon
Phylum VII: Thermomicrobia
Class I: Thermomicrobia
Order I: Thermomicrobiales
Family I: Thermomicrobiaceae
Thermomicrobium
Phylum VIII: Nitrospira
Class I: Nitrospira
Order I: Nitrospirales
Family I: Nitrospiraceae
Leptospirillum
Magnetobacterium
Nitrospira
Thermodesulfovibrio
Phylum IX: Deferribacteres
Class I: Deferribacteres
Order I: Deferribacterales
Family I: Deferrivacter
Deferribacter
Denitrovibrio
Flexistipes
Geovibrio
Genera incertae sedis
Synergistes
Phylum X: Cyanobacteria
Class I: Cyanobacteria
SubSection I
Chamaesiphon
Chroococcus
Cyanobacterium
Cyanobium
Cyanothece

Dactylococcopsis
Gloeobacter
Gloeocapsa
Gloeothece
Microcystis
Prochlorococcus
Prochloron
Synechococcus
Synechocystis
SubSection II
Chroococcidiopsis
Cyanocystis
Dermocarpella
Myxosarcina
Pleurocapsa
Stanieria
Xenococcus
SubSection III
Arthrospira
Borzia
Crinalium
Geitlerinema
Halospirulina
Leptolyngbia
Limnothrix
Lyngbya
Microcoleus
Oscillatoria
Planktothrix
Prochlorothrix
Pseudoanabaena
Spirulina
Starria
Symploca
Trichodesmium
Tychonema
Subsection IV
Anabaena
Anabaenopsis
Aphanizomenon
Calothrix
Cyanospira
Cylindrospermum
Nodularia
Nostoc
Rivularia
Scytonema
Tolypothrix
SubSection V
Chlorogloeopsis
Fischerella
Geitleria
Iyengariella
Nostochopsis
Stigonema
Phylum XI: Chlorobi
Class I: Chlorobia

Order I: Chlorobiales
Family I: Chlorobiaceae
Ancalochloris
Chlorobium
Chloroherpeton
Pelodictyon
Prosthecochloris
Phylum XII: Proteobacteria
Class I: Alpha-Proteobacteria
Order I: Rhodospirillales
Family I: Rhodosprillaceae
Azospirillum
Magnetospirillum
Phaeospirillum
Rhodocista
Rhodospira
Rhodospirillum
Rhodothalassium
Rhodovibrio
Roseospira
Shermanella
Family II: Acetobacteraceae
Acetobacter
Acidiphilium
Acidosphaera
Acidocella
Acidomonas
Asaia
Craurococcus
Gluconacetobacter
Gluconobacter
Paracraurococcus
Rhodopila
Roseococcus
Stella
Zavarzinia
Order II: Rickettsiales
Family I: Rickettsiaceae
Orientia
Rickettsia
Wolbachia
Family II: Ehrlichiaceae
Aegyptianella
Anaplasma
Cowdria
Ehrlichia
Neorickettsia
Xenohaliotis
Family III: Holosporaceae
Caedibacter
Holospora
Lyticum
Odyssella
Polynucleobacter
Pseudocaedibacter
Symbiotes
Tectibacter

Order III: Rhodobacterales
  Family I: Rhodobacteraceae
    Ahrensia
    Amaricoccus
    Antarctobacter
    Gemmobacter
    Hirschia
    Hyphomonas
    Maricaulis
    Methylarcula
    Octadecabacter
    Paracoccus
    Rhodobacter
    Rhodovulum
    Roseobacter
    Roseibium
    Roseinatronobacter
    Roseovarius
    Roseovivax
    Rubrimonas
    Ruegeria
    Sagittula
    Staleya
    Stappia
    Sulfitobacter
Order IV: Sphingomonadales
  Family I: Sphingomonodaceae
    Blastomonas
    Erythrobacter
    Erythromicrobium
    Erythromonas
    Porphyrobacter
    Rhizomonas
    Sandaracinobacter
    Sphingomonas
    Zymomonas
Order V: Caulobacterales
  Family I: Caulobacteraceae
    Asticcacaulis
    Brevundimonas
    Caulobacter
    Phenylobacterium
Order VI: Rhizobiales
  Family I: Rhizobiaceae
    Agrobacterium
    Carbophilus
    Chelatobacter
    Ensifer
    Rhizobium
    Sinorhizobium
  Family II: Bartonellaceae
    Bartonella
  Family III: Brucellaceae
    Brucella
    Mycoplana
    Ochrobactrum
  Family IV: Phyllobacteriaceae

    Allorhizobium
    Aminobacter
    Aquamicrobium
    Defluvibacter
    Mesorhizobium
    Phyllobacteriaceae
    Pseudaminobacter
  Family V: Methylocystaceae
    Methylocystis
    Methylopila
    Methylosinus
  Family VI: Beijerinckiaceae
    Beijerinckia
    Chelatococcus
    Derxia
  Family VII: Bradyrhizobiaceae
    Afipia
    Agromonas
    Blastobacter
    Bosea
    Bradyrhizobium
    Nitrobacter
    Oligotropha
    Rhodopseudomonas
  Family VIII:
  Hyphomicrobiaceae
    Ancalomicrobium
    Ancylobacter
    Angulomicrobium
    Aquabacter
    Azorhizobium
    Blastochloris
    Devosia
    Dichotomicrobium
    Filomicrobium
    Gemmiger
    Hyphomicrobium
    Labrys
    Methylorhabdus
    Pedomicrobium
    Prosthecomicrobium
    Rhodomicrobium
    Rhodoplanes
    Seliberia
    Starkya
    Xanthobacter
  Family IX:
  Methylobacteriaceae
    Methylobacterium
    Protomonas
    Roseomonas
  Family X: Rhodobiaceae
    Rhodobium
Class II: Beta-Proteobacteria
Order I: Burkholderiales
  Family I: Burkholderiaceae
    Burkholderia

    Cupriavidus
    Lautropia
    Pandoraea
    Thermothrix
  Family II: Ralstoniaceae
    Ralstonia
  Family III: Oxalobacteraceae
    Duganella
    Herbaspirillum
    Janthinobacterium
    Massilia
    Oxalobacter
    Telluria
  Family IV: Alcaligenaceae
    Achromobacter
    Alcaligenes
    Bordetella
    Pelistega
    Sutterella
    Taylorella
  Family V: Comamonadaceae
    Acidovorax
    Aquabacterium
    Brachymonas
    Comamonas
    Delftia
    Hydrogenophaga
    Ideonella
    Leptothrix
    Polaromonas
    Rhodoferax
    Rubrivivax
    Sphaerotilus
    Tepidimonas
    Thiomonas
    Variovorax
Order II: Hydrogenophilales
  Family I: Hydrogenophilus
    Hydrogenophaga
    Thiobacillus
Order III: Methylophilales
  Family I: Methylophilaceae
    Methylobacillus
    Methylophilus
    Methylovorus
Order IV: Neisseriales
  Family I: Neisseriaceae
    Alysiella
    Aquaspirillum
    Catenococcus
    Chromobacterium
    Eikenella
    Formivibrio
    Iodobacter
    Kingella
    Microvirgula
    Neisseria

Prolinoborus
Simonsiella
Vitreoscilla
Vogesella
Order V: Nitrosomonadales
  Family I: Nitrosomonadaceae
   Nitrosomonas
   Nitrosospira
  Family II: Spirillaceae
   Spirillum
  Family III: Gallionellacea
   Gallionella
Order VI: Rhodocyclales
  Family I: Rhodocyclaceae
   Azoarcus
   Azonectus
   Azospira
   Azovibrio
   Propionibacter
   Propionivibrio
   Rhodocyclus
   Thauera
   Zoogloea
Class III: Gammaproteobacteria
  Order I: Chromatiales
   Family I: Chromatiaceae
    Allochromatium
    Amoebobacter
    Chromatium
    Halochromatium
    Halothiobacillus
    Isochromatium
    Lamprobacter
    Lamprocystis
    Marichromatium
    Nitrosococcus
    Pfennigia
    Rhabdochromatium
    Thermochromatium
    Thioalkalicoccus
    Thiocapsa
    Thiococcus
    Thiocystis
    Thiodictyon
    Thiohalocapsa
    Thiolamprovum
    Thiopedia
    Thiorhodococcus
    Thiorhodovibrio
    Thiospirillum
   Family II:
   Ectothiorhodospiraceae
    Arhodomonas
    Ectothiorhodospira
    Halorhodospira
    Nitrococcus
    Thiorhodospira

Order II: Acidothiobacillales
  Family I: Acidothiobacillaceae
   Acidothiobacillus
Order III: Xanthomonadales
  Family I: Xanthomonadaceae
   Frateuria
   Lureimonas
   Lyssobacter
   Nevskia
   Pseudoxanthomonas
   Rhodanobacer
   Stenotrophomonas
   Xanthomonas
   Xylella
Order IV: Cardiobacteriales
  Family I: Cardiobacteriaceae
   Cardiobacterium
   Dichelobacter
   Suttonella
Order V: Thiotrichales
  Family I: Thiotrichaceae
   Achromatium
   Beggiatoa
   Leucothrix
   Macromonas
   Thiobacterium
   Thiomargarita
   Thioploca
   Thiospira
   Thiothrix
  Family II: Piscirickettsiaceae
   Cycloclasticus
   Hydrogenovibrio
   Methylophaga
   Piscirickettsia
   Thiomicrospira
  Family III: Francisellaceae
   Francisella
Order VI: Legionellales
  Family I: Legionellaceae
   Legionella
  Family II: Coxiellaceae
   Coxiella
   Rickettsiella
Order VII: Methylococcales
  Family I: Methylococcaceae
   Methylobacter
   Methylocaldum
   Methylococcus
   Methylomicrobium
   Methylomonas
   Methylosphaera
Order VIII: Oceanospirillales
  Family I: Oceanospirillaceae
   Balneatrix
   Fundibacter
   Marinomonas

Marinospirillum
Neptunomonas
Oceanospirillum
  Family II: Halomonadaceae
   Alcanivorax
   Carnimonas
   Chromohalobacter
   Deleya
   Halomonas
   Zymobacter
Order IX: Pseudomonadales
  Family I: Pseudomonadaceae
   Azomonas
   Azotobacter
   Cellvibrio
   Chryseomonas
   Flavimonas
   Lampropedia
   Mesophilobacter
   Morococcus
   Oligella
   Pseudomonas
   Rhizobacter
   Rugamonas
   Serpens
   Thermoleophilum
   Xylophilus
  Family II: Moraxellaceae
   Acinetobacter
   Moraxella
   Psychrobacter
Order X: Alteromonadales
  Family I: Alteromonadaceae
   Allishewanella
   Alteromonas
   Colwellia
   Ferrimonas
   Idiomarina
   Marinobacter
   Marinobacterium
   Microbulbifer
   Moritella
   Pseudoalteromonas
   Shewanella
Order XI: Vibrionales
  Family I: Vibrionaceae
   Allomonas
   Enhydrobacter
   Listonella
   Photobacterium
   Salvinivibrio
   Vibrio
Order XII: Aeromonadales
  Family I: Aeromonadaceae
   Aeromonas
   Oceanomonas
   Tolumonas

Family II: Succinivibrionaceae
Anaerobiospirillumm
Ruminobacter
Succinomonas
Succinivibrio
Order XIII: Enterobacteriales
Family I: Enterobacteriaceae
Alterococcus
Arsenophonus
Brenneria
Buchnera
Budvicia
Buttiauxella
Calymmatobacterium
Cedecea
Citrobacter
Edwardsiella
Enterobacter
Erwinia
Escherichia
Ewingella
Hafnia
Klebsiella
Kluyvera
Leclercia
Leminorella
Moellerella
Morganella
Obesumbacterium
Pantoea
Pectobacterium
Photorhabdus
Plesiomonas
Pragia
Proteus
Providencia
Rahnella
Saccharobacter
Salmonella
Serratia
Shigella
Sodalis
Tatumella
Trabulsiella
Wigglesworthia
Xenorhabdus
Yersinia
Yokenella
Order XIV: Pasteurellales
Family I: Pasteurellaceae
Actinobacillus
Haemophilus
Lonepinella
Pasteurella
Mannheimia
Phocoenobacter
Class VI: Deltaproteobacteria

Order I: Desulfurales
Family I: Desulfurellaceae
Desulfurella
Hippea
Order II: Desulfovibrionales
Family I: Desulfovibrionaceae
Bilophila
Desulfovibrio
Lawsonia
Family II: Desulfomicrobiaceae
Desulfomicrobium
Family III: Desulfonalobiaceae
Desulfohalobium
Desulfomonaas
Desulfonatronovibrio
Order III: Desulfobacterales
Family I: Desulfobacteraceae
Desulfobacter
Desulfobacterium
Desulfobacula
Desulfococcus
Desulfofaba
Desulfofrigus
Desulfonema
Desulfosarcina
Desulfospira
Desulfocella
Desulfotalea
Desulfotignum
Family II: Desulfobulbaceae
Desulfobulbus
Desulfocapsa
Desulfofustis
Desulforhopalus
Family III: Nitrospinaceae
Nitrospina
Desulfobacca
Desulfomonile
Order IV: Desulfuromonadales
Family I: Desulfomonadaceae
Desulfuromonas
Desulfuromusa
Family II: Geobacteraceae
Geobacter
Family III: Pelobacteriaceae
Pelobacter
Melonomonas
Trichlorobacter
Order V: Syntrophobacterales
Family I: Syntrophobacteraceae
Desulfacinum
Syntrophobacter
Desulforhabdus
Desulfovirga
Thermodesulforhabdus
Family II: Syntrophaceae
Smithella

Syntrophus
Order VI: Bdellovibrionales
Family I: Bdellovibrionaceae
Bacteriovorax
Bdellovibrio
Micavibrio
Vampirovibrio
Order VII: Myxococcales
Family I: Myxococcaceae
Angiococcus
Myxococcus
Family II: Archangiaceae
Archangium
Family III: Cystobacteraceae
Cystobacter
Melittangium
Stigmatella
Family IV: Polyangiaceae
Chondromyces
Nannocystis
Polyangium
Class V: Epsilonproteobacteria
Order I: Campylobacterales
Family I: Campylobacteraceae
Arcobacter
Campylobacter
Sulfurospirillum
Thiovulum
Family II: Helicobacteraceae
Helicobacter
Wolinella
Phylum XIII: Firmicutes
Class I: Clostridia
Order I: Clostridiales
Family I: Clostridiaceae
Acetivibrio
Acidaminobacter
Anaerobacter
Caloramator
Clostridium
Coprobacillus
Natronineola
Oxobacter
Sarcina
Sporobacter
Thermobrachium
Thermohalobacter
Tindallia
Family II: Lachnospiraceae
Acetitomaculum
Anaerofilum
Butyrivibrio
Catenibacterium
Catonella
Coprococcus
Johnsonella
Lachnospira

Pseudobutyrivibrio
Roseburia
Ruminococcus
Sporobacterium
Family III:
Peptostreptococcace
Filifactor
Fusibacter
Helcococcus
Micromonas
Peptostreptococcus
Tissierella
Family IV: Eubacteriaceae
Acetobacterium
Anaerovorax
Eubacterium
Mogibacterium
Pseudoramibacter
Family V: Peptococcaceae
Anaeroaarcus
Anaerosinus
Anaerovibrio
Carboxydothermus
Centipeda
Dehalobacter
Dendrosporobacter
Desulfitobacterium
Desulfonispora
Desulfosporosinus
Desulfotomaculum
Mitsuokella
Peptococcus
Propionispira
Succinispira
Syntrophobotulus
Thermoterrabacterium
Family VI: Heliobacteriaceae
Heliobacterium
Heliobacillus
Heliophilum
Heliorestis
Family VII:
Acidaminocaccaceae
Acetonema
Anaeromonas
Acidaminococcus
Dialister
Megasphaera
Papillibacter
Pectinatus
Phascolarctobacterium
Quinella
Schwartzia
Selenomonas
Sporomusa
Succiniclasticum
Veillonella

Zymophilus
Family VIII:
Syntrophomonadaceae
Acetogenium
Aminobacterium
Aminomonas
Anaerobaculum
Anaerobranca
Caldicellulosiruptor
Dethiosulfovibrio
Pelospora
Syntrophomonas
Syntrophospora
Syntrophothermus
Thermoaerobacter
Thermoaerovibrio
Thermohydrogenium
Thermosyntropha
Order II: Thermoanaerobactriales
Family I:
Thermoanaerobacteriaceae
Ammonifex
Carboxydobrachium
Coprothermobacter
Moorella
Sporotomaculum
Thermacetogenium
Thermoanaerobacter
Thermoanaerobacterium
Thermoanaerobium
Order III: Haloanaerobiales
Family I: Haloanaerobiaceae
Haloanaerobium
Halocella
Halothermothrix
Natroniella
Family II: Halobacteroidaceae
Acetohalobium
Haloanaerobacter
Halobacteroides
Orenia
Sporohalobacter
Class II: Mollicutes
Order I: Mycoplasmatales
Family I: Mycoplasmataceae
Mycoplasma
Eperythrozoon
Haemobartonella
Ureaplasma
Order II: Entoplasmatales
Family I: Entoplasmatales
Entomoplasma
Mesoplasma
Family II: Spiroplasmataceae
Spiroplasma
Order III: Acholeplasmataceae
Family I: Anaeroplasmataceae

Acholeplasma
Order IV: Anaeroplasmatales
Anaeroplasma
Asteroleplasma
Order V: Incertae sedis
Family I: Erysipelothrichaeceae
Bulleidia
Erysipelothrix
Holdemania
Solobacterium
Class III: Bacilli
Order I: Bacillales
Family I: Bacillaceae
Amphibacillus
Anoxybacillus
Bacillus
Exiguobacterium
Gracilibacillus
Halobacillus
Saccharococus
Salibacillus
Virgibacillus
Family II: Planococcaceae
Filibacter
Kurthia
Planococcus
Sporosarcina
Family III: Caryophanaceae
Caryophanon
Family IV: Listeriaceae
Brochothrix
Listeria
Family V: Staphylococcaceae
Gemella
Macrococcus
Salinicoccus
Staphylococcus
Family VI:
Sporolactobacillaceae
Marinococcus
Sporolactobacillus
Family VII: Paenibacillaceae
Ammoniphilus
Aneurinibacillus
Brevibacillus
Oxalophagus
Paenibacillus
Thermicanus
Thermobacillus
Family VIII: Alicyclobacillaceae
Alicyclobacillus
Pasteuria
Sulfobacillus
Family IX:
Thermoactinomycetaceae
Thermoactinomyces
Order II: Lactobacillales

Family I: Lactobacillaceae
Lactobacillus
Paralactobacillus
Pediococcus
Family II: Aerococcaceae
Abiotrophia
Aerococcus
Dolosicoccus
Eremococcus
Facklamia
Globicatella
Tetragenococcus
Ignavigranum
Family III: Carnobacteriaceae
Agitococcus
Alloiococcus
Carnobacterium
Desemzia
Dolosigranulum
Granulicatella
Lactosphaera
Trichococcus
Family IV: Enterococcaceae
Atopobacter
Enterococcus
Melissococcus
Tetragenococcus
Vagococcus
Family V: Leuconostocaceae
Leuconostoc
Oenococcus
Weissela
Family VI: Streptococcaceae
Lactococcus
Streptococcus
Family VII: Incertae sedis
Acetoanaerobium
Oscillospira
Syntrophococcus

Phylum XIV: The Actinobacteria
Class I: Actinobacteria
Order I: Acidimicrobiales
Family I: Acidimicrobiaceae
Acidimicrobium
Order II: Rubrobacterales
Family I: Rubrobacteraceae
Rubrobacter
Order III: Coriobacteridae
Family I: Coriobacteriaceae
Atophobium
Collinsella
Coriobacterium
Cryptobacterium
Denitrobacterium
Eggerthella
Slakia
Order IV: Sphaerobacterales

Family I: Sphaerobacteraceae
Sphaerobacter
Order V: Actinomycetales
Suborder: Actinomycineae
Family I: Actinomycetaceae
Actinobaculum
Actinomyces
Arcanobacterium
Mobiluncus
Suborder: Micrococcineae
Family I: Micrococcaceae
Arthrobacter
Kocuria
Micrococcus
Nesterenkonia
Renibacterium
Rothia
Stomatococcus
Family II: Bogoricellaceae
Bogoriella
Family III: Rarobacteriaceae
Rarobacter
Family IV: Sanguibacteraceae
Sanguibacter
Family V: Brevibacteriaceae
Brevibacterium
Family VI: Cellulomonadaceae
Cellulomonas
Oerskovia
Family VII: Dermabacteraceae
Brachybacterium
Dermabacter
Family VIII: Dermatophilaceae
Dermatophilus
Family IX: Dermacoccaceae
Dermacoccus
Demetria
Kytococcus
Family X: Intrasporangiaceae
Intrasporangium
Janibacter
Ornithinicoccus
Ornithinimicrobium
Terrabacter
Terracoccus
Tetrasphaera
Family XI: Jonesiaceae
Jonesia
Family XII: Microbacteriaceae
Agrococcus
Agromyces
Aureobacterium
Clavibacter
Cryobacterium
Curtobacterium
Frigoribacterium
Leifsonia

Leucobacter
Microbacterium
Rathayibacter
Subtercola
Family XIII: Beutenbergiaceae
Beutenbergia
Family XIV:
Promicromonosporaceae
Promicromonospora
Suborder: Corynebacterineae
Family I: Corynebacteriaceae
Corynebacterium
Family II: Dietziaceae
Dietzia
Family III: Gordoniaceae
Gordonia
Skermania
Family IV: Mycobacteriaceae
Mycobacterium
Family V: Nocardiaceae
Nocardia
Rhodococus
Family VI: Tsukamurellaceae
Tsukamurella
Family VII: Williamsiaceae
Williamsia
Suborder: Micromonosporaceae
Family I: Micromonosporaceae
Actinoplanes
Catellatospora
Catenuloplanes
Couchioplanes
Dactylosporangium
Micromonospora
Pilimelia
Spirilliplanes
Verrucosispora
Suborder: Propionibacterineae
Family I:
Propionibacteriaceae
Luteococcus
Microlunatus
Propionibacterium
Propioniferax
Tessaracoccus
Family II: Nocardioidaceae
Aeromicrobium
Friedmanniella
Hongia
Kribbella
Micropruina
Marmoricola
Nocardioides
Suborder: Pseudonocardineae
Family I: Pseudonocardiaceae
Actinoalloteichus
Actinopolyspora

Amycolatopsis
Kibdelosporangium
Kutzneria
Prauserella
Pseudonocardia
Saccharomonospora
Saccharopolyspora
Streptoalloteichus
Thermobispora
Thermocrispum
Family II:
Actinosynnemataceae
Actinokineospora
Actinosynnema
Lentzea
Saccharothrix
Suborder: Streptomycineae
Family I: Streptomycetaceae
Streptomyces
Streptoverticillium
Suborder: Streptosporangineae
Family I: Streptosporangiaceae
Acrocarpospora
Herbidospora
Microbispora
Microtetraspora
Nonomuraea
Planobispora
Planomonospora
Planopolyspora
Planotetraspora
Streptosporangium
Family II: Nocardiopsaceae
Nocardiopsis
Thermobifida
Family III:
Thermomonosporaceae
Actinomadura
Spirillospora
Thermomonospora
Suborder: Frankineae
Family I: Frankiaceae
Frankia
Family II:
Geodermatophilaceae
Blastococcus
Geodermatophilus
Modestobacter
Family III: Microsphaeraceae
Microsphaera
Family IV: Sporichthyaceae
Sporichthya
Family V: Acidothermaceae
Acidothermus
Family VI: Kineosporaceae
Cryptosporangium
Kineococcus

Kineosporia
Suborder XI: Glycomycineae
Family I: Glycomycetaceae
Glycomyces
Order VI: Bifidobacteriales
Family I: Bifidobacteriaceae
Bifidobacterium
Falcivibrio
Gardnerella
Family II: Unknown Affiliation
Actinobispora
Actinocorallia
Excellospora
Pelczaria
Turicella
Phylum XV: Planctomycetes
Order I: Planctomycetales
Family I: Planctomycetaceae
Gemmata
Isosphaera
Pirellula
Planctomyces
Phylum XVI: Chlamydiae
Order I: Chlamydiales
Family I: Chlamydiaceae
Chlamydia
Chlamydophila
Family II: Parachlydiaceae
Parachlamydia
Family III: Simkaniaceae
Simkania
Family IV: Waddliaceae
Waddlia
Phylum XVII: Spirochaetes
Class I: Spirochaetes
Order I: Spirochaetales
Family I: Spirochaetaceae
Borrelia
Brevinema
Clevelandina
Cristispira
Diplocalyx
Hollandina
Pillotina
Spirochaeta
Treponema
Family II: Serpulinaceae
Brachyspira
Serpulina
Family III: Leptospiraceae
Leptonema
Leptospira
Phylum XVIII: Fibrobacteres
Class I: Fibrobacteres
Family I: Fibrobacteraceae
Fibrobacter
Phylum XIX: Actinobacter

Family I: Acidobacteriaceae
Acidobacterium
Geothrix
Holophaga
Phylum XX: Bacteroidetes
Class I: Bacteroidetes
Order I: Bacteroidales
Family I: Bacteroidaceae
Acetofilamentum
Acetomicrobium
Acetothermus
Anaerorhabdus
Bacteroides
Megamonas
Family II: Rikenellaceae
Marinilabilia
Rikenella
Family III:
Porphyromonadaceae
Dysgonomonas
Porphyromonas
Family IV: Prevotellaceae
Prevotella
Class II: Flavobacteria
Order I: Flavobacteriales
Family I: Flavobacteriaceae
Bergeyella
Capnocytophaga
Cellulophaga
Chryseobacterium
Coenonia
Empedobacter
Flavobacterium
Gelidibacter
Ornithobacterium
Polaribacter
Psychroflexus
Psychroserpens
Riemerella
Saligentibacter
Weeksella
Family II: Myroideaceae
Myroides
Psychromonas
Family III: Blattabacteriaceae
Blattabacterium
Class III: Sphingobacteria
Order I: Sphingobacteriales
Family I: Sphingobacteriaceae
Pedobacter
Sphingobacterium
Family II: Saprospiraceae
Haliscomenobacter
Lewinella
Saprospira
Family III: Flexibacteraceae
Cyclobacterium

Cytophaga
Dyadobacter
Flectobacillus
Flexibacter
Hymenobacter
Meniscus
Microscilla
Runella
Spirosoma
Sporocytophaga
Family IV: Flammeovirgaceae
Flammeovirga
Flexithrix
Persicobacter
Thermonema

Family V: Crenotrichaceae
Chitinophaga
Crenothrix
Rhodothermus
Toxothrix
Phylum XXI: Fusobacteria
Class I: Fusobacteria
Order I: Fusobacteriales
Family I: Fusobacteriaceae
Fusobacterium
Ilyobacter
Leptotrichia
Propionigenium
Sebaldella
Streptobacillus

Genera incertae sedis
Cetobacterium
Phylum XXII: Verrucomicrobia
Class I: Verrucomicrobiae
Order I: Verrucomicrobiales
Family I: Verrucomicrobiaceae
Prosthecobacter
Verrucomicrobium
Family II:
Xiphinematobacteriaceae
Xiphinematobacter
Phylum XXIII: Dictyoglomas
Class I: Dictyoglomi
Order I: Dictyoglomales
Family I: Dictyoglomaceae
Dictyoglomus

APPENDIX D: Major Microbial Agents of Disease by Body System Affected

Table D.1 **Microbial Disease Agents Affecting the Skin, Mucous Membranes, and Eyes**

Bacterial Disease Agent	Target of Infection	Disease(s) Caused	Mode of Transmission/Acquisition	Text Reference:
Staphylococcus aureus	Moist skin folds	Scalded skin syndrome; impetigo; folliculitis; furuncles (boils); carbuncles; toxic shock syndrome	Direct contact between individuals; via fomites; historically, use of highly absorbent tampons	Ch. 19 (pp. 535–537)
Streptococcus pyogenes	Skin and deeper tissues	Pyoderma; erysipelas; necrotizing fasciitis	Respiratory droplets; close contact between individuals	Ch. 19 (pp. 538–540)
Bacillus anthracis	Skin	Cutaneous anthrax	Inhalation of spores; inoculation of spores into a body via break in skin; ingestion of spores	Ch. 19 (pp. 545–547)
Clostridium perfringens	Skin, muscle and connective tissue	Gas gangrene	Physical trauma resulting in introduction of spores into the body	Ch. 19 (pp. 547–548)
Mycobacterium leprae	Peripheral nerve endings; skin cells	Leprosy (Hansen's disease)	Person-to-person contact; inhalation of respiratory droplets; via breaks in skin	Ch. 19 (pp. 559–560)
Propionibacterium acnes	Sebaceous (oil) glands of skin	Acne	Normal microbiota on skin	Ch. 19 (pp. 560–561)
Actinomyces spp.	Mucous membranes	Actinomycosis	Via breaks in mucous membranes resulting from trauma or infection by other pathogens	Ch. 19 (p. 562)
Pasteurella multocida	Skin and local lymph nodes	Pasteurellosis	Animal bites or scratches; inhalation of aerosols from animals	Ch. 20 (p. 582)
Bartonella henselae	Skin and local lymph nodes	"Cat-scratch disease"	Cat scratches and bites	Ch. 20 (p. 584)
Neisseria gonorrhoeae	Cornea	Ophthalmia neonatorum	Gonococcal infection during birth	Ch. 20 (p. 569)
Chlamydia trachomatis	Conjunctiva	Trachoma	Infection during birth; bacteria from genitalia introduced to eyes via fomites or fingers	Ch. 21 (p. 604)

Table D.1 (continued)

Fungal Disease Agent	Target of Infection	Disease(s) Caused	Mode of Transmission/Acquisition	Text Reference:
Blastomyces dermatitidis	Lungs, skin	Cutaneous blastomycosis	Inhalation of dust carrying fungal spores, resulting in dissemination from lungs to skin	Ch. 22 (p. 628)
Trichophyton spp. Microsporum spp. Epidermophyton floccosum	Cutaneous tissues (skin, nails; *Trichophyton* also infects hair)	Dermatophytoses (e.g. Tinea pedis, or "athlete's foot"; Tinea cruris, or "jock-itch")	Spores; fungal elements shed in the environment by infected individuals	Ch. 22 (pp. 638–640)
Malassezia furfur	Skin	Pityriasis; folliculitis; seborrheic dermatitis; dandruff	Normal microbiota on skin	Ch. 22 (p. 639)
Fonsecaea pedrosoi; F. compacta; Phialophora verrucosa; Cladophialophora carrionii	Subepidermal tissues in skin	Chromoblastomycoses	Traumatic introduction of fungal elements to subepidermal tissues	Ch. 22 (p. 640)
Madurella, Pseudallescheria, Exophalia, Acremonium	Skin, fascia (lining of muscles), bones of the hands and feet	Mycetomas	Prick wounds or scrapes caused by twigs, thorns, or leaves contaminated with fungi	Ch. 22 (p. 642)
Sporothrix schenckii	Subcutaneous tissues, usually in arms or legs	Sporotrichosis ("rose-gardener's disease")	Thorn pricks, wood splinters	Ch. 22 (p. 642)
Aspergillus spp.	Eyelids, conjunctiva, eye socket	Aspergillosis	Contamination from spores	Ch. 22 (p. 632)

Parasitic Disease Agent	Target of Infection	Disease(s) Caused	Mode of Transmission/Acquisition	Text Reference:
Acanthamoeba	Eye	Keratitis	Through the conjunctiva via abrasions from contact lenses or trauma; inhalation of contaminated water	Ch. 23 (p. 652)
Leishmania	Skin; mucous membranes of the mouth, nose, or soft palate	Cutaneous leishmaniasis; mucocutaneous leishmaniasis	Bite from infected sand fly	Ch. 23 (pp. 654–655)

Table D.1   *(continued)*

Viral Disease Agent	Target of Infection	Disease(s) Caused	Mode of Transmission/Acquisition	Text Reference:
Smallpox virus (Variola)	Skin	Smallpox	Inhalation of droplets or dried crusts	Ch. 24 (pp. 680–682)
Molluscipoxvirus	Skin	Molluscum contagiosum	Person-to-person contact	Ch. 24 (p. 682)
Herpes simplex virus 1	Lips, skin	Fever blisters (cold sores); ocular herpes; whitlow	Via mucous membranes; cut or break in skin	Ch. 24 (pp. 684–687)
Herpes simplex virus 2	Genitals	Genital herpes; whitlow	Sexual transmission	Ch. 24 (pp. 684–687)
Varicella-zoster virus	Respiratory tract, eyes, blood, skin	Chickenpox; shingles	Person-to-person contact	Ch. 24 (pp. 687–689)
Human herpesvirus 6	Skin, lymph nodes	Roseola	Contact with droplets	Ch. 24 (pp. 692–693)
Human papillomavirus	Epithelium of skin or mucous membranes	Warts, verruca	Via direct contact and fomites; sexual transmission	Ch. 24 (pp. 693–695)
Parvovirus B19	Skin (cheeks, arms, thighs, buttock, trunk)	Erythema infectiosum (Fifth disease)	Contact with droplets	Ch. 24 (pp. 699–700)
Coxsackie A virus	Mouth, pharynx, skin	Herpangina; hand-foot-and-mouth-disease; acute hemorrhagic conjunctivitis; viral meningitis	Via fecal-oral route	Ch. 25 (pp. 707–708)
Rubella virus	Skin, respiratory tract, lymph nodes, blood	Rubella	Via respiratory route	Ch. 25 (p. 713)
Measles virus	Respiratory tract, lymph, blood, conjunctiva, urinary tract, blood vessels, lymphatics, central nervous system	Measles, rubeola	Via respiratory droplets	Ch. 25 (pp. 725–727)

Table D.2   Microbial Disease Agents Affecting the Nervous System

Bacterial Disease Agent	Target of Infection	Disease(s) Caused	Mode of Transmission/Acquisition	Text Reference:
Clostridium botulinum	Neuromuscular junction	Botulism	Consumption of toxin from canned foods; ingestion of endospores	Ch. 19 (pp.549–550)
Clostridium tetani	Inhibitory neurons	Tetanus	Introduction of endospores to break in skin, resulting in formation of toxin	Ch. 19 (pp. 550–551)
Mycobacterium leprae	Peripheral nerve endings	Leprosy (Hansen's disease)	Person-to-person contact; inhalation of respiratory droplets	Ch. 19 (pp.559–560)
Neisseria meningitidis	Brain and spinal cord membranes	Meningococcal meningitis; meningococcal septicemia	Person-to-person contact; inhalation of respiratory droplets	Ch. 20 (pp. 569–570)
Haemophilus influenzae	Brain, meninges	Meningitis; infantile arthritis; epiglottitis; subcutaneous tissue inflammation	Person-to-person contact	Ch. 20 (pp. 582)

Fungal Disease Agent	Target of Infection	Disease(s) Caused	Mode of Transmission/Acquisition	Text Reference:
Cryptococcus neoformans	Central nervous system	Cryptococcal meningitis	Inhalation of spores and/or dried yeast forms from contaminated pigeon or chicken droppings	Ch. 22 (p. 635)

Parasitic Disease Agent	Target of Infection	Disease(s) Caused	Mode of Transmission/Acquisition	Text Reference:
Acanthamoeba spp.	Brain	Amoebic encephalitis	Introduction via cuts or scrapes in the skin; inhalation of contaminated water	Ch. 23 (p. 652)
Echinococcus granulosus	Brain, liver, other organs	Hydatid disease	Consumption of food or water contaminated with eggs shed in dogs' feces.	Ch. 23 (p. 666)
Naegleria spp.	Nasal mucosa; brain	Amoebic meningoencephalitis	Inhalation of contaminated water	Ch. 23 (p. 652)
Trypanosoma brucei	Lymphatic, circulatory and central nervous systems	African sleeping sickness	Bite of infected tsetse fly	Ch. 23 (p. 653)
Trypanosoma cruzi	Blood, lymph, spinal fluid, other body cells	Chagas' disease	Feces from infected *Triatoma* bug introduced via bite wound	Ch. 23 (p. 655)

Table D.2   (continued)

Viral Disease Agent	Target of Infection	Disease(s) Caused	Mode of Transmission/Acquisition	Text Reference:
JC polyomaviruses	Kidney cells, blood, brain, central nervous system	Progressive multifocal leukoencephalopathy	Via respiratory route	Ch. 24 (p. 695)
Poliovirus	Pharyngeal and intestinal cells, meninges, central nervous system, spinal cord, brain	Poliomyelitis	Contaminated water, food; via fecal-oral route	Ch. 25 (pp. 706–707)
Togavirus	Blood; occasionally brain, liver, skin, and blood vessels	Eastern equine encephalitis (EEE); Western equine encephalitis (WEE); Venezuelan equine encephalitis (VEE)	Arthropod vectors	Ch. 25 (pp. 709–710)
Flavivirus	Blood; occasionally brain, liver, skin, and blood vessels	St. Louis, Japanese, West Nile, and Russian spring-summer encephalitis; Dengue fever; yellow fever	Arthropod vectors	Ch. 25 (pp. 710–713)
Rabies virus	Skeletal muscle cells; neurons; spinal cord; brain; salivary glands	Rabies	Via saliva of infected animals; introduction of virus through break in skin	Ch. 25 (pp. 728–730)
Bunyavirus	Blood, brain, kidney, liver, blood vessels	California encephalitis; Rift Valley fever	Arthropod vectors	Ch. 25 (pp. 735–736)
Human immunodeficiency virus (HIV)	T-helper cells, macrophage cells; smooth muscle cells; dendritic cells	AIDS	Sexual transmission; intravenous drug use; blood transfusion; contact with infected bodily fluid	Ch. 25 (pp. 718–724)
Prions	Brain	Spongiform encephalopathies (Creutzfeldt-Jakob disease, Kuru)	Consumption of infected tissue	Ch. 13 (p. 399)

| Table D.3 | | Microbial Disease Agents Affecting the Cardiovascular and Lymphatic Systems | | |

Bacterial Disease Agent	Target of Infection	Disease(s) Caused	Mode of Transmission/Acquisition	Text Reference:
Staphylococcus aureus	Lining of heart valves; vagina	Endocarditis; staphylococcal toxic shock syndrome	Introduction of pathogen via skin wound or contaminated medical device; historically, use of highly absorbent tampons	Ch. 19 (pp. 535–536)
Streptococcus pyogenes (Group A streptococcus)	Blood, skin, heart valves, muscle	Septicemia; pericarditis; rheumatic fever; scarlet fever; streptococcal toxic shock syndrome	Person-to-person contact; infection of skin or mucous membrane leads to invasion of deeper tissues and organs	Ch. 19 (pp. 538-541)
Bacillus anthracis	Skin, blood, lungs	Anthrax	Inoculation of spores into body via break in skin; ingestion of spores; inhalation of spores	Ch. 19 (p. 545)
Yersinia pestis	Blood	Plague	Bite of infected rat flea	Ch. 20 (pp. 580–581)
Bartonella henselae	Skin, lymph nodes	"Cat-scratch" disease	Infected cat scratches and bites; bites from infected fleas	Ch. 20 (p. 584)
Brucella spp.	Mucous membranes	Brucellosis (undulant fever)	Consumption of contaminated dairy products; contact with animal blood, urine, and placentas via breaks in mucous membranes	Ch. 20 (p. 585)
Francisella tularensis	Skin, lymph nodes	Tularemia	Bite of infected tick or contact with infected animals such as rabbits and muskrats (bacteria can pass through apparently unbroken skin)	Ch. 20 (pp. 588–589)
Rickettsia ricketsii	Endothelial cells lining blood vessels	Rocky Mountain spotted fever	Bite of infected tick	Ch. 21 (p. 599)
Rickettsia prowazekii	Blood vessels	Epidemic typhus	Bite of infected human body louse	Ch. 21 (p. 600)
Ehrlichia chafeensis; Ehrlichia equi	Blood, leukocytes	Ehrlichiosis	Bite of infected tick, including the dog tick and deer tick	Ch. 21 (pp. 600–601)
Borrelia burgdorferi	Blood	Relapsing fever; Lyme disease	Bite of infected Ixodes tick	Ch. 21 (pp. 610–611)

Parasitic Disease Agent	Target of Infection	Disease(s) Caused	Mode of Transmission/Acquisition	Text Reference:
Trypanoma cruzi	Blood	Chagas' disease (results in inflammation and damage of heart muscle)	Bite of infected Triatoma bugs	Ch. 23 (p. 655)
Leishmania spp.	Liver, spleen, bone marrow, lymph nodes	Leishmaniasis	Bite of infected female sand flies	Ch. 23 (p. 655)
Plasmodium spp.	Blood, liver, kidney	Malaria	Bite of infected female Anopheles mosquito	Ch. 23 (pp. 658–659)
Toxoplasma gondii	Lymph nodes	Toxoplasmosis	Ingestion of contaminated undercooked meat; ingestion or inhalation of contaminated soil; contact with infected cats	Ch. 23 (p. 661)
Schistosoma spp.	Blood; liver, lungs, brain	Schistosomiasis; swimmer's itch	Skin contact with contaminated water	Ch. 23 (p. 667)
Wuchereria bancrofti	Lymphatic system and subcutaneous tissues	Filariasis (elephantiasis)	Infected mosquitoes	Ch. 23 (p. 671)

Table D.3 *(continued)*

Viral Disease Agent	Target of Infection	Disease(s) Caused	Mode of Transmission/Acquisition	Text Reference:
Epstein Barr virus	Epithelial cells of the pharynx and parotid salivary glands; blood; B lymphocytes	Burkitt's lymphoma; infectious mononucleosis	Via saliva	Ch. 24 (pp. 689–691)
Cytomegalovirus	Lymph glands, liver	Birth defects	Via bodily secretions such as saliva, mucus, milk, urine, feces, semen, cervical secretions; sexual transmission; birth; blood transfusions; organ transplants; infected needles	Ch. 24 (pp. 691–692)
Coxsackie B virus	Heart muscle, skin	Myocarditis; pericardial infections; pleurodynia ("devil's grip"); viral meningitis	Via fecal-oral route	Ch. 25 (pp. 707–708)
HTLV-1, HTLV-2	Lymphocytes	Adult acute T-cell lymphocytic leukemia (HTLV-1); Hairy-cell leukemia (HTLV-2)	Sexual transmission; blood transfusion, infected needles	Ch. 25 (p. 717)
Marburg virus, Ebola virus	Multiple body cells, especially macrophages and liver cells	Viral hemorrhagic fevers	Person-to-person via contaminated body fluids, primarily blood	Ch. 25 (pp. 730–731)
Hantavirus	Blood, lungs	Hantavirus pulmonary syndrome	Inhalation of virions in dried deer-mouse urine or feces	Ch. 25 (p. 735)

Table D.4   Microbial Disease Agents Affecting the Respiratory System

Bacterial Disease Agent	Target of Infection	Disease(s) Caused	Mode of Transmission/Acquisition	Text Reference:
Streptococcus pyogenes	Pharynx; tongue	Strep throat; scarlet fever	Inhalation of respiratory droplets	Ch. 19 (p. 539)
Streptococcus pneumoniae	Pharynx; lungs; alveoli	Pneumococcal pneumonia	Inhalation of pneumococci from the pharynx into damaged lungs	Ch. 19 (p. 543)
Bacillus anthracis	Lungs	Inhalation anthrax	Inhalation of spores; inoculation of spores into a body via break in skin; ingestion of spores	Ch. 19 (pp. 545–547)
Corynebacterium diphtheriae	Respiratory membranes	Diphtheria	Inhalation of respiratory droplets	Ch. 19 (pp. 554–555)
Mycobacterium tuberculosis	Lungs	Tuberculosis	Inhalation of respiratory droplets	Ch. 19 (pp. 556–557)
Haemophilus influenzae	Lungs, pharynx	Bronchitis; pneumonia	Inhalation of respiratory droplets	Ch. 20 (p. 583)
Bordetella pertussis	Ciliated epithelial cells of the trachea	Whooping cough	Inhalation of respiratory droplets	Ch. 20 (p. 585)
Legionella pneumophila	Lungs	Legionellosis	Inhalation of aerosols such as those produced by showers, misters, hot tubs, vaporizers, whirlpools, air conditioning systems	Ch. 20 (p. 590)
Coxiella burnetti	Lungs, liver	Q fever	Inhalation of infected bodies that become airborne when released from dried tick and animal feces /urine; consumption of contaminated milk	Ch. 20 (p. 590)
Mycoplasma pneumoniae	Lungs; mucous membranes of the respiratory and urinary tracts	Mycoplasmal pneumonia	Nasal secretions among people in close contact	Ch. 21 (pp. 596–597)
Chlamydia pneumonia	Lungs	Bronchitis, pneumonia, sinusitis	Inhalation of respiratory droplets	Ch. 21 (p. 605)
Chlamydia psittaci	Lungs	Pneumonia caused by psittacosis (ornithosis, or "parrot fever")	Inhalation of respiratory droplets	Ch. 21 (p. 606)

Fungal Disease Agent	Target of Infection	Disease(s) Caused	Mode of Transmission/Acquisition	Text Reference:
Blastomyces dermatitidis	Lungs	Pulmonary blastomycosis	Inhalation of spores from dust	Ch. 22 (p. 628)
Coccidioides immitis	Alveolar spaces of lungs	Coccidioidomycosis	Inhalation of spores	Ch. 22 (p. 629)
Histoplasma capsulatum	Alveolar macrophages in lungs	Histoplasmosis	Inhalation of spores near bird droppings	Ch. 22 (p. 630)
Pneumocystis carinii	Lungs	*Pneumocystis* pneumonia	Inhalation of spores	Ch. 22 (p. 636)

Viral Disease Agent	Target of Infection	Disease(s) Caused	Mode of Transmission/Acquisition	Text Reference:
Adenovirus	Respiratory tract	Common cold	Inhalation of respiratory droplets	Ch. 24 (p. 696)
Rhinovirus	Upper respiratory tract	Common cold	Inhalation of respiratory droplets; via fomites; direct person-to-person contact	Ch. 25 (pp. 704–705)
Coronavirus	Respiratory tract	Severe acute respiratory syndrome (SARS); common cold	Inhalation of respiratory droplets	Ch. 25 (p. 715)
Mumps virus	Upper respiratory cells	Mumps	Respiratory secretions	Ch. 25 (pp. 727–728)
Respiratory syncytial virus	Respiratory tract	Respiratory syncytial disease	Via fomites, hands; especially in child day-care centers	Ch. 25 (p. 728)
Influenzavirus	Lungs, respiratory tract	Influenza	Inhalation of airborne viruses; via fomites	Ch. 25 (pp. 731–734)

Table D.5   Microbial Disease Agents Affecting the Digestive System

Bacterial Disease Agent	Target of Infection	Disease(s) Caused	Mode of Transmission/Acquisition	Text Reference:
Staphylococcus aureus	Lower digestive system	Staphylococcal food poisoning	Consumption of contaminated food, such as potato salad, processed meats, ice cream	Ch. 19 (p. 535)
Viridans streptococci	Mouth	Dental caries	Normal microbiota	Ch. 19 (p. 542)
Clostridium perfringens	Lower digestive system	Gastroenteritis	Consumption of contaminated food	Ch. 19 (p. 547)
Escherichia coli	Lower digestive system	Gastroenteritis	Consumption of contaminated food	Ch. 20 (p. 575)
Salmonella enterica Dublin	Lower digestive system	Salmonellosis	Consumption of contaminated food, such as chicken eggs, or inadequately pasteurized milk	Ch. 20 (p. 577)
Salmonella typhi	Lower digestive system	Typhoid fever	Ingestion of contaminated food or water	Ch. 20 (p. 578)
Shigella	Lower digestive system	Shigellosis	Poor personal hygiene; ingestion of bacteria on hands; consumption of contaminated foods	Ch. 20 (p. 579)
Yersinia enterocolitica	Lower digestive system	Gastroenteritis	Consumption of contaminated food or water	Ch. 20 (p. 580)
Vibrio cholerae	Lower digestive system	Cholera	Ingestion of contaminated food or water	Ch. 21 (p. 615)
Vibrio parahemolyticus	Lower digestive system	Gastroenteritis	Ingestion of contaminated shellfish	Ch. 21 (p. 616)
Vibrio vulnificus	Lower digestive system; blood	Gastroenteritis; septicemia	Consumption of contaminated shellfish	Ch. 21 (p. 616)
Campylobacter jejuni	Lower digestive system	Gastroenteritis	Consumption of contaminated poultry	Ch. 21 (p. 616)
Helicobacter pylori	Lower digestive system	Peptic ulcers	Not known for certain; possibly fecal-oral path of infection; cats may transmit disease to humans	Ch. 21 (p. 618)

Fungal Disease Agent	Target of Infection	Disease(s) Caused	Mode of Transmission/Acquisition	Text Reference:
Aspergillus flavus	Liver	Aflatoxin poisoning	Consumption of toxin in food crops; ingestion of contaminated milk	Ch. 22 (p. 643)

Table D.5   (continued)

Parasitic Disease Agent	Target of Infection	Disease(s) Caused	Mode of Transmission/Acquisition	Text Reference:
Entamoeba histolytica	Lumen of the intestine, colon, intestinal mucosa	Amoebic dysentery (amoebiasis)	Drinking of water contaminated with feces containing cysts; fecal contamination of hands or food; oral-anal intercourse	Ch. 23 (p. 651)
Giardia intestinalis	Lumen of the intestine; intestinal mucosa; colon	Giardiasis	Ingestion of cysts in contaminated water; contaminated raw foods; contact with feces	Ch. 23 (p. 657)
Cryptosporidium parvum	Intestine	Cryptosporidiosis	Drinking contaminated water; fecal-oral transmission; poor hygiene	Ch. 23 (p. 662)
Cyclospora cayetanensis	Intestine	Cyclospora diarrheal infection	Ingestion of oocytes on contaminated foods and water	Ch. 23 (p. 663)
Taenia saginata	Intestine	Tapeworm (beef)	Ingestion of cysticerci in raw or undercooked beef	Ch. 23 (p. 665)
Taenia solium	Intestine	Tapeworm (pork)	Ingestion of cysticerci in raw or undercooked pork	Ch. 23 (p. 665)
Echinococcus granulosus	Intestine, circulatory system, liver	Hydatid disease	Consumption of food or water contaminated with eggs shed in dogs' feces	Ch. 23 (p. 666)
Necator americanus; Ancylostoma duodenale	Small intestine; intestinal mucosa	Hookworm	Burrow through skin	Ch. 23 (p. 670)
Ascaris lumbricoides	Small intestine; circulatory and lymphatic systems; lungs	Ascariasis	Consumption of contaminated food or water	Ch. 23 (p. 670)
Enterobius vermicularis	Anus	Pinworms	Ingestion of eggs from fingers or food	Ch. 23 (p. 671)

Viral Disease Agent	Target of Infection	Disease(s) Caused	Mode of Transmission/Acquisition	Text Reference:
Hepatitis B	Liver	Serum hepatitis; hepatic cancer	Sexual transmission; contact with infected bodily fluid; contaminated syringes	Ch. 24 (pp. 696–699)
Hepatitis A	Liver cells	Infectious hepatitis	Fecal-oral route	Ch. 25 (p. 709)
Hepatitis C	Liver cells	Chronic hepatitis	Sexual transmission; contaminated syringes	Ch. 25 (p. 709)
Hepatitis D	Liver cells	Delta agent hepatitis	Sexual transmission; contaminated syringes	Ch. 25 (p. 709)
Hepatitis E	Liver cells	Enteric hepatitis	Fecal-oral route	Ch. 25 (p. 709)
Rotavirus	Respiratory and digestive tracts	Infantile gastroenteritis; diarrhea	Fecal-oral route	Ch. 25 (p. 737)
Cytomegalovirus	Lymph glands, liver	Birth defects	Via bodily secretions such as saliva, mucus, milk, urine, feces, semen, cervical secretions; sexual transmission; birth; blood transfusions; organ transplants	Ch. 24 (pp. 691–692)

Table D.6   Microbial Disease Agents Affecting the Urinary and Reproductive Systems

Bacterial Disease Agent	Target of Infection	Disease(s) Caused	Mode of Transmission/Acquisition	Text Reference:
Staphylococcus aureus	Vagina, blood	Toxic shock syndrome	Historically, use of highly absorbent tampons	Ch. 19 (pp. 535–537)
Neisseria gonorrhoeae	Epithelial cells of mucous membranes lining the genital, urinary and digestive tracts	Gonorrhea; pelvic inflammatory disease (PID)	Sexual transmission	Ch. 20 (pp. 568–569)
Escherichia coli	Urethra, bladder	Cystitis (urinary bladder infection); kidney infection	Normal microbiota	Ch. 20 (p. 575)
Haemophilus ducreyi	Genitals	Chancroid	Sexual transmission	Ch. 20 (p. 582)
Mycoplasma genitalium; Ureaplasma urethritis	Urinary and genital tracts	Nongonococcal urethritis (NGU)	Sexual transmission	Ch. 20 (p. 598)
Chlamydia trachomatis	Mucous membranes of urethra, uterus, uterine tubes, anus, rectum; lymph nodes	Lymphogranuloma venereum	Sexual transmission	Ch. 20 (p. 603)
Treponema pallidum	Genitals	Syphilis	Sexual transmission	Ch. 20 (p. 607)
Leptospira interrogans	Skin, mucous membranes, blood, central nervous system, kidneys	Leptospirosis (kidney infection)	Direct contact with urine of infected animals; indirectly via contact with spirochetes in contaminated environments	Ch. 20 (p. 613)

Fungal Disease Agent	Target of Infection	Disease(s) Caused	Mode of Transmission/Acquisition	Text Reference:
Candida albicans	Vagina, urinary tract	Candidiasis	Normal microbiota; sexual contact; childbirth	Ch. 22 (p. 633)

Parasitic Disease Agent	Target of Infection	Disease(s) Caused	Mode of Transmission/Acquisition	Text Reference:
Trichomonas vaginalis	Vagina	Vaginitis	Sexual transmission	Ch. 23 (p. 658)

Viral Disease Agent	Target of Infection	Disease(s) Caused	Mode of Transmission/Acquisition	Text Reference:
Herpes simplex virus 2	Genitals	Genital herpes; whitlow	Sexual transmission	Ch. 24 (pp. 684–687)
Human papillomavirus	Epithelium of skin or mucous membranes	Warts, verruca	Via direct contact and fomites; sexual transmission	Ch. 24 (pp. 693–694)
Hepatitis B	Liver	Serum hepatitis; hepatic cancer	Sexual transmission; contact with infected bodily fluid; contaminated syringes	Ch. 24 (pp. 696–699)
Human immunodeficiency virus (HIV)	T-helper cells, macrophage cells; smooth muscle cells; dendritic cells	AIDS	Sexual transmission; intravenous drug use; blood transfusion; contact with infected bodily fluid	Ch. 25 (pp. 718–724)

Glossary

A site In a ribosome, a binding site which accommodates tRNA delivering an amino acid.

Abscess An isolated site of infection such as a pimple, boil, or pustule.

Abyssal zone In marine habitats, the zone of water beneath the benthic zone, virtually devoid of life except around hydrothermal vents.

Acellular Noncellular.

Acetyl-CoA Combination of two carbon atoms and co-enzyme A.

Acid Compound that dissociates into one or more hydrogen ions and one or more anions.

Acid-fast stain In microscopy, a differential stain used to penetrate waxy cell walls.

Acidic dye In microscopy, an anionic chromophore used to stain alkaline structures. Works most effectively in acidic environments.

Acidophile Microorganism requiring acidic pH.

Acne Skin disorder characterized by presence of whiteheads, blackheads, and in severe cases, cysts; caused by infection with *Propionibacterium acnes*.

Acquired (secondary) immunodeficiency diseases Any of a group of immunodeficiency diseases that develop in older children, adults, and the elderly as a direct consequence of some other recognized cause, such as infectious disease.

Acquired immunodeficiency syndrome (AIDS) Cluster of characteristic signs and symptoms that develop several years after infection with the human immunodeficiency virus (HIV), which destroys helper T cells.

Actinomycetes High G+C Gram-positive bacteria that form branching filaments and produce spores, thus resembling fungi.

Activation energy The amount of energy needed to trigger a chemical reaction.

Active site Functional site of an enzyme, the shape of which is complementary to the shape of the substrate.

Active transport The movement of a substance against its electrochemical gradient via carrier proteins and requiring cell energy from ATP.

Acute anaphylaxis Condition in which the release of inflammatory mediators overwhelms the body's coping mechanisms.

Acute disease Any disease that develops rapidly but lasts only a short time, whether it resolves in convalescence or death.

Acute inflammation Type of inflammation that develops quickly, is short-lived, and is usually beneficial.

Adenine Ring-shaped nitrogenous base found in nucleotides of DNA and RNA.

Adenosine triphosphate (ATP) The primary short-term, recyclable energy molecule fueling cellular reactions.

Adherence Process by which phagocytes attach to microorganisms through the binding of complementary chemicals on the cell membranes.

Adhesion The attachment of microorganisms to host cells.

Adhesion factors A variety of structures or chemical receptors by which microorganisms attach to host cells.

Adjuvant Chemical added to a vaccine to increase its ability to stimulate active immunity.

Aerobic respiration Type of cellular respiration requiring oxygen atoms as final electron acceptors.

Aerosol A cloud of small droplets and solid particles suspended in the air.

Aerotolerant anaerobe Microorganism which prefers anaerobic conditions but can tolerate exposure to low levels of oxygen.

African sleeping sickness Potentially fatal disease caused by a bite from a tsetse fly carrying *Trypanosoma brucei* and characterized by chancre formation at the site of the bite, followed by parasitemia and central nervous system invasion.

Agar Gel-like polysaccharide isolated from red algae and used as thickening agent.

Agglutination Clumping caused when antibodies bind to two antigens, hindering the activity of pathogenic microorganisms and increasing the chance that they will be phagocytized.

Agglutination test In serology, a procedure in which antiserum is mixed with a sample that potentially contains its target antigen.

Agranulocyte Type of leukocyte having a uniform cytoplasm lacking large granules.

Agroterrorism The use of microbes to terrorize humans by destroying the livestock and crops that constitute their food supply.

Airborne transmission Spread of pathogens to the respiratory mucous membranes of a new host via an *aerosol*.

Alcohol Intermediate-level disinfectant that denatures proteins and disrupts cell membranes.

Aldehyde Compound containing terminal –CHO groups used as a high-level disinfectant because it cross-links organic functional groups in proteins and DNA.

Algae Eukaryotic unicellular or multicellular photosynthetic organisms with simple reproductive structures.

Alkalinophile Microorganism requiring alkaline pH environments.

Allergen An antigen that stimulates an allergic response.

Allergic contact dermatitis Type of delayed hypersensitivity reaction in which chemically modified skin proteins trigger a cell-mediated immune response.

Allergy An immediate hypersensitivity response against an antigen.

Allograft Type of graft in which tissues are transplanted from a donor to a genetically dissimilar recipient.

Alpha-interferons Interferons secreted by virally infected monocytes, macrophages, and some lymphocytes within hours after infection.

Alphaproteobacteria Class of aerobic bacteria in the phylum Proteobacteria capable of growing at very low nutrient levels.

Alternation of generations Method of sexual reproduction in which diploid thalli alternate with haploid thalli.

Alveolar macrophage Fixed macrophage of the lungs.

Alveolates Protozoa with small membrane-bound cavities called *alveoli* beneath their cell surfaces.

Ames test Method for screening mutagens that is commonly used to identify potential carcinogens.

Amination Reaction involving the addition of an amine group to a metabolite to make an amino acid.

Amino acid A monomer of protein.

Aminoglycoside Antimicrobial agent that inhibits protein synthesis by changing the shape of the 30S ribosomal subunit.

Ammonification Process by which microorganisms disassemble proteins in soil wastes into their amino acids, which are then converted to ammonia.

Amoeba Protozoan that moves and feeds by pseudopodia, lacks mitochondria, and reproduces via binary fission.

Amoebiasis A mild to severe dysentery that, if invasive, can cause the formation of lesions in the liver, lungs, brain, and other organs; caused by infection with *Entamoeba histolytica*.

Amoebic encephalitis Often fatal inflammation of the brain characterized by headache, altered mental state, and neurological deficit; caused by infection with *Acanthamoeba*.

Amoebic meningoencephalitis Often fatal inflammation of the brain characterized by headache, vomiting, fever, and destruction of neurological tissue; caused by infection with *Naegleria*.

Amphibolic reaction A reversible metabolic reaction; that is, a reaction which can be catabolic or anabolic.

Amphitrichous Term used to describe a cell having flagella at both ends.

Amphotericin B Medication that is widely used to treat mycoses.

Anabolism All of the synthesis reactions in an organism taken together.

Anaerobic respiration Type of cellular respiration not requiring oxygen atoms as final electron acceptors.

Analytical epidemiology Detailed investigation of a disease, including analysis of data to determine the probable cause, mode of transmission, and possible means of prevention.

Anaphase Third stage of mitosis, during which sister chromatids separate and move to opposite poles of the spindle to form chromosomes. Also used for the comparable stage of meiosis.

Anaphylactic shock Condition in which the release of inflammatory mediators overwhelms the body's coping mechanisms, causing suffocation, edema, smooth muscle contraction, and often death.

Anion A negatively charged ion.

Anthrax Gastrointestinal, cutaneous, or pulmonary disease that is usually fatal without aggressive treatment; caused by ingestion, inoculation, or inhalation of spores of *Bacillus anthracis*.

Antibiotic Antimicrobial agent that is produced naturally by an organism.

Antibody (*immunoglobulin*) Proteinaceous antigen-binding molecule secreted by plasma cells.

Antigen Molecule that triggers a specific immune response.

Antigen presenting cell (APC) Monocytes, macrophages, and their close relatives found in lymphoid organs and under the surface of the skin and characterized by many long, thin cytoplasmic processes called *dendrites*.

Antigen-binding site Site formed by the variable regions of a heavy and light chain of an antibody.

Antigenic determinant (*epitope*) The three-dimensional shape of a region of an antigen that is recognized by the immune system.

Antigenic drift Phenomenon that occurs every 2–3 years when a single strain of influenza mutates within a local population.

Antigenic shift Major antigenic change that occurs on average every 10 years and results from the reassortment of genomes from different influenzavirus strains within host cells.

Antihistamines Drugs which specifically neutralize histamine.

Antimicrobial agent Chemotherapeutic agent used to treat microbial infection.

Antimicrobic Any compound used to treat infectious disease; may also function as intermediate-level disinfectant.

Antisepsis The inhibition or killing of microorganisms on skin or tissue by the use of a chemical antiseptic.

Antiseptic Chemical used to inhibit or kill microorganisms on skin or tissue.

Antiserum In serology, blood fluid containing antibodies which binds to the antigens that triggered their production.

Antitoxin Antibodies formed by the host that bind to and protect against toxins.

Antiviral proteins Proteins triggered by alpha- and beta-interferons that prevent viral replication.

Apicomplexans In protozoan taxonomy, group of pathogenic alveolate protozoa characterized by the complex of special intracellular organelles located at the apices of the infective stages of these microbes.

Apoenzyme The protein portion of protein enzymes that is inactive unless bound to one or more cofactors.

Apoptosis Cell suicide.

Applied microbiology Branch of microbiology studying the commercial use of microorganisms.

Arachnid Group of arthropods distinguished by the presence of eight legs, such as spiders, ticks, and mites.

Arborviruses Group of enveloped, positive ssRNA viruses which are transmitted by arthropods.

Archaea (*Archaeon*, sing.) In Woese's taxonomy, domain which includes all prokaryotic cells having archaeal rRNA sequences.

Arenaviruses Group of segmented, negative ssRNA viruses which cause zoonotic diseases.

Arthropod Animal with a segmented body, hard exoskeleton, and jointed legs, including arachnids and insects.

Artificial wetlands Use of successive ponds, marshes, and meadowlands to remove wastes from sewage as the water moves through the wetlands.

Artificially acquired active immunity Type of immunity which occurs when the body receives antigens by injection, as with vaccinations, and mounts a specific immune response.

Artificially acquired passive immunity Type of immunity which occurs when the body receives via injection preformed antibodies in antitoxins or antisera, which can destroy fast-acting and potentially fatal antigens such as rattlesnake venom.

Ascariasis Symptomatic but not typically fatal disease caused by infection with the nematode *Ascaris lumbricoides*.

Ascomycota Division of fungi characterized by the formation of haploid ascospores within sacs called asci.

Ascospore Haploid germinating structure of ascomycetes.

Ascus Sac in which haploid ascospores are formed and from which they are released in fungi of the division Ascomycota.

Aseptate Lacking crosswalls.

Aseptic Characteristic of an environment or procedure that is free of contamination by pathogens.

Aspergillosis Term for several localized and invasive diseases caused by infection with *Aspergillus* species.

Assembly In virology, fourth stage of the lytic replication cycle, in which new virions are assembled in the host cell.

Asthma Hypersensitivity reaction affecting the lungs and characterized by bronchial constriction and excessive mucous production.

Astroviruses Group of small, round enteric viruses which cause diarrhea, typically in children.

Asymptomatic Characteristic of disease that may go unnoticed because of absence of symptoms, even though clinical tests may reveal signs of disease.

Attachment In virology, first stage of the lytic replication cycle, in which the virion attaches to the host cell.

Atom The smallest chemical unit of matter.

Atomic force microscope Type of probe microscope that uses a pointed probe to traverse the surface of a specimen. A laser beam detects vertical movements of the probe which a computer translates to reveal the atomic topography of the specimen.

Atomic mass (*atomic weight*) The sum of the masses of the protons, neutrons, and electrons in an atom.

Atomic number The number of protons in the nucleus of an atom.

ATP synthase Enzyme that phosphorylates ATP in oxidative phosphorylation and photophosphorylation.

Attenuation The process of reducing vaccine virulence.

Attenuated vaccine Inoculum in which pathogens are weakened so that, theoretically, they no longer cause disease; residual virulence can be a problem.

Autoantigens Antigens on the surface of normal body cells.

Autoclave Device that uses steam heat under pressure to sterilize chemicals and objects that can tolerate moist heat.

Autograft Type of graft in which tissues are moved to a different location within the same patient.

Autoimmune disease Any of a group of diseases which result when an individual begins to make autoantibodies or cytotoxic T cells against normal body components.

Autoimmune hemolytic anemia Disease resulting when an individual produces antibodies against his or her own red blood cells.

Avirulent Harmless.

Axial filament In cell morphology, structure composed of rotating endoflagella that allows a spirochete to "corkscrew" through its medium.

B cell receptor (BCR) Antibody integral to the cytoplasmic membrane and expressed by B lymphocytes.

B lymphocyte (*B cell*) Lymphocyte which arises in the red bone marrow in adults and is found primarily in the spleen, lymph nodes, red bone marrow, and Peyer's patches of the intestines and which secretes antibodies.

Bacillus Rod-shaped prokaryotic cell.

Bacitracin Antimicrobial which blocks NAG and NAM secretion from the cytoplasm, prompting cell lysis.

Bacteremia The presence of bacteria in the blood; often caused by infection with *Staphylococcus aureus* or *Streptococcus pneumoniae*.

Bacteria Prokaryotic microorganisms typically having cell walls composed of peptidoglycan. In Woese's taxonomy, domain which includes all prokaryotic cells having bacterial rRNA sequences.

Bacteriophage (*phage*) Virus that infects and usually destroys bacterial cells.

Bacteriorhodopsin Purple protein synthesized by *Halobacterium* that absorbs light energy to synthesize ATP.

Bacteroid Diverse group of Gram-negative

microbes that have similar rRNA nucleotide sequences, e.g., *Bacteroides.*

Balantidiasis A mild gastrointestinal illness caused by *Balantidium coli.*

Barophile Microorganism requiring the extreme hydrostatic pressure found at great depth below the surface of water.

Bartonellosis Often fatal disease characterized by fever, severe anemia, headache, muscle and joint pain, and chronic skin infections; caused by infection with *Bartonella bacilliformis.*

Base Molecule that binds with hydrogen ions when dissolved in water.

Base-excision repair Mechanism by which enzymes excise a section of a DNA strand containing an error, and then DNA polymerase fills in the gap.

Base pair A complementary arrangement of nucleotides in a strand of DNA or RNA. For example, in both DNA and RNA, guanine and cytosine always pair.

Basic dye In microscopy, a cationic chromophore used to stain acidic structures. Works most effectively in alkaline environments.

Basidiocarp Fruiting body of basidiomycetes; includes mushrooms, puffballs, stinkhorns, jelly fungi, bird's nest fungi, and bracket fungi.

Basidiomycota Division of fungi characterized by production of basidiospores and basidiocarps, fruiting bodies such as mushrooms, puffballs, and bracket fungi.

Basophil Type of granulocyte which stains blue with the basic dye methylene blue.

Bejel An oral disease caused by infection with *Treponema pallidum endemicum.*

Benign tumor Mass of neoplastic cells that remains in one place and is not generally harmful.

Benthic zone Bottom zone of freshwater or marine water, devoid of light and nutrients.

Beta-interferons Interferons secreted by virally infected fibroblasts within hours after infection.

Beta-lactam Antimicrobial whose functional portion is composed of beta-lactam rings, which inhibit peptidoglycan formation by irreversibly binding to the enzymes that cross-link NAM subunits.

Beta-lactamase Bacterial enzyme which breaks the beta-lactam rings of penicillin and similar molecules, rendering them inactive.

Beta-oxidation A catabolic process in which enzymes split pairs of hydrogenated carbon atoms from a fatty acid and join them to coenzyme A to form acetyl-CoA.

Betaproteobacteria Diverse group of Gram-negative bacteria in the phylum Proteobacteria capable of growing at very low nutrient levels.

Binary fission The most common method of asexual reproduction of prokaryotes, in which the parental cell disappears with the formation of progeny.

Binomial nomenclature The classification method used in the Linnaean system of taxonomy, which assigns each species both a genus name and a specific epithet.

Biochemical oxygen demand (BOD) A measure of the amount of oxygen aerobic bacteria require to metabolize organic and inorganic nitrogenous wastes in water.

Biochemistry Branch of chemistry which studies the chemical reactions of living things.

Bioconversion The transformation of organic materials into fuels via the use of microorganisms.

Biodiversity The number of species living in a given ecosystem.

Biofilm A slimy, organized system of bacteria on an environmental surface.

Biogeochemical cycling The movement of elements and nutrients from unusable forms to useable forms due to the activities of microorganisms.

Biological vector Biting arthropod or other animal that transmits pathogens and serves as host for the multiplication of the pathogen during some stage of the pathogen's life cycle.

Biomass The quantity of all organisms in a given ecosystem.

Bio-mining The science involved in searching natural populations of soil microbes for useful organisms.

Bioremediation The use of microorganisms to metabolize toxins in the environment to reclaim soils and waterways.

Bioreporter Type of biosensor composed of microbes with innate signaling capabilities.

Biosensor Device that combines bacteria or microbial products such as enzymes with electronic measuring devices to detect other bacteria, bacterial products, or chemical compounds in the environment.

Biosphere The region of Earth inhabited by living organisms.

Biotechnology Branch of microbiology in which microbes are manipulated to manufacture useful products.

Bioterrorism The use of microbes or their toxins to terrorize human populations.

Black piedra Benign scalp disease caused by infection with *Piedraea hortae.*

Blastomycosis Pulmonary disease found in the southeastern United States caused by infection with *Blastomyces dermatidis.*

Blood group antigens The surface molecules of red blood cells.

Bodily fluid transmission Spread of pathogenic microorganisms via blood, urine, saliva, or other bodily fluids.

Botulism Potentially fatal intoxication with botulism toxin; three types include foodborne botulism, infant botulism, and wound botulism.

Broad-spectrum drug Antimicrobial that works against many different kinds of pathogens.

Broth A liquid, nutrient-rich medium used for cultivating microorganisms.

Broth-dilution test Test for determining the minimum inhibitory concentration in which a standardized amount of bacteria is added to serial dilutions of antimicrobial agents in tubes or wells containing broth.

Bruton-type agammaglobulinemia An inherited disease in which affected babies cannot make immunoglobulins and experience recurrent bacterial infections.

Bubonic plague Severe systemic disease, fatal if untreated in 75% of patients, and characterized by fever, tissue necrosis, and the presence of buboes; caused by infection with *Yersinia pestis.*

Budding In prokaryotes, reproductive process in which an outgrowth of the parent cell receives a copy of the genetic material, enlarges, and detaches. In virology, extrusion of enveloped virions through the host's cell membrane.

Buffer A substance, such as a protein, that prevents drastic changes in internal pH.

Bulbar poliomyelitis Infection of the brain stem and medulla resulting in paralysis of muscles in the limbs or respiratory system; caused by infection with poliovirus.

Bunyaviruses Group of zoonotic pathogens that have a segmented genome of three −ssRNA molecules and are transmitted to humans via arthropods.

Burkitt's lymphoma Infectious cancer of the jaw caused by infection with Epstein-Barr virus.

Caliciviruses Group of small, round enteric viruses which cause diarrhea, nausea, and vomiting.

Calor Heat.

Calvin-Benson cycle Stage of photosynthesis in which atmospheric carbon dioxide is fixed and reduced to produce glucose.

Cancer Disease characterized by the presence of one or more malignant tumors.

Candidiasis Term for several opportunistic diseases caused by infection with *Candida* species.

Capnophile Microorganism which grows best with high levels of carbon dioxide in addition to low levels of oxygen.

Capsid A protein coat surrounding the nucleic acid core of a virion.

Capsomere A proteinaceous subunit of a capsid.

Capsule Glycocalyx composed or repeating units of organic chemicals firmly attached to the cell surface.

Capsule stain (see *negative stain*)

Carbohydrate Organic macromolecule consisting of atoms of carbon, hydrogen, and oxygen.

Carbon cycle Biogeochemical cycle in which carbon is cycled in the form of organic molecules.

Carbon fixation The attachment of atmospheric carbon dioxide to ribulose 1,5-bisphosphate (RuBP).

Carbuncle The coalescence of several furuncles extending deep into underlying tissues; caused by infection with *Staphylococcus aureus.*

Carcinogen Chemical capable of causing genetic mutations that result in cancer.

Caries (*cavities*) Tooth decay; caused by viridans streptococci.

Carotenoid Plant pigment which acts as an antioxidant.

Carrageenan Gel-like polysaccharide isolated from red algae and used as thickening agent.

Carrier In human pathology, continuous asymptomatic human source of infection.

Catabolism All of the decomposition reactions in an organism taken together.

Catarrhal phase In pertussis, initial phase lasting 1–2 weeks and characterized by signs and symptoms resembling those of a common cold.

Cation A positively charged ion.

Cat-scratch disease Common and occasionally serious infection in children; characterized by fever and malaise plus localized swelling; caused by infection with *Bartonella henselae*.

CD95 pathway In cell-mediated cytotoxicity, pathway involving CD95 protein which triggers apoptosis of infected cells.

Cell culture Cells isolated from an organism and grown on the surface of a medium or in broth. Viruses can be grown in a cell culture.

Cell wall In prokaryotic and some eukaryotic cells, structural boundary composed of polysaccharide chains that provides shape and support against osmotic pressure.

Cell-mediated immune response Immune response used by T cells to fight intracellular pathogens and abnormal body cells.

Cellular respiration Metabolic process that involves the complete oxidation of substrate molecules and production of ATP via a series of redox reactions.

Cellular slime mold Individual haploid myxamoeba that phagocytizes bacteria, yeasts, dung, and decaying vegetation.

Central dogma In genetics, fundamental description of protein synthesis which states that genetic information is transferred from DNA to RNA to polypeptides, which function alone or in conjunction as proteins.

Centrioles Nonmembranous organelles in animal cells which appear to function in the formation of flagella and cilia and in cell division.

Centrosome Region of a cell containing centrioles.

Cesspool Home equivalent of primary wastewater treatment in which wastes enter a series of porous concrete rings buried underground and are digested by microbes.

Cestodes (*tapeworms*) Group of helminths that are long, flat, and segmented and lack digestive systems.

Chagas' disease Potentially fatal disease caused by a bite from a kissing bug carrying *Trypanosoma cruzi* and characterized by the formation of swellings called *chagomas* at the site of the bite, followed by fever, swollen lymph nodes, myocarditis, organ enlargement, and eventually congestive heart failure.

Chancre Painless red lesion that appears at the site of infection with *Treponema pallidum*, the agent of syphilis.

Chemical bond An interaction between atoms in which electrons are either shared or transferred in such a way as to fill their valence shells.

Chemical fixation In microscopy, a technique which uses methyl alcohol or formalin to attach a smear to a slide.

Chemical reaction The making or breaking of a chemical bond.

Chemiosmosis Use of ion gradients to generate ATP.

Chemoautotroph Microorganism which uses carbon dioxide as a carbon source and catabolizes organic molecules for energy.

Chemoheterotroph Microorganism which uses organic compounds for both energy and carbon.

Chemokine An immune system cytokine which signals leukocytes to rush to the site of inflammation or infection and activate other leukocytes.

Chemotactic factors Peptides derived from complement and chemicals called cytokines which are released by leukocytes at a site of infection and attract phagocytic cells.

Chemotaxis Cell movement that occurs in response to chemical stimulus.

Chemotherapeutic agent Chemical used to treat disease.

Chemotherapy A branch of medical microbiology in which chemicals are studied for their potential to destroy pathogenic microorganisms.

Chitin Strong, flexible nitrogenous polysaccharide found in fungal cell walls and in the exoskeletons of insects and other arthropods.

Chlamydia Any of the Gram-negative pathogenic cocci having both DNA and RNA and growing and reproducing within the cells of mammals, birds, and a few invertebrates.

Chlorophyll Pigment molecule which captures light energy for use in photosynthesis.

Chlorophyta Green-pigmented division of algae that have chlorophylls *a* and *c*, store sugar and starch as food reserves, have cell walls composed of cellulose, and have rRNA sequences similar to plants. Considered the progenitors of plants.

Chloroplast Light-harvesting organelle found in photosynthetic eukaryotes.

Cholera Disease contracted through the ingestion of food and water contaminated with *Vibrio cholerae* and characterized by vomiting and watery diarrhea.

Cholera toxin Exotoxin produced by *Vibrio cholerae* which causes the movement of water out of the intestinal epithelium.

Chromatin Thread-like mass of DNA and associated histone proteins that becomes visible during mitosis as chromosomes.

Chromatin fiber An association of nucleosomes and proteins found within the chromosomes of eukaryotic cells.

Chromoblastomycosis Cutaneous and subcutaneous disease characterized by lesions that can disseminate internally; caused by traumatic introduction of dematiaceous fungi into the skin.

Chromosome In prokaryotes, a typically circular molecule of DNA associated with protein and RNA molecules and localized in a region of the cytosol called the nucleoid. In eukaryotes, chromosomes are thread-like and are most visible during mitosis and meiosis.

Chronic disease Any disease that develops slowly, usually with less severe symptoms, and is continual or recurrent.

Chronic granulomatous disease Primary immunodeficiency disease in which children have recurrent infections characterized by the development of large masses of inflammatory cells in lymph nodes, lungs, bones, and skin.

Chronic inflammation Type of inflammation that develops slowly, lasts a long time, and can cause damage (even death) to tissues, resulting in disease.

Chrysophyta Division of algae including the golden algae, yellow-green algae, and diatoms.

Cilia Short, hairlike, rhythmically motile projections covering some eukaryotic cells.

Ciliate In protozoan taxonomy, group of alveolate protozoa characterized by the presence of cilia in their trophozoite stages.

Class Taxonomical grouping of similar orders of organisms.

Clinical specimen Sample of human material, such as feces or blood, that is examined or tested for the presence of microorganisms.

Clonal deletion Process by which cells with receptors that respond to autoantigens are selectively killed via apoptosis.

Coccidioidomycosis Pulmonary disease found in the southwestern United States caused by infection with *Coccidioides immitis*.

Coccobacillus A prokaryotic cell intermediate in shape between a sphere and a rod, e.g., an elongated coccus.

Coccus Spherical prokaryotic cell.

Codon Triplet of mRNA nucleotides that codes for specific amino acids. For example, AAA is a codon for lysine.

Coenocyte Multinucleate cell resulting from repeated mitosis but postponed or absent cytokinesis.

Coenzyme Organic cofactor.

Cofactor Inorganic ions or organic molecules that are essential for enzyme action.

Cold enrichment Incubation of a specimen in a refrigerator to enhance the growth of cold-tolerant species.

Coliforms Enteric Gram-negative bacteria that ferment lactose to gas and are found in the intestinal tracts of animals and humans.

Colony Visible population of microorganisms living in one place.

Colony-forming unit A single cell or group of related cells used to obtain a pure culture.

Colorado tick fever Mild disease characterized by fever and chills and caused by infection with a coltivirus.

Coltiviruses Group of reoviruses that cause Colorado tick fever, a mild disease involving fever and chills.

Commensalism Symbiotic relationship in which one member benefits without significantly affecting the other.

Communicable disease Any infectious disease that comes either directly or indirectly from another host.

Competence   Ability of a prokaryotic cell to take up DNA from the environment.

Competitive inhibitor   Inhibitory substance that blocks enzyme activity by blocking active sites.

Complement fixation test   A complex assay used to determine the presence of specific antibodies by the ability of the antibodies to nullify the action of complement on sheep red blood cells.

Complement system   Set of proteins that act as chemotactic attractants, trigger inflammation and fever, and ultimately effect the destruction of foreign cells.

Complementary DNA (cDNA)   DNA synthesized from an mRNA template using reverse transcriptase.

Complex medium   Culturing medium that contains nutrients released by the partial digestion of yeast, beef, soy, or other proteins; thus, the exact chemical composition is unknown.

Complex transposon   Transposon containing inverted repeats, transposase, and other genes not connected with transposition.

Compound   A molecule containing atoms of more than one element.

Compound microscope   Microscope using a series of lenses for magnification.

Concentration gradient   The difference in concentration of a chemical on the two sides of a membrane. Also called a *chemical gradient*.

Condenser lens   In a compound microscope, a lens which directs light through the specimen as well as one or more mirrors which deflect its path.

Condyloma acuminata   Large, cauliflower-like genital warts caused by infection with a papillomavirus.

Confocal microscope   Type of light microscope which uses ultraviolet lasers to illuminate fluorescent chemicals in a single plane of the specimen.

Congenital syphilis   Disease characterized by mental retardation, organ malformation, and in some cases, death of the fetus of a woman infected with *Treponema pallidum*, the agent of syphilis.

Conjugation   Method of horizontal gene transfer in which a bacterium containing a fertility plasmid forms a conjugation pilus that attaches and transfers plasmid genes to a recipient.

Conjugation pilus   Proteinaceous, rodlike structure extending from the surface of a cell; mediates conjugation.

Contagious disease   A communicable disease that is easily transmitted from a reservoir or patient.

Continuous cell culture   Type of cell culture created from tumor cells.

Contrast   The difference in intensity between two objects, or between an object and its background.

Convalescence   In the infectious disease process, final stage during which the patient recovers from the illness and tissues and systems are repaired and return to normal.

Convalescent phase   In pertussis, final phase lasting 3–4 weeks during which the ciliated lining of the trachea grows back and frequency of coughing spells diminishes, but secondary bacterial infections may ensue.

Cord factor   A cell-wall component of the cell walls of *Mycobacterium tuberculosis* that produces strands of daughter cells, inhibits migration of neutrophils, and is toxic to body cells.

Coronaviruses   Group of enveloped ssRNA viruses which cause colds as well as gastroenteritis in children.

Corticosteroids   Immunosuppressive agents, including prednisone and methylprednisone, which suppress the response of T cells.

Counterstain   In a Gram stain, red stain which provides contrasting color to the primary stain, causing Gram-negative cells to appear pink.

Covalent bond   The sharing of a pair of electrons by two atoms.

Coxsackieviruses   Group of enteroviruses which cause mild fever, muscle aches, and malaise in humans.

Cristae   Folds within the inner membrane of mitochondria that increase its surface area.

Cross resistance   Phenomenon in which resistance to one antimicrobial drug confers resistance to similar drugs.

Crossing over   Process in which portions of homologous chromosomes are recombined during the formation of gametes.

Croup   Inflammation and swelling of the larynx, trachea, and bronchi and a "seal bark" cough caused by infection with a parainfluenza virus.

Cryptococcoses   Two diseases often resulting in meningitis and caused by infection with *Cryptococcus neoformans*.

Cryptosporidiosis   A gastrointestinal disease caused by infection with *Cryptosporidium parvum*; in humans, characterized by diarrhea and fluid and weight loss; may be fatal in HIV-positive patients.

Culture   Act of cultivating microorganisms or the microorganisms that are cultivated.

Cyanobacteria   Gram-negative phototrophic bacteria that vary greatly in shape, size, and method of reproduction.

Cyclic photophosphorylation   Return of electrons to the original reaction center after passing down an electron transport chain.

Cycloserine   Semisynthetic antibiotic used to treat infections with Gram-positive bacteria.

Cyst   In protozoan morphology, the hardy resting stage characterized by a thick capsule and a low metabolic rate.

Cytokines   Chemicals released by leukocytes at a site of infection.

Cytokinesis   Division of a cell's cytoplasm.

Cytoplasm   General term used to describe the semiliquid, gelatinous material inside a cell.

Cytoplasmic membrane   Membrane found in prokaryotic and eukaryotic cells, and composed of a fluid mosaic of phospholipids and proteins.

Cytosine   Ring-shaped nitrogenous base found in nucleotides of DNA and RNA.

Cytoskeleton   Internal network of fibers contributing to the basic shape of eukaryotic and rod-shaped prokaryotic cells.

Cytosol   The liquid portion of the cytoplasm.

Cytotoxic drugs   Group of drugs which inhibit mitosis and cytokinesis.

Cytotoxic T cell (*CD8 cell*)   In cell-mediated immune response, type of cell characterized by CD8 cell-surface glycoprotein which secretes perforins and granzymes that destroy infected or abnormal body cells.

Dark-field microscope   Microscope used for studying pale or small specimens; deflects light rays so that they miss the objective lens.

Dark repair   Mechanism by which enzymes cut damaged DNA sections from a molecule, creating a gap that is repaired by DNA polymerase and DNA ligase.

Deamination   Process in which amine groups are split off from amino acids.

Death phase   Phase in a growth curve in which the organisms are dying more quickly than they are being replaced by new organisms.

Decimal reduction time   The time required to destroy 90% of the microbes in a sample.

Decline   In the infectious disease process, period in which the body gradually returns to normal as the patient's immune response and any medical treatments vanquish the pathogens.

Decolorizing agent   In a Gram stain, a solution which breaks down the wall of Gram-negative cells, allowing the primary stain to be washed away.

Decomposition reaction   A chemical reaction in which the bonds of larger molecules are broken to form smaller atoms, ions, and molecules.

Deep freezing   Long-term storage of cultures at temperatures ranging from –50°C to –95°C.

Deeply branching bacteria   Prokaryotic autotrophs with rRNA sequences and growth characteristics thought to be similar to those of earliest bacteria.

Defensins   Small peptide chains that act against a broad range of pathogens.

Defined medium (synthetic medium)   Culturing medium of which the exact chemical composition is known.

Definitive host   In the life cycle of parasites, host in which mature and sometimes sexual forms of the parasite are present and usually reproducing.

Degerming   The removal of microbes from a surface by scrubbing.

Dehydration synthesis   Type of synthesis reaction in which two smaller molecules are joined together by a covalent bond, and a water molecule is formed.

Delayed hypersensitivity reaction (*type IV hypersensitivity*)   T cell-mediated inflammatory reaction that takes 24–72 hours to reach maximal intensity.

Deletion   Type of mutation in which a nucleotide base pair is deleted.

Deltaproteobacteria   Group of proteobacteria that includes *Desulfovibrio*, *Bdellovibrio*, and myxobacteria.

Denaturation   Process by which the bonds within an enzyme or other protein break, disrupting the protein's three-dimensional structure.

Dendritic cells (*Langerhans cells*)   Cells of the epidermis that devour pathogens.

Dengue fever   Self-limiting but extremely painful disease caused by a flavivirus transmitted by *Aedes* mosquitoes.

Dengue hemorrhagic fever   Potentially fatal disease involving a hyperimmune response to reinfection with the dengue virus and causing ruptured blood vessels, internal bleeding, and shock.

Denitrification   The conversion of nitrate into nitrogen gas by anaerobic respiration.

Deoxyribonucleic acid (**DNA**)   Nucleic acid consisting of nucleotides made up of phosphoric acid, a deoxyribose pentose sugar, and an arrangement of the bases adenine, guanine, cytosine, and thymine.

Dermatophytoses   Any of a variety of superficial skin, nail, and hair infections caused by dermatophytes.

Dermis   The layer of the skin deep to the epidermis and containing hair follicles, glands, and nerve endings.

Descriptive epidemiology   The careful recording of data concerning a disease.

Dessication   Inhibition of microbial growth by drying.

Deuteromycetes   Informal grouping of fungi having no known sexual stage.

Diapedesis (*emigration*)   Process whereby leukocytes leave intact blood vessels by squeezing between lining cells.

Diatom   Type of algae in the division Chrysophyta which has cell walls made of silica arranged in nesting halves called frustules.

Dichotomous key   In taxonomy, method of identifying organisms in which information is arranged in paired statements, only one of which applies to any particular organism.

Differential interference contrast microscope   Type of phase microscope which uses prisms to split light beams, giving images a three-dimensional appearance.

Differential medium   Culturing medium formulated such that either the presence of visible changes in the medium or differences in the appearances of colonies helps microbiologists differentiate among different kinds of bacteria growing on the medium.

Differential stain   In microscopy, a stain using more than one dye so that different structures can be distinguished. The Gram stain is the most commonly used.

Differential white blood cell count   Lab technique that indicates the relative numbers of leukocytes.

Diffusion   The net movement of a chemical down its concentration gradient.

Diffusion-susceptibility test (Kirby-Bauer test)   Simple, inexpensive test widely used to reveal which drug is most effective against a particular pathogen. Procedure involves inoculating a Petri plate uniformly with a standardized amount of the pathogen in question and arranging on the plate disks soaked in the drugs to be tested.

DiGeorge anomaly   Failure of the thymus to develop, and thus, absence of T cells.

Dimorphic   Having two forms, e.g., dimorphic fungi have both yeast-like and mold-like thalli.

Dinoflagellate   In protozoan taxonomy, group of unicellular alveolate protozoa characterized by photosynthetic pigments such as carotene and chlorophylls.

Diphtheria   Mild to potentially fatal respiratory disease caused by diphtheria toxin following infection with *Corynebacterium diphtheriae.*

Diplococcus   A pair of cocci.

Diploid   A nucleus with two copies of each chromosome.

Diploid cell culture   Type of cell culture created from embryonic animal, plant, or human cells that have been isolated and provided appropriate growth conditions.

Direct contact transmission   Spread of pathogens from one host to another involving body contact between the hosts.

Direct fluorescent antibody test   Immune test allowing direct observation of the presence of antigen in a tissue sample flooded with labeled antibody.

Disaccharide   Carbohydrate consisting of two monosaccharide molecules joined together.

Disease   Any adverse internal condition severe enough to interfere with normal body functioning.

Disease process   Definite sequence of events following contamination and infection.

Disinfectant   Physical or chemical agent used to inhibit or destroy microorganisms on inanimate objects.

Disinfection   The use of physical or chemical agents to inhibit or destroy microorganisms on inanimate objects.

Disk-diffusion method   Method of evaluating the effectiveness of a disinfectant or antiseptic in which a zone of inhibition is visible around disks soaked in effective agents. Also called *Kirby-Bauer method.*

Disseminated intravascular coagulation (DIC)   The formation of blood clots within blood vessels throughout the body; triggered by lipid A.

Dolor   Pain.

Domain   Any of three basic types of cell groupings distinguished by Carl Woese, containing or replacing the Linnaean taxon of *kingdom.*

Donor cell   In horizontal gene transfer, a cell which contributes part of its genome to a recipient.

Droplet transmission   Spread of pathogens from one host to another via droplet nuclei that exit the body during exhaling, coughing, and sneezing.

E site   In translation, site at which tRNA exits from the ribosome.

Eastern equine encephalitis (EEE)   Potentially fatal infection of the brain caused by a togavirus.

Ebola virus   Type of hemorrhagic fever caused by a filamentous virus and fatal in 90% of cases.

Echoviruses (*enteric cytopathic human orphan viruses*)   Group of enteroviruses which cause viral meningitis and colds.

Ecosystem   All of the organisms living in a particular habitat, and the relationships between the two.

Electrical gradient   Voltage across a membrane created by the electrical charges of the chemicals on either side.

Electrochemical gradient   The chemical and electrical gradients across a cell membrane.

Electrolyte   Any hydrated cation or anion; can conduct electricity through a solution.

Electron   A negatively charged subatomic particle.

Electron transport chain   Series of redox reactions that pass electrons from one membrane-bound carrier to another, and then to a final electron acceptor.

Electronegativity   The attraction of an atom for electrons.

Elek test   Immunodiffusion assay used to detect the presence of diphtheria toxin in a fluid sample.

Element   Matter that is composed of a single type of atom.

Elementary bodies   Infectious stage in the life cycle of chlamydias.

Elephantiasis   Enlargement and hardening of tissues, especially in the lower extremities, where lymph has accumulated following infection with *Wuchereria bancrofti.*

Empyema   In patients with staphylococcal pneumonia, the presence of pus in the alveoli of the lungs.

Encystment   In the life cycle of protozoa, stage in which cysts form in host tissues.

Endemic   In epidemiology, a disease that occurs at a relatively stable frequency within a given area or population.

Endemic relapsing fever   Disease characterized by recurrent episodes of fever and septicemia separated by symptom-free intervals; caused by infection with any of several types of *Borrelia* carried by ticks in the genus *Ornithodoros.*

Endocarditis   Potentially fatal inflammation of the endocardium; typically caused by infection with *Staphylococcus aureus* or *Streptococcus pneumoniae.*

Endocytosis   Active transport process, used by some eukaryotic cells, in which pseudopodia surround a substance and move it into the cell.

Endogenous antigen   Antigen produced by microorganisms that multiply inside the cells of the body.

Endogenous infection   An infection arising within the patient from opportunistic pathogens.

Endoflagellum   A special flagellum of spirochetes that spirals tightly around a cell, rather than protruding from it.

Endoplasmic reticulum   Netlike arrangement of hollow tubules continuous with the outer membrane of the nuclear envelope and functioning as a transport system.

Endospore   Environmentally resistant structure produced by the transformation of a vegetative cell of the Gram-positive genus *Bacillus* or *Clostridium*.

Endosymbiotic theory Theory proposing that eukaryotes formed from the phagocytosis of small aerobic prokaryotes by larger anaerobic prokaryotes.

Endothermic reaction Any chemical reaction that requires energy.

Endotoxin (*lipid A*) Potentially fatal toxin released from the cell wall of dead and dying Gram-negative bacteria.

Enrichment culture Technique used to enhance the growth of less abundant microorganisms by using a selective medium.

Enterobacteriaceae (enteric bacteria) Family of oxidase negative Gram-negative bacteria which can be pathogenic.

Enteroviruses Group of picornaviruses that are transmitted via the fecal-oral route but cause disease in any of a variety of target organs.

Entner-Doudoroff pathway Series of reactions that catabolize glucose to pyruvic acid using different enzymes from those used in either glycolysis or the pentose phosphate pathway.

Entry In virology, second stage of the lytic replication cycle, in which the virion or its genome enters the host cell.

Envelope In virology, membrane surrounding the viral capsid.

Environmental microbiology Branch of microbiology studying the role of microorganisms in soils, water, and other habitats.

Environmental specimen Sample of material taken from such sources as ponds, soil, and air and tested for the presence of microorganisms.

Enzyme An organic catalyst.

Enzyme-linked immunosorbent assays (ELISAs) A family of simple immune tests that use an enzyme as a label and that can be readily automated and read by machine.

Eosinophil Type of granulocyte which stains red to orange with the acidic dye eosin.

Epidemic In epidemiology, a disease that occurs at a greater than normal frequency for a given area or population.

Epidemic relapsing fever Disease characterized by recurrent episodes of fever and septicemia separated by symptom-free intervals; caused by infection with *Borrelia recurrentis*.

Epidemic typhus Disease characterized by high fever, mental and physical depression, and rash following exposure to lice carrying *Rickettsia prowazekii*.

Epidemiology Study of the occurrence, distribution, and spread of disease in humans.

Epidermis The outermost layer of the skin.

Epsilonproteobacteria Group of Gram-negative rods, vibrios, or spirals in the phylum Proteobacteria.

Erysipelas Pyoderma occurring on the faces of children accompanied by pain and inflammation and caused by infection with group A streptococci.

Erythema infectiosum (*fifth disease*) Harmless red rash occurring in children and caused by infection with B19 virus.

Erythrocyte Red blood cell.

Erythrocytic cycle In the lifecycle of *Plasmodium*, stage during which merozoites infect and cause lysis of erythrocytes.

Etest Test for determining minimum inhibitory concentration in which a plastic strip containing a gradient of the antimicrobial agent being tested is placed on a plate inoculated with the pathogen of interest.

Ethambutol Antimicrobial which blocks mycolic acid synthesis in the walls of mycobacteria.

Etiology The study of the causation of disease.

Euglenoids Euglenozoan protozoans that store food as paramylon, lack cell walls, and have eyespots used in positive phototaxis.

Eukarya In Woese's taxonomy, domain which includes all eukaryotic cells.

Eukaryote Any organism made up of cells containing a nucleus composed of genetic material surrounded by a distinct membrane. Classification includes animals, plants, algae, fungi, and protozoa.

Eutrophication The overgrowth of microorganisms in aquatic systems.

Exchange reaction Type of chemical reaction in which atoms are moved from one molecule to another by means of the breaking and forming of covalent bonds.

Excystment In the life cycle of protozoa, stage following ingestion by the host, in which cysts become trophozoites.

Exocytosis Active transport process, used by some eukaryotic cells, in which vesicles fuse with the cytoplasmic membrane and export their substances from the cell.

Exoerythrocytic cycle In the lifecycle of *Plasmodium*, stage during which infected mosquito infects sporozoites into the blood.

Exogenous antigen Antigen produced by microorganisms that multiply outside the cells of the body.

Exogenous infection An infection caused by pathogens acquired from the health care environment.

Exon Coding sequence of mRNA. Exons are connected to produce a functional mRNA molecule.

Exothermic reaction Any chemical reaction that releases energy.

Exotoxin Toxin secreted by a pathogenic microorganism into its environment.

Experimental epidemiology The testing of hypotheses resulting from analytical epidemiology concerning the cause of a disease.

Extracellular state In virology, state in which a virus exists outside of a cell and is called a virion.

Extremophile Microbe in the domain Archaea that requires extreme conditions of temperature, pH, and/or salinity to survive.

F (fertility) plasmid (F factor) Molecule of DNA coding for conjugation pili. Cells that contain an F plasmid are called F^+ cells and serve as donors during conjugation.

Facilitated diffusion Movement of substances across a cell membrane via protein channels.

Facultative anaerobe Microorganism which can live with or without oxygen.

Family Taxonomical grouping of similar genera of organisms.

Fats Compounds composed of three fatty acid molecules linked to a molecule of glycerol.

F_C region The stem region of an antibody.

Fecal-oral infection Spread of pathogenic microorganisms in feces to the mouth, such as results from drinking sewage-contaminated water.

Feedback inhibition (negative feedback) Method of controlling the action of enzymes in which the end-product of a series of reactions inhibits an enzyme in an earlier part of the pathway.

Fermentation The partial oxidation of sugar to release energy using an organic molecule rather than an electron transport chain as the final electron acceptor.

Fever Body temperature above 37°C.

Fever blisters (*cold sores*) Painful, itchy lesions on the lips; characteristic of infection with herpes simplex virus type 1.

Filariasis Infection of the lymphatic system caused by a filarial nematode.

Filtration The passage of air or liquid through a material that traps and removes microbes.

Fimbriae Sticky, nonmotile, proteinaceous extensions of some bacterial cells that function to adhere cells to one another and to environmental surfaces.

Firmicutes Phylum of bacteria that includes clostridia, mycoplasmas, and low G+C Gram-positive bacilli and cocci.

Flagellates Group of protozoa that possess at least one long flagellum, generally used for movement.

Flagellum A long, whiplike structure, composed of a basal body, hook, and filament, protruding from a cell.

Flavine adenine dinucleotide (FAD) Important vitamin-derived electron carrier molecule.

Fluid mosaic model Model describing the arrangement and motion of the proteins within the cytoplasmic membrane.

Fluorescent microscope Type of light microscope which uses an ultraviolet light source to fluoresce objects.

Folliculitis Infection of a hair follicle by *Staphylococcus aureus*.

Fomite Object inadvertently used to transfer pathogens to new hosts, e.g., needle, coin, etc.

Food microbiology The use of microorganisms in food production and the prevention of food-borne illnesses.

Food-borne transmission Spread of pathogenic microorganisms in or on foods that are poorly processed, undercooked, or improperly refrigerated.

Formed elements Cells and cell fragments suspended in blood plasma.

Frameshift mutation Type of mutation in which nucleotide triplets subsequent to an insertion or deletion are displaced, creating new sequences of codons that result in vastly altered polypeptide sequences.

Fungi Eukaryotic organisms lacking cell walls which obtain food from other organisms.

Functional group An arrangement of atoms common to all members of a class of organic molecules, such as the amine group found in all amino acids.

Furuncle A large, painful, nodular extension of folliculitis into surrounding tissue; caused by infection with *Staphylococcus aureus*.

Gametocyte In sexual reproduction of protozoa, cell which fuses with another gametocyte to form a diploid zygote.

Gamma-interferons Interferon produced by T lymphocytes and NK lymphocytes which activates macrophages and neutrophils days after an infection.

Gammaproteobacteria Largest and most diverse class of proteobacteria, including purple sulfur bacteria, methane oxidizers, pseudomonads, and others.

Gaseous agent High-level disinfecting gas used to sterilize heat-sensitive equipment and large objects.

Gel electrophoresis Technique used in recombinant DNA technology to separate molecules by size, shape, and electrical charge.

Gene A specific sequence of nucleotides that code for polypeptides or RNA molecules.

Gene library Collection of bacterial or phage clones, each of which carries a fragment of an organism's genome.

Gene therapy The use of recombinant DNA technology to insert a missing gene or repair a defective gene in human cells.

Generation time Time required for a bacterial cell to grow and divide.

Genetic engineering The manipulation of genes via recombinant DNA technology for practical applications.

Genetic fingerprinting (DNA fingerprinting) Technique which identifies unique sequences of DNA to: determine paternity; connect blood, semen, or skin cells to suspects in criminal investigations; or identify pathogens.

Genetic mapping Application of recombinant DNA technology in which genes are located on a nucleic acid molecule.

Genetic recombination The exchange of segments, typically genes, between two DNA molecules that are composed of homologous nucleotide sequences.

Genetic screening Procedure by which laboratory tests are used to screen patient and fetal DNA for mutant genes.

Genetics The study of inheritance and inheritable traits as expressed in an organism's genetic material.

Genome The sum of all the genetic material in a cell or virus.

Genomics The sequencing, analysis, and comparison of genomes.

Genotype Actual set of genes in an organism's genome.

Genus Taxonomical grouping of similar species of organisms.

Germ theory of disease Hypothesis formu-lated by Pasteur in 1857 that microorganisms are responsible for disease.

Giardiasis A mild to severe gastrointestinal illness caused by ingestion of cysts of *Giardia intestinalis*.

Glomerulonephritis Deposition of immune complexes in the walls of the glomeruli, networks of minute blood vessels in the kidneys; may result in kidney failure; typically caused by infection with Group A streptococcus.

Glycocalyx Sticky external sheath of prokaryotic and eukaryotic cells.

Glycolysis (Embden-Meyerhof pathway) First step in the catabolism of glucose via respiration and fermentation.

Goblet cells Mucous-secreting cells in the epithelium of the respiratory and gastrointestinal tracts.

Golgi body In eukaryotic cells, a series of flattened, hollow sacs surrounded by phospholipid bilayers and functioning to package large molecules for export in secretory vesicles.

Gonorrhea A common sexually transmitted disease caused by infection with *Neisseria gonorrhoeae*.

gp41 Antigenic HIV glycoprotein that promotes fusion of the viral envelope to a target cell.

gp120 Antigenic glycoprotein that is the primary attachment molecule of HIV.

Grafts Tissues or organs that have been transplanted, whether between sites on the same patient or between a donor and a recipient.

Graft-versus-host disease Disease resulting when donated bone marrow cells mount an immune response against the recipient's cells.

Gram-negative cell Bacterial cell having a wall composed of a thin layer of peptidoglycan, an external membrane of lipopolysaccharide, and a periplasmic space between and appearing pink after the Gram staining procedure.

Gram-positive cell Bacterial cell having a wall composed of a thick layer of peptidoglycan containing teichoic acids and retaining the crystal violet dye used in the Gram staining procedure.

Gram stain Technique for staining microbial samples by applying a series of dyes which leave some microbes purple and others pink. Developed by Christian Gram in 1884.

Granulocyte Type of leukocyte having large granules in the cytoplasm.

Granzyme Protein molecule in the cytoplasm of cytotoxic T cells which causes an infected cell to undergo apoptosis.

Graves' disease Production of autoantibodies that stimulate pituitary cells to secrete thyroid-stimulating hormone, which elicits excessive production of thyroid hormone and growth of the thyroid gland.

Group A streptococcus (*S. pyogenes*) A coccus that produces protein M and a hyaluronic acid capsule, both of which contribute to the pathogenicity of the species.

Group B streptococcus (*S. agalactiae*) A Gram-positive coccus that normally resides in the lower GI, genital, and urinary tracts but can cause disease in newborns.

Group translocation Active process, occurring in some prokaryotes, by which a substance being actively transported across a cell membrane is chemically changed during transport.

Growth An increase in size.

Growth curve Graph that plots the number of bacteria growing in a population over time.

Growth factor Organic chemical such as a vitamin required in very small amounts for microbial metabolism. In immunology, an immune system cytokine that stimulates stem cells to divide, ensuring that the body is supplied with sufficient leukocytes of all types.

Guanine Ring-shaped nitrogenous base found in nucleotides of DNA and RNA.

Gumma Syphilitic lesion that occurs in bones, nervous tissue, or on skin in patients with tertiary syphilis.

Habitat The physical localities in which organisms are found.

Halogen One of the four very reactive, nonmetallic chemical elements: iodine, chlorine, bromine, and fluorine. Used in disinfectants and antiseptics.

Halophile Microorganism requiring a saline environment (greater than 9% NaCl).

Hantavirus pulmonary syndrome Rapid, severe, and often fatal pneumonia caused by infection with a hantavirus.

Hantaviruses Group of bunyaviruses that are transmitted to humans via inhalation of virions in dried deer-mouse urine and cause hantavirus pulmonary syndrome.

Haploid A nucleus with a single copy of each chromosome.

Haustoria Modified hyphae which penetrate the tissue of the host to withdraw nutrients.

Hay fever Allergic reaction localized to the upper respiratory tract and characterized by nasal discharge, sneezing, itchy throat and eyes, and excessive tear production.

Heat fixation In microscopy, a technique which uses the heat from a flame to attach a smear to a slide.

Heavy metal ions The ions of high-molecular-weight metals such as arsenic that are used as antimicrobial agents because they denature proteins. They have largely been replaced because they are also toxic to human cells.

Helminths Multicellular eukaryotic worms, some of which are parasitic.

Helper T cell (T$_H$ cell) In cell-mediated immune response, type of cell characterized by CD4 cell-surface glycoprotein which regulates the activity of B cells and cytotoxic T cells during an immune response.

Hemagglutinin (HA) Component of glycoprotein spikes in the lipid envelope of influenza viruses that help them attach to pulmonary epithelial cells.

Hemolytic disease of the newborn Disease that results when antibodies made by the mother of an Rh negative woman cross the placenta and destroy the red blood cells of an Rh positive fetus.

Hepatitis B Inflammatory condition of the

liver caused by infection with hepatitis B virus and characterized by jaundice.

Hepatitis C   Chronic disease that can cause liver damage or hepatic cancer; caused by infection with hepatitis C virus.

Hepatitis D   Disease that can cause liver damage or hepatic cancer; caused by infection with hepatitis D virus simultaneously with hepatitis B virus.

Hepatitis E (*enteric hepatitis*)   Infection of the liver which is fatal in 20% of infected pregnant women.

Herpes zoster (*shingles*)   Extremely painful skin rash caused by reactivation of latent varicella-zoster virus.

Heterocyst   Thick-walled nonphotosynthetic cell of cyanobacteria which transports reduced nitrogen to neighboring cells in exchange for glucose.

Hfr (high frequency of recombination) cell   Cell containing an F plasmid that is integrated into the prokaryotic chromosome. Hfr cells form conjugation pili and transfer cellular genes more frequently than normal F$^+$ cells.

Histamine   Inflammatory chemical released from damaged cells that causes vasodilation of capillaries.

Histone   Globular protein found in eukaryotic chromosomes.

Histoplasmosis   Pulmonary, cutaneous, ocular, or systemic disease found in the Ohio River Valley and caused by infection with *Histoplasma capsulatum*.

Holoenzyme   The combination of an apoenzyme and its cofactors.

Horizontal gene transfer   Process in which a donor cell contributes part of its genome to a recipient cell which may be a different species or genus from the donor.

Human granulocytic ehrlichiosis   Disease caused by infection with *Ehrlichia equi* transmitted by ticks, and characterized by leukopenia, rash, malaise, and death in about 10% of untreated cases.

Human immunodeficiency viruses   Group of retroviruses which destroy the immune system.

Human monocytic ehrlichiosis   Disease caused by infection with *Ehrlichia chaffeensis* transmitted by ticks, and characterized by leukopenia, rash, malaise, and death in about 5% of untreated cases.

Human T-lymphocyte viruses   Group of oncogenic retroviruses associated with cancer of lymphocytes.

Humoral immune response   The antigen-antibody immune response used by B cells to fight exogenous antigens.

Hybridomas   Tumor cells created by fusing antibody-secreting plasma cells with cancerous plasma cells called *myelomas*.

Hydatid   Fluid-filled.

Hydatid disease   Potentially fatal disease caused by infection with the canine tapeworm *Echinococcus granulosus* and characterized by the presence of fluid-filled cysts in the liver and other tissues.

Hydrogen bond   The electrical attraction between a partially charged hydrogen atom and a full or partial negative charge on a different region of the same molecule or another molecule. Hydrogen bonds confer unique properties to water molecules.

Hydrolysis   A decomposition reaction in which a covalent bond is broken, and the ionic components of water are added to the products.

Hydrophobia   Literally a fear of water; symptom caused by painful swallowing characteristic of rabies infection.

Hydrophobic   Insoluble in water.

Hydrophilic   Attracted to water.

Hydrothermal vent   Vent in marine abyssal zone that spews super-heated, nutrient-rich water.

Hydroxyl radical   Most reactive of the toxic forms of oxygen.

Hypersensitivity   Any immune response against a foreign antigen that is exaggerated beyond the norm.

Hypersensitivity pneumonitis   Type of pneumonia resulting from repeated inhalation of mold spores and other antigens deep into the lungs, stimulating the formation of immune complexes.

Hyperthermophile   Microorganism requiring temperatures above 80°C.

Hypertonic   Characteristic of a solution having a higher concentration of solutes than another.

Hyphae   Long, branched, tubular filaments in the thalli of molds.

Hypotonic   Characteristic of a solution having a lower concentration of solutes than another.

Iatrogenic infections   A subset of nosocomial infections that are the direct result of a medical procedure or treatment, such as surgery.

Illness   In the infectious disease process, the most severe stage, in which signs and symptoms are most evident.

Immune complexes   Antigen-antibody complexes.

Immune thrombocytopenic purpura   Disease resulting when drugs bound to platelets bind antibodies and complement, causing the platelets to lyse.

Immunochromatographic assay   Rapid modification of ELISA test in which an antigen solution flows through a porous strip, encountering labeled antibody; used for pregnancy testing and for rapid identification of infectious agents.

Immunoelectrophoresis   Type of immune test used to analyze complex antigen mixtures in which electrophoresis is used to separate antigens before precipitation testing is performed.

Immunofiltration assay   Rapid modification of ELISA test using membrane filters rather than polystyrene plates.

Immunoglobulin (Ig)   (see *antibody*)

Immunoglobulin A (IgA)   The antibody class most commonly associated with various body secretions, including tears and milk. IgA pairs with a secretory component to form *secretory IgA*.

Immunoglobulin D (IgD)   A membrane-bound antibody molecule found in some animals as a B cell receptor.

Immunoglobulin E (IgE)   Signal antibody molecule which triggers the inflammatory response, particularly in allergic reactions and infections by parasitic worms.

Immunoglobulin G (IgG)   The predominant antibody class found in the bloodstream and the primary defender against invading bacteria.

Immunoglobulin M (IgM)   The second most common antibody class and the predominant antibody produced first during a primary humoral immune response.

Immunology   Study of the body's specific defenses against pathogens.

Immunophilins   Immunosuppressive drugs, such as cyclosporine, which inhibit T cell function.

Impetigo   Presence of red, pus-filled vesicles on the face and limbs of children; caused by infection with *Staphylococcus aureus*.

Inactivated polio vaccine (IPV)   Inoculum developed by Jonas Salk in 1955 for vaccination against poliovirus.

Inactivated vaccine   Inoculum containing either whole agents or subunits and often adjuvants.

Incidence   In epidemiology, the number of new cases of a disease in a given area or population during a given period of time.

Inclusion   Deposited substance such as a lipid, gas vesicle, or magnetite, stored within the cytosol of a prokaryote.

Inclusion bodies   In the life cycle of chlamydias, phagosomes full of reticulate bodies.

Incubation period   Stage in infectious disease process between infection and occurrence of the first symptoms or signs of disease.

Index case   In epidemiology, the first of the disease in a given area or population.

Indirect contact transmission   Spread of pathogens from one host to another via inanimate objects called *fomites*.

Indirect fluorescent antibody test   Immune test allowing observation through a fluorescent microscope of the presence of antigen in a tissue sample flooded with labeled antibody.

Inducible operon   Type of operon that is not normally transcribed and must be activated by inducers.

Induction   In virology, excision of a prophage from the host chromosome, at which point the prophage reenters the lytic phase.

Industrial microbiology   Branch of microbiology in which microbes are manipulated to manufacture useful products.

Infection   Successful invasion of the body by a pathogenic microorganism.

Infection control   Branch of microbiology studying the prevention and control of infectious disease.

Infectious mononucleosis (*mono*)   Disease characterized by sore throat, fever, fatigue, and enlargement of the spleen and liver; caused by infection with Epstein-Barr virus.

Influenza (*flu*)   Infectious disease caused by two species of orthomyxoviruses and characterized by fever, malaise, headache, and myalgia; certain strains can be fatal.

Initial body (*reticulate body*)   Reproductive

structure of chlamydias that undergoes repeated binary fissions until the host cell is filled.

Innate (*inborn*) **resistance** Resistance to a pathogen conferred by cellular or physiological incompatibility with the pathogen.

Inoculum Sample of microorganisms.

Inorganic chemical Molecule lacking carbon.

Insertion Type of mutation in which a base pair is inserted into a genetic sequence.

Insertion sequence A simple transposon consisting of no more than two inverted repeats and a gene that encodes the enzyme transposase.

Integrase Enzyme carried by the virions of HIV which allows integration into a human chromosome.

Interferons Protein molecules that inhibit the spread of viral infections.

Interleukin An immune system cytokine that signals among leukocytes.

Interleukin-1 A pyrogen released by phagocytes that have phagocytized bacteria.

Intermediate host In the life cycle of parasites, host in which immature forms of the parasite are present and undergoing various stages of maturation.

Intracellular state In virology, state in which a virus exists solely as nucleic acid within a cell.

Intron Noncoding sequence of mRNA which is removed to make functional mRNA.

In-use test Method of evaluating the effectiveness of a disinfectant or antiseptic which tests efficacy under specific, real-life conditions.

Inverted repeat Palindromic sequence found at each end of a transposon.

Ion An atom or group of atoms that has either a full negative charge or a full positive charge.

Ionic bond A type of bond formed from the attraction of opposite electrical charges. Electrons are not shared

Ionizing radiation Form of radiation that uses wavelengths shorter than 1 nm to strike molecules, creating ions by ejecting electrons from atoms.

Isograft Type of graft in which tissues are moved between genetically identical individuals (identical twins).

Isoniazid (INH) Antimicrobial which blocks mycolic acid synthesis in the walls of mycobacteria.

Isotonic Characteristic of a solution having the same concentration of solutes and water as another.

Isotopes Atoms of a given element that differ only in the number of neutrons they contain.

Ixodes Genus of ticks which transmit Lyme disease.

Kinetoplastid Euglenozoan protozoan with a single large mitochondrion that contains an apical region of mitochondrial DNA and is called a *kinetoplast*.

Kingdom Taxonomical grouping of similar phyla of organisms.

Kinins Powerful inflammatory chemicals released by mast cells.

Koch's postulates A series of steps, elucidated by Robert Koch, that must be taken to prove the cause of any infectious disease.

Koplik's spots Mouth lesions characteristic of measles.

Krebs cycle Series of eight enzymatically catalyzed reactions that transfer stored energy from acetyl-CoA to coenzymes NAD^+ and FAD.

Küpffer cells Fixed macrophages of the liver.

Lag phase Phase in a growth curve in which the organisms are adjusting to their environment.

Lagging strand Daughter strand of DNA synthesized in short segments that are later joined. Synthesis of the lagging strand always moves away from the replication fork, and lags behind synthesis of the leading strand.

Lassa hemorrhagic fever Infectious disease characterized by hemorrhage and caused by infection with an arenavirus.

Latency In virology, process by which an animal virus is not incorporated into the chromosomes of the cell and remains inactive in the cell, possibly for years.

Latent disease Any disease in which a pathogen remains inactive for a long period of time before becoming active.

Latent virus (see *provirus*) An animal virus that remains inactive in a host cell.

Leading strand Daughter strand of DNA synthesized continuously toward the replication fork as a single long chain of nucleotides.

Legionnaires' disease Severe pneumonia caused by infection with *Legionella pneumophila*.

Leishmaniasis Any of three clinical syndromes caused by a bite from a sand fly carrying *Leishmania* and ranging from painless skin ulcers to disfiguring lesions to visceral leishmaniasis, which is systemic and fatal in 95% of untreated cases.

Leprosy Disease caused by infection with *Mycobacterium leprae* that produces either a nonprogressive tuberculoid form or a progressive lepromatous form which destroys tissues, including facial features, digits, and other structures.

Leptospirosis Zoonotic disease contracted by humans upon exposure to infected animals; characterized by pain, headache, and liver and kidney disease; caused by infection with *Leptospira interrogans*.

Leukocyte White blood cell.

Leukotrienes Inflammatory chemical released from damaged cells that increase vascular permeability.

Lichen Organism composed of a fungus living in partnership with photosynthetic microbes, either green algae or cyanobacteria.

Light-dependent reaction Reaction of photosynthesis requiring light.

Light-independent reaction Reaction synthesizing glucose from carbon dioxide and water and not requiring light.

Light repair Mechanism by which prokaryotic DNA photolyase enzyme breaks the bonds between adjoining thymine nucleotides, restoring the original DNA sequence.

Limnetic zone Sunlit, upper layer of freshwater or marine water away from the shore.

Littoral zone Shoreline zone of freshwater or marine water.

Lipid Any of a diverse group of organic macromolecules not composed of monomers and insoluble in water.

Lipid A The lipid component of lipopolysaccharide which is released from dead Gram-negative bacterial cells and can trigger shock and other symptoms in human hosts.

Lipopolysaccharide (**LPS**) Molecule composed of lipid A and polysaccharide found in the external membrane of Gram-negative cell walls.

Lithotroph Microorganism which acquires electrons from inorganic sources.

Log phase Phase in a growth curve in which the population is most actively growing.

Logarithmic (exponential) growth Increase in size of a microbial population in which the number of cells doubles in a fixed interval of time.

Lophotrichous Term used to describe a cell having a tuft of flagella at one end.

Lyme disease Disease carried by ticks infected with *Borrelia burgdorferi* and characterized by a "bull's eye" rash, neurologic and cardiac dysfunction, and severe arthritis.

Lymphatic system Body system composed of lymphatic vessels and lymphoid tissues and organs.

Lymphocyte Type of small agranulocyte which originates and matures in the red bone marrow and has nuclei that nearly fill the cell.

Lymphocytic choriomeningitis (LCM) Type of hemorrhagic fever resulting in flu-like symptoms and caused by infection with an arenavirus.

Lymphogranuloma venereum Sexually transmitted disease caused by infection with *Chlamydia trachomatis* and leading in some cases to proctitis or, in women, pelvic inflammatory disease.

Lyophilization Removal of water from a frozen culture or other substance by means of vacuum pressure. Used for the long-term preservation of cells and foods.

Lysogenic replication cycle (*lysogeny*) Process of viral replication in which a bacteriophage enters a bacterial cell, inserts into the DNA of the host, but remains inactive. The phage is then replicated every time the host cell replicates its chromosome.

Lysosome Vesicle in animal cells that contains digestive enzymes.

Lysozyme Antibacterial protein secreted in sweat.

Lytic replication cycle Process of viral replication consisting of five stages ending with lysis of and release of new virions from the host cell.

Macrolide Antimicrobial agent that inhibits protein synthesis by inhibiting the ribosomal 50S subunits.

Macrophage Mature form of monocyte, and phagocyte of bacteria, fungi, spores, and dust, as well as dead cell bodies.

Macule   Any flat, reddened skin lesion; characteristic of early infection with a poxvirus.

Magnification   The apparent increase in size of an object viewed via microscopy.

Major histocompatibility complex (MHC)   A cluster of genes, located on each copy of chromosome 6 in humans, that codes for membrane-bound glycoproteins called major histocompatibility antigens.

Malaria   A mild to potentially fatal disease caused by a bite from an Anopheles mosquito carrying any of four species of *Plasmodium*; characterized by fever, chills, hemorrhage, and potential destruction of brain tissue.

Malignant tumor   Mass of neoplastic cells that can invade neighboring tissues and may metastasize to cause tumors in distant organs or tissues.

Marburg virus   Type of hemorrhagic fever caused by a filamentous virus and fatal in 90% of cases.

Margination   Process by which leukocytes stick to the walls of blood vessels at the site of infection.

Mast cells   Specialized cells located in connective tissue that release histamine when they are exposed to complement.

Matter   Anything that takes up space and has mass.

Measles (*rubeola*)   Contagious disease characterized by fever, sore throat, headache, dry cough, conjunctivitis, and lesions called Koplik's spots; caused by infection with *Morbillivirus*.

Mechanical vector   Housefly, cockroach, or other animal that passively carries pathogens to new hosts on its feet or other body parts and is not infected by the pathogens it carries.

Medium   A collection of nutrients used for cultivating microorganisms.

Meiosis   Nuclear division of diploid eukaryotic cells resulting in four haploid nuclei.

Membrane attack complexes (MAC)   The end products of the complement cascade, which form circular holes in a pathogen's membrane.

Membrane filters   Thin circles of nitrocellulose or plastic containing specific pore sizes, some small enough to trap viruses.

Membrane filtration   Direct method of estimating population size in which a large sample is poured through a filter small enough to trap cells.

Memory B cell   B lymphocyte that migrates to lymphoid tissues to await a subsequent encounter with the same antigen.

Memory T cell   Type of T cell that persists in lymphoid tissues for months or years awaiting subsequent contact with an antigenic determinant matching its TCR, at which point it produces cytotoxic T cells.

Memory response   The rapid and enhanced immune response to a subsequent encounter with a familiar antigen.

Mesophile   Microorganism requiring temperatures ranging from 20°C to about 40°C.

Messenger RNA (mRNA)   Form of ribonucleic acid which carries genetic information from DNA to a ribosome.

Metabolism   The sum of all chemical reactions, both anabolic and catabolic, within an organism.

Metachromatic granules   Inclusions of *Corynebacteria* that store phosphate and stain differently from the rest of the cytoplasm.

Metaphase   Second stage of mitosis, during which chromosomes line up and attach to microtubules of the spindle. Also used for the comparable stage of meiosis

Metastasis   The spreading of malignant cancer cells to nonadjacent organs and tissues, where they produce new tumors.

Methane oxidizer   Any Gram-negative bacterium that utilizes methane both as a carbon and as an energy source.

Methanogen   Obligate anaerobe that produces methane gas.

Methicillin-resistant *Staphylococcus aureus* (MRSA)   Strain of *Staphylococcus aureus* that is resistant to many common antimicrobial drugs, and has emerged as a major nosocomial problem.

Methylation   Process in which a cell adds a methyl group to one or two bases that are part of specific nucleotide sequences.

Microaerophile   Microorganism which requires low levels of oxygen.

Microbe   An organism too small to be seen without a microscope.

Microbial antagonism (*microbial competition*)   Normal condition in which established microbiota use up available nutrients and space, reducing the ability of arriving pathogens to colonize.

Microbial contamination   The presence of microorganisms in or on the body.

Microbial death   Permanent loss of reproductive capacity of a microorganism.

Microbial death rate   A measurement of the efficacy of an antimicrobial agent.

Microbial ecology   The study of the interactions of microorganisms among themselves and their environment.

Microglia   Fixed macrophages of the nervous system.

Micrograph   A photograph of a microscopic image.

Microorganism   An organism too small to be seen without a microscope.

Microscopy   The use of light or electrons to magnify objects.

Minimum bactericidal concentration (MBC) test   An extension of the MIC test in which samples taken from clear MIC tubes are transferred to plates containing a drug-free growth medium and monitored for bacterial replication.

Minimum inhibitory concentration (MIC)   The smallest amount of a drug that will inhibit a pathogen.

Mismatch repair   Mechanism by which enzymes scan newly synthesized, nonmethylated DNA for mismatched bases, removes them, and replaces them.

Missense mutation   A substitution in a nucleotide sequence resulting in a codon that specifies a different amino acid: what is transcribed makes sense, but not the right sense.

Mitochondria   Spherical to elongated structures found in most eukaryotic cells that produce most of the ATP in the cell.

Mitosis   Nuclear division of a eukaryotic cell resulting in two nuclei with the same ploidy as the original.

Mold   A typically multicellular fungus that grows as long filaments called *hyphae* and reproduces by means of spores.

Molecular biology   Branch of biology combining aspects of biochemistry, cell biology, and genetics to explain cell function at the molecular level, particularly via the use of genome sequencing.

Molecular mimicry   Process in which microorganisms with antigenic determinants similar to self-antigens trigger autoimmune tissue damage.

Molecule   Two or more atoms held together by chemical bonds.

Molluscum contagiosum   Infectious disease caused by *Molluscipoxvirus* which results in tumor-like growths on the skin.

Monoclonal antibodies   Identical antibodies secreted by a cell line originating from a single plasma cell.

Monocyte   Type of agranulocyte which has slightly lobed nuclei.

Monomer   A subunit of a macromolecule such as a protein.

Mononuclear phagocytic system   (see *reticuloendothelial system*)

Monosaccharide (*simple sugar*)   A monomer of carbohydrate, such as a molecule of glucose.

Monotrichous   Term used to describe a cell having a single flagellum.

Morbidity   Any change from a state of health.

Mordant   In microscopy, a substance that binds to a dye and makes it less soluble.

Most probable number (MPN) method   Statistical estimation of the size of a microbial population based upon the dilution of a sample required to eliminate microbial growth.

Multiple sclerosis   Autoimmune disease in which cytotoxic T cells attack and destroy the myelin sheath that insulates neurons.

Mumps   Disease caused by infection with the mumps virus and characterized by fever, parotitis, pain in swallowing, and in some cases, meningitis or deafness.

Murine typhus (*endemic typhus*)   Disease transmitted by fleas and characterized by high fever, headache, chills, muscle pain, and nausea; caused by infection with *Rickettsia typhi*.

Mutagen   Physical or chemical agent that introduces a mutation that changes the microbe's phenotype.

Mutant   A cell with an unrepaired genetic mutation, or any of its descendents.

Mutation   In genetics, change in the nucleotide base sequence of a genome.

Mutualism   Symbiotic relationship in which both members benefit from their interaction.

Mycelium   Tangled mass of hyphae.

Mycetismus   Mushroom poisoning.

Mycetoma Destructive, tumor-like infection of the skin, fascia, and/or bones of the hands or feet caused by mycelial fungi of several genera in the division Ascomycota.

Mycolic acid Long carbon-chain waxy lipid found in the walls of cells in the genus *Mycobacterium* that makes them resistant to dessication and staining with water-based dyes.

Mycology The field of medicine concerned with the diagnosis, management, and prevention of fungal diseases.

Mycosis Fungal disease.

Mycotoxicosis Poisoning caused by eating food contaminated with fungal toxins.

Mycotoxins Secondary metabolites produced by fungi and toxic to humans.

Mycoplasmas Class of low G+C bacteria which lack cytochromes, enzymes of the Krebs cycle, and cell walls and are pleomorphic.

Myxobacteria Gram-negative, aerobic, soil-dwelling bacteria with a unique life cycle including a stage of differentiation into fruiting bodies containing resistant myxospores.

Narrow-spectrum drug Antimicrobial that works against only a few kinds of pathogens.

Natural killer (NK) lymphocyte Type of nonspecific, defensive leukocyte that secretes toxins onto the surfaces of virally infected cells and neoplasms.

Naturally acquired active immunity Type of immunity which occurs when the body responds to exposure to antigens by mounting specific immune responses.

Naturally acquired passive immunity Type of immunity which occurs when a fetus, newborn, or child receives antibodies across the placenta or within breastmilk.

Necrotizing fasciitis Potentially fatal condition marked by toxemia, organ failure, and destruction of muscle and fat tissue following infection with group A streptococci

Negative feedback (see *feedback inhibition*)

Negative selection (*indirect selection*) Process by which auxotrophic mutants are isolated and cultured.

Negative stain (*capsule stain*) In microscopy, a staining technique used primarily to reveal bacterial capsules and involving application of an acidic dye which leaves the specimen colorless and the background stained.

Negative strand RNA (-RNA) Viral single-stranded RNA transcribed from the +ssRNA genome by viral RNA polymerase.

Negri bodies Aggregates of virions in the brains of rabies patients.

Neisseria Genus of pathogenic Gram-negative cocci which has fimbriae, a polysaccharide capsule, and lipooligosaccharide in the cell walls.

Nematodes Group of round, unsegmented helminths with pointed ends that have a complete digestive tract.

Neoplasia Uncontrolled cell division in a multicellular animal.

Neuraminidase (NA) Component of glycoprotein spikes in the lipid envelope of influenza viruses that provides access to cell surfaces by hydrolyzing mucus in the lungs.

Neutralization Antibody function in which the action of a toxin or attachment of a pathogen is blocked.

Neutralization test Immune test that measures the ability of antibodies to neutralize the biological activity of pathogens and toxins.

Neutron An uncharged subatomic particle.

Neutrophil Type of granulocyte which stains lilac with a mixture of acidic and basic dyes.

Neutrophile Microorganism requiring neutral pH.

Nicotinamide adenine dinucleotide (NAD$^+$) Important vitamin-derived electron carrier molecule.

Nicotinamide adenine dinucleotide phosphate (NADP$^+$) Important vitamin-derived electron carrier molecule.

Nitrification The process by which bacteria convert reduced nitrogen compounds such as ammonia into nitrate, which is more available to plants.

Nitrifying bacteria Chemoautotrophic bacteria that derive electrons from the oxidation of nitrogenous compounds.

Nitrogen cycle Biogeochemical cycle involving nitrogen fixation, ammonification, nitrification, and denitrification.

Nitrogen fixation The conversion of atmospheric nitrogen to ammonia.

Noncommunicable disease An infectious disease that arises from outside of hosts or from normal microbiota.

Noncompetitive inhibitor Inhibitory substance that blocks enzyme activity by binding to an allosteric site located elsewhere on the enzyme.

Noncyclic photophosphorylation The production of ATP by noncyclic electron flow.

Nonionizing radiation Electromagnetic radiation with a wavelength greater than 1 nm.

Nonliving reservoir of infection Soil, water, food, and inanimate object that is a continuous source of infection.

Nonsense mutation A substitution in a nucleotide sequence which causes an amino acid codon to be replaced by a stop codon.

Nonspecific defense Type of defense that protects the body against a wide variety of pathogens and that operates in a generalized way independent of the nature of the intruder.

Normal microbiota Microorganisms that colonize the surfaces of the human body without normally causing disease. They may be resident or transient.

Noroviruses Group of caliciviruses that cause diarrhea.

Nosocomial disease A disease acquired in a health care facility.

Nosocomial infection An infection acquired in a health care facility.

Nuclear envelope Double membrane composed of phospholipid bylayers surrounding a cell nucleus.

Nuclear pores Spaces in the nuclear envelope that function to control the transport of substances through it.

Nucleoid Region of the prokaryotic cytosol containing the cell's chromosome(s).

Nucleolus Specialized region in a cell nucleus where RNA is synthesized.

Nucleoplasm The semiliquid matrix of a cell nucleus.

Nucleosome Bead of DNA bound to histone in a eukaryotic chromosome.

Nucleotide Monomer of a nucleic acid.

Nucleotide analog Compound structurally similar to a normal nucleotide that can be incorporated into DNA, resulting in mismatched base pairing.

Nucleus Spherical to ovoid membranous organelle containing a eukaryotic cell's genetic material.

Numerical aperture The ability of a lens to gather light.

Nutrient Any chemical such as carbon, hydrogen, etc., required for growth of microbial populations.

Objective lens In microscopy, the lens immediately above the object being magnified.

Obligate aerobe Microorganism which requires oxygen as the final electron acceptor of the electron transport chain.

Obligate anaerobe Microorganism which cannot tolerate oxygen and uses a final electron acceptor other than oxygen.

Obligate halophile Microorganism requiring high osmotic pressure.

Ocular herpes (*ophthalmic herpes*) Disorder characterized by conjunctivitis, a gritty feeling in the eye, and pain, and characteristic of latent herpes simplex virus type 1.

Ocular lens In microscopy, the lens closest to the eyes. May be single (*monocular*) or paired (*binocular*).

Operator Regulatory element in an operon where repressor protein binds to stop transcription.

Operon A series of genes, a promoter, and an operator sequence controlled by one regulatory gene. The operon model explains gene regulation in prokaryotes.

Opportunistic pathogens Microorganisms that cause disease when the immune system is suppressed, when microbial antagonism is reduced, or when introduced into an abnormal area of the body.

Opsonin Antimicrobial protein which readies a pathogen for phagocytosis.

Opsonization The coating of pathogens by proteins called *opsonins*, making them more vulnerable to phagocytes.

Optimum growth temperature Temperature at which a microorganism's metabolic activities produce the highest growth rate.

Oral polio vaccine (OPV) Inoculum developed by Albert Sabin in 1961 for vaccination against poliovirus.

Order Taxonomical grouping of similar families of organisms.

Organelle Internal, membrane-bound structure found in eukaryotic cells that acts as a tiny organ to carry out one or more cell functions.

Organic compounds Molecules that contain both carbon and hydrogen atoms.

Organotroph Microorganism which acquires electrons from organic sources.

Ornithosis (*parrot fever*) A respiratory disease of birds that can be transmitted to humans and is caused by infection with *Chlamydia psittaci.*

Osmosis The diffusion of water molecules across a selectively permeable membrane.

Osmotic pressure The pressure exerted across a selectively permeable membrane by the solutes in a solution on one side of the membrane. The osmotic pressure exerted by high-salt or high-sugar solutions can be used to inhibit microbial growth in certain foods.

Osteomyelitis Inflammation of the bone marrow and surrounding bone; caused by infection with *Staphylococcus.*

Otitis media Inflammation of the middle ear; typically caused by *Streptococcus pneumoniae.*

Oxidation-reduction reaction (**redox reaction**) Any metabolic reaction involving the transfer of electrons from an electron donor to an electron acceptor; reactions in which electrons are accepted are called *reduction* reactions, whereas reactions in which electrons are donated are *oxidation* reactions.

Oxidation lagoons Successive wastewaster treatment areas (lagoons) used by farmers and ranchers to treat animal wastes and from which water is released into natural water systems.

Oxidative phosphorylation The use of energy from redox reactions to attach inorganic phosphate to ADP.

Oxidizing agent Antimicrobial agent which releases oxygen radicals.

P site In a ribosome, a binding site which holds a tRNA and the growing polypeptide.

Palisade arrangement In cell morphology, a folded arrangement of bacilli.

Pandemic In epidemiology, the occurrence of an epidemic on more than one continent simultaneously.

Papilloma (*wart*) Benign growth of the epithelium of the skin or mucous membranes.

Papule Any raised, reddened skin lesion that progresses from a macule; characteristic of infection with a poxvirus.

Parabasalid Group of single-celled, animal-like microorganisms that contain a Golgi-like parabasal body.

Paracoccidioidomycosis Pulmonary disease found from southern Mexico to South America caused by infection with *Paracoccidioides brasiliensis.*

Parainfluenza viruses Group of enveloped, negative ssRNA viruses which cause respiratory disease, particularly in children.

Parasite A microbe that derives benefit from its host while harming it or even killing it.

Parasitism Symbiotic relationship in which one organism derives benefit while harming, or even killing, its host.

Parasitology The study of protozoan and helminthic parasites of animal and humans.

Parenteral route A means by which pathogenic microorganisms can be deposited directly into deep tissues of the body, as in puncture wounds and surgery.

Paroxysmal phase In pertussis, second phase lasting 2–4 weeks and characterized by exhaustive coughing spells.

Parvoviruses Group of extremely small, pathogenic ssDNA viruses.

Pasturellaceae Family of gammaproteobacteria, two genera of which—Pasteurella and Haemophilus—are pathogenic.

Pasteurization The use of heat to kill pathogens and reduce the number of spoilage microorganisms in food and beverages.

Pathogen A microorganism capable of causing disease.

Pathogenicity A microorganism's ability to cause disease.

Pelvic inflammatory disease Infection of the uterus and uterine tubes; may be caused by infection with any of several bacteria.

Pentose phosphate pathway Enzymatic formation of phosphorylated pentose sugars from glucose 6-phosphate.

Peptic ulcer Erosion of the mucous membrane of the stomach or duodenum and caused by infection with *Helicobacter pylori.*

Peptide bond A covalent bond between amino acids in proteins.

Peptidoglycan Large, interconnected polysaccharide composed of chains of two alternating sugars and crossbridges of four amino acids. Main component of bacterial cell walls.

Perforin Protein molecule in the cytoplasm of cytotoxic T cells which forms channels (perforations) in an infected cell's membrane.

Periplasmic space In Gram-negative cells, the space between the cell membrane and the outer membrane containing peptidoglycan and periplasm.

Peritrichous Term used to describe a cell having flagella covering the cell surface.

Peroxide Toxic form of oxygen which is detoxified by catalase or peroxidase.

Peroxisome Vesicle found in all eukaryotic cells that degrades poisonous metabolic wastes.

Pertussis (*whooping cough*) Pediatric disease characterized by development of copious mucus, loss of tracheal cilia, and deep "whooping" cough; caused by infection with *Bordetella pertussis.*

Petri plate Dish filled with solid medium used in culturing microorganisms.

pH scale A logarithmic scale used for measuring the concentration of hydrogen ions in a solution.

Phaeohyphomycosis Cutaneous and subcutaneous disease characterized by lesions that can disseminate internally; caused by traumatic introduction of dematiaceous fungi into the skin.

Phaeophyta Brown-pigmented division of algae having cell walls composed of cellulose and alginic acid, a thickening agent.

Phage (see *bacteriophage.*)

Phage typing Method of classifying microorganisms in which unknown bacteria are identified by observing plaques.

Phagocytes Cells such as leukocytes which are capable of phagocytosis.

Phagocytosis Type of endocytosis in which solids are moved into the cell.

Phagolysosome Digestive vesicle formed by the fusing of a lysosome with a phagosome.

Phagosome A sac formed by a phagocyte's pseudopodia.

Pharyngitis (*strep throat*) Inflammation of the throat caused by infection with group A streptococci.

Phase microscope Type of microscope used to examine living microorganisms or fragile specimens.

Phase-contrast microscope Type of phase microscope that produces sharply defined images in which fine structures can be seen in living cells.

Phenol coefficient Method of evaluating the effectiveness of a disinfectant or antiseptic which compares the agent's efficacy to that of phenol.

Phenolic Compound derived from phenol molecules that have been chemically modified to denature proteins and disrupt cell membranes in a wide variety of pathogens.

Phenotype The physical features and functional traits of an organism expressed by genes in the genotype.

Phospholipid Phosphate-containing lipid made up of molecules with two fatty acid chains.

Phospholipid bilayer Bipolar structure of a cytoplasmic membrane.

Photoautotroph Microorganism which requires light energy and uses carbon dioxide as a carbon source.

Photoheterotroph Microorganism which requires light energy and gains nutrients via catabolism of organic compounds.

Photophosphorylation The use of energy from light to attach inorganic phosphate to ADP.

Photosynthesis Process in which light energy is captured by chlorophylls and transferred to ATP and metabolites.

Photosystem Network of light-absorbing chlorophyll molecules and other pigments held within a protein matrix on thylakoids.

Phototaxis Cell movement that occurs in response to light stimulus.

Phylum Taxonomical grouping of similar classes of organisms.

Picornaviruses Family of viruses that contain positive single-stranded RNA and naked polyhedral capsids; many are human pathogens.

Pilus A long, hollow, nonmotile tube of protein that connects two bacterial cells and mediates the transfer of DNA. Also called *conjugation pilus.*

Pinocytosis Type of endocytosis in which liquids are moved into the cell.

Pinta A disfiguring skin disease caused by infection with *Treponema carateum.*

Pityriasis Condition characterized by depigmented or hyperpigmented patches of scaly skin resulting from infection with *Malassezia furfur.*

Plaque In phage typing, the clear region within the bacterial lawn where growth is inhibited.

Plaque assay   Technique for estimating phage numbers in which each plaque corresponds to a single phage in the original bacterium/virus mixture.

Plasma   The liquid portion of blood.

Plasma cells   B cells that are actively fighting against exogenous antigens and secreting antibodies.

Plasmid   A small, circular molecule of DNA that replicates independently of the chromosome. Each carries genes for its own replication and often for one or more nonessential functions such as resistance to antibiotics.

Plasmodial slime mold   (acellular slime mold) Streaming, coenocytic, colorful filaments of cytoplasm that phagocytize organic debris and bacteria.

Plasmolysis   Shriveling of a cell's cytoplasm caused by osmotic pressure.

Platelet   Cell fragments involved in blood clotting.

Pleomorphic   In cell morphology, term used to describe a variably-shaped prokaryotic cell.

***Pneumocystis* pneumonia (PCP)**   Debilitating fungal pneumonia that is the leading cause of death in AIDS patients and is caused by opportunistic infection with *Pneumocystis jiroveci*.

Pneumonia   Inflammation of the lungs; typically caused by infection with a strain of *Staphylococcus* or *Streptococcus pneumoniae*.

Pneumonic plague   Fever and severe respiratory distress caused by infection of the lungs with *Yersinia pestis*; fatal if untreated in nearly 100% of cases.

Point mutation   A genetic mutation affecting only one or a few base pairs in a genome. Point mutations include substitutions, insertions, and deletions.

Polar   In cell morphology, pertaining to either end of a cell, e.g., polar flagella.

Polar covalent bond   Type of bond in which there is unequal sharing of electrons between atoms with opposite electrical charges.

Poliomyelitis   (*polio*) Infection of varied degrees of severity from asymptomatic to crippling and caused by infection with poliovirus.

Polluted   Containing microorganisms or chemicals in excess of acceptable values.

Polyenes   Group of antimicrobial drugs such as amphotericin B which disrupt the cytoplasmic membrane of targeted cells by becoming incorporated into the membrane and damaging its integrity.

Polymer   Repeating chains of covalently linked monomers found in macromolecules.

Polymerase chain reaction (PCR)   Technique of recombinant DNA technology that allows researchers to produce a large number of DNA molecules rapidly *in vitro*.

Polyomaviruses   Group of opportunistic viruses that infect the kidneys of most people but cause tumors in immunocompromised patients.

Polysaccharide   Carbohydrate polymer composed of several to thousands of covalently linked monosaccharides.

Polyunsaturated fat   Triglyceride with several double bonds between adjacent carbon atoms in its fatty acids.

Portal of entry   Entrance site of pathogenic microorganisms, including the skin, mucous membranes, and placenta.

Portal of exit   Exit site of pathogenic microorganisms, including the nose, mouth, and urethra.

Positive selection   Process by which mutants are selected by eliminating wild type phenotypes.

Positive strand RNA (+RNA)   Viral single-stranded RNA that can act directly as mRNA.

Potable   Fit to drink.

Pour plate   Method of culturing microorganisms in which colony-forming units are separated from one another using a series of dilutions.

Pox   (*pocks*; see *pustule*)

Precursor metabolite   Any of twelve molecules typically generated by a catabolic pathway and essential to the synthesis of organic macromolecules in many bacterial cells.

Prevalence   In epidemiology, the total number of cases of a disease in a given area or population during a given period of time.

Primary atypical pneumonia   So-called *walking pneumonia* characterized by mild respiratory symptoms that last for several weeks; caused by infection with *Mycoplasma pneumoniae*.

Primary immunodeficiency diseases   Any of a group of diseases detectable near birth and resulting from a genetic or developmental defect.

Primary response   The slow and limited immune response to a first encounter with an unfamiliar antigen.

Primary stain   In a Gram stain, the initial stain, which colors all cells purple.

Prion   Proteinaceous infectious particle that lacks nucleic acids and replicates by converting similar normal proteins into new prions.

Privileged sites   The thymus, the cornea of the eye, and other sites at which grafts are not likely to be rejected.

Probe   Nucleic acid molecule with a specific nucleotide sequence that has been labeled with a radioactive or fluorescent chemical so that its location can be detected.

Prodromal period   In the infectious disease process, short time of generalized, mild symptoms that precedes illness.

Products   The atoms, ions, or molecules that remain after a chemical reaction is complete.

Profundal zone   Zone of freshwater or marine water beneath the limnetic zone and above the benthic zone.

Proglottids   Body segments of a tapeworm that grow continuously as long as the worm remains attached to its host.

Progressive multifocal leukoencephalopathy (PML)   Irreversible and typically fatal neurological disease caused by infection with JC virus in immunocompromised patients.

Prokaryote   Any unicellular microorganism that lacks a nucleus. Classification includes bacteria and archaea.

Promoter   Region of DNA where transcription begins.

Prophage   An inactive bacteriophage.

Prophase   First stage of mitosis, during which DNA condenses into chromatids and the spindle apparatus forms. Also used for the comparable stage of meiosis.

Prostaglandins   Inflammatory chemicals released from damaged cells that increase vascular permeability.

Protease   Enzyme secreted by microorganisms that digests proteins into amino acids outside a microbe's cell wall; in inflammatory reactions, chemicals released by mast cells that activate the complement system; in virology, an internal viral enzyme that makes HIV virulent.

Protein   A complex macromolecule consisting of carbon, hydrogen, oxygen, nitrogen, and sulfur and important to many cell functions.

Proteobacteria   Phylum of prokaryotes that includes five classes (designated alpha, beta, gamma, delta, and epsilon) of Gram-negative bacteria sharing common 16S rRNA nucleotide sequences.

Proton   A positively charged subatomic particle.

Proton gradient   Electrochemical gradient of hydrogen ions across a membrane.

Protozoa   Single-celled eukaryotes that lack a cell wall and are similar to animals in their nutritional needs and structure.

Provirus   Viral DNA that has become incorporated into a host chromosome.

Pseudomonad   Any Gram-negative, aerobic, rod-shaped bacterium in the class Gammaproteobacteria that catabolizes carbohydrates by the Entner-Doudoroff and pentose phosphate pathways.

Pseudopodia   Movable extensions of the cytoplasm and membrane of some eukaryotic cells.

Psychrophile   Microorganism requiring very cold temperatures (below 20°C).

Pure culture (axenic culture)   Culture containing cells of only one species of microorganism.

Purple sulfur bacteria   Group of gammaproteobacteria which are obligate anaerobes and oxidize hydrogen sulfide to sulfur.

Pustule   (*pox*) Any raised, pus-filled skin lesion; characteristic of infection with a poxvirus.

Pyoderma   Any confined, pus-producing lesion on the exposed skin of the face, arms, or legs; often caused by infection with group A streptococci.

Pyrogen   Chemical that triggers the hypothalamic "thermostat" to reset at a higher temperature, inducing fever.

Q fever   Disease characterized by high fever, severe headache, chills, and mild pneumonia, and caused by infection with *Coxiella burnetii*.

Quaternary ammonium compound (quat)   Detergent antimicrobial which is harmless to humans.

Quorum sensing   Process by which bacteria respond to changes in microbial density by utilizing signal and receptor molecules.

Rabies   Neuromuscular disease characterized by hydrophobia, seizures, hallucinations, and paralysis, and fatal if untreated, caused by infection with the rabies virus.

Radial immunodiffusion Type of immunodiffusion test in which an antigen solution is allowed to diffuse into agar containing specific concentrations of antibodies, causing a ring of precipitate to form.

Radiation The release of high-speed subatomic particles or waves of electromagnetic energy from atoms.

Reactants The atoms, ions, or molecules that exist at the beginning of a chemical reaction.

Reaction center chlorophyll In a photosystem, a chlorophyll molecule in which electrons excited by light energy are passed to an acceptor molecule which is the initial carrier of an electron transport chain.

Recipient cell In horizontal gene transfer, a cell that receives part of the genome of a donor cell.

Recombinant Cell resulting from genetic recombination between donated and recipient DNA.

Recombinant DNA technology Type of biotechnology in which scientists change the genotypes and phenotypes of organisms to benefit humans in various medical and agricultural applications.

Red tide Abundance of red-pigmented dinoflagellate protozoa in marine water.

Reducing medium Special culturing medium containing compounds that combine with free oxygen and remove it from the medium.

Refrigeration Short-term storage of cultures at 5°C.

Release In virology, final stage of the lytic replication cycle, in which the new virions are released from the host cell, which lyses.

Reoviruses Group of naked, segmented, dsRNA viruses that cause respiratory and gastrointestinal disease.

Repressible operon Type of operon that is continually transcribed until deactivated by repressors.

Reproduction An increase in number.

Reservoir of infection Living or nonliving continuous source of infectious disease.

Resolution The ability to distinguish between objects that are close together.

Responsiveness An ability to respond to environmental stimuli.

Restriction enzyme Enzyme which cuts DNA at specific nucleotide sequences and is used to produce recombinant DNA molecules.

Reticulate bodies (*initial bodies*) Noninfectious stage in the life cycle of chlamydias.

Reticuloendothelial system (*mononuclear phagocytic system*) Collectively, macrophages of all types and monocytes attached to endothelial cells.

Retrovirus Any +ssRNA virus that uses the enzyme reverse trasncriptase carried within its capsid to transcribe DNA from its RNA.

Reverse transcriptase Complex enzyme that allows retroviruses to make dsDNA from RNA templates.

Revolving nosepiece Portion of a compound microscope on which several objective lenses are mounted.

Rh antigens Cytoplasmic membrane proteins common to the red blood cells of 85% of humans as well as rhesus monkeys.

Rheumatic fever A complication of untreated group A streptococcal pharyngitis in which inflammation leads to damage of the heart valves and muscle.

Rheumatoid arthritis A crippling, systemic autoimmune disease, resulting from a type III hypersensitivity reaction, in which B cells in the joints produce autoantibodies against collagen.

Rhinoviruses Group of picornaviruses that cause upper respiratory tract infection and the "common cold."

Rhodophyta Red algae containing the pigment phycoerythrin, the storage molecule floridean starch, and cell walls of agar or carrageenan.

Ribonucleic acid (RNA) Nucleic acid consisting of nucleotides made up of phosphoric acid, a ribose pentose sugar, and an arrangement of the bases adenine, guanine, cytosine, and uracil.

Ribosomal RNA (rRNA) Form of ribonucleic acid which, together with polypeptides, makes up the structure of ribosomes.

Ribosome Nonmembranous organelle found in prokaryotes and eukaryotes that is composed of protein and ribosomal RNA and functions to make proteins.

Ribozyme RNA molecule functioning as an enzyme.

Rickettsias Group of extremely small, Gram-negative, obligate intracellular parasites that appear almost wall-less.

RNA polymerase Enzyme that synthesizes RNA by linking RNA nucleotides that are complementary to genetic sequences in DNA.

Rocky Mountain spotted fever (RMSF) Serious illness caused by infection with *Rickettsia rickettsii* transmitted by ticks and characterized by rash, malaise, petechiae, encephalitis, and death in 5% of cases.

Roseola Endemic illness of children characterized by an abrupt fever, sore throat, enlarged lymph nodes, and faint pink rash; caused by infection with human herpesvirus 6.

Rotaviruses Group of reoviruses that cause a potentially fatal infantile gastroenteritis.

Rough endoplasmic reticulum (RER) Type of endoplasmic reticulum that has ribosomes adhering to its outer surface which produce proteins for transport throughout the cell.

R-plasmid Extrachromosomal piece of DNA containing genes for resistance to antimicrobial drugs.

Rubella (*German measles*) Disease caused by infection with *Rubivirus* resulting in characteristic rash lasting about three days; mild in children but potentially teratogenic to fetuses of infected women.

Rubor Redness

Salmonellosis A serious diarrheal disease usually resulting from consumption of food contaminated with the enteric bacterium *Salmonella*.

Salt A crystalline compound formed by ionic bonding of elements.

Sanitization The process of disinfecting surfaces and utensils used by the public.

Saprobe Fungus that absorbs nutrients from dead organisms.

Sarcina A cuboidal packet of cocci.

Satellite virus A virus, such as hepatitis D virus, that requires glycoproteins coded by another virus to complete its replication cycle.

Saturated fat A triglyceride in which all but the terminal carbon atoms are covalently linked to two hydrogen atoms.

Scalded skin syndrome Reddening and blistering of the skin caused by infection with *Staphylococcus aureus*.

Scanning electron microscope (SEM) Type of electron microscope that uses magnetic fields within a vacuum tube to scan a beam of electrons across a specimen's metal-coated surface.

Scanning tunneling microscope Type of probe microscope in which a tungsten probe passes slightly above the surface of the specimen, revealing surface details at the atomic level.

Scarlet fever Diffuse rash and sloughing of skin caused by infection with group A streptococci.

Schaeffer-Fulton endospore stain In microscopy, staining technique that uses heat to drive a malachite green primary stain into an endospore.

Schistosomiasis A potentially fatal disease caused by infection with a blood fluke in the genus *Schistosoma*; may cause tissue damage in the liver, lungs, brain, or other organs.

Schizogony Special type of asexual reproduction in which the protozoan *Plasmodium* undergoes multiple mitoses to form a multinucleate schizont.

Schizont Multinucleate body formed by the protozoan *Plasmodium* which undergoes cytokinesis to release merozoites.

Scientific method Process by which scientists attempt to prove or disprove hypotheses through observations of the outcomes of carefully controlled experiments.

Scolex Small attachment organ that possesses suckers and/or hooks used to attach a tapeworm to host tissues.

Scrub typhus Disease caused by infection with *Orienta tsutsugamushi* transmitted by mites and characterized by fever, headache, muscle pain, and in a few cases, heart failure and death.

Sebum Oily substance secreted by the sebaceous glands of the skin that lowers pH.

Secretory IgA The combination of IgA and a secretory component, found in tears, mucous membrane secretions, and breastmilk, where it agglutinates and neutralizes antigens.

Secretory vesicle In eukaryotic cells, vesicles containing secretions packaged by the Golgi body which fuse to the cytoplasmic membrane and then release their contents outside the cell via exocytosis.

Selective medium Culturing medium containing substances that either favor the growth of particular microorganisms or inhibit the growth of unwanted ones.

Selective toxicity   Principle by which an effective antimicrobial agent must be more toxic to a pathogen than to the pathogen's host.

Selectively permeable   In cell physiology, characteristic of a membrane that allows some substances to cross while preventing the crossing of others.

Semisynthetic antibiotic   Antibiotic that has been chemically altered.

Septate   Characterized by the presence of crosswalls.

Septic tank   The home equivalent of primary wastewater treatment, consisting of a sealed concrete holding tank in which solids settle to the bottom and the effluent flows into a leach field that acts as a filter.

Serology   The study and use of immunologic tests to diagnose and treat disease or identify antibodies or antigens.

Severe combined immunodeficiency disease (SCID)   Primary immunodeficiency disease in children which affects both T cells and B cells and causes recurrent infections.

Shiga toxin   Exotoxin secreted by *Shigella dysenteriae* which stops proteins synthesis in host cells.

Shigellosis   A severe form of dysentery caused by any of four species of *Shigella*.

Shingles   (see *herpes zoster*)

Signs   In human pathology, objective manifestations of a disease that can be observed or measured by others.

Silent mutation   Mutation produced by base pair substitution that does not change the amino acid sequence due to the redundancy of the genetic code.

Simple microscope   Microscope containing a single magnifying lens.

Simple stain   In microscopy, a stain composed of a single dye such as crystal violet.

Singlet oxygen   Toxic form of oxygen which is neutralized by pigments called carotenoids.

Sinusitis   Inflammation of the nasal sinuses; typically caused by *Streptococcus pneumoniae*.

Slant tube (slant)   Test tube containing agar media that solidified while the tube was resting at an angle.

Slime layer   Loose, water-soluble glycocalyx.

Slime mold   Eukaryotic microbe resembling a filamentous fungus but lacking a cell wall and phagocytizing rather than absorbing nutrients.

Sludge   After primary treatment of wastewater, the heavy material remaining at the bottom of settling tanks.

Smallpox   Infectious disease eradicated in nature by 1980 and characterized by high fever, malaise, delirium, pustules, and death in about 20% of untreated cases.

Smear   In microscopy, the thin film of organisms on the slide.

Smooth endoplasmic reticulum (SER)   Type of endoplasmic reticulum that lacks ribosomes and plays a role in lipid synthesis and transport.

Snapping division   A variation of binary fission in Gram-positive prokaryotes in which the

parent cell's outer wall tears apart with a snapping movement to create the daughter cells.

SOS response   Mechanism by which prokaryotic cells with extensive DNA damage use a variety of processes to induce DNA polymerase to copy the damaged DNA.

Southern blot   Technique used in recombinant DNA technology that allows researchers to stabilize specific DNA sequences from an electrophoresis gel and then localize them using DNA dyes or probes.

Species   Taxonomical category of organisms that can successfully interbreed.

Specific epithet   In taxonomy, latter portion of the descriptive name of a species of organisms, describing that species' specific characteristics.

Specific immunity   The ability of a vertebrate to recognize and defend against distinct species or strains of invaders.

Spectrum of action   The number of different kinds of pathogens a drug acts against.

Spiral   In cell morphology, a spiral-shaped prokaryotic cell.

Spirillus   A stiff spiral-shaped prokaryotic cell.

Spirochetes   Group of helical, Gram-negative bacteria with axial filaments that cause the organism to corkscrew, enabling it to burrow into a host's tissues.

Spoilage   Any unwanted change to a food.

Spontaneous generation   The theory that living organisms can arise from nonliving matter.

Sporadic   In epidemiology, characteristic of a disease that occurs in only a few scattered cases within a given area or population during a given period of time.

Spore   Reproductive cell of actinomycetes.

Sporogonic cycle   In the lifecycle of *Plasmodium*, stage during which sporozoites are produced in the mosquito's digestive tract, and migrate into the mosquito's salivary glands.

Sporotrichosis   Subcutaneous infection usually limited to the arms and legs in which lesions form around the site of infection with *Sporothrix schenckii*.

Staining   Coloring microscopy specimens with stains called *dyes*.

Staphylococcus   A cluster of cocci.

Starter culture   Group of known microorganisms that carry out specific and reproducible fermentation reactions.

Stationary phase   Phase in a growth curve in which new organisms are being produced at the same rate at which older organisms are dying.

Stem cells   Generative cells capable of dividing to form daughter cells of a variety of types.

Sterile   Free of microbial contamination.

Sterilization   The eradication of all microorganisms, including bacterial endospores and viruses, although not prions, in or on an object.

Steroid   Lipids consisting of four fused carbon rings attached to various side chains and functional groups.

Streak plate   Method of culturing microorganisms in which a sterile inoculating loop is used to spread an inoculum across the surface of a solid medium in Petri dishes.

Streptococcal toxic shock syndrome   Rare but

potentially fatal syndrome characterized by fever, vomiting, diarrhea, pain, and organ failure, and caused by systemic infection with group A streptococci.

Streptococcus   A long chain of cocci.

Structural analog   Antimicrobial drug that competes with a structurally similar molecule, such as para-aminobenzoic acid (PABA), that is critical to the pathogen's metabolism.

Sty   Infection of a hair follicle at the base of an eyelid by *Staphylococcus aureus*.

Subacute disease   Any disease that has a duration and severity that lies somewhere between acute and chronic.

Subacute sclerosing panencephalitis (SSPE)   Slow, progressive disease of the central nervous system that results in memory loss, muscle spasms, and death several years after infection with a defective measles virus.

Subclinical   (see *asymptomatic*).

Substitution   Type of mutation in which a nucleotide base pair is replaced.

Substrate   The molecule upon which an enzyme acts.

Substrate-level phosphorylation   The transfer of phosphate to ADP from another phosphorylated organic compound.

Subunit vaccine   Type of immunization developed using recombinant DNA technology which exposes the recipient's immune system to a pathogen's antigens but not the pathogen itself.

Sulfonamide   Antimetabolic drug that is a structural analog of para-aminobenzoic acid (PABA).

Superoxide radical   Toxic form of oxygen which is detoxified by superoxide dismutase.

Surfactant   Chemical which acts to reduce the surface tension of solvents such as water by decreasing the attraction among solvent molecules.

Symbiosis   A close association between two or more organisms that is often but not always mutually beneficial.

Symptoms   Subjective characteristics of a disease that can be felt by the patient alone.

Syncytium   Giant, multinucleated cell formed by fusion of virally infected cell to neighboring cells.

Syndrome   A group of symptoms and signs that collectively characterizes a particular disease or abnormal condition.

Synergism   Interplay between drugs that results in efficacy that exceeds the efficacy of either drug alone.

Synthesis   In virology, third stage of the lytic replication cycle, in which the host cell's metabolic machinery is used to synthesize new nucleic acids and viral proteins.

Synthesis reaction   A chemical reaction involving the formation of larger, more complex molecules.

Synthetic drug   Antimicrobial that has been completely synthesized in a laboratory.

Syphilis   Sexually transmitted disease caused by infection with *Treponema pallidum* and characterized by a painless red lesion called a *chancre*.

Systemic lupus erythematosus (SLE) (*lupus*) A systemic autoimmune disease in which the individual produces autoantibodies against numerous antigens, including nucleic acids.

T cell receptor (TCR)   Antigen receptor generated in the cytoplasmic membrane of T lymphocytes.

T lymphocyte (*T cell*)   Lymphocyte that matures in the thymus and acts directly against endogenous antigens in cell-mediated immune responses.

Taxa   Nonoverlapping groups of organisms sorted on the basis of mutual similarities.

Taxis   Cell movement that occurs as a positive or negative response to light or chemicals.

Taxonomic system   A system for naming and grouping similar organisms together.

Taxonomy   The science of classifying and naming organisms.

Telophase   Final stage of mitosis, during which nuclear envelopes form around the daughter nuclei. Also used for the comparable stage of meiosis.

Temperate phage (*lysogenic phage*; see *prophage*) A bacteriophage that enters a host cell but remains inactive.

Teratogenic   Characterized by an ability to cause birth defects.

Terminator   Region of DNA where transcription ends.

Tetanus   Potentially fatal infection with the endospores of *Clostridium tetani*, which produce tetanus toxin, a potent neurotoxin.

Tetracycline   Antimicrobial agent that inhibits protein synthesis by blocking the tRNA docking site.

Thallus   Nonreproductive body of a filamentous mold or yeast.

Thermal death time   The time it takes to completely sterilize a particular volume of liquid at a set temperature.

Thermal death point   The lowest temperature that kills all cells in a broth in ten minutes.

Thermophile   Microorganism requiring temperatures above 45°C.

Thylakoid   In photosynthetic cells, portion of cellular membrane containing light-absorbing photosystems.

Thymine   Ring-shaped nitrogenous base found in nucleotides of DNA.

Thymine dimer   Mutation caused by nonionizing radiation in the form of ultraviolet light in which adjacent thymine bases covalently bond to one another.

Titer   The highest dilution of blood serum giving a positive reaction during titration.

Titration   Serial dilution of blood serum to test for agglutination activity.

Total magnification   A multiple of the magnification achieved by the objective and ocular lenses of a compound microscope.

Toxemia   Presence in the blood of poisons called *toxins*.

Toxic shock syndrome   Potentially fatal syndrome characterized by fever, vomiting, red rash, low blood pressure, and loss of sheets of skin, and caused by systemic infection with strains of *Staphylococcus* that produce TSS toxin.

Toxin   Chemicals produced by pathogens that either harm tissues or trigger host immune responses that cause damage.

Toxoid vaccine   Inoculum using modified toxins to stimulate antibody-mediated immunity.

Toxoplasmosis   A disease affecting cats and other animals and caused by infection with *Toxoplasma gondii*; in humans, characterized by mild, febrile symptoms; may be fatal in AIDS patients; transplacental transmission may result in miscarriage, stillbirth, or severe birth defects.

Trace element   Element required in very small amounts for microbial metabolism.

Trachoma   Serious eye disease caused by *Chlamydia trachomatis*.

Transamination   Reaction involving transfer of an amine group from one amino acid to another.

Transcription   Process in which the genetic code from DNA is copied as RNA nucleotide sequences.

Transducing phage   Virus that transfers bacterial DNA from one bacterium to another.

Transduction   Method of horizontal gene transfer in which DNA is transferred from one cell to another via a replicating virus.

Transfer RNA (tRNA)   Form of ribonucleic acid which carries amino acids to the ribosome.

Transformation   Method of horizontal gene transfer in which a recipient cell takes up DNA from the environment.

Transgenic organism   Plant or animal that has been genetically altered by the inclusion of genes from other organisms.

Translation   Process in which the sequence of genetic information carried by mRNA is used by ribosomes to construct polypeptides with specific amino acid sequences.

Transmission electron microscope (TEM) Type of electron microscope which generates a beam of electrons that passes through the specimen and produces an image on a fluorescent screen.

Transport medium   A special type of medium used to move clinical specimens from one location to another while preserving the relative abundance of organisms and preventing contamination of the specimen or environment.

Transposon   Segment of DNA found in many prokaryotes, eukaryotes, and viruses that codes for the enzyme transposase and can move from one location in a DNA molecule to another location in the same or different molecule.

Trematodes (*flukes*)   Group of helminths that are flat, leaf-shaped, have incomplete digestive systems, and have oral and ventral suckers.

Trench fever   Fever, headache, and pain in the long bones caused by infection with *Bartonella quintana*.

Trophozoite   The motile feeding stage of all protozoa.

Tuberculin response   Type of delayed hypersensitivity reaction in which the skin of an individual exposed to tuberculosis or tuberculosis vaccine reacts to a subcutaneous injection of tuberculin.

Tuberculin skin test   Test for a delayed hypersensitivity reaction to a subcutaneous injection of tuberculin.

Tuberculosis (TB)   A respiratory disease caused by infection with *Mycobacterium tuberculosis*; its disseminated form can result in wasting away of the body and death.

Tularemia   Zoonotic disease causing fever, chills, malaise, and fatigue, and caused by infection with *Francisella tularensis*.

Tumor   In the pathology of cancer, a mass of neoplastic cells. In inflammation, a symptom of swelling (*edema*).

Tumor necrosis factor   An immune system cytokine secreted by macrophages and T cells to kill tumor cells and to regulate immune responses and inflammation.

Type I diabetes mellitus   Immunologic attack on the islets of Langerhans cells resulting in the inability to produce the hormone insulin.

Typhoid fever   Fever, headache, and malaise produced by infection with *Salmonella typhi*; severe infections may cause peritonitis.

Uncoating   In animal viruses, the removal of a viral capsid within a host cell.

Unsaturated fat   A triglyceride with at least one double bond between adjacent carbon atoms, and thus at least one carbon atom bound to only a single hydrogen atoms.

Uracil   Ring-shaped nitrogenous base found in nucleotides of RNA.

Urticaria   Hives.

Use-dilution test   Method of evaluating the effectiveness of a disinfectant or antiseptic against specific microbes in which the most effective agent is the one that entirely prevents microbial growth at the highest dilution.

Vaccination   Active immunization.

Vaccine   The inoculum used in active immunization.

Vacuole   General term for membranous sac that stores or carries a substance in a cell.

Vaginitis   A sexually transmitted disease caused by infection with *Trichomonas vaginalis* and characterized by discharge, vaginal and cervical lesions, and pain.

Valence   The combining capacity of an atom.

Vancomycin   Antibiotic obtained from *Streptomyces orientalis* and used to treat infections with Gram-positive bacteria.

Varicella (*chickenpox*)   Highly infectious disease characterized by fever, malaise, and skin lesions, and caused by infection with varicella-zoster virus.

Variola   Common name for the smallpox virus.

Variola major   Variant of smallpox virus which causes severe disease with a mortality rate of 20% or higher.

Variola minor   Variant of smallpox virus which causes less severe disease and mortality rate of less than 1%.

Vector   Nucleic acid molecule such as a viral genome, transposon, or plasmid, that is used to deliver a gene into a cell. In epidemiology, an animal (typically an arthropod) that transmits disease from one host to another.

Vehicle transmission   Spread of pathogens via air, drinking water, and food, as well as bodily fluids being handled outside the body.

Venezuelan equine encephalitis (VEE)   Potentially fatal infection of the brain caused by a togavirus.

Vesicle   General term for membranous sac that stores or carries a substance in a cell; in human pathology, any raised skin lesion filled with clear fluid.

Viable plate count   Estimation of the size of a microbial population based upon the number of colonies formed when diluted samples are plated onto agar media.

Vibrio   A slightly curved rod-shaped prokaryotic cell.

Viral hemagglutination inhibition test   Immune test commonly used to detect antibodies against influenza, measles, and other viruses that naturally agglutinate red blood cells.

Viremia   The presence of virions in the blood.

Viridans streptococci   Group of alpha-hemolytic streptococci which produce a green pigment when grown on blood media and normally inhabit the mouth, throat, GI, genital, and urinary tract.

Virion   A virus outside of a cell, consisting of a proteinaceous capsid surrounding a nucleic acid core.

Viroid   Extremely small, circular piece of RNA that is infectious and pathogenic in plants.

Virulence   A measure of pathogenicity.

Virulence factors   Enzymes, toxins, and other factors that affect the relative ability of a pathogen to infect and cause disease.

Virus   Tiny infectious acellular agent with nucleic acid surrounded by proteinaceous capsomeres that form a coat called a capsid.

Wandering macrophage   Type of macrophage that leaves the blood via diapedesis to travel to distant sites of infection.

Wastewater (*sewage*)   Any water that leaves homes or businesses after being used for washing or flushed from toilets.

Water mold   Eukaryotic microbe resembling a filamentous fungus but having tubular cristae in their mitochondria, cell walls of cellulose, two flagella, and true diploid thalli.

Water-borne transmission   Spread of pathogenic microorganisms via water.

Wavelength   The distance between two corresponding points of a wave.

Wax   Alcohol-containing lipid made up of molecules with one fatty-acid chain.

Western blot test   Variation of an ELISA test which can detect the presence of antibodies against multiple antigens; used to verify the presence of antibodies against HIV in the serum of individuals who have tested positive by ELISA.

Western equine encephalitis (WEE)   Potentially fatal infection of the brain caused by a togavirus.

White piedra   Benign scalp disease caused by infection with *Trichosporon beigelii*.

Whitlow   Inflamed blister that may result from infection with herpes simplex viruses via a cut or break in the skin.

Wild type cell   A cell normally found in nature (in the wild); a non-mutant.

Xenograft   Type of graft in which tissues are transplanted between individuals of different species.

Xenotransplant   Technique involving recombinant DNA technology in which human genes are inserted into animals to produce cells, tissues, or organs that are then introduced into the human body.

Yaws   A skin disease caused by infection with *Treponema pallidum pertenue*.

Yeast   A unicellular, typically oval or round fungus that usually reproduces asexually by budding.

Yellow fever   Often fatal hemorrhagic disease contracted through a mosquito bite carrying a flavivirus.

Zone of inhibition   In a diffusion susceptibility test, a clear area surrounding the drug-soaked disk where the microbe does not grow.

Zoonoses (*zoonotic diseases*)   Diseases that are naturally spread from usual animal host to humans.

Zygomycoses   Opportunistic fungal infections involving the brain and caused by various genera of fungi classified in the division Zygomycota.

Zygomycota   Division of fungi including coenocytic molds called zygomycetes. Most are saprobes.

Zygospore   Thick, black, rough-walled sexual structure of zygomycetes that can withstand desiccation and other harsh environmental conditions.

Zygote   In sexual reproduction of protozoa, diploid cell formed by the union of gametocytes.

ILLUSTRATION CREDITS

Chapter 1: Figures 1.10, 1.12, 1.14, 1.15 J.B. Woolsey Associates, LLC; Figure 1.13 Precision Graphics.

Chapter 2: 2.1–2.12, 2.15–2.27 Precision Graphics; 2.13, 2.14 J.B. Woolsey Associates, LLC.

Chapter 3: 3.5, 3.7, 3.12, 3.13, 3.14, 3.23, 3.30, 3.31, 3.32, 3.33, 3.36, 3.37 Kenneth Probst; 3.8, 3.11, 3.19, 3.20, 3.29 Precision Graphics; 3.15, 3.16, 3.18, 3.35 Darwen Hennings/Precision Graphics; 3.24 Darwen Hennings.

Chapter 4: 4.1, 4.3, 4.26 Precision Graphics; 4.2, 4.4, 4.5, 4.6, 4.7, 4.8, 4.14, 4.17 J.B. Woolsey Associates, LLC; 4.25, 4.31 Darwen Hennings.

Chapter 5: 5.1–5.20, 5.22–5.25, 5.29–5.34, Highlight Box, Precision Graphics; 5.21, 5.26, 5.27, 5.28 Kenneth Probst.

Chapter 6: 6.1, 6.4a, 6.5, 6.7a, 6.17a, 6.18, 6.19, 6.20 Precision Graphics; 6.3, 6.8a, 6.9a, 6.16, 6.21, 6.22a, 6.23, 6.24, 6.25 J.B. Woolsey Associates, LLC; 6.15 Cassio Lynm.

Chapter 7: 7.3– 7.24, 7.28, 7.30–7.33, 7.35, Highlight Box, Closer Look Box Precision Graphics; 7.25–7.27, 7.29 J.B. Woolsey Associates, LLC.

Chapter 8: 8.1–8.6, 8.8–8.11 Precision Graphics.

Chapter 9: 9.1–9.3, 9.5, 9.6, 9.13, 9.15 Precision Graphics; 9.8, 9.10a, 9.11, J.B. Woolsey Associates, LLC.

Chapter 10: 10.2 Ken Probst; 10.3– 10.8, 10.13, 10.16 Precision Graphics; 10.12, 10.15 J.B. Woolsey Associates, LLC.

Chapter 11: 11.1, 11.5, 11.21, 11.26 Kenneth Probst; 11.2, 11.6–11.8, 11.10, 11.25 Precision Graphics.

Chapter 12: 12.1, 12.3, 12.4 Precision Graphics; 12.6, 12.7, 12.10, 12.12a, 12.14c, 12.18, 12.19, 12.21, 12.23, 12.24, 12.26, 12.30, 12.32 Kenneth Probst.

Chapter 13: 13.1, 13.4, 13.6, 13.7, 13.8, 13.11, 13.12, 13.13, 13.17 Kenneth Probst; 13.9, 13.14, 13.15, 13.21 Precision Graphics.

Chapter 14: Table 14.2, 14.3, 14.4, 14.11 Kenneth Probst; 14.5, 14.8, 14.9, 14.10, 14.13–14.19 Precision Graphics; 14.7 J.B. Woolsey Associates, LLC.

Chapter 15: 15.1, 15.12, 15.13, 15.15, 15.16 Precision Graphics; 15.3, 15.4, 15.5, 15.9, 15.10, 15.18, 15.19, 15.20 Kenneth Probst.

Chapter 16: 16.1, 16.3b, 16.5–16.17 Precision Graphics; 16.2, 16.4 Kenneth Probst.

Chapter 17: 17.1, 17.2, 17.3, 17.5, 17.6, 17.12, 17.14, 17.16, 17.17 Precision Graphics; 17.4, 17.7–17.11 J.B. Woolsey Associates, LLC.

Chapter 18: 18.1, 18.7, 18.8 Precision Graphics; 18.6, 18.9 Cassio Lynm; 18.12 J.B. Woolsey Associates, LLC.

Chapter 19: 19.5, 19.14, 19.17 Precision Graphics; 19.16a,b, 19.24, Cassio Lynm; 19.21 Kenneth Probst.

Chapter 20: 20.2, 20.5, 20.7, 20.13, 20.14, 20.15, 20.22, 20.23, 20.25, Precision Graphics; 20.8, 20.16, 20.18 Kenneth Probst.

Chapter 21: 21.4, 21.6, 21.7, 21.8, 21.12, 21.14, 21.20, 21.21, 21.24, 21.25, 21.27 Precision Graphics; 21.18 Kenneth Probst.

Chapter 22: 22.1 Kenneth Probst; 22.2 Precision Graphics.

Chapter 23: 23.1, 23.3, 23.5, 23.10, 23.11, 23.14, 23.17 Kenneth Probst; 23.9, 23.13 Precision Graphics.

Chapter 24: 24.2, 24.5, 24.10, 24.23, 24.24b Kenneth Probst; 24.12, 24.19b, 24.24 Precision Graphics.

Chapter 25: 25.3, 25.10, 25.11, 25.14, 25.21, 25.23, 25.24, 25.28, 25.30, 25.32, 25.35 Precision Graphics; 25.8, 25.12, 25.16, 25.17, 25.19, 25.20, 25.25, 25.38b, 25.39, 25.41 Kenneth Probst; 25.9 Darwen Hennings.

Chapter 26: 26.1, 26.2, 26.3, 26.8, 26.10, 26.11, 26.12, 26.15–26.20 Precision Graphics.

PHOTO CREDITS

Chapter Opener 1   M.I. Walker / Photo Researchers, Inc.; Highlight 1   Center for Disease Control; 1.1   Courtesy of Pfizer, Inc.; 1.2   Alan Shinn; 1.3   Rich Robison, Brigham Young University; 1.4 a   Manfred Kage/Peter Arnold; 1.4 b   John D. Cunningham/Visuals Unlimited; 1.5 a   *Microbiology; A Photographic Atlas for the Laboratory;* Benjamin Cummings 2001; 1.5 b   M. Abbey/ Visuals Unlimited; 1.5 c, d   Center for Disease Control; 1.6 a   M.I. Walker/Photo Researchers, Inc.; 1.6 b   Lester V. Berman/ Corbis; 1.7, 1.8   *Microbiology; A Photographic Atlas for the Laboratory;* Benjamin Cummings 2001; 1.9   Lee D. Simon/Photo Researchers, Inc.; 1.11   Louis Pasteur in his library: Oil on canvas, 1889, by Albert Edelfelt / The Granger Collection; 1.16   Culver Pictures, Inc.; 1.17, 1.18   *Microbiology; A Photographic Atlas for the Laboratory;* Benjamin Cummings 2001; 1.19   The Granger Collection, French Colored Engraving, 1883; 1.21   C. James Webb / Phototake.

Chapter Opener 2   NASA/Mark Marten/Photo Researchers, Inc.; Highlight 1   Topham / The Image Works; 2.12   Herman Eisenbeiss Photo Researchers, Inc.

Chapter Opener 3   D. Balkwill, D. Maratea; Highlight 1   NIBSC / Science Photo Library / Photo Researchers, Inc.; Highlight 2   Esther R. Angert / Phototake; New Frontiers 1   David Scharf/ Peter Arnold, Inc.; 3.1 a   Dr. Gopal Murti / Photo Researchers, Inc.; 3.1 b   Dennis Kunkel/ Phototake; 3.1 c   Cabisco/Visuals Unlimited; 3.1 d   VU/ Science VU/Visuals Unlimited; 3.4 a   Dr. Immo Rantala / Science Photo Library / Photo

Researchers, Inc.; 3.4 b   Omikron / Photo Researchers, Inc.; 3.6 a   Hans Glederbion / Stone; 3.6 b   Dr. K.S.Kim / Peter Arnold, Inc.; 3.6 c   Ed Reschke/ Peter Arnold, Inc.; 3.6 d   A.B. Dowsett / Science Photo Library / Photo Researchers, Inc.; 3.7 a   Chris Bjornberg/ Photo Researchers. Inc.; 3.9. 3.10   Dennis Kunkel/ Phototake; 3.21   P. S. Poole; 3.22   Ruth Carallido-Lopez; 3.23 a, b   *Microbiology; A Photographic Atlas for the Laboratory;* Benjamin Cummings 2001; 3.23 c   Omnikron / Science Library / Photo Researchers, Inc.; 3.25   Robert W. Bauman; 3.26   Don. W. Fawcett / Visuals Unlimited; 3.27 a–d   Mike Abbey/ Visuals Unlimited; 3.28   *Microbiology; A Photographic Atlas for the Laboratory;* Benjamin Cummings 2001; 3.29 b   Albert Tousson/ Phototake; 3.30 a   David M. Phillipis / Visuals Unlimited; 3.31   D.W. Fawcett / Photo Researchers, Inc.; 3.32   R. Bolender, D. Fawcett / Visuals Unlimited; 3.33   Biophoto Associates / Photo Researchers, Inc.; 3.34   George B. Chapman & Peter K. Chan/Visuals Unlimited; 3.36   D.W. Fawcett / Photo Researchers, Inc.; 3.37   G. Chapman / Visuals Unlimited.

Chapter Opener 4   Robert Bauman; U1   Phillipe Fauchet, Phd /ASM News Volume 68 No.2, 2002; 4.4   Rich Robison, Brigham Young University; 4.4   Charles D. Winter / Photo Researchers, Inc.; 4.6 a–d   Rich Robison, Brigham Young University; 4.9 a, b   Rich Robison, Brigham Young University; 4.10   Fred Marsik / Visuals Unlimited; 4.11   David Becker / SPL / Photo Researchers, Inc.; 4.12   WM Omerod/Visuals Unlimited; 4.13   Microworks Color / Phototake; 4.15 a   Stanley Flegler/Visuals Unlimited; 4.15 b   David M. Phillips/Visuals Unlimited; 4.15 c   Oliver Meckes/Photo Researchers, Inc.; 4.15 d   Stanley Flegler/Visuals Unlimited; 4.16 a, b   © 1996 Digital Instruments; 4.17   *Microbiology; A Photographic Atlas for the Laboratory;* Benjamin Cummings 2001; 4.18 a–c   Rich Robison, Brigham Young University; 4.19 a–e   Rich Robison, Brigham Young University; 4.20 a, b   *Microbiology; A Photographic Atlas for the Laboratory;* Benjamin Cummings 2001; 4.21. 4.22   Rich Robison, Brigham Young University; 4.23, 4.24 a–c, 4.27 a, b, 4.28 a, b   *Microbiology; A Photographic Atlas for the Laboratory;* Benjamin Cummings 2001; 4.29 a   Raymond B. Otero / Visuals Unlimited; 4.30   Courtesy of the Microbial Diseases Laboratory, Berkeley, CA.

Chapter Opener 5   Randall Hyman; Highlight 1   Photo Researchers, Inc.; New Frontiers 1   Dr. Morley Read / Science Photo Library, Photo Researchers, Inc.; 5.6   Ken Eward Biografx; 5.26   eos Wanner Photo Researchers, Inc.

Chapter Opener 6   Wayne Armstrong; N1   NASA; U1   Douglas Faulkner Photo Researchers, Inc.; 6.2, 6.4 b   *Microbiology; A Photographic Atlas for the Laboratory;* Benjamin Cummings 2001; 6.6 a   Gerald & Buff Corsi /Visuals Unlimited; 6.6 b   Dr. Ron Hoham; 6.7   *Microbiology; A Photographic Atlas for the Laboratory;* Benjamin Cummings 2001; 6.8 b   Raymond B. Otero/Visuals Unlimited; 6.9 b   *Microbiology; A Photographic Atlas for the Laboratory;* Benjamin Cummings 2001; 6.10, 6.11   *Microbiology; A Photographic Atlas for the Laboratory;* Benjamin Cummings 2001; 6.12   Rich Robison, Brigham Young University; 6.13, 6.14 a–c   *Microbiology; A Photographic Atlas for the Laboratory;* Benjamin Cummings 2001; 6.15b   Coy Laboratories, Inc.; 6.17 b   Alfred Pasieka / Peter Arnold, Inc.; 6.22 b   Pall/Visuals Unlimited; 6.22 c   Jack Bosrtak/Visuals Unlimited; 6.25 a   Fundamental Photographs; 6.25 b   Courtesy of TOPAC.

New Frontiers 1   CDC / Janice Carr; 7.2 a   Visuals Unlimited; 7.2 b   Huntington Potter and David Dressler; 7.3 a   Courtesy Barbara Hamkalo; 7.3 b   Courtesy of J. R. Paulsen and U. K. Laemmli, *Cell*, Copyright Cell Press.; 7.3 c   G.F. Bahr / Armed Forces Institute of Pathology; 7.3 d   G.F. Bahr / Armed Forces Institute of Pathology; 7.16   Visuals Unlimited; 7.31 a   Dennis Kunkel / Phototake; 7.34   John A. Anderson Earth Scenes.

Chapter Opener 8   David Nance / ARS / NSDA; Highlight 1   Kent Loeffler; New Frontiers 1   Michael Gadomski/Animal Animals; New Frontiers 3   Michael Gadomski/Animal Animals; 8.7 b   M. Schroder, Argus Fotoachiv / Peter Arnold; 8.9 d   Sovereign / Phototake; 8.12   D. Parker / Science Photo Library; 8.13   SCIENCE VU/MONSANTO/VU.

Chapter Opener 9   Dianne Bodareff / POOL / AP Wide World Photo; U1   Lois Ellen Frank / Corbis; 9.4   *Microbiology; A Photographic Atlas for the Laboratory;* Benjamin Cummings 2001; 9.7 a   Custom Medical; 9.8   Ricahrd Megna/Fundamental Photographs; 9.9   Jonathon Blair/CORBIS; 9.10 b   Pall / Visuals Unlimited; 9.12 a   Richard Megra / Fundamental Photographs; 9.12 b   Courtesy Iowa State University Food Safety Project; 9.14   Richard Megra / Fundamental Photographs; 9.16   Talaro / Visuals Unlimited.

Chapter Opener 10   Dr. Kari Lounatmaa / Science Photo Library; F1   AP Wideworld Photo; Highlight 1   Leonard Lessin/Photo Researchers, Inc.; 10.1   Rich Robison, Brigham Young University; 10.3   Lennart Nilsson / *The Body Victorious,* Delacourt Publishing; 10.3   Lennart Nilsson; 10.9   *Microbiology; A Photographic Atlas for the Laboratory;* Benjamin Cummings 2001; 10.10   Dr. Craig Warner / Albert Einstein College of Medicine; 10.11   *Microbiology; A Photographic Atlas for the Laboratory;* Benjamin Cummings 2001; 10.14 a   Barts Medical Library / Phototake; 10.14 b   Ken Greer / Visuals Unlimited; 10.17   E. Vercauteren, Department of Microbiology, University Hospital, Antwerp, Belgium.

Chapter Opener 11   Dr. H.W. Jannasch, Woods Hole Oceanographic Institute; Highlight 1   Joseph Janovic, MD. Director Parkinson's Disease Center and Movement Disorders Clinic, Baylor College of Medicine; 11.3 a, b   T. A. Krulwich; 11.4   Yoko Takahashi, Kitasato Institute for Life Sciences, Kitasato University, Tokyo; 11.6 a–e   *Microbiology; A Photographic Atlas for the Laboratory;* Benjamin Cummings 2001; 11.7 a–d   *Microbiology; A Photographic Atlas for the Laboratory;* Benjamin Cummings 2001; 11.9 a, b   Rich Robison, Brigham Young University; 11.11 a   Dr. Reinhard Rachel; 11.11 b   Dr. Reinhard Rachel and K.O. Stetter; 11.12   Stephen Fuller/Peter Arnold, Inc; 11.13   Science VU/NASAGSFC/Visuals Unlimited; 11.14 a   Phillip Sze/ Visuals Unlimited; 11.14 b   M. Abbey / Visuals Unlimited; 11.14 c   Eichelberger / Visuals Unlimited; 11.15 c   Paul Johnson / Biological Photo Service; 11.16   Michael Gabridge / Visuals Unlimited; 11.17   J.R. Adams / Visuals Unlimited; 11.18   *Microbiology; A Photographic Atlas for the Laboratory;* Benjamin Cummings 2001; 11.19   Dr. Yves Brun, University of Indiana; 11.20   Wally Eberhart/Visuals Unlimited; 11.22   Leonard Lessin/Peter Arnold, Inc; 11.23   J. Ott; 11.24

INDEX

NOTE: A *t* following a page number indicates tabular material, an *f* following a page number indicates a figure, and a *b* following a page number indicates a boxed feature.

A. *See* Adenine
Abdominal infections, *Bacteroides fragilis* causing, 591
Abiogenesis (spontaneous generation), 7–10
ABO system, 515–516, 516*f*, 517*t*
Abscesses, 456
   in group A streptococcal pharyngitis, 539
   Streptococcus anginosus causing, 542
Absidia genus/spp., 636–637
Abyssal zone, 771, 771*f*
Acanthamoeba genus/spp., 353, 652–653, 664*t*
Acceptor stem, transfer RNA, 214, 215*f*
Accessory photosynthetic pigments, 366
Acellular agents, 378. *See also* Prions; Viroids; Viruses
Acellular fermentation, Buchner's demonstration of, 11–12, 13*f*
Acellular obligatory parasites, viruses as, 6
Acellular (plasmodial) slime molds, 370–371, 372*f*
Acetic acid, microbial fermentation producing, 150*f*, 747
Acetic acid bacteria, 747
Acetobacter genus/spp., 332, 334*t*
   fermentation products of, 150*f*, 747
Acetone
   in Gram staining, 109, 109*f*
   microbial fermentation producing, 150*f*
Acetylcholine, botulism toxins affecting, 549, 549*f*
Acetyl-CoA (acetyl-coenzyme A), 160*t*, 162*f*
   in fat catabolism, 151, 151*f*
   synthesis of, 142, 142*f*, 149*t*, 160*t*, 162*f*
Acidanus, 321, 322*t*
Acid-base balance, 38
Acid-fast bacteria, 67, 110–111
Acid-fast rods, 111
Acid-fast stain, 110–111, 111*f*, 114*t*, 329
Acidic dyes, 108
Acid mine drainage, 762–763, 763*f*, 764*f*
Acidophiles, pH range tolerated by, 38, 174–176
Acids, 37–39, 38*f*
   pH of, 38, 38*f*
Acid-tolerant microbes, pH range tolerated by, 174
Acinetobacter genus/spp., 588
Acne, 532, 561–562, 561*f*
Acquired immunity, 482–483, 483*t*
   artificial. *See also* Immunization; Vaccines
      active/active immunization, 483, 483*t*, 488, 489–494, 490*t*, 491*f*, 492*t*, 493*f*, 494*b*
      passive/passive immunization, 483, 483*t*, 494–496, 495*f*
   natural
      active, 482, 483*t*
      passive, 483, 483*t*
Acquired immunodeficiency, 528. *See also* HIV infection/AIDS
Acquired immunodeficiency syndrome (AIDS), 528, 718–724, 719*t*, 725*b*. *See also* HIV infection/AIDS
Acremonium genus/spp., 641

Acridine, as frameshift mutagen, 224
Actin, in eukaryotic cytoskeleton, 81, 82*f*
Actinomyces genus/spp., 329*t*, 330, 562–563, 563*f*
Actinomyces israelii, 330
Actinomycetes, 329–330, 329*t*, 330*f*
   reproduction in, 315, 316*f*
Actinomycosis, 562–563, 563*f*
Activated charcoal filters, in water treatment, 756–758, 757*f*
Activated sludge system, wastewater treatment and, 759, 760*f*
Activation energy, enzymes affecting, 130, 131, 131*f*
Active immunity
   artificially acquired/active immunization, 483, 483*t*, 488, 489–494, 490*t*, 491*f*, 492*t*, 493*f*, 494*b*. *See also* Immunization; Vaccines
   naturally acquired, 482, 483*t*
Active site
   chlorophyll, 153, 153*f*
   enzyme, 130, 131, 132*f*
Active transport, 73–74, 75*f*, 76*t*
Acute disease, definition of, 423, 425*t*
Acute hemorrhagic conjunctivitis, 707
Acute inflammation, 455
ACV. *See* Acyclovir
Acyclovir, 296*f*, 308*t*
   spectrum of action of, 293*f*, 308*t*
Adaptation, microbial, survival and, 764–765
Adenine (A), 50, 50*f*
Adenosine arabinoside, 296*f*, 308*t*
Adenosine diphosphate. *See* ADP
Adenosine monophosphate. *See* AMP
Adenosine triphosphate. *See* ATP
Adenoviruses (*Adenoviridae*), 385*t*, 696, 696*f*, 697*f*, 698*b*
Adenylate cyclase toxin, 585
Adherence, in phagocytosis, 448–449, 448*f*
Adhesins, 411
   Bordetella producing, 411, 585
   Enterobacteriaceae producing, 573, 574*f*
   pneumococcal, 543
   Pseudomonas aeruginosa producing, 587
   Yersinia producing, 580
Adhesion, in infection, 410–411
Adhesion disks, 410–411
Adhesion factors, 410–411
Adjuvants, for inactivated vaccines, 490*t*, 491
ADP, 52, 52*f*, 162*f*
   in active transport, 74, 75*f*
   phosphorylation of, 128, 129*f*. *See also* Phosphorylation
      in chemiosmosis, 147
      in glycolysis, 138*f*, 139
Adult acute T-cell lymphocytic leukemia, HTLV-1 causing, 717
Aedes mosquitoes, as disease vectors, 424*t*
   for dengue fever, 711, 711*t*
   for filariasis (*Wuchereria bancrofti*), 671
Aerobes, 147, 169, 171
   obligate, 169, 171
Aerobic respiration, 147, 149*t*, 150*t*
Aerosols, airborne transmission of disease and, 422
Aerotolerant anaerobes, 171
   culturing, 186

Aflatoxicosis, 643
Aflatoxins, 643
   as mutagens, 224
African sleeping sickness, 653–654, 654*f*
   discovery of cause of, 17*t*
AFRs. *See* Acid-fast rods
Agammaglobulinemia, Bruton-type, 527, 528*t*
Agar, 14
   in algal cell wall, 79, 367
   for cultures, 181, 181*f*
Agaricus genus/spp., 363
Agarose, for gel electrophoresis, 248, 250*f*
Agglutination, antibody, 469, 469*f*
Agglutination tests, 120, 120*f*, 499–500, 499*f*, 500*f*, 506*t*
Agranulocytes, 444, 445*f*
Agranulocytosis, 518
Agricultural crops, assessing biological threats to, 773–774
Agricultural microbiology, 20*t*, 755, 756*t*
   recombinant DNA technology in, 257–258, 258*b*, 258*f*, 755
Agricultural wastes, microbes in treatment of, 760–761
Agrobacterium genus/spp., 332, 332*f*, 334*t*
Agrobacterium tumefaciens
   chromosomes of, 201–202
   soil-borne disease caused by, 770*t*
Agroterrorism, 772, 773–774, 775. *See also* Bioterrorism
AIDS (acquired immunodeficiency syndrome), 528, 718–724, 719*t*, 725*b*. *See also* HIV infection/AIDS
AIDS "cocktail," 724
   drug resistance and, 303
Airborne transmission of disease, 421, 422, 422*t*
Akinetes, of cyanobacteria, 323, 324*f*
Alcoholic beverages, microbial production of, 13*t*, 150*f*, 746–747, 746*f*, 747*f*, 748*t*
Alcohols. *See also* Ethanol
   antimicrobial actions of, 278–279, 282*t*
Aldehydes, antimicrobial action of, 281, 282*t*
Algae, 5, 6*f*, 366–369, 371*t*
   brown, 369, 369*f*, 370*f*, 371*t*
   cell walls of, 79, 79*f*
   classification of, 348, 367–369, 371*t*
   distribution of, 366
   golden, 367–369, 371*t*
   green, 367, 371*t*
   morphology of, 366
   red, 367, 369*f*, 371*t*
   reproduction in, 367, 368*f*
   in soil, 769
   yellow-green, 367–369, 371*t*
Algin, in algal cell wall, 79
Alginic acid/alginate, 369
Alkalinophiles, pH range tolerated by, 176
Allergens, 510
   fungal, 643, 644
   in hay fever, 512, 513*f*
   initial exposure to, 510, 511*f*, 513*b*
   testing for, 514, 514*f*
Allergic contact dermatitis, 520–521, 521*f*
Allergic reactions/allergies, 510–515, 512*f*, 523*t*
   to antimicrobial drugs, 301
   clinical signs of
      in localized reaction, 512, 513*f*
      in systemic reactions, 512–514
   degranulation/degranulating cells in, 510–511, 511*f*, 512*f*, 514*t*

Allergic reactions/allergies, *Cont.*
  diagnosis of, 514, 514*f*
  to fungi, 627
  initial exposure/sensitization and, 510,
    511*f*, 513*b*
  poison ivy as, 509, 520, 521*f*
  prevention of, 514–515, 515*b*
  treatment of, 515
  vaccines causing, 494
Allicin, in garlic, food preservation and, 751
Allografts, 521, 521*f*
Allosteric (excitatory) activation, 135, 135*f*
Allosteric (noncompetitive) inhibition, 135,
  135*f*, 136*f*
Allosteric site, 135, 135*f*, 136*f*
Alpha-amanitin, 644
Alpha- (α-) carbon, amino acid, 47, 47*f*
Alpha chains, 468
Alpha- (α-) helices, 48–50, 49*f*
Alpha-hemolysis, by *Streptococcus pneumoniae*,
  183, 183*f*
Alpha-hemolytic streptococci (viridans
    streptococci), 542
Alpha- (α-) interferon, 453, 454*f*, 455*t*
Alpha- (α-) ketoglutaric acid, 159*f*, 160*t*
Alphaproteobacteria, 325*t*, 330–333, 331*f*, 332*f*,
    334*t*, 583, 584
  pathogenic, 331
Alphaviruses, as biological weapons, 774*t*
Alternaria genus/spp., 640
Alternate (properdin) pathway of complement
    activation, 450, 451, 451*f*
Alternation of generations, 367, 368*f*
Alternative fuels, microbial production of,
    754–755, 754*f*, 755*b*, 756*t*
Aluminum, Bohr diagram of, 30*f*
Alveolar macrophages, 444
  in reticuloendothelial system, 446*f*
Alveolata (kingdom), 348, 348*f*
Alveolates, 351, 356*t*
Alveoli, of protozoa, 351, 351*f*
Amanita muscaria, 363, 364*f*, 644
Amanita phalloides, 644, 644*f*
Amantadine, 308*t*
  for influenza, 735
  mechanism of action of, 295, 308*t*
Amastigotes
  in *Leishmania* life cycle, 656, 656*f*
  in *Trypanosoma cruzi* life cycle, 654*f*, 655
Ambylomma ticks, as disease vectors, for
    ehrlichiosis (*Ehrlichia* species), 601,
    602*t*
Amebiasis. *See* Amoebiasis
Ames, Bruce, 228
Ames test, 228, 229*f*
Amikacin, 306*t*
  mechanism of action of, 292, 306*t*
Amination, amino acid, 159, 159*f*
Amino acids, 47–48, 47*f*, 48*f*
  amination of, 159, 159*f*
  biosynthesis of, 159–160, 159*f*, 162*f*
  deamination of, 151
  essential, 159
  for microbial growth, 173*t*
  peptide bonds of, 48, 49*f*
  in protein catabolism, 151, 152*f*
  sequence of (primary structure of protein),
    48, 49*f*
  transamination of, 159*f*, 160
Aminoglycosides, 306*t*
  mechanism of action of, 292, 293*f*, 306*t*
  toxicity of, 301, 306*t*
Amino group, 40*t*
  in amino acids, 47, 47*f*
Ammonia, 151

Ammonification, 767, 767*f*
Ammonium ion, 151
Amoebae, 352–353, 353*b*, 353*f*, 356*t*, 650–653,
    664*t*
  free-living, 353, 356*t*, 652–653, 664*t*
Amoebiasis, 651–652
Amoebic dysentery, 651–652, 773*t*
Amoebic encephalitis, *Acanthamoeba*
    causing, 652
Amoebic meningoencephalitis, *Naegleria*
    causing, 652
Amoeboid action, 80, 81*f*
Amoxicillin, 305*t*
AMP, 52, 52*f*
Amphibolic reactions, 157
Amphitrichous flagellum, 63, 64*f*
Amphotericin B, 309*t*, 627
  mechanism of action of, 292, 293*f*, 309*t*
Ampicillin, 305*t*
Amplifying hosts, for *Yersinia pestis*, 580, 581*f*
Amylase, microbial production of, 753, 756*t*
Amylopectin, 45, 46*f*
Amylose, 45, 46*f*
Anabaena genus/spp., 324
Anabolism/anabolic pathways, 36, 126,
    126–127, 126*f*, 156–157, 160*t*, 162*f*
  amino acid biosynthesis, 159–160,
    159*f*, 162*f*
  carbohydrate biosynthesis, 157–158,
    158*f*, 162*f*
  integration and regulation of, 160–161, 162*f*
  lipid biosynthesis, 158, 159*f*, 162*f*
  nucleotide biosynthesis, 160, 161*f*, 162*f*
  photosynthesis, 152–156, 153*f*, 154*f*, 155*f*,
    156*t*, 157*f*
Anaerobes, 147, 169, 171
  aerotolerant, 171
    culturing, 186
  culturing, 183–185, 185*f*
  facultative, 171
    glycolytic, 334*t*, 335, 336*t*
    yeasts as, fermentation and, 11, 12*f*
  obligate, 169, 171
Anaerobic bacteria, in fermentation, 11, 12*f*
Anaerobic media, 183–185, 185*f*
Anaerobic respiration, 147, 149*t*, 150*t*
Analytical epidemiology, 429
Anaphase
  meiosis, 345*f*, 346
  mitosis, 344, 345*f*
Anaphylactic shock, 512–514
  antimicrobial drugs causing, 301
  vaccines causing, 494
Anaphylaxis, 512–514. *See also* Anaphylactic
    shock
Ancylostoma duodenale, 670–671, 670*f*, 673*t*
Anemia, hemolytic, 518, 524–525
Angiomatosis, bacillary, 584
Animal bites, *Pasteurella* infecting, 582
Animal cultures, 185
  viral, 396
Animal feed (silage), fermentation in
    production of, 745, 748*t*
Animalia (kingdom), 115*f*, 116, 116*f*, 348*f*
Animal reservoirs, 420, 421*t*
Animal viruses
  latency of, 394, 394*t*, 683
  replication of, 390–394, 391*f*, 392*t*,
    393*f*, 394*t*
    attachment in, 390, 394*t*
    entry and uncoating in, 390, 391*f*, 394*t*
    synthesis in, 390–392, 392*t*, 394*t*
Anions, 34, 37, 38*f*
  of salts, 39
  in stain dyes, 108

Anopheles mosquitoes, as disease vectors, 424*t*
  for filariasis (*Wuchereria bancrofti*), 671
  for malaria (*Plasmodium* species), 659,
    659*f*, 661
Anoxygenic organisms
  green and purple bacteria as, 324
  noncyclic photophosphorylation in, 155
Antacids, bacteria in stomach affected by, 39, 39*b*
Antagonism, microbial
  adaptation/survival and, 764
  drug resistance and, 304
  in host defense, 440
  lactobacilli and, 327
  microbial growth affected by, 177
  opportunistic infections and, 407–409
  superinfection and, 297
Anthrax, 327, 421*t*, 545, 545–546, 547*b*, 547*f*. *See
    also Bacillus anthracis*
  bioterrorism and, 22, 545, 546, 547*b*, 774*t*
  ciprofloxacin for, 297*b*
  cutaneous, 327, 546, 547*f*
  discovery of cause of, 14, 17*t*
  inhalation, 546, 705*t*
    manifestations of, 705*t*
  irradiation systems in control of, 262
  soil-borne disease transmission of, 769, 770*t*
  vaccine against, 546
Anthrax toxin, 545
Anthropophilic dermatophytes, 638
Antibacterial drugs, 305–307*t*. *See also*
    Antibiotics
Antibacterial soap, 283*b*
Antibiotics, 282–283, 288, 305–307*t*. *See also
    specific agent and* Antimicrobial drugs
  microbial production of, 13*t*, 288, 289*b*, 289*t*,
    755, 756*t*
  spectrum of action of, 298*f*
Antibodies, 47, 443, 468–471. *See also*
    Immunoglobulins
  antitoxins as, 416
  classes of, 468, 469–470, 470*f*
  deficient, 527
  fluorescent dye linked to, 101, 101*f*
  function of, 468–469, 469*f*
  monoclonal, in passive immunization,
    495, 495*f*
  in passive immunization, 494–496, 495*f*
  plasma cell secretion of, 477, 478*f*
  serological tests for, 120, 120*f*
  structure of, 468, 468*f*
Antibody sandwich ELISA, 503, 504*f*
Antibody titers, 500, 500*f*
Anticodons, 214–215, 215*f*, 216, 216*f*
Antifungal drugs, 298*f*, 309*t*, 627
Antigen-antibody complexes, in
    hypersensitivity, 518–520, 519*f*, 523*t*
Antigen-binding groove, on MHC molecules,
    474, 475*f*
Antigen-binding sites, 468, 468*f*
  antigenic determinant (epitope)
    complementarity and, 468
Antigenic determinants (epitopes), 463, 464*f*
  antigen-binding site complementarity
    and, 468
Antigenic drift, influenza virus, 732, 734*f*
Antigenic shift, influenza virus, 732, 734*f*
Antigen-presenting cells (APC), 474, 475*f*
  in humoral immunity, 477, 478*f*
Antigens, 463–465, 464*f*
  B cell receptors recognizing, 471, 471*f*
  dye-tagged antibodies binding to, 101, 101*f*
  of Enterobacteriaceae, 572–573, 574*f*
  presentation of, in humoral immunity, 477,
    478*f*. *See also* Antigen-presenting cells
  processing, 475–477, 475*f*, 476*f*

properties of, 463–464, 464*f*
serological tests and, 120, 120*f*
T cell receptors recognizing, 471–472
T-dependent, 471
T-independent, 475, 475*f*
types of, 464–465, 464*f*
in vaccines, recombinant DNA technology in introduction of, 254, 255*b*, 492, 493*f*
Antigen wells
in enzyme-linked immunosorbent assay (ELISA), 502–503, 504*f*
in immunodiffusion, 496, 497, 497*f*, 498*f*
Antihelminthic drugs, 298*f*, 309*t*
Antihistamines, for type I hypersensitivity, 515
Antilymphocyte globulin, 523
Antimeasles immunoglobulin, 487
Antimetabolic agents/antimetabolites, 290*f*, 294–295, 294*f*, 307*t*
Antimicrobial drugs/antimicrobials, 287–312, 305–310*t*
allergies to, 301
cell wall synthesis affected by, 290–292, 290*f*, 291*f*, 305–306*t*
clinical considerations in prescribing, 296–301
cytoplasmic membranes affected by, 290*f*, 292–294, 293*f*, 307*t*, 309*t*
distribution of, 300
efficacy of, 297–300, 298*f*, 299*f*, 300*f*
history of, 288, 288*f*, 289*b*, 289*t*
indiscriminate use of, 304
mechanisms of action of, 289–295, 290*f*, 296*f*, 297*b*
metabolic pathways affected by, 290*f*, 294–295, 294*f*, 307*t*
microbial production of, 13*t*, 288, 289*b*, 289*t*, 755, 756*t*
normal microbiota affected by, 301
nucleic acid synthesis affected by, 290*f*, 295, 296*f*, 307*t*, 308*t*
protein synthesis affected by, 290*f*, 292, 293*f*, 306*t*
resistance to, 287, 301–304, 303*b*. *See also* Resistance
routes of administration of, 300, 300*f*
safety/side effects of, 300–301, 301*f*
sources of, 289*t*
spectrum of action of, 297, 298*f*
toxicity of, 301, 301*f*
virus attachment affected by, 290*f*, 295, 308*t*
Antimicrobial methods, 262–286. *See also specific type and* Antimicrobial drugs
action of, 264–265
chemical, 277–283, 282*t*
desiccation, 272, 273*f*, 277*t*
environmental conditions and, 266–267, 267*f*
factors affecting efficacy of, 265–267, 266*f*, 267*f*
filtration, 274, 274*f*, 275*f*, 275*t*, 277*t*
heat-related, 268–272, 269*f*, 270*f*, 271*f*, 271*t*, 272*b*, 277*t*
lyophilization, 273, 277*t*
microbial susceptibility and, 266, 266*f*
osmotic pressure, 274–275, 277*t*
physical, 268–277, 277*t*
radiation, 262, 275–277, 276*f*, 277*t*
refrigeration/freezing, 272, 273*b*, 277*t*
selection of, 265–268
treatment site and, 266
Antimicrobics, 282–283, 282*t*. *See also* Antibiotics
Antimony sodium gluconic agent, 310*t*
Antiparallel strands, DNA, 51*f*, 52, 200, 201*f*
Antiphagocytic chemicals, 417*f*, 418, 449
Antiphagocytic factors, 417*f*, 418

Antiports, 73–74, 75*f*
Antiprotozoan drugs, 298*f*, 310*t*
Anti-RBC antibodies, in complement fixation test, 501, 501*f*
Anti-Rh serum, 518
Antisepsis/antiseptics, 19*f*, 263, 264*t*. *See also* Hygiene
chemical agents used for, 277–283, 282*t*
evaluation of efficacy of, 267–268
in infection control/prevention, discovery of importance of, 17–18, 19*f*
Antiserum (antisera), 120
passive immunization and, 483, 494–496, 494*f*
Antitoxins, 416
passive immunization and, 483, 495
Antiviral drugs, 308*t*
attachment affected by, 290*f*, 295, 308*t*
viral metabolism affected by, 295, 308*t*
Antiviral proteins (AVPs), 453, 454*f*
Antioxidants, 171
APC. *See* Antigen presenting cells
Apicomplexans, 351–352, 356*t*, 658–663, 664*t*
Apoenzymes, 130, 130*f*
Apoptosis
in Epstein-Barr virus infection, 691
in self-tolerance, 473–474, 474*f*
Apple juice, unpasteurized, *E. coli* contamination of, 272*b*
Applied microbiology, 20*t*, 744
food microbiology, 13*t*, 19*f*, 20*t*, 150*f*, 744, 744–752, 748*t*, 749*t*, 752*t*
industrial microbiology/biotechnology, 11, 13*t*, 19*f*, 20*t*, 744, 752–763
recombinant DNA technology in, 242, 243*f*
Aquaspirillum magnetotacticum, 56*f*
Aquatic microbiology, 770–772, 771*f*, 773*t*
aquatic habitats and, 770–772, 771*f*
water-borne illnesses and, 422, 422*t*, 772, 773*t*
water pollution and, 756, 772
Aquifex genus/spp., 323
Aquificae (phylum), 320*f*, 323
Arachnids, as disease vectors, 673–674, 674*f*
Araezoa, 354–355
Arboviruses, 710
bunyaviruses as, 735
coltiviruses as, 737–738
togaviruses and flaviviruses as, 710–713, 711*t*
Archaea, 5–6, 58, 115*f*, 117–118, 318, 319–322, 320*f*. *See also specific type*
cell walls of, 5, 68–69, 319–320
classification of, 115*f*, 117–118, 319–320, 320*f*
nucleic acid sequencing and, 21
cytoplasmic membranes of, 70
as domain, 115*f*, 117–118, 318
temperature range of, 174, 321, 322*t*
Arenaviruses (*Arenaviridae*), 385*t*, 736–737, 737*f*, 738*t*
as biological weapon, 774*t*
Argon, Bohr diagram of, 30*f*
Arildone, 308*t*
spectrum of action of, 293*f*, 308*t*
virus attachment affected by, 295, 308*t*
Aristotle, 8
Arithmetic growth, 187, 188*f*
Armillaria genus/spp., 356
Armillaria ostoyae, 342
Arsenic, antimicrobial action of, 280–281, 282*t*
Arthritis, rheumatoid, 526, 526*f*
Arthrobacter genus/spp., 320*f*
Arthroconidia, 358, 629
Arthropods, 672
Arthropod vectors, 423, 424*t*, 672–675, 674*f*
arbovirus diseases transmitted by, 710, 711*t*, 712*f*

bunyavirus diseases transmitted by, 735, 736
rickettsial diseases transmitted by, 598
Arthrospores, 629
Artifacts, in transmission electron micrographs, 102–104
Artificial bioremediation, 761
Artificially acquired immunity. *See also* Immunization; Vaccines
active/active immunization, 483, 483*t*, 488, 489–494, 490*t*, 491*f*, 492*t*, 493*f*, 494*b*
passive/passive immunization, 483, 483*t*, 494–496, 495*f*
Artificial sweetener, microbial production of, 13*t*
Artificial wetlands, in wastewater treatment, 761, 762*f*
Ascariasis, 670, 673*t*
Ascaris lumbricoides, 670, 673*t*
Asci, 360, 361, 361*f*, 362*f*
Ascocarps, 360, 361*f*
Ascomycetes, 360–362, 361*f*, 362*f*
Ascomycota (division), 359, 360–362, 366*t*, 628, 632
Ascospores, 360, 361, 361*f*, 362*f*, 366*t*
Aseptate hyphae, 356, 357*f*
Aseptic/aseptic technique, 263, 264*t*
pure cultures and, 179
Aseptic packaging, in food preservation, 751
Asexual reproduction, 57
in fungi, 358, 359*f*
Asexual spores, 4, 358, 359*f*. *See also* Spores
A site, ribosomal, 216
Aspartic acid
in amination and transamination, 159, 159*f*
in nucleotide biosynthesis, 159, 161*f*
Aspergilloma, noninvasive, 632, 632*f*
Aspergillosis, 632–633, 632*f*
Aspergillus genus/spp., 632–633, 632*f*
aflatoxins produced by, 643
mutagenic effects of, 224
Aspergillus flavus, 632, 643
Aspergillus fumigatus, 632, 637
Aspergillus niger, 632
Aspergillus nomius, 643
Aspergillus oryzae
in amylase production, 753
in sake production, 747
in vegetable fermentation, 744
Aspergillus parasiticus, 643
Aspirin, Reye's syndrome and, 689, 691*b*, 735
Assembly
animal virus, 393, 394*t*
bacteriophage, in lytic replication, 385, 386*f*, 387–388, 389*f*, 394*t*
in HIV replication, 720*f*, 722
Asthma, allergic, 512
Astroviruses (*Astroviridae*), 385*t*, 704, 709, 738*t*
Asymptomatic infection, 412, 425*t*
Atherosclerosis, *Chlamydia pneumoniae* infection and, 605, 605*b*
Athlete's foot (tinea pedis), 623, 640*t*, 641*b*
Atomic force microscopy/microscopes, 105, 106*f*, 107*f*
Atomic mass/atomic weight, 27, 28*t*
Atomic mass unit (dalton), 27
Atomic number, 27, 28*t*
Atomic structure, 27, 27*f*, 28*t*
Atoms, 27–29, 27*f*, 28*f*, 28*t*, 29*f*
Atopobium genus/spp., 320*f*
ATP, 52–53, 52*f*, 57, 126, 127–128, 162*f*
in active transport, 73–74, 75*f*
in Calvin-Benson cycle, 156, 157*f*, 162*f*
catabolic pathways producing, 127, 127–128, 129*f*
in cellular respiration, 141–148, 149*t*, 150*t*
in chemiosmosis, 144, 147–148

ATP, *Cont.*
in electron transport chain, 144–147, 144*f*, 145*f*, 149*t*
energy storage and, 127–128, 129*f*
in Entner-Doudoroff pathway, 141
in glycolysis, 137, 138*f*, 139, 149*t*
in Krebs cycle, 142, 143*f*, 149*t*
in pentose phosphate pathway, 140*f*, 141
in photophosphorylation, 154, 155*f*
ATP synthases (ATPases), 147
Attachment
in animal virus replication, 390, 394*t*
antimicrobial drugs affecting, 290*f*, 295, 308*t*
bacteriophage, in lytic replication, 385, 386*f*, 387, 389*f*, 394*t*
in HIV replication, 720*f*, 721, 721*f*
Attachment antagonists, 308*t*
mechanism of action of, 290*f*, 295, 308*t*
Attachment proteins, 411
Attenuated (live) vaccines, 489–490
Attenuation, in vaccine preparation, 489
recombinant DNA technology in, 492, 493*f*
Atypical pneumonia, primary, *Mycoplasma pneumoniae* causing, 597
Auramine O dye, 100, 100*f*
Autoantibodies
molecular mimicry and, 524
in systemic lupus erythematosus, 525
Autoantigens, 464, 464*f*
lymphocyte editing (clonal deletion) and, 473–474, 474*f*
Autoclave/autoclaving, in microbial control, 269–270, 269*f*, 270*f*, 271*f*, 277*t*
Autograft, 521, 521*f*
Autoimmune diseases, 474, 523–526, 526*f*. *See also specific type*
causes of, 523–524
single tissue, 524–525
systemic, 525–526, 526*f*
Autoimmune hemolytic anemia, 524–525
Autoinoculation, in papillomavirus infections, 693
Autotrophs, 168–169
lithotropic (chemolithoautotrophs), 169, 169*b*
Auxotroph, identification of, 227–228, 228*f*
Avery, Oswald, 20
Avirulent organisms, 411
attenuation for vaccines and, 489–490
AVPs. *See* Antiviral proteins
Axenic culture, 178. *See also* Pure culture
Axenic environment, 406
defense mechanisms affected by, 441
Axial filament, 64, 65*f*
Azathioprine, 523
Azidothymidine (AZT/zidovudine), 296*f*, 308*t*, 719*t*
in AIDS "cocktail," 303*b*
Azithromycin, 306*t*
Azoles, 309*t*, 627
spectrum of action of, 293*f*, 309*t*
Azomonas genus/spp., 334*t*, 335
Azospirillum genus/spp., 331, 334*t*
Azotobacter genus/spp., 334*t*, 335
nitrogen fixation by, 767
AZT. *See* Azidothymidine
Aztreonam, 291*f*

B. *See* Boron
B19 parvovirus, 700, 700*f*
Babesia genus/spp., 352
Bacillary angiomatosis, 584
Bacillary peliosis hepatitis, 584
Bacillus (bacilli), 314, 314*f*, 329*t*
arrangements of, 315–316, 317*f*
Gram-negative, 566–594. *See also specific agent*
Gram-positive, 327–328. *See also specific agent*

Bacillus genus/spp., 327, 327*f*, 329*t*, 545–546
antimicrobials produced by, 288, 289*t*, 755
endospores of, 316, 318
as sterilization indicator, 270, 271*f*
susceptibility of, 266
fermentation products of, 150*f*
nitrogen fixation by, 767
Bacillus anthracis, 327, 545, 545*f*, 546, 547*b*. *See also* Anthrax
bioterrorism and, 22, 327, 545, 547*b*, 774*t*
ciprofloxacin affecting, 297*b*
discovery of, 17*t*
irradiation affecting, 262
isothiocyanate dye staining, 100
soil-borne disease caused by, 769, 770*t*
Bacillus cereus, endospores of, 327
Bacillus licheniformis, antimicrobials produced by, 289*t*, 305*t*, 327
Bacillus polymyxa, antimicrobials produced by, 289*t*, 292, 307*t*, 327
Bacillus popilliae, in pest control, 755
Bacillus sphaericus, in mosquito control, 755
Bacillus stearothermophilus, as sterilization indicator, 270
Bacillus thuringiensis, Bt-toxin produced by, 257–258, 258*b*, 327, 327*f*, 755
Bacitracin, 305*t*
mechanism of action of, 292, 305*t*
Bacteremia
Bacteroides fragilis causing, 591
pneumococcal, 543
staphylococcal, 536
Bacteria, 5–6, 58, 318, 320*f*, 323–338, 400*t*. *See also specific type*
acid-fast, 67
antimicrobials produced by, 289*b*, 289*t*
cardiovascular disease and, D-6
cell walls of, 5, 66–69, 67*f*, 68*f*, 69*f*
peptidoglycan in, 5, 45, 66, 68*f*
chromosomes of, 87*t*, 200–202, 202*f*
classification of, 115*f*, 117–118, 320*f*, 323, C-1 to C-9
nucleic acid sequencing and, 21, 117–118, 117*b*
colony formation by, 14, 15*f*, 177, 178*f*
conjugation in, 231–232, 233*f*, 234*f*, 235*f*, 235*t*
culturing viruses in, 396, 396*f*
deeply branching, 320*f*, 323
as domain, 115*f*, 117–118, 318
endospores of. *See* Endospores
enteric. *See* Enterobacteriaceae
eye infections caused by, D-1
in fermentation, 11, 12*f*, 149, 150*f*
gastrointestinal infections caused by, D-9
genomes of, 200, 200–202, 202*f*, 204*t*
Gram-negative, 16, 16*f*, 320*f*, 330–338, 334–335*t*. *See also* Gram-negative bacteria
Gram-positive, 16, 16*f*, 320*f*, 325–330, 329*t*, 532–565. *See also* Gram-positive bacteria
horizontal genetic transfer in, 229–232, 235*t*
lymphatic system infections caused by, D-6
magnetosomes in (magnetobacteria), 56, 56*f*, 75
nervous system infections caused by, D-4
nitrifying, 331
opportunistic infections caused by, in HIV infection/AIDS, 718*t*
pathogenic, 5
phototrophic, 323–325, 325*t*
reproductive system infections caused by, D-11
respiratory infections caused by, D-8
skin infections caused by, D-1

in soil, 769
disease caused by, 769, 770*t*
urinary tract infections caused by, D-11
viruses/viroids/prions compared with, 400*t*
water-borne illnesses caused by, 773*t*
zoonoses caused by, 421*t*
Bacterial viruses, 379, 380*f*
Bactericides, 264
Bacteriochlorophyll a, 153
Bacteriochlorophylls, green and purple bacteria using, 324
Bacteriocidal drug, 299
Bacteriocin factors, 202
Bacteriocins, 47, 202
Bacteriology, 20*t*. *See also* Bacteria
Bacteriophages (phages), 377, 377*f*, 379
bacterial pathogens affected by, 381*b*
high-pressure, 387, 387*b*
lysogenic, 388
replication of, 394*t*
lysogenic, 388–390, 389*f*
lytic, 385, 385–388, 386*f*, 388*f*, 389*f*
shapes of, 382, 383*f*
in taxonomy/microbial classification, 120, 121*f*
temperate, 388
in transduction, 230–231, 232*f*, 235*t*
Bacteriorhodopsins, 322
Bacteriostatic drug, 299
Bacteroides genus/spp., 335*t*, 339, 591, 591*f*
Bacteroides fragilis, 591, 591*f*
Bacteroidetes (phylum), 320*f*, 335*t*
Bacteroids, 339
Balantidiasis, 650
Balantidium genus/spp., 351
Balantidium coli, 650, 664*t*
*Bam*HI restriction enzyme, 245*t*
in genome mapping, 252, 252*t*, 253*f*
Bang's disease. *See* Brucellosis
Barophiles, 176
Bartonella genus/spp., 583–584, 584*f*
Bartonella bacilliformis, 584
Bartonella henselae, 584
Bartonella quintana, 584
Bartonellosis, 584
Basal body, of bacterial flagellum, 62, 63, 63*f*
Base-excision repair, 225, 226*f*
Base number, in logarithm, B-1
Base pairs (bp), 200, 201*f*
point mutations affecting, 221–222, 222*f*, 223*t*
in taxonomy/microbial classification, 325, 329*t*
Bases, 37–39, 38*f*
nucleotide, 50–51, 50*f*, 51
in DNA, 50–51, 51, 51*f*, 52, 200, 201*f*
in RNA, 50–51, 51, 51*f*, 200, 201*f*
pH of, 38, 38*f*
Basic dyes, 108
Basidia, 363
Basidiocarps, 363, 363*f*
Basidiomycetes, 363
Basidiomycota (division), 359, 363, 366*t*
Basidiospores, 363, 366*t*
Basophils, 444, 445*f*
in type I (immediate) hypersensitivity, 511, 511*f*, 512*f*
Batch method, for pasteurization, 271, 271*t*
Batch production, in industrial fermentation, 753
B cell receptors, 471, 471*f*
B cells. *See* B lymphocytes
BCG vaccine
for leprosy, 560
for tuberculosis, 559
BCR. *See* B cell receptors
Bdellovibrio genus/spp., 334*t*, 336, 337*f*
Be. *See* Beryllium

Beadle, George, 20, 355
"Beaver fever" (giardiasis), 657–658
Bedrock, 768, 768f
Beef tapeworm (Taenia saginata), 649, 665–666, 673t
Beer, fermentation in production of, 747, 747f, 748t
Bee stings, anaphylaxis caused by, 512–514
Beggiatoa genus/spp., in sulfur cycle, 768
von Behring, Emil, 19f, 22
Beijerinck, Martinus, 16, 17t, 19f, 21, 186
Bejel, 609
Benign tumor, 395
Benthic zone, 771, 771f
Benzalkonium chloride, 280, 280f
Benzoic acid, in food preservation, 751
Benzopyrene, as frameshift mutagen, 224
Bergey's Manual of Determinative Bacteriology, 118
Bergey's Manual of Systematic Bacteriology, 118, 319, C-1 to C-9
Beryllium (Be), Bohr diagram of, 30f
Beta-carotene, gene for, insertion of into rice, 258
Betadine, antimicrobial action of, 279, 279f
Beta-hemolysis, by Streptococcus pyogenes, 183, 183f, 546t
Beta-hemolytic streptococci, 542
Beta- (β-) interferon, 453, 454f, 455t
Beta-lactam antibiotics, 305t
mechanism of action of, 290, 291f, 305t
Beta-lactamase, 302, 303f, 534
penicillin and, 221b, 302, 303f
Beta-lactam rings, 290, 291f, 302, 303f
Beta-oxidation, of fatty acids, 151, 151f
Beta- (β-) pleated sheets, 48–50, 49f
Beta-propiolactone, for gas sterilization, 281–282, 281f, 282t
Betaproteobacteria, 325t, 333, 334t, 567
Beverages, microbial production of, 13t, 19f, 20t, 150f, 746–747, 746f, 747f, 748t
Bidirectionality, of DNA replication, 207, 207f
Bilayers, phospholipid, 41, 43f, 70, 70f
cholesterol in, 42, 44f
Bile, in host defense, 443t
Bilirubin, accumulation of in hemolytic disease of newborn, 518
Binary fission, 186–187, 187f, 314, 315f
in amoebae, 352, 353b
Binocular microscope, 97
Binomial nomenclature, 114–116
Biochemical oxygen demand (BOD), reduction of in wastewater treatment, 759
Biochemistry, 12, 19–20
practical applications of research in, 20
in taxonomy/microbial classification, 118–120, 119f
Bioconversion, in microbial production of fuels, 754
Biodiversity, 764
Biofilms, 67b, 177, 177b, 411, 412f
confocal microscopes for examination of, 101
fimbriae in, 65
in-use test of organisms in, 268
Biogeochemical cycles, microorganisms in, 765–768, 766f, 767f
Biological vectors, in disease transmission, 423, 424t
Biological warfare, 772–777. See also Bioterrorism
Biological weapons, 772. See also Bioterrorism
Bioluminescence, 146b
Biomass, 764
microbes in reclamation of, 761
Bio-mining, 769
Bioreceptors, microbes as, 756
Bioremediation, 20t, 21, 761, 763b
Biosensors, microbes as, 756

Biosphere, 764, 765f
Biot, Jean Baptiste, 47
Biotechnology/industrial microbiology, 11, 13t, 19f, 20t, 744, 752–763
agricultural products of, 755, 756t
alternative fuel production and, 754–755, 754f, 755b, 756t
enzyme production and, 753–754, 754b, 756t
fermentations, 752–753, 753f
pesticide production and, 755, 756t
pharmaceutical products of, 755, 756t
products of, 753–756, 756t
recombinant DNA technology in, 242, 243f
Bioterrorism, 22, 22f, 772–776
animal pathogens and, 774–775
anthrax and, 22, 545, 546, 547b
assessing microorganisms' potential and, 772–774
defense against, 775, 775f
definition of, 772
Francisella and, 589
human pathogens used in, 774, 774t
known microbial threats and, 774–775, 774t
plant pathogens and, 775
recombinant genetic technology and, 775–776, 776b
smallpox and, 681, 683b
Bisphenolics, 278, 278f, 282t
Bites, Pasteurella infecting, 582
BK virus, 695
"Black Death," 580
"Black hairy tongue," metronidazole causing, 301, 301f
Black piedra, 638, 638f
Blackwater fever, 660
Bladder infections
discovery of cause of, 17t
E. coli causing, 575
Blades, of algal thalli, 366
Blastoconidia, 358
Blastomyces dermatitidis, 624, 628, 628f
geographic distribution of, 626, 628
soil-borne disease caused by, 770t
Blastomycosis, 628–629, 628f, 770t
Blattella species, as disease vectors, 424t
Bleach. See Sodium hypochlorite
Blepharospasm, botulinum toxin in management of, 326b
Blood. See also specific component
defensive components of, 443–447, 444f, 445f, 446f, 447b
Blood agar, 182, 184t
as differential medium, 182–183, 183f, 184t
for Neisseria culture, 567
Blood cells. See also specific type
autoimmunity affecting, 524–525
defensive, 443–447, 444f, 445f
Blood flow, in host defense, 443t
Blood flukes (Schistosoma species), 667–668, 668f, 673t
water-borne transmission of, 773t
Blood group antigens, 515. See also ABO system
Blood poisoning. See Septicemia
Blood specimen, collection of, 179t
Blood typing, hemagglutination tests in, 499–500, 499f
Blood vessels, dilation/increased permeability of, in inflammation, 455–456, 455f, 456f
Bloom, 768
intoxications caused by, 772
Blotting, in western blot test, 504, 505f
Blue dye, microbial production of, 754, 754b
Blue-green bacteria/"blue-green algae." See Cyanobacteria
Blunt ends, of restriction enzymes, 244, 245f

B lymphocytes/B cells, 465, 467f, 468, 473t
activation of, 477, 478f
deficient, 527, 527b, 528t
editing of, 473–474, 474f
memory, 477, 477–480, 478f, 479f
receptors on, 471, 471f
immunological memory and, 477–480, 479f
BOD. See Biochemical oxygen demand
Body fluids, disease transmission via, 423, 423b
Body temperature, hypothalamus in regulation of, 458, 458f
Bohr, Niels, 27
Bohr diagrams, 29, 29f, 30f
Boiling, in microbial control, 269, 277t
Bone marrow
B lymphocyte maturation in, 465
lymphocyte editing in, 473, 474f
red, lymphoid cells produced in, 465, 466f, 467f
Bone marrow transplants
graft-versus-host disease and, 522
for severe combined immunodeficiency (SCID), 527b
Bordetella genus/spp., 333, 334t, 585–587
adhesins produced by, 411, 585
Bordetella parapertussis, 585
Bordetella pertussis, 585–587, 586f, 587f
Bordet-Genou medium, for Bordetella pertussis culture, 586
Boron, 28t
Bohr diagram of, 30f
Borrelia genus/spp., 335t, 339, 610–613
Borrelia burgdorferi, 610–612, 610f
motility of, 64
Borrelia recurrentis, 613
Botryococcus barunii, hydrocarbons produced by, 755
Botulism, 326b, 549–550, 549f
as bioterrorist threat, 774t
Botulism (botulinum) toxins, 326b, 549, 549f
Bound coagulase, 534
Bovine growth hormone (BGH), 258
Bovine spongiform encephalitis (BSE/mad cow disease), 48, 399
bp. See Base pairs
Bradyzoites, in Toxoplasma life cycle, 662, 662f
Branhamella catarrhalis. See Moraxella catarrhalis
Brazilian purpuric fever, 583
Bread, microbial production of, 13t, 744, 748t
Breakbone fever, 711
Brequinar sodium, 523
Bright-field microscopy/microscopes, 95, 96–98, 97f, 98f, 99f, 107t
compound, 96–98, 97f, 98f
simple, 96
Brill-Zinsser disease, 600
Broad-spectrum drugs, 297, 298f
Bromine, antimicrobial action of, 279, 282t
Bronchitis, manifestations of, 705t
Bronchodilators, for asthma, 515
Broth dilution test, 299, 299f
Broths, 177, 181
turbid, estimating microbial growth and, 193–194, 194f
Broth tubes, carbohydrate utilization, 183, 183f
Brown algae, 369, 369f, 370f, 371t
Bruce, David, 17t
Brucella genus/spp., 331–332, 334t, 584–585, 585f
as biological weapon, 774t
Brucella abortus, 585
Brucella canis, 585
Brucella melitensis, 584–585
discovery of, 17t
pasteurization in control of, 271
Brucella suis, 585

Brucellosis (undulant fever), 331, 584–585, 585f
  as bioterrorist threat, 774t
  discovery of cause of, 17t
Bruton-type agammaglobulinemia, 527, 528t
BSE. See Bovine spongiform encephalitis
Bt-toxin, 257–258, 258b, 327, 327f, 755, 756t
Buboes
  in bubonic plague, 566, 580, 581f
  in lymphogranuloma venereum, 603
Bubonic plague, 421t, 566, 580–581, 581f
  discovery of cause of, 17t
Buchner, Eduard, experiments in fermentation by, 11–12, 13f, 19f
Budding, 315, 316f
  fungus reproduction by, 358
  virus release by, 393, 393f
    influenza virus and, 732, 735f
  yeast reproduction by, 4
Buffers, 38
Bulbar poliomyelitis, 706
"Bull's eye" rash, in Lyme disease, 612, 612f
Bunyaviruses (Bunyaviridae), 385t, 735–736, 736f, 738t
  diseases caused by, 711t, 735–736
Burkholderia genus/spp., 333, 334t
Burkholderia mallei, as biological weapon, 774t
Burkitt, Denis, 690
Burkitt's lymphoma, 690, 690f
Burn wounds, Pseudomonas contamination of, 588
Bursa of Fabricius, 468
Burst size, 388, 388f
Burst time, 388, 388f
Buttermilk, fermentation in production of, 745, 748t

C. See Carbon; Cytosine
C3, 451, 451f, 452f
C3a, 451, 451f, 452f
  in inflammation
    blood vessels affected by, 455, 455f
    phagocyte migration and, 457
C3b, 451, 451f, 452f
C5, 451, 451f, 452f
C5a, in inflammation
    blood vessels affected by, 455, 455f
    phagocyte migration and, 457
C5a peptidase, 538
C5b, 451, 451f, 452f
^{12}C (carbon-12), 28, 28f
^{13}C (carbon-13), 28, 28f
^{14}C (carbon-14), 28, 28f
Ca. See Calcium
Cadmium, cycling of, 768
Calcium, 28t
  Bohr diagram of, 30f
  for microbial growth, 170t, 172
Calcium hypochlorite, antimicrobial action of, 279
Caliciviruses (Caliciviridae), 385t, 704, 709, 738t
California encephalitis, 735
Calor, in inflammation, 455
Calvin-Benson cycle, 156, 157f, 162f, A-7
Campylobacter genus/spp., 334t, 336
Campylobacteria, 320f
Campylobacter jejuni, 616–617, 617b, 752t
  water-borne transmission of, 616, 752t, 773t
Cancer, 395
  adenoviruses in treatment of, 698b
  genital warts and, 693
  T cell therapy for, 482b
  viruses in, 394–395, 395f
    HIV infection/AIDS and, 718t
    retroviruses, 717, 718f
Candida genus/spp., 624, 633–635, 633f

Candida albicans, 4, 633, 637
  opportunistic infection caused by, 407–409, 633–635, 633f, 637
  reproduction in, 358
  superinfection caused by, 301
Candidemia, 634t
Candidiasis, 633–635, 633f, 634t
Candle jars, 186, 186f
Canning, industrial, in food preservation, 750, 750f
Capillaries, lymphatic, 465, 466f
Capnophiles, culturing, 186
Capsids, viral, 378, 378f, 380
  of enveloped positive single-stranded RNA viruses, 710, 710f
Capsomeres/capsomers, 380
Capsular antigens, 572, 574f
Capsule (glycocalyx)
  of Enterobacteriaceae, 573, 574f
  prokaryotic, 62, 62f
    as antiphagocytic factor, 417f, 418
Capsule (negative) stains, 112, 112f, 114t
Carbohydrates, 42–45, 44f, 45f, 46f, 47b
  biosynthesis of, 157–158, 158f, 162f
  catabolism of, 136–149, 137f, 149t, 162f
    cellular respiration in, 136–137, 137f, 141–148, 149t, 150t
    electron transport chain in, 137, 142–147, 144f, 145f
    Entner-Doudoroff pathway in, 141, 141f
    fermentation in, 137, 137f, 148–149, 148f
    glycolysis in, 137, 137f, 138f, 139, 139f
      alternatives to, 139–141, 140f, 141f
    Krebs cycle in, 137, 142, 143f
    pentose phosphate pathway in, 139–141, 140f
Carbohydrate utilization broth tubes, 183, 183f
Carbohydrate utilization test, 119f
Carbolfuchsin
  in flagellar staining, 112
  in Ziehl-Neelsen acid-fast staining, 111
Carbolic acid. See Phenol
Carbon, 28t, 32
  alpha (α), amino acid, 47, 47f
  atomic number of, 27, 28t
  Bohr diagram of, 30f
  isotopes of, 28, 28f
  for microbial growth, 168–169, 168f, 170t
Carbon-12, 28, 28f
Carbon-13, 28, 28f
Carbon-14, 28, 28f
Carbon chains, 32
Carbon cycle, 766, 766f
Carbon dioxide (CO_2)
  fixation of, 766, 766f
    in light-independent photosynthesis, 156, 157f
  microbial fermentation producing, 150f
  production of, 766, 766f
Carbon dioxide incubators, 185–186
Carbonyl group, 40t
Carboxyl group, 40t
  in amino acids, 47, 47f
Carbuncle, 535
carcino- (prefix), definition of, 413t
Carcinogens, 395
  identification of, 228, 229f
Cardiovascular system
  in host defense, 443t
  microbial disease agents affecting, D-6 to D-7
Caries, dental, viridans streptococci causing, 542
Carotene, in chrysophytes, 368
β-Carotene, gene for, insertion of into rice, 258
Carotenoids, singlet oxygen affected by, 170

Carrageenan, in algal cell wall, 79, 367
Carrier proteins
  in active transport, 73
  in electron transport chains, 144–146
  in facilitated diffusion, 72
Carriers (disease), human, 420, 420b
  of Entamoeba histolytica, 651
Cascade, complement, 451–452, 451f, 452f
Caseous necrosis, in tuberculosis, 556, 557f
Catabolism/catabolic pathways, 36, 126, 126–127, 126f, 162f
  carbohydrate/glucose, 136–149, 137f, 149t
  integration and regulation of, 160–161, 162f
  lipid, 149–151, 151f
  protein, 151, 152b, 152f, 162f
Catalase, 171, 171f, 533
  hydrogen peroxide affected by, 279
  in peroxisomes, 84
Catalysts, 45
Catarrhal phase, of pertussis, 586, 587f
Cat fleas. See Fleas
Cathode ray machines, in microbial control, 276
Cations, 34, 37, 38f
  of salts, 39
  in stain dyes, 108
Cats, in toxoplasmosis, 661–662, 662f
Cat-scratch disease, 584, 584f
Caulobacter genus/spp., 332, 332f, 334t
Cavities (dental), viridans streptococci causing, 542
CCR5, HIV replication and, 721
CD (cluster of differentiation) designations, 465
CD3, monoclonal antibodies against, 523
CD4 receptors, HIV replication and, 720f, 721, 721f
CD4 T cells, in HIV infection/AIDS, 472b, 528
  attachment/penetration/uncoating and, 720f, 721, 721f
  course of infection and, 722, 722f
CD8 T cells, in HIV infection/AIDS, 528
CD95 cytotoxic pathway, 481, 481f
CDC. See Centers for Disease Control and Prevention
cDNA. See Complementary DNA
Cefaclor, 305t
Cell. See Cells
Cell counter, for estimating microbial growth, 192, 192f
Cell cultures, 185
  viral, 396–397, 397f
Cell death, programmed, in Epstein-Barr virus infection, 691
Cell division, centrioles in, 81
Cell-free coagulase, 534
Cell-mediated (type IV) hypersensitivity, 520–522, 520f, 521f, 523t
Cell-mediated immune response, 472, 480–482, 480f, 482b
Cell (cytoplasmic/plasma) membranes
  antimicrobial drugs affecting, 290f, 292–294, 293f, 307t, 309t
  electrical potential of, 71, 71f
  eukaryotic, 60f, 79–80, 79f, 80f, 81f, 81t, 87t
  movement across, 71–74, 76t
    active processes in, 73–74, 75f, 76t
    passive processes in, 71–73, 72f, 73f, 74f, 76t
  permeability of, 71
  phospholipid bilayers forming, 41, 43f, 70, 70f
    cholesterol in, 42, 44f
  prokaryotic, 60f, 70–74, 70f, 87t
    electron transport chains in, 144, 145f
    function of, 71–74, 71f, 72f, 73f, 74f, 75f
Cell plate, 346

Cells, 58
  structure/function of, 56–91
    cell walls. *See also* Cell walls
      of eukaryotes, 77–79, 79*f*
      of prokaryotes, 60*f*, 66–69, 67*f*, 68*f*, 69*f*
    cytoplasm
      of eukaryotes, 60*f*, 80–85, 82*f*, 83*f*, 84*f*,
        85*f*, 86*f*, 87*t*
      of prokaryotes, 60*f*, 74–76, 75*f*, 76*f*, 77*f*
    cytoplasmic membranes. *See also*
        Cytoplasmic (cell/plasma)
        membranes
      of eukaryotes, 60*f*, 79–80, 79*f*, 80*f*, 81*f*, 81*t*
      of prokaryotes, 60*f*, 70–74, 70*f*, 71*f*, 72*f*,
        73*f*, 74*f*, 75*f*
    endosymbiotic theory and, 85–86
    in eukaryotic versus prokaryotic cells,
        58–61, 59*b*, 59*f*, 60*f*, 61*f*, 87*t*
    external structures
      of eukaryotes, 60*f*, 76–77, 78*f*, 79*f*
      of prokaryotes, 60*f*, 61–66
    life processes and, 57–58, 57*t*
    in microbes, 57*t*
  types of, 58, 59*f*
Cellular inclusions
  in eukaryotic cells, 87*t*
  in prokaryotic cells, 60*f*, 75, 76*f*, 87*t*
Cellular PrP, 398–399
  amino acid substitutions in, 48, 399
Cellular respiration, 136–137, 137*f*, 141–148,
    149*t*, 150*t*
  acetyl-CoA synthesis in, 142, 142*f*, 149*t*
  aerobic, 147, 149*t*, 150*t*
  anaerobic, 147, 149*t*, 150*t*
  chemiosmosis and, 144, 147–148
  electron transport in, 142–147, 144*f*, 145*f*, 149*t*
  Krebs cycle in, 142, 143*f*, 149*t*
Cellular slime molds, 371, 372*f*
Cellulose, 45, 46*f*
  indigestibility of wood and, 47*b*
  in plant cell walls, 79
Cell walls
  absence of in some bacteria, 68
  antimicrobial agents affecting, 265
  antimicrobial drugs affecting, 290–292, 290*f*,
    291*f*, 305–306*t*
  of eukaryotic cells, 77–79, 79*f*, 87*t*
  of prokaryotic cells, 60*f*, 66–69, 87*t*
    archaeal, 5, 68–69, 319–320
    bacterial, 5, 66–69, 67*f*, 68*f*, 69*f*
      Gram-negative, 67, 68, 69*f*
      Gram-positive, 67, 69*f*
      peptidoglycan in, 5, 45, 66, 68*f*
Centers for Disease Control and Prevention
    (CDC), 426
  data sharing and, 432
  Morbidity and Mortality Weekly Report
      published by, 426, 428*f*
  notifiable diseases listed by, 426, 429*f*
Centimeter (cm), 93, 93*t*
Central dogma, 208, 209*b*
  retrovirus transcription and, 716
Central vacuole, 84, 85*f*
Centrioles, 60*f*, 81, 82*f*, 87*t*
Centromere, 344, 345*f*
Centrosome, 81, 82*f*
Cepacol. *See* Cetylpyridinium chloride
Cephalexin, 305*t*
Cephalosporins, 305*t*
  mechanism of action of, 290, 291*f*, 305*t*
Cephalosporium genus/spp., antimicrobials
    produced by, 288, 289*t*, 305*t*
Cephalothin, 291*f*, 305*t*
Cercariae, in fluke life cycle, 667, 668*f*
Cerebral malaria, 660
Cerebral phaeohyphomycosis, 641

Cerebrospinal fluid, specimen collection
    and, 179*t*
Cesspools, 759–760
Cestodes (tapeworms), 421*t*, 649, 663, 664–666,
    665*f*, 673*t*
  beef (*Taenia saginata*), 649, 665–666, 673*t*
  Echinococcus granulosus, 666, 666*f*, 673*t*
  pork (*Taenia solium*), 649, 665–666, 673*t*
Cetylpyridinium chloride, 280, 280*f*
CFU. *See* Colony-forming unit
CGD. *See* Chronic granulomatous disease
Chagas, Carlos, 655
Chagas' disease, 654*f*, 655
Chagomas, 655
Chancre
  soft (chancroid), 583, 583*f*
  syphilitic, 608–609, 608*f*
Chancroid (soft chancre), 583, 583*f*
Channel proteins
  in active transport, 73
  in facilitated diffusion, 72
Charcoal, activated, in water treatment,
    756–758, 757*f*
Cheese, microbial production of, 13*t*, 150*f*,
    745–746, 745*f*, 748*t*
Chemical bonds, 29–35, 35*t*
  definition of, 29
  hydrogen, 34–35, 34*f*, 35*t*
  ionic, 33–34, 33*f*, 34*f*, 35*t*
  nonpolar covalent, 30–32, 31*f*, 32*f*, 35*t*
  polar covalent, 32, 33*f*, 35*t*
Chemical elements, 27, 28*t*, 30*f*
  for microbial growth, 168–172, 168*f*, 170*t*
  periodic table of, 30*f*
Chemical fixation, 108
Chemical (concentration) gradient, 71
Chemical mutagens, 223–224, 224*f*
Chemical reactions, 19–20, 35–36. *See also*
      Chemistry
  enzymes affecting, 130–136, 131*f*,
      132*f*, 133*f*
    inhibition and, 134–136, 135*f*, 136*f*
    pH and, 132–133, 134*f*
    substrate concentration and,
      133–134, 134*f*
    temperature and, 132, 134*f*
Chemical requirements, for microbial growth,
    168–172, 168*f*, 170*t*, 173*t*
Chemiosmosis, 144, 147–148
Chemistry, 26–55. *See also under Chemical*
  atoms and, 27–29, 27*f*, 28*f*, 28*t*, 29*f*
  chemical bonds and, 29–35, 35*t*
  chemical reactions and, 19–20, 35–36
  inorganic substances and, 37–39
  organic substances and, 39–53
Chemoautotrophs, 168*f*, 169
  in carbon cycle, 766, 766*f*
Chemoheterotrophs, 168*f*, 169
Chemokine receptors, HIV replication
    and, 721
Chemokines, 448, 473. *See also* Cytokines
  in specific immunity, 473
Chemolithoautotrophs, 169, 169*b*
Chemolithotrophs, 144
Chemostat, 189
Chemotactic factors, 448
Chemotaxis, 64
  in phagocytosis, 448, 448*f*
Chemotherapy/chemotherapeutic agents, 19,
    19*f*, 20*t*, 288. *See also specific type and*
      Antimicrobial drugs
Chemotrophs, 169
Chicken eggs, embryonated, culturing viruses
    in, 396, 397*f*
Chickenpox (varicella), 687–689, 687*f*
  vaccine against, 689

Chiggers (mites), as disease vectors, 424*t*, 601*b*,
    673, 674*f*
  for scrub typhus (*Orienta tsutsugamushi*),
    600, 602*t*
Childbirth (puerperal) fever, streptococcal,
    17, 541
Chitin, in fungal cell walls, 79, 355
Chlamydia genus/spp./chlamydias, 320*f*, 334*t*,
    337, 602–606, 606*t*
  animal/cell cultures for growth of, 185
  antimicrobial agents affecting, 298*f*
  comparison of with other small
      microbes, 606*t*
Chlamydiae (phylum), 320*f*, 334*t*, 602
Chlamydia pneumoniae, 605, 605*b*
Chlamydia psittaci, 605–606, 606*f*
  as biological weapon, 774*t*
Chlamydia trachomatis, 602–605, 604*f*, 605*f*
  LGV strain of, 603
Chlamydomonas nivalis, 175*f*
Chlamydospores, 358, 359*f*
Chloramine
  antimicrobial action of, 279
  phenol coefficient of, 267
Chloramphenicol, 306*t*
  mechanism of action of, 292, 293*f*, 306*t*
Chlorine, 28*t*
  antimicrobial action of, 279, 282*t*
  Bohr diagram of, 30*f*
  for microbial growth, 170*t*
  in sodium chloride formation,
    33–34, 33*f*
  in wastewater treatment, 759, 759*f*
  for water disinfection, 757*f*, 758
Chlorine dioxide, antimicrobial action of, 279
Chlorobi (phylum), 320*f*, 324, 325*t*
Chloroflexi (phylum), 320*f*, 324, 325*t*
Chlorophylls, 153, 153*f*
  cyanobacteria using, 323
  reaction center, 154, 154*f*
Chlorophyta (division), 367, 371*t*
Chloroplasts, 85, 86*f*, 87*t*
  endosymbiotic theory and, 85–86
Chloroquine, 310*t*
  for malaria prophylaxis, 661
Chocolate, fermentation in production of,
    125, 744
Chocolate agar, for *Neisseria* culture, 567
Cholera, 615–616, 615*f*, 773*t*
  discovery of cause of, 17*t*
  incubation period for, 419*t*
Cholera toxin, 615–616, 615*f*
Cholesterol, 42, 44*f*
  in eukaryotic membranes, 79
  for microbial growth, 173*t*
Chondrus genus/spp., 367
Choriomeningitis, lymphocytic, 736
Chromatids, 344, 345*f*
Chromatin/chromatin fibers, 83, 83*f*, 203,
    203*f*, 343
Chromatium genus/spp., 325*f*
Chromoblastomycosis, 640–641, 641*f*, 642*f*
Chromophore, 108
Chromosomes, 83, 87*t*, 344, 345*f*
  eukaryotic, 83, 87*t*, 203–204, 203*f*, 204*t*, 343
  prokaryotic, 87*t*, 200–202, 202*f*, 204*t*
Chronic disease, definition of, 423, 425*t*
Chronic fatigue syndrome, Epstein-Barr virus
    and, 690, 690*f*
Chronic granulomatous disease, 527, 528*t*
  gamma-interferon deficiency in, 453, 454
Chronic inflammation, 455
Chrysolaminarin, 367
Chrysophyta (division), 367–369, 371*t*
Chrysophytes, 367–369, 371*t*
-cide/-cidal (suffix), definition of, 263, 264*t*

Ciguatoxin (*Gambierdiscus* genus/spp.), water-borne illness caused by, 772, 773*t*
Cilia, 60*f*, 77, 78*f*, 79*f*, 87*t*
    protozoan, 4, 5*f*
Ciliates, 351, 351*f*, 356*t*, 650, 664*t*
Ciliophora, 351
Ciprofloxacin, 307*t*
    for anthrax, 546
    mechanism of action of, 295, 297*b*, 307*t*
Citric acid (Krebs) cycle, 137, 142, 143*f*, 149*t*, 162*f*, A-6
Citrobacter genus/spp., 576
CJD. *See* Creutzfeldt-Jakob disease
Cl. *See* Chlorine
Cladophialophora genus/spp., 640
Cladophialophora bantiana, 641
Cladophialophora carrionii, 640
Clarithromycin, 306*t*
Class I major histocompatibility complex proteins, 474, 475*f*
Class II major histocompatibility complex proteins, 474–475, 475*f*
Classes, 114, 115*f*, C-1 to C-9
Classical pathway of complement activation, 450, 451, 451*f*, 452*f*, 453*f*
Claviceps purpurea, 361, 643
Clavulanic acid, synergistic effects of, 304
Clindamycin, 306*t*
    mechanism of action of, 292, 306*t*
Clinical specimens, for microbial culture, 178, 179*t*
Clofazimine, 307*t*
    mechanism of action of, 295, 307*t*
Clonal deletion, 473–474, 474*f*
Clonal selection, in humoral immunity, 477, 478*f*
Clones
    in gene libraries, 246–247, 248*f*
    selection of, 248
Clostridium genus/spp./clostridia, 320*f*, 325–326, 329*t*, 547–551
    endospores of, 316, 318
        susceptibility of, 266
    fermentation products of, 150*f*
    food contaminated by, 750, 752*t*
    nitrogen fixation by, 767
    in pectinase production, 753
Clostridium botulinum, 326, 326*b*, 548–550, 549*f*, 752*t*
    as biological weapon, 774*t*
    growth of, heat-related control methods for, 269, 550
Clostridium difficile, 326, 548
    pseudomembranous colitis (superinfection) caused by, 301, 431, 548
Clostridium perfringens, 326, 547–548, 548*f*
    discovery of, 17*t*
Clostridium tetani, 326, 550–551, 550*f*. *See also* Tetanus
    discovery of, 17*t*
    soil-borne disease caused by, 770*t*
    vaccination against, 551
Clouds (electron shells), 29, 29*f*
Cluster of differentiation (CD) designations, 465
cm. *See* Centimeter
CMV. *See* Cytomegalovirus genus/spp.
Co. *See* Cobalt
Coagulase, 534
    as virulence factor, 416, 417*f*
Coagulation, in host defense, 443*t*
Cobalt, 28*t*
    for microbial growth, 170*t*
Coccidioides immitis, 356, 624, 629–630, 629*f*
    disseminated, 629, 629*f*
    geographic distribution of, 626*f*, 629
    soil-borne disease caused by, 770*t*

Coccidioidomycosis, 356, 629–630, 629*f*, 630*f*, 770*t*
Coccobacillus, 314, 314*f*
Coccus (cocci), 314, 314*f*
    arrangements of, 315, 317*f*
    Gram-negative, 566–594. *See also specific agent*
    Gram-positive, 327–328, 532–565. *See also specific agent*
Cockroaches, as disease vectors, 424*t*
Codium genus/spp., 367
Codons, 212, 214*f*
Coefficient, in scientific notation, B-1
Coenocytes, 346
    molds, zygomycetes, 359–360, 361*f*, 366*t*
Coenzyme A
    in acetyl-CoA synthesis, 142, 142*f*
    as cofactor in enzymatic activity, 131*t*
    in Krebs cycle, 142, 143*f*
Coenzyme Q, 146
Coenzymes, 130, 130*f*, 131*t*
    ATP in formation of, 53
Cofactors, 130, 130*f*, 131*t*
Coffee, fermentation in production of, 744
Cohesiveness, of water, 37, 37*f*
col-/colo- (prefix), definition of, 413*t*
Cold (common), 705*t*
    adenoviruses causing, 696
    coronaviruses causing, 715
    manifestations of, 705*t*
    rhinoviruses causing, 704–705, 704*f*, 705*t*
    vaccine for, 490*b*
Cold enrichment, 186
    Listeria detected by, 553
Cold sores (oral herpes), 684–685, 684*f*, 685*f*
Cold vaccine, 490*b*
Coliforms, 574–576
    canned food contaminated by, 750
    fecal, as indicator organisms, 758, 758*f*
    nosocomial infection caused by, 571*t*
Colitis, pseudomembranous, *Clostridium difficile* causing, 301, 431, 548
Collagen, in host defense, 438–439
Collagenase, as virulence factor, 416, 417*f*
Colony, microbial, 168, 177, 178*f*
    bacteria and fungi forming, 14, 15*f*, 178*f*
    characteristics of, 177, 178*f*
Colony-forming units, 178
    microbial growth estimation affected by membrane filtration and, 191
    viable plate counts and, 191
Colorado tick fever, 711*t*, 737–738
Coltiviruses, 737–738
Combustion, in carbon cycle, 766*f*
Commensalism, 405, 406*t*
Commercial sterilization, 263
Common antigen, 572, 574*f*
Common cold, 705*t*
    adenoviruses causing, 696
    coronaviruses causing, 715
    manifestations of, 705*t*
    rhinoviruses causing, 704–705, 704*f*, 705*t*
    vaccine for, 490*b*
Common-source epidemics, food poisoning and, 752
Communicable diseases, 424, 425*t*
Communities, 764, 765*f*
Competent cells/competency, 230
Competition, microbial
    adaptation/survival and, 764
    food spoilage and, 749, 749*t*
    opportunistic pathogens and, 407–409
Competitive inhibition/inhibitors, 134–135, 135*f*

Complement/complement system, 47, 450–452, 451*f*, 452*f*, 453*f*, 459*t*
    activation of, 451, 451*f*, 452*f*
    deficiency of, in systemic lupus erythematosus, 526
    in host defense, 450–452, 451*f*, 452*f*, 453*f*, 459*t*
    in immune-complex (type III) hypersensitivity, 518–519, 519*f*
    inactivation of, 451–452
    in inflammation, blood vessels affected by, 455, 455*f*
Complementarity, DNA, replication and, 205
Complementary DNA (cDNA), reverse transcriptase in synthesis of, 243, 243*f*, 255*t*
Complement fixation test, 501–502, 501*f*, 506*t*
Complement proteins, 443, 450. *See also* Complement/complement system
Complete blood count, 447, 447*b*
Complex media, 182, 184*t*
Complex transposons, 235, 236*f*
Complex viruses, 382, 382*f*
Compound, chemical bonds in formation of, 29–30
Compound microscopes, 96–98, 97*f*, 98*f*
Compromised host. *See* Immunocompromised host
Computational subtraction, detection of infectious agents and, 254*b*
Concentration (chemical) gradient, 71
Condenser lenses, of light microscope, 97*f*, 98
Condoms, in genital wart prevention, 695
Condylomata acuminata, 693, 695*f*
Confocal microscopy/microscopes, 95, 101, 102*f*, 107*t*
Congenital diseases, 412, 415*t*
Congenital rubella, 713
Congenital syphilis, 609
Conidiophores, 358
Conidiospores/conidia, 358, 359*f*
Conjugation, 66
    bacterial, 231–232, 233*f*, 234*f*, 235*f*, 235*t*
    fertility (F) factors and, 202, 232, 233*f*
    in protozoa, 349–350, 350*f*
Conjugation (sex) pili, 66, 66*f*, 231–232, 233*f*
Conjunctiva, of eye, as portal of entry, 409*f*, 410
Conjunctivitis
    adenoviruses causing (pinkeye), 696, 697*f*
    coxsackie A virus causing (acute hemorrhagic conjunctivitis), 707
Constant region, immunoglobulin, 468*f*
Consumption, in biogeochemical cycling, 765
Consumption (tuberculosis), 557. *See also* Tuberculosis
Contact dermatitis, allergic, 520–521, 521*f*
Contact lens wear, *Acanthamoeba* eye infection and, 652
Contact transmission of disease, 421, 422*t*
Contagious diseases, 424, 425*t*
Contamination
    microbial, 409
    of water, 772
Continuous cell cultures, for viruses, 397
Continuous flow production, in industrial fermentation, 753
Contractile vacuoles, in protozoa, 349, 349*f*
Contrast, 95
Contrast microscopy/microscopes, differential interference (Nomarski), 92, 98–100, 99*f*, 107*t*
Control groups, in scientific method, 9–10, 11*f*
Convalescence, 419, 419*f*
Convalescent phase, of pertussis, 586, 587*f*
Cooperation, microbial, adaptation/survival and, 764

Copper, 28t
  antimicrobial action of, 280–281, 282t
  cycling of, 768
  for microbial growth, 170t, 172
Cord factor, in *Mycobacterium tuberculosis* cell
    wall, 555, 556f
Corepressor, tryptophan as, 220
Core temperature, hypothalamus in regulation
    of, 458, 458f
Corn, genetically-engineered, 241
  monarch butterflies and, 258b
Coronaviruses (*Coronaviridae*), 385t, 709, 710,
    715f, 738f
  diseases caused by, 715–716
  envelope of, 710, 710f, 715f
  SARS caused by, 715b, 716
Corticosteroids
  for asthma, 515
  immunosuppressive actions of, 522, 524t
Cortinarius gentilis, 644
Cortinellis genus/spp., 363
Corynebacterium genus/spp., 320f, 328, 329t,
    554–555, 554f, 555f
Corynebacterium diphtheriae, 328, 554–555, 554f.
    See also Diphtheria
  discovery of, 17t
  reproduction in, 315–316, 554, 554f
Coulter counter, for estimating microbial
    growth, 192
Counterstain
  in Gram staining, 109
  in Schaeffer-Fulton endospore staining, 111
  in Ziehl-Neelsen acid-fast staining, 111
Covalent bonds, 30, 31f
  definition of, 30
  nonpolar, 30–32, 31f, 32f, 35t
  polar, 32, 33f, 35t
Cowpox, 680
  smallpox prevention and, 18, 488, 681
Coxiella genus/spp., 334t, 335, 590–591, 590f
Coxiella burnetii, 590–591, 590f
  as biological weapon, 774t
Coxsackieviruses (coxsackie A and coxsackie B
    viruses), 707–708, 708f
Crenarchaeota (phylum), 319, 320f
Creutzfeldt-Jakob disease, 48, 399
Crimean–Congo hemorrhagic fever,
    711t, 736
Cristae, 85, 86f
  electron transport chains in, 144, 145f
Crops, assessing biological threats to,
    773–774
Crop yields, recombinant DNA technology
    affecting, 258
Crossing over, 344, 345f
  in recombination, 229, 230f
Cross resistance, 303
Croup, 727
Crust, in poxvirus infection, 680, 681f
Crustose lichens, 365, 365f
Cryptic plasmids, 202
Cryptococcoma, 635
Cryptococcosis, 635
Cryptococcus neoformans, 363, 635, 637
Cryptococcus neoformans gattii, 635
Cryptococcus neoformans neoformans, 635
Cryptosporidiosis, 662–663, 663f
Cryptosporidium genus/spp., 662–663, 663f
Cryptosporidium parvum, 662–663, 663f, 664t
Crystal violet, 114t
  in Gram staining, 109, 109f
Crytpococcosis, 635
Ctenocephalides felis, as disease vector, in murine
    (endemic) typhus (*Rickettsia typhi*),
    600, 602t
Cu. *See* Copper

Culex mosquitoes, as disease vectors
  for equine encephalitis, 712f
  for filariasis (*Wuchereria bancrofti*), 671
Culture (microbial), 177–186
  animal, 185
  for biochemical testing, 118
  cell, 185
  clinical sampling for, 178, 179t
  definition of, 177
  enrichment, 186
  environmental sampling for, 177–178
  low-oxygen, 185–186, 186f
  media for, 177, 181–185, 181f. *See also*
    Culture media
  preserving, 186
  pure, 178–181, 179f, 180f, 181f
  special techniques in, 185–186
  viral, 395–397, 396f, 397f
Culture media, 177, 181–185, 181f
  anaerobic, 183–185, 185f
  complex, 182, 184t
  defined (synthetic), 181–182, 182t
  differential, 182–183, 183f, 184f, 184t
  selective, 182, 182t
  selective and differential, 183, 184t
  transport, 178, 185
Cuprum. *See* Copper
Curds, in cheese production, 745–746, 745f, 748t
Cutaneous anthrax, 327, 546, 547f
Cutaneous aspergillosis, 633
Cutaneous blastomycosis, 628, 628f
Cutaneous candidiasis, 634, 634t
Cutaneous cryptococcosis, 635
Cutaneous diphtheria, 554
Cutaneous histoplasmosis, 630
Cutaneous leishmaniasis, 656
Cutaneous mycoses, 640–643, D-2
Cutaneous paracoccidioidomycosis, 631, 631f
Cutaneous sporotrichosis, fixed, 642
Cutaneous zygomycosis, 637
Cuticle
  roundworm, 669
  tapeworm, 664
CXCR4, HIV replication and, 721, 721f
Cyanobacteria (blue-green bacteria), 320f,
    323–324, 324f, 325t
  nitrogen fixation by, 172, 323–324
  pH range tolerated by, 38
  reproduction in, 315, 323, 324f
Cyclic photophosphorylation, 154, 155f
Cyclophosphamide, 523
Cycloserine, 305t
  mechanism of action of, 292, 305t
Cyclospora cayetanensis, 663, 664t
Cyclosporine, 355, 523
Cystic acne, 561, 561f
Cysticerci, tapeworm, 665, 665f
Cystic fibrosis, *Pseudomonas* infections
    and, 588
Cyst stage, of protozoa, 349, 349f, 650
  in *Balantidium coli* infection, 650
  in *Entamoeba histolytica* infection, 651, 652
  in *Giardia* infection, 657
Cytochrome oxidase, in *Neisseria*, 567, 567f
Cytochromes, in electron transport chains, 146
Cytokine network, 473
Cytokines, 448. *See also specific type*
  immune system, 472, 473
Cytokinesis (cytoplasmic division), 346, 346f
  centrioles in, 81
Cytolytic toxins, 534
Cytomegalovirus genus/spp., 683, 684,
    691–692, 692f
Cytomegalovirus mononucleosis, 692
Cytometry, flow, for estimating microbial
    growth, 192

Cytopathic effect, viral neutralization tests
    and, 500
Cytophaga genus/spp., 335t, 339
Cytoplasm. *See also* Cytosol; Inclusions;
    Organelles
  division of (cytokinesis), 346, 346f
  in eukaryotic cells, 60f, 80–85, 82f, 83f, 84f,
    85f, 86f, 87t
  in prokaryotic cells, 60f, 74–76, 75f, 76f, 77f
Cytoplasmic (cell/plasma) membranes
  antimicrobial agents affecting, 290f, 292–294,
    293f, 307t, 309t
  electrical potential of, 71, 71f
  eukaryotic, 60f, 79–80, 79f, 80f, 81f, 81t, 87t
  movement across, 71–74, 76t
    active processes in, 73–74, 75f, 76t
    passive processes in, 71–73, 72f, 73f,
      74f, 76t
  permeability of, 71
  phospholipid bilayers forming, 41, 43f, 70, 70f
    cholesterol in, 42, 44f
  prokaryotic, 60f, 70–74, 70f, 87t
    electron transport chains in, 144, 145f
    function of, 71–74, 71f, 72f, 73f, 74f, 75f
Cytosine (C), 50, 50f
Cytoskeleton, 87t
  of eukaryotic cells, 60f, 81, 82f, 87t
  of prokaryotic cells, 76, 77f, 87t
Cytosol
  in eukaryotic cells, 80, 87t
  in prokaryotic cells, 76, 87t
Cytotoxic (type II) hypersensitivity,
    515–518, 523t
  ABO system/transfusion reactions and,
    515–516, 516f, 517t
  drug-induced, 518, 518f
  Rh system/hemolytic disease of newborn
    and, 517–518, 517f
Cytotoxic drugs, immunosuppressive actions
    of, 522–523, 524t
Cytotoxic T cells, 472, 473t
  in cell-mediated immune response, 480–481,
    480f, 481f
Cytotoxins, 416, 417f, 480, 481f

d4T. *See* Stavudine
D. *See* Decimal reduction time
DADPH, 127
Dairy products. *See also* Cheese; Milk
    Brucella contamination of, 585
  fermented, 745–746, 745f, 748t
Dalton (atomic mass unit), 27
Dandruff, *Malassezia furfur* causing, 639
Dane particles/bodies, hepatitis B virus
    producing, 698, 699f
Dark-field microscopy/microscopes, 95, 98,
    99f, 107t
Dark-field stop, 98, 99f
Dark reactions. *See* Light-independent
    reactions
Dark repair, DNA, 225, 226f
Data sharing, by public health agencies, 432
DC-SIGN, HIV replication and, 721
ddC. *See* Dideoxycytidine
ddI. *See* Dideoxyinosine
Deamination, 151, 767, 767f
"Death-cap mushroom," 363, 644, 644f
Death (decline) phase, of microbial growth,
    189, 189f
Decarboxylation, in acetyl-CoA synthesis,
    142, 142f
Decimal reduction time (D), 269, 269t
Decimeter (dm), 93, 93t
Decline phase of disease, 419, 419f
Decline (death) phase of microbial growth,
    189, 189f

Decolorizing agent
　in Gram staining, 109
　in Schaeffer-Fulton endospore staining, 111
　in Ziehl-Neelsen acid-fast staining, 111
Decomposition, in biogeochemical cycling, 765
　carbon cycle, 766f
Decomposition reactions, 36, 36f
Deep-freezing, for preserving cultures, 186
Deeply branching bacteria, 320f, 323
Deeps, 181
Deer ticks. *See also* Ixodes ticks
　as disease vectors
　　for ehrlichiosis (*Ehrlichia* species), 601, 602t
　　for Lyme disease (*Borrelia burgdorferi*),
　　　610–611, 611f
Defecation, in host defense, 443t
Defensins, 454–455, 459t
Defined (synthetic) media, 181–182, 182t
Definitive host, 650
　for flukes, 667, 668f
　for tapeworms, 664, 665f
Degenerative diseases, 412, 415t
Degerming, 263, 264t
Degranulation/degranulating cells, in type I
　(immediate) hypersensitivity,
　510–511, 511f, 512f, 514t
Dehydration synthesis, 35, 36f
　of sucrose, 45, 45f
Dehydrogenation, 127. *See also* Oxidation
Deinococcus genus/spp., 323
Deinococcus-thermus (phylum), 320f, 323
Delayed (type IV) hypersensitivity, 520–522,
　520f, 521f, 523t
Deletions, in mutations, 221, 222f, 223t
Delta agent (hepatitis D virus), 709t, 737
　hepatitis B virus coinfection and, 697, 737
Delta chains, 468
Deltaproteobacteria, 334t, 335–336
　electricity produced by, 755b
Deltavirus genus/spp., 717, 737. *See also*
　Hepatitis D virus
Denaturation
　DNA, in polymerase chain reaction,
　　247, 249f
　protein, 50
　　antimicrobial action and, 265
　protein enzyme, 132, 134f
Dendritic (Langerhans) cells, 438
　HIV infection of, 721
Dengue, 711–712, 711t, 713f
Dengue hemorrhagic fever, 711–712, 711t, 714f
Denitrification, 767, 767f
Dental caries, viridans streptococci causing, 542
Dental plaque, as biofilm, 67b, 177, 411, 412f, 542
Deoxyribonucleases, 538
Deoxyribonucleic acid. *See* DNA
Deoxyribonucleotide, 51, 200, 201f
Deoxyribose, 44
Dermacentor ticks, as disease vectors, 424t
　for ehrlichiosis (*Ehrlichia* species), 601, 602t
　for Rocky Mountain spotted fever (*Rickettsia
　　rickettsii*), 599, 602t
Dermatiaceous fungi, 640–641
Dermatitis
　allergic contact, 520–521, 521f
　seborrheic, *Malassezia furfur* causing, 639
dermato- (prefix), definition of, 413t
Dermatomes, shingles lesions following,
　688, 689f
Dermatophytes, 624, 638–639, 640t
　as emerging pathogens, 626
Dermatophytoses, 638–639, 640t
Dermis, 438, 439f
Dermonecrotic toxin, 585
Descriptive epidemiology, 426–429, 430f

Desiccation
　in food preservation, 750
　in microbial control, 272, 273f, 277t
Desulfovibrio genus/spp., 334t, 335–336
　in sulfur cycle, 768
Desulfuromonas acetoxidans, electricity produced
　by, 755b
Detergents, antimicrobial action of, 280,
　280f, 282t
Deuteromycetes, 359, 363, 366t
Deuteromycota (division), 359, 363. *See also*
　Ascomycota
Devil's grip (pleurodynia), coxsackie B virus
　causing, 708
Dextran, viridans streptococci producing, 542
D forms, 48
DHAP. *See* Dihdyroxyacetone phosphate
Diabetes mellitus
　coxsackie B virus infection and, 708
　type I, autoimmunity in, 525
Diapedesis, 444, 445f, 456
Diaper rash, candidal, 633f
Diarrhea. *See also* Dysentery; Food-borne
　　illnesses/food poisoning;
　　Gastroenteritis
　astroviruses causing, 709
　caliciviruses causing, 709
　Campylobacter jejuni causing, 616–617,
　　617b, 752t
　in cholera, 615
　in cryptosporidiosis, 662
　Cyclospora causing, 663
　in giardiasis, 657
　infectious, treatment of, 573
　rotaviruses causing, 737
　in shigellosis, 579
　　discovery of cause of, 17t
　traveler's, discovery of cause of, 17t
　Vibrio parahaemolyticus causing, 616
Diatomaceous earth, 369
　microbial production of, 13t
Diatoms, 5, 367–369, 369f, 371t
DIC. *See* Disseminated intravascular
　coagulation
Dichotomous keys, 121–122, 121f
Dictyostelium genus/spp., 371
Dideoxycytidine (ddC), 296f, 719t
Dideoxyinosine (ddI), 296f, 719t
Didinium genus/spp., 351, 351f
Diethylcarbamazine, 309t
Differential interference contrast (Nomarski)
　microscopy/microscopes, 92,
　98–100, 99f, 107t
Differential media, 182–183, 183f, 184f, 184t
Differential stains, 108–112, 114t. *See also specific
　type*
Differential white blood cell count, 447, 447b
Diffusion, 71–72, 72f, 76t
　facilitated, 72, 72f, 76t
Diffusion susceptibility (Kirby-Bauer) tests,
　297–298, 298f
DiGeorge anomaly, 527, 528t
Digestion, in phagocytosis, 448f, 449
Digestive system
　in host defense, 443t
　microbial disease agents affecting, D-9 to
　　D-10
　normal microbiota of, 408t
Digestive vesicles (phagolysosomes), 448f, 449
Dihdyroxyacetone phosphate (DHAP), in
　glycolysis, 138f, 139
Dihydrofolic acid, antimicrobial drugs
　affecting, 294, 295f
Dikaryon, 358, 360f
Dilution tests, 299, 299f

Dimorphic fungi, 356, 357f, 625, 625f, 628
Dinoflagellates, 352, 352f, 356t, 371f
　blooms of, intoxications caused by, 772
Dioecious helminths, 663
Dipeptide, 48, 49f
Diphtheria, 554–555, 555f
　discovery of cause of, 17t
　vaccination against, 555
Diphtheria toxin, 554
Diphtheria toxoid, 555
Diplobacilli, 317f
Diplococci, 315, 316f
Diploid cell cultures, for viruses, 397
Diploid nucleus, 343
Diplomonadida (kingdom), 348, 348f,
　354, 356t
Diplomonads, 354, 356t
Direct contact transmission of disease,
　421, 422f
Direct fluorescent antibody tests, 502, 506t
　for *Chlamydia trachomatis*, 604, 605f
Directly Observed Treatment, Shortcouse
　(DOTS), for tuberculosis, 558–559
Disaccharides, 44–45, 45f
Disease. *See also* Infection/infectious diseases
　biofilm manipulation in prevention of, 177b
　categories of, 412–413, 415t
　causes of, 13–16
　defenses against, 21–22, 22f. *See also* Host
　　defenses
　　nonspecific, 437–461
　　specific, 438, 439f, 462–486. *See also*
　　　Immune response
　definition of, 412
　endemic, 425, 427f
　epidemic, 425–426, 427f
　germ theory of, 14, 413
　terminology of, 413t
Disease process, 418, 419f
Disinfection/disinfectants, 263, 264t
　chemical agents, 277–283, 282t
　evaluation of efficacy of, 267–268
　in wastewater treatment, 759, 760f
　in water treatment, 757f, 758
Disk-diffusion (Kirby-Bauer) method,
　267–268, 268f
Disseminated intravascular coagulation
　(DIC), 567
Dissociation (ionization), 34, 34f
Distilled spirits, fermentation in production of,
　747, 748t
Distribution, antimicrobial drug, 300
Disulfide bridges, 50
Divisions
　classification of algae into, 367–369
　classification of fungi into, 359–366
dm. *See* Decimeter
DNA, 50, 51–52, 52t
　autoantibodies against, in systemic lupus
　　erythematosus, 525
　complementary (cDNA), reverse
　　transcriptase in synthesis of, 243,
　　243f, 255t
　double helix structure of, 51f, 52
　double-stranded, 200, 201f
　　in viruses, 379, 680. *See also* Double-
　　　stranded DNA viruses
　in eukaryotic genomes, 203–204, 204t
　extranuclear, 204
　function of, 52
　insertion of into cells, 251–252, 251f, 255t
　　vaccine development and, 492, 493f
　mitochondrial, 85
　nucleotide bases in, 50–51, 51, 51f, 52,
　　200, 201f

polymerase chain reaction in replication of, 247–248, 249f, 255t
in prokaryotic genomes, 200–202, 202f, 204t
repair of, 224–226, 226f
  methylation affecting, 208
replication of, 204–208, 205f, 206f, 207f, 218t
  antimicrobial drugs affecting, 295, 296f, 297b, 307t, 308t
  polymerase chain reaction for, 247–248, 249f, 255t
single-stranded, in viruses, 379, 680. See also Single-stranded DNA viruses
structure of, 200, 201f
synthesis of, 160, 161f
viral, 58b, 379
DNA base composition, 120
DNA fingerprinting, 256, 257f
  Southern blots for, 249–251
DNA gyrase, quinolones/fluoroquinolones affecting, 295, 297b, 307t
DNA helicases, 205, 206f
DnaK (heat shock protein), defensins affecting, 455
DNA ligase, in lagging strand synthesis, 206f, 207
DNA photolyase, 225
DNA polymerase, 205, 206f
DNA primer, in polymerase chain reaction, 247, 249f
DNA probes, synthetic nucleic acids in synthesis of, 244, 255t
DNA sequencing. See Genomes, sequencing
DNA viruses, 385t, 679–702. See also specific type
adenoviruses (Adenoviridae), 385t, 682t, 696, 696f, 697f, 698b
genomes of, 200
hepadnaviruses (Hepadnaviridae), 385t, 682t, 696–699
herpesviruses (Herpesviridae), 385t, 682t, 683–693, 684f
papillomaviruses (Papillomaviridae), 385t, 682t, 693–695, 694f, 695b, 695f
parvoviruses (Parvoviridae), 385t, 682t, 699–700, 700f
pathogenic, 679–702
polyomaviruses (Polyomaviridae), 385t, 682t, 695
poxviruses (Poxviridae), 385t, 680–683, 680f, 681f, 682t
Dog tapeworms (Echinococcus granulosus), 666, 666f, 673t
Dog ticks. See also Dermacentor ticks
as disease vectors for ehrlichiosis (Ehrlichia species), 601, 602t
Dolor, in inflammation, 455
Domagk, Gerhard, 22, 288
Domains, 115f, 117–118, 318, C-1 to C-9
Domestic water, 770
Donor cell, in horizontal gene transfer, 229
Donor-recipient matching, prevention of graft rejection and, 522
DOTS. See Directly Observed Treatment, Shortcouse
Double covalent bond, 30, 31f
Double helix, of DNA, 51f, 52
Double immunodiffusion (Ouchterlony test), 496–497, 497f
Double-stranded DNA, 200, 201f
Double-stranded DNA viruses, 379, 680, 682t
synthesis of, 390–392, 392t
Double-stranded RNA viruses, 379
naked segmented, 737–738, 738f
synthesis of, 392, 392t
Doxycycline, 306t
DPT vaccine, 555, 587
Drain opener, microbial production of, 13t

Drinking (potable) water, 432
microbes in production of, 756–758, 757f
Droplet nuclei, 421
Droplet transmission of disease, 421, 422f, 422t
Drug administration, routes of, 300, 300f
Drug allergies, 301
Drug distribution, 300
Drug-induced cytotoxic (type II) hypersensitivity, 518, 518f
Drug resistance. See Resistance
Dry heat, in microbial control, 271–272, 277t
Drying (desiccation)
in food preservation, 750
in microbial control, 272, 273f, 277t
Dry weight, for estimating microbial growth, 193
dsDNA. See Double-stranded DNA
dsDNA viruses. See Double-stranded DNA viruses
dsRNA viruses. See Double-stranded RNA viruses
Duffy antigens, absence of, malaria resistance and, 660
Dutton, Joseph, 17t
Dyes, for specimen preparation. See Staining
Dysentery. See also Diarrhea
amoebic, 651–652, 773t
Shigella causing, 579, 773t
Dyspnea, in respiratory syncytial virus infection, 728

Ear infections, Prevotella causing, 592
Eastern equine encephalitis, 711, 711t
Eberth, Carl, 17t
Ebola virus, 703, 730–731, 731f, 732f
EBV. See Epstein-Barr virus
Echinococcus granulosus, 666, 666f, 673t
Echoviruses, 708
E. coli. See Escherichia coli
Ecological associations, microbial, growth affected by, 177
Ecological microbiology, 19f, 763–765, 765f
adaptation/survival and, 764–765
levels of microbial associations and, 764, 765f
EcoRI restriction enzyme, 244, 245t
in genome mapping, 252, 252t, 253f
EcoRII restriction enzyme, 244, 245t
Ecosystems, 764, 765f
aquatic
  freshwater, 770, 770–771, 771f
  marine, 770, 771–772, 771f
  specialized, 772
microbes in maintenance of, 763
Ectopic pregnancy, gonococcal PID and, 568
Edema, in inflammation, 455
Edwardsiella genus/spp., 577
EEE. See Eastern equine encephalitis
Efficacy, of antimicrobial drugs, 297–300, 298f, 299f, 300f
Eflornithine, for Trypanosoma brucei, 654
Eggs
culturing viruses in, 396, 397f
salmonella contamination of, 577, 752t
Ehrlich, Paul, 18–19, 19f, 288
Ehrlichia genus/spp., 332, 334t, 600–601, 601f, 602t
Ehrlichia chaffeensis, 600
Ehrlichia equi, 600
Ehrlichiosis, 332, 600–601, 601f, 602t
Elastase, Pseudomonas aeruginosa producing, 587
Electrical gradient, 71, 71f
Electricity, microbial production of, 755b
Electrochemical gradient, 71
in chemiosmosis, 147
Electrolytes, 34, 39

Electromagnetic radiation. See also Ultraviolet light
in microbial control, 276–277, 277t
Electromagnetic spectrum, microscopy and, 94, 94f
Electron beams, in microbial control, 276, 277t
Electron carrier molecules, 127
Electron configurations, 28–29, 29f, 30f
Electronegativity, 30–31, 32f
Electronic counters, for estimating microbial growth, 192
Electron microscopy/microscopes, 101–104, 107t
invention of, 102b
scanning, 104, 104f, 105f, 107t
staining specimens for, 112
transmission, 102–104, 103f, 107t
Electrons, 27, 27f
atomic mass of, 27
in microbial growth, 168–169
Electron shells, 29, 29f
Electron transport/electron transport chain, 137, 142–147, 144f, 145f, 149t, 162f
light produced by, 146b, 147
Electrophoresis, 248–249, 250f, 255t
in immunoelectrophoresis, 499, 499f
in western blot test, 504, 505f
Electroporation, 251, 251f, 255t
Elek test, in diphtheria, 555
Elementary bodies
Chlamydia forming, 337, 602, 603f
Ehrlichia forming, 601, 601f
Elements (chemical), 27, 28t, 30f
for microbial growth, 168–172, 168f, 170t
periodic table of, 30f
Elephantiasis, 672, 672f, 673t
Elimination, in phagocytosis, 448f, 449–450
ELISA (enzyme-linked immunosorbent assay), 120, 502–503, 504f, 506t
antibody sandwich, 503, 504f
in western blot test, 504, 505f
Elongation
RNA transcript, 209–210, 210f
in translation, 216–217, 217f, 218f
Elongation factor, diphtheria toxin affecting, 554
EMB (eosin methylene blue) agar, 184t
Embden-Meyerhof pathway (glycolysis), 137, 137f, 138f, 139, 139f, A-2 to A-3
alternatives to, 139–141, 140f, 141f
Embryonated chicken eggs, culturing viruses in, 396, 397f
Emerging disease
ehrlichiosis as, 600
Nipah encephalitis as, 726b
-emia (suffix), definition of, 413t
Emigration (diapedesis), 444, 445f, 456
Empty magnification, 95
Empyema, staphylococcal, 536
Encephalitis
amoebic, Acanthamoeba causing, 652
arboviral (flavivirus/togavirus), 711, 711t, 712f
as bioterrorist threat, 774t
Encephalopathies, spongiform, 399
Encystment, 650
Endemic disease, 425, 427f
Endemic relapsing fever, 613, 613f
Endemic (murine) typhus, 600, 602t
Endergonic reactions, 127
endo- (prefix), definition of, 413t
Endocarditis
candidal, 634t
pneumococcal, 543
staphylococcal, 536

Endocrine (hormonal) diseases, 412, 415t
  autoimmunity and, 525
Endocytosis, 79–80, 80f, 81t
Endoflagella, 64, 65f
Endogenous antigens, 464, 464f
  processing, 476f, 477
Endogenous nosocomial infections, 430
Endoplasmic reticulum, 60f, 80, 83, 84f, 87t
  rough, 60f, 83, 83f, 84f
  smooth, 60f, 83, 84f, 85f
Endospores, 316–318, 318f, 319f. See also specific
        organisms
  locations of, 317, 319f
  in soil-borne diseases, 769
  staining, 111, 111f, 114t
  as sterilization indicator, 270, 271f
  survivors of death phase of growth and, 189,
        317–318
  susceptibility of, 266, 266f, 317–318
Endosymbiotic theory, 85–86
Endothelial cells, in reticuloendothelial system,
        446f, 447
Endothermic reactions, 36
Endotoxin (lipid A), in Gram-negative cell
        walls, 68, 69f, 416–417, 417f, 418t
  Enterobacteriaceae, 572, 574f
  Neisseria, 567
  Pseudomonas aeruginosa, 587
End-product inhibition (feedback
        inhibition/negative feedback),
        135–136, 136f
Energy
  activation, enzymes affecting, 130, 131, 131f
  for microbial growth, 168–169, 168f
  storage of
    in ATP, 53, 57
    in fats, 41
  synthesis reactions requiring, 35–36
Energy parasites, chlamydias as, 602
Enriched media, 182
Enrichment culture, 186
Entamoeba genus/spp., 353, 651–652, 652f
Entamoeba histolytica, 651–652, 652f, 664t
  water-borne transmission of, 651, 773t
Enteric bacteria. See Enterobacteriaceae
Enteric cytopathic human orphan virus
        (echovirus), 708
Enterobacter genus/spp., 336t, 576
  sites of infection caused by, 582f
Enterobacteriaceae, 335, 336t. See also specific
        types
  coliforms, 574–576
    nosocomial infection caused by, 571f
  culturing, 573, 575f
  dichotomous key for identification of,
        572, 572f
  diseases caused by, 573–574, 575f, 582f
  noncoliforms, 576–577
    nosocomial infection caused by, 571f
  opportunistic, 574–577
  oxidase test in detection of, 570, 571f
  pathogenic, 577–582
  pathogenicity of, 572–573, 574f
  site of infection caused by, 582f
  structure and physiology of, 571–572,
        571f, 572f
Enterobius vermicularis (pinworm), 671,
        671f, 673t
Enterococcus genus/spp./enterococci, 328,
        544–545, 544f, 546t
  superbug (resistant) strains of, 303
Enterococcus faecalis, 544, 544f, 546t
  culture of, 183, 183f
  Entner-Doudoroff pathway used by, 140
Enterococcus faecium, 544, 546t

Enterotoxins, 416
  clostridial, 547
  in E. coli gastroenteritis, 575
  Shigella producing, 579
  staphylococcal, 534, 535, 752t
Enteroviruses (Enterovirus genus/spp.),
        705–708
Entner-Doudoroff pathway, 141, 141f, 587, A-5
Entry, virus
  in animal virus replication, 390, 391f
  in lytic replication, 385, 386f, 387, 389f
Envelope, viral/enveloped viruses/virions,
        378, 382–384, 383f
  antimicrobial action and, 265
  budding in release and, 393
  for coronaviruses, 710, 710f
  susceptibility and, 266, 266f
  for togaviruses and flaviviruses,
        709–710, 710f
Environment, antimicrobial efficacy affected
        by, 266–267, 267f
Environmental microbiology, 19f, 20t, 21, 744,
        763–772
  aquatic microbiology and, 770–772,
        771f, 773t
  biogeochemical cycles and, 765–768,
        766f, 767f
  microbial ecology and, 19f, 763–765, 765f
  soil microbiology and, 768–770, 768f, 770t
Environmental specimens, for microbial
        culture, 177–178
Enzyme cofactors, 130, 130f, 131t
Enzyme-dependent termination, of
        transcription, 210f, 212
Enzyme-linked immunosorbent assay (ELISA),
        120, 502–503, 504f, 506t
  antibody sandwich, 503, 504f
  in western blot test, 504, 505f
Enzymes
  active site of, 130, 131, 132f
  activity of, 130–136, 131f, 132f, 133f
    inhibition of, 134–136, 135f, 136f
    pH affecting, 132–133, 134f
    substrate concentration affecting,
        133–134, 134f
    temperature affecting, 132, 134f
  catalytic, 45
  definition of, 12
  denaturation of, 132, 134f
  extracellular, as virulence factors, 416, 417f
  in fermentation, Buchner's demonstration
        of, 12, 13f
  induced fit model of substrate interaction
        and, 131, 132f
  makeup of, 130, 130f, 131t
  in metabolism, 126, 128–136, 130t, 131t
  microbial production of, 753–754, 754b, 756t
  naming/classification of, 129, 130t
  restriction, in recombinant DNA technology,
        244, 245f, 245t, 255t
    gene library production and, 247, 248f
    genome mapping and, 252, 252t, 253f
  staphylococcal, 534
Enzyme-substrate complex, 131, 132f, 133f
Enzyme-substrate specificity, 131, 132f
Eosin methylene blue (EMB) agar, 184t
Eosinophilia, 450, 511
Eosinophils, 444, 445f, 459t
  in type I (immediate) hypersensitivity, 511,
        511f, 512f
Epidemic cholera. See Cholera
Epidemic disease, 425–426, 427f
  common-source, food poisoning and, 752
Epidemic relapsing (louse-born) fever, 613, 613f
Epidemic (louse-born) typhus, 599–600, 602t

Epidemiology, 18, 19f, 20t, 424–433, 431b, 432b
  analytical, 429
  definition of, 424
  descriptive, 426–429, 430f
  disease frequency and, 424–426, 425f, 426f,
        427f, 428f, 429t
  exceptions to Koch's postulates and, 415
  experimental, 429
  hospital, 430–432, 432f. See also Nosocomial
        infections
  public health and, 432–433
  research studies used in, 426–429
Epidermis, 438, 439f, 440b
Epidermophyton floccosum, 623, 639, 640t
Epiglottitis, Haemophilus influenzae causing, 582
Epimastigote
  in Trypanosoma brucei life cycle, 653, 654f
  in Trypanosoma cruzi life cycle, 654f, 655
Epinephrine, for anaphylactic shock/type I
        hypersensitivity, 514, 515
Epithelium, mucous membrane, in host
        defense, 440, 441f
Epithet, specific, 114
Epitopes (antigenic determinants), 463, 464f
  antigen-binding site complementarity
        and, 468
Epsilon chains, 468
Epsilonproteobacteria, 334t, 336
Epstein-Barr virus (EBV), 684, 689–691, 690f
Epulopiscium genus/spp., 326, 329t
Epulopiscium fishelsoni, 59b
Equine encephalitis, 711, 711t, 712f
ER. See Endoplasmic reticulum
Ergometrine, 643
Ergosterol, in fungal membrane, antifungal
        drugs affecting, 627
  amphotericin B, 292, 293f
Ergot alkaloids, 643
Ergotamine, 643
Erysipelas, 540, 540f
Erythema infectiosum (fifth disease), 700, 700f
Erythrocytes (red blood cells), 59f, 443
  autoantibodies to, 524–525
  in Plasmodium life cycle/malaria, 659,
        659f, 660
  sheep, in complement fixation test,
        501, 501f
Erythrocytic cycle, in Plasmodium life cycle,
        659–660, 659f
Erythrogenic toxins, 538
Erythromycin, 306t
  mechanism of action of, 292, 293f, 306t
  spectrum of action of, 297, 298f, 306t
Erythropoietin, recombinant DNA in
        production of, 256t
Erythrose 4-phosphate, 160t
Erythrovirus genus/spp., 700, 700f
Eschar, in cutaneous anthrax, 546
Escherich, Theodore, 17t
Escherichia genus/spp., 336t
  sites of infection caused by, 582f
Escherichia coli, 59f, 335, 336t, 575
  conjugation in, 231–232, 233f, 234f, 235f, 235t
  discovery of, 17t
  fermentation products of, 150f
  fimbriae and flagella of, 66f
  food-borne illness caused by, 752t
  genes/genome of, 199, 200, 235f
  as indicator organism, 758, 758f
  in indigo production, 754b
  insulin produced by, 743, 755
  juice contaminated by, 272b
  lac operon of, 219–220, 219f
  pasteurization in control of, 271
  pili of, 66f

replication of, 385–388, 386f, 388f
water-borne transmission of, 773t
Escherichia coli O157:H7, 575
E site, ribosomal, 216
Esophageal candidiasis, 634t
Essential amino acids, 159
Ester group, 40t
Esters, in fat synthesis, 40, 41f
Etest, 299, 299f
Ethambutol, 306t
  mechanism of action of, 292
Ethanol (ethyl alchol)
  antimicrobial action of, 278–279
  in Gram staining, 109, 109f
  microbial fermentation producing, 150f,
    754, 756t
Ether group, 40t
Ethidium bromide, as frameshift mutagen, 224
Ethyl alcohol. *See* Ethanol
Ethylene oxide, for gas sterilization, 281–282,
    281f, 282t
Etiology, 14, 19f, 20t, 412–415, 414f, 415t
  definition of, 413
Eubostrichus parasitiferus, 406b
Euchromatin, 203f, 204
Euglena genus/spp., 354f
Euglenoid movement, 353
Euglenoids, 353, 354f, 356t, 371t
Euglenophyta, 371t. *See also* Euglenoids
Euglenozoa (kingdom), 348, 348f, 353, 354f, 356t
Eukarya (domain), 115f, 117–118, 318
Eukaryotes/eukaryotic cells, 58–61, 59b, 59f, 60f,
    61f, 87t, 342–376. *See also specific type*
  cell walls of, 77–79, 79f
  characteristics of, 343–347
  chromosomes of, 83, 87t, 203–204, 203f, 343
  classification of, 347–348, 348f
    nucleic acid sequencing and, 21
  cytoplasmic membrane of, 60f, 79–80, 79f,
    80f, 81f, 81t. *See also* Cytoplasmic
    (cell/plasma) membranes
  cytoplasm of, 60f, 80–85, 82f, 83f, 84f, 85f,
    86f, 87t
  endosymbiotic theory of formation of, 85–86
  external structures of, 60f, 76–77, 78f, 79f
  fungi, 4
  genomes of, 202–204, 203f, 204t
  life processes in, 57t
  prokaryotic cells compared with, 87t
  protozoa, 4
  reproduction in, 343–347, 345f, 346f, 347f, 347t
  size of, 58, 59b, 61f
Eukaryotic ribosomes, 60f, 80, 83f, 84f, 87t,
    215–216, 215f
Eupenicillium genus/spp., 362
Euryarchaeota (phylum), 319, 320f
Eutrophication, phosphorus and, 768
Exchange (transfer) reactions, 36
Excystment, 650
Exergonic reactions, 127
Exfoliative toxins, 534
Exocytosis, 80, 81f, 81t
  in phagocytosis, 448f, 449–450
Exoenzyme S, *Pseudomonas aeruginosa*
    producing, 587
Exoerythrocytic cycle, in *Plasmodium* life cycle,
    659, 659f
Exogenous antigens, 464, 464f
  processing, 475–477, 476f
Exogenous nosocomial infections, 430
Exons, 214, 214f, 698b
Exophiala genus/spp., 640, 641, 642
Exothermic reactions, 36
Exotoxin A, *Pseudomonas aeruginosa*
    producing, 587

Exotoxins, 416, 417f, 418t
  Enterobacteriaceae producing, 573, 574f
Experimental epidemiology, 429
Experimental groups, in scientific method,
    9–10, 11f
Exponential (logarithmic) growth, 187, 188f,
    B-1
  growth curves of, 188, 189f
Exponential (scientific) notation, B-1
Exponential power, in scientific notation, B-1
Extension, in polymerase chain reaction,
    247, 249f
Extracellular killing, 450, 459t
Extraintestinal amoebiasis, invasive, 652
Extranuclear DNA, 204
Extreme thermophiles (hyperthermophiles),
    132, 174, 174f, 175b, 321, 321f
Extremophiles, 167, 320–322, 322t, 764
Eye
  lacrimal apparatus of, in host defense,
    441, 442f
  normal microbiota of, 408t
  as portal of entry, 409f, 410
Eye infections, organisms causing, D-1 to D-3
  Chlamydia trachomatis causing, 603,
    604, 604f

F$_{ab}$ regions, 468, 468f
Fabricius, bursa of, 468
Facilitated diffusion, 72, 72f, 76t
Factor VIII, recombinant DNA in production
    of, 256t
Factor B (properdin), 451, 451f
Factor D (properdin), 451, 451f
Factor P (properdin), 451, 451f
Facultative anaerobes, 171
  glycolytic, 334t, 335, 336t
  yeasts as, fermentation and, 11, 12f
Facultative halophiles, 176
FAD (flavine adenine dinucleotide), 127, 130,
    131t, 162f
  in Krebs cycle, 142, 143f, 162f
FADH$_2$, 127, 130, 131t, 162f
  in beta-oxidation, 151, 151f
  in chemiosmosis, 147–148
  in electron transport chain, 144f,
    145f, 146
  in Krebs cycle, 142, 143f, 149t, 162f
Families, 114, 115f, C-1 to C-9
Farmer's lung, 519
Fasciitis, necrotizing, 540, 541f
Fasciola genus/spp. (liver flukes), 669, 673t
Fasciola gigantica, 669, 673t
Fasciola hepatica, 669, 673t
Fastidious organisms, for growth factor assays,
    181–182
Fats, 40–41, 41f, 42t
  biosynthesis of, 158, 159f
  catabolism of, 149–151, 151f
Fatty acids
  beta-oxidation of, 151, 151f
  in fat catabolism, 151, 151f, 162f
  in fats, 40–41, 41f, 42t
  in phospholipids, 41, 43f
F$^+$ cells, 232, 233f, 234f
F$^-$ cells, 232, 233f, 234f
Fc region, 468, 468f
Fe. *See* Iron
Fecal coliforms, as indicator organisms,
    758, 758f
Fecal-oral infection
  food-borne transmission in, 422–423
  water-borne transmission in, 422
Feedback inhibition (negative feedback/end-
    product inhibition), 135–136, 136f

Fermentation/fermentation pathways, 125,
    148–149, 148f, 150t
  acellular, Buchner's demonstration of,
    11–12, 13f
  in alcoholic beverage production, 13t, 150f,
    746–747, 746f, 747f, 748t
  in bread production, 744, 748t
  in carbohydrate/glucose catabolism, 137,
    137f, 148–149, 148f
  causes of, 10–12, 12f, 13f
  in chocolate production, 125, 744
  of dairy products, 745–746, 745f, 748t
  definition of, 11, 148
  in meat product preparation, 745, 748t
  products of, 149, 150f, 744–748, 748t
  in vegetable preparation, 744–745, 748t
Fermentations, industrial, 752–753, 753f
Ferroplasma acidarmarnus, 763, 764f
Ferrum. *See* Iron
Fetus
  prevention of rejection of, 521–522
  Toxoplasma infection of, 662
Fever, 457–458, 458f, 459t
Fever blisters (oral herpes), 684–685, 684f, 685f
Fever of Crete. *See* Brucellosis
F (fertility) factors/plasmids, 202, 232,
    233f, 234f
Fibrobacteres (phylum), 320f
Fibroblasts, in tissue repair, 457
Fifth disease (erythema infectiosum),
    700, 700f
Filament, of bacterial flagellum, 62–63, 63f
Filamentous hemagglutinin, *Bordetella pertussis*
    producing, 585
Filamentous particles, hepatitis B virus
    producing, 698, 699f
Filarial nematodes (*Wuchereria bancrofti*), 670,
    671–672, 672f, 673t
Filariasis, 671–672, 672f, 673t
Filoviruses (*Filoviridae*), 385t, 724–725, 730–731,
    731f, 738t
  as biological weapons, 774t
Filterable viruses, 274
Filtration
  in microbial control, 274, 274f, 275f,
    275t, 277t
  in water treatment, 756–758, 757f
Fimbriae, 65, 66f, 87t
  of Enterobacteriaceae, 573
  of *Escherichia coli*, 66f
  of *Neisseria*, 65, 567, 568
  of *Pseudomonas aeruginosa*, 587
Final electron acceptor, 144, 144f, 147
Fingerprinting (DNA/genetic), 256, 257f
  Southern blots for, 249–251
Firmicutes (phylum), 320f, 325, 329t, 533
Fish
  fermentation of, 745, 748t
  raw, antimicrobial control in preparation of,
    273b
Fission, binary, 186–187, 187f, 314, 315f
5-day fever, 584
Five-kingdom taxonomic scheme, Whitaker's,
    116–117, 116f, 348f
Fixation
  carbon, 766f
    in light-independent photosynthesis,
      156, 157f
  chemical, 108
  heat, 108
  nitrogen, 172, 767, 767f
    by alphaproteobacteria, 331, 331f
Fixed cutaneous sporotrichosis, 642
Flaccid paralysis, botulism toxins
    causing, 549

Flagella
of brown alga, 369, 370f
of eukaryotic cells, 77, 78f, 79f, 87t
of prokaryotic cells, 60f, 62–64, 63f, 64f, 65f, 87t
of protozoa, 4, 5f
staining of, 112, 113f, 114t
Flagellar antigens, 572, 574f
Flagellates, 653–658, 664t
Flagellin, 63–64
Flagyl. See Metronidazole
"Flashlight fish," 146b
Flash pasteurization, 271, 271t
Flat warts, 693, 694f
Flavine adenine dinucleotide (FAD), 127, 130, 131t, 162f
in Krebs cycle, 142, 143f, 162f
Flavin mononucleotide (FMN), 146, 146b
Flaviviruses (Flaviviridae), 385t, 709, 738t
discovery of, 17t
diseases caused by, 711, 711–713, 711t, 713f
envelope of, 710, 710f
Flavoproteins, in electron transport chains, 146
Fleas, as disease vectors, 424t, 674, 674f
for murine (endemic) typhus (Rickettsia typhi), 600, 602t
in plague (Yersinia pestis) transmission, 580, 581f, 674
Fleming, Alexander, 19f, 22, 288
"Flesh-eating strep" (necrotizing fasciitis), 540, 541f
Flies, as disease vectors, 424t, 674–675, 674f
for African sleeping sickness (Trypanosoma brucei), 653, 654, 654f, 675
for leishmaniasis, 655, 656f
Flocculation, in water treatment, 756, 757f
Flocs, betaproteobacteria forming, 333
Floridean starch, 367
Flow cytometry, for estimating microbial growth, 192
Flu. See Influenza
Fluconazole, 627
Fluid mosaic model, of membrane structure, 70
Flukes (trematodes), 663, 666–669, 667f, 668f, 669b, 673t
blood (Schistosoma species), 667–668, 668f, 673t
liver (Fasciola species), 669, 673t
Fluorescein, for fluorescent antibody tests, 502, 502f
Fluorescein isothiocyanate, 100
Fluorescence, 95
Fluorescent antibody tests, 502, 502f, 503f
for Chlamydia pneumoniae, 605
for Chlamydia trachomatis, 604, 605f
Fluorescent microscopy/microscopes, 95, 100–101, 100f, 101f, 107t
Fluorine
antimicrobial action of, 279, 282t
Bohr diagram of, 30f
5-Fluorocytosine, 309t, 627
Fluoroquinolones, 307t
mechanism of action of, 295, 307t
Flu vaccines, 735
FMD. See Foot-and-mouth disease
FMN (flavin mononucleotide), 146, 146b
Focal infection, 425t
Focal point, light refraction and image magnification and, 95, 95f
Folic acid
antimicrobial agents affecting synthesis of, 294f, 295
in nucleotide biosynthesis, 160, 161f
Foliose lichens, 365, 365f
Folliculitis, 535
Malassezia furfur causing, 639
staphylococcal, 535

Fomites, 421
Fonsecaea compacta, 640
Fonsecaea pedrosoi, 640, 641f, 642f
Food
microbes in production of, 13t, 19f, 20t, 150f, 744–748, 748t
nutritional value of, recombinant DNA technology affecting, 258
processing of, spoilage prevention and, 750–751, 750f, 751f
safety of, 432. See also Food-borne illnesses/food poisoning
spoilage of, causes/prevention of, 748–751, 749f, 750f, 751f. See also Food spoilage
Food additives
microbial production of, 753, 756t
for preservation, 751
Food allergies, 512, 514, 515b
Food-borne illnesses/food poisoning, 422–423, 422t, 751–752, 752t. See also Gastroenteritis
as bioterrorist threat, 774t
botulism, 550, 752t
Brucella causing, 585
Campylobacter jejuni causing, 616, 752t
clostridial, 547, 548, 752t
Cyclospora causing, 663
infections, 752
intoxications, 752
Listeria causing, 272, 327, 554, 751, 752t
mycotoxins causing, 643
roundworms (nematodes) causing, 669
salmonellae causing, 421t, 577–578, 578f, 579b, 752t
discovery of cause of, 17t
incubation period for, 419t
staphylococcal, 535, 752t
Vibrio causing, 615, 752t
Yersinia causing, 580, 752t
Food microbiology, 13t, 19f, 20t, 150f, 744–752, 748t, 749t, 752t
food-borne illnesses and, 422–423, 422t, 751–752, 752t
food production and, 13t, 19f, 20t, 150f, 744–748, 748t
food spoilage and, 748–751, 749t, 750f, 751f. See also Food spoilage
Food spoilage, 744, 748–751, 749t, 750f, 751f
causes of, 748–749, 749t
extrinsic factors in, 748, 749, 749t
intrinsic factors in, 748, 748–749, 749t
potential for, food classified by, 750
prevention of, 750–751, 750f, 751f
Food supplements, microbial production of, 753, 756t
Food vesicle, 80, 85f
Foot-and-mouth disease (FMD), as biological threat, 775
Foraminifera, 352, 353f, 356t
Forde, Robert, 17t
Formaldehyde
antimicrobial action of, 281, 282t
covalent bonds in formation of, 31f, 32
in vaccine preparation, 490
Formalin
antimicrobial action of, 281
for specimen preparation, 108
Formed elements, 443
Fortified foods, intrinsic spoilage factors and, 749
Fracastoro, Girolamo, 14
Fraenkel, Albert, 17t
Fragments, in complement activation, 451
Frameshift mutations, 221, 222f, 223t
chemicals causing, 224, 225f
Francisella genus/spp., 588–589

Francisella tularensis, 588–589
as biological weapon, 774t
Free-living amoebae, 353, 356t, 652–653, 664t
Free-living mycoplasmas, 596
Free-living nitrogen fixers, 767
Freeze-drying (lyophilization)
for food preservation, 750, 751f
for microbial control, 273, 277t
for preserving cultures, 186
Freeze resistance, recombinant DNA technology and, 257
Freezing
in food preservation, 751
for microbial control, 272, 273b, 277t
for preserving cultures, 186
Freshwater ecosystems, 770, 770–771, 771f
Fructose, 44, 44f
Fructose 1,6-bisphosphate
catabolic lysis of, 131, 133f
in glycolysis, 138f, 139
Fructose 6-phosphate, 160t
Fruiting bodies
of molds, 356
of myxobacteria, 336, 338f
Frustules, 368, 369, 369f
Fruticose lichens, 365, 365f
Fuels, alternative, microbial production of, 754–755, 754f, 755b, 756t
Functional groups, 39–40, 40t
Fungal allergens, 643, 644
Fungal toxins, 643–644
Fungal viruses, 379, 380f
Fungi, 4, 4f, 355–366, 366t, 623–648. See also specific organism
allergies to, 644, 645b
antimicrobial agents affecting, 298f, 309t. See also Antifungal drugs
antimicrobials produced by, 289t, 355
cell walls of, 79, 355
classification of, 348, 359–366, 366t
colony formation by, 14, 15f, 177, 178f
diseases/infections caused by, 626. See also Mycoses
cutaneous/subcutaneous, 640–643, D-2
epidemiology of, 624
gastrointestinal involvement and, D-9
intoxications/allergies, 626, 627, 643–644, 645b
nervous system involvement and, D-4
respiratory system involvement and, D-8
superficial, 638–640
systemic
opportunistic, 631–637
pathogenic, 628–631
urinary tract/reproductive system involvement and, D-11
zoonoses, 421t
morphology of, 355–356, 357f, 358f
nutrition of, 356–357, 358f
opportunistic, 624–626, 625t, 631–637
in HIV infection/AIDS, 718t
pathogenic, 624–626, 625f, 626f, 628–631
pH range tolerated by, 38
reproduction in, 357–359, 359f, 360f
significance of, 355
in soil, 769
disease caused by, 770t
Fungi (kingdom), 115f, 116, 116f, 348, 348f, 355
Fungicides, 264
Furuncle, 535
Fusarium genus/spp., 637
Fusarium oxysporum, soil-borne disease caused by, 770t
Fusobacteria, 320f

G + C ratio, 325, 329t, 533
  in gram-positive bacteria
    high, 320f, 328–329, 329t, 533
    low, 320f, 325–328, 329t, 533
G3P. See Glyceraldehyde 3-phosphate
G. See Guanine
Gallo, Robert, 717
Galls, 332, 332f
Gambierdiscus genus/spp., blooms of,
    intoxications caused by, 772, 773t
Gametes, 343
  protozoan, 349
    in Plasmodium life cycle, 659f, 660
    in Toxoplasma life cycle, 661
Gametocytes
  in Plasmodium life cycle, 659–660, 659f
  in protozoa, 349
Gamma chains, 468
Gamma-hemolysis, by Enterococcus faecalis,
    183, 183f
Gamma- (γ-) interferon (macrophage activation
    factor), 453, 455t
Gammaproteobacteria, 325t, 333–335, 334t,
    336t, 570, 587
Gamma rays
  for food preservation, 750–751
  in microbial control, 276, 276f, 277t
  as mutagens, 223
Ganciclovir, 296f, 308t
Gangrene, gas, 548, 548f
  discovery of cause of, 17t
Gartner, A. A., 17t
Gases, antimicrobial action of, 281–282, 282t
Gas gangrene, 548, 548f
  discovery of cause of, 17t
Gasohol, 754
Gastric acid, in host defense, 443t
Gastroenteritis. See also Diarrhea; Food-borne
    illnesses/food poisoning
  astroviruses causing, 709
  caliciviruses causing, 709
  Campylobacter jejuni causing, 616–617, 617b,
    752t, 773t
  E. coli causing, 575
  organisms causing, D-9 to D-10
  rotaviruses causing, 737
  Vibrio cholerae causing, 615
  Vibrio parahaemolyticus causing, 616
Gastroferritin, in host defense, 443t
Gastrointestinal anthrax, 546
Gastrointestinal candidiasis, 634t
Gastrointestinal system
  in host defense, 443t
  microbial disease agents affecting, D-9 to D-10
  normal microbiota of, 408t
Gastrointestinal zygomycosis, 637
Gas vesicles, 75
Gated channels, in active transport, 73
Gel electrophoresis, 248–249, 250f, 255t
Gelidium genus/spp., 367
Gene activation, in cholera, 615
Gene expression, control of, in microbial
    metabolism, 161
Gene gun, for DNA insertion into cells, 251f,
    252, 255t
  vaccine development and, 492, 493f
Gene libraries, in recombinant DNA
    technology, 246–247, 248f, 255t
Genera. See Genus
Generation time, 188, B-1
Genes, 200. See also Genetics
  expression of
    control of, 218–221, 219f, 220f,
      221b, 221t
    genotype and phenotype and, 208
    methylation affecting, 207

function of, 208–221
  insertion of into cells, 251–252, 251f, 255t
    vaccine development and, 492, 493f
  locating, recombinant DNA technology for,
    252, 253f
  mutations of, 221–228, 229f. See also Mutations
Gene sequencing. See Genomes, sequencing
-genesis (suffix), definition of, 413t
Gene therapy, 21, 256
  adenoviruses in, 698b
Genetic code, 212, 213f, 213t
  synthetic nucleic acids in elucidation of, 244
Genetic engineering/recombinant DNA
    technology, 19f, 20t, 21, 241–261, 243f
  applications of, 252–258, 256t
  in biotechnology, 242, 243f
  bioterrorism and, 775–776, 776b
  for DNA fingerprinting, 249–251, 256, 257f
  ethics of, 259
  for gene therapy, 256
  for genetic screening, 256
  goals of, 242
  hyperthermophiles in, 175b, 321
  medical uses of, 254–257, 256t
  for protein synthesis, 254
  safety of, 258b, 259
  techniques of, 247–252, 255t
  tools of, 242–247, 255t
  in vaccine production, 254–256, 255b,
    491–492, 493f
  for xenotransplants, 257
Genetic fingerprinting, 256, 257f
  Southern blots for, 249–251
Genetic information, transfer of, 208, 209b. See
    also Transcription; Translation
Genetic maps, 232, 233f. See also Genomics
  recombinant DNA technology and, 252,
    253f, 255t
Genetic recombination/transfer, 229–236, 230f
  horizontal gene transfer and, 229–232, 231f,
    232f, 233f, 234f, 235f, 235t
  mutations caused by, 223
  transposons/transposition and, 232–236,
    235f, 236f
Genetics, 19f, 20–21, 200
  autoimmune disease and, 524
  microbial, 20–21, 20t, 199–240. See also
    specific aspect
    DNA replication and, 204–208, 205f,
      206f, 207f
    eukaryotic genomes and, 202–204, 203f, 204t
    gene function and, 208–221
    genome structure and replication and,
      200–208
    mutations and, 221–228, 229f
    nucleic acid structure and, 200, 201f
    prokaryotic genomes and, 200–202,
      202f, 204t
    recombination and transfer and, 229–236
Genetic screening, recombinant DNA
    technology for, 256
Gene transfer, horizontal, 229–232, 235t
  bacterial conjugation, 231–232, 233f, 234f,
    235f, 235t
  transduction, 230–231, 232f, 235t
  transformation, 229–230, 231f, 235t
gen-/-gen (prefix or suffix), definition of, 413t
Genital herpes, 684f, 685, 685f
  incubation period for, 419t
Genital warts, 693, 693–694, 694f, 695f
Genomes, 200–208. See also specific organism
  bacterial/prokaryotic, 200–202, 202f, 204t
  DNA replication and, 204–208, 205f,
    206f, 207f
  eukaryotic, 202–204, 203f, 204t
  influenza virus, 732

mapping, 252–254, 253f, 255t
  nucleic acid structure and, 200, 201f
  sequencing, 21, 199, 253–254, 253f, 255t
    detection of infectious agents and, 254b
    microbial classification and, 21, 117–118,
      117b, 120–121, 318–319, 320f, 323
  viral, 200, 378, 379, 379f
    segmented, in negative single-stranded
      RNA viruses, 731
  as vectors, 245–246, 246f, 255t
Genomics (genome/nucleotide sequencing),
    21, 199, 253–254, 253f, 255t
  detection of infectious agents and, 254b
  microbial classification and, 21, 117–118,
    117b, 120–121, 318–319, 320f, 323
Genotype, 208
Gentamicin, 306t
  mechanism of action of, 292, 306t
Genus (genera), 114, 115f, C-1 to C-9
  binomial nomenclature and, 114
Genus name, 114
Geophilic dermatophytes, 638
German measles (rubella), 713, 714f, 727t
  vaccine against. See
    Measles/mumps/rubella (MMR)
    vaccine
Germicides, 264
  effectiveness of, 266
Germinal centers, lymph node, 465, 467f
Germistatic chemicals, in food preservation, 751
Germ theory of disease, 14, 413
Ghon complexes, 556
Giardia genus/spp., 354, 657–658, 657f
Giardia intestinalis/lamblia, 657–658, 657f, 664t
  water-borne transmission of, 657, 658, 773t
Giardiasis, 657–658, 773t
Giemsa stain, for rickettsias, 598, 598f
Gimenez stain, for rickettsias, 598
Gingivitis, gonococcal, 569
Gingivostomatitis, herpetic, 685
Glanders, as bioterrorist threat, 774t
Gliding, cyanobacterial movement by, 323
Glomeruli
  immune-complex (type III) hypersensitivity
    affecting, 519–520
  streptococcal infection affecting, 540
Glomerulonephritis, 519–520
  poststreptococcal, 540
Glossina flies, as disease vectors, 424t
  for African sleeping sickness (Trypanosoma
    brucei), 653, 654f
Glucomannan, in fungal cell walls, 79
Gluconeogenesis, 158, 158f, 162f
Gluconobacter genus/spp., 332, 334t
  in vinegar production, 747
Glucose, 43, 44, 44f
  biosynthesis of, 157–158, 158f, 162f
  catabolism of, 136–149, 137f, 149t, 162f
    cellular respiration in, 136–137, 137f,
      141–148, 149t, 150f
    electron transport chain in, 137, 142–147,
      144f, 145f
    Entner-Doudoroff pathway in, 141, 141f
    fermentation in, 137, 137f, 148–149, 148f
    glycolysis in, 137, 137f, 138f, 139, 139f
      alternatives to, 139–141, 140f, 141f
    Krebs cycle in, 137, 142, 143f
    pentose phosphate pathway in,
      139–141, 140f
  group translocation of, accumulation in
    bacterial cell and, 74, 75f
  polysaccharides composed of, 45, 46f
Glucose 6-phosphate, 160t, 162f
  in glycolysis, 138f, 160t, 162f
  in pentose phosphate pathway, 139,
    140f, 162f

Glucose-6-phosphate dehydrogenase deficiency, malaria resistance and, 660
Glutamine, in nucleotide biosynthesis, 160, 161f
Glutaraldehyde, antimicrobial action of, 281, 282t
Glyceraldehyde 3-phosphate (G3P), 160t, 162f
    in Calvin-Benson cycle, 156, 157f
    in glycolysis, 138f, 139, 160t, 162f
Glycerol
    in fat catabolism, 151, 151f, 162f
    in fat synthesis, 40, 41f, 158, 159f, 162f
Glycocalyx (glycocalyces)
    of eukaryotic cell, 77, 87t
    of prokaryotic cell, 60f, 62, 62f, 87t
Glycogen, 43, 45, 46f, 158
Glycolysis, 137, 137f, 138f, 139, 139f, 149t, 162f, A-2 to A-3
    alternatives to, 139–141, 140f, 141f
Glycolytic facultative anaerobes, 334t, 335, 336t
Glycoproteins, 50, 70f, 71
Glyphosate, recombinant DNA technology and, 257
Goblet cells, in host defense, 440, 441f
Goiter, 525
Gold, microbial reduction of, 128b
Golden algae, 367–369, 371t
Golgi body/apparatus, 60f, 83–84, 84f, 85f, 87t
Gonococcus. See Gonorrhea; Neisseria gonorrhoeae
Gonorrhea, 568–569, 569f. See also Neisseria gonorrhoeae
    discovery of cause of, 17t
    indirect fluorescent antibody test in diagnosis of, 502
Gonyaulax genus/spp., 352
    blooms of, intoxications caused by, 772, 773t
gp41, 716f, 719, 719t
gp120, 716f, 718–719, 719t
    in HIV replication, 720f, 721, 721f
Graft rejection, 521–522, 521f
Graft-versus-host disease, 522
Gram, Hans Christian, 16, 19f, 109
Gram-negative bacteria, 16, 16f, 320f, 330–338, 334–335t, 566–594. See also specific agent
    aerobic bacilli, 583–591
    anaerobic bacilli, 591–592
    antimicrobial agents affecting, 298f
    cell walls of, 67, 68, 69f, 566
    cocci (Neisseria species), 567–570
    curved bacilli, 614–618
    facultatively anaerobic bacilli, 570–583
    pathogenic, 566–594
    sensor in identification of, 110b
    staining, 109–110, 109f, 114t
Gram-positive bacteria, 16, 16f, 320f, 325–330, 329t, 532–565. See also specific agent
    antimicrobial agents affecting, 298f
    cell walls of, 67, 69f
    high G + C, 320f, 328–330, 329t, 533
    low G + C, 320f, 325–328, 329t, 533
    pathogenic, 532–565
    sensor in identification of, 110b
    staining, 109–110, 109f, 110f, 114t
Gram stain, 16, 16f, 67, 109–110, 109f, 110f, 114t
Grana, chloroplast, 86f, 153
Granulocytes, 444, 445f
Granzymes, 480, 481f
Graves' disease, autoimmunity and, 525
Gravid proglottids, 664, 665f
Green algae, 367, 371t
Greenhouse gas
    carbon dioxide as, 766
    methane as, 322, 766
Green nonsulfur bacteria, 320f, 324, 325t

Green sulfur bacteria, 320f, 324, 325t
Griffith, Frederick, 229
Griseofulvin, 309t, 627
Gross mutations, 222
Ground itch, 671
Group A streptococcus (Streptococcus pyogenes), 538–541, 546t
    culture of, 183, 183f
    diseases caused by, 539–541, 539f, 540f, 541f
    epidemiology of, 538–539, 539b
    pathogenicity of, 538
Group B streptococcus (Streptococcus agalactiae), 541–542, 546t
Group C streptococcus, 546t
Group D streptococcus, 544, 546t
Group F streptococcus, 546t
Group G streptococcus, 546t
Group translocation, 74, 75f, 76t
Growth
    as characteristic of life, 57
    microbial, 57t, 167–198. See also Microbial growth
Growth curve, 188–189, 189f
Growth factors, 172, 173t
    fastidious organisms for identification of, 181–182
    in immune response, 473
Growth hormone
    bovine (BGH), 258
    human
        microbial production of, 13t, 756t
        recombinant DNA in production of, 256t
GTP, in Krebs cycle, 142, 143f
Guanine (G), 50, 50f
Guanosine triphosphate (GTP), in Krebs cycle, 142, 143f
Guilds, 764, 765f
Gummas, in syphilis, 608f, 609
Gymnodinium genus/spp., 352
Gyrase, DNA, quinolones/fluoroquinolones affecting, 295, 297b, 307t
Gyromitra esculenta, 644

H. See Hydrogen
H₂S. See Hydrogen sulfide
HA. See Hemagglutinin
Habitats, microbial, 763, 765f. See also Environmental microbiology
    aquatic, 770–772, 771f
Haemophilus genus/spp., 336t, 582–583, 583f
Haemophilus aphrophilus, 582
Haemophilus ducreyi, 582, 583, 583f
Haemophilus influenzae, 582–583, 583f
    strain aegyptius, 583
    vaccine against, 582
    in viral influenza patients, 732
Hafnia genus/spp., 576
Hairy-cell leukemia, HTLV-2 causing, 717, 718f
Hairy leukoplakia, oral, Epstein-Barr virus and, 690, 690f
Haloarcula genus/spp., 313
Halobacterium genus/spp./halobacteria, 167, 176, 322
Halobacterium salinarium, 322, 772
Halogens, antimicrobial action of, 279, 279f, 282t
Halophiles, 176, 320f, 321–322, 322f
Hand-foot-and-mouth disease, 707, 708f
Hand washing, in infection prevention/control, 432
    discovery of importance of, 17
Hansen, Gerhard, 559
Hansen's disease. See Leprosy
Hantaviruses, 735
    as biological weapons, 774t
    soil-borne disease caused by, 769, 770t

Hantavirus pulmonary syndrome, 421t, 735, 769, 770t
    as bioterrorist threat, 774t
H antigens, 572, 574f
    of E. coli, 575
Haploid nucleus, 343
Hapten, 463
Haustoria, 357
HAV. See Hepatitis A virus
Hay fever, 512, 513f
HBV. See Hepatitis B virus
HCV. See Hepatitis C virus
HDCV. See Human diploid cell vaccine
Heart disease, Chlamydia pneumoniae infection and, 605, 605b
Heat, localized, in inflammation, 455
Heat fixation, 108
Heat-related microbial control methods, 268–272, 269f, 270f, 271f, 271t, 272b, 277t. See also specific type
    in food preservation, 751
Heat shock protein, defensins affecting, 455
Heavy chains, immunoglobulin, 468, 468f
Heavy metals, antimicrobial action of, 280–281, 281f, 282t, 310t
HeLa cell culture, 397
Helical viruses, 382, 382f
Helicases, DNA, 205, 206f
Helicobacter genus/spp., 334t, 336
Helicobacter pylori, 39, 617–618, 617f, 618f
    pH range tolerated by, 39, 176
Helium, Bohr diagram of, 30f
Helix
    α, 48–50, 49f
    double, of DNA, 51f, 52
Helminths, 6, 7f, 371–373, 421t, 663–672, 673t. See also specific type
    antimicrobial agents affecting, 298f, 309t. See also Antihelminthic drugs
    zoonoses caused by, 421t
Helper T lymphocytes (T_H cells), 472, 473t
    in antigen processing, 475, 477, 478f
    in cell-mediated immunity, 480, 480f
    in HIV infection/AIDS, 472, 472b
        attachment/penetration/uncoating and, 720f, 721, 721f
        course of infection and, 722, 722f
    in humoral immunity, 477, 478f
    type 1 (T_H1), 472, 473t
    type 2 (T_H2), 472, 473t
        in humoral immunity, 477, 478f
Hemagglutination, 499–500, 499f. See also Viral hemagglutination
Hemagglutinin
    filamentous, Bordetella pertussis producing, 585
    influenza virus, 732, 733f
Hematopoiesis, 443, 444f
Heme
    cytochromes associated with, in electron transport chains, 146
    for microbial growth, 173t
Hemoglobin C, malaria resistance and, 660
Hemoglobin S, malaria resistance and, 660
Hemolysins
    Enterobacteriaceae producing, 573, 574f
    Streptococcus agalactiae producing, 541
Hemolysis, in transfusion reaction, 516, 516f
Hemolytic anemia, 518, 524–525
Hemolytic disease of newborn, 517–518, 517f
Hemolytic uremic syndrome, E. coli O157:H7 causing, 575
Hemorrhagic fevers
    arenavirus, 736
    as bioterrorist threats, 774t
    bunyavirus, 735

Crimean–Congo, 711*t*, 736
dengue, 711–712, 711*t*, 714*f*
filovirus, 730–731, 731*f*
HEPA (high-efficiency particulate air) filters, in microbial control, 274, 275*f*
Hepadnaviruses (*Hepadnaviridae*), 384, 385*t*, 682*t*, 696–699. *See also under Hepatitis*
genomes of, 697, 697*f*
hepat- (prefix), definition of, 413*t*
Hepatic cancer, hepatitis and, 699, 700*f*
Hepatic candidiasis, 634*t*
Hepatitis, 708, 709*f*
type A, 708, 709*t*, 773*t*
type B, 697–699, 700*f*
incubation period for, 419*t*
type C (non-A/non-B), 709*t*, 714–715
type D, 709*t*, 737
type E (enteric), 709, 709*t*
Hepatitis A virus, 708, 709*t*
water-borne transmission of, 773*t*
Hepatitis B virus, 696, 697, 697–699, 699*f*, 709*t*
genome of, 697, 697*f*, 699*f*
hepatic cancer and, 699, 700*f*
hepatitis D virus coinfection and, 697, 737
Hepatitis B virus vaccine, 698, 699
recombinant DNA in production of, 256*t*
Hepatitis C virus, 709*t*, 714–715
Hepatitis D virus, 709*t*, 737
hepatitis B virus coinfection and, 697, 737
Hepatitis E virus, 709, 709*t*
Hepatovirus genus/spp., 708
Hepevirus genus/spp., 709
Hepeviruses (*Hepeviridae*), 708, 709, 738*t*
Herbicide resistance, recombinant DNA technology and, 257
Herd immunity, 490
Hereditary diseases, 412, 415*t*
Herpangina, 707
Herpes gladiatorum, 686
Herpes simplex viruses, 684–687, 687*t*
genital infection caused by, 684*f*, 685, 685*f*
incubation period for, 419*t*
in HIV infection/AIDS, 722, 723*f*
latency/recurrent infection and, 684, 684*f*
neonatal infections caused by, 685
ocular/ophthalmic infection caused by, 684*f*, 685, 686*f*
oral infection caused by, 684–685, 684*f*, 685*f*
type 1, 684, 684–685, 686, 687*t*
type 2, 684, 685, 686, 687*t*
whitlow caused by, 684*f*, 685, 686*f*
Herpesviruses (*Herpesviridae*), 384, 385*t*, 682*t*, 683–693, 684*f*
Herpes zoster, 688, 688*f*, 689, 689*b*, 689*f*
Herpetic gingivostomatitis, 685
Heterochromatin, 203*f*, 204
Heterocysts
in cyanobacteria, 324, 324*f*
of nitrogen-fixing cyanobacteria, 767
Heterotrophs, 169
in carbon cycle, 766
Hexachlorophene, 278, 278*f*
Hexoses, 44
Hfr (high frequency of recombination) cells, 232, 234*f*
HGE. *See* Human granulocytic ehrlichiosis
HHV-1, 684. *See also* Herpes simplex viruses, type 1
HHV-2, 684. *See also* Herpes simplex viruses, type 2
HHV-3, 684. *See also* Varicella-zoster virus
HHV-4, 684, 689. *See also* Epstein-Barr virus
HHV-5, 684. *See also Cytomegalovirus* genus/spp.
HHV-6, 684, 692. *See also Roseolovirus* genus/spp.

HHV-7, 684
HHV-8, 684, 693, 693*f*
Hib vaccine, 582
Hierarchy, phylogenetic, 117
High-efficiency particulate air (HEPA) filters, in microbial control, 274, 275*f*
High-energy bonds, ATP, energy release and, 52*f*, 53
High frequency of recombination (Hfr) cells, 232, 234*f*
High G + C Gram-positive bacteria, 320*f*, 328–330, 329*t*, 533
High-level germicides, 266
High power/high dry objective lens, of light microscope, 97
*Hind*II restriction enzyme, 244, 245*t*
*Hind*III restriction enzyme, 244, 245*t*
in genome mapping, 252, 252*t*, 253*f*
*Hinf*I restriction enzyme, 245*t*
Hinge region, immunoglobulin, 468, 468*f*
Histamine
in inflammation, blood vessels affected by, 455–456, 455*f*, 456*f*
in type I (immediate) hypersensitivity, 511, 511*f*, 514*t*
Histiolytic toxins, clostridial, 547
Histones, 83, 203, 203*f*, 204*f*
Histoplasma capsulatum, 356, 363, 624, 630, 630*f*
geographic distribution of, 626*f*, 630
soil-borne disease caused by, 769, 770*t*
strain *capsulatum*, 630
strain *duboisii*, 630
Histoplasmosis, 363, 630, 769, 770*t*
HIV (human immunodeficiency virus), 716, 716*f*, 718, 719*t*. *See also* HIV infection/AIDS
origin of, 719
pathogenicity of, 719*t*
as provirus, 394, 722
replication of, 719–722, 720*f*, 721*f*
resistance to, 302, 303*b*
structure of, 716*f*, 718–719
vaccine development and, 724
as vector, 245–246, 246*f*
HIV-1, 718
HIV-2, 718
Hives (urticaria), in type I (immediate) hypersensitivity, 512, 513*f*
HIV infection/AIDS, 528, 718–724, 719*t*, 725*b*. *See also* HIV
antibody sandwich ELISA in diagnosis of, 503
aspergillosis in, 633, 637
blastomycosis in, 628
candidiasis in, 633*f*, 634–635, 637
cryptococcosis in, 635, 637
cryptosporidiosis in, 662–663
Cyclospora infection in, 663
cytomegalovirus infection in, 692
diagnosis of, 723–724
drug resistance in, 303
epidemiology of, 722–723, 723*f*, 724*f*
helper T lymphocytes in, 472, 472*b*, 720*f*, 721, 721*f*
HHV-6 and, 692
incubation period for, 419*t*
leishmaniasis in, 656
mycobacterial infections in, 559, 560
opportunistic infections/tumors in, 407, 718*f*. *See also specific type*
fungal infections, 637, 718*t*
pathogenesis of, 722, 722*f*, 723*f*
Pneumocystis pneumonia in, 636
prevention of, 724
terminology of, 719*t*
toxoplasmosis in, 662
transmission of, 723, 724*f*

treatment of, 724
tuberculosis in, 559
vaccine development and, 724
western blot test in diagnosis of, 504, 505*f*
HLA. *See* Human leukocyte antigens
HME. *See* Human monocytic ehrlichiosis
Hodgkin's lymphoma, Epstein-Barr virus and, 690
Holdfasts, of algal thalli, 366
Holoenzymes, 130, 130*f*
Homologous sequences, in genetic recombination, 229*f*, 230
Honey, infant botulism and, 550
Honey mushroom (*Armillaria ostoyae*), 342
Hook, of bacterial flagellum, 62, 63, 63*f*
Hookworms (*Ancylostoma* and *Necator*), 670–671, 670*f*, 673*t*
Hopanoids, 70
Hops, in beer production, 747, 747*f*
Horizontal gene transfer, 229–232, 235*t*
bacterial conjugation, 231–232, 233*f*, 234*f*, 235*f*, 235*t*
transduction, 230–231, 232*f*, 235*t*
transformation, 229–230, 231*f*, 235*t*
Hospital epidemiology, 430–432, 432*f*. *See also* Nosocomial infections
Hospital microbiology, 19*f*
Host defenses, 21–22, 22*f*, 438
nonspecific, 437–461, 439*f*, 459*t*
blood, 443–447, 444*f*, 445*f*, 446*f*, 447*b*. *See also specific blood component*
chemical, 450–455, 450*b*
complement/complement system, 450–452, 451*f*, 452*f*, 453*f*, 459*t*
defensins, 454–455, 459*t*
extracellular killing, 450, 459*t*
fever, 457–458, 458*f*, 459*t*
inflammation, 455–457, 455*f*, 456*f*, 457*f*, 458*f*, 459*t*
interferons, 452–454, 453*f*, 455*t*, 459*t*
lacrimal apparatus of eye, 441, 442*f*
mucous membranes, 437, 439–440, 439*f*, 441*f*, 442*t*, 459*t*
normal microbiota, 440–441
phagocytosis, 448–450, 448*f*, 449*b*, 449*f*, 459*t*
secretions, 441–442, 443*t*
skin, 438–439, 439*f*, 440*b*, 442*t*, 459*t*
specific, 438, 439*f*, 462–486. *See also* Immune response
Hosts
movement of microbes into, 409–411, 409*f*, 410*f*, 411*f*, 411*t*, 412*f*. *See also* Infection; Portals of entry
movement of microbes out of, 419, 420*f*
of parasite, 406, 650
symbiotic relationships with microbes and, 405–409, 405*f*, 406*b*, 407*f*, 408*t*
of viruses, 379, 380*f*, 381*b*
Hot air, in microbial control, 271–272, 277*t*
Houseflies, as disease vectors, 424*t*
*Hpa*I restriction enzyme, 245*t*
HPS. *See* Hantavirus pulmonary syndrome
HRIG. *See* Human rabies immune globulin
HSV. *See* Herpes simplex viruses
HTLV-1. *See* Human T-lymphotrophic virus 1
HTLV-2. *See* Human T-lymphotrophic virus 2
HTLV-5. *See* Human T-lymphotrophic virus 5
Human diploid cell vaccine, for rabies, 729
Human genome, 200
detection of infectious agents and, 254*b*
Human granulocytic ehrlichiosis (HGE), 600, 601, 601*f*, 602*t*
Human growth hormone
microbial production of, 13*t*, 756*t*
recombinant DNA in production of, 256*t*

Human herpesviruses, 385t, 682t, 683–693, 684f. *See also specific type under HHV*
Human immunodeficiency virus (HIV), 716, 716f, 718, 719t. *See also* HIV infection/AIDS
  origin of, 719
  pathogenicity of, 719t
  as provirus, 394, 722
  replication of, 719–722, 720f, 721f
  resistance to, 302, 303b
  structure of, 716f, 718–719
  type 1, 718
  type 2, 718
  vaccine development and, 724
  as vector, 245–246, 246f
Human insulin, microbial production of, 13t, 756t
  by *Escherichia coli*, 743, 755
  recombinant DNA in, 256t
Human leukocyte antigens (HLA), 474
Human monocytic ehrlichiosis (HME), 600, 601, 601f, 602t
Human rabies immune globulin (HRIG), 729
Human T-lymphotrophic virus 1 (HTLV-1), 717
Human T-lymphotrophic virus 2 (HTLV-2), 717
Human T-lymphotrophic virus 5 (HTLV-5), 717
Humoral immune response, 468, 477–480, 478f, 479f
Humus, in topsoil, 768
Hyaluronic acid capsule, of group A streptococci, 538
Hyaluronidase, 534
  microbial production of, 753
  as virulence factor, 416, 417f
Hybridization, nucleic acid, 120
Hybridomas, in passive immunization, 495, 495f
Hydatid cysts/hydatid disease, 666, 666f, 673t
Hydrocarbons, microbial production of, 755, 756t
Hydrochloric acid, in Ziehl-Neelsen acid-fast staining, 111
Hydrogen, 28t, 31–32
  as alternative fuel, microbial production of, 755, 756t
  atomic number of, 27, 28t
  Bohr diagram of, 30f
  covalent bonds in formation of, 31–32, 31f, 32. *See also* Hydrogen bonds
  for microbial growth, 169, 170t
Hydrogen bonds, 34–35, 34f, 35t
Hydrogen ions, 37, 38f
  in acid-base balance, 38, 38f
Hydrogen peroxide
  antimicrobial action of, 279
  peroxide anion in, 171
Hydrogen sulfide (H_2S), production of, 757f, 768
Hydrogen sulfide (H_2S) test, 118, 119f
Hydrolases, 129, 130t
Hydrolysis, 36, 36f
  enzymes in, 129, 130t
  of sucrose, 45, 45f
Hydrophilic molecules, in phospholipids, 41, 43f, 70, 70f
Hydrophobia, in rabies, 729
Hydrophobic molecules, 40
  in phospholipids, 43f, 70, 70f
Hydrostatic pressure, microbial growth affected by, 176
Hydrothermal vents, 771–772
Hydroxyl group, 40t
Hydroxyl ions, 37, 38f
Hydroxyl radicals, 171

Hygiene, infection control/prevention and, 16
  Lister's contribution and, 17–18
  Nightingale's contribution and, 18
  Semmelweis' contribution and, 17
  Snow's contribution/epidemiology and, 18
Hypersensitivity, 510–522, 523t
  type I (immediate), 510–515, 511f, 512f, 513b, 513f, 514f, 514t, 515b, 523t. *See also* Allergic reactions
    fungal allergens causing, 644
  type II (cytotoxic), 515–518, 516f, 517f, 517t, 518f, 523t
  type III (immune-complex mediated), 518–520, 519f, 523t
    fungal allergens causing, 644
  type IV (delayed/cell-mediated), 520–522, 520f, 521f, 523t
Hypersensitivity aspergillosis, 632
Hypersensitivity pneumonitis, 519
Hyperthermophiles, 132, 174, 174f, 175b, 321, 321f
Hypertonic solutions, 73, 74f, 176
  microbial growth affected by, 275, 277t
Hyphae, 4, 4f, 356, 357f, 624, 625f
Hypnozoites, in *Plasmodium* life cycle, 659
Hypothalamus, body temperature controlled by, 458, 458f
Hypothesis, in scientific method, 9–10, 11f
Hypotonic solutions, 73, 74f, 176

I. *See* Iodine
Iatrogenic diseases, 415t
  infection, 431
Ibotenic acid, 644
ICTV. *See* International Committee on Taxonomy of Viruses
idio- (prefix), definition of, 413t
Idiopathic diseases, 415t
IFNs. *See* Interferons
Ig. *See* Immunoglobulins
IgE, in type I (immediate) hypersensitivity, 510, 511f, 513b
Il-2R. *See* Interleukin 2 receptor
Illness phase of disease, 419, 419f
IM. *See* Intramuscular (IM) drug administration
Immediate (type I) hypersensitivity, 510–515, 512f, 523t
  to antimicrobial drugs, 301
  clinical signs of
    in localized reaction, 512, 513f
    in systemic reactions, 512–514
  degranulation/degranulating cells in, 510–511, 511f, 512f, 514t
  diagnosis of, 514, 514f
  fungal allergens causing, 644
  initial exposure/sensitization and, 510, 511f, 513b
  prevention of, 514–515, 515b
  treatment of, 515
  vaccines causing, 494
Immersion oil, resolution/resolving power affected by, 97, 98f
Immune-complex mediated (type III) hypersensitivity, 518–520, 519f, 523t
  fungal allergens causing, 644
Immune globulin, human rabies, 729
Immune hypersensitivities. *See* Hypersensitivity
Immune response (immunity), 438, 439f, 462–486
  acquired, 482–483, 483t
  antigen processing and, 475–477, 476f
  antigens in, 463–465, 464f
  body's preparation for, 473–477
  cell-mediated, 472, 480–482, 480f, 482b

  cytokines in, 473
  elements of, 463–473
  herd, 490
  humoral, 468, 477–480, 478f, 479f
  lymphatic system in, 465–467, 466f, 467f
  lymphocyte editing in, 473–474, 474f
  major histocompatibility complex and, 474–475, 475f
Immune serum. *See* Antiserum
Immune suppression, opportunistic pathogens and, 407
Immune testing, 496–505, 506b, 506t. *See also specific test*
  agglutination tests in, 499–500, 499f, 500f, 506t
  complement fixation test in, 501–502, 501f, 506t
  labeled antibody tests in, 502–504, 502f, 503f, 504f, 505f, 506t
  neutralization tests in, 500, 506t
  precipitation tests in, 496–499, 497f, 498f, 499f, 506t
  recent developments in, 505, 506b, 506t
Immune thrombocytopenic purpura, 518, 518f
Immunization, 18, 488–496. *See also* Vaccines
  active, 488, 489–494, 490t, 491f, 492t, 493f, 494b, 495f
  artificially acquired immunity and, 483
  CDC recommendations for, 491, 491f, 492t
  history of, 488, 489f, 490b
  humoral immune response in, 478–480, 479f
  passive, 488, 494–496, 495f
  toxoids in, 416, 491
Immunochromatographic assays, 505
Immunocompromised host
  candidiasis in, 633f, 634–635
  cryptococcosis in, 635
  cytomegalovirus infection in, 692
  Listeria infection in, 553
  opportunistic pathogens and, 407–409. *See also specific type*
    fungal, 637
  Pneumocystis pneumonia in, 636
  zygomycoses in, 636–637
Immunodeficiency diseases, 526–528, 527b, 528t
  acquired (secondary), 527, 528. *See also* HIV infection/AIDS
  primary, 526–528, 527, 527b, 528t
  severe combined, 527, 527b, 528t
Immunodiffusion, 496–498, 506t
  double (Ouchterlony test), 496–497, 497f
  radial, 497–498, 498f
Immunoelectrophoresis, 498–499, 499f, 506t
Immunofiltration assays, 505
Immunofluorescence, 101, 101f
Immunoglobulin A (IgA), 468, 470, 470f, 471t
  deficiency of, 527
  secretory, 470, 470f
  in pneumococcal infection, 543
Immunoglobulin D (IgD), 468, 470, 470f, 471t
Immunoglobulin E (IgE), 468, 470, 470f, 471t
Immunoglobulin G (IgG), 468, 469, 470f, 471t
Immunoglobulin M (IgM), 468, 469–470, 470f, 471t
Immunoglobulins, 443, 468–471, 471t. *See also* Antibodies
  classes of, 468, 469–470, 470f
  deficient, 527
  structure of, 468, 468f
Immunological diseases, 412, 415t
Immunological memory, 477–480, 479f
Immunological synapse, 482
Immunology, 18, 19f, 20t, 22. *See also* Immune response
Immunophilins, 523, 524t
Immunosuppressive drugs, 522–523, 524t

Immunotherapy, for type I hypersensitivity, 514–515
Impetigo
 staphylococcal, 535, 536f
 streptococcal, 539–540
Inactivated (killed) vaccines, 490–491
 adjuvants for, 490t, 491
Inactivated polio vaccine (IPV/Salk vaccine), 707, 707t
Inborn (innate) resistance, 438
Incidence of disease, 425, 425f, 426f
Incinceration, in microbial control, 272, 277t
Inclusion bodies
 chlamydias forming, 602, 603f
 in cytomegalovirus infection, 692, 692f
Inclusions, cellular
 in eukaryotic cells, 87t
 in prokaryotic cells, 60f, 75, 76f, 87t
Incubation, in streak plate method of isolation, 179
Incubation period, 418, 419f, 419t
 of pertussis, 586, 587f
Index case, 426
Indicator organisms, in water quality testing, 758, 758f
Indiginous microbiota, 406–409, 407f, 408t. See also Normal microbiota
Indigo, microbial production of, 754, 754b, 756t
Indirect contact transmission of disease, 421, 422t
Indirect fluorescent antibody tests, 502, 503f, 506t
Indirect (negative) selection, in mutant identification, 227–228, 228f
Indoleamine 2,3-dioxygenase, fetal placental production of, 522
Induced fit model, 131, 132f
Inducers, in operon, 219, 219f
Inducible operons, 219, 219f, 221t
 lactose (lac) operon as, 219–220, 219f
Induction, in lysogeny, 389f, 390
Industrial canning, in food preservation, 750, 750f
Industrial fermentations, 752–753, 753f
Industrial microbiology/biotechnology, 11, 13t, 19f, 20t, 744, 752–763
 agricultural products of, 755, 756t
 alternative fuel production and, 754–755, 754f, 755b, 756t
 enzyme production and, 753–754, 754b, 756t
 fermentations and, 752–753, 753f
 pesticide production and, 755, 756t
 pharmaceutical products of, 755, 756t
 products of, 753–756, 756t
 recombinant DNA technology in, 242, 243f
Infant botulism, 550
Infection/infectious diseases, 409–411, 411–419, 415t
 adhesion in, 410–411
 causes (etiology) of, 412–415, 414f, 415t
 classification of, 423–424, 424t, 425t
 contamination and, 409
 definition of, 409, 412
 epidemiology of, 424–433, 425f, 426f, 427f, 428f, 429f, 429t, 430f, 431b, 432b
 germ theory of, 14, 413
 iatrogenic, 431
 manifestations of, 412, 412t
 nature of, 411–419
 nosocomial, 16, 430–432, 432f
 notifiable, 426, 429t
 persistent, enveloped viruses causing, 393, 393f
 portals of entry and, 409–410, 409f, 410f
 prevention of, 16–19
 reservoirs of, 420, 421t

 sources of, 419–421, 421t
 stages of, 418–419, 419f, 419t
 terminology of, 413t
 transmission of, 421–423, 422f, 422t, 423b
  public health agencies in interruption of, 432–433
 virulence factors and, 415–418, 416f, 417f, 418t
Infection control, 18, 19f, 20t
 nosocomial infections and, 431–432
Infectious mononucleosis (mono)
 cytomegalovirus causing, 692
 Epstein-Barr virus causing, 690, 690f
Inflammation, 455–457, 455f, 456f, 457f, 458t, 459t
 as host defense, 455–457, 455f, 456f, 457f, 458t, 459t
Influenza, 705t, 731–735
 diagnosis/treatment/prevention of, 735
 epidemiology of, 733
 incubation period for, 419t
 manifestations of, 705t, 732–733
 vaccine against, 735
Influenza viruses, 731–735, 733f, 734f, 735f
 pandemics caused by, 731–732, 732f
Ingestion, in phagocytosis, 448f, 449
INH. See Isoniazid
Inhalation anthrax, 327, 546, 705t
 bioterrorism and, 22, 545, 546, 547b, 774t
 manifestations of, 705t
Inhibition/inhibitors
 allosteric, 135, 135f, 136f
 competitive, 134–135, 135f
 enzyme, 134–136, 135f, 136f
 feedback, 135–136, 136f
 noncompetitive, 135, 135f
Initial (reticulate) bodies
 Chlamydia forming, 337, 602, 603f
 Ehrlichia forming, 601, 601f
Initiation
 transcription, 209, 210f, 211f
 translation, 216, 216f
Initiation complex, 216, 216f
Injection
 for DNA insertion into cells, 251f, 252, 255t
  vaccine development and, 492, 493f
 intramuscular (IM)/intravenous (IV), for drug administration, 300, 300f
Innate (inborn) resistance, 438
Inoculum/inocula, 177
 clinical sampling and, 178, 179t
 environmental sampling and, 177–178
 sources of, 177–178
Inorganic chemicals, 37–39. See also specific type
 acids, 37
Insecticides, microbial production of, 13t, 755, 756t
Insects, as disease vectors, 674–675, 674f
Insertions, in mutations, 221, 222f, 223t
Insertion sequences (IS), 235, 236f
Insulin
 autoimmunity affecting production of, 525
 microbial production of, 13t, 756t
  by Escherichia coli, 743, 755
  recombinant DNA in, 256t
Integral proteins, 70–71, 70f
 in electron transport chains, 146
Integrase, 716, 716f
 HIV, 716f, 719t
 latency and, 722
Interference (differential interference contrast/Nomarski microscopy/microscopes), 92, 98–100, 99f, 107t

Interferons, 452–454, 453f, 455t, 459t
 alpha (α), 453, 454f, 455t
 beta (β), 453, 454f, 455t
 clinical therapy with, 453–454
 gamma (γ) (macrophage activation factor), 453, 455t
 in host defense, 452–454, 453f, 455t, 459t
 in immune response, 473
 recombinant DNA in production of, 256t
Interleukin 1, in fever, 458, 458f
Interleukin 2 receptor, monoclonal antibodies against, 523
Interleukins
 in immune response, 473
  cell-mediated, 480, 480f
 recombinant DNA in production of, 256t
Intermediate filaments, in eukaryotic cytoskeleton, 81, 82f
Intermediate hosts, 650
 for flukes, 667, 668f
 for helminths, 663
 for tapeworms, 664, 665f
Intermediate-level germicides, 266
Internal carbonyl group, 40t
International Committee on Taxonomy of Viruses (ICTV), 384
Intestinal secretions, in host defense, 443t
Intoxications
 food. See Food-borne illnesses/food poisoning
 fungi causing, 626, 627, 643–644, 645b
  mushroom poisoning, 363, 364f, 643–644, 644, 644f
Intracellular pathogens, 335
Intramuscular (IM) drug administration, 300, 300f
Intravenous (IV) drug administration, 300, 300f
Introns, 214, 214f, 698b
In-use test, 268
Invasive extraintestinal amoebiasis, 652
Inverted repeat (IR), 235
Iodine, 28t
 antimicrobial action of, 279, 282t
 in Gram staining, 109, 109f
Iododeoxyuridine, 296t
Iodophors, antimicrobial action of, 279, 279f
Ionic bonds, 33–34, 33f, 34f, 35t
Ionization (dissociation), 34, 34f
Ionizing radiation. See also Radiation
 in microbial control, 275–276, 276f, 277t
 as mutagen, 223
IPV. See Inactivated polio vaccine
IR. See Inverted repeat
Iron, 28t
 cycling of, 768
 for microbial growth, 170t, 172
Iron-binding compounds, of Enterobacteriaceae, 573, 574f
Irradiation. See Radiation
IS. See Insertion sequences
Islets of Langerhans, autoimmunity affecting, 525
Isografts, 521, 521f
Isolation, for pure culture, 178–181, 179f, 180f, 181f
Isomerases, 129, 130t
Isoniazid, 305t
 mechanism of action of, 292, 305t
 spectrum of action of, 298f, 305t
Isopropanol
 antimicrobial actions of, 278–279
 microbial fermentation producing, 150f
Isotonic solutions, 73, 74f
Isotopes, 27–28, 28f
 radioactive, 28

-itis (suffix), definition of, 413t
Itraconazole, 627
IV. See Intravenous (IV) drug administration
Ivanowski, Dmitri, 16, 17t, 19f, 380
Ivermectin, 309t
Ixodes ticks, as disease vectors, 424t
   for ehrlichiosis (Ehrlichia species), 601, 602t
   for Lyme disease (Borrelia burgdorferi),
      610–611, 611f

Jacob, Francois, 218
Japanese encephalitis, 711, 711t
Jaundice
   in hemolytic disease of newborn, 518
   in hepatitis A, 708
   in hepatitis B, 697
   in malaria, 660
J (joining) chain, 470, 470f
JC virus, 695
Jenner, Edward, 18, 19f, 488, 681
"Jock itch" (tinea cruris), 640t
Joining (J) chain, 470, 470f
Juice, unpasteurized, E. coli contamination
      of, 272b
"Jumping genes," 233, 235f
Junin hemorrhagic fever, 736

K. See Potassium
Kala-azar (visceral leishmaniasis), 656
Kalium. See Potassium
Kanamycin, 306t
K antigens, 572, 574f
   E. coli producing, 575
   Haemophilus influenzae producing, 582
Kaposi's sarcoma
   HHV-8 causing, 693, 693f
   in HIV infection/AIDS, 722, 723f
Kelps, 5
   giant, 369, 369f
Keratin, superficial mycoses and, 638
Keratitis, Acanthamoeba, 652
Ketoconazole, 309t, 627
α-Ketoglutaric acid, 159f, 160t
Kidneys
   immune-complex (type III) hypersensitivity
      affecting, 519–520
   streptococcal infection affecting, 540
Kimchi, fermentation in production of, 744, 748t
Kinases, as virulence factors, 416, 417f
Kinetoplast, 354
Kinetoplastids, 353–354, 354f, 356t
Kingdoms, 114, 115f, 348f
   Witaker's, 116–117, 116f, 348f, 351
Kinins, in type I (immediate) hypersensitivity,
      511, 511f, 514t
Kirby-Bauer (disk-diffusion) method,
      267–268, 268f
Kirby-Bauer (diffusion susceptibility) tests,
      297–298, 298f
Kissing bugs (Triatoma), as disease vectors,
      424t, 674f, 675
   for Chagas' disease (Trypanosoma cruzi), 654f,
      655, 675
Kitasato, Shibasaburo, 17t, 19f, 22
Klebs, Edwin, 16, 17t
Klebsiella genus/spp., 336t, 575, 576f
   sites of infection caused by, 582f
Klebsiella pneumoniae, 575, 576f
   capsules of, 62
   negative (capsule) staining of, 112
Kluyver, Albert, 19
Koch, Robert, 14–16, 14f, 17t, 19f, 413
Koch's postulates, 15–16, 19f, 413, 414f
   exceptions to, 414–415
Koplik's spots, in measles, 726, 726f
Krebs cycle, 137, 142, 143f, 149t, 162f, A-6

Küpffer cells, 444, 445f, 446f
Kuru, 399, 400b

Labeled antibody tests, 502–504, 502f, 503f,
      504f, 505f. See also specific type
lac operator, 219f, 220
lac operon, 219–220, 219f
Lacrimal apparatus, in host defense, 441, 442f
LaCrosse encephalitis, 711t, 735
β-Lactam antibiotics, 305t
   mechanism of action of, 290, 291f, 305t
β-Lactamase, 302, 303f, 534
   penicillin and, 221b, 302, 303f
β-Lactam rings, 290, 291f, 302, 303f
Lactic acid, microbial fermentation
      producing, 150f
   vegetable fermentation and, 744
Lactobacilli, 320f
Lactobacillus genus/spp., 327–328, 327b, 329t
   fermentation products of, 150f,
      327–328, 328b
      vegetable fermentation and, 744
Lactobacillus bulgaricus, in yogurt
      production, 745
Lactococcus genus/spp., in vegetable
      fermentation, 744
Lactococcus lactis
   in cheese production, 745
   subspecies cremoris, in buttermilk
      production, 745
Lactose, 45
Lactose (lac) operon, 219–220, 219f
Lagging strand, in DNA replication, 205,
      206f, 207
Lagoons, oxidation, in wastewater treatment,
      760–761
Lag phase, of microbial growth, 188, 189f
Lamellae, photosynthetic, 85
   in phototrophic bacteria, 323
Laminarin, 369
Lamivudine, 296f
Lancefield, Rebecca, 537
Lancefield antigens, 537
Lancefield groups, 537, 546t
Landfills
   biomass reclamation/bioremediation
      and, 761
   microbial production of methane and, 754,
      754f
Langerhans (dendritic) cells, 438, 444
   HIV infection of, 721
Langerhans, islets of, autoimmunity
      affecting, 525
Larvae, tapeworm, 665, 665f
Lassa hemorrhagic fever, 736
Latency
   animal virus, 394, 394t, 683
   herpes simplex virus, 684, 684f
   HIV, 720f, 721–722
   varicella-zoster virus, 688, 688f. See also
      Herpes zoster
Latent diseases, definition of, 424, 425t
Latent phase, in syphilis, 609
Latent viruses (proviruses), 394
   herpesviruses as, 683
      herpes simplex virus, 684, 684f
      varicella-zoster virus, 688, 688f
   HIV as, 394, 722
Laundry enzymes, microbial production of,
      13t, 756t
Laveran, Charles, 16, 17t
Laws/theories, in scientific method, 9, 11f
LCM. See Lymphocytic choriomeningitis
Leach fields, 759, 761f
Leading strand, in DNA replication, 205,
      205–207, 206f

van Leeuwenhoek, Antoni, 2–3, 2f, 3f, 6–7, 19f,
      93–94, 95, 96
Leflunomide, 523
Legionella genus/spp., 334t, 335, 589–590, 590f
Legionella pneumophila, 431b, 589–590, 590f
Legionnaires' disease (legionellosis), 589, 590
   epidemiology of, 431b, 590
Leishmania genus/spp., 655–656, 656f, 657f, 664t
Leishmaniasis, 655–656, 656f, 657f
Lenses
   in compound microscope, 96–98, 97f, 98f
   numerical aperture of, resolution affected
      by, 95
      immersion oil and, 97, 98f
Lentivirus genus/spp., 717
Lepromatous leprosy, 559, 559f
Leprosy, 559–560, 559f, 560b
   incubation period for, 419t
Leptospira genus/spp., 613–614, 614f
Leptospira interrogans, 613–614, 614f
Leptospirosis, 613–614
Leptotrombidium species, as disease vectors, 424t
   for scrub typhus (Orienta tsutsugamushi),
      600, 602t
Leuconostoc genus/spp., in vegetable
      fermentation, 744
Leuconostoc citrovorum, in buttermilk
      production, 745
Leukemia
   adult acute T-cell lymphocytic, HTLV-1
      causing, 717
   hairy-cell, HTLV-2 causing, 717, 718f
Leukocidins, 418
Leukocytes, 443
   autoantibodies to, 524
   extracellular killing by, in host defense,
      450, 459f
   in host defense, 443–444, 444f, 445f, 447, 447f
   lab analysis of, 447, 447b
   polymorphonuclear. See Neutrophils
Leukoencpehalopathy, progressive multifocal
      (PML), 695
Leukopenia, in ehrlichiosis, 332, 601
Leukotrienes
   in inflammation, blood vessels affected by,
      455–456, 456f
   in type I (immediate) hypersensitivity, 511,
      511f, 514t
Levy, Stuart, 283b
L forms, 48
Li. See Lithium
Librarian's lung, 519
Libraries, in recombinant DNA technology,
      246–247, 248f, 255t
Lice, as disease vectors, 424t, 674, 674f
   for epidemic relapsing fever (Borrelia
      recurrentis), 613
   for epidemic typhus (Rickettsia prowazekii),
      599–600, 602t
   for trench fever (Bartonella quintana), 584
Lichens, 362, 363–365, 365f
Life processes, 57–58, 57t
   in microbes, 57–58, 57t, 58b
      viruses, 57t, 58, 58b, 400
   possibility of on other planets, 26, 169b, 313
Ligands, as adhesion factors, 411, 411f
Ligases, 129, 130t
   DNA, in lagging strand synthesis, 206f, 207
Light
   electron transport chain producing, 146b
   for microbial growth, 168f, 169
   refraction of, image magnification and,
      94–95, 95f
      oil immersion lens affecting, 97, 98f
   wavelengths of, microscopy and, 94, 94f
Light chains, immunoglobulin, 468, 468f

Light-dependent reactions, 153, 154–156, 154*f*, 155*f*, 156*t*
Light-independent reactions, 153, 156, 157*f*
Light microscopy/microscopes, 95–101, 107*t*
    bright-field, 95, 96–98, 97*f*, 98*f*, 99*f*, 107*t*
    dark-field, 95, 98, 99*f*, 107*t*
Light reactions. *See* Light-dependent reactions
Light repair, DNA, 225, 226*f*
Limiting nutrient, 169
Limnetic zone, 770, 771*f*
Linnaean system of taxonomy, 11, 113–117, 115*f*
Linnaeus, Carolus, 3, 19*f*, 113, 114, 117, 348*f*, 351
Linoleic acid, 42*t*
Lipases, 534
    in fat catabolism, 149, 151*f*
Lipid A (endotoxin), in Gram-negative cell walls, 68, 69*f*, 416–417, 417*f*, 418*t*
    Enterobacteriaceae, 572, 574*f*
    Neisseria, 567
    Pseudomonas aeruginosa, 587
Lipids, 40–42, 41*f*, 42*t*, 43*f*, 44*f*. *See also specific type*
    biosynthesis of, 158, 159*f*, 162*f*
    catabolism of, 149–151, 151*f*, 162*f*
Lipooligosaccharide (LOS), in *Neisseria* cell walls, 567
Lipopolysaccharide (LPS), in Gram-negative cell walls, 68, 69*f*
Lipoproteins, 50
Lipoteichoic acids, in Gram-positive cell walls, 67, 69*f*
Lister, Joseph, 17–18, 19*f*
Listeria genus/spp., 327, 329*t*, 551–554, 553*f*
    growth of, in refrigerated foods, 272, 327, 554, 751, 752*t*
Listeria monocytogenes, 327, 551–554, 752*t*
Listeriolysin O, 553
Lithium (Li), Bohr diagram of, 30*f*
Lithotrophs, 169, 169*b*
Littoral zone, 770, 771, 771*f*
Liver
    cancer of, hepatitis B and, 699, 700*f*
    hepatitis A affecting, 708
    hepatitis B affecting, 697, 698
Liver flukes (*Fasciola* species), 669, 673*t*
Livestock, assessing biological threats to, 773, 774–775
Local drug administration, 300
Local infection, 425*t*
Lockjaw, 551. *See also* Tetanus
Loeffler's syndrome, 670, 671
Loffler's medium, for *Corynebacterium diphtheriae* culture, 555
Logarithmic (exponential) growth, 187, 188*f*, B-1
    growth curves of, 188, 189*f*
Logarithms, B-1
Log phase, of microbial growth, 188–189, 189*f*
Lone Star tick, as disease vector, for ehrlichiosis (*Ehrlichia* species), 601, 602*t*
Long-term nonprogressors, in HIV infection, 724
Lophotrichous flagellum, 63, 64*f*
LOS. *See* Lipooligosaccharide
Loss of function, in inflammation, 455
Louse-borne (epidemic) relapsing fever, 613, 613*f*
Louse-borne (epidemic) typhus, 599–600, 602*t*
Lovely, Derek, 128*b*
Lowenstein-Jensen agar, for *Mycobacterium tuberculosis* culture, 555, 556*f*
Low G + C Gram-positive bacteria, 320*f*, 325–328, 329*t*, 533
Low-level germicides, 266
Low-oxygen cultures, 185–186, 186*f*

Low power objective lens, of light microscope, 97
LPS. *See* Lipopolysaccharide
LSD (lysergic acid diethylamide), 361
Luciferase, 146*b*
Luminal amoebiasis, 651
Luminescence, electron transport chain producing, 146*b*
Lungs, specimen collection from, 179*t*
Lupus erythematosus, systemic, 525–526, 526*f*
Lutzomyia flies, leishmaniasis transmitted by, 655, 656*f*
Lyases, 129, 130*t*
Lyme disease, 421*t*, 610–612, 611*f*, 612*f*
Lymph, 465, 466*f*
Lymphatic capillaries, 465, 466*f*
Lymphatic ducts, 465, 466*f*
Lymphatic filariasis, 671–672, 672*f*
Lymphatic system, 465–467, 466*f*, 467*f*
    microbial disease agents affecting, D-6 to D-7
Lymphatic vessels, 465, 466*f*, 467*f*
Lymph nodes, 465, 467*f*
Lymphocryptovirus genus/spp., 683, 689. *See also* Epstein-Barr virus
Lymphocutaneous sporotrichosis, 642, 643*f*
Lymphocyte-depleting therapies, 523, 524*t*
Lymphocytes, 444, 445*f*, 465, 467*f*. *See also* B lymphocytes; T lymphocytes
    editing, by clonal deletion, 473–474, 474*f*
    natural killer, 450
Lymphocytic choriomeningitis, 736
Lymphogranuloma venereum, 337, 603, 604*f*
Lymphoid cells, 465, 467*f*
Lymphoma
    Burkitt's, Epstein-Barr virus in, 690, 690*f*
    Hodgkin's, Epstein-Barr virus in, 690
Lyophilization (freeze-drying)
    for food preservation, 750, 751*f*
    for microbial control, 273, 277*t*
    for preserving cultures, 186
Lysergic acid diethylamide (LSD), 361
Lysogenic phages, 388. *See also* Lysogeny/lysogenic replication
Lysogeny/lysogenic replication, 388–390, 389*f*
Lysol. *See* Orthophenylphenol
Lysosomes, 60*f*, 84, 85*f*, 87*t*
    in phagocytosis, 448*f*, 449
Lysozyme
    in host defense, 439, 459*t*
    in lytic replication, 387
Lyssavirus genus/spp., 728
Lytic replication, 385, 385–388, 386*f*, 388*f*, 389*f*

m. *See* Meter
MAC. *See* Membrane attack complexes
MacConkey agar, 182, 184*f*, 184*t*
    as selective and differential medium, 183, 184*f*
Machupo hemorrhagic fever, 736
MacLeod, Colin, 20
Macrocystis genus/spp., 369, 369*f*
Macrolides, 306*t*
    mechanism of action of, 292, 293*f*, 306*t*
Macromolecules, 40. *See also specific type*
Macronucleus, in protozoa, 349
    reproduction and, 350, 350*f*
Macrophage activation factor (gamma-interferon), 453, 455*t*
Macrophage colony stimulating factor, recombinant DNA in production of, 256*t*
Macrophages, 444
    HIV infection of, 721, 721*f*
    in reticuloendothelial system, 446*f*, 447

Macules, in poxvirus infection, 680, 681*f*
Mad cow disease, 48, 399
Madurella genus/spp., 641
Madurella mycetomatis, 642*f*
"Magic bullet," 18–19, 22
Magnesium, 28*t*
    Bohr diagram of, 30*f*
    in chlorophyll, 153
    as cofactor in enzymatic activity, 131*t*
    cycling of, 768
    for microbial growth, 170*t*, 172
Magnetite, cellular inclusions of, 75
Magnetobacteria, 56, 56*f*, 75
Magnetosomes, 56, 56*f*
Magnification, 94–95, 95*f*
    determination of, for compound microscope, 97–98
    empty, 95
Major histocompatibility complex, 474–475, 475*f*
Major histocompatibility complex proteins, 473, 474–475, 475*f*
    prevention of graft rejection and, 522
Malachite green, in Schaeffer-Fulton endospore staining, 111
Malaria, 421*t*, 659–661, 659*f*
    discovery of cause of, 17*t*
    vaccine against, 661, 661*b*
Malassezia furfur, 639–640, 639*f*
Malignant tumor, 395. *See also* Cancer
Mallon, Mary, 579*b*
Malta fever. *See* Brucellosis
Maltose, 45
    in beer production, 747, 747*f*
Manganese, 28*t*
    for microbial growth, 170*t*, 172
Mannheimia haemolytica, vaccine against, 255*b*
Marburg virus, 730–731, 731*f*, 732*f*
Margination, 456
Marine ecosystems, 770, 771–772, 771*f*
Mast cells
    in inflammation, 455
    in type I (immediate) hypersensitivity, 511, 511*f*, 512*f*
Mastigophora, 351
Matter, definition of, 27
Maximum growth temperature, 173–174
MBC test. *See* Minimum bactericidal concentration (MBC) test
McCarty, Maclyn, 20
McClintock, Barbara, 233
MDR *M tuberculosis*. *See* Multidrug resistant (MDR) *Mycobacterium tuberculosis*
Measles (rubeola), 725–727, 726*f*, 727*f*, 727*t*
    vaccine against, 726–727, 727*f*. *See also* Measles/mumps/rubella (MMR) vaccine
Measles/mumps/rubella (MMR) vaccine, 487, 726–727, 727*f*, 728, 728*f*
Measurement, units of, 93, 93*t*
Meat/pork
    fermentation in preparation of, 745, 748*t*
    tapeworm infestation of, 649, 665, 665*f*
Mebendazole, 309*t*
Mechanical vectors, in disease transmission, 423, 424*t*
Mediators, inflammatory, 458*t*
Medium (culture), 177, 181–185, 181*f*. *See also* Culture media
Medulla, lymph node, 465, 467*f*
Megakaryocytes, 443
Meiosis, 344–346, 345*f*, 347*t*
Melioidosis, as bioterrorist threat, 774*t*
Membrane attack complexes (MAC), 451, 452*f*, 453*f*
    complement fixation test and, 501

Membrane filters/filtration
    in estimating microbial growth, 191, 191*f*
    in microbial control, 274, 274*f*, 275*t*
    in water quality testing, 758, 758*f*
    in water treatment, 756, 757*f*
Membrane fusion, in animal virus replication, 390, 391*f*
Membrane proteins, 70–71, 70*f*
    in electron transport chains, 144–146
Membranes. *See also* Cytoplasmic (cell/plasma) membranes
    phospholipid bilayers forming, 41, 43*f*, 70, 70*f*
        cholesterol in, 42, 44*f*
Membranous organelles, 81–85, 87*t*
Memory B cells, 477, 477–480, 478*f*, 479*f*
Memory responses, 480
Memory T cells, 480, 480*f*, 481
Meningeal candidiasis, 634*t*
Meningitis
    in blastomycosis, 628
    cryptococcal, 635
    Haemophilus influenzae causing, 582
    Listeria causing, 553
    meningococcal, 569–570
        discovery of cause of, 17*t*
    pneumococcal, 543
    viral, coxsackieviruses causing, 708
Meningococcal meningitis, 569–570
    discovery of cause of, 17*t*
Meningococcal septicemia, 570, 570*f*
Meningococcus. *See Neisseria meningitidis*
Meningoencephalitis
    in African sleeping sickness (*Trypanosoma brucei*), 653
    amoebic, *Naegleria* causing, 652
Menstrual flow, in host defense, 443*t*
Mental disease, 412, 415*t*
Mercury, antimicrobial action of, 280–281, 282*t*
Merismopedia genus/spp., 324
Merozoites, in *Plasmodium* life cycle, 347, 347*f*, 659, 659*f*
Mesophiles, 174, 174*f*
Messenger RNA (mRNA), 208, 213–214, 214*f*
    in translation, 216–217, 216*f*, 218*f*
Metabolic activity, for estimating microbial growth, 193
Metabolic expression, control of, 161
Metabolism, 12, 19*f*, 20*t*, 36, 57, 57*t*, 125–166, 126*f*, 162*f*, A-1 to A-7. *See also specific compound*
    amino acid biosynthesis, 159–160, 159*f*, 162*f*
    antimicrobial drugs affecting, 290*f*, 294–295, 294*f*, 307*t*
        resistance and, 303
    ATP production/energy storage and, 127–128, 129*f*
    carbohydrate biosynthesis, 157–158, 158*f*, 162*f*
    carbohydrate catabolism, 136–149, 137*f*, 149*t*
    catabolism and anabolism and, 36, 126–127, 126*f*
    as characteristic of life, 57
    chemical reactions underlying, 126–136
    enzymes in, 126, 128–136, 130*t*, 131*t*. *See also* Enzymes
    integration/regulation of functions of, 160–161, 162*f*
    lipid biosynthesis, 158, 159*f*, 162*f*
    lipid catabolism, 149–151, 151*f*, 162*f*
    nucleotide biosynthesis, 160, 161*f*, 162*f*
    oxidation-reduction reactions and, 127, 127*f*, 128*b*
    pathways for, 126, A-1 to A-7. *See also specific type*

photosynthesis and, 152–156, 153*f*, 154*f*, 155*f*, 156*t*, 157*f*
protein catabolism, 151, 152*b*, 152*f*, 162*f*
rate of, estimating microbial growth and, 193
resistance and, 303
Metabolites
    precursor, 126, 127, 156, 160*t*
        for amino acid conversion, 159–160
        Entner-Doudoroff pathway in production of, 141, 141*f*
        pentose phosphate pathway in production of, 139–141, 140*f*
    primary and secondary, of industrial fermentation, 753
Metacercariae, in fluke life cycle, 667, 668*f*
Metachromatic granules, in corynebacteria, 328
Metal-containing proteins, in electron transport chains, 146
Metal ions, cycling of, 768
Metalloproteins, 50
Metaphase
    meiosis, 344–346, 345*f*
    mitosis, 344, 345*f*
Metastasis, 395
Meter (m), 93, 93*t*
Methane
    as alternative fuel source, microbial production of, 754, 754*f*, 756*t*
    covalent bonds in formation of, 31*f*, 32
    methanogens producing, 322, 754, 754*f*
Methane oxidizers, 334*t*, 335
Methanobacterium genus/spp., 322
Methanogens, 320*f*, 322
    methane from landfills produced by, 754, 754*f*
Methanol (methyl alcohol), for specimen preparation, 108
Methanopyrus genus/spp., 322
Methicillin, 291*f*, 305*t*, 537
    mechanism of action of, 291*f*, 292, 305*t*
    Staphylococcus aureus resistance and, 287, 537, 538*b*
Methicillin-resistant *Staphylococcus aureus* (MRSA), 287, 537, 538*b*
Methotrexate, 523
Methyl alcohol. *See* Methanol
Methylation, in DNA replication, 207–208
Methylene blue, 114*t*
    in Ziehl-Neelsen acid-fast staining, 111
4-Methylumbilliferyl-β-D-glucoronide (MUG), in water quality testing, 758
Metric units of measurement, 93, 93*t*
Metronidazole, 310*t*
    toxicity of, 301, 301*f*, 310*t*
Mg. *See* Magnesium
MHA-TP (microhemagglutination assay against *T. pallidum*), 609
MHC. *See* Major histocompatibility complex
Micelles, 41, 43*f*
Miconazole, 309*t*
Microaerophiles, 171
Microbes/microorganisms, 1
    adaptation in survival of, 764–765
    in biogeochemical cycles, 765–768, 766*f*, 767*f*
    as biological weapons, 772–774, 774*t*. *See also* Bioterrorism
    classification of, 3–7, 112–122, C-1 to C-9
    death of, 264. *See also* Microbial death rate
    discovery of, 2–3, 3*f*
    genetics of. *See* Genetics, microbial
    growth of, 57*t*, 167–198. *See also* Microbial growth
    identification of, 113, 118–122. *See also* Taxonomy/microbial classification
    industrial uses of, 11, 13*t*, 19*f*, 744, 752–763. *See also* Industrial microbiology

isolation of, in pure cultures, 178–181, 179*f*, 180*f*, 181*f*
life on other planets and, 26, 169*b*, 313
life processes in, 57–58, 57*t*, 58*b*
microscopes for observation of, 2–3, 2*f*, 93–105, 107*t*. *See also* Microscopy
movement of into hosts, 409–411, 409*f*, 410*f*, 411*f*, 411*t*, 412*f*. *See also* Infection; Portals of entry
movement of out of hosts, 419, 420*f*
normal (normal flora), 406–409, 407*f*, 408*t*
pathogenic, 14, 21. *See also specific type*
pH range tolerated by, 38–39, 39*b*
requirements of for existence, 26, 167, 169*b*. *See also* Chemistry; Microbial growth
sites of infection of, D-1 to D-11
in soil, 768–770, 768*f*, 770*t*
spontaneous generation (abiogenesis) of, 7–10
susceptibility of to antimicrobial agents, 266, 266*f*
symbiotic relationships with hosts and, 405–409, 405*f*, 406*b*, 407*f*, 408*t*
Microbial adaptation, survival and, 764–765
Microbial antagonism
    adaptation/survival and, 764
    drug resistance and, 304
    in host defense, 440
    lactobacilli and, 327
    microbial growth affected by, 177
    opportunistic infections and, 407–409
    superinfection and, 297
Microbial competition
    adaptation/survival and, 764
    food spoilage and, 749, 749*f*
    opportunistic pathogens and, 407–409
Microbial control methods. *See* Antimicrobial drugs; Antimicrobial methods
Microbial cooperation, adaptation/survival and, 764–765
Microbial cultures. *See* Culture (microbial)
Microbial death rate, 264, 265*f*
    susceptibility of organisms and, 266
    thermal methods of microbial control and, 268–269, 269*f*
Microbial ecology, 19*f*, 763–765, 765*f*
    adaptation/survival and, 764–765
    levels of microbial associations and, 764, 765*f*
Microbial genetics. *See* Genetics, microbial
Microbial growth, 57*t*, 167–198
    arithmetic, 187, 188*f*
    control of, 262–286, 287–312
        basic principles of, 263–265
        in body, 287–312. *See also* Antimicrobial drugs
        chemical methods for, 277–283, 282*t*, 287–312
        in environment, 262–286. *See also* Antimicrobial methods
        physical methods for, 268–277, 277*t*
        selection of methods for, 265–268
        terminology of, 263–264, 264*t*
    in culture, 177–186. *See also* Culture (microbial)
    generation time and, 188, B-1
    logarithmic (exponential), 187, 188*f*, B-1
        growth curves of, 188, 189*f*
    mathematical considerations in, 187–188, 188*f*, B-1
    measuring, 189–194
        direct methods for, 190–193, 190*f*, 191*f*, 192*f*, 193*f*, 195*t*
        indirect methods for, 193–194, 194*f*
    phases of, 188–189, 189*f*
    requirements for, 168–177, 168*f*, 170*t*

chemical, 168–172, 168f, 170t, 173t
ecological associations and, 177, 177b
physical, 172–177
Microbial growth curve, 188–189, 189f
Microbial load, chlorine disinfection of water and, 758
Microbial metabolism, 12, 19f, 20t, 57t, 125–166, 126f, 162f, A-1 to A-7. See also specific compound and Metabolism
amino acid biosynthesis, 159–160, 159f, 162f
ATP production/energy storage and, 127–128, 129f
carbohydrate biosynthesis, 157–158, 158f, 162f
carbohydrate catabolism, 136–149, 137f
catabolism and anabolism and, 36, 126–127, 126f
as characteristic of life, 57
chemical reactions underlying, 126–136
enzymes in, 128–136, 130t, 131t. See also Enzymes
integration/regulation of functions of, 160–161, 162f
lipid biosynthesis, 158, 159f, 162f
lipid catabolism, 149–151, 151f, 162f
nucleotide biosynthesis, 160, 161f, 162f
oxidation-reduction reactions and, 127, 127f, 128b
pathways for, 126, A-1 to A-7
photosynthesis and, 152–156, 153f, 154f, 155f, 156t, 157f
protein catabolism, 151, 152b, 152f, 162f
Microbial morphology, 19f
Microbial resistance. See Resistance
Microbiology
agricultural, 20t, 755, 756t
recombinant DNA technology in, 257–258, 258b, 258f, 755
applied, 20t, 744
food microbiology, 13t, 19f, 20t, 150f, 744, 744–752, 748t, 749t, 752t
industrial microbiology/biotechnology, 11, 13t, 19f, 20t, 744, 752–763
aquatic, 770–772, 771f, 773t
bioterrorism and, 22, 22f, 772–776. See also Bioterrorism
chemistry of, 26–55
ecological, 19f, 763–765, 765f
environmental, 19f, 20t, 21, 744, 763–772
fields of, 20t. See also specific type
history of, 1–25
early years, 2–7
future, 23
Golden Age, 7–19, 19f
modern age, 19–22, 20t
hospital, 19f
industrial (biotechnology), 11, 13t, 19f, 20t, 752–763
recombinant DNA technology in, 242, 243f
mathematical considerations in, B-1
medical, 20t
pharmaceutical, 19f, 20t, 755, 756t
public health, 20t
soil, 768–770, 768f, 770t
Microbiota, normal. See Normal microbiota
Microcephaly, maternal toxoplasmosis causing, 662
Microfilaments, in eukaryotic cytoskeleton, 81, 82f
Microfilariae, in Wuchereria bancrofti infection, 671, 672, 672f
Microglia/microglial cells, 444
HIV infection of, 721, 721f
in reticuloendothelial system, 446f
Micrographs, 98
Microhabitats, 764, 765f

Microhemagglutination assay against T. pallidum (MHA-TP), 609
Microinjection, for DNA insertion into cells, 251f, 252, 255t
Micrometer (μm), 93, 93t
Micromonospora purpurea, antimicrobials produced by, 289t, 306t
Micronucleus, in protozoa, 349
reproduction and, 350, 350f
Microorganisms. See Microbes/microorganisms
Micropipettes
for DNA insertion into cells, 251f, 252, 255t
for microbial isolation, 181
Microscopy counts, for estimating microbial growth, 192, 192f
Microscopy/microscopes, 93–105, 107t. See also specific type
bright-field, 95, 96–98, 97f, 98f, 99f, 107t
confocal, 95, 101, 102f, 107t
dark-field, 95, 98, 99f, 107t
definition of, 93
electron, 101–104, 102b, 103f, 104f, 105f, 107t
fluorescent, 95, 100–101, 100f, 101f, 107t
Leeuwenhoek's development of, 2–3, 3f, 93–94, 95, 96
light, 95–101, 97f, 98f, 99f, 107t
phase, 98–100, 99f, 100f, 107t
principles of, 94–95, 94f, 95f, 96f
probe, 104–105, 106f, 107t
Microsporidia, 354, 356f
Microsporidium genus/spp., 354
Microsporum genus/spp., 639, 640t
Microtubules
triplets of, in centrioles, 81, 82f
tubulin, in eukaryotic cytoskeleton, 81, 82f
MIC test. See Minimum inhibitory concentration (MIC) test
Milk
Campylobacter jejuni contamination of, 616, 617b, 752t
fermentation of, 745–746, 745f, 748t
pasteurization of, 271, 271t, 432–433
safety of, 432–433
Milk curds, in cheese production, 745–746, 745f
Millimeter (mm), 93, 93t
Mimicry, molecular, 524
Mineral soil (subsoil), 768, 768f
Minimum bactericidal concentration (MBC) test, 299–300, 300f
Minimum growth temperature, 173
Minimum inhibitory concentration (MIC) test, 298–299, 299f
Miracidia, in fluke life cycle, 668f
Mismatch repair, 225, 226f
Miso, fermentation in production of, 748t
Missense mutations, 222, 222f, 223t
Mites (chiggers), as disease vectors, 424t, 601b, 673, 674f
for scrub typhus (Orienta tsutsugamushi), 600, 602t
Mitochondria, 60f, 84–85, 84f, 86f, 87t
endosymbiotic theory and, 85–86
in protozoa, 349
Mitochondrial DNA, 85
Mitochondrial membranes, 84–85, 86f
electron transport chains in, 144, 145f
Mitochondrial ribosomes, 85, 86f
Mitosis, 83, 187, 204, 344, 345f, 347t
centrioles in, 81
chromatin fibers during, 204
mm. See Millimeter
MMR. See Measles/mumps/rubella (MMR) vaccine
MMWR. See Morbidity and Mortality Weekly Report
Mn. See Manganese

Mo. See Molybdenum
Modified live vaccines (attenuated/live vaccines), 489–490
Modified Thayer-Martin medium, for Neisseria culture, 567, 569
Moist heat, in microbial control, 269–271, 269f, 270f, 271f, 271t, 272b, 277t
Moisture, in soil, microbial survival and, 769
Molds, 4, 4f. See also Fungi
adverse reactions to, 644, 645b
coenocytic, zygomycetes, 359–360, 361f, 366t
slime, 369, 370–371
cellular, 371, 372f
plasmodial (acellular), 370–371, 372f
thalli of, 356, 357f
water, 369, 370, 370f
Molecular biology, 21
Molecular formula, 32
Molecular mimicry, 524
Molecule, chemical bonds in formation of, 29–30
Mollicutes, 329t, 596
Molluscipoxvirus, 682, 683f
Molluscum contagiosum, 682, 683f
Molybdenum, 28t
for microbial growth, 170t
Monarch butterflies, genetically modified corn affecting, 258b
Monera (Prokaryotae) (kingdom), 116, 116f
Monkeypox, 680, 683
Monobactams, 305t
mechanism of action of, 291f, 292, 305t
Monoclonal antibodies
immunosuppressive actions of, 523
in passive immunization, 495, 495f
Monocular microscope, 97
Monocytes, 444, 445f
HIV infection of, 721, 721f
Monod, Jacques, 218
Monoecious helminths, 663
Monomers, 40
Mononuclear phagocytic system, 446f, 447
Mononucleosis, infectious (mono)
cytomegalovirus causing, 692
Epstein-Barr virus causing, 690, 690f
Monosaccharides, 44, 44f
Monotrichous flagellum., 63, 64f
Monounsaturated fatty acids, 42t
Moraxella genus/spp., 588
Moraxella catarrhalis, 588
Morbidity, 412. See also Disease
Morbidity and Mortality Weekly Report (MMWR), 426, 428f
Morbillivirus genus/spp. (measles virus), 725–727
Morchella esculenta, 361f
Mordants, 109
in flagellar staining, 112
Morganella genus/spp., 577
sites of infection caused by, 582f
Morula, Ehrlichia forming, 601, 601f
Mosquitoes, as disease vectors, 424t, 674f, 675
for arboviruses, 710, 711t, 712f
dengue fever and, 711, 711t
encephalitis and, 711, 711t
West Nile virus and, 711, 711t, 715b
for filarial nematodes, 670, 671, 672
for malaria (Plasmodium species), 659, 659f, 661
Most probable number (MPN), for estimating microbial growth, 192–193, 193f, 195t
in water quality testing, 758
Motility, eukaryotic versus prokaryotic, 87t
MPN. See Most probable number
M protein, 418, 449
mRNA. See Messenger RNA
MRSA. See Methicillin-resistant Staphylococcus aureus

MS. *See* Multiple sclerosis
*Msp*I restriction enzyme, 245*t*
Mu chains, 468
Mucocutaneous candidiasis, 634*t*
Mucocutaneous leishmaniasis, 656, 657*f*
Mucor genus/spp., 636–637
Mucosa-associated lymphatic tissue (MALT), 465–466, 466*f*
Mucous membranes
    in host defense, 437, 439–440, 439*f*, 441*f*, 442*t*, 459*t*
    as portals of entry, 409–410, 409*f*
    specimen collection from, 179*t*
Mucus, in host defense, 437, 459*t*
MUG, in water quality testing, 758
Muller, Hermann, 223
Multidrug resistant (MDR) *Mycobacterium tuberculosis*, 558–559
Multiple resistance, 303
Multiple sclerosis (MS)
    autoimmunity in, 525
    HHH-6 and, 692
Mumps, 727–728, 727*f*, 728*f*
    vaccine against, 728, 728*f*. *See also* Measles/mumps/rubella (MMR) vaccine
Mumps virus, 727–728
Municipal wastewater treatment/municipal sewer systems, 759, 760*f*
Murine (endemic) typhus, 600, 602*t*
Musca genus/spp., as disease vectors, 424*t*
Muscimol, 644
Muscle contraction, in tetanus, 551, 552*f*
Mushroom grower's lung, 519
Mushroom poisoning, 363, 364*f*, 643–644, 644, 644*f*
Mushrooms, honey (*Armillaria ostoyae*), 342
Mutagens, 223–224, 223*f*, 224*f*, 225*f*
    chemical, 223–224, 224*f*
    frameshift, 224, 225*f*
    identification of, 226–228, 227*f*, 228*f*, 229*f*
    radiation as, 223, 223*f*
    in recombinant DNA technology, 242, 255*t*
Mutants
    definition of, 226
    identification of, 226–228, 227*f*, 228*f*
Mutations, 221–228, 229*f*
    agents (mutagens) causing, 223–224, 223*f*, 224*f*
    definition of, 221
    DNA repair and, 224–226, 226*f*
    effects of, 222, 222*f*, 223*t*
    frameshift, 221, 222*f*
    frequency of, 224
    gross, 222
    identification of
        mutants/mutagens/carcinogens and, 226–228, 227*f*, 228*f*, 229*f*
    point, 221–222, 222*f*, 223*t*
    spontaneous, 223
Mutualism, 405, 405*f*, 406*b*, 406*t*
Mycelium/mycelia, 356, 358*f*
Mycetismus, 643–644, 644*f*
Mycetomas
    fungal, 641–642, 642*f*
    Nocardia causing, 562, 562*f*
Mycetozoa (kingdom), 348, 348*f*
Mycobacterium genus/spp./mycobacteria, 320*f*, 328–329, 329*t*, 555–560
    acid-fast staining of, 110–111, 112*f*
    antimicrobial agents affecting, 292, 298*f*
    mycolic acid in cell walls of, 67, 158, 329, 555
    susceptibility of, 266, 266*f*
Mycobacterium avium-intracellulare, 560

Mycobacterium bovis
    acid-fast staining of, 112*f*
    BCG vaccine using, 559
    pasteurization in control of, 271
Mycobacterium leprae, 329, 559–560
    animal culture for growth of, 185
    clofazimine affecting, 295
Mycobacterium tuberculosis, 329, 555–559, 556*f*. *See also* Tuberculosis
    auramine O dye staining, 100, 100*f*
    direct fluorescent antibody test in detection of, 502
    discovery of, 15, 17*t*
    resistant strains of, 558–559
    rifampin affecting, 295
    susceptibility of, 266
    tuberculin test in detection of, 520, 520*f*, 558
Mycolic acid, in *Mycobacterium* cell walls, 67, 158, 329, 555
Mycology/mycologists, 20*t*, 355. *See also* Fungi
    medical, 624–648. *See also* Mycoses
        antifungal therapies and, 627
        categories of fungal agents and (pathogenic/opportunistic), 624–626, 625*f*, 625*t*, 626*f*
        clinical manifestations of fungal disease and, 626–627
        diagnosis of fungal disease and, 627
        epidemiology of mycoses and, 624
Mycophenolate mofetil, 523
Mycoplasma genus/spp., 329*t*, 596–597
Mycoplasma genitalium, 597–598
Mycoplasma hominis, 597–598
Mycoplasma pneumoniae, 596, 597, 597*f*
    absence of cell wall and, 68
Mycoplasmas, 320*f*, 326–327, 327*f*, 533, 596–598, 596*f*, 597*f*, 606*t*
    comparison of with other small microbes, 606*t*
Mycorrhizae, 355
Mycoses, 355
    clinical manifestations of, 626–627
    cutaneous/subcutaneous, 640–643, D-2
    diagnosis of, 627
    epidemiology of, 624
    infections, 626
    opportunistic, 631–637
        aspergillosis, 632–633, 632*f*
        candidiasis, 633–635, 633*f*
        in HIV infection/AIDS, 718*t*
        Pneumocystis pneumonia, 636, 636*f*
        predisposing factors and, 625–626, 625*t*
    pathogenic, 628–631
        blastomycosis, 628–629, 628*f*
        coccidioidomycosis, 629–630, 629*f*, 630*f*
        cryptococcosis, 635
        histoplasmosis, 630
        paracoccidioidomycosis, 631, 631*f*
        zygomycoses, 636–637
    superficial, 638–640
    systemic
        opportunistic, 631–637
        pathogenic, 628–631
    treatment of, 627
Mycotoxicosis, 643
Mycotoxins, 643
Myelin sheaths, autoimmune destruction of, in multiple sclerosis, 525
Myelomas, in hybridoma production, 495, 495*f*
Myocardial candidiasis, 634*t*
Myocarditis, coxsackie B virus causing, 707
Myonecrosis (gas gangrene), 548, 548*f*
    discovery of cause of, 17*t*

Myxamoebae
    of acellular slime molds, 371, 372*f*
    of cellular slime molds, 371, 372*f*
Myxobacteria, 320*f*, 334*t*, 336, 338*f*
Myxospores, 336, 338*f*

N. *See* Nitrogen
NA. *See* Neuraminidase; Sodium
N-acetylglucosamine (NAG), 66, 67*f*, 68*f*
N-acetylmuramic acid (NAM), 66, 67*f*, 68*f*
    antimicrobial agents affecting, 290–292, 291*f*
NAD⁺ (nicotinamide adenine dinucleotide), 127, 130, 131*t*, 162*f*
    in fermentation, 148*f*, 149
    in Krebs cycle, 142, 143*f*, 162*f*
NADH, 127, 162*f*
    acetyl-CoA synthesis producing, 142, 142*f*, 149*t*
    in beta-oxidation, 151, 151*f*
    in chemiosmosis, 147–148
    in electron transport chain, 144*f*, 145*f*, 146
    glycolysis producing, 138*f*, 139, 149*t*
    in Krebs cycle, 142, 143*f*, 162*f*
    for microbial growth, 173*t*
    pentose phosphate pathway producing, 140*f*, 141
NADP⁺ (nicotinamide adenine dinucleotide phosphate), 127, 130, 131*t*
    pentose phosphate pathway producing, 140*f*, 141
    in photophosphorylation, 154, 155*f*
NADPH
    in Calvin-Benson cycle, 156, 157*f*, 162*f*
    Entner-Doudoroff pathway in production of, 141, 141*f*
    in noncyclic photophosphorylation, 154, 155*f*
NADPH oxidase, defective, in primary immunodeficiency, 527
Naegleria genus/spp., 353, 652–653, 664*t*
NAG. *See* *N*-acetylglucosamine
Nail infections, dermatophytes causing, 638–639, 640*t*
Naked (nonenveloped) viruses/virions, 382
    positive ssRNA, 704–709, 738*t*
    release of, 394
    susceptibility of, 266, 266*f*
Nalidixic acid, 307*t*
NAM. *See* *N*-acetylmuramic acid
Nanometer (nm), 93, 93*t*
Narrow-spectrum drugs, 297, 298*f*
Nasopharyngeal cancer, Epstein-Barr virus and, 690, 690*f*
Natamycin, 283
Natrium. *See* Sodium
Natural bioremediation, 761
Natural killer lymphocytes (NK cells), 450, 459*t*
Naturally acquired immunity
    active, 482, 483*t*
    passive, 483, 483*t*
Necator americanus, 670–671, 673*t*
Necrotizing fasciitis, 540, 541*f*
Needham, Tuberville, experiments in spontaneous generation by, 8
Neelsen, Friedrich, 111
Negative feedback (feedback/end-product inhibition), 135–136, 136*f*
Negative (indirect) selection, in mutant identification, 227–228, 228*f*
Negative (capsule) stains, 112, 112*f*, 114*t*
Negative strand RNA, 392
Negative strand RNA viruses
    segmented, 731–737, 738*t*
    synthesis of, 392, 392*f*
    unsegmented, 724–731, 738*t*
Negri bodies, in rabies, 729, 730*f*

Neisser, Albert, 17*t*
Neisseria genus/spp./Neisserias, , 320*f*, 333, 334*t*, 567–570, 573*b*
structure and physiology of, 567, 567*f*
Neisseria gonorrhoeae (gonococcus), 568–569. *See also* Gonorrhea
adhesins produced by, 411
culturing, 186, 569
discovery of, 17*t*
fimbriae on, 65, 568
indirect fluorescent antibody test in detection of, 502, 503*f*
resistant strains of, 569
Neisseria meningitidis (meningococcus), 569–570
discovery of, 17*t*
vaccine against, 570
Nematodes (roundworms), 663, 669–672, 669*f*, 673*t*
Ascaris, 670, 673*t*
filarial (*Wuchereria bancrofti*), 670, 671–672, 672*f*, 673*t*
hookworms (*Ancylostoma* and *Necator*), 670–671, 670*f*, 673*t*
pinworms (*Enterobius*), 671, 671*f*, 673*t*
Neon, Bohr diagram of, 30*f*
Neoplasia/neoplasms, 395, 412, 415*t*
Nephrons, streptococcal infection affecting, 540
Nervous system/nervous tissue
autoimmunity affecting, 525
in host defense, 443*t*
microbial disease agents affecting, D-4 to D-5
Neuraminidase
influenza virus, 732, 733*f*
Pseudomonas aeruginosa fimbriae affected by, 587
Neuraminidase inhibitors, for influenza, 735
Neurologic disease, organisms causing, D-4 to D-5
Neuromuscular junction, botulism toxins affecting, 549, 549*f*
Neurospora genus/spp., 355, 362
Neurospora crassa, microbial genetics research and, 20
Neurotoxins, 416
clostridial, 547, 752*t*
dinoflagellates producing, 352
tetanus toxin, 551, 552*f*
Neutralization, antibody, 469, 469*f*, 500
Neutralization tests, 500, 506*t*
Neutrons, 27, 27*f*
atomic mass of, 27
Neutrophiles, pH range tolerated by, 174
Neutrophils, 59*f*, 444, 445*f*
Nevinic acids, 355
Newborn
candidiasis in, 634, 634*t*
gonococcal infection in, ophthalmia neonatorum caused by, 569
group B streptococcal infection in, 541
prevention of, 542
hemolytic disease of, 517–518, 517*f*
hepatitis in, 698
herpes simplex infection in, 685
New variant Creutzfeldt-Jakob disease, 48, 399
NGU. *See* Nongonococcal urethritis
NH_3. *See* Ammonia
NH_4^+. *See* Ammonium ion
Niacin (nicotinic acid), for microbial growth, 173*t*
Nickel, for microbial growth, 170*t*
Niclosamide, 309*t*
spectrum of action of, 293*f*, 309*t*
Nicotinamide adenine dinucleotide (NAD^+), 127, 130, 131*t*, 162*f*
in fermentation, 148*f*, 149
in Krebs cycle, 142, 143*f*, 162*f*

Nicotinamide adenine dinucleotide phosphate ($NADP^+$), 127, 130, 131*t*
in pentose phosphate pathway, 140*f*, 141
Nicotinic acid. *See* Niacin
van Niel, C.B., 19
Nightingale, Florence, 18, 18*f*, 19*f*
Nipah virus, 726*b*
as biological weapon, 774*t*
Nisin, 283
Nitrification, 331, 767, 767*f*
Nitrifying bacteria, 320*f*, 331
Nitrobacter genus/spp., 331, 334*t*
in nitrification, 767
Nitrogen, 28*t*
Bohr diagram of, 30*f*
for microbial growth, 170*t*, 171–172
Nitrogenase, in nitrogen fixation, 767
Nitrogen cycle, 766–767, 767*f*
Nitrogen fixation, 172, 767, 767*f*
by alphaproteobacteria, 331, 331*f*
Nitrogenous wastes, 151
Nitroimidazoles, 310*t*
Nitrosomonas genus/spp., 333, 334*t*
in nitrification, 767
Nitrous acid, as mutagen, 224
NK cells. *See* Natural killer lymphocytes
nm. *See* Nanometer
Nocardia genus/spp., 320*f*, 329*t*, 330, 562, 562*f*
acid-fast staining of, 110–111
Nocardia asteroides, 562
Nomarski (differential interference contrast) microscopy/microscopes, 92, 98–100, 99*f*, 107*t*
Nomenclature, 113. *See also* Taxonomy/microbial classification
binomial, 114–116
Noncoliform Enterobacteriaceae, 576–577
nosocomial infection caused by, 571*f*
Noncommunicable diseases, 424, 425*t*
Noncompetitive inhibition/inhibitors, 135, 135*f*
Noncyclic photophosphorylation, 154–155, 155*f*
Nonenveloped viruses/virions, 382
release of, 394
susceptibility of, 266, 266*f*
Nongonococcal urethritis (NGU)
Chlamydia trachomatis causing, 604
mycoplasmas causing, 598
Noninvasive aspergillomas, 632, 632*f*
Nonionizing radiation. *See also* Radiation
in microbial control, 276–277, 277*t*
as mutagen, 223, 223*f*
Nonliving reservoirs, 420–421
Nonmembranous organelles, 87*t*
in eukaryotic cells, 80–81, 82*f*, 87*t*
in prokaryotic cells, 75–76, 77*f*, 87*t*
Nonmunicipal wastewater treatment, 759–760, 761*f*
Nonparalytic polio, 706
Nonperishable food, 750
Nonpolar covalent bonds, 30–32, 31*f*, 32*f*, 35*t*
Nonsense mutations, 222, 222*f*, 223*t*
Nonspecific lines of defense, 437–461. *See also* Host defenses, nonspecific
Nonsulfur bacteria
green, 320*f*, 324, 325*t*
purple, 320*f*, 324, 325*t*, 331, 334*t*
Norfloxacin, 307*t*
Normal microbiota (normal flora), 406–409, 407*f*, 408*t*
antimicrobial drugs affecting, 301
coliforms, 574–576
in cultures, 178

disease caused by, 407–409. *See also* Opportunistic pathogens
in host defense, 440–441
noncoliform Enterobacteriaceae, 576–577
Noroviruses, 709
Northern blot, 249
Norwalk virus, water-borne transmission of, 773*t*
Nosema genus/spp., 354
Nosepiece, revolving nosepiece, of light microscope, 96–97
Nosocomial diseases. *See also* Nosocomial infections
cost of, 430
Nosocomial infections, 16, 430–432, 432*f*
causes of, 571*f*
control of, 431–432
factors affecting, 431, 432*f*
Pseudomonas aeruginosa causing, 588
Nuclear division, 343–346, 345*f*
Nuclear envelope, 60*f*, 83, 83*f*, 203
Nuclear pores, 60*f*, 83, 83*f*
Nucleic acid hybridization, 120
Nucleic acid probes, synthetic nucleic acids in synthesis of, 244, 255*t*
Nucleic acids, 50–53, 50*f*, 51*f*, 52*f*, 52*t*, 204*t*. *See also* DNA; Nucleotides; RNA
antimicrobial agents affecting, 265
antimicrobial drugs affecting, 295, 296*f*, 297*b*, 307*t*, 308*t*
autoantibodies against, in systemic lupus erythematosus, 525
function of, 52
structure of, 51–52, 51*f*, 200
synthetic, in recombinant DNA technology, 243–244, 255*t*
Nucleic acid sequencing, 21, 253–254, 253*f*, 255*t*
detection of infectious agents and, 254*b*
microbial classification and, 21, 117–118, 117*b*, 120–121, 318–319, 320*f*, 323
Nucleocapsid, 378, 378*f*, 380
Nucleoid, in prokaryotic cells, 60*f*, 76, 201, 202*f*
Nucleoli/nucleolus, 81–83, 83*f*
eukaryotic, 60*f*
Nucleoplasm, 81, 83*f*
Nucleoproteins, 50
shape–function relationship and, 50
Nucleosomes, 203*f*, 204
Nucleotide analogs, 296*f*, 308*t*
clinical use of, 223
mechanism of action of, 295, 296*f*, 308*t*
as mutagens, 223, 224*f*
Nucleotide bases, 50–51, 50*f*
in DNA, 50–51, 51, 51*f*, 52, 200, 201*f*
in RNA, 50–51, 51, 51*f*, 200, 201*f*
Nucleotides, 50–51, 50*f*
biosynthesis of, 160, 161*f*, 162*f*
chemicals altering, mutagenic effects of, 224
Nucleotide sequencing, 253–254, 253*f*, 255*t*
detection of infectious agents and, 254*b*
Nucleus. *See also under* Nuclear
atomic, 27, 27*f*
cell, 81–83, 83*f*, 87*t*, 203
Numerical aperture of lens, resolution affected by, 95
immersion oil and, 97, 98*f*
Nursing, Florence Nightingale and, 18
Nutrient agar, 181
Nutrient broth, 181, 182
Nutrients
food spoilage and, 749, 749*t*
limiting, 169
for metabolism, 126. *See also* Cytoplasmic (cell/plasma) membranes, movement across

Nutrients, *Cont.*
  for microbial growth, 168–172, 168*f*, 170*t*, 173*t*
  in soil, microbial diversity and, 769
Nutrition, microbial, 167–198. *See also* Microbial growth
Nutritional diseases, 412, 415*t*
Nuttall, George, 17*t*
Nystatin, 309*t*

O₂. *See* Oxygen
O₃. *See* Ozone
O antigen, 572, 574*f*
  of *E. coli*, 575
Objective lens, of light microscope, 96–97, 97*f*
Obligate acidophiles, pH range tolerated by, 174
Obligate aerobes, 169, 171
Obligate anaerobes, 169, 171
  culturing, 183–185, 185*f*
Obligate halophiles, 176
Ocular candidiasis, 633*f*, 634*t*
Ocular/ophthalmic herpes, 684*f*, 685, 686*f*
Ocular histoplasmosis, 630
Ocular infections, *Chlamydia trachomatis* causing, 603, 604, 604*f*
Ocular lenses, of light microscope, 97, 97*f*
OH. *See* Hydroxyl radical
Oil immersion objective lens, of light microscope, 97, 98*f*
Oil spills, bioremediation in clean up of, 761, 763*b*
Okazaki fragments, 206*f*, 207
Oleic acid, 42*t*
Oligoadenylate synthetase, 453
-oma (suffix), definition of, 413*t*
Oncogenes, 395, 395*f*
ONPG (*O*-nitrophenyl-β-D-galactopyranoside), in water quality testing, 758
Onychomycosis
  Candida species causing, 633*f*
  dermatophytes causing (tinea unguium), 640*t*
Onyx-015, cancer cells destroyed by, 698*b*
Oocysts
  in *Cryptosporidium* life cycle, 662
  in *Cyclospora* life cycle, 663
  in *Plasmodium* life cycle, 559*f*, 660
  in *Toxoplasma* life cycle, 661–662, 662*f*
Ookinetes, in *Plasmodium* life cycle, 660
Operator, in operon, 219, 219*f*
  lac, 219*f*, 220
Operons/operon model, 218–221, 219*f*, 220*f*, 221*t*
Ophthalmia neonatorum, gonococcal, 569
Ophthalmic/ocular herpes, 684*f*, 685, 686*f*
Opportunistic mycoses, 631–637
  aspergillosis, 632–633, 632*f*
  candidiasis, 633–635, 633*f*
  cryptococcosis, 635
  in HIV infection/AIDS, 718*t*
  Pneumocystis pneumonia, 636, 636*f*
  predisposing factors and, 625–626, 625*t*
  zygomycoses, 636–637
Opportunistic pathogens (opportunists), 407–409. *See also specific organism or disease*
  Cryptococcus species, 635
  enteric
    coliform, 574–576
    noncoliform, 576–577
    nosocomial infections caused by, 571*f*
  fungi, 624–626, 625*t*, 631–637
    in HIV infection/AIDS, 718*t*
  Pseudomonas aeruginosa, 587–588, 588*f*

Opsonins, 449, 469, 469*f*
Opsonization, 449, 469, 469*f*
Optical density, light refraction and image magnification and, 94–95, 95*f*
Optimum growth temperature, 174
OPV. *See* Oral polio vaccine
Oral drug administration, 300, 300*f*
Oral hairy leukoplakia, Epstein-Barr virus and, 690, 690*f*
Oral herpes (fever blisters/cold sores), 684–685, 684*f*, 685*f*
Oral polio vaccine (OPV/Sabin vaccine), 490, 494, 707
  safety of, 490, 494, 707
Orders, 114, 115*f*, C-1 to C-9
Orf, 680
Organelles, 58, 87*t*. *See also specific type*
  membranous, 81–85, 87*t*
  nonmembranous, 87*t*
    in eukaryotic cells, 80–81, 82*f*, 87*t*
    in prokaryotic cells, 75–76, 77*f*, 87*t*
Organic compounds, 32, 39–53. *See also specific type*
  acids, 37
  ATP, 52–53, 52*f*
  carbohydrates, 42–45, 44*f*, 45*f*, 46*f*, 47*b*
  functional groups of, 39–40, 40*t*
  lipids, 40–42, 41*f*, 42*t*, 43*f*, 44*f*
  for microbial growth, 168*f*
  monomers, 40
  nucleic acids, 50–53, 50*f*, 51*f*, 52*f*, 52*t*
  proteins, 45–50, 47*f*, 48*f*, 49*f*
Organic phosphate group, 40*t*
Organotrophs, 169
Orienta genus/spp., 600
Orienta tsutsugamushi, 600, 602*t*
Origin, in DNA replication, 205
Ornithodoros ticks, as disease vectors, for epidemic relapsing fever (*Borrelia recurrentis*), 613
Ornithosis, 605–606, 606*f*
Oropharyngeal candidiasis (thrush), 633*f*, 634, 634*t*, 635
  antimicrobial drug use and, 301
Orphan viruses
  HHV-7 as, 684
  reoviruses and, 737
Orthohepadnavirus genus/spp., 696
Orthomyxoviruses (*Orthomyxoviridae*), 385*t*, 731–734, 738*t*
Orthophenylphenol, 278, 278*f*
Orthopoxvirus genus/spp., 682. *See also* Poxviruses
Oscillatoria genus/spp., 324
Oseltamivir, for influenza, 735
-osis (suffix), definition of, 413*t*
Osmosis, 72–73, 72*f*, 73*f*, 76*t*, 176
Osmotic pressure, 176
  microbial growth affected by, 176, 274–275, 277*t*
    antimicrobial drug effect on cell walls and, 290–292, 291*f*
Osteoarticular blastomycosis, 628
Osteoarticular candidiasis, 634*t*
Osteomyelitis, staphylococcal, 536
Otitis media, pneumococcal, 543
Ouchterlony test (double immunodiffusion), 496–497, 497*f*
Owl's eye cell, in cytomegalovirus infection, 692, 692*f*
Oxaloacetate/oxaloacetic acid, 159, 159*f*, 160*t*
  in amination and transamination, 159, 159*f*
  in Krebs cycle, 142, 143*f*, 160*t*

Oxidase
  in *Neisseria*, 567, 567*f*
  in pasteurellae, 570, 571*f*
  in peroxisomes, 84
Oxidase test, 570, 571*f*
Oxidation, 127, 127*f*
  beta, of fatty acids, 151, 151*f*
Oxidation lagoons, in wastewater treatment, 760–761
Oxidation-reduction (redox) reactions, 127, 127*f*, 128*b*
Oxidative phosphorylation, 147, 156*t*
  chemiosmosis and, 147–148
Oxidizing agents, antimicrobial action of, 279–280, 282*t*
Oxidoreductases, 129, 130*t*
Oxygen, 28*t*
  atomic number of, 27, 28*t*
  Bohr diagram of, 30*f*
  covalent bonds in formation of, 31*f*
  as final electron acceptor, 147
  forms of, 170–171
  microbial cultures and, 185–186, 186*f*
  for microbial growth, 169–171, 170*t*, 171*f*, 172*f*
  singlet, 170
  in soil, microbial survival and, 769
Oxygen demand, biochemical (BOD), reduction of in wastewater treatment, 759
Oxygenic organisms, noncyclic photophosphorylation in, 154
Oxygen radicals, 170–171
Ozone, antimicrobial action of, 279, 280, 282*t*

P. *See* Phosphorus
PABA. *See* Para-aminobenzoic acid
Pain, in inflammation, 455
Palindromes/palindromic sequences
  in restriction enzymes, 244, 245*f*
  in transposons, 234–235
Palisades, 316, 317*f*
  in *Corynebacterium*, 554
Palmitic acid, 42*t*
Pandemic, 426, 427*f*
Panencephalitis, subacute sclerosing, 726
Papanicolaou (Pap) smear, 694
Papillomas, 693–695, 694*f*, 695*b*, 695*f*
Papillomaviruses (*Papillomaviridae*), 385*t*, 693–695, 694*f*, 695*b*, 695*f*
Papules, in poxvirus infection, 680, 681*f*
Para-aminobenzoic acid (PABA)
  for microbial growth, 173*t*
  structural analogs of, antimicrobial action of, 294–295, 294*f*
Parabasala (kingdom), 348, 348*f*, 354, 355*f*, 356*t*
Parabasal body, 354
Parabasalids, 354, 356*t*
Paracoccidioides brasiliensis, 624, 631, 632*f*
  geographic distribution of, 626*f*, 631
Paracoccidioidomycosis, 631, 631*f*
Parainfluenza virus, 727
Paralysis, flaccid, botulism toxins causing, 549
Paralytic polio, 706, 707*f*
Paralytic shellfish poisoning, 773*t*
Paramecium genus/spp., 59*f*
  morphology of, 349, 349*f*
Paramylon, of euglenoids, 353
Paramyxoviruses (*Paramyxoviridae*), 385*t*, 724–725, 725–728, 725*f*, 738*t*
  as biological weapons, 774*t*
Paranasal sinus phaeohyphomycosis, 641
Pararosanaline, in flagellar staining, 112
Parasitemia, in African sleeping sickness (*Trypanosoma brucei*), 653

Parasites, 405–406
    acellular obligatory, viruses as, 6
    cardiovascular disease and, D-6
    energy, chlamydias as, 602
    gastrointestinal disease caused by, D-10
    hosts of, 406, 650
    lymphatic system infections caused
        by, D-6
    nervous system infections caused by, D-4
    protozoan, 649, 650–663, 664f
    skin/eye infections caused by, D-2
    transmission of to humans, 650, 651f
    urinary tract/reproductive system infections
        caused by, D-11
    worms (parasitic helminths). See Helminths
Parasitism, 405–406, 406t
Parasitology, 20t, 649. See also Parasites
Parenteral route, as portal of entry, 410
Parotitis, in mumps, 727, 727f
Paroxysmal phase, of pertussis, 586, 587f
Parrot fever (ornithosis), 605–606, 606f
Parvovirus B19, 700, 700f
Parvoviruses (Parvoviridae), 385t, 682t,
        699–700, 700f
Passive immunity
    artificially acquired/artificial immunization,
        483, 483t, 494–496, 495f
    naturally acquired, 483, 483t
Passive transport processes, in prokaryotic
        cells, 71–73, 72f, 73f, 74f, 76t
Pasteur, Louis, 9, 9f, 19f, 270, 488
    experiments in fermentation by, 11, 12f
    experiments in spontaneous generation by,
        9, 10f
Pasteurella genus/spp., 582
Pasteurella multocida, 582
Pasteurellaceae, 336t, 582–583
    oxidase test in detection of, 570, 571f
Pasteurization, 11, 19f, 263, 264t, 270–271, 272b,
        277t, 750
patho (prefix/suffix), definition of, 413t
Pathogenicity, 415. See also Virulence
Pathogens, 14, 21. See also specific type
    animal, in bioterrorism, 774–778
    bacterial, 5
    human, in bioterrorism, 774, 774t
    parasitic, 406t
    plant, in bioterrorism, 775
Pauling, Linus, 21, 32f
Pauling scale, 32f
PCP. See Pneumocystis pneumonia
PCR. See Polymerase chain reaction
PEAS. See Possible estuary-associated
        syndrome
Pectinase, microbial production of, 753
Pediculus genus/spp. (lice), as disease vectors,
        424t, 674, 674f
    for epidemic relapsing fever (Borrelia
        recurrentis), 613
    for epidemic typhus (Rickettsia prowazekii),
        599–600, 602t
    for trench fever (Bartonella quintana), 584
Pediculus humanus, as disease vector
    for epidemic relapsing fever (Borrelia
        recurrentis), 613
    for epidemic typhus (Rickettsia prowazekii),
        599–600, 602t
Peliosis hepatitis, bacillary, 584
Pellicle, estimations of microbial growth
        affected by, 194
Pelvic inflammatory disease (PID)
    Chlamydia trachomatis causing, 604
    gonococcal, 568
    Mycoplasma hominis causing, 598
Penciclovir, 296f

Penetration
    direct, in animal virus replication, 390,
        391f, 394t
    in HIV replication, 720f, 721
Penicillin, 291f, 305t
    allergic reaction to, 301
    beta-lactamase and, 221b, 302, 303f
    discovery of, 22, 288
    mechanism of action of, 290, 291f, 305t
    Penicillium producing, 4, 288, 289t,
        305t, 755
    spectrum of action of, 297, 298f, 305t
Penicillinase (beta-lactamase), 302, 303f, 534
    penicillin and, 221b, 302, 303f
Penicillium genus/spp.
    conidiospores of, 359f
    penicillin produced by, 4, 288, 289t,
        305t, 755
Penicillium chrysogenum, 4, 4f, 288, 288f, 289t
Penicillium griseofulvin, 289t, 309t
Penicillium marneffei, 637
Pentose phosphate pathway, 139–141, 140f,
        162f, A-4
Pentoses, 44
    in nucleotides, 50, 50f
Pentothenic acid, for microbial growth, 173t
PEP. See Phosphoenolpyruvic acid
Peptic ulcers, Helicobacter pylori infection and,
        617–618, 618f
Peptide bonds, 48, 49f
Peptidoglycan
    in bacterial cell wall, 5, 45, 66, 68f
        antimicrobial agents affecting, 290, 291f
    structure of, 66, 67f, 68f
Peracetic acid, antimicrobial action of, 279,
        280, 282t
Percentage of transmission/percentage of
        absorbance, spectrophotometer
        measuring, 193–194, 194f
Perforin-granzyme cytotoxic pathway,
        480–481, 481f
Perforins, 480, 481f
Pericardial infection
    in candidiasis, 634t
    coxsackie B virus causing, 707
Perinuclear space, 60f
Periodic table of elements, 30f
Periodontal infections, Prevotella causing, 592
Peripheral proteins, 70f, 71
Periplasm, 68
Periplasmic space, in Gram-negative cell walls,
        68, 69f
Perishable foods, 750
Peristalsis, in host defense, 443t
Peritonitis
    candidal, 634t
    in typhoid fever, 578
Peritrichous flagellum, 64, 64f
Permeability
    blood vessel, in inflammation, 455–456, 456f
    cell membrane, selective, 71
Permeases, in facilitated diffusion, 72
Peroxidase, 171
Peroxide anion, 171
Peroxides, antimicrobial action of, 279, 282t
Peroxisomes, 84, 87t
Perplaneta genus/spp., as disease vectors, 424t
Persistent infections, enveloped viruses
        causing, 393, 393f
Pertussis, 585–587, 586f, 587f
Pertussis toxin, 585
Pesticides, microbial production of, 13t,
        755, 756t
Pest resistance, recombinant DNA technology
        and, 257

Petechiae
    in meningococcal septicemia, 570, 570f
    in Rocky Mountain spotted fever, 599, 599f
Petri dish, 179, 181
Petri plates, 181
    in anaerobic cultures, 183, 185f
Petroff-Hauser counting chamber, for estimating
        microbial growth, 192, 192f
Peyer's patches, 466, 466f
Pfiesteria genus/spp., 343, 352
Pfiesteria toxin, 352
pH
    enzymatic activity affected by, 132–133, 134f
    of food, spoilage and, 749, 749t
    microbial growth affected by, 174–176
    microorganism tolerance of changes in,
        38–39, 39b
    soil, microbial population affected by, 769
Phaeohyphomycosis, 640–641
Phaeophyta (division), 369, 371t
Phages (bacteriophages), 377, 377f, 379
    bacterial pathogens affected by, 381b
    high-pressure, 387, 387b
    lysogenic, 388
    replication of, 394t
        lysogenic, 388–390, 389f
        lytic, 385, 385–388, 386f, 388f, 389f
    shapes of, 382, 383f
    in taxonomy/microbial classification,
        120, 121f
    temperate, 388
    in transduction, 230–231, 232f, 235t
Phage typing, in taxonomy/microbial
        classification, 120, 121f
Phagocytes, 448, 448f, 456–457, 459t. See also
        Macrophages; Phagocytosis
    migration of, in inflammation, 456–457
Phagocytosis, 80, 81t, 448–450, 448f, 449b,
        449f, 459t
    in animal virus replication, 390, 391f
    evasion of, virulence factors and, 417f, 418
    in host defense, 448–450, 448f, 449b,
        449f, 459t
Phagolysosomes, 448f, 449
Phagosome, 448f, 449
Phalloidin, 644
Pharmaceutical microbiology, 19f, 20t, 755, 756t
Pharyngitis
    gonococcal, 569
    herpetic, 685
    streptococcal
        group A streptococci causing, 539
        Streptococcus equisimilis causing, 542
Phase (light), contrast affected by, 95
Phase-contrast microscopy/microscopes, 98,
        99f, 107t
Phase microscopy/microscopes, 98–100,
        100f, 107t
    differential interference contrast (Nomarski),
        92, 98–100, 99f, 107t
    phase-contrast, 98, 99f, 107t
pH buffers, 38
Phenol/phenolics, 278, 278f, 282t
Phenol coefficient, 267
Phenotype, 208
phi29, pressure in capsid of, 387, 387b
Phialophora verrucosa, 640
Phlebotomus flies
    bartonellosis transmitted by, 584
    leishmaniasis transmitted by, 655, 655f, 674
Phosphate
    in ATP production (phosphorylation),
        127–128, 129f
    in nucleotides, 50, 50f
    in phosphorus cycle, 768

Phosphate bonds, ATP, energy release and, 52–53, 52f
Phosphate (organic) group, 40t
Phosphoenolpyruvic acid (PEP), 131, 160t
  in glycolysis, 138f, 139, 160t
3-Phosphoglyceric acid, 160t, 162f
  in Calvin-Benson cycle, 156, 157f
Phospholipid bilayers, 41, 43f, 70, 70f, 79
  cholesterol in, 42, 44f
Phospholipids, in membranes, 41, 43f, 70f, 79
Phosphoric acid, in nucleotides, 50, 50f
Phosphorus, 28t
  Bohr diagram of, 30f
  for microbial growth, 170t, 172
Phosphorus cycle, 768
Phosphorylation, 127–128, 129f, 156t. See also
    Photophosphorylation
  oxidative, 147, 156t
    chemiosmosis and, 147–148
  substrate-level, 138f, 139, 156t
    in Krebs cycle, 142, 143f
Phosphorylcholine, in Streptococcus pneumoniae
    cell wall, 543
Photic zone, algae in, 366
Photoautotrophs, 168f, 169
  in carbon cycle, 766, 766f
Photobacterium genus/spp., electron transport
    chain of, 146b
Photoblepharon palpebratus, 146b
Photoheterotrophs, 168f, 169
Photolyase, DNA, 225
Photophosphorylation, 154–156, 155f, 156t
  chemiosmosis and, 147
  cyclic, 154, 155f
  noncyclic, 154–155, 155f
Photosynthesis, 152–156
  by algae, 5, 6f
  in carbon cycle, 766, 766f
  chemicals and structures in, 153, 153f
  chloroplasts in, 85, 86f
  by cyanobacteria, 323
  light-dependent reactions in, 153, 154–156,
    154f, 155f, 156t
  light-independent reactions in, 153, 156, 157f
  pentose phosphate pathway and, 139, 140f
Photosynthetic lamellae, 85
  in phototrophic bacteria, 323
Photosynthetic organs/structures, 153, 153f. See
    also Photosynthesis
  of algae, 5, 6f
Photosynthetic pigments, accessory, 366
Photosystem I, 153, 154, 154f, 155f
Photosystem II, 153, 154, 155f
Photosystems, 153, 153f
Phototaxis, 64
Phototrophic bacteria/phototrophs, 169,
    323–325, 325t, 331
pH scale, 38, 38f
Phycoerythrin, 366, 367
Phycology/phycologists, 20t, 366
Phylogenetic hierarchy, 117
Phylum (phyla), 114, 115f, 318, C-1 to C-9
Phytophthora genus/spp., 370
Phytophthora cinnamomi, soil-borne disease
    caused by, 770t
Phytoplankton, 368–369
Pickles, fermentation in production of,
    744–745, 748t
Pickling, for food preservation/flavoring,
    744–745
Picornaviruses (Picornaviridae), 384, 385t,
    704–708, 738t
PID. See Pelvic inflammatory disease
Piedra, black and white, 638, 638f
Piedraia hortae, 638, 638f
Pigeon breeder's lung, 519

Pili, 66, 66f, 87t
  conjugation (sex), 66, 66f, 231–232, 233f
Pilin, 66
Pinkeye, 696, 697f
Pinocytosis, 80, 81t
Pinta, 609
Pinworm (Enterobius vermicularis), 671,
    671f, 673t
Piperazine, 309t
Pityriasis, 639, 639f
Placenta, as portal of entry, 409f, 410
Plague
  as bioterrorist threat, 774t
  bubonic, 421t, 566, 580–581, 581f
    discovery of cause of, 17t
  pneumonic, 580, 581, 581f
Plankton, 348
Plantae (kingdom), 115f, 116, 116f, 348, 348f
Plantar warts, 693, 694f
Plantomycetes (phylum), 320f
Plants
  culturing viruses in, 396
  diseases of
    in biological warfare, 773–774, 775
    soil-borne, 769, 770t
  Plant viruses, 379, 380f
Plaque (dental), as biofilm, 67b, 177, 411,
    412f, 542
Plaque assay, 396
Plaques, 396, 396f
  in phage typing, 120, 121f
Plasma, in host defense, 443
Plasma cells, 468, 477, 478f
Plasmalemma, 79
Plasma membrane. See Cell
    (cytoplasmic/plasma) membranes
Plasmids, 202, 202f, 204t
  F (fertility)/F factors, 202, 232, 233f, 234f
  R (resistance)/R factors, 202, 235, 236f, 302
  as vectors, 245–246, 246f, 255t
Plasmodial (acellular) slime molds,
    370–371, 372f
Plasmodium (slime mold thallus), 370–371, 372f
Plasmodium genus/spp., 343, 352, 659–661,
    659f, 661b, 664t
  discovery of, 17t
  nonmotile, 4
  reproduction in, 347, 659–660, 659f
  superbug strains of, 303
Plasmodium falciparum, 659, 660
  genome sequencing of, 253
Plasmodium malariae, 659, 660
Plasmodium ovale, 659, 660
Plasmodium vivax, 659, 660
Plasmolysis, 73, 176
Plastics, microbial production of, 754, 756t
Platelets, 443
  autoantibodies to, 524
  destruction of, in immune thrombocytopenic
    purpura, 518, 518f
Pleomorphic prokaryotes, 314, 314f. See also
    specific type
  mycoplasmas, 596, 596f
Pleurodynia, coxsackie B virus causing, 708
PML. See Progressive multifocal
    leukoencephalopathy
Pneumococcal meningitis, 543
Pneumococcal pneumonia, 543
Pneumococcal vaccine, 544
Pneumococcus. See Streptococcus pneumoniae
Pneumocystis genus/spp., 624, 636
Pneumocystis jiroveci (Pneumocystis carinii),
    636, 636f
Pneumocystis pneumonia, 636, 636f
Pneumocysts, of algal thalli, 366
Pneumolysin, 543

Pneumonia
  bacterial, manifestations of, 705t
  Chlamydia pneumoniae causing, 605
  cryptococcal, 635
  discovery of cause of, 17t
  Klebsiella pneumoniae causing, 575
  Mycoplasma pneumoniae causing (primary
    atypical pneumonia), 597
  Nocardia causing, 562
  pneumococcal, 543
  Pneumocystis, 636, 636f
  staphylococcal, 536
  viral, 705t
    manifestations of, 705t
Pneumonic plague, 580, 581, 581f
Pneumonitis, hypersensitivity, 519
Pneumovirus genus/spp., 728
PO₄²⁻. See Phosphate
Pocks (pustules), in poxvirus infection,
    680, 681f
Point mutations, 221–222, 222f, 223t
Point-source infections, 772
Poisonings. See Toxicoses
Poison ivy, 509, 520, 521f
Polar covalent bonds, 32, 33f, 35t
Polar flagella, 64
Poliomyelitis (polio), 706–707, 706f, 707f, 773t
Polio vaccine, 706, 707, 707t
  inactivated (IPV/Salk), 707, 707t
  oral (OPV/Sabin), 707, 707t
  safety of, 490, 494, 707
Poliovirus, 706
  as biological weapon, 776, 776b
  water-borne transmission of, 773t
Poliovirus vaccine. See Polio vaccine
Pollution, water, 756, 772. See also Water-borne
    illnesses
Polyenes, 309t
  mechanism of action of, 292, 309t
Polyhedral viruses, 382, 382f
Polymerase chain reaction (PCR), 120, 247–248,
    249f, 255t
  hyperthermophiles in, 175b, 321
Polymerases, 129, 130t
  DNA, in DNA replication, 205, 206f
  RNA
    antimicrobial drugs affecting, 295
    in transcription, 209, 210f
Polymerization, in metabolism, 126, 162f
Polymers, 40
Polymorphonuclear leukocytes. See
    Neutrophils
Polymyxa genus/spp., soil-borne disease
    caused by, 770t
Polymyxin, 307t
  mechanism of action of, 292–294, 307t
  spectrum of action of, 298f, 307t
  toxicity of, 301, 307t
Polyomaviruses (Polyomaviridae), 385t, 682t, 695
Polypeptides, 48
Polyribosome, 217, 218f
Polysaccharides, 45, 46f
Polyunsaturated fatty acids, 41, 42t
Pontiac fever, 590
Population, microbe, 764, 765f
  growth of, 57t, 167, 186–194, 195t. See also
    Microbial growth
  generation time and, 188, B-1
  mathematical considerations in, 187–188,
    188f, B-1
  measuring, 189–194
    direct methods for, 190–193, 190f, 191f,
      192f, 193f, 195t
    indirect methods for, 193–194, 194f
  phases of, 188–189, 189f
Pores, nuclear, 60f, 83, 83f

Porins
in Gram-negative cell walls, 68, 69f
in resistance, 302
Pork. *See* Meat/pork
Pork tapeworm (*Taenia solium*), 649, 665–666, 673t
Portals of entry, 409–410, 409f, 410f
Portals of exit, 419, 420f
Ports, in active transport, 73
Positive selection, in mutant identification, 227, 227f
Positive strand RNA, 392
Positive strand RNA viruses
enveloped, 709–716, 738t
with reverse transcriptase, 716–724, 738t
naked, 704–709, 738t
synthesis of, 392, 392t
Possible estuary-associated syndrome, 352
Postpartum fever, *Mycoplasma hominis* causing, 598
Postpolio syndrome, 706
Potable water, 432
microbes in production of, 756–758, 757f
Potassium, 28t
Bohr diagram of, 30f
for microbial growth, 170t
Potassium alum, in flagellar staining, 112
Potassium chloride, ionic bonds in formation of, 34
Poultry
assessing biological threats to, 773
Campylobacter jejuni contamination of, 616, 752t
salmonella contamination of, 577
Pour plates, 180, 180f
Pox (pustules), in poxvirus infection, 680, 681f
Poxviruses (*Poxviridae*), 385t, 680–683, 680f, 681f, 682t
Praziquantel, 309t
spectrum of action of, 293f, 309t
Precipitation tests, 496–499, 497f, 498f, 499f, 506t
Precursor metabolites, 126, 127, 156, 160t
for amino acid conversion, 159–160
Entner-Doudoroff pathway in production of, 141, 141f
pentose phosphate pathway in production of, 139–141, 140f
Pregnancy
antimicrobial drug use during, 301
ectopic, gonococcal PID and, 568
HIV infection/AIDS during, 723
Listeria infection during, 553
rubella during, 713
toxoplasmosis during, 662
Pre-messenger RNA, 213–214, 214f
Preservatives, food, 751
Pressure cooking, in microbial control, 269–270, 269f, 270f, 271f, 277t
Prevalence of disease, 425–426, 425f, 427f
Prevotella genus/spp., 591–592
Primaquine, 310t
Primary atypical pneumonia, *Mycoplasma pneumoniae* causing, 597
Primary humoral immune response, 479, 479f
Primary immunodeficiency diseases, 526–528, 527, 527b, 528t
Primary infection, 425t
Primary metabolites, of industrial fermentation, 753
Primary producers, in carbon cycle, 766, 766f
Primary stain
in Gram staining, 109, 109f
in Schaeffer-Fulton endospore staining, 111
in Ziehl-Neelsen acid-fast staining, 111

Primary structure of protein, 48, 49f
Primase
in lagging strand synthesis, 206f, 207
in leading strand synthesis, 205, 206f
Priming, in polymerase chain reaction, 247, 249f
Prion PrP, 398–399, 399f
Prions, 398–399, 398f, 399f, 400b, 400t
amino acid substitutions and, 48, 399
control of growth of, 263
Privileged sites, 521
Probe microscopy/microscopes, 104–105, 106f, 107t
atomic force, 105, 106f, 107t
scanning tunneling, 104, 106f, 107t
Probes, synthetic nucleic acids in synthesis of, 244, 255t
Probiotics, 328b
Proctitis
Chlamydia trachomatis causing, 604
gonococcal, 569
Prodromal period, 418–419, 419f
Production, in biogeochemical cycling, 765
Products, 35
in decomposition reactions, 36
in industrial microbiology, 753–756, 756t
in synthesis reactions, 35
Profundal zone, 770–771, 771f
Proglottids, tapeworm, 664, 665f
Programmed cell death, in Epstein-Barr virus infection, 691
Progressive multifocal leukoencephalopathy (PML), 695
Proguanil, for malaria prophylaxis, 661
Prokaryotae (Monera) (kingdom), 116, 116f
Prokaryotes/prokaryotic cells, 5–6, 6f, 58–61, 59b, 59f, 60f, 61f, 87f, 314–341. *See also* Archaea; Bacteria
arrangements of, 315–316, 317f
cell walls of, 60f, 66–69
characteristics of, 314–318
chromosomes of, 87t, 200–202, 202f
classification of, 318–319, 320f
cytoplasmic membrane of, 60f, 70–74, 70f. *See also* Cytoplasmic (cell/plasma) membranes
cytoplasm of, 60f, 74–76, 75f, 76f, 77f
eukaryotic cells compared with, 87t
external structures of, 60f, 61–66
genomes of, 200–202, 202f, 204t
life processes in, 57t
morphology of, 314, 314f
reproduction in, 314–315, 315f, 316f
size of, 58, 59b, 61f
Prokaryotic operons, 218–221, 219f, 220f, 221t
Prokaryotic ribosomes, 60f, 75–76, 87t, 215f, 216
Promastigotes, in *Leishmania* life cycle, 656, 656f
Promoter
in operon, 219
in transcription, 209, 210f, 211f
Properdin (alternate) pathway of complement activation, 450, 451, 451f
Prophage, 388, 389f
Prophase
meiosis, 344, 345f
mitosis, 344, 345f
Propionibacterium genus/spp., 560–562
fermentation products of, 150f
Propionibacterium acnes, 38, 532, 561
pH range tolerated by, 38–39
Propionic acid, microbial fermentation producing, 150f
Propylene oxide, for gas sterilization, 281–282, 281f, 282t

Prostaglandins
in inflammation, blood vessels affected by, 455–456, 456f
in type I (immediate) hypersensitivity, 511, 511f, 514t
Prostate gland, secretions of, in host defense, 443t
Prostate specific antigen, immunoelectrophoresis in measurement of, 499
Prosthecae, 330, 331f
Protease inhibitors, 308t, 719t
in AIDS "cocktail," 303b
Proteases, 716, 716f
HIV, 716f, 719t
release and, 722
microbial production of, 753, 756t
in protein catabolism, 151, 152f
secretory IgA, 543
Streptococcus agalactiae producing, 541
in type I (immediate) hypersensitivity, 511, 511f, 514t
Protein A, 534
Protein adhesin. *See* Adhesins
Protein-dependent termination, of transcription, 210f, 212
Protein kinase, interferon action and, 453
Protein M, 538
Proteins, 45–50, 47f, 48f, 49f. *See also specific type*
catabolism of, 151, 152b, 152f, 162f
complement, 443
membrane, 70–71, 70f
regulatory, 45
structure of, 48–50, 49f
antimicrobial agents affecting, 265
Protein synthesis
antimicrobial drugs affecting, 80, 289, 290f, 293f, 306t
recombinant DNA technology in, 254
in ribosomes, 75, 80
Proteobacteria (phylum), 320f, 324, 325t, 334t
α, 325t, 330–333, 331f, 332f, 334t, 583, 584
pathogenic, 331
β, 325t, 333, 334t, 567
δ, 334t, 335–336
electricity produced by, 755b
ε, 334t, 336
γ, 325t, 333–335, 334t, 336t, 570, 587
Gram-negative, 330–338, 334t–335t. *See also* Gram-negative bacteria
Proteus genus/spp., 336t, 576–577, 577f
sites of infection caused by, 582f
Proteus mirabilis, 577
Protheca genus/spp., 367
Protista (kingdom), 116, 116f, 347, 348f, 351
Proton gradient
in electron transport chain, 144, 147
halobacteria establishing, 322
Proton motive force, 147
Protons, 27, 27f
atomic mass of, 27
Protooncogenes, 395, 395f
Protoplasts
for electroporation, 251
fusion of, for DNA insertion into cells, 251–252, 251f, 255t
Protoscoleces, of hydatid cysts, 666, 666f
Protozoa, 4, 5f, 348–355, 356t, 650–663, 664t
antimicrobial agents affecting, 298f, 310t
classification of, 348, 348f, 350–355, 356t
cysts of, susceptibility of, 266, 266f
distribution of, 348
locomotive structure of, 4, 5f
morphology of, 348–349, 349f
nutrition of, 349

Protozoa, *Cont.*
opportunistic infections caused by, in HIV infection/AIDS, 718*t*
reproduction in, 349–350, 350*f*
in soil, 769
water-borne illness caused by, 773*t*
zoonoses caused by, 421*t*
Protozoology/protozoologists, 20*t*, 348. *See also* Protozoa
Providencia genus/spp., 577
Proviruses (latent viruses), 394
herpesviruses as, 683
herpes simplex virus, 684, 684*f*
varicella-zoster virus, 688, 688*f*
HIV as, 394, 722
PrP, 398–399, 399*f*
Prusiner, Stanley, 398
Pseudallescheria genus/spp., 641, 642
Pseudocysts
in Chagas' disease, 655
in *Toxoplasma* life cycle, 662, 662*f*
Pseudohypha, 358
Pseudomembrane, in diphtheria, 554, 555*f*
Pseudomembranous colitis, *Clostridium difficile* causing, 301, 431, 548
Pseudomonads, 320*f*, 335, 587–588
Pseudomonas genus/spp., 334*t*, 335, 335*f*
degradation of crude oil by, 761, 763*b*
superbug strains of, 303
Pseudomonas aeruginosa, 587–588, 588*f*
biofilm formed by, 177, 588
Entner-Doudoroff pathway used by, 140, 587
in use-dilution test, 267
Pseudomonas putida, in biofilm, 67*b*
Pseudomonas syringae, industrial applications for, 755
Pseudoplasmodium, 371, 372*f*
Pseudopodia, 80, 81*f*
in phagocytosis, 448, 448*f*, 449, 449*f*
of protozoa, 4, 5*f*, 352, 353*f*
Psilocybe cubensis, 644
Psilocybin, 644
P site, ribosomal, 216
Psittacosis (ornithosis), 605–606, 606*f*
as bioterrorist threat, 774*t*
Psychrophiles, 174, 174*f*, 175*f*
Public health
assessing biological threats and, 773
epidemiology and, 432–433
Public health education, 433
Public health microbiology, 20*t*
Public perception, assessing biological threats and, 773
Puerperal (childbirth) fever, streptococcal, 17, 541
Pulmonary anthrax, 327, 546, 705*t*
bioterrorism and, 22, 545, 546, 547*b*
manifestations of, 705*t*
Pulmonary aspergillosis, acute invasive, 632
Pulmonary blastomycosis, 628
Pulmonary candidiasis, 634*t*
Pulmonary coccidioidomycosis, 629
Pulmonary cryptococcosis, 635
Pulmonary histoplasmosis, 630
Pulmonary infections. *See also* Respiratory system
organisms causing, D-8
Pulmonary paracoccidioidomycosis, 631
Pulmonary tuberculosis, 556, 557*f*
Pulmonary zygomycosis, 637
Pure culture, 178–181, 179*f*, 180*f*
Purines, 50, 50*f*
biosynthesis of, 160, 161*f*
for microbial growth, 173*t*
Purple nonsulfur bacteria, 320*f*, 324, 325*t*, 331, 334*t*

Purple sulfur bacteria, 324, 325*f*, 325*t*, 333, 333*f*, 334*t*
Purpura, immune thrombocytopenic, 518, 518*f*
Purulent abscesses, in group A streptococcal pharyngitis, 539
Pus, 456
Pustules, in poxvirus infection, 680, 681*f*
Pyelonephritis
E. coli causing, 575
Mycoplasma hominis causing, 598
Pyocyanin, *Pseudomonas aeruginosa* producing, 587
Pyoderma, 539–540
Pyogenic lesions, staphylococcal, 535
Pyogenic toxins, 538
Pyridoxal phosphate
as cofactor in enzymatic activity, 131*t*
in transamination, 159*f*, 160
Pyrimidines, 50, 50*f*
biosynthesis of, 160, 161*f*
for microbial growth, 173*t*
Pyrite, acid mine drainage and, 763, 764*f*
Pyrodictium, 321, 321*f*, 322*t*
temperature range of, 174, 321, 322*t*
Pyrogenic toxins, 538
Pyrogens, 458
Pyroxidine (vitamin B_6), for microbial growth, 173*t*
Pyrrhophyta, 371*t*. *See also* Dinoflagellates
Pyruvic acid, 160*t*, 162*f*
in acetyl-CoA synthesis, 142, 142*f*, 162*f*
in glycolysis, 138*f*, 139, 160*t*, 162*f*

Q fever, 591
as bioterrorist threat, 774*t*
Quarternary structure of protein, 49*f*, 50
Quaternary ammonium compounds (quats), antimicrobial action of, 280, 280*f*
Quellung reaction, 544
Quinines, 310*t*
Quinolones, 307*t*
mechanism of action of, 295, 307*t*
Quorum sensing, 177, 177*b*

Rabbit fever, 589
Rabies, 421*t*, 728–730, 729*f*, 730*b*, 730*f*
Rabies vaccine, 729
for treatment, 729–730, 730*b*
Rabies virus, 728–730, 729*f*
direct fluorescent antibody test in detection of, 502, 502*f*
Radial immunodiffusion, 497–498, 498*f*
Radiation
for food preservation, 750–751
in microbial control, 262, 275–277, 276*f*, 277*t*
as mutagen, 223, 223*f*
DNA repair and, 225, 226*f*
wavelengths of
microbial control and, 275, 277*t*
microscopy and, 94, 94*f*
Radioactive decay, 28
Radioactive isotopes, 28
Radiolarians, 352, 353*f*, 356*t*
Ralstonia solanacearum, soil-borne disease caused by, 770*t*
Rapid sand filters, in water treatment, 756
Rash
in Lyme disease, 612, 612*f*
in measles, 726, 726*f*
in rubella, 713, 714*f*
in syphilis, 608*f*, 609
Rat fleas. *See* Fleas
Reactants, 35
in decomposition reactions, 36
in synthesis reactions, 35

Reaction center chlorophyll, 154, 154*f*
Recalcitrant substances, biomass reclamation and, 761
Receptor alteration, in resistance, 302
Recipient cell, in horizontal gene transfer, 229
Recombinant DNA technology/genetic engineering, 19*f*, 20*t*, 21, 241–261, 243*f*
in agriculture, 257–258, 258*b*, 258*f*, 755
applications of, 252–258, 256*t*
in biotechnology, 242, 243*f*
bioterrorism and, 775–776, 776*b*
for DNA fingerprinting, 249–251, 256, 257*f*
ethics of, 259
for gene therapy, 256
for genetic screening, 256
goals of, 242
hyperthermophiles in, 175*b*, 321
medical uses of, 254–257, 256*t*
for protein synthesis, 254
safety of, 258*b*, 259
techniques of, 247–252, 255*t*
tools of, 242–247, 255*t*
in vaccine production, 254–256, 255*b*, 491–492, 493*f*
for xenotransplants, 257
Recombinants, 229
selection of, 248
Recombination (genetic), 229–236, 230*f*. *See also* Recombinant DNA technology
horizontal gene transfer and, 229–232, 231*f*, 232*f*, 233*f*, 234*f*, 235*f*, 235*t*
mutations caused by, 223
transposons/transposition and, 232–236, 235*f*, 236*f*
Red algae, 367, 369*f*, 371*t*. *See also* Rhodophyta
Red blood cells. *See* Erythrocytes
Red bone marrow, lymphoid cells produced in, 465, 466*f*, 467*f*
Redi, Francesco, experiments in spontaneous generation by, 8, 8*f*
Red measles, 725–727, 726*f*, 727*f*, 727*t*
Red mites, as disease vectors, for scrub typhus (*Orienta tsutsugamushi*), 600
Redness, in inflammation, 455
Redox (oxidation-reduction) reactions, 127, 127*f*, 128*b*
Red tide, 352, 772
Reducing chemicals, for anaerobic cultures, 183, 185*f*
Reducing media, 183
Reduction, 127, 127*f*, 128*b*
Reed, Walter, 16, 17*t*
Refraction, image magnification and, 94–95, 95*f*
oil immersion lens affecting, 97, 98*f*
Refrigeration
in food preservation, 751
for microbial control, 272, 277*t*
for preserving cultures, 186
Regulatory proteins, 45
Rejection, transplant (graft), 521–522, 521*f*
Relapsing fever
endemic, 613, 613*f*
epidemic (louse-borne), 613, 613*f*
Relaxin, recombinant DNA in production of, 256*t*
Release
animal virus, 393, 394*t*
bacteriophage, in lytic replication, 385, 386*f*, 388, 388*f*, 389*f*, 394*t*
in HIV replication, 720*f*, 722
Release factors, in translation, 217
Rennin, in curdling process, 745
Reoviruses (*Reoviridae*), 385*t*, 737–738, 738*t*
diseases caused by, 711*t*, 737–738

Repair
  DNA, 224–226, 226f
    methylation affecting, 208
  tissue, 457, 457f
Replica plating, in negative (indirect) selection, 227, 228f
Replication
  DNA, 204–208, 205f, 206f, 207f, 218t
    polymerase chain reaction for, 247–248, 249f, 255t
  viral, 384–394. See also Viral replication
Replication fork, 205, 206f
Repressible operons, 219, 220f, 221t
  tryptophan (trp) operon as, 220–221, 220f
Repressor protein/repressors, in operon, 219, 219f, 220, 220f
Reproduction
  as characteristic of life, 57
  eukaryotic, 343–347, 345f, 346f, 347f, 347t
  fungal, 357–359, 359f, 360f
  microbial, 57t, 168. See also Microbial growth
    by binary fission, 186–187, 187f, 314, 315f
  prokaryotic, 314–315, 315f, 316f
  protozoal, 349–350, 350f
Reproductive spores, 4, 4f, 315, 316f
Reproductive system
  in host defense, 443t
  microbial disease agents affecting, D-11
  normal microbiota of, 408t
RER. See Rough endoplasmic reticulum
Reservoirs of infection, 420, 421t
Resident microbiota, 406, 408t. See also Normal microbiota
Residual body, 449
Resistance, 283, 283b, 287, 301–304, 303b. See also specific organism
  cross, 303
  development of, 301–302, 302f
  diffusion susceptibility tests in determination of, 297–298, 298f
  multiple, 303
  retarding, 304, 304f
  synergism and, 304, 304f
Resistance (disease), innate/inborn, 438
Resistance (R) factors/plasmids, 202, 235, 236f, 302
Resolution/resolving power, 95, 96f
  immersion oil affecting, 97, 98f
Respiration
  in carbon cycle, 766f
  cellular. See Cellular respiration
Respiratory syncytial virus, 728
Respiratory system. See also under Pulmonary
  adenovirus infection of, 696
  anthrax affecting, 327, 546
    bioterrorism and, 22, 545, 546, 547b
  aspergillosis affecting, 632
  blastomycosis affecting, 628
  candidal infection of, 634t
  coccidioidomycosis affecting, 629
  histoplasmosis affecting, 630
  in host defense, 437, 439–440, 439f, 441f, 442t
  manifestations of infections of, 705t
  microbial disease agents affecting, D-8
  normal microbiota of, 408t
  paracoccidioidomycosis affecting, 631
  as portal of entry, 409f, 410
  rhinovirus infection of, 704–705, 705t
  tuberculosis affecting, 556, 557f
  type III hypersensitivity affecting, 519
  zygomycosis affecting, 637
Responsiveness
  as characteristic of life, 57
  in microbes, 57t
    viruses, 58b

Restriction enzymes, in recombinant DNA technology, 244, 245f, 245t, 255t
  gene library production and, 247, 248f
  genome mapping and, 252, 252t, 253f
Restriction fragmentation, for genome mapping, 252, 252t, 253f
Restriction sites, 244, 245f, 245t
Reticulate (initial) bodies
  Chlamydia forming, 337, 602, 603f
  Ehrlichia forming, 601, 601f
Reticuloendothelial system, 446f, 447
Reticulum, in phagocytosis, 444, 445f
Retrospective studies, analytical studies as, 429
Retroviruses (Retroviridae), 385t, 716–724, 716f, 738t
  oncogenic, 717, 718f
  synthesis of, 392, 392t
Reverse transcriptase, 716–717, 716f, 717f. See also Retroviruses
  in cDNA synthesis, 243, 243f, 255t
  in HIV, 716f, 719t
  in retrovirus synthesis, 392, 392t, 716, 717f
Reverse transcriptase inhibitors, in AIDS "cocktail," 303b
Reverse transcriptase polymerase chain reaction, for hantavirus detection, 735
Revolving nosepiece, of light microscope, 96–97
Reye, Ralph, 691b
Reye's syndrome, 689, 691b, 735
R (resistance) factors/plasmids, 202, 235, 236f, 302
R group, in amino acids, 47, 47f
Rhabdoviruses (Rhabdoviridae), 385t, 724–725, 728–730, 729f, 738t
Rh (Rhesus) antigen, 517. See also Rh system
Rheumatic fever, streptococcal infection and, 540
Rheumatoid arthritis, 526, 526f
Rhinocerebral zygomycosis, 637
Rhinorrhea, in common cold, 704
Rhinoviruses, 704–705, 704f, 705t
  common cold caused by, 704–705, 704f, 705t
Rhizobium genus/spp., 320f, 331, 331f, 334t
  nitrogen fixation by, 172, 331, 331f, 767
Rhizopus genus/spp., 359, 361f, 636–637
Rhizopus nigricans, 359
Rhodophyta (division/kingdom), 348, 348f, 367, 371t
Rhodospirilla, 320f
Rhogam, 518
Rh system, 517–518, 517f
Ribavirin, 296f, 308t
  for Crimean–Congo hemorrhagic fever, 736
  for respiratory syncytial virus infection, 728
  spectrum of action of, 293f, 308t
Riboflavin (vitamin B₂), for microbial growth, 173t
Ribonucleic acid. See RNA
Ribonucleotide, 51
Ribose 5-phosphate, 160t
Ribosomal RNA (rRNA), 208, 215–216, 215f
  microbial taxonomy and, 117–118, 117b, 318–319, 320f, 323
Ribosomes, 87t, 215–216, 215f
  eukaryotic, 60f, 80, 83f, 84f, 87t, 215–216, 215f
  mitochondrial, 85, 86f
  prokaryotic, 60f, 75–76, 87t, 215f, 216
  translation in, 212, 216–218, 216f, 217f, 218f
    antimicrobial drugs affecting, 80, 289, 290f, 293f
Ribozymes, 130
  in transcription, 214, 214f
  in translation, 216

Ribulose 1,5-bisphosphate (RuBP), in Calvin-Benson cycle, 156, 157f
Rice
  Bacillus cereus endospores contaminating, 327
  nutritional value of, recombinant DNA technology affecting, 258
Rice-water stool, in cholera, 615
Ricketts, Howard, 598
Rickettsia genus/spp., 331, 334t, 598–600
Rickettsia prowazekii, 598, 599–600, 602t
  as biological weapon, 774t
Rickettsia rickettsii, 598, 599, 599f, 602t
Rickettsia tsutsugamushi. See Orienta tsutsugamushi
Rickettsia typhi, 598, 600, 602t
Rickettsias, 320f, 337, 598–602, 602t, 606t
  animal/cell cultures for growth of, 185
  antimicrobial agents affecting, 298f
  comparison of with other small microbes, 606t
Rifampin, 307t
  mechanism of action of, 295, 307t
Rift Valley fever, 711t, 735, 736
Rimantadine, 308t
  mechanism of action of, 295, 308t
Ringworm, 421t, 638–639, 640t. See also Dermatophytes
RMSF. See Rocky Mountain spotted fever
RNA, 50, 52t
  DNA compared with, 52t
  function of, 52
  messenger (mRNA), 208, 213–214, 214f
    in translation, 216–217, 216f, 218f
  negative strand, 392
  nucleotide bases in, 50–51, 51, 51f, 200, 201f
  positive strand, 392
  ribosomal (rRNA), 215–216, 215f
    microbial taxonomy and, 117–118, 117b, 318–319, 320f, 323
  structure of, 200, 201f
  synthesis of, 160, 161f
    in nucleoli, 83
  transfer (tRNA), 208, 214–215, 215f
    in translation, 216–217, 216f
  viral, 58b, 379
RNA-dependent RNA transcriptase, 392
RNA polymerase
  antimicrobial drugs affecting, 295
  in transcription, 209, 210f, 211f
RNA primer, in leading strand synthesis, 205, 206f
RNA probes, synthetic nucleic acids in synthesis of, 244, 255t
RNA transcript, elongation of, 209–210, 210f
RNA viruses, 385t, 701–741, 738t. See also specific type
  enveloped
    positive ssRNA, 709–716, 738t
      with reverse transcriptase, 716–724, 738t
    segmented negative ssRNA, 731–737, 738t
    unsegmented negative ssRNA, 724–731, 738t
  genomes of, 200
  segmented, 704
  naked
    positive ssRNA, 704–709, 738t
    segmented dsRNA, 737–738, 738t
  pathogenic, 703–741
Rochalimaea genus/spp.. See Bartonella genus/spp.
Rock fever of Gibraltar. See Brucellosis
Rocky Mountain spotted fever (RMSF), 599, 599f, 602t
Root nodules, in nitrogen fixation, 767

Rose-gardener's disease (sporotrichosis), 642–643, 643f
Roseola, 692, 693f
Roseolovirus genus/spp., 683, 684, 692, 693f
RotaShield, 737
Rotaviruses, 737, 738f
Rough endoplasmic reticulum, 60f, 83, 83f, 84f
Roundup. *See* Glyphosate
Roundworms (nematodes), 663, 669–672, 669f, 673t
  Ascaris, 670, 673t
  filarial (*Wuchereria bancrofti*), 670, 671–672, 672f, 673t
  hookworms (*Ancylostoma* and *Necator*), 670–671, 670f, 673t
  pinworms (*Enterobius*), 671, 671f, 673t
Rous, F. Peyton, 395
rRNA. *See* Ribosomal RNA
RSV. *See* Respiratory syncytial virus
Rubbing alcohol. *See* Isopropanol
Rubella (German/three-day measles), 713, 714f, 727t
  vaccine against, 713, 714f. *See also* Measles/mumps/rubella (MMR) vaccine
Rubella virus, 713, 714f
Rubella virus vaccine, 713, 714f
Rubeola (measles), 725–727, 726f, 727f, 727t
Rubivirus genus/spp., 713
Rubor, in inflammation, 455
"Runs," in bacterial motility, 64, 65f
Rupture event scanning, 506b
Russian spring–summer encephalitis, 711, 711t
Rusts, 363

S. *See* Sulfur; Svedbergs
Sabin vaccine (oral polio vaccine/OPV), 707, 707t
  safety of, 490, 494, 707
Sabouraud dextrose agar, 182, 182f
Saccharomyces genus/spp., 355, 362
  fermentation products of, 150f, 362
Saccharomyces cerevisiae, 4, 4f
  in beer production, 747, 747f
  in hepatitis B vaccine production, 755
  plasmids of, 204
  in wine and spirits production, 746–747, 746f
Safranin
  in Gram staining, 109, 109f
  in Schaeffer-Fulton endospore staining, 111
Sake, fermentation in production of, 747, 748t
Saliva, in host defense, 443t
Salk vaccine, 707, 707t
Salmonella genus/spp., 336t, 577–579, 578f, 579b, 752t
  in Ames test, 228, 229f
  discovery of, 17t
  sites of infection caused by, 582f
  water-borne transmission of, 577–578, 773t
Salmonella cholerasuis, in use-dilution test, 267
Salmonella enterica, 577
  biofilm formed by, 177
Salmonella paratyphi, 578
Salmonella typhi, 578–579, 578f, 579b
  discovery of, 17t
  growth of, chloramine affecting, 267
Salmonellosis, 421t, 578, 578f, 752t, 773t
  discovery of cause of, 17t
  incubation period for, 419t
Salt. *See* Sodium
Salts, 34, 39
Salt tolerance, in plants, recombinant DNA technology and, 257
Salvarsan, 310t
Sand filters, slow and rapid, in water treatment, 756, 757f

Sand flies
  bartonellosis transmitted by, 584
  leishmaniasis transmitted by, 655, 655f, 674
Sand fly fever, 711t
Sanitization, 263, 264t
Saprobes, fungi as, 356
Sarcinae (arrangement), 315, 317f
Sarcodina, 351
Sarcoma, Kaposi's, HHV-8 causing, 693, 693f
Sarcomastigophora, 351
SARS (severe acute respiratory syndrome), 404, 715b, 716
Satellite virus, hepatitis D virus as, 737
Saturated fats, 41
Saturated fatty acids, 40, 41f, 42t
Sauerkraut, fermentation in production of, 744, 748t
Saxitoxin (*Gonyaulax* genus/spp.), water-borne illness caused by, 772, 773t
Scalded skin syndrome, 535, 535f
Scanning electron microscopy/microscopes, 104, 104f, 105f, 107t
Scanning objective lens, of light microscope, 97
Scanning tunneling microscopy/microscopes, 104, 106f, 107t
Scarlet fever (scarlatina), 539
Scars, in poxvirus infection, 680, 681f
Schaeffer-Fulton endospore stain, 111, 111f, 114t
Schistosoma genus/spp. (blood flukes), 667–668, 668f, 673t
  water-borne transmission of, 773t
Schistosoma haematobium, 667, 673t
Schistosoma japonicum, 667, 673t
Schistosoma mansoni, 667, 668f, 673t
Schistosomiasis, 667–668, 673t, 773t
Schizogony, 343, 347, 347f, 658
Schizonts, 347, 347f, 658
Schleiden, Matthias, 58
Schwann, Theodor, 58
SCID. *See* Severe combined immunodeficiency disease
Science, definition of, 2
Scientific method, 9–10, 11f
Scientific (exponential) notation, B-1
Sclerotic bodies, in chromoblastomycosis, 641, 642f
Scolex, tapeworm, 664, 665f
Scrapie, 48, 399
Scrub typhus, 600, 602t
Seaweeds, 5
Seborrheic dermatitis, *Malassezia furfur* causing, 639
Sebum, in host defense, 439
Secondary cultures, in commercial food/beverage production, 744
Secondary humoral immune response, 479, 479f
Secondary infection, 425t
Secondary metabolites
  antibiotics as, 289
  of industrial fermentation, 753
Secondary structure of protein, 48–50, 49f
Second-generation drugs, 304
Secretory component, 470, 470f
Secretory IgA, 470, 470f
  in pneumococcal infection, 543
Secretory IgA protease, 543
Secretory vesicles, 60f, 83–84, 84f, 85f
Sediment, estimations of microbial growth affected by, 194
Sedimentation, in water treatment, 756, 757f
Seed warts, 693, 694f
Selective media, 182, 182f
Selective permeability, of cell membrane, 71
Selective toxicity, 289
Selenium, for microbial growth, 170t

Self-termination, of transcription, 210f, 211–212
Self-tolerance, 464
  autoimmune diseases and, 524
  lymphocyte editing (clonal deletion) and, 473–474, 474f
SEM. *See* Scanning electron microscopy/microscopes
Semiconservative DNA replication, 205, 207
Semilogarithmic graph, 188, 189f
Semi-perishable food, 750
Semisynthetics, 282, 288, 289t, 305t. *See also* Antibiotics; Antimicrobial drugs
Semmelweis, Ignaz, 17
Sensitization, in type I (immediate) hypersensitivity, 510, 511f, 513b
septi- (prefix), definition of, 413t
Septate hyphae, 356, 357f
Septicemia
  meningococcal, 570, 570f
  Vibrio vulnificus causing, 595, 616
Septic tanks, 759, 761f
SER. *See* Smooth endoplasmic reticulum
Serology, 19f, 20t, 22, 488, 496. *See also* Immune testing
  in taxonomy/microbial classification, 120, 120f
Serovars, 63
Serratia genus/spp., 336t, 576, 576f
Serratia marcescens, 576, 576f
Serum, 443
  for passive immunization, 494
Severe acute respiratory syndrome (SARS), 404, 715b, 716
Severe combined immunodeficiency disease (SCID), 527, 527b, 528t
Sewage, treatment of, microbes in, 759–761, 760f, 761f, 762f
Sex (conjugation) pili, 66, 66f, 231–232, 233f
Sexually transmitted diseases. *See also specific disease*
  Chlamydia trachomatis causing, 603, 603–604
  Trichomonas vaginalis causing, 658
Sexual reproduction, 57
  in fungi, 358, 360f
Sexual spores, 4, 358, 360f. *See also* Spores
Sheath, of cyanobacteria, 323, 324f
Sheep red blood cells, in complement fixation test, 501, 501f
Shiga, Kiyoshi, 17t
Shiga-like toxin, *E. coli* O157:H7 producing, 575
Shiga toxin, 579
Shigella genus/spp., 336t, 579, 580f, 752t
  sites of infection caused by, 582f
  water-borne transmission of, 773t
Shigella boydii, 579
Shigella dysenteriae, 579
  discovery of, 17t
Shigella flexneri, 579
Shigella sonnei, 579
Shigellosis, 579, 580f, 752t, 773t
  discovery of cause of, 17t
Shingles, 688, 688f, 689, 689b, 689f
"Shipping fever," vaccine against, 254, 255b
Shock
  anaphylactic, antimicrobial drugs causing, 301
  in toxic shock syndrome, 536
Si. *See* Silicon
Sickle-cell trait, malaria resistance and, 660
Sigma factor, in transcription, 209, 210f, 211f
Signs of disease, 412, 412t
Silage, fermentation in production of, 745, 748t
Silent mutations, 222, 222f, 223t
Silica, dinoflagellate plates made of, 352
Silicates, in algal cell wall, 79

Silicon, 28t
  Bohr diagram of, 30f
  for microbial growth, 170t
Silver, antimicrobial action of, 280–281, 282t
Silver nitrate, 281
Simian immunodeficiency virus (SIV), 719
Simple microscope, 96
Simple stains, 108, 108f, 114t
Simple sugars (monosaccharides), 44, 44f
Simplexvirus genus/spp., 683, 684. See also
      Herpes simplex viruses
Simulium flies, as disease vectors, 424t
Single-stranded DNA viruses, 379, 680, 682t
  synthesis of, 392, 392t
Single-stranded RNA viruses, 379
  enveloped
      positive, 709–716, 738t
          with reverse transcriptase, 716–724, 738t
          segmented negative, 731–737, 738t
          unsegmented negative, 724–731, 738t
  naked, positive, 704–709, 738t
  synthesis of, 392, 392t
Singlet oxygen, 170
Sinusitis
  pneumococcal, 543
  Prevotella causing, 592
SIV. See Simian immunodeficiency virus
Skin
  diseases/infections of. See also specific type
      and under Cutaneous
      acne, 532, 561–562, 561f
      actinomycosis, 562–563, 563f
      bacterial, D-1
      fungal (dermatophytoses), 638–639, 640t,
          D-2
      mycetoma (Nocardia), 562, 562f
      organisms causing, D-1 to D-3
      parasitic, D-2
      staphylococcal, 535, 535f
      viral, D-3
  in host defense, 438–439, 439f, 440b, 442t, 459t
  normal microbiota of, 405, 407, 408t
  as portal of entry, 409, 409f, 410f
  specimen collection from, 179t
Slant tubes/slants, 181, 181f
SLE. See Systemic lupus erythematosus
Sleeping sickness, African, 653–654, 654f
  discovery of cause of, 17t
Slime layer, 62, 62f
  Staphylococcus aureus, 534
Slime molds, 369, 370–371, 372f
  cellular, 371, 372f
  classification of, 348, 370
  plasmodial (acellular), 370–371, 372f
Slow sand filter, in water treatment, 756
Slow viruses. See Prions
Sludge, wastewater treatment and, 759, 760f
Slug (pseudoplasmodium), 371, 372f
SmaI restriction enzyme, 244, 245t
Smallpox, 680–682, 682f, 683b
  as bioterrorist threat, 774t
Smallpox vaccination, 681
  discovery of, 18, 488
  for general public, 494b
Smallpox virus (variola), 680–682, 683b
  as biological weapon, 774t
"Smart" bandages, 110b
Smears, preparation of, 106–107
Smooth endoplasmic reticulum, 60f, 83, 84f, 85f
Smooth muscle cells, HIV infection of, 721, 721f
Smuts, 363
Snails, in fluke life cycle, 667, 668f
Snapping division, 314–315, 315f
  in Corynebacterium, 554
Snow, John, 18, 19f

Soap
  antibacterial, 283b
  antimicrobial action of, 280, 282t
Sodium (salt), 28t
  Bohr diagram of, 30f
  in food preservation, 751
  ionic bonds in formation of, 33–34, 33f
  microbial growth affected by, 176, 274–275
      halobacteria and, 167, 176
      as nutrient, 170t
  in sodium chloride formation, 33–34, 33f
  in sweat, in host defense, 439, 459t
Sodium hypochlorite, antimicrobial action of, 279
Soft chancre (chancroid), 583, 583f
Soil-borne diseases, 769, 770t
Soil microbiology, 768–770, 768f, 770t
Solvent, water as, 37
Soredia, 364
SOS response, in DNA repair, 226
Sour cream, microbes in production of, 13t
Southern, Ed, 249
Southern blot test, 249, 250f, 255t
Soy sauce, fermentation in production of, 13t,
      150f, 744, 748t
Spallanzani, Lazzaro, experiments in
      spontaneous generation by, 8–9
Species, 113, 115f
Specific epithet, 114
Specific immunity, 438, 439f, 462–486. See also
      Immune response
Specimen collection, for culture
  clinical sampling, 178, 179t
  environmental sampling, 177–178
Spectrophotometer, turbidity measured by, for
      estimating microbial growth,
      193–194, 194f
Spectrum of action, of antimicrobial drug,
      297, 298f
Spermatia, 367
Sphaerotilus genus/spp., 333, 334t
Spherical particles, hepatitis B virus producing,
      698, 699f
Spherule, Coccidioides forming, 629, 630f
Spindle, 344, 345f
Spiral-shaped prokaryotes, 314, 314f
Spirillum (spirilla), 314, 314f
Spirillum genus/spp., 333, 334t
Spirillum minus, 333
Spirits, fermentation in production of, 747, 748t
Spirochaetes (phylum), 320f, 335t, 607
Spirochetes, 64, 314, 314f, 320f, 337–338, 607–614
  endoflagella/axial filaments of, 64, 65f
Spirogyra genus/spp., 6f
Spleen, 465–466
Splenic candidiasis, 634t
Spongiform encephalopathies, 399
Spontaneous generation (abiogenesis), 7–10
Sporadic disease, 425, 427f
Sporangiophores, 358, 359f
Sporangiospores, 358, 359f
Sporangium/sporangia, 336, 338f, 358, 359f
Spores. See also Endospores
  of microsporidia, 354
  reproductive, 4, 4f, 315, 316f, 329
      asexual, 4, 358, 359f
      sexual, 4, 359, 360f
Sporogonic cycle, in Plasmodium life cycle,
      659f, 660
Sporothrix schenckii, 642–643, 643f
Sporotrichosis, 642–643, 643f
Sporozoa, 351, 658. See also Apicomplexans
Sporozoites
  in Plasmodium life cycle, 659, 659f, 660
  in Toxoplasma life cycle, 661–662, 662f
Sporulation, 316–317, 318f, 319f

Spotted fevers, Rocky Mountain (RMSF), 599,
      599f, 602t
ssDNA viruses. See Single-stranded DNA
      viruses
SSPE. See Subacute sclerosing panencephalitis
ssRNA viruses. See Single-stranded RNA viruses
Stab culture, 183
Stachybotrys chartarum, 645b
Staining, 106–112, 114t
  acid-fast stain and, 110–111, 111f, 114t
  differential stains and, 108–112, 114t
  for electron microscopy, 112
  endospore stain and, 111, 111f, 114t
  flagellar stain and, 112, 113f, 114t
  fluorescent dyes for, 100
  Gram stain and, 16, 16f, 67, 109–110, 109f,
      110f, 114t
  for light microscopy, 106–112, 114t
  negative (capsule) stain and, 112, 112f, 114t
  principles of, 108
  simple stains and, 108, 108f, 114t
  special stains and, 112, 114t
  specimen preparation for, 106–108, 106f
Stanley, Wendell, 380
Staphylococci (arrangement), 315, 317f
Staphylococcus genus/spp./staphylococci, 328,
      329t, 533–537, 533f
  diseases caused by, 535–536, 535f, 536f,
      537f, 752t
  resistant strains of, 537
  on skin, as commensal/normal flora, 405, 407
  superbug strains of, 303
  virulence factors of, 534, 535t
Staphylococcus aureus, 328, 533, 533f, 534
  diseases caused by, 535–536, 535f, 536f,
      537f, 752t
  epidemiology of, 534
  growth of
      biofilm formation and, 177
      chloramine affecting, 267
      osmotic pressure affecting, 176
      temperature affecting, 173f
  resistant strains of, 287, 537, 538b
  toxins of, 534
  in use-dilution test, 267
  vaccine against, 537
  virulence factors of, 534, 535t
Staphylococcus epidermidis, 533, 534
  epidemiology of, 534
  on skin, 533, 534
      acquisition of, 407
  virulence factors of, 534, 535t
Staphylokinase, 534
  as virulence factor, 416
Starch, 43
Starter cultures, in commercial food/beverage
      production, 744
-stasis/-static (suffix), definition of, 263, 264t
Stationary phase, of microbial growth, 189, 189f
Stavudine (d4T), 296f
Steam, in autoclaving, 269–270, 269f, 270f, 277t
Stearic acid, 42t
Stem cells, 465
  cytomegalovirus infection of, teratogenic
      effects of, 692
  in epithelial cell replacement, 440
Stereoisomers, 48, 48f
Sterility (infertility), gonococcal PID and, 568
Sterilization, 263, 264t
  with autoclave, 269–270, 269f, 270f,
      271f, 277t
  commercial, 263
  gas, 281–282, 282t
  pure cultures and, 179
  ultrahigh-temperature, 271, 271t, 277t

Steroids, 42, 44f. *See also* Corticosteroids
  in eukaryotic membranes, 79
  immunosuppressive actions of, 522, 524t
Sterols, 42, 44f
  in eukaryotic membranes, 79
  in mycoplasma membranes, 326
Sticky ends, of restriction enzymes, 244, 245f
Stipes, of algal thalli, 366
St. Louis encephalitis, 711
Stomach, specimen collection from, 179t
Stomach acid, in host defense, 443t
Strabismus, botulinum toxin in management of, 326b
Strain R bacteria, transformation and, 229–230, 231f
Strains, 113
Strain S bacteria, transformation and, 229–230, 231f
Stramenopila (kingdom), 348, 348f
Streak plates, 179–180, 179f
"Strep throat," 539, 705t
  manifestations of, 705t
Streptobacilli, 317f
Streptococci (arrangement), 315, 317f
Streptococcus genus/spp./streptococci, 6f, 328, 329t, 537–544, 546t
  alpha-hemolytic, 542
  classification of, 537
  fermentation products of, 150f
    vegetable fermentation and, 744
  group A (*Streptococcus pyogenes*), 538–541, 546t
    culture of, 183, 183f
    diseases caused by, 539–541, 539f, 540f, 541f
    epidemiology of, 538–539, 539b
    pathogenicity of, 538
  group B (*Streptococcus agalactiae*), 541–542, 546t
  group C, 546t
  group D, 544, 546t
  group F, 546t
  group G, 546t
  Lancefield groups of, 537
  puerperal fever caused by, 17
  superbug (resistant) strains of, 303, 328
  in vegetable fermentation, 744
  viridans group, 542
Streptococcus agalactiae (group B streptococcus), 541–542, 546t
Streptococcus anginosus, 542, 546t
Streptococcus equisimilis, 542, 546t
Streptococcus mitis, 542
Streptococcus mutans, 546t
Streptococcus pneumoniae, 542–544, 542f, 546t
  capsules of, 62
  culture of, 182–183, 183f
  discovery of, 17t
  diseases caused by, 543–544
  epidemiology of, 543
  microbial genetics research and, 20
  pathogenicity of, 543
  transformation in, 229–230, 231f
  vaccine against, 544
Streptococcus pyogenes (group A streptococcus), 538–541, 546t
  culture of, 183, 183f
  diseases caused by, 539–541, 539f, 540f, 541f
  epidemiology of, 538–539, 539b
  pathogenicity of, 538
Streptococcus sanguis, 542
Streptococcus thermophilus, in yogurt production, 745
Streptokinases, 416, 417f, 538
  microbial production of, 753, 756t

Streptolysins, 538
Streptomyces genus/spp., 320f, 329t, 330
  antimicrobials produced by, 288, 289b, 289t, 306t, 330, 755
Streptomyces aureofaciens, 289t
Streptomyces avermitilis, 289t
Streptomyces erythraeus, 289t, 306t
Streptomyces fradiae, 289t
Streptomyces griseus, 289t
Streptomyces mediterranei, 289t, 307t
Streptomyces nodosus, 289t, 309t
Streptomyces orientalis, 289t, 305t
Streptomyces scabies, soil-borne disease caused by, 770t
Streptomyces venezuelae, 289t
Streptomycin, 306t
  mechanism of action of, 292, 306t
  spectrum of action of, 297, 298f, 306t
Strip-mining, acid mine damage and, 762–763, 763f
Strobila, 664
Stroma, chloroplast, 85, 86f, 153
Structural formula, 31–32
Sty, 535
Subacute disease, definition of, 423–424, 425t
Subacute sclerosing panencephalitis, 726
Subclinical infection, 412
Subcutaneous mycoses, 640–643
Subsoil (mineral soil), 768, 768f
Substitutions, in mutations, 221, 223t
Substrate, 129, 132f
  concentration of, enzyme activity affected by, 133–134, 134f
  induced fit model of enzyme interaction and, 131, 132f
Substrate-level phosphorylation, 138f, 139, 156t
  in Krebs cycle, 142, 143f
Subunit vaccines, 254, 490
Succinyl-CoA, 160t
Sucrose, 45, 45f
Sugars
  in food preservation, 751
  microbial growth affected by, 274–275
  simple (monosaccharides), 44, 44f
Sulfa drugs, discovery of, 22
Sulfamethoxazole, 294f
Sulfanilamide, 294f
  as competitive inhibitor, 134–135, 134f
  history of use of, 288
  mechanism of action of, 294–295, 294f
Sulfate-reducing microbes, 326
Sulfhydryl group, 40t
Sulfisoxazole, 294f
Sulfolobus, 321, 321f, 322t
Sulfonamides, 294f, 307t, 310t
  mechanism of action of, 294–295, 294f, 307t, 310t
  spectrum of action of, 293f, 307t, 310t
Sulfur, 28t
  Bohr diagram of, 30f
  for microbial growth, 170t, 172
Sulfur bacteria, 333
  green, 320f, 324, 325f, 325t
  purple, 324, 325f, 325t, 333, 333f, 334t
Sulfur cycle, 767–768, 767f
Sulfur dissimilation, 767f, 768
Supella genus/spp., as disease vectors, 424t
Superbugs, 303
Superinfections, 297, 301, 431
Superoxide dismutases, 171
Superoxide radicals, 170–171
Suppressor T cell, 472
Suramin, 309t
Surface tension, of water, 37, 37f

Surfactants, antimicrobial action of, 280, 280f, 282t
Susceptibility, 266, 266f
  diffusion susceptibility tests in determination of, 297–298, 298f
Sushi, antimicrobial control in preparation of, 273b
Svedbergs (S), 76
Sweat, in host defense, 439, 459t
Swelling, in inflammation, 455
Swimmer's itch, 667
Symbiosis/symbiotic relationships, 405–409, 405f, 406b, 406t, 407f, 408t
  microbial growth affected by, 177
Symbiotic nitrogen fixers, 767
Symports, 74, 75f
Symptoms of disease, 412, 412t
Synaptic cleft, botulism toxins affecting, 549, 549f
Syncytia
  HIV replication and, 721
  paramyxovirus infection and, 725, 725f
  respiratory syncytial virus infection and, 728
Syndrome
  AIDS as, 718
  definition of, 412
  terminology of, 413t
Synergistic relationship/synergism
  microbial growth affected by, 177
  resistance and, 304, 304f
Synthesis
  animal virus, 390–392, 392t, 394t
  bacteriophage, in lytic replication, 385, 386f, 387, 389f, 394t
  in HIV replication, 720f, 721–722
Synthesis reactions, 35–36, 36f
  sucrose formed in, 45, 45f
Synthetase enzymes, in feedback inhibition, 136, 136f
Synthetic (defined) media, 181–182, 182t
Synthetics, 282, 288. *See also* Antibiotics; Antimicrobial drugs
Syphilis, 607–609, 608f
  congenital, 609
  incubation period for, 419t
  primary, 608–609
  secondary, 609
  tertiary, 609
Systemic aspergillosis, 633
Systemic histoplasmosis, 630
Systemic infection, 425t
  fungal
    opportunistic fungi causing, 631–637
    pathogenic fungi causing, 628–631
Systemic lupus erythematosus, 525–526, 526f

T. *See* Thymine
Taenia saginata (beef tapeworm), 649, 665–666, 673t
Taenia solium (pork tapeworm), 649, 665–666, 673t
Tannic acid, in flagellar staining, 112
Tapeworms (cestodes), 421t, 649, 663, 664–666, 665f, 673t
  beef (*Taenia saginata*), 649, 665–666, 673t
  Echinococcus granulosus, 666, 666f, 673t
  pork (*Taenia solium*), 649, 665–666, 673t
Tatum, Edward, 20, 355
Taxa, 112, 114, 318. *See also* Taxonomy/microbial classification
Taxis, 57, 64
Taxol, microbial production of, 756t
  recombinant DNA and, 256t
Taxonomic keys, 121–122, 121f

Taxonomists, 318
Taxonomy/microbial classification, 3–7, 19f,
    112–122, 117b, 318–319, 320f
    bacterial classification and, 115f, 117–118,
        320f, 323
        nucleic acid sequencing in, 21,
            117–118, 117b
    biochemical tests in, 118–120, 119f
    categories in, 113–117, 115f, 116f
    characteristics in, 118–121
    historical basis of, 113–117, 115f, 116f
    nucleic acid sequencing in, 21, 117–118, 117b,
        120–121, 318–319, 320f, 323
    phage typing in, 120, 121f
    physical characteristics in, 118
    serological tests in, 120, 120f
    viral classification and, 118, 384, 385t
TB. See Tuberculosis
T$_c$ cells. See Cytotoxic T cells
T cell receptor, 471, 472f
    in cell-mediated immunity, 480, 481f
T cells. See T lymphocytes
TCR. See T cell receptor
T-dependent antigens, 471, 475–476, 476f
Td vaccine, 555
Tears, in host defense, 443t
Teichoic acids, in Gram-positive cell walls, 67, 69f
Telophase
    meiosis, 345f, 346
    mitosis, 344, 345f
TEM. See Transmission electron
        microscopy/microscopes
TEM images. See Transmission electron
        micrographs
Temperate phages, 388. See also
        Lysogeny/lysogenic replication
Temperature
    antimicrobial efficacy affected by,
        266–267, 267f
    core body
        elevated (fever), 457–458, 458f, 459t
        hypothalamus in regulation of, 458, 458f
    enzymatic activity affected by, 132, 134f
    during food processing/storage, spoilage
        and, 751
    microbial growth affected by, 172–174, 173f,
        174f, 175b, 175f
Temperature range
    for microbial growth, 174, 174f, 175b
    thermophiles and, 321, 322t
terato- (prefix), definition of, 413t
Teratogen, cytomegalovirus as, 692
Terminal carbonyl group, 40t
Termination
    transcription, 210f, 211–212
    translation, 217
Terminator sequence, 210f, 211–212
Tertiary structure of protein, 49f, 50
Tetanus, 551, 552f, 770t
    discovery of cause of, 17t
    incubation period for, 419t, 551
    vaccination against, 551
Tetanus toxin, 551, 552f
    immunological memory and, 479, 479f
Tetanus toxoid, 551
    immunological memory and, 477–479, 479f
Tetracyclines, 306t
    fetal toxicity of, 301, 301f
    mechanism of action of, 292, 293f, 306t
    spectrum of action of, 298f, 306t
Tetrads, 315, 317f
    chromosomal, 344, 345f
Tetrahydrofolate/tetrahydrofolic acid (THF)
    antimicrobial drugs affecting, 294f, 295
    as cofactor in enzymatic activity, 131t

T$_H$1 cells. See Helper T lymphocytes (T$_H$ cells),
        type 1
T$_H$2 cells. See Helper T lymphocytes (T$_H$ cells),
        type 2
Thallus/thalli, 356, 357f, 358, 625
    of algae, 366
    slime mold (plasmodium), 370–371, 372f
Thayer-Martin medium, modified, for Neisseria
        culture, 567, 569
T-helper cells. See Helper T lymphocytes
Theories/laws, in scientific method, 9, 11f
Thermal death point, 268
Thermal death time, 268
Thermococcus litoralis, 175b
Thermocycler, for polymerase chain
        reaction, 248
Thermoduric organisms, 271
Thermophiles, 174, 174f, 320f, 321, 321f, 322t
    extreme (hyperthermophiles), 132, 174, 174f,
        175b, 321, 321f
Thermophilic organisms, 271. See also
        Thermophiles
Thermus aquaticus, 175b
THF. See Tetrahydrofolate/tetrahydrofolic acid
Thiabendazole, 309t
Thiamine (vitamin B$_1$), for microbial
        growth, 173t
Thiamine pyrophosphate, as cofactor in
        enzymatic activity, 131t
Thimerosal, 281
Thiobacillus genus/spp., 333, 334t
    in sulfur cycle, 768
Thiomargarita namibiensis, 59b
Third-generation drugs, 304
Three-day measles (rubella), 713, 714f, 727t
Thrush (oropharyngeal candidiasis), 633f, 634,
        634t, 635
    antimicrobial drug use and, 301
Thylakoids, 85, 86f, 153, 153f
    in phototrophic bacteria, 323
Thymine (T), 50, 50f
Thymine dimers
    in microbial control, 276
    mutations caused by, 223, 223f
    repair of, 225, 226f
Thymus
    lymphocyte editing in, 473
    T lymphocyte maturation in, 465, 467f, 471
Thyroid-stimulating hormone, autoimmunity
        affecting production of, 525
Tick fever, 589
Ticks, as disease vectors, 424t, 673, 674f
    for ehrlichiosis (Ehrlichia species), 601, 602t
    for epidemic relapsing fever (Borrelia
        recurrentis), 613
    for Lyme disease (Borrelia burgdorferi),
        610–611, 611f, 612
    for rickettsial diseases, 598, 602t
    for Rocky Mountain spotted fever (Rickettsia
        rickettsii), 599, 602t
Tinctures, 278
T-independent antigens, 475, 475f
Tinea, 638–639, 640t. See also Dermatophytes
Tinea capitis, 640t
Tinea corporis, 640t
Tinea cruris ("jock itch"), 640t
Tinea pedis (athlete's foot), 623, 640t, 641b
Tinea unguium (onychomycosis), 640t
Tinea versicolor, 639, 639f
Tissue, diseased, specimen collection from, 179t
Tissue cultures. See Cell cultures
Tissue plasminogen activating factor,
        recombinant DNA in production
        of, 256t
Tissue repair, 457, 457f

Tissue typing, prevention of graft rejection
        and, 522
Titer, antibody, 500, 500f
Titration, 500, 500f
T lymphocytes/T cells, 465, 467f, 471–472,
        472b, 472f, 473t
    in cell-mediated immune response, 472,
        480–482, 480f, 482b
    cytotoxic, 472, 473t
        in cell-mediated immune response,
            480–481, 480f, 481f
    deficient, 527, 527b, 528, 528t
    editing of, 473–474
    helper, 472, 473t. See also Helper T
        lymphocytes
    memory, 480, 480f, 481
    receptors on, 471, 472f
        in cell-mediated immunity, 480, 481f
    regulation of, 481–482
    suppressor, 472
TMA. See Trimethylamine
TMAO. See Trimethylamine oxide
TNF. See Tumor necrosis factor
Toadstools, 363
Tobacco mosaic disease, discovery of cause
        of, 17t
Tobacco mosaic virus
    discovery of, 17t
    soil-borne disease caused by, 770t
Tobamovirus (tobacco mosaic virus), discovery
        of, 17t
Tobramycin, 306t
    mechanism of action of, 292, 306t
Togaviruses (Togaviridae), 385t, 709, 710, 738t
    diseases caused by, 711, 711t, 712f
    envelope of, 709–710, 710f
Tolerance, to autoantigens. See Self-tolerance
Tonsils, 466, 466f
Topical drug administration, 300
Topoisomerase II. See DNA gyrase
Topsoil, 768, 768f
Total magnification, 97–98
tox- (prefix), definition of, 413t
Toxemia, 416
    in necrotizing fasciitis, 540
Toxicity, antimicrobial drug, 301, 301f
    selective, 289
Toxicodendron radicans (poison ivy), allergic
        reaction to, 509, 520, 521f
Toxicoses (poisonings)
    food. See Food-borne illnesses/food
        poisoning
    fungi causing, 626, 627, 643–644, 645b
        mushroom poisoning, 363, 364f, 643–644,
            644, 644f
Toxic shock syndrome, 536, 536f, 537f
    streptococcal, 540
Toxic-shock syndrome (TSS) toxin, 534, 536
Toxins. See also Endotoxin; Exotoxins
    adenylate cyclase, 585
    anthrax, 545
    as biological weapons, 774t
    botulism/botulinum, 326b, 549, 549f
    Bt, 257–258, 258b, 327, 327f, 755, 756t
    cholera, 615–616, 615f
    clostridial, 547, 752t
        as biological weapon, 774t
    dermonecrotic, 585
    diphtheria, 554
    erythrogenic, 538
    fungal, 643–644
    pertussis, 585
    pyrogenic, 538
    staphylococcal, 534
    tetanus, 479, 479f, 551, 552f

Toxins, *Cont.*
  as virulence factors, 416–417, 417f, 418t
  water-borne illnesses and, 773t
Toxoid vaccines, 416, 491
  diphtheria, 555
  tetanus, 477–479, 479f, 551
Toxoplasma genus/spp., 352, 661–662, 662f
Toxoplasma gondii, 661–662, 662f, 664t, 752t
Toxoplasmosis, 421t, 662, 662f, 752t
Trace elements, for microbial growth, 170t, 172
Tracheal cytotoxin, 585
Trachoma, 603, 604, 604f
Transamination, amino acid, 159f, 160
Transcription, 208–212, 209b, 210f, 211f, 218t
  antimicrobial drugs affecting, 295, 296f
  control of, 218–221, 219f, 220f, 221b, 221t. *See also* Operons
  in retroviruses, 716
Transcription factors, 209
Transducing phages, 231, 232f
Transduction, 230–231, 232f, 235t, 388
  generalized, 231
  specialized, 231
Transferases, 129, 130t
Transfer (exchange) reactions, 36
Transferrin, in host defense, 443t, 450, 450b
Transfer RNA (tRNA), 208, 214–215, 215f
  in translation, 216–217, 216f
Transformation, 229–230, 231f
Transfusion reactions, ABO system and, 515–516, 516f, 517t
Transgenic organisms, in agricultural microbiology, 257, 755
  safety of, 259
Transient microbiota, 406–407. *See also* Normal microbiota
Translation, 209b, 212–218, 216f, 217f, 218f, 218t
  antimicrobial drugs affecting, 80, 289, 290f, 293f
  genetic code and, 212, 213f, 213t
  participants in, 213–216, 214f, 215f
Translocation, group, 74, 75f, 76t
Transmission electron micrographs, 102, 103f
Transmission electron microscopy/microscopes, 102–104, 103f, 107t
Transovarian transmission, in Rocky Mountain spotted fever (*Rickettsia rickettsii*), 599
Transplacental transmission, in toxoplasmosis, 662
Transplantation
  disease transmission and, 423, 423b
  rejection and, 521–522, 521f
Transportation proteins, 47
Transport media, 178, 185
Transport vesicles, 60f, 85f
Transposase, 235, 236f
Transpositions, in mutations, 221–222, 232–236, 235f, 236f
Transposons, 232–236, 235f, 236f
  complex, 235, 236f
  simple (insertion sequences), 235, 236f
  as vectors, 245–246, 246f, 255t
Traveler's diarrhea, discovery of cause of, 17t
Trebouxia genus/spp., 367
Trematodes (flukes), 663, 666–669, 667f, 668f, 669b, 673t
  blood (*Schistosoma* species), 667–668, 668f, 673t
  liver (*Fasciola* species), 669, 673t
Trench fever, 584
Treponema genus/spp., 335t, 339, 607–610
Treponema carateum/Treponema pallidum carateum, 609
Treponema pallidum endemicum, 609

Treponema pallidum pallidum (T. pallidum), 607–609, 607f, 608f
  animal culture for growth of, 185
  motility of, 64, 65f
  tests for detection of, 609
Treponema pallidum pertenue/Treponema pertenue, 609
Treponemal diseases, nonvenereal, 609–610, 610f
Triatoma genus/spp., as disease vectors, 424t, 674f, 675
  in Chagas' disease (*Trypanosoma cruzi*), 654f, 655, 675
Tricarboxylic acid (TCA/Krebs) cycle, 137, 142, 143f, 149t, 162f, A-6
Trichinosis, 421t
Trichomonas genus/spp., 354, 658, 658f
Trichomonas vaginalis, 658, 658f, 664t
Trichophyton genus/spp., 363, 639, 640t
Trichosporon beigelii, 637, 638, 638f
Trickle filter system, in wastewater treatment, 759, 760f
Triclosan, 278, 278f, 283b
Trifluridine, 296f
Triglycerides (fats), 40–41, 41f, 42f
Trimethoprim, 307t, 310t
  mechanism of action of, 295, 307t, 310t
Trimethylamine (TMA), 152b
Trimethylamine oxide (TMAO), 151, 152b
Trimocycline, 306t
Triphosphate deoxyribonucleotides, in DNA replication, 204–205, 205f
Triple sugar iron agar, 184t
Triplets, of microtubules, in centrioles, 81, 82f
tRNA. *See* Transfer RNA
Trophozoites, 349, 349f, 650
  in *Acanthamoeba* infection, 652
  in *Balantidium coli* infection, 650
  in *Entamoeba histolytica* infection, 651, 652
  in *Giardia* infection, 657–658
  in *Naegleria* infection, 652
  in *Plasmodium* infection, 659, 659f
  in *Trichomonas* infection, 658
trp operon, 220–221, 220f
Truffles, 362, 363b
Trychonympha genus/spp., 354, 355f
Trypanosoma genus/spp., 354, 354f, 653–654, 654f, 655f
Trypanosoma brucei, 653–654, 654f, 655f, 664t
Trypanosoma brucei gambiense, 653
  discovery of, 17t
Trypanosoma brucei rhodesiense, 653
Trypanosoma cruzi, 354f, 653, 654f, 655, 664t
Trypomastigote
  in *Trypanosoma brucei* life cycle, 653, 654f
  in *Trypanosoma cruzi* life cycle, 654f, 655
Trypticase soy agar, 182
Tryptophan auxotroph, identification of, 227–228, 228f
Tryptophan (*trp*) operon, 220–221, 220f
Tsetse flies, as disease vectors, 674–675, 674f
  for African sleeping sickness (*Trypanosoma brucei*), 653, 654f, 675
TSH. *See* Thyroid-stimulating hormone
Tubercles, 556, 557f
Tuberculin skin test/tuberculin response, 520, 520f, 558, 558f
  BCG vaccine affecting, 559
Tuberculoid leprosy, 559, 560b
Tuberculosis, 329, 555–559, 556f, 557f. *See also* *Mycobacterium tuberculosis*
  diagnosis/treatment/prevention of, 558–559, 558f
  discovery of cause of, 15, 17t
  disseminated, 557

  epidemiology of, 557–558
  pathogenesis of, 556–557, 557f
  primary, 556, 557f
  secondary/reactivated, 556, 557f
  vaccine against, 559
Tuberculous cavity, 556
Tuber genus/spp., 362
Tuber magnatum pico, 363b
Tuber melanosporum, 363b
Tubulin microtubules, in eukaryotic cytoskeleton, 81, 82f
Tularemia, 589
  as bioterrorist threat, 774t
"Tumbles," in bacterial motility, 64, 65f
Tumbling motility, in *Listeria*, 553–554
Tumor, 395, 415t. *See also* Neoplasia
Tumor (edema), in inflammation, 455
Tumor necrosis factor
  in immune response, 473
  recombinant DNA in production of, 256t
Tungsten, for microbial growth, 170t
Tunneling current, 104
Turbidity, for estimating microbial growth, 193–194, 194f
Type III secretion system, of Enterobacteriaceae, 573, 574f
  E. coli O157:H7, 575
  Yersinia, 580
Typhoid fever, 578, 578f, 579b, 773t
  discovery of cause of, 17t
"Typhoid Mary," 579b
Typhus, 421t
  as bioterrorist threat, 774t
  epidemic (louse-borne), 599–600, 602t
  murine (endemic), 600, 602t
  scrub, 600, 602t

U. *See* Uracil
Ubiquinones, in electron transport chains, 146
Ulcers
  in gastrointestinal zygomycosis, 637
  peptic, *Helicobacter pylori* infection and, 617–618, 618f
Ultrahigh-temperature pasteurization, 271, 271t
Ultrahigh-temperature sterilization, 271, 271t, 277t
Ultramicrotome, 102
Ultrastructure, electron microscopsy in study of, 102
Ultraviolet light. *See also* Light
  fluorescent microscopes using, 100
  in food preservation, 751
  in microbial control, 276–277, 277t
  as mutagen, 223, 223f
  DNA repair and, 225, 226f
Ulva genus/spp., alternation of generations in, 368f
Uncoating
  in animal virus replication, 390, 391f, 394t
  antiviral drugs affecting, 308t
  in HIV replication, 720f, 721
Undulant fever (brucellosis), 331, 584–585, 585f
  as bioterrorist threat, 774t
  discovery of cause of, 17t
Uniports, 73, 75f
Units of measurement, 93, 93t
Universal precautions, 178, 431
Unsaturated fatty acids, 40–41, 41f, 42t
Uracil (U), 50, 50f
Ureaplasma genus/spp., 596
Ureaplasma urealyticum, 597–598
Urease, *Proteus* producing, 577
Urethritis, nongonococcal (NGU)
  Chlamydia trachomatis causing, 604
  mycoplasmas causing, 598

Urinary tract
in host defense, 443*t*
microbial disease agents affecting, D-11
normal microbiota of, 408*t*
Urinary tract infections
candidal, 634*t*
E. coli causing, 575
organisms causing, D-11
Proteus mirabilis causing, 577
Urine, in host defense, 443*t*
Urine specimen, collection of, 179*t*
Urticaria, in type I (immediate)
hypersensitivity, 512, 513*f*
Urushiol, allergic reaction to (poison ivy), 509,
520, 521*f*
Use-dilution test, 267
UTIs. *See* Urinary tract infections
UV light. *See* Ultraviolet light

Vaccination, 18, 488–496. *See also*
Immunization; Vaccines
Vaccines, 489–494, 490*b*, 494*b*. *See also specific*
type and specific organism or disease
artificially acquired immunity and, 483
attenuated (live), 489–490
CDC recommendations for, 491, 491*f*, 492*t*
discovery of, 18, 488
egg cultures in preparation of, 396
inactivated (killed), 490–491
adjuvants for, 490*t*, 491
recombinant DNA technology in production
of, 254–256, 255*b*, 491–492, 493*f*
safety of, 492–494
subunit, 254, 490
toxoid, 416, 491
whole agent, 490
Vaccinia virus, smallpox vaccine and, 18,
488, 681
Vacuoles, 84, 85*f*, 87*t*
contractile, in protozoa, 349, 349*f*
Vacuum filtration, in microbial control,
274, 274*f*
Vaginal secretions, in host defense, 443*t*
Vaginitis
candidal, 633, 634*t*, 635
antimicrobial drug use and, 301
Trichomonas, 658
Valence, definition of, 29
Valence electrons, 29
Valence shell, 29
Vanadium, for microbial growth, 170*t*
Vancomycin, 305*t*
mechanism of action of, 292, 305*t*
Staphylococcus aureus resistance and,
287, 537
van Leeuwenhoek, Antoni, 2–3, 2*f*, 3*f*, 6–7, 19*f*,
93–94, 95, 96
van Niel, C.B., 19
Variable region, immunoglobulin, 468, 468*f*
Variant Creutzfeldt-Jakob disease, 48, 399
Varicella (chickenpox), 687–689, 687*f*
vaccine against, 689
Varicella-zoster virus (VZV), 687–689, 687*f*
latency/recurrent infection and, 688, 688*f*,
689, 689*b*, 689*f*
Varicellovirus genus/spp., 683, 684, 687. *See also*
Varicella-zoster virus
Variola (smallpox) virus, 680–682, 683*b*
as biological weapon, 774*t*
Variola major, 682
as biological weapon, 774*t*
Variola minor, 682
Variolation, 488
Vasodilation, in inflammation, 455–456,
455*f*, 456*f*

Vectors, 373
in disease transmission, 423, 424*t*. *See also*
specific disease and specific vector
arthropod, 423, 424*t*, 672–675, 674*f*
in recombinant DNA technology, 245–246,
246*f*, 255*t*
VEE. *See* Venezuelan equine encephalitis
Vegetables, fermented, 744–745, 748*t*
Vegetative cells, 316, 318*f*
Vehicle transmission of disease, 422, 422*t*
Veillonella genus/spp., 326, 329*t*
Venezuelan equine encephalitis, 711, 711*t*
Vesicles, 84, 85*f*, 87*t*
food, 80, 85*f*
gas, 75
in poxvirus infection, 680, 681*f*
secretory, 60*f*, 83–84, 84*f*, 85*f*
transport, 60*f*, 85*f*
Viable plate counts, for estimating microbial
growth, 190–191, 190*f*
Vi antigens, 572, 574*f*
Vibrio genus/spp., 336*t*, 614–616,
614*f*, 615*f*
Vibrio cholerae, 17*t*, 614*f*, 615–616, 615*f*. *See also*
Cholera
chromosomes of, 202
culturing, 186
pH range tolerated by, 176
strain O1 *el Tor*, 615, 615*f*
strain O139 Bengal, 615
water-borne transmission of, 615, 773*t*
Vibrio parahaemolyticus, 616
Vibrio vulnificus, 595, 616, 752*t*
antacids affecting, 39*b*
Vibrionaceae, 336*t*
Vibrios, 314, 314*f*, 614
Vinegar, microbial production of, 13*t*, 150*f*,
747, 748*t*
Viral envelope, 378, 382–384, 383*f*. *See also*
Viruses, enveloped
Viral hemagglutination, 500
Viral hemagglutination inhibition test,
500, 506*t*
Viral meningitis, coxsackieviruses
causing, 708
Viral neutralization, 500, 506*t*
Viral pneumonia, manifestations of, 705*t*
Viral proteins, antiviral drugs affecting, 308*t*
Viral replication, 384–394, 394*t*
of animal viruses, 390–394, 391*f*, 392*t*,
393*f*, 394*f*
of bacteriophages, 394*t*
lysogenic, 388–390, 389*f*
lytic, 385, 385–388, 386*f*, 388*f*, 389*f*
Viremia
in arbovirus infection, 710
in Epstein-Barr virus infection, 691
Viricides, 264
Viridans streptococci, 542
Virions, 378, 378*f*. *See also* Viruses
enveloped, 382–384, 383*f*
budding in release of, 393, 393*f*
nonenveloped (naked), 382
shapes of, 382, 382*f*, 383*f*
Viroid-like agents, 398
Viroids, 397–398, 398*f*, 400*t*
Virology, 16, 19*f*, 20*t*. *See also* Viruses
characteristics of, 384*f*
Virulence, 415–418, 416*f*, 417*f*, 418*t*. *See also*
specific organism
residual, vaccines and, 490, 494
Virulence factors, 415–418, 416*f*, 417*f*, 418*t*. *See*
also specific organism
of Enterobacteriaceae, 573, 574*f*
Virulence plasmids, 202

Viruses, 6, 7*f*, 16, 377–397, 400*t*, 606*t*. *See also*
specific type and under Viral
animal/cell cultures for growth of, 185
antimicrobial agents affecting, 298*f*. *See also*
Antiviral drugs
attachment of
in animal virus replication, 390, 394*t*
antimicrobial drugs affecting, 290*f*, 295, 308*t*
in lytic replication, 385, 386*f*, 387, 389*f*, 394*t*
autoimmune diseases and, 524
bacteria compared with, 400*t*
in cancer, 394–395, 395*f*
HIV infection/AIDS and, 718*t*
characteristics of, 378–384, 378*f*
classification of, 118, 384, 385*t*
comparison of with other small
microbes, 606*t*
culturing, 395–397, 396*f*, 397*f*
definition of, 378
discovery of, 16
diseases caused by, 378
cardiovascular system/lymphatic system
involvement and, D-7
gastrointestinal involvement and, D-10
nervous system involvement and, D-5
respiratory system involvement and, D-8
skin/eye involvement and, D-3
urinary tract/reproductive system
involvement and, D-11
DNA, 679–702
enveloped, 378, 382–384, 383*f*
antimicrobial action and, 265
budding in release of, 393, 393*f*
susceptibility of, 266, 266*f*
filtration in control of, 274
genomes of, 200, 378, 379, 379*f*
segmented, 731
as vectors, 245–246, 246*f*, 255*t*
high-pressure, 387, 387*b*
hosts of, 379, 380*f*, 381*b*
life processes in, 57*t*, 58, 58*b*, 400
methylation affecting, 207–208
nonenveloped, susceptibility of, 266, 266*f*
opportunistic infections caused by, in HIV
infection/AIDS, 718*t*
replication of, 384–394
resistance to, 302, 303*b*
shapes of, 382, 382*f*, 383*f*
sizes of, 380, 381*f*
slow. *See* Prions
in soil, 769
disease caused by, 770*t*
uncoating of
in animal virus replication, 390, 391*f*, 394*t*
antiviral agents affecting, 308*t*
water-borne illness caused by, 773*t*
zoonoses caused by, 421*t*
Visceral leishmaniasis (kala-azar), 656
Vitamin B₁ (thiamine), for microbial
growth, 173*t*
Vitamin B₂ (riboflavin), for microbial
growth, 173*t*
Vitamin B₆ (pyroxidine), for microbial
growth, 173*t*
Vitamins
for microbial growth, 173*t*
microbial production of, 13*t*
Volvox, 59*f*
Vomiting, in host defense, 443*t*
von Behring, Emil, 19*f*, 22
Vorticella genus/spp., 351
Vulvovaginal infection, candidal, 633,
634*t*, 635
antimicrobial drug use and, 301
VZV. *See* Varicella-zoster virus

Walking pneumonia, *Mycoplasma pneumoniae* causing, 597
Walsby, Tony, 313
Wandering macrophages, 444
Wangiella genus/spp., 640
Warts, 693–695, 694*f*, 695, 695*b*, 695*f*
Wasserman test, 502
Wastewater treatment, microbes in, 759–761, 760*f*, 761*f*, 762*f*
Water, 37, 37*f. See also* Aquatic microbiology
  bond angle for, 32, 33*f*
  diseases transmitted by. *See* Water-borne illnesses
  domestic, 770
  physical effects of, microbial growth and, 176
  polar covalent bonds in formation of, 32, 33*f*
  pollution of, 756, 772. *See also* Water-borne illnesses
  potable, 432
    microbes in production of, 756–758, 757*f*
  in Schaeffer-Fulton endospore staining, 111
  testing quality of, microbes in, 758–759, 758*f*
Water activity, of food, spoilage and, 749, 749*t*
Water-borne illnesses, 422, 422*t*, 772, 773*t*
  Acanthamoeba causing, 652
  as bioterrorist threats, 774*t*
  Campylobacter jejuni causing, 616, 752*t*, 773*t*
  Cryptosporidium causing, 662
  Cyclospora causing, 663
  Entamoeba causing, 651, 773*t*
  Giardia causing, 657, 658, 773*t*
  Legionella causing, 589, 590
  Naegleria causing, 652
  roundworms (nematodes) causing, 669
  Salmonella causing, 577–578, 773*t*
  Vibrio causing, 615, 773*t*
  Yersinia causing, 580
Water molds, 369, 370, 370*f*
Wavelengths of radiation
  microbial control and, 275, 277*t*
  microscopy and, 94, 94*f*
Waxes, 41–42
WEE. *See* Western equine encephalitis
Weichselbaum, Anton, 17*t*
Wells, antigen
  in enzyme-linked immunosorbent assay (ELISA), 502–503, 504*f*
  in immunodiffusion, 496, 497, 497*f*, 498*f*

Welsh, William, 17*t*
Western blot test, 120, 503–504, 505*f*, 506*t*
Western equine encephalitis, 711, 711*t*
West Nile encephalitis, 711, 711*t*, 715*b*
West Nile virus, 711, 711*t*, 715*b*
  transmission of, 423*b*, 715*b*
Wetlands, artificial, in wastewater treatment, 761, 762*f*
Wheat soil-borne mosaic virus, disease caused by, 770*t*
Whey, in cheese production, 745, 745*f*
White blood cells. *See* Leukocytes
White piedra, 638, 638*f*
Whitlow, herpetic, 684*f*, 685, 686*f*
Whittaker, Robert, 116–117, 116*f*, 348*f*, 351
Whole agent vaccines, 490
Whooping cough (pertussis), 585–587, 586*f*, 587*f*
Wild type cells, 226–227
Wine, fermentation in production of, 746–747, 746*f*, 748*t*
Winogradsky, Sergei, 19*f*, 21
Woese, Carl, 21, 117, 117*b*
Working distance, oil immersion lens and, 97
Worms. *See* Helminths
Wort, in beer production, 747, 747*f*
Wound, open, specimen collection from, 179*t*
Wound botulism, 550
Wound infection, *Vibrio vulnificus* causing, 616
Wuchereria bancrofti, 670, 671–672, 672*f*, 673*t*

Xanthophylls, 369
Xenodiagnosis, in Chagas' disease (*Trypanosoma cruzi*), 655
Xenografts, 521, 521*f*
Xenopsylla fleas, as disease vectors, 424*t*
  for murine (endemic) typhus (*Rickettsia typhi*), 600, 602*t*
Xenotransplants, recombinant DNA technology and, 257
X-rays
  in microbial control, 276, 277*t*
  as mutagens, 223

Yaws, 609, 610*f*
Yeasts, 4, 4*f*
  in fermentation, 11, 12*f*
    wine and spirits production and, 746–747, 746*f*
  thalli of, 356, 357*f*

Yellow fever, 421*t*, 711*t*, 712–713
  discovery of cause of, 17*t*
Yellow fever virus, discovery of, 17*t*
Yellow-green algae, 367–369, 371*t*
Yersin, Alexandre, 17*t*
Yersinia genus/spp., 336*t*, 580–581, 581*f*
  growth of, in refrigerated products, 272
  sites of infection caused by, 582*f*
Yersinia enterocolitica, 580, 752*t*
Yersinia pestis, 566, 580–581, 581*f*
  as biological weapon, 774*t*
  discovery of, 17*t*
Yersinia pseudotuberculosis, 580
Yogurt, fermentation in production of, 13*t*, 150*f*, 745

Zanamivir, for influenza, 735
Zephiran. *See* Benzalkonium chloride
Zidovudine (azidothymidine/AZT), 296*f*, 308*t*, 719*t*
  in AIDS "cocktail," 303*b*
Ziehl, Franz, 111
Ziehl-Neelsen acid-fast stain, 111, 111*f*, 114*t*
Zinc, 28*t*
  antimicrobial action of, 280–281, 282*t*
  cycling of, 768
  for microbial growth, 170*t*
Zn. *See* Zinc
Zone of inhibition
  in diffusion susceptibility test, 298, 298*f*
  in disk-diffusion test, 267–268, 268*f*
Zoogloea genus/spp., 333, 334*t*
Zoonoses/zoonotic diseases, 420, 421*t*
  arbovirus infections and, 710, 711*t*
  arenavirus infections and, 736
  rabies, 729, 729*f*, 730*b*
Zoophilic dermatophytes, 638
Zoster, 688, 688*f*, 689, 689*b*, 689*f*
Zygomycetes, 359–360, 362*f*
Zygomycoses, 636–637
Zygomycota (division), 359, 359–360, 366*t*, 636
Zygospores, 359–360, 361*f*, 366*t*
Zygotes, 343
  protozoan
    in *Plasmodium* life cycle, 659*f*, 660
    in *Toxoplasma* life cycle, 661

Pronunciations of Selected Microorganisms

Fonsecaea pedrosoi (fon-sē-sē'ă pe-drō'sō-ē)
Francisella tularensis (fran'si-sel'ă too-lă-ren'sis)
Fusarium (fū-zā'rē-ŭm)

Gambierdiscus (gam'bē-er-dis-kŭs)
Gelidium (jel-li'dē-ŭm)
Giardia (jē-ar'dē-ă)
Gluconobacter (gloo-kon'ō-bak-ter)
Gonyaulax (gon-ē-aw'laks)
Gymnodinium (jīm-nō-din'ē-ŭm)
Gyromitra esculenta (gī-rō-mē'tră es-kū-len'tă)

Haemophilus aphrophilus (hē-mof'i-lŭs a-frō-fil'ŭs)
Haemophilus ducreyi (hē-mof'i-lŭs doo-krā'ē)
Haemophilus influenzae (hē-mof'i-lŭs in-flū-en'zē)
Hafnia (haf'nē-ă)
Halobacterium salinarium (hā'lō-bak-tēr'ē-ŭm sal-ē-nar'ē-ŭm)
Helicobacter pylori (hel'ĭ-kō-bak'ter pī'lō-rē)
Histoplasma capsulatum (his-tō-plaz'mă kap-soo-lā'tŭm)

Isabella abottae (iz-ă-bel'ă ab'ot-tē)

Klebsiella pneumoniae (kleb-sē-el'ă nū-mō'nē-ī)

Lactobacillus bulgaricus (lak-tō-bă-sil'ŭs bul-gā'ri-kŭs)
Lactococcus lactis (lak-tō-kok'ŭs lak'tis)
Legionella pneumophila (lē-jŭ-nel'lă noo-mō'fi-lă)
Leishmania (lēsh-man'ē-ă)
Lentivirus (len'ti-vī-rŭs)
Leptospira interrogans (lep'tō-spī'ră in-ter'ră-ganz)
Leuconostoc citrovorum (loo'kō-nos-tŏk sit-rō-vō'rum)
Listeria monocytogenes (lis-tēr'ē-ă mo-nō-sī-to'je-nēz)
Lyssavirus (lis'ă-vī-rŭs)

Madurella (mad'ū-rel'ă)
Malassezia furfur (mal-ă-sā'zē-ă fur'fur)
Methanobacterium (meth'a-nō-bak-tēr'ē-ŭm)
Methanopyrus (meth'a-nō-pī'rŭs)
Microsporidium (mī-krō-spor-i'dē-ŭm)
Microsporum (mī-kros'po-rŭm)
Moraxella catarrhalis (mōr'ak-sel'ă ka-tar-hal'is)
Morganella (mōr'gan-el'ă)
Mucor (mū'kōr)
Mycobacterium avium-intracellulare (mī-kō-bak-tēr'ē-ŭm ā'vē-ŭm in'tra-sel-yu-la'rē)
Mycobacterium bovis (mī-kō-bak-tēr'ē-ŭm bō'vis)
Mycobacterium leprae (mī-kō-bak-tēr'ē-ŭm lep'rī)
Mycobacterium tuberculosis (mī-kō-bak-tēr'ē-ŭm tū-ber-kyū-lō'sis)
Mycoplasma genitalium (mī'kō-plaz-mă jen-ē-tal'ē-ŭm)
Mycoplasma hominis (mī'kō-plaz-mă ho'mi-nis)
Mycoplasma pneumoniae (mī'kō-plaz-mă nū-mō'nē-ī)

Naegleria (nā-glē'rē-ă)
Neisseria gonorrhoeae (nī-se'rē-ă go-nor-re'-ī)
Neisseria meningitidis (nī-se'rē-ă me-nin-ji'ti-dis)
Neurospora crassa (nū-ros'pōr-ă kras'ă)
Nitrosomonas (nī-trō-sō-mō'nas)
Nocardia asteroides (nō-kar'dē-ă as-ter-oy'dēz)
Nosema (nō-sē'mă)

Orienta tsutsugamushi (ōr-ē-en'tă tsoo-tsoo-gă-mū'chē)

Paracoccidioides brasiliensis (par'ă-kok-sid-ē-oy'dēz bră-sil-ē-en'sis)
Paramecium (par-ă-mē'sē-ŭm)
Pasteurella haemolytica (pas-ter-el'ă hē-mō-lit'i-kă)
Pasteurella multocida (pas-ter-el'ă mul-tŏ'si-da)
Penicillium chrysogenum (pen-i-sil'ē-ŭm krī-so'jĕn-ŭm)
Penicillium marneffei (pen-i-sil'ē-ŭm mar-nef-ē'ī)
Pfiesteria (fes-te'rē-ă)
Phialophora verrucosa (fī-ă-lof'ŏ-ră ver-ū-kō'să)
Physarum (fī-sa'-rŭm)
Phytophthora (fī-tof'tho-ră)
Plasmodium falciparum (plaz-mō'dē-ŭm fal-sip'ar-ŭm)
Plasmodium malariae (plaz-mō'dē-ŭm mă-lār'ē-ī)
Plasmodium ovale (plaz-mō'dē-ŭm ō-vă'lē)
Plasmodium vivax (plaz-mō'dē-ŭm vī'vaks)
Pneumocystis jiroveci (nū-mō-sis'tis jē-rō-vĕt'zē)
Prevotella (prev'ō-tel'-ă)
Propionibacterium acnes (prō-pē-on-i-bak-tēr'ē-ŭm ak'nēz)
Proteus mirabilis (prō'tē-ŭs mi-ra'bi-lis)
Prototheca (prō-tō-thē'kă)
Providencia (prov'i-den'sē-ă)
Pseudallescheria (sood'al-es-kē-rē-ă)
Pseudomonas aeruginosa (soo-dō-mō'nas ā-rū-ji-nō'să)
Pseudomonas putida (soo-dō-mō'nas pyoo'ti-dă)
Pseudomonas syringae (soo-dō-mō'nas sēr'in-jē)
Psilocybe cubensis (sī-lō-sī'bē kū-benz'is)
Pyrodictium (pī-rō-dik'tē-um)

Rhizobium (rī-zō'bē-ŭm)
Rhizopus nigricans (rī-zō'pŭs ni'gri-kanz)
Rickettsia prowazekii (ri-ket'sē-ă prō-wă-ze'kē-ē)
Rickettsia rickettsii (ri-ket'sē-ă ri-ket'sē-ē)
Rickettsia typhi (ri-ket'sē-ă tī'fē)

Saccharomyces cerevisiae (sak-ă-rō-mī'sēz se-ri-vis'ē-ē)
Salmonella cholerasuis (sal'mŏ-nel'ă kol-er-a-su'is)
Salmonella enterica (sal'mŏ-nel'ă en-ter'i-kă)
Salmonella paratyphi (sal'mŏ-nel'ă par'a-tī'fē)
Salmonella typhi (sal'mŏ-nel'ă tī'fē)
Salmonella typhimurium (sal'mŏ-nel'ă tī'fē-mur-ē-ŭm)
Sarcina (sar'si-nă)
Schistosoma haematobium (skis-tō-sō'mă hē'mă-tō'bē-ŭm)